Volume 3

Australian Beetles

Polyphaga (Part 2)

Editors: Hermes E. Escalona and Adam Ślipiński

PUBLISHING

NATIONAL LIBRARY OF AUSTRALIA

A catalogue record for this book is available from the National Library of Australia

ISBN: 9781486314539 (hbk)
ISBN: 9781486314546 (epdf)
ISBN: 9781486314553 (epub)

How to cite:
Escalona HE, Ślipiński A (Eds) (2025) *Australian Beetles: Polyphaga (Part 2)*. Vol. 3. CSIRO Publishing, Melbourne.

Published by:

CSIRO Publishing
36 Gardiner Road, Clayton VIC 3168
Private Bag 10, Clayton South VIC 3169
Australia

Telephone: +61 3 9545 8400
Email: publishing.sales@csiro.au
Website: www.publish.csiro.au
Sign up to our email alerts: publish.csiro.au/earlyalert

Front cover: (top to bottom, left to right) *Trigonoptera vittata* (Gestro), *Rhytiphora pardalis* (Newman), *Euops rudis* Lea, *Car* sp. (South Australia), *Leaoanthribus crucifer* (Lea), *Platynotocis pyriformis* Lea, *Rhinotia princeps* Zimmerman, *Ipsichora desiderabilis* Lea, *Chrysolopus spectabilis* (Fabricius), *Merimnetes celmisiae* Lea, *Schizoeupsalis promissa* (Pascoe) male, *Syrphax bovilli* (Blackburn), *Agriochaeta crinita* Pascoe, *Misophricoides* sp. (Western Australia), *Ochrometa amoena* Pascoe (photographs by Debbie Jennings, except *T. vittata* and *R. pardalis* by Cate Lemann).
Back cover: *Myllocerus* cf. *M. australis* Boisduval (photograph by Debbie Jennings).

Cover design by Cath Pirret
Typeset by Envisage Information Technology
Index by Indexicana
Printed in Australia by Ligare Book Printers

CSIRO acknowledges the Traditional Owners of the lands that we live and work on across Australia and pays its respect to Elders past and present. CSIRO recognises that Aboriginal and Torres Strait Islander peoples have made and will continue to make extraordinary contributions to all aspects of Australian life including culture, economy and science. CSIRO is committed to reconciliation and demonstrating respect for Indigenous knowledge and science. The use of Western science in this publication should not be interpreted as diminishing the knowledge of plants, animals and environment from Indigenous ecological knowledge systems.

MIX
Paper | Supporting responsible forestry
FSC www.fsc.org FSC® C011613

The paper this book is printed on is in accordance with the standards of the Forest Stewardship Council®. The FSC® promotes environmentally responsible, socially beneficial and economically viable management of the world's forests.

Jan25_01

CONTENTS

PREFACE

It was not envisioned by John F. Lawrence and Adam Ślipiński, when preparing the first volume of this series in 2013, that the planned three-volume series *Australian Beetles* will include weevils. However, over the years it became clear that the continuation of the *Australian Weevils* series, as it was originally planned by E. C. Zimmerman in 1991, was impractical. Therefore, the decision was made to include the generic level treatment of these beetles in the *Australian Beetles* series.

This volume of *Australian Beetles* deals with two phytophagous beetle groups: the weevils, comprising seven families, and part of the longhorns. Both are of broad interest to biosecurity and as agricultural pests but also play a significant role in ecosystems as pollinators, biocontrol of weeds and decomposing plant material, among others.

Weevils belong in the superfamily Curculionoidea and represent about 20% or more of all the beetle species occurring in Australia. The approximate number of genera and species is difficult to estimate, with thousands of undescribed taxa accumulated in entomological collections. The chapters of the different weevil families presented here are, therefore, by no means definitive. Our aim has been to provide a new synthesis that, together with the monumental work of Zimmerman (1991, 1992, 1993, 1994a, b), Oberprieler & Zimmerman (2020) and the Australian Weevil Catalogue (Pullen *et al.* 2014), will facilitate the study of this fascinating fauna. The bark beetles (Platypodinae, Scolytinae) constitute two separate chapters with unprecedented details for the Australian fauna, thanks to the efforts of James R. M. Bickerstaff and collaborators. The weevil chapters were enriched by unpublished taxonomical, distributional and biological information as well as several artistic illustrations accumulated by the late E. C. Zimmerman and carefully conserved by Rolf Oberprieler. Zimmerman's colossal endeavour and long-term vision, not only regarding weevil taxonomical research but also about the role of Natural History Collections in Australian society, is a lesson to the scientific community.

The second part of the book is dedicated to longhorn beetles, the family Cerambycidae. Our previous work (Ślipiński & Escalona 2013, 2016; Ślipiński *et al.* 2023) has covered the Australian fauna in detail, and we present here a practical synopsis with some minor updates. These chapters only include the introduction and the treatment of Lamiinae; the other subfamilies will be included in the next volume.

Australian Beetles Volume 4 will include the remaining families of Polyphaga and a new chapter on the classification of the Australian beetles to reflect ongoing changes in the higher classification of Coleoptera.

The following are abbreviations throughout the text and keys for Australian states and territories: ACT (Australian Capital Territory), NSW (New South Wales), NT (Northern Territory), QLD (Queensland), SA (South Australia), VIC (Victoria) and WA (Western Australia).

References

Oberprieler RG, Zimmerman EC (2020) *Australian Weevils (Coleoptera: Curculionoidea). Volume VI. Obituary: Elwood Curtin Zimmerman 1912–2004, Curculionidae: Entiminae Part I, Literature Consulted II.* CSIRO Australia, Clayton, xiv + 650 pp.

Pullen K, Jennings D, Oberprieler RO (2014) Annotated catalogue of Australian weevils (Coleoptera: Curculionoidea). *Zootaxa* **3896**(1), 1–481. doi:10.11646/zootaxa.3896.1.1

Ślipiński A, Escalona HE (2013) *Australian Longhorn Beetles (Coleoptera). Volume 1. Introduction and Subfamily Lamiinae.* CSIRO Publishing, Collingwood, Victoria.

Ślipiński A, Escalona HE (2016) *Australian Longhorn Beetles (Coleoptera). Volume 2: Subfamily Cerambycinae.* CSIRO Publishing, Collingwood, Victoria.

Ślipiński A, de Keyzer R, Jin M (2023) *Australian Longhorn Beetles (Coleoptera: Cerambycidae). Volume 3. Subfamily Prioninae of the Australo-Pacific Region.* CSIRO Publishing and Australian Biological Resources Study, Clayton South, Victoria, and Canberra, ACT.

Zimmerman EC (1991) *Australian Weevils (Coleoptera: Curculionoidea). Volume V. Colour Plates 1–304.* CSIRO Australia, Melbourne, x + 633 pp., 303 pls.

Zimmerman EC (1992) *Australian Weevils (Coleoptera: Curculionoidea). Volume VI. Colour Plates 305–632.* CSIRO Australia, Melbourne, viii + 707 pp., 328 pls.

Zimmerman EC (1993) *Australian Weevils (Coleoptera: Curculionoidea). Volume III: Nanophyidae, Rhynchophoridae, Erirhinidae, Curculionidae: Amycterinae, Literature Consulted.* CSIRO Australia, Melbourne, x + 854 pp.

Zimmerman EC (1994a) *Australian Weevils (Coleoptera: Curculionoidea). Volume I. Orthoceri. Anthribidae to Attelabidae: The Primitive Weevils.* CSIRO Australia, Melbourne, xxxii + 741 pp.

Zimmerman EC (1994b) *Australian Weevils (Coleoptera: Curculionoidea). Volume II. Brentidae, Eurhynchidae, Apionidae and a Chapter on Immature Stages by Brenda May.* CSIRO Australia, Melbourne, x + 755 pp.

DEDICATION

This volume is dedicated to the memory of

GEOFFREY B. MONTEITH

January 12, 1942 – October 23, 2024

Geoff's passion for entomology, and for natural history more broadly, has made substantial contributions to the knowledge of Australian invertebrates and beyond. He played a major role in developing exceptional entomological collections at the University of Queensland and the Queensland Museum (QM) in Brisbane, now forming one collection at QM. Geoff has also been pivotal in promoting entomological activities among natural history enthusiasts and community groups, particularly through his long term engagement with organisations like the Entomological Society of Queensland. Geoff's generosity, along with his infectious enthusiasm, charisma, and never-ending fascination with insects and animals, big and small, have indeed sparked the interest of generations of students and amateur naturalists. He has provided colleagues overseas with critical and often much needed support to navigate and familiarise themselves with a country like no other. Domestic scientists have also benefited from his endless knowledge and incomparable memory for the most minute details, whether that was the page where to find certain information in a publication or the rock where to turn to find the exact site you need to visit.

Geoff is indeed one of the most extraordinary field entomologists and collectors and has conducted entomological fieldwork across Australia, Papua New Guinea and New Caledonia, but he has explored his native Queensland more extensively, with only a few corners left unexplored. Geoff holds a special place among Australian entomologists for his selfless generosity, ingenuity, pragmatic approach and his sharp sense of humour. He has devoted his life to documenting Australia's biodiversity, driven by a fascination with its creatures and deep concern for the growing destruction of its natural habitats.

His own research has focussed over the years on different groups, starting from Hemiptera, with a PhD unravelling the systematics of Australian Aradidae. Never losing sight of true bugs, he then moved onto beetles that have been the centrepiece of his scientific production for many years, particularly Australian native dung beetles and Carabidae. The collections and discoveries resulted from his endeavours have represented the backbone of many successful museum exhibitions and continue to serve as a valuable resource for entomologists worldwide studying our rich but still largely unknown fauna.

Federica Turco
ANIC-CSIRO

ACKNOWLEDGEMENTS

We gratefully acknowledge The Australian National Insect Collection (CSIRO) and its extraordinary staff, who, under the guidance of David Yeates and Federica Turco, have provided continuous support, an exceptional work environment and cutting-edge facilities. The Zimmerman Trust is acknowledged for supporting H. Escalona, D. Jennings and R. Oberprieler. The support from the Australian Biological Resources Study (ABRS) and the Department of Agriculture, Fisheries and Forestry have been critical for our research on longhorns the last 15 years.

Our sincere thanks are due to the curators of different entomological collections for their permanent collaboration attending to our specimen requests and visits. In particular, we would like to thank: Geoff Monteith, Christine Lambkin and Karin Koch (Queensland Museum, QM, Brisbane), Richard Leschen (Landcare Research, New Zealand), Nik Tatarnic (Western Australian Museum, WAM, Perth), Ben Parslow (South Australia Museum, SAM, Adelaide), Ken Walker and Simon Hinkley (Museum Victoria, MV, Melbourne), Chris Reid and Derek Smith (Australian Museum, AM, Sydney), Justin Bartlett (Queensland Department of Agriculture and Forestry, QDAF), Chris Lyal, Max Barclay and Michael Geiser (Natural History Museum London, NHML); Lourdes Chamorro (Smithsonian Institution–National Museum of Natural History, NMNH), Harald Schillhammer (Naturhistorisches Museum Wien), as well as Paul M Hutchinson (Perth), Jack Hasenpusch (Innisfail), Allen Sundholm (Sydney) and Roger de Keyzer (Sydney).

We also thank Bruno Bell, Nick Monaghan, Dash Huang, Nick Volpe, Jean and Fred Hort, Nicholas J. Fisher, Jessa Thurman, Bernhard Jacobi, Owen Kelly, Jiří & Marie Lochman (Lochman Transparencies), Reiner Richter, Owen Kelly, Jack Hasenpusch, James Tweed and Nick Porch for their wonderful photographs of beetles in their natural habitat.

The tireless dedication and talented work of Debbie Jennings (ANIC-CSIRO) made the weevil chapters possible; her stunning photographs and SEM images in those chapters are a major highlight in this volume. Cate Lemann and Lingzi Zhou (ANIC) have provided support and assistance in many ways.

H.E. thanks the staff at the Museo de Zoologia Agricola de la UCV (MIZA), particularly to L.J. Joly and Vilma Savini, for the many years of entomological training and encouragement, as well as the extraordinary Carlos Bordon A. (1921–2012) for stirring his teenage entomological enthusiasm and introducing him to the wonderful world of weevils. Our dear friend Richard Leschen ('Chiquito') has been a permanent support for many years.

We have been fortunate to have had John F. Lawrence's mentorship, unwavering support and friendship for several years. This volume is also part of his legacy.

Finally, thanks to our families for their understanding, patience and love.

CONTRIBUTORS TO VOLUME 3

Editors: Hermes E. Escalona and Adam Ślipiński
Australian National Insect Collection
CSIRO
GPO Box 1700
Canberra, ACT 2601, Australia

Lauren G. Ashman
Australian National Insect Collection
CSIRO

James R. M. Bickerstaff
Australian National Insect Collection
CSIRO

Anthony I. Cognato
Department of Entomology
Michigan State University
288 Farm Lane, 243 Natural Science Bldg.
East Lansing, MI 48824, USA

Debbie Jennings
Australian National Insect Collection
CSIRO

Deborah S. Kent
Brierfield, NSW 2454
Australia

Roger de Keyzer
Australian Museum
1 William Street
Sydney NSW 2010, Australia

Rolf G. Oberprieler
Australian National Insect Collection
CSIRO

Sarah M. Smith
Department of Entomology
Michigan State University
288 Farm Lane, 243 Natural Science Bldg.
East Lansing, MI 48824, USA

Marek Wanat
Muzeum Przyrodnicze
Uniwersytet Wroclawski
Ul. Sienkiewicza 21
50-335 Wroclaw, Poland

37. SUPERFAMILY CURCULIONOIDEA LATREILLE, 1802

Rolf G. Oberprieler, Debbie Jennings and Hermes E. Escalona

Introduction. The superfamily Curculionoidea, informally also called Rhynchophora, is among the largest groups of insects, with more than 62 000 described species worldwide (Oberprieler *et al.* 2007) and more than 4200 in Australia (Pullen *et al.* 2014), but this is just a fraction of the total number of species estimated to exist both in the world and in Australia. Beetles in this group are generally called weevils or 'snout beetles', but some derived forms have their own common name, for example the bark beetles (subfamilies Platypodinae and Scolytinae). The morphological signature of Curculionoidea is the projection of the anterior part of the head to form a rostrum, which is regarded as a key evolutionary adaptation that enabled the group to take full advantage of plants as a source of food (Anderson 1993, 1995; Farrell 1998; Oberprieler *et al.* 2007, 2014). This singular characteristic also allowed weevils to be identified as a natural group in the early stages of the classification of the Coleoptera, but other morphological traits associated with the rostrum have also been used in their classification, mainly the structure of the antennae (straight vs geniculate or 'elbowed') and the prementum, forming phanerognathous (Fig. 37.12L) vs adelognathous (Figs 37.10B, 37.12N) mouthparts. The division of the superfamily into natural families and subfamilies has been challenging for a long time but was gradually resolved in the 19th Century mainly by the works of Crowson (1955), Morimoto (1962b), Thompson (1992), Zimmerman (1993, 1994a, b) and Kuschel (1995), especially by the last applying phylogenetic methods to morphological characters and delimiting a smaller number of families that have since been generally corroborated by molecular data as well. These studies have resulted in the development of a more stable classification system that is now widely used in the literature, including in general reviews and catalogues (Alonso-Zarazaga & Lyal 1999, 2002, 2009; Oberprieler *et al.* 2007, 2014; Pullen *et al.* 2014; Alonso-Zarazaga *et al.* 2017, 2023).

Several authors have contributed relatively recently to the current understanding of weevil systematics with comparative morphological work of different structures, in particular the head (Morimoto 1962a; Lyal 1995; Morimoto & Kojima 2003; Davis 2017) and mandibles (Thompson 1992), elytro-tergal stridulation systems (Lyal & King 1996), sclerolepidia (Lyal *et al.* 2006), tibial apices (Thompson 1992; Oberprieler & Zimmerman 2020), hindwings (Zherikhin & Gratshev 1995; Lawrence *et al.* 2022), abdomen and terminalia (Morimoto 1962a; Kuschel 1971; Thompson 1992; Wanat 2007), ovipositor (Howden 1995), internal anatomy (Kasap & Crowson 1977; Calder 1989, 1990) and larvae (May 1993, 1994; Marvaldi 1997).

Modern phylogenetic studies based on morphological and molecular data contributed significantly to the development of a natural classification and the reconstruction of the evolutionary history of the superfamily (e.g. Marvaldi 1997; Marvaldi *et al.* 2002, 2006, 2009; Marvaldi & Morrone 2000; McKenna *et al.* 2009; Gunter *et al.* 2016; Shin *et al.* 2018), mostly confirming phylogenetic hypotheses of earlier workers at the family level but not yet successful at resolving relationships at subfamily and lower levels, largely due to limitations of taxon and data sampling and the challenges of analysing large-scale datasets. In the current classification, the Curculionoidea comprise eight extant families (Cimberididae, Nemonychidae, Anthribidae, Belidae, Attelabidae, Caridae, Brentidae and Curculionidae) and one extinct one (Mesophyletidae), of which all but Cimberididae and Mesophyletidae are represented in Australia.

The fossil record of weevils is long and rich, extending back to the Upper Jurassic, approximately 160 Ma ago, but it is much more extensive in the Northern than the Southern Hemisphere. Only a single authentic weevil fossil has been described from Australia so far, a Jurassic nemonychid named *Talbragarus averyi* (Oberprieler & Oberprieler 2012), but weevil fossils of Miocene age are also known from Cape York amber (Soriano *et al.* 2010) and McGraths Flat in New South Wales (McCurry *et al.* 2022), though remaining undescribed. Important accounts of the fossil record of weevils have been published by Davis *et al.* (2012), Oberprieler *et al.* (2014) and Clarke *et al.* (2019).

The main food source of weevils are plants, and host specialisation is a common phenomenon in the superfamily. Associations with conifers are prevalent in Nemonychidae, Belidae and Caridae, but the host associations of Anthribidae, Attelabidae, Brentidae and Curculionidae are more diverse and their life-styles include mycetophagy (many Anthribidae and Scolytinae), coprophagy (the curculionid genus *Tentegia*) and myrmecophily (some Brentinae). Female weevils typically use their rostrum to drill holes into plant tissue, into which they then lay their eggs using their extendable ovipositor. Weevil larvae feed on virtually all structures of the plants, generally preferring tissues with weakened chemical defences, such as damaged, sick or dying organs, although some develop in healthy, growing parts of stems, leaves, buds, fruits and seeds and also pollen (Nemonychidae). Sophisticated symbiotic associations between weevils and fungi occur mainly in Scolytinae and Platypodinae but also in some other groups, such as Attelabidae, in which they are, however, much more poorly understood. These specialised relationships are often facilitated by the weevils possessing one or more kinds of exoskeletal cavities, which have generally been interpreted as spore-carrying organs (mycetangia, mycangia) in Scolytinae. Unique among weevils is the development of a haplo-diploidy sex determination system, predominant in the Dryocoetini and Xyleborini tribes of Scolytinae.

The Curculionoidea are widespread in Australia and occur in all habitats, including the inhospitable ones on the remote subantarctic Heard and McDonald Islands (in Ectemnorhinini). Their largest diversity occurs in wet tropical forests along the eastern side of the continent, to which some groups, such as the whole family Nemonychidae, are restricted, but weevils are also very numerous and diverse in arid regions including sandy deserts and in the floristically rich southwestern parts of Western Australia.

Because of their close association with particular species or genera of plants, weevils are an important group of beetles economically. Many species are regarded as major or minor pests of agricultural and sylvicultural crops and ornamental plants. On the other hand, several species have been successfully employed as biological control agents of various noxious weeds in Australia. Weevils are also important pollinators of plants, in particular cycads, palms and the more basal lineages of dicotyledonous angiosperms.

The scientific exploration and description of the Australian weevil fauna commenced with the earliest European naturalists at Botany Bay in 1770 and proceeded in exponential fashion through the 19th and early 20th Centuries, before tapering off after the death of the prolific weevil taxonomist Arthur Lea (1868–1932), who described almost half of all weevil species in the country. The discovery and documentation of the Australian weevil fauna and the contributions made to it by various taxonomists over the last two-and-a-half centuries were reviewed by Pullen *et al.* (2014), who also analysed the composition of the fauna and catalogued all described genera (then 832) and species (then 4110, now more than 4200). Large parts of the Curculionoidea were taxonomically treated and illustrated in the monograph series *Australian Weevils* (Zimmerman 1991, 1992, 1993, 1994a, b; Oberprieler & Zimmerman 2020).

Morphology

General characteristics of adults and larvae. The key morphological feature of weevils is the elongation of the anterior part of the head to form a muzzle, or rostrum, of varying length (Fig. 37.9), with concomitant modifications of the mouthparts, such as fusion of the labrum to the frontoclypeus (except in Nemonychidae and Anthribidae, Figs 37.11A, C, D, 37.12E), reduction of the size of the mandibles, fusion of galea and lacinia of the maxillae (except Sitonini) and the short, rigid maxillary and labial palps. The pseudotetramerous tarsi (secondarily reduced in some groups) and a reduced type of cucujiform aedeagus, in which the tegmen forms a ring, in the adult and the loss of a mandibular mola in the larva are features shared with Chrysomeloidea, but the mouthpart modifications of the adult combined with the hypopharyngeal bracon in the larva distinguish weevils from chrysomeloids. Adult weevils usually have a more or less rigid body, with posteriorly closed procoxal cavities, mesocoxal cavities laterally closed by the junction of the meso- and metaventrite (except some Belidae), globular procoxae with concealed trochantins (barely exposed in some Belidae) and connation of two or more ventrites (free in Caridae and Belidae). The antennae are almost always distinctly clubbed and usually have a long scape that may fit into a groove (scrobe) along the sides of the rostrum. The larvae of weevils are almost always legless, but minute legs occur in some Nemonychidae, many Anthribidae and Brentidae and well-developed legs are present in Caridae. Phytophagous habits are almost universal in the superfamily, with notable exceptions in Australia being the mycetophagous habits of most Anthribidae, Platypodinae and Scolytinae and the coprophagous ones of *Tentegia* (Curculionidae).

The terminology of morphological structures in the Curculionoidea chapter is that of Lawrence & Ślipiński (2013) and Oberprieler *et al.* (2014). Adults and larvae of each family are described in detail in each chapter, and here only additional comments on structures that are of general use and broad interest in weevil systematics are provided. The body length given is the standard length (SL), measured from the anterior pronotal margin to the elytral apex in lateral view, as in Fig. 37.10J (Zimmerman 1994a).

Adults. The head capsule (Figs 37.10H, 37.11A–Y) is typically hypognathous, with the frontoclypeal region produced forward forming a short muzzle (Fig. 37.11V) or longer rostrum (Fig. 37.11K) and the antennal insertions and the frons displaced towards the apex of the rostrum (DuPorte 1960). Sulci separating the frons from the epifrons (anteromedian part of the vertex), genae and clypeus are absent, and distinct frons and clypeus sclerites are not discernible. In forms with a shorter rostrum (e.g. Entiminae, Fig. 37.11S) three well-marked areas are usually distinguishable in both dorsal and lateral view (Fig. 37.11S): a distal epistome, a triangular frons extending to between the antennal insertions and a proximal epifrons extending to the anterior margin of the eyes. When a long rostrum is present (Fig. 37.1C), the region in front of the antennal insertions (anatomically the frontoclypeus) is often referred to as the prorostrum (or distirostrum) and the region behind the insertions (anatomically the epifrons) as the metarostrum (or basirostrum). The sockets of the antennae usually open posteriorly into a groove or scrobe along the sides of the rostrum, which may be short, shallow and posteriorly open or long, deep and posteriorly closed and sometimes curving or bending onto the underside of the rostrum (Fig. 37.4E).

The labium (Fig. 37.10C, E, F) has only a single posterior sclerite, the elongate postmentum (Figs 37.4F, 37.12F–R), also termed submentum but its homology unclear (Lyal 1995), but its lateral borders, the subgenal sutures, are usually obliterated (but discernible in many Attelabidae), and often also the outer occipital sutures separating the genae from the postgenae are partly to completely obsolete. The gula is distinct in Nemonychidae and Belidae, with each gular suture ending anteriorly in the posterior tentorial pit, but in the other families the gular sutures when present (e.g. apparently absent in Anthribidae) are usually confluent, forming a single gular suture that may end anteriorly in a single posterior tentorial pit (Fig. 37.12A–D). In the Attelabidae the tentorial pit is usually faint and situated further forward and the gular suture joins the subgenal sutures on the ventral side of the rostrum.

The antennae (Fig. 37.13A–T) are normally 11-segmented but often appear as 12-segmented when the last segment is divided into two parts (Fig. 37.13F; see also Clarke *et al.* 2019), which may be clearly distinct even in compact terminal clubs (Fig. 37.13H, J, M). In many weevils, however, the number of antennal segments is reduced, up to five in Platypodinae. In some Anthribidae the antennae are much longer than the body, especially in males. The first antennal segment, the scape, is usually elongated and the second, the pedicel, small and very similar to the following six segments, which together form the

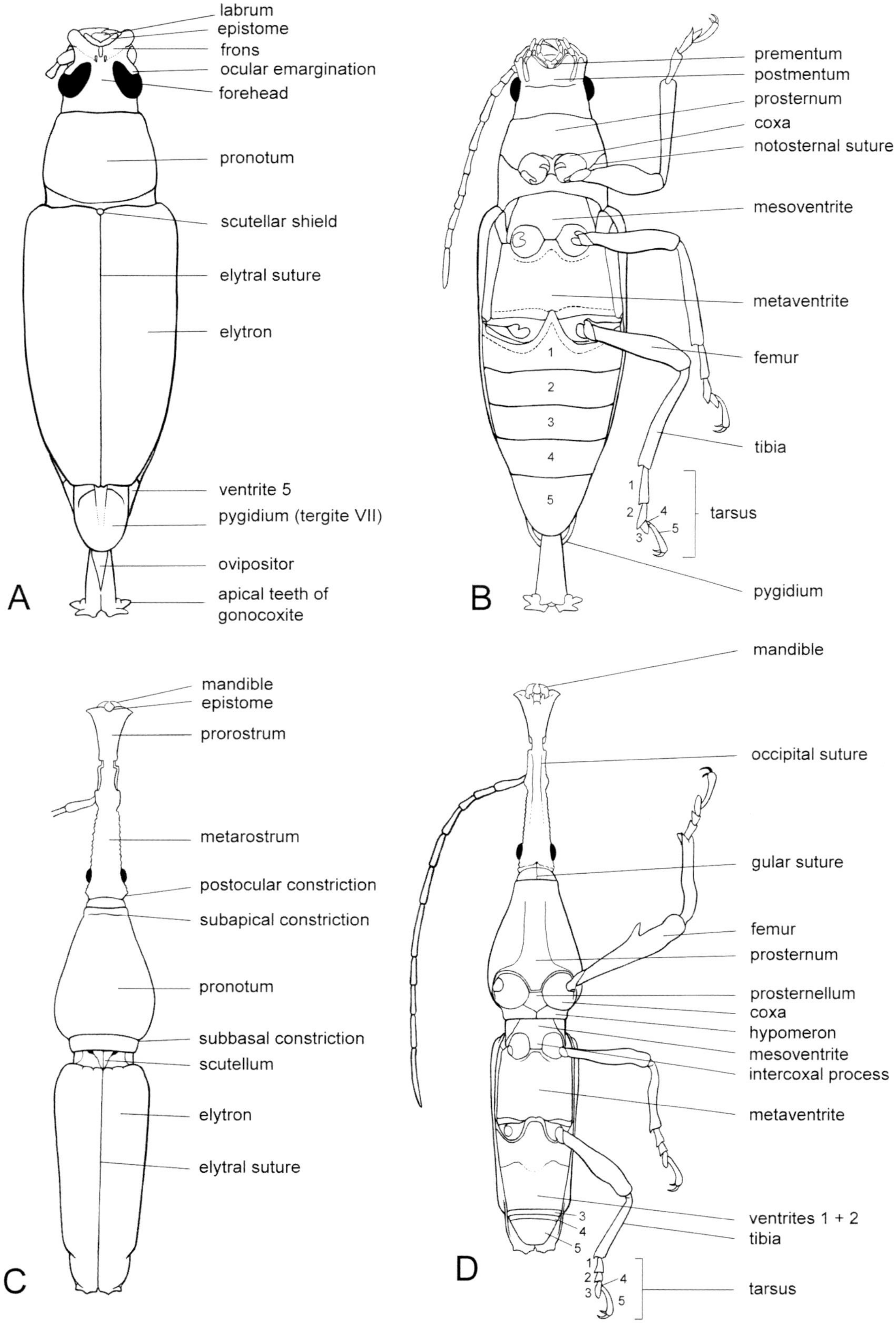

Fig. 37.1. Structural details and terms for Curculionoidea, adult. A–B, *Xenocerus* Schoenherr habitus: A, dorsal; B, ventral; C–D, *Ectocemus* Pascoe habitus: C, dorsal; D, ventral. A–B, after Zimmerman (1994a); C–D, after Zimmerman (1994b).

portion referred to as the funicle, leaving the last three or four segments forming a more or less distinct club, whose segments may be flexible (Fig. 37.13A, C, D, F, H, L, N) or compacted and immovable from each other (Fig. 37.13B, E, G, H, I, J, K, O–T). The scape is often longer and wider than the pedicel and the latter inserted in a subterminal position at the apex of the scape so that the funicle forms an angle with the scape, resulting in antennae referred to as geniculate or 'elbowed'. However, a sharp bend in the scape may result in a superficially similar condition and form a subgeniculate antenna.

The prothorax of weevils (Fig. 37.14A–R) is characterised by the meeting to solid fusion of the posterior ventral projections of the propleura, or hypomera, medially behind the procoxae (Fig. 37.14G, H, Q, R). The prosternum extends into a process between the procoxae when these are separate, and when the procoxal cavities become secondary confluent the posterior part of this process remains as a distinct median sclerite behind the procoxae, referred to as the prosternellum (Fig. 37.14O), although this is highly unlikely to be homologous to a similar structure in other insect orders. In the metathorax, a row of scale-like structures called sclerolepidia (Lyal *et al.* 2006) occurs along the metanepisternal sutures (which separate the metanepisterna from the metaventrite, Fig. 37.15O), but their function and taxonomic significance remain unclear. The elytra generally have distinctive punctures aligned into longitudinal striae, typically ten in number but sometimes fewer or more, and in the families Nemonychidae, Anthribidae, Belidae and Attelabidae there is an additional abbreviated inner stria, at the base next to the median elytral suture, termed the scutellary striole. In some species, however, the punctures are irregularly scattered over the elytra, and they may also be reduced to obsolescent or absent.

The tibiae often carry a pair of socketed spurs at the inner angle of the apex, located between the fringing setae and the tarsal socket and directed distad. Their number can vary between the pro-, meso- and metatibiae, yielding a spur formula (e.g. 1-1-2) that can characterise genera or species. In the more derived taxa, however, these spurs are mostly completely absent. Between and above the spurs the ring of fringing setae may be interrupted by a short, fixed spine (the mucro) directed more or less perpendicular to the tibial axis, but in some groups this spine has become dislocated to the outer edge of the apex and modified into a stronger, sharper uncus directed at varying angles to the tibial axis (see Thompson 1992), sometimes differently between males and females (e.g. in *Melanterius*). Additional shorter named spines or projections occur in some species (Fig. 37.10I). In many terricolous weevils the meta- and sometimes the mesotibiae also possess a flat, elliptical apical surface that seemingly aids the weevil in pushing off the ground. When this plane is formed by an inward flexion of the outer edge of the tibial apex, it usually carries a complete ring of setae and is covered with setae or scales and termed a (true) corbel, but when it is formed between the fringing setae and the tarsal socket, it carries no setae along its inner edge and its surface is bare and it is termed a false corbel (Oberprieler & Zimmerman 2020). True corbels are only known to occur in the curculionid subfamilies Brachycerinae and Entiminae, but false corbels occur sporadically in other subfamilies too. In the tarsus, the fourth segment (tarsite) is strongly reduced and concealed in the cleft of the bilobed third segment, yielding a seemingly 4-segmented condition termed pseudotetramerous (Fig. 37.17L), which also occurs in the superfamily Chrysomeloidea and is one of the characters identifying Curculionoidea and Chrysomeloidea as sister-groups. The terminal (fifth) tarsal segment (termed onychium) apically typically carries a pair of stout claws, which can be simple or appendiculate (dentate or bifid) in shape and in direction vary from parallel to divergent to divaricate (spread out into a plane perpendicular to the tarsal axis) (Fig. 37.17P–V). In several groups of Curculionidae, the tarsal claws are reduced, sometimes unequally so, and may be fused into a single claw or absent altogether, and also the onychium may be reduced to completely lost (Fig. 37.17M).

The hindwings of Curculionoidea generally possess a rather complete venation, which was studied in detail by Zherikhin & Gratshev (1995), but the terminology of the veins used in this work differs from the one currently in use for Coleoptera as a whole (Lawrence *et al.* 2022). The wing venation of representative families is shown in Fig. 37.18A–I.

The weevil aedeagus (Fig. 37.17A–H) is a reduced version of the cucujoid (ring) type, consisting of a more or less cylindrical penis, a ring-like tegmen with posterior parameral plate or separate parameres on the dorsal side and a median apodeme (the manubrium) extending anteriad on the ventral side, and an inflatable endophallus that everts during copulation and lodges in the bursa copulatrix of the female. In the plesiomorphic, pedotectal type of weevil aedeagus, the penis comprises a ventral portion (the pedon) and a dorsal one (the tectum), the latter extending over the dorsal opening of the penis (the ostium) (Fig. 37.2A), but the majority of weevils have the pedal type of aedeagus, in which the tectum of the penis is absent and the ostium exposed on the dorsal membrane of the pedon (Fig. 37.3G). At the anterior end the penis is extended into a pair of apodemes (the temones), which in the pedotectal type are each connected to the tectum by a dorsal arm and to the pedon by a ventral arm (Fig. 37.3A) but in the pedal type only by a single arm connected to the pedon (Fig. 37.3G). The tegmen also differs between the pedotectal and pedal types of aedeagus, in the former being hinged laterally and dorsally comprising a large basal plate with short, broad, setose parameral lobes (Fig. 37.3C) but in the latter being entire laterally and dorsally consisting of two narrow, separate parameres (Fig. 37.3F). The dorsal parts of the tegmen may, however, be reduced to varying degrees to being completely absent. The pedotectal aedeagus occurs in the families Nemonychidae, Anthribidae, Belidae, Attelabidae, Caridae, Brentidae and in the basal subfamilies of Curculionidae (e.g. Brachycerinae), whereas the derived pedal aedeagus is characteristic of all 'higher' Curculionidae. The membranous endophallus, which everts through the ostium of the penis, usually extends anteriad out of the penis body and in its lumen carries various internal sclerites that are extruded during copulation and anchor in the bursa copulatrix of the female. Such endophallic sclerites plesiomorphically consist of a long, single or double hollow flagellum, which enters the spermathecal duct of the female during copulation, but in most 'higher' weevils the endophallus instead usually has a complex, paired or asymmetrical copulatory sclerite as well as fields of spines or fine asperities that anchor the everted endophallus in the bursa copulatrix of the female during copulation. The sternites of segments VIII and IX also differ between the pedotectal and pedal types of aedeagus; in the former sternite VIII is more or less complete with a short basal median apodeme (the spiculum relictum) and sternite IX has a large basal plate, whereas in the pedal type sternite VIII is generally divided into two crescentic hemisternites without any apodeme and the basal plate of sternite IX is usually narrowly bifurcate.

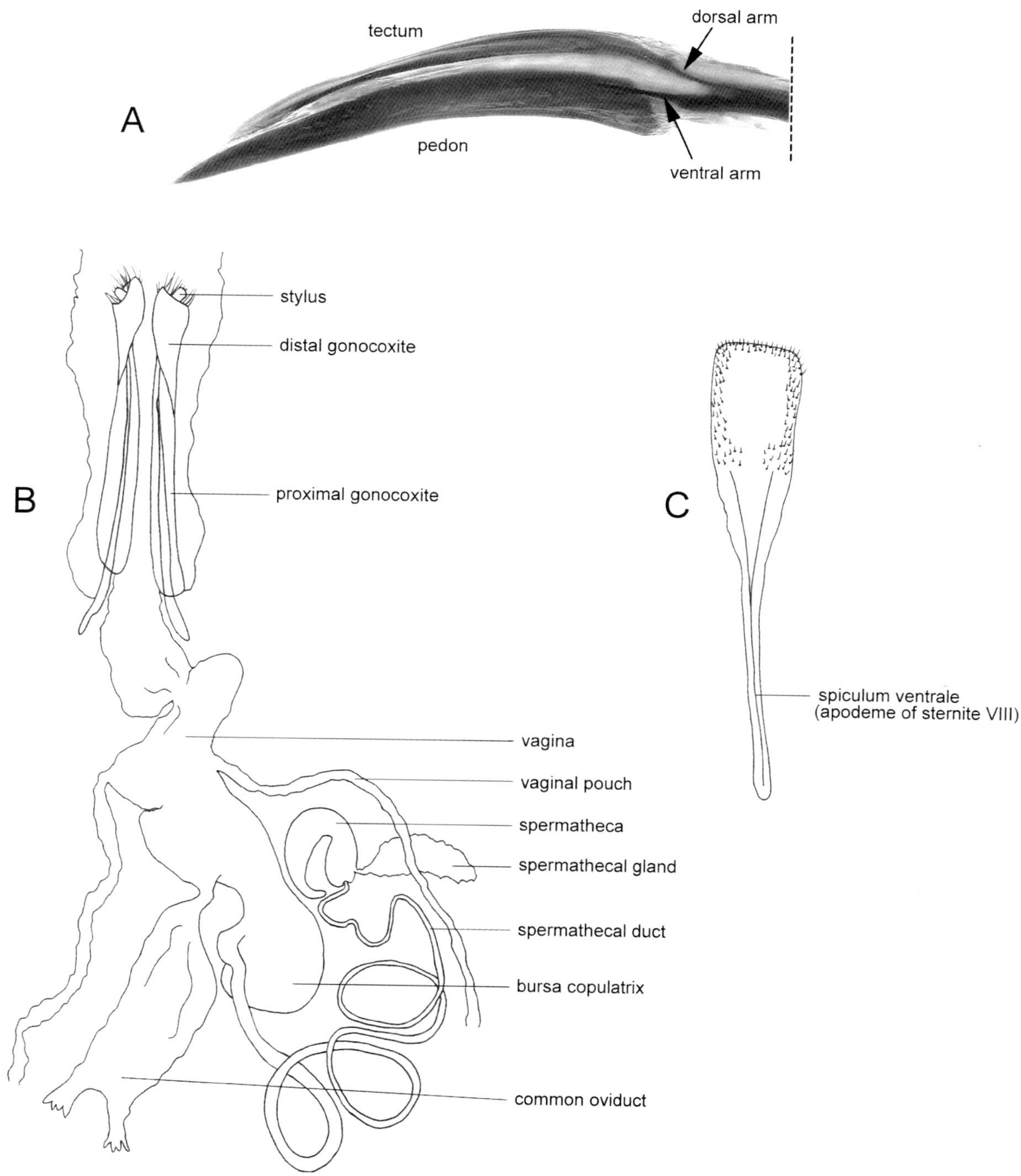

Fig. 37.2. Structural details and terms of Curculionoidea genitalia. A, *Car* sp., pedotectal aedeagus, lateral; B–C, *Ectocemus decemmaculatus* (Montrouzier): B, female genital tract, dorsal; C, sternite VIII, dorsal. B–C, after Zimmerman (1994b).

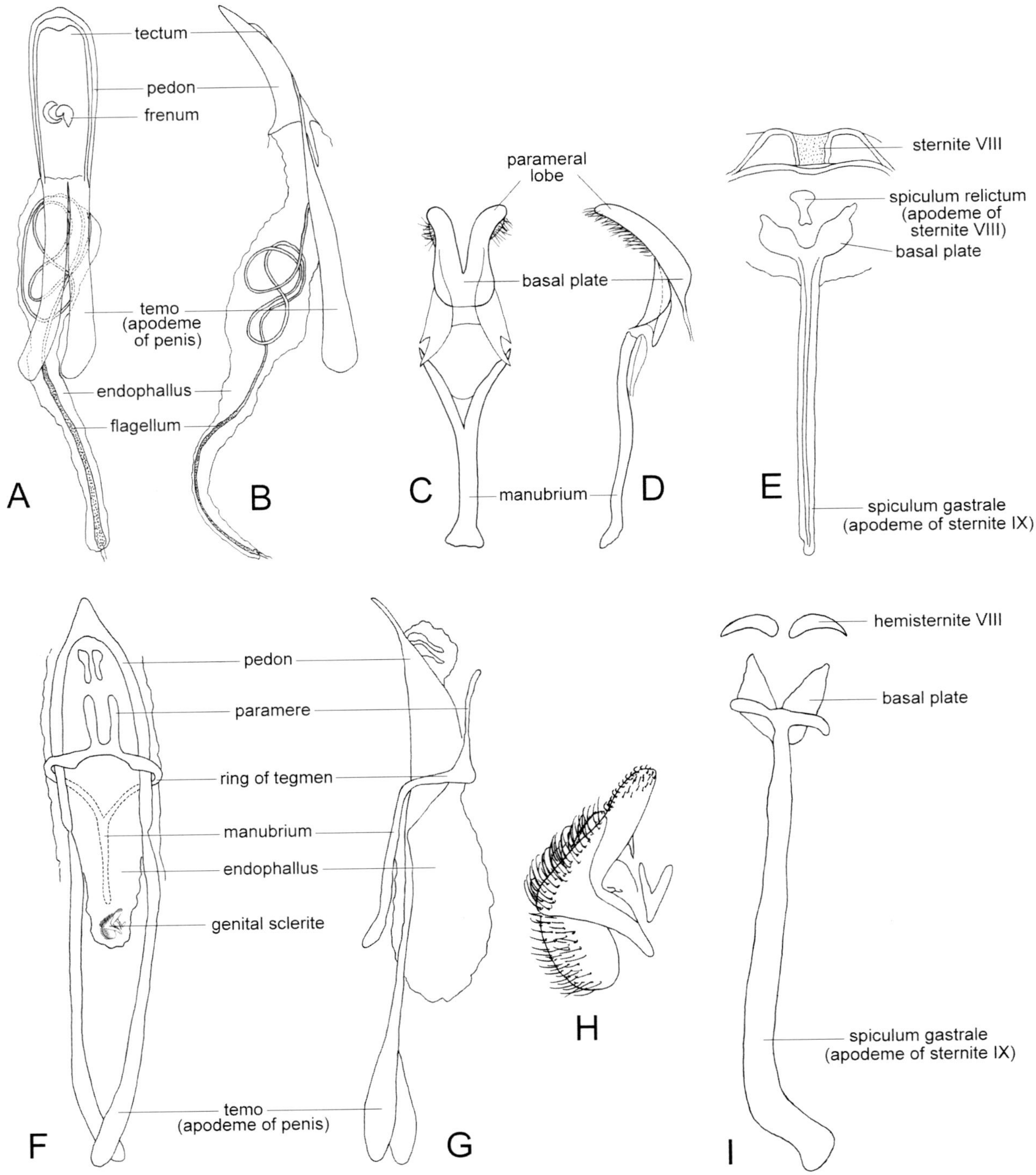

Fig. 37.3. Structural details and terms of male genitalia. A–E, *Ectocemus* Pascoe, pedotectal aedeagus: A, aedeagus, dorsal; B, aedeagus lateral; C, tegmen, dorsal; D, tegmen lateral; E, sternites VIII, dorsal; F–I, *Phlyctinus* Schoenherr, pedal aedeagus: F, aedeagus, dorsal; G, aedeagus, lateral; H, genital sclerite, dorsal; I, sternites VIII and IX, dorsal. A–E, after Zimmerman (1994b); F–I, after Oberprieler *et al.* (2014).

The external female genitalia of weevils are relatively simple, usually consisting of a pair of sclerotised gonocoxites divided into a distal and a proximal part, with the former each carrying an apical or subapical, setose stylus (Figs 37.2B, 37.17I). In some taxa, however (e.g. Anthribidae), also the more proximal paraprocts are discernible, often with stiffened rods (baculi) in their walls, and the apex of the distal gonocoxites is mostly strongly sclerotised and toothed. The internal female genital

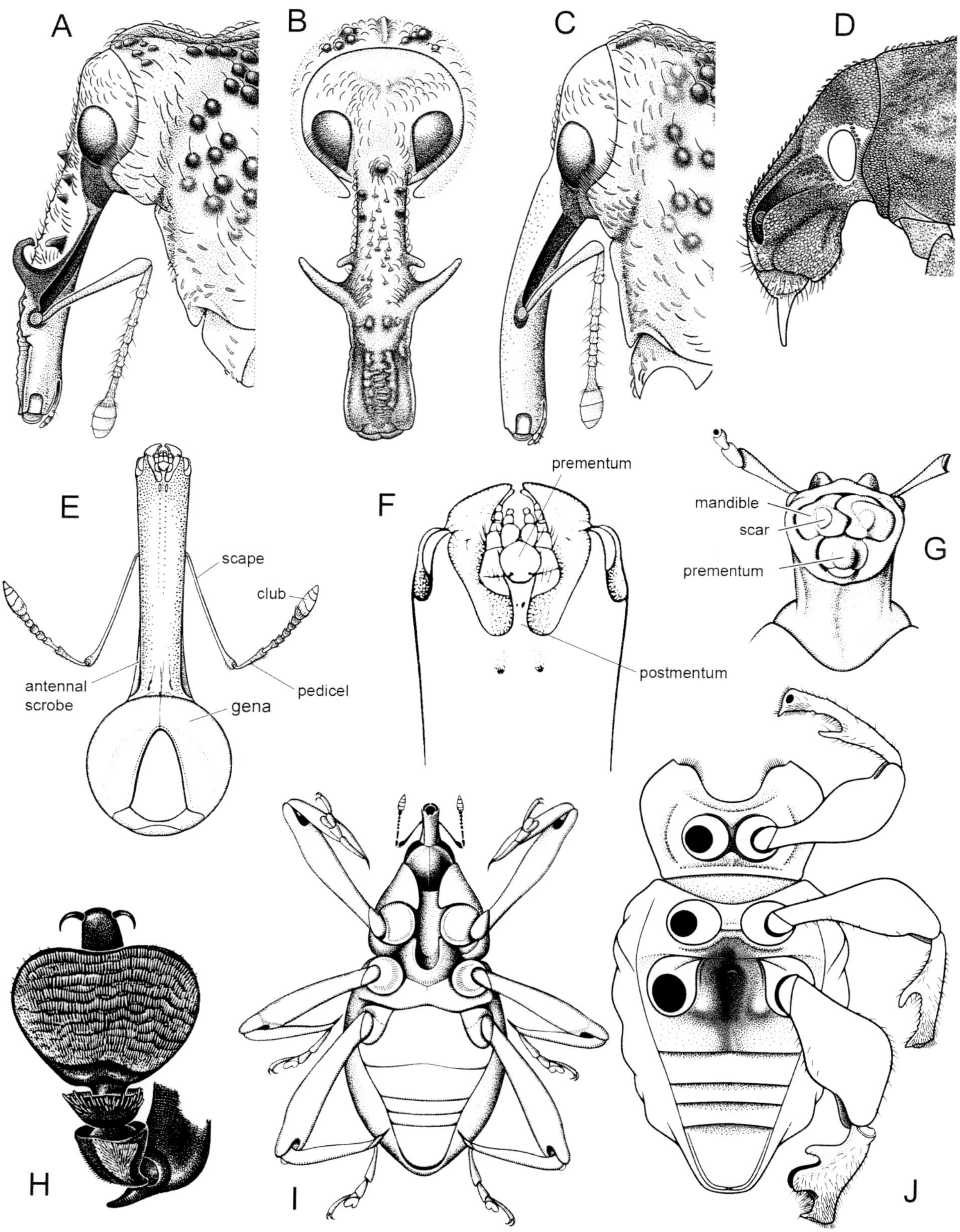

Fig. 37.4. General structural details of weevils. A–C, *Glochinorhinus doubledayi* Waterhouse, head and rostrum: A, ♂, lateral; B, ♂, dorsal; C, ♀, lateral; D, *Onesorus armipennis* Lea, head and rostrum, lateral; E–F, *Orthorhinus cylindrirostris* (Fabricius): E, head, ventral; F, rostral apex, ventral; G, *Leptopius* sp., rostral apex, ventral; H, *Diathetes morio* Pascoe, tarsus, ventral; I, *Paleticus subereus* (Pascoe), habitus, ventral; J, *Mandalotus variabilis* Lea, thorax and abdomen, ventral. E–I, after Lawrence & Ślipiński (2013).

track comprises a membranous vagina and bursa copulatrix, the latter sometimes with one or more simple sclerotised rods in its walls and usually divided into several, often complex lobes, as well as a sclerotised spermatheca, which is usually crescentic in shape but can be long and thin and variously twisted. The duct of the spermatheca can be short to very long and coiled, and it enters the bursa at its apex or near its junction with the common oviduct, and the spermathecal

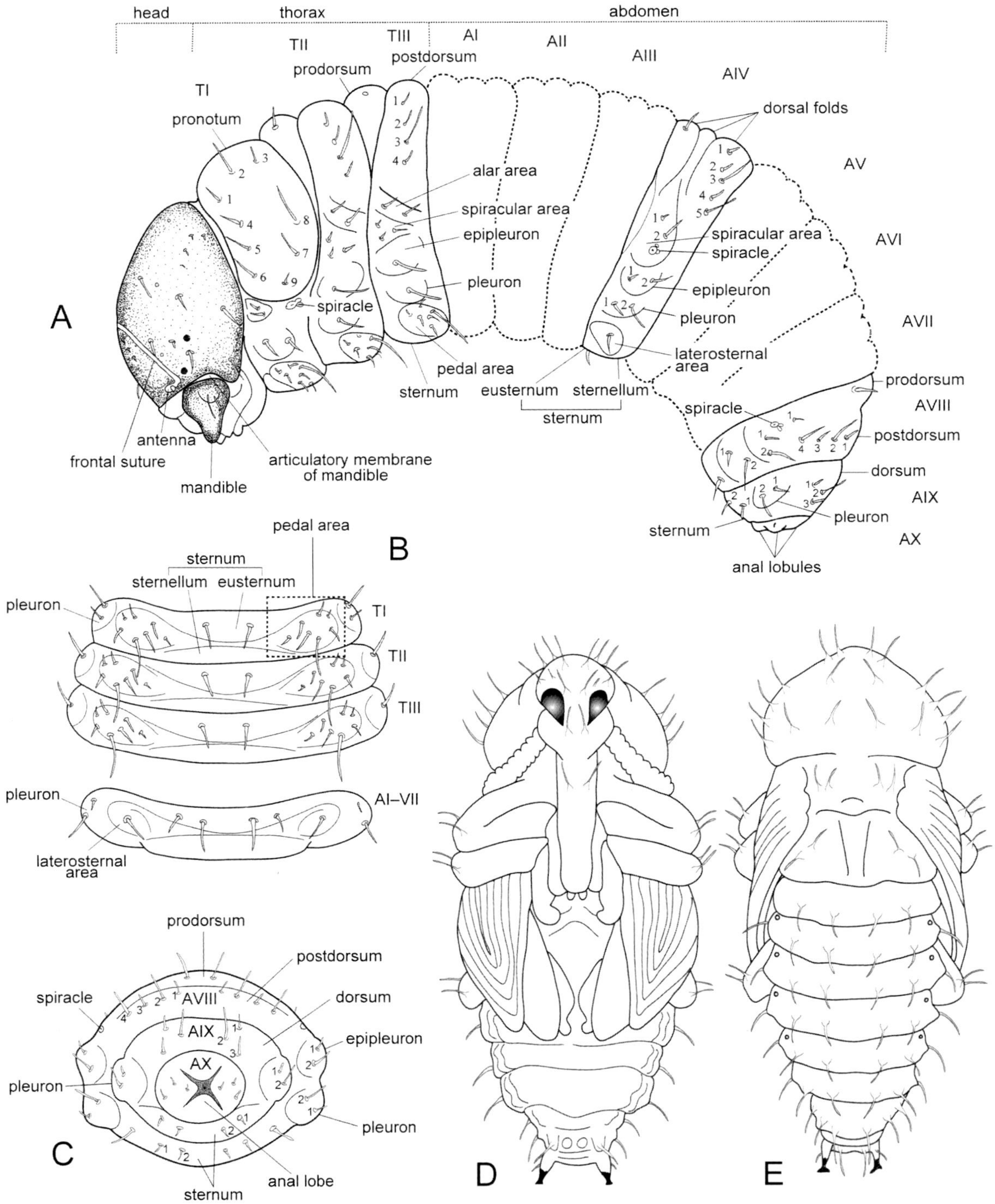

Fig. 37.5. Structural details of generalised Curculionoidea larva and pupa. A–C, larva: A, habitus, lateral; B, thorax and abdomen, ventral; C, caudal; D–E, pupa: D, habitus, dorsal; E, ventral. A–C, after Marvaldi (1999); D–E, after May (1994).

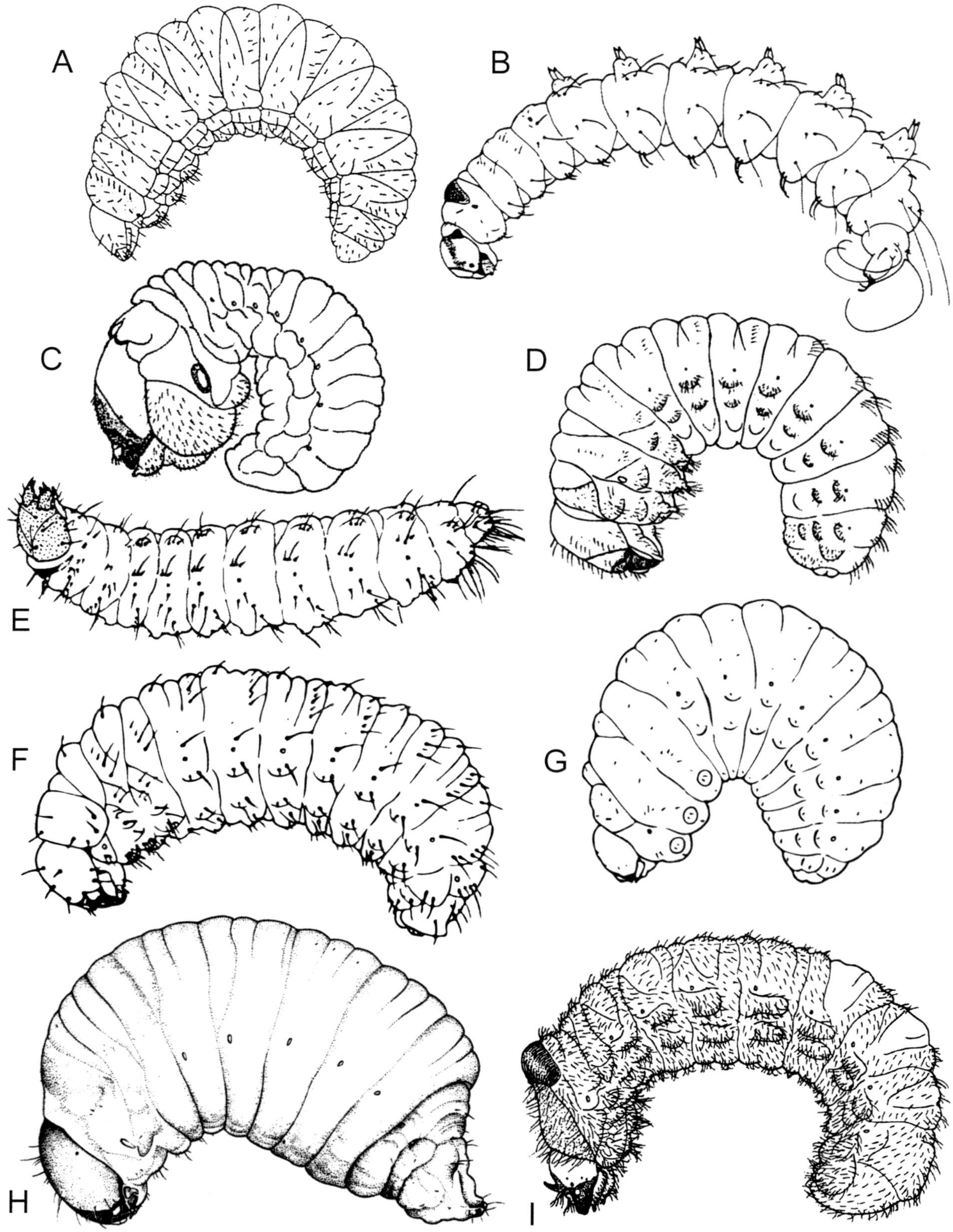

Fig. 37.6. Curculionoidea larvae, lateral habitus. A, *Auletobius uniformis* (Roelofs); B, *Cyrtobagous singularis* Hustache; C, *Phloeobius gigas* (Fabricius); D, *Cacephatus huttoni* (Sharp); E, *Notomacer araucariae* Kuschel; F, *Mandalotus miricollis* (Broun); G, *Exapion ulicis* (Forster); H, *Trigonotarsus calandroides* Guérin-Méneville; I, *Rhinotia bidentata* (Donovan). A–G, I after May (1994).

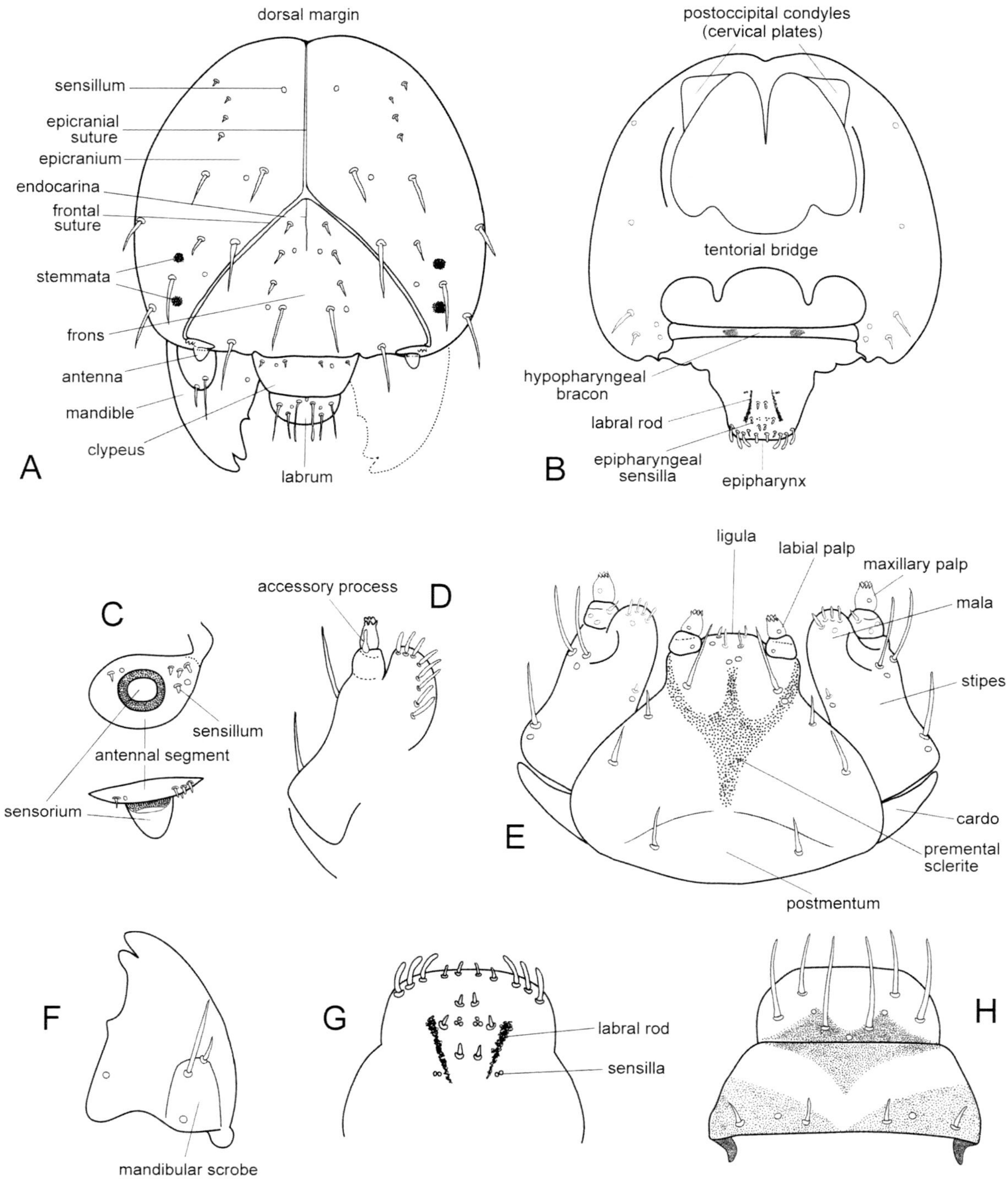

Fig. 37.7. Structural details of a generalised hypognathous head of Curculionoidea larva. A–B, head capsule: A, anterior; B, posterior; C, antenna, ventral and anterior; D, maxilla, anterior; E, maxillae and labium, posterior; F, mandible, anterior; G, epipharynx, posterior; H, clypeus and labrum, anterior. After Marvaldi (1999).

gland varies from round and bulbous to more elongate and cylindrical.

Larvae. The body (Figs 37.5, 37.6) is soft, generally slightly curved but varying from stout and crescentic to elongate and almost straight, the abdominal segments are unsclerotised and dorsally divided into transverse folds and the abdominal apex lacks urogomphi. The head is well sclerotised and the mandibles are generally strong. The legs are absent or vestigial, and locomotion is achieved by crawling, often with the aid of dorsal, ventral or lateral ampullae or weals, or by

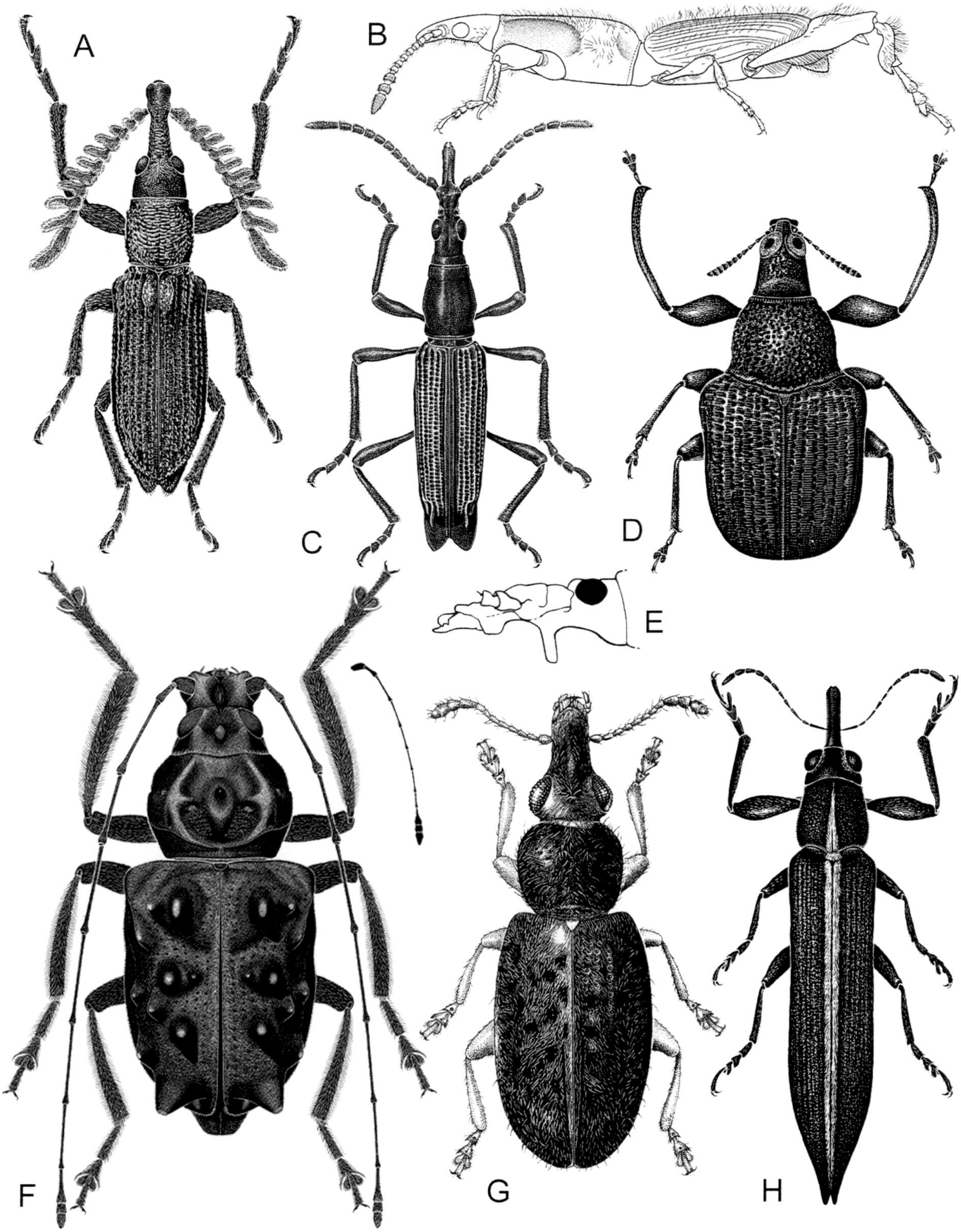

Fig. 37.8. Orthocerous weevils (with non-geniculate antennae). A, *Ctenaphides porcellus* Pascoe (Brentidae: Eurhynchinae); B, *Catagogus diorymerus* (Lea) (Brentidae: Brentinae); C, *Cordus acutipennis* Power (Brentidae: Brentinae); D, *Euops falcatus* (Guérin-Méneville) (Attelabidae: Attelabinae); E, *Ankleineella subscripta* (Damoiseau) (Brentidae: Brentinae); F, *Commista baccula* Jordan (Anthribidae: Anthribinae); G, *Bunyaeus monteithi* Kuschel (Nemonychidae: Rhinorhynchinae); H, *Rhinotia suturalis* (Macleay) (Belidae: Belinae).

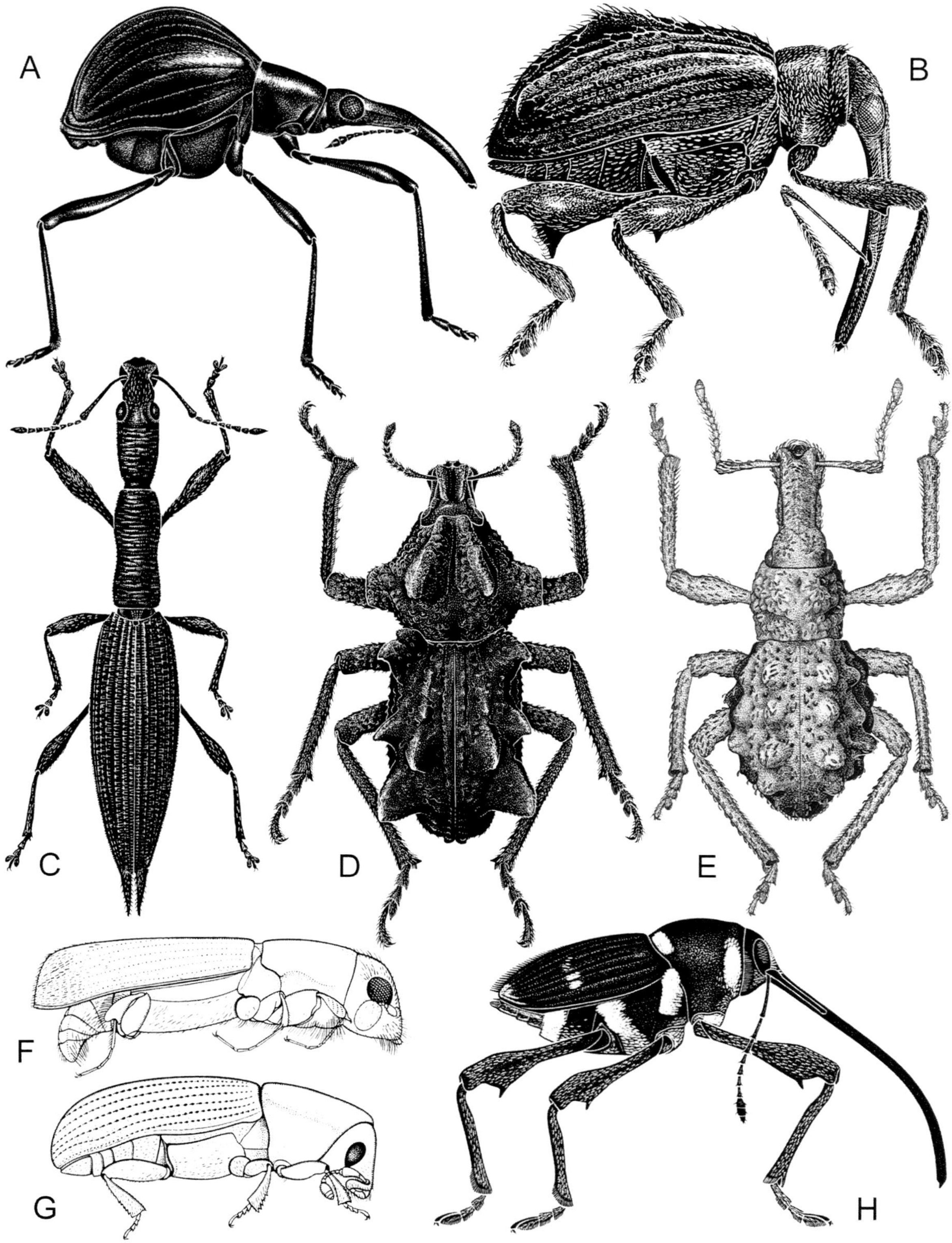

Fig. 37.9. Orthocerous and gonatocerous weevils (with non-geniculate and geniculate antennae). A, *Pseudopiezotrachelus convexipennis* (Lea) (Brentidae: Apioninae); B, *Meriphus tuberculatus* Lea (Curculionidae: Curculioninae); C, *Rhadinosomus lacordairei* Pascoe (Curculionidae: Cyclominae); D, *Tetralophus excursus* Pascoe (Curculionidae: Cyclominae); E, *Amisallus phymatodis* (Lea) (Curculionidae: Entiminae); F, *Platypus subgranosus* Schedl (Curculionidae: Platypodinae); G, *Hylastes ater* (Paykull) (Curculionidae: Scolytinae); H, *Curculio simulator* Zimmerman & Oberprieler (Curculionidae: Curculioninae).

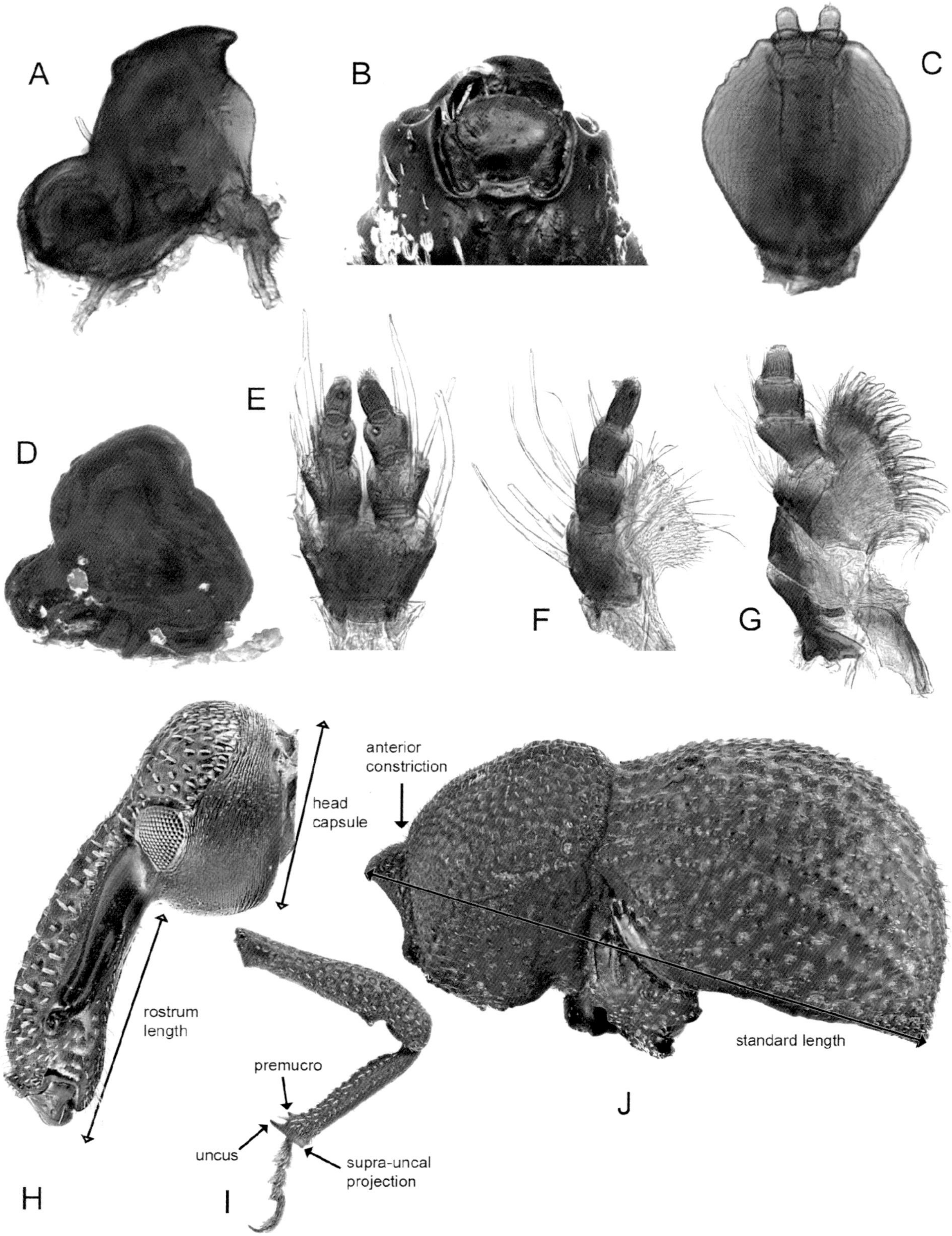

Fig. 37.10. Mouthparts and measurement details. A–C, *Undarobius howarthi* Escalona & Oberprieler (Curculionidae: Entiminae): A, mandible, dorsal; B, mouthparts, ventral; C, labium, ventral; D–G, *Tentegia amplipennis* Lea (Curculionidae: Molytinae): D, mandible, dorsal; E, prementum, ventral; F, prementum, lateral; G, maxilla, ventral; H–J, *Tentegia* spp.: H, head; I, leg; J, thorax and elytra, lateral.

peristaltic contractions of the body (typically in Curculionidae: Dryophthorinae). The different body parts and main structures and their chaetotaxy (setae and sensilla in modal numbers) are illustrated in Fig. 37.5.

The head of weevil larvae is typically hypognathous and fully exposed (Fig. 37.5A), though sometimes partially retracted into the prothorax (deeply and permanently so in Belidae). The epicranial and frontal sutures are usually distinct and form an inverted Y (Fig. 37.7A), but the frontal sutures are sometimes effaced or absent. Larvae of Curculionidae are characterised by the frontal sutures being 'incomplete', or separated from the articulatory membrane of the mandibles by a fronto-epicranial bracon, whereas in other families, when distinct, they are 'complete', or extending to the mandibular membrane. The median endocarina can be present, vestigial or absent. The antennae (Fig. 37.7C) are reduced and usually 1-segmented, consisting of a basal membranous segment bearing a conspicuous apical sensorium and several smaller sensilla, but 2-segmented antennae occur in Belinae (Belidae) and Rhynchitinae (Attelabidae). The stemmata vary from two pairs to none in Curculionidae and most Brentidae, but up to six pairs can be present in larvae of other families. The clypeus is separated from the frons by a distinct suture (Fig. 37.7A), which is effaced in Nemonychidae and many Anthribidae. The labrum (Fig. 37.7H) usually bears three pairs of setae in Curculionidae but four pairs in other families. Posteroventrally the head capsule has a hypopharyngeal bracon (Fig. 37.7B). A hypopharyngeal sclerome (consisting of a single sclerite) occurs only in larvae of Nemonychidae and Anthribidae. In the mandibles (Fig. 37.7F) the mola is usually absent, except in Nemonychidae and Anthribidae. Labral tormae (lateral sclerites extending from the posterolateral portions of the labrum) occur in Nemonychidae, Anthribidae and Belidae, whereas the epipharynx in the other families has a pair of labral rods (or epipharyngeal rods), which are paramedian, usually subparallel sclerites not attached to the posterolateral portions of the labrum (Fig. 37.7G) (Lawrence & Ślipiński 2013). The maxillae have an undivided cardo, and the lacinia and galea are usually fused into one lobe, the mala (Fig. 37.7D), but larvae of Nemonychidae and Anthribidae can have a lacinial lobe or spine. The labium (Fig. 37.7E) consists of a prementum and a postmentum, the former usually featuring a median premental sclerite that is trident-shaped in most Curculionidae. Also characteristic of weevil larvae is the absence of a gular region, the labium being contiguous with the cervical membrane (Lawrence *et al.* 2011).

The prothorax bears a single dorsal plate or pronotum, whereas the other two segments are dorsally divided into pro- and postdorsal folds (Fig. 37.5A). Legs are absent in larvae of Curculionidae, Attelabidae and Belidae, but they are usually present, although often vestigial or with faint segmentation, in those of the remaining families, and a tarsal claw occurs in *Car* (Caridae). The single pair of thoracic spiracles is located on the mesothorax or in the membrane between meso- and prothorax, but in Curculionidae it is placed on the prothorax.

The abdomen comprises ten segments, the first seven or eight dorsally divided into transverse folds (Fig. 37.5A). In all families other than Curculionidae, the typical abdominal segments (usually AII to AVII) have two such dorsal folds, but in Curculionidae there are three or four. Segments I–VIII each carry a pair of annular lateral spiracles with usually two accessory chambers or airtubes, the bicameral condition. The airtubes are often more strongly developed in the earlier larval instars and can be vestigial or even absent in mature larvae. Segment X consists of the anal lobes (variable in number, usually four) surrounding the anal opening.

Pupae. Weevil pupae (Fig. 37.5D, E) are exarate. The integument is usually creamy white. Setae are usually placed on tubercles on the rostrum, head, pronotum, femoral apex and dorsum of the abdominal segments. The head is deflexed beneath the prothorax and the rostrum onto the legs. The rostral apex bears the mandibular thecae (sheaths), and the wings are represented by pterothecae. The abdomen comprises nine segments. Functional spiracles are usually present on the first four or five, whereas those on segments VI–VIII are often reduced and non-functional. The terminal abdominal segment usually carries a pair of posterior processes ('pupal urogomphi'), which are often sclerotised and furnished with an apical spine (May 1994).

Key to the Australian families of Curculionoidea (adults)

1. Labrum free, articulated (Figs 37.11A, C, D, 37.12E), separated from frontoclypeus by a membrane; antennae straight, non-geniculate (Fig. 37.13A, C) 2

– Labrum fused to frontoclypeus (Fig. 37.11B, F, J); antennae geniculate (Fig. 37.13B, G, I) or not (Fig. 37.13L) 3

2(1). Prothorax without lateral carinae; scutellum with stridulatory files (Fig. 37.15D); abdominal ventrites free; tibial apex with one or two spurs .... **NEMONYCHIDAE** (Fig. 37.8G)

– Prothorax almost always with lateral bracteate carinae, these sometimes very short or indistinct; scutellum without stridulatory files; abdominal ventrites 1 to 4 or 1 to 5 connate; tibial spurs absent .... **ANTHRIBIDAE** (Fig. 37.8F)

3(1). Antennae straight, non-geniculate (Figs 37.1D, 37.13D, F, L) 4

– Antennae geniculate ('elbowed') (Figs 37.4E, 37.13B, G, I, K, M) 7

4(3). Gular sutures short, separate; abdominal ventrites free .... **BELIDAE**

– Gular sutures fused or confluent (Fig. 37.12A–C); abdominal ventrites free or connate 5

5(4). Antennal insertions ventral (*Car*, Fig. 37.12A) or lateral (*Carodes*); abdominal ventrites free (Fig. 37.16D, N) .... **CARIDAE**

– Antennal insertions variable; abdomen with two or more connate ventrites (Fig. 37.16A, B, C, E–J) 6

6(5). Mandibles exodont (Fig. 37.12R); antennal insertions dorsal, exposed in dorsal view (Fig. 37.11G) .... **ATTELABIDAE**

– Mandibles not exodont (Fig. 37.12C); antennal insertions lateral or ventral, not exposed in dorsal view .... **BRENTIDAE: Brentinae, Eurhynchinae**

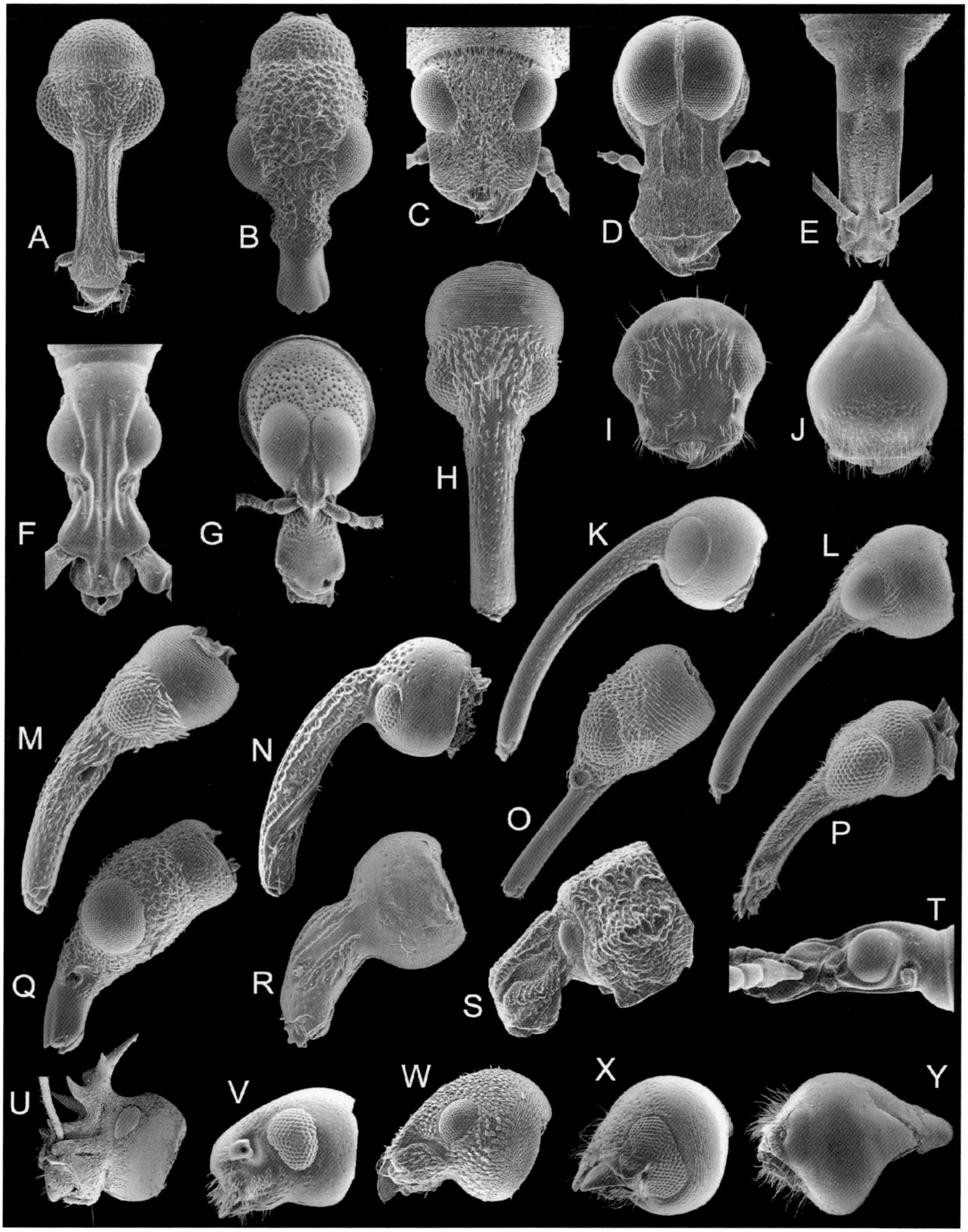

Fig. 37.11. Structural details of weevil heads, dorsal and lateral. A, *Aragomacer uniformis* Kuschel, ♀; B, *Sphinctobelus cinereus* (Blanchard), ♂; C, *Basitropis peregrina* Pascoe; D, *Litocerus variegatus* Jordan; E, *Matesia maculata* Lea; F, *Cordus hospes* Schoenherr; G, *Euops falcatus* (Guérin-Méneville); H, *Apiotherium* sp.; I, *Dinoplatypus pseudocupulatus* (Schedl), ♀; J, *Cnestus solidus* (Eichhoff), ♀; K, *Myctides imberbis* Lea, ♀; L, *Car* sp., ♀; M, *Apiotherium carpophagum* (Lea), ♂; N, *Achelocis rudis* Lea, ♂; O, *Rhinorhynchidius* sp., ♂; P, *A. uniformis* Kuschel, ♀; Q, *S. cinereus* (Blanchard), ♂; R, *Notiomimetes pascoei* Wollaston, ♂; S, *Leptopius echidna* (W. S. Macleay); T, *Cordus hospes* Schoenherr; U, *Acantholophus niveovittatus* Blackburn; V, *Notoplatypus elongatus* Lea; W, *Peliocis subcylindricus* Lea; X, *Beaverium insulindicus* (Eggers), ♀; Y, *C. solidus* (Eichhoff), ♀.

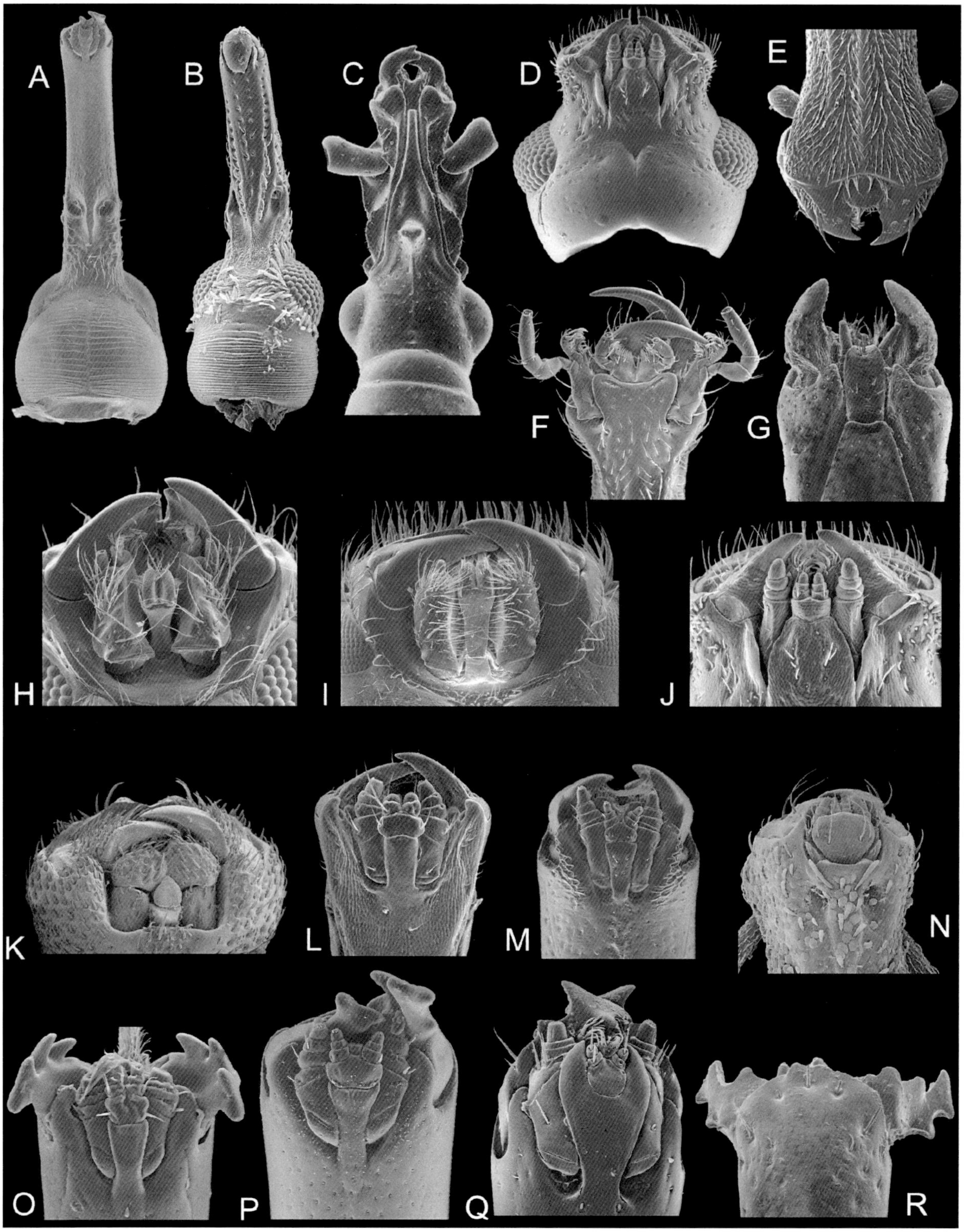

Fig. 37.12. Structural details of weevil heads and mouthparts, ventral. A, *Car* sp., ♀; B, *Apiotherium carpophagum* (Lea), ♂; C, *Kleineella barbata* (Kleine), ♂; D, *Notoplatypus elongatus* Lea; E, *Telala* sp.; F, *Aragomacer uniformis* Kuschel, ♀; G, *Cordus occidentalis* Damoiseau; H, *Beaverium insulindicus* (Eggers), ♀; I, *Cnestus solidus* (Eichhoff), ♀; J, *N. elongatus* Lea; K, *Euomus stephensii* (Gyllenhal); L, *Decilaus* sp., ♂; M, *Rhinotia bimaculata* (Pascoe), ♀; N, *Merimnetes* sp.; O, *Car pini* Lea, ♀; P, *Car* sp., ♂; Q, *Metopum* sp.; R, *Metopum filirostre* (Pascoe).

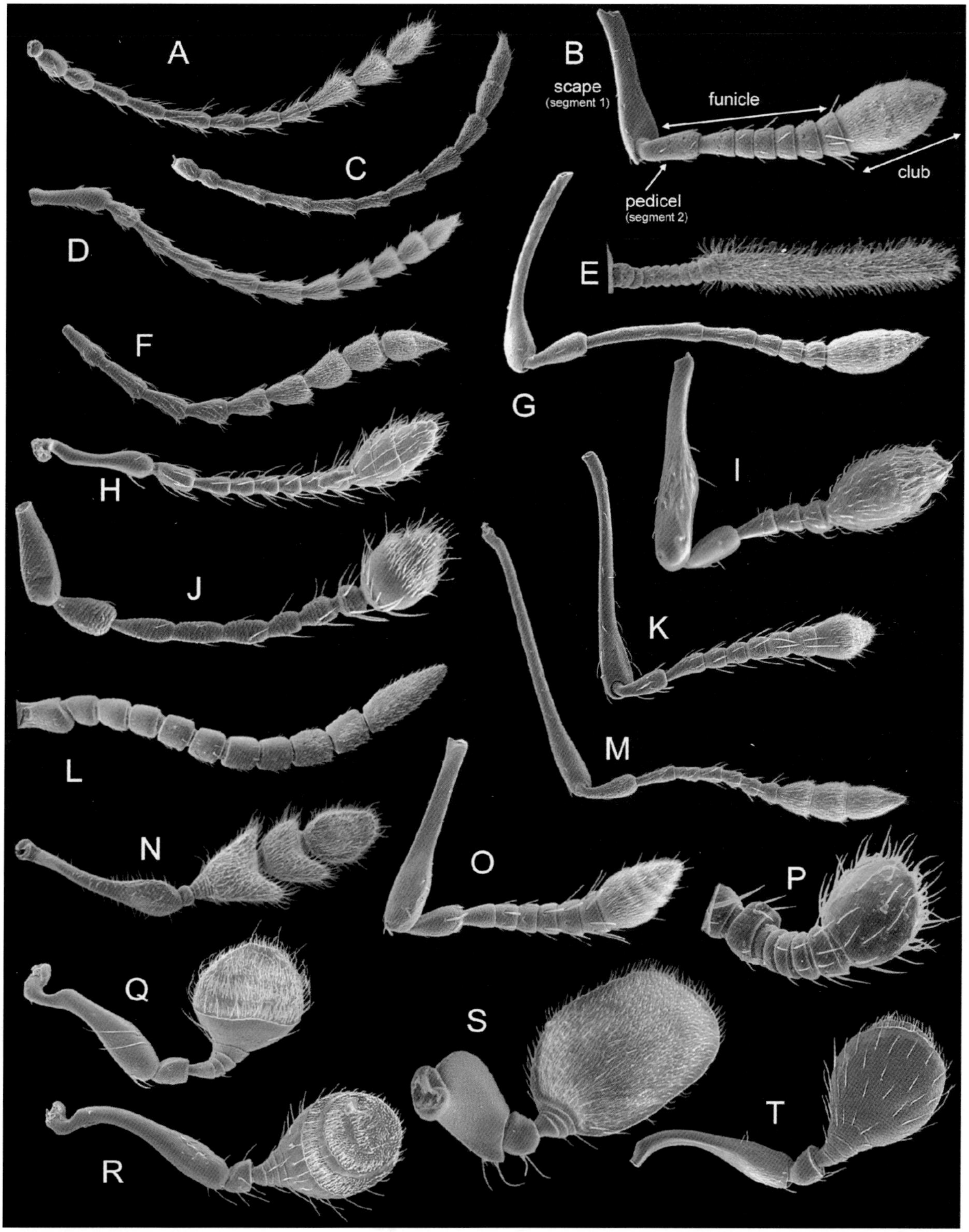

Fig. 37.13. Structural details of weevil antennae, dorsal. A, *Aragomacer uniformis* Kuschel, ♀; B, *Baris leucospila* Pascoe, ♂; C, *Sphinctobelus cinereus* (Blanchard), ♀; D, *Car* sp., ♀; E, *Cylas formicarius* (Fabricius), ♂; F, *Rhinotia bimaculata* (Pascoe), ♂; G, *Agametis bifasciata* Lea; H, *Apiotherium carpophagum* (Lea), ♂; I, *Notiomimetes pascoei* Wollaston, ♂; J, *Hypurus bertrandi* Perris; K, *Achelocis rudis* Lea, ♂; L, *Cordus hospes* Schoenherr; M, *Meriphus longirostris* Pascoe, ♀; N, *Aricerus eichhoffi* Blandford; O, *Myctides imberbis* Lea, ♂; P, *Coptocorynus* sp.; Q, *Beaverium insulindicus* (Eggers), ♀; R, *Dinoplatypus pseudocupulatus* (Schedl); S–T, *Cnestus solidus* (Eichhoff), ♀: S, ventral; T, dorsal.

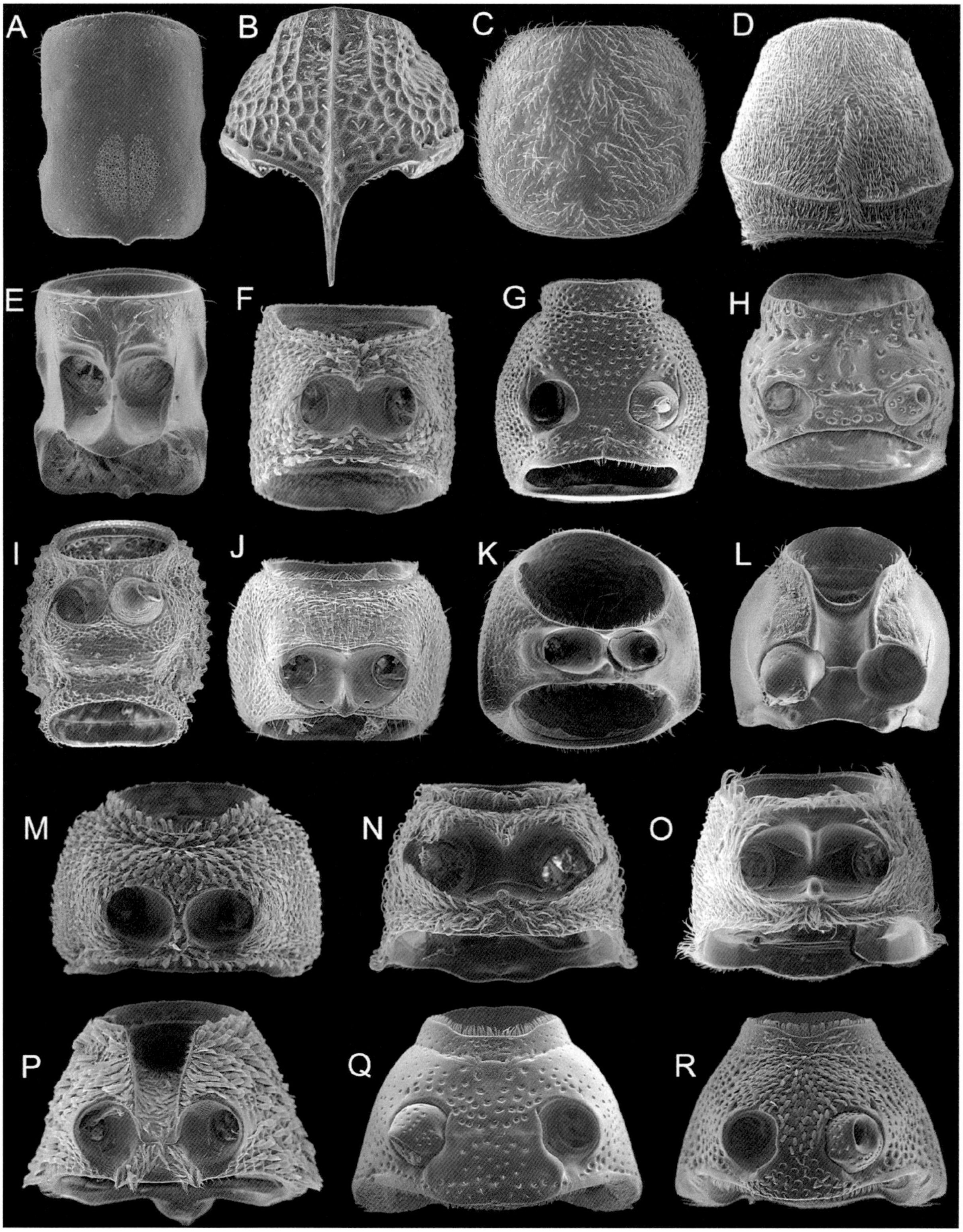

Fig. 37.14. Structural details of weevil prothoraxes. A, *Dinoplatypus pseudocupulatus* (Schedl), ♀; B, *Mecysmoderes* sp., ♂; C, *Bunyaeus monteithi* Kuschel, ♀; D, *Telala* sp.; E, *Dinoplatypus pseudocupulatus* (Schedl), ♀; F, *Peliocis subcylindricus* Lea; G, *Arecophaga* sp., ♂; H, *Notiomimetes pascoei* Wollaston, ♂; I, *Tasmanica* sp., ♂; J, *Basiliorhinus araucariae* Kuschel, ♀; K, *Beaverium insulindicus* (Eggers), ♀; L, *Bepharus* sp., ♀; M, *Orichora trivirgata* Pascoe, ♀; N, *Sphinctobelus cinereus* (Blanchard), ♂; O, *Car* sp., ♀; P, *Lobotrachelus stigma* Pascoe, ♀; Q, *Ipsichora desiderabilis* Lea; R, *Myctides imberbis* Lea, ♂.

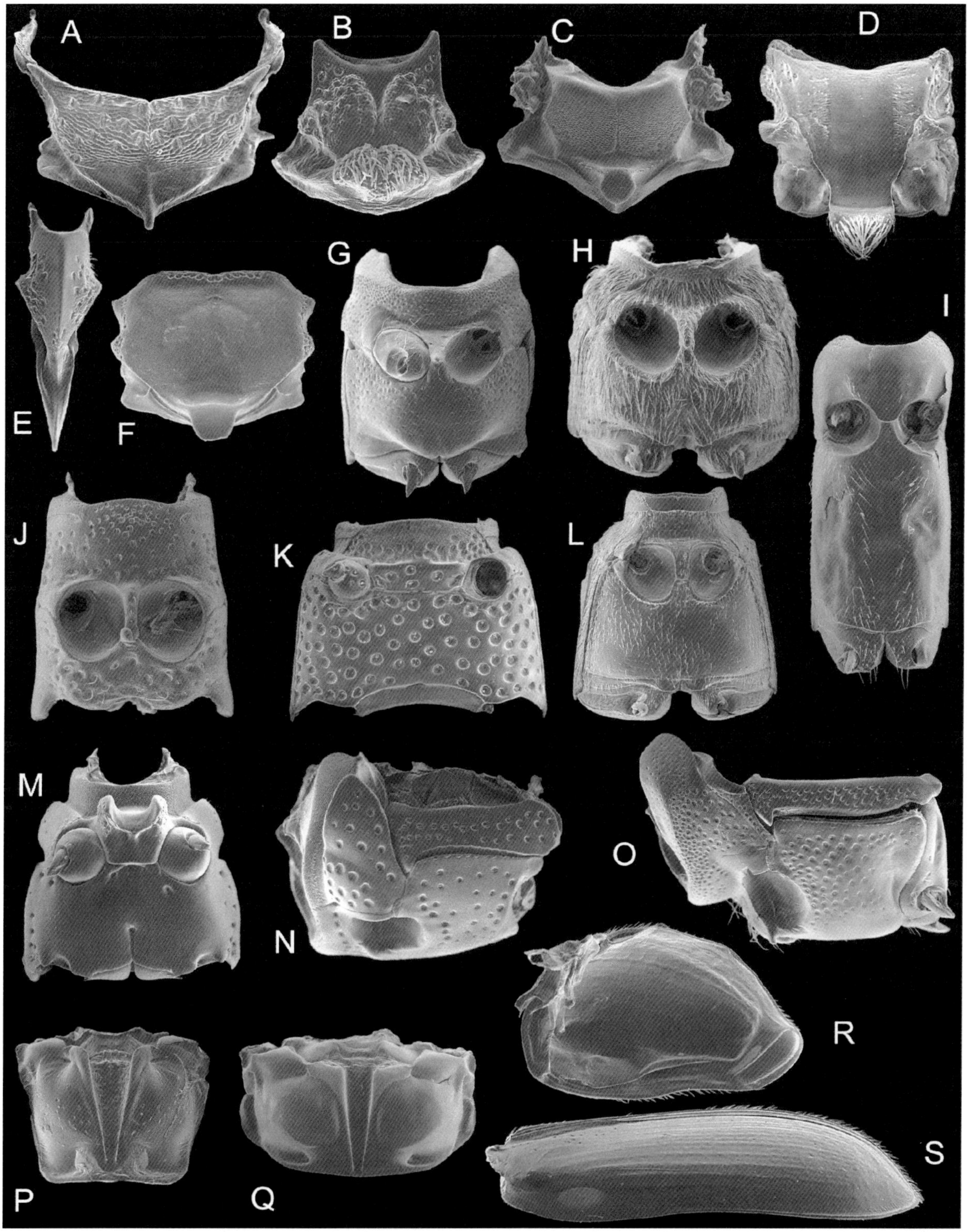

Fig. 37.15. Structural details of weevil pterothoraxes, scutella and elytra. A–F, scutellum, dorsal: A, *Hypurus bertrandi* Perris; B, *Sphinctobelus cinereus* (Blanchard), ♂; C, *Ipsichora desiderabilis* Lea; D, *Aragomacer uniformis* Kuschel; E, *Notoplatypus elongatus* Lea; F, *Xylosandrus crassiusculus* (Motschulsky); G–M, pterothorax, ventral: G, *Beaverium insulindicus* (Eggers), ♀; H, *Car* sp., ♀; I, *Dinoplatypus pseudocupulatus* (Schedl), ♀; J, *Syagrius* sp., ♂; K, *Notiomimetes pascoei* Wollaston, ♂; L, *A. uniformis*, ♂; M, *Bepharus* sp., ♀; N–O, pterothorax, lateral: N, *I. desiderabilis;* O, *B. insulindicus*, ♀; P–Q, metanotum, dorsal: P, *Rhinotia bimaculata* (Pascoe); Q, *Car* sp., ♀; R–S, elytron, inner lateral: R, *Lobotrachelus stigma* Pascoe, ♀; S, *A. uniformis*, ♂.

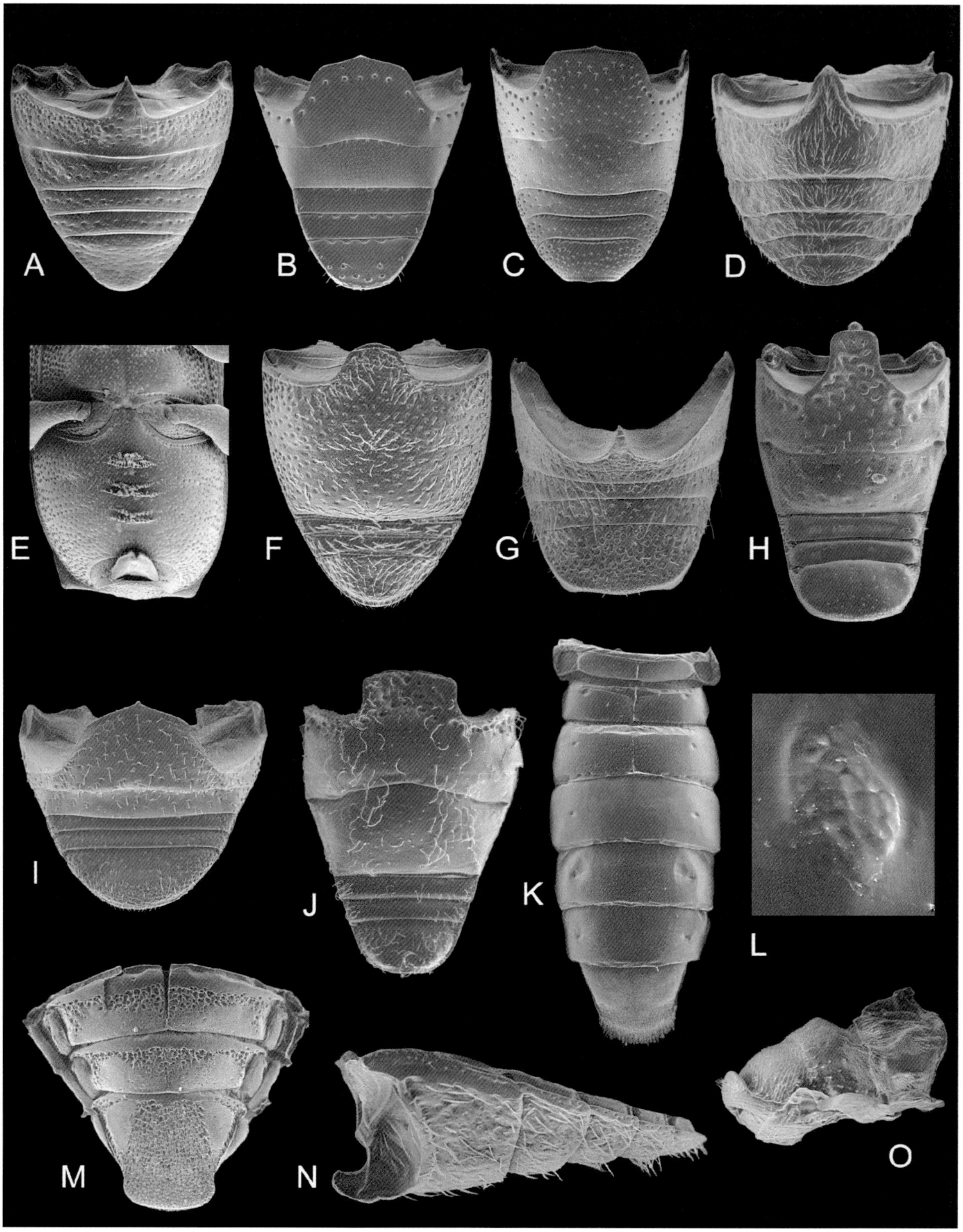

Fig. 37.16. Structural details of weevil abdomens. A–J, abdominal ventrites: A, *Beaverium insulindicus* (Eggers), ♀; B, *Bepharus* sp., ♀; C, *Ipsichora desiderabilis* Lea; D, *Car* sp., ♀; E, *Euops falcatus* (Guérin-Méneville); F, *Apiotherium* sp.; G, *Platypus omnivorus* (Lea); H, *Syagrius* sp., ♂; I, *Psepholax* sp., ♂; J, *Mallixus sylvicola* (Oke), ♂; K–L, *Sphinctobelus cinereus* (Blanchard), ♂: K, tergites; L, pore, tergite detail; M, *I. desiderabilis*, tergites; N, *Car pini* Lea, ♀, abdominal ventrites, lateral; O, *Apiotherium carpophagum* (Lea), ♂, pygidium, lateral.

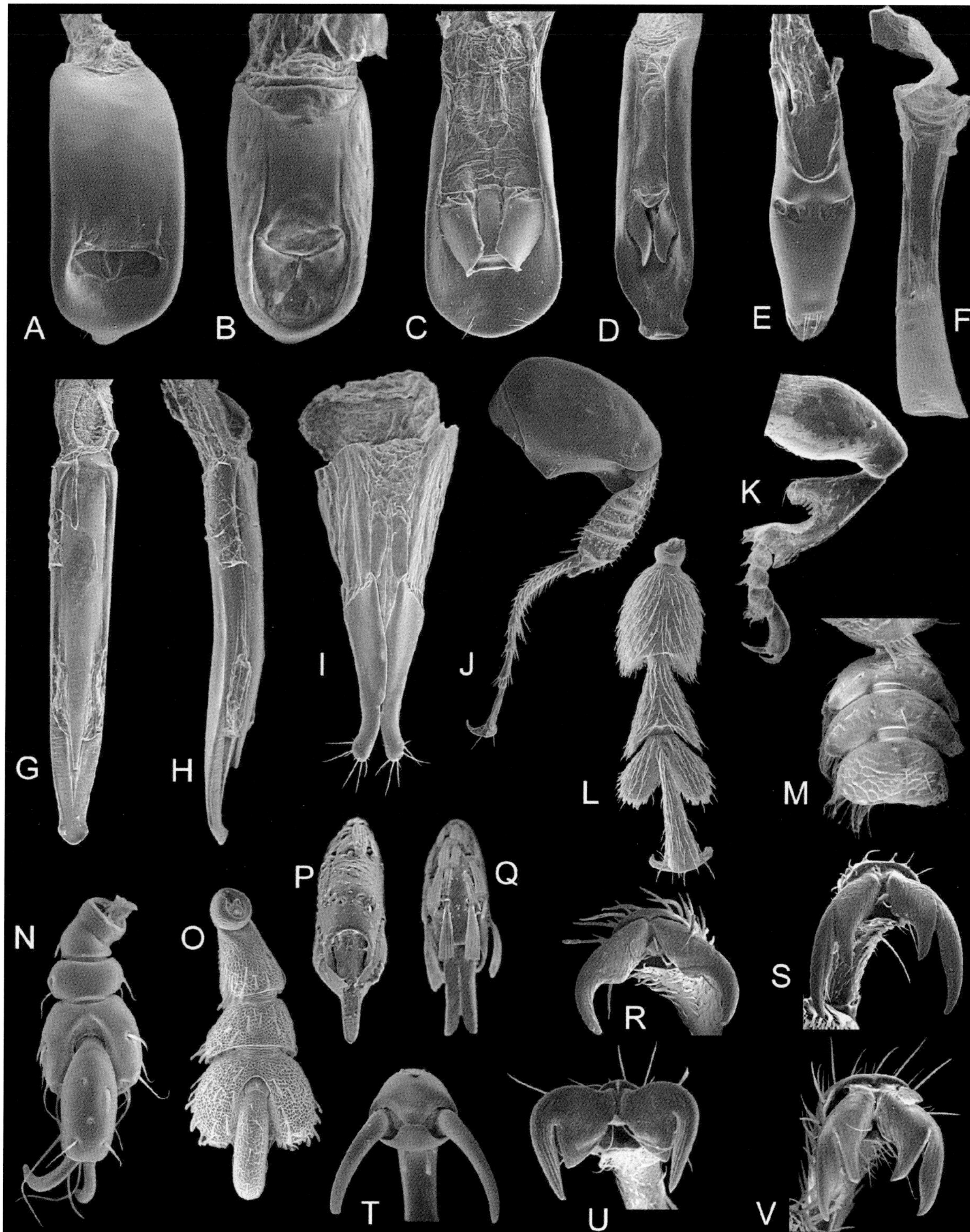

Fig. 37.17. Structural details of weevil genitalia, legs and tarsal claws. A–H, penis, dorsal: A, *Myctides imberbis* Lea; B, *Orichora trivirgata* Pascoe; C, *Psepholax* sp.; D, *Mandalotina bicolor* Oke; E, *Pseudapion philanthum* (Lea); F, *Austroplatypus incompertus* (Schedl); G–H, *Aragomacer uniformis* Kuschel dorsal and lateral; I, *A. uniformis*, ovipositor, dorsal; J–K, foreleg, lateral: J, *Dinoplatypus pseudocupulatus* (Schedl), ♀; K, *Cerobates laevipennis* Senna; L–O, tarsi, dorsal: L, *Car* sp.; M, *Geochus* sp., ♀; N, *Notiomimetes* sp., ♀; O, *Orchidophilus aterrimus* Waterhouse, ♀; P–V, claws: P, *Haplonyx spencei* Gyllenhal; Q, *Menechirus parryi* Jennings & Oberprieler; R, *Car* sp.; S, *Telala* sp.; T, *Diathetes signaticollis* Faust; U, *Meriphus fimbriatus* Lea, ♂; V, *A. uniformis*, ♀.

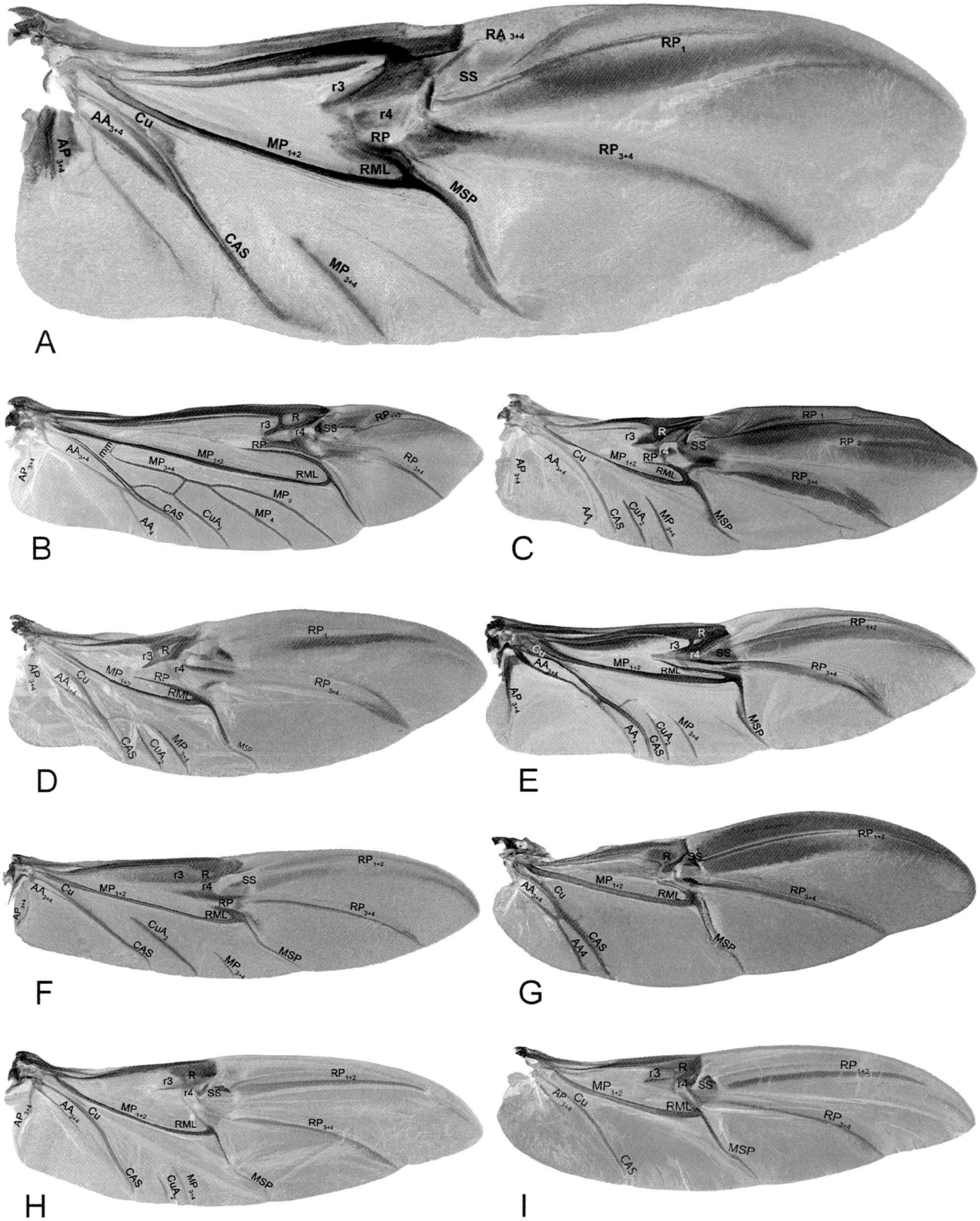

Fig. 37.18. Structural details of weevil hindwings with veins. A, *Araecerus palmaris* (Pascoe); B, *Rhinotia bidentata* (Donovan); C, *Euops falcatus* (Guérin-Méneville); D, *Car condensatus* Balckburn; E, *Eurhynchus laevior* (Kirby); F, *Tracheloschizus dichrous* (Lacordaire); G, *Rhynchophorus ferrugineus* (Olivier); H, *Tranes lyterioides* (Pascoe); I, *Sympiezoscelus spencei Waterhouse*. After Lawrence *et al.* (2022).

7(3). Antennae 9-segmented with loose, 3-segmented antennal club .... **BRENTIDAE: Nanophyinae**
– Antennae usually 11-segmented with 3-segmented antennal club, if 9-segmented then club compact and 1- or 2-segmented (some Scolytinae) **8**
8(7). Maxillary palps 2-segmented; labial palps 1-segmented; trochanters often elongate, trochantero-femoral joints not or only slightly oblique so that femora not reaching coxae; tarsal claws often dentate .... **BRENTIDAE: Apioninae**
– Maxillary palps usually 3-segmented; labial palps usually 2-segmented; trochantero-femoral joints strongly oblique so that femora reaching or almost reaching coxae; tarsal claws dentate or simple .... **CURCULIONIDAE**

Acknowledgements

We thank James Bickerstaff (ANIC-CSIRO) for allowing the use of some of his scanning electron microscope (SEM) images of Platypodinae and Scolytinae and Adriana Marvaldi (CONICET, Argentina) for the use of figures 37.5A–C and 37.7A–H.

References

Alonso-Zarazaga MA, Lyal CHC (1999) *A World Catalogue of Families and Genera of Curculionoidea (Insecta: Coleoptera) (Excepting Scolytidae and Platypodidae)*. Entomopraxis, Barcelona, 315 pp.

Alonso-Zarazaga MA, Lyal CHC (2002) Addenda and corrigenda to a 'World Catalogue of Families and Genera of Curculionoidea (Insecta: Coleoptera)'. *Zootaxa* **63**, 1–37. doi:10.5281/zenodo.155588

Alonso-Zarazaga MA, Lyal CHC (2009) A catalogue of family and genus group names in Scolytinae and Platypodinae with nomenclatural remarks (Coleoptera: Curculionidae). *Zootaxa* **2258**, 1–134. doi:10.11646/zootaxa.2258.1.1

Alonso-Zarazaga MA, Barrios H, Borovec R, Bouchard P, Caldara R, *et al.* (2017) Cooperative catalogue of Palaearctic Coleoptera Curculionoidea. *Monografías electrónicas S.E.A.* **8**, 1–729.

Alonso-Zarazaga MA, Barrios H, Borovec R, Bouchard P, Caldara R, *et al.* (2023) Cooperative catalogue of Palaearctic Coleoptera Curculionoidea, 2nd edition. *Monografías electrónicas S.E.A.* **14**, 1–780.

Anderson RS (1993) Weevils and plants: phylogenetic versus ecological mediation of evolution of host plant associations in Curculionidae (Curculioninae). *Memoirs of the Entomological Society of Canada* **165**, 197–232. doi:10.4039/entm125165197-1

Anderson RS (1995) An evolutionary perspective on diversity in Curculionoidea. *Memoirs of the Entomological Society of Washington* **14**, 103–114.

Calder AA (1989) The alimentary canal and nervous system of Curculionoidea (Coleoptera): gross morphology and systematic significance. *Journal of Natural History* **23**(6), 1205–1265. doi:10.1080/00222938900770671

Calder AA (1990) Gross morphology of the soft parts of the male and female reproductive systems of Curculionoidea (Coleoptera). *Journal of Natural History* **24**(2), 453–505. doi:10.1080/00222939000770341

Clarke DJ, Limaye A, McKenna DD, Oberprieler RO (2019) The weevil fauna preserved in Burmese amber – snapshot of a unique, extinct lineage (Coleoptera: Curculionoidea). *Diversity* **11**(1), 1. doi:10.3390/d11010001

Crowson RA (1955) *The Natural Classification of the Families of Coleoptera*. Nathaniel Lloyd & Co., London, 214 pp.

Davis SR (2017) The weevil rostrum (Coleoptera: Curculionoidea): internal structure and evolutionary trends. *Bulletin of American Museum of Natural History* **416**(416), 1–76. doi:10.1206/0003-0090-416.1.1

Davis SR, Engel MS, Legalov A, Ren D (2012) Weevils of the Yixian Formation, China (Coleoptera: Curculionoidea): phylogenetic considerations and comparison with other Mesozoic faunas. *Journal of Systematic Palaeontology* **11**(4), 399–429. doi:10.1080/14772019.2012.691906

DuPorte M (1960) Evolution of cranial structure in adult Coleoptera. *Canadian Journal of Zoology* **38**(3), 655–675. doi:10.1139/z60-069

Farrell BD (1998) "Inordinate Fondness" explained: why are there so many beetles? *Science* **281**(5376), 555–559. doi:10.1126/science.281.5376.555

Gunter NL, Oberprieler RG, Cameron SL (2016) Molecular phylogenetics of Australian weevils (Coleoptera: Curculionoidea): exploring relationships in a hyperdiverse lineage through comparison of independent analyses. *Austral Entomology* **55**(2), 217–233. doi:10.1111/aen.12173

Howden AT (1995) Structures related to oviposition in Curculionoidea. *Memoirs of the Entomological Society of Washington* **14**, 53–100.

Kasap H, Crowson RA (1977) A comparative study of the internal anatomy and abdominal structures in Curculionoidea. *Hacettepe Bulletin of Natural Sciences and Engineering* **6**, 35–86.

Kuschel G (1971) Entomology of the Aucklands and other islands south of New Zealand: Coleoptera: Curculionidae. *Pacific Insects Monographs* **27**, 225–259.

Kuschel G (1995) A phylogenetic classification of Curculionoidea to families and subfamilies. *Memoirs of the Entomological Society of Washington* **14**, 5–33.

Lawrence JF, Ślipiński SA (2013) *Australian Beetles. Volume 1. Morphology, Classification and Keys*. CSIRO Publishing, Collingwood, Victoria, viii + 561 pp.

Lawrence JF, Ślipiski A, Seago AE, Thayer MK, Newton AF, *et al.* (2011) Phylogeny of the Coleoptera based on morphological characters of adults and larvae. *Annales Zoologici* **61**(1), 1–217. doi:10.3161/000345411X576725

Lawrence JF, Zhou YL, Lemann C, Sinclair B, Ślipiński A (2022) The hind wing of Coleoptera (Insecta): morphology, nomenclature and phylogenetic significance: part 2. Further discussion, Histeroidea, Bostrichoidea to Curculionoidea. *Annales Zoologici* **72**(3), 433–755. doi:10.3161/00034541ANZ2022.72.3.004

Lyal CHC (1995) The ventral structures of the weevil head (Coleoptera: Curculionoidea). *Memoirs of the Entomological Society of Washington* **14**, 35–51.

Lyal CHC, King T (1996) Elytro-tergal stridulation in weevils (Insecta: Coleoptera: Curculionoidea). *Journal of Natural History* **30**(5), 703–773. doi:10.1080/00222939600770391

Lyal CHC, Douglas DA, Hine SJ (2006) Morphology and systematic significance of sclerolepidia in the weevils (Coleoptera: Curculionoidea). *Systematics and Biodiversity* **4**(2), 203–241. doi:10.1017/S1477200005001714

Marvaldi AE (1997) Higher level phylogeny of Curculionidae (Coleoptera: Curculionoidea) based mainly on larval characters, with special reference to broad-nosed weevils. *Cladistics* **13**(4), 285–312. doi:10.1111/j.1096-0031.1997.tb00321.x

Marvaldi AE (1999) Morfología larval en Curculionidae (Insecta: Coleoptera). *Acta Zoologica Lilloana* **45**(1), 7–24.

Marvaldi AE, Morrone JJ (2000) Phylogenetic systematics of weevils (Coleoptera: Curculionoidea) a reappraisal based on larval and adult morphology. *Insect Systematics and Evolution* **31**, 43–58.

Marvaldi AE, Duckett CN, Kjer KM, Gillespie JJ (2009) Structural alignment of 18S and 28S rDNA sequences provides insights into phylogeny of Phytophaga (Coleoptera: Curculionoidea and Chrysomeloidea). *Zoologica scripta* **38**(1), 63–77. doi:10.1111/j.1463-6409.2008.00360.x

Marvaldi AE, Sequeira AS, O'Brien CW, Farrell BD (2002) Molecular and morphological phylogenetics of weevils (Coleoptera, Curculionoidea): do niche shifts accompany diversification? *Systematic Biology* **51**(5), 761–785. doi:10.1080/10635150290102465

Marvaldi AE, Lyal CHC, Oberprieler RG, Bradbury T, Anderson RS (2006) Phylogeny of the Oxycoryninae *s. l.* (Coleoptera: Belidae) and evolution of host-plant associations. *Invertebrate Taxonomy* **20**(4), 447–476. doi:10.1071/IS05059

May BM (1993) *Fauna of New Zealand. Number 28. Larvae of Curculionoidea (Insecta: Coleoptera) a Systematic Overview.* Manaaki Whenua Press, Lincoln, New Zealand, 226 pp.

May BM (1994) An introduction to the immature stages of Australian Curculionoidea. In *Australian Weevils (Coleoptera: Curculionoidea). Volume II–Brentidae, Eurhynchidae, Apionidae and a Chapter on Immature Stages by Brenda May.* (Ed. EC Zimmerman) pp. 365–728. CSIRO, Canberra, Australia, x + 755 pp.

McCurry MR, Cantrill DJ, Smith PM, Beattie R, Dettmann M, *et al.* (2022) A Lagerstätte from Australia provides insight into the nature of Miocene mesic ecosystems. *Science Advances* **8**(1), eabm1406. doi:10.1126/sciadv.abm1406

McKenna DD, Sequeira AS, Marvaldi AE, Farrell BD (2009) Temporal lags and overlap in the diversification of weevils and flowering plants. *Proceedings of the National Academy of Sciences of the United States of America* **106**(17), 7083–7088. doi:10.1073/pnas.081061810

Morimoto K (1962a) Comparative morphology and phylogeny of the superfamily Curculionoidea of Japan (Comparative morphology, phylogeny and systematics of the superfamily Curculionoidea of Japan. I). *Journal of the Faculty of Agriculture, Kyushu University* **11**(4), 331–373. doi:10.5109/22690

Morimoto K (1962b) Key to families, subfamilies, tribes and genera of the superfamily Curculionoidea of Japan excluding Scolytidae, Platypodidae and Cossoninae (Comparative morphology, phylogeny and systematics of the superfamily Curculionoidea of Japan. III). *Journal of the Faculty of Agriculture, Kyushu University* **12**(1), 21–66. doi:10.5109/22693

Morimoto K, Kojima H (2003) Morphologic characters of the weevil head and phylogenetic implications (Coleoptera, Curculionoidea). *Esakia* **43**, 133–169. doi:10.5109/2677

Oberprieler RG, Oberprieler SK (2012) *Talbragarus averyi* gen. et sp. n., the first Jurassic weevil from the southern hemisphere (Coleoptera: Curculionoidea: Nemonychidae). *Zootaxa* **3478**(3478), 256–266. doi:10.11646/zootaxa.3478.1.25

Oberprieler RG, Zimmerman EC (2020) *Australian Weevils (Coleoptera: Curculionoidea). Volume VI. Obituary: Elwood Curtin Zimmerman 1912–2004, Curculionidae: Entiminae Part I, Literature Consulted II.* CSIRO Australia, Clayton, xiv + 650 pp.

Oberprieler RG, Anderson RS, Marvaldi AE (2014) 3. Curculionoidea Latreille, 1820: Introduction, Phylogeny. In *Handbook of Zoology. Arthropoda: Insecta. Coleoptera, Beetles. Volume 3: Morphology and Systematics (Phytophaga).* (Eds RAB Leschen and RG Beutel) pp. 285–300. Walter de Gruyter, Berlin, Germany/Boston, USA, xii + 675 pp.

Oberprieler RG, Marvaldi AE, Anderson RS (2007) Weevils, weevils, weevils everywhere. *Zootaxa* **520**(1668), 491–520. doi:10.11646/zootaxa.1668.1.24

Pullen KR, Jennings D, Oberprieler RO (2014) Annotated catalogue of Australian weevils (Coleoptera: Curculionoidea). *Zootaxa* **3896**(1), 1–481. doi:10.11646/zootaxa.3896.1.1

Shin S, Clarke DJ, Lemmon AR, Moriarty Lemmon E, Aitken AL, *et al.* (2018) Phylogenomic data yield new and robust insights into the phylogeny and evolution of weevils. *Molecular Biology and Evolution* **35**(4), 823–836. doi:10.1093/molbev/msx324

Soriano C, Archer M, Azar D, Creaser Ph, Delclòs X, *et al.* (2010) Synchrotron X-ray imaging of inclusions in amber. *Comptes Rendus Palevol* **9**(6–7), 361–368. doi:10.1016/j.crpv.2010.07.014

Thompson RT (1992) Observations on the morphology and classification of weevils (Coleoptera: Curculionoidea) with a key to the major groups. *Journal of Natural History* **26**(4), 835–891. doi:10.1080/00222939200770511

Wanat M (2007) Alignment and homology of male terminalia in Curculionoidea and other Coleoptera. *Invertebrate Systematics* **21**(2), 147–171. doi:10.1071/IS05055

Zherikhin VV, Gratshev VG (1995) A comparative study of the hind wing venation of the superfamily Curculionoidea, with phylogenetic implications. In *Biology, Phylogeny and Classification of Coleoptera. Vol. 2, Papers Celebrating the 80th Birthday of Roy A. Crowson.* (Eds J Pakaluk and SA Ślipiński) pp. 633–777. Muzeum i Instytut Zoologii Polska Akademia Nauk, Warszawa, 1092 pp.

Zimmerman EC (1991) *Australian Weevils (Coleoptera: Curculionoidea). Volume V. Colour Plates 1–304.* CSIRO Australia, Melbourne, x + 633 pp.

Zimmerman EC (1992) *Australian Weevils (Coleoptera: Curculionoidea). Volume VI. Colour Plates 305–632.* CSIRO Australia, Melbourne, vii + 707 pp.

Zimmerman EC (1993) *Australian Weevils (Coleoptera: Curculionoidea). Volume III. Nanophyidae, Rhynchophoridae, Erirhinidae, Curculionidae: Amycterinae, Literature Consulted.* CSIRO Australia, Melbourne, x + 854 pp.

Zimmerman EC (1994a) *Australian Weevils (Coleoptera: Curculionoidea), Volume I. Orthoceri, Anthribidae to Attelabidae: The Primitive Weevils.* CSIRO Australia, Melbourne, xxxii + 741 pp.

Zimmerman EC (1994b) *Australian Weevils (Coleoptera: Curculionoidea). Volume II. Brentidae, Eurhynchidae, Apionidae and a Chapter on Immature Stages by Brenda May.* CSIRO Australia, Melbourne, x + 755 pp.

38. NEMONYCHIDAE BEDEL, 1882

Hermes E. Escalona, Debbie Jennings and Rolf G. Oberprieler

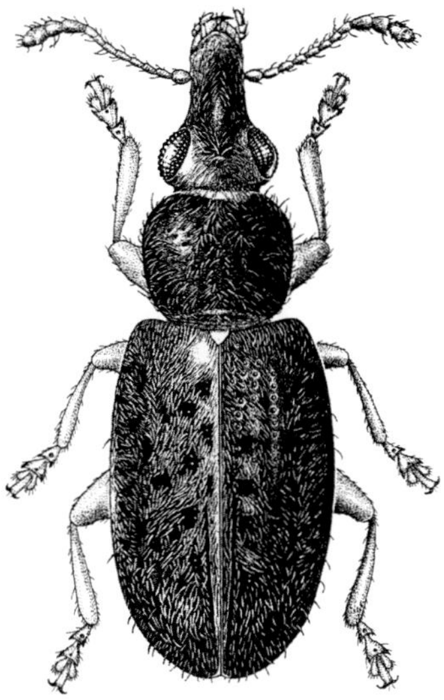

Fig. 38.1. *Bunyaeus monteithi* Kuschel.

Introduction. Nemonychidae are a small group with 16 extant species in 8 genera in Australia. The family is now divided into two subfamilies, Rhinorhynchinae, which includes all the Australian species and it is distributed mainly in the Southern Hemisphere, and Nemonychinae, which is restricted to the Palaeartic region. A former third subfamily, Cimberidinae, is now recognised as a distinct family, Cimberididae (Shin *et al.* 2018; Alonso-Zarazaga *et al.* 2023). The major contributions to the systematics and biology of the group were made by the late Guillermo Kuschel and collaborators, who revised the world fauna of the family (Kuschel 1959, 1993, 1994, 2003; Kuschel & May 1997; Kuschel & Leschen 2011). The family is one of major interest in Curculionoidea due to its primitive morphological features, rich fossil record, host specificity and predominant diet of pollen for both adults and larvae. The monophyly and phylogenetic position of Nemonychidae in Curculionoidea have been a subject of considerable debate (Crowson 1985a, b; Kuschel 1989, 1994, 1995; Marvaldi & Morrone 2000). Recent molecular analysis with extensive gene sampling (Shin *et al.* 2018) revealed a close relation of Nemonychidae with Anthribidae, as also evident from some morphological data (Marvaldi & Morrone 2000), but to the exclusion of the former subfamily Cimberidinae, which was recovered as the most basal lineage of Curculionoidea and afforded family status again. It is expected that new molecular analysis with better taxon and gene sampling, together with morphological data, will provide a clearer picture of the relationships and placement of the group. Nemonychidae are well represented in the Mesozoic fossil record, as summarised by Anderson *et al.* (2014). The Nemonychidae of Australia, New Caledonia and New Guinea were comprehensively treated by Kuschel (1994) and Kuschel & Leschen (2011). Almost all the extant genera and species occurring in Australia were described in these two publications, except for two species earlier described by Lea (1926), and May (1994) described the larvae of all Australian Nemonychidae as known to date. A single Australian nemonychid fossil has been described, *Talbragarus averyi*† Oberprieler & Oberprieler, 2012, tentatively assigned to the subfamily Rhinorhynchinae but of uncertain tribal placement. The most recent addition to the taxonomy of the Australian Nemonychidae was made by Legalov (2017), who proposed a new tribe (Zimmiellini) and new subtribe (Bunyaeina) for the rhinorhynchine genera *Zimmiellus* and *Bunyaeus*, respectively. We here follow the classification of Kuschel & Leschen (2011), which is based on a detailed morphological and cladistic analysis of the Rhinorhynchinae.

Biology. Nemonychidae are predominately associated with conifers, only a few South American species living on angiosperms (Nothofagaceae). The Rhinorhynchinae of the Australian region have been found associated only with Araucariaceae and Podocarpaceae and the Australian species so far only with the former family, mostly with *Araucaria cunninghamii* (recorded for the genera *Aragomacer, Bunyaeus, Eutactobius, Notomacer, Basiliogeus* and *Basiliorhinus*) but also with *Araucaria bidwillii, Agathis robusta* and *Agathis atropurpurea*. Like their hosts, the Australian nemonychid taxa are restricted to eastern Australia, from northern New South Wales to northern Queensland. Current knowledge of their host associations is largely due to the collecting and rearing efforts of G. Kuschel and Geoff Monteith conducted in the late 1970s and early 1980s, but the hosts of several species remain unknown, particularly those of *Pagomacer* and *Zimmiellus*. Their type localities are in the distribution ranges of *Araucaria* and *Agathis*, respectively, but Podocarpaceae (*Dacrycarpus* sp.) have been reported as hosts for one species of *Basiliogeus* in New Guinea and could also host Nemonychidae in Australia. Nemonychids are elusive and rarely collected. The most effective way to obtain specimens is by rearing them from male cones of Araucariaceae, as described in detail by Kuschel (1994). Adult Nemonychidae feed on the pollen of their host plants and larvae on pollen and structures of the male cones. Larval development is rapid, and mature larvae drop to the ground and pupate in the soil, from which adults emerge after several months (Kuschel 1994).

Characteristics. Adults. Body length 2.0–4.3 mm (SL). Body about 2.2–2.8 × as long as wide and more or less parallel-sided (Fig. 38.2), slightly flattened or convex, cuticle yellow to dark brown in colour, sometimes with patterns of dark markings on elytra and occasionally pronotum; clothed with fine, decumbent hairs and sparser, scattered erect ones, without scales. Head not or slightly declined at base, but vertex in *Zimmiellus* strongly declined so that rostrum extends ventrally forming an acute angle with forehead (Pl. 26J); occipital region without transverse ridge or stridulatory file. Rostrum generally expanded apically, slightly to distinctly curved (*Zimmiellus*), base continuous with or gently to deeply saddled against forehead. Eyes slightly to strongly protuberant, not emarginate, finely or coarsely facetted, without or with short, fine interfacettal setae. Antennal insertions lateral, concealed or barely exposed from above, usually located close to apex of rostrum but near middle in *Basiliorhinus*, *Basiliogeus* and *Zimmiellus*; scrobes (subantennal grooves) well-developed or indistinct, shallow, generally concealed in dorsal view. Vertex (epifrons) produced forward to form rostrum, which may be only slightly longer than wide (*Bunyaeus*) but is usually much longer than wide (as much as 9 × as long as wide in *Basiliogeus*). Frontoclypeal suture absent; labrum free, exposed, on same plane as and fitting tightly against epistome (frontoclypeus), slightly transverse, triangular to subquadrate, broadly rounded but sometimes finely denticulate at apex; generally with 3 pairs of dorsal setae and 0–6 pairs of peg-like apical setae. Antennae 11-segmented, not geniculate, scape elongate but not much longer than pedicel, funicle 7-segmented, subequal in length in both sexes to slightly longer in female, club 3-segmented, loosely articulated, elongate with weak club. Mandibles usually moderately elongate, slender and distinctly falcate, unidentate; incisor edge simple or with single tooth on one mandible; molae reduced and sometimes asymmetrical; reduced, setose prostheca sometimes present on one mandible; inner edge of mandibular base with moderately short pubescent process. Maxillae with short, setose galea and lacinia, each with inner tuft of setae; maxillary palps moderately well-developed, 4-segmented, apical segment fusiform. Prementum slightly transverse, apically emarginate, fused at base; ligula bilobed; labial palps 3-segmented, attached ventrally near base, apical segment fusiform. Gular sutures short and widely separate. Corpotentorium absent; anterior tentorial arms not expanded mesally; cervical sclerites reduced and transverse.

Pronotum about 0.8–1.15 × as long as wide, widest at about middle; base narrower than combined elytral bases; sides slightly curved; posterior angles right or rounded; posterior edge straight or evenly curved, finely margined; disc simple, flat or slightly convex, without impressions or groove. Prosternum in front of procoxae about as long or slightly longer than shortest diameter of coxal cavity, flat to slightly convex. Prosternal process complete but interrupted at middle, intercoxal process strongly concave, the apex expanded and truncate or broadly angulate. Notosternal sutures incomplete. Procoxae projecting below prosternum, with short concealed lateral extensions; trochantins concealed. Procoxal cavities subcircular, contiguous, externally broadly closed, without narrow lateral extensions, internally broadly closed; postcoxal (hypomeral) processes meeting disjunct apex of prosternal process (sternellum).

Scutellar shield not abruptly elevated, anteriorly simple, posteriorly broadly rounded; mesoscutum with stridulatory file, which may be divided or undivided and sometimes asymmetrical (e.g. *Basiliogeus*). Elytra about 1.5–1.9 × as long as wide and 2.3–2.75 × as long as pronotum, with 10 complete striae (puncture rows) and short scutellary striole; elytral apices meeting or almost meeting at suture and not exposing any abdominal tergites; epipleura narrow and incomplete; inner subcostal flange absent or present (e.g. *Aragomacer*). Mesoventrite separated by complete sutures from mesanepisterna, which are distinctly separated at midline; anterior edge on same plane as metaventrite, without paired procoxal rests but with broadly transverse prothoracic rest. Mesocoxae projecting, with concealed trochantins. Mesocoxal cavities narrowly separated, subcircular, not or slightly oblique, closed laterally by meeting of mesoventrite and metaventrite. Mesoventral process narrowly rounded at apex; mesometaventral junction monocondylic. Metaventrite slightly convex; transverse (metakatepisternal) suture long but close to posterior edge of ventrite; discrimen about half as long as ventrite excluding anterior intercoxal process; postcoxal lines absent; exposed portion of metanepisterna moderately elongate; metepimera slightly exposed. Metacoxae narrowly separated, almost reaching elytral margin. Metendosternite with long stalk and lateral arms, well-developed but slender laminae and very short anterior process bearing widely separated anterior tendons. Hindwings with apical field about half as long as total wing length, containing distinct sclerite just beyond radial cell and vaguely indicated remnants of RP; radial cell moderately well-developed, slightly longer than wide with base more or less straight and inner posterobasal angle right; cross-vein r_3 short and moderately oblique, arising well away from r_4, this weakly sclerotised but more or less complete; basal portion of RP moderately long; R-M loop forming acute angle; medial spur ending before wing margin; medial field with 3 or 4 free veins and no medial fleck; MP_{3+4} without base; anal lobe moderately well-developed with notch-like embayment. Legs moderately long and slender; trochanterofemoral joints strongly oblique with base of femur separate from coxa; femora slightly inflated; tibiae with outer proximal angle not to slightly extended, inner edge straight to incurved, usually ending in a small to medium-sized mucro or mucro absent; tibial spurs apical, 2-2-2 in Mecomacerini, 2-2-1 in Rhinorhynchini. Tarsi 5-5-5-segmented in both sexes, pseudotetramerous (segment 4 reduced) and segment 3 distinctly bilobed; tarsal claws bifid or with lamina (laminate), without ventrobasal seta; empodium small.

Abdomen with 5 free ventrites subequal in length; intercoxal process acute or narrowly rounded; apex of ventrite 5 broadly rounded or slightly emarginate. Abdominal terga lightly pigmented, except for tergite VII, which bears two patches of wing-folding asperities; functional spiracles on segments I–VII, those on I–VI in pleural membrane, those on VII at lateral edges of tergite. Tergite VIII in male partly exposed and sclerotised; sternite VIII reduced and lightly sclerotised, with short median strut. Anterior edge of segment IX in male with distinct spiculum gastrale; tergite X membranous. Aedeagus cucujiform, of pedotectal type (with separate pedon and tectum), usually symmetrical; tegmen forming a ring, anterior edge ventrally with moderately long apodeme, sometimes expanded at apex; parameres fused to phallobase and partly or entirely fused together. Penis flattened with dorsal and ventral lobes (tectum and pedon) and with paired anterior apodemes (temones) dorsally connected at bases by apodemal bridge; endophallus with complex armature. Sternite VIII in female more or less membranous, with long, fixed spiculum ventrale (apodeme). Ovipositor elongate and lightly sclerotised, except

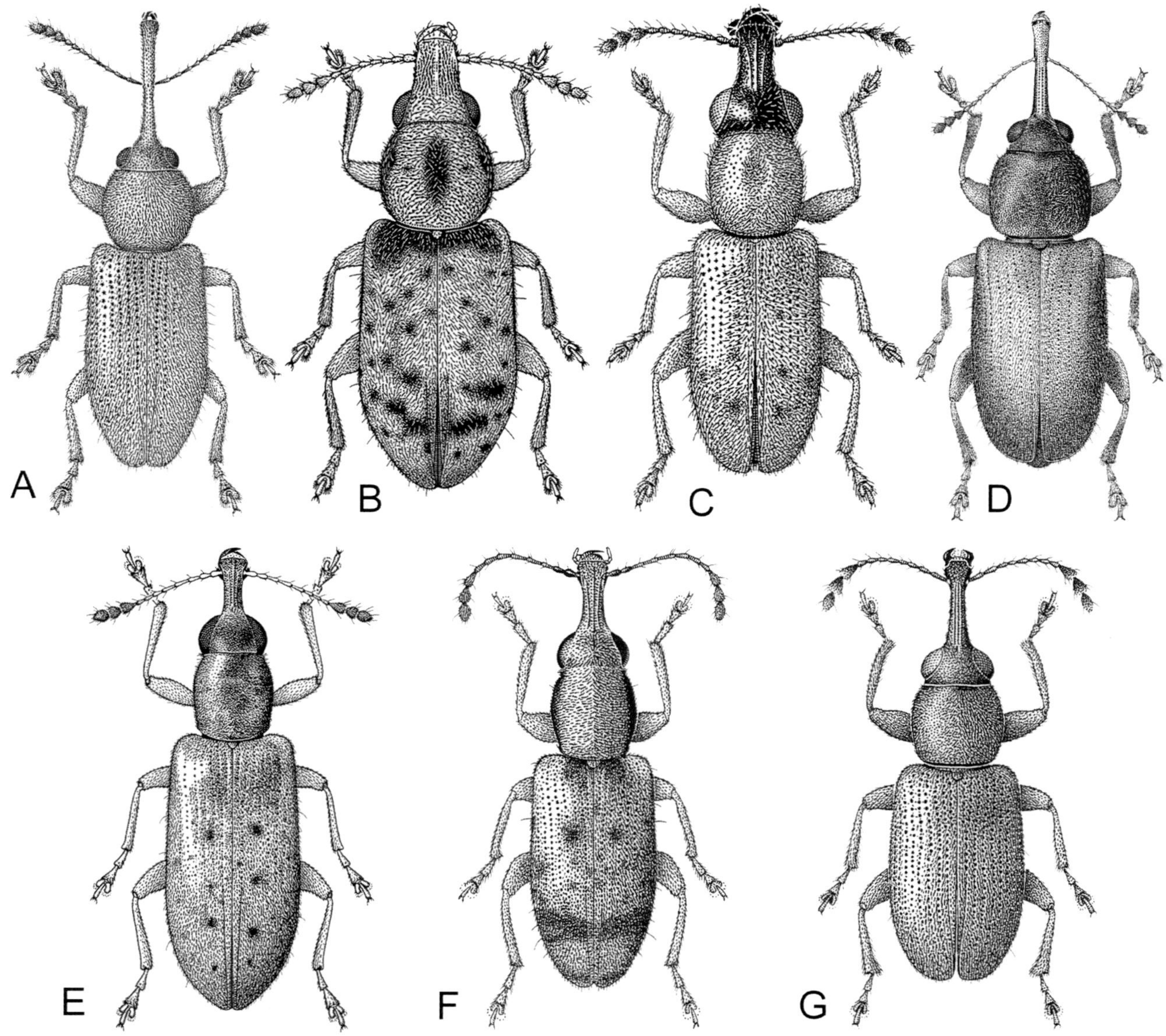

Fig. 38.2. Nemonychidae, dorsal habitus. A, *Basiliogeus prasinus* Kuschel, ♂; B, *Bunyaeus monteithi* Kuschel, ♀; C, *Eutactobius puellus* Kuschel, ♂; D, *Basiliorhinus araucariae* Kuschel, ♀; E, *Notomacer araucariae* Kuschel, ♀; F, *Notomacer reginae* Kuschel, ♂; G, *Pagomacer deceptus* Kuschel, ♂. After Kuschel (1994).

for baculi; paraprocts usually subequal to or slightly shorter than gonocoxites, with longitudinal baculi; gonocoxites located distally of paraprocts, not or weakly divided, with narrow distal parts bearing well-developed, terminal styli. Bursa copulatrix membranous with strongly curved, sclerotised capsular spermatheca and spermathecal gland attached at or near its base. Proventriculus without blades. (Kuschel 1959, 1983, 1989, 1993, 1994, 1995, 2003; Crowson 1985a; Thompson 1992; Kuschel & May 1997; Anderson 2002; Kuschel & Leschen 2011; Lawrence & Ślipiński 2013; Anderson *et al.* 2014).

Larvae (Figs 38.3A, F–J, K–O). Body elongate and more or less parallel-sided or slightly wider at middle and narrowing anteriorly and posteriorly, somewhat flattened (Fig. 38.3F); lightly pigmented except for head capsule and prothorax; vestiture consisting of scattered long and short, simple setae. Head hypognathous or slightly declined, not or partially retracted into prothorax; posterior edge of head capsule not emarginate; epicranial stem moderately long; frontal arms V-shaped, contiguous at base and reaching articulatory membrane of mandibles; median endocarina extending almost to clypeal region. Stemmata on each side usually 4, one stemma with lens, and as many as 3 vestigial eyespots. Antennae very short, 1-segmented, with terminal, conical sensorium. Frontoclypeal suture absent; labrum short and free, but without tormae or labral rods. Mandibles more or less symmetrical,

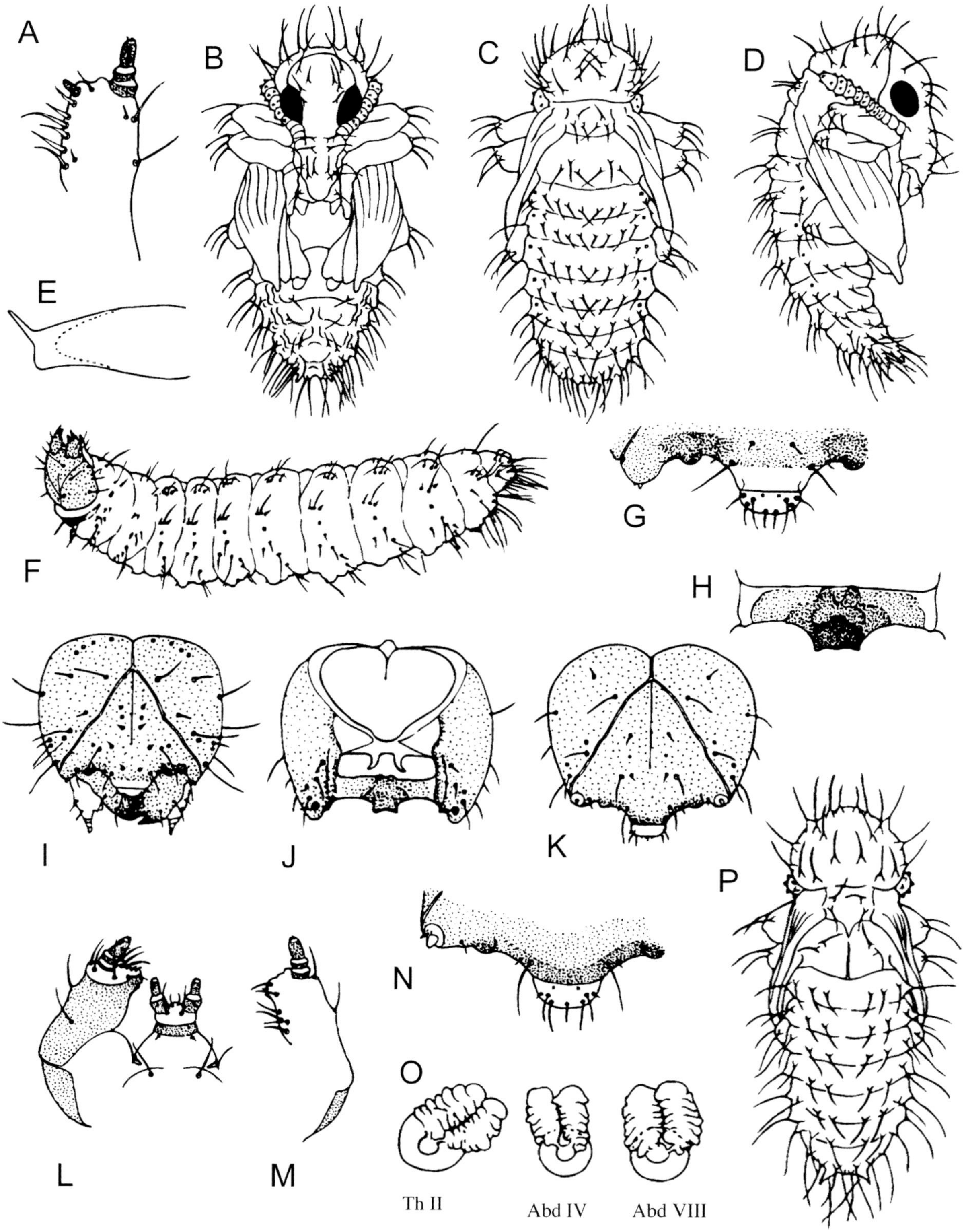

Fig. 38.3. Nemonychidae, structural details of immature stages. A, *Bunyaeus monteithi* Kuschel, larva, mala, dorsal; B–D, *Notomacer caledonicus* Kuschel, pupa: B, ventral; C, dorsal; D, lateral; E, primary pterotheca, apex; F–J, *Notomacer araucariae* Kuschel, larva: F, habitus, lateral; G, anterior portion of the head, labrum and antenna; H, hypopharyngeal bracon and sclerome; I, head, dorsal; J, head, ventral; K–O, *Basiliorhinus araucariae* Kuschel, larva: K, head, dorsal; L, mala and labium, ventral; M, mala, dorsal; N, anterior portion of pseudoclypeus, labrum and antenna; O, spiracles, showing alignment; P, *Basiliorhinus araucariae* Kuschel, pupa, dorsal. After May (1994).

short and broad, obliquely bidentate; accessory ventral process absent; incisor edge with or without one subapical tooth, mola reduced and sub-basal; prostheca absent. Ventral mouthparts retracted; maxillae with articulating area present, cardo moderately strongly oblique, undivided; stipes about as long as wide; mala fixed, apically truncate with a few setae; inner apical angle with slight tooth; maxillary palp 3-segmented without distinct palpifer. Labium consisting of prementum and postmentum; ligula absent; labial palps 2-segmented, separated by distance equal to width of first segment. Hypopharyngeal bracon and sclerome present. Hypostomal rods short and slightly diverging; ventral epicranial ridges absent. Gular region very short.

Prothoracic tergum about as long as meso- or metatergum, with or without lightly pigmented plate. Legs absent. Abdomen more than twice as long as thorax; abdominal terga not extending laterally beyond edges of sterna; meso- and metaterga and abdominal terga I–VIII each with transverse ambulatory ampulla; segment IX subequal to or slightly shorter than VIII; completely dorsal, without urogomphi; sternum 9 simple; segment X much shorter than IX, terminal, without pygopods; anal region posteriorly oriented; anal opening T-shaped. Spiracles annular-biforous with circular peritreme and moderately long accessory chambers. Alimentary canal with proventriculus short; mycetomes lacking around cardiac valve; gastric caecae absent. Malpighian tubules 4 or 6, purple in colour and visible through cuticle; cryptonephridium weakly developed. Rectal bracon not visible. (Anderson 1991; May 1993, 1994; Anderson 2002; Marvaldi 2003; Kuschel & Leschen 2011; Lawrence & Ślipiński 2013; Anderson *et al.* 2014).

Pupae (Fig. 38.3B–E, P). Terminal abdominal segment with paired posterior processes ('pupal urogomphy') small, acute, shorter than associated setae (May 1994).

Key to the Australian genera

Modified after Kuschel (1994) and Kuschel & Leschen (2011).

1. Eyes elongate (Fig. 38.5B); antennal insertions next to mandibles (Fig. 38.5A); antennal cavities not visible in dorsal view (Fig. 38.5A) .... **MECOMACERINI** ... 2

– Eyes oval or rounded (Fig. 38.6B); antennal insertions well separated from mandibles (Fig. 38.6A); antennal cavities visible in dorsal view (Fig. 38.6A) .... **RHINORHYNCHINI** ... 5

2(1). Forehead about as wide as apex of rostrum (Fig. 38.5A); scutellum with two stridulatory files (Fig. 38.4O) 3

– Forehead distinctly wider than apex of rostrum (Fig. 38.7A); scutellum with one stridulatory file (Fig. 38.7I) 4

3(2). Prothorax usually considerably longer than broad (Fig. 38.2E, F), finely and shallowly punctate and partially smooth on midline; endophallus without long baculiform sclerites; spermatheca thick and strongly curved; QLD, New Caledonia. Hosts: *Araucaria* and *Agathis* .... ***Notomacer*** **Kuschel**

– Prothorax about as long as wide (Pl. 26A), coarsely and deeply punctate, not partially smooth on midline; endophallus with 1 or 2 pairs of long, baculiform sclerites; spermatheca relatively flat and weakly curved; Southeast QLD, Papua New Guinea. Hosts: *Araucaria* and *Agathis* .... ***Aragomacer*** **Kuschel**

4(2). Rostrum in lateral view discontinuous with forehead, with distinct depression at base; mesotibiae in male with mucro; QLD (Canungra, Cainbable). Host: *Araucaria* .... ***Eutactobius*** **Kuschel**

– Rostrum in lateral view continuous with forehead (Fig. 38.8B), lacking a depression at base; Mesotibiae in male without mucro; QLD (Cooyar, Yarraman, Benarkin, Canungra, Mt. Nebo, Cainbable, Blackall Range, Bunya Mt.). Host: *Araucaria* .... ***Bunyaeus*** **Kuschel**

5(1). Head with vertex strongly declined; rostrum, in lateral view, emerging from ventral surface of head (Pl. 26J). QLD (nr. Paluma Range). Host: unknown .... ***Zimmiellus*** **Kuschel**

– Head with vertex not strongly declined; rostrum, in lateral view, emerging from the middle of head (Fig. 38.6B) 6

6(5). Pubescence on dorsal surface of rostrum directed anteriad 7

– Pubescence on dorsal surface of rostrum directed posteriad; north QLD (Bellenden Ker Range). Host: unknown .... ***Pagomacer*** **Kuschel**

7(6). Rostrum forming shallow sinus with forehead; scutellum with two symmetrical stridulatory files (Fig. 38.6K); QLD (Bunya Mt.). Host: *Araucaria* .... ***Basiliorhinus*** **Kuschel**

– Rostrum forming deep sinus with forehead; scutellum with one stridulatory file or with two strongly asymmetrical files; QLD (Coen River, Paradise Park, Bunya Mt., Atherton Tablelands) and New Guinea. Hosts: *Araucaria* spp., Podocarpaceae in New Guinea .... ***Basiliogeus*** **Kuschel**

Classification of the Australian genera

Subfamily Rhinorhynchinae Voss, 1922

Tribe Mecomacerini Kuschel, 1994

Aragomacer Kuschel, 1994 (Pl. 26A; Figs 38.4, 38.5)

Type species, by original designation: *Aragomacer leai* Kuschel, 1994.

Australian species. *A. leai* Kuschel, 1994, *A. uniformis* Kuschel, 1994.

Distribution. Australia (QLD), Papua New Guinea.

Notes. Two of the five species of the genus occur in Australia. May (1994) provided a generic diagnosis of the larvae and described the larva and pupa of *A. leai*. Kuschel (1994) noted that pollen has been found in the gut of *A. papuae* Kuschel (1994), whereas the gut of A. *uniformis* contained fungi but not pollen. The recorded hosts in Australia are *Agathis robusta*, *Agathis atropurpurea* and *Araucaria cunninghamii* and in Papua New Guinea *A. cunninghamii* and *Araucaria hunsteinii.*

References. Kuschel (1994); May (1994); Kuschel & Leschen (2011)

Bunyaeus Kuschel, 1994 (Pl. 26B; Figs 38.2B, 38.3A, 38.7, 38.8)

Type species, by original designation: *Bunyaeus monteithi* Kuschel, 1994.
Australian species. *B. eutactae* Kuschel, 1994, *B. monteithi* Kuschel, 1994.
Distribution. Australia (southern QLD).
Notes. May (1994) described the larvae and pupae of *B. monteithi* and *B. eutactae*. The known hosts are *Araucaria bidwillii* and *A. cunninghamii*. The larvae feed mainly in male sporophylls and parts of the strobili (Kuschel 1994).
References. May (1994); Kuschel (1994); Kuschel & Leschen (2011); Legalov (2017).

Eutactobius Kuschel, 1994 (Pl. 26C; Fig. 38.2C)

Type species, by original designation: *Eutactobius puellus* Kuschel, 1994.
Australian species. *E. puellus* Kuschel, 1994.
Distribution. Australia (southern QLD).
Notes. May (1994) described the larva of *E. puellus*. The only known host is *Araucaria cunninghamii*.
References. May (1994); Kuschel (1994).

Notomacer Kuschel, 1994 (Pl. 26D; Figs 38.2E, F, 38.3B–J)

Type species, by original designation: *Notomacer araucariae* Kuschel, 1994.
Australian species. *N. australiae* (Lea, 1926), *N. brittoni* Kuschel, 1994, *N. eximius* Kuschel, 1994, *N. reginae* Kuschel, 1994, *N. zimmermani* Kuschel, 1994.
Distribution. Australia (QLD), New Caledonia.
Notes. May (1994) provided a generic diagnosis, descriptions and illustrations of the larvae of the Australian species *N. australiae* and *N. reginae* and the New Caledonian *N. caledonicus* Kuschel and *N. araucariae* Kuschel. The recorded hosts are *Araucaria cunninghamii*, *Agathis robusta* and *Agathis atropurpurea* in Australia and *Araucaria biramulata* and *Araucaria columnaris* in New Caledonia (Kuschel 1994).
References. May (1994); Kuschel (1994); Kuschel & Leschen (2011).

Tribe Rhinorhynchini Voss, 1922

Basiliogeus Kuschel, 1994 (Pl. 26E; Fig. 38.2A)

Type species, by original designation: *Basiliogeus prasinus* Kuschel, 1994.
Australian species. *B. inops* Kuschel, 2011 (*in* Kuschel & Leschen 2011), *B. prasinus* Kuschel, 1994, *B. striatopunctatus* (Lea, 1926).
Distribution. Australia (QLD), Papua New Guinea, Malaysia (Sabah).
Notes. May (1994) provided a generic diagnosis for the larvae and described the larva of *B. prasinus*. The known host plants are *Araucaria bidwillii* and *A. cunninghamii* in Australia and *Dacrycarpus* sp. (Podocarpaceae) for *B. dacrycarpi* Kuschel & Riedel (2011) in New Guinea. In addition, a putative larva has been found on *Agathis dammara* in Papua New Guinea (May 1994).
References. May (1994); Kuschel (1994); Kuschel & Leschen (2011).

Basiliorhinus Kuschel, 1994 (Pl. 26F; Figs 38.2D, 38.3K–P, 38.6)

Type species, by original designation: *Basiliorhinus araucariae* Kuschel, 1994.
Australian species. *B. araucariae* Kuschel, 1994.
Distribution. Australia (southern QLD).
Notes. The recorded host is *Araucaria bidwillii*.
References. May (1994); Kuschel (1994); Kuschel & Leschen (2011).

Pagomacer Kuschel, 1994 (Pl. 26G, I; Fig. 38.2G)

Type species, by original designation: *Pagomacer deceptus* Kuschel, 1994.
Australian species. *P. deceptus* Kuschel, 1994.
Distribution. Australia (northeast QLD).
Notes. The only species of this genus is known from a single specimen collected in a Malaise trap in the Bellenden Ker Range in North QLD Kuschel (1994) suggested that the host could be *Agathis atropurpura* or even a *Podocarpus* species.
References. Kuschel (1994); Kuschel & Leschen (2011).

Zimmiellus Kuschel, 2011 (Pl. 26H, J)

Type species, by original designation: *Zimmiellus fronto* Kuschel, 2011.
Australian species. *Z. fronto* Kuschel, 2011 (*in* Kuschel & Leschen 2011).
Distribution. Australia (north-eastern QLD).
Notes. The most distinctive feature of *Zimmiellus* is the unusual configuration of the rostrum, emerging from the underside of the head, as it also does in the New Caledonian *Idiomacer* Kuschel 2011. Another remarkable character is the exodont mandibles. The only known species of the genus, *Z. fronto*, is known from a single female collected in a Malaise trap in the Paluma Range in northern QLD.
References. Kuschel & Leschen (2011); Legalov (2017).

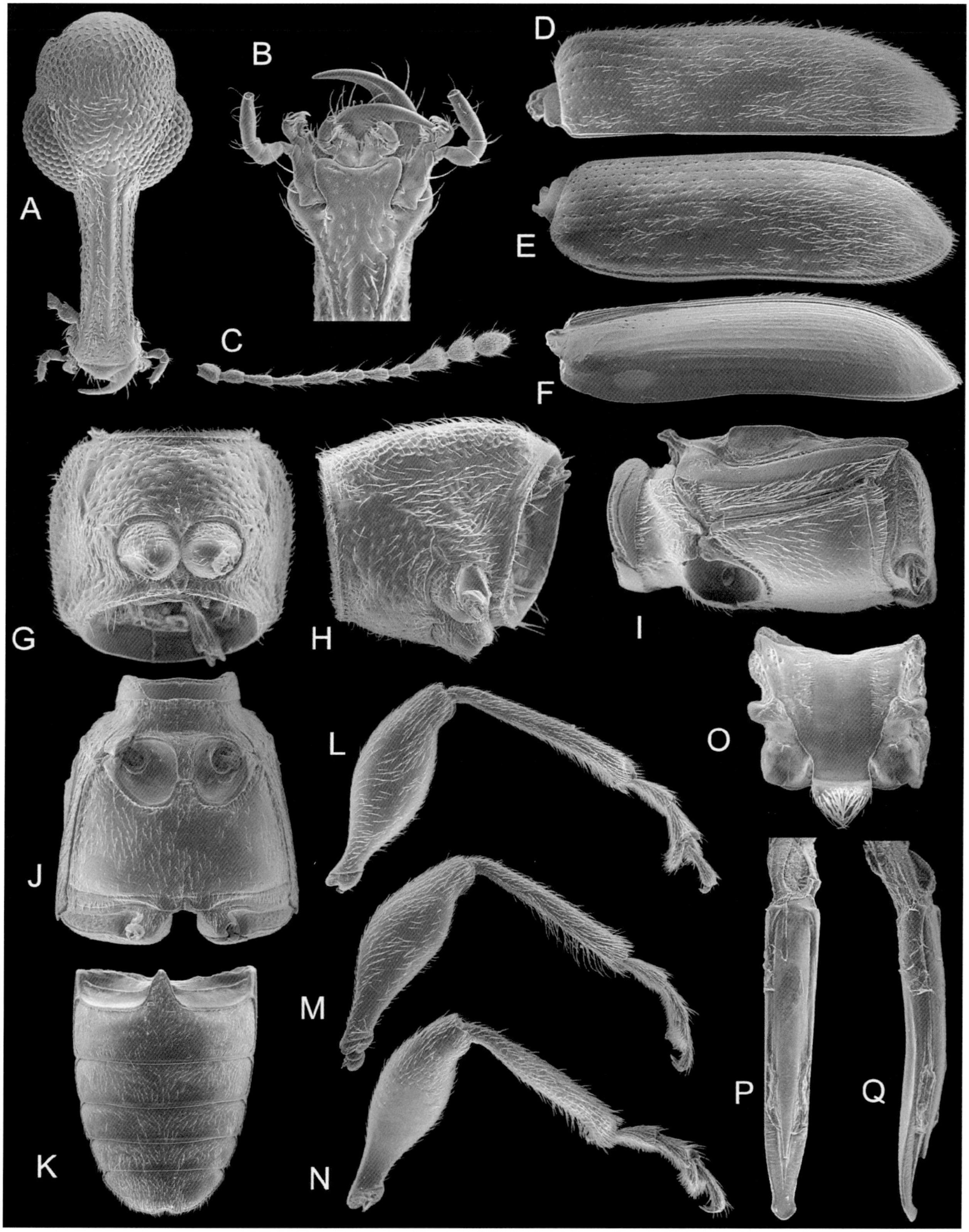

Fig. 38.4. Nemonychidae, structural details, *Aragomacer uniformis* Kuschel, ♂. A, head, dorsal; B, mouthparts, ventral; C, antenna; D–F, elytron: D, dorsal; E, lateral; F, inner lateral; G–H, prothorax: G, ventral; H, lateral; I–J, pterothorax: I, lateral; J, ventral; K, abdominal ventrites; L, foreleg; M, middle leg; N, hindleg; O, scutellum; P–Q, penis: P, dorsal; Q, lateral.

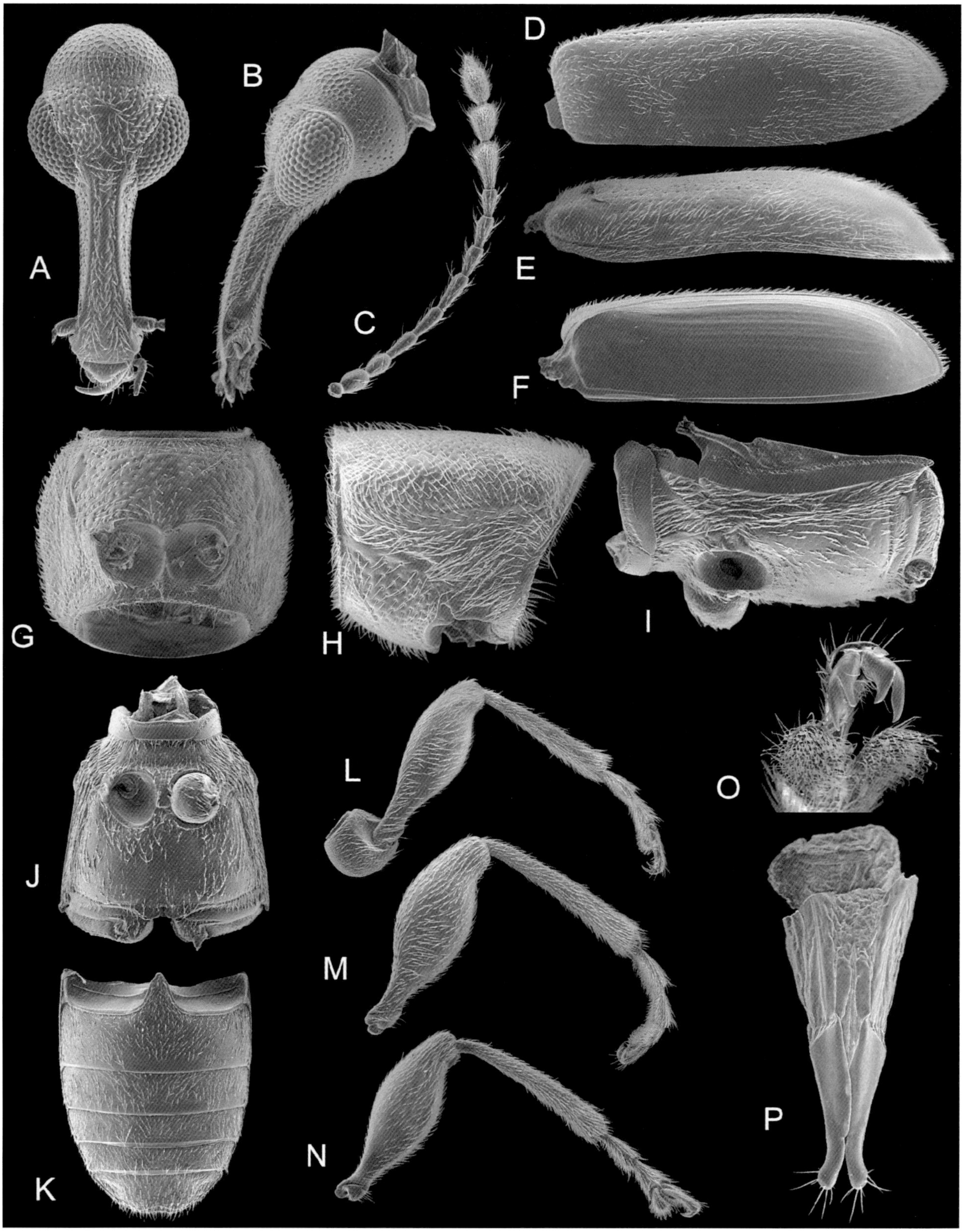

Fig. 38.5. Nemonychidae, structural details, *Aragomacer uniformis* Kuschel, ♀. A–B, head: A, dorsal; B, lateral; C, antenna; D–F, elytron: D, dorsal; E, lateral; F, inner lateral; G–H, prothorax: G, ventral; H, lateral; I–J, pterothorax: I, lateral; J, ventral; K, abdominal ventrites; L, middle leg; M, foreleg; N, hindleg; O, tarsal claws; P, ovipositor.

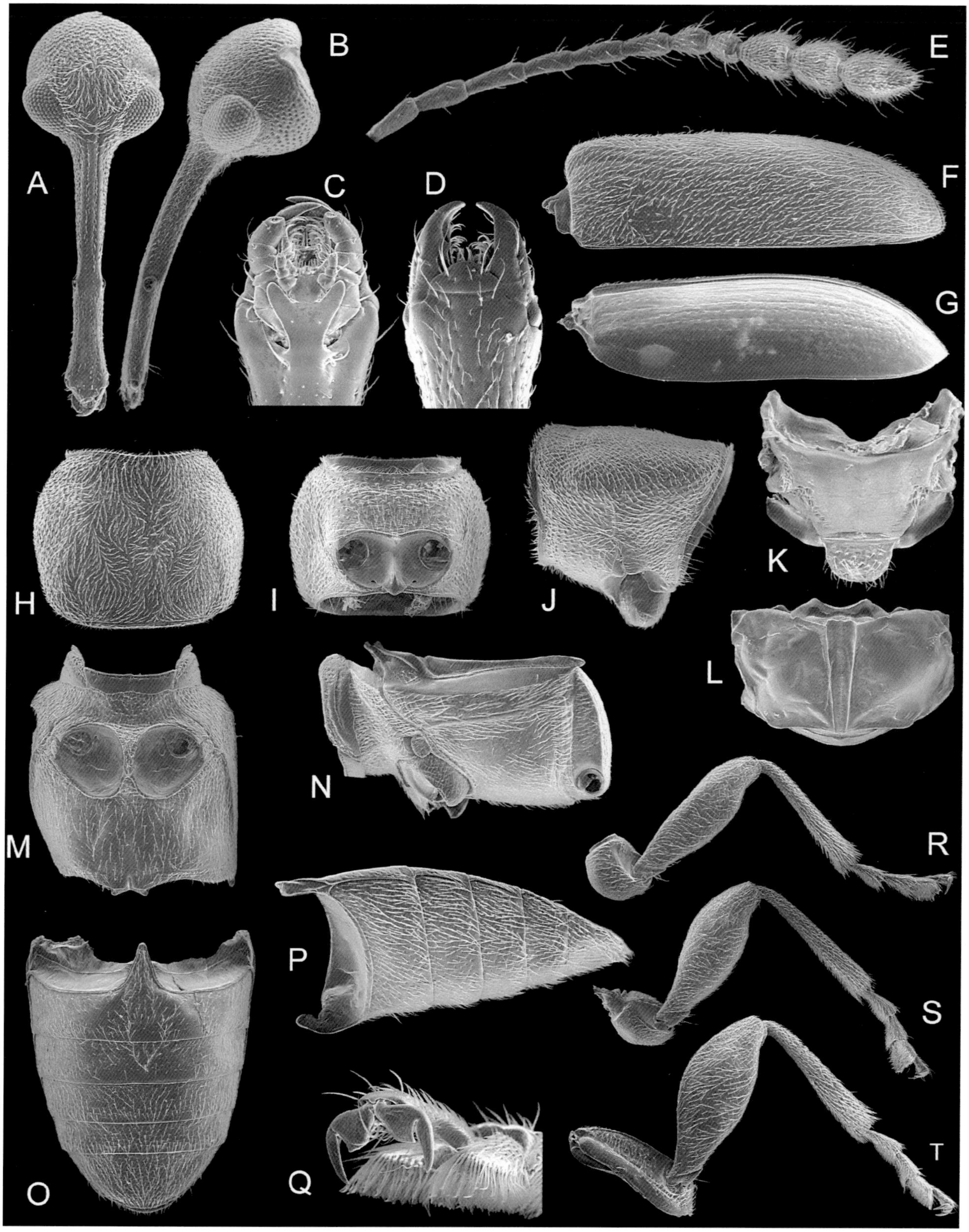

Fig. 38.6. Nemonychidae, structural details, *Basiliorhinus araucariae* Kuschel, ♀. A–B, head: A, dorsal; B, lateral; C–D, mouthparts: C, ventral; D, dorsal; E, antenna; F–G, elytron: F, dorsal; G, inner lateral; H–J, prothorax: H, dorsal; I, ventral; J, lateral; K, scutellum; L, metanotum; M–N, pterothorax: M, ventral; N, lateral; O–P, abdominal ventrites: O, ventral; P, lateral; Q, tarsal claws; R, foreleg; S, middle leg; T, hindleg.

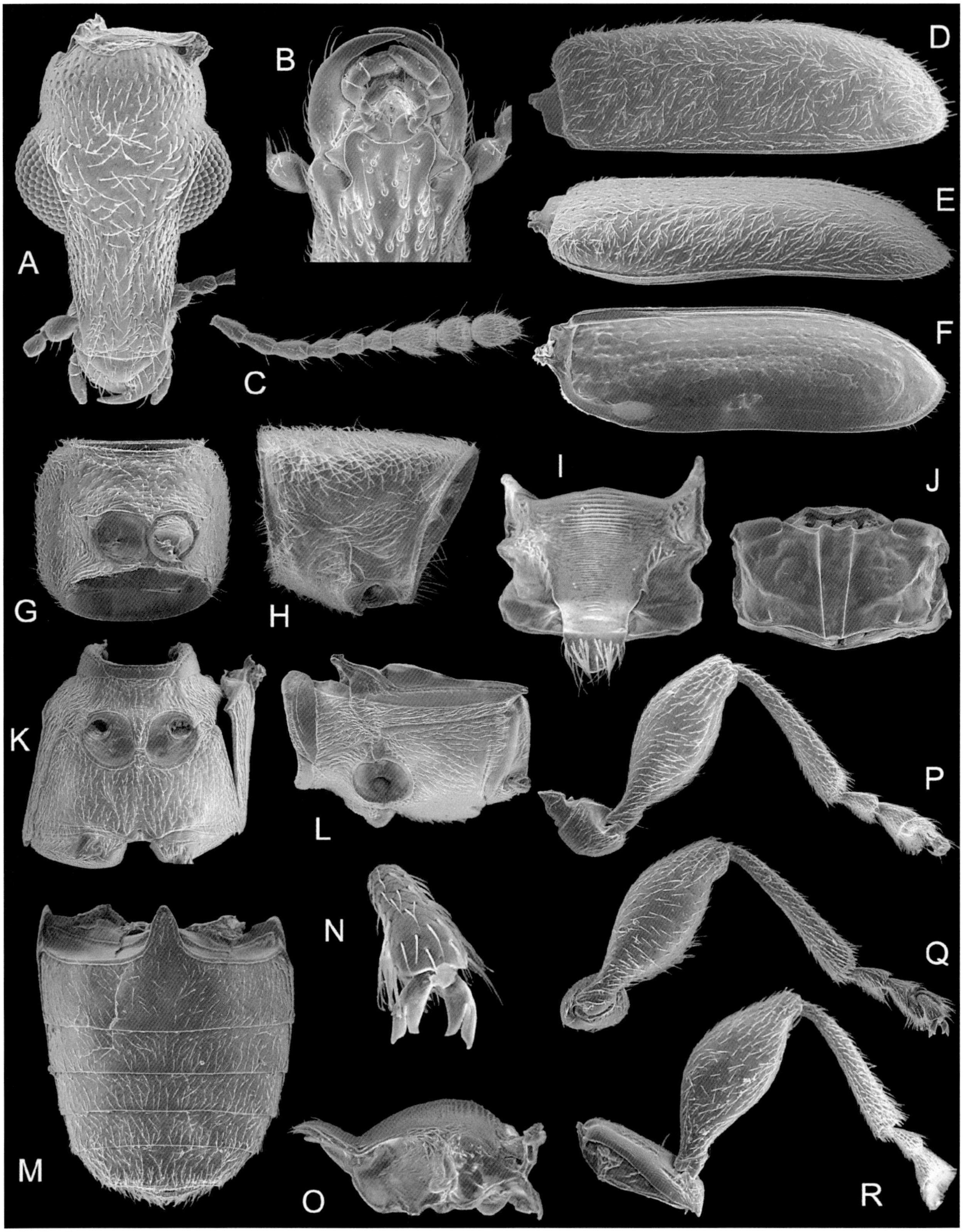

Fig. 38.7. Nemonychidae, structural details, *Bunyaeus monteithi* Kuschel, ♂. A, head, dorsal; B, mouthparts; C, antenna; D–F, elytron: D, dorsal; E, lateral; F, inner lateral; G–H, prothorax: G, ventral; H, lateral; I, scutellum; J, metanotum; K–L, pterothorax: K, ventral; L, lateral; M, abdominal ventrites; N, tarsal claws; O, scutellum, lateral; P, foreleg; Q, middle leg; R, hindleg.

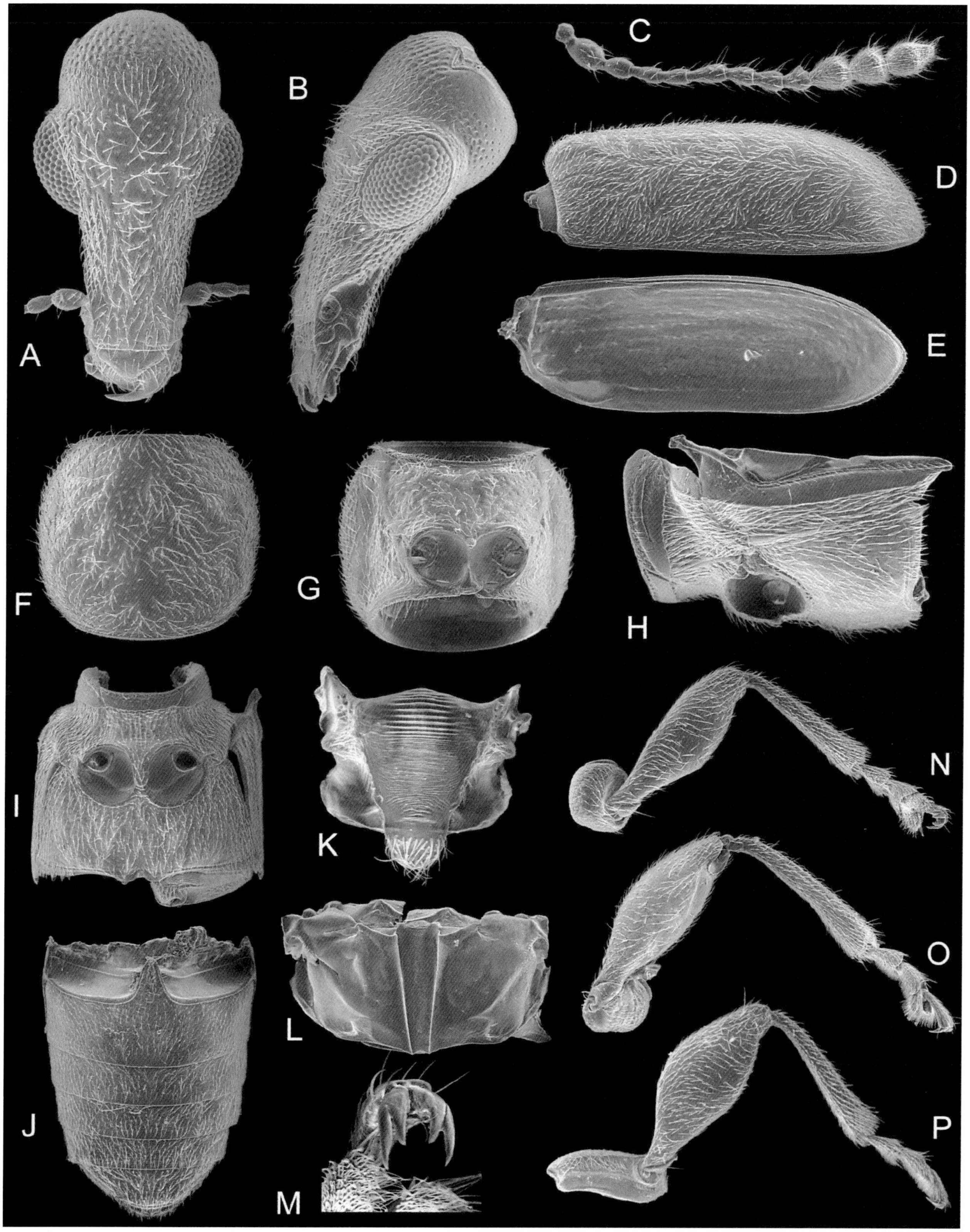

Fig. 38.8. Nemonychidae, structural details, *Bunyaeus monteithi* Kuschel, ♀. A–B, head: A, dorsal; B, lateral; C, antenna; D–E, elytron: D, dorsal; E, inner lateral; F–G, prothorax: F, dorsal; G, ventral; H–I, pterothorax: H, lateral; I, ventral; J, abdominal ventrites; K, scutellum; L, metanotum; M, tarsal claws; N, middle leg; O, foreleg; P, hindleg.

References

Alonso-Zarazaga MA, Barrios H, Borovec R, Bouchard P, Caldara R, *et al.* (2023) Cooperative catalogue of Palaeartic Coleoptera Curculionoidea. 2nd Edition. *Monografías electrónicas S.E.A.* **14**, 1-780.

Anderson DM (1991) The nemonychid weevils. In *Immature Insects. Vol. 2.* (Ed. FW Stehr) pp. 585–586. Kendall/Hunt Publishing Company, Dubuque, Iowa.

Anderson RS (2002) Chapter 125. Nemonychidae. In *American Beetles Volume 2. Polyphaga: Scarabaeoidea through Curculionoidea* (Eds RH Arnett, MC Thomas, P Skelley and JH Frank) pp. 692–694. CRC Press, Boca Raton, FL, 880 pp.

Anderson RS, Oberprieler RG, Marvaldi AE (2014) 3.1 Nemonychidae Bedel, 1882. In *Handbook of Zoology, Coleoptera, Beetles Volume 3: Morphology and Systematics (Phytophaga)* (Eds RAB Leschen and RG Beutel) pp. 301–398. De Gruyter, Berlin/Boston, xii + 675 pp.

Crowson RA (1985a) The systematic position of *Nemonyx* Redtenbacher (Col.). *Entomologist's Monthly Magazine* **121**, 144.

Crowson RA (1985b) Comments of Insecta of the Rhynie chert. *Entomologia Generalis* **11**(1–2), 97–98. doi:10.1127/entom.gen/11/1985/97

Kuschel G (1959) Nemonychidae, Belidae y Oxycorynidae de la fauna Chilena, con algunas consideraciones biogeograficas. *Investigaciones zoológicas chilenas* **5**, 229–271.

Kuschel G (1983) Past and present of the relict family Nemonychidae (Coleoptera, Curculionoidea). *Geojournal* **7**(6), 499–504.

Kuschel G (1989) The Nearctic Nemonychidae (Coleoptera: Curculionoidea). *Entomologica Scandinavica* **20**(2), 121–171. doi:10.1163/187631289X00276

Kuschel G (1993) The Palaearctic Nemonychidae (Coleoptera: Curculionoidea). *Annales de la Société entomologique de France* **29**(1), 23–46. doi:10.1080/21686351.1993.12277645

Kuschel G (1994) Nemonychidae of Australia, New Guinea and New Caledonia. In *Australian Weevils. Volume I. Orthoceri. Anthribidae to Attelabidae. The Primitive Weevils.* (Ed. EC Zimmerman) pp. 563–637. CSIRO, Australia, xxxii + 741 pp.

Kuschel G (1995) A phylogenetic classification of Curculionoidea to families and subfamilies. *Memoirs of the Entomological Society of Washington* **14**, 5–33.

Kuschel G (2003) Nemonychidae, Belidae, Brentidae (Insecta: Coleoptera: Curculionoidea). *Fauna of New Zealand* **45**, 1–100. doi:10.7931/J2/FNZ.45

Kuschel G, Leschen RAB (2011) Phylogeny and taxonomy of the Rhinorhynchinae (Coleoptera: Nemonychidae). *Invertebrate Systematics* **24**(6), 573–615. doi:10.1071/is09027

Kuschel G, May BM (1997) A new genus and species of Nemonychidae (Coleoptera) associated with *Araucaria araucana* in Brazil. *New Zealand Entomologist* **20**, 15–22. doi:10.1080/00779962.1997.9722664

Lawrence JF, Ślipiński SA (2013) *Australian Beetles. Vol. 1. Morphology, Classification and Keys*. CSIRO Publishing, Collingwood, Victoria, viii + 561 pp.

Lea AM (1926) On some Australian Curculionidae. *Proceedings of the Linnean Society of New South Wales* **51**(3), 327–362.

Legalov AA (2017) Contribution to the knowledge of the family Nemonychidae (Coleoptera) with descriptions of new taxa. *Ukrainian Journal of Ecology* **7**(2), 64–87. doi:10.15421/2017_22

Marvaldi AE (2003) Key to larvae of the South American subfamilies of weevils (Coleoptera, Curculionoidea). *Revista Chilena de Historia Natural* **76**(4), 603–612. doi:10.4067/S0716-078X2003000400005

Marvaldi AE, Morrone JJ (2000) Phylogenetic systematics of weevils (Coleoptera: Curculionoidea) A reappraisal based on larval and adult morphology. *Insect Systematics & Evolution* **31**, 43–58.

May BM (1993) *Fauna of New Zealand. Number 28. Larvae of Curculionoidea (Insecta: Coleoptera) a Systematic Overview*. Manaaki Whenua Press, Lincoln, New Zealand, 226 pp.

May BM (1994) An Introduction to the Immature Stages of Australian Curculionoidea. In *Australian Weevils. Volume II. Brentidae, Eurhynchidae and Apionidae and a chapter on Immature Stages by Brenda May*. (Ed. EC Zimmerman) pp. 365–728. CSIRO, Australia, x + 751 pp.

Oberprieler RG, Oberprieler, SK (2012) *Talbragarus averyi* gen. et sp. n., the first Jurassic weevil from the southern hemisphere (Coleoptera: Curculionoidea: Nemonychidae). *Zootaxa* **3478**(1), 256–266. doi:10.11646/zootaxa.3478.1.25

Shin S, Clarke DJ, Lemmon AR, Moriarty Lemmon E, Aitken AL, *et al.* (2018) Phylogenomic data yield new and robust insights into the phylogeny and evolution of weevils. *Molecular Biology and Evolution* **35**(4), 823–836. doi:10.1093/molbev/msx324

Thompson RT (1992) Observations on the morphology and classification of weevils (Coleoptera: Curculionoidea) with a key to the major groups. *Journal of Natural History* **26**, 835–891. doi:10.1080/00222939200770511

39. ANTHRIBIDAE BILLBERG, 1820

Hermes E. Escalona, Debbie Jennings and Rolf G. Oberprieler

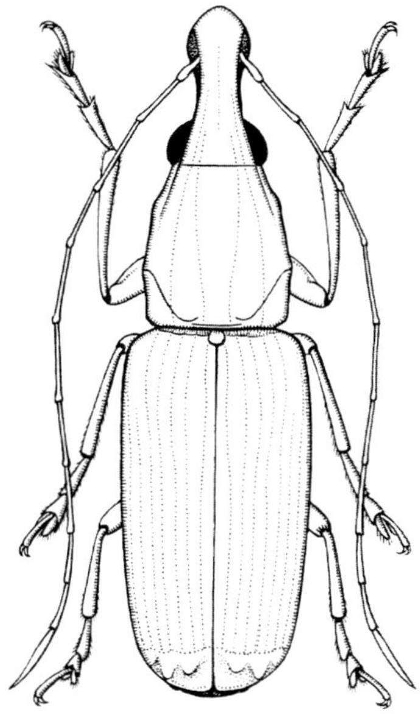

Fig. 39.1. *Ancylotropis waterhousei* Jekel.

Introduction. Anthribidae are a monophyletic basal family of Curculionoidea, as corroborated by morphological and molecular data (Kuschel 1995; Wink & Rheinheimer 1997; Marvaldi & Morrone 2000; Marvaldi *et al.* 2002; Shin *et al.* 2018). The family comprises more than 3800 species in about 369 genera and is usually divided into three subfamilies, Anthribinae, Choraginae and Urodontinae, the former two containing most of the species and with a worldwide distribution and the latter occurring only in southern Europe, northern Africa, the Near East, central Asia and in southern Africa, where it is most diverse (Louw 1993). The family is delineated by nine apomorphic morphological characters (Marvaldi *et al.* 2002; Mermudes & Leschen 2014). Although its monophyly and phylogenetic position in Curculionoidea seem to be settled, its internal classification and relationships require considerable further study, as the subfamily Anthribinae is indicated to be paraphyletic with respect to Choraginae (Kuschel 1995; McKenna *et al.* 2009) and most currently recognised tribes are poorly defined and likely artificial constructs, so that a tribal classification is mostly only used in catalogues (e.g. Alonso-Zarazaga & Lyal 1999, 2002; Lyal & Alonso-Zarazaga 2006; Bouchard *et al.* 2011; Alonso-Zarazaga *et al.* 2017, 2023).

A catalogue of the entire family was compiled by Rheinheimer (2004), and Pullen *et al.* (2014) published a checklist of the 48 genera and 85 species then recorded from Australia. The Australian fauna was revised by Zimmerman (1991, 1992, 1994), who provided a key to the 44 genera and their species as then known from the country, together with numerous photographs, drawings and extensive notes, and he also enumerated and illustrated another 28 genera recorded from Australia but not identified. The larvae of the few Australian species known, in the genera *Araecerus*, *Cacephatus*, *Euciodes*, *Eucorynus* and *Phloeobius*, were described by May (1994). The New Zealand fauna of Anthribidae was revised by Holloway (1982), the New Caledonian one by Kuschel (1998) and the Oriental one by Morimoto (1972, 1978, 1979, 1980, 1981). The males of some anthribid species bear setose patches on the sides of the abdominal ventrites, as in *Dendropemon albopictus* (Pascoe) (Zimmerman 1991, pl. 10, Fig. 5) and some species of *Taburnus* Zimmerman, but such sex patches can also occur on other parts of the body in the family (Holloway 1985; Kuschel 1998).

Anthribidae are largely mycetophagous, feeding on fungi in decaying wood as adults and larvae, but some are phytophagous, feeding on live stems, fruits and seeds, and a few are even adephagous, both adults and larvae preying on scale insects like the Eurasian genus *Anthribus* Geoffroy (Kosztarab & Kozar 1983). Little is known about the biology of anthribids in Australia. Most species likely also develop on fungi in association with dead trees (Zimmerman 1994; Kuschel 1995), but the larvae of *Euciodes suturalis* Pascoe develop in grass stems, feeding on the pith (Penman 1978), and those of some species of *Araecerus*, such as *A. fasciculatus* (De Geer), *A. lutatus* (Fairmaire), *A. palmaris* (Pascoe) and *Araecerus* sp. 2 (Zimmerman 1994) develop in fruits and seeds (often dried and mummified), to such an extent that some may on occasion be considered as pests of stored food products. *Araecerus palmaris* also commonly develops in galls produced by the rust fungus *Uromycladium tepperianum* on acacia stems (Zimmerman 1994; Bashford 2002), and *A. bicristatus* Blackburn has been reared from rotten conifer and cycad cones. The females of *Streneoderma* have the ovipositor modified into a fruit-piercing organ (Zimmerman 1994) and the larvae evidently develop in seeds, as adults of *S. contemptum* Blackburn have been reared from seeds of *Cryptocarya* (Lauraceae).

Characteristics. Adults (Figs 39.7–39.10). Body length 1.4–20.1 mm (SL) and about 1.45–2.65 × as long as wide, usually with elytra parallel-sided and pronotum at least slightly rounded or with sides of pronotum and elytra independently rounded, rarely with sides evenly rounded (e.g. *Cisanthribus*), slightly flattened to strongly convex (e.g. *Cisanthribus*), usually yellowish brown to dark brown or black, occasionally red, but often bicolored, with pattern of dark markings on elytra and occasionally pronotum; almost always densely clothed with decumbent, thickened setae usually forming a pattern, but sometimes with suberect or erect and usually fine hairs, rarely subglabrous (e.g. *Cisanthribus*). Vestiture directed forwards on the head and pronotum and backwards on the elytra. Stridulatory files on head, thorax and abdomen absent. Head not to moderately strongly declined at base, or with vertex strongly declined so that rostrum directed ventrad (e.g. *Hucus*,

Litocerus, *Mucronianus*); with short to moderately long occipital endocarina; rostrum mostly short and strongly transverse but also elongate (Figs 39.8 and 39.9), more than twice as long as wide, often more or less flattened, often with one or more longitudinal carinae dorsally; frontoclypeal suture absent, frontoclypeal apex usually truncate to slightly emarginate; labrum free, exposed, much narrower than frontoclypeus, slightly transverse and broadly rounded. Eyes slightly to strongly protuberant, usually well separate but sometimes almost contiguous (e.g. *Litocerus*, *Mucronianus*), margin entire or anteriorly shallowly to deeply emarginate, finely or coarsely facetted, without interfacettal setae. Antennal insertions exposed dorsally and located between eyes (Choraginae) or more or less concealed in dorsal view and located laterally in front of eyes (Anthribinae), sockets foveiform or produced into well-developed scrobes. Antennae non-geniculate, 11-segmented, short (not reaching beyond base of pronotum) to very long (Fig. 39.1), sometimes extending well beyond elytral apices, especially in males (e.g. *Ancylotropis*, *Antribisomus*, *Exillis*, *Mecocerinopis*, *Xenocerus*), very slender to moderately stout, usually with relatively weak club; scapes elongate but not much longer than pedicel, often somewhat inflated; clubs loose, 3-segmented (appearing 4-segmented in *Eucorynus*), occasionally stronger or more compact. Mandibles short and broad, unidentate or bidentate, incisor edge with or without tooth; with well-developed or reduced mola but no prostheca; inner edge of mandibular base with short to moderately long, pubescent process; outer side usually with deep setiferous groove. Maxillae with slender, articulated, setose galea and slightly wider setose lacinia, the latter with or without uncus; maxillary palps moderately well-developed, 4-segmented, apical segment fusiform. Prementum slightly transverse, apically emarginate, more or less fused at base; ligula bilobed; labial palps 3-segmented, apical segment fusiform. Gular region usually concave or transversely grooved; gular sutures absent. Corpotentorium absent; anterior tentorial arms not expanded mesally; cervical sclerites absent.

Pronotum about 0.40–1.0 × as long as wide, usually widest posteriorly or in middle; base not or only slightly narrower than base of elytra, sides usually slightly curved; posterior angles usually right or obtuse, occasionally acute; posterior edge straight, evenly curved or occasionally weakly sinuate, usually weakly marginate; disc slightly convex, almost always with distinct, transverse, finely crenulate bracteate carina located just anteriorly of basal pronotal margin (subbasal) or slightly further in front (antebasal), carina evenly rounded or sinuate, occasionally interrupted in middle, laterally usually curved and extended anteriad into a lateral carina on either side, this usually not reaching beyond middle of pronotum but sometimes almost to anterior edge (e.g. *Basitropis*, *Dendropemon*), occasionally strongly reduced or absent, from corner of transverse and lateral carina often also a short carinule extending to posterior corner of pronotum; pronotum behind antebasal transverse carina flat or slightly declined, behind subbasal carina steeply declined or vertical so that carina abutting anterior edges of elytra. Prosternum shorter to distinctly longer than shortest diameter of procoxal cavity, flat to slightly convex, not carinate; prosternal process complete but interrupted in middle, intercoxal process strongly concave or apparently divided by transverse suture, apex expanded and truncate and more or less fused to postcoxal (notal) projections; notosternal sutures probably absent but a suture sometimes extending dorsolaterad from lateral edge of procoxal cavity to lateral edge of pronotal disc. Procoxae not or slightly projecting below prosternum, with long concealed lateral extensions; trochantins concealed. Procoxal cavities subcircular, contiguous to moderately widely separate but joined by transverse suture, externally broadly closed, without narrow lateral extensions, internally broadly closed; postcoxal (hypomeral) processes fused to disjunct apex of prosternal process.

Scutellar shield well-developed, highly reduced or absent, abruptly elevated, anteriorly simple, posteriorly acute to broadly rounded or truncate. Elytra about 0.85–2.0 × as long as wide and 1.75–3.0 × as long as pronotum; anterior edges often sharp or with raised carina; humeri well-developed or not; sides usually subparallel to weakly rounded and conjointly rounded to slightly truncate at apex; pygidium often exposed, strongly declined or vertical; usually with 10 distinct striae (puncture rows) and short scutellary striole, sometimes with various elevations or tubercles, with posteriadly directed horns in *Commista*; apical end of elytral suture with internal interlocking ridge; epipleura usually narrow and incomplete. Mesoventrite partly fused to or separated by complete sutures from mesanepisterna, these distinctly separate at midline; anterior edge on different plane than metaventrite, with broadly transverse prothoracic rest. Mesoventral cavity absent. Mesocoxae projecting, with concealed trochantins. Mesocoxal cavities moderately to very widely separate, subcircular, not or slightly oblique, closed laterally by meeting of mesoventrite and metaventrite. Mesoventral process narrowly rounded to broadly rounded or truncate at apex; mesometaventral junction usually formed by an abutting or overlapping of ventrites. Metaventrite slightly convex; transverse (metakatepisternal) suture long but close to posterior edge of ventrite; discrimen usually complete to base of anterior intercoxal process, occasionally short or absent; postcoxal lines usually absent, occasionally arched and strongly recurved; metaventrite reduced in wingless forms; exposed portion of metanepisterna moderately elongate; metepimera slightly exposed. Metacoxae transverse, globular in *Cisanthribus*, narrowly to widely separate, not extending laterad to reach elytra. Metendosternite usually with long stalk and lateral arms, well-developed laminae, slender and very short or barely indicated anterior process bearing narrowly separate anterior tendons. Hindwings with apical field at least half as long as total wing length, containing distinct sclerite just beyond radial cell and two RP remnants; radial cell small, short and broad or absent; inner posterobasal angle right; cross-vein r_3 moderately long and oblique to absent, sometimes fused basally with r_4; r_4 sometimes with apical spur; basal portion of RP short to moderately long; R-M loop relatively narrow; medial spur usually extending almost to wing margin; medial field with three or fewer free veins and no medial fleck; MP_{3+4} usually without base; wedge cell absent; anal lobe usually well-developed without or with weak embayment; wings sometimes absent. Legs short and stout to moderately long and slender;

trochanterofemoral joints strongly oblique with base of femur sometimes meeting coxa; femora not to slightly inflated; tibial spurs absent. Tarsi 5-5-5 in both sexes but pseudotetramerous (segment 4 strongly reduced), segment 3 distinctly bilobed; tarsal claws usually mostly bifid, occasionally simple, without ventrobasal seta; empodium not observed.

Abdomen with 5 ventrites, first 4 or all 5 connate; ventrite 1 at its shortest point subequal in length to 2, usually without postcoxal lines (occasionally with recurved lines just behind edges of coxal cavities); intercoxal process acute to broadly rounded or rarely truncate; ventrites 2, 3 and 4 laterally with pairs of large sex patches in males of *Dendropemon*. Abdominal terga usually lightly pigmented except for tergite VII (pygidium), this with basal median notch interlocking with elytral apices; tergites III–VI in *Araecerus* with moderately sclerotised plates and numerous rounded microsclerites; patches of wing-folding asperities sometimes present on tergites V and VI; functional spiracles on segments I–VII, those on I–VI in pleural membrane, those on VII at lateral edges of tergite (sometimes reduced). Tergite VIII in male partly concealed beneath pygidium and relatively weakly sclerotised; sternite VIII somewhat reduced, without median spicule. Anterior edge of segment IX in male with distinct spiculum gastrale; tergite X membranous. Aedeagus of pedotectal type, usually symmetrical; anterior edge of tegmen with moderately long manubrium; parameres fused together and sometimes to phallobase. Penis flattened and usually curved, with distinct tectum and pedon and long, paired anterior apodemes (temones); endophallus with complex armature. Segment VIII in female more or less membranous, with short, fixed spiculum ventrale. Ovipositor elongate and weakly sclerotised, except for baculi and distal gonocoxites; proctigers elongate, with paired baculi; paraprocts more or less subequal in length to gonocoxites, with longitudinal baculi; gonocoxites divided into elongate proximal part with sclerotised longitudinal baculum and short, broad, distal part, the latter heavily sclerotised with lateradly directed teeth and subterminal, sclerotised styli. Internal tract with strongly curved, sclerotised capsular spermatheca. (Valentine 1960, 1999, 2002; Morimoto 1972, 1978, 1979, 1980, 1981; Holloway 1982, 1984, 1985; Thompson 1992; Zimmerman 1994; Lawrence & Ślipiński 2013; Mermudes & Leschen 2014).

Larvae (Figs 39.3–39.5). Body elongate (Figs 39.3A, 39.4J), usually more or less parallel-sided and circular in cross-section, sometimes with enlarged prothorax; all surfaces lightly pigmented except for head capsule or mouth-frame and mandibles; vestiture consisting of scattered fine, simple setae, sometimes more abundant on thorax. Head hypognathous, usually transverse, protracted and more or less circular in outline or partially retracted, posterolaterally compressed and narrowed behind; posterior edge not emarginate, occasionally produced and acute; postoccipital condyles absent; ecdysial lines often obscure; epicranial stem long; frontal arms V-shaped, contiguous at base; median endocarina extending anteriorly between frontal arms. Stemmata 3 or fewer on each side, one stemma with lens, and 1 or 2 additional eye spots. Antennae very short, 1-segmented, with terminal, conical sensorium. Frontoclypeal suture distinct or indistinct; labrum free, with four pairs of primary setae and lateral tormae. Mandibles

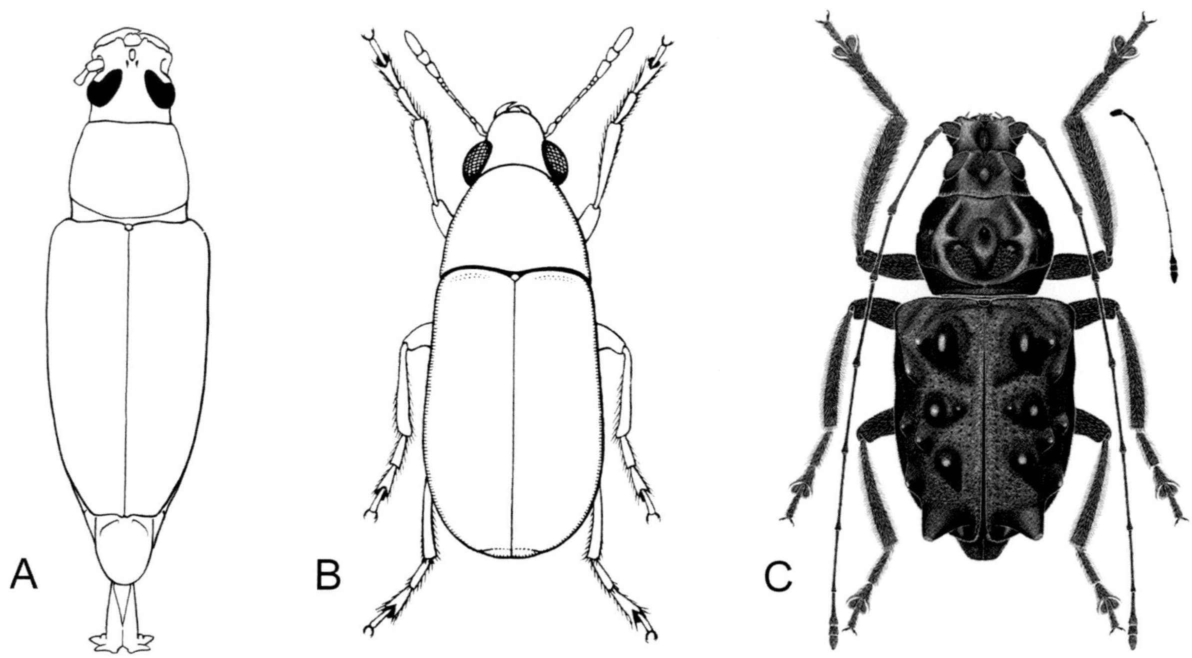

Fig. 39.2. Anthribidae, habitus, dorsal. A, *Xenocerus* sp.; B, *Euparius lateripictus* (Jordan); C, *Commista baccula* Jordan. A and C after Zimmerman (1994); B after Lawrence & Ślipiński (2013).

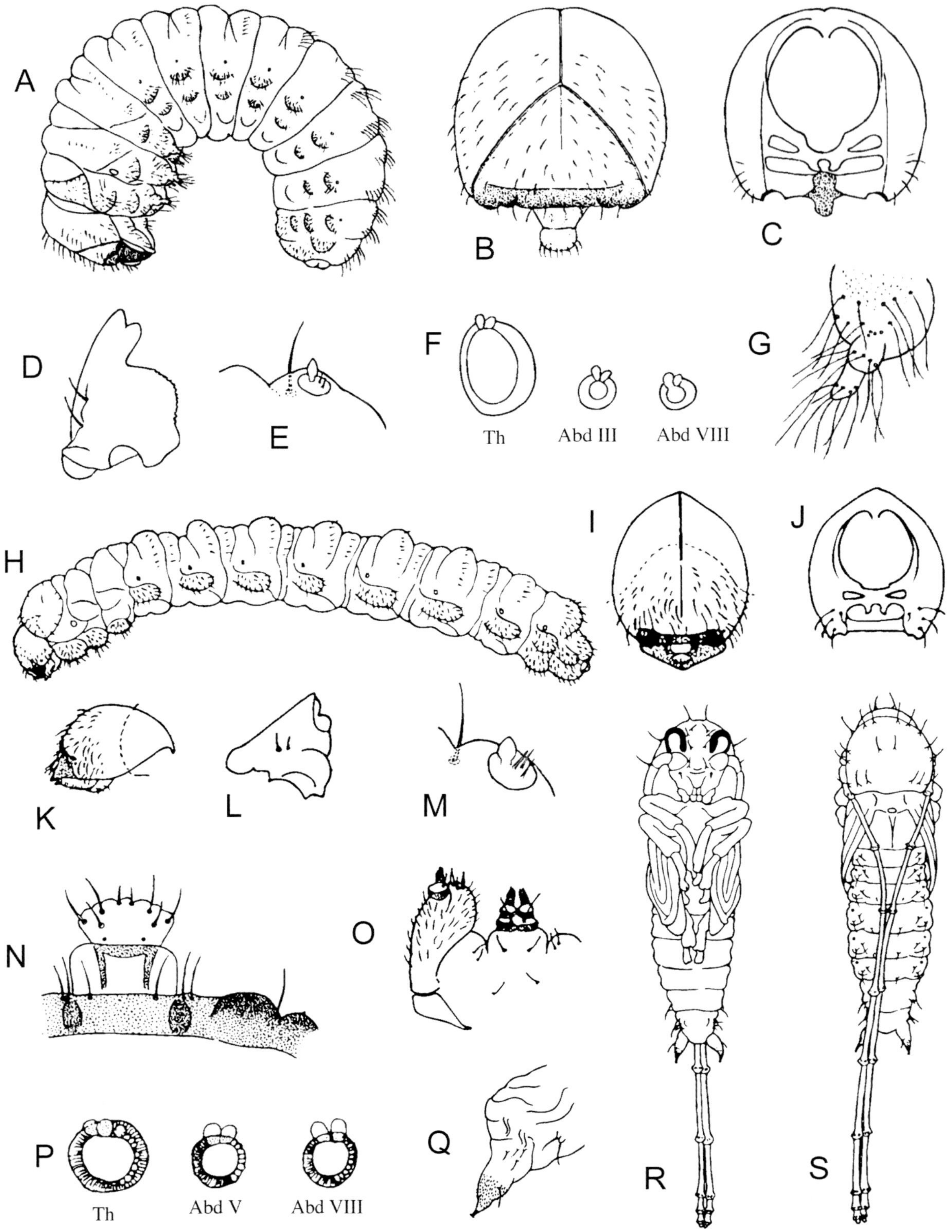

Fig. 39.3. Anthribinae, structural details of immature stages. A–G, *Cacephatus huttoni* (Sharp), larva from New Zealand: A, habitus, lateral; B, head, dorsal; C, head, ventral; D, mandible; E, antenna; F, spiracles, showing alignment; G, pedal papilla; H–P, *Euciodes suturalis* Pascoe, larva from New Zealand: H, habitus, lateral; I, head, dorsal; J, head, ventral; K, head, lateral; L, mandible; M, antenna; N, labrum, clypeus, epistoma and antenna; O, mala and labium, ventral; P, spiracles, showing alignment; Q–S, *Euciodes suturalis* Pascoe, pupa from New Zealand: Q, pseudocercus; R, ♂, ventral; S, ♂, dorsal. After May (1994).

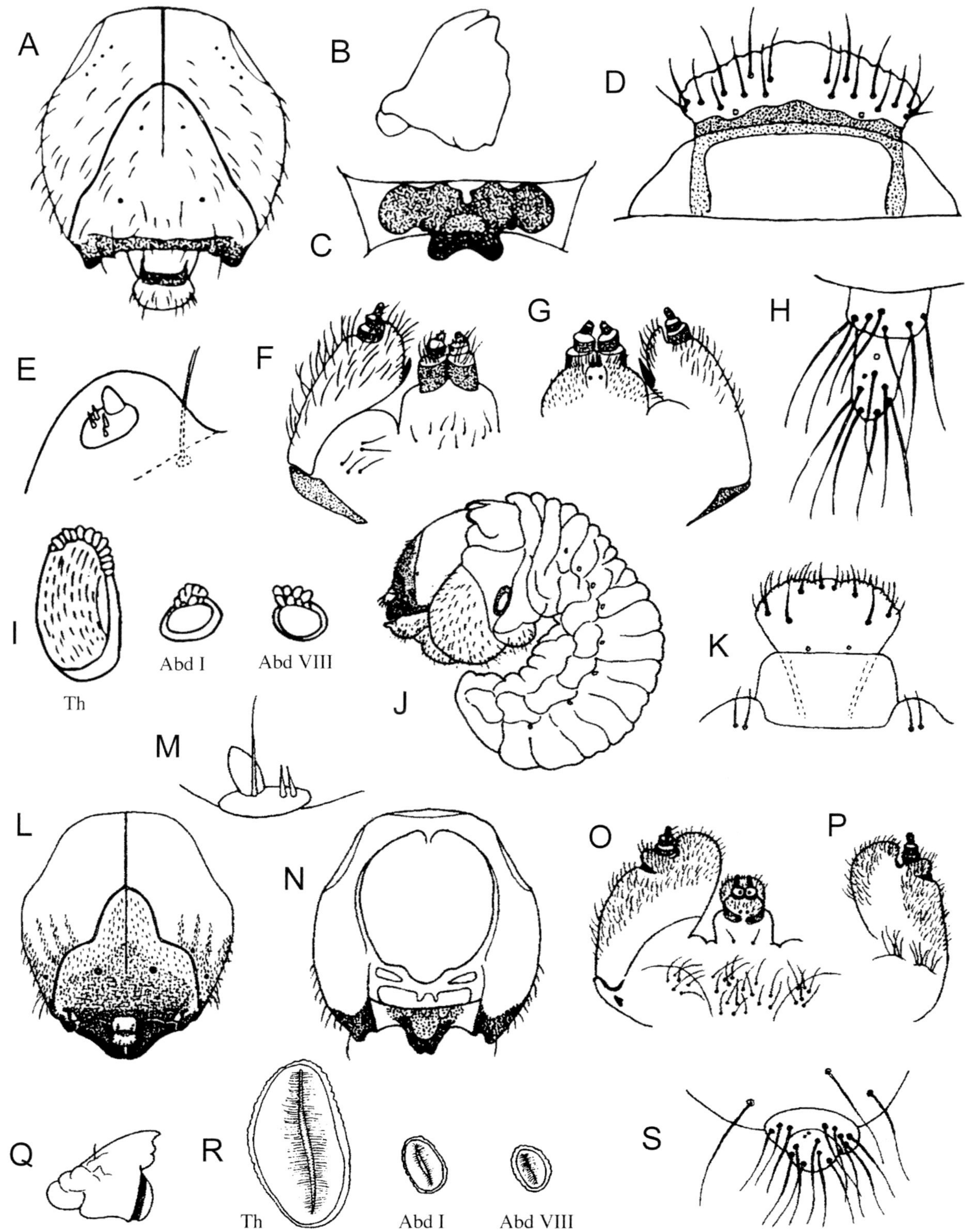

Fig. 39.4. Anthribinae, structural details of larvae. A–I, *Eucorynus mastersi* Blackburn: A, head, dorsal; B, mandible; C, hypopharyngeal sclerome, D; labrum and clypeus; E, antenna, on ventral surface; F, mala and labium, ventral; G, mala and labium, dorsal, H; pedal papilla; I, spiracles, showing alignment; J–S, *Phloeobius gigas* (Fabricius): J, lateral habitus; K, labrum and clypeus; L, head, dorsal; M, antenna; N, head, ventral; O, mala and labium, ventral; P, mala, dorsal; Q, mandible, showing smooth dorsal face; R, spiracles, showing alignment; S, pedal papilla. After May (1994).

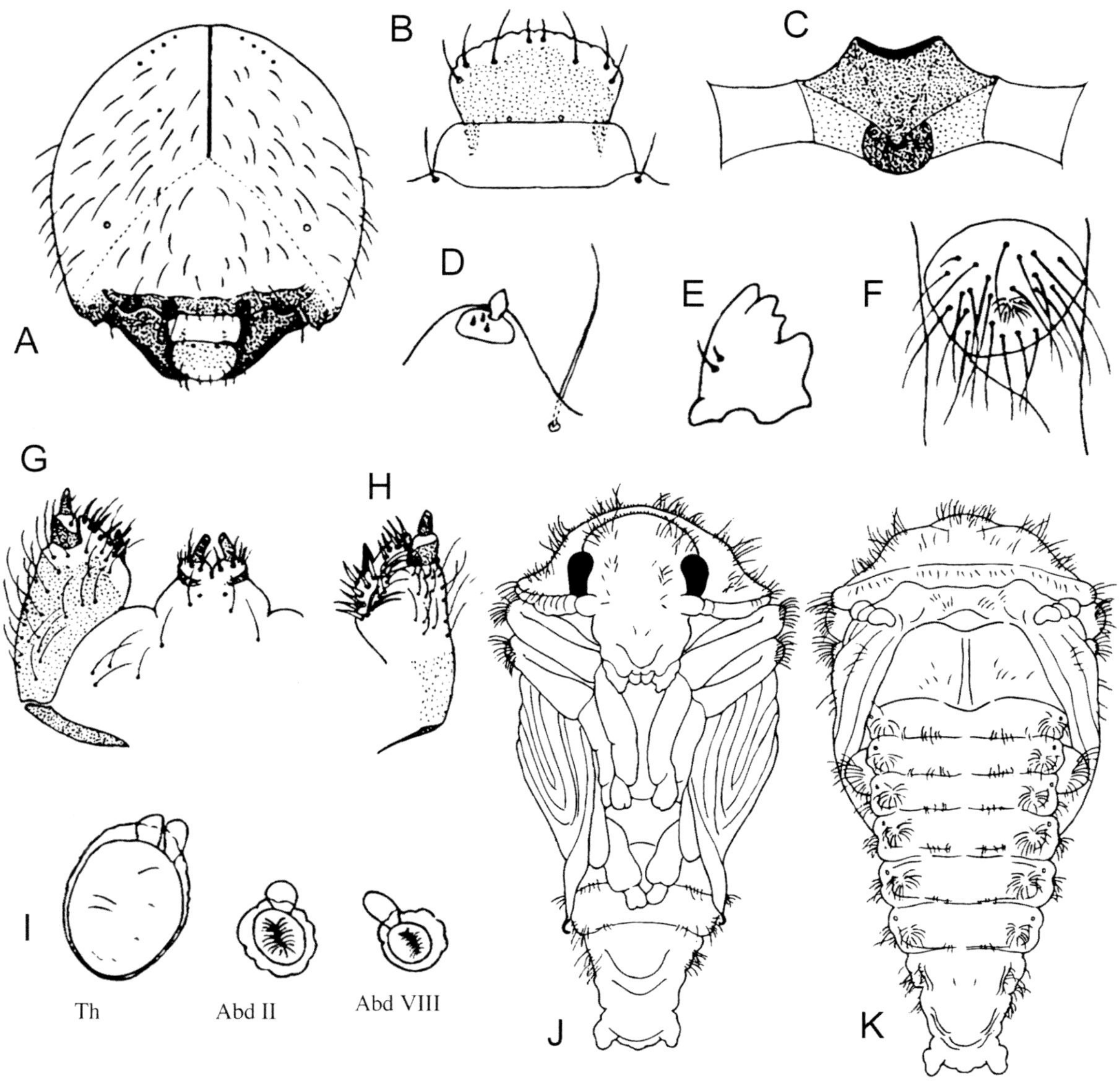

Fig. 39.5. Choraginae, structural details of immature stages. A–I, *Araecerus palmaris* (Pascoe) larva from New Zealand: A, head, dorsal; B, labrum and clypeus; C, hypopharyngeal sclerome; D, antenna; E, mandible; F, pedal area; G, mala and labium, ventral; H, mala, dorsal; I, spiracles, showing alignment; J–K, *Araecerus palmaris* (Pascoe) pupa from New Zealand: J, ventral; K, dorsal. After May (1994).

short and broad, obliquely bidentate; incisor edge usually with subapical tooth, mola usually well-developed (sometimes reduced); prostheca absent. Ventral mouthparts retracted; maxillary articulating area present; cardo slightly to strongly oblique, undivided; stipes longer than wide; usually with fixed, rounded, setose mala and sclerotised laciniar spine (absent in *Euciodes* and *Phloeobius*); maxillary palps 2- or 3-segmented, occasionally with distinct palpifer. Labium consisting of prementum and postmentum; ligula absent; labial palps usually 2-segmented (1-segmented in *Araecerus*), narrowly separated. Hypopharyngeal bracon present; hypopharyngeal sclerome usually present (absent in *Euciodes*). Hypostomal rods usually short and subparallel or converging; ventral epicranial ridges absent. Gular region very short or absent. Tentorial bridge divided on either side of midline.

Prothoracic tergum usually about as long as meso- or metatergum (as long as the two combined in *Phloeobius*), without pigmented plate. Legs often present (absent in *Euciodes*), usually very short and 1- or 2-segmented, without claw-like pretarsus. Abdomen more than twice as long as thorax (more than 4 × as long in *Euciodes*); abdominal terga not extending laterally beyond edges of sterna; most abdominal terga with 2 to 4 plicae; terga I–VI each with transverse ambulatory ampulla in *Euciodes*; segment IX usually shorter than VIII; completely dorsal; sternum IX simple; segment X much shorter than IX, terminal; anal region orientated posteriad; anal opening X-shaped (4-lobed). Spiracles annular-biforous, annular-uniforous, annular-multiforous or annular; thoracic spiracles intersegmental or on mesothorax. (Anderson 1947; May 1993, 1994; Valentine 2002; Marvaldi 2003).

Key to the Australian subfamilies and genera

Modified after Zimmerman (1994).

1. Antennal insertions lateral or dorsal (Fig. 39.8A–B); eyes not or deeply emarginate (Figs 39.7B, 39.9E); pronotum mostly with transverse carina close of basal pronotal margin (subbasal) or well separated from basal pronotal margin (antebasal, Fig. 39.6H, K) ..... **ANTHRIBINAE** ... 2

– Antennal insertions visible in dorsal view (Fig. 39.6A); eyes barely emarginate; pronotum with transverse carina close of basal pronotal margin (subbasal) ..... **CHORAGINAE** ... 48

2(1). Rostrum transverse and wider than head capsule (Pl. 27C); antennae filiform and at least twice the length of the body; Norfolk Island ..... ***Antribisomus* Perroud**

– Rostrum transverse or not, not wider than head capsule; antennae often capitate and less than twice the length of the body 3

3(2). Pronotum with or without a transverse subbasal carina 4

– Pronotum with a transverse antebasal carina 17

4(3). Procoxal cavities separated by more than width of procoxal cavity (Fig. 39.10G) 5

– Procoxal cavities separated by less than width of procoxal cavity (Fig. 39.7J) 8

5(4). Body length <4.0 mm; tarsal claws bifid (Fig. 39.7U) ..... ***Streneoderma* Blackburn**

– Body length >5.0 mm; tarsal claws simple (Fig. 39.10H) 6

6(5). Body length >10 mm; pronotum and elytra tuberculate; scrobes with posterior margin well separate from eye (Fig. 39.9I) ..... ***Platystomos* Hellwig**

– Body length <10.0 mm; pronotum and elytra not tuberculate; scrobes with posterior margin in contact with eye 7

7(6). Eyes very large, deeply emarginate (Fig. 39.6B); vertex narrower than eye; lateral carinae of prothorax incomplete; winged species, widely distributed ..... ***Phloeobius* Schoenherr**

– Eyes small, subcircular, not emarginate; vertex several times as wide as eye (Fig. 39.9D); lateral carinae of prothorax complete; flightless, restricted to Lord Howe Island ..... ***Howeanthribus* Zimmerman**

8(4). Lateral carinae of pronotum extending nearly to apex of pronotum 9

– Lateral carinae of pronotum not extending beyond middle of pronotum 10

9(8). Antennae not reaching base of prothorax; scrobes ventral (Fig. 39.8D) ..... ***Basitropis* Jekel**

– Antennae extending beyond middle of elytra; scrobes lateral (Fig. 39.9J) ..... ***Protaedus* Pascoe** (part)

10(8). Scrobes distinctly separate from eyes 11

– Scrobes contiguous or subcontiguous with eyes (Fig. 39.7B) 13

11(10). Body length 9–14 mm; posterior angles of pronotum projecting posteriad (Fig. 39.6C) ..... ***Taburnus* Zimmerman**

– Body length <5 mm; posterior angles of pronotum not projecting posteriad (Fig. 39.6D) 12

12(11). Eyes emarginated anteriorly; distance between eyes distinctly narrower than rostrum width ..... ***Plesiobasis* Jordan**

– Eyes entire, not emarginated anteriorly; distance between eyes about as wide as rostrum width ..... ***Erichsonocis* Zimmerman**

13(10). Antennae reaching beyond base of elytra 14

– Antennae not reaching beyond base of elytra 16

14(13). Antennae reaching only to about middle of elytra ..... ***Mauia* Blackburn**

– Antennae reaching beyond middle of elytra 15

15(14). Eyes slightly emarginate (Fig. 39.9J) ..... ***Protaedus* Pascoe** (part)

– Eyes deeply emarginate (Fig. 39.9B) ..... ***Exillis* Pascoe** (part)

16(13). Scutellar shield invisible ..... ***Ozotomerus* Perroud**

– Scutellar shield visible ..... ***Euparius* Schoenherr**

17(3). Eyes with anterior margin strongly emarginate 18

– Eyes with anterior margin entire or barely emarginate 22

18(17). Rostrum with tubercles surrounding antennal insertions (Fig. 39.9J) ..... ***Protaedus* Pascoe** (part)

– Rostrum without tubercles surrounding antennal insertions 19

19(18). Scrobes with posterior margins well distant from anterior margin of eye ..... ***Notoecia* Blackburn**

– Scrobes almost touching or slightly invading ocular emargination 20

20(19). Body length >10.0 mm; with pattern of white lines on brown background ..... ***Xenocerus* Schoenherr**

– Body length <8.0 mm; with mixed maculation 21

21(20). Lateral pronotal carinae very short, not reaching notosternal suture (Fig. 39.7J) ..... ***Euciodes* Pascoe**

– Lateral pronotal carinae reaching notosternal suture ..... ***Exillis* Pascoe** (part)

22(17). Scutellar shield mostly or entirely concealed (Fig. 39.6E) ..... ***Teratanthribus* Zimmerman**

– Scutellar shield exposed 23

23(22). Scrobes at least partly visible in dorsal view 24

– Scrobes not visible in dorsal view 37

24(23). Eyes horizontally elongate, ovate 25

– Eyes subcircular in outline 29

25(24). Anterior angles of pronotum conspicuously acutely produced outward and downward (Fig. 39.6F) ..... ***Genethila* Pascoe**

– Anterior angles of pronotum obtuse, not produced 26

26(25). Pronotum with an unusual, additional, medially interrupted transverse carina in front of antebasal carina (Fig. 39.1) ..... ***Ancylotropis* Jekel**

– Pronotum with a single transverse antebasal carina 27

27(26). Antennae not or hardly extending caudad of base of prothorax; elytral declivity tuberculate (Fig. 39.6G) ..... ***Telala* Jordan** (part)

– Antennae extending distinctly caudad of base of elytra; elytra not tuberculate 28

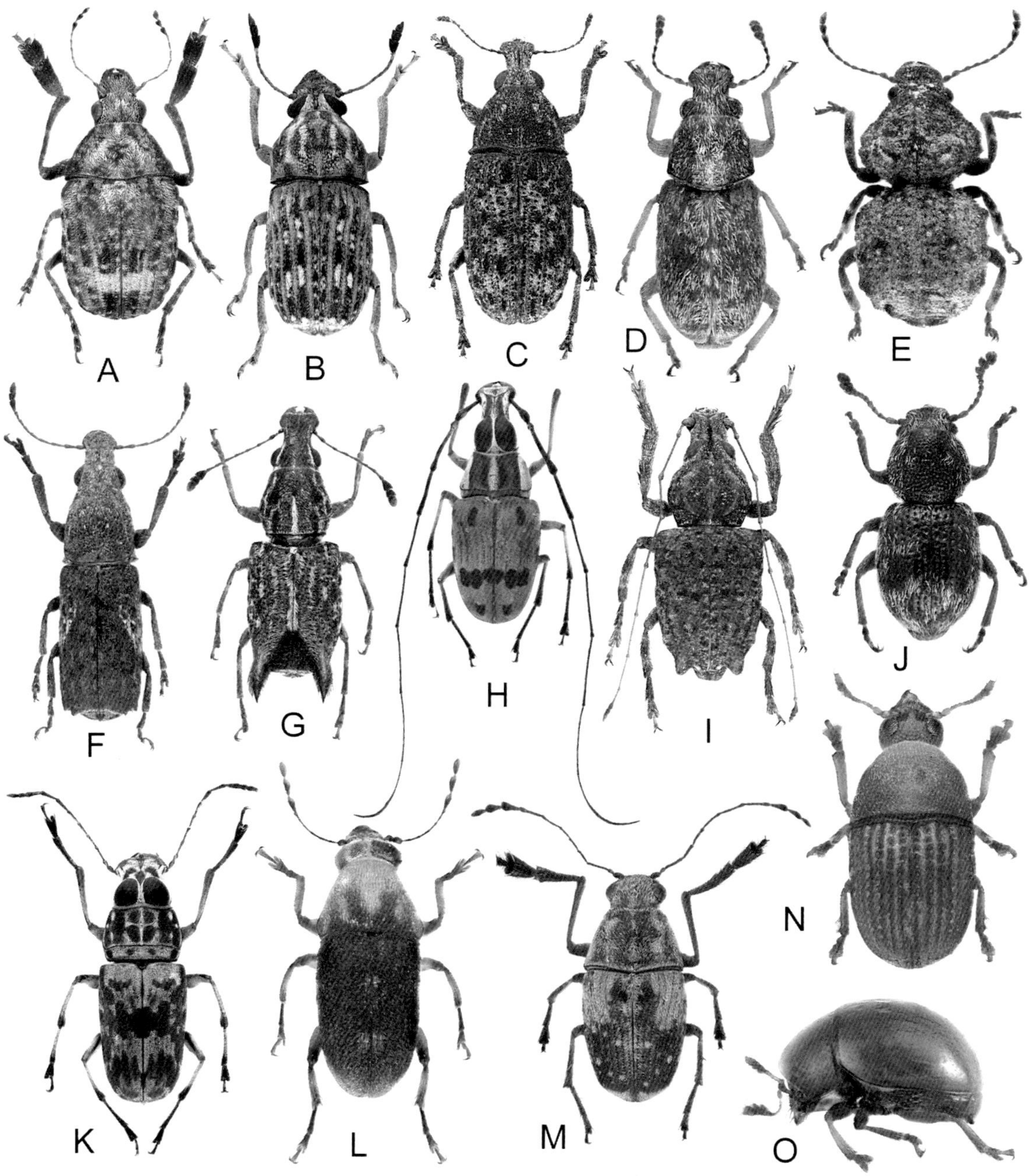

Fig. 39.6. Anthribidae, characters for key to genera, habitus. A, *Araecerus palmaris*, ♂ (Pascoe); B, *Phloeobius gigas* (Fabricius), ♀; C, *Taburnus neglectus* Zimmerman; D, *Erichsonocis alboginosus* (Erichson); E, *Teratanthribus* sp.; F, *Genethila retusa* Pascoe; G, *Telala* sp.; H, *Mecocerinopis amabilis* (Pascoe), ♂; I, *Commista baccula* Jordan; J, *Xynotropis micans* Blackburn; K, *Mucronianus rufipes* Jordan; L, *Deropygus* sp.; M, *Misthosima* sp., ♂; N, *Stenorhis* sp.; O, *Cisanthribus* sp.

28(27). Metatibiae notably compressed and widened laterally .... ***Eczesaris* Pascoe**

– Metatibiae subcylindrical .... ***Cacephatus* Blackburn** (males)

29(24). Antennal sockets sulciform, long (scrobes), almost reaching front of eye and underside of rostrum **30**

– Antennal sockets foveiform or sulciform but not close to eyes **31**

30(29). Antennal funicles 6-segmented and clubs 4-segmented; lateral carinae of pronotum not extending anteriad beyond middle .... ***Eucorynus* Schoenherr**

– Antennal funicles 7-segmented and clubs 3-segmented; lateral carinae of pronotum extending anteriad beyond middle .... ***Dendropemon* Schoenherr**

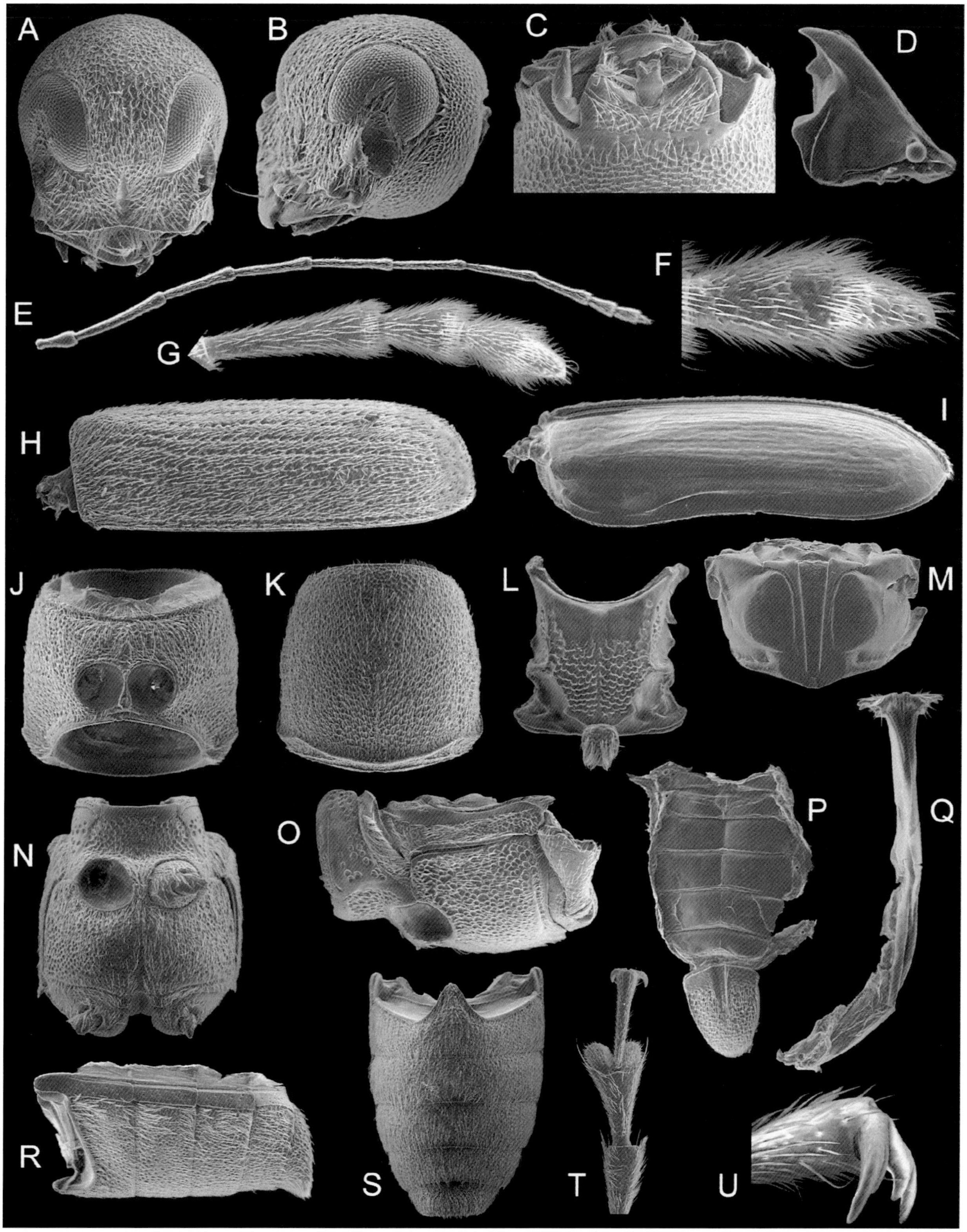

Fig. 39.7. Anthribinae, structural details, *Euciodes* sp., ♀. A–B, head: A, anterior; B, lateral; C, mouthparts; D, mandible; E–G, antenna: E, segments 2–11 (without scape); F, club segment 3; G, club; H–I, elytron: H, dorsal; I, inner lateral; J–K, prothorax: J, ventral; K, dorsal; L, scutellum; M, metanotum; N–O, pterothorax: N, ventral; O, lateral; P, tergites, dorsal; Q, ovipositor; R–S, abdominal ventrites: R, lateral; S, ventral; T, tarsus, dorsal; U, bifid tarsal claws, lateral.

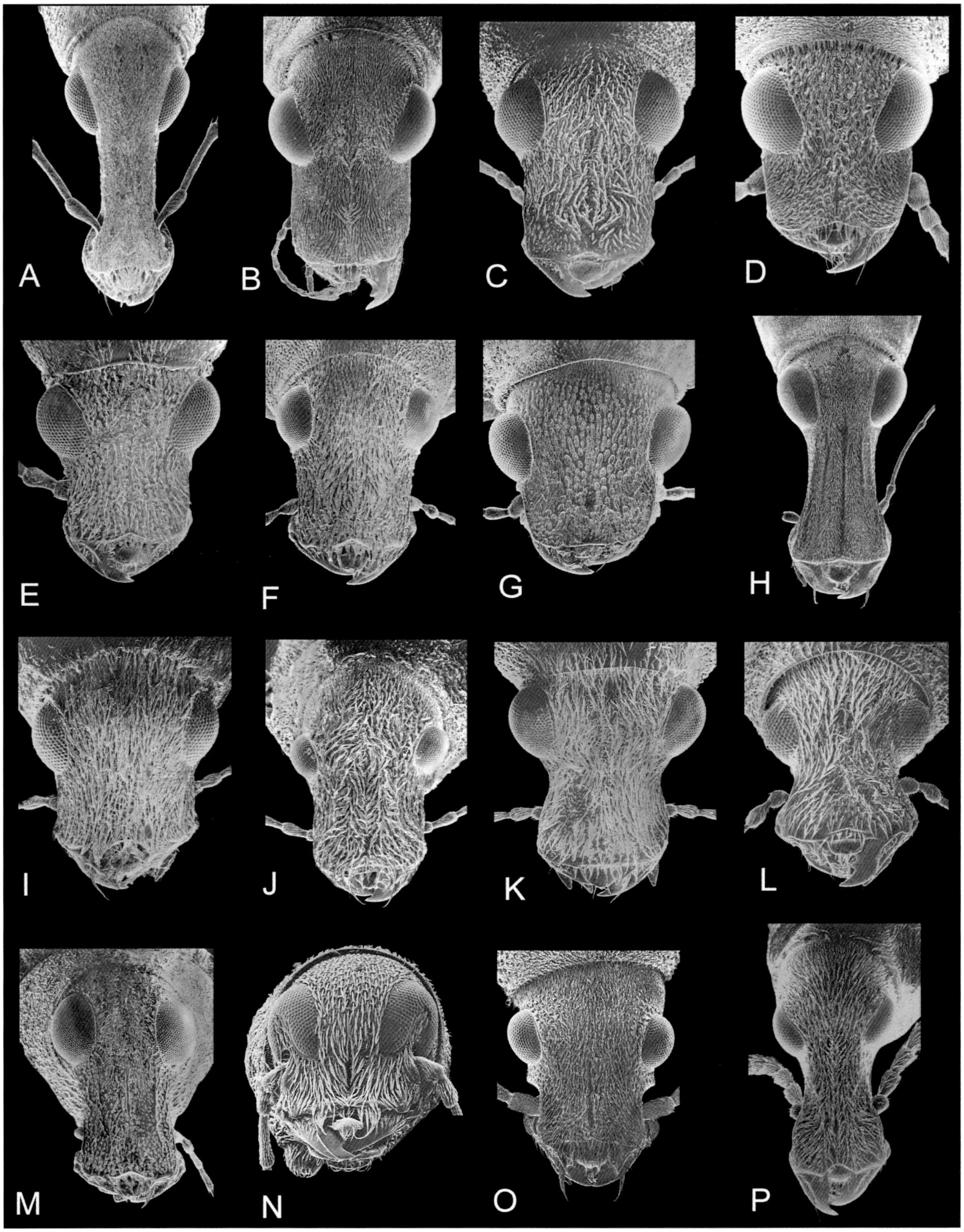

Fig. 39.8. Anthribinae, structural details of heads, dorsal or anterior. A, *Ancylotropis waterhousei* Jekel; B, *Apatenia insignis* Jordan; C, *Atropideres evanescens* (Blackburn); D, *Basitropis peregrina* Pascoe; E, *Cacephatus sericeus* Blackburn; F, *Commista grisea* Jordan; G, *Dendropemon albopictus* (Pascoe); H, *Eczesaris atomaria* Pascoe; I, *Enspondus bigibbosus* Blackburn; J, *Entromus dorsoplagiatus* Blackburn; K, *Epargemus marmoratus* Blackburn; L, *Erichsonocis albuginosus* (Erichson); M, *Erichsonocis albuginosus* (Erichson); N, *Euciodes suturalis* Pascoe; O, *Eucorynus crassicornis* (Fabricius); P, *Eupanteos ornatus* Jordan. After Zimmerman (1994).

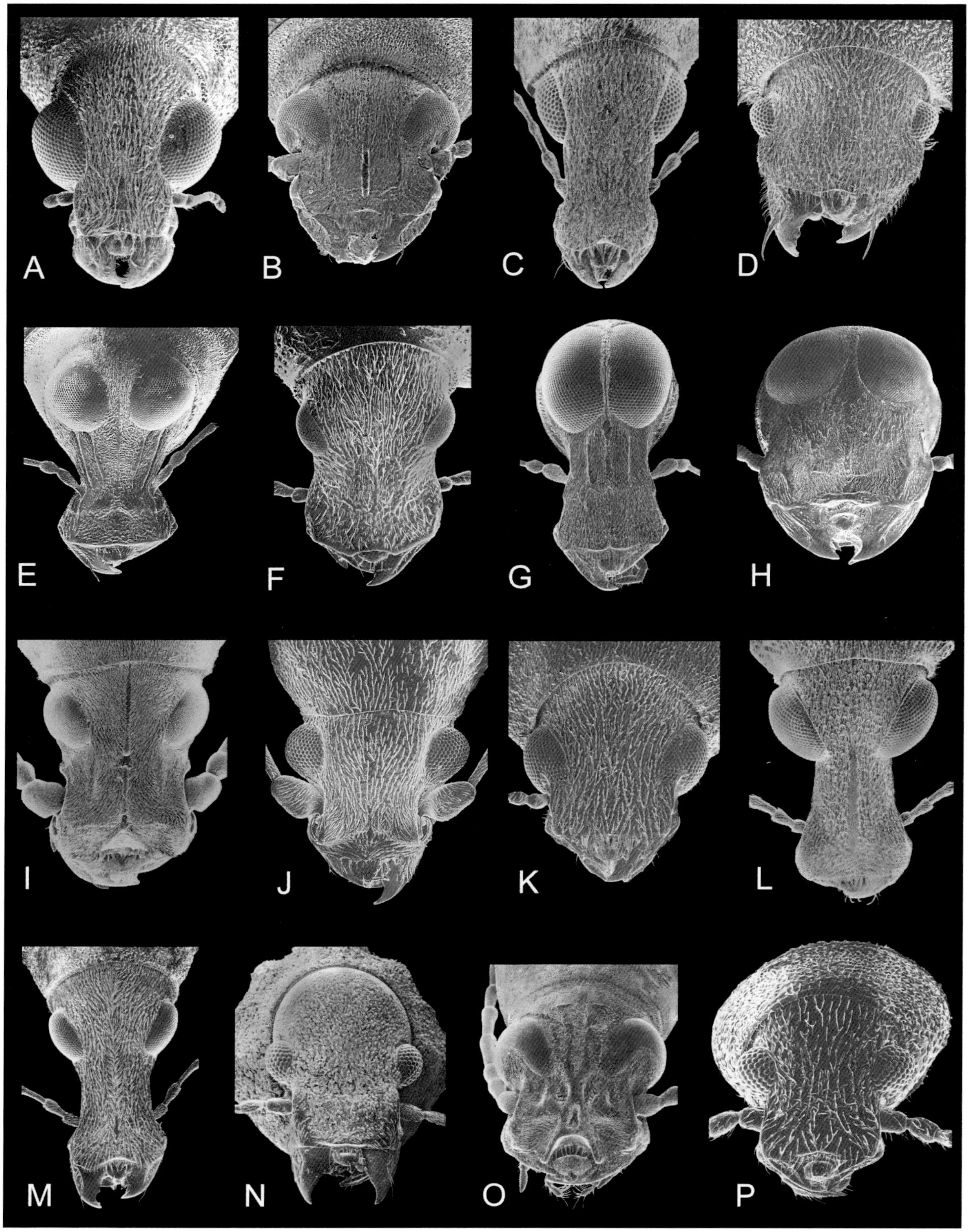

Fig. 39.9. Anthribinae, structural details of heads, dorsal or anterior. A, *Euparius didymus* (Jordan); B, *Exillis australis* Frieser; C, *Genethila retusa* Pascoe; D, *Howeanthribus bufo* Zimmerman; E, *Hucus persimilis* Jordan; F, *Leaoanthribus crucifer* (Lea); G, *Litocerus variegatus* Jordan; H, *Mucronianus rufipes* Jordan; I, *Platystomos wallacei* (Pascoe); J, *Protaedus bryanti* Jordan; K, *Streneoderma contemptum* Blackburn; L, *Taburnus neglectus* Zimmerman; M, *Telala armigera* Jordan; N, *Teratanthribus monstrosus* Zimmerman; O, *Xenocerus speracerus* Montrouzier; P, *Xynotropis micans* Blackburn. After Zimmerman (1994).

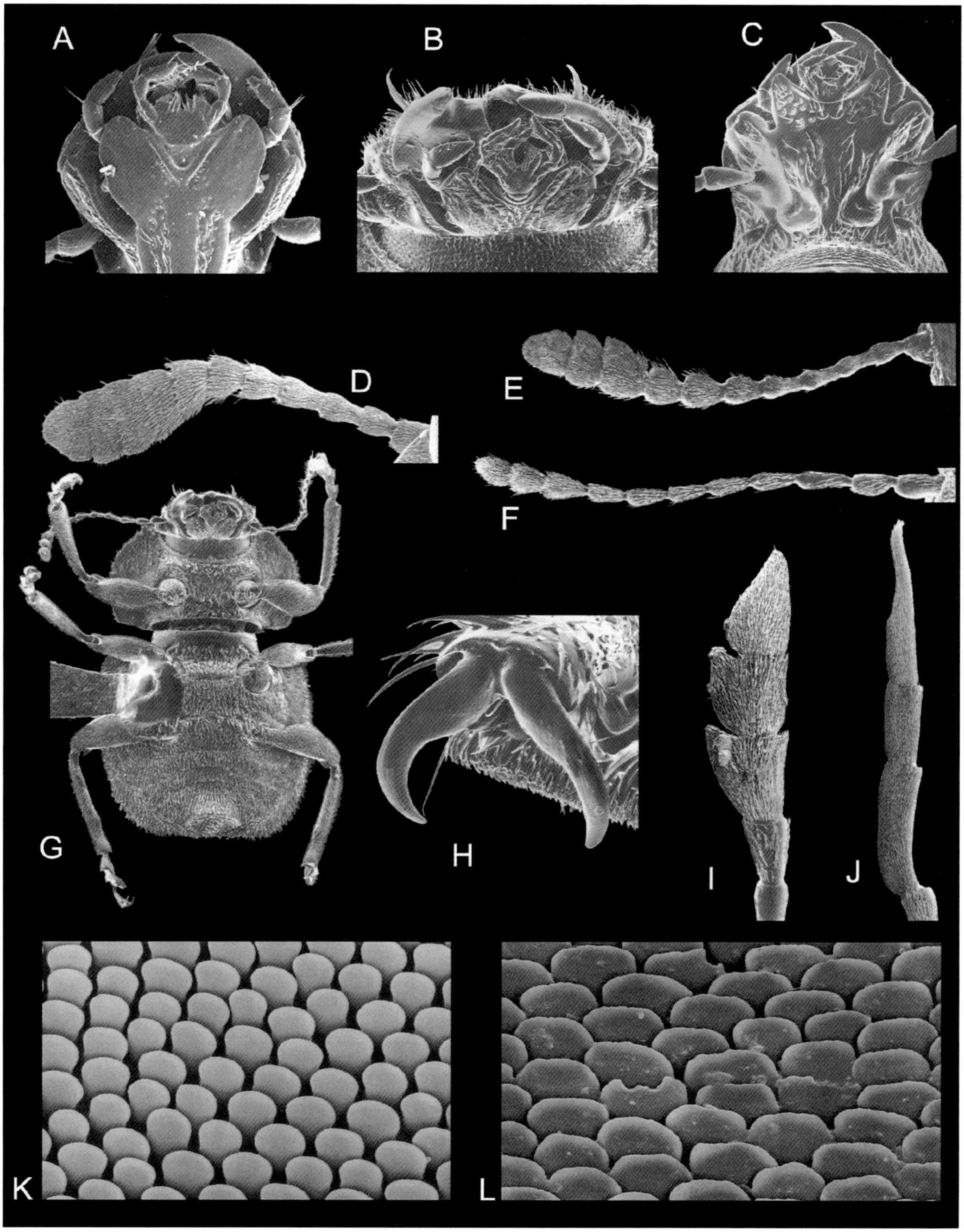

Fig. 39.10. Anthribinae, structural details. A–C, mouthparts, ventral: A, *Taburnus tessellatus* Zimmerman; B, *Howeanthribus bufo* Zimmerman; C, *Leaoanthribus crucifer* (Lea); D–F, antenna: D, *Basitropis peregrina* Pascoe; E, *Dendropemon marmoratus* (Montrouzier); F, *Teratanthribus monstrosus* Zimmerman; G, *H. bufo* Zimmerman, ventral habitus; H, *H. bufo* Zimmerman, tarsal claws; I–J, antennal club *Phloeobius gigas* (Fabricius): I, ♀; J, ♂; K–L, wing-folding pad processes: K, *Ozotomerus bipunctatus* (Montrouzier); L, *P. gigas* (Fabricius). After Zimmerman (1994).

31(29). Body length <3.0 mm; antennae inserted close to eye and scrobes almost in contact with eye (Fig. 39.9P) ..... ***Xynotropis* Blackburn** (part)
– Body length >3.0 mm; antennae well separate from eyes **32**
32(31). Scrobes with ventral margin forming oblique sulcus ..... ***Telala* Jordan** (part)
– Scrobes with ventral margin entire, without oblique sulcus **33**
33(32). Lateral pronotal carinae straight, directed toward eye **34**
– Lateral pronotal carinae oblique, directed ventrad **35**
34(33). Rostrum dorsally with 5 carinae (Fig. 39.9E); distance between eye and scrobe much greater than width of forehead ..... ***Hucus* Pascoe**
– Rostrum dorsally without carinae (Fig. 39.8E); distance between eye and scrobe much less than width of forehead ..... ***Cacephatus* Blackburn** (females)
35(33). Head in dorsal view with interocular distance less than eye width (Fig. 39.6H) ..... ***Mecocerinopis* Zimmerman**
– Head in dorsal view with interocular distance more than eye width **36**
36(35). Pronotal surface even, disc convex; elytra without tubercles ..... ***Ethneca*** Pascoe
– Pronotal surface irregular, disc with longitudinal depressions on each side of midline or fasciculate; elytra tuberculate (Fig. 39.6I) ..... ***Commista* Jordan**
37(23). Metaventrite very short, distance between meso- and metacoxa ~0.3 × width of mesocoxa; body length <3.0 mm (Fig. 39.6J) ..... ***Xynotropis* Blackburn** (part)
– Metaventrite longer, distance between meso- and metacoxa greater than width of mesocoxa; body length variable **38**
38(37). Scrobes entirely lateral, not extending beneath rostrum, as close as or closer to eye than to base of mandible ..... ***Enspondus* Blackburn**
– Scrobes not so constructed **39**
39(38). Lateral pronotal carinae prominent, forming conspicuous angle with transverse carina (Fig. 39.6K) **40**
– Lateral pronotal carinae reduced or not well differentiated from transverse carina or not distinctly reaching middle of prothorax, often forming a continuous curve with transverse carina **43**
40(39). Eyes as much frontal as lateral; interocular area subparallel-sided **41**
– Eyes lateral; interocular area V-shaped **42**
41(40). Rostrum much broader than long; antennal insertions subapical, closer to mandible than to eye (Fig. 39.9H) ..... ***Mucronianus* Jordan**
– Rostrum longer than broad; antennal insertions about midway between eye and mandible (Fig. 39.9G) ..... ***Litocerus* Schoenherr**
42(40). Rostrum ~2 × as wide as long ..... ***Phaulimia* Pascoe**
– Rostrum slightly wider than long (Fig. 39.8B) ..... ***Apatenia* Pascoe**
43(39). Rostrum subparallel-sided **44**
– Rostrum narrowed in front of antennal insertions **46**
44(43). Dorsum with raised tufts of vestiture and low fasciculate tubercles; rostrum tricarinate (Fig. 39.8J); antennal clubs with segment 1 slightly longer than broad, ~2 × as long as segment 7 of funicle ..... ***Entromus* Blackburn**
– With different combination of characters **45**
45(44). Lateral pronotal carinae distinct, reaching to ca. one-third length of side, partly visible dorsally ..... ***Cacephatus* Blackburn** (females)
– Lateral pronotal carinae not visible dorsally ..... ***Atropideres* Zimmerman**
46(43). Rostrum not carinate (Fig. 39.8K) ..... ***Epargemus* Blackburn**
– Rostrum longitudinally carinate or carinulate **47**
47(46). Rostrum broader than long (Fig. 39.9F); antennae inserted close to eye ..... ***Leaoanthribus* Zimmerman**
– Rostrum much longer than broad (Fig. 39.8P); antennae inserted in apical third of rostrum ..... ***Eupanteos* Jordan**
48(1). Body length <2.0 mm; body convex (Fig. 39.6O), shiny black; prothorax with complete lateral carinae; scutellar shield invisible; metacoxae not much wider than mesocoxae ..... ***Cisanthribus* Zimmerman**
– Without the above combination; metacoxae strongly transverse and extending laterad to well-exposed metanepisterna **49**
49(48). Lateral pronotal carinae extending to anterior margin of pronotum, forming acute, downward projecting angle with transverse carina (Fig. 39.6N) ..... ***Stenorhis* Jordan**
– Lateral pronotal carinae not reaching much beyond middle of pronotum **50**
50(49). Antennae inserted above eye level ..... ***Araeocerodes* Blackburn**
– Antennae inserted below eye level **51**
51(50). Eyes notably large and transverse (Fig. 39.6L) ..... ***Deropygus* Sharp**
– Eyes circular or oval **52**
52(51). Eyes circular; metatarsi with segment 1 as long as or longer than remainder of tarsus and half or more as long as metatibia (Fig. 39.6M) ..... ***Misthosima* Pascoe**
– Eyes oval; metatarsi with segment 1 variable in length ..... ***Araecerus* Schoenherr**

Classification of the Australian genera

Subfamily Anthribinae Billberg, 1820

Ancylotropis Jekel, 1855 (Pl. 27A; Figs 39.1, 39.8A)

Type species, by monotypy: *Ancylotropis waterhousei* Jekel, 1855.

Australian species. *A. niveinasus* (Motschulsky, 1875), *A. scabrata* (Motschulsky, 1875), *A. waterhousei* Jekel, 1855.
Distribution. Australia, New Guinea.

Antribisomus Perroud, 1865 (Pl. 27B, C)

Type species, by monotypy: *Antribisomus tessellatus* Perroud, 1865 (*in* Perroud & Montrouzier 1865).

Australian species. *A. latifrons* Kuschel, 1998.
Distribution. Australia (Norfolk Island), New Caledonia.
Notes. The single species recorded from Australia was originally named 'Genus B, new species 1' by Zimmerman (1991: 36, pl. 18; 1994: 186, Fig. 128) but later identified by Kuschel (1998). It was described from New Caledonia.

Apatenia Pascoe, 1859 (Pl. 27D; Fig. 39.8B)

Type species, by monotypy: *Apatenia viduata* Pascoe, 1859b.
Australian species. *A. apicalis* Gahan, 1900, *A. insignis* Jordan, 1895c.
Distribution. Australia, Pacific region, tropical Africa.
Notes. *Apatenia apicalis* is only known from the Australian territory of Christmas Island.

Atropideres Zimmerman, 1994 (Pl. 32I, J; Fig. 39.8C)

Type species, by original designation: *Tropideres evanescens* Blackburn, 1900.
Australian species. *A. evanescens* (Blackburn, 1900).
Distribution. Australia.
Notes. The genus is monotypic, comprising only a single species.

Basitropis Jekel, 1855 (Pl. 27E, F, G; Figs 39.8D, 39.10D)

Type species, by original designation: *Basitropis nitidicutis* Jekel, 1855.
Australian species. *B. ingrata* Pascoe, 1859b, *B. pallida* Blackburn, 1900, *B. peregrina* Pascoe, 1859b, *B. relicta* Blackburn, 1900, *B. solitaria* Pascoe, 1860a.
Distribution. Australia, Asia, Madagascar.
Notes. The genus contains several undescribed species in Australia.

Cacephatus Blackburn, 1900 (Pl. 27H, I; Figs 39.3A–G, 39.8E)

Type species, by monotypy: *Cacephatus sericeus* Blackburn, 1900.
Australian species. *C. sericeus* Blackburn, 1900.
Distribution. Australia (including Lord Howe Island, Norfolk Island), New Zealand, New Caledonia.
Notes. May (1994) provided a generic diagnosis for the pupa and larva based on the New Zealand species *C. aucklandicus* (Brookes), *C. huttoni* (Sharp) and *C. inornatus* (Sharp), with illustrations of larvae and pupae of *C. huttoni* (Sharp), e.g. Fig. 39.3A–G. The larvae develop in dry wood and pupate in situ (May 1994). There are several undescribed species in Australia.

Commista Jordan, 1895 (Pl. 27K, L; Figs 39.2C, 39.8F)

Type species, by original designation: *Commista latifrons* Jordan, 1895b.
Australian species. *C. baccula* Jordan, 1924b, *C. bispina* (Erichson, 1842), *C. grisea* Jordan, 1924b, *C. latifrons* Jordan, 1895b.
Distribution. Australia.
Notes. The genus contains several undescribed species in Australia.

Dendropemon Schoenherr, 1839 (Pl. 28A, B; Figs 39.8G, 39.10E)

Type species, by original designation: *Anthribus perfolicornis* Fabricius, 1801.
Australian species. *D. albinasus* (Jordan, 1923a), *D. albopictus* (Pascoe, 1860a), *D. marmoratus* (Montrouzier, 1855), *D. subfasciatus* (Fåhraeus, 1839) (*in* Schoenherr 1839), *D. virgatus* (Jordan, 1923a).
Distribution. Australia, New Caledonia, Asia.
Notes. There is at least one undescribed species in Australia.

Eczesaris Pascoe, 1859 (Pl. 28K; Fig. 39.8H)

Type species, by monotypy: *Eczesaris atomaria* Pascoe, 1859a.
Australian species. *E. atomaria* Pascoe, 1859a, *E. australis* Frieser, 1983.
Distribution. Australia, New Guinea, Melanesia.
Notes. The genus appears to only contain two species in Australia.

Enspondus Blackburn, 1900 (Pl. 28C; Fig. 39.8I)

Type species, by monotypy: *Enspondus bigibbosus* Blackburn, 1900.
Australian species. *E. bigibbosus* Blackburn, 1900.
Distribution. Australia.
Notes. The genus is monotypic and apparently restricted to Australia.

Entromus Blackburn, 1900 (Pl. 28D; Fig. 39.8J)

Type species, by monotypy: *Entromus dorsoplagiatus* Blackburn, 1900.
Australian species. *E. dorsoplagiatus* Blackburn, 1900.
Distribution. Australia.
Notes. The genus is monotypic and apparently restricted to Australia.

Epargemus Blackburn, 1900 (Pl. 28E; Fig. 39.8K)

Type species, by monotypy: *Epargemus marmoratus* Blackburn, 1900.

Australian species. *E. marmoratus* Blackburn, 1900.
Distribution. Australia.
Notes. The genus is monotypic and apparently restricted to Australia.

Erichsonocis Zimmerman, 1994 (Pl. 28I, J; Fig. 39.8L)

Type species, by original designation: *Tropideres albuginosus* W. F. Erichson, 1842.
Australian species. *E. albuginosus* (Erichson, 1842).
Distribution. Australia.
Notes. The genus is apparently restricted to Australia, but there are several undescribed species.

Ethneca Pascoe, 1860 (Pl. 28F; Fig. 39.8M)

Type species, by monotypy: *Ethneca bakewelli* Pascoe, 1860b.
Australian species. *E. bakewellii* Pascoe, 1860b.
Distribution. Australia.
Notes. The genus is monotypic and apparently restricted to Australia.

Euciodes Pascoe, 1866 (Pl. 28G, H; Figs 39.3H–S, 39.7A–U, 39.8N)

Type species, by monotypy: *Euciodes suturalis* Pascoe, 1866.
Australian species. *E. suturalis* Pascoe, 1866.
Distribution. Australia, New Zealand (introduced).
Notes. The larvae and pupae of *E. suturalis* were described and illustrated by May (1994) (in part Fig. 39.3H–S), and Penman (1978) and May (1994) recorded the life history in New Zealand. Oviposition occurs into the flower stalks of grasses, and the larvae feed inside the stalks, where pupation also occurs. All larval hosts recorded are from New Zealand, namely *Arrhenatherum elatius*, *Bromus*, *Dactylis glomerata*, *Festuca arundinacea* and *Holcus lanatus*. The adults apparently feed on fungi such as *Puccinia* (Holloway 1982). There are several undescribed species in Australia.

Eucorynus Schoenherr, 1823 (Pl. 28L; Figs 39.4A–I, 39.8O)

Type species, by original designation: *Anthribus crassicornis* Fabricius, 1801.
Australian species. *E. crassicornis* (Fabricius, 1801).
Distribution. Australia, New Guinea, Asia.
Notes. May (1994) described and illustrated the larva of *E. crassicornis* from specimens collected in rotten wood.

Eupanteos Jordan, 1923 (Pl. 29A–D; Fig. 39.8P)

Type species, by original designation: *Eupanteos ornatus* Jordan, 1923b.
Australian species. *E. bifasciatus* (Lea, 1926), *E. montanus* (Oke, 1934), *E. ornatus* Jordan, 1923b.
Distribution. Australia.
Notes. The genus is restricted to eastern Australia, and there are several undescribed species.

Euparius Schoenherr, 1823 (Pl. 29E; Figs 39.2B, 39.9A)

Type species, by monotypy: *Anthribus lunatus* Fabricius, 1801.
Australian species. *E. didymus* (Jordan, 1924b), *E. lateripictus* (Jordan, 1895a).
Distribution. Australia, Melanesia, Asia, North and Central America.

Exillis Pascoe, 1860 (Pl. 29F, G; Fig. 39.9B)

Type species, by monotypy: *Exillis longicornis* Pascoe, 1860b.
Australian species. *E. australis* Frieser, 1981, *E. biimpressus* Frieser, 2009, *E. conicollis* Frieser, 2009, *E. decleratus* Frieser, 2009, *E. scabrosus* Frieser, 2009.
Distribution. Australia (including Norfolk Island), Polynesia, Asia.
Notes. There are several undescribed species in Australia. Some of the ones recorded by Zimmerman (1994) may be conspecific with species subsequently described by Frieser (2009).

Genethila Pascoe, 1860 (Pl. 29H; Fig. 39.9C)

Type species, by monotypy: *Genethila retusa* Pascoe, 1860b.
Australian species. *G. retusa* Pascoe, 1860b.
Distribution. Australia, New Guinea.
Notes. There are some undescribed species in Australia.

Howeanthribus Zimmerman, 1994 (Pl. 29J, K; Figs 39.9D, 39.10B, 39.10G, H)

Type species, by original designation: *Howeanthribus bufo* Zimmerman, 1994.
Australian species. *H. bufo* Zimmerman, 1994.
Distribution. Australia (Lord Howe Island).
Notes. The genus is monotypic and apparently restricted to Lord Howe Island.

Hucus Pascoe, 1859 (Pl. 29L; Fig. 39.9E)

Type species, by monotypy: *Hucus melanostoma* Pascoe, 1859b.
Australian species. *H. persimilis* Jordan, 1903.
Distribution. Australia, Melanesia, Asia.

Notes. There is apparently only one species of this large tropical genus in Australia.

Leaoanthribus Zimmerman, 1994 (Pl. 30F; Figs 39.9F, 39.10C)

Type species, by original designation: *Epargemus crucifer* Lea, 1926.
Australian species. *L. crucifer* (Lea, 1926).
Distribution. Australia.
Notes. The genus is monotypic and apparently restricted to Australia.

Litocerus Schoenherr, 1833 (Pl. 30I; Fig. 39.9G)

Type species, by original designation: *Litocerus histrio* Gyllenhal, 1833 (*in* Schoenherr, 1833).
Australian species. *L. jordani* Gahan, 1900, *L. macrophthalmus* (Montrouzier, 1855), *L. variegatus* Jordan, 1894.
Distribution. Australia, Melanesia, Asia, Africa.
Notes. There may be more species of this very large tropical genus in Australia.

Mauia Blackburn, 1885 (Pl. 30A, B)

Type species, by monotypy: *Mauia satelles* Blackburn, 1885 (*in* Blackburn & Sharp, 1885).
Australian species. *M. subnotata* (Boheman, 1859).
Distribution. Australia, Pacific region, Southeast Asia.
Notes. Apart from the widespread *M. subnotata*, which may be introduced to Australia, there are several undescribed species in the country.

Mecocerinopis Zimmerman, 1994 (Pl. 30C–E)

Type species, by original designation: *Litocerus balli* Olliff, 1889.
Australian species. *M. amabilis* (Pascoe, 1859a), *M. balli* (Olliff, 1889).
Distribution. Australia (incl. Lord Howe Island), New Guinea.
Notes. The genus appears to only contain these two species.

Mucronianus Jordan, 1894 (Pl. 30L; Fig. 39.9H)

Type species, by original designation: *Mucronianus rufipes* Jordan, 1894.
Australian species. *M. lunulatus* Frieser, 2001, *M. rufipes* Jordan, 1894.
Distribution. Australia, New Guinea, Southeast Asia, Japan.
Notes. The presence of the widespread *M. rufipes* in Australia requires confirmation, as all specimens seen represent *M. lunulatus*. An undescribed species occurs on the Cape York peninsula.

Notoecia Blackburn, 1900 (Pl. 31A)

Type species, by monotypy: *Notoecia reticulata* Blackburn, 1900.
Australian species. *N. reticulata* Blackburn, 1900.
Distribution. Australia.
Notes. The genus is monotypic and apparently restricted to Australia.

Ozotomerus Perroud, 1853 (Pl. 30G, H; Fig. 39.10K)

Type species, by monotypy: *Ozotomerus maculosus* Perroud, 1853.
Australian species. *O. bipunctatus* (Montrouzier, 1855).
Distribution. Australia, New Guinea, Asia (to Japan and Siberia).
Notes. The single species known from Australia is widespread in the Pacific region.

Phaulimia Pascoe, 1859 (Pls 1B, 30J, K)

Type species, by monotypy: *Phaulimia ephippiata* Pascoe, 1859b.
Australian species. *P. albirostris* (Frieser, 1985), *P. australiaca* (Jordan, 1928b), *P. punctata* (Montrouzier, 1855).
Distribution. Australia, Asia (to Japan and Siberia).
Notes. There are several undescribed species of this large and widespread genus in Australia.

Phloeobius Schoenherr, 1823 (Pl. 31B, C; Figs 39.4J–S, 39.10J)

Type species, by original designation: *Anthribus griseus* Fabricius, 1792.
Australian species. *P. gigas* (Fabricius, 1775).
Distribution. Australia, Oriental, Africa.
Notes. There appears to be only one species of this large and widespread genus in Australia. Its larva and pupa were described by May (1994) from specimens found in a dry branch.

Platystomos Schneider, 1791 (Pl. 31E, F; Fig. 39.9I)

Type species, by subsequent designation (Pierce, 1916: 463): *Curculio albinus* Linnaeus, 1758.
Australian species. *P. wallacei* (Pascoe, 1860b).
Distribution. Australia, New Guinea, Oriental, Afrotropical and Palaearctic regions.
Notes. Only one widespread Oriental species of this large genus appears to occur in Australia.

Plesiobasis Jordan, 1939 (Pl. 31D)

Type species, by original designation: *Plesiobasis monera* Jordan, 1939.
Australian species. *Plesiobasis* sp.
Distribution. Australia, Melanesia (Fiji, Vanuatu).
Notes. A single species of this small Melanesian genus was recorded from Australia by Zimmerman (1992: 26, pl. 317; 42, pl. 325; 1994: 189) as 'Genus G, new species 1'.

Protaedus Pascoe, 1860 (Pl. 31G; Fig. 39.9J)

Type species, by monotypy: *Protaedus moerens* Pascoe, 1860b.
Australian species. *P. bryanti* Jordan, 1923b.
Distribution. Australia (QLD), Melanesia, Southeast Asia.
Notes. Only one species of this large Oriental genus has been recorded from Australia, but a few apparently undescribed ones occur in the country as well.

Rhaphitropis Reitter, 1916

Type species, by subsequent designation (Jordan, 1925: 257): *Anthribus marchicus* Herbst, 1797.
Australian species. *R. basiplaga* Frieser, 1989.
Distribution. Australia, Oriental, Afrotropical and Palaearctic regions.
Notes. This large and very widespread genus has only been recorded from Australia with a single species, which was described from New Guinea but reported to occur in Australia as well (Rheinheimer 2004). As we were unable to locate any specimens to confirm its presence in Australia, we have not illustrated and included this genus in the key.

Streneoderma Blackburn, 1900 (Pl. 31H–J; Fig. 39.9K)

Type species, by subsequent designation (Zimmerman, 1994: 159): *Streneoderma planatum* Blackburn, 1900.
Australian species. *S. contemptum* Blackburn, 1900, *S. planatum* Blackburn, 1900.
Distribution. Australia, Southeast Asia (?).
Notes. The genus presently contains two described and several undescribed species in Australia and one described species in Southeast Asia (Zimmerman 1994; Rheinheimer 2004), but it remains to be ascertained whether the last truly belongs in the genus. The genus is unusual in having a fruit-piecing ovipositor and carpophagous (seed-feeding) larvae.

Taburnus Zimmerman, 1994 (Pl. 32C, D; Figs 39.9L, 39.10A)

Type species, by original designation: *Taburnus neglectus* Zimmerman, 1994.
Australian species. *T. breviceps* Zimmerman, 1994, *T. neglectus* Zimmerman, 1994, *T. setosus* Zimmerman, 1994, *T. tessellatus* Zimmerman, 1994.
Distribution. Australia.
Notes. The genus appears restricted to Australia. The adults may graze on bark (Zimmerman 1994).

Telala Jordan, 1895 (Pls 1C, 32A, B; Fig. 39.9M)

Type species, by original designation: *Telala armigera* Jordan, 1895b.
Australian species. *T. armigera* Jordan, 1895b, *T. musiva* (Erichson, 1842).
Distribution. Australia.
Notes. The genus is restricted to eastern Australia and contains several undescribed species.

Teratanthribus Zimmerman, 1994 (Pl. 32E, H; Figs 39.9N, 39.10F)

Type species, by original designation: *Teratanthribus monstrosus* Zimmerman, 1994.
Australian species. *T. monstrosus* Zimmerman, 1994.
Distribution. Australia.
Notes. This unusual genus is restricted to Australia and contains several undescribed species.

Xenocerus Schoenherr, 1833 (Pl. 32F, G; Figs 39.2A, 39.9O)

Type species, by original designation: *Xenocerus saperdoides* Gyllenhal, 1833.
Australian species. *X. nativitatis* Gahan, 1900, *X. speracerus* Montrouzier, 1855.
Distribution. Australia, Melanesia, Oriental region.
Notes. This huge and characteristic Oriental genus is represented in Australia by only two species, and one of them, *X. nativitatis*, only occurs on Christmas Island.

Xynotropis Blackburn, 1900 (Pls 1A, 32K; Fig. 39.9P)

Type species, by monotypy: *Xynotropis micans* Blackburn, 1900.
Australian species. *X. micans* Blackburn, 1900, *X. tasmanica* Frieser, 2001.
Distribution. Australia.
Notes. The genus is restricted to Australia and contains at least two undescribed species.

Anthribinae not confirmed to occur in Australia

The Australian species named 'Genus U, new species 1' by Zimmerman (1992: 32, 44, pls 320, 326; 1994: 193) was tentatively placed in *Helmoreus* by Kuschel (1998) and, following this work, by Pullen *et al.* (2014), but this assignment is not correct and it is considered here as genus undetermined (Pl. 29I). We noticed that the type species of *Helmoreus*, *H. sharpi* (Broun, 1880), has an antebasal carina on the pronotum as well as a long rostrum versus a subbasal carina and short rostrum in 'Genus U, new species 1', furthermore Kuschel (1998) key to genera of Anthribidae in New Caledonia and Vanuatu does not lead to *Helmoreus*.

Subfamily Choraginae Kirby, 1819

Araecerus Schoenherr, 1823 (Pl. 33A–C; Figs 39.5A–K)

Type species, by original designation: *Anthribus coffeae* Fabricius, 1801.

Australian species. *A. asperulus* Blackburn, 1900, *A. bicristatus* Blackburn, 1900, *A. eurous* (Jordan, 1895b), *A. fasciculatus* (De Geer, 1775), *A. lindensis* Blackburn, 1900, *A. lutatus* (Fairmaire, 1849), *A. palmaris* (Pascoe, 1882).

Distribution. Cosmopolitan.

Notes. The genus contains several species of economic importance in the dried-fruit and stored-product industry, particularly the cosmopolitan *A. fasciculatus*, which is introduced in Australia. Several undescribed species occur in Australia. Descriptions of the larva and pupa of *A. fasciculatus*, *A. palmaris* and *A. lutatus* were given by May (1994).

Araeocerodes Blackburn, 1900 (Pl. 33D, E)

Type species, by monotypy: *Araeocerodes lilliputanus* Blackburn, 1900.

Australian species. *A. difficilis* (Jordan, 1924a), *A. lilliputanus* Blackburn, 1900.

Distribution. Australia, Melanesia, Southeast Asia, Japan.

Notes. The genus is large and widespread in the Pacific and Oriental regions if it includes *Melanopsacus* Jordan, 1924a (Zimmerman 1994; Pullen *et al.* 2014), but this wider concept is not always accepted (Rheinheimer 2004). In addition to the two named species in Australia there are many undescribed ones.

Cisanthribus Zimmerman, 1938 (Pl. 33F, G)

Type species, by original designation: *Cisanthribus convexus* Zimmerman, 1938.

Australian species. *Cisanthribus sp.*

Distribution. Australia (Lord Howe Island, Norfolk Island), Pacific region, Japan.

Notes. The genus is poorly known but contains several species in the Pacific region, including at least two in Australia, one on Lord Howe Island (Zimmerman 1994: 234) and one in northern Queensland, both undescribed.

Deropygus Sharp, 1891 (Pl. 33I, J)

Type species, by subsequent designation (Morimoto, 1978: 23): *Deropygus histrio* Sharp, 1891.

Australian species. *Deropygus sp.*

Distribution. Australia, New Guinea, Melanesia, Polynesia, Southeast Asia, Japan.

Notes. The genus is thus far only recorded from Australia by one undescribed species (Zimmerman 1992: 4, pl. 306; 36, pl. 322; 1994: 227).

Misthosima Pascoe, 1859 (Pl. 33K, L)

Type species, by subsequent designation (Zimmerman, 1994: 228): *Misthosima mera* Pascoe, 1859b.

Australian species. *M. dorsonotata* Blackburn, 1900.

Distribution. Australia, New Guinea, Melanesia, Asia (to Japan).

Notes. This large genus contains some undescribed species in Australia.

Stenorhis Jordan, 1928 (Pl. 33H)

Type species, by original designation: *Stenorhis ampedus* Jordan, 1928a.

Australian species. *Stenorhis spp.*

Distribution. Australia (incl. Lord Howe and Norfolk Islands), Pacific region, Southeast Asia, Asia (to Japan).

Notes. The genus contains several undescribed species in Australia (Zimmerman 1992: 20, pl. 314; 40, pl. 324; 1994: 230–231).

Acknowledgements

Thanks are due to Nick Monaghan and Nick Porch for the photographs of Anthribidae in their natural habitat.

References

Alonso-Zarazaga MA, Lyal CHC (1999) *A World Catalogue of Families and Genera of Curculionoidea (Insecta: Coleoptera) (Excepting Scolytidae and Platypodidae)*. Entomopraxis, Barcelona, 315 pp.

Alonso-Zarazaga MA, Lyal CHC (2002) Addenda and corrigenda to a "World Catalogue of Families and Genera of Curculionoidea (Insecta: Coleoptera)". *Zootaxa* **63**(1), 1–37. doi:10.11646/zoootaxa.63.1.1

Alonso-Zarazaga MA, Barrios H, Borovec R, Bouchard P, Caldara R, *et al.* (2017) The cooperative catalogue of Palaearctic Coleoptera Curculionoidea. *Monografias Electronicas S.E.A.* **8**, 5–729.

Alonso-Zarazaga MA, Barrios H, Borovec R, Caldara R, Colonnelli E, *et al.* (2023) The cooperative catalogue of Palaearctic Coleoptera Curculionoidea. *Monografias Electronicas S.E.A.* **14**, 5–780.

Anderson WH (1947) Larvae of some genera of Anthribidae (Coleoptera). *Annals of the Entomological Society of America* **40**(3), 480–517. doi:10.1093/aesa/40.3.489

Bashford R (2002) The insect fauna inhabiting *Uromycladium* (Uredinales) rust galls on silver wattle (*Acacia dealbata*) in Tasmania. *The Australian Entomologist* **29**(3), 81–95.

Billberg GJ (1820) *Enumeratio Insectorum in Museo Gust. Joh. Billberg*. Typis Gadelianis, Stockholm, 138 pp. doi:10.5962/bhl.title.49763

Blackburn T (1900) Further notes on Australian Coleoptera, with descriptions of new genera and species. Part XXVII. *Transactions of the Royal Society of South Australia* **24**, 113–169.

Blackburn T, Sharp D (1885) Memoirs on the Coleoptera of the Hawaiian Islands. *Scientific Transactions of the Royal Dublin Society* (Series 2) **3**, 119–289, 300, pls. 4–5.

Boheman CH (1859) Coleoptera. In *Kongliga Svenska Fregatten Eugenies Resa omkring Jorden under befäl af C. A. Virgin Åren 1851–1853. Vetenskapliga iakttagelser på H. M. Konung Oscar Den Förstes befallning*. (Ed. CAK Virgin) pp. 1–217. Svenska Vetenskaps Akademien. P. A. Norstedt & Söner, Stockholm. doi:10.5962/bhl.title.2467

Bouchard P, Bousquet Y, Davies AE, Alonso-Zarazaga MA, Lawrence JF, *et al*. (2011) Family-group names in Coleoptera (Insecta). *Zookeys* **88**(88), 1–972. doi:10.3897/zookeys.88.807

Broun T (1880) *Manual of the New Zealand Coleoptera. Part 1*. Colonial Museum and Geological Survey Department, Wellington, pp. 1–651.

De Geer C (1775) *Mémoires pour servir à l'histoire des Insectes. Tome cinquième*. P. Hesselberg, Stockholm, vi + 448 pp. 16 pls.

Erichson WF (1842) Conspectus Insectorum Coleopterorum, quae in Republics Peruana observata sunt. *Archiv für Naturgeschichte* **13**(1), 67–185.

Fabricius JC (1775) *Systema entomologiae, sistens insectorum classes, ordines, genera, species, adiectis synonymis, locis, descriptionibus, observationibus*. Officina Libraria Kortii, Flensburgi et Lipsiae, xxxii + 832 pp. doi:10.5962/bhl.title.36510

Fabricius JC (1792) *Entomologia systematica emendata et aucta. Secundum classes, ordines, genera, species adjectis synonymis, locis, observationibus, descriptionibus. Tomus I. Pars II*. C. G. Proft, Hafniae, xx + 538 pp.

Fabricius JC (1801) *Systema eleutheratorum secundum ordines, genera, species: adiectis synonymis, locis, observationibus, descriptionibus. Tomus II*. Bibliopolii Academici Novi, Kiliae, 687 pp.

Fairmaire L (1849) Essai sur les Coléoptères de la Polynésie. *Revue et Magasin de Zoologie Pure et Appliquée* Series 2, **1**, 445–460.

Fairmaire L (1895) Diagnoses de quelques Coléoptères de Madagascar et des îles Seychelles et notes synonymiques. *Annales de la Société Entomologique de France* **64**, 278–282.

Fairmaire L (1903) Matériaux pour la faune coléoptérique malgache (15e note). *Revue d'Entomologie, Caen* **22**, 13–46.

Frieser R (1981) Neue tropische Anthribiden (Coleptera, Anthribidae). *Annales Historico-Naturales Musei Nationalis Hungarici (Zoologica)* **73**, 207–216.

Frieser R (1983) Zur Kenntnis der Anthribidenfauna Neuguineas (Coleoptera: Anthribidae). *Folia Entomologica Hungarica* **44**(1), 47–62.

Frieser R (1985) Neue Anthribiden, meist aus Fernost (Coleoptera, Anthribidae). *Mitteilungen der Münchner Entomologischen Gesellschaft* **75**, 83–101.

Frieser R (1989) Ein weiterer Bericht zur Anthribiden-fauna Papua-Neuguineas (Coleoptera: Anthribidae). *Beiträge zur Entomologie* **39**(1), 105–123. doi:10.21248/contrib.entomol.39.1.105-123

Frieser R (2001) Neue Anthribiden von Australien, Neuguinea und aus der indomalaiischen Region (Coleoptera: Anthribidae). *Acta Coleopterologica* **12**(2), 39–67.

Frieser R (2009) Beitrag zur Kenntnis der Anthribiden (Coleoptera: Anthribidae). *Acta Coleopterologica (Munich)* **25**(1), 46–70.

Gahan CJ (1900) Order 4. Coleoptera. In *A Monograph of Christmas Island (Indian Ocean): Physical Features and Geology; with Descriptions of the Fauna and Flora by numerous Contributors*. (Ed. CW Andrews) pp. 89–127. British Museum (Natural History), London. doi:10.5962/bhl.title.25555

Herbst JFW (1797) *Natursystem aller bekannten in- und ausländischen Insekten; als eine Fortsezzung der von Büffonschen Naturgeschichte. Der Käfer siebenter Theil. Mit 26 illuminirten Kupfertafeln*. Pauli, Berlin, xi + 346 pp. pl. XCVI–CXVI.

Holloway BA (1982) Anthribidae (Insecta: Coleoptera). *Fauna of New Zealand* **3**, 1–264.

Holloway BA (1984) Morphology of pronotal carinae and associated surface structures in New Zealand Anthribidae (Insecta: Coleoptera). *New Zealand Journal of Zoology* **11**(4), 445–450. doi:10.1080/03014223.1984.10428259

Holloway BA (1985) Comparative morphology of setiferous sex patches in male Anthribidae (Insecta: Coleoptera). *New Zealand Journal of Zoology* **12**(2), 159–167. doi:10.1080/03014223.1985.10428275

Jekel H (1855) Coleoptera. Curculionides. - Part I. In *Insecta Saundersiana: Or Characters of Undescribed Insects in the Collection of William Wilson Saunders, Esq. Vol. II*. (Ed. WW Saunders) pp. 1–153. John van Voorst, London.

Jordan K (1894) On Anthribidae in the museum of the Honourable Walter Rothschild. *Novitates Zoologicae* **1**(4), 591–651.

Jordan K (1895a) Beitrag zur Kenntniss der Anthribidae. II. *Stettiner Entomologische Zeitung* **56**, 122–204.

Jordan K (1895b) Zur Kenntniss der Anthribidae. III. *Stettiner Entomologische Zeitung* **56**, 245–265.

Jordan K (1895c) Zur Kenntniss der Anthribidae. IV. *Stettiner Entomologische Zeitung* **56**, 369–401.

Jordan K (1903) New oriental Anthribidae. *Novitates Zoologicae* **10**, 415–434.

Jordan K (1923a) New Anthribidae from the eastern hemisphere. *Novitates Zoologicae* **30**, 216–221.

Jordan K (1923b) New Eastern Anthribidae. *Novitates Zoologicae* **30**(1), 167–185. doi:10.5962/bhl.part.28983

Jordan K (1924a) New Anthribidae allied to *Choragus*. *Annals and Magazine of Natural History* Series 9, **13**(78), 606–620. doi:10.1080/00222932408633088

Jordan K (1924b) New Anthribidae. *Novitates Zoologicae* **31**, 231–255.

Jordan K (1925) New Anthribidae collected by J.C. Corporaal on Sumatra. *Annals and Magazine of Natural History* Series 9, **16**(92), 257–264.

Jordan K (1928a) Anthribidae. *Insects of Samoa and Other Samoan Terrestrial Arthropoda* **4**(2), 161–172.

Jordan K (1928b) New Anthribidae from the Old World. *Novitates Zoologicae* **34**, 105–128.

Jordan K (1939) Descriptions of Anthribidae, mostly from Fiji. *Novitates Zoologicae* **41**, 423–432.

Kosztarab M, Kozar F (1983) Introduction of *Anthribus nebulosus* (Coleoptera: Anthribidae) in Virginia for control of scale insects: a review. *Virginia Journal of Science* **34**(4), 223–236.

Kuschel G (1995) A phylogenetic classification of Curculionoidea to families and subfamilies. *Memoirs of the Entomological Society of Washington* **14**, 5–33.

Kuschel G (1998) The subfamily Anthribinae in New Caledonia and Vanuatu (Coleoptera: Anthribidae). *New Zealand Journal of Zoology* **25**(4), 335–408. doi:10.1080/03014223.1998.9518163

Lawrence JF, Ślipiński SA (2013) *Australian Beetles. Vol. 1. Morphology, Classification and Keys*. CSIRO Publishing, Collingwood, Victoria, viii + 561 pp.

Lea AM (1926) Notes on some miscellaneous Coleoptera, with descriptions of new species. Part VI. *Transactions of the Royal Society of South Australia* **50**, 45–84.

Linnaeus C (1758) *Systema naturae per regna tria naturae, secundem classes, ordines, genera, species, cum characteribus, differentis, synonymis, locis. Editio decimal, reformata. Tom. I.* Laurentii Salvii, Holmiae, 824 pp. doi:10.5962/bhl.title.542

Louw S (1993) Systematics of the Urodontidae (Coleoptera: Curculionoidea) of southern Africa. *South Africa Department of Agriculture Entomology Memoir* **87**, 1–92.

Lyal CHC, Alonso-Zarazaga MA (2006) Addenda and corrigenda to "A world catalogue of families and genera of Curculionoidea (Insecta: Coleoptera)". 2. *Zootaxa* **1202**(1), 21–31. doi:10.11646/ZOOTAXA.1202.1.2

Marvaldi AE (2003) Key to larvae of the South American subfamilies of weevils (Coleoptera, Curculionoidea). *Revista Chilena de Historia Natural* **76**(4), 603–612.

Marvaldi AE, Morrone JJ (2000) Phylogenetic systematics of weevils (Coleoptera: Curculionoidea): a reappraisal based on larval and adult morphology. *Insect Systematics and Evolution* **31**, 43–58.

Marvaldi AE, Sequeira AS, O'brien CW, Farrell BD (2002) Molecular and Morphological Phylogenetics of weevils (Coleoptera, Curculionoidea): do niche shifts accompany diversification? *Systematic Biology* **51**(5), 761–785. doi:10.1080/10635150290102465

May BM (1993) *Fauna of New Zealand. Number 28. Larvae of Curculionoidea (Insecta: Coleoptera) a Systematic Overview.* Manaaki Whenua Press, Lincoln, New Zealand, 226 pp.

May BM (1994) An introduction to the immature stages of Australian Curculionoidea. In *Australian Weevils (Coleoptera: Curculionoidea). Volume II, Brentidae, Eurhynchidae, Apionidae and a Chapter on Immature Stages by Brenda May.* (Ed. EC Zimmerman) pp. 365–728. CSIRO, Canberra, Australia, 755 pp.

McKenna DD, Sequeira AS, Marvaldi AE, Farrell BD (2009) Temporal lags and overlap in the diversification of weevils and flowering plants. *Proceedings of the National Academy of Sciences of the USA* **106**(17), 7083–7088. doi:10.1073/pnas.081061810

Mermudes JRM, Leschen RAB (2014) Anthribidae Billberg, 1820. In *Handbuch der Zoologie/Handbook of Zoology. Band/Volume IV Arthropoda: Insecta Teilband/Part 38. Coleoptera, Beetles. Volume 3. Morphology and Systematics (Polyphaga).* (Eds RAB Leschen and RG Beutel) pp. 309–315. W. DeGruyter, Berlin.

Montrouzier X (1855) Essai sur la faune l'île de Woodlark ou Moiou. *Annales des Sciences Physiques et Naturelles, d'Agriculture et d'Industrie, Société impériale d'Agriculture (etc.) de Lyon*, Series 2, **7**(1), 1–114.

Morimoto K (1972) A key to the genera of oriental Anthribidae (Coleoptera). *Bulletin of the Forest Experiment Station, Meguro, Tokyo* **246**, 35–54.

Morimoto K (1978) The family Anthribidae of Japan (Coleoptera). Part 1. *Esakia* **12**, 17–47.

Morimoto K (1979) The family Anthribidae of Japan (Coleoptera). Part 2. *Esakia* **14**, 1–23.

Morimoto K (1980) The family Anthribidae of Japan (Coleoptera). Part 3. *Esakia* **15**, 11–47.

Morimoto K (1981) The family Anthribidae of Japan (Coleoptera). Part 4. *Esakia* **17**, 53–107.

Motschulsky V de (1875) Enumération des nouvelles espèces de Coléoptères rapportés de ses voyages. 14ième article. *Bulletin de la Société Impériale des Naturalistes de Moscou* **48**(4), 226–242.

Oke C (1934) On some Australian Curculionoidea. Part 2. *Proceedings of the Royal Society of Victoria (New Series)* **46**(2), 250–263.

Olliff AS (1889) The insect fauna of Lord Howe Island. *Memoirs of the Australian Museum* **2**(4), 75–98. doi:10.3853/j.0067-1967.2.1889.482

Olliff AS (1890) Additions to the insect-fauna of Lord Howe Island, and descriptions of two new Australian Coleoptera. *Records of the Australian Museum* **1**(3), 72–76. doi:10.3853/j.0067-1975.1.1890.1230

Pascoe FP (1859a) On some new Anthribidae. *Annals and Magazine of Natural History* Series 3, **4**(23), 327–333.

Pascoe FP (1859b) On some new Anthribidae. *Annals and Magazine of Natural History* Series 3, **4**(24), 431–439.

Pascoe FP (1860a) Notices of new or little-known genera and species of Coleoptera. Part I. *Journal of Entomology* **1**(1), 36–64. doi:10.5962/bhl.title.24970

Pascoe FP (1860b) On some new Anthribidae. *Annals and Magazine of Natural History* Series 3, **5**(25), 35–48.

Pascoe FP (1866) Notices of new or little known genera and species of Coleoptera. Part V. *Journal of Entomology* **2**(14), 443–493. doi:10.5962/bhl.title.24970

Pascoe FP (1882) Notes on Coleoptera, with descriptions of new genera and species. Part IV. *Annals and Magazine of Natural History* Series 5, **9**(49), 25–37. doi:10.1080/00222938209458985

Penman DR (1978) Biology of *Euciodes suturalis* (Coleoptera, Anthribidae) infesting cocksfoot in Canterbury. *New Zealand Entomologist* **6**(4), 421–425. doi:10.1080/00779962.1978.9722309

Perroud BP, Montrouzier X (1865) Essai sur la faune entomologique de Kanala (Nouvelle Calédonie) et descriptions de quelques espèces nouvelles ou peu connues. *Annales de la Société Linnéenne de Lyon* **11**, 46–257.

Pierce WD (1916) Studies on weevils (Rhynchophora) with descriptions of new genera and species. *Proceedings of the United States National Museum* **51**(2159), 461–473. doi:10.5479/si.00963801.51-2159.461

Pullen K, Jennings D, Oberprieler RO (2014) Annotated catalogue of Australian weevils (Coleoptera: Curculionoidea). *Zootaxa* **3896**(1), 1–481. doi:10.11646/zootaxa.3896.1.1

Reitter E (1916) *Fauna Germanica. Die Käfer des Deutschen Reiches. Nach der analytischen Methode bearbeitet. Band V.* K. G. Lutz, Stuttgart, 343 pp.

Rheinheimer J (2004) Illustrierter Katalog und Bibliographie der Anthribidae der Welt (Insecta: Coleoptera). *Mitteilungen Entomologischer Verein Stuttgart* **39**(1–2), 2–243.

Schneider DH (1791) Nachrichten von neu angenommenen Gattungen (Generibus) im entomologischen System. *Neuestes Magazin für die Liebhaber der Entomologie* **1**(1), 11–89.

Schoenherr CJ (1823) Curculionides [Tabula synoptica familiae Curculionidum]. *Isis von Oken* **1823**(10), columns 1132–1146.

Schoenherr CJ (1833) *Genera et species Curculionidum, cum synonymia hujus familiae. Species novae aut hactenus minus cognitae, descriptionibus a Dom. Leonardo Gyllenhal, C. H. Boheman, et entomologis aliis illustratae. Tomus primus. –Pars prima.* Roret, Parisii, pp. i–xv, 1–381 + 3 pp. Corrigenda.

Schoenherr CJ (1839) *Genera et species curculionidum, cum synonymia hujus familiae. Species novae aut hactenus minus cognitae, descriptionibus a Dom. L. Gyllenhal, C. H. Boheman, O. J. Fahraeus et entomologis aliis illustratae. Tomus quintus. –Pars prima. Supplementum continens.* Roret, Parisii/Fred. Fleischer, Lipsia, pp. i–viii, 1–456. doi:10.5962/bhl.title.8952

Sharp D (1891) The rhynchophorous Coleoptera of Japan. Part II. Apionidae and Anthribidae. *Transactions of the Royal Entomological Society of London* **39**(2), 293–328. doi:10.1111/j.1365-2311.1891.tb01652.x

Shin S, Clarke DJ, Lemmon AR, Moriarty Lemmon E, Aitken AL, *et al.* (2018) Phylogenomic data yield new and robust insights into the phylogeny and evolution of weevils. *Molecular Biology and Evolution* **35**(4), 823–836. doi:10.1093/molbev/msx324

Thompson RT (1992) Observations on the morphology and classification of weevils (Coleoptera, Curculionoidea) with a key to major groups. *Journal of Natural History* **26**(4), 835–891. doi:10.1080/00222939200770511

Thomson J (1858) Voyage au Gabon. Historie Naturelle des Insectes et des Arachnides recueillis pendant un voyage fait au Gabon en 1856 et en

1857 par M. Henry C. Deyrolle sous les auspices de M. M. Le Comte de Mniszech et James Thomson précédée de l'histoire du voyage; Arachnides par H. Lucas. *Archives Entomologiques* **2**, 1–469.

Valentine BD (1960) The genera of the weevil family Anthribidae north of Mexico (Coleoptera). *Transactions of the American Entomological Society* **86**(1), 41–85.

Valentine BD (1999) A Review of Nearctic and some related Anthribidae (Coleoptera). *Insecta Mundi* **12**(3/4), 251–296.

Valentine BD (2002) 126. Anthribidae Billberg, 1820. In *American Beetles. Vol. 2. Polypphaga: Scarabaeoidea through Curculionoidea.* (Eds RH Arnett, MC Thomas, PE Skelley and JH Frank) pp. 695–700. CRC Press LLC, Boca Raton, FL, xiv + 861pp.

Wink M, Rheinheimer J (1997) Phylogenetic relationships in weevils (Coleoptera: Curculionoidea) inferred from nucleotide sequences of mitochondrial 16S rDNA. *Naturwissenschaften* **84**(7), 318–321. doi:10.1007/s001140050404

Zimmerman EC (1938) Anthribidae of southeastern Polynesia (Coleoptera). *Occasional Papers of the Bernice P. Bishop Museum* **14**(13), 219–250.

Zimmerman EC (1991) *Australian Weevils (Coleoptera: Curculionoidea). Volume V. Colour Plates 1–304.* CSIRO Australia, Melbourne, x + 633 pp. 303 pls.

Zimmerman EC (1992) *Australian Weevils (Coleoptera: Curculionoidea). Volume VI. Colour Plates 305–632.* CSIRO Australia, Melbourne, viii + 707 pp. 328 pls.

Zimmerman EC (1994) *Australian Weevils (Coleoptera: Curculionoidea). Volume I. Orthoceri. Anthribidae to Attelabidae. The Primitive Weevils.* CSIRO Australia, Melbourne, xxxii + 741 pp.

40. BELIDAE SCHOENHERR, 1826

Hermes E. Escalona, Debbie Jennings and Rolf G. Oberprieler

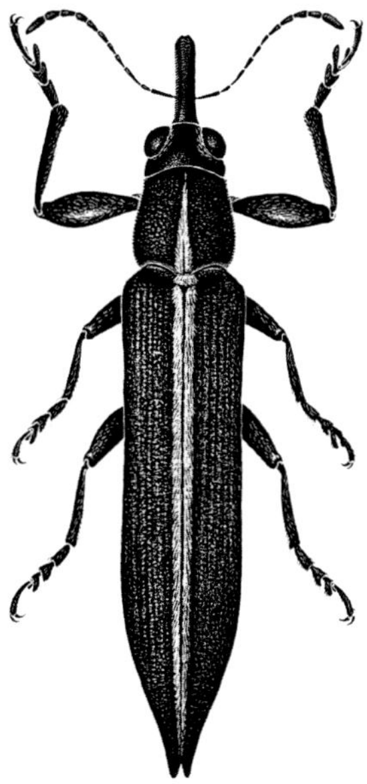

Fig. 40.1. *Rhinotia suturalis* (Macleay).

Introduction. The small family Belidae, with approximately 40 genera and 360 species worldwide, is an ancient and relictual lineage of weevils (Shin *et al.* 2018). It comprises two subfamilies, Belinae and Oxycoryninae, the former containing 25 genera and 145 species and restricted to the Southern Hemisphere (South America, Australia-New-Guinea (Meganesia) and New Zealand) and the latter, with 13 genera and approximately 200 species, distributed over both the Northern and the Southern Hemispheres but in a fragmented pattern, i.e. southern and central America, southern Africa, Southeast Asia and the Pacific region as far south as New Caledonia and New Zealand but absent from Australia (Lawrence & Ślipiński 2013; Marvaldi & Ferrer 2014). With 18 genera and 116 species of Belinae, Australia harbours almost half the generic diversity of the family and a third of the number of species, the latter largely due to the result of a species radiation, to approximately 87, in the genus *Rhinotia* Kirby (Marvaldi & Ferrer 2014; Pullen *et al.* 2014). The subfamily Belinae is divided into three tribes, Agnesiotidini, Belini and Pachyurini, which all have their greatest diversity in Australia. The monophyly of Belidae is supported by both morphological and molecular data (Kuschel 1995; Marvaldi & Morrone 2000; Marvaldi *et al.* 2002, 2009; Marvaldi & Ferrer 2014), and Belidae are generally recognised as constituting the sister-group of all other weevil families except Cimberididae, Nemonychidae and Anthribidae (e.g. Kuschel 1995; Oberprieler *et al.* 2007; McKenna *et al.* 2009, 2019; Li *et al.* 2023). Phylogenetic analyses of the tribes and genera of Belinae using morphological characters (Kuschel & Leschen 2003; Marvaldi 2005; Marvaldi *et al.* 2006) have not been able to resolve the tribal relationships unequivocally, but a recent molecular approach (Li *et al.* 2024) indicates that Agnesiotidini form the sister-group of Belini and Pachyurini. The fossil record of Belidae is poor and controversial, as nearly all fossils described as belids do not show the critical characters needed for an unambiguous assignment to the family. The only Mesozoic ones that do are a few specimens from the Crato Formation in Brazil, of Upper Aptian age (~113 Ma), described as *Preclarusbelus vanini* (Santos *et al.* 2007) and *Cratonemonyx martinsnetoi* (Gratshev & Legalov 2014) and both representing the subfamily Oxycoryninae (based on features such as double gular sutures, short maxillary palps, protibiae with grooming brush, short tarsites 1, elongate tarsites 5, short antennal segments 3, basal to subbasal antennal insertions). No authentic fossils of Belinae are known to date. The purportedly oldest representative of Belidae, *Sinoeuglypheus daohugouensis* (Yu *et al.* 2019), from the Tiaojishan Formation at Daohugou in China, of Upper Jurassic age (~165 Ma), does not display any explicit and indisputable features enabling its assignment to Belidae; it more likely is a representative of the family Cimberididae instead.

The life histories and larval hostplants of the Australian Belidae are also poorly known, and records of adults collected on certain plants often do not signify the hosts of their larvae. Most species of Pachyurini are associated with conifers (*Agathis, Araucaria, Callitris*), their larvae developing in the branches, stems and sometimes cones, but the larvae of Agnesiotidini and Belini mostly develop in dying or dead twigs and branches of angiosperms, specifically acacias (*Rhinotia*) and eucalypts (*Cyrotyphus, Isacantha*). The adult weevils often fly actively during the day, with specimens of *Rhinotia* sometimes relatively commonly found on the foliage of acacias but also of several other plants that are not necessarily their larval hosts. Some species of *Rhinotia* and *Rhinotiodes* appear to be mimics of Lycidae.

Characteristics. Adults (Fig. 40.1). Body length 5–23 mm (SL), elongate, about 1.9–6.4 × as long as wide, elytra parallel-sided or slightly wider near apex, pronotum slightly to strongly rounded laterally, slightly flattened to moderately convex, colour usually yellowish-brown to reddish-brown, dark brown or black, occasionally bicoloured on elytra or pronotum and elytra; almost always densely clothed with decumbent, thickened setae, variously coloured and often forming a pattern, sometimes interspersed with suberect finer setae, thickened setae sometimes limited to small patches. Head not to slightly declined at base, rarely constricted behind eyes (e.g. *Sphinctobelus*, Fig. 40.8A); occipital region without transverse ridge or stridulatory file, with moderately long occipital endocarina; rostrum inserted ventrally and porrect to gently arcuate, usually elongate (up to 8 × as long as wide) but rarely very short (e.g. *Arhinobelus*), sexually dimorphic, i.e. shorter and thicker in males, longer in females, subcylindrical to slightly flattened, not or only slightly widened apically, without distinct longitudinal dorsal carinae;

epistomal apex often notched. Labrum absent, completely fused to clypeus, frontoclypeal suture also absent. Eyes protuberant to strongly protuberant, hemispherical, separated by at least width of rostrum, entire and finely facetted, without or with sparse, fine interfacetal setae. Antennal insertions behind middle of rostrum, rarely about middle; sockets lateral or ventral, concealed from above, foveiform or produced into well-developed scrobes. Antennae non-geniculate, 11-segmented (but see Clarke *et al.* 2019), moderately long, slender, incrassate or with weak, 3- to 5-segmented club; scapes at least slightly elongate and slightly to much longer than pedicel. Mandibles very short and broad, bidentate or tridentate; incisor edge without subapical teeth, mola or prostheca; inner edge of mandibular base with moderately long, pubescent process. Maxillae with short, broad, fixed, setose or spinose galea and lacinia, the latter without uncus; maxillary palps moderately short to very short, 4-segmented, apical segment cylindrical. Prementum elongate, apically truncate or emarginate, fused at base to gulamentum; ligula undivided; labial palps short, 3-segmented, apical segment cylindrical. Pregula usually very long and narrow; gular sutures separate, short, sometimes meeting anteriorly in a single tentorial pit; gula longer than wide. Corpotentorium absent; anterior tentorial arms not expanded mesally; cervical sclerites absent.

Pronotum about 0.75–1.1 × as long as wide, widest behind middle, often constricted anteriorly; base not to distinctly narrower than combined elytral bases; sides usually slightly curved or sinuate, lateral pronotal carinae absent; anterior angles not produced; posterior angles rounded to slightly acute; posterior edge slightly to strongly sinuate or bi-emarginate, usually weakly margined but with distinct ventral interlocking ridge; disc sometimes with median longitudinal groove and endocarina, base often with paired, broad basal impressions into which anterior elytral projections fit, but weak to absent in *Agnesiotis*. Prosternum slightly to much shorter than shortest diameter of procoxal cavity, flat to somewhat declined; prosternal process narrowly or broadly interrupted at middle, intercoxal process strongly concave in middle, apex slightly expanded and then narrowed to form an apically acute or angulate sternellum; notosternal sutures incomplete, extending dorsolaterad from edge of procoxal cavity and then abruptly curving anteriad but never reaching anterior edge of prothorax. Procoxae projecting well below prosternum, with short concealed lateral extensions; trochantins concealed or barely exposed; procoxal cavities subcircular, contiguous or narrowly separate, externally broadly closed, with or without narrow lateral extensions, internally broadly closed; postcoxal (hypomeral) processes meeting prosternal process or meeting one another behind prosternal process.

Scutellar shield well-developed, abruptly elevated or not, anteriorly simple, posteriorly acute to broadly rounded or truncate; mesoscutum without stridulatory file. Elytra about 1.5–5.3 × as long as wide and 3.3–5.4 × as long as pronotum; anterior edge often sharp or carinate; humeri well-developed; sides usually subparallel, sometimes diverging to posterior third, apically usually conjointly broadly to narrowly rounded or subacute but individually acutely pointed in some Belini; always fully covering abdominal tergites; disc irregularly punctate, without distinct striae, with protuberances on either side of scutellum in *Cyrotyphus*, *Pachybelus* and *Isacantha*; lateral portion with internal interlocking ridge just anteriorly of middle; epipleura very narrow but sometimes complete. Mesoventrite separated by complete sutures from mesanepisterna; anterior edge on at least slightly different plane than metaventrite, with broadly transverse prothoracic rest. Mesoventral cavity absent. Mesocoxae not projecting, trochantins exposed or not. Mesocoxal cavities narrowly separate, subcircular, not or slightly oblique, closed laterally by meeting of meso- and metaventrite or partly closed by mesepimeron. Mesoventral process broadly rounded or truncate at apex; mesometaventral junction monocondylic. Metaventrite moderately convex; transverse (metakatepisternal) suture long but not crossing midline; discrimen usually complete to base of anterior intercoxal process; metanepisterna with moderately elongate exposed portion; metepimera slightly exposed. Metacoxae narrowly separate, not extending laterally to reach elytra. Metendosternite with long stalk, long lateral arms, well-developed laminae and very short anterior process bearing widely separated anterior tendons. Hindwings with apical field approximately 0.3 × as long as whole wing, containing distinct sclerite just beyond radial cell and two RP remnants; radial cell slightly elongate, with inner posterobasal angle right or obtuse; cross-vein r_3 usually moderately long and oblique, sometimes absent; r_4 without apical spur; basal portion of RP moderately long; R-M loop relatively broad; medial spur straight, extending to wing margin, there with distinct embayment; medial field usually with 5 free veins and no medial fleck; MP_{3+4} usually with basal cross-vein; wedge cell absent; anal lobe usually well-developed without embayment. Legs usually short and stout, forelegs longer than other two pairs; trochanterofemoral joint oblique but with base of femur usually separate from coxa; all tibiae, but particularly protibiae, with row of teeth along inner edge; tibial spur formula usually 2-2-2, with spurs articulated, but protibiae often with single spur combined with mucro at internal apical angle. Tarsi 5-5-5-segmented in both sexes but pseudotetramerous (segment 4 reduced), segments 3 distinctly bilobed; tarsal claws simple; empodium present (Fig. 40.5O).

Abdomen with 5 free ventrites; ventrite 1 at its shortest point subequal in length to 2; intercoxal process acute; ventrite 5 (sternite VII) emarginate exposing a portion of sternite VIII in some *Rhinotia* males. Abdominal tergites II–VI as darkly pigmented as tergite VII, the latter with a pair of wing-folding patches; functional spiracles on segments I–VII, all in pleural membrane. Tergite VIII in male usually concealed beneath tergite VII, exposed in some *Rhinotia*; sternite VIII with short median spicule. Anterior edge of segment IX in male with distinct spiculum gastrale; tergite X membranous. Aedeagus of pedotectal type, symmetrical; anterior edge of tegmen with moderately long manubrium; parameres partly or entirely fused together and fused to phallobase; penis with tectum and pedon distinct but sometimes partially or largely fused into a tube in Belini, with paired temones; endophallus with complex armature. Segment VIII in female with sclerotised tergite and sternite and long, fixed spiculum ventrale. Ovipositor

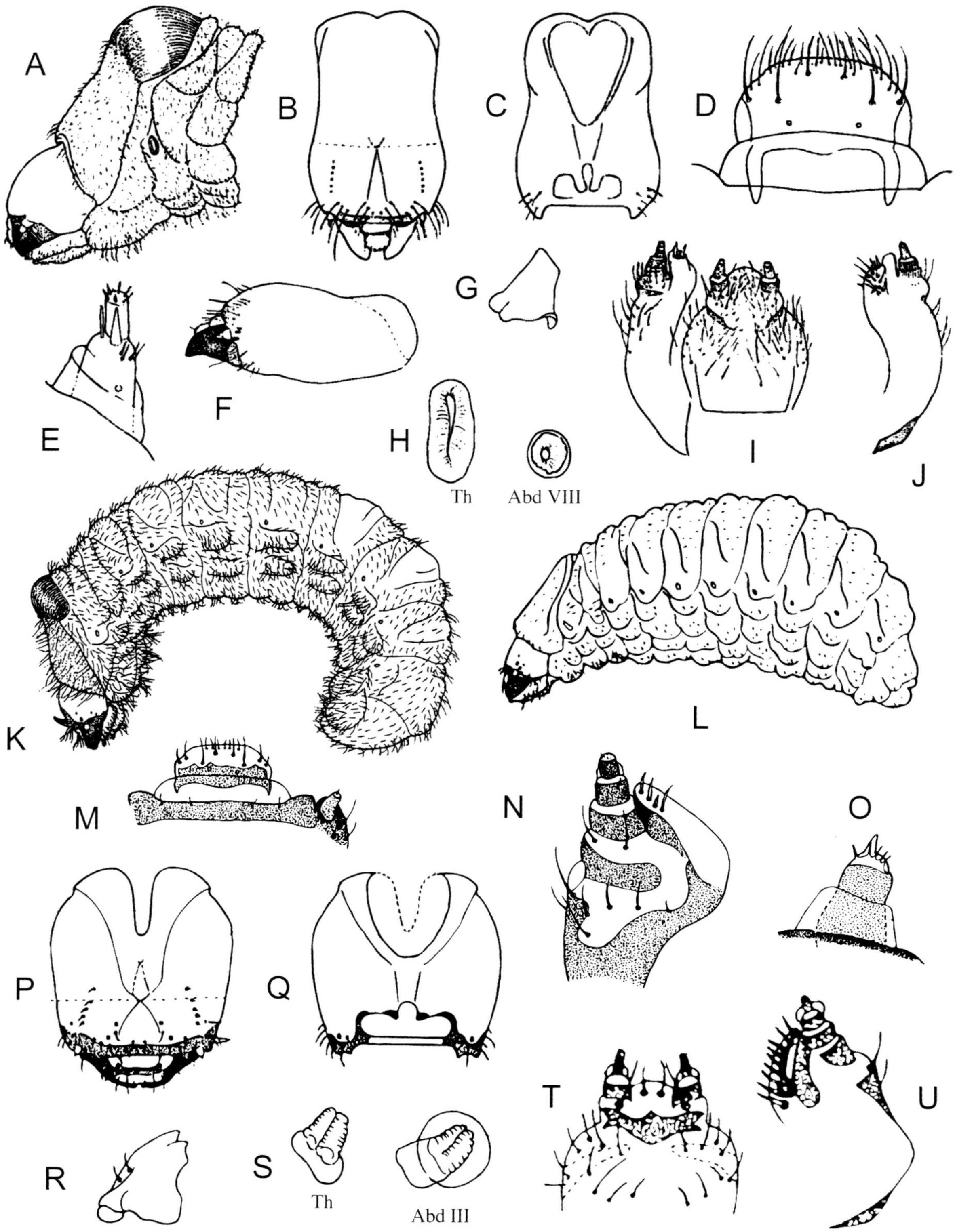

Fig. 40.2. Belidae, structural details of larvae. A, *Rhinotia* sp., larva, head and thorax, lateral; B–J, *Rhinotia haemoptera* Kirby, larva: B, head, dorsal; C, head, ventral; D, labrum and clypeus; E, antenna; F, head, lateral; G, mandible; H, spiracles, showing alignment; I, mala and labium, ventral; J, mala, dorsal; K, *Rhinotia bidentata* (Donovan) larva, lateral habitus; L–U, *Hadrobelus undulatus* (Zimmerman), larva: L, lateral habitus; M, labrum, epistoma, antenna and primary ocelli; N, mala, dorsal; O, antenna; P, head, dorsal; Q, head, ventral; R, mandible; S, spiracles, showing alignment; T, mala, ventral; U, labium, ventral. After May (1994).

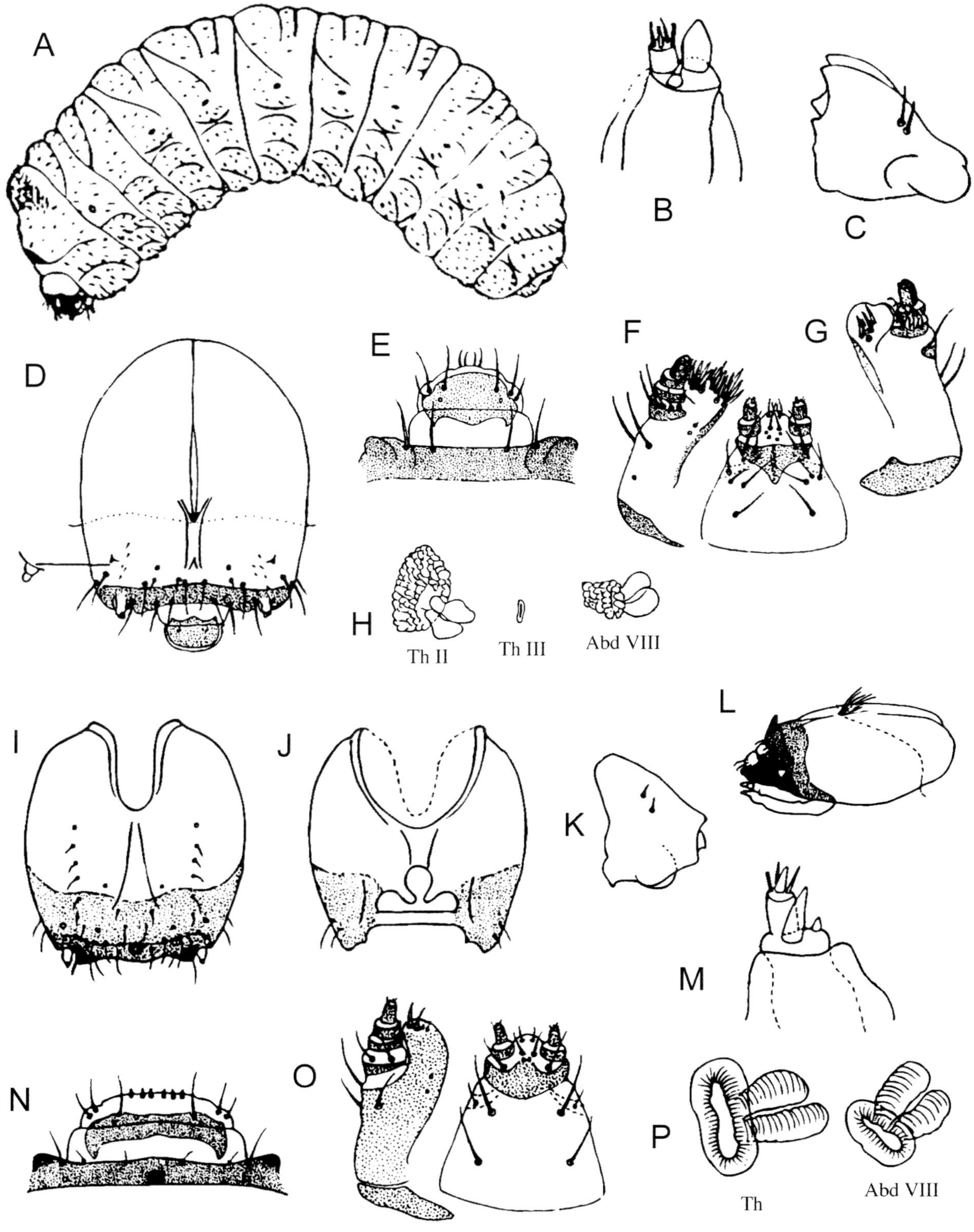

Fig. 40.3. Belidae, structural details of larvae. A–H, *Cyrotyphus blandus* (Faust), larva: A, lateral habitus; B, antenna; C, mandible; D, head, dorsal; E, labrum, clypeus and epistome; F, mala and labium, ventral; G, mala, dorsal; H, spiracles, showing alignment; I–P, *Sphinctobelus niger* Zimmerman, larva: I, head, dorsal; J, head, ventral; K, mandible, L; head, lateral; M, antenna; N, labrum, clypeus and epistome; O, mala and labium, ventral; P, spiracles, showing alignment. After May (1994).

moderately elongate and lightly sclerotised except for baculi, distal gonocoxites and styli; proctiger with paired baculi; paraprocts more or less subequal in length to coxites and lateral to them, with longitudinal baculi; gonocoxites divided into elongate basal part with long sclerotised baculum and short, heavily sclerotised distal part with terminal sclerotised stylus. Internal tract with spermathecal gland several times as large as sclerotised spermathecal capsule. Proventriculus with 8 or more rows of spinuliform setae. Hind gut with 6 cryptonephritic Malpighian tubules. Nervous system and female reproductive tract as in Calder (1989, 1990). (Kuschel 1959, 1995, 2003; Vanin 1976; Calder 1989, 1990; Thompson 1992; Anderson 2002, 2005; Marvaldi *et al.* 2006; Oberprieler *et al.* 2007, 2010; Marvaldi & Ferrer 2014).

Larvae (Figs 40.2, 40.3). Body usually elongate, slightly to strongly C-shaped, more or less parallel-sided, circular in cross-section, but shorter, broader and less curved in *Hadrobelus* (Fig. 40.2L); all surfaces lightly pigmented except for mouth-frame and mandibles and sometimes posterior portion of protergum (*Rhinotia*); vestiture consisting of simple, sometimes dense setae. Head prognathous and strongly retracted, at least slightly to clearly elongate and somewhat flattened; posterior edge deeply emarginate (*Hadrobelus*, *Sphinctobelus*) or not (*Cyrotyphus*, *Rhinotia*), postoccipital condyles absent; ecdysial lines often obscure; epicranial stem long; frontal arms V-shaped, contiguous at base; endocarina V-shaped. Stemmata 1–6 pairs (3 as vestigial eyespots). Antennae very short, usually 2-segmented, with conical sensorium on basal segment as long as or longer than apical segment (1-segmented with terminal sensorium in *Hadrobelus*). Frontoclypeal suture present; labrum free, with 4 pairs of primary setae; tormae and epipharyngeal rods absent. Mandibles short and broad, unidentate or bidentate, with or without additional tooth on cutting edge; mola and prostheca absent. Ventral mouthparts retracted; maxillary articulating area present; cardo slightly oblique, undivided; stipes longer than wide; usually with fixed, rounded, setose mala, usually cleft at apex, without laciniar spine; maxillary palps 3-segmented, with distinct palpifer. Labium consisting of prementum and postmentum; ligula rounded, longer than basal palp segment; labial palps 2-segmented and widely separate. Hypopharyngeal bracon present; hypopharyngeal sclerome absent. Hypostomal rods and ventral epicranial ridges absent. Gular region absent. Tentorial bridge wide, not divided on either side of midline.

Prothoracic tergum usually longer than meso- and metatergum combined, usually with posterior sclerotised elevation, this sometimes longitudinally striate or asperate (e.g. *Rhinotia*, Fig. 40.2A). Legs absent. Abdomen more than twice as long as thorax; abdominal terga not extending laterally beyond edges of sterna; most abdominal terga with 2 or 3 plicae, sometimes with transverse ambulatory ampullae; segment IX shorter than VIII; completely dorsal, without urogomphi; sternum IX simple; segment X much shorter than IX, terminal, without pygopods; anal region posteriorly or posteroventrally orientated; anal opening Y-shaped. Spiracles annular-biforous or annular; thoracic spiracles on mesothorax. (May 1993, 1994; Anderson 2002; Marvaldi 2003, 2005; Marvaldi & Ferrer 2014).

Pupae. Terminal abdominal segment with paired posterior processes (pupal urogomphy) short, acute, curved dorsolaterad. (May 1994).

Key to the Australian genera

Modified after Zimmerman (1994).

1. Antennae with segment 1 slightly to distinctly longer than 2, segments 1 and 2 shorter than 3 (Fig. 40.8C); antennal sockets often foveiform (Fig. 40.7B) **2**

– Antennae with segment 1 often longer than 2, segments 1 and 2 as long as or longer than 3 (Fig. 40.6B); antennal sockets variable but often elongate (sulciform) (Fig. 40.6A) **15**

2(1). Rostrum wider than long (Fig. 40.4C); antennal insertions distant from base of mandible by about length of antennal segment 1; QLD .... ***Arhinobelus* Zimmerman**

– Rostrum longer than wide; antennal insertions distant from base of mandible by more than length of antennal segment 1 .... **3**

3(2). Head with distinct longitudinal sulci on vertex, usually bordering inner margin of eye and extending to base of rostrum **4**

– Head without sulci on vertex **10**

4(3). Femora simple, without ventral tooth; meso- and metatibiae without crenulate carina on outside **5**

– Femora with ventral tooth; meso- and metatibiae with crenulate carina on outside **9**

5(4). Temples as long as or longer than eye **6**

– Temples shorter than eye **7**

6(5). Elytra widest behind middle (Fig. 40.4H); vertex about as wide as rostrum; QLD, NSW, VIC, TAS, SA, WA .... ***Pachyura* Hope** (part)

– Elytra narrowest behind middle (Fig. 40.4B); vertex narrower than rostrum; QLD .... ***Macrobelus* Lea**

7(5). Left mandible tilted upwards; QLD .... ***Agathobelus* Zimmerman**

– Left mandibles not tilted upwards **8**

8(7). Procoxal cavities separate; QLD .... ***Apagobelus* Zimmerman**

– Procoxal cavities contiguous; QLD .... ***Basiliobelus* Zimmerman** (part)

9(4). Body covered with long, fine, erect hairs (Fig. 40.4M); pronotum and elytra without elevations or tubercles; QLD, NSW, VIC .... ***Agnesiotis* Pascoe**

– Body not covered with fine long hairs on dorsum and elytra (Fig. 40.4J); pronotum and elytra with conspicuous carinae or tubercles; NSW, VIC, TAS, SA, WA .... ***Cyrotyphus* Pascoe**

10(3). Antennae inserted near middle of rostrum (Fig. 40.4H); tibial spur formula 2-2-2 .... **11**

– Antennae inserted near base of rostrum; tibial spur formula 1-2-2 or 2-2-2 .... **12**

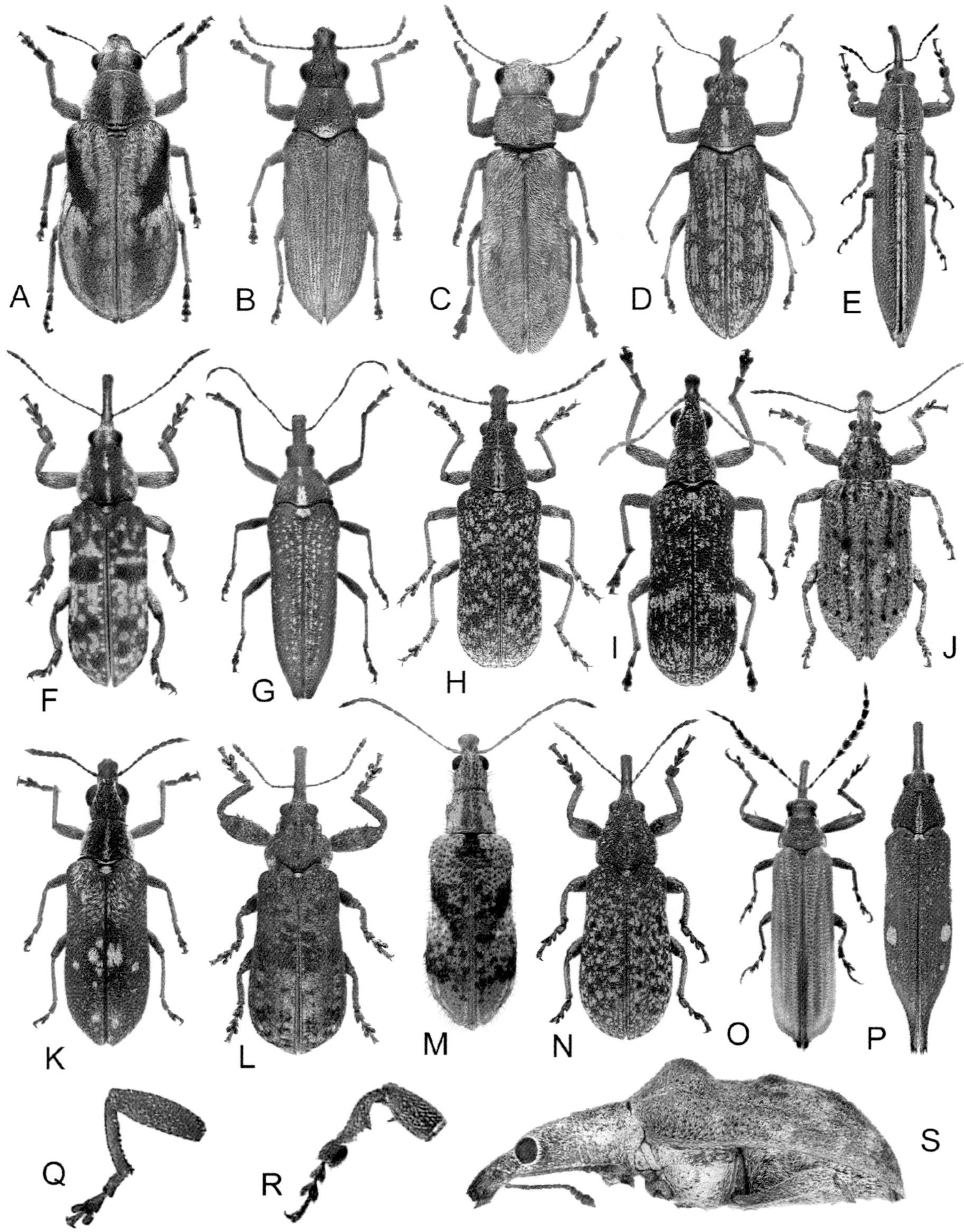

Fig. 40.4. Belidae, characteristics for key to genera. A, *Agathobelus bivittatus* Zimmerman; B, *Apagobelus brevirostris* (Lea); C, *Arhinobelus agathophagus* Zimmerman; D, *Basiliobelus flavovittatus* Zimmerman; E, *Stenobelus tibialis* (Blackburn); F, *Hadrobelus undulatus* (Zimmerman); G, *Macrobelus insignis* (Lea); H, *Pachyura australis* Hope; I, *Sphinctobelus cinereus* (Blanchard); J, *Cyrotyphus fascicularis* Pascoe; K, *Sphinctobelus quadrimaculatus* (Lea); L, *Isacantha dermestiventris* (Boisduval); M, *Agnesiotis pilosula* Pascoe; N, *Isacanthodes monilis* (Newman); O, *Rhinotiodes spinipennis* (Lacordaire); P, *Rhinotia bidentata* (Donovan); Q–R, forelegs: Q, *Araiobelus filum* (Jekel); R, *S. tibialis* (Blackburn); S, *Pachybelus tuberculatus* Zimmerman, lateral habitus.

11(10). Elytra not tuberculate; QLD, NSW, VIC, TAS, SA, WA .... ***Pachyura* Hope** (part)
– Elytra tuberculate (Fig. 40.4S); QLD, NSW .... ***Pachybelus* Zimmerman**
12(10). Head with postocular constriction (Fig. 40.7A); mandibles when closed not visible from above; small species (<7.0 mm SL); QLD, NSW, VIC, TAS, SA .... ***Sphinctobelus* Zimmerman**
– Head without postocular constriction; mandibles when closed visible from above **13**
13(12). Prosternum longer than procoxal diameter; QLD .... ***Habrobelus* Zimmerman** (part)
– Prosternum about as long as procoxal diameter **14**
14(13). Temples about as long as eye diameter; elytral surface uneven, with many conspicuous depressions; QLD, NSW .... ***Habrobelus* Zimmerman** (part)
– Temples shorter than eye diameter; elytral surface even .... ***Basiliobelus* Zimmerman** (part)
15(1). Metafemora reaching apex of ventrite 2 when folded back **16**
– Metafemora reaching about middle of ventrite 2 when folded back **20**
16(15). Elytral apices extended into long, narrow processes (Fig. 40.5D–F) .... ***Rhinotia* Kirby** (part)
– Elytral apices not extended **17**
17(16). Prosternum longer than procoxal diameter .... ***Rhinotia* Kirby** (part)
– Prosternum shorter than procoxal diameter **18**
18(17). Profemora without ventral tooth (Fig. 40.4N); protibiae neither serrate not denticulate on inside, with 2 spurs and no mucro .... ***Isacanthodes* Zimmerman**
– Profemora with ventral tooth; protibiae denticulate or serrate on inside, with 1 spur and mucro **19**
19(18). Base of elytra around scutellum projected over base of pronotum (Fig. 40.4L) and above level of scutellar shield .... ***Isacantha* Hope**
– Base of elytra not elevated above scutellar shield .... ***Rhinotia* Kirby** (part)
20(15). Apex of elytra produced upwards as tuberculiform, setose process (Fig. 40.4O); femora with ventral tooth .... ***Rhinotiodes* Zimmerman**
– Apex of elytra not produced; femora toothed or not **21**
21(20). Protibiae with 1 spur at base of mucro-like process .... ***Rhinotia* Kirby** (part)
– Protibiae with 2 spurs **22**
22(21). Profemora without ventral tooth (Fig. 40.4Q); protibiae on inside partly serrate (Fig. 40.4Q) .... ***Araiobelus* Zimmerman**
– Profemora with ventral tooth (Fig. 40.4R); protibiae on inside with strong tooth near base (Fig. 40.4R) .... ***Stenobelus* Zimmerman**

Classification of the Australian genera

Subfamily Belinae Schoenherr, 1826

Tribe Agnesiotidini Zimmerman, 1994

Agnesiotis Pascoe, 1870 (Pls 2A, 34A, B; Fig. 40.4M)

Type species, by monotypy: *Agnesiotis pilosula* Pascoe, 1870b.
Australian species. *A. pilosula* Pascoe, 1870b.
Distribution. Australia (QLD, NSW, VIC).
Notes. Adults of *A. pilosula* have been collected on *Acacia melanoxylon* (Fabaceae), but the hostplants of the larvae are unknown.

Cyrotyphus Pascoe, 1870 (Pls 2E, 34C; Figs 40.3A–H, 40.4J)

Type species, by monotypy: *Cyrotyphus fascicularis* Pascoe, 1870a.
Australian species. *C. blandus* (Faust, 1892), *C. fascicularis* Pascoe, 1870a, *C. variegatus* Lea, 1919, *C. vestitus* (Pascoe, 1873).
Distribution. Australia (QLD, NSW, VIC, TAS, SA, WA).
Notes. In addition to the four described species, Zimmerman (1994) identified another two undescribed ones in Australia, and he also subsumed the monotypic New Zealand genus *Agathinus* Broun, 1880 into *Cyrotyphus*, but this action (and the synonymy of their names) was rejected by Kuschel (2003). The life histories of the genus are poorly known. Specimens of *C. blandus* have been reared from dead, decaying *Eucalyptus* logs (May 1994; Zimmerman 1994), but adults of *C. fascicularis* and *C. vestitus* have been collected from *Callitris* foliage (Cupressaceae) and it is possible that their larvae develop in branches or trunks of this plant genus.

Tribe Belini Schoenherr, 1826

Araiobelus Zimmerman, 1994 (Pl. 34D; Fig. 40.4Q)

Type species, by original designation: *Belus acicularis* Pascoe, 1872b.
Australian species. *A. acicularis* (Pascoe, 1872b), *A. filum* (Jekel, 1860), *A. floccosus* (Lea, 1917), *A. maculipennis* (Lea, 1925).
Distribution. Australia (NSW, VIC, TAS, SA, WA).
Notes. *Araiobelus* includes only four species of very narrow, reddish belids that are difficult to distinguish from similar narrow species of *Rhinotia*, and the concept of the genus is in need of verification as it lacks stable differences from *Rhinotia*, such as the varying length of its metafemora and its variable spur formula, which can differ not only intraspecifically but even in single specimens (Zimmerman 1994).

Isacantha Hope, 1834 (Pl. 34E; Fig. 40.4L)

Type species, by monotypy: *Isacantha rhinotioides* Hope, 1834.

Australian species. *I. dermestiventris* (Boisduval, 1835), *I. inculta* Olliff, 1889, *I. interrupta* (Lea, 1917), *I. punctirostris* (Lea, 1908b), *I. rhinotioides* Hope, 1834, *I. serrata* Lea, 1908b.

Distribution. Australia (QLD, NSW, VIC, TAS, SA, Lord Howe Is.).

Notes. In addition to the six described species, Zimmerman (1994) identified another two undescribed ones. The adults of *I. dermestiventris* appear to mimic bird droppings, and the species is widespread and very variable in size and colour pattern. Its larvae have been found developing in dead limbs and branches of *Eucalyptus* (Myrtaceae) and *Nothofagus cunninghamii* (Nothofagaceae). Adults of *I. serrata* and one of the undescribed species have been collected from *Casuarina* (Casuarinaceae), but it is unknown whether their larvae also develop in branches of this genus.

Isacanthodes Zimmerman, 1994 (Pls 2C, 34F; Fig. 40.4N)

Type species, by original designation: *Pachyura monilis* Newman, 1838.

Australian species. *I. ganglionicus* (Pascoe, 1873), *I. monilis* (Newman, 1838).

Distribution. Australia (QLD, NSW, VIC, NT, WA).

Notes. The genus contains two described and another five undescribed species (Zimmerman 1994). The life histories of its species are almost completely unknown, except that adults of *I. ganglionicus* and *I. monilis* have been collected from *Callitris* trees (Cupressaceae) and those of one of the undescribed species from foliage of *Hakea* (Proteaceae).

Rhinotia Kirby, 1819 (Pls 2F–H, 34G–J; Figs 40.1, 40.2A–K, 40.4P, 40.5, 40.6)

Type species, by monotypy: *Rhinotia haemoptera* Kirby, 1819.

Australian species. *R. acaciae* (Lea, 1899), *R. acanthoptera* (Lea, 1910), *R. acrobeles* (Olliff, 1889), *R. acutipennis* (Lea, 1917), *R. adelaidae* (Blackburn, 1893), *R. affinis* (Perroud, 1853), *R. amplicollis* (Jekel, 1860), *R. anguinea* (Pascoe, 1872b), *R. angustata* (Lea, 1917), *R. angustula* (Germar, 1848), *R. aphthosa* (Pascoe, 1872b), *R. apicalis* Zimmerman, 1994, *R. bassiae* (Marshall, 1936), *R. bidentata* (Donovan, 1805), *R. bimaculata* (Pascoe, 1871), *R. bispinosa* (Perroud, 1853), *R. brevipes* (Lea, 1908a), *R. brunnea* (Guérin-Méneville, 1838), *R. centralis* (Pascoe, 1872a), *R. corallina* Pascoe, 1872b, *R. cristata* (Lea, 1908a), *R. cruenta* Pascoe, 1870b, *R. cylindrica* (Lea, 1917), *R. difficilis* (Blackburn, 1893), *R. distincta* (Blackburn, 1893), *R. divisa* (Pascoe, 1885), *R. edentula* (Lea, 1899), *R. elegans* (Blackburn, 1893), *R. exigua* (Pascoe, 1873), *R. exilis* (Lea, 1917), *R. farinaria* (Pascoe, 1872b), *R. filiformis* (Germar, 1848), *R. flindersi* (Blackburn, 1893), *R. frater* (Blackburn, 1893), *R. granicollis* (Lea, 1908a), *R. haemoptera* Kirby, 1819, *R. halmaturina* (Lea, 1917), *R. helmsi* (Blackburn, 1893), *R. hemisticta* (Germar, 1848), *R. inconstans* (Lea, 1908a), *R. insipida* (Blackburn, 1889), *R. irrorata* (Jekel, 1860), *R. lacustris* (Lea, 1917), *R. lineata* (Donovan, 1805), *R. marginella* Boheman, 1839, *R. melanocephala* (Boheman, 1839), *R. mimica* (Lea, 1917), *R. multimaculata* (Lea, 1917), *R. niveopilosa* (Lea, 1908a), *R. orthodoxa* (Lea, 1917), *R. parallela* (Pascoe, 1872b), *R. parva* Lea, 1908a, *R. perplexa* (Blackburn, 1893), *R. phoenicoptera* (Germar, 1848), *R. pica* (Jekel, 1860), *R. pictirostris* (Lea, 1908a), *R. plagiata* (Pascoe, 1870b), *R. podagrosa* (Lea, 1917), *R. poverus* (Lea, 1917), *R. princeps* Zimmerman, 1994, *R. pruinosa* Pascoe, 1871, *R. pudica* (Lea, 1899), *R. pulverulenta* (Lea, 1908a), *R. regalis* (Blackburn, 1893), *R. ruficornis* (Lea, 1908b), *R. scalaris* (Germar, 1848), *R. semipunctata* (Fabricius, 1775), *R. serpens* (Pascoe, 1870b), *R. simplicipennis* Lea, 1908a, *R. sparsa* (Germar, 1848), *R. subparallela* (Jekel, 1860), *R. subsuturalis* (Lea, 1908b), *R. suturalis* (Macleay, 1826), *R. tenuis* (Lea, 1899), *R. ursus* (Lea, 1910), *R. variabilis* (Lea, 1917), *R. venusta* Pascoe, 1872c, *R. vetusta* (Pascoe, 1870c), *R. villosa* (Lea, 1917).

Distribution. Australia (incl. Lord Howe Is.), New Guinea, Solomon Is., Indonesia (Aru Is., Mysol Is., Morotai Is.).

Notes. *Rhinotia* is by far the most speciose genus of Belidae. As pointed out by Zimmerman (1994), its species-level taxonomy is complicated due to the large variation among morphological characters, and the seven species groups he used in his key to assist with species identifications do not constitute natural (monophyletic) groups. Many undescribed species are known to exist, and Zimmerman's (1994) treatise of the genus was only a preliminary one. The larval hosts and life histories are only known for few species. Most host records of the Australian species pertain to the genus *Acacia* (Fabaceae) (Hawkeswood *et al.* 1994; Zimmerman 1994), but adults have also been collected on several other plants, in the genera *Betula* (Betulaceae), *Cassia* (Fabaceae), *Casuarina* (Casuarinaceae), *Atriplex*, *Bassia* and *Salicornia* (Chenopodiaceae), *Cyathodes* (Epacridaceae), *Loranthus* (Loranthaceae), *Prunus* (Rosaceae) and *Argyrodendron* (Malvaceae) (Zimmerman 1994). Detailed life-history accounts are only provided by Froggatt (1893, 1902, 1910, 1923) and Hawkeswood *et al.* (1994), and May (1994) published descriptions and illustrations of the larvae of *R. haemoptera* Kirby and *R. bidentata* (Donovan). It appears that the females generally lay their eggs in live branches and twigs but that these die off as the larvae develop inside them and may fall to the ground, and the larvae pupate in the dead stems (Froggatt 1893, 1923).

Rhinotiodes **Zimmerman, 1994 (Pls 3A, 34K; Fig. 40.4O)**

Type species, by original designation: *Rhinotia spinipennis* Lacordaire, 1863.
Australian species. *R. spinipennis* (Lacordaire, 1863).
Distribution. Australia (QLD, NSW).
Notes. The genus contains a single but variable species, whose colouration appears to mimic that of Lycidae (Zimmerman 1994) and perhaps other beetle families with similar aposematic colour patterns, such as Pyrochroidae. Nothing appears to be known about its larval hosts and life history.

Stenobelus **Zimmerman, 1999 (Pl. 34L; Fig. 40.4F)**

Type species, by original designation: *Belus tibialis* Blackburn, 1893.
Australian species. *S. testaceus* (Waterhouse, 1839), *S. tibialis* (Blackburn, 1893).
Distribution. Australia (QLD, SA, WA).
Notes. The genus was originally described as *Leptobelus* Zimmerman, 1994, but as this name is a junior homonym, it was renamed as *Stenobelus* by Zimmerman in Alonso-Zarazaga & Lyal (1999). *Stenobelus* was revised by Oberprieler *et al.* (2010), who found that the name of one of the two species included in *Leptobelus* by Zimmerman (1994), *L. linearis* (Pascoe), is a junior synonym of *Belus testaceus* Waterhouse, 1839, a name based on a single specimen collected by Charles Darwin in 1836 in Western Australia but since forgotten in the literature. Species of *Stenobelus* can be differentiated from those of other beline genera by the distinctive clamping device of the forelegs (Fig. 40.4R). Nothing is known about the larval hosts and life histories of the genus, but the habitats of both species appear to be acid swamplands dominated by *Melaleuca* trees and sedges (Oberprieler *et al.* 2010).

Tribe Pachyurini Kuschel, 1959

Agathobelus **Zimmerman, 1994 (Pl. 35A; Fig. 40.4A)**

Type species, by original designation: *Agathobelus bivittatus* Zimmerman, 1994.
Australian species. *A. bivittatus* Zimmerman, 1994.
Distribution. Australia (East Coast QLD).
Notes. A specimen of *A. bivittatus* has been reared from a log of *Agathis robusta*, and another two in the ANIC are labelled as having been reared from Hoop Pine (*Araucaria cunninghamii*) (Zimmerman 1994), but the latter record needs confirmation.

Apagobelus **Zimmerman, 1994 (Pl. 35B; Fig. 40.4B)**

Type species, by original designation: *Pachyura brevirostris* Lea, 1917.
Australian species. *A. brevirostris* (Lea, 1917).
Distribution. Australia (QLD, NSW).
Notes. Specimens of *A. brevirostris* have been reared from billets, female cones and crown galls of *Araucaria cunninghamii* as well as cones of *Agathis* (Zimmerman 1994; James Nicholls pers. comm. 2024).

Arhinobelus **Zimmerman, 1994 (Pl. 35C; Fig. 40.4C)**

Type species, by original designation: *Arhinobelus agathophagus* Zimmerman, 1994.
Australian species. *A. agathophagus* Zimmerman, 1994.
Distribution. Australia (QLD).
Notes. As indicated by its name, *Arhinobelus* does not have a definite rostrum, thus does not appear like a weevil at all. The only three known specimens of *A. agathophagus* were reared from *Agathis* (presumably logs) (Zimmerman 1994).

Basiliobelus **Zimmerman, 1994 (Pl. 35D; Fig. 40.4D)**

Type species, by original designation: *Basiliobelus flavovittatus* Zimmerman, 1994.
Australian species. *B. alveatus* Zimmerman, 1994, *B. flavovittatus* Zimmerman, 1994, *B. lepidus* Zimmerman, 1994.
Distribution. Australia (QLD, NSW).
Notes. All three species of the genus are only known from very few specimens, with one female of *B. flavovittatus* having been reared from an immature (female) cone of *Araucaria cunninghamii* (Zimmerman 1994).

Habrobelus **Zimmerman, 1994 (Pl. 35J, K; Fig. 40.4N)**

Type species, by original designation: *Habrobelus maculatus* Zimmerman, 1994.
Australian species. *H. maculatus* Zimmerman, 1994.
Distribution. Australia (QLD).
Notes. The genus and its single species are only known from the holotype male specimen of *H. maculatus* (Zimmerman 1994). Nothing is known about its hosts and life history, but it is likely to be associated with *Araucaria* and/or *Agathis* as well.

Hadrobelus **Zimmerman, 1999 (Pl. 35E; Figs 40.2L–U, 40.4F)**

Type species, by original designation: *Brachybelus undulatus* Zimmerman, 1994.
Australian species. *H. undulatus* (Zimmerman, 1994).
Distribution. Australia (QLD).
Notes. The genus was originally described as *Brachybelus* Zimmerman, 1994, but as this name is a junior homonym, it was renamed as *Hadrobelus* by Zimmerman in

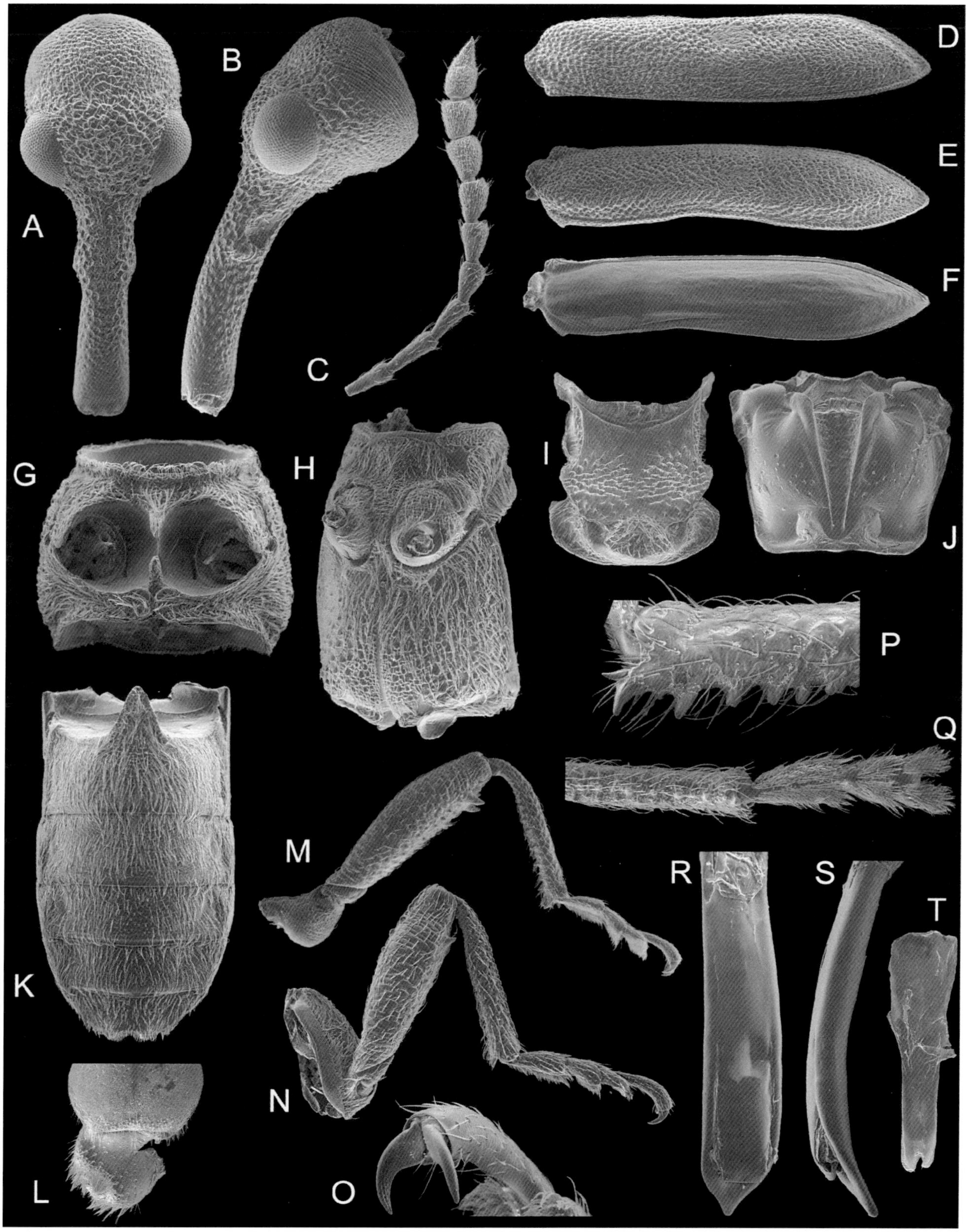

Fig. 40.5. Belini, structural details, *Rhinotia bimaculata* (Pascoe), ♂. A–B, head: A, dorsal, B; lateral; C, antenna; D–F, elytron: D, dorsal; E, lateral; F, inner lateral; G, prothorax, ventral; H, pterothorax, ventral; I, scutellum; J, metanotum; K, abdominal ventrites; L, tergites VIII–IX, dorsal; M, foreleg; N, hindleg; O, tarsal claws; P, protibia, apex; Q, tarsi, ventral; R–S, penis: R, dorsal; S, lateral; T, tegmen.

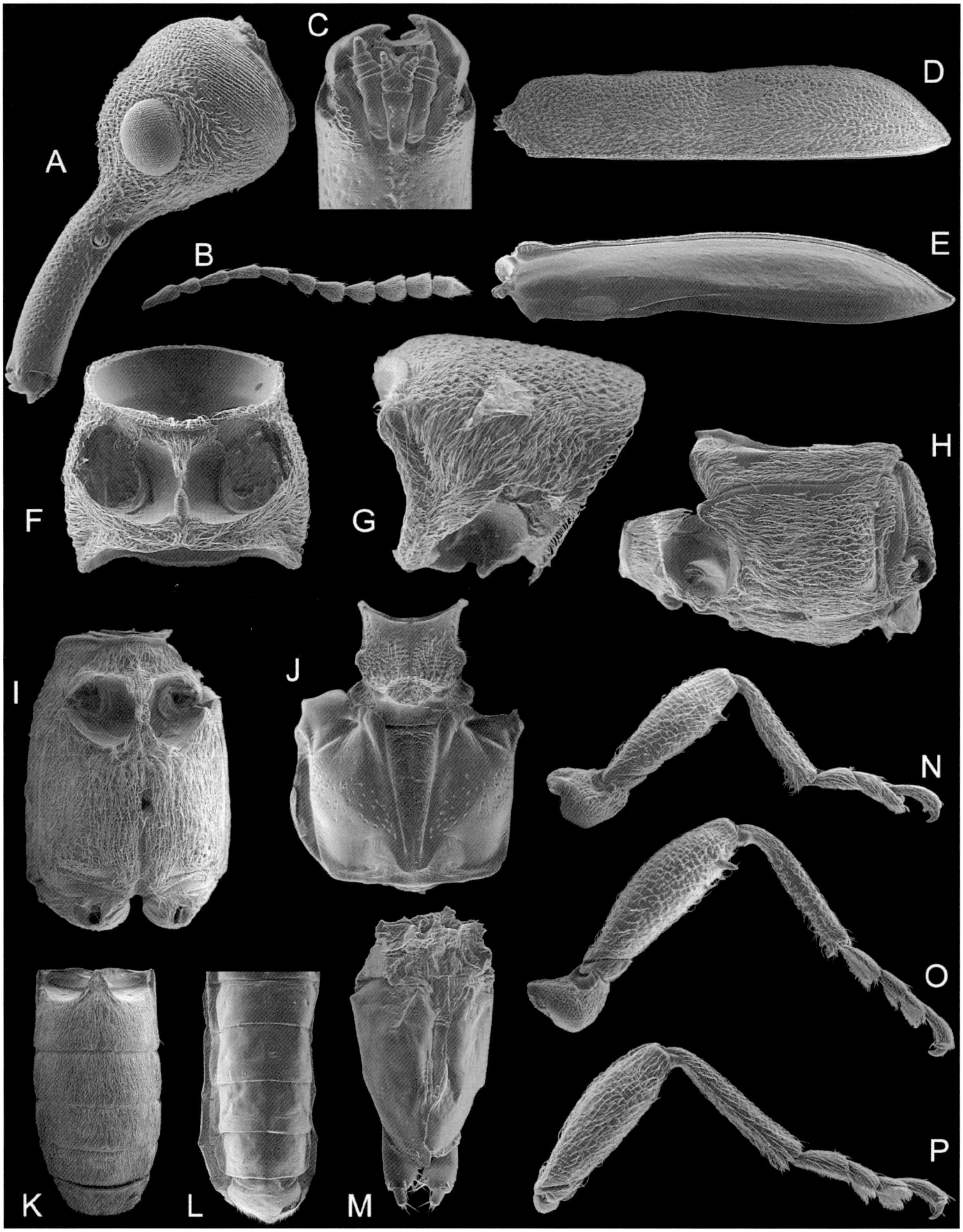

Fig. 40.6. Belini, structural details, *Rhinotia bimaculata* (Pascoe), ♀. A, head, lateral; B, antenna; C, mouthparts; D–E, elytron: D, dorsal; E, inner lateral; F–G, prothorax: F, ventral; G, lateral; H–I, pterothorax: H, lateral; I, ventral; J, scutellum and metanotum; K, abdominal ventrites; L, abdominal tergites; M, ovipositor; N, middle leg; O, foreleg; P, hindleg.

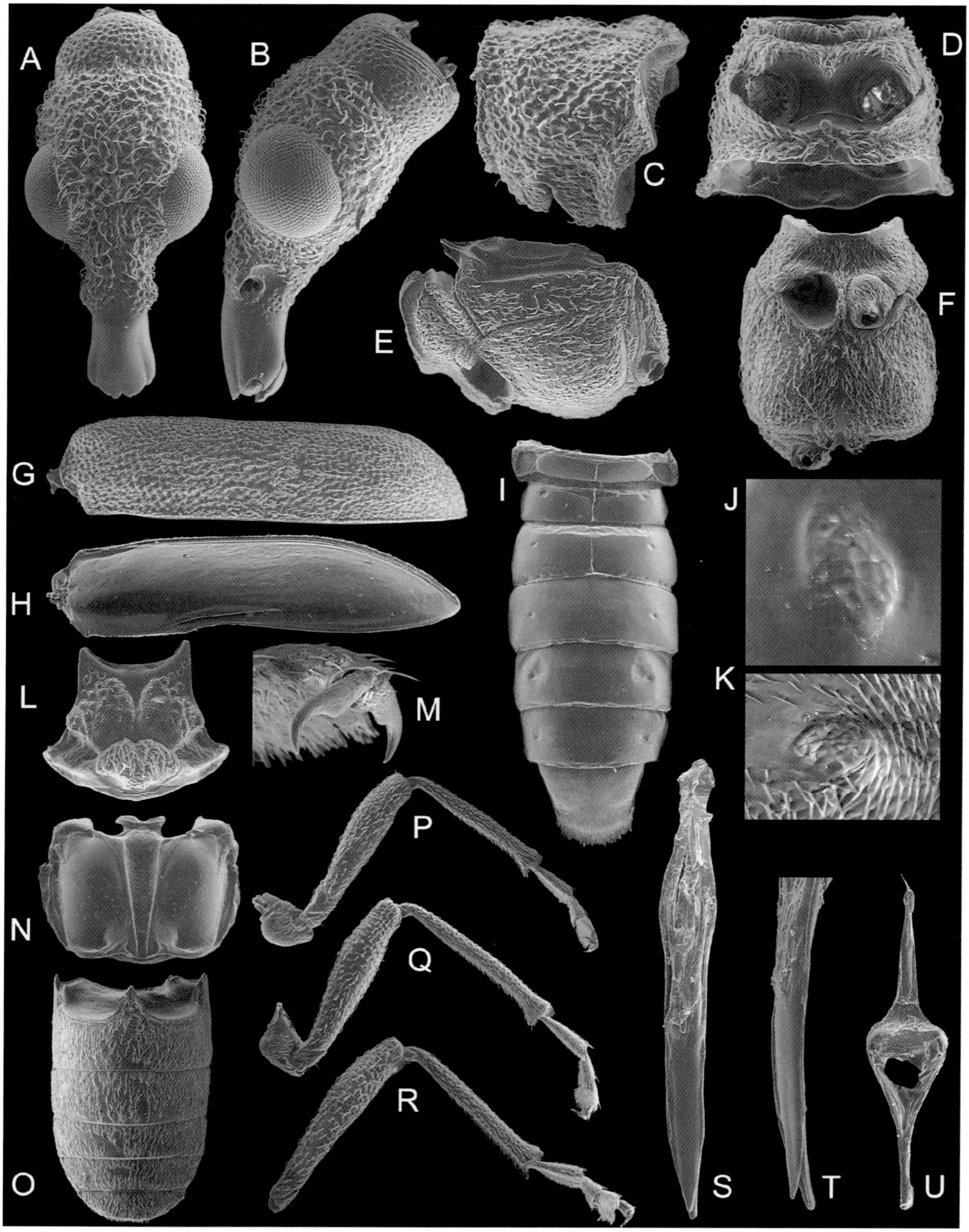

Fig. 40.7. Pachyurini, structural details, *Sphinctobelus cinereus* (Blanchard), ♂. A, head, dorsal; B, head, lateral; C–D, prothorax: C, lateral; D, ventral; E–F, pterothorax: E, lateral; F, ventral; G–H, elytron: G, dorsal; H, inner lateral; I, tergites; J, pore on tergite II; K, pore on tergite IV; L, scutellum; M, tarsal claws; N, metanotum; O, abdominal ventrites; P, middle leg; Q, foreleg; R, hindleg; S–T, penis: S, dorsal; T, lateral; U, tegmen, dorsal.

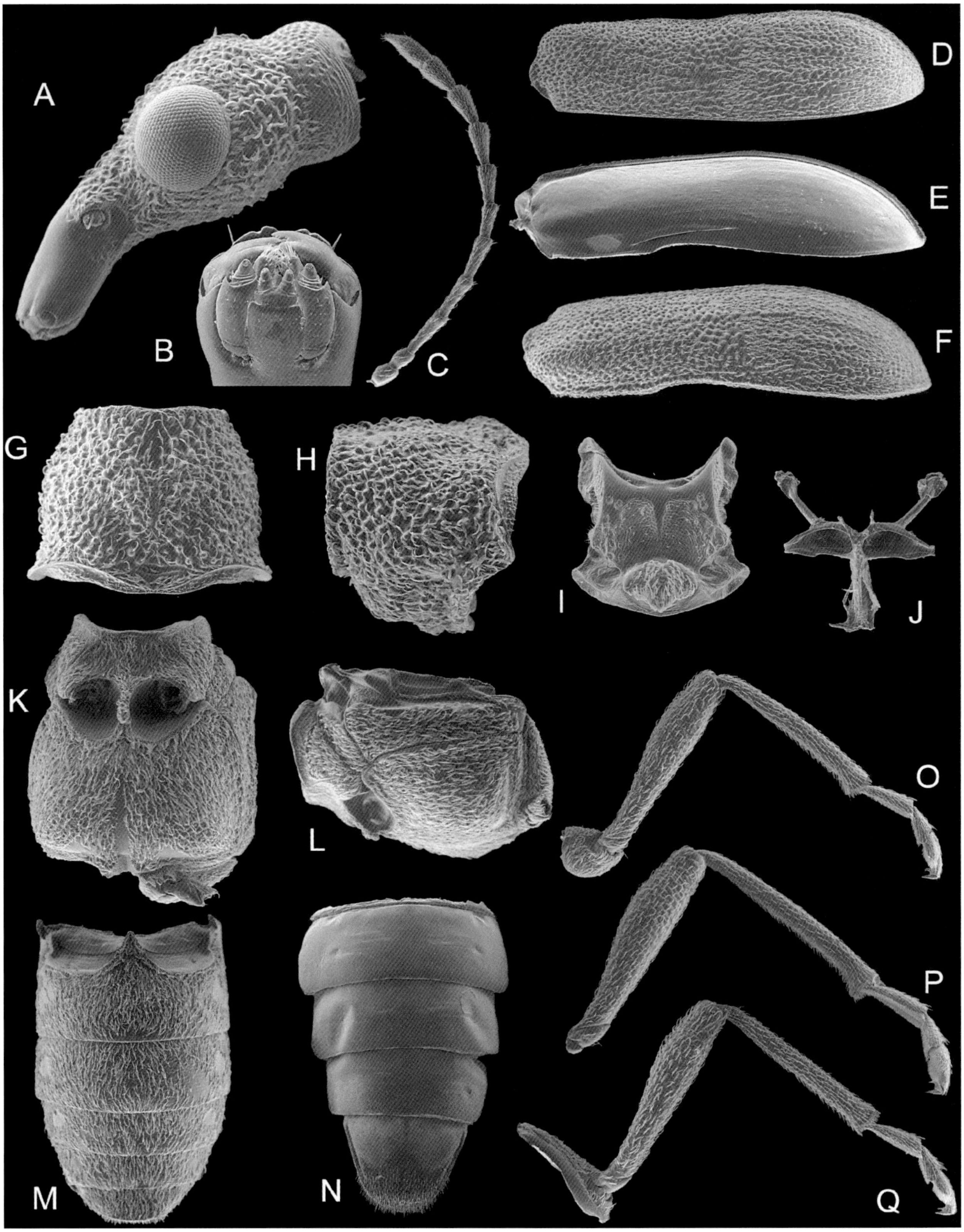

Fig. 40.8. Pachyurini, structural details, *Sphinctobelus cinereus* (Blanchard), ♀. A, head; B, mouthparts; C, antenna; D–F, elytron: D, dorsal; E, inner lateral; F, lateral; G–H, prothorax: G, dorsal; H, lateral; I, scutellum; J, metendosternite; K–L, pterothorax: K, ventral; L, lateral; M, abdominal ventrites; N, abdominal tergites; O, middle leg; P, foreleg; Q, hindleg.

Alonso-Zarazaga & Lyal (1999). Specimens of the single species have been reared from fallen branchlets of *Araucaria cunninghamii*, in which the larvae tunnel and pupate (May 1994; Zimmerman 1994). The larva and pupa were described by May (1994).

Macrobelus Lea, 1909 (Pl. 35F; Fig. 40.4G)

Type species, by original designation: *Belus insignis* Lea, 1909.
Australian species. *M. insignis* (Lea, 1909) (*in* Lea & Bovie 1909).
Distribution. Australia (QLD).
Notes. The single species is only known from a few specimens, and nothing is known about its hostplant(s) and life history (Zimmerman 1994).

Pachybelus Zimmerman, 1994 (Pl. 35L, M; Fig. 40.4S)

Type species, by original designation: *Pachybelus tuberculatus* Zimmerman, 1994.
Australian species. *P. tuberculatus* Zimmerman, 1994.
Distribution. Australia (QLD, NSW).
Notes. The single species of *Pachybelus* is only known from two specimens, one of them, the male, partly destroyed (Zimmerman 1994). Nothing is known about its hostplant(s) and life history.

Pachyura Hope, 1834 (Pls 2D, 35G; Fig. 40.4H)

Type species, by monotypy: *Pachyura australis* Hope, 1834.
Australian species. *P. australis* Hope, 1834.
Distribution. Australia (QLD, NSW, VIC, TAS, SA).
Notes. The only species now retained in *Pachyura*, *P. australis*, is widely distributed in Australia. Specimens have been collected from species of *Hakea* (Proteaceae) (Zimmerman 1994), but whether these constitute the larval hosts is unknown.

Sphinctobelus Zimmerman, 1994 (Pls 2B, 35H, I; Figs 40.3I–P, 40.4I, K, 40.7–8)

Type species, by original designation: *Pachyura cinerea* Blanchard, 1853.
Australian species. *S. ater* Zimmerman, 1994, *S. cinereus* (Blanchard, 1853), *S. niger* Zimmerman, 1994, *S. pyriatra* (Lea, 1911), *S. quadrimaculatus* (Lea, 1917), *S. rufibeccus* Zimmerman, 1994.
Distribution. Australia (QLD, NSW, VIC, TAS, SA).
Notes. The six species of *Sphinctobelus* have been recorded from various plant taxa, but not all may be the larval hosts. Three species appear associated with conifers, specimens of *S. niger* having been reared from male and female cones of *Araucaria cunninghamii* (May 1994; Zimmerman 1994) and specimens of *S. pyriatra* and *S. quadrimaculatus* collected from *Callitris* (ANIC records), whereas the other three species appear associated only with angiosperms, *S. cinereus* frequently found on *Eucalyptus* (particularly *E. pauciflora*) and specimens reared from *Apiomorpha* (Eriococcidae) galls on *Eucalyptus* and rust (*Puccinia*) galls on *Acacia* and specimens of *S. ater* and *S. rufibeccus* collected from *Acacia melanoxylon* and of the latter reared from twigs of *A. binervata* (Zimmerman 1994; ANIC records).

Acknowledgements

We thank Jiri Lochman, Nick Monaghan and Nick Porch for kindly providing photographs of live Belidae in their natural habitats and James Nicholls (CSIRO-ANIC) for sharing his observations on the host and life history *Apagobelus brevirostris*.

References

Alonso-Zarazaga MA, Lyal CHC (1999) *A World Catalogue of Families and Genera of Curculionoidea (Insecta: Coleoptera) (Excepting Scolytidae and Platypodidae)*. Entomopraxis, Barcelona, 315 pp.

Anderson RS (2002) Chapter 131. Curculionidae. In *American Beetles Polyphaga: Scarabaeoidea through Curculionoidea, Volume 2.* (Eds RH Arnett, Jr., MC Thomas, PE Skelley and JH Frank) pp. 722–815. CRC Press, Boca Raton, FL.

Anderson RS (2005) New Oxycoryninae from Central and South America: phylogenetic and biogeographical implications (Coleopter: Belidae). *Systematic Entomology* **30**(4), 644–652. doi:10.1111/j.1365-3113.2005.00304.x

Blackburn T (1889) Further notes on Australian Coleoptera, with descriptions of new genera and species. *Proceedings of the Linnean Society of New South Wales* Series 2, **3**, 1387–1506.

Blackburn T (1893) Coleoptera (continued) [Scientific results of the Elder Exploring Expedition]. *Transactions of the Royal Society of South Australia* **16**(2), 177–202.

Blanchard E (1853) Description des insectes. In *Voyage au Pole Sud et dans l'Océanie sur les corvettes l'Astrolabe et la Zélée; exécuté par ordre du Roi pendant les années 1837–1828–1839–1840, sous le commandement de M. J. Dumont d'Urville, capitaine de vaisseau; publié par ordre du government, sous la direction supérieure de M. Jacquinot, capitaine de vasseau, commandant de la Zélée. Zoologie. Tome quatrième.* (Eds JB Hombron and H Jacquinot) pp. 1–422. Gide & Baudry, Paris.

Boheman CH (1839) [Taxonomic descriptions]. In *Genera et species curculionidum, cum synonymia hujus familiae. Species novae aut hactenus minus cognitae, descriptionibus a Dom. Leonardo Gyllenhal, C.H. Boheman, et entomologis aliis illustratae.* Vol 5(1). (Ed. CJ Schoenherr) pp. 1–456. Fleischer, Lipsia.

Boisduval JA (1835) *Voyage de découvertes de l'Astrolabe, exécuté par ordre du Roi, pendant les années 1826–1827–1828–1829, sous le commandement de M. J. Dumont d'Urville. Faune entomologique de l'Océan Pacifique, avec l'illustration des insectes nouveaux recueillis pendant le voyage. Deuxième Partie. Coléoptères et autres Ordres.* J. Tastu, Paris, vii + 716 pp.

Broun T (1880) *Manual of the New Zealand Coleoptera. Part 1.* Colonial Museum and Geological Survey Department, Wellington, 651 pp.

Calder AA (1989) The alimentary canal and nervous system of Curculionoidea (Coleoptera): gross morphology and systematic significance. *Journal of Natural History* **23**(6), 1205–1265. doi:10.1080/00222938900770671

Calder AA (1990) Gross morphology of the soft parts of the male and female reproductive systems of Curculionoidea (Coleoptera). *Journal of Natural History* **24**(2), 453–505. doi:10.1080/00222939000770341

Clarke DJ, Limaye A, Mckenna DD, Oberprieler RG (2019) The weevil fauna preserved in Burmese amber – snapshot of a unique, extinct lineage (Coleoptera: Curculionoidea). *Diversity* **11**(1), 1. doi:10.3390/d11010001

Donovan E (1805) *An Epitome of the Natural History of the Insects of New Holland, New Zealand, New Guinea, Otaheite, and other Islands in the Indian, Southern, and Pacific Oceans: Including Figures and Descriptions of One Hundred and Fifty-three Species of the more Splendid, Beautiful, and Interesting Insects, hitherto Discovered in those Countries, and which for the most Part have not appeared in the Works of any Preceding Author.* Rivington, London, iv + [167] pp., [41] pls.

Fabricius JC (1775) *Systema entomologiae, sistens insectorum classes, ordines, genera, species, adiectis synonymis, locis, descriptionibus, observationibus.* Officina Libraria Kortii, Flensburgi et Lipsiae, xxxii + 832 pp. doi:10.5962/bhl.title.36510

Faust J (1892) Fünf neue Curculioniden von Australien. *Stettiner Entomologische Zeitung* **53**, 179–184.

Froggatt WW (1893) On the life-histories of Australian Coleoptera. Part 1. *Proceedings of the Linnean Society of New South Wales* **8**, 27–42.

Froggatt WW (1902) Insects of wattle trees. *Agricultural Gazette of New South Wales* **13**, 701–720.

Froggatt WW (1910) Insects which damage saltbush. *Agricultural Gazette of New South Wales* **21**, 465–470.

Froggatt WW (1923) *Forest Insects of Australia.* Government Printer, Sydney, 171pp.

Germar EF (1848) Beiträge zur Insektenfauna von Adelaide. *Linnaea Entomologica* **3**, 153–247.

Gratshev VG, Legalov AA (2014) The Mesozoic Stage of Evolution of the Family Nemonychidae (Coleoptera, Curculionoidea). *Paleontogical Journal* **48**(8), 851–944. doi:10.1134/S0031030114080012

Guérin-Méneville FE (1838) Histoire naturelle des crustacés, arachnides et insectes, recueillis dans le voyage autour du monde de la corvette de sa Majesté, La Coquille, exécuté pendant les années 1822, 1823, 1824 et 1825, sous le commandement du Capitaine Duperrey. 1e division (Crustacés, arachnides et insectes). In Voyage *autour du Monde, Exécuté par Ordre du Roi, sur la Corvette de Sa Majesté,* La Coquille, *pendant les années 1822, 1823, 1824 et 1825. Zoologie. Tome Second. 2e Partie.* (Ed. LI Duperrey) pp. 9–319. Arthus Bertrand, Paris, pp. iv–xii, 9–319.

Hawkeswood TJ, Turner JR, LeBreton M (1994) The biology and host plants of the Australian weevil *Rhinotia haemoptera* (Kirby). *Spixiana* **17**(3), 237–245.

Hope FW (1834) Characters and descriptions of several new genera and species of coleopterous insects. *Transactions of the Zoological Society of London* **1**(2), 91–112. doi:10.1111/j.1096-3642.1835.tb00607.x

Jekel H (1860) Coleoptera. Fam. Curculionides. – Part II. In *Insecta Saundersiana: or Characters of Undescribed Insects in the Collection of William Wilson Saunders, Esq. Vol. II.* (Ed. WW Saunders) pp. 155–244. John van Voorst, London.

Kirby W (1819) A century of insects, including several new genera described from his cabinet. *Transactions of the Linnean Society of London* **12**(2), 375–453. doi:10.1111/j.1095-8339.1817.tb00239.x

Kuschel G (1959) Nemonychidae, Belidae y Oxycorynidae de la fauna chilena, con algunas consideraciones biogeográficas. *Investigaciones Zoológicas Chilenas* **5**, 229–271.

Kuschel G (1995) A phylogenetic classification of Curculionoidea to families and subfamilies. *Memoirs of the Entomological Society of Washington* **14**, 5–33.

Kuschel G (2003) Nemonychidae, Belidae, Brentidae (Insecta: Coleoptera: Curculionoidea). *Fauna of New Zealand* **45**, 1–100.

Kuschel G, Leschen RAB (2003) Appendix 1. Phylogenetic relationships of the genera of Belinae. In *Nemonychidae, Belidae, Brentidae (Insecta: Coleoptera: Curculionoidea),* Fauna of New Zealand **45**. (Ed. G Kuschel) pp. 48–55. Manaaki Whenua Press, Lincoln, Canterbury, New Zealand, 1–100.

Lacordaire T (1863) *Histoire naturelle des Insectes. Genera des Coléoptères ou exposé méthodique et critique de tous les genres proposés jusqu'ici dans cet ordre d'insectes. Tome sixième.* Roret, Paris, 637 pp. doi:10.5962/bhl.title.67686

Lawrence JF, Ślipiński SA (2013) *Australian Beetles. Vol. 1. Morphology, Classification and Keys.* CSIRO Publishing, Collingwood, Victoria, viii + 561 pp.

Lea AM (1899) Descriptions of new species of Australian Coleoptera. Part V. *Proceedings of the Linnean Society of New South Wales* **23**(4), 521–645.

Lea AM (1908a) Descriptions of Australian Curculionidae, with notes on previously described species. Part VI. *Transactions of the Royal Society of South Australia* **32**, 203–251.

Lea AM (1908b) Notes on the Australian Curculionidae in the Belgian Museum with descriptions of new species. Part I. *Mémoires de la Société Entomologique de Belgique* **16**, 127–186.

Lea AM (1910) Notes on Australian Curculionidae in the collection of the German Entomological National Museum of Berlin (Col.). *Deutsche Entomologische Zeitschrift* **1910**, 153–172, 505–526.

Lea AM (1911) Descriptions of Australian Curculionidae, with notes on previously described species. Part IX. *Transactions of the Royal Society of South Australia* **35**, 62–88.

Lea AM (1917) Insecta. [Results of the South Australian Museum expedition to Strzelecki and Cooper Creeks, September and October, 1916, (n).] *Transactions of the Royal Society of South Australia* **41**, 489–630, pls. 33–37.

Lea AM (1919) Descriptions of new species of Australian Coleoptera. Part XIV. *Proceedings of the Linnean Society of New South Wales* **43**(4), 715–746.

Lea AM (1925) Descriptions of new species of Australian Coleoptera. Part XVIII. *Proceedings of the Linnean Society of New South Wales* **50**(4), 414–431.

Lea AM, Bovie A (1909) Coleoptera. Fam. Curculionidae. Subfam. Belinae. In *Genera Insectorum. 91[me] Fascicule.* (Ed. P Wytsman) pp. 1–13. P. Wytsman, Bruxelles.

Li YD, Engel MS, Tihelka E, Cai C (2023) Phylogenomics of weevils revisited: data curation and modelling compositional heterogeneity. *Biology Letters* **19**(9), 20230307. doi:10.1098/rsbl.2023.0307.

Li X, Marvaldi AE, Oberprieler RG, Clarke D, Farrell BD, *et al.* (2024) The evolutionary history of the ancient weevil family Belidae (Coleoptera: Curculionoidea) reveals the marks of Gondwana breakup and major floristic turnovers, including the rise of angiosperms. *eLife* **13**, RP97552. doi:10.7554/eLife.97552.1

Macleay WS (1826) Annulosa. Catalogue of Insects, collected by Captain King, R. N. Appendix B. Containing a list and description of the subjects of natural history collected during Captain King's survey of the intertropical and western coasts of Australia. In *Narrative of a Survey of the Intertropical and Western Coasts of Australia, Performed between the Years 1818 and 1822. Vol. II.* (Ed. PP King) pp. 438–469, pl. B. John Murray, London, 637 pp.

Marshall GAK (1936) New Australian Curculionidae (Col.). *Proceedings of the Royal Entomological Society of London (B)* **5**(10), 189–195. doi:10.1111/j.1365-3113.1936.tb01309.x

Marvaldi AE (2003) Key to larvae of the South American subfamilies of weevils (Coleoptera, Curculionoidea). *Revista Chilena de Historia Natural* **76**(4), 603–612.

Marvaldi AE (2005) Larval morphology and biology of oxycorynine weevils, and the higher phylogeny of Belidae (Coleoptera, Curculionoidea). *Zoologica Scripta* **34**(1), 37–48. doi:10.1111/j.1463-6409.2005.00169.x

Marvaldi AE, Ferrer MS (2014) 3.1 Belidae Schoenherr, 1826. In *Handbook of Zoology, Coleoptera, Beetles Volume 3: Morphology and Systemat-*

ics (Phytophaga). (Eds RAB Leschen and RG Beutel) pp. 316–328. De Gruyter, Berlin/Boston, xii + 675 pp.

Marvaldi AE, Morrone JJ (2000) Phylogenetic systematics of weevils (Coleoptera: Curculionoidea): a reappraisal based on larval and adult morphology. *Insect Systematics and Evolution* **31**(1), 43–58. doi:10.1163/187631200X00309

Marvaldi AE, Duckett CN, Kjer KM, Gillespie JJ (2009) Structural alignment of 18S and 28S rDNA sequences provides insights into phylogeny of Phytophaga (Coleoptera: Curculionoidea and Chrysomeloidea). *Zoologica Scripta* **38**(1), 63–77. doi:10.1111/j.1463-6409.2008.00360.x

Marvaldi AE, Sequeira AS, O'Brien CW, Farrell BD (2002) Molecular and morphological phylogenetics of weevils (Coleoptera, Curculionoidea): do niche shifts accompany diversification? *Systematic Biology* **51**(5), 761–785. doi:10.1080/10635150290102465

Marvaldi AE, Oberprieler RG, Lyal CHC, Bradbury T, Anderson RS (2006) Phylogeny of the Oxycoryninae *s. l.* (Coleoptera Phytophaga) and evolution of plant-weevil interactions. *Invertebrate Systematics* **20**(4), 447–476.

May BM (1993) Larvae of Curculionoidea (Insecta: Coleoptera): a systematic overview. *Fauna of New Zealand* **28**, 1–1221.

May BM (1994) An introduction to the immature stages of Australian Curculionoidea. In *Australian Weevils (Coleoptera: Curculionoidea). Vol. II. Brentidae, Eurhynchidae, Apionidae and a Chapter on Immature Stages by Brenda May*. (Ed. EC Zimmerman) pp. 365–728. CSIRO Publishing, Melbourne, xxxii + 741 pp.

McKenna DM, Sequeira AS, Marvaldi AE, Farrell BD (2009) Temporal lags and overlap in the diversification of weevils and flowering plants. *Proceedings of the National Academy of Sciences of the United States of America* **106**(17), 7083–7088. doi:10.1073/pnas.0810618106

McKenna DD, Shin S, Ahrens D, Balke M, Beza-Beza C, *et al.* (2019) The evolution and genomic basis of beetle diversity. *Proceedings of the National Academy of Sciences* **116**(49), 24729–24737. doi:10.1073/pnas.1909655116

Newman E (1838) Entomological notes. *Entomological Magazine* **5**, 168–181.

Oberprieler RO, Marvaldi AE, Anderson RS (2007) Weevils, weevils, weevils everywhere. *Zootaxa* **520**(1668), 491–520. doi:10.11646/zootaxa.1668.1.24

Oberprieler RG, Thompson RT, Peterson M (2010) Darwin's forgotten weevil. *Zootaxa* **2675**(1), 33–46. doi:10.11646/zootaxa.2675.1.3

Olliff AS (1889) The insect fauna of Lord Howe Island. *Memoirs of the Australian Museum* **2**(4), 75–98. doi:10.3853/j.0067-1967.2.1889.482

Pascoe FP (1870a) Contributions towards a knowledge of the Curculionidae [part]. *Journal of the Linnean Society of London, Zoology* **10**(47), 434–458. doi:10.1111/j.1096-3642.1870.tb00671.x

Pascoe FP (1870b) [Contributions towards a knowledge of the Curculionidae] [concl.]. *Journal of the Linnean Society of London, Zoology* **10**(48), 459–493. doi:10.1111/j.1096-3642.1870.tb00672.x

Pascoe FP (1870c) Descriptions of some genera and species of Australian Curculionidae. *Transactions of the Entomological Society of London* **1870**(2), 181–209. doi:10.1111/j.1365-2311.1870.tb01873.x

Pascoe FP (1871) Additions to the Australian Curculionidae. Part I. *Annals and Magazine of Natural History* Series 4, **8**(44), 89–99. doi:10.1080/00222937108696441

Pascoe FP (1872a) Additions to the Australian Curculionidae. Part III. *Annals and Magazine of Natural History* Series 4, **10**(56), 84–101. doi:10.1080/00222937208696651

Pascoe FP (1872b) Contributions towards a knowledge of the Curculionidae. Part III. *Journal of the Linnean Society of London, Zoology* **11**(55), 440–492. doi:10.1111/j.1096-3642.1872.tb01668.x

Pascoe FP (1872c) Additions to the Australian Curculionidae. Part II. *Annals and Magazine of Natural History* Series 5, **9**(50), 132–142.

Pascoe FP (1873) Additions to the Australian Curculionidae. Part V [continued]. *Annals and Magazine of Natural History* Series 4, **12**(70), 278–286. doi:10.1080/00222937308680768

Pascoe FP (1885) List of the Curculionidae of the Malay Archipelago collected by Dr. Odoardo Beccari, L. M. d'Albertis, and others. *Annali del Museo Civico di Storia Naturale Genova* **22**, 201–332.

Perroud BP (1853) Descriptions de quelques Coléoptères nouveaux ou peu connus. *Annales de la Société Linnéenne de Lyon* Series 2, **1**, 389–528.

Pullen K, Jennings D, Oberprieler RO (2014) Annotated catalogue of Australian weevils (Coleoptera: Curculionoidea). *Zootaxa* **3896**(1), 1–481. doi:10.11646/zootaxa.3896.1.1

Santos MFA, Mermudes JRM, Fonseca VMM (2007) Description of a new genus and species of Belinae (Belidae, Curculionoidea, Coleoptera) from the Santana Formation (Crato member, Lower Cretaceous) of the Araripe basin, northeastern Brazil. In *Paleontology: Life Scenarios, v.1, Interscience*. (Eds IS Carvalho, RCT Cassab and C Schwanke) pp. 449–455. Rio de Janeiro, Brazil.

Shin S, Clarke DJ, Lemmon AR, Moriarty Lemmon E, Aitken AL, *et al.* (2018) Phylogenomic data yield new and robust insights into the phylogeny and evolution of weevils. *Molecular Biology and Evolution* **35**(4), 823–836. doi:10.1093/molbev/msx324

Thompson RT (1992) Observations on the morphology and classification of weevils (Coleoptera, Curculionoidea) with a key to major groups. *Journal of Natural History* **26**(4), 835–891. doi:10.1080/00222939200770511

Vanin SA (1976) Taxonomic revision of the South American Belidae (Coleoptera). *Arquivos de Zoologia, S. Paulo* **28**(1), 1–75. doi:10.11606/issn.2176-7793.v28i1p1-75

Waterhouse GR (1839) Descriptions of some new species of exotic insects. *Transactions of the Entomological Society of London* Series 1, **2**(3), 188–196. doi:10.1111/j.1365-2311.1836.tb00323.x

Yu Y, Davis SR, Shih C, Ren D, Pang H (2019) The earliest fossil record of Belidae and its implications for the early evolution of Curculionoidea (Coleoptera). *Journal of Systematic Palaeontology* **17**(24), 2105–2117. doi:10.1080/14772019.2019.1588401

Zimmerman EC (1994) *Australian Weevils (Coleoptera: Curculionoidea). Volume 1. Orthoceri. Anthribidae to Attelabidae. The Primitive Weevils.* CSIRO Australia, Melbourne, xxxii + 741 pp.

41. ATTELABIDAE BILLBERG, 1820

Hermes E. Escalona, Debbie Jennings and Rolf G. Oberprieler

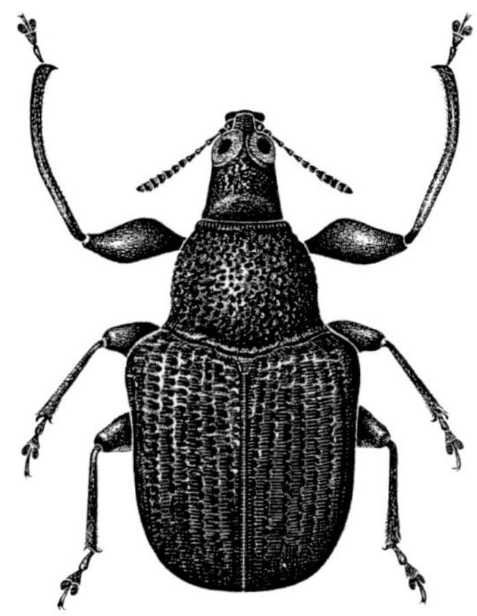

Fig. 41.1. *Euops falcatus* (Guérin-Méneville).

Introduction. The Attelabidae are a relatively small family of Curculionoidea, with approximately 2500 species in 120 genera world-wide, and the attelabid fauna in Australia is even smaller, comprising only three genera and approximately 85 species (Pullen *et al.* 2014). The family is divided into two subfamilies, Attelabinae and Rhynchitinae, which are both represented in Australia, the former by one genus, *Euops* Schoenherr with about 29 species, and the latter by two genera, *Australotobius* Legalov with one species and *Metopum* Agassiz with about 55 species. The Attelabidae are a monophyletic group and form the sister-group of Caridae, Brentidae and Curculionidae, as based on both morphological characters (Kuschel 1995; Marvaldi & Morrone 2000; Marvaldi *et al.* 2002, 2009) and molecular data and phylogenomic analysis (McKenna *et al.* 2009; Shin *et al.* 2018). However, only few imaginal and larval autapomorphies have been identified for the family (Kuschel 1995; Marvaldi & Morrone 2000; Riedel 2014), and although the subfamily Attelabinae is well supported as a monophyletic group, the subfamily Rhynchitinae is not and likely to be paraphyletic in its current concept (Kobayashi *et al.* 2012; Riedel 2014). The classification of the subfamilies into tribes (Riedel 2014) is also unsettled and requires inputs from molecular-phylogenetic analyses. The taxonomy of the Attelabidae has suffered significantly from recent works of Legalov (2003, 2007 and others) – see Alonso-Zarazaga (2011) for an assessment – and requires considerable efforts to untangle. A comprehensive overview of the taxonomy, morphology and biology of the family is provided by Riedel (2014). The fossil record of Attelabidae is poor and similarly contentious, with the earliest authentic representative (*Sayrevilleus* Gratshev & Zherikhin, 2000) dating to the early Upper Cretaceous and others (mostly Rhynchitinae) only to the Cenozoic Era (Riedel 2014).

Attelabidae life histories include widespread parental care to varying degrees culminating in the construction of elaborate leaf-rolls, or nidi, and also in kleptoparasitism, and mutualistic associations with symbiotic fungi, most of which have been documented in other regions (Wasmann 1884; Prell 1924; Kôno 1927; Velbinger 1950; Lengerken 1954; Sakurai 1985; Sawada & Morimoto 1986; Riedel 2002, 2014). The hostplants of the larvae are mostly dicotyledons spanning many families (Riedel 2014). In Australia, *Euops* species appear to be mostly associated with Myrtaceae such as *Eucalyptus*, whose leaves the females roll into cigar-shaped structures that fall to the ground and are often collected and distributed by ants (McAreavey 1938; Zimmerman 1994). The leaf-rolling behaviour of *Euops falcatus* (Guérin-Méneville) was described in detail by McAreavey (1938). *Euops* females have transverse double rows (combs) of setae on the underside of the abdomen as well as mycetangia behind the metacoxae (in the space between metathorax and abdomen), and during construction of the leaf roll they brush spores from the mycetangia, mixed with a secretion from abdominal glands opening between the rows of setae, onto the leaf surface (Sakurai 1985; Sawada & Morimoto 1986; Riedel 2014). The fungi appear to be *Penicillium* species, which not only assist in the breakdown of the leaf tissue but also suppress other microbes that can be harmful to the larvae developing inside the nidi (Kobayashi *et al.* 2008; Li *et al.* 2012). Nothing is known of the oviposition behaviour and life histories of the rhynchitine genera *Australotobius* and *Metopum* in Australia, but the females are likely to bite leaves or flower or seed buds with their mandibles, causing the structures to whither or drop and then serve as feeding substrate for the larvae as in other Rhynchitinae (Sawada 1993). *Australotobius* and *Metopum* females possess pairs of long tubes at the opening of the vagina (Zimmerman 1994), possibly for storing secretions from collaterial glands (Riedel 2014), which may play a role in the preparation of the plant substrates used as food for the larvae.

Characteristics. Adults (Fig. 41.2). Body length 1.3–7 mm (SL), about 1.65–2.5 × as long as wide, elytra subparallel-sided or weakly rounded, pronotum independently rounded and usually narrowed anteriorly, moderately convex; colour yellowish-brown to red, dark brown or black, rarely bicoloured; subglabrous (*Euops*) or clothed with fine, decumbent (*Metopum*) or long, erect (*Australotobius*) setae. Head not to moderately declined at base; temples short (Fig. 41.2A, C, D) or long (Fig. 41.2B), occipital region with moderately long median endocarina in *Metopum*; forehead abruptly declined in *Euops*, without median groove; rostrum in *Euops* slightly longer than wide, sharply declined, somewhat flattened and distinctly expanded apically, in *Australotobius* and *Metopum* more than 3 × as long as wide, not or only slightly declined or expanded. Labrum completely fused to clypeus; frontoclypeal suture also absent. Eyes moderately to strongly protuberant, dorsally widely separate (*Australotobius*, *Metopum*) to subcontiguous (*Euops*), not emarginate, without interfacetal setae. Antennal insertions barely concealed from above, located relatively close to base of rostrum, laterally (*Metopum*, *Australotobius*) or dorsally on either side of a median elevation (*Euops*); sockets extended into short, weakly defined scrobes. Antennae non-geniculate, 11-segmented, moderately long, with weak 3-segmented club but terminal segment subdivided; scapes elongate, only slightly longer than pedicel. Mandibles short, broad to slightly elongate, exodont in *Australotobius* and some *Metopum* species, with mesally and laterally directed teeth at apex, non-exodont and unidentate in *Euops*; incisor edge with 1 or more subapical teeth; mola subbasal, weakly developed or absent, prostheca absent; inner edge of mandibular base with long pubescent process; in some non-Australian Rhynchitinae with deciduous mandibular process. Maxillae with short, broad, densely setose galea and narrower or reduced lacinia bearing a few spines; maxillary palps short and inflexible, extending only slightly beyond malar apex, 4-segmented, apical segment cylindrical. Prementum elongate and fused at base in *Metopum*, shorter, broader and basally articulated in *Euops*; ligula simple and articulated (*Metopum*) or fused with prementum (*Euops*) and often with pointed projections; labial palps 3-segmented in *Metopum*, absent in *Euops*. Gular sutures fused, long, extending onto underside of rostrum. Corpotentorium present; cervical sclerites absent.

Pronotum about 0.8–1.05 × as long as wide, widest at base in *Euops* and *Australotobius*, at about middle in *Metopum*, base distinctly narrower than combined elytral bases; sides at least slightly curved, without lateral pronotal carinae; anterior angles not produced; posterior angles rounded in *Australotobius* and *Metopum*, right to obtuse in *Euops*; posterior edge straight or weakly sinuate; disc with weak, subbasal transverse shallow groove (obscured mesally in *Metopum*), with narrow anterior collar in *Euops*. Prosternum about half as long as shortest diameter of procoxal cavity, flat to slightly convex in *Euops*, strongly reduced in *Australotobius* and indiscernible in *Metopum*; prosternal process in *Euops* broadly interrupted in middle with anterior section short, acute and posterior section fused to postcoxal processes, in *Australotobius* present only anteriorly, in *Metopum* anteriorly absent but posteriorly present as minute sternellum between postcoxal processes; notosternal sutures transverse and incomplete. Procoxae projecting below prosternum, with short concealed lateral extensions; trochantins concealed; procoxal cavities subcircular, contiguous, externally broadly closed, postcoxal (hypomeral) processes meeting disjunct apex of prosternal process in *Australotobius* or fused with prosternal process in *Euops*, without narrow lateral extensions, internally broadly closed.

Scutellar shield not abruptly elevated, anteriorly simple, posteriorly broadly rounded. Elytra about 1.15–1.7 × as long as wide and 1.7–2.5 × as long as pronotum, weakly rounded laterally, apices meeting or almost meeting at suture and barely exposing abdominal tergites in *Australotobius* and *Metopum*, subparallel laterally and broadly truncate exposing pygidium in *Euops*; disc irregularly punctate in *Australotobius* and *Metopum*, punctostriate in *Euops* with 9 complete striae and short scutellary striole; epipleura narrow to very narrow and incomplete; with internal longitudinal interlocking ridge (postscutellar locking device) laterally at about middle in *Euops*. Mesoventrite separated by complete sutures from mesanepisterna, but partly fused to mesanepisterna in *Metopum*; anterior edge on different plane than metaventrite, usually with narrow transverse prothoracic rest continuing onto mesanepisterna. Mesocoxae projecting, with concealed trochantins. Mesocoxal cavities moderately narrowly separate in *Euops* and *Australotobius*, contiguous in *Metopum*, subcircular, not or slightly oblique, often closed laterally by meeting of mesoventrite and metaventrite, open in *Australotobius*. Mesoventital process emarginate at apex and mesometaventral junction dicondylic in *Euops* and *Australotobius*, process absent in *Metopum*. Metaventrite moderately convex; transverse (metakatepisternal) suture long but not crossing midline; discrimen complete to base of anterior intercoxal process; metanepisterna with exposed portion about 3 × as long as wide in *Metopum*, <2 × in *Euops*; metepimera exposed. Metacoxae narrowly separated, not extending laterally to reach elytra; posteriorly (in cavity between metathorax and abdomen) with pair of mycetangia in *Euops* females. Metendosternite with long stalk, long lateral arms, short, broad laminae and very short anterior process bearing narrowly to widely separated anterior tendons. Hindwings with apical field more than half as long as total wing, containing distinct sclerite just beyond radial cell and remnants of RP; radial cell moderately well-developed, slightly longer than wide with base more or less straight and inner posterobasal angle acute; cross-vein r_3 moderately long and oblique, arising close to base of r_4, this sometimes incomplete; basal portion of RP moderately long, its apical portion extending to wing margin; R-M loop moderately broad; medial spur sinuate and extending almost to wing margin; medial field with 3 free veins in *Euops*, 1 or 2 in *Metopum*, and no medial fleck; MP_{3+4} without base; wedge cell absent; anal lobe moderately well-developed or somewhat reduced, without embayment. Legs in *Australotobius* and *Metopum* slender with femora not or only slightly inflated and tibial apices not expanded, with fringe of spines and paired spurs, legs in *Euops* with femora distinctly inflated, especially in male, in which forelegs much longer than posterior pairs, meso- and metafemora sometimes with tufts of setae and protibiae very long, slender, curved and lined with teeth internally; tibiae at apex with distinct uncus or premucro and no spurs; meso- and metatibiae crenulate in *Euops* and *Australotobius*. Tarsi pseudotetramerous (5-5-5, segment 4 reduced), segments 3 distinctly bilobed. 1–3, but especially 3, densely setose beneath; tarsal claws in *Euops* simple and basally connate, in *Australotobius* and *Metopum* cleft (bifid) and free at base; empodium not apparent.

Abdomen with 5 ventrites, ventrites 1–2 (in *Metopum* males), 1–3 (in *Metopum* females) and 1–4 (in *Euops*) connate; 1 at shortest point not longer than 2; 1–3 with double rows of modified setae in *Euops* females; intercoxal process acute in *Euops*, broadly rounded in *Australotobius* and *Metopum*. Abdominal tergites in *Metopum* relatively lightly pigmented except for more or less horizontal tergite VII, with patches of wing-folding asperities on tergites V–VI and functional spiracles in pleural membrane of I–VI, in *Euops* moderately heavily sclerotised and accompanied by laterotergites, with pygidium strongly sclerotised and vertical, patches of wing-folding patches on IV–VII and spiracles on tergites. Tergite VIII in male more or less concealed beneath VII but well sclerotised; sternite VIII without median spicule. Anterior edge of segment IX in male with distinct spiculum gastrale; tergite X membranous. Aedeagus of pedotectal type, symmetrical; tegmen anteriorly with long manubrium, slightly expanded and truncate at apex, body with complete ring and elongate parameres fused together and to phallobase; penis with distinct tectum and pedon and paired basal temones; endophallus with complex sclerites. Segment VIII in female more or less membranous, with long, fixed spiculum ventrale. Ovipositor reduced, with proctiger and paraprocts membranous and not clearly defined, without baculi; gonocoxites undivided, without baculi in *Metopum*, not clearly delimited in *Euops*; styli absent or highly reduced. Internal genital tract with well sclerotised, curved spermathecal capsule, spermathecal duct and gland close to each other. Proventriculus with 8 primary folds densely covered with setae or spicules or with sclerotised gnathal ridges. Hindgut with 6 cryptonephritic Malpighian tubules. Nervous system, testis and ovary as in Calder (1989, 1990). (Thompson 1992; Zimmerman 1994; Hamilton 2002; Lawrence & Ślipiński 2013; Riedel 2014).

Larvae (Fig. 41.3). Body moderately elongate but robust, subcylindrical or widest at middle and ventrally curved (C-shaped); all surfaces lightly pigmented except for mandibles and mouth-frame; vestiture consisting of scattered to moderately abundant, short, simple setae or sometimes asperities. Head prognathous and retracted, longer than wide; posterior edge of head capsule not emarginate, posteriorly produced to form narrow, flattened extension; epicranial stem long; frontal arms V-shaped, usually indistinct, contiguous at base and extending to mandibles; median endocarina extending between frontal arms but not reaching clypeal region. Stemmata variable, usually with 3 pairs and 3 pairs of vestigial eyespots but sometimes fewer. Antennae very short, 2-segmented, with terminal, conical sensorium. Frontoclypeal suture present; labrum free, with 4 pairs of setae in *Auletobius*, 3 in *Euops*; epipharynx with very long and slender epipharyngeal rods. Mandibles short and broad, obliquely bidentate; accessory ventral process absent; mola and prostheca absent. Ventral mouthparts retracted; cardo slightly oblique, undivided; stipes elongate; galea and lacinia present in *Auletobius*, apically rounded, with few stout spines at apex of each; in *Euops* with undivided mala bearing distinct lacinial spine at inner apical angle; maxillary palps 2-segmented (*Auletobius*), 3-segmented (Rhynchitinae) with distinct palpifer. Labium consisting of prementum and postmentum in *Auletobius*, a single sclerite (prementum fused to postmentum) in *Euops*; premental sclerite absent, ligula simple, longer than basal segment; labial palps 2-segmented, separated by about width of basal segment in *Euops*, more widely so in *Auletobius*. Hypopharyngeal bracon narrow in *Auletobius*, absent in *Euops*; sclerome absent. Hypostomal rods moderately long, diverging; ventral epicranial ridges absent. Gular region absent.

Prothoracic tergum subequal in length to meso- or metatergum, without or with very lightly pigmented plate. Legs absent. Abdomen more than twice as long as thorax; abdominal terga not extending laterally beyond edges of sterna; abdominal terga I–VIII each with transverse groove forming 2 plicae but posterior plica (postdorsum) subdivided laterally in *Auletobius*; abdominal terga strongly humped in middle in *Euops*; segment IX distinctly shorter than VIII and dorsal; sternum IX simple; segment X reduced and more or less fused to IX; anal region posteriorly or posteroventrally orientated; anal opening bordered by 4 lobes in *Euops*, transverse in *Auletobius*. Spiracles annular-biforous with very long accessory chambers; thoracic spiracles located on mesothorax or intersegmental fold in *Euops*. (Gardner 1934; van Emden 1938; Lee & Morimoto 1988a, b; May 1993; Hamilton 2002; Lawrence & Ślipiński 2013; Riedel 2014).

The larvae of Australian Attelabidae remains unknown and undescribed, this family description is based on overseas material.

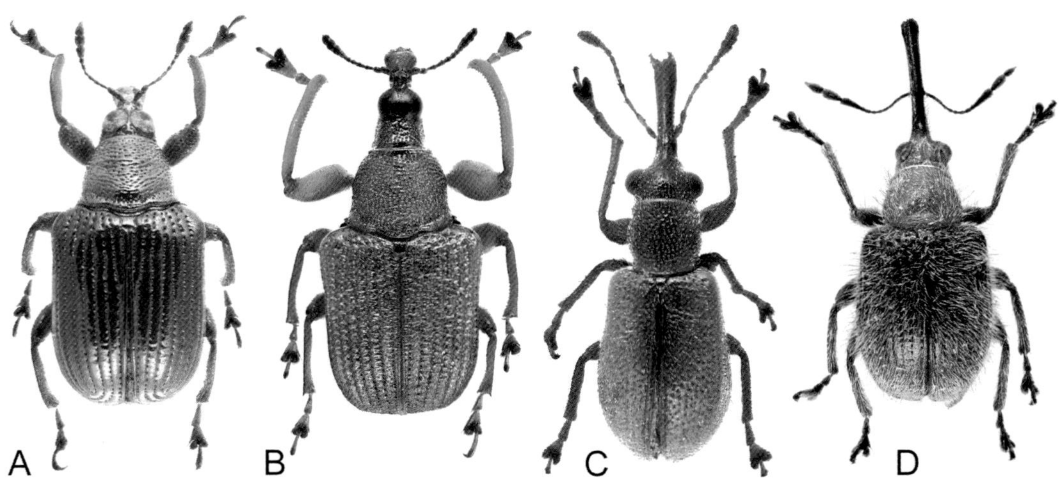

Fig. 41.2. Attelabidae, dorsal habitus. A, *Euops bakewelli* Jekel; B, *Euops falcatus* (Guérin-Méneville); C, *Metopum densum* (Lea); D, *Australotobius zimmermani* (Legalov).

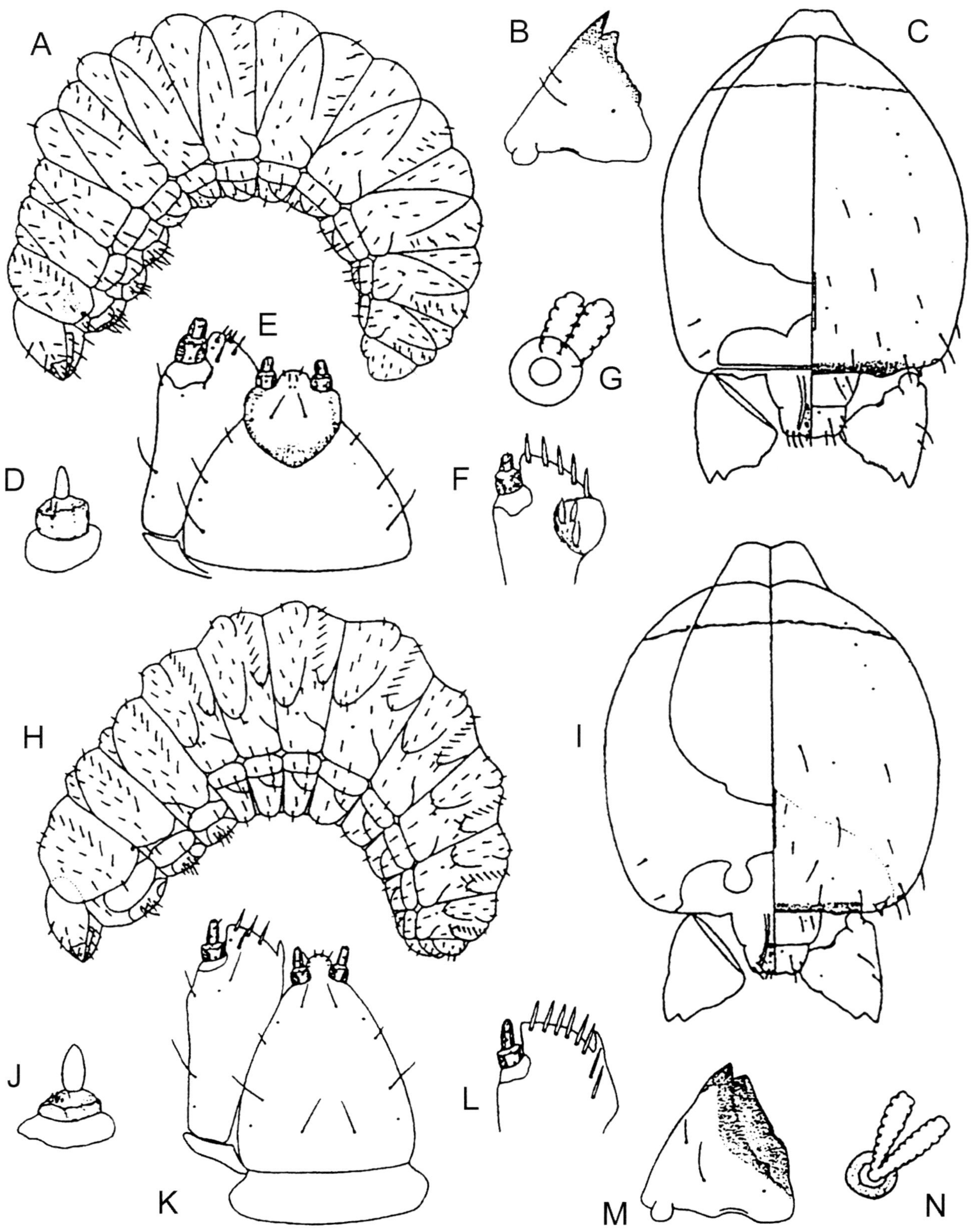

Fig. 41.3. Attelabidae, structural details of larvae. A–G, *Auletobius uniformis* (Roelofs), larva from Japan: A, habitus, lateral; B, mandible; C, head, dorsal; D, antenna; E, mala and labium; F, mala, dorsal; G, spiracle; H–N, *Euops splendidus* Voss, larva from Japan: H, habitus, lateral; I, head, dorsal, J; antenna; K, mala and labium; L, mala, dorsal; M, mandible; N, spiracle. After Lee & Morimoto (1988a, b).

Key to the Australian genera

1. Eyes in dorsal view widely separate (Fig. 41.4A); elytra irregularly multipunctate (Fig. 41.2C, D) .... **RHYNCHITINAE** ...2
– Eyes in dorsal view subcontiguous or narrowly separate (Fig. 41.4M); elytra regularly punctostriate (Fig. 41.2A, B) .... **ATTELABINAE** ... ***Euops* Schoenherr**

2(1). Antennal segments 9 about as long as wide (Fig. 41.2C); antennae inserted near basal fourth of rostrum; prosternum not visible; widespread .... ***Metopum* Agassiz**
– Antennal segments 9 longer than wide (Fig. 41.2D); antennae inserted near basal half of rostrum; prosternum distinctly visible; North QLD .... ***Australotobius* Legalov**

Classification of the Australian genera

Subfamily Rhynchitinae Gistel, 1848

Tribe Auletini Desbrochers des Loges, 1908

Metopum Agassiz, 1846 (Pls 1E, 36F–G; Figs 41.2C, 41.4A–L)

Type species, by monotypy: *Metopon suturalis* Waterhouse, 1842.

Australian species. *M. adelaidae* (Legalov, 2003), *M. aeneiceps* (Voss, 1933), *M. aeneum* (Voss, 1922b), *M. albipilosum* (Voss, 1922b), *M. anthracinum* (Lea, 1926), *M. aterrimum* (Lea, 1910), *M. brevirostre* (Lea, 1899), *M. bryophagum* (Lea, 1910), *M. calceatum* (Pascoe, 1874a), *M. cariniceps* (Lea, 1926), *M. castor* (Lea, 1926), *M. decipiens* (Lea, 1910), *M. densum* (Lea, 1899), *M. erythroderes* (Lea, 1926), *M. eucalypti* (Lea, 1899), *M. filirostre* (Pascoe, 1874a), *M. flavipenne* (Lea, 1926), *M. imitator* (Lea, 1910), *M. incanum* (Lea, 1926), *M. inconstans* (Lea, 1910), *M. inflaticolle* (Lea, 1926), *M. insigne* (Lea, 1899), *M. kingi* (Lea, 1908), *M. laterirostre* (Lea, 1926), *M. latipenne* (Lea, 1926), *M. leai* (Voss, 1922b), *M. leucotrichum* (Lea, 1926), *M. lineatopunctatum* (Lea, 1926), *M. melaleucae* (Lea, 1899), *M. melanocephalum* (Erichson, 1842), *M. melanostethum* (Lea, 1926), *M. meridianum* (von Dalla Torre & Voss, 1937), *M. minus* (Lea, 1899), *M. monticola* (Voss, 1937), *M. nigritarse* (Pascoe, 1874a), *M. obscurum* (Lea, 1926), *M. orientale* (Lea, 1926), *M. orthorrhinum* (Lea, 1926), *M. pallipes* (Lea, 1899), *M. pilosum* (Lea, 1899), *M. pollux* (Lea, 1926), *M. postscutellare* (Lea, 1926), *M. psilorrhinum* (Lea, 1926), *M. puncticolle* (Lea, 1910), *M. punctipenne* (Lea, 1910), *M. rhyparochromum* (Lea, 1926), *M. rubricolle* (Voss, 1922b), *M. semicrudum* (Lea, 1899), *M. sobrinus* (Lea, 1910), *M. subcalceatum* (Lea, 1910), *M. sulcibasis* (Lea, 1926), *M. suturale* (Waterhouse, 1842), *M. turbidum* (Pascoe, 1874a), *M. variicolle* (Lea, 1910), *M. variipenne* (Lea, 1909a).

Distribution. Australia (QLD, NSW, VIC, TAS, SA, NT, WA), New Guinea, New Caledonia.

Notes. The genus *Metopum* was separated from *Auletobius* by Pullen *et al.* (2014) and encompasses the Austro-Pacific *suturalis* species-group formerly included in *Auletobius*, which differs from *Auletobius* in several characters (Kuschel 2008). The precise concept and composition of the genus requires further study, however, in particular in regard to several generic names applied to New-Guinean species by A. Legalov (Pullen *et al.* 2014).

Biology. Only few plants have been recorded as hosts of *Metopum* species, i.e. *Dillwynia* (Fabaceae) for *M. brevirostre*, *Eucalyptus* (Myrtaceae) for *M. calceatum* and *M. turbidum*, a dwarf *Melaleuca* species for *M. melaleucae* and *Ficus* (Moraceae) for *M. minor* (Zimmerman 1994), but it is unknown whether any of them constitute true (larval) hosts for these species.

Tribe Rhynchitini Gistel, 1848

Australotobius Legalov, 2007 (Pl. 36H; Fig. 41.2D)

Type species, by original designation: *Elautobius zimmermani* Legalov, 2003.

Australian species. *A. zimmermani* (Legalov, 2003).

Distribution. Australia (northern QLD).

Notes. A single specimen of this genus was reported and illustrated by Zimmerman (1992: Pl 328, 3–6, 1994: Fig. 354), tentatively identified as a new species of *Involvulus* Schrank, 1798. Based solely on Zimmerman's photographs, Legalov (2003) first described the species as *Elautobius zimmermani* but later placed it in its own genus, *Australotobius* (Legalov 2007). Both genus and species still require proper redescription. Only one additional specimen appears to have been found, at the Normanby River in the Lakefield N.P. in July 1998 by R. G. Oberprieler (ANIC).

Subfamily Attelabinae Billberg, 1820

Tribe Euopini Voss, 1925

Euops Schoenherr, 1839 (Pls 1D, F, G, 36A–E; Figs 41.1, 41.2A, B, 41.3H–N, 41.4M–N)

Type species, by original designation: *Euops australasiae* Fåhraeus (*in* Schoenherr, 1839).

Australian species. *E. affinis* Voss, 1924, *E. bakewelli* Jekel, 1860, *E. castaneipennis* Voss, 1924, *E. clavigerus* Pascoe, 1874b, *E. contactus* Lea, 1929, *E. corrugatus* Lea, 1899, *E. coxalis* Lea, 1929, *E. effulgens* Lea, 1909b, *E. episternalis* Lea, 1929, *E. eucalypti* Pascoe, 1874b, *E. falcatus* (Guérin-Méneville, 1833), *E. flavomaculatus* Lea, 1909b, *E. hoppla*

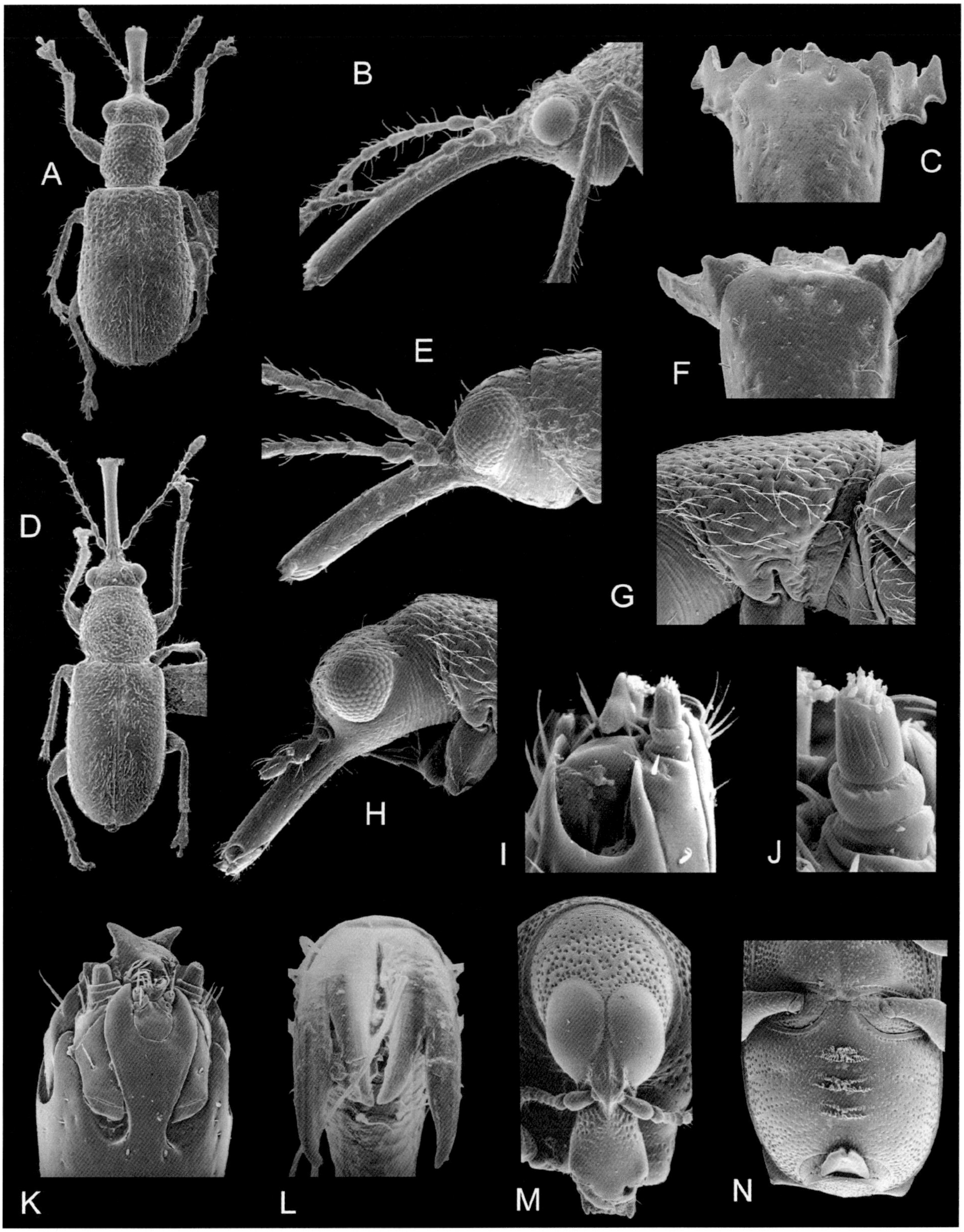

Fig. 41.4. Attelabidae, structural details. A, *Metopum calceatum* (Pascoe), habitus, dorsal; B–D, *Metopum filirostre* (Pascoe), holotype, NHML: B, head, lateral; C, mandibles; D, habitus, dorsal; E–F, *Metopum nigritarse* (Pascoe), holotype, NHML: E, head, lateral; F, mandibles; G–L, *Metopum* sp.: G, notosternal suture; H, head, lateral; I, mouthparts, lateral; J, maxillary palp; K, mouthparts, ventral; L, claws; M–N, *Euops falcatus* (Guérin-Méneville): M, head, dorsal; N, abdominal ventrites, ♀. After Zimmerman (1994).

Riedel, 2001b, *E. impuncticollis* Lea, 1909b, *E. lateralis* Lea, 1909b, *E. leai* Voss, 1922a, *E. micans* Lea, 1929, *E. montanus* Voss, 1924, *E. oberprieleri* Riedel, 2001b, *E. paraniger* Legalov, 2002, *E. parvoarmatus* Lea, 1929, *E. pulchellus* Pascoe, 1875, *E. quadrifasciculatus* Lea, 1929, *E. reidi* Riedel, 2001b, *E. rudis* Lea, 1909b, *E. strigiventris* Lea, 1898, *E. suturalis* Lea, 1899, *E. tuberculatus* Lea, 1921, *E. victoriensis* Blackburn, 1894.

Distribution. Australia (QLD, NSW, VIC, SA, NT), New Guinea, Oriental, eastern Palaearctic and Afrotropical regions.

Notes. The Australian fauna of *Euops* remains poorly studied, although Riedel (1999, 2001a, c) made several fine contributions to the knowledge of the fauna of the Australo-Pacific region. The hostplants and life histories of the genus in Australia are also very poorly known, the sole exception being *E. falcatus*, whose leaf-rolling behaviour on *Eucalyptus macrorrhyncha* was described by McAreavey (1938) and which has also been collected on *Angophora intermedia* and seemingly on other *Eucalyptus* species (Zimmerman 1994).

Acknowledgements

We thank Bruno Bell, Nick Monaghan and Nick Porch for kindly supplying photographs of Attelabidae in their natural habitats.

References

Agassiz L (1846) Nomenclatoris zoologici index universalis, continens nomina systematica classium, ordinum, familiarum et generum animalium omnium, tam viventium quam fossilium, secundum ordinem alphabeticum unicum disposita, adjectis homonymiis plantarum, nec non variis adnotationibus et emendationibus. In *Nomenclator Zoologicus, continens nomina systematica generum animalium tam viventium quam fossilium, secundum ordinem alphabeticum disposita, adjectis auctoribus, libris, in quibus reperiuntur, anno editionis, etymologia et familiis, ad quas pertinent, in singulis classibus*. Fasc. 12. Agassiz, L. (1842–1846). Jent & Gassmann, Soloduri, pp. i–viii, 1–393. doi:10.5962/bhl.title.49761

Alonso-Zarazaga MA (2011) Comments. Rhynchitidae. In *Catalogue of Palaearctic Coleoptera. Volume 7*. (Eds I Löbl and A Smetana) pp. 66–73. Apollo Books, Stenstrup, 373 pp.

Blackburn T (1894) Notes on Australian Coleoptera, with descriptions of new species. Part XIV. *Proceedings of the Linnean Society of New South Wales* Series 2, **8**, 245–286.

Calder AA (1989) The alimentary canal and nervous system of Curculionoidea (Coleoptera): gross morphology and systematic significance. *Journal of Natural History* **23**(6), 1205–1265. doi:10.1080/00222938900770671

Calder AA (1990) Gross morphology of the soft parts of the male and female reproductive systems of Curculionoidea (Coleoptera). *Journal of Natural History* **24**(2), 453–505. doi:10.1080/00222939000770341

Erichson WF (1842) Contribution to the insect fauna of Van Diemen's Land (now Tasmania), with particular reference to the geographical distribution of insects. (translated by Miss S. A. Fogg). *Transactions and Proceedings of the Royal Society of Tasmania* **3**(2), 298–338.

Gardner JCM (1934) Immature stages of Indian Coleoptera (14) (Curculionidae). *Indian Forest Records* **20**(2), 1–48.

Gratshev V, Zherikhin VV (2000) The weevils from the Late Cretaceous New Jersey amber (Coleoptera, Curculionoidea). In *Studies on fossils in amber, with particular reference to the Cretaceous of New Jersey*. (Ed. DA Grimaldi) pp. 241–254. Backhuys Publishers Leiden, Leiden, the Netherlands.

Guérin-Méneville FE (1833) *Iconographie du règne animal de G. Cuvier, ou représentation d'après nature de l'une des espèces les plus remarquables, et souvent non figurées, de chaque genre d'animaux, avec un texte descriptif mis au courant de la science. Ouvrage pouvant servir d'atlas a tous les traitès de zoologie. Tome VII. Insectes*. Baillière, Paris, 110 pls. [Plates 36–39 bis published in 1833, plate 40 in 1835, after Cowan, 1971; text only completed in 1844].

Hamilton RW (2002) Attelabidae Billberg 1820. In *American Beetles. Volume II. Polyphaga: Scarabaeoidea through Curculionoidea*. (Eds RH Arnett, MC Thomas, PE Skelley and JH Frank) pp. 703–710. CRC Press, Boca Taron, FL. 861pp.

Jekel H (1860) Coleoptera. Curculionides. – Part II. In *Insecta Saundersiana: or Characters of Undescribed Insects in the Collection of William Wilson Saunders, Esq. Vol. II*. (Ed. WW Saunders) pp. 155–244. John van Voorst, London, 244 pp.

Kobayashi C, Fukasawa Y, Hirose D, Kato, M (2008) Contribution of symbiotic mycangial fungi to larval nutrition of a leaf-rolling weevil. *Evolutionary Ecology* **22**(6), 711–722. doi:10.1007/s10682-007-9196-2

Kobayashi C, Okuyama Y, Kawazoe K, Kato M (2012) The evolutionary history of maternal plant-manipulation and larval feeding behaviours in attelabid weevils (Coleoptera; Curculionoidea). *Molecular Phylogenetics and Evolution* **64**(2), 318–330. doi:10.1016/j.ympev.2012.04.006

Kôno H (1927) Beitrag zur Kenntnis der Attelabiden-Fauna Japans. *Insecta Matsumurana* **2**(1), 34–45.

Kuschel G (1995) A phylogenetic classification of Curculionoidea to families and subfamilies. *Memoirs of the Entomological Society of Washington* **14**, 5–33.

Kuschel G (2008) Curculionoidea (weevils) of New Caledonia and Vanuatu: Basal families and some Curculionidae. In *Zoologia Neocaledonica 6. Biodiversity studies in New Caledonia*. (Ed. P Grandcolas) pp. 99–249. *Mémoires du Muséum national d'Histoire naturelle* **197**.

Lawrence JF, Ślipiński SA (2013) *Australian Beetles. Vol. 1. Morphology, Classification and Keys*. CSIRO Publishing, Collingwood, Victoria, viii + 561 pp.

Lea AM (1898) Descriptions of new species of Australian Coleoptera. Part IV. *Proceedings of the Linnean Society of New South Wales* **22**(3), 584–638. doi:10.5962/bhl.part.12734

Lea AM (1899) Descriptions of new species of Australian Coleoptera. Part V. *Proceedings of the Linnean Society of New South Wales* **23**(4), 521–645.

Lea AM (1908) The Coleoptera of King Island, Bass Strait. *Proceedings of the Royal Society of Victoria (New Series)* **20**(2), 143–207.

Lea AM (1909a) Curculionidae. In *Die Fauna Südwest-Australiens. Ergebnisse der Hamburger südwest-australischen Forschungsreise 1905. Band II* (Eds W. Michaelsen and R. Hartmeyer) pp. 215–232. Gustav Fischer, Jena, 493 pp. doi:10.5962/bhl.title.7416

Lea AM (1909b) Descriptions of Australian Curculionidae, with notes on previously described species. Part VII. *Transactions of the Royal Society of South Australia* **33**, 145–196.

Lea AM (1910) Descriptions of Australian Curculionidae, with notes on previously described species. Part VIII. *Transactions of the Royal Society of South Australia* **34**, 13–58.

Lea AM (1921) On Coleoptera, mostly from Queensland. *Memoirs of the Queensland Museum* **7**(3), 182–240.

Lea AM (1926) On some Australian Curculionidae. *Proceedings of the Linnean Society of New South Wales* **51**(3), 327–362.

Lea AM (1929) Descriptions of new species of Australian Coleoptera. XX. *Proceedings of the Linnean Society of New South Wales* **54**(5), 519–549.

Lee C, Morimoto K (1988a). Larvae of the weevil family Attelabidae of Japan. Part 1. Subfamily Attelabinae (Insecta: Coleoptera). *Journal of the Faculty of Agriculture, Kyushu University* **32**, 215–237. doi:10.5109/23878

Lee C, Morimoto K (1988b). Larvae of the weevil family Attelabidae of Japan. Part 2. Subfamily Rhynchitinae (Insecta: Coleoptera). *Journal of the Faculty of Agriculture, Kyushu University* **32**, 239–254. doi:10.5109/23879

Legalov AA (2002) New taxonomic names of the leaf-rolling weevils (Coleoptera, Curculionoidea: Rhynchitidae, Attelabidae). *Evraziatskii Entomologicheskii Zhurnal* **1**(1), 91–92.

Legalov AA (2003) *Taksonomiya, klassifikatsiya i filogeniya rinkhitid i trubkovertov (Coleoptera, Rhynchitidae, Attelabidae) mirovoi fauny.* [*Taxanomy [sic], Classification and Phylogeny of the Leaf-rolling Weevils (Coleoptera, Rhynchitidae, Attelabidae) of the World Fauna*]. CD-ROM. Novosibirsk, 733 pp. [in Russian]

Legalov AA (2007) *Leaf-rolling Weevils (Coleoptera, Rhynchitidae, Attelabidae) of the World Fauna.* Agro-Siberia, Novosibirsk, 523 pp.

Lengerken Hv (1954) *Die Brutfürsorge und Brutpflege instinkte der Käfer*, 2 Aufl. Geest & Portig, Leipzig 383 pp.

Li X, Guo W, Ding J (2012) Mycangial fungus benefits the development of a leaf-rolling weevil, *Euops chinensis*. *Journal of Insect Physiology* **58**(6), 867–873. doi:10.1016/j.jinsphys.2012.03.011

Marvaldi AE, Morrone JJ (2000) Phylogenetic systematics of weevils (Coleoptera: Curculionoidea): a reappraisal based on larval and adult morphology. *Insect Systematics & Evolution* **31**(1), 43–58.

Marvaldi AE, Duckett CN, Kjer KM, Gillespie JJ (2009) Structural alignment of 18S and 28S rDNA sequences provides insights into phylogeny of Phytophaga (Coleoptera: Curculionoidea and Chrysomeloidea). *Zoologica scripta* **38**(1), 63–77. doi:10.1111/j.1463-6409.2008.00360.x

Marvaldi AE, Sequeira AS, O'Brien CW, Farrell BD (2002) Molecular and morphological phylogenetics of weevils (Coleoptera: Curculionoidea): do niche shifts accompany diversification? *Systematic Biology* **51**(5), 761–785. doi:10.1080/10635150290102465

May BM (1993) *Fauna of New Zealand. Number 28. Larvae of Curculionoidea (Insecta: Coleoptera) a Systematic Overview.* Manaaki Whenua Press, Lincoln, New Zealand, 226 pp.

McAreavey J (1938) On the so-called "ant letters". *Victorian Naturalist* **54**, 123–127.

McKenna DD, Sequeira AS, Marvaldi AE, Farrell BD (2009) Temporal lag and overlap in the diversification of weevils and flowering plants. *Proceedings of the National Academy of Sciences of the United States of America* **106**(17), 7083–7088. doi:10.1073/pnas.0810618106

Pascoe FP (1874a) Additions to the Australian Curculionidae. Part VI. *Annals and Magazine of Natural History* Series 4, **12**(77), 383–389. doi:10.1080/00222937408680885

Pascoe FP (1874b) Contributions towards a knowledge of the Curculionidae. Part IV. *Journal of the Linnean Society of London, Zoology* **12**, 1–99. doi:10.1111/j.1096-3642.1875.tb02572.x

Pascoe FP (1875) Additions to the Australian Curculionidae. Part VIII. *Annals and Magazine of Natural History* Series 4, **16**(91), 55–67. doi:10.1080/00222937508681122

Prell H (1924) Die biologische Gruppe der deutschen Rhynchitiden. *Zoologischer Anzeiger* **61**(7/8), 153–170.

Pullen K, Jennings D, Oberprieler RO (2014) Annotated catalogue of Australian weevils (Coleoptera: Curculionoidea). *Zootaxa* **3896**(1), 1–481. doi:10.11646/zootaxa.3896.1.1

Riedel A (1999) The *spinosus*-group of *Euops* Schoenherr (Coleoptera: Curculionoidea, Attelabidae), weevils with humeral spines from New Guinea. *Entomologica scandinavica* **30**(1), 75–117.

Riedel A (2001a) The *pygmaeus*-group of *Euops* Schoenherr (Coleoptera, Curculionoidea, Attelabidae), weevils associated with *Nothofagus* in New Guinea. *Journal of Natural History* **35**(8), 1173–1237.

Riedel A (2001b) Revision of the *Euops quadrifasciculatus*-group (Coleoptera: Curculionoidea: Attelabidae) from the Australian region, with a discussion of shifts between *Nothofagus* and *Eucalyptus* host plants. *Invertebrate Taxonomy* **15**(4), 551–587.

Riedel A (2001c) Revision of the *simulans*-group of *Euops* Schoenherr (Coleoptera, Curculionoidea, Attelabidae) from the Papuan region. *Deutsche Entomologische Zeitschrift* **48**(2), 139–221. doi:10.1002/dez.200100016

Riedel A (2002) *Taxonomy, phylogeny, and zoogeography of the weevil genus Euops (Insecta: Coleoptera: Curculionoidea) in the Papuan Region.* Dissertation zur Erlangung des Doktorgrades der Fakultät für Biologie der Ludwig-Maximilians-Universität, München, Teil 1, 216 pp.

Riedel A (2014) 3.4 Attelabidae Billberg, 1820. In *Handbook of Zoology, Coleoptera, Beetles, Volume 3.* (Eds RAB Leschen and RG Beutel) pp. 328–355. De Gruyter, Göttingen, 675 pp.

Sakurai K (1985) An attelabid weevil (*Euops splendida*) cultivates fungi. *Journal of Ethology* **3**, 151–156. doi:10.1007/BF02350306

Sawada Y (1993) A systematic study of the family Rhynchitidae of Japan (Coleoptera: Curculionoidea). *Humans and Nature* **2**, 1–93. doi:10.24713/hitotoshizen.2.0_1

Sawada Y, Morimoto K (1986) The mycetangia and the mode of the fungus transmission in the weevil genus *Euops* (Coleoptera: Attelabidae). *Scientific Bulletin of the Faculty of Agriculture, Kyushu University* **40**(4), 197–205.

Schoenherr CJ (1839) *Genera et species curculionidum, cum synonymia hujus familiae. Species novae aut hactenus minus cognitae, descriptionibus a Dom. Leonardo Gyllenhal, C.H. Boheman, et entomologis aliis illustratae.* Roret, Paris, Vol. 5(1), 1–456.

Shin S, Clarke DJ, Lemmon AR, Moriarty Lemmon E, Aitken AL, *et al.* (2018) Phylogenomic data yield new and robust insights into the phylogeny and evolution of weevils. *Molecular Biology and Evolution* **35**(4), 823–836. doi:10.1093/molbev/msx324

Thompson RT (1992) Observations on the morphology and classification of weevils (Coleoptera, Curculionoidea) with a key to major groups. *Journal of Natural History* **26**(4), 835–891. doi:10.1080/00222939200770511

van Emden F (1938) On the taxonomy of the Rhynchophora larvae (Coleoptera). *Transactions of the Royal Entomological Society of London* **87**(1), 1–37. doi:10.1111/j.1365-2311.1938.tb01800.x

Velbinger HH (1950) Beitrag zur biologie und Bekämpfung der Apfelfruchtsecher *Rhynchites aequatus* L. und Rhynchites bacchus L.) (Col. Cucrcul.). *Intenational Congress of Entomology* **8**, 958–973.

Von Dalla Torre KW, Voss E (1937) Curculionidae: Mesoptiliinae, Rhynchitinae I. In *Coleopterorum Catalogus. Pars 158* (Eds W. Junk and S. Schenkling) W. Junk, 's-Gravenhage, 56 pp.

Voss E (1922a) Kurze Bemerkungen über Rüsselkäfer. *Deutsche Entomologische Zeitschrift* **1922**(1), 174. doi:10.1002/mmnd.192219220117

Voss E (1922b) Monographische Bearbeitung der Unterfamilie Rhynchitinae (Curc.). I. Teil: Nemonychini-Auletini. *Archiv für Naturgeschichte*, Abteilung A **88**(8), 1–113.

Voss E (1924) Die Gattung *Euops* Sch. (Col. Curc.). *Deutsche Entomologische Zeitschrift* **1924**(1), 33–64. doi:10.1002/mmnd.192419240103

Voss E (1933) Monographie der Rhynchitinen-Tribus Auletini. III. Teil der Monographie der Rhynchitinae-Pterocolinae. *Stettiner Entomologische Zeitung* **94**(1), 108–136.

Voss E (1937) Über Arten und Gattungen der Unterfamilien Belinae, Rhynchitinae und Attelabinae. *Stettiner Entomologische Zeitung* **98**, 199–209.

Wasmann E (1884) *Der Trichterwickler. Eine naturwissenschaftliche Studie über den Thierinstinkt.* Münster, Verlag der Aschendorff'schen Buchhandlung. VII + 264 pp.

Waterhouse GR (1842) Description of two new genera of Curculionidae, closely allied to *Rhynchites*. *Annals and Magazine of Natural History* Series 1, **10**(62), 68.

Zimmerman EC (1992) *Australian Weevils (Coleoptera: Curculionoidea). Volume* VI. Colour Plates 305–632. CSIRO Australia, Melbourne, viii + 707 pp. 328 pls.

Zimmerman EC (1994) *Australian Weevils (Coleoptera: Curculionoidea). Volume 1. Orthoceri. Anthribidae to Attelabidae. The Primitive Weevils.* CSIRO Australia, Melbourne, xxxii + 741 pp.

42. CARIDAE THOMPSON, 1992

Hermes E. Escalona, Debbie Jennings and Rolf G. Oberprieler

Fig. 42.1. *Car pini* Lea.

Introduction. The Caridae represent the smallest family of extant weevils. The included genera *Car* Blackburn and *Caenominurus* Voss were earlier variously placed in Nemonychidae, Belidae, Attelabidae and Brentidae but later recognised as representing a discrete family-group, which was first treated as a subfamily of Belidae (Thompson 1992) or Brentidae (Kuschel 1992, 1995; May 1993) but subsequently recognised as constituting a distinct family (e.g. Zimmerman 1994; Morrone 1997; Marvaldi & Morrone 2000; Oberprieler *et al.* 2007). Caridae form the sister-group of Brentidae and Curculionidae, as based on both morphological characters (Marvaldi *et al.* 2002) and molecular data (Marvaldi *et al.* 2009; McKenna *et al.* 2009; Shin *et al.* 2018). The family includes only four described genera, *Car* and *Carodes* Zimmerman in Australia and *Caenominurus* and *Chilecar* Kuschel in South America (Argentina and Chile), but an undescribed genus with several species is known to occur in New Guinea (Oberprieler 2014). All genera except *Car* are monotypic. The fossil record of Caridae is confused and controversial, as none of the fossils placed in the family can be unequivocally assigned to it (Oberprieler 2014; Clarke *et al.* 2018).

The described species of Caridae are associated with Cupressaceae, those of *Car* with *Callitris* in Australia and *Caenominurus topali* Voss with *Austrocedrus chilensis* and *Chilecar pilgerodendri* Kuschel with *Pilgerodendron uviferum* in South America (Kuschel 1992; Oberprieler 2014). The New Guinean species have been collected on *Dacrycarpus* and *Podocarpus* (Podocarpaceae) (Oberprieler 2014), but it is unknown whether their larvae develop in tissues of these taxa. The hostplant of *Carodes* Zimmerman remains unknown. The known larvae of *Car* develop in green female strobili (cones) of *Callitris*, feeding in the sporophylls and sometimes also the immature seeds, and drop to the ground for pupation (May 1994; Oberprieler 2014). However, whether all Caridae develop in the female cones of their hosts is unknown.

Characteristics. Adults (Fig. 42.3). Body length 2.0–5.5 mm (SL) and 1.55–1.9 × as long as wide, pronotum and elytra independently rounded, moderately convex; colour yellow to dark brown or black, sometimes bicolored with pattern of dark markings on elytra and occasionally pronotum; clothed with short decumbent hairs, some sometimes thickened and dense to form a pattern (*Carodes*), and usually with scattered, fine erect sensory hairs. Head not or slightly declined at base, temples shorter than or as long as eyes, forehead and vertex flat, the latter with small median pit or linear fovea near posterior margin of eyes but no groove extending to occiput; occipital region without transverse ridge or stridulatory files; rostrum evenly cylindrical but sometimes apically slightly flattened, in male about as long as pronotum, in females up to twice as long; slightly evenly downcurved; at base dorsally very faintly set off from head in profile; glabrous, with fine sparse punctures or faintly shagreened in front of antennal insertions, sparsely setose behind insertions; epistomal area not differentiated, sometimes laterally demarcated by a pair of setae. Labrum absent (completely fused to frontoclypeus). Eyes slightly convex, not emarginate, finely facetted, without interfacettal setae. Antennal insertions at about basal fourth of rostrum, situated ventrally close to midline (*Car*) or more laterally (*Carodes*); sockets foveiform to reniform, scrobes absent or weakly defined. Antennae non-geniculate, 11-segmented (but see Clarke *et al.* 2018), about as long as head and rostrum; scapes elongate, slightly expanded apically and about twice as long as pedicel, apically clavate and largely glabrous; funicles 7-segmented; clubs very weak, loose, 3- or 4-segmented. Mandibles asetose, moderately elongate, internally bluntly tridentate, exodont (Fig. 42.3E, *Car*) or not (*Carodes*); incisor edge with single, large tooth; mola and prostheca absent; inner edge of mandibular base with long pubescent process. Maxillae with short, broad, densely setose mala (galea fused with lacinia); maxillary palps 3-segmented, extending only slightly beyond malar apex, apical segment cylindrical. Prementum elongate, fused at base; ligula simple, about as long as first 2 labial palp segments; labial palps 3-segmented, separated by less than width of first segment, apical segment cylindrical. Gular sutures fused (confluent). Corpotentorium absent; cervical sclerites absent (*Car*).

Pronotum about 0.80–0.87 × as long as wide, widest at base, there distinctly narrower than combined elytral bases; in dorsal view subrectangular to trapezoidal with sides slightly curved; anterior angles not produced; posterior angles rounded; posterior edge straight or weakly sinuate, not margined; disc simple. Prosternum about 0.25 × as long as shortest diameter of procoxal cavity, flat to slightly convex, without carinae; prosternal process more or less complete but broadly interrupted at middle, intercoxal process strongly concave, apex expanded and truncate; notosternal sutures transverse and incomplete, directed dorsad and then bent anteriad. Procoxae projecting below prosternum, with short concealed lateral extensions; trochantins concealed; procoxal cavities subcircular, contiguous, externally broadly closed, without narrow lateral extensions; postcoxal (hypomeral) processes meeting behind disjunct apex of prosternal process; cavities broadly closed internally.

Scutellar shield abruptly elevated, anteriorly simple, posteriorly broadly rounded with obtuse median groove in *Carodes*; densely covered with stout setae. Elytra about 1.03–1.37 × as long as wide and 1.93–2.63 × as long as pronotum, weakly rounded laterally, lateral edge gently sinuate, apically meeting or almost meeting at suture, not exposing any abdominal tergites; puncto-striate with 10 complete striae, without scutellary striole, interstriae generally flat but in *Carodes* 3, 5 and 7 slightly costate-tuberculate; epipleura absent; with internal longitudinal interlocking ridge laterally at about middle. Mesoventrite separated by complete sutures from mesanepisterna, these distinctly separated at midline; anterior edge on different plane than metaventrite, with narrow transverse prothoracic rest continuing onto mesanepisterna; mesanepisterna and mesepimera both covered with dense yellowish setae. Mesoventral cavity absent. Mesocoxae not conical and projecting, with concealed trochantins. Mesocoxal cavities narrowly separated, subcircular, not or slightly oblique, closed laterally by meeting of meso- and metaventrite. Mesoventral process subtruncate at apex; mesometaventral junction monocondylic. Metaventrite slightly convex; transverse (metakatepisternal) suture absent; discrimen complete to base anterior intercoxal process; postcoxal lines absent; exposed portion of metanepisternum about 3 × as long as wide; metanepisternal sutures without sclerolepidia; metepimera concealed. Metacoxae narrowly separate, not extending laterally to reach elytra. Metendosternite with long stalk, long lateral arms, well-developed but slender laminae and very short anterior process bearing widely separated anterior tendons. Hindwings with apical field more than half as long as whole wing, with distinct sclerite just beyond radial cell and vaguely indicated remnants of RP; radial cell moderately well-developed, slightly longer than wide with base more or less straight and inner posterobasal angle right; cross-vein r_3 moderately long and strongly oblique, arising close to base of r_4, this weakly sclerotised but more or less complete; basal portion of RP moderately long, its apical portion extending to wing margin; R-M loop forming acute angle; medial spur sinuate and extending almost to wing margin; medial field with 1 (*Car pini*) or 3 (*Car condensatus*) free veins and no medial fleck; MP_{3+4} without base; wedge cell absent; anal lobe moderately well-developed with highly reduced (*C. pini*) or shallow (*C. condensatus*) embayment. Legs short and stout; femora slightly inflated; outer edges of meso- and metatibiae in *Carodes* with crenulate (beaded) carina in distal half; tibiae with apices truncate, tibial spur formula 2-2-2, inner spur in *Carodes* curved and non-articulated (mucro-like); meso- and metatibiae slightly expanded at apex, with fringe of spines. Tarsi pseudotetramerous (5-5-5), segments 3 distinctly bilobed, 1–3 densely setose beneath; tarsal claws divaricate, with slight basal tooth and ventrobasal seta; empodium bisetose.

Abdomen with 5 free ventrites at same level (Figs 42.3N, 42.4H); ventrite 1 at its shortest point about 1.5 × as long as 2, 2–4 subequal, 5 × as long as 4 in *Carodes*; intercoxal process acute; apex of ventrite 5 broadly rounded. Abdominal tergites lightly pigmented except for VII and VIII, IV–VII with patches of wing-folding asperities; functional spiracles on segments I–V located in pleural membrane, on VI reduced and on VII absent. Tergite VIII in male more or less concealed beneath VII but well sclerotised; sternite VIII moderately sclerotised, subdivided in *Carodes*, with short articulated median apodeme (spiculum relictum). Anterior edge of segment IX in male with distinct spiculum gastrale; tergite X membranous. Aedeagus of pedotectal type, symmetrical; tegmen with long manubrium, slightly expanded and truncate at apex; body of tegmen elongate; parameres short, broad, separate and articulated; penis with distinct pedon and tectum (Fig. 42.4P) and paired basal temones; endophallus with flagellum, in *Carodes* also with short, paired sclerite in base of aedeagal body. Segment VIII in female more or less membranous, with long, fixed spiculum ventrale, slightly expanded at apex. Ovipositor slightly elongate and lightly sclerotised; proctiger and paraprocts membranous and not clearly defined, without baculi; gonocoxites undivided, without basal baculi; styli moderately well-developed, slender, terminal. Internal genital tract with sclerotised spermathecal complex, duct short and curled in *Car condensatus* group but long to very long and coiled in other species. Nervous system, testes and ovaries as in Calder (1989: Fig. 109; 1990: figs 15, 143, respectively), summarised by Oberprieler (2014). Alimentary canal with foregut without crop, hindgut with six cryptonephritic Malpighian tubules and rectal ring (Calder 1989: Fig. 14). (Voss 1932, 1965; Kuschel 1992, 1995; Thompson 1992; Zimmerman 1994; Lawrence & Ślipiński 2013; Oberprieler 2014).

Larvae (Fig. 42.2A–K). Body moderately elongate and robust, up to 10 mm in length, somewhat flattened and ventrally curved (C-shaped); all surfaces lightly pigmented except for head capsule and prothorax; vestiture consisting of scattered long and short, simple setae. Head strongly declined (hypognathous), not retracted; convex, subcircular in diameter; colour reddish-brown, frontal region paler except for dark epistome; posterior edge of head capsule not emarginate; epicranial stem long; frontal arms V-shaped, contiguous at base and separated from mandibular membranes by broad closure of frons and epicranium (not by frontoepicranial bracon); median endocarina extending between frontal arms but not reaching clypeal region. Stemmata 4 pairs, all with distinct lenses. Antennae very short, 1-segmented, with terminal, conical sensorium and several papillae. Frontoclypeal suture present; labrum fused with clypeus, triangular, anteriorly weakly trilobate, with 4 pairs of setae and 3 sensilla; epipharynx with relatively short, broad epipharyngeal rods, tormae parallel. Mandibles more or less symmetrical, short and broad, obliquely

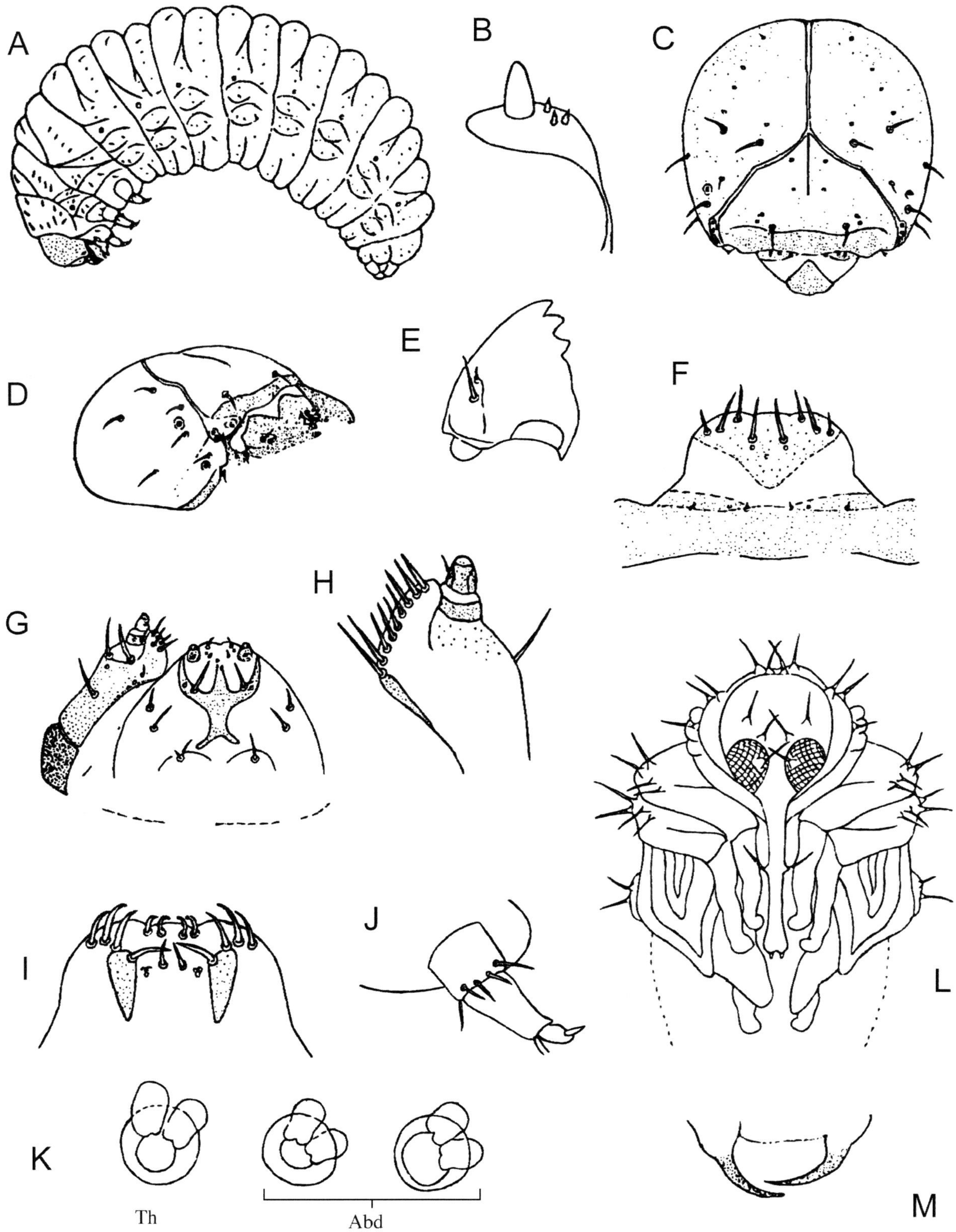

Fig. 42.2. Caridae, structural details of immature stages, *Car condensatus* Blackburn. A–K, larva: A, habitus, lateral; B, antenna; C, head, dorsal; D, head, frontolateral view showing incomplete frontal suture; E, mandible; F, labrum, clypeus and epistoma; G, mala and labium, ventral; H, mala, ventral; I, epipharyngeal lining; J, leg; K, spiracles, showing alignment; L–M, pupa: L, habitus, ventral; M, pseudocerci of abdominal segment IX. After May (1994).

bidentate; without accessory ventral process; incisor edge with 2 subapical teeth, mola and prostheca absent. Ventral mouthparts retracted; maxillary articulating area absent; maxillae with cardo slightly oblique, undivided; stipes elongate; mala fixed, apically rounded with row of stout setae along inner edge; palp 2-segmented, without distinct palpifer. Labium consisting of prementum and postmentum; premental sclerite with acute anterior extension and bifurcate posterior extension; ligula absent; labial palps 1-segmented, widely separated. Hypopharyngeal bracon present; sclerome absent. Hypostomal rods moderately long, subparallel; ventral epicranial ridges absent. Gular region absent.

Prothoracic tergum about as long as meso- or metatergum, with or without lightly pigmented plate. Legs well-developed, 3-segmented, including claw-like pretarsus bearing single seta. Abdomen >2 × as long as thorax; abdominal terga with 2 transverse folds, not extending laterally beyond edges of sterna; terga I–VIII each with transverse groove forming 2 plicae; segment IX distinctly shorter than VIII; completely dorsal, without urogomphi, sternum simple; segment X much shorter than IX, terminal, without pygopods; anal region posteriadly orientated; anal opening bordered by 4 lobes. Spiracles minute, annular-biforous with circular peritreme and moderately large accessory chambers; thoracic spiracle located on mesothorax. (May 1994; Oberprieler 2000, 2014; Lawrence & Ślipiński 2013).

Pupae (Fig. 42.2L–M). Length about 5 mm. Abdomen with terminal segment (IX) tapering, sclerotized, with pointed urogomphi (May 1994).

Key to the Australian genera and species

Modified after Zimmerman (1994).

1. Antennae inserted ventrally (Fig. 42.3D), sockets narrowly separated, area between them ridged (Fig. 42.3D); mandibles exodont (Fig. 42.3F); meso- and metatibiae dorsally not crenulate **2**
– Antennae inserted laterally, sockets widely separate, surface between them flat; mandibles externally smoothly rounded; meso- and metatibiae dorsally strongly crenulate from base to apex (Fig. 42.3H–I) .... ***Carodes revelatus* Zimmerman**

2(1). Antennal clubs indistinct (Fig. 42.5C); large species (SL > 5 mm) .... ***Car condensatus* Blackburn**
– Antennal clubs distinct (4-segmented) (Fig. 42.6C); small species (SL < 3.5 mm) **3**

3(2). Disc of elytra with decumbent setae in lateral view .... ***Car intermedius* Lea**
– Disc of elytra with fine erect setae mixed with coarse, decumbent setae in lateral view .... ***Car pini* Lea**

Classification of the Australian genera

Car Blackburn, 1897 (Pls 3B–D, 36I–K; Figs 42.1; 42.2; 42.3A–B, D–G, J–N; 42.4–42.6)

Type species, by monotypy: *Car condensatus* Blackburn, 1897.

Australian species. *C. condensatus* Blackburn, 1897, *C. intermedius* Lea, 1926, *C. pini* Lea, 1911.

Distribution. Australia (QLD, NSW, VIC, TAS, SA, NT, WA).

Notes. In addition to the three described species, *Car* contains approximately another eight undescribed ones, especially in the drier and more poorly sampled northern regions, and especially *C. condensatus* is a complex of several very similar species. The species fall into two broad groups, which prompted Legalov (2013) to divide the genus into two genera (*Car* and *Crowsonicar* Legalov, 2013), but the alleged distinguishing characters do not hold, in particular with respect to *C. intermedius*, and Pullen *et al.* (2014) therefore synonymised the name *Crowsonicar* with that of *Car*. *Car* species have been collected from several species of *Callitris* (*C. canescens, C. collumnellaris, C. endlicheri, C. glaucophylla, C. preissii* and *C. rhomboidea*) and in Tasmania one specimen also from *Diselma archeri* (also Cupressaceae) (O. Bell pers. comm.), but it is unknown whether the latter is also a larval host. Often one species of the *C. condensatus* group and one of the *C. pini* group live together on the same host tree. The larvae of the *C. condensatus* group develop in green female cones (strobili), feeding in the sporophylls and sometimes also the seeds (May 1994; Oberprieler 2014), but it is not known whether the larvae of the smaller species of the *C. pini* group do so as well. The larva of *C. condensatus* or a species near it was described by May (1994).

Carodes Zimmerman, 1994 (Pl. 36L; Fig. 42.3C, H–I)

Type species, by original designation: *Carodes revelatus* Zimmerman, 1994.

Australian species. *C. revelatus* Zimmerman, 1994.

Distribution. Australian (northern QLD).

Notes. The single species of the genus remains known from only two specimens (Zimmerman 1994), and its host-plant(s) and life history have also remained elusive.

Acknowledgements

We thank Bruno Bell for his photographs of Caridae species in their natural habitat and Otto Bell for sharing his notes on host records.

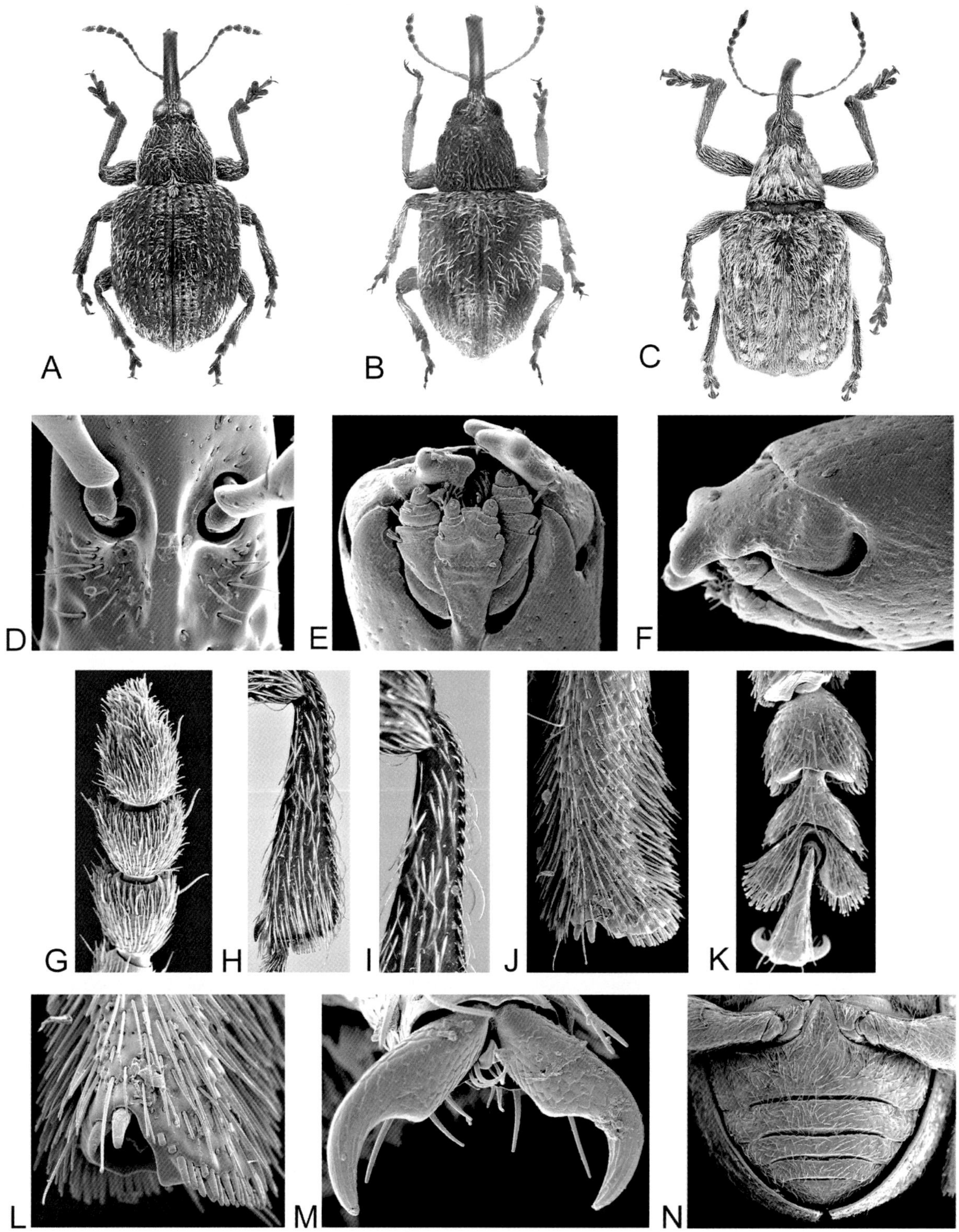

Fig. 42.3. Caridae, dorsal habitus (A–C) and structural details (D–N). A, *Car condensatus* Blackburn; B, *Car pini* Lea; C, *Carodes revelatus* Zimmerman; D–G, *Car condensatus* Blackburn: D, rostrum with ventral antennal insertions; E, mouthparts, ventral; F, left mandible, lateral; G, loosely 3-segmented antennal club; H–I, *C. revelatus*: H, metatibia with upper fused, mucro-like spur; I, metatibia showing outer crenulate ridge; J–N, *C. condensatus*: J, metatibia with two inner spurs; K, tarsus with broadened basal segment; L, metatibial spurs; M, tarsal claws showing basal swelling and ventrobasal seta; N, abdominal ventrites. D–N, after Oberprieler (2014).

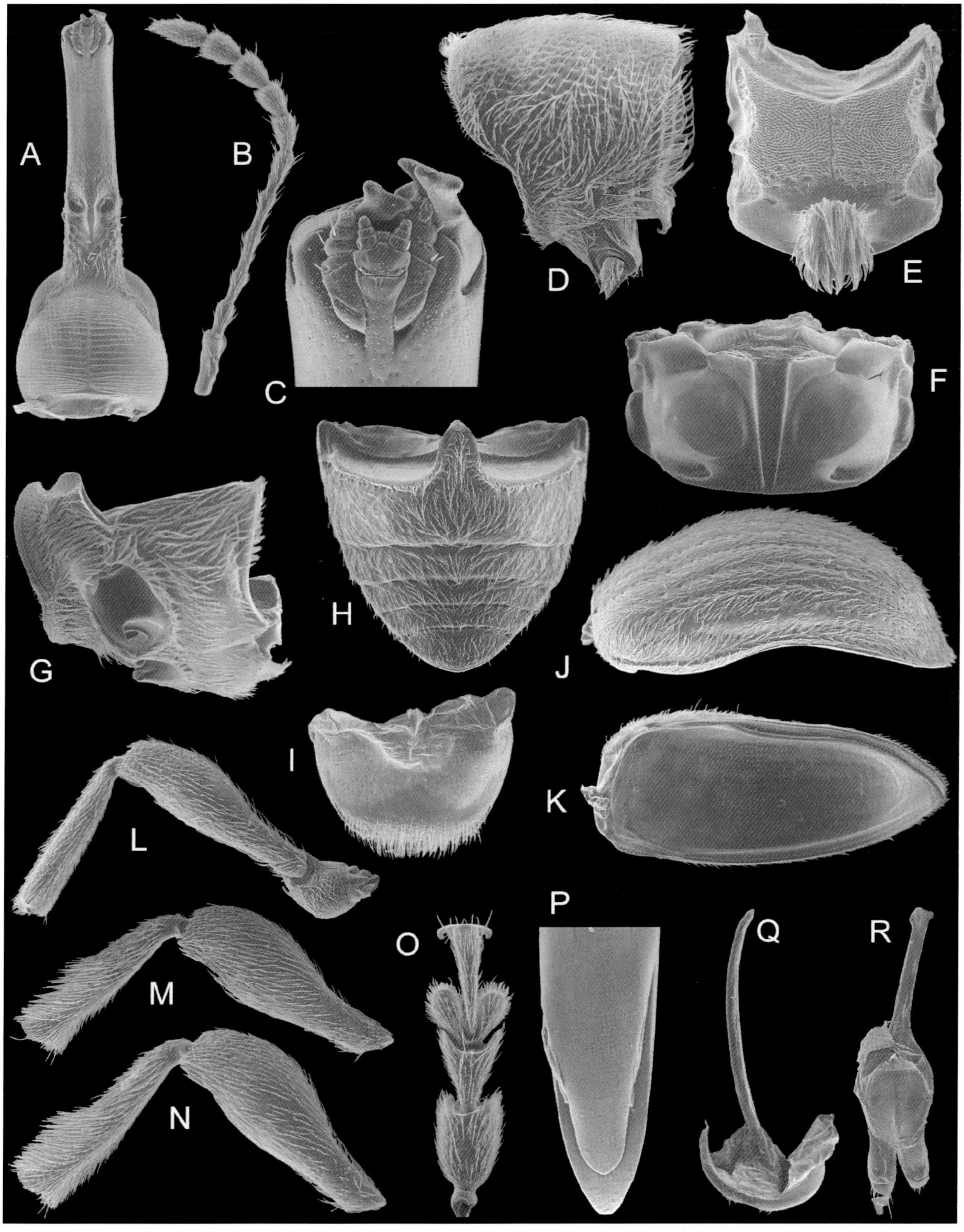

Fig. 42.4. Caridae, structural details, *Car condensatus* Blackburn, ♂. A, head, ventral; B, antenna; C, mouthparts, ventral; D, prothorax, lateral; E, scutellum; F, metanotum; G, pterothorax, lateral; H, abdominal ventrites; I, pygidium; J–K, elytron: J, lateral; K, inner lateral; L, foreleg; M, middle leg; N, hindleg; O, tarsus, dorsal; P, penis, apex, dorsal; Q, sternites VIII and IX; R, tegmen.

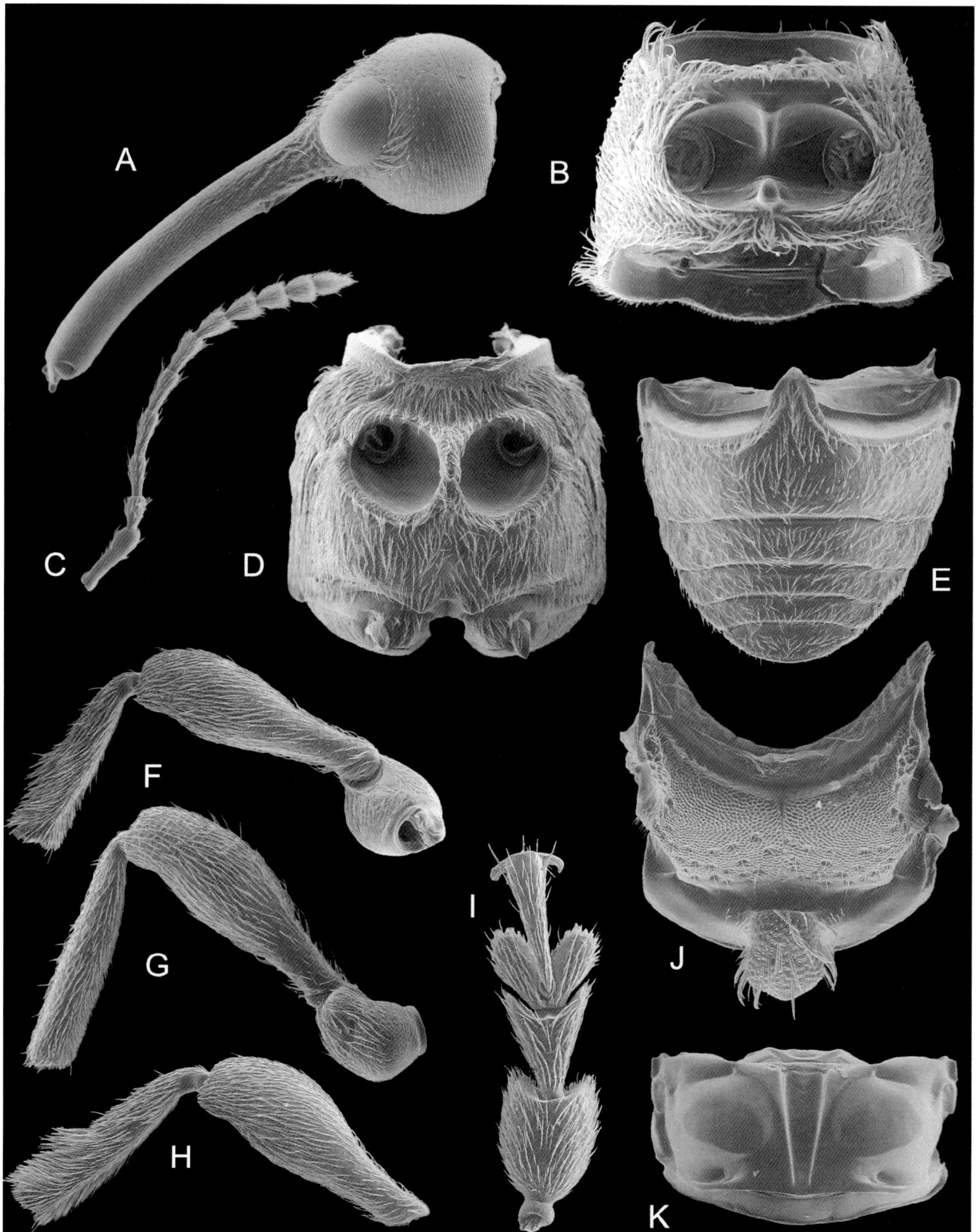

Fig. 42.5. Caridae, structural details, *Car condensatus* Blackburn, ♀. A, head, lateral; B, prothorax, ventral; C, antenna; D, pterothorax, ventral; E, abdominal ventrites; F, foreleg; G, middle leg; H, hindleg; I, tarsus, dorsal; J, scutellum; K, metanotum.

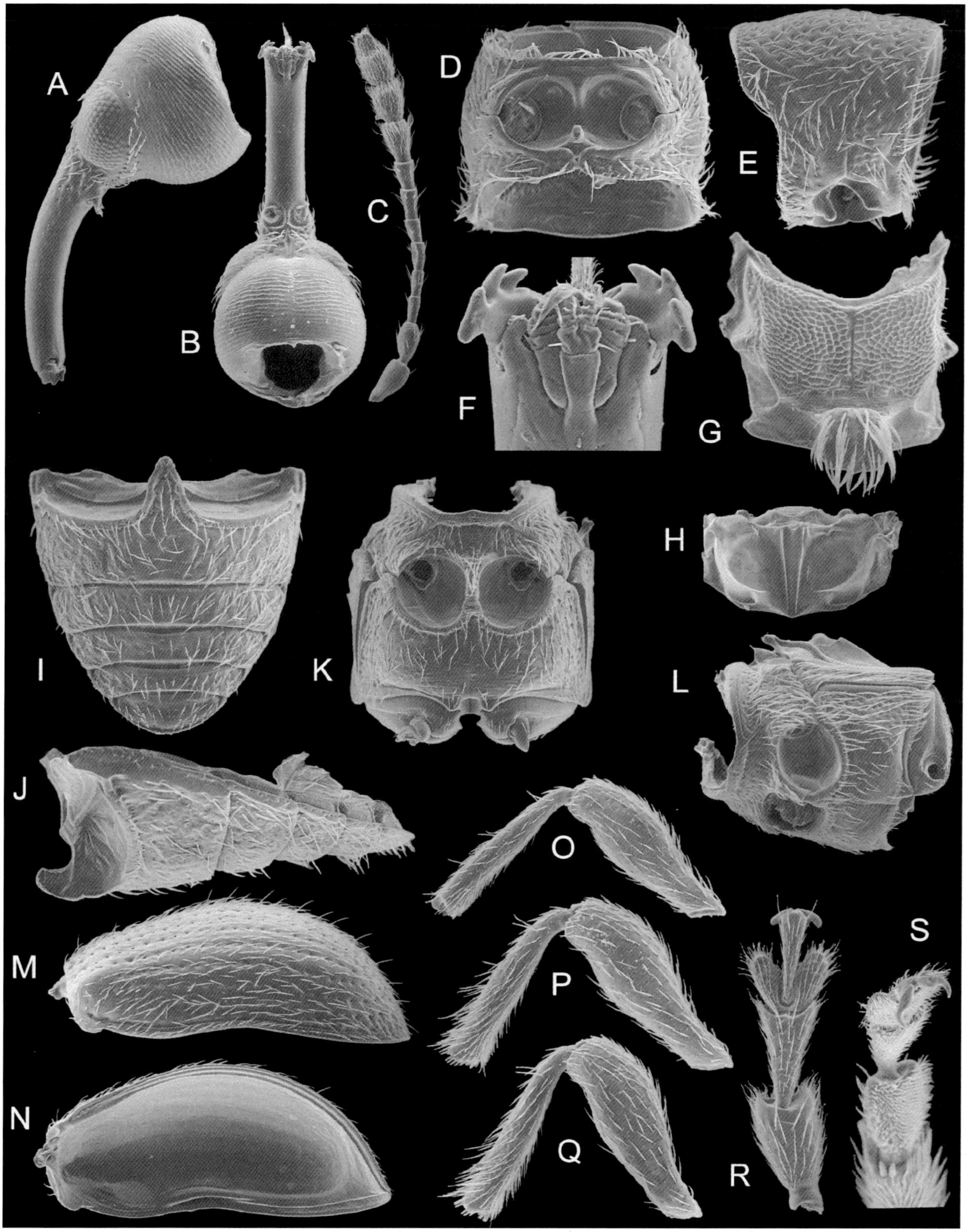

Fig. 42.6. Caridae, structural details, *Car pini* Lea. A, head, lateral; B, head, ventral; C, antenna; D–E, prothorax: D, ventral; E, lateral; F, mouthparts, ventral; G, scutellum; H, metanotum; I–J, abdominal ventrites: I, ventral; J, lateral; K–L, pterothorax: K, ventral; L, lateral; M–N, elytron: M, lateral; N, inner lateral; O, foreleg; P, middle leg; Q, hindleg; R, tarsus, dorsal; S, tarsus with claws, ventrolateral.

References

Blackburn T (1897) Further notes on Australian Coleoptera, with descriptions of new genera and species. Part XXI. *Transactions of the Royal Society of South Australia* **21**, 28–39.

Calder AA (1989) The alimentary canal and nervous system of Curculionoidea (Coleoptera): gross morphology and systematic significance. *Journal of Natural History* **23**(6), 1205–1265. doi:10.1080/00222938900770671

Calder AA (1990) Gross morphology of the soft parts of the male and female reproductive systems of Curculionoidea (Coleoptera). *Journal of Natural History* **24**(2), 453–505. doi:10.1080/00222939000770341

Clarke DJ, Limaye A, Mckenna DD, Oberprieler RG (2018) The weevil fauna preserved in Burmese amber – snapshot of a unique, extinct lineage (Coleoptera: Curculionoidea). *Diversity* **11**(1), 1–233. doi:10.3390/d11010001

Kuschel G (1995) A phylogenetic classification of curculionoidea to families and subfamilies. *Memoirs of the Entomological Society of Washington* **14**, 5–33.

Kuschel W (1992) Reappraisal of the Baltic amber Curculionoidea described by E. Voss. *Mitteilungen aus dem Geologisch-Paläntologischen Institut der Universität Hamburg* **73**, 191–215.

Lawrence JF, Ślipiński SA (2013) *Australian Beetles. Vol. 1. Morphology, Classification and Keys*. CSIRO Publishing, Collingwood, Victoria, viii + 561 pp.

Lea AM (1911) Notes on Australian Curculionidae in the Belgian Museum, with descriptions of new species. Part II. *Mémoires de la Société entomologique de Belgique* **18**, 61–127.

Lea AM (1926) On some Australian Curculionidae. *Proceedings of the Linnean Society of New South Wales* **51**(3), 327–362.

Legalov AA (2013) A new genus of the subfamily Carinae (Coleoptera, Ithyceridae) from Australia. *Amurian Zoological Journal* **5**(1), 23–26. doi:10.33910/1999-4079-2013-5-1-23-26

Marvaldi AE, Morrone JJ (2000) Phylogenetic systematics of weevils (Coleoptera: Curculionoidea): a reappraisal based on larval and adult morphology. *Insect Systematics and Evolution* **31**, 43–58.

Marvaldi AE, Duckett CN, Kjer KM, Gillespie JJ (2009) Structural alignment of 18S and 28S rDNA sequences provides insights into phylogeny of Phytophaga (Coleoptera: Curculionoidea and Chrysomeloidea). *Zoologica Scripta* **38**(1), 63–77. doi:10.1111/j.1463-6409.2008.00360.x

Marvaldi AE, Sequeira AS, O'Brien CW, Farrell BD (2002) Molecular and morphological phylogenetics of weevils (Coleoptera, Curculionoidea): do niche shifts accompany diversification? *Systematic Biology* **51**(5), 761–785. doi:10.1080/10635150290102465

May BM (1993) *Fauna of New Zealand. Number 28. Larvae of Curculionoidea (Insecta: Coleoptera) a Systematic Overview*. Manaaki Whenua Press, Lincoln, New Zealand, 226 pp.

May BM (1994) An introduction to the immature stages of Australian Curculionoidea. In *Australian Weevils (Coleoptera: Curculionoidea). Volume II – Brentidae, Eurhynchidae, Apionidae and a Chapter on Immature Stages by Brenda May*. (Ed. EC Zimmerman) pp. 365–721. CSIRO Australia, Melbourne, vii + 755 pp.

McKenna DD, Sequeira AS, Marvaldi AE, Farrell BD (2009) Temporal lag and overlap in the diversification of weevils and flowering plants. *Proceedings of the National Academy of Sciences* **106**(17), 7083–7088. doi:10.1073/pnas.0810618106

Morrone JJ (1997) The impact of cladistics on weevil classification, with a new scheme of families and subfamilies (Coleoptera: Curculionoidea). *Trends in Entomology* **1**, 129–136.

Oberprieler RG (2000) The larvae of the weevil tribe Eurhynchini and the phylogeny of the Brentidae (Coleoptera: Curculionoidea). *Invertebrate Taxonomy* **14**(6), 755–770. doi:10.1071/IT00032

Oberprieler RG (2014) Caridae Thompson, 1992. In *Handbook of Zoology. Arthropoda: Insecta. Coleoptera, Beetles. Volume 3: Morphology and Systematics (Phytophaga)*. (Eds RAB Leschen and RG Beutel) pp. 356–363. Walter de Gruyter, Berlin/Boston, xxi + 675 pp.

Oberprieler RG, Marvaldi AE, Anderson RS (2007) Weevils, weevils, weevils everywhere*. *Zootaxa* **1668**(1), 491–520. doi:10.11646/zootaxa.1668.1.24

Pullen KR, Jennings D, Oberprieler RO (2014) Annotated catalogue of Australian weevils (Coleoptera: Curculionoidea). *Zootaxa* **3896**(1), 1–481. doi:10.11646/zootaxa.3896.1.1

Shin S, Clarke DJ, Lemmon AR, Lemmon EM, Aitken AL, *et al.* (2018) Phylogenomic data yield new and robust insights into the phylogeny and evolution of weevils. *Molecular Biology and Evolution* **35**(4), 823–836. doi:10.1093/molbev/msx324

Thompson RT (1992) Observations on the morphology and classification of weevils (Coleoptera, Curculionoidea) with a key to the major groups. *Journal of Natural History* **26**(4), 835–891. doi:10.1080/00222939200770511

Voss E (1932) Monographie der Rhynchitinen-Tribus Rhynchitini (V. i. Teil der Monographie der Rhymchitinae-Pterocolinae.). *Koleopterologische Rundschau* **18**, 153–189.

Voss E (1965) The Zoological results of Gy. Topal's collectings in south Argentina. 17. Attelabidae (Coleoptera). *Annales Historico-Naturales Musei Nationalis Hungarici, Pars Zoologica* **57**, 329–332.

Zimmerman EC (1994) *Australian Weevils (Coleoptera: Curculionoidea). Volume 1. Orthoceri. Anthribidae to Attelabidae. The Primitive Weevils*. CSIRO Australia, Melbourne, xxxii + 741 pp.

43. BRENTIDAE BILLBERG, 1820

Hermes E. Escalona, Marek Wanat, Debbie Jennings and Rolf G. Oberprieler

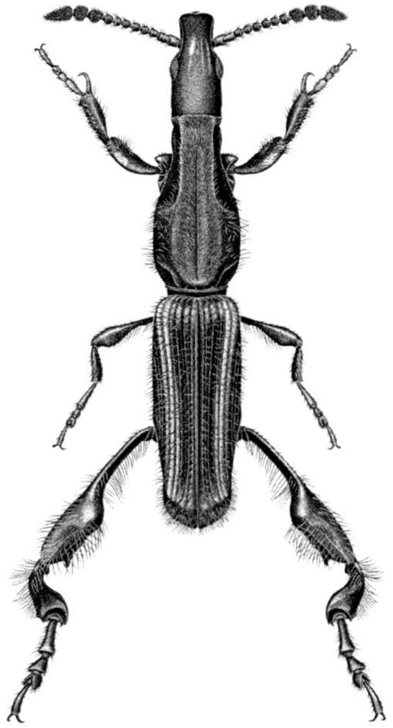

Fig. 43.1. *Catagogus diorymerus* (Lea).

Introduction. The modern concept of the family Brentidae (Oberprieler *et al.* 2007; Oberprieler 2014) was first developed by Crowson (1955) and Morimoto (1962, 1976) and subsequently modified and consolidated by Thompson (1992), May (1993), Kuschel (1995) and Oberprieler (2000). In this modern concept, which was corroborated by a recent molecular-phylogenetic analysis (Shin *et al.* 2018), the family comprises six extant subfamilies (Apioninae, Brentinae, Eurhynchinae, Ithycerinae, Microcerinae and Nanophyinae) and approximately 540 genera and 4400 species (Oberprieler 2014), most of the taxa belonging to Apioninae and Brentinae. With the monotypic Ithycerinae restricted to North America (Sanborne 1981) and Microcerinae to Africa (Louw 2004), the Australian fauna consists of only four subfamilies. These were revised by Zimmerman (1993, 1994) and catalogued by Pullen *et al.* (2014), and the Brentinae were also included in the world-wide brentid catalogue of Sforzi & Bartolozzi (2004). The New Zealand Brentidae were revised by Kuschel (2003), accompanied by notes on their systematics and biology, and the Australo-Pacific genera of some Apioninae were studied by Wanat (2001). The family was comprehensively reviewed by Oberprieler (2014), Sforzi *et al.* (2014), Alonso-Zarazaga & Wanat (2014) and Alonso-Zarazaga (2014).

In the modern concept, Brentidae are indicated to be a monophyletic lineage forming the sister-group of Curculionidae, as corroborated by morphological characters (Kuschel 1995; Oberprieler 2000, 2014) and a recent phylogenomic analysis (Shin *et al.* 2018), although Kuschel (1995) also included Caridae in Brentidae and Shin *et al.* (2018) did not sample Nanophyinae in their study. Previous phylogenetic analyses, based on both morphological and molecular characters or molecular data only (e.g. Marvaldi *et al.* 2002; Hunt *et al.* 2007; McKenna *et al.* 2009; Gillett *et al.* 2014; Gunter *et al.* 2016), were more limited in data and did not include all brentid lineages, in particular Ithycerinae and Microcerinae. The phylogeny estimate of Shin *et al.* (2018) resolved Ithycerinae and Eurhynchinae as basal lineages and Brentinae plus Microcerinae forming the sister-group of Apioninae, but larger sampling of taxa and also support by morphological characters are required to derive at a more comprehensive and robust internal classification. Oberprieler *et al.* (2007) and Oberprieler (2014) listed a series of morphological characters that typically occur in Brentidae, although not shared across all the subfamilies, such as the eyes covered by a corneal lens (Zimmerman 1994), the labial palps with only one or two segments and sunk into grooves, more exposed in Apioninae and Nanophyinae, the abdominal ventrites 1 and 2 on a lower level than 3–5, tergite VIII of the males pouch-like inflated with an inflexed median, tongue-like process, usually accompanied by a corresponding single or paired sclerite in the ventral membrane of genital sheath, the penis with a pair of crescentic frenal sclerites and the ring of the tegmen laterally constricted or articulated. The fossil record of Brentidae is poor, authentically stretching back to the Lower Cretaceous, with *Axelrodiellus ruptus* described from the Crato Formation in Brazil (125–112 Ma) (Zherikhin & Gratshev 2004) and *Orapaeus cretaceus* Kuschel & Oberprieler, 1994 (*in* Kuschel *et al.* 1994) from the Orapa kimberlite diatreme in Botswana (94–88 Ma) (Kuschel *et al.* 1994), both assigned to Eurhynchinae (Oberprieler 2014). These dates are congruent with the projected age of Eurhynchinae (Shin *et al.* 2018). The large and more derived subfamilies Apioninae and Brentinae are evidently younger, dating to the Upper Cretaceous (Shin *et al.* 2018), which accords with their absence in the large mid-Cretaceous weevil fauna discovered in Burmese amber (~100 Ma) (Clarke *et al.* 2018).

Characteristics. Adults (Figs 43.2–43.4, 43.9–43.34). Body length 1.5–40 mm (SL), length vs width ratio 3–8 (Brentinae) to 2.1–3 (Eurhynchinae) and 1.6–2.5 (Apioninae), some Brentinae present size-related sexual dimorphism. Body elongate and subcylindrical (Brentinae, Eurhynchinae) to convex or globose (*Cylas*, Apioninae, Nanophyinae); pronotum and elytra usually independently rounded or with elytra more or less parallel-sided; yellowish brown to red, dark brown or black, rarely metallic blue (some *Aporhina* and Apioninae), bicolored with red head and prothorax and black elytra in *Cylas* or with pattern of yellow markings on dark background in some Brentinae; surface commonly microreticulate in Apioninae and with apophyses, carinae, grooves and pubescent areas in myrmecophilous Brentinae. Vestiture when present (most Eurhynchinae, *Austronanodes*, *Rhinorhynchidius*, most Apionini) consisting of decumbent hairs, some thickened to almost squamose, forming a pattern, or of only scattered, fine hairs or body glabrous

(Brentinae, Myrmacicelini and Rhadinocybini of Apioninae). Head prognathous, not declined at base, in Brentinae often slightly constricted and then widened behind eyes, forming a broad neck (Fig. 43.3I); with long temples in Eurhynchinae and some Brentinae; occipital region without transverse ridge or stridulatory file, with median occipital endocarina and short frontal groove in *Eurhynchus*, forehead (region between eyes) usually slightly depressed to deeply, shortly sulcate; vertex evenly convex, smooth; rostrum shorter (*Brachycybus, Cyphagogus*) or several times longer than head capsule, porrect but slightly downcurved, apically usually slightly narrower than base, subcircular to subquadrate in cross-section, inserted in about middle of head in profile and dorsally forming shallow to abrupt sinus with outline of head, usually longer and thinner in females; length and shape sometimes sexually dimorphic, in males of some Brentinae robust, widened or with protuberances; frontoclypeal suture absent, frontoclypeal area in Apioninae often indicated by lateral incisions bearing 1–2 setae, labrum absent, epistome not demarcated from clypeus but anteriorly with pronounced median lobe. Eyes not to strongly protuberant, not emarginate, finely or coarsely facetted, without interfacettal setae; medially separated by width of rostrum or slightly less, rarely subcontiguous (*Austronanodes*) or contiguous (*Ommatocybus*); with facetted corneal layer covering central ommatidia as well as thin dark outer ring in Eurhynchinae and Brentinae. Antennal insertions lateral or ventral (several Apionini) with a raised transverse keel visible in profile as subocular tooth (some Brentinae), concealed from above, placed closer to base in females than in males; scrobes usually short, absent in *Rhinorhynchidius* (Apioninae) and some Brentinae, long in *Austronanodes*. Antennae almost always non-geniculate, geniculate in Nanophyinae (*Austronanodes*), almost always 11-segmented but 10-segmented in *Cylas* and 9-segmented in *Austronanodes* and *Neosebus*; moderately long, usually longer in males, filiform, moniliform, incrassate, rarely pectinate (males of *Ctenaphides*) or capitate with relatively weak club, this almost always 3-segmented (1-segmented in *Cylas*); scapes elongate (in Eurhynchinae longer and laterally compressed in females), but rarely more than 2 × as long as pedicel, more than 4 × as long in *Austronanodes*; funicles 7-segmented, proximal segments elongate but distal ones gradually shorter, glabrous to finely setose but densely shortly finely hirsute in males of *Ctenaphides* and some species of *Eurhynchus*; in *Ctenaphides* males each segment with a flat or cylindrical ventral process, in females only segments 4–7 (*C. porcellus* Pascoe) or 6–7 (*C. maculatus* Pascoe) with apex slightly produced ventrally; clubs loosely 3-segmented. Mandibles usually short and broad, occasionally slightly elongate, unidentate to obliquely tridentate (Eurhynchinae, Apioninae, Nanophyinae) or bidentate with large subapical tooth on incisor edge, ventrally directed in Cyphagogini; mola and prostheca absent; inner edge of mandibular base with long pubescent process. Maxillae with short and broad to elongate, setose mala, without uncus in Brentinae; maxillary palps 3-segmented in Eurhynchinae, Brentinae and Nanophyinae, 2-segmented in Apioninae, extending only slightly beyond malar apex and usually with distinct palpifer; apical segment cylindrical. Postmentum usually elongate, fused at base; prementum usually elongate; ligula simple or absent (Eurhynchinae); labial palps 1- or 2-segmented (*Austronanodes*) and more or less concealed in pits in Eurhynchinae and Brentinae, 2-segmented and exposed in Nanophyinae and 1-segmented and exposed in Apioninae; apical segment cylindrical to fusiform. Gular sutures fused, ending anteriorly in posterior tentorial pit in most Apioninae. Corpotentorium broad or narrow (*Austronanodes*, Apioninae) to absent; cervical sclerites absent.

Prothorax often elongate to slightly transverse. Pronotum about 1.2–2.4 × (Brentinae), 0.95–1.1 × (Eurhynchinae) or 0.5–1.4 × (Apioninae and Nanophyinae) as long as wide, usually widest in about middle (at base in *Austronanodes* and some Apionini); base distinctly narrower than combined elytral bases in Eurhynchinae and some Brentinae and Apioninae, basal edge with setal fringe in some Apioninae and crenulate in *Austronanodes*; sides slightly to strongly curved, lateral pronotal carinae absent; anterior angles not produced; posterior angles rounded, barely projected outwards in some Apioninae; posterior edge usually straight or weakly rounded, gradually narrowing posteriorly in some Brentinae, medially slightly emarginate in Eurhynchinae; disc with median longitudinal groove in some Brentinae and Eurhynchinae, with a small prescutellar fovea or pit in most Apionini, constricted at base and sometimes with posterior margin or collar in some Brentinae, *Rhynolaccus* and some Rhadinocybini, anteriorly raised in *Cyphagogus*; sides of prothorax with anterior lateral impressions for profemora in Cyphagogini (Brentinae); in many Apionini basal margin of pronotum with scales arranged parallel to body axis (basal flange). Prosternum much shorter than shortest diameter of procoxal cavity in *Austronanodes*, Apioninae and Eurhynchinae, about as long in *Cylas* and much longer in most Brentinae, usually flat to slightly convex, strongly concave in *Austronanodes* and *Rhinorhynchidius*. Prosternal process usually more or less complete but broadly interrupted at middle, intercoxal process strongly concave, the apex usually represented by a disjunct sclerite (sternellum or prosternellum) behind procoxae but often very small or completely fused to postcoxal processes. Notosternal sutures incomplete or absent, present in Nanophyinae, vertical above procoxae in Eurhynchinae, in most Apioninae with pit at anterior end. Procoxae usually projecting below prosternum (not so in some Brentinae), with short concealed lateral extensions; trochantins concealed. Procoxal cavities subcircular, usually contiguous (*Cylas*, Apioninae, Nanophyinae) but well separated in some other Brentinae, externally broadly closed, without narrow lateral extensions; postcoxal (hypomeral) processes meeting behind sternellum, fused or partially fused (Eurhynchinae) between them and sometimes to prosternal process; cavities broadly closed internally.

Scutellar shield distinct but often reduced, some flightless Apioninae, or absent (Brentinae, Nanophyinae), usually abruptly elevated, anteriorly simple, posteriorly acute to broadly rounded, in some Apioninae with two pointed tubercles at base and raised apex. Elytra moderately elongate, narrow (Eurhynchinae, Brentinae) and flat to oval and convex (Apioninae, *Austronanodes*); about 2.1–5.6 × (Brentinae), 1.75–

2.4 × (Eurhynchinae) and 1.5–2.2 × (Apioninae) as long as wide; about 1.3–2.65 × (Brentinae), 2.9–4.0 × (Eurhynchinae) and 2.0–3.8 × (Apioninae) longer than pronotum; sides subparallel (most Brentinae) to weakly rounded laterally, elytral apices usually conjointly rounded or projected in some Brentinae (*Hoplopisthius*), abdominal tergites not exposed, apical third in lateral view gently declivous to strongly convex towards apex, apices sometimes truncate or independently rounded or acute; base crenulate in *Austronanodes*; with 9 (most Apionini) or 10 (Eurhynchinae, some Apioninae) distinct striae (puncture rows), scutellary striole absent; punctures often coarse or very faint (*Ithystenus*, some *Aporhina* and Apioninae); interstriae sometimes raised and coloured red or yellow in Brentinae; interstriae 2 often anteriorly elevated into multituberculate to multispinose callus or ridge, in *Eurhynchus* 3 and 5 each often in middle with short to stout median spine, in *Aporhina* 4 usually in middle with very large, sharp spine; anterior margins together broadly V-shaped in Eurhynchinae and each extended anteriad into oblique ridge (*Ctenaphides*) to strong lobe (*Aporhina*) fitting over dorsolateral margins of prothorax; epipleura very narrow or absent (Eurhynchinae); sometimes with internal longitudinal interlocking ridge laterally near base, in Apioninae with multituberculate internal interlocking area near base and often with apical interlocking area (absent in Rhadinocybini, Myrmacicelini, *Diselmobius* and *Rhinorhynchidius*). Mesanepisterna large; mesepimera narrowly triangular. Mesoventrite in Eurhynchinae and *Austronanodes* separated by complete sutures from mesanepisterna, which are distinctly separated at midline, partly or completely fused to mesanepisterna in Brentinae and Apioninae; anterior edge usually on same or different (e.g. Apioninae) plane than metaventrite, paired procoxal rests absent but usually with narrow transverse prothoracic rest continuing onto mesanepisterna; sides of mesothorax deeply emarginate in *Cylas*; prosternal rest broadly transverse, absent in Brentinae. Mesocoxae usually not conical and projecting (distinctly projecting in some Apioninae), with concealed trochantins. Mesocoxal cavities usually narrowly separated, contiguous in *Rhinorhynchidius, Diselmobius* and Rhadinocybini, widely separated in *Austronanodes*, subcircular, not or slightly oblique, closed laterally by meeting of meso- and metaventrite. Mesoventral intercoxal process usually narrowly rounded to subtruncate at apex (process absent in *Rhinorhynchidius*); mesometaventral junction, if present, often monocondylic (absent due to complete fusion of meso- and metaventrite in *Austronanodes*). Metaventrite with disc moderately to strongly convex (Eurhynchinae), in some Apioninae with median depression; transverse (metakatepisternal) suture absent or very short; discrimen short or absent; exposed portion of metanepisterna moderately to extremely long and narrow; metepimera usually concealed but shortly exposed in Eurhynchinae; metanepisternal sutures without sclerolepidia. Metacoxae narrowly to moderately widely separate, sometimes extending laterad to reach elytra (not in Brentinae). Metendosternite with long stalk and lateral arms, well-developed, broad laminae in Eurhynchinae, but reduced or absent in most groups, anterior tendons close together in middle (absent in *Cylas*). Hindwings with apical field more than half as long as total wing, with 2 remnants of RP; radial cell usually highly reduced (radial window in most Apioninae) or absent, moderately well developed and triangular in Eurhynchinae, with inner posterobasal angle acute; cross-vein r_3 usually absent, joined basally with r_4 in Eurhynchinae; r_4 short or absent; basal portion of RP short or absent; R-M loop forming narrowly acute angle; medial spur sinuate and apically curved; medial field usually with 3, 4 (Eurhynchinae) or fewer free veins and no medial fleck; MP_{3+4} without base; wedge cell absent; anal lobe moderately well-developed without embayment, reduced with notch-like embayment in Apioninae and some Brentinae. Legs usually moderately long and slender (very long in *Ithystenus* and *Mesetia*), short and stout in some Cyphagogini; procoxae usually elongate and protruding, mesocoxae slightly elongate in *Aporhina*, flat in *Ctenaphides* and *Eurhynchus*; metacoxae shortly transverse, not quite reaching elytra, separated by about metacoxal width; trochanters short in Eurhynchinae and Brentinae, elongate in Nanophyinae and most Apioninae (except *Rhinorhynchidius*); trochanterofemoral joint slightly oblique in Nanophyinae, *Cylas* and most Apioninae; trochanteral cavities with prominent posterior peg limiting posterior articulation of trochanters in Eurhynchinae; femora usually not or only slightly inflated, with ventral spines in *Austronanodes*, without ventral spine in Apioninae, strongly inflated in some Brentinae, profemora with small to stout subapical ventral spine in some *Aporhina* (sometimes only in males) and some Brentinae; tibiae subcylindrical, usually not apically expanded or spinose, very short and stout in some Brentinae, protibiae with subapical antenna cleaner in some Brentinae, protibial apex in Eurhynchinae expanded and dorsally deeply notched on inside to allow tarsi to rotate up into ~90° angle, spur formula 2-2-2 in most Brentinae, 1-2-2 in Eurhynchinae, 1-2-2 or 1-1-1 in various Brentinae, 0-0-0 in *Cylas*, *Austronanodes* and Apioninae (spurs absent); mucro present in some males of Apioninae, usually on meso- and metatibiae. Tarsi 5-5-5-segmented in both sexes but pseudotetramerous (segment 4 strongly reduced), segment 3 usually distinctly bilobed but not or barely lobed in Cyphagogini, 1 toothed in some Apioninae, 1–3 or 3 only usually densely setose beneath (except Cyphagogini); tarsal claws simple and free in Eurhynchinae (divergent in *Ctenaphides* and *Aporhina*, divaricate in *Eurhynchus*) and most Brentinae, simple and connate in *Cylas* and *Austronanodes*, dentate in most Apioninae; empodium present in Apioninae and Eurhynchinae, absent in Brentinae.

Abdomen with 5 ventrites, 1 and 2 enlarged and connate, together longer than 3–5 and on different plane (lower) than 3–5, clearly separated by suture in Eurhynchinae but with suture incomplete or vaguely indicated in most other groups; ventrite 1 at shortest point subequal in length to 2, more or less equal in length to 5 and longer than 3 and 4 combined, intercoxal process narrowly to broadly rounded; ventrite 5 in Eurhynchinae roundly triangular to trapezoidal, apex emarginate in males of *E. laevior* and *C. maculatus*, more or less truncate and often with marked sexual dimorphism in males of Apioninae. Abdominal tergites I–VI or VII lightly to moderately pigmented, VIII and sometimes VII more heavily pigmented, usually with patches of wing-folding asperities on tergites IV–VII; functional spiracles on segments I–VII in Eurhynchinae and Brentinae and on I–VI in Nanophyinae and Apioninae, vestigial on VIII, anterior spiracles located in pleural membrane, those on segment VII on tergite. Tergite VIII in male usually heavily sclerotised, often forming capsulate pygidium in males of Apioninae and with apical margin medially inflexed into a tongue-like process in Eurhynchinae, Brentinae, Nanophyinae and Apioninae (excluding Apionini); sternite VIII reduced, with short median spicule (spiculum relictum) in males of Eurhynchinae; between tergites VII and VIII with membranous pouch in Eurhynchinae and several Apioninae (Myrmacicelini, Rhinorhynchidiini, Rhadinocybini except *Nanomyrmacyba*); anterior edge of segment IX in male with distinct apodeme (spiculum gastrale); tergite X membranous. Aedeagus of pedotectal type, symmetrical; tegmen laterally articulated, body elongate, anterior edge with short or long apodeme (manubrium); parameres short and broad (*Higonius*) to moderately long and narrow, separate and articulated or at least partly separated but fused to phallobase or fused together and to phallobase; penis with distinct tectum, pedon and paired anterior apodemes (temones); endophallus often with pair of crescentic sclerites (frena) and flagellum (some Brentinae, *Aporhina*, Myrmacicelini), only flagellum (some Rhadinocybini) or simple sclerites and denticles, spinulae and microplates arranged irregularly or in rows (some Apionini). Tergite VIII in female more or less membranous in most groups but sclerotised laterally in Eurhynchinae and forming heavily sclerotised toothed process in at least some Brentinae, usually with a pair of glands (tergal pouches) in Brentinae; sternite VIII with long, fixed spiculum ventrale, with sclerotised posterior process in Eurhynchinae. Ovipositor (gonocoxites and styli) elongate and lightly sclerotised; gonocoxites undivided, with longitudinal baculi; styli moderately well-developed or short (Eurhynchinae, some Brentinae), slender and apical (*Austronanodes*) or subapical (Eurhynchinae). Internal genital tract with heavily sclerotised spermathecal capsule; in Eurhynchinae insertions of duct and gland on spermatheca separated by at least width of duct, not on bulbus, bursa copulatrix often without sclerites (rarely present in Apioninae). (Morimoto 1962, 1976; Kissinger 1968; Alonso-Zarazaga 1990; Thompson 1992; Zimmerman 1993, 1994; Kuschel 1995; Oberprieler 2000, 2004, 2014; Wanat 2001; Anderson & Kissinger 2002; Lawrence & Ślipiński 2013; Alonso-Zarazaga 2014; Alonso-Zarazaga & Wanat 2014; Sforzi *et al.* 2014).

Larvae (Figs 43.5–43.8). Body elongate, C-shaped (Figs 43.5A–43.6A), cylindrical to shorter, broader and wider at middle; lightly pigmented, usually with exception of head capsule and protergum (pronotal plate); vestiture consisting of scattered long and short, simple setae; legs present in some Brentinae and 2-segmented in Eurhynchinae. Head moderately to strongly declined (hypognathous), protracted or slightly retracted; posterior edge of head capsule not emarginate, subglobose to oval in cross-section and strongly sclerotised; epicranial stem long; frontal arms U- or V-shaped, contiguous at base, extending to mandibular membrane; median endocarina present, weak in *Austronanodes*, located beneath epicranial stem in most Brentinae, extending between frontal arms in *Cylas*, Eurhynchinae, Nanophyinae and Apioninae. Stemmata in 3 (Eurhynchinae), 2 (*Cylas*), 1 or 2 (Apioninae), 1 (*Austronanodes*) pairs or absent (most Brentinae); postoccipital condyles absent. Antennae very short, 1-segmented, often with terminal conical sensorium (absent in Brentinae), papillae present in Eurhynchinae and Apioninae. Frontoclypeal suture present; clypeus pigmented and transverse; clypeolabral suture distinct; labrum free and broadly roundly triangular, with 2–3 pairs of setae; epipharynx with relatively short and broad to longer and narrow epipharyngeal rods (tormae). Mandibles robust, more or less symmetrical, short and broad, obliquely bidentate; accessory ventral process absent; incisor edge with or without subapical teeth, mola and prostheca absent. Ventral mouthparts retracted; maxillary articulating area often absent, present in Brentinae; cardo slightly oblique, undivided; stipes elongate; mala fixed, broad and without lacinial tooth, apically rounded with row of stout setae along inner edge; maxillary palp 1 (some Apioninae) or 2-segmented (Brentinae) without distinct palpifer, basal segment broad, subquadrate, with three sensilla, apical segment narrowly conical with accessory process in basal membrane. Labium consisting of prementum and postmentum; prementum with anterior extension in Eurhynchinae and *Austronanodes*; ligula simple and rounded or absent (*Cylas*); labial palps usually 1-segmented (2-segmented in Eurhynchinae and *Cylas*) and widely separated. Hypopharyngeal bracon present; sclerome absent. Hypostomal rods short to moderately long and diverging; ventral epicranial ridges often absent, present in Eurhynchinae. Gular region absent. Tentorial bridge sometimes laterally subdivided in Brentinae.

Prothoracic tergum about as long as meso- or metatergum, with pigmented plate in Eurhynchinae and some Brentinae. Mesocoxae relatively widely separated in Brentinae. Legs absent or very short (non-articulated lobes) and sparsely setose, distinctly 2-segmented in Eurhynchinae, in *Cylas* with 2 disc-like segments, without terminal claw in Eurhynchinae and some Apioninae, *Austronanodes* and *Cylas*. Abdomen usually more than twice as long as thorax (less so in some Apioninae); terga not extending laterally beyond edges of sterna, with patches of asperities in Eurhynchinae (spinules in *Eurhynchus* or carinulae in *Aporhina*) and some Brentinae; abdominal terga I–VIII each with transverse groove forming 2 plicae in Apioninae, *Austronanodes* and *Cylas*, with 3 or 4 folds in Eurhynchinae and most Brentinae; segment IX distinctly shorter than VIII; completely dorsal, without urogomphi; sternum 9 simple; segment X much shorter than IX, terminal, without pygopods; anal region posteriadly or postoventradly orientated; anal opening bordered by 4 lobes in Eurhynchinae, *Austronanodes* and *Cylas*, 3 lobes in at least some Brentinae and 2 lobes (with transverse anal opening) in Apioninae. Spiracles usually annular-biforous, annular in some Brentinae and Apioninae; abdominal spiracles annular in *Austronanodes*; spiracles absent (rarely vestigial) on segment VIII in Apioninae. Alimentary canal with 4 Malpighian tubules. (Gardner 1935; van Emden 1938; May 1993, 1994; Oberprieler 2000, 2004; Anderson and Kissinger 2002; Lawrence & Ślipiński 2013; Alonso-Zarazaga & Wanat 2014; Sforzi *et al.* 2014).

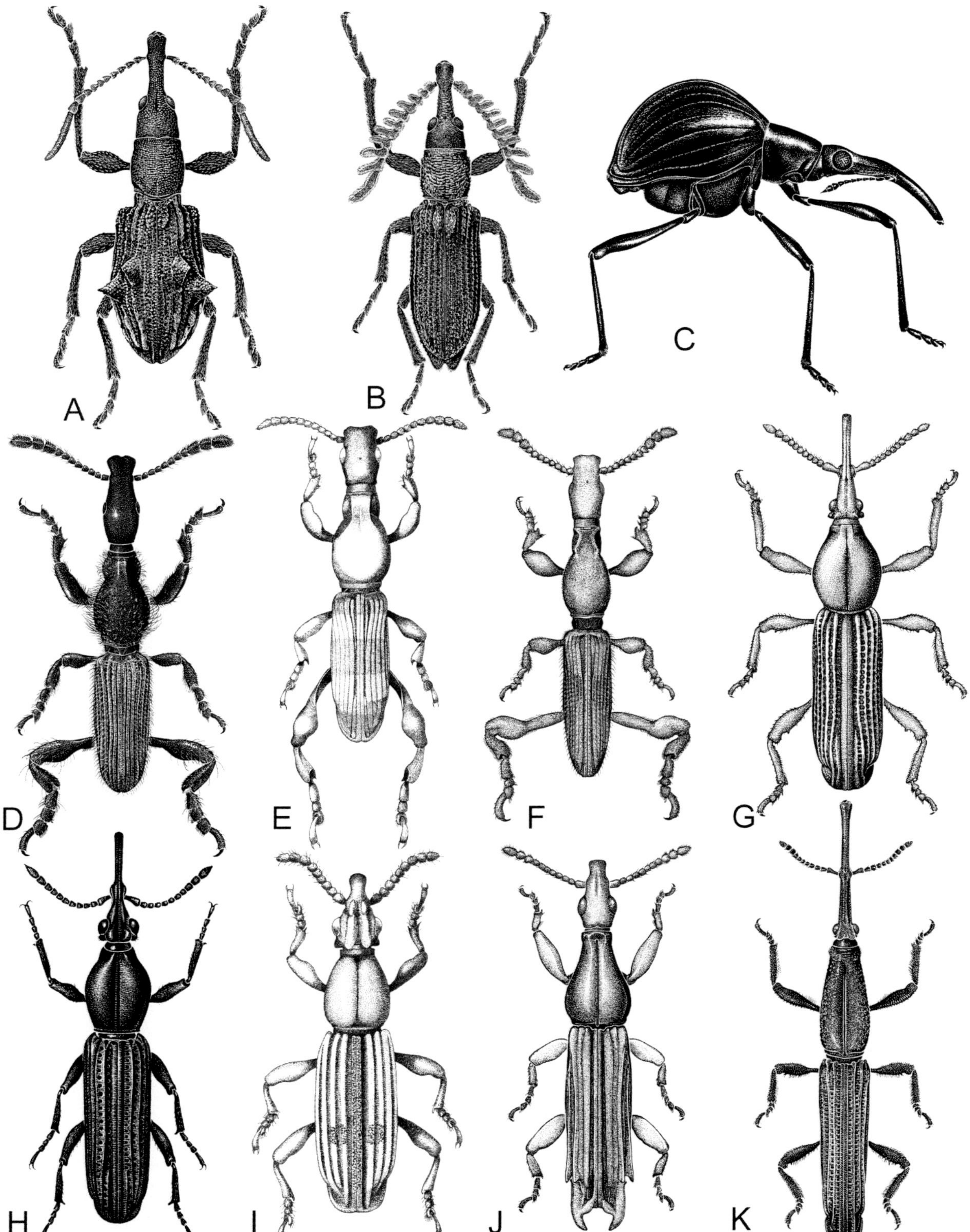

Fig. 43.2. Eurhynchinae (A, B), Apioninae (C), Brentinae (D–K), dorsal habitus. A, *Eurhynchus acanthopterus* Boisduval; B, *Ctenaphides porcellus* Pascoe; C, *Pseudopiezotrachelus convexipennis* (Lea); D, *Cyphagogus nigraustralis* Damoiseau; E, *Cyphagogus tristriatus* Kleine; F, *Cyphagogus crassitarsus* Damoiseau; G, *Euschizus dictatorius* Kleine; H, *Euschizus internatus* (Pascoe); I, *Higonius novenarius* Damoiseau; J, *Hoplopisthius kolbei* Senna; K, *Hormocerus reticulatus* (Lund). After Zimmerman (1994), except C after Lawrence & Britton (1994).

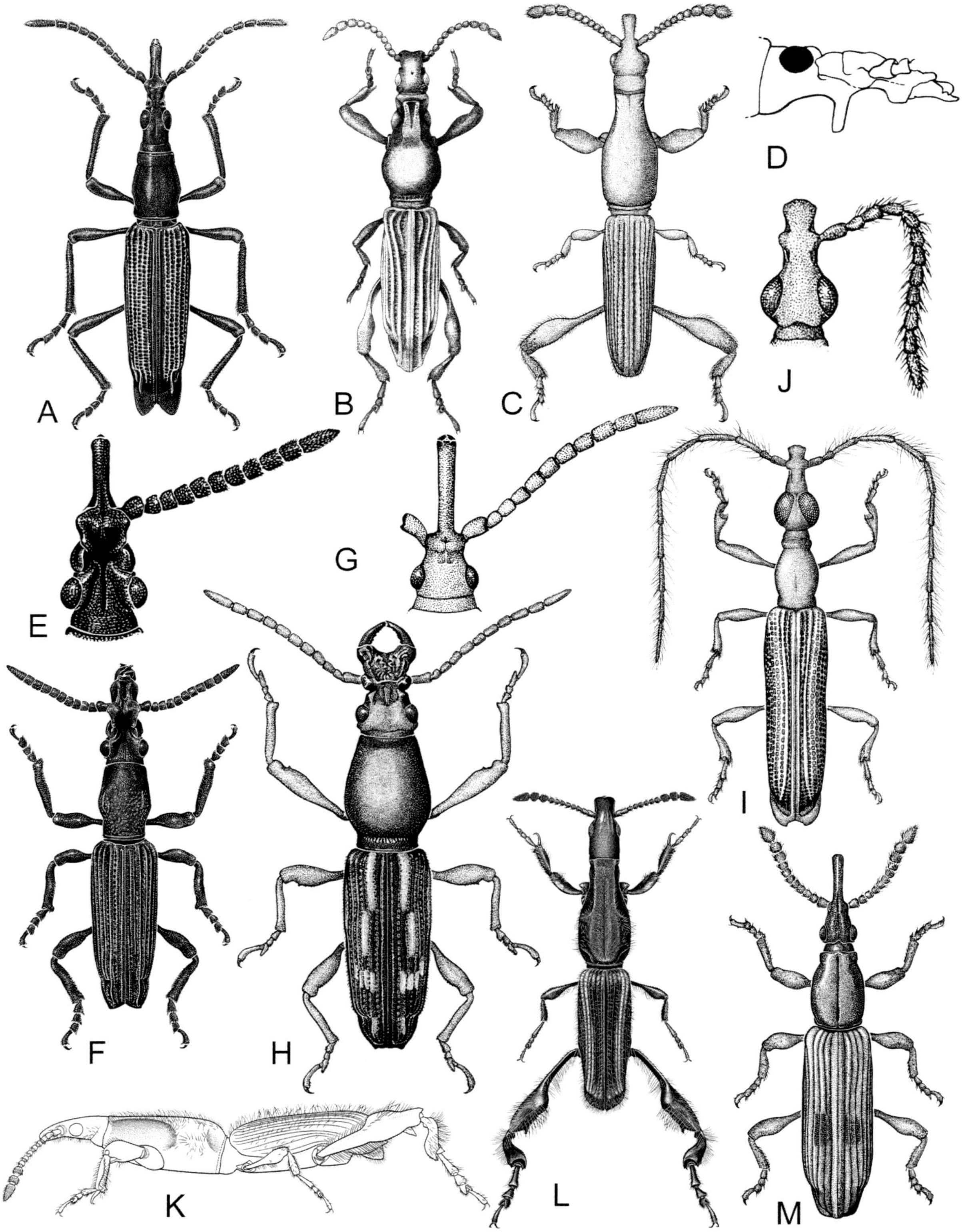

Fig. 43.3. Brentinae, habitus and head structures. A, *Cordus acutipennis* Power; B, *Allaeometrus breviceps* Senna; C, *Ancylobrentus australicola* Damoiseau; D–F, *Ankleineella subscripta* (Damoiseau): D, head, ♂; E, head, ♀; F, habitus, ♂; G–H, *Baryrhynchus lineicollis* Power: G, head, ♀; H, habitus, ♂; I–J, *Cerobates ophthalmicus* (Pascoe): I, habitus ♂; J, head, ♀; K–L, *Catagogus diorymerus* (Lea), lateral and dorsal; M, *Araiorrhinus howittii* (Pascoe). After Zimmerman (1994).

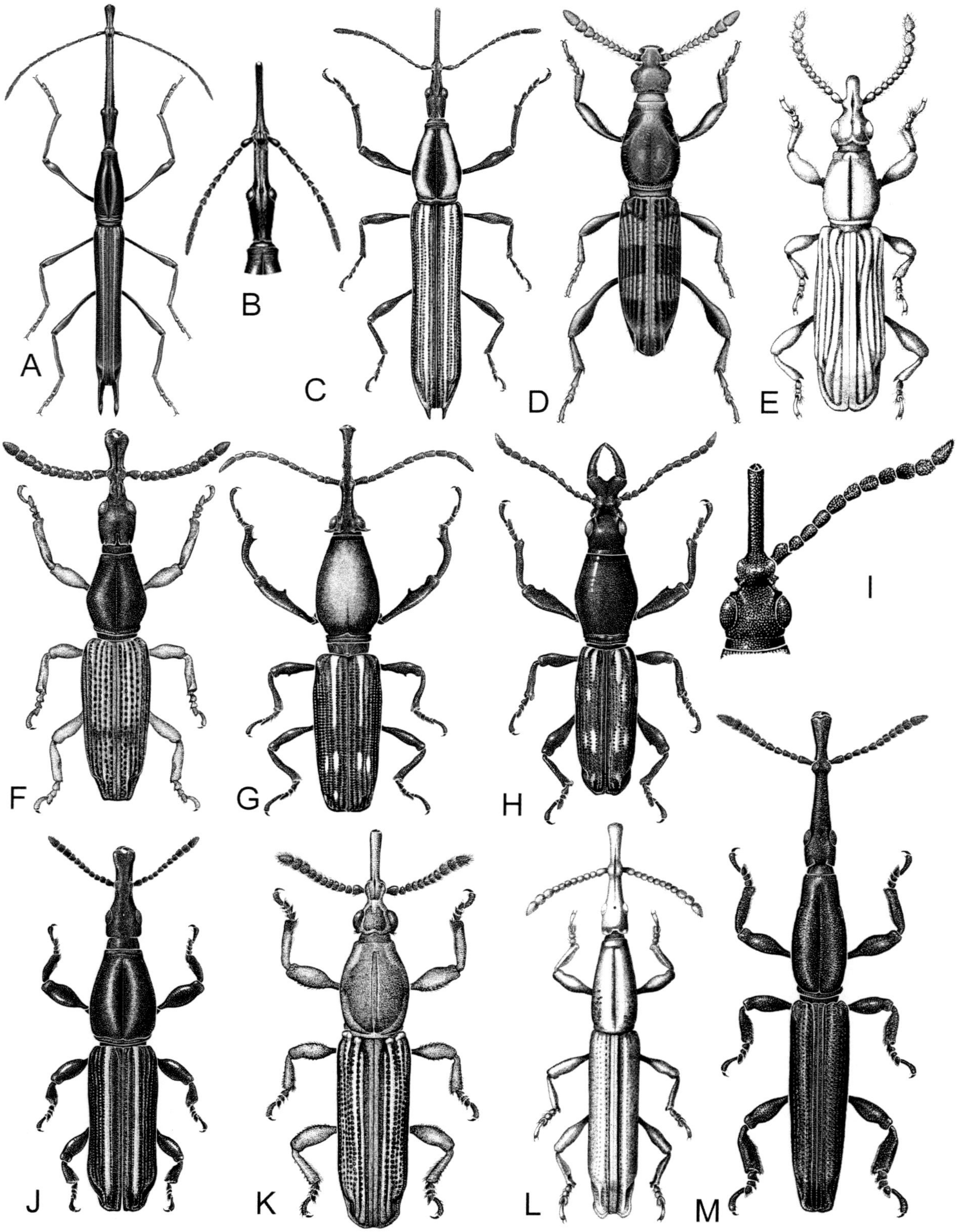

Fig. 43.4. Brentinae, habitus and head structures. A–B, *Ithystenus hollandiae* (Boisduval): A, habitus; B, head, ♀; C, *Mesetia amoena* Blackburn; D, *Mesoderes guttatus* (Kleine); E, *Microtrachelizus montrouzieri* Senna; F, *Miolispa Australiana* Senna; G, *Orychodes digramma* (Boisduval); H–I, *Schizoeupsalis promissa* (Pascoe): H, habitus, ♂; I, head, ♀; J, *Stenobrentus lineatus* Damoiseau; K, *Trachelizus bisulcatus* (Lund); L, *Tracheloschizus angulaticeps* (Senna); M, *Tracheloschizus dichrous* (Lacordaire). After Zimmerman (1994).

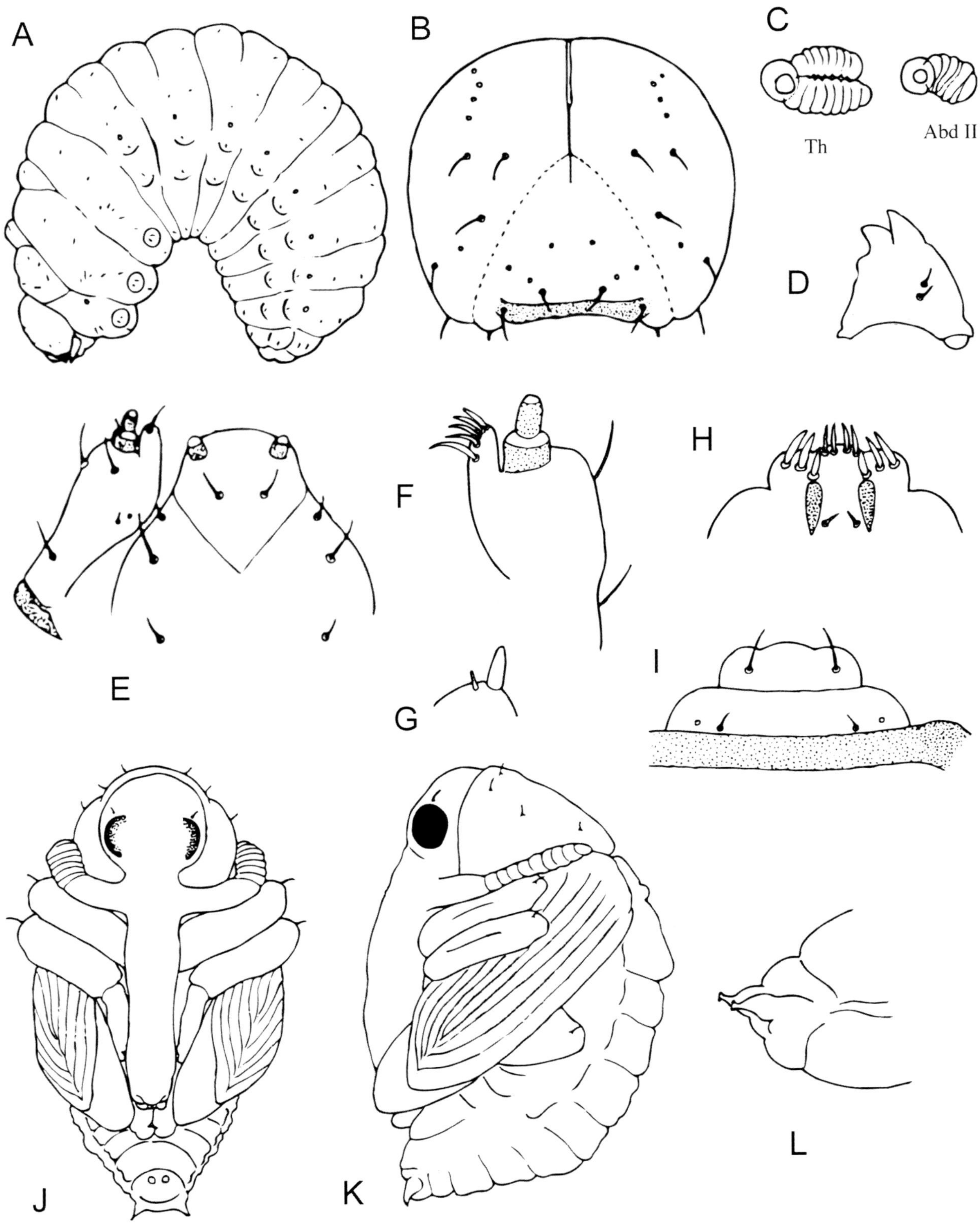

Fig. 43.5. Apioninae, structural details of immature stages, *Exapion ulicis* (Forster). A–I, larva: A, habitus, lateral; B, head, dorsal; C, spiracles, showing alignment; D, mandible; E, mala and labium, ventral; F, mala, dorsal; G, antenna; H, epipharyngeal lining; I, labrum, clypeus and epistoma; J–L, pupa: J, habitus, ventral; K, habitus, lateral; L, terminal segments, lateral. After May (1994).

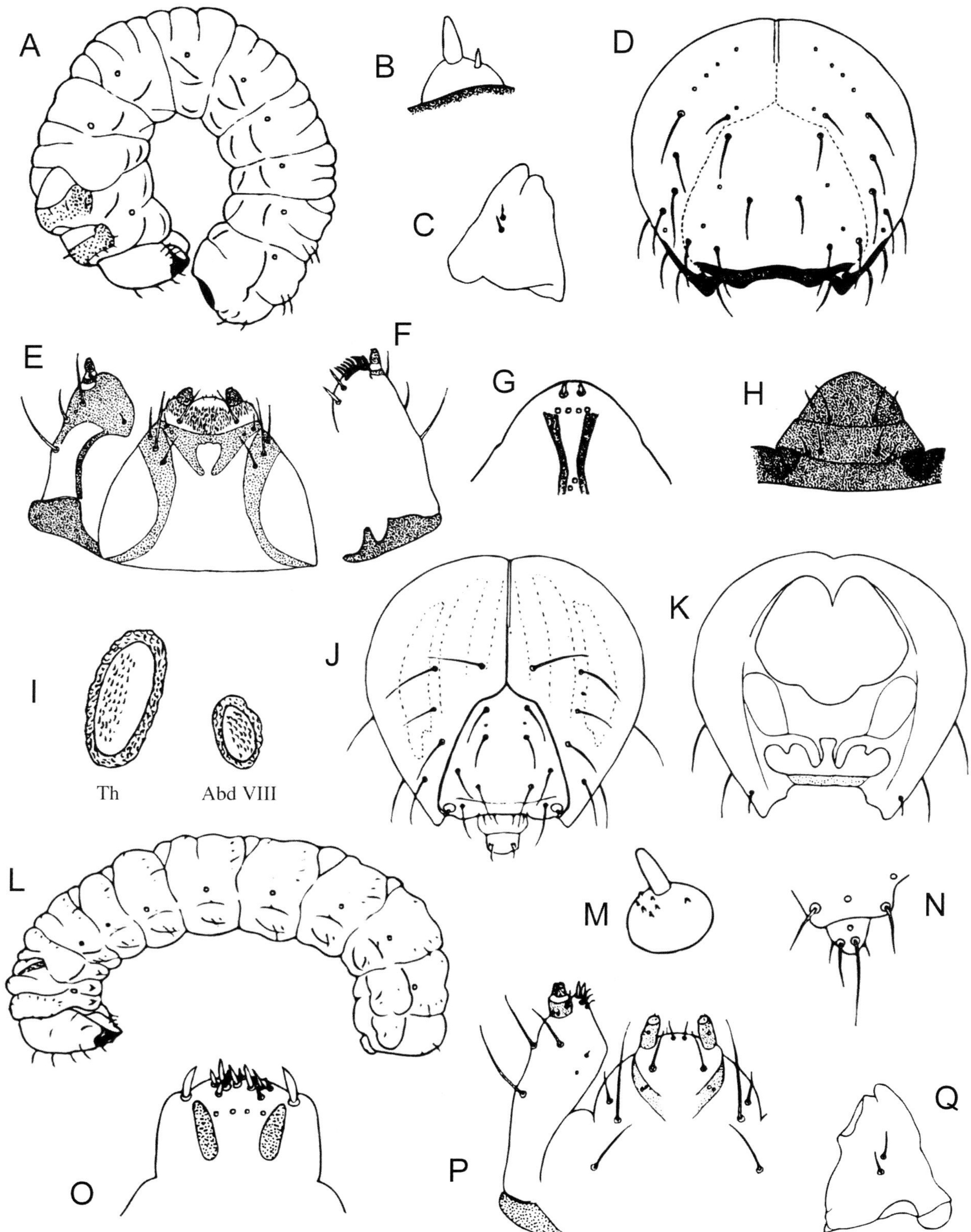

Fig. 43.6. Brentinae, structural details of larva. A–I, *Allaeometrus breviceps* Senna: A, habitus, lateral; B, antenna; C, mandible; D, head, dorsal; E, mala and labium, ventral; F, mala, dorsal; G, epipharyngeal lining; H, labrum, clypeus and epistome; I, spiracles, showing alignment; J–Q, *Cerobates ophthalmicus* (Pascoe): J, head, dorsal; K, head, ventral; L, habitus, lateral; M, antenna; N, leg; O, epipharyngeal lining; P, mala and labium, ventral; Q, mandible. After May (1994).

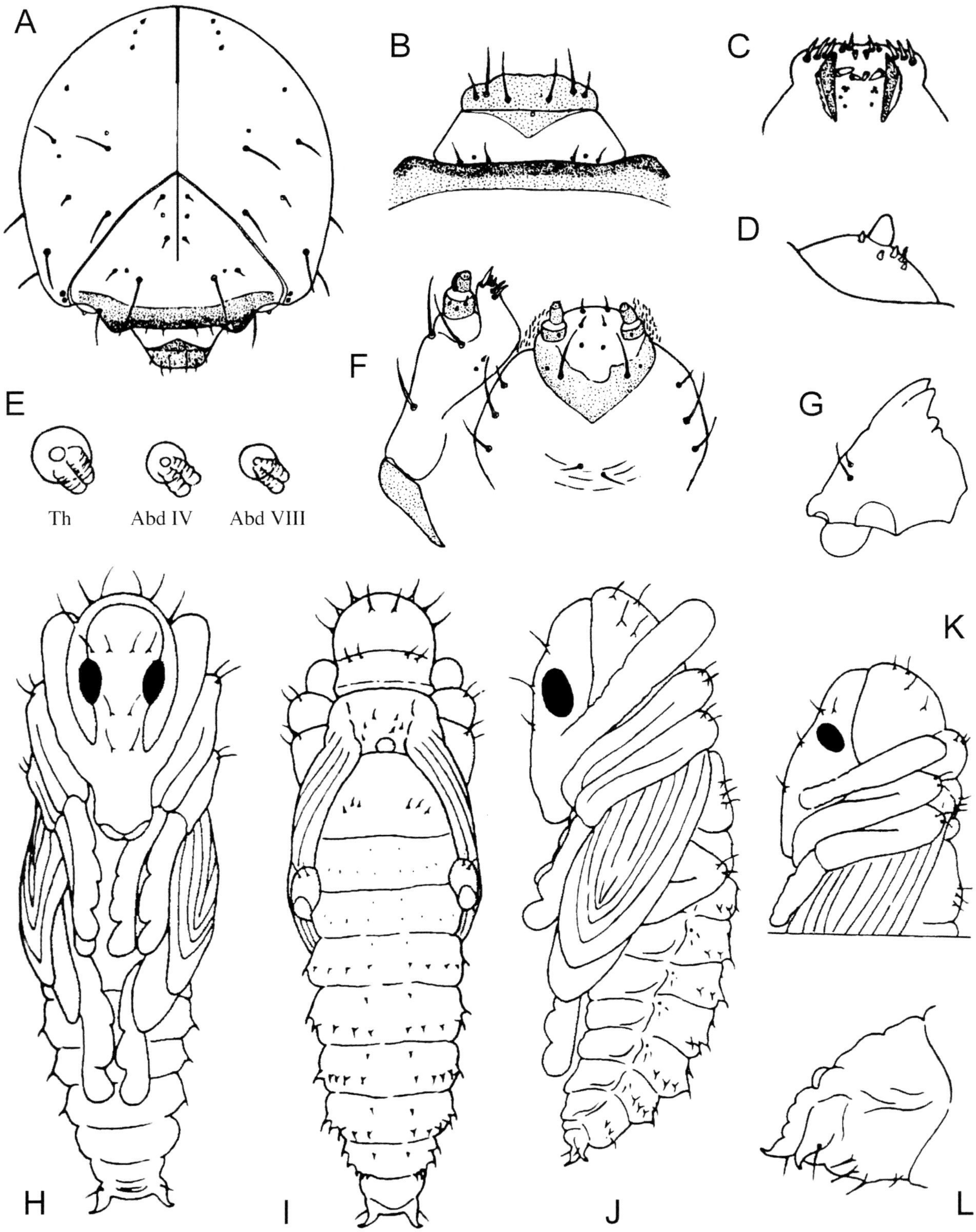

Fig. 43.7. Brentinae, structural details of immature stages, *Cylas formicarius* (Fabricius). A–G, larva: A, head, dorsal; B, labrum, clypeus and epistome; C, epipharyngeal lining; D, antenna; E, spiracles, showing alignment; F, mala and labium, ventral; G, mandible; H–L, pupa: H, female, ventral habitus; I, female, dorsal dorsal; J, male, lateral habitus; K, female, lateral habitus, anterior part; L, terminal segments, lateral. After May (1994).

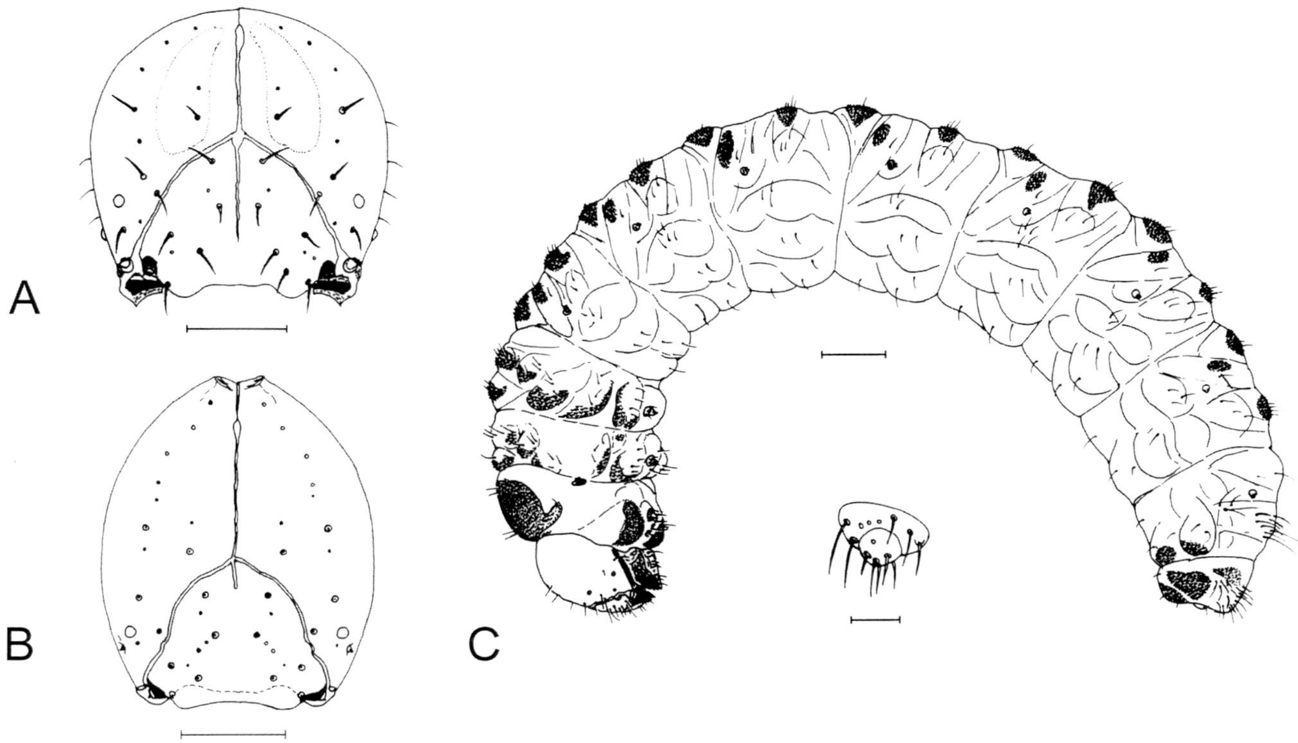

Fig. 43.8. Eurhynchinae, structural details of larva. A, *Eurhynchus laevior* (Kirby), head, dorsal; B, *Aporhina australis* (Heller), head, dorsal; C, *Eurhynchus laevior* (Kirby), habitus, lateral. After Oberprieler (2014).

Key to the Australian genera

Eurhynchinae and Brentinae modified after Zimmerman (1994).

1. Antennae 9-segmented, geniculate (Fig. 43.9A); scape >4 × as long as pedicel; scrobes long ...... **NANOPHYINAE** ... ***Austronanodes* Zimmerman**

– Antennae 10- or 11-segmented, non-geniculate (Fig. 43.9B–L); scape <3 × as long as pedicel; scrobes short ... **2**

2(1). Protibiae with apex expanded and dorsally deeply notched (Fig. 43.9P); mesoventrite separated by complete sutures from mesanepisterna ... **EURHYNCHINAE ... 4**

– Protibiae with apex not expanded or dorsally deeply notched; mesoventrite fused to mesanepisterna or separated from it by incomplete sutures ... **3**

3(2). Maxillary palps 3-segmented; labial palps 1- or 2-segmented, more or less concealed in pits; trochanters very short, trochanterofemoral joints strongly oblique; tibial spurs present; tarsal claws simple ... **BRENTINAE ... 6**

– Maxillary palps 2-segmented; labial palps 1-segmented, exposed; trochanters usually elongate, trochanterofemoral joints not or only slightly oblique; tibial spurs absent; tarsal claws often dentate ... **APIONINAE ... 42**

EURHYNCHINAE

4(2). Body vestiture forming patches on glabrous areas (Fig. 43.9J); rostrum strongly arcuate; profemora with subapical tooth on ventral surface; elytra each with single, large spine ... ***Aporhina* Boisduval**

– Body covered with vestiture, without large glabrous areas (Fig. 43.9H, K); rostrum slightly arcuate; profemora without tooth on ventral surface; elytra each with two spines or none ... **5**

5(4). Prosternum shorter than hypomeral lobes behind procoxae; elytra without spines or each with two spines near middle of interstriae 3 and 5 ... ***Eurhynchus* Schoenherr**

– Prosternum subequal in length to hypomeral lobes behind procoxae; elytra without spines ... ***Ctenaphides* Pascoe**

BRENTINAE

6(3). Elytra never blue; antennae 11- or (rarely) 9-segmented; procoxal cavities narrowly to broadly separate; procoxae globular; tarsal claws free ... **7**

– Elytra steely blue in colour; antennae 10-segmented (Fig. 43.9M); procoxal cavities confluent; procoxae conical, prominent; tarsal claws connate at base ... **CYLADINI ... *Cylas* Latreille**

7(6). Rostrum without tubercles between antennal insertions and eyes ... **8**

– Rostrum with tubercles between antennal insertions and eyes (Fig. 43.9N) ... **BRENTINI (part) ... 11**

8(7). Protibiae with large, deep, setose emargination on inner edge (Fig. 43.9U), base of emargination projected ..... ***Cerobates*** **Schoenherr**
– Protibiae not emarginate **9**
9(8). Prothorax laterally not emarginate for reception of femora when folded **10**
– Prothorax laterally emarginate for reception of femora when folded ..... **CYPHAGOGINI** ... **21**
10(9). Midline of pronotum without groove; procoxae widely separated; tarsal segment 3 deeply bilobed ..... **BRENTINI** (part) ... **11**
– Midline of pronotum with groove (Fig. 43.9F) or not, if absent, tarsal segment 3 not deeply bilobed or procoxae narrowly separated ..... **TRACHELIZINI** ... **30**
11(10). Tarsal segment 3 not bilobed **12**
– Tarsal segment 3 bilobed, usually distinctly wider than tarsal segment 2 ..... **14**
12(11). Rostrum with ventral surface without sulcus on each side of medial line (Fig. 43.9X) ..... ***Ankleineella*** **Zimmerman**
– Rostrum with ventral surface sulcate on each side of medial line (Fig. 43.9O) **13**
13(12). Head with ventral sulci confined to rostrum, curving upward to reach base of rostrum in front of eyes (Fig. 43.9W) ..... ***Kleineella*** **Strand**
– Head with ventral sulci continued behind eyes (Fig. 43.9O) ..... ***Cordus*** **Schoenherr**
14(11). Rostrum expanded towards apex and depressed (Fig. 43.9C), males **15**
– Rostrum parallel-sided and tubular (Fig. 43.9E), females **18**
15(14). Rostrum short and broad; mandibles widely separated at base, long and arcuate, projecting beyond apex of rostrum **16**
– Rostrum long and slender; mandibles small and inconspicuous **17**
16(15). Rostrum on each side of base with large gland opening in front of eye; elytra glabrous ..... ***Schizoeupsalis*** **Kleine**
– Rostrum on each side of the base without such gland opening; elytra with vestiture ..... ***Baryrhynchus*** **Lacordaire**
17(15). Antennae not extending beyond base of prothorax (Pl. 37K, L); head with a spine behind each eye; elytra glabrous ..... ***Orychodes*** **Pascoe**
– Antennae extending beyond base of prothorax (Fig. 43.9I); head lacking a spine behind each eye; elytra covered with vestiture ..... ***Ectocemus*** **Pascoe**
18(14). Elytra glabrous **19**
– Elytra with vestiture **20**
19(18). Rostrum with base at each side with a single, large gland opening ..... ***Schizoeupsalis*** **Kleine**
– Rostrum with base at each side with two gland openings ..... ***Orychodes*** **Pascoe**
20(18). Antennal segments notably elongate, segments 4 to 8 each as long or longer than 1 ..... ***Ectocemus*** **Pascoe**
– Antennal segments not so elongate, segments 4 to 8 each shorter than 1 ..... ***Baryrhynchus*** **Lacordaire**
21(9). Elytra with apex rounded, lacking caudal processes **22**
– Elytra with apex projected, forming pointed processes (Fig. 43.9D) ..... ***Hoplopisthius*** **Senna**
22(21). Antennae 9-segmented ..... ***Neosebus*** **Senna**
– Antennae 11-segmented **23**
23(22). Metafemora reaching elytral apices **24**
– Metafemora surpassing elytral apices **28**
24(23). Elytra with 9 or 10 distinct striae **25**
– Elytra appearing to have only 6 or 8 striae **27**
25(24). Head without ventral tubercles; mesocoxae separated by a distance smaller than width of mesofemur ..... ***Opisthenoxys*** **Kleine**
– Head with a tooth or spine between eye and scrobe; mesocoxae separated by a distance greater than width of mesofemur **26**
26(25). Pronotum somewhat flattened, with median groove; head and pronotum with coarse setae and scales; body length less than 4.0 mm ..... ***Isomorphus*** **Kleine**
– Pronotum strongly convex, medial line absent or shallow; head and pronotum with fine setae ..... ***Mesoderes*** **Senna**
27(24). Head long; eyes separated from postocular constriction for more than an eye length; prothorax forming a long neck; profemora not reaching anterior margin of prothorax; elytra each with apex emarginate; tarsal segment 3 deeply bilobed; tarsal segment 1 of meso- and metatarsi compressed laterally ..... ***Callipareius*** **Senna**
– Head short; eyes separated from postocular constriction by less than an eye length; prothorax not forming a long neck; profemora reaching anterior margin of prothorax; elytra with apex entire, convex; tarsal segment 3 not bilobed; tarsal segment 1 of meso- and metatarsi not compressed laterally ..... ***Allaeometrus*** **Senna**
28(23). Head barely constricted behind eyes; metafemora with large subapical ventral tooth (e.g. Fig. 43.9S, V); protarsi with tarsal segment 1 long and slender, as long as 2 and 3; prosternum shorter than hypomeral lobes behind procoxae ..... ***Catagogus*** **Kleine**
– Head clearly constricted behind eyes, forming a distinct neck; metafemora not dentate ventrally; protarsi with tarsal segment 1 stout, shorter than 2 and 3; prosternum about as long as hypomeral lobes behind procoxae **29**
29(28). Rostrum broad; postocular constriction separated from eye for about an eye length ..... ***Cyphagogus*** **Parry**
– Rostrum slender; postocular constriction separate from eye by less than an eye length ..... ***Ancylobrentus*** **Damoiseau**
30(10). Antennal segments long and slender; elytra each with a pointed projection **31**
– Antennal segments 4 to 8 short and broad; elytra without pointed projection **32**
31(30). Antennae inserted near middle of rostrum (Fig. 43.10A); metafemora exceeding apex of ventrite 5 ..... ***Ithystenus*** **Pascoe**
– Antennae inserted in apical third of rostrum (Fig. 43.10B); metafemora not exceeding apex of ventrite 2 ..... ***Mesetia*** **Blackburn**
32(30). Small slender species, body length <7.0 mm; tibial spur formula 1-2-2, protibial spur developed into a long, strong, arcuate process that extends towards tarsomere 1 ..... **33**
– Without the above combination **35**
33(32). Each side of head and base of rostrum elevated and covered with scales; antennae 9-segmented ..... ***Higonius*** **Lewis**
– Head not as above; antennae 11-segmented **34**
34(33). Rostrum in front of antennal insertions longer than head behind antennal insertions ..... ***Araiorrhinus*** **Senna**
– Rostrum in front of antennal insertions shorter than head behind antennal insertions ..... ***Microtrachelizus*** **Senna**

35(32). Body surface with vestiture, colour dark to black; tibial spur formula 2-2-2 .... **36**
– Body surface mostly glabrous, colour castaneous, at most with small velutinous areas; tibial spur formula 1-1-1, 1-2-2 or 2-2-2 .... **37**
36(35). Elytra with apex broadly rounded; postocular constriction shorter than an eye length .... ***Hormocerus* Schoenherr**
– Elytra each with apex projected as a caudiform process; postocular constriction longer than an eye length .... ***Uropteroides* Kleine**
37(35). Head with a tubercle behind eyes; elytra with interstria 2 only developed on apical third .... ***Euschizus* Kleine**
– Head without tubercle behind eyes; elytra with interstria 2 complete throughout length **38**
38(37). Elytra with interstria 2 absent at basal margin of elytra **39**
– Elytra with interstria 2 present at basal margin of elytra **40**
39(38). Forehead as wide as base of rostrum; median groove of pronotum deep and wide, extending beyond middle; elytral humeri protuberant, disc flat .... ***Trachelizus* Schoenherr**
– Forehead narrower than base of rostrum; median groove of pronotum shallow, not reaching middle, elytral humeri absent (Fig. 43.10C), disc concave .... ***Howeius* Mantilleri**
40(38). Elytra with striae partly or weakly impressed, marked by lines of minute punctures on middle section **41**
– Elytra with striae clearly impressed, punctures coarse .... ***Miolispa* Pascoe**
41(40). Elytra with interstriae 2 and 3 on same level and of similar width, interstriae 3 yellow, contrasting with castaneous elytra; apices of meso- and metatibiae with wide, concave, shovel-like corbel-like area .... ***Stenobrentus* Damoiseau**
– Elytra with interstriae 2 narrower than 3 and depressed below level of 3, this of same colour as rest of elytra; apices of tibiae not modified as above .... ***Tracheloschizus* Damoiseau**

APIONINAE

42(3). Eyes flat, dorsal (Fig. 43.10G); pronotum pedunculate (Figs 43.9L, 43.10J); trochanters ca. half as long as femur (e.g. Fig. 43.10R) .... **MYRMACICELINI** ... ***Rhynolaccus* Guérin-Méneville**
– Eyes convex, lateral (Fig. 43.10H, I); pronotum with different shape; trochanters 0.2 × as long as femur **43**
43(42). Elytra with suture at apex straight, simple, with dovetailed lock (Fig. 43.10M); interstriae 10 present (except in *Diselmobius*), striae 9 and 10 well developed, joining far from apical corner of elytron, thus interstriae 10 wedge-shaped; mesocoxae contiguous (Fig. 43.10L); spermatheca with smooth surface **44**
– Elytra with suture at apex sinuous, with pocket-like lock (Fig. 43.10N, P); interstriae 10 absent (Fig. 43.10N, P); striae 10, when present, only between apices of equally arranged striae 1 and 2; mesocoxae separated by joining processes of meso- and metaventrite (Fig. 43.10K); spermatheca with transversely wrinkled surface .... **APIONINI** ... **57**
44(43). Body with distinct vestiture of unordered piliform scales or setae; elytra with 5 striae evident at base, between scutellum and humeral callus; elytral striae impressed and punctate **45**
– Body glabrous (Fig. 43.10D–F); elytra with 4 striae evident at base, between scutellum and humeral callus; inner striae (1–4) scratch-like, impunctate and not impressed, or striae partially reduced .... **RHADINOCYBINI** ... **46**
45(44). Body covered with long, appressed, whitish piliform scales; rostrum tooth-like expanded at base, around antennal insertions (Fig. 43.10H), scrobes absent .... **RHINORHYNCHIDIINI** ... ***Rhinorhynchidius* Voss**
– Body vestiture inconspicuous, consisting of short, curved, hair-like setae; rostrum subcylindrical in basal third and around antennal insertions, scrobes distinct, lateroventral .... ***Diselmobius* Wanat**
46(44). Rostrum expanded at base, thin and subparallel towards apex; antennal insertions basal or subbasal (Fig. 43.10F) **47**
– Rostrum cylindrical or conical; antennal insertions at basal third of rostrum **48**
47(46). Eyes very large, almost contiguous (Fig. 43.10F), ommatidia enlarged and convex; elytral striae largely reduced; parameres without projecting apical lobes .... ***Ommatocybus* Wanat**
– Eyes not enlarged, well separated, ommatidia of normal size; elytral striae complete; parameres apically with sclerotised rounded outer lobes and a pair of membranous thin inner processes .... ***Nanomyrmacyba* Wanat**
48(46). Abdominal ventrite 2 with declining margin broadly hollowed .... ***Coeliapion* Wanat**
– Abdominal ventrite 2 with declining margin simple **49**
49(48). Rostrum as long as or shorter than head, at most 2.1 × as long as wide (Fig. 43.10S), with deep subgenal sulci below antennal cavities in lateral view .... ***Brachycybus* Wanat**
– Rostrum longer than head and >2.5 × as long as wide, without subgenal sulci below antennal cavities **50**
50(49). Antennae short, funicle segments 2–7 triangular in outline; scutellar shield 1.5–3.0 × as long as wide, in minute species isodiametric; elytral striae 1 broadly outcurved, limiting course of striae 2 and reaching base of elytron at level of striae 3; inner elytral striae complete .... ***Notapion* Zimmerman**
– Antennae longer, funicle segments 2–7 rounded to subcylindrical; scutellar shield always subisodiametric; elytral striae 1 and 2 outcurved and joining at elytral base, or inner striae vestigial **51**
51(50). Body elongate, subcylindrical (Fig. 43.10D); elytra 1.9–2.2 × as long as wide, slightly convex; metaventrite >2 × as long as mesocoxae .... ***Myrmecapion* Wanat**
– Body more robust (Fig. 43.10E); elytra 1.3–1.8 × as long as wide, distinctly convex (Fig. 43.10S); metaventrite <2 × as long as mesocoxae **52**
52(51). Pronotum at least 1.2 × as long as wide with equally deep subapical and subbasal constrictions **53**
– Pronotum at most 1.1 × as long as wide, with weak subapical constriction **55**
53(52). Rostrum 1.20–1.25 × as long as head, inserted at lower part of head in profile; head transverse .... ***Micronotapion* Wanat**
– Rostrum at least 1.5 × as long as head, unless head distinctly elongate, inserted at upper or middle part of head in profile; head as long as wide or elongate **54**

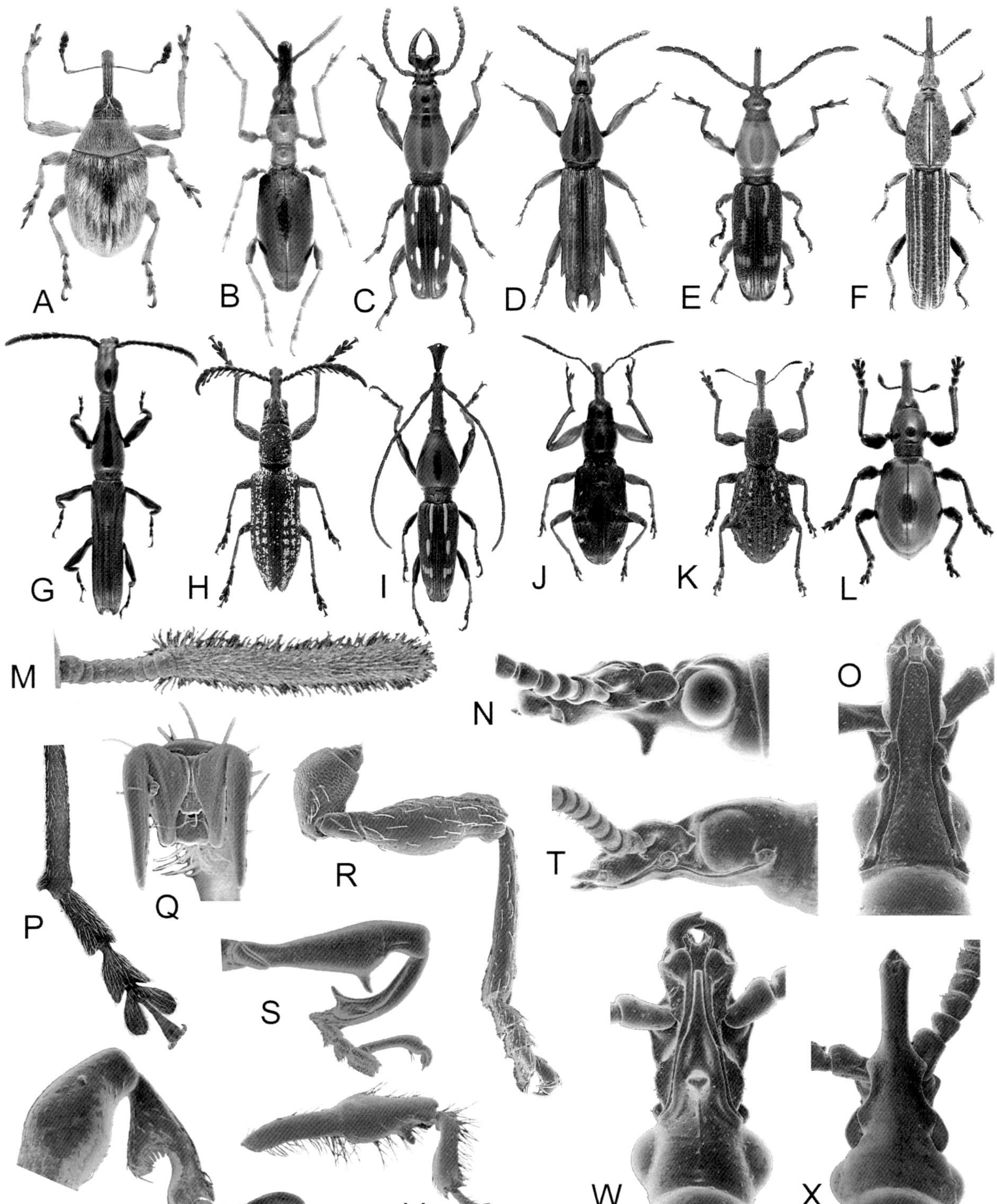

Fig. 43.9. Brentidae, characteristics for key to genera. A, *Austronanodes nigrovarius*, (Lea); B, *Cylas formicarius* (Fabricius); C, *Schizoeupsalis promissa* (Pascoe); D, *Hoplopisthius kolbei* Senna; E, *Baryrhynchus lineicollis* Power; F, *Hormocerus reticulatus* (Lund); G, *Callipareius planitarsus* (Perroud & Montrouzier); H, *Ctenaphides porcellus* Pascoe; I, *Ectocemus decemmaculatus* (Montrouzier); J, *Aporhina australis* (Heller); K, *Eurhynchus* sp.; L, *Rhynolaccus exsertus* (Pascoe); M, *Cylas* sp., antenna; N, *Ankleineella* sp., head, lateral; O, *Cordus* sp., head, ventral; P, *Eurhynchus* sp., emarginate tibial apex and tarsus; Q, *Apiotherium* sp., tarsal claws; R, *Apiotherium* sp., leg; S, *Orychodes* sp., leg; T, *Cordus* sp., head, lateral; U, *Cerobates* sp., leg; V, *Catagogus* sp., leg; W, *Kleineella* sp., head, ventral; X, *Ankleineella* sp., head, ventral.

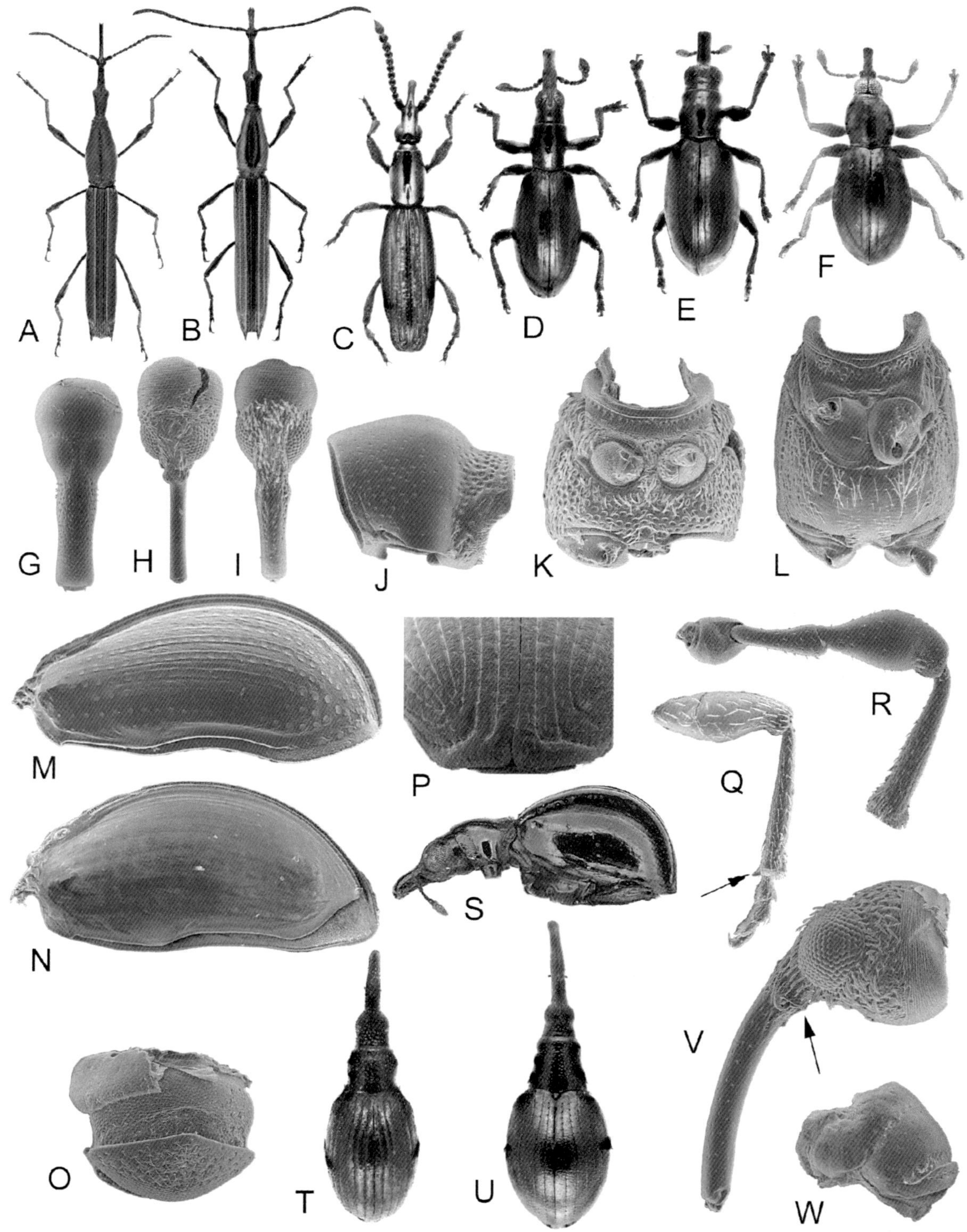

Fig. 43.10. Brentinae and Apioninae, characteristics for key to genera. A, *Ithystenus hollandiae* (Boisduval); B, *Mesetia amoena* Blackburn; C, *Howeius micropterus* Mantilleri; D, *Myrmecapion australicum* Wanat; E, *Micronotapion gibbiceps* Wanat; F, *Ommatocybus* sp.; G–I, head, dorsal: G, *Rhynolaccus* sp.; H, *Rhinorhynchidius* sp.; I, *Pseudapion* sp.; J, *Rhynolaccus* sp., prothorax, lateral; K, *Pseudapion* sp.,♀, pterothorax, ventral; L–M, *Rhinorhynchidius* sp.: L, pterothorax, ventral; M, elytron, inner lateral; N, *Pseudapion* sp., elytron, inner lateral, ♂; O, *Pseudopiezotrachelus* sp., pygidium; P, *Ceratapion onopordi* (Kirby), specimen from Europe, elytral apex, showing left-curved suture course; Q, *Apiotherium* sp., leg; R, *Rhynolaccus* sp., leg; S, *Brachycybus* sp., habitus, lateral; T, *Conapium* sp., habitus, dorsal; U, *Pseudopiezotrachelus* sp., habitus, dorsal; V, *Exapion* sp., head, lateral; W, *Apiotherium* sp., pygidium, lateral. P, after Wanat (2001).

54(53). Rostrum thin, porrect, widened between base and antennal insertions; antennae inserted at basal 0.20–0.35 part of rostrum; scapes at least 1.2 × as long as rostrum width at antennal insertions and slightly longer than two basal funicle segments combined .... ***Allonotapion*** **Wanat**

– Rostrum thicker, slightly to distinctly curved, slightly widened around antennal insertions, antennae inserted at basal 0.37–0.42 part of rostrum; scapes at most 1.1 × as long as rostrum width at antennal insertions and as long as 3–4 basal funicle segments combined .... ***Ganoapion*** **(*Ganoapion*) Wanat**

55(52). Rostrum with subgenal sulci invisible in profile, ventrally with numerous short erect setae between antennal cavities and scrobes; scapes orange-testaceous, well contrasting with almost black funicles and clubs; endophallus without flagellum .... ***Ganoapion*** **(*Proganoapion*) Wanat**

– Rostrum usually with punctate subgenal sulci visible in profile of prorostrum, ventrally without erect setae between antennal cavities and scrobes; antennae unicolorous; endophallus with flagellum **56**

56(55). Antennae thick; mesepimeral sulcus partially or completely reduced; abdominal ventrite 5 in male rounded apically as in female, pygidium entirely concealed; endophallus with short, subconical flagellum receiving strongly inflated ejaculatory duct; spermatheca very long and thin, with tail part irregularly curved in different plains or even looped, openings of duct and gland on corpus distant from each other .... ***Lissapion*** **Zimmerman**

– Antennae thinner; mesepimeral sulcus distinctly edged, at most slightly shortened; abdominal ventrite 5 of male broadly truncate apically, pygidium partially exposed; endophallus with long filiform flagellum, ejaculatory duct not inflated; spermatheca simply c-shaped, openings of duct and gland close together .... ***Alissapion*** **Wanat**

57(43). Elytral striae 1 and 9 confluent apically, stria 2 joining 8 or free **58**

– Elytral striae 1, 2 and 9 confluent apically **59**

58(57). Body vestiture on dorsum vestigial; prorostrum tapering, weakly arched; mesocoxae contiguous; elytral striae 2 and 8 disjunct apically; parameres with raised median crista at base .... ***Pseudorhinapion*** **Voss**

– Body vestiture on dorsum distinct, consisting of short, sparse, white piliform scales; prorostrum subcylindrical, distinctly curved; mesocoxae narrowly separate; elytral striae 2 and 8 confluent apically; parameres lacking raised crista .... ***Pseudoconapion*** **Voss**

59(57). Integument with distinct aureocupreous shine; mesocoxae contiguous .... ***Perapion*** **Wagner**

– Integument testaceous to black, lacking metallic shine; mesocoxae separate by joined processes of meso- and metaventrite **60**

60(59). Rostrum long, thin, cylindrical, base with sharp tooth-like expansion behind antennal cavities (Fig. 43.10V) .... ***Exapion*** **Bedel**

– Rostrum without tooth-like basal expansion, antennae not inserted at extreme base of rostrum **61**

61(60). Basal tarsomere of pro- and mesotarsi in male with spine, absent on metatarsi; elytra light reddish-castaneous with darkened suture and outer intervals .... ***Coelocephalapion*** **Wagner**

– Pro- and mesotarsi in male unarmed or, if armed then entire elytra black **62**

62(61). Body black, vestiture vestigial, if piliform scales present beneath eyes or on procoxae (one species) then all basal tarsal segments in male with ventral spine; hindwings with small radial window; pygidium in male half-exposed, with complete transverse sulcus (Fig. 43.10O); parameres with sclerotised median keel or crista at base **63**

– Body vestiture distinct, aggregated on thoracic epimera and episterna, often forming various spots and patterns on elytra; hindwings without radial window; pygidium in male concealed or at most one-fifth exposed; parameres different **64**

63(62). Pronotum campaniform in dorsal view (Fig. 43.10T) .... ***Conapium*** **Motschulsky**

– Pronotum trapeziform in dorsal view (Fig. 43.10U) .... ***Pseudopiezotrachelus*** **Wagner**

64(62). Scutellar shield conspicuous, 2–3 × as long as wide, with 3 pointed tubercles prominent over elytra in lateral view **65**

– Scutellar shield variable, isodiametric to almost 2 × as long as wide, flat or with blunt tubercles not visible above elytra in lateral view **66**

65(64). Scales on anterior pronotal margin obliquely appressed; elytra with condensed white piliform scales around scutellar shield, derived from expanded and joined basal spots on interstriae 3 .... ***Lopatinapion*** **Friedman**

– Scales on anterior pronotal margin directed anteriad; elytra without scales around scutellar shield, those at bases of interstriae 3 inconspicuous .... ***Pseudaspidapion*** **Wanat**

66(64). Rostrum in male mostly orange-testaceous, paler than in female; tibiae in male unarmed or meso- and metatibiae with minute apical spine, hardly projecting from fringing setae, or metatibiae with dark lamelliform apical process, eventually accompanied by spine on basal metatarsomere; scutellar shield small, isodiametric, without tubercles; pygidium in male concealed .... ***Pseudapion*** **Schilsky**

– Rostrum usually concolorous in male and female; meso- and metatibiae in male always with clearly visible mucro differing in shape (Fig. 43.10Q), tarsal segments without ventral spine; scutellar shield isodiametric or elongate, with 2 or 3 blunt tubercles; pygidium in male with narrowly exposed ridge separated by transverse sulcus (Fig. 43.10W) .... ***Apiotherium*** **Beguin-Billecocq**

Classification of the Australian genera

Subfamily Eurhynchinae Lacordaire, 1863

Characteristics. Adults with protibiae with emarginate or notched apex; labial palps inserted submedially on ental surface of labium; membrane surrounding ovipositor with strip-like lateral sclerites. Larvae with corrugated head surface and densely setose maxillary malae and apex of prementum. (Zimmerman 1994; Kuschel 1995; Oberprieler 2000, 2004; Wanat 2001).

Distribution. The Eurhynchinae are a unique element of the Australian and New Guinea brentid fauna, comprising three genera (*Aporhina, Ctenaphides* and *Eurhynchus*) with 29 described species (Pullen *et al.* 2014). The classification, morphology and biology of the subfamily were treated by Zimmerman (1994), Oberprieler (2000, 2004, 2014) and Wanat (2001).

Aporhina **Boisduval, 1835 (Pls 3F, 40H; Figs 43.8B, 43.9J)**

Type species, by monotypy: *Aporhina bispinosa* Boisduval, 1835.

Australian species. *A. australis* (Heller, 1896), *A. pulchra* Oberprieler, 2004.

Distribution. Australia (eastern QLD, northeastern NSW), New Guinea.

Notes. *Aporhina pulchra* was recorded as *A. splendida* (Blackburn, 1894) by Zimmerman (1994), but the name *splendida* Blackburn is a junior homonym and the species was renamed as *A. pulchra* by Oberprieler (2004). In addition to the two species in Australia, *Aporhina* comprises 20 in New Guinea (Oberprieler 2004, 2014; Setliff 2007), many of them brightly metallically coloured. The Australian species occur in coastal habitats at elevations up to 1000 m above sea level. Only the larva of *A. australis* is known, which develops in twig galls on *Litsea leefiana* (Lauraceae) (Oberprieler 2000, 2014). A New-Guinean species has also been collected on *Litsea* and another on *Cryptocarya* (Oberprieler 2000, 2014), suggesting that Lauraceae may be the main or only hostplants of the genus.

Ctenaphides **Pascoe, 1870 (Pls 3E, 40I; Figs 43.2B, 43.9H)**

Type species, by monotypy: *Ctenaphides porcellus* Pascoe, 1870.

Australian species. *C. maculatus* (Pascoe, 1873), *C. porcellus* Pascoe, 1870.

Distribution. Australia (southwestern WA).

Notes. The characteristic feature of the genus are the conspicuously pectinate antennae of the males. In addition to the two described species, at least one undescribed one is known (Zimmerman 1994). The adults have been collected on *Banksia* and *Dryandra* (Proteaceae), but the larvae are still unknown (Oberprieler 2000, 2004, 2014).

Eurhynchus **Kirby, 1828 (Pl. 40J, K; Figs 43.2A, 43.8A, C, 43.9K, P)**

Type species, by original designation: *Eurhinus laevior* Kirby, 1819.

Australian species. *E. acanthopterus* Boisduval, 1835, *E. laevior* (Kirby, 1819), *E. quadridens* Erichson, 1842, *E. quadrituberculatus* Boheman, 1839 (*in* Schoenherr, 1839), *E. scabrior* (Kirby, 1819).

Distribution. Australia (QLD, NSW, VIC, SA, TAS).

Notes. In addition to the five currently recognised species (Zimmerman 1994), at least another six unnamed ones are known (Oberprieler 2000, 2014). The species are somewhat heterogeneous, as *E. laevior* is more similar in habitus, genitalia and larval hosts to *Ctenaphides* than to the other *Eurhynchus* species (Oberprieler 2014), and the genus thus requires revision. The larva is only known for *E. laevior*, which develops in elongate galls in woody branches of *Persoonia* (Proteaceae) (Froggatt 1896; Oberprieler 2000, 2004), and adults can be found on *Persoonia* as well as on *Lambertia* (also Proteaceae). Adults of the other species are generally collected on *Eucalyptus* (Myrtaceae), but their larvae and precise hostplants remain unknown.

Reference. Kirby & Spence (1828).

Subfamily Brentinae Billberg, 1820

Characteristics. Eye with facetted corneal layer covering central ommatidia, antennae with funicle and club of about the same width, prosternum longer than hypomeral lobes behind procoxae, scutellar shield not exposed, tibial spurs present, tarsal segment 5 longer than 2 and 3; claws divergent. (Kuschel 1995; Sforzi & Bartolozzi 2004; Sforzi *et al.* 2014).

Notes. Brentinae have a worldwide distribution in tropical and subtropical regions and are also well represented in Australia, with 31 genera and 79 species named (Pullen *et al.* 2014) but several additional species recognised (Zimmerman 1994). The classification of the subfamily into tribes is unsettled, as it was previously often treated as a family and then divided into several subfamilies and tribes. The Australian fauna is classified into four tribes, Brentini (with 28 species), Cyladini (1), Cyphagogini (33) and Trachelizini (19). Adult brentines are narrowly elongate with usually a slender rostrum, but the rostrum can be strongly sexually dimorphous and much shorter to almost absent in males of some Cyphagogini. The larvae generally bore in dead or dying wood of dicotyledonous angiosperms, but some have more specialised lifestyles such as mycetophagy, myrmecophily or predation. Adults of Cyphagogini have cylindrical, compact bodies with excavations for the retraction of the legs, which allow them to enter the tunnels of platypodine and scolytine ambrosia beetles, where both adults and larvae can prey on the ambrosia beetles and their brood (Sforzi & Bartolozzi 2004; Sforzi *et al.* 2014). The adults of *Ankleineella*, *Cordus* and *Kleineella* are often found in ant nests and have adaptations for a myrmecophilous lifestyle in possessing extraordinarily sculptured heads with ventral or lateral funnels for disseminating glandular fluids that appear to appease the host ants (Zimmerman 1994), but despite several observations on their natural history their larvae and life histories remain unknown. The only species of Cyladini in Australia, *Cylas formicarius* (Fabricius), is an introduction from the Oriental Region and often regarded as a serious pest of cultivated Sweetpotato (*Ipomoea batatas*).

A comprehensive overview of the biology and ecology of Brentinae is provided by Sforzi & Bartolozzi (2004).

Tribe Brentini Billberg, 1820

Ankleineella Zimmerman, 1994 (Pl. 37A, B; Fig. 43.11)

Type species, by original designation: *Amorphocephalus sulcicollis* Pascoe, 1872.

Australian species. *A. australis* (Lacordaire, 1865), *A. mniszechi* (Power, 1879a), *A. subscripta* (Damoiseau, 1979), *A. sulcicollis* (Pascoe, 1872).

Distribution. Australia (QLD, NSW, VIC, SA, NT, WA), New Guinea, Solomon Is., Indonesia.

Notes. Zimmerman (1994) provided a differential diagnosis between *Ankleineella* and *Kleineella* and a key to the species of the former. The species are probably associated with ants, but the larvae are unknown.

Baryrhynchus Lacordaire, 1865 (Pl. 37C, D; Figs 43.3G, H, 43.9E)

Type species, by subsequent designation (Kleine, 1938: 91): *Arrhenodes latirostris* Gyllenhal, 1833.

Australian species. *B. lineicollis* Power, 1879b, *B. poweri* Roelofs, 1879.

Distribution. Australia, Melanesia, Southeast Asia.

Notes. The genus is represented in Australia by the subgenus *Eupsalomimus* Kleine, 1916a, with apparently two species (*B. poweri* recorded by Sforzi & Bartolozzi (2004) but in need of confirmation). The larvae are wood-borers.

Cordus Schoenherr, 1847 (Pl. 37E, F, G; Figs 43.3A, 43.9O, T, 43.12)

Type species, by original designation: *Cordus hospes* Schoenherr, 1847.

Australian species. *C. acutipennis* Power, 1879a, *C. demarzi* Damoiseau, 1980, *C. festus* Damoiseau, 1980, *C. firmus* Damoiseau, 1980, *C. ganglbaueri* Senna, 1919, *C. hospes* Schoenherr, 1847, *C. laevis* Damoiseau, 1980, *C. occidentalis* Damoiseau, 1980, *C. pascoei* Power, 1879a, *C. plumipennis* Damoiseau, 1980, *C. queenslandicus* Senna, 1919, *C. recticornis* Damoiseau, 1980, *C. rostralis* Damoiseau, 1980, *C. schoenherri* Power, 1879a, *C. ultimus* Damoiseau, 1980, *C. vermiculatus* Damoiseau, 1980, *C. vicinus* Senna, 1919.

Distribution. Australia (QLD, NSW, VIC, SA, NT, WA, TAS), New Guinea, Solomon Is., Borneo.

Notes. *Cordus* is the largest brentine genus in Australia with 17 described species and another seven unnamed ones (Zimmerman 1994). *Cordus ganglbaueri*, *C. hospes* and *C. pascoei* are widely distributed in Australia but other species known from a few localities only. Specimens are often collected at light. Zimmerman (1994) illustrated the heads of several species and commented on their remarkable modifications, which include three ventrolateral gland openings and long channels on the underside, probably related to their association with ants. Associations with ants are recorded for *C. ganglbaueri* (*Iridomyrmex nitidus*), *C. hospes* (*Aphaenogaster longiceps*, *Camponotus*, *Crematogaster laeviceps*, *Iridomyrmex nitidus*), *C. schoenherri* (*Iridomyrmex nitidus*) and *Cordus* 'new species 1' (*Iridomyrmex conifer*). The larvae of the genus remain unknown, however.

Ectocemus Pascoe, 1862 (Pl. 37I, J; Fig. 43.9I)

Type species, by monotypy: *Ectocemus wallacei* Pascoe, 1862.

Australian species. *E. decemmaculatus* (Montrouzier, 1855).

Distribution. Australia (QLD), New Caledonia, Melanesia, India, Southeast Asia.

Notes. Of the five species of the genus, only the widely distributed *E. decemmaculatus* occurs in Australia. The larva of *E. cinnamomi* (Herbst, 1783) was described by Gardner (1935).

Kleineella Strand, 1918 (Pl. 37H; Fig. 43.14E–L)

Type species, by monotypy: *Mastax barbata* Kleine, 1918.

Australian species. *K. barbata* (Kleine, 1918).

Distribution. Australia (QLD), New Guinea, Solomon Is, Polynesia, Indonesia.

Notes. The name of the genus has been confused, given either as *Kleineella* Strand, 1918 (Zimmerman 1994, Alonso-Zarazaga & Lyal 1999, Pullen *et al.* 2014) or *Kleinella* Kleine, 1917a (Sforzi & Bartolozzi 2004). *Kleinella* Kleine, 1917a has priority over *Kleineella* Strand, 1918 but is a junior homonym of *Kleinella* Adams, 1860 (Mollusca), making *Kleineella* Strand, 1918 (a replacement name for the junior homonym *Mastax* Kleine, 1918) the valid name. The genus includes six species, but only *K. barbata* occurs in Australia. The ventral side of its head is modified in a similar way as in *Cordus*, with a pair of convergent grooves extending from the anterior margin of the eyes towards the postmentum (Zimmerman 1994).

Orychodes Pascoe, 1862 (Pl. 37K, L; Figs 43.4G, 43.9S, 43.13N)

Type species, by subsequent designation (Lacordaire, 1865: 433): *Brentus serrirostris* Lund, 1800.

Australian species. *O. andrewsi* Gahan, 1900, *O. digramma* (Boisduval, 1835).

Distribution. Australia (QLD, NSW, Christmas Is.), Melanesia, Polynesia, Asia.
Notes. *Orychodes digramma* is widespread in the New Guinea region. Its larvae bore in wood, but no specific hostplants have been recorded for it (Zimmerman 1994).

Schizoeupsalis **Kleine, 1917 (Pls 3G, 38A, B; Figs 43.4H, I, 43.9C)**

Type species, by subsequent designation (Zimmerman, 1994: 124): *Eupsalis promissa* Pascoe, 1872.
Australian species. *S. promissa* (Pascoe, 1872).
Distribution. Australia (QLD, NT), New Guinea, Indonesia.
Notes. The genus only includes the widely distributed and highly variable species *S. promissa*.
Reference. Kleine, 1917b.

Tribe Cyladini Schoenherr, 1823

Cylas **Latreille, 1802 (Pl. 38C; Figs 43.7, 43.9B, M, 43.13D–M)**

Type species, by monotypy: *Brentus brunneus* Olivier, 1790.
Australian species. *C. formicarius* (Fabricius, 1798).
Distribution. Australia (QLD, NT, NSW), New Zealand, Asia, Africa, Central and South America.
Notes. *Cylas* currently includes 24 species, mostly in Africa but with a few in the Oriental region or with a tropicopolitan distribution (Wolfe 1991; Alonso-Zarazaga 2004). Its unique morphological characteristics have resulted in a chequered classificatory history, but its closest affinities have been shown to lie with Brentinae (Morimoto 1976; Wanat 2001; Shin *et al.* 2018), in which it is now usually classified as a monotypic tribe (Pullen *et al.* 2014; Sforzi *et al.* 2014). The most notorious species is *C. formicarius*, the Sweetpotato Weevil, which is regarded as a serious pest of Sweetpotato (*Ipomoea batatas*) and has been widely dispersed throughout the tropics, including to Australia (Wolfe 1991; Brookes *et al.* 2019). Both its adults and larvae feed in living sweetpotato stems and tubers and can cause major crop losses.

Tribe Cyphagogini Kolbe, 1892

Allaeometrus **Senna, 1903 (Pl. 38E, Figs 43.3B, 43.6A–I)**

Type species, by original designation: *Allaeometrus breviceps* Senna, 1903.
Australian species. *A. breviceps* Senna, 1903.
Distribution. Australia (QLD), New Guinea, Southeast Asia including India and China.
Notes. The genus includes only two species, the widely distributed *A. breviceps* and *A. setosus* (Schönfeldt 1906), the latter known only from New Guinea. The larvae develop in wood, with *Shorea robusta* (Dipterocarpaceae) recorded as a larval host in India (Beeson 1941).

Ancylobrentus **Damoiseau, 1965 (Pl. 38D; Fig. 43.3C)**

Type species, by monotypy: *Ancylobrentus australicola* Damoiseau, 1965.
Australian species. *A. australicola* Damoiseau, 1965.
Distribution. Australia (QLD).
Notes. This monotypic genus is only known from Australia and from a few specimens (Zimmerman 1994).

Araiorrhinus **Senna, 1893 (Pls 4G, 39E; Fig. 43.3M)**

Type species, by subsequent designation (Kleine, 1938: 71): *Araiorrhinus longirostris* Senna, 1893b.
Australian species. *A. howittii* (Pascoe, 1872), *A. zimmermani* Mantilleri, 2011.
Distribution. Australia (QLD, NSW, Lord Howe Is., Norfolk Is., VIC), New Guinea, Southeast Asia.
Notes. Mantilleri (2011) described a second species of the genus from Australia, and Mantilleri (2012) subsequently revised the genus and provided a key to its ten species. The larvae have been reported to bore in dead trees.

Callipareius **Senna, 1892 (Pl. 38F; Fig. 43.9G)**

Type species, by monotypy: *Callipareius feae* Senna, 1892.
Australian species. *C. planitarsus* (Perroud & Montrouzier, 1865).
Distribution. Australia (QLD), New Caledonia, Vanuatu, Asia, Africa.
Notes. The genus includes 14 species (Sforzi & Bartolozzi 2004), but only one occurs in Australia, first recorded from there by Zimmerman (1994). Nothing is known about its hostplants and life history, but specimens of *C. japonicus* (Nakane 1963) have been reported developing in (or emerging from) dead wood of *Ardisia sieboldii* (Primulaceae), *Machilus thunbergii* (Lauraceae), *Mallotus japonicus* (Euphorbiaceae), *Morus australis* (Moraceae) and *Stewartia pseudocamellia* (Theaceae), in some cases collected nearby or from galleries of other weevils (Toki & Yoshitake 2012).

Catagogus **Kleine, 1926 (Pl. 38G; Figs 43.1, 43.3K, L, 43.9V, 43.13Q)**

Type species, by original designation: *Catagogus desidiosus* Kleine, 1926.
Australian species. *C. diorymerus* (Lea, 1899).

Distribution. Australia (QLD, NSW).

Notes. The genus only includes one Australian species as known (Zimmerman 1994). Its cylindrical body with retractable legs is adapted to invade the galleries of ambrosia beetles.

Cerobates Schoenherr, 1840 (Pls 4C, 38H; Figs 43.3I, J, 43.6J–Q, 43.9U, 43.13A–C, P)

Type species, by original designation: *Brentus tristriatus* Lund, 1800.

Australian species. *C. grouvellei* Senna, 1893b, *C. laevipennis* Senna, 1896, *C. ophthalmicus* (Pascoe, 1872), *C. sexsulcatus* Motschulsky, 1858, *C. tristriatus* (Lund, 1800).

Distribution. Australia (QLD, NSW, Lord Howe Is.), Melanesia, Asia, Africa.

Notes. The genus contains more than 50 species with remarkable sexual dimorphism. *Cerobates laevipennis, C. sexsulcatus* and *C. tristriatus* are widely distributed from India to Melanesia and Australia. Gardner (1925, 1935) described and illustrated the larva of *C. tristriatus*, and Beeson (1941) provided a detailed description of the galleries constructed by the larvae of *C. tristriatus* and *C. crematus* Lacordaire, 1865.

Cyphagogus Parry, 1849 (Pl. 38I, J; Figs 43.2D–F, 43.13O)

Type species, by subsequent designation (Lucas, 1920: 219): *Cyphagogus westwoodi* Parry, 1849.

Australian species. *C. bipunctatus* Senna, 1897, *C. crassitarsus* Damoiseau, 1964, *C. delicatus* Lea, 1899, *C. eggersi* Kleine, 1921, *C. modiglianii* Senna, 1893c, *C. nigraustralis* Damoiseau, 1966b, *C. odewahnii* Pascoe, 1864, *C. tristriatus* Kleine, 1916b.

Distribution. Australia (QLD, NT, NSW, VIC, TAS), Melanesia, Indonesia, Asia (Japan, Korea), Madagascar.

Notes. The genus includes 64 species, and the eight recorded from Australia may be introductions (Zimmerman 1994). Based on observations by several authors (e.g. Beeson 1941; Browne 1961; H. Roberts in Thompson 1996), the adults invade the galleries of ambrosia beetles (Platypodinae and Scolytinae), where they and their larvae may prey on the original occupants.

Higonius Lewis, 1883 (Pl. 39H; Figs 43.2I, 43.14A–D)

Type species, by subsequent designation (Lucas, 1920: 329): *Higonius poweri* Lewis, 1883.

Australian species. *H. novenarius* Damoiseau, 1987.

Distribution. Australia (QLD, NSW), New Guinea, Southeast Asia.

Notes. After Mantilleri's (2009) revision of *Higonius*, the genus contains 17 species, of which only one occurs in Australia. Zimmerman (1994) erected a new genus, *Higonodes*, for this species, but from his phylogenetic analysis of morphological characters Mantilleri (2009, 2012) sunk this taxon to subgenus of *Higonius*.

Hoplopisthius Senna, 1892 (Pl. 38K; Figs 43.2J, 43.9D)

Type species, by monotypy: *Hoplopisthius trichemerus* Senna, 1892.

Australian species. *H. kolbei* Senna, 1893a.

Distribution. Australia (QLD), Asia.

Notes. The genus is represented in Australia only by one confirmed species, *H. kolbei*, which was placed in *Carcinopisthius* Kolbe by Zimmerman (1994), but in a review of these genera Mantilleri (2010) downgraded *Carcinopisthius* to a subgenus of *Hoplopisthius* (thus synonymising their names). The presence of *H. doriae* Senna in Australia, as reported by Schedl (1959) and followed by Sforzi & Bartolozzi (2004) and Pullen *et al.* (2014), was refuted by Zimmerman (1994) and also not confirmed by Mantilleri (2010).

Howeius Mantilleri, 2011 (Pl. 39I, Fig. 43.10C)

Type species, by original designation: *Howeius micropterus* Mantilleri, 2011.

Australian species. *H. micropterus* Mantilleri, 2011.

Distribution. Australia (Lord Howe Is.).

Notes. The only known species of the genus, *H. micropterus*, was illustrated by Zimmerman (1994) as 'Lord Howe Genus A'. It has partially fused elytra and was collected from leaf litter, pit falls and pan traps.

Isomorphus Kleine, 1916 (Pl. 38L)

Type species, by original designation: *Isomorphus unicolor* Kleine, 1916b.

Australian species. Two unnamed species.

Distribution. Australia (QLD, NT), New Guinea, New Caledonia, Indonesia, Taiwan.

Notes. Zimmerman (1994) recorded two *Isomorphus* species from Australia but could not name them; he suspected them to be introductions from Indonesia. The species from Queensland was illustrated by Zimmerman (1991).

Mesoderes Senna, 1898 (Pl. 39B, C; Figs 43.4D, 43.14O)

Type species, by subsequent designation (Lucas, 1920: 406): *Mesoderes sexnotatus* Senna, 1898.

Australian species. *M. guttatus* (Kleine, 1916b), *M. maculatus* Senna, 1898, *M. sexnotatus* Senna, 1898.

Distribution. Australia (QLD), New Guinea to India.

Notes. Three of the eight species of the genus occur in Australia, *M. guttatus* being the most common one. Nothing is known about their hostplants and life histories.

Microtrachelizus Senna, 1893 (Pl. 40A; Figs 43.4E, 43.14M, N)

Type species, by subsequent designation (Lucas, 1920: 417): *Trachelizus lyratus* Perroud & Montrouzier, 1865.

Australian species. *M. altostriatus* Mantilleri, 2011, *M. australicus* Mantilleri, 2011, *M. montrouzieri* Senna, 1903, *M. occultus* Kleine, 1935, *M. queenslandicus* Damoiseau, 1987.

Distribution. Australia (QLD), New Guinea, New Caledonia, Southeast Asia, Africa.

Notes. Zimmerman (1994) recorded seven species of *Microtrachelizus* from Australia, all but one, *M. montrouzieri* (as *M. laevis* Damoiseau), unnamed. In his revision of the genus, Mantilleri (2011) identified two of the latter (as *M. occultus* and *M. queenslandicus*) and described three as new (*Araiorrhinus zimmermani*, *M. altostriatus*, *M. australicus*). The genus now contains 42 species, for which Mantilleri (2011, 2012) provided identifications keys. The life histories of the Australian species are unknown, but the Indian *M. beneficus* Kleine, 1925 has been reported using the prepupal tunnels of a species of Cerambycidae for oviposition and larval development (Beeson 1941; Zimmerman 1994).

Reference. Senna, 1893b.

Neosebus Senna, 1903 (Pl. 39A)

Type species, by monotypy: *Neosebus bidentatus* Senna, 1903.

Australian species. *Neosebus* sp.

Distribution. Australia (QLD), Sumatra, Borneo.

Notes. *Neosebus* contains three described species and is one of the few genera of Australian Brentidae with 9-segmented antennae. Zimmerman (1992, 1994) reported and illustrated one specimen from the Iron Range as possibly belonging in this genus.

Tribe Trachelizini Lacordaire, 1865

Euschizus Kleine, 1922 (Pl. 39F, G; Fig. 43.2G, H)

Type species, by original designation: *Euschizus alatus* Kleine, 1922c.

Australian species. *E. dictatorius* Kleine, 1926, *E. internatus* (Pascoe, 1872), *E. tenuitarsis* (Pascoe, 1872).

Distribution. Australia (QLD, NSW, Norfolk Is.).

Notes. The genus only contains three Australian species, for which Zimmerman (1994) provided a key. Specimens of *E. internatus* have been associated with borings of *Eurhamphus fasciculatus* Shuckard (Molytinae: Orthorhinini) in trunks of *Araucaria cunninghamii* (Araucariaceae) (Zimmerman 1994).

Hormocerus Schoenherr, 1823 (Pl. 39J; Figs 43.2K, 43.9F)

Type species, by original designation: *Brentus reticulatus* Lund, 1800.

Australian species. *H. compressitarsus* (Senna, 1895), *H. fossulatus* Blackburn, 1896, *H. reticulatus* (Lund, 1800).

Distribution. Australia (QLD), Polynesia, Southeast Asia.

Notes. This widely distributed genus contains three species, all of which occur in Australia. A key to them was provided by Zimmerman (1994).

Ithystenus Pascoe, 1862 (Pls 4B, 39K; Figs 43.4A, B, 43.10A)

Type species, by monotypy: *Brenthus angustatus* Guérin-Méneville, 1831.

Australian species. *I. barbirostris* Kleine, 1920, *I. frontalis* Pascoe, 1862, *I. hollandiae* (Boisduval, 1835), *I. sabulosus* Kleine, 1919.

Distribution. Australia (QLD), Melanesia, Polynesia, Indonesia.

Notes. The genus comprises 44 species, most of them in New Guinea (Sforzi & Bartolozzi 2004). The differentiation of *I. barbirostris* and *I. sabulosus* (both also described from Australia) from *I. hollandiae* is unclear, Zimmerman (1994) having been unable to distinguish more than one species in Australia.

Mesetia Blackburn, 1896 (Pls 39L, 40A; Figs 43.4C, 43.10B)

Type species, by monotypy: *Mesetia amoena* Blackburn, 1896.

Australian species. *M. amoena* Blackburn, 1896.

Distribution. Australia (northern NSW).

Notes. Specimens of *M. amoena* are uncommon in collections. A specimen has been collected on blossoms of *Leptospermum* (Zimmerman 1994).

Miolispa Pascoe, 1862 (Pl. 40B; Fig. 43.4F)

Type species, by monotypy: *Miolispa suturalis* Pascoe, 1862.

Australian species. *M. Australiana* Senna, 1897, *M. novaguineensis* (Guérin-Méneville, 1831).

Distribution. Australia (QLD, NSW, VIC), New Guinea, Melanesia, Southeast Asia, Japan.

Notes. *Miolispa* is one of the largest genera of Brentinae, with 74 described species (Sforzi & Bartolozzi 2004; Goossens 2011). Zimmerman (1994) provided a key to the two Australian species. Nothing is known about their life histories.

Stenobrentus Damoiseau, 1966 (Pl. 40C, D; Fig. 43.4J)

Type species, by original designation: *Stenobrentus lineatus* Damoiseau, 1966a.

Australian species. *S. lineatus* Damoiseau, 1966a.

Distribution. Australia (QLD).

Notes. The genus is monotypic, containing only one Australian species as known.

Trachelizus Dejean, 1834 (Pl. 40E; Fig. 43.4K)

Type species, by subsequent designation (Schoenherr, 1840: 490): *Brentus bisulcatus* Lund, 1800.

Australian species. *T. bisulcatus* (Lund, 1800).

Distribution. Australia (QLD), New Guinea, Solomon Is., Southeast Asia, Japan.

Notes. The genus contains 15 described species (Sforzi & Bartolozzi 2004), only one widespread and variable species occurring in Australia (Zimmerman 1994). Its larva bores in wood (Beeson 1941).

Tracheloschizus Damoiseau, 1966 (Pl. 40F; Fig. 43.4L, M)

Type species, by original designation: *Schizotrachelus dichrous* Lacordaire, 1865.

Australian species. *T. altilis* (Kleine, 1922c), *T. angulaticeps* (Senna, 1899), *T. dichrous* (Lacordaire, 1865).

Distribution. Australia (QLD, NSW), Southeast Asia, Melanesia.

Notes. The genus includes 38 described species (Sforzi & Bartolozzi 2004), three of them in Australia together with four unnamed ones identified by Zimmerman (1994). *Tracheloschizus dichrous* appears to be restricted to Australia, but the other two species also occur further north in New Guinea and the Solomon Islands (*T. altilis*) and widely in Southeast Asia (*T. angulaticeps*). Nothing is recorded about their life histories.

Reference. Damoiseau, 1966b.

Uropteroides Kleine, 1922 (Pl. 40G)

Type species, by original designation: *Brenthus douei* Montrouzier, 1861.

Australian species. *U. gestroi* (Senna, 1894)

Distribution. Australia (QLD), New Guinea, New Caledonia, Polynesia.

Notes. *Uropteroides* contains two species, of which only one is recorded from Australia but also occurs widely New Guinea and surrounding islands (Zimmerman 1994). *U. douei* (Montrouzier 1861) have been also recorded from Australia by Sforzi & Bartolozzi (2004) but we have not been able to confirm it. The adults are frequently found on dead trees in rainforest.

Reference. Kleine, 1922b.

Brentinae not confirmed to occur in Australia

A species of *Stratiorrhina* Pascoe, 1872, *Stratiorrhina xiphias* (Westwood 1847), was reported as occurring in Australia by Sforzi & Bartolozzi (2004) and, following this work, by Pullen *et al.* (2014), but this is evidently an error caused by confusion of the type locality of *S. xiphias*, the Prince of Wales' Island (Penang Island) in Malaysia, with the Prince of Wales Island (Muralag) in the Torres Strait Archipelago of Australia. Calabresi (1921) in her revision of *Stratiorrhina* correctly cited Penang ('Is. Pr. Di Gales') as the distribution of *S. xiphias*.

A species of *Opisthenoxys* Kleine, 1922a, *O. ochraceus* Kleine, 1922a, was also reported as occurring in Australia by Sforzi & Bartolozzi (2004) and Pullen *et al.* (2014), apparently based on Zimmerman's (1994) note that a specimen of *O. ochraceus* had been intercepted in Adelaide with timber imported from Indonesia and that it may occur in Australia because of its wide distribution in Southeast Asia and New Guinea and the ease of its dispersal by trade in timber. We have, however, not found any specimen of the species taken in Australia and therefore do not regard it as occurring in the country. We nonetheless include the genus in the key to the genera of Brentinae and illustrate it (Pl. 39D) to allow its recognition.

Subfamily Apioninae Schoenherr, 1823

Apioninae are a cosmopolitan subfamily of approximately 205 genera and 2200 species (Alonso-Zarazaga & Wanat 2014). The Australian fauna is currently deemed to comprise 23 genera with 70 described species (Pullen *et al.* 2014; Wanat 2023), but a recent reclassification (Wanat & Escalona 2024) provides an estimate of more than 150 described and undescribed species, including those recorded by Zimmerman (1991, 1992, 1994). As with Brentinae, the classification of Apioninae is unsettled as the subfamily was also previously often treated as a family and then variously divided into subfamilies, supertribes and tribes. Four tribes are recognised as being represented in Australia (Pullen *et al.* 2014). Some genera of Apionini occurring in Australia (*Apiotherium*, *Conapium*, *Pseudaspidapion*, *Pseudoconapion*, *Pseudopiezotrachelus*, *Pseudorhinapion*) are widely distributed in New Guinea, the Pacific region and tropical Asia and Africa, but the tribes Myrmacicelini, Rhadinocybini and Rhinorhynchidiini are geographically much more restricted, the first to Australia and New Guinea, the second to the Australo-Pacific region

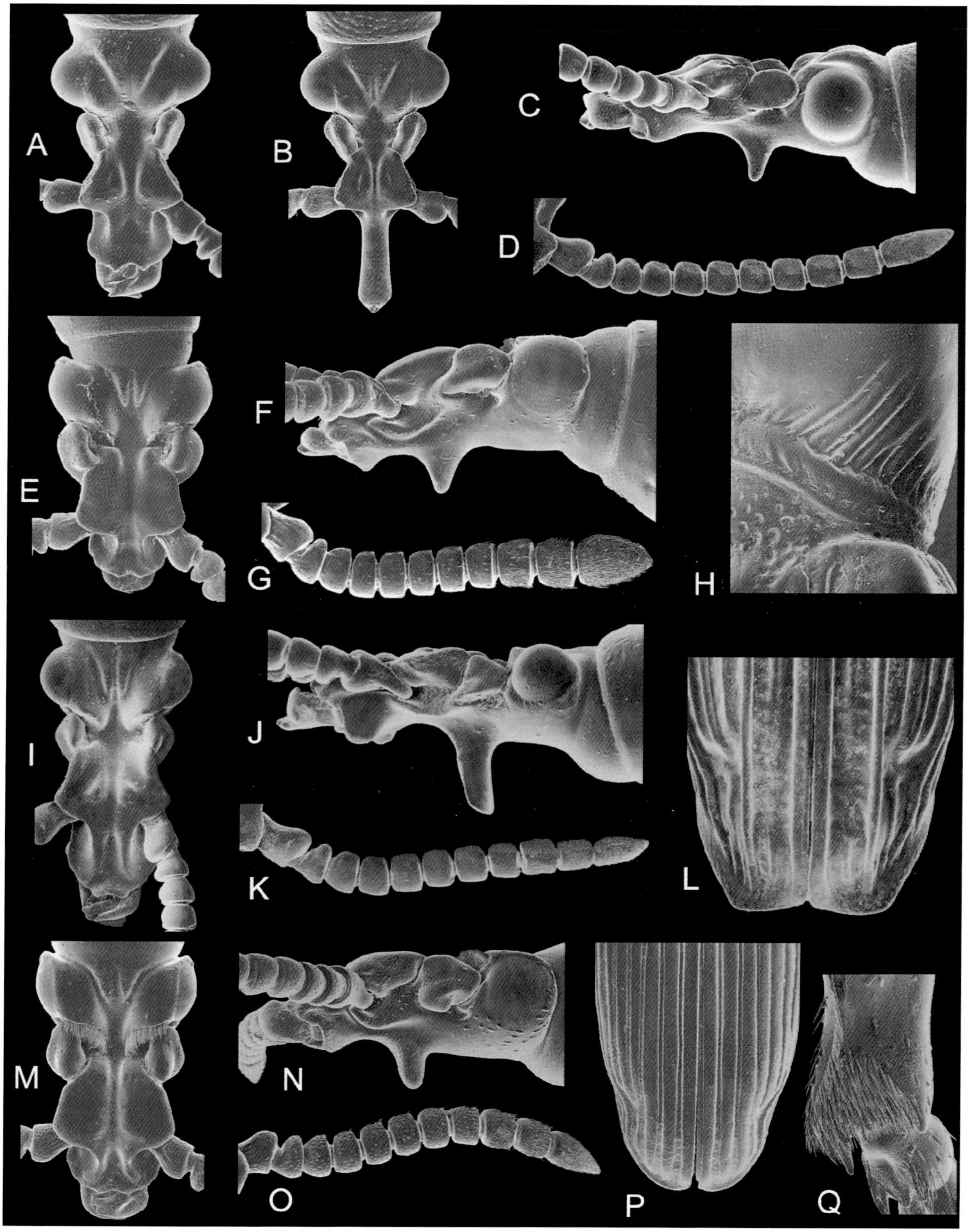

Fig. 43.11. Brentinae, structural details, *Ankleineella Zimmerman*. A–D, *A. australis* (Lacordaire): A, head, ♂, dorsal; B, head, ♀, dorsal; C, head, ♂, lateral; D, antenna, ♂; E–H, *A. mniszechi* (Power), ♂: E, head, dorsal; F, head, lateral; G, antenna; H, oblique pronotal sulci; I–L, *A. subscripta* (Damoiseau), ♂: I, head, dorsal; J, head, lateral; K, antenna; L, elytral apices; M–Q, *A. sulcicollis* (Pascoe), ♂: M, head, dorsal; N, head, lateral; O, antenna; P, elytral apices; Q, right protibia with antennal cleaner and spur, apicoventral.

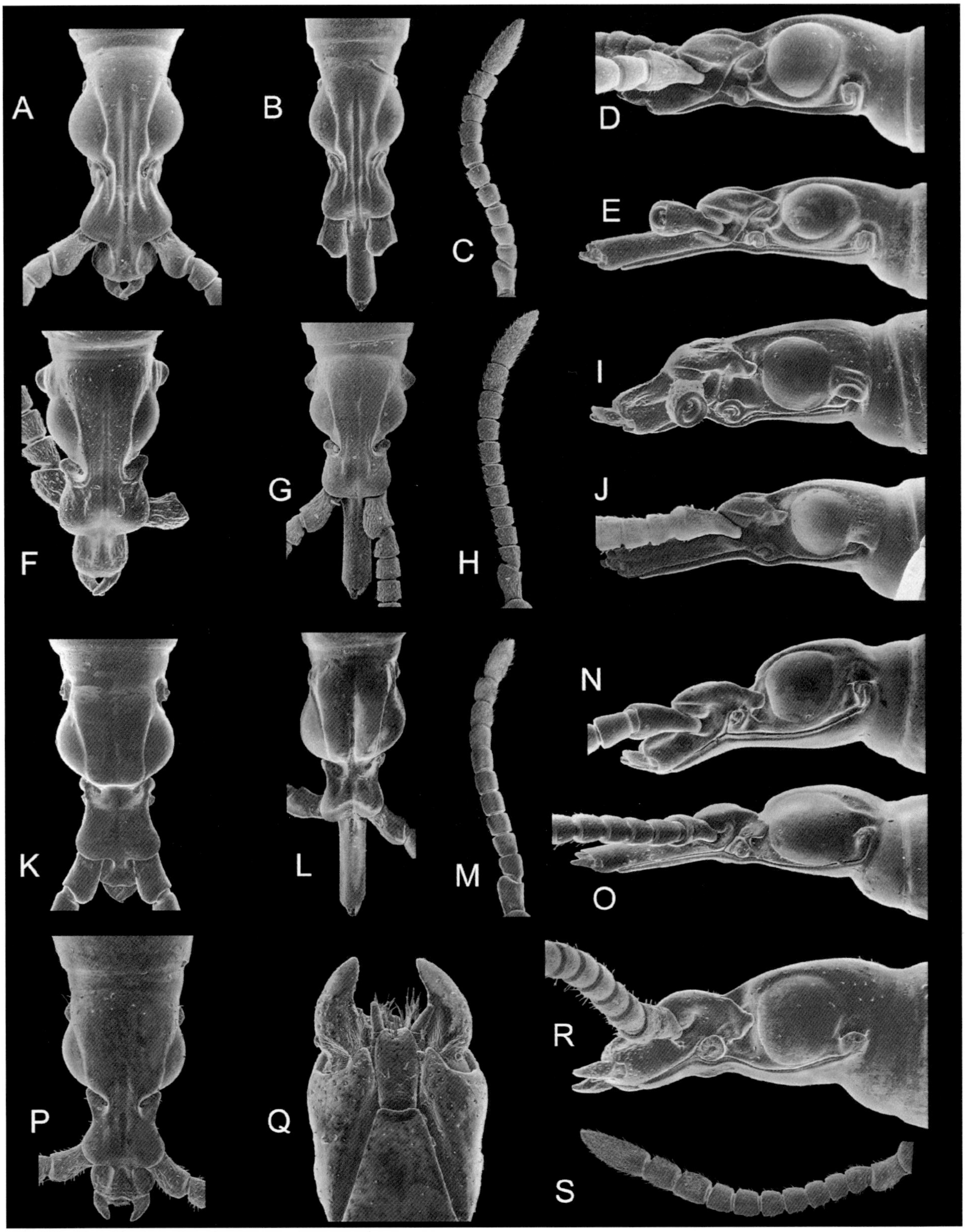

Fig. 43.12. Brentinae, structural details, *Cordus* Schoenherr. A–E, *C. hospes* Schoenherr: A, head, ♂, dorsal; B, head, ♀, dorsal; C, antenna; D, head, ♂, lateral; E, head, ♀, lateral; F–J, *C. queenslandicus* Senna: F, head, ♂, dorsal; G, head, ♀, dorsal; H, antenna; I, head, ♂, lateral; J, head, ♀, lateral; K–O, *C. vicinus* Senna: K, head, ♂, dorsal; L, head, ♀, dorsal; M, antenna; N, head, ♂, lateral; O, head, ♀, lateral; P–S, *C. occidentalis* Damoiseau, ♂: P, head, dorsal; Q, mouthparts; R, head, lateral; S, antenna.

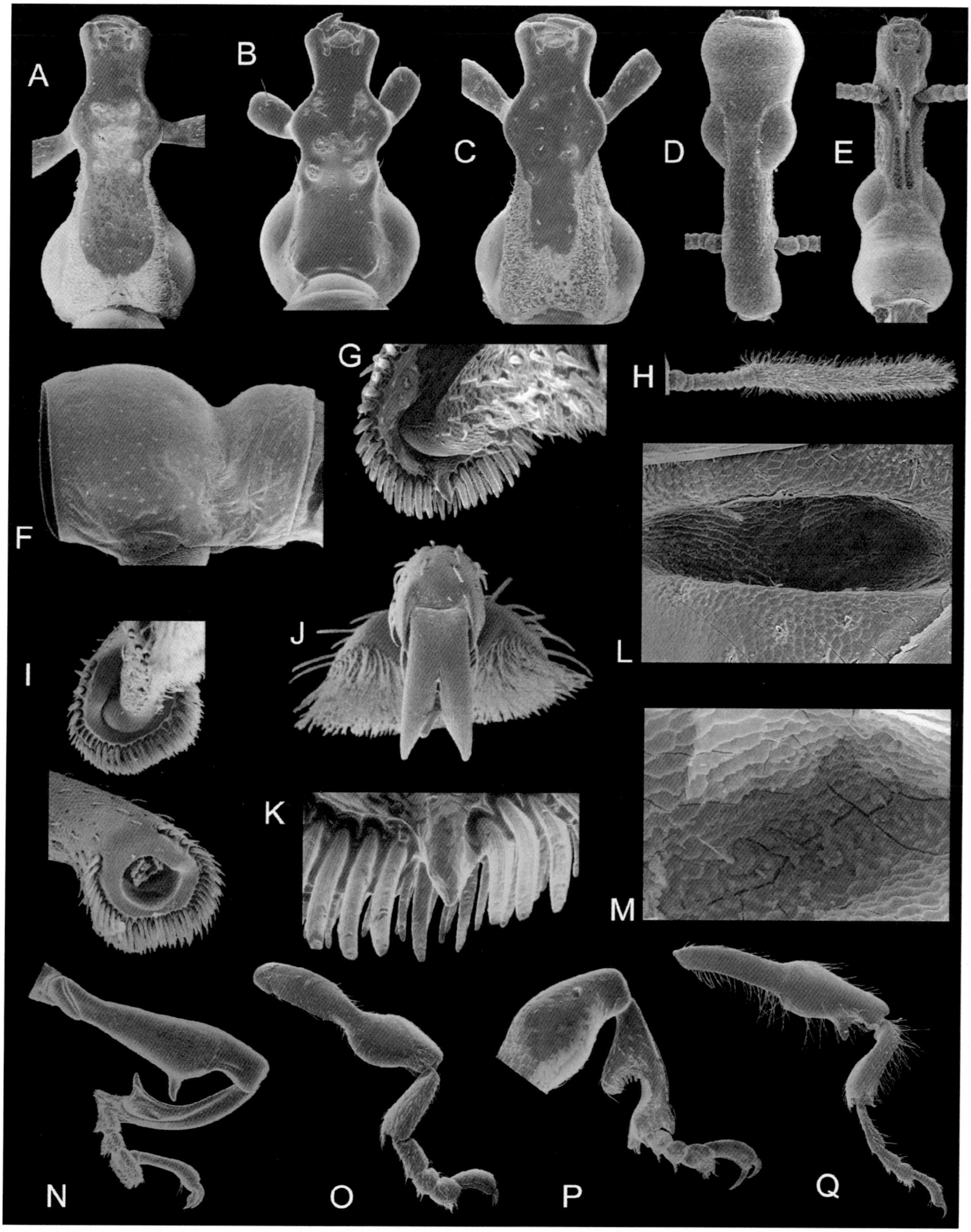

Fig. 43.13. Brentinae, structural details. A, *Cerobates laevipennis* Senna, head, ventral; B, *Cerobates sexsulcatus* Motschulsky, head, ventral; C, *Cerobates tristriatus* (Lund), head, ventral; D–M, *Cylas formicarius* (Fabricius): D, head, dorsal; E, head, ventral; F, prothorax, lateral; G, metatibial apex; H, antenna, ♂; I, tibial apices; J, tarsal claws; K, metatibial spur; L, mesothoracic concavity; M, mesothoracic pit; N, *Orychodes digramma* (Boisduval), foreleg; O, *Cyphagogus crassitarsus* Damoiseau, hindleg; P, *Cerobates laevipennis*, foreleg; Q, *Catagogus diorymerus* (Lea), hindleg.

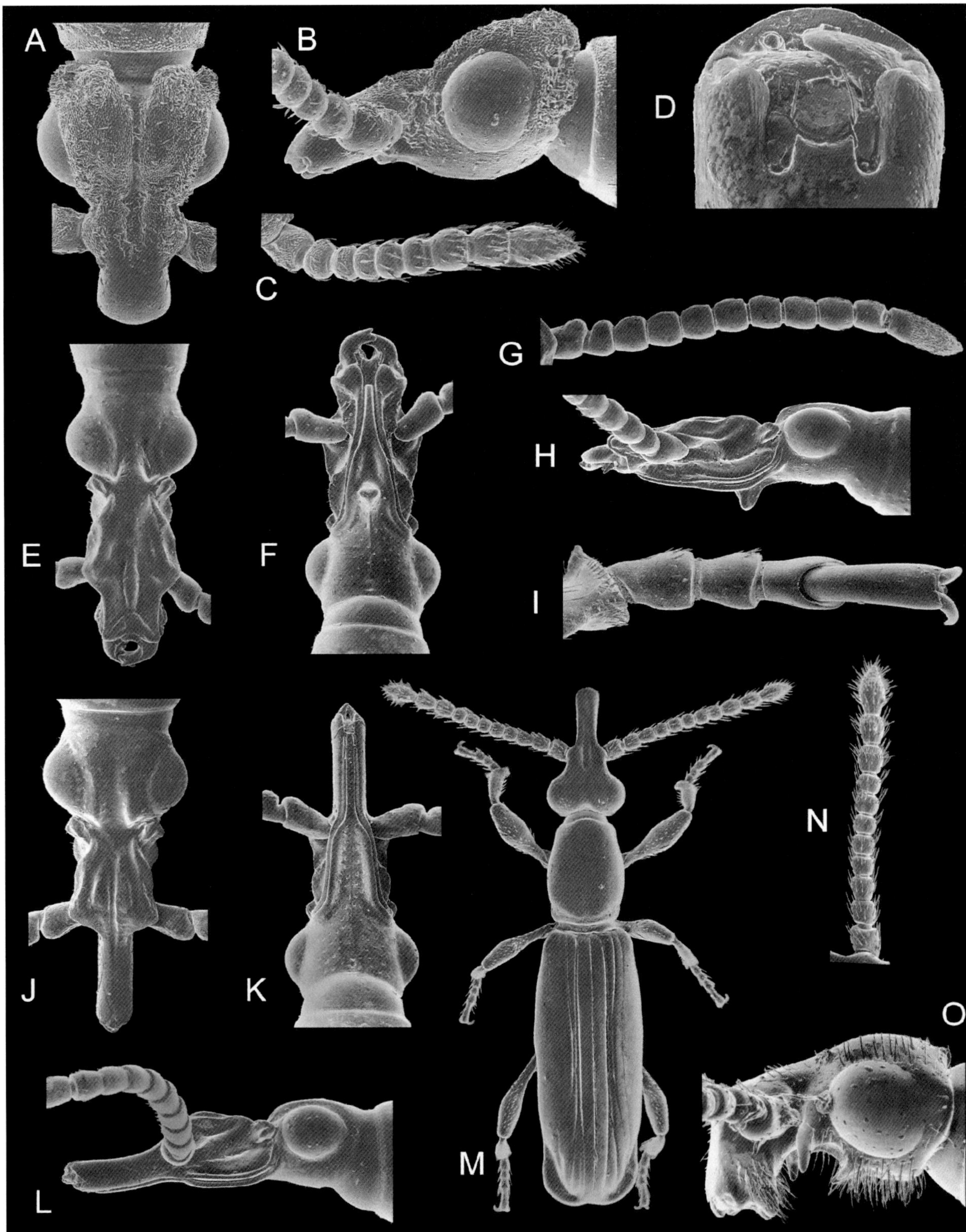

Fig. 43.14. Brentinae, structural details. A–D, *Higonius novenarius* Damoiseau: A, head, dorsal; B, head, lateral; C, antenna; D, mouthparts; E–I, *Kleineella barbata* (Kleine), ♂: E, head, dorsal; F, head, ventral; G, antenna; H, head, lateral; I, tarsi and claws, dorsal; J–L, *K. barbata* (Kleine), ♀: J, head, dorsal; K, head, ventral; L, head lateral; M–N, *Microtrachelizus* sp.: M, habitus, dorsal; N, antenna; O, *Mesoderes guttatus* (Kleine), head, lateral.

and the last to only Australia. The larvae of Apioninae develop mainly in stems but also in other green plant tissues, such as galls, fruits and seeds, of closely related plant species. In Australia their main hosts are angiosperms, predominantly Malvaceae, but two genera, *Rhinorhynchidius* and *Diselmobius*, are associated with the conifers *Callitris* and *Diselma* (both Cupressaceae), respectively. The monotypic Myrmacicelini develop in the galls induced by pteromalid wasps or fungi on various species of *Acaccia*. Several species of Apionidae are of economic importance, either as pests or as biological control agents. No Australian species is thus far regarded as a major pest, but a species of *Coelocephalapion* imported from South America has been used for the control of *Mimosa pigra* (Fabaceae) (Forno *et al.* 1994). Major contributions to the systematics of Apioninae were made by Kissinger (1968, 1992), Alonso-Zarazaga (1990) and Wanat (1995, 2001), and Alonso-Zarazaga & Wanat (2014) provided a comprehensive summary of the taxonomy, biology, morphology and phylogeny of the subfamily.

Tribe Apionini Schoenherr, 1823

Characteristics. The Apionini can be recognized by the presence of an internal interlocking area at the apex of the elytral suture, externally indicated by a left turn of the elytral suture in its apical part, and by the elytral striae 10 being reduced to an apical portion linking apices of 1 and 2. In addition, the male pygidium lacks inflection of apical margin, genital sheath lacks sclerite(s) between sternites VIII and IX or pouch-like invagination between tergites VII and VIII, and the endophallus is short, largely situated in the penis in repose and without flagellum, frena or any composite sclerotised transfer apparatus. In females the spermatheca has a thin wall (collapsing in glycerol or resins during preparation) with more or less dense transverse wrinkles or strigae on its surface.

This tribe comprises more than 85% of the world fauna and is considered a derived group of Apioninae (Wanat 2001). It is also the largest tribe of the subfamily in Australia, with 43 named species but many undescribed represented in collections (Pullen *et al.* 2014; Wanat & Escalona 2024). The Australian fauna is related to Oriental and Wallacean groups, often comprising genera or even species groups that are widespread in palaeotropical regions.

Apiotherium Beguin-Billecocq, 1905 (Pls 41D, 42G, 44A–E; Figs 43.9Q, R, 43.10Q, W, 43.15–43.22)

Type species, by subsequent designation (Kissinger, 1968: 29): *Apion suturale* Beguin-Billecocq, 1905.

Australian species. *A. amabile* (Lea, 1899), *A. anthidium* (Lea, 1899), *A. astri* (Lea, 1926), *A. binotatum* (Lea, 1899), *A. carpophagum* (Lea, 1899), *A. commersoniae* (Rheinheimer, 1997), *A. comosum* (Pascoe, 1874), *A. condensatum* (Lea, 1899), *A. congestum* (Lea, 1910a), *A. congruum* (Lea, 1899), *A. foveicolle* (Lea, 1899), *A. fraserianum* (Rheinheimer, 1997), *A. fuscosuturale* (Lea, 1899), *A. immundum* (Lea, 1899), *A. integricolle* (Lea, 1899), *A. melvillense* (Lea, 1926), *A. microscopicum* (Lea, 1910a), *A. nigrosuturale* (Lea, 1910a), *A. nigroterminale* (Lea, 1915b), *A. niveodispersum* (Lea, 1910a), *A. parvocastaneum* (Lea, 1926), *A. pilistriatum* (Lea, 1910a), *A. pseudofraserianum* (Legalov, 2000), *A. pulicare* (Pascoe, 1874), *A. quadricolor* (Lea, 1926), *A. rivulare* (Lea, 1926), *A. symbolum* (Faust, 1897), *A. tasmanicum* (Lea, 1910a), *A. teretirostre* (Lea, 1899), *A. turbidum* (Lea, 1910b), *A. vertebrale* (Lea, 1910a).

Distribution. Australia, palaeotropical (Oriental and Australo-Pacific regions incl. New Guinea and New Caledonia, southeastern Palaearctic and Afrotropical regions incl. Madagascar and Arabian Peninsula).

Notes. Most of the species previously classified in *Apion* as well as those in *Sterculapion* Rheinheimer (Zimmerman 1991, 1994; Pullen *et al.* 2014) were recently transferred to this genus by Wanat & Escalona (2024). Most are associated with Malvaceae but several also with Rhamnaceae (Rheinheimer 1989), the larvae probably developing in the reproductive organs. The genus occurs in both arid habitats and rainforest canopies.

Coelocephalapion Wagner, 1914 (Pl. 41G, H)

Type species, by subsequent designation (Kissinger, 1968: 29): *Apion bryanti* Wagner, 1914.

Australian species. *C. pigrae* Kissinger, 1992.

Distribution. Australia (NT; introduced), North, Central and South America.

Notes. Two species of the genus were introduced in northern Australia for the biological control of the weed *Mimosa pigra* (Fabaceae), *C. aculeatum* (Fall, 1898) from Mexico in 1992 and *C. pigrae* from Venezuela in 1994, but only the latter established (Paynter 2004; Ostermeyer & Grace 2007; Heard & Paynter 2009). Its larvae develop in flower buds of the host but can also feed on its leaves.

Conapium Motschulsky, 1866 (Pl. 41I, J; Fig. 43.10T)

Type species, by monotypy: *Apion gracile* Gerstaecker, 1854.

Australian species. *C. argutulum* (Pascoe, 1874).

Distribution. Australia (QLD, NT, NSW, WA), paleotropical (Oriental and Australo-Pacific regions incl. New Guinea, Loyalty Is. of New Caledonia, southern Palaearctic region in China, Himalayas, Afrotropical region incl. Madagascar and Arabian Peninsula).

Notes. At least two species of the genus occur in Australia, one undescribed (Wanat & Escalona 2024). The host plants are mainly Fabaceae.

Exapion **Bedel, 1887 (Pl. 41K, L; Figs 43.5, 43.10V, 43.23, 43.24)**

Type species, by subsequent designation (Schilsky, 1901: 38F): *Curculio fuscirostris* Fabricius, 1775.

Australian species. *E. ulicis* (Forster, 1771).

Distribution. Australia (NSW, VIC, SA, TAS; introduced).

Notes. The single species in Australia was introduced from New Zealand to Tasmania in 1939 for the biological control of the invasive gorse *Ulex europaeus* (Fabaceae) (Evans 1942; Ireson *et al.* 2004; Ireson & Davies 2012). It occurs is areas where the hostplant is abundant, the adults feeding on the foliage and the larvae developing and pupating in the pods (Davies *et al.* 2008). The species is native to western Europe and has been introduced to several countries in the world.

Lopatinapion **Friedman, 2013 (Pl. 44F, G)**

Type species, by original designation: *Lopatinapion simsari* Friedman, 2013.

Australian species. *Lopatinapion* sp.

Distribution. Australia (QLD, WA), New Guinea.

Notes. One undescribed species was illustrated by Zimmerman (1992: 52) as *Apion* 'new species 8', standing in collections (South Australian Museum Adelaide, Natural History Museum London) under the manuscript name *Apion bimaculialbum* Lea. The Australian species previously included in this genus (Pullen *et al.* 2014) was transferred to *Apiotherium* by Wanat & Escalona (2024), as *A. niveodispersum* (Lea).

Perapion **Wagner, 1907 (Pl. 42A, B)**

Type species, by subsequent designation (Hustache, 1931: 19): *Apion curtirostre* Germar, 1817.

Australian species. *P. antiquum* (Gyllenhal, 1833).

Distribution. Australia (SA, WA; introduced), Holarctic Region, South Africa.

Notes. The single species in Australia was introduced from South Africa several times between 1974 and 1988 for the biological control of *Emex australis* (Polygonaceae) but has not had a major impact on the weed (Yeoh *et al.* 2012). The adults feed on the foliage and the larvae bore in the stems and roots.

Reference. Wagner, 1907a.

Pseudapion **Schilsky, 1906 (Pls 41B, C, E, 42E; Figs 43.10I, K, N, 43.25–43.29)**

Type species, by original designation: *Apion fulvirostre* Gyllenhal, 1833.

Australian species. *P. aemulum* (Lea, 1899), *P. hoblerae* (Lea, 1915a), *P. philanthum* (Lea, 1899).

Distribution. Australia (QLD, NT, NSW, VIC, SA, WA), western Palaearctic region.

Notes. The genus previously only included one species in Australia, *P. aemulum* (Rheinheimer 1998; Pullen *et al.* 2014), but Wanat & Escalona (2024) transferred another two from *Apion* to it. The hosts of the genus as known are Malvaceae, the larvae feeding on developing seeds.

Pseudaspidapion **Wanat, 1990 (Pl. 44I, J; Figs 43.30, 43.31)**

Type species, by original designation: *Apion spadiceum* Wagner, 1908.

Australian species. *P. subopacum* (Lea, 1910a).

Distribution. Australia (QLD, NT, WA), Afrotropical, Oriental, and southern Palaearctic regions.

Notes. The single Australian species of the genus was transferred to it from *Apion* by Wanat & Escalona (2024). The hosts of the genus as known are Malvaceae, the larvae feeding on the seeds.

Pseudoconapion **Voss, 1959 (Pls 41F, 44H)**

Type species, by original designation: *Conapion sparsum* Faust, 1899.

Australian species. *P. pudicum* (Lea, 1899), *P. terraereginae* (Blackburn, 1892).

Distribution. Australia (QLD, NT, NSW, WA), New Guinea, palaeotropical (Oriental, southern Palaearctic and Afrotropical regions.

Notes. The genus includes two species in Australia, one of them was only recently transferred to it by Wanat & Escalona (2024). The hosts of the genus as known are Fabaceae, including the widespread genus *Indigofera*, also present in Australia; the larvae developing in the seeds.

Pseudopiezotrachelus **Wagner, 1907 (Pls 41A, 42C, D; Figs 43.2C, 43.10O, U, 43.32)**

Type species, by original designation: *Apion probum* Faust, 1899.

Australian species. *P. macleayensis* (Lea, 1926), *P. basiinflatus* (Lea, 1926), *P. convexipennis* (Lea, 1910a).

Distribution. Australia (QLD, NT, NSW, WA), New Guinea, paleotropical (Oriental and Australo-Pacific, southern and eastern Palaearctic and Afrotropical regions.

Notes. The genus previously only included one species in Australia, *P. macleayensis* (Pullen *et al.* 2014), but Wanat

& Escalona (2024) transferred another two from *Apion* to it. The hosts of the genus as known are Fabaceae.

Reference. Wagner, 1907b.

Pseudorhinapion Voss, 1959 (Pl. 42F)

Type species, by original designation: *Apion severini* Wagner, 1907b.

Australian species. *P. agonis* (Lea, 1899).

Distribution. Australia (QLD, NT, NSW, WA), Afrotropical region.

Notes. The genus is represented by a single species in Australia (Pullen *et al.* 2014; Wanat & Escalona 2024), which was earlier placed in the genus *Leaoapion* by Zimmerman (1994) but transferred to *Pseudorhinapion* by Wanat (2001), the name *Leaoapion* synonymised with the older name *Pseudorhinapion*. The hosts of the genus as known are Fabaceae.

Tribe Myrmacicelini Zimmerman, 1994

Characteristics. The single genus placed in this tribe, *Rhynolaccus*, possesses a unique combination of characters: eyes located on dorsal side of head; antennal clubs connate; trochanters about half as long as femora; tarsal claws separated by apical processes of tarsite 5; elytral suture simply dove-tailing apically; mesocoxae separated by long process of mesoventrite; endophallus with filiform flagellum and frena. Myrmacicelini also share a set of characters with some South African apionine genera considered basal as well, although the relationships among these taxa are still uncertain: tongue-like internal process of pygidium in male and often an additional sclerite (rarely two) in membrane between sternites VIII and IX; tegminal plate simple; spermatheca with thick wall and smooth surface (Wanat 2001).

Rhynolaccus Guérin-Méneville, 1831 (Pls 4G, 42H; Figs 43.9L, 43.10G, J, R, 43.33)

Type species, by monotypy: *Rhynolaccus formicarius* Guérin-Méneville, 1831.

Australian species. *R. exsertus* (Pascoe, 1872), *R. formicarius* Guérin-Méneville, 1831, *R. puerulus* (Lea, 1899).

Distribution. Australia (QLD, NSW, ACT., VIC, SA, WA), New Guinea.

Notes. The genus was long known as *Myrmacicelus* (e.g. Zimmerman 1994), but Alonso-Zarazaga & Lyal (1999) established that the name is a junior synonym of *Rhynolaccus*. In addition to the three described species, the genus contains several undescribed in Australia and is also represented in New Guinea. The larvae develop as inquilines in galls on *Acacia* caused by *Trichilogaster* wasps (Pteromalidae), and adults have also been found in *Acacia* galls induced by the fungus *Uromycladium* (Noble 1940; Zimmerman 1994).

Tribe Rhadinocybini Alonso-Zarazaga, 1992

Characteristics. Rhadinocybini are distinguishable by the presence of four elytral striae on the disc reaching the base between scutellar shield and humeral calli, a glabrous body and contiguous mesocoxae (except some species of *Nanomyrmacyba*), and other common features of the group are the apically straight elytral suture with a simple dove-tailing lock and smooth surface of the spermatheca. The tribe is considered one of the putative basal groups of Apioninae (Wanat 2001) and contains dozens of genera and several hundreds of species in the Australo-Pacific region, in contrast to the other basal groups of Apioninae, i.e. Antliarhinini, Mecolenini, Apiomorphini, Setapiini, Tanaini, Rhynchitapiini, Apodytapiini and Lepanomiini (Wanat 2001, 2021), which include only few genera and species and occur mainly in Africa. The Australian fauna of Rhadinocybini currently comprises 20 described species, occurring largely in rainforests of Queensland and New South Wales, and is markedly less diverse compared with the New-Guinean and New-Caledonian faunas, although several undescribed species are expected to exist in Australia.

Alissapion Wanat, 2001 (Pl. 42I)

Type species, by original designation: *Alissapion zimmermani* Wanat, 2001.

Australian species. *A. zimmermani* Wanat, 2001.

Distribution. Australia (QLD, NSW), New Guinea.

Notes. The genus contains another described species in New Guinea and another undescribed one in Australia (Wanat 2001). Its hosts and life histories are unknown.

Allonotapion Wanat, 2001 (Pl. 42J)

Type species, by original designation: *Allonotapion subulirostre* Wanat, 2001.

Australian species. *A. subulirostre* Wanat, 2001.

Distribution. Australia (QLD, NSW), New Guinea, Indonesia (Sulawesi).

Notes. The genus contains another described species in New Guinea and another seven undescribed ones, three of those in Australia (Wanat 2001). Its hosts and life histories are unknown.

Brachycybus Wanat, 2001 (Pl. 43I; Fig. 43.10S)

Type species, by original designation: *Brachycybus silvaticus* Wanat, 2001.
Australian species. *Brachycybus* sp. (Wanat, 2001: 247)
Distribution. Australia (QLD), New Guinea.
Notes. The genus contains two described species in New Guinea and another eight undescribed ones, the single one from Australia apparently known only from a single male (Wanat 2001). Its hosts and life histories are unknown.

Coeliapion Wanat, 2001 (Pl. 43G, H)

Type species, by original designation: *Coeliapion bilocalis* Wanat, 2001.
Australian species. *C. mednonstriatum* (Lea, 1926).
Distribution. Australia (QLD, NSW), New Guinea.
Notes. In addition to the single Australian species, the genus contains another described and eight undescribed species in New Guinea (Wanat 2001). Its hosts and life histories are unknown.

Ganoapion Zimmerman, 1994 (Pl. 42K)

Type species, by original designation: *Apion clavicorne* Lea, 1926.
Australian species. *G. clavicorne* (Lea, 1926), *G. rhadinocyboides* Wanat, 2001.
Distribution. Australia (QLD, NSW).
Notes. The genus was described as a subgenus of *Notapion* by Zimmerman (1994) but raised to genus level by Wanat (2001). It appears to be endemic to Australia and contains two species in distinct subgenera (*Ganoapion, Proganoapion*); another two undescribed species of *Ganoapion* (*s. str.*) are known in collections (Wanat 2001). Its hosts and life histories are unknown.

Lissapion Zimmerman, 1994 (Pl. 42L)

Type species, by original designation: *Apion varistriatum* Lea, 1926.
Australian species. *L. albertisii* (Pascoe, 1885), *L. sculpticeps* (Lea, 1926), *L. varistriatum* (Lea, 1926).
Distribution. Australia (QLD, NSW), New Guinea.
Notes. In addition to the three described Australian species, the genus contains three undescribed ones in Australia and another five to seven in New Guinea (Wanat 2001). Its hosts and life histories are unknown.

Micronotapion Wanat, 2001 (Pl. 43A; Fig. 43.10E)

Type species, by original designation: *Micronotapion gibbiceps* Wanat, 2001.
Australian species. *M. gibbiceps* Wanat, 2001.
Distribution. Australia (QLD, NSW, Lord Howe Is.), New Guinea.
Notes. From NSW recorded by Wanat & Escalona (2024); ocurrence in the Lord Howe Is. is derived from the photographs in Zimmerman (1992: 55). The genus contains another described species in New Guinea, *M. magnicolle* Wanat, 2001. Its hosts and life histories are unknown.

Myrmecapion Wanat, 2001 (Pl. 43B; Fig. 43.10D)

Type species, by original designation: *Myrmecapion australicum* Wanat, 2001.
Australian species. *M. australicum* Wanat, 2001.
Distribution. Australia (QLD).
Notes. The genus comprises a single species as known (Wanat 2001). Its hosts and life histories are unknown.

Nanomyrmacyba Wanat, 1990 (Pl. 43C)

Type species, by original designation: *Nanomyrmacyba minuta* Wanat, 1990.
Australian species. *N. illawarrensis* (Lea, 1926), *N. inflaticollis* (Lea, 1926), *N. stilba* (Lea, 1910a).
Distribution. Australia (QLD, NSW), New Guinea, Indonesia, Philippines, Taiwan, Vietnam.
Notes. The three Australian species of the genus were placed in *Kelainapion* by Zimmerman (1994), but this name was synonymised with *Nanomyrmacyba* by Wanat (2001). The genus contains another described species in Sumatra, *N. minuta* Wanat, 1990, and approximately 40 additional undescribed species outside of Australia (Wanat 2001). Its hosts and life histories are unknown.

Notapion Zimmerman, 1994 (Pl. 43D)

Type species, by original designation: *Apion trilobicolle* Lea, 1926.
Australian species. *N. mediopunctum* (Lea, 1926), *N. striatipenne* (Lea, 1926), *N. tenuistriatum* (Lea, 1910a), *N. trilobicolle* (Lea, 1926), *N. varirostre* (Lea, 1926).
Distribution. Australia (QLD, NSW), New Guinea.
Notes. The genus contains another described species in New Guinea, *N. riedeli* Wanat, 2001, and another undescribed one each in Australia and New Guinea (Wanat 2001). Adults of *N. striatipenne* and *N. tenuistriatum* have been found to be common on *Doryphora sassafras* (Monimiaceae) (Rheinheimer 1989).

Ommatocybus Wanat, 2001 (Pl. 43J, K; Fig. 43.10F)

Type species, by original designation: *Ommatocybus simplex* Wanat, 2001.

Australian species. *Ommatocybus* sp.

Distribution. Australia (QLD), New Guinea, Sulawesi, Malaysia.

Notes. The genus has not been recorded from Australia before (Wanat & Escalona 2024). It contains one described species in New Guinea and undescribed ones in New Guinea (4–5), Sulawesi (2) and Malaysia (1) (Wanat 2001). Its hosts and life histories are unknown.

Tribe Rhinorhynchidiini Zimmerman, 1994

Characteristics. This monotypic and morphologically distinct tribe is endemic to Australia and characterised by the absence of an apical pocket-like lock of the elytral suture, the smooth surface of spermatheca, the well-developed body vestiture of long piliform scales and the basally dentate rostrum with the antennae inserted basally in foveiform sockets not extended into scrobes (Zimmerman 1994; Wanat 2001).

Rhinorhynchidius Voss, 1922 (Pl. 43E, F; Figs 43.10H, L, M, 43.34)

Type species, by monotypy: *Rhinorhynchidius cylindrirostris* Voss, 1922.

Australian species. *R. australasiae* (Lea, 1910b), *R. migueli* Pullen, Jennings & Oberprieler, 2014

Distribution. Australia (NSW, ACT, SA, TAS, WA).

Notes. The genus currently contains two described species and at least three undescribed ones (Zimmerman 1994). *R. migueli* was renamed from *R. meridionalis* (Lea) as used by Zimmerman (1994) because of a complex situation of homonymy among the names applied to the species (Pullen *et al*. 2014). The larvae induce twig and foliage galls on *Callitris* (Cupressaceae), some resulting in characteristic deformities referred to as 'pineapple-top' galls (Moore 1972; Zimmerman 1993, 1994).

Tribe *incertae sedis*

Diselmobius Wanat, 2023 (Pl. 44K)

Type species, by original designation: *Diselmobius lechi* Wanat, 2023.

Australian species. *D. lechi* Wanat, 2023.

Distribution. Australia (TAS).

Notes. This monotypic endemic Australian genus was only described recently, and its tribal placement is uncertain. Similarly to the New Zealand genera *Cecidophyus* Kuschel, *Neocyba* Kissinger, *Strobilobius* Kuschel and *Zelapterus* Kuschel, its lacks the key diagnostic characters of Apionini, Rhadinocybini and Rhinorhynchidiini, differing from the first in its elytral suture having a simply dove-tailed apex and its spermatheca a smooth surface (suture apex left-turned and spermathecal surface wrinkled in Apionini), from the second in having five striae reaching the base of elytral disc and a distinctly pilose body (four striae and glabrous body in Rhadinocybini) and from the last in having shallow scrobes (absent in Rhinorhynchidiini), among other diagnostic characters (Wanat 2023). The single species is only known from four females collected on *Diselma archeri* (Cupressaceae), which is likely to be its larval host.

Subfamily Nanophyinae Gistel, 1848

Tribe Nanophyini Gistel, 1848

Characteristics. Eyes subcontiguous; funicles 5-segmented, segment 4 asymmetrical; basal margins of pronotum and elytra crenulate-carinate; scutellar shield absent; prosternum reduced to small, transverse bar; notosternal sutures short, without anterior pit; tarsal claws connate; pygidium of male uniform, without grooves; male sternite IX forming asymmetrical fork, endophallus usually with frena and tubular flagellum; both male ejaculatory duct and female spermathecal duct with spiral or annular sclerotized reinforcements (Alonso-Zarazaga 1989, 2014).

Notes. Nanophyinae are represented in Australia by a single genus with five described species, which live on aquatic plants such as *Ludwigia* (Onagraceae), *Myriophyllum* (Haloragaceae) and *Nymphoides* (Menyanthaceae) (Zimmerman 1993). In the Holarctic, Oriental and Afrotropical regions the subfamily is far more diverse, with about 33 genera and more than 300 described species and hosts belonging to several plant families (Alonso-Zarazaga 2014). Very little is known about the life history of *Austronanodes*, except that the larvae of some species develop in the stems (Zimmerman 1993; May 1994). The current classification of Nanophyinae follows Alonso-Zarazaga (1989).

Austronanodes Zimmerman, 1993 (Pls 4D, 40L; Fig. 43.9A)

Type species, by original designation: *Nanophyes nigrovarius* Lea, 1907.

Australian species. *A. alleni* (Lea, 1907), *A. maurus* (Pascoe, 1875), *A. nigrovarius* (Lea, 1907) *A. pallidicornis* (Lea, 1907), *A. v-notatus* (Lea, 1907).

Distribution. Australia (QLD, NT, NSW, VIC, SA).

Notes. The genus contains some undescribed species in Australia, and similar unidentified species has been found in New Caledonia (Wanat pers. com.). The larva of *A. nigrovarius* was described by May (1994), it develops in the stems of *Ludwigia*.

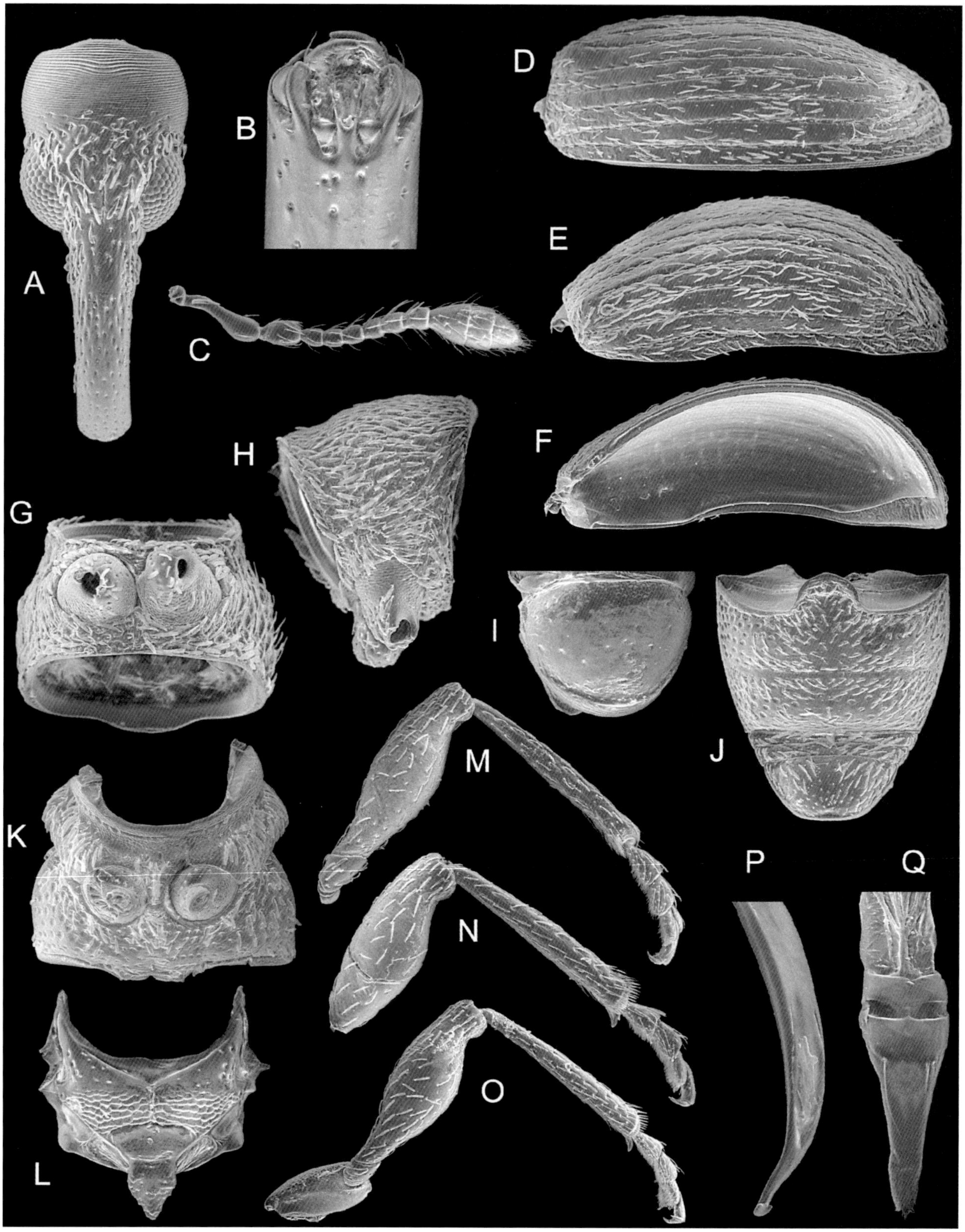

Fig. 43.15. Apioninae, structural details, *Apiotherium anthidium* (Lea), ♂. A, head, dorsal; B, mouthparts; C, antenna; D–F, elytron: D, dorsal; E, lateral; F, inner lateral; G–H, prothorax: G, ventral; H, lateral; I, pygidium, dorsal; J, abdominal ventrites; K, pterothorax, ventral; L, scutellum; M, foreleg; N, middle leg; O, hindleg; P–Q, aedeagus: P, penis, distal part, dorsolateral; Q, parameres, dorsal.

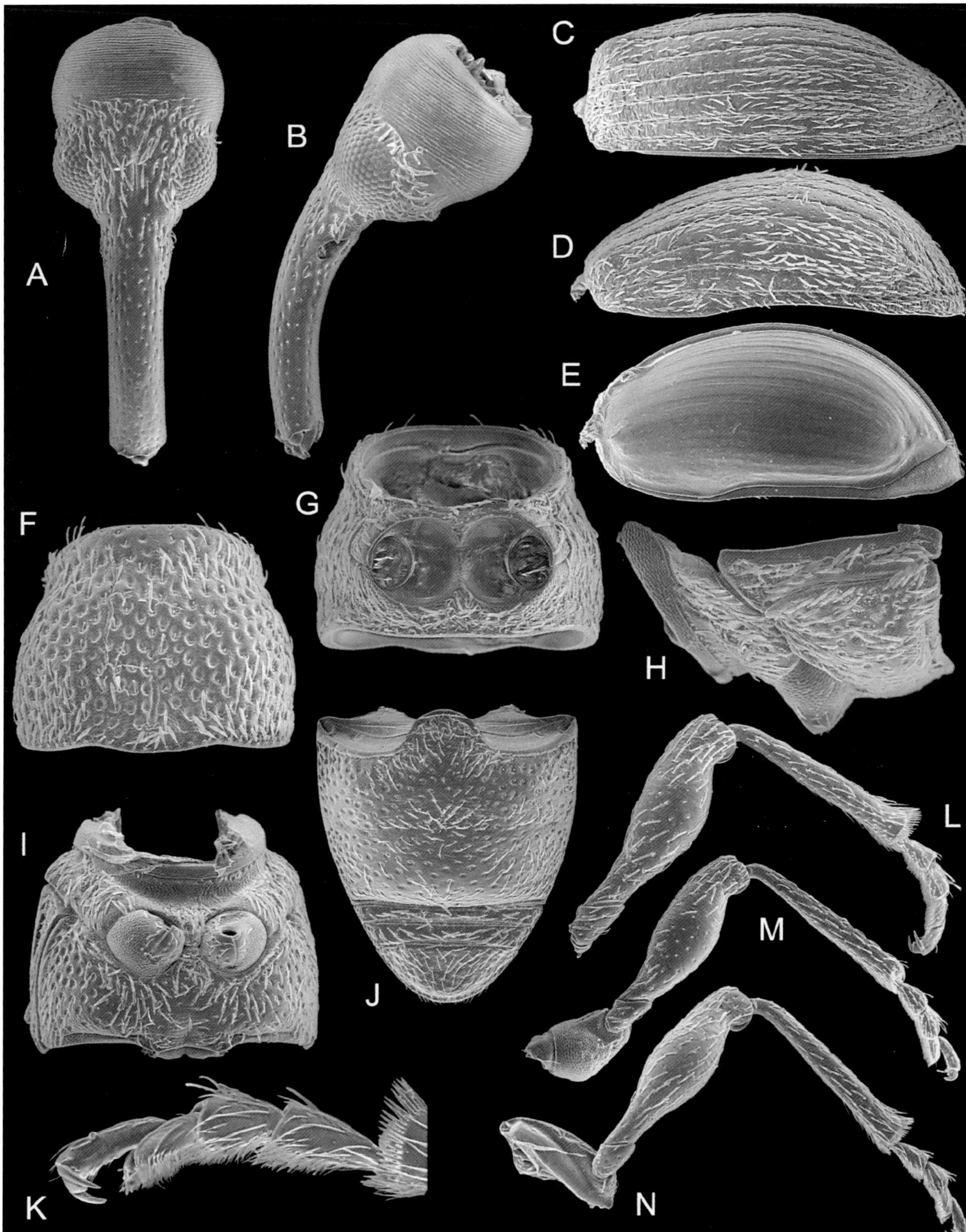

Fig. 43.16. Apioninae, structural details, *Apiotherium anthidium* (Lea), ♀. A–B, head: A, dorsal; B, lateral; C–E, elytron: C, dorsal; D, lateral; E, inner lateral; F–G, prothorax: F, dorsal; G, ventral; H–I, pterothorax: H, lateral; I, ventral; J, abdominal ventrites; K, tarsus with claws, lateral; L, foreleg; M, middle leg; N, hindleg.

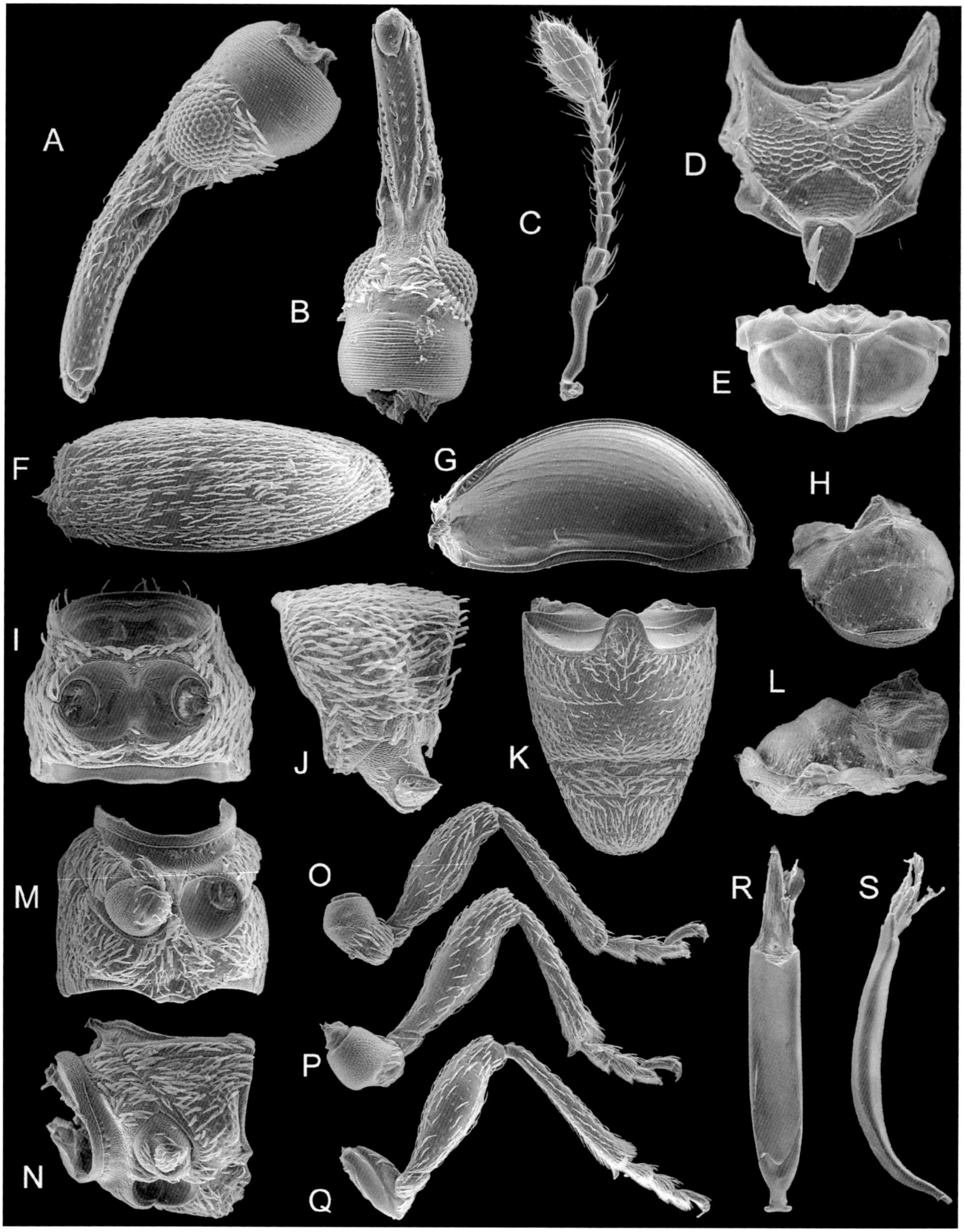

Fig. 43.17. Apioninae, structural details, *Apiotherium carpophagum* (Lea), ♂. A–B, head: A, lateral; B ventral; C, antenna; D, scutellum; E, metanotum; F–G: elytron: F, dorsal; G, inner lateral; H, pygidium, dorsal; I–J, prothorax: I, ventral; J, lateral; K, abdominal ventrites; L, pygidium, lateral; M–N, pterothorax: M, ventral; N, lateral; O, foreleg; P, middle leg; Q, hindleg; R–S, penis, dorsal and lateral.

Fig. 43.18. Apioninae, strutural details, *Apiotherium carpophagum* (Lea), ♀. A, head, lateral; B, antenna; C, metanotum; D, scutellum; E–F, elytron: E, lateral; F, inner lateral; G, prothorax, ventral; H, pterothorax, lateroventral; I, tarsus and claws, lateral; J, ovipositor extruded; K, abdominal ventrites; L, foreleg; M, middle leg; N, hindleg.

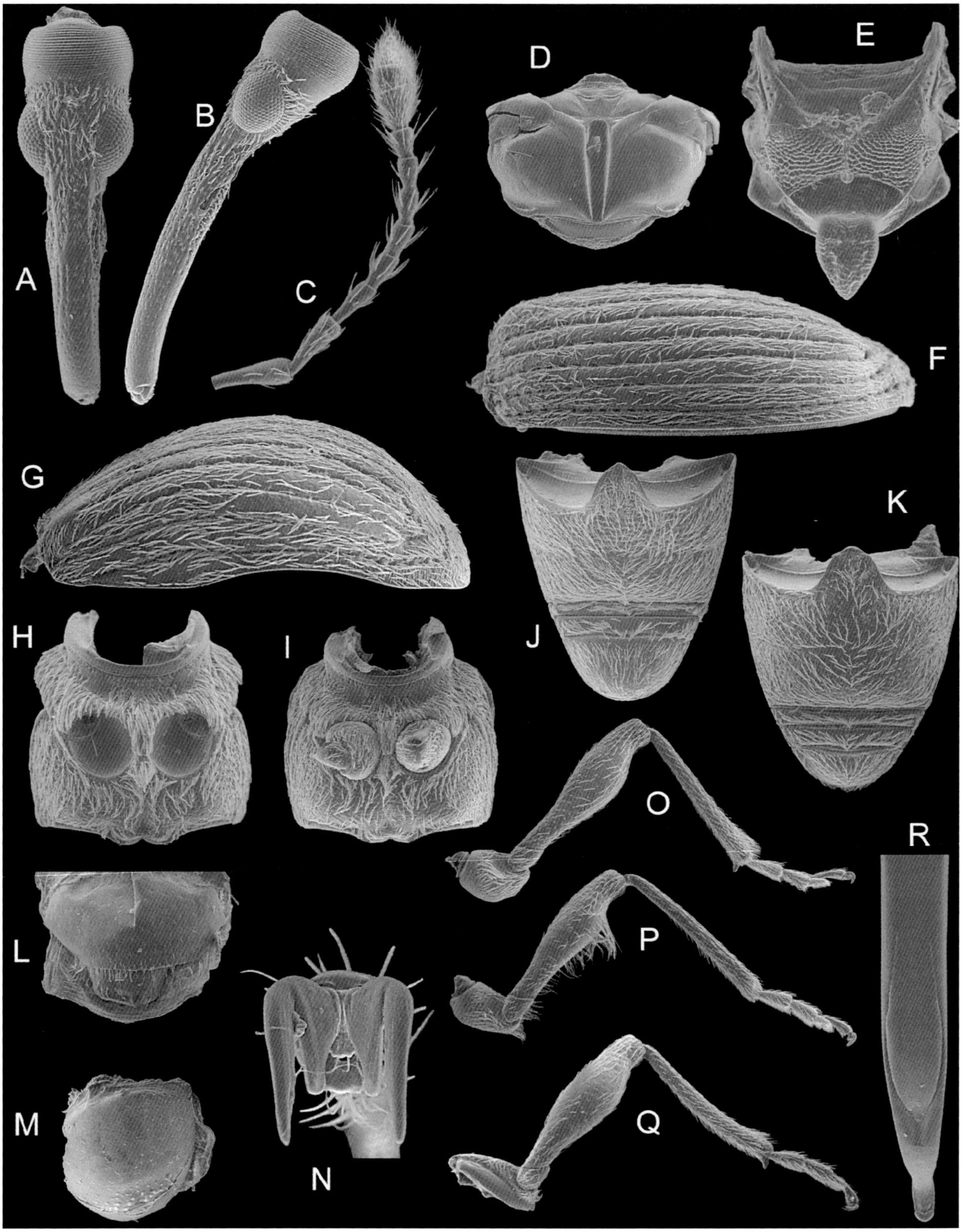

Fig. 43.19. Apioninae, structural details, *Apiotherium commersoniae* (Rheinheimer). A–B, head: dorsal, ♂; B, lateral, ♀; C, antenna; D, metanotum; E, scutellum; F–G, elytron: F, dorsal; G, lateral; H–I, pterothorax, ventral: H, ♂; I, ♀; J–K, abdominal ventrites: J, ♂; K, ♀; L, terminalia, dorsal, ♀, tergite VII and VIII (part); M, pygidium, ♂; N, dentate tarsal claws; O, middle leg; P, foreleg; Q, hindleg; R, penis, distal part, dorsal.

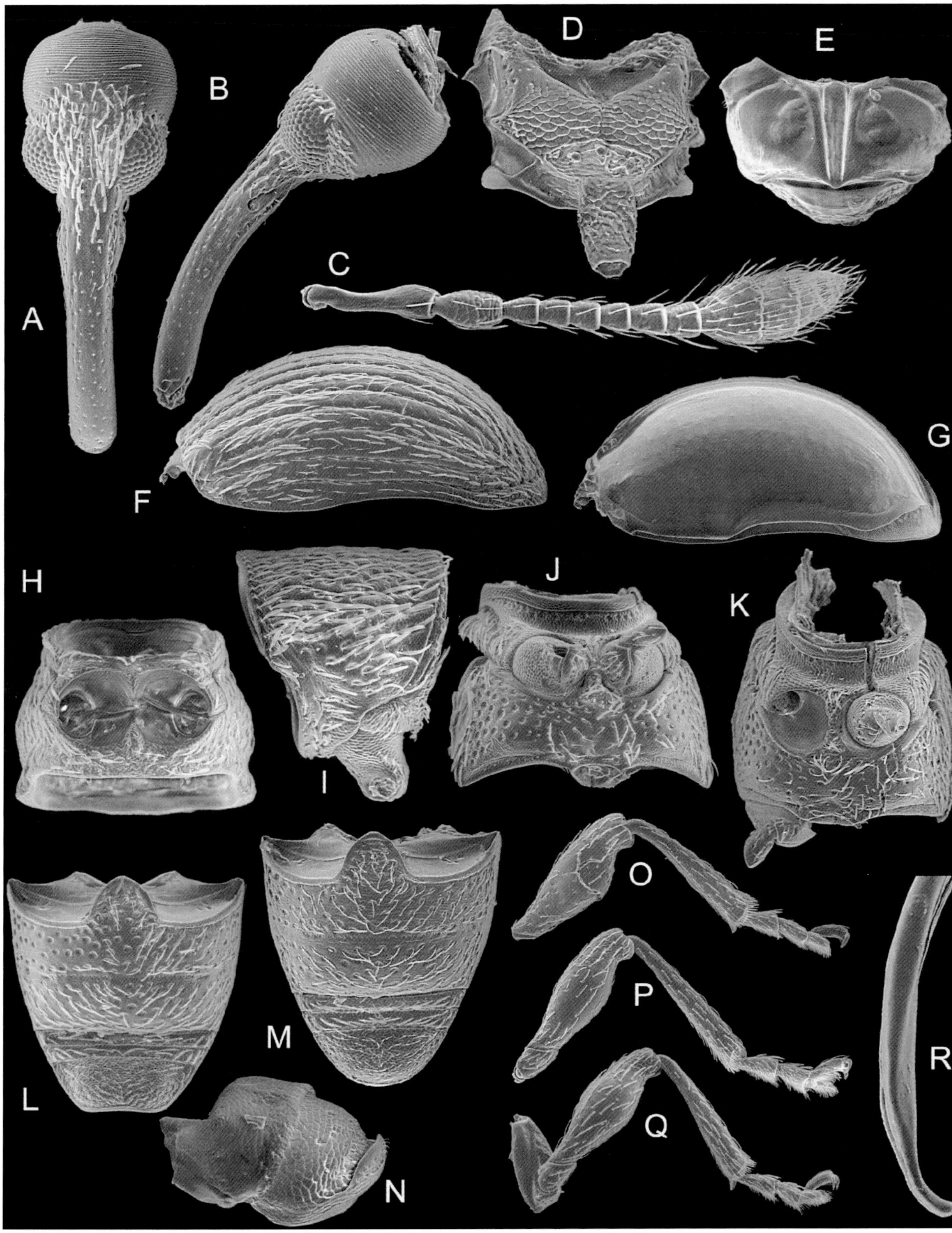

Fig. 43.20. Apioninae, structural details, *Apiotherium fuscosuturale* (Lea). A–B, head: A, dorsal, ♂; B, lateral, ♀; C, antenna; D, scutellum; E, metanotum; F–G, elytron: F, lateral; G, inner lateral; H–I, prothorax: H, ventral; I, lateral; J–K, pterothorax: J, ventral, ♂; K, ventral, ♀; L–M, abdominal ventrites: L, ♂; M, ♀; N, pygidium, ♂, lateral; O, middle leg, ♂; P, foreleg, ♂; Q, hindleg, ♂; R, penis, lateral.

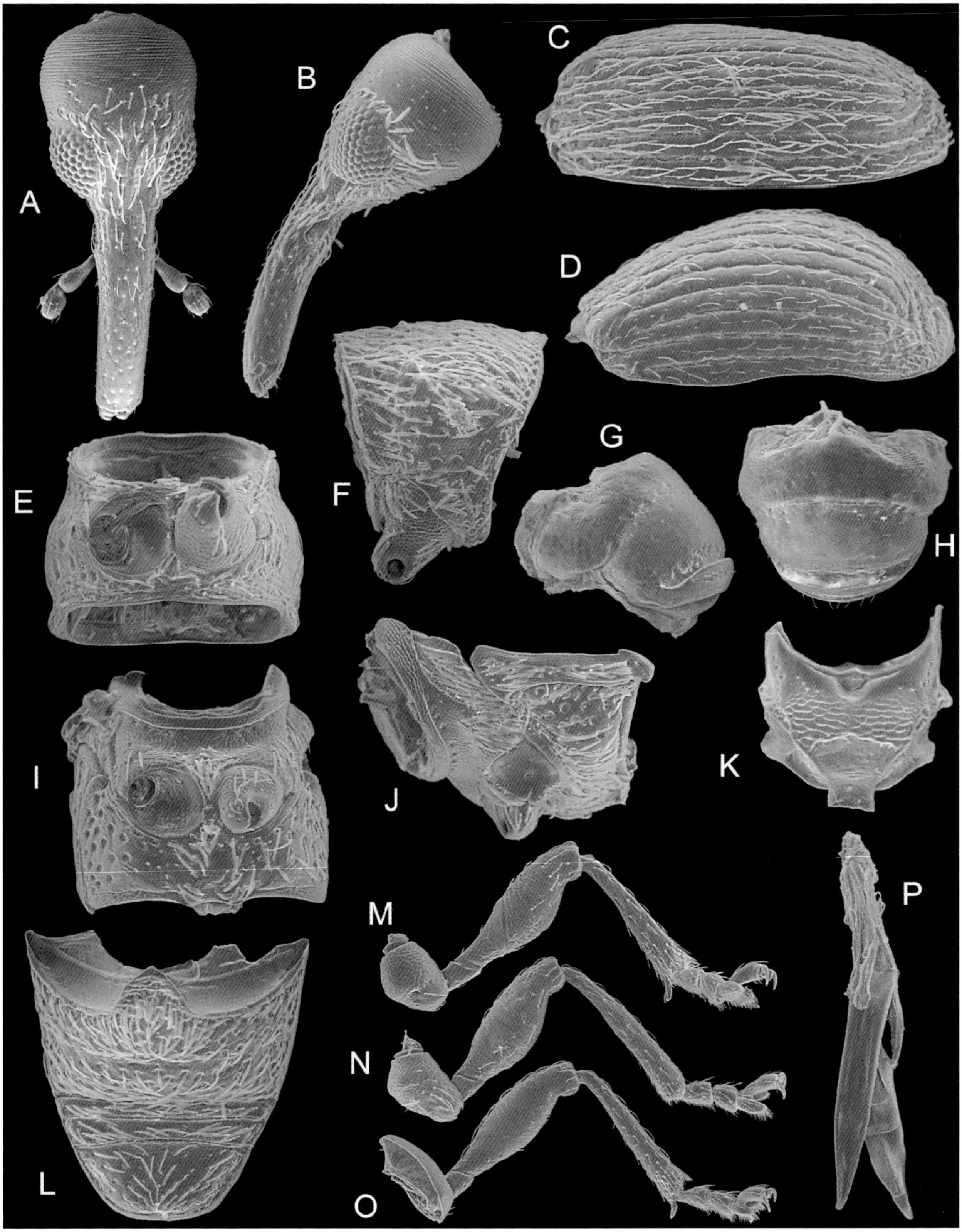

Fig. 43.21. Apioninae, structural details, *Apiotherium microscopicum* (Lea), ♂. A–B, head: A, dorsal; B, lateral; C–D, elytron: C, dorsal; D, lateral; E–F: prothorax: E, ventral; F, lateral; G–H, pygidium: G, lateral; H, dorsal; I–J: pterothorax: I, ventral; J, lateral; K, scutellum; L, abdominal ventrites; M, middle leg; N, foreleg; O, hindleg; P, aedeagus.

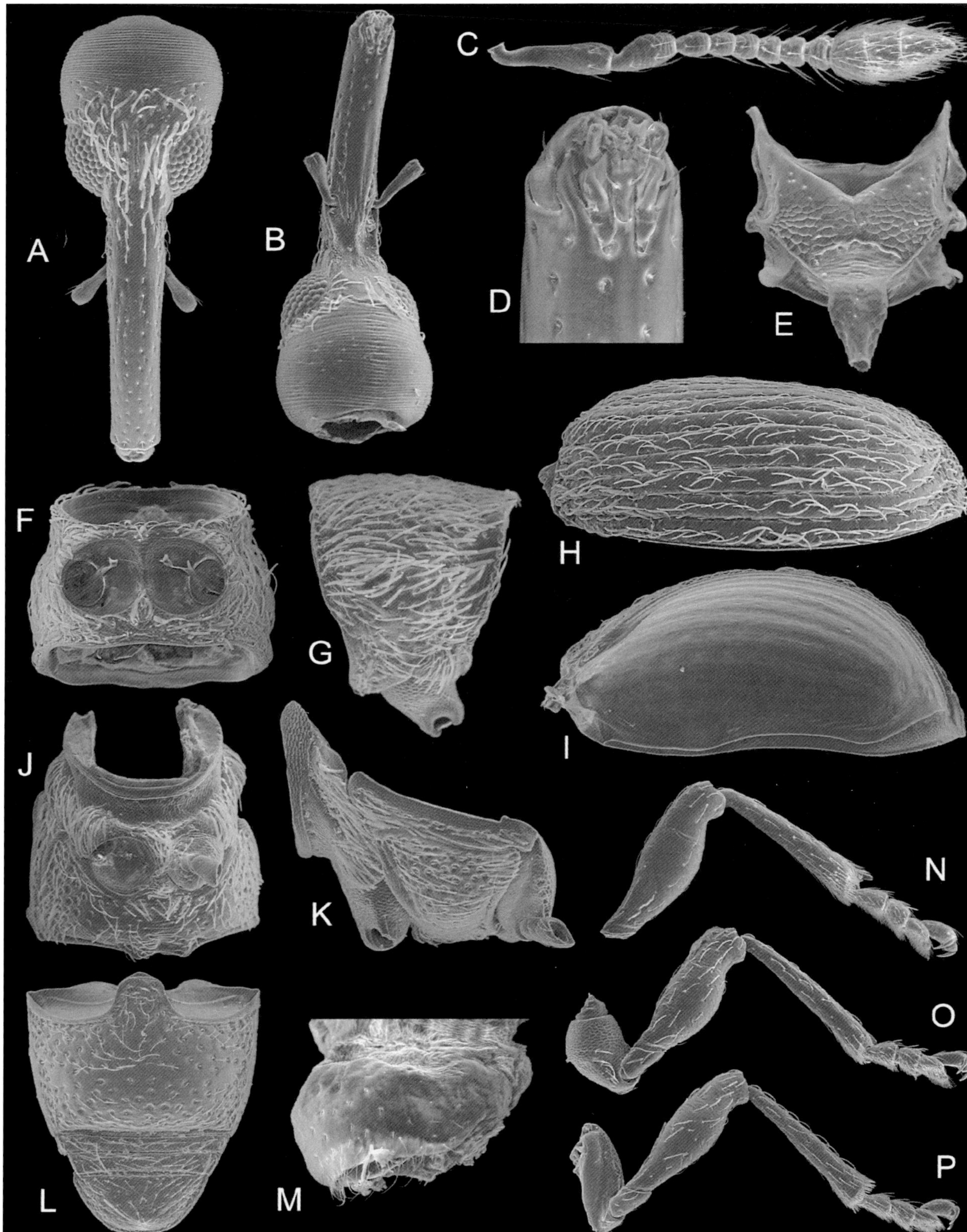

Fig. 43.22. Apioninae, structural details, *Apiotherium microscopicum* (Lea), ♀. A, head, dorsal; B, head, ventral; C, antenna; D, mouthparts; E, scutellum; F–G, prothorax: F, ventral; G, lateral; H–I, elytron: dorsal; I, inner lateral; J–K, pterothorax: J, ventral; K, lateral; L, abdominal ventrites; M, tergite VII, dorsolateral; N, foreleg; O, middle leg; P, hindleg.

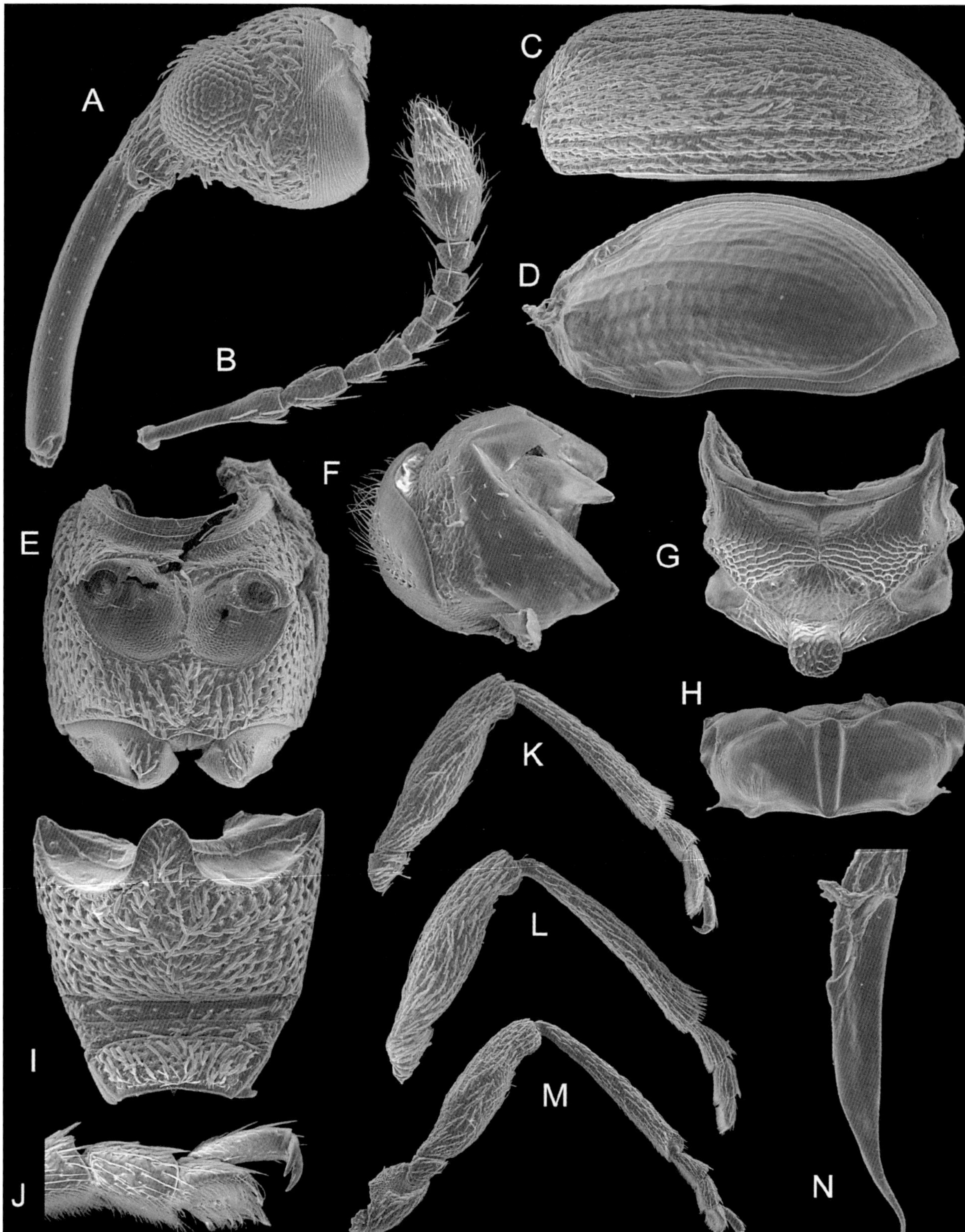

Fig. 43.23. Apioninae, structural details, *Exapion ulicis* (Forster), ♂. A, head, lateral; B, antenna; C–D, elytron: C dorsal; D, inner lateral; E, pterothorax, ventral; F, pygidium, laterodorsal; G, scutellum; H, metanotum; I, abdominal ventrites; J, tarsus with claws, lateral; K, foreleg; L, middle leg; M, hindleg; N, penis, lateral.

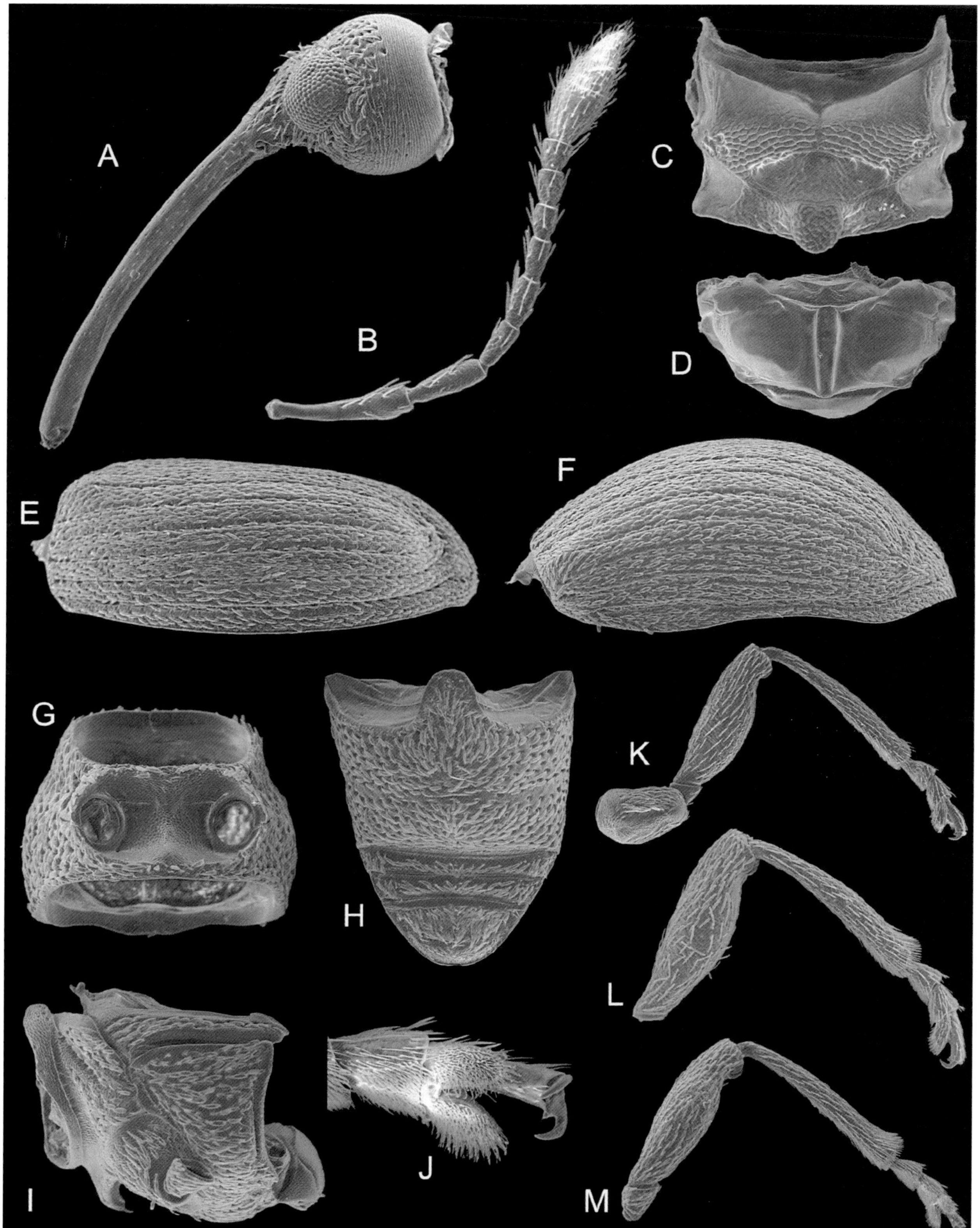

Fig. 43.24. Apioninae, structural details, *Exapion ulicis* (Forster), ♀. A, head, lateral; B, antenna; C, scutellum; D, metanotum; E, elytron, dorsal; F, elytron, lateral; G, prothorax, ventral; H, abdominal ventrites; I, pterothorax, lateral; J, tarsal segments 3–5 and tarsal claws, ventral; K, foreleg; L, middle leg; M, hindleg.

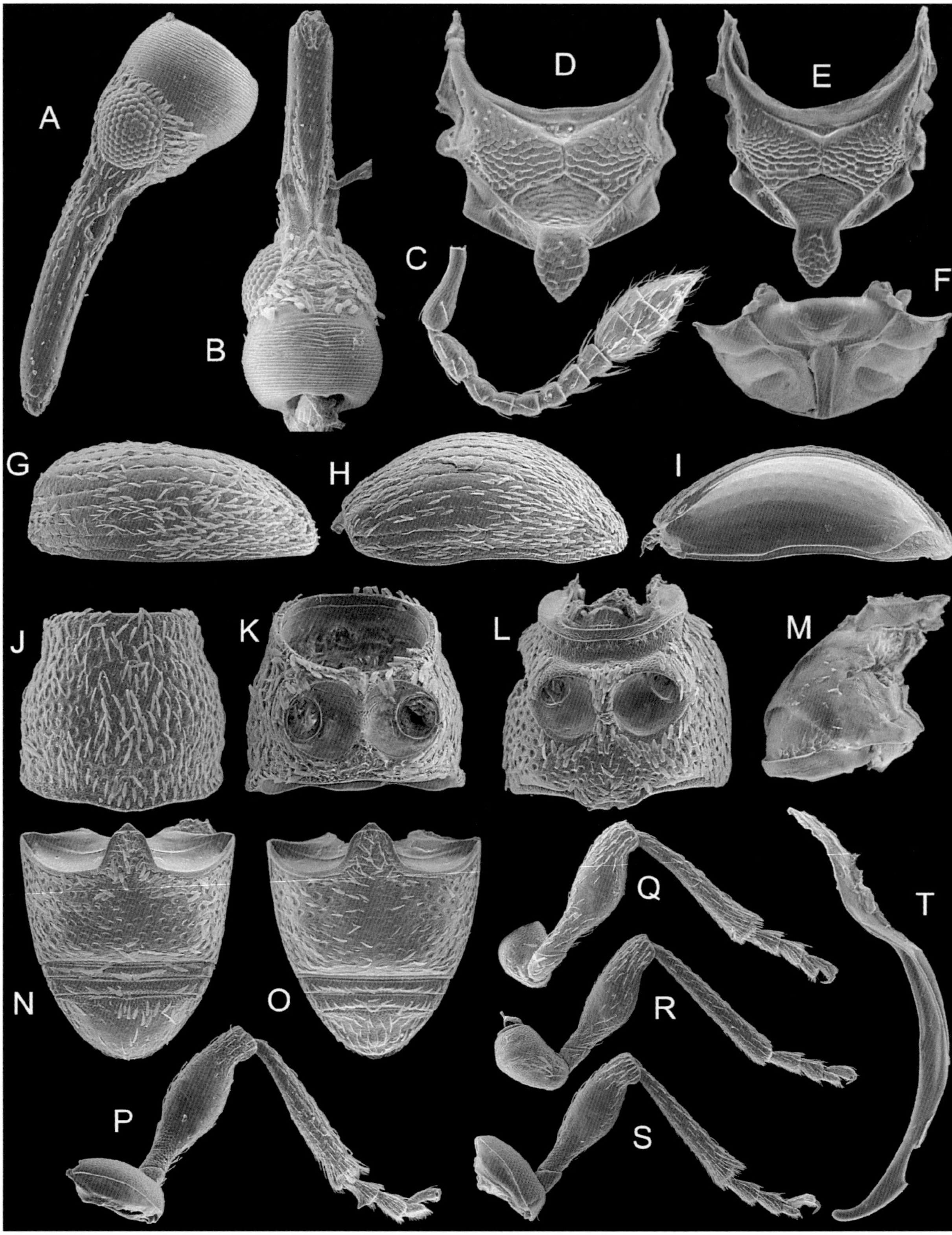

Fig. 43.25. Apioninae, structural details, *Pseudapion aemulum* (Lea). A, head, ♀, lateral; B, head, ♀, ventral; C, antenna; D, scutellum, ♂; E, scutellum, ♀; F, metanotum; G, elytron, dorsal; H, elytron, lateral; I, elytron, inner lateral; J, prothorax, dorsal; K, prothorax, ventral; L, pterothorax, ventral; M, pygidium, ♂, lateral; N–O, abdominal ventrites: N, ♂; O, ♀; P, hindleg, ♂; Q, foreleg, ♀; R, middle leg, ♀; S, hindleg, ♀; T, penis, lateral.

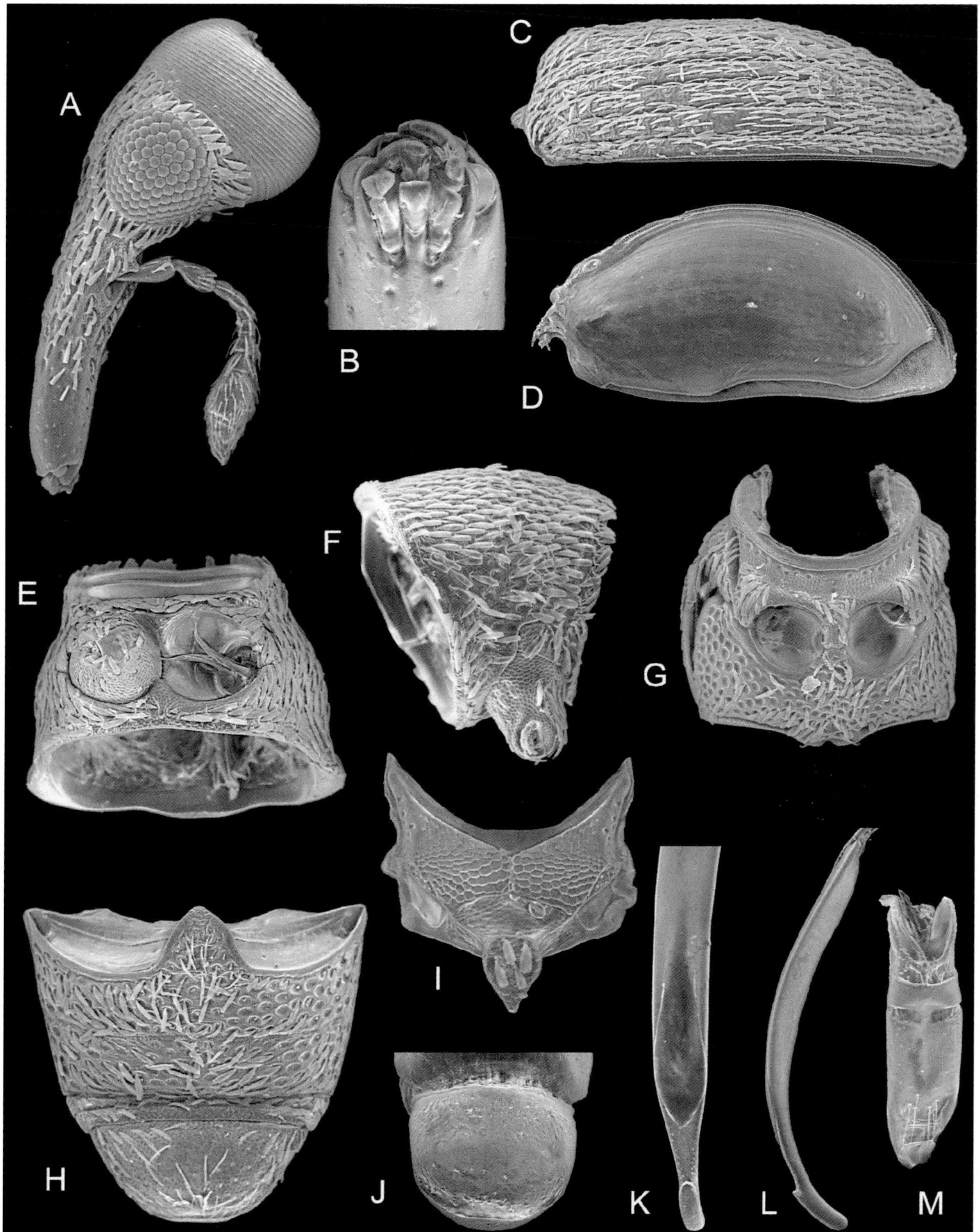

Fig. 43.26. Apioninae, structural details, *Pseudapion hoblerae* (Lea), ♂. A, head, lateral; B, mouthparts; C, elytron, dorsal; D, elytron, inner lateral; E, prothorax, ventral; F, prothorax, lateral; G, pterothorax, ventral; H, abdominal ventrites; I, scutellum; J, pygidium, dorsal; K–L, penis: K, dorsal; L, lateral; M, parameres, dorsal.

Fig. 43.27. Apioninae, structural details, *Pseudapion hoblerae* (Lea), ♀. A–B, head: A, dorsal; B, lateral; C, antenna; D–E, elytron: D, lateral; E, inner lateral; F–G, prothorax: F, dorsal; G, ventral; H–I, pterothorax: H, ventral; I, lateral; J, scutellum; K, abdominal ventrites; L, metanotum; M, foreleg; N, middle leg; O, hindleg.

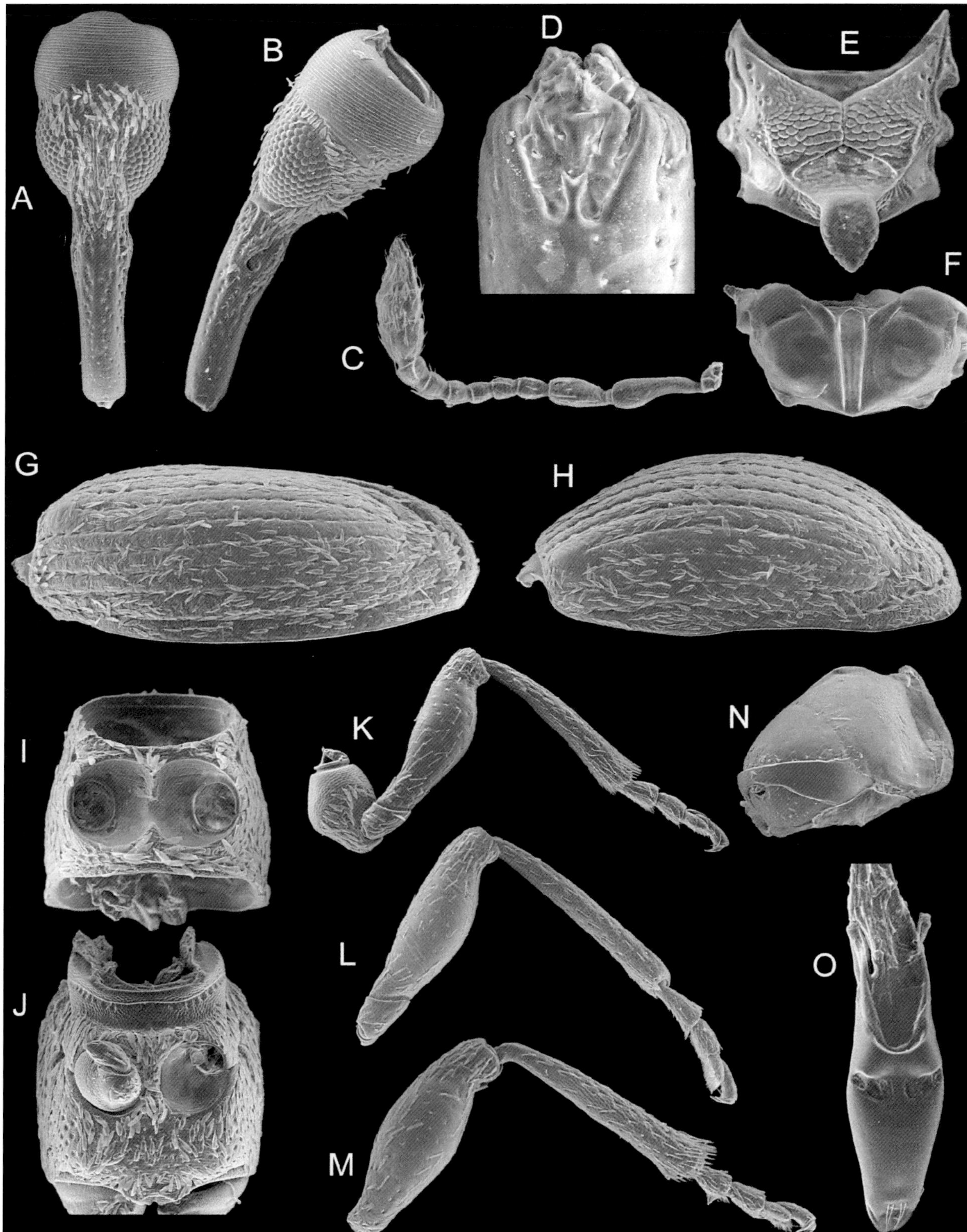

Fig. 43.28. Apioninae, structural details, *Pseudapion philanthum* (Lea), ♂. A–B, head: A, dorsal; B, lateral; C, antenna; D, mouthparts; E, scutellum; F, metanotum; G–H, elytron: G, dorsal; H, lateral; I, prothorax, ventral; J, pterothorax, ventral; K, foreleg; L, middle leg; M, hindleg; N, pygidium, lateral; O, parameres, dorsal.

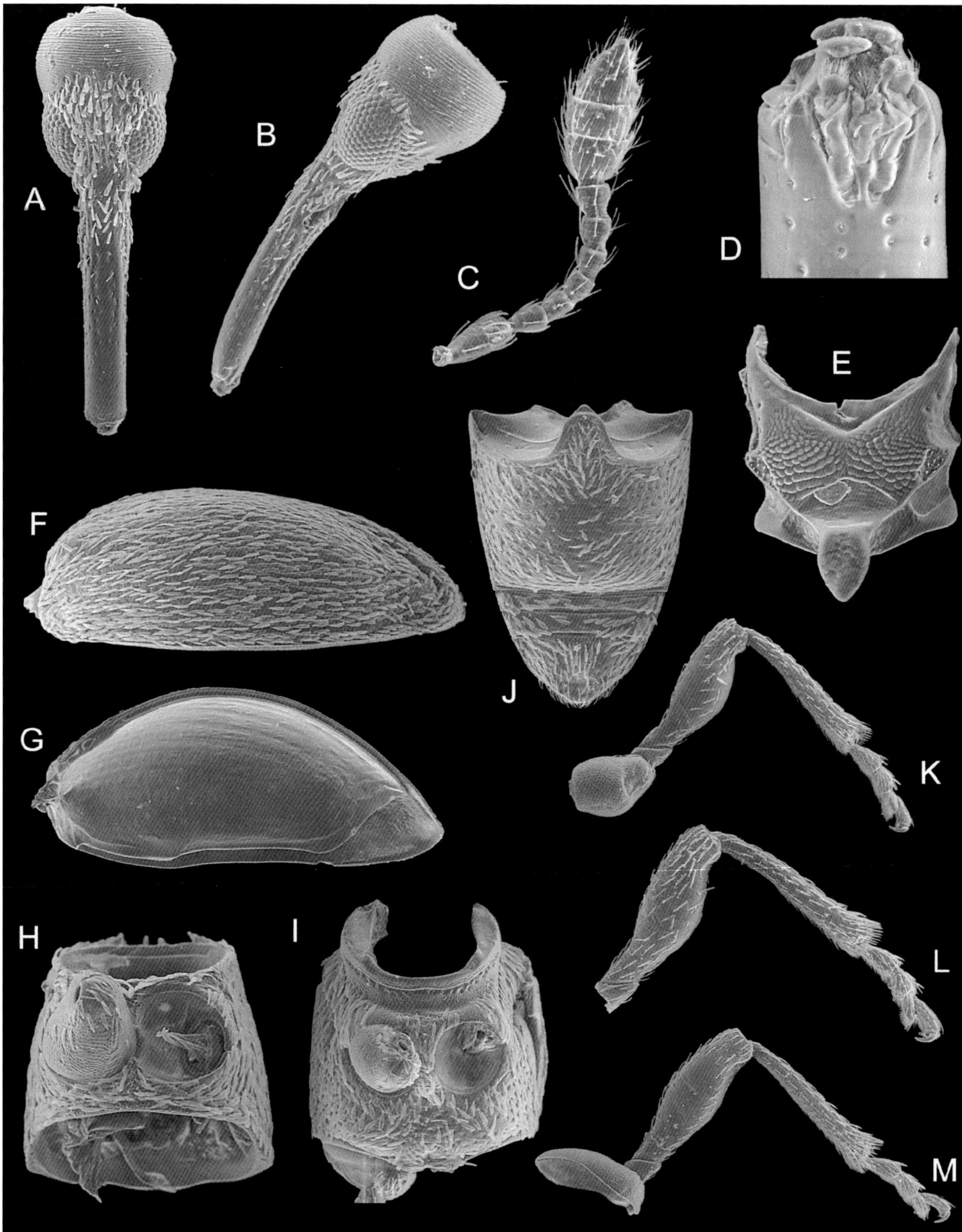

Fig. 43.29. Apioninae, structural details, *Pseudapion philanthum* (Lea), ♀. A–B, head: A, dorsal; B, lateral; C, antenna; D, mouthparts; E, scutellum; F–G, elytron: F, dorsal; G, inner lateral; H, prothorax, ventral; I, pterothorax, ventral; J, abdominal ventrites; K, foreleg; L, middle leg; M, hindleg.

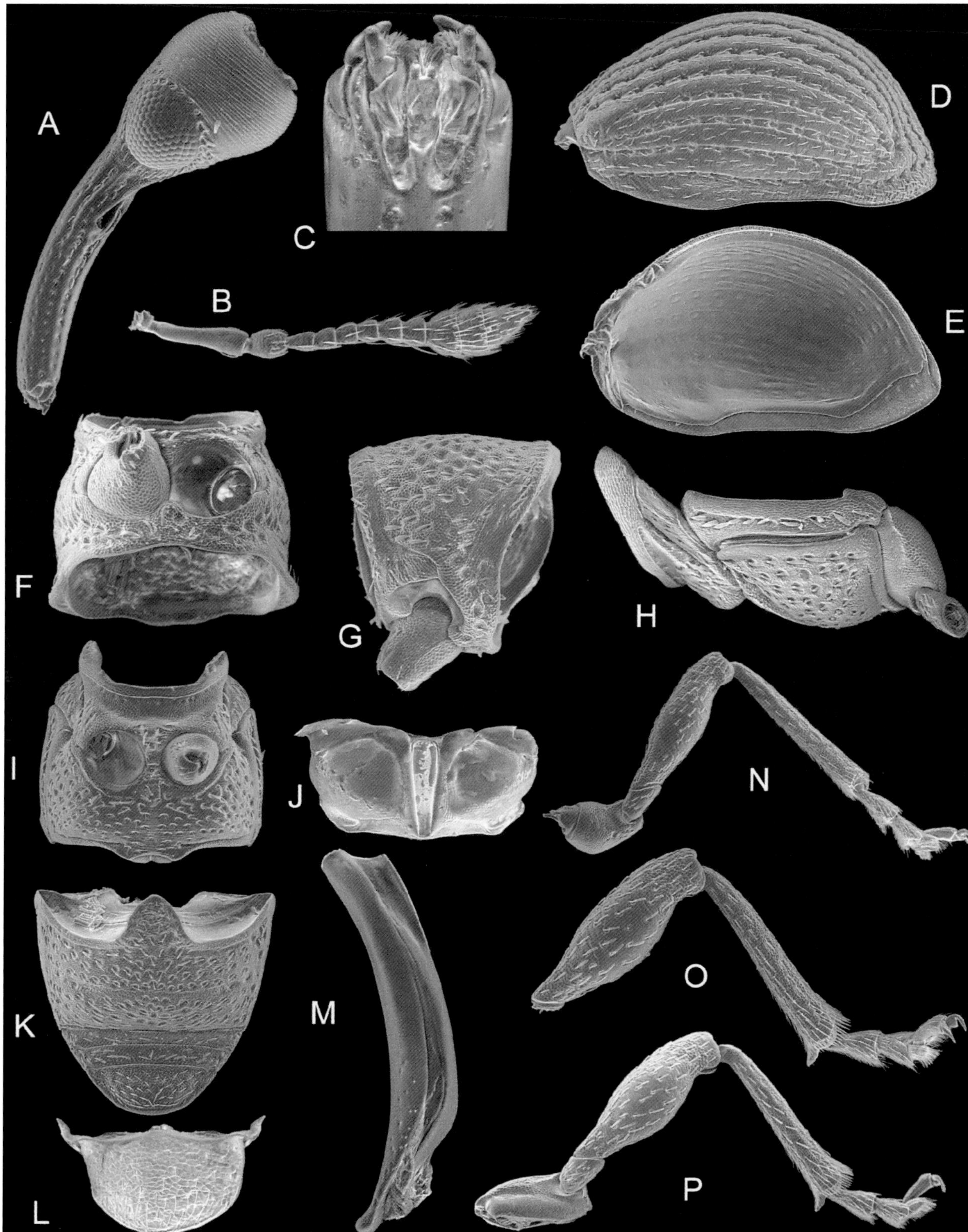

Fig. 43.30. Apioninae, structural details, *Pseudaspidapion subopacum* (Lea), ♂. A, head, lateral; B, antenna; C, mouthparts; D–E, elytron: D, lateral; E, inner lateral; F–G, prothorax; F, ventral; G, lateral; H–I, pterothorax: H, lateral; I, ventral; J, metanotum; K, abdominal ventrites; L, pygidium, dorsal; M, penis, dorsolateral; N, foreleg; O, middle leg; P, hindleg.

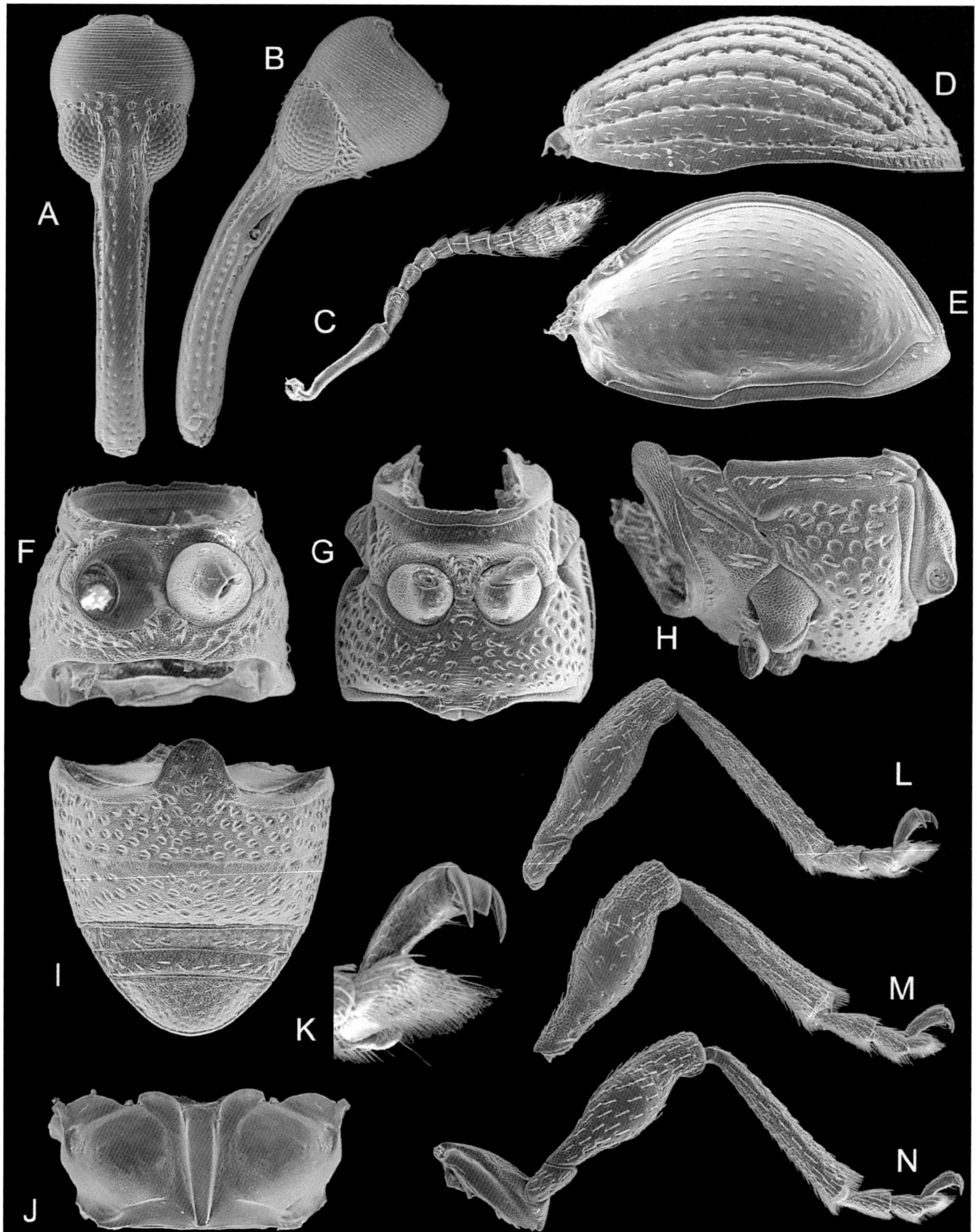

Fig. 43.31. Apioninae, structural details, *Pseudaspidapion subopacum* (Lea), ♀. A–B, head: A, dorsal; B, lateral; C, antenna; D–E, elytron: D, lateral; E, inner lateral; F, prothorax, ventral; G–H, pterothorax; G, ventral; H, lateral; I, abdominal ventrites; J, metanotum; K, tarsal claws; L, foreleg; M, middle leg; N, hindleg.

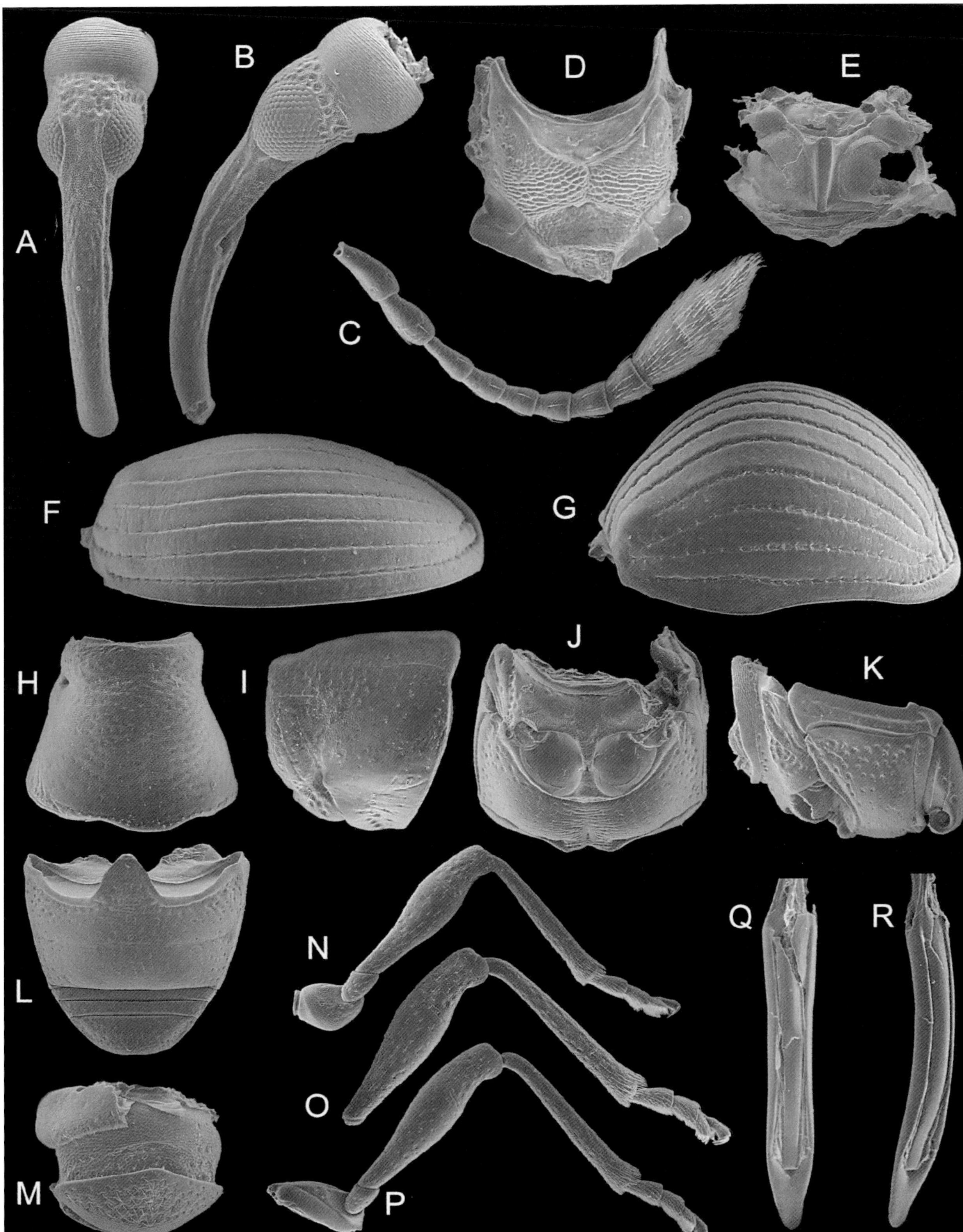

Fig. 43.32. Apioninae, structural details, *Pseudopiezotrachelus macleayensis* (Lea), ♂. A–B, head: A, dorsal; B, lateral; C, antenna; D, scutellum; E, metanotum; F–G, elytron: F, dorsal; G, lateral; H–I, prothorax: H, dorsal; I, lateral; J–K, pterothorax: J, ventral; K, lateral; L, abdominal ventrites; M, pygidium, dorsal; N, foreleg; O, middle leg; P, hindleg; Q –R: penis; Q, dorsal; R, dorsolateral.

Fig. 43.33. Apioninae, structural details, *Rhynolaccus exsertus* (Pascoe). A–B, head: A, dorsal; B, ventral; C, antenna; D, mouthparts; E, meso- and metanotum; F, scutellum; G–I, elytron: G, dorsal; H, lateral; I, inner lateral; J–K, prothorax: J, lateral; K, ventral; L, pterothorax, ventral; M, pygidium, ♂, dorsal; N–O, abdominal ventrites: N, ♂; O, ♀; P, tarsal claws, apical; Q, middle leg; R, foreleg; S, hindleg; T, penis, lateral; U, tegmen, dorsal.

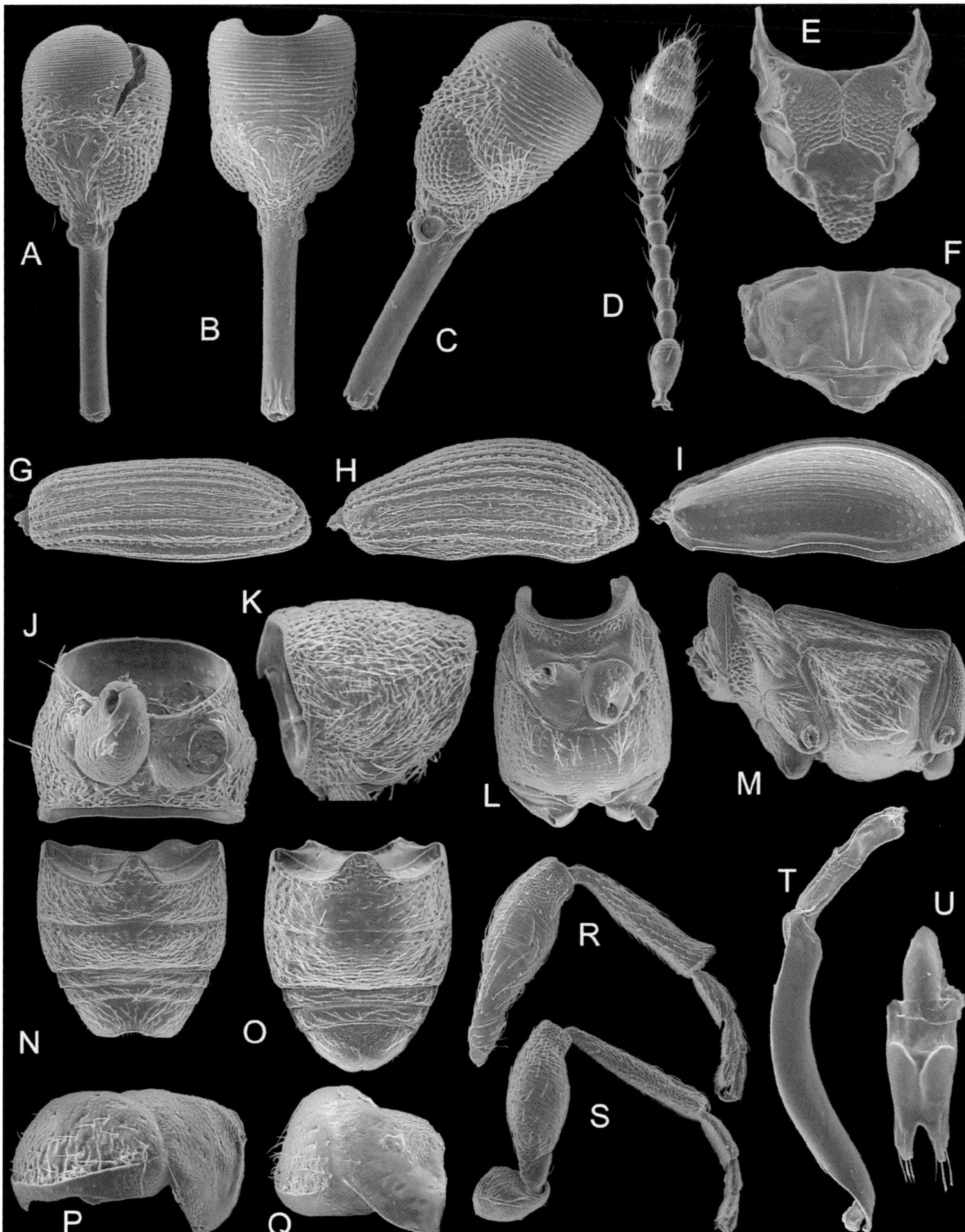

Fig. 43.34. Apioninae, structural details, *Rhinorhynchidius* sp. A–C, head: A, ♀, dorsal; B, ♂, dorsal; C, ♂, lateral; D, antenna; E, scutellum; F, metanotum; G–I, elytron: G, dorsal; H, lateral; I, inner lateral; J–K, prothorax; J, ventral; K, lateral; L–M, pterothorax: L, ventral; M, lateral; N–O, abdominal ventrites: N, ♂; O, ♀; P–Q, pygidium, ♂: P, dorsolateral; Q, lateral; R, middle leg; S, foreleg; T, penis, lateral; U, parameres, dorsal.

Acknowledgements

We thank Bruno Bell, Jiri Lochman, Nick Monaghan, Nick Porch and Nick Volpe for their photographs of live brentids and Adam Ślipiński for his support during of MW's visits to the ANIC during 2004 and 2008.

References

Adams A (1860) On some new genera and species of Mollusca from Japan. *Annals and Magazine of Natural History* (3) **5**(35), 299–303. doi:10.1080/00222936008697336

Alonso-Zarazaga MA (1989) Revision of the supraspecific taxa in the Palaearctic Apionidae Schoenherr, 1823. 1. Introduction and subfamily Nanophyinae Seidlitz, 1891 (Coleoptera, Curculionoidea). *Fragmenta Entomologica* **21**(2), 205–262.

Alonso-Zarazaga MA (1990) Revision of the supraspecific taxa in the Palaearctic Apionidae Schoenherr, 1823 (Coleoptera, Curculionoidea). 2. Subfamily Apioninae Schoenherr, 1823: introduction, keys and descriptions. *Graellsia* **46**, 19–156.

Alonso-Zarazaga MA (2004) Cyladinae Schoenherr, 1823 (Coleoptera, Curculionoidea). In *Brentidae of the World (Coleoptera, Curculionoidea)*. (Eds A Sforzi and L Bartolozzi) pp. 855–871. Monografia del Museo Regionale di Scienze Naturali, Torino **38**, 1–971.

Alonso-Zarazaga MA (2014) 3.6.4. Nanophyinae Gistel, 1848. In *Handbook of Zoology. Arthropoda: Insecta. Coleoptera, Beetles. Volume 3: Morphology and Systematics (Phytophaga)*. (Eds RAB Leschen and RG Beutel) pp. 416–423. Walter de Gruyter, Berlin/Boston, xii + 675 pp.

Alonso-Zarazaga MA, Lyal CHC (1999) *A World Catalogue of Families and Genera of Curculionoidea (Insecta: Coleoptera) (excepting Scolytidae and Platypodidae)*. Entomopraxis, Barcelona, 315 pp.

Alonso-Zarazaga MA, Wanat M (2014) 3.6.3. Apioninae Schoenherr, 1823. In *Handbook of Zoology. Arthropoda: Insecta. Coleoptera, Beetles. Volume 3: Morphology and Systematics (Phytophaga)*. (Eds RAB Leschen and RG Beutel) pp. 395–415. Walter de Gruyter, Berlin/Boston, xii + 675 pp.

Anderson RS, Kissinger DG (2002) Family 129. Brentidae: Billberg 1820. In *American Beetles. Vol. 2. Polyphaga: Scarabaeoidea through Curculionoidea*. (Eds RH Arnett, M Thomas, PE Skelley and JH Frank) pp. 711–719. CRC Press, Boca Raton, FL, 880 pp.

Beeson CFC (1941) *The Ecology and Control of the Forest Insects of India and the Neighbouring Countries*. Forest Research Institute, India, 1007 pp.

Beguin-Billecocq L (1905) Diagnoses d'espèces nouvelles d'Apionidae [Col.] provenant de la région Malgache et recueillies par MM. Alluaud, Decorse, Mocquerys, Perrier et Sicard. *Annales de la Société entomologique de France* **74**, 134–159.

Billberg GJ (1820) *Enumeratio Insectorum in Museo Gust. Joh. Billberg*. Typis Gadelianis, Stockholm, 138 pp. doi:10.5962/bhl.title.49763

Blackburn T (1892) Notes on Australian Coleoptera with descriptions of new species. Part XI. *Proceedings of the Linnean Society of New South Wales (Series 2)* **7**, 65–151.

Blackburn T (1894) Further notes on Australian Coleoptera with descriptions of new genera and species. Part XV. *Transactions of the Royal Society of South Australia* **18**, 139–168.

Blackburn T (1896) Further notes on Australian Coleoptera, with descriptions of new genera and species. Part XIX. *Transactions of the Royal Society of South Australia* **20**, 35–109.

Boisduval JA (1835) *Voyage de découvertes de l'Astrolabe, Exécuté par ordre du Roi, pendant les années 1826–1827–1828–1829, sous le commandement de M. J. Dumont d'Urville. Faune entomologique de l'Océan Pacifique, avec l'illustration des insectes nouveaux recueillis pendant le voyage. Deuxième Partie. Coléoptères et autres Ordres*. J. Tastu, Paris, vii + 716 pp.

Brookes DR, Hereward JP, Walter GH, Furlong MJ (2019) Origins, divergence, and contrasting invasion history of the sweet potato weevil pests *Cylas formicarius* (Coleoptera: Brentidae) and *Euscepes batatae* (Coleoptera: Curculionidae) in the Asia-Pacific. *Journal of Economic Entomology* **112**(6), 2931–2939.doi:10.1093/jee/toz198

Browne FG (1961) The biology of Malayan Scolytidae and Platypodidae. *Malyan Forest Records* **22**, 1–255.

Calabresi E (1921) Revisione del genere *Stratiorrhina* Pascoe (Coleop. Brentidae). *Bollettino della Società Entomologica Italiana* **53**, 42–52.

Clarke D, Limaye A, McKenna D, Oberprieler R (2018) The Weevil Fauna Preserved in Burmese Amber – Snapshot of a Unique, Extinct Lineage (Coleoptera: Curculionoidea). *Diversity* **11**(1). doi:10.3390/d11010001

Crowson RA (1955) *The Natural Classification of the Families of Coleoptera*. N. Lloyd, London, 187 pp.

Damoiseau R (1964) Contribution à la connaissance des Brentidae (Coleoptera-Phytophagoidea). 18. Genres *Calodromus* Guérin et *Cyphagogus* Parry: notes synonymiques et description d'espèces nouvelles. *Bulletin et Annales de la Société Royale Belgique d'Entomologie* **100**(36), 453–500.

Damoiseau R (1965) Contribution à la connaissance des Brentidae (Coleoptera-Curculionoidea). 22. Révision des Calodrominae palaeotropicaux et description d'espèces nouvelles. *Bulletin de l'Institut Royal des Sciences Naturelles de Belgique* **41**(34), 1–28.

Damoiseau R (1966a) Nouveaux Brentidae de l'Ancien Monde (Coleoptera-Curculionoidea). *Bulletin et Annales de la Société Royale Belgique d'Entomologie* **102**(9), 139–154.

Damoiseau R (1966b) Nouveaux Brentidae du Museum Frey (Coleoptera: Phytophagoidea). *Entomologische Arbeiten aus dem Museum G. Frey* **17**, 7–24.

Damoiseau R (1979) Les Amorphocephalini (Coleoptera-Brentidae). 1. *Amorphocephala* Schoenherr et sa parentèle. *Bulletin Institut Royal des Sciences Naturelles de Belgique* **51**(10), 1–35.

Damoiseau R (1980) Les Amorphocephalini (Coleoptera-Brentidae). 2. *Cordus* Schoenherr et les genres voisins. *Bulletin de l'Institut Royal des Sciences Naturelles de Belgique* **52**(21), 1–32.

Damoiseau R (1987) Contribution à la systématique et corrections à la nomenclature des Calodrominae (Coleoptera-Brentidae) II. Tribus des Hoplopisthiini, Atopobrentini et Stereodermini. *Bulletin de l'Institut Royal des Sciences Naturelles de Belgique* **57**, 31–91.

Davies JT, Ireson JE, Allen GR (2008) The phenology and impact of the gorse seed weevil, *Exapion ulicis*, on gorse, *Ulex europaeus*, in Tasmania. *Biological Control* **45**(1), 85–92. doi:10.1016/j.biocontrol.2008.01.009

Erichson WF (1842) Beitrag zur Insecten-Fauna von Vandiemensland, mit besonderer Berücksichtigung der geographischen Verbreitung der Insecten. *Archiv für Naturgeschichte* **8**(1), 83–287. doi:10.5962/bhl.part.21656

Evans JW (1942) The gorse weevil. *Tasmanian Journal of Agriculture* **13**, 15–18.

Fabricius JC (1775) *Systema entomologiae, sistens insectorum classes, ordines, genera, species, adiectis synonymis, locis, descriptionibus, observationibus*. Officina Libraria Kortii, Flensburgi et Lipsiae, xxxii + 832 pp. doi:10.5962/bhl.title.36510

Fabricius JC (1798) *Supplementum entomologiae systematicae*. Proft et Storch, Hafniae, ii + 573 pp.

Fall HC (1898) Revision of the species of Apion of America North of Mexico. *Transactions of the American Entomological Society (1890-)* **25**, 105–184.

Faust J (1897) Neue Gattungen und Arten in der Celeuthetiden-Gruppe. *Entomologische Zeitung (Stettin)* **58**, 229–298.

Faust J (1899) Viaggio di Lamberto Loria nella Papuasia Orientale xxiii. Curculionidae (Curculionidae von Neu Guinea). *Annali del Museo Civico di Storia Naturale Genova* **40**, 1–130.

Forno W, Heard TA, Day MD (1994) Host specificity and aspects of the biology of *Coelocephalapion aculeatum* (Coleoptera: Apionidae), a potential biological control agent of *Mimosa pigra* (Mimosaceae). *Environmental Entomology* **23**(1), 147–153. doi:10.1093/ee/23.1.147

Forster JR (1771) *Novæ species insectorum. Centuria I.* T. Davies & B. White, London, viii + 100 pp.

Friedman AL (2013) A new genus and species of Malvapiini (Coleoptera: Curculionoidea: Apionidae) from Papua New Guinea. *Caucasian Entomological Bulletin* **9**(1), 130–136. doi:10.23885/1814-3326-2013-9-1-130-136

Froggatt WF (1896) Life-histories of Australian Coleoptera. Part III. *Proceedings of the Linnean Society of New South Wales (series 2)* **10**, 325–336. doi:10.5962/bhl.part.24348

Gahan CJ (1900) Order 4. Coleoptera. In *A Monograph of Christmas Island (Indian Ocean): Physical Features and Geology; with Descriptions of the Fauna and Flora by Numerous Contributors.* (Ed. CW Andrews) pp. 89–127. British Museum (Natural History), London, xiv + 337 pp. doi:10.5962/bhl.title.25555

Gardner JCM (1925) Description of (a) the larva of *Cerobates tristriatus* Lund, (b) the larva of *Cerobates sexsulcatus* Motsch., (c) the pupa of *Cyphagogus corporaali* Kleine. *Indian Forest Records Calcutta* **11**, 189–194.

Gardner JCM (1935) Immature stages of Indian Coleoptera (18) (Brenthidae). *The Indian Forest Records (New Series)* **1**, 139–148.

Germar EF (1817) W. Kirby's Monographie der Gattung Apion Herbst. Aus dem Englischen übersetzt, mit Bemerkungen, und eingeschalteten Beschreibungen neuer Arten. *Magazin der Entomologie* **2**, 114–265.

Gerstaecker CEA (1854) Beschreibung neuer Arten der Gattung Apion Herbst. *Entomologische Zeitung (Stettin)* **15**, 234–261.

Gillett C, Crampton-Platt A, Timmermans MJ, Jordal BH, Emerson BC, *et al.* (2014) Bulk *de novo* mitogenome assembly from pooled total DNA elucidates the phylogeny of weevils (Coleoptera: Curculionoidea). *Molecular Biology and Evolution* **31**(8), 2223–2237. doi:10.1093/molbev/msu154

Gistel JNFX (1848) *Naturgeschichte des Thierreichs. Fur hohere Schulen.* Hoffmann, Stuttgart, xvi + 216 + 32 pls.

Goossens J (2011) Description d'une nouvelle espèce du genre *Miolispa* Pascoe. *Bulletin Société Royale Belge d'Entomologie* **147**, 137–141.

Guérin-Méneville FE (1831) Histoire naturelle, Zoologie. Atlas. In *Voyage autour du Monde, Exécuté par Ordre du Roi, sur la Corvette de Sa Majesté, La Coquille, pendant les années 1822, 1823, 1824 et 1825. Vol. 2.* (Ed. LI Duperrey). 21 pls. Arthus Bertrand, Paris. [for text see Guérin-Méneville, 1838] doi:10.5962/bhl.title.59769

Gunter NL, Oberprieler RG, Cameron SL (2016) Molecular phylogenetics of Australian weevils (Coleoptera: Curculionoidea): exploring relationships in a hyperdiverse lineage through comparison of independent analyses. *Austral Entomology* **55**(2), 217–233. doi:10.1111/aen.12173

Gyllenhal L (1833) [Taxonomic descriptions]. In *Genera et species curculionidum, cum synonymia hujus familiae; a C.J. Schoenherr. Species novae aut hactenus minus cognitae, descriptionibus a Dom. Leonardo Gyllenhal, C. H. Boheman, et entomologis aliis illustratae. Tomus primus. – Pars prima.* (Ed. CJ Schoenherr) 1833. Roret, Parisii, i–xv + 1–381.

Heard TA, Paynter Q (2009) *Mimosa pigra* L. (Leguminosae). In *Biological Control of Tropical Weeds Using Arthropods.* (Eds R Muniappan, GVP Reddy and A Raman) pp. 256–273. Cambridge University Press, Cambridge, UK, 495 pp.

Heller KM (1896) Remarks on the curculionid genus *Chalcocybebus. Notes from the Leyden Museum* **18**(1), 19–24.

Herbst JFW (1783) Kritisches Verzeichniss meiner Insektensammlunf. *Archiv der Insectengeschichte* **4**(1), 1–72.

Hustache A (1931) *Tableaux analytique des Coléoptères de la faune Franco-Rhénane (France, Hollande, Belgique, Région rhénane, Valais). Famille LXXXIX. Curculionidae. Tribu: Apioninae.* Publications des Miscellanea Entomologica, Toulouse, 286 pp.

Hunt T, Bergsten J, Levkanicova Z, Papadopoulou A, St. John O, *et al.* (2007) A comprehensive phylogeny of beetles reveals the evolutionary origins of a superradiation. *Science* **318**(5858), 1913–1916. doi:10.1126/science.1146954

Ireson JE, Davies JT (2012) *Ulex europaeus* L. – Gorse. In *Biological Control of Weeds in Australia.* (Eds M Julien, R McFadyen and J Cullen) pp. 581–590. CSIRO Publishing, Melbourne.

Ireson JE, Kwong RM, Gourlay H, Davies JT, Holloway RJ, *et al.* (2004) Progress on the biological control of gorse (*Ulex europaeus*) in Australia. In *Proceedings of the XI International Symposium on Biological Control of Weeds* (2004). (Eds JM Cullen, DT Briese, DJ Kriticos, WM Lonsdale, L Morin and JK Scott) pp. 415–418, CSIRO Entomology, Canberra, ACT.

Kirby W (1819) A century of insects, including several new genera described from his cabinet. *Transactions of the Linnean Society of London* **12**(2), 375–453. doi:10.1111/j.1095-8339.1817.tb00239.x

Kirby W, Spence W (1828) *An Introduction to Entomology: Or Elements of the Natural History of Insects: with Plates. Vol. III.* Longman, Rees, Orme, Brown and Green, London, 731 pp. doi:10.5962/bhl.title.8131

Kissinger DG (1968) *Curculionidae Subfamily Apioninae of North and Central America.* Taxonomic Publications, South Lancaster, MA, 559 pp.

Kissinger DG (1992) Apionidae from North and Central America. Part 4. Generic classification and introduction to the genus *Coelocephalapion* Wagner, with new species from Mexico and Venezuela (Coleoptera). *Insecta Mundi* **6**(2), 65–77.

Kleine R (1916a) Die Gattung *Baryrrhynchus* und ihr Verwandschaftskreis. *Entomologische Blätter* **12**, 121–137, 150–190.

Kleine R (1916b) Neue Taphroderini (Brenthidae, Col.). *Entomologische Mitteilungen* **5**, 1–92.

Kleine R (1917a) Beiträge zur Kenntnis der *Amorphocephalus*-Verwandtschaft. *Zoologische Mededelingen* **3**, 275–278.

Kleine R (1917b) Die Gattung *Eupsalis* und ihr Verwandtschaftskreis. *Archiv für Naturgeschichte* **82A**(4), 55–150.

Kleine R (1918) *Mastax*, ein neues Brenthidengenus aus Queensland. *Archiv für Naturgeschichte* **82A**(12), 162–168.

Kleine R (1919) Die Gattung *Ithystenus* Pascoe. *Archiv für Naturgeschichte* **83A**(7), 40–136.

Kleine R (1920) 1. Nachtrag zu meiner Arbeit über die Gattung *Ithystenus* Pascoe. *Archiv für Naturgeschichte* **85A**(5), 69–71.

Kleine R (1921) Die Gattung *Cyphagogus* Parry. *Archiv für Naturgeschichte* **87A**(6), 282–328.

Kleine R (1922a) Beitrag zur Kenntnis der Brenthidenfauna der Philippinen. *Archiv für Naturgeschichte* **87A**(10), 24–37.

Kleine R (1922b) Brethidenstudien. II. Folge. Archiv für Naturgeschichte **88A**, 201–222.

Kleine R (1922c) Neue Brenthiden aus dem Dahlemer Museum. *Archiv für Naturgeschichte* **88A** (3), 223–228.

Kleine R (1925) Fauna Buruana. Coleoptera Fam Brenthidae. *Treubia Buitenzorg* **7**, 14–37.

Kleine R (1926) Neue Gattungen und Arten aus der Familie Brenthidae. *Stettiner Entomologische Zeitung* **87**, 354–372.

Kleine R (1935) Bericht über die von Miss Cheesman in British Neu-Guinea gesammelten Brenthiden und Lyciden. *Nova Guinea (Zoologie)* **17**, 303–322.

Kleine R (1938) Coleoptera. Fam. Brenthidae. *Genera Insectorum. Fascicule CCVII* **207**, 1–197.

Kolbe HJ (1892) Beitrage zur Kenntniss der Brenthiden. I. Die Gruppe Cyphagoginae. *Stettiner Entomologische Zeitung* **53**(4/6), 162–175.

Kuschel G (1995) A phylogenetic classification of Curculionoidea to families and subfamilies. *Memoirs of the Entomological Society of Washington* **14**, 5–33.

Kuschel G (2003) Nemonychidae, Belidae, Brentidae (Insecta: Coleoptera: Curculionoidea). *Fauna of New Zealand* **45**, 1–100.

Kuschel G, Oberprieler RG, Rayner RJ (1994) Cretaceous weevils from southern Africa, with description of a new genus and species and phylogenetic and zoogeographical comments (Coleoptera: Curculionidae). *Entomologica Scandinavica* **25**, 137–149.

Lacordaire JT (1863) *Histoire Naturelle des Insectes. Genera des Coléoptères ou exposé méthodique et critique de tous les genres proposés jusqu'ici dans cet ordre d'insectes. Tome sixième, contenant les familles des Curculionides*. Libraire Encyclopédique de Roret, Paris, 637 pp.

Lacordaire T (1865[1866]) *Histoire naturelle des Insectes. Genera des Coléoptères ou exposé méthodique et critique de tous les genres proposés jusqu'ici dans cet ordre d'insectes. Tome septième*. Roret, Paris, 620 pp. doi:10.5962/bhl.title.67686

Latreille PA (1802) *Histoire Naturelle, générale et particulière, des Crustacés et des Insectes. Ouvrage faisant suite aux Oeuvres de Leclercq de Buffon, et partie du Cours complet d'Histoire naturelle rédigé par C. S. Sonnini, membre des plusieurs Sociétés savantes. Tome troisième*. F. Dufart, Paris, xii +467 pp. doi:10.5962/bhl.title.15764

Lawrence JF, Britton EB (1994) *Australian Beetles*. Melbourne University Press, Melbourne, 192 pp.

Lawrence JF, Ślipiński SA (2013) *Australian Beetles. Volume 1. Morphology, Classification and Keys*. CSIRO Publishing, Collingwood, Victoria, viii + 561 pp.

Lea AM (1899) Descriptions of new species of Australian Coleoptera. Part V. *Proceedings of the Linnean Society of New South Wales* **23**(4), 521–645. doi:10.5962/bhl.part.18131

Lea AM (1907) Descriptions of Australian Curculionidae, with notes on previously described species. Part V. *Transactions of the Royal Society of South Australia* **31**, 129–168.

Lea AM (1910a) Descriptions of Australian Curculionidae, with notes on previously described species. Part VIII. *Transactions of the Royal Society of South Australia* **34**, 13–58.

Lea AM (1910b) Notes on Australian Curculionidae in the collection of the German Entomological National Museum of Berlin (Col.). *Deutsche Entomologische Zeitschrift* **1910**, 153–172, 505–526.

Lea AM (1915a) Descriptions of new species of Australian Coleoptera. Part X. *Proceedings of the Linnean Society of New South Wales* **39**(4), 650–694.

Lea AM (1915b) On some Australian Malacodermidae and Curculionidae collected by Mr. G. E. Bryant [concluded]. *Annals and Magazine of Natural History (Series 8)* **15**(89), 452–481. doi:10.1080/00222931508693660

Lea AM (1926) On some Australian Curculionidae. *Proceedings of the Linnean Society of New South Wales* **51**(3), 327–362.

Legalov AA (2000) Eine neue Art der Gattung *Sterculapion* Rheinheimer, 1997 (Coleoptera: Apionidae) aus dem nordöstlichen Australien. *Russian Entomological Journal* **9**(3), 261–262.

Lewis G (1883) On Japan Brenthidae, and notes on their habits. *Journal of the Linnean Society of London* **17**, 295–302.

Lucas R (1920) Catalogus alphabeticus generum et subgenerum Coleopterorum orbis terrarum totius (famil., trib., subtrib., sect. incl.). Pars I. *Archiv für Naturgeschichte* **84A**(1), i–xxi, 1–128; (2), 129–288; (3), 289–448.

Lund NT (1800) Nogle arter af slaegten *Brentus*. *Skrivter af Naturhistorie-Selskabet* **5**(2), 54–91.

Louw S (2004) Microcerini Lacordaire, 1863 (Coleoptera, Curculionoidea). In *Brentidae of the World (Coleoptera, Curculionoidea)*. (Eds A Sforzi and L Bartolozzi) pp. 905–935. Monografia del Museo Regionale di Scienze Naturali, Torino **38**, 1–971.

Mantilleri A (2009) Révision des *Higonius* Lewis, 1883, et genres apparentés (Insecta, Coleoptera, Brentidae). *Zoosystema* **31**(4), 899–943. doi:10.5252/z2009n4a7

Mantilleri A (2010) A taxonomic review and phylogeny of the genera *Hoplopisthius* Senna, 1892 and *Carcinopisthius* Kolbe, 1892 (Coleoptera: Brentidae, Hoplopisthiini). *Zootaxa* **2516**(1), 1–26. doi:10.5281/zenodo.196132

Mantilleri A (2011) Revision of the tribe Microtrachelizini Zimmerman, 1994 from Australia: new taxa and records (Insecta: Coleoptera, Brentidae). *Records of the Australian Museum* **63**(1), 89–98. doi:10.3853/j.0067-1975.63.2011.1555

Mantilleri A (2012) Taxonomy, nomenclature and phylogeny of the tribes Hoplopisthiini Senna & Calabresi, 1919 and Microtrachelizini Zimmerman, 1994 (Coleoptera, Brentidae). *Zoosystema* **34**(3), 561–633. doi:10.5252/z2012n3a5

Marvaldi AE, Sequeira AS, O'Brien CW, Farrell BD (2002) Molecular and morphological phylogenetics of weevils (Coleoptera, Curculionoidea): do niche shifts accompany diversification? *Systematic Biology* **51**(5), 761–785. doi:10.1080/10635150290102465

May BM (1993) Larvae of Curculionoidea (Insecta: Coleoptera): a systematic overview. *Fauna of New Zealand* **28**, 1–223. doi:10.7931/J2/FNZ.28

May BM (1994) An introduction to the immature stages of Australian Curculionoidea. In *Australian Weevils (Coleoptera: Curculionoidea). Volume II. Brentidae, Eurhynchidae, Apionidae and a Chapter on Immature Stages by Brenda May*. (Ed. EC Zimmerman) pp. 365–728. CSIRO Australia, Melbourne, x + 755 pp.

McKenna DD, Sequeira AS, Marvaldi AE, Farrell BD (2009) Temporal lag and overlap in the diversification of weevils and flowering plants. *Proceedings of the National Academy of Sciences of the United States of America* **106**(17), 7083–7088. doi:10.1073/pnas.0810618106

Montrouzier X (1855) Essai sur la faune l'île de Woodlark ou Moiou. *Annales des Sciences Physiques et Naturelles, d'Agriculture et d'Industrie, Société impériale d'Agriculture (etc.) de Lyon (Series 2)* **7**(1), 1–114.

Montrouzier Z (1861) Essai sur la faune entomologique de la Nouvelle-Calédonie (Balade) et des îles des Pins, Art, Lifu, etc. (Suite). *Annales de la Société Entomologique de France (Series 3)* **8**, 867–916.

Moore KM (1972) Observations on some Australian forest insects. 27. Some insects attacking *Callitris hugelii* (White Cypress Pine). *Australian Zoologist* **17**(1), 40–46.

Morimoto K (1962) Key to families, subfamilies, tribes and genera of the superfamily Curculionoidea of Japan excluding Scolytidae, Platypodidae and Cossoninae. (Comparative morphology, phylogeny and systematics of the superfamily Curculionoidea of Japan. III). *Journal of the Faculty of Agriculture, Kyushu University* **12**(1), 21–66.

Morimoto K (1976) Notes on the family characters of Apionidae and Brentidae (Coleoptera), with key to the related families. *Kontyû* **44**(4), 469–476.

Motschulsky V de (1858) Insectes des Indes Orientales. *Études Entomologiques* **7**, 20–122.

Motschulsky V de (1866) Essai d'un catalogue des insectes de l'île de Ceylan. Supplément. *Bulletin de la Société des Naturalistes de Moscou* **39**(1), 393–446.

Nakane T (1963) New or little-known Coleoptera from Japan and its adjacent regions. XVI–XXII. *Fragmenta Coleopterologica Kyoto* **4–12**, 18–48.

Noble NS (1940) *Trichilogaster acaciae-longifoliae* (Froggatt) (Hymenopt., Chalcidoidea), a wasp causing galling of the flower-buds of *Acacia longifolia* Willd., *A. floribunda* Sieber and *A. sophorae* R. Br. *Transactions of the Entomological Society of London* **90**(2), 13–38. doi:10.1111/j.1365-2311.1940.tb02688.x

Oberprieler RG (2000) The larvae of the weevil tribe Eurhynchini and the phylogeny of the Brentidae (Coleoptera: Curculionidae). *Invertebrate Taxonomy* **14**(6), 755–770. doi:10.1071/IT00032

Oberprieler RG (2004) Eurhynchinae Lacordaire, 1863 (Coleoptera, Curculionoidea). In *Brentidae: of the World (Coleoptera: Curculionoidea)*. (Eds A Sforzi and L Bartolozzi) pp. 873–935. Monografia del Museo Regionale di Scienze Naturali, Torino **38**, 1–971.

Oberprieler RG (2014) 3.6 Brentidae Billberg, 1820. In *Handbook of Zoology. Arthropoda: Insecta. Coleoptera, Beetles. Volume 3: Morphology and Systematics (Phytophaga)*. (Eds RAB Leschen and RG Beutel) pp. 363–423. Walter de Gruyter, Berlin/Boston, xii + 675 pp.

Oberprieler RG, Marvaldi AE, Anderson RS (2007) Weevils, weevils, weevils everywhere. *Zootaxa* **520**(1668), 491–520. doi:10.11646/zootaxa.1668.1.24

Olivier AG (1790) *Entomologie, ou histoire naturelle des insectes, avec leurs caractères génériques et spécifiques, leur description, leur synonymie, et leur figure enluminée. Coléoptères, genera 9-34.* Baudouin, Paris, Vol. 2, 485 pp.

Ostermeyer N, Grace BS (2007) Establishment, distribution and abundance of *Mimosa pigra* biological control agents in northern Australia: implications for biological control. *BioControl* **52**(5), 703–720. doi:10.1007/s10526-006-9054-0

Parry JFS (1849) Descriptions of some new species of Coleoptera. *Transactions of the Entomological Society of London* **5**, 179–185. doi:10.1111/j.1365-2311.1849.tb01715.x

Pascoe FP (1862) Notes on the Brenthidae. *Journal of Entomology* **1**(5), 388–394.

Pascoe FP (1864) A new *Atractocerus* and a new *Cyphagogus*, and…notes respecting them. *Proceedings of the Entomological Society of London* **3**(2), 45–46.

Pascoe FP (1870) Contributions towards a knowledge of the Curculionidae [concl.]. *Journal of the Linnean Society of London, Zoology* **10**(48), 459–493. doi:10.1111/j.1096-3642.1870.tb00672.x

Pascoe FP (1872) Notes on Coleoptera, with descriptions of new genera and species. Part II. *Annals and Magazine of Natural History (Series 4)* **10**(59), 317–326. doi:10.1080/00222937208696709

Pascoe FP (1873) Additions to the Australian Curculionidae. Part V [continued]. *Annals and Magazine of Natural History (Series 4)* **12**(70), 278–286. doi:10.1080/00222937308680768

Pascoe FP (1874) Additions to the Australian Curculionidae. Part VI. *Annals and Magazine of Natural History (Series 4)* **12**(77), 383–389. doi:10.1080/00222937408680885

Pascoe FP (1875) Additions to the Australian Curculionidae. Part VIII. *Annals and Magazine of Natural History (Series 4)* **16**(91), 55–67. doi:10.1080/00222937508681122

Pascoe FP (1885) List of the Curculionidae of the Malay Archipelago collected by Dr. Odoardo Beccari, L. M. d'Albertis, and others. *Annali del Museo Civico di Storia Naturale Genova* **22**, 201–332.

Paynter Q (2004) Evaluating *Mimosa pigra* biological control in Australia. In *Research and Management of Mimosa pigra.* (Eds M Julien, G Flanagan, T Heard, B Hennecke, Q Paynter and C Wilson) pp. 141–148. CSIRO Entomology, Canberra, 173 pp.

Perroud BP, Montrouzier X (1865) Essai sur la faune entomologique de Kanala (Nouvelle Calédonie) et description de quelques espèces nouvelles ou peu connues. *Annales de la Société Linnéenne de Lyon (Series 2)* **11**, 46–257.

Power G (1879a) Notes pour servir à la monographie des Brenthides. *Annales de la Société Entomologique de France (Series 5)* **8**(4), 477–496.

Power G (1879b) Diagnoses de nouvelles espèces de Brenthides. *Petites Nouvelles Entomologiques* **2**(213), 297–298.

Pullen K, Jennings D, Oberprieler RG (2014) Annotated catalogue of Australian weevils (Coleoptera: Curculionoidea). *Zootaxa* **3896**(1), 1–481. doi:10.11646/zootaxa.3896.1.1

Rheinheimer J (1989) Notes on the host plants of some adult Australian weevils (Coleoptera: Curculionidae). *Australian Entomological Magazine* **16**(1), 27–31.

Rheinheimer J (1997) *Sterculapion*, eine neue Apioniden-Gattung aus Australien (Coleoptera: Apionidae). *Koleopterologische Rundschau* **67**, 257–262.

Rheinheimer J (1998) Biology and worldwide distribution of some weevil genera (Coleoptera: Curculionoidea). In *Taxonomy, Ecology and Distribution of Curculionoidea (Coleoptera: Polyphaga).* Proceedings of a Symposium (28 August 1996, Florence, Italy) XX International Congress of Entomology (Eds E Colonnelli, S Louw and G Osella) pp. 287–294. Atti del Museo Regionale di Scienze Naturali, Torino, 1–294.

Roelofs W (1879) Diagnoses de nouvelles espèces de Curculionides, Brenthides, Anthribides et Bruchides du Japon. *Annales de la Société Entomologique de Belgique (Comptes-rendus)* **22**, 53–55. doi:10.1002/mmnd.48018790218

Sanborne M (1981) Biology of *Ithycerus noveboracensis* (Forster) (Coleoptera) and weevil phylogeny. *Evolutionary Monographs* **4**, 1–80.

Schedl KE (1959) Brenthidae of the Australian Region, I. *Memoirs of the National Museum, Melbourne* **24**, 59–65. doi:10.24199/j.mmv.1959.24.05

Schilsky J (1901) *Die Käfer Europa's. Nach der Natur beschrieben von Dr. H. C. Küster und Dr. G. Kraatz, fortgesetzt von J. Schilsky. Heft 38.* Bauer und Raspe, Nürnberg, i–vi + A–K + 100 nrs.

Schilsky J (1906) *Die Käfer Europa's. Nach der Natur beschrieben von Dr. H. C. Küster und Dr. G. Kraatz, fortgesetzt von J. Schilsky. Heft 43.* Bauer und Raspe, Nürnberg, i–v + 1–119 (as I–CXIX) + 30 nrs.

Schoenherr CJ (1823) Curculionides [Tabula synoptica familiae Curculionidum]. *Isis von Oken* **1823**(10), columns 1132–1146.

Schoenherr CJ (1839) *Genera et species curculionidum, cum synonymia hujus familiae. Species novae aut hactenus minus cognitae, descriptionibus a Dom. Leonardo Gyllenhal, C.H. Boheman, et entomologis aliis illustratae.* Roret, Paris, Vol. 5(1) pp. 1–456.

Schoenherr CJ (1840) *Genera et species curculionidum, cum synonymia hujus familiae. Species novae aut hactenus minus cognitae, descriptionibus a Dom. L. Gyllenhal, C. H. Boheman, O. J. Fahraeus et entomologis aliis illustratae. Tomus quintus. Pars secunda. Supplementum continens.* Roret, Parisii/Fred. Fleischer, Lipsia, pp. 465–970 (+ 4 pp. Corrigenda).

Schoenherr CJ (1847) *Mantissa secunda familiae Curculionidum. Seu descriptiones novorum quorundam generum Curculionidum.* Norstedt et filii, Holmia, 86 pp. doi:10.5962/bhl.title.5794

Schönfeldt H von (1906) Brenthidae. Nova Guinea, Resultats de l'Expedition scientifique Neerlandaise a la Nouvelle-Guinee en 1903. sous les auspices de Arthur Wichmann. *Chef de l'Expedition* **5**(1), 35–36.

Senna A (1892) Viaggio di Leonardo Fea in Birmania e regioni vicine: Brentidi. *Annali del Museo Civico di Storia Naturale Genova (Series 2)* **12**, 429–494.

Senna A (1893a) Contribuzioni allo studio dei Brentidi. Nota XIII. Diagnosi di alcune nuove specie del genere *Hoplopisthius. Bollettino della Società Entomologica Italiana* **24**, 253–256.

Senna A (1893b) Contribuzioni allo studio dei Brentidi. Nota XXII. Su alcuni Brentidi importati col tabacco secco di Sumatra. *Bulletino della Société Entomologica Italiana* **25**, 288–330.

Senna A (1893c) Contribuzioni allo studio dei Brentidi. Nota XVI. Brentidi raccolti nell'Isola di Engano dal Dott. Elio Modigliani. *Annali del Museo Civico di Storia Naturale Genova (Series 2)* **13**, 256–284.

Senna A (1894) Viaggio di Lamberto Loria nella Papuasia orientale. *Annali di Museo Civico di Storia Naturale di Genova (Series 2)* **14**, 554–564.

Senna A (1895) Description of two new Brenthidae. *Notes from the Leyden Museum* **17**(1/3), 57–62.

Senna A (1896) On the genus *Cerobates* Schh. and description of some new species. *Notes from the Leyden Museum* **17**(4), 209–224.

Senna A (1897) Descriptions de quelques espèces nouvelles de brenthides du Musée Royal de Belgique et de ma collection et notes diverses. *Annales de la Société Entomologique de Belgique (Comptes-rendus)* **41**, 227–243.

Senna A (1898) On a new genus and several new species of Brenthidae. *Notes from the Leyden Museum* **20**(1), 65–76.

Senna A (1899) Aggiunte alla fauna brenthologica di Celebes. *Bollettino della Société Entomologica Italiana* **31**, 300–311.

Senna A (1903) Brentidi delle regione di Batacchi Indipendenti (Sumatra). *Bulletino della Société Entomologica Italiana* **34**, 152–178.

Senna A (1919) Nuove specie di *Cordus*, Schh. *Bollettino della Società Entomologica Italiana* **50**, 78–83.

Setliff GP (2007) Annotated checklist of weevils from the Papuan region (Coleoptera, Curculionidae). *Zootaxa* **1536**(1536), 1–296. doi:10.11646/zootaxa.1536.1.1

Sforzi A, Bartolozzi L (2004) *Brentidae of the World (Coleoptera, Curculionoidea).* Museo Regionale di Scienzia Naturali, Torino, 976 pp.

Sforzi A, Bartolozzi L, Leschen RAB (2014) 3.6.2 Brentinae Billberg, 1820. In *Handbook of Zoology. Arthropoda: Insecta. Coleoptera, Beetles. Volume*

3: Morphology and Systematics (Phytophaga). (Eds RAB Leschen and RG Beutel) pp. 384–395. Walter de Gruyter, Berlin/Boston, xii + 675 pp.

Shin S, Clarke DJ, Lemmon AR, Moriarty Lemmon E, Aitken AL, *et al.* (2018) Phylogenomic data yield new and robust insights into the phylogeny and evolution of weevils. *Molecular Biology and Evolution* **35**(4), 823–836. doi:10.1093/molbev/msx324

Thompson RT (1992) Observations on the morphology and classification of weevils (Coleoptera, Curculionoidea) with a key to major groups. *Journal of Natural History* **26**(4), 835–891. doi:10.1080/00222939200770511

Thompson RT (1996) The species of *Phaenomerus* Schönherr (Coleoptera: Curculionidae: Zygopinae) of the Australian Region. *Invertebrate Systematics* **10**(5), 937–993. doi:10.1071/IT9960937

Toki W, Yoshitake H (2012) Additional records and preliminary data on the biology of *Callipareius* (*Callipareius*) *japonicus* (Nakane, 1963) (Coleoptera, Brentidae). *Elytra* **2**(2), 147–149.

Van Emden F (1938) On the taxonomy of Rhynchophora larvae (Coleoptera). *Transactions of the Royal Entomological Society of London* **87**(1), 1–37. doi:10.1111/j.1365-2311.1938.tb01800.x

Voss E (1922) Monographische Bearbeitung der Unterfamilie Rhynchitinae (Curc.). I. Teil: Nemonychini-Auletini. *Archiv für Naturgeschichte, Abteilung A* **88**(8), 1–113.

Voss E (1959) Ein Beitrag zur Kenntnis der Apioniden-Fauna Zentral-Afrikas (Coleoptera Curculionoidea). *Annales du Musée Royal du Congo Belge, Zoologie* **76**, 7–119.

Wagner H (1907a) Beitrag zur Kenntnis der südafrikanischen Apioniden. *Mitteilungen der Schweizerischen Entomologischen Gesellschaft* **11**(6), 259–265.

Wagner H (1907b) Neue Apioniden aus Afrika aus dem Königl. Naturh. Museum zu Brüssel. *Annales de la Société Entomologique de Belgique* **51**, 271–279.

Wagner H (1908) Neue central-und sudafrikanische Apionen. *Stettiner Entomologische Zeitung* **69**, 59–92.

Wagner H (1914) Beitrag zur Kenntnis der Apion-Fauna Zentral- und Sud-Amerikas. Beschreibungen neuer Arten. *Archiv für Naturgeschichte* **79**, 137–164.

Wanat M (1990) Apionidae (Coleoptera, Curculionoidea) of the Arabian Peninsula. *Fauna of Saudi Arabia* **11**, 55–81.

Wanat M (1995) Systematics and phylogeny of the tribe Ceratapiini (Coleoptera: Curculionoidea: Apionidae). *Genus* **(Suppl)**, 1–406.

Wanat M (2001) *Genera of Australo-Pacific Rhadinocybinae and Myrmacicelinae, with Biogeography of the Apionidae (Coleoptera: Curculionoidea) and Phylogeny of the Brentidae (s. lato)*. Mantis, Olsztyn, 432 pp.

Wanat M (2021) New basal taxa of South African Apioninae (Coleoptera: Curculionoidea: Brentidae). *Zootaxa* **5035**(1), 1–60. doi:10.11646/zootaxa.5035.1.1

Wanat M (2023) *Diselmobius*, a new basal genus of apionine weevil (Coleoptera: Brentidae: Apioninae) from Tasmania. *Annales Zoologici* **73**(3), 389–396. doi:10.3161/00034541ANZ2023.73.3.003

Wanat M, Escalona HE (2024) Notes on systematics and distribution of Australian Apioninae (Coleoptera: Brentidae). *Annales Zoologici* **74**(2), 337–349. doi:10.3161/00034541ANZ2024.74.2.007

Westwood JO (1847) *The Cabinet of Oriental Entomology; Being a Selection of Some of the Rarer and More Beautiful Species of Insects, Natives of India and the Adjacent Islands, the Greater Portion of which are now for the First Time Described and Figured*. William Smith, London, 88 pp. doi:10.5962/bhl.title.34273

Wolfe GW (1991) The origin and dispersal of the pest species of *Cylas* with a key to the pest species groups of the world. In *Sweet Potato Pest Management* (Eds RK Jansson and KV Raman) pp. 13–43. CRC Press, 458 pp. doi:10.1201/9780429308109

Yeoh PB, Julien MH, Scott JK (2012) *Emex australis* Steinheil – doublegee and *Emex spinosa* (L.) Campdera – Lesser Jack. In *Biological control of weeds in Australia*. (Eds MH Julien and MH McFadyen) pp. 238–255. CSIRO Publishing, Melbourne, 648 pp.

Zherikhin VV, Gratshev VG (2004) Fossil Curculionoid Beetles (Coleoptera, Curculionoidea) from the Lower Cretaceous of Northeastern Brazil. *Paleontological Journal* **38**(5), 528–537.

Zimmerman EC (1991) *Australian Weevils (Coleoptera: Curculionoidea). Volume V. Colour Plates 1-304*. CSIRO Australia, Melbourne, x + 633 pp.

Zimmerman, EC (1992) *Australian Weevils (Coleoptera: Curculionoidea). Volume VI. Colour Plates 305-632*. CSIRO Australia, Melbourne, viii + 707 pp.

Zimmerman EC (1993) *Australian Weevils (Coleoptera: Curculionoidea). Volume III. Nanophyidae, Rhynchophoridae, Erirhinidae, Curculionidae: Amycterinae, Literature Consulted*. CSIRO Australia, Melbourne, x + 854 pp.

Zimmerman EC (1994) *Australian Weevils (Coleoptera: Curculionoidea). Volume II. Brentidae, Eurhynchidae, Apionidae and a Chapter on Immature Stages by Brenda May*. CSIRO Australia, Melbourne, x + 755 pp.

44. CURCULIONIDAE LATREILLE, 1802

Hermes E. Escalona, Debbie Jennings and Rolf G. Oberprieler

Fig. 44.1. *Curculio simulator* Zimmerman & Oberprieler.

Introduction. The Curculionidae are the largest family of Curculionoidea, with about 53 000 described species in about 4800 genera worldwide (Oberprieler *et al.* 2007). Although the concept of the family and its position in Curculionoidea as sister-group of Brentidae have now been well established by morphological (Kuschel 1995; Marvaldi & Morrone 2000) and molecular (McKenna *et al.* 2009, 2019; Shin *et al.* 2018) data, its internal classification is still a work in progress, with no resolution and general agreement on the number and constitution of its subfamilies and tribes. The only consistent pattern emerging from all recent phylogenetic analyses is that the family comprises a number of more basal lineages, classified as subfamilies Brachycerinae (including Erirhininae), Dryophthorinae, Platypodinae and Bagoinae, and two sister clades of 'higher' Curculionidae, termed CEGH and CCCMS by Gunter *et al.* (2016) and roughly corresponding to the broad-snouted and long-snouted groups sometimes recognised (e.g. Thompson 1992). The CEGH clade includes the current subfamilies Cyclominae, Entiminae and Hyperinae and the unplaced tribes Gonipterini, Phrynixini and Viticiini, whereas the CCCMS clade contains the current subfamilies Curculioninae, Conoderinae, Cossoninae, Molytinae and Scolytinae. However, a recent molecular analysis of Curculioninae found several genera (particularly Australian ones) traditionally placed in this subfamily (thus in CCCMS) to belong in the CEGH clade instead (Haran *et al.* 2023), and analyses sampling more taxa of the CCCMS clade (McKenna *et al.* 2009; Gillett *et al.* 2014; Riedel *et al.* 2016; Mugu *et al.* 2018; Shin *et al.* 2018; Hsiao *et al.* 2023) have also failed to recover plausible subfamilies in this clade except, to a degree, Scolytinae. As a result, the composition and classification of both the CEGH and CCCMS clades remain highly unsettled and the concepts of nearly all traditional subfamilies of Curculionidae in doubt. The Australian fauna includes representatives of all the current subfamilies but in uneven proportions; an analysis of their numbers was conducted by Pullen *et al.* (2014). The subfamilies with the highest numbers of genera and species in Australia are Molytinae (mainly the tribe Cryptorhynchini) and Entiminae (mainly the tribe Leptopiini), and, as in other regions of the world, these groups also include the largest numbers of undescribed genera and species. The only endemic Australian tribe are the Amycterini (Cyclominae), which include several genera and species adapted to dry environments, especially in Western Australia (Zimmerman 1993).

The host associations and life histories of Curculionidae are very diverse, in some cases being generalist and opportunistic but in others much more specific. **Brachycerinae** are mostly associated with monocotyledonous angiosperms, including aquatic and semi-aquatic taxa (Tanysphyrini), but the larvae of a few appear to develop in tissues of dicotyledons (*Aonychus*, *Tadius*), including roots in the soil (Myrtonymini). **Bagoinae** also live on aquatic monocotyledons, with the apparently related tribe Notiomimetini occurring along beaches of the southern Australian coast in association with seagrasses. **Dryophthorinae** are closely associated with terrestrial monocotyledons, many with palms but others with bananas, pandans, grasstrees and grasses and a few with dead wood and stored plant products such as grain. Most **Platypodinae** are ambrosia beetles that cultivate and feed on yeast-like fungi. Little is known about the hosts and habits of the Australian **Hyperinae**, except that the larvae of some genera (*Gerynassa* and relatives) mine in leaves and others may have ectophytic larvae feeding on leaves of various plant taxa; pupation of all genera is likely to occur in meshed cocoons as typical for this subfamily (Oberprieler *et al.* 2014b). The diverse and taxonomically poorly resolved **Cyclominae** have diverse host associations and life histories, the larvae of some developing in underground bulbs, tubers and rhizomes (Amycterini), of others boring in shoots, stems and roots (Aterpini), tunnelling in stems or roots of terrestrial and aquatic plants (Listroderini) or feeding ectophytically on leaves (Gonipterini, Rhythirrinini). Related taxa of as yet uncertain status include the Phrynixini, which live among leaf litter and probably have wood-boring larvae, and the Viticiini, whose larvae mine in leaves, both living and dead in the case of *Geochus* (May 1992; Leschen *et al.* 2022). **Entiminae** feed on leaves of a multitude of plants as adults and on roots in the soil as larvae, with several introduced species of the genera *Naupactus*, *Otiorhynchus*, *Phlyctinus* and *Sitona* regarded as agricultural and horticultural pests (Oberprieler & Zimmerman 2020). The larvae of **Curculioninae** develop mainly in flower buds, fruits and seeds of many various plants and also include leaf miners, and several species are known or indicated to play a key role in the pollination of their hosts. **Conoderinae**, another taxonomically poorly resolved subfamily, are only poorly represented in Australia, and little is known about their life histories, except that their larvae develop in seeds, fruits and stems of various plants. The larvae of **Cossoninae** mostly develop in old and rotten wood and are easily dispersed in this medium over large distances, but some develop in seeds and even in cycad cones (*Zimmiodes*); the taxonomy of the Australian species is also poorly understood. **Molytinae** are by and large borers in living and dead wood, including stems and branches as well as fruits, seeds and leaf litter, with the tribe Cryptorhynchini being particularly diverse and speciose and several species of the *Tranes* group being cycad pollinators. **Scolytinae** or bark and ambrosia beetles construct galleries in the phloem layer beneath the bark of living or dead tree trunks and include ambrosia beetles (Xyleborini), with several species of economic relevance introduced into Australia.

Characteristics. Adults. Body length 0.8–60 (usually 1.5–40) mm (SL). Body about 1.2–5.7 × as long as wide, most elongate in some Aterpini, Cossoninae and Platypodinae, usually with pronotum and elytra independently rounded or with elytra more or less parallel-sided, usually slightly flattened to strongly convex (strongly flattened in a few Cossoninae), colour usually brown to black, rarely bicolored (e.g. some species of *Cossonus*) or metallic (e.g. *Ipsichora*); vestiture usually consisting of fine hairs to stout setae or scales, varying from white to brown or black in colour but usually in part metallic green or blue (e.g. Entiminae, Hyperinae) often forming distinct patterns, rarely obsolete to absent (e.g. some Cossoninae and Conoderinae). Head usually more or less prognathous but sometimes variously declined (Cryptorhynchini, some Conoderinae), behind eyes rarely abruptly constricted posteriorly to form temples and neck, slightly constricted in *Rhadinosomus*; occipital region without transverse ridge or stridulatory file but with longitudinal stridulatory ridge (plectrum) in some females of Scolytinae (Bedoya *et al.* 2021), with median occipital endocarina in most Platypodinae and Scolytinae; anterior part typically extending forward to form long, narrow, subcylindrical to somewhat flattened rostrum, but also rostrum short and broad (e.g. most Entiminae, some Cossoninae) and virtually absent in most Scolytinae and Platypodinae; frontoclypeal suture absent but frons often more or less distinct in some groups (e.g. Entiminae). Eyes usually entire, finely or coarsely facetted, without interfacettal setae; not or only slightly protuberant but prominent in some taxa (e.g. some Aterpini and Entiminae), enlarged and dorsally approximate or contiguous in others (e.g. *Mecopus*, *Orchestes*, *Phaenomerus*), shallowly to deeply emarginate in some Scolytinae, sometimes reduced (microphthalmic condition) (e.g. some Notiomimetini, Phrynixini, several undescribed subterranean Molytinae and Cyclominae, Pls 57H–L, 98E) to entirely absent (anophthalmic condition) (e.g. *Halorhynchus*, *Myrtonymus*, *Undarobius*, *Tasmanica*; see Escalona & Oberprieler 2021). Antennal insertions often concealed from above but partly to fully exposed dorsally in many Entiminae and some Platypodinae, located at various positions on rostrum; often extending posteriad into long, narrow and deep scrobe but sometimes only into short, broad and shallow depression (some Entiminae, Platypodinae, Scolytinae). Antennae 11- or 12-segmented, about as long as or longer than rostrum and head; scapes elongate, usually >3 × as long as pedicel, apically clavate; funicles usually 7-segmented but 6-segmented in Amycterini, Tanysphyrini and most Dryophthorinae, 5-segmented in many Cossoninae, 4-segmented in *Dryophthoroides*, *Dryophthorus*, *Stenommatus*, some Tanysphyrini and most Platypodinae, 3-segmented in *Notoplatypus* and 3- to 6-segmented in some Scolytinae; clubs compact, usually 3- or 4-segmented but 1-segmented (segments fused) in Dryophthorinae and Platypodinae. Mandibles usually short and broad but strongly reduced when rostrum long and narrow, sometimes large when rostrum very short (e.g. Amycterini); on outside usually with few long stiff setae (paucisetose condition) but sometimes with many setae or scales (plurisetose, plurisquamose), in most Entiminae apically or apicoventrally also with dehiscent cusp leaving characteristic scar when discarded; on inside more or less truncate with one to three teeth (unidentate, bidentate, tridentate conditions), mola and prostheca absent, inner edge of mandibular base usually with long pubescent process (short or absent in Platypodinae and Scolytinae). Maxillae with galea and lacinia fused into mala (but divided in Sitonini), mala short and broad to moderately elongate, setose or spinose; maxillary palp usually 3-segmented with distinct palpifer and cylindrical apical segment extending slightly beyond malar apex. Postmentum usually elongate, fused at base; prementum variable; ligula present or absent; labial palp usually 2-segmented with distinct palpiger, narrowly separated; apical segment cylindrical. Gular sutures fused. Corpotentorium narrow, incomplete or absent; cervical sclerites absent.

Pronotum about 0.45–2.65 × as long as wide, base not to distinctly narrower than combined elytral bases; sides straight to strongly curved or sinuate; lateral pronotal carinae almost always absent but weakly defined and incomplete in Platypodinae and some Scolytinae (associated with crural impressions); anterior angles not produced forward; posterior angles rounded to obtuse or right; posterior edge usually straight or weakly rounded, sometimes sinuate or lobed; disc usually simple, rarely with basal impressions, median longitudinal groove or sublateral carinae, sometimes with anterior collar. Prosternum shorter or longer than shortest diameter of procoxal cavity, usually flat to slightly convex or slightly concave, sometimes strongly concave forming a canal to receive rostrum in repose (e.g. Aedemonini, Cryptorhynchini, *Aonychus*, *Lobotrachelus*, *Hypurus*) or with a canal in front of the procoxae, also termed pectoral canal (Figs 44.94F, 44.102C, e.g. several Storeini and Cleogonini); prosternal process more or less complete but sometimes broadly interrupted at middle, intercoxal process then strongly concave and apex represented by disjunct but often small sclerite (sternellum), sometimes completely fused to postcoxal processes, suture between prosternal and postcoxal (hypomeral) processes distinct or not; notosternal sutures usually incomplete or absent. Procoxae more or less globular but usually projecting at least slightly below prosternum, with short or occasionally long concealed lateral extensions; trochantins always concealed; procoxal cavities subcircular, internally broadly closed, externally usually contiguous (mesally open) but well separated (mesally closed) in taxa with broad prosternal and postcoxal processes (e.g.

flattened forms such as some Cossoninae) or in taxa with rostral canal; postcoxal (hypomeral) processes joined behind procoxal cavities or sternellum, junction often obliterated but sometimes distinct, sometimes extending into single or bifurcate spine (some Entiminae: Cyphicerini).

Scutellum without stridulatory files; scutellar shield often reduced and sometimes absent, usually abruptly elevated, anteriorly simple, posteriorly acute to broadly rounded. Elytra about 0.7–4.2 × as long as wide and 1.15–4.6 × as long as pronotum; sides subparallel to broadly rounded laterally, with elytral apices usually conjointly rounded, exposing no abdominal tergites, but occasionally independently rounded (e.g. *Ectemnorhinus*), independently acute (e.g. *Methypora*, *Rhadinosomus*) or truncate exposing pygidium from above (e.g. *Sitophilus*, some Curculioninae, Mesoptiliini); underside with subapical stridulatory file in Platypodinae, Listroderini, Orthorhinini, Cossoninae, Scolytinae, Ceutorhynchini and a few Curculioninae, or apical file in some Amycterini; surface usually simple, occasionally tuberculate or dentate, with 9 or 10 striae but no scutellary striole; epipleura very narrow or absent. Mesoventrite usually separated by complete sutures from mesanepisterna, but partly or completely fused to mesanepisterna in various groups; anterior edge usually on same plane as metaventrite but on different plane in taxa with mesothoracic extension of the rostral groove (e.g. Cryptorhynchini), without paired procoxal rests but usually with narrow transverse prothoracic rest continuing onto mesanepisterna. Mesoventral cavity absent but mesoventrite sometimes deeply concave forming continuation of rostral canal. Mesocoxae usually not or only slightly projecting, with concealed trochantins; mesocoxal cavities narrowly to very widely separated, subcircular and closed laterally by meeting of mesoventrite and metaventrite. Mesoventral process narrowly rounded to broadly truncate; mesometaventral junction simple or complex, often a straight or curved line, occasionally absent due to fusion of ventrites. Metaventrite moderately to strongly convex; transverse (metakatepisternal) suture absent or not extending across midline; discrimen long to short or absent; exposed portion of metanepisterna usually long and narrow, with modified scales (sclerolepidia) lining inner edge in various Conoderinae, Cossoninae, Molytinae and Scolytinae; metepimera usually concealed. Metacoxae usually moderately to widely separated but narrowly (*Peliocis*) to very narrowly so or contiguous in Platypodinae and some Scolytinae, sometimes extending laterad to reach elytra. Metendosternite usually with long stalk, long lateral arms, reduced or no laminae and anterior tendons located on arms. Hindwings with apical field more than half as long as entire wing, with 2 remnants of RP; radial cell usually highly reduced or absent; cross-vein r_3 usually absent and r_4 short or absent; basal portion of RP short or absent; R-M loop forming narrowly acute angle; medial spur often sinuate; medial field with 3 or fewer free veins and no medial fleck; MP_{3+4} without base; wedge cell absent; anal lobe moderately well-developed without embayment, occasionally reduced with notch-like embayment (e.g. some Scolytinae).

Legs short and stout to long and slender; trochanterofemoral joints usually strongly oblique, sometimes with femur abutting coxa; femora often somewhat inflated in middle or at apex, sometimes enlarged and ventrally dentate (e.g. Ochyromerini, Trachodini), especially metafemora (Campyloscelini, Eugnomini, Rhamphini); tibiae often expanded and obliquely truncate at apex to form ovate corbel (Oberprieler & Zimmerman 2020), either on outside above row of fringing setae (true corbel; Fig. 44.58I) in many Entiminae and some Brachycerinae or on inside of fringing setae (false corbel; Fig. 44.58K–L) in various groups; inner apical angle often with tooth-like process (mucro) or outer apical angle with inwardly curved spine (uncus), sometimes with additional inner preapical tooth (premucro; e.g. *Tentegia*, Fig. 37.10I), subapically sometimes expanded with outer edge ending in a large tooth (e.g. *Psepholax*), teeth or socketed spines in most Scolytinae (not *Scolytus*), some Platypodinae and a few Cossoninae; tibial spurs almost always reduced or absent but present in some Cyclominae (Myrtonymini, Listroderini, some Aterpini; Fig. 44.57G–I) and some Entiminae. Tarsi usually 5-5-5-segmented (pseudotetramerous) but sometimes 4-4-4 (Myrtonymini, Stromboscerini, *Cryptoplus*) or 3-3-3 (terminal two segments lost) (e.g. *Achelocis, Aonychus*, *Cenchrena*, *Cisolea*, *Geochus*, *Syarbis*, *Zeopus*), segment 3 distinctly lobed beneath but basal three segments not or only weakly lobed in some groups (e.g. Dryophthorini, Platypodinae, Scolytinae, some Cossoninae); tarsal claws usually simple but sometimes dentate or bifid, mostly divergent but sometimes divaricate or connate to reduce to one claw (e.g. some Cryptoplini), absent when onychium lost; empodium present or not.

Abdomen with 5 ventrites; ventrites 1 and 2 connate, together usually longer than 3 and 4 combined (not in Platypodinae), ventrite 1 rarely with postcoxal lines (e.g. males of *Orpha*; Fig. 44.78L); intercoxal process usually broadly rounded to truncate but narrowly rounded to acute in some Platypodinae and Scolytinae; ventrite 5 with apex usually broadly rounded to truncate. Abdominal tergites I–VI or I–VII weakly to moderately sclerotised but VIII and sometimes VII more heavily so, IV–VII usually with patches of wing-folding asperities; I–VII with functional spiracles but spiracles on VII sometimes reduced (non-functional) or absent, VIII occasionally with non-functional spiracles (Thompson 1992), anterior spiracles located in pleural membrane, those on VII sometimes on tergite; tergite VIII in males usually heavily sclerotised to form pygidium, sometimes exposed beyond elytral apices but concealed beneath VII in males of Dryophthorinae, Platypodinae, some Brachycerinae and some Scolytinae, in female weakly sclerotised and concealed beneath VII. Sternite VIII reduced in males, anteriorly sometimes with remnant of apodeme (spiculum relictum), in females larger with stout anterior apodeme (spiculum ventrale). Tergite X membranous; sternite X in males consisting of basal plate and usually long, stout anterior apodeme (spiculum gastrale). Aedeagus of pedotectal (Brachycerinae, some Dryophthorinae) or pedal type (all other groups); tegmen ring-like, dorsally with basal plate bearing parameral lobes or with pair of separate parameres (sometimes reduced to absent, e.g. Platypodinae, many Cossoninae), ventrally with long anterior apodeme (manubrium); penis with paired anterior apodemes (temones); endophallus various in shape, often with flagellum or smaller copulatory sclerites and fields of spines or asperities in wall. Ovipositor highly variable, short and broad to narrowly elongate; gonocoxites usually undivided, styli moderately well-developed to reduced or absent. Internal genital tract of females usually with heavily sclerotised spermathecal capsule. (Zimmerman 1991, 1992, 1993; Thompson 1992; Howden 1995; Kuschel 1995; Lyal & King 1996; Anderson 2002; Marvaldi *et al.* 2002; Lyal *et al.* 2006; Oberprieler *et al.* 2007; Oberprieler 2010, 2014a; Lawrence & Ślipiński 2013).

Larvae. Body elongate and cylindrical to shorter, broader and wider at middle, usually slightly to strongly curved ventrally (C-shaped); ventrally flattened in Gonipterini and probably body flattened in leaf miners as well (*Austrocis*, *Geochus*, *Gerynassa*); all surfaces weakly sclerotised, usually except mouth frame and mandibles or entire head capsule, sometimes protergum and rarely tergum IX; vestiture usually consisting of scattered long and short, simple setae, occasionally dense. Head almost always protracted and strongly declined (hypognathous), rarely retracted; posterior edge of head capsule not emarginate; epicranial stem long; frontal arms V-shaped, contiguous at base, not extending to mandibular membrane; median endocarina located beneath epicranial stem and usually extending forward between frontal arms. Postoccipital condyles usually present. Stemmata sometimes 1 or 2 pairs, often absent. Antennae very short, 1-segmented, with terminal, conical sensorium. Frontoclypeal suture present or absent; labrum free, with 3 pairs of setae (fewer in Platypodinae); epipharynx with relatively short and broad to longer and narrow epipharyngeal rods. Mandibles more or less symmetrical, short and broad, obliquely bidentate; accessory ventral process absent; incisor edge with or without subapical teeth, mola and prostheca absent, dorsal pseudomola present in Platypodinae and some Scolytinae. Ventral mouthparts retracted; maxillae with articulating area absent, cardo slightly oblique, undivided, stipes short and broad to moderately elongate, mala fixed, apically rounded with row of stout setae along inner edge, palp usually 2-segmented without distinct palpifer (1-segmented in Platypodinae). Labium consisting of prementum and postmentum; prementum often with trident-shaped premental sclerite, ligula simple and rounded or absent, palp usually 2-segmented and well separated (1-segmented in Platypodinae and Bagoinae). Hypopharyngeal bracon present; sclerome absent. Hypostomal rods present or absent; ventral epicranial ridges absent. Gular region absent.

Thoracic segments usually more or less equal in length and not much longer than anterior abdominal segments, but thorax greatly enlarged in a few groups (e.g. *Neolaemosaccus*, *Strongylorhinus*); prothoracic tergum usually about as long as meso- or metatergum, with or without sclerotised plate, with transverse row of setiferous rings in Platypodinae; upper surfaces of thorax and abdomen usually covered with dark tubercles or conical protuberances in Gonipterini. Legs absent. Abdomen usually more than twice as long as thorax; abdominal terga not extending laterally beyond edges of sterna, terga without patches or rows of asperities; abdominal terga I–VIII each with at least two transverse grooves forming three plicae (three grooves and four plicae in some Cossoninae). Segment VIII mostly subequal in length to VII but in most Brachycerinae enlarged and modified, forming tergal disc or plate, IX shorter than VIII, simple and subterminal, X shorter than or more or less fused to IX and terminal, IX and X ventral in most Brachycerinae. Spiracles annular-biforous, annular-uniforous or annular; thoracic spiracles usually located on prothorax or pro-mesothoracic junction; abdominal spiracles usually lateral, those on segments II–VII in *Cyrtobagous*, I–VII in *Echinocnemus* and VIII in *Neochetina* associated with blade-like process (scalpellum); spiracles on segments I and VIII vestigial in *Cyrtobagous*; those on VIII dorsally placed in *Bagous*, *Echinocnemus* and *Neohydronomus*. (Gardner 1932, 1934, 1938; Emden 1938, 1952; Browne 1972; May 1993, 1994; Kuschel 1995; Anderson 2002; Marvaldi *et al.* 2002; Oberprieler *et al.* 2007; Lawrence & Ślipiński 2013; Oberprieler 2014b).

The distribution of the genera follows largely Alonso-Zarazaga & Lyal (1999) with updates and additions from more recent work.

Key to the Australian subfamilies, tribes and genera

1. Mesanepisterna ascending and visible dorsally between prothorax and elytra (Fig. 44.2V); abdomen with small, exposed pygidium .... **CONODERINAE**

– Mesanepisterna not ascending or visible dorsally between prothorax and elytra (Fig. 44.2U); abdomen with or without small, exposed pygidium **2**

2(1). Mandibles with deciduous process or cusp and distinguishing scar (Fig. 44.2K); mouthparts often with most of maxillae concealed by large prementum (adelognathous condition, Fig. 44.2K) .... **ENTIMINAE** (part)

– Mandibles without deciduous process, cusp or distinguishing scar; mouthparts with most of maxillae exposed (phanerognathous condition, Fig. 44.2J) **3**

3(2). Vestiture consisting of sparse setae only; mandibles with 3–4 long and a few small setae (paucisetose); scrobes absent; tarsal claws with long, cylindrical ventrobasal seta; restricted to subantarctic Heard and McDonald Islands .... **ENTIMINAE: Ectemnorhinini**

– Not with this combination of characters **4**

4(3). Rostrum well developed, antennae 8-segmented with 1-segmented club and finely densely pubescent (spongiform) apex (Fig. 44.2M); prementum concealed from below by postmentum; tarsal claws separated by dorsal and ventral lobes of tarsal apex (Fig. 44.2X) .... **DRYOPHTHORINAE**

– Rostrum variable in length, antennae rarely 8-segmented with 1-segmented club (e.g. *Coptocorynus*, Cossoninae); prementum not concealed from below by postmentum; tarsal claws rarely separated by dorsal and ventral lobes (e.g. *Elaeagna*, Cryptorhynchini) **5**

5(4). Body length commonly 0.9–2.5 mm, rarely up to 9.5 mm; head without rostrum (Fig. 44.2G, a) or with very short rostrum and 1-segmented antennal clubs (*Notoplatypus*, Fig. 44.2F) or with short rostrum and prothorax transverse with base emarginate at sides, well defined elytral striae and female with preapical transverse groove on pronotal disc (*Dactylipalpus*) **6**
– Body length 0.8–60 mm, commonly 1.5–40 mm; head with rostrum (Fig. 44.2C–E), if rostrum short clubs not 1-segmented, or without above character combination **7**
6(5). Eyes entire in outline (Fig. 44.2F); tibiae short, surface spiculate or ridged; prothorax longer than wide with hypomera excavated to accommodate profemora; tarsi slender and long, often longer than tibiae (Fig. 44.2Y), tarsal segments 3 without ventral lobes .... **PLATYPODINAE**
– Eyes often emarginate in outline (Fig. 44.2G); tibiae with surface not spiculate or ridged; prothorax longer than wide with sides subparallel; hypomera not excavated; tarsi shorter than tibiae (Fig. 44.2Z), tarsal segments 3 often with ventral lobes .... **SCOLYTINAE**
7(5). Rostrum robust, short to very short; mandibles with more than 10 setae (plurisetose) (Fig. 44.2H) **8**
– Rostrum variable; mandibles with fewer than 10 long setae (paucisetose) **9**
8(7). Mouthparts with mandibles blade-like elongate (Fig. 44.2H) and usually with single subapical tooth; maxillae coarsely sculptured, with palpifer covering short 3-segmented palp (Fig. 44.2L); labial palps concealed in ventral view .... **CYCLOMINAE: Amycterini**
– Mouthparts with mandibles broad with two subapical teeth; maxillae not coarsely sculptured, with palpifer not covering palp; labial palps exposed in ventral view .... **CYCLOMINAE: Aterpini (*Aesiotes*)**
9(7). Size small, body length 0.9–1.7 mm; eyeless **10**
– Size small to large, body length 0.8–60 mm; eyes present **11**
10(9). Pronotum and elytra with shallow, small punctures; tibiae with two spurs .... **BRACHYCERINAE: Myrtonymini**
– Pronotum and elytra with deep, coarse and large punctures; tibiae without spurs .... **CYCLOMINAE: Listroderini (*Tasmanica*)**
11(9). Body length 1.4–5.3 mm; aedeagus of pedotectal type, penis with distinct tectum; rostrum apicolaterally often with oblique groove running to below antennal insertions and carrying a row of long erect setae giving rostrum a 'whiskered' appearance (Fig. 44.2C) .... **BRACHYCERINAE**
– Body length 0.9–60 mm; aedeagus of pedal type, penis without tectum; rostrum apicolaterally without oblique groove running to below antennal insertions and without row of long erect setae **12**
12(11). Body with aquifuge plastron; antennal scapes setose around the margin; prosternum with canal; tibiae distally arcuate towards apex, uncus prominent; tarsal segments 3 not bilobed, about as wide as tarsal segments 2 .... **BAGOINAE**
– Not with this combination of characters **13**
13(12). Tibiae with apical spurs, sometimes only on metatibiae .... **CYCLOMINAE** (part)
– Tibiae without apical spurs **14**
14(13). Body length 1.4–5.6 mm, shape elongate (Fig. 44.2d); rostrum short to moderately long; coxae widely separated; meso- and metaventrite at same level; outer apical angle of tibiae extending into long, curved uncus; protibiae with ventral grooming comb near apex; meso- and metatibiae without distal comb of setae .... **COSSONINAE**
– Body length 0.8–60 mm, shape and rostrum variable; coxae rarely broadly separated; meso- and metaventrite often at different levels; tibiae with uncus when present variably formed, all with or without distal comb of setae **15**
15(14). Body length 1.1–2.6 mm; metafemora enlarged for jumping (Fig. 44.2c) .... **CURCULIONINAE: Rhamphini**
– Body length 0.8–60 mm; metafemora not enlarged or modified for jumping **16**
16(15). Prosternum with canal-like excavation in front of procoxae (Fig. 44.2N) **17**
– Prosternum without canal-like excavation in front of procoxae (Fig. 44.2P–R), rarely slightly concave (e.g. *Elleschodes*) **23**
17(16). Prosternal canal deep and narrow, extending to between procoxae and terminating in receptacle on mesoventrite (Fig. 44.2S) .... **MOLYTINAE** (part)
– Prosternal canal shallower and broader, rarely extending to between procoxae (e.g. *Psepholax*) and without receptacle on mesoventrite (Fig. 44.2T); procoxae contiguous or separated by prosternal process, rarely by the canal (e.g. *Psepholax*) **18**
18(17). Walls of prosternal canal in front of procoxae with narrow cleft to large hole, often lined with dense setae and lying over eyes when rostrum folded down .... **CURCULIONINAE: Storeini s. str.**
– Walls of prosternal canal in front of the coxae without cleft or hole **19**
19(18). Procoxae contiguous and prosternal process interrupted to form posterior sternellum (Fig. 44.2O) **20**
– Procoxae separate, prosternal process not interrupted (Fig. 44.2N) **22**
20(19). Elytral interstriae costate or to some degree carinate .... **MOLYTINAE: Cleogonini**
– Elytral interstriae flat **21**
21(20). Antennal insertions lateral; scrobes not reaching eyes .... **CURCULIONINAE: Smicronychini**
– Antennal insertions dorsal; scrobes reaching eyes .... **Notiomimetini (*Psaldus*)**
22(19). Tibiae with well-developed uncus (Fig. 44.2f) .... **MOLYTINAE** (part)
– Tibiae with uncus poorly developed or absent or present only in one leg pair .... **CURCULIONINAE** (part)
23(16). Mouthparts with labial palps covered by prementum (Fig. 44.2I) .... **MOLYTINAE ... (*Eristinus*, *Thaumastophasis*)**
– Mouthparts with labial palps visible, not covered by prementum (Fig. 44.2J) **24**
24(23). Rostrum shorter than pronotum (Fig. 44.2b; except *Prophaesia*); mesoventrite steeply declivous in front of mesocoxae, mesoventral process separating mesocoxal cavities often produced into tubercle fitting into posterior notch between procoxae when the prothorax is bent down .... **Gonipterini**
– Rostrum variable in length; mesoventrite rarely declivous in front of mesocoxae, mesoventral process not produced into tubercle **25**
25(24). Procoxae contiguous, prosternal process interrupted (Fig. 44.2Q), forming small sclerite (sternellum) behind procoxae **26**
– Procoxae separate, prosternal process not interrupted (Fig. 44.2R) **35**
26(25). Pygidium laterally with stridulatory files (Fig. 44.2W); body covered with scales; body length approximately 2.3 mm .... ***Ethadomorpha* Blackburn**
– Pygidium laterally without stridulatory files; body with vestiture variable; body length 0.8–50 mm **27**

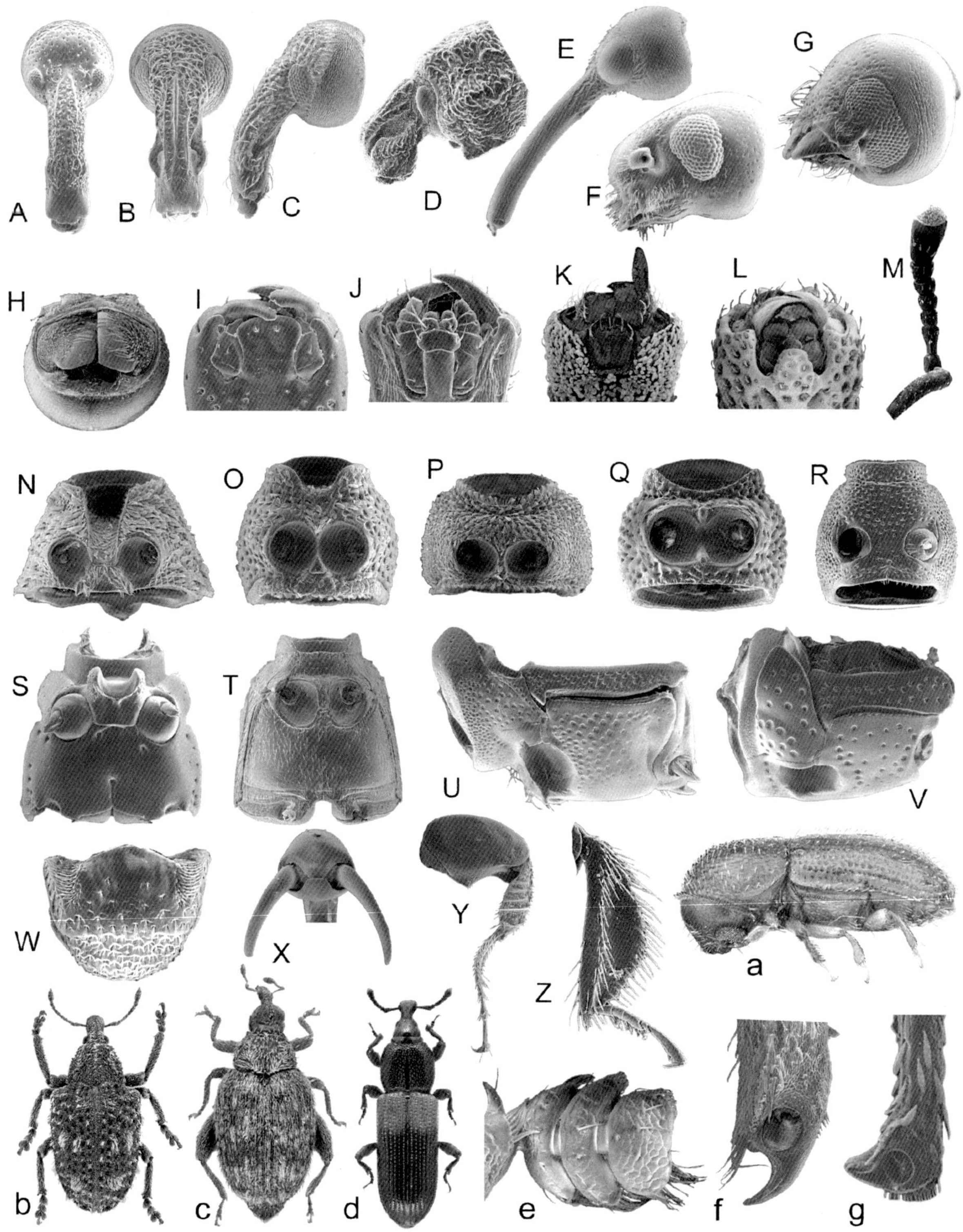

Fig. 44.2. Characters for key to subfamilies. A–B, head, dorsal: A, *Crossixus*; B, *Mandalotina*; C–G, head, lateral: C, *Athor*; D, *Leptopius*; E, *Car*; F, *Notoplatypus*; G, *Beaverium*; H–L, mouthparts, ventral: H, *Amycterus*; I, *Thaumastophasis*; J, *Decilaus*; K, *Onesorus*; L, *Atychoria*; M, antenna, *Rhabdoscelus*; N–T, prothorax, ventral: N, *Lobotrachelus*; O, *Smicronyx*; P, *Orichora*; Q, *Mandalotina*; R, *Arecophaga*; S, *Bepharus*; T, *Aragomacer*; U–V, pterothorax, lateral: U, *Beaverium*; V, *Ipsichora*; W, pygidium, *Ethadomorpha*; X, tarsal claws, *Diathetes*; Y, middle leg, *Dinoplatypus*; Z, hindleg, *Amasa*; a–d, habitus: a, *Cosmoderes*; b, *Oxyops*; c, *Orchestes*; d, *Cossonus*; e, tarsus, *Geochus*; f–g, tibial apex: f, *Chimades*; g, *Smicronyx*.

27(26). Tibiae with well-developed uncus (Fig. 44.2f) 28

– Tibiae with uncus absent or poorly developed or with mucro (Fig. 44.2g) 30

28(27). Antennal insertions dorsal (Fig. 44.2B) .... **CYCLOMINAE** (part)

– Antennal insertions lateral 29

29(28). Body surface with soil incrustation between setae or scales; eyes small and protuberant (Fig. 44.2A) or flattened; tibiae mucronate (Fig. 44.2g) .... **Phrynixini**

– Body surface without soil incrustations between setae or scales; eyes small and protuberant or flattened; tibiae uncinate .... **MOLYTINAE** (part)

30(27). Rostrum shorter than head capsule .... **Viticiini** (part)

– Rostrum longer than head capsule 31

31(30). Body length about 7.0 mm; pronotum along sides, on narrow median strip and elytral sides with white scales; rostrum apex, beyond antennal insertions, mostly glabrous with sparse long setae; prothorax with anterior ocular lobes; tarsal claws divaricate .... ***Hypsomus* Schoenherr**

– Not with this combination of characters 32

32(31). Body length 2.4–3.4 mm; body covered by appressed scales, cuticle not visible; prothorax with incomplete anterior collar, only visible in front of procoxae; prothorax without postocular lobes .... ***Symbothynus* Blackburn**

– Not with this combination of characters 33

33(32). Body length about 2.0 mm; vestiture consisting of scattered scale-like setae, without erect hairs; cuticle brown; prothorax without ocular lobes; tarsal claws absent .... ***Thechia* Pascoe**

– Not with this combination of characters 34

34(33). Surface of pronotum and elytra with blunt ridges or protuberances .... **HYPERINAE** (part)

– Surface of pronotum and elytra without ridges or protuberances .... **CURCULIONINAE** (part)

35(25). Procoxae separated by about diameter of procoxa (Fig. 44.2R) 36

– Procoxae separated by less than half diameter of procoxa (Fig. 44.2P) 38

36(35). Eyes reduced; scrobes open to eyes; uncus absent .... **Notiomimetini**

– Eyes of normal size; scrobes not open to eyes; uncus present 37

37(36). Rostrum in lateral view more than twice as long as head capsule; pterothorax longer than wide; tarsal claws often present .... **MOLYTINAE** (part)

– Rostrum in lateral view about as long as head capsule; pterothorax transverse; tarsal claws absent (Fig. 44.2e) .... **Viticiini** (***Geochus***)

38(35). Tibiae with well-developed uncus (Fig. 44.2f) .... **MOLYTINAE** (part)

– Tibiae with uncus absent (Fig. 44.2g) or poorly developed 39

39(38). Surface of pronotum and elytra with blunt ridges or protuberances .... **HYPERINAE** (part)

– Surface of pronotum and elytra without ridges or protuberances .... **CURCULIONINAE** (part)

Curculionidae: Dryophthorinae Schoenherr, 1826

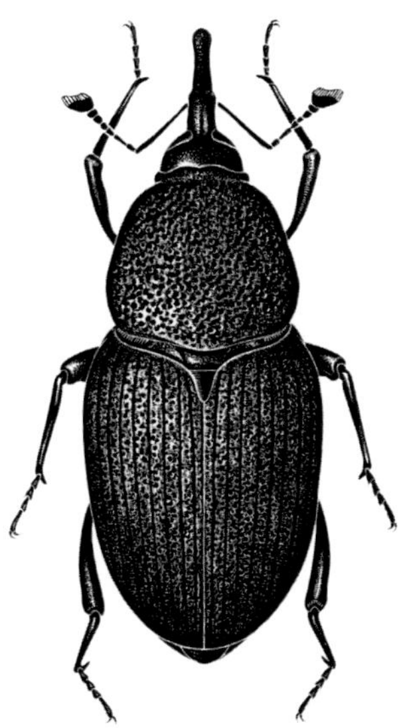

Fig. 44.3. *Trigonotarsus calandroides* Guérin-Méneville.

Introduction. Dryophthorinae are a diverse subfamily in the tropical regions of the world, particularly the Indo-Pacific Region, but are only poorly represented in Australia, mainly by species belonging to widely distributed genera and several of them introduced through human trade. The Australian fauna was revised by Zimmerman (1991, 1993) under the name Rhynchophoridae, presenting keys to the tribes, genera and species, colour photographs and habitus drawings of several species and their genitalia and innumerable notes on classification and biology. May (1994) described the larvae of a few species, and some of her figures are reproduced here (Figs 44.4, 44.5). Four of the usually recognised five tribes of Dryophthorinae are represented in Australia, Dryophthorini, Orthognathini, Rhynchophorini and Stromboscerini, Dryophthorini by only a few species of *Dryophthorus* and an undescribed one of *Stenommatus*, Orthognathini by only the widely distributed *Sipalinus gigas* (Fabricius), Rhynchophorini by ten genera with approximately 16 species and Stromboscerini by single undescribed species of *Dryophthoroides* and *Allaeotes*. Nearly all taxa have strong affinities to the Melanesian and Oriental faunas, the only autochthonous element being the monotypic genus *Trigonotarsus*. Most species are associated with monocotyledonous plants such as Arecaceae (palms), Musaceae (bananas), Orchidaceae (orchids), Pandanaceae (screw-palms), Poaceae (grasses and bamboos) and Asphodelaceae: Xanthorrhoeoideae (grasstrees), their larvae mostly developing in the stems and roots. *Rhabdoscelus obscurus* (Boisduval) and *Cosmopolites sordidus* (Germar) are pests of sugarcane and bananas, respectively, whereas the three species of *Sitophilus* are imported stored-product pests. The genus *Rhynchophorus*, with its notorious species of palm pests, is currently not considered present in Australia

(Pullen *et al.* 2014; Oberprieler & Jennings 2016). (Lawrence & Ślipiński 2013; Anderson & Marvaldi 2014).

According to the results of recent phylogenetic studies, the Dryophthorinae are one of the basal lineages of Curculionidae and form the sister-group of Platypodinae, supported by evidence from both morphological characters and DNA data (Marvaldi 1997; Marvaldi *et al.* 2002; McKenna *et al.* 2009, 2019; Shin *et. al.* 2018). The phylogenetic relationships among the dryophthorine genera were analysed by Chamorro *et al.* (2021), resolving four main clades broadly corresponding with the current tribes but with some differences in their concepts.

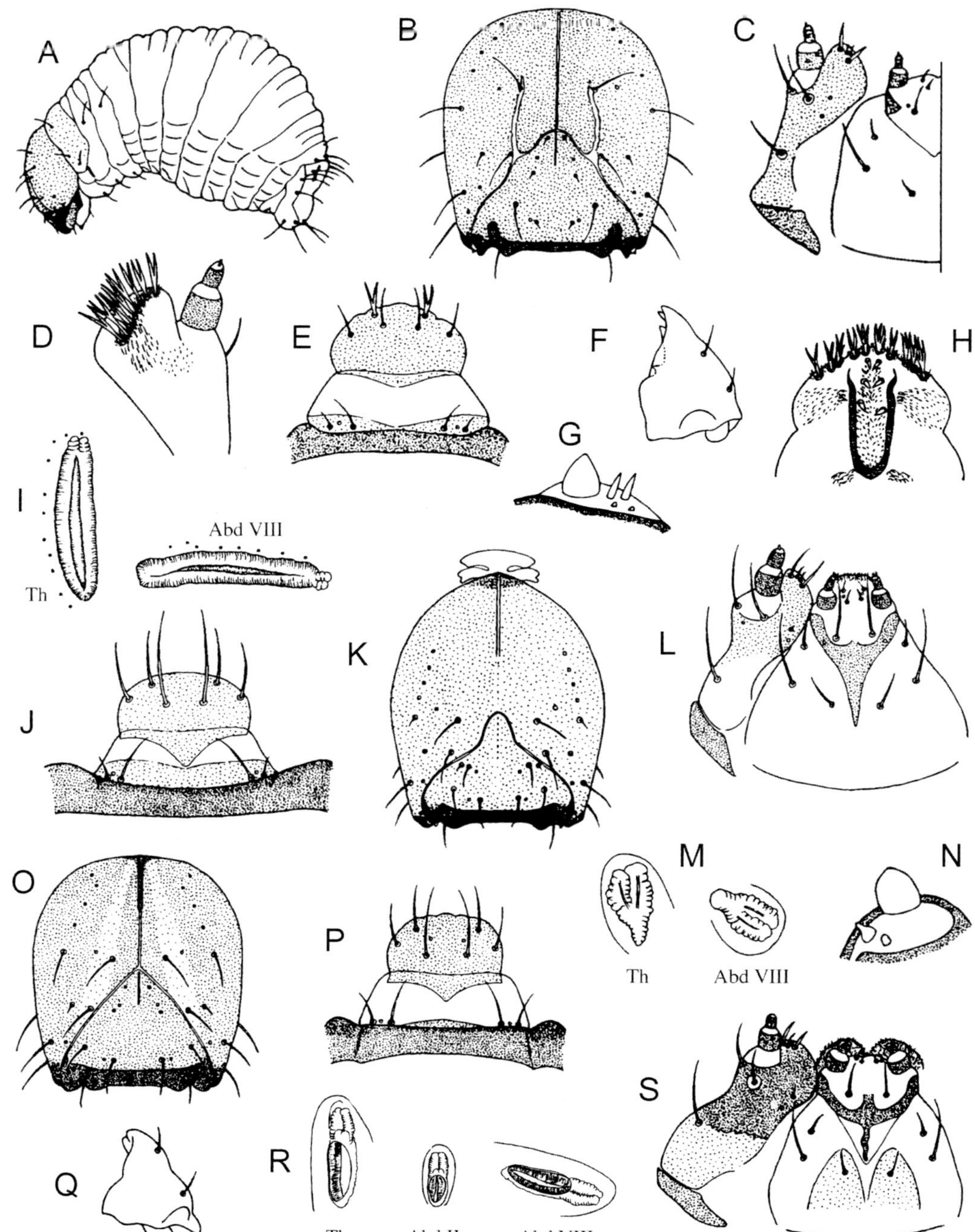

Fig. 44.4. Dryophthorinae, structural details of larvae (Rhynchophorini). A–I, *Cosmopolites sordidus* (Germar): A, habitus, lateral; B, head, anterior; C, mala and labium, posterior; D, mala, anterior; E, labrum, clypeus and epistome, anterior; F, mandible; G, antenna; H, epipharyngeal lining; I, spiracles, showing alignment; J–N, *Diocalandra frumenti* (Fabricius): J, labrum, clypeus and epistoma, anterior; K, head, anterior; L, mala and labium, posterior; M, spiracles, showing alignment; N, antenna; O–S, *Rhabdoscelus obscurus* (Boisduval): O, head, anterior; P, labrum, clypeus and epistoma, anterior; Q, mandible; R, spiracles, showing alignment; S, mala and labium, posterior. After May (1994).

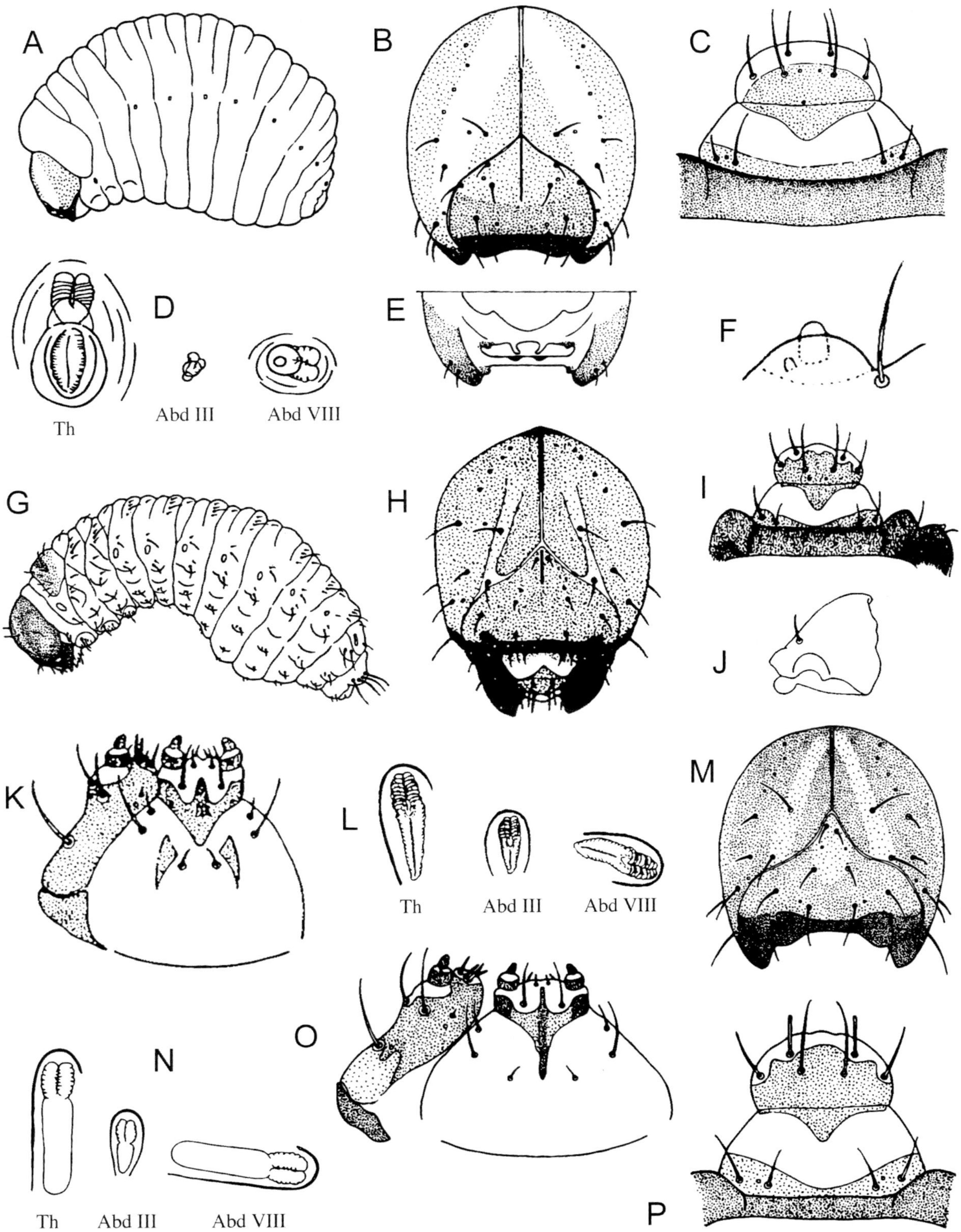

Fig. 44.5. Dryophthorinae, structural details of larvae (Rhynchophorini). A–F, *Sitophilus oryzae* (Linnaeus): A, habitus, lateral; B, head, anterior; C, labrum, clypeus and epistoma, anterior; D, spiracles, showing alignment; E, head, posterior; F, antenna; G–L, *Sphenophorus brunnipennis* (Germar): G, habitus, lateral; H, head, anterior; I, labrum, clypeus, epistoma; J, mandible; K, mala and labium, posterior; L, spiracles, showing alignment; M–P, *Trigonotarsus calandroides* Guérin-Méneville: M, head, anterior; N, spiracles, showing alignment; O, mala and labium, posterior; P, labrum, clypeus and epistome, anterior. After May (1994).

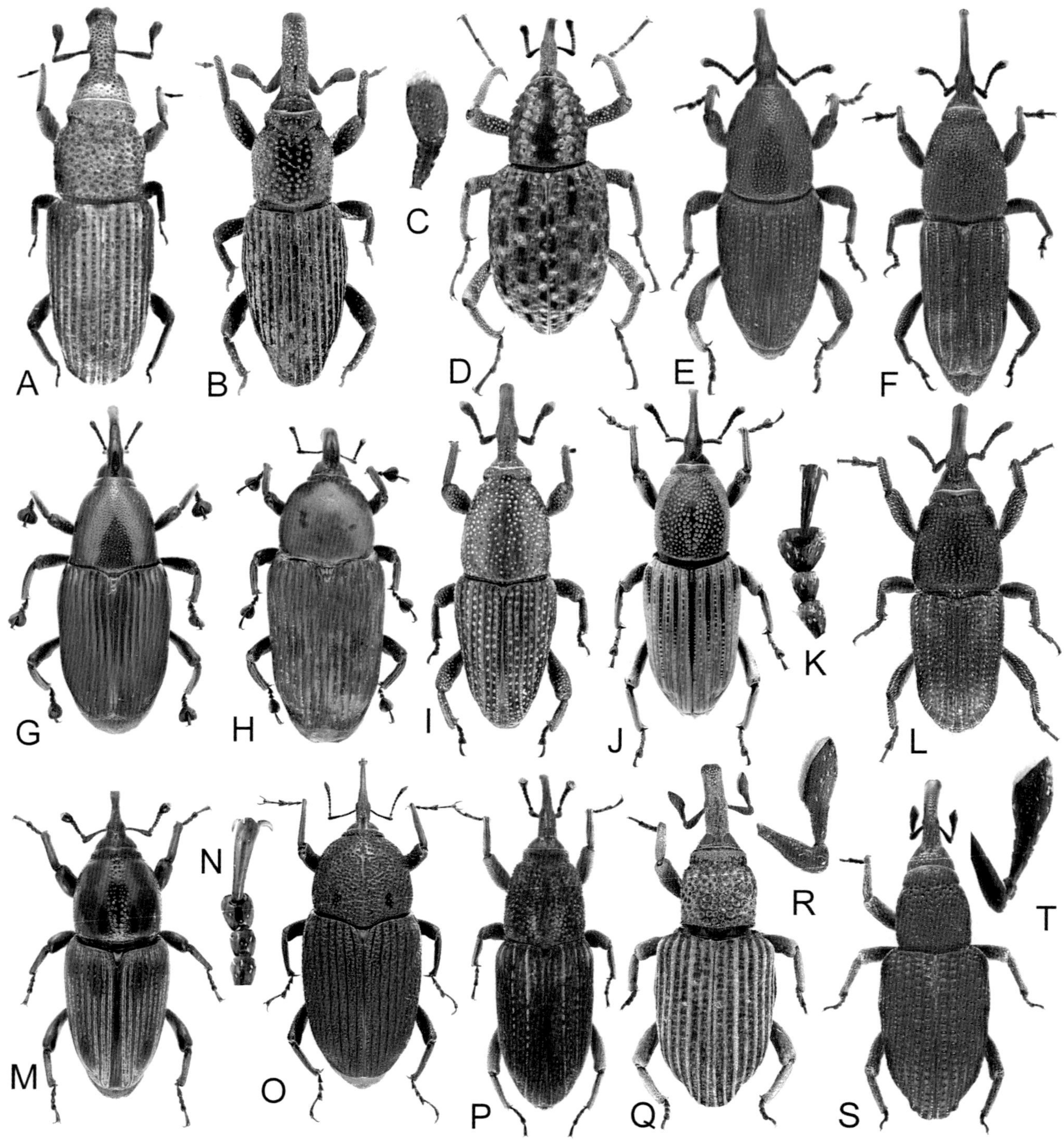

Fig. 44.6. Dryophthorinae, characters for key to genera. A, *Dryophthorus assimilis* Gahan; B–C, *Stenommatus*, habitus and antenna; D, *Sipalinus gigas* (Fabricius); E, *Cosmopolites sordidus* (Germar); F, *Diocalandra frumenti* (Fabricius); G, *Diathetes morio* Pascoe; H, *Iphthimorhinus australasiae* Roelofs; I, *Polytus mellerborgii* (Boheman); J–K, *Rhabdoscelus interstitialis* (Boheman), habitus and tarsi; L, *Sitophilus oryzae* (Linnaeus); M–N, *Sphenophorus brunnipennis* (Germar), habitus and tarsus; O, *Trigonotarsus calandroides* Guérin-Méneville; P, *Trochorhopalus strangulatus* (Gyllenhal); Q–R, *Allaeotes*, habitus and antenna; S–T, *Dryophthoroides*, habitus and antenna.

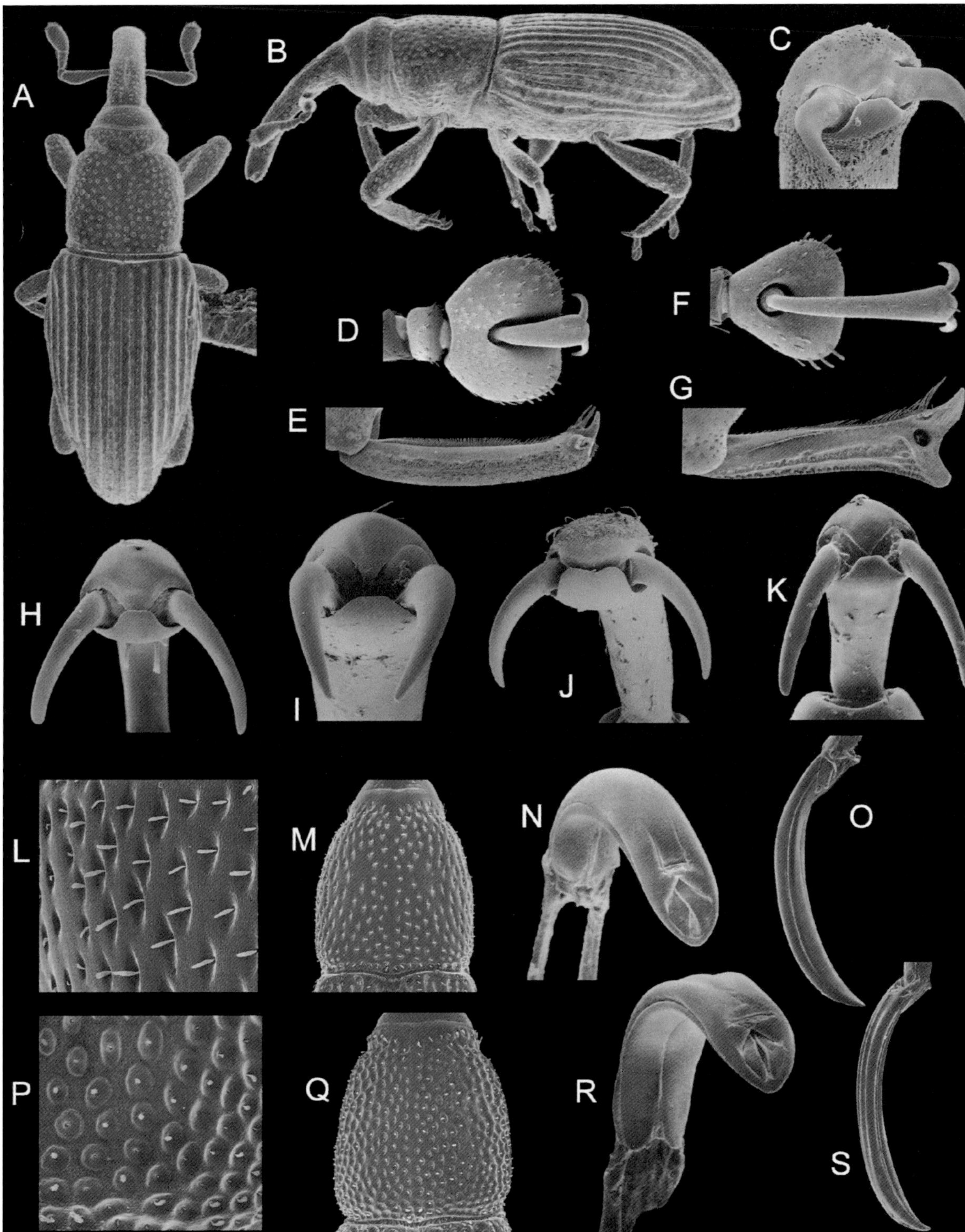

Fig. 44.7. Dryophthorinae, structural details. A–B, *Stenommatus* sp.: A, dorsal; B, lateral habitus; C, *Cosmopolites sordidus* (Germar), tarsal claws; D–E, *Diathetes morio* Pascoe: D, protarsus; E, protibia, lateral; F–G, *Diathetes signaticollis* Faust: F, metatarsus (apical part); G, protibia, lateral; H–K, tarsal claws: H, *Diathetes signaticollis* Faust; I, *Iphthimorhinus australasiae* Roelofs; J, *Sipalinus gigas* (Fabricius); K, *Sphenophorus brunnipennis* (Germar); L–M, *Sitophilus granarius* (Linnaeus): L, pronotal vestiture; M, pronotum, dorsal; N–Q, *Sitophilus oryzae* (Linnaeus): N, penis, apicodorsal; O, penis, lateral; P, pronotal vestiture; Q, pronotum, dorsal; R–S, *Sitophilus zeamais* Motschulsky, penis: R, apicodorsal; S, lateral. After Zimmerman (1993).

Key to the Australian genera of Dryophthorinae

Modified after Zimmerman (1993).

1. Pygidium concealed beneath elytra (Fig. 44.6D) or apex narrowly exposed **2**

– Pygidium clearly exposed (Fig. 44.6F) **6**

2(1). Funicles 6-segmented; scutellum exposed; pygidium without longitudinal groove; body length >10 mm; habitus as Fig. 44.6D ..... ***Sipalinus* Marshall**

– Funicles 4-segmented; scutellum concealed; pygidium with longitudinal groove; body length <6 mm **3**

3(2). Antennal clubs symmetrical (Fig. 44.6C), pilose apical part broadly convex or conical; procoxae separated; tarsi with segment 4 exposed beyond 3 ..... **4**

– Antennal clubs strongly asymmetrical (Fig. 44.6R, T), pilose apical part oblique; procoxae contiguous; tarsi with segment 4 hidden between lobes of 3 ..... **5**

4(3). Body robust (Fig. 44.6A); rostrum shorter than pronotum; eyes well separated ventrally ..... ***Dryophthorus* Schoenherr**

– Body slender (Fig. 44.6B); rostrum about as long as or longer than pronotum (Fig. 44.7B); eyes contiguous or subcontiguous ventrally ..... ***Stenommatus* Wollaston**

5(3). Antennal clubs with pilose area not extending to basal half (Fig. 44.6T); habitus as Fig. 44.6S ..... ***Dryophthoroides* Roelofs**

– Antennal clubs with pilose area extending to basal fourth (Fig. 44.6R); habitus as Fig. 44.6Q ..... ***Allaeotes* Pascoe**

6(1). Pygidium with basal longitudinal groove where elytral sutural margins lock; body length approximately 2.8 mm (Fig. 44.6L) ..... ***Sitophilus* Schoenherr**

– Pygidium without basal longitudinal groove; body length variable **7**

7(6). Pygidium with transverse irregular groove at about middle of length, area below groove densely pilose (Fig. 44.6P) ..... ***Trochorhopalus* Kirsch**

– Pygidium without transverse groove at middle of length **8**

8(7). Tarsi with segment 3 at apex emarginate and bilobed ..... ***Diocalandra* Faust**

– Tarsi with segment 3 not bilobed, apical margin entire (Fig. 44.7D, F) **9**

9(8). Body length <5 mm; tibiae with two strong apical hooks (apical uncus and preapical premucro); metanepisterna narrow, not wider than lateral elytral intervals; habitus as Fig. 44.6I ..... ***Polytus* Faust**

– Body length >5 mm; tibiae with single apical hook (uncus), premucro small to rudimentary, flanked by pair of setal tufts (Fig. 44.7E); metanepisterna broader than lateral elytral interval **10**

10(9). Elytra with stria 10 abbreviated, not continued beyond metanepisternum **11**

– Elytra with stria 10 complete, continued above abdomen to elytral apex **13**

11(10). Tarsi with segment 3 much broader than 1 or 2 (Fig. 44.6K); segment 4 inserted near middle of 3 (Fig. 44.6J) ..... ***Rhabdoscelus* Marshall**

– Tarsi with segment 3 not or slightly wider than 1 and 2 (Fig. 44.6N); segment 4 inserted subapically **12**

12(11). Tarsi smooth and shiny dorsally; funicle segments 2 shorter than 3+4; scutellar shield narrowed towards apex (Fig. 44.6M) ..... ***Sphenophorus* Schoenherr**

– Tarsi densely pruinose dorsally; funicle segments 2 about as long as 3+4; scutellar shield widened towards apex (Fig. 44.6E) ..... ***Cosmopolites* Chevrolat**

13(10). Prothorax and elytra unusually coarsely and densely punctate (Fig. 44.6O); tarsi with segment 4 inserted subapically in 3 (Fig. 44.6O) ..... ***Trigonotarsus* Schoenherr**

– Prothorax and elytra minutely punctate (Fig. 44.6G, H); tarsi with segment 4 inserted in about middle of 3 ..... **14**

14(13). Antennal clubs with setose apex almost circular; elytra mostly red with black maculae; body length 20–35 mm or more (Fig. 44.6H) ..... ***Iphthimorhinus* Roelofs**

– Antennal clubs with setose apex elongate and ovate; elytral colouration and body length variable (Fig. 44.6G) ..... ***Diathetes* Pascoe**

Classification of the Australian genera of Dryophthorinae

Tribe Dryophthorini Schoenherr, 1825

Dryophthorus Germar, 1824 (Pl. 45A)

Type species, by monotypy: *Curculio lymexylon* Fabricius, 1792.

Australian species. *D. assimilis* Gahan, 1900, *Dryophthorus* spp. (Zimmerman, 1993: 126).

Distribution. Australia (QLD, NSW, VIC, TAS), Pacific region, Oriental region, Palaearctic region, Americas, Africa.

Notes. The genus contains more than 60 species and is widely distributed in the tropics of the world. Several unnamed or undescribed species occur in Australia. The larvae develop in decaying and rotten wood.

Stenommatus Wollaston, 1873 (Pl. 45B; Fig. 44.7A, B)

Type species, by monotypy: *Stenommatus fryi* Wollaston, 1873b.

Australian species. *Stenommatus* sp. (Zimmerman, 1993: 128).

Distribution. Australia (QLD), Pacific region, Oriental region, Central and South America.

Notes. The genus contains more than 25 described species. Only a single, undescribed one is recorded from Australia, but it is unknown whether it is native or introduced. Additional species may occur in the country. The Australian species is likely associated with bananas (*Musa* spp.) and perhaps related Musaceae.

Tribe Orthognathini Lacordaire, 1865

Sipalinus Marshall, 1943 (Pl. 45C; Fig. 44.7J)

Type species, by original designation: *Calandra granulata* Fabricius, 1801.

Australian species. *S. gigas* (Fabricius, 1775).

Distribution. Australia (QLD), New Guinea, Oriental, eastern Palaearctic and Afrotropical regions.

Notes. The genus contains seven species but only one in the Australo-Pacific and Oriental regions (Vaurie 1971), the large and widespread *S. gigas*, which may be introduced in Australia (Zimmerman 1993). The larvae tunnel communally in diseased or dying tree trunks of various kinds (Chamorro 2019).

Tribe Rhynchophorini Schoenherr, 1833

Cosmopolites Chevrolat, 1885 (Pl. 45D; Figs 44.4A–I, 44.7C)

Type species, by original designation: *Calandra sordida* Germar, 1824.

Australian species. *C. sordidus* (Germar, 1824).

Distribution. Australia (QLD, NSW), tropical regions worldwide (widely introduced).

Notes. The genus contains only two species, of which the widespread *C. sordidus* has been introduced in Australia. It is a major pest of bananas (*Musa* spp.), the larvae (May 1994) developing in the stems and rootstocks.

Diathetes Pascoe, 1874 (Pl. 45F; Fig. 44.7D–H)

Type species, by subsequent designation (Zimmerman, 1993: 61): *Diathetes ruficollis* Pascoe, 1874c.

Australian species. *D. morio* Pascoe, 1874c, *D. schoenherri* (Gyllenhal, 1838) (*in* Schoenherr 1838), *D. signaticollis* Faust, 1899.

Distribution. Australia (QLD, NT, northern WA), New Guinea area (including Aru Is., Solomon Is., Maluku Is.), Pacific region (Fiji, Samoa, Vanuatu, New Caledonia), Oriental region (Philippines).

Notes. The genus contains several species, but its taxonomy requires revision. Two undescribed species occur in Australia in addition to the three described ones. The larvae as known develop in stems and flower stalks of *Pandanus* (Zimmerman, 1993) and seemingly of *Freycinetia* (May 1994).

Diocalandra Faust, 1894 (Pl. 45E; Fig. 44.4J–N)

Type species, by monotypy: *Calandra frumenti* Fabricius, 1801.

Australian species. *D. frumenti* (Fabricius, 1801).

Distribution. Australia (QLD, NT, northern WA), New Guinea, Pacific, Oriental Region, east Palaearctic and African regions, also introduced to Ecuador, Canada and USA.

Notes. The genus contains approximately 10 species, of which only the widespread *D. frumenti* has been introduced to Australia. It is often confused in the literature and in collections with the similar Polynesian species *D. taitensis* (Guérin-Méneville 1834). Its larvae (May 1994) develop in sick or injured parts of palms, particularly of coconut palms (*Cocos nucifera*).

Iphthimorhinus Roelofs, 1892 (Pl. 45G; Fig. 44.7I)

Type species, by monotypy: *Iphthimorhinus australasiae* Roelofs, 1892.

Australian species. *I. australasiae* Roelofs, 1892.

Distribution. Australia (QLD, NT), New Guinea.

Notes. The genus contains only a single species, which is native to northern Australia and New Guinea. The female has a peculiarly shaped rostrum, resembling a flamingo bill. Its larval hosts are unknown but likely palms or pandans.

Polytus Faust, 1894 (Pl. 45H)

Type species, by original designation: *Sitophilus mellerborgii* Boheman, 1838.

Australian species. *P. mellerborgii* (Boheman, 1838) (*in* Schoenherr 1838).

Distribution. Australia (QLD, northern NSW), New Guinea, tropical regions worldwide (widely introduced).

Notes. The genus contains only one species, which is native to Southeast Asia but has been widely dispersed throughout the tropics, including to Australia. Both adults and larvae (May 1994) bore in decaying corms and stems of bananas (*Musa* spp.).

Rhabdoscelus Marshall, 1943 (Pl. 45I; Fig. 44.4O–S)

Type species, by subsequent designation (Faust, 1894: 348): *Sphenophorus maculatus* Gyllenhal, 1838 (*in* Schoenherr 1838).

Australian species. *R. interstitialis* (Boheman, 1859), *R. obscurus* (Boisduval, 1835).

Distribution. Australia (QLD, NT, NSW, VIC), New Guinea, Pacific and Oriental regions to Japan.

Notes. The genus comprises about 10 species but is in need of taxonomic revision. Two species have been introduced to Australia, of which *R. obscurus* is more common as it is a serious pest of sugarcane (*Saccharum*), but its larvae (May 1994) also attack a broad range of palms and occasionally other plants (Zimmerman 1993).

Sitophilus Schoenherr, 1838 (Pl. 45J; Figs 44.5A–F, 44.7L–S)

Type species, by original designation: *Curculio oryzae* Linnaeus, 1763.
Australian species. *S. granarius* (Linnaeus, 1758), *S. linearis* (Herbst, 1795), *S. oryzae* (Linnaeus, 1763), *S. zeamais* Motschulsky, 1855.
Distribution. Australia (throughout), cosmopolitan.
Notes. The genus contains about 15 species, but its taxonomy is poorly resolved. The four species in Australia are all introduced and well-known pests of stored grains. Zimmerman (1993) compiled a comprehensive review of the Australian species, with a key and illustration of diagnostic characters (e.g. Fig. 44.7L–M, N–Q, R–S), and May (1994) provided diagnoses of the larvae.

Sphenophorus Schoenherr, 1838 (Pl. 45K; Figs 44.5G–L, 44.7K)

Type species, by original designation: *Curculio abbreviatus* Fabricius, 1787.
Australian species. *S. brunnipennis* (Germar, 1824).
Distribution. Australia (QLD, NSW, VIC, SA, WA), native to the New World.
Notes. The genus contains numerous species, of which one has been introduced to Australia from South America and is a pest of cultivated grasses, in particular couch grass (*Cynodon dactylon*) and kikuyu (*Pennisetum clandestinum*), and also of sedges (Zimmerman 1993; May 1994).

Trigonotarsus Guérin-Méneville, 1833 (Pl. 45L; Figs 44.3, 44.5M–P)

Type species, by monotypy: *Trigonotarsus calandroides* Guérin-Méneville, 1833.
Australian species. *T. calandroides* Guérin-Méneville, 1833.
Distribution. Australia (QLD, NSW).
Notes. The genus is monotypic and native to Australia. The single species was referred to as *T. rugosus* (Boisduval) in the past (e.g. Zimmerman 1993). Its larvae develop in the trunks of grasstrees (*Xanthorrhoea* spp.) (May 1994; Chamorro 2019), especially in plants damaged or weakened by fire or some other calamity, but infected plants usually survive.

Trochorhopalus Kirsch, 1877 (Pl. 45M)

Type species, by monotypy: *Sphenophorus strangulatus* Gyllenhal, 1838.
Australian species. *T. strangulatus* (Gyllenhal, 1838).
Distribution. Australia (Christmas Is.), New Guinea, Oriental region (Indonesia, Cambodia, southern China, India, Myanmar, Philippines, Sri Lanka, Thailand), Indian Ocean islands (Mauritius, Reunion, Seychelles).
Notes. The genus contains about 15 species, mostly in the Oriental region. Only the widespread *T. strangulatus* has been recorded from the Australian territory, occurring on Christmas Is.

Tribe Stromboscerini Lacordaire, 1865

Allaeotes Pascoe, 1885 (Pl. 45N)

Type species, by monotypy: *Allaeotes griseus* Pascoe, 1885.
Australian species. *Allaeotes* sp. (Pullen *et al.* 2014: 74).
Distribution. Australia (QLD), Maluku Is., Sri Lanka, China, Japan, Vietnam, Cuba (introduced).
Notes. The genus contains five known species (only two described). The species in Australia was not included in Zimmerman's (1993) treatment of the Australian dryophthorines but only recorded later by Pullen *et al.* (2014).

Dryophthoroides Roelofs, 1879 (Pl. 45O)

Type species, by monotypy: *Dryophthoroides sulcatus* Roelofs, 1879.
Australian species. *Dryophthoroides* sp. n. (Zimmerman, 1993: 131).
Distribution. Australia (QLD), New Guinea, Maluku Is., Bangladesh, India, China, Japan.
Notes. The genus comprises six described species and more undescribed ones, including one in Australia (Zimmerman 1993). The weevils inhabit dead wood in moist environments.

Curculionidae: Bagoinae Thomson, 1859

Introduction. The Bagoinae, included as a tribe in Brachycerinae by Pullen *et al.* (2014) but treated as a tribe of uncertain placement by Oberprieler *et al.* (2014b), have since been shown by phylogenetic analyses to represent a different lineage more closely related to the 'higher' weevils (the CEGH and CCCMS clades) (Gillett *et al.* 2014; Shin *et al.* 2018; Baird *et al.* 2021; Haran *et al.* 2023) and are again treated as a subfamily (Shin *et al.* 2018).

Key to the Australian genera of Bagoinae

1. Forehead with longitudinal fovea between eyes; prosternum with deep canal to receive rostrum .... ***Bagous* Germar**
– Forehead without longitudinal fovea between eyes; prosternum with shallow depression to receive rostrum; habitus as Fig. 44.11C .... ***Azollaebagous* Caldara *et al.***

Classification of the Australian genera of Bagoinae

Azollaebagous Caldara, O'Brien & Meregalli, 2017 (Pl. 47A, B)

Type species, by original designation: *Bagous clarenciensis* Blackburn, 1894.

Australian species. *A. clarenciensis* (Blackburn, 1894a).

Distribution. Australia (QLD, NSW, VIC), Oriental region.

Notes. The formerly large genus *Bagous* was divided into four genera by Caldara *et al.* (2017), among them *Azollaebagous*. The genus contains five species, four in the Oriental region and one in Australia. The larvae develop in floating plants of the aquatic fern *Azolla* (Salviniaceae).

Bagous Germar, 1817 (Pl. 47C, D; Fig. 44.9A–I)

Type species, by subsequent designation (Schoenherr, 1826: 21): *Curculio binodulus* Herbst, 1795.

Australian species. *B. adelaidae* Blackburn, 1894c, *B. apicalis* O'Brien, 1992, *B. australasiae* Blackburn, 1894c, *B. blyxae* O'Brien, 1992, *B. blyxodes* O'Brien, 1992, *B. brittoni* O'Brien, 1992, *B. callosus* O'Brien, 1992, *B. dostinei* O'Brien, 1992, *B. dubius* O'Brien, 1992, *B. femoralis* O'Brien, 1992, *B. fornoae* O'Brien, 1992, *B. hydrillae* O'Brien, 1992, *B. infrequens* O'Brien, 1992, *B. josephi* O'Brien, 1992, *B. matthewsi* O'Brien, 1992, *B. meridionalis* O'Brien, 1992, *B. natator* O'Brien, 1992, *B. occiduus* O'Brien, 1992, *B. pauxillulus* O'Brien, 1992, *B. propinquus* O'Brien, 1992, *B. proximus* O'Brien, 1992, *B. purcelli* O'Brien, 1992, *B. rieki* O'Brien, 1992, *B. simulans* O'Brien, 1992, *B. tarsalis* O'Brien, 1992, *B. utriculariae* O'Brien, 1992.

Distribution. Australia, cosmopolitan except for Central and South America, mainly Europe, USA, India.

Notes. The genus contains more than 300 described species worldwide and 26 in Australia, but more are likely to occur in both these realms. The Australian fauna was revised by O'Brien & Askevold (1992). The larvae develop in many different aquatic plants (see O'Brien & Askevold (1992) for details). The biology, ecology, morphology and taxonomy of the subfamily were reviewed by Oberprieler *et al.* (2014b), and a phylogenetic analysis based on morphological characters was undertaken by Caldara *et al.* (2017).

Curculionidae: Brachycerinae Billberg, 1820

Introduction. Brachycerinae in the current, wide sense (Oberprieler *et al.* 2007; Oberprieler 2014b) are a heterogeneous group, encompassing approximately 95 genera and 1350 species worldwide. The Australian fauna is small, comprising 13 genera with 54 described species (Pullen *et al.* 2014), though several further genera and species remain to be described. It was revised by Zimmerman (1991, 1993) (as a family Erirhinidae), the treatment including keys to the tribes, genera and species and numerous illustrations and notes on the various taxa. May (1994) described the larvae of a few species, and some of her figures are reproduced here (Figs 44.9, 44.10). Only three of the seven tribes recognised by Oberprieler (2014b) occur in Australia, the Erirhinini, Myrtonymini and Tanysphyrini, Erirhinini with four genera (*Aonychus*, '*Bagoopsis*', *Echinocnemus*, *Tadius*), Myrtonymini with two (*Hexonymus*, *Myrtonymus*) and Tanysphyrini with eight, the species of four of them (*Cyrtobagous*, *Hydrotimetes*, *Neochetina*, *Neohydronomus*) introduced for biological control of aquatic weeds. Of the native genera, '*Bagoopsis*', *Echinocnemus* and *Niphobolus* also occur in aquatic to subaquatic habitats, whereas *Athor* and *Baeosomus* are associated with mosses and the minute species of *Hexonymus* and *Myrtonymus* live in log litter and among fine roots in the soil. Several undescribed genera and species are known in all these groups (e.g. Fig. 44.12L–T).

The checkered history of the concept and classification of the Brachycerinae was reviewed by Oberprieler (2014b). In its current, wide sense the subfamily is evidently a paraphyletic construct (defined by the plesiomorphic pedotectal type of aedeagus), but its recognition prevents acceptance of a polyphyletic subfamily Erirhininae, whose different lineages and their relationships remain unclear. The relationships and natural taxa among the various curculionid groups with a pedotectal aedeagus will likely only become apparent from more comprehensive molecular phylogenetic analyses.

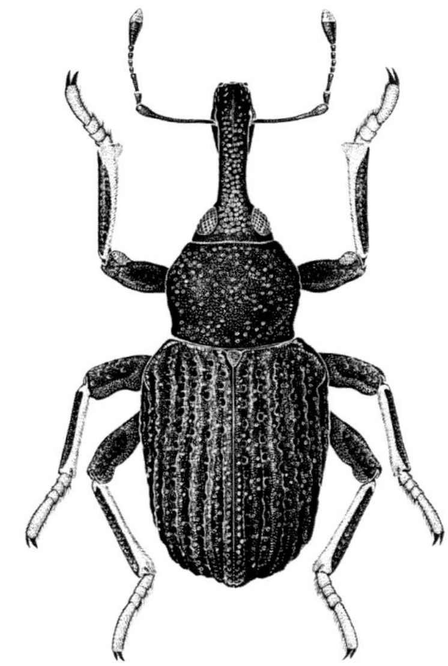

Fig. 44.8. *Cyrtobagous salviniae* Calder & Sands.

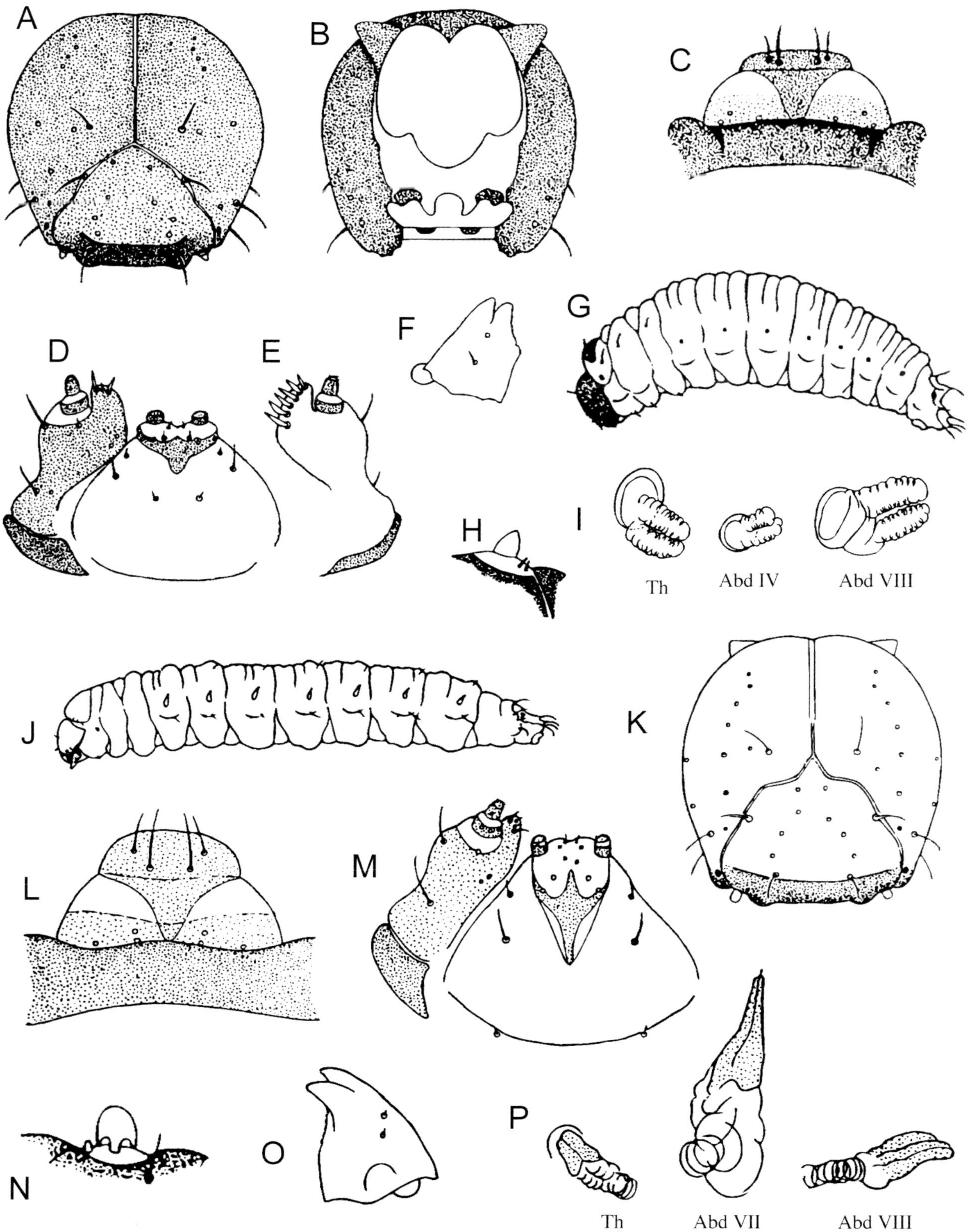

Fig. 44.9. Bagoinae and Brachycerinae, structural details of larvae (Erirhinini). A–I, *Bagous australasiae* Blackburn: A, head, anterior; B, head, posterior; C, labrum, clypeus and epistoma, anterior; D, mala and labium, posterior; E, mala, anterior; F, mandible; G, habitus, lateral; H, antenna; I, spiracles, showing alignment; J–P, *Echinocnemus* sp.: J, habitus, lateral; K, head, anterior; L, labrum, clypeus and epistoma, anterior; M, mala and labium, posterior; N, antenna; O, mandible; P, spiracles, showing alignment and modification on AVII. After May (1994).

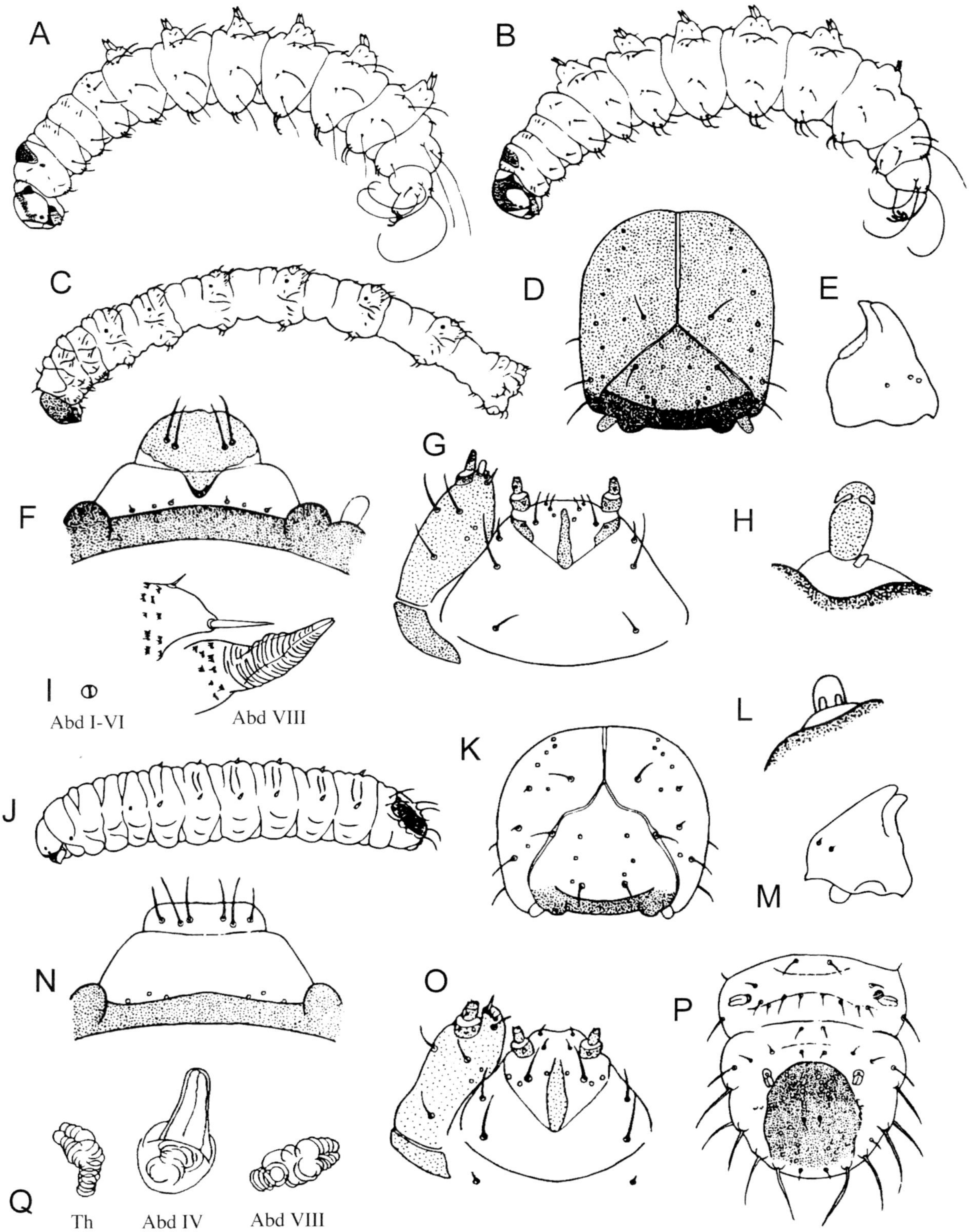

Fig. 44.10. Brachycerinae, structural details of larvae (Tanysphyrini). A–B, *Cyrtobagous*: A, *Cyrtobagous singularis* Hustache, habitus, lateral; B, *Cyrtobagous salviniae* Calder & Sands, habitus, lateral; C–I, *Neochetina eichhorniae* Warner: C, habitus, lateral; D, head, anterior; E, mandible; F, labrum, clypeus, and epistoma, anterior; G, mala and labium, posterior; H, antenna; I, spiracles showing alignment and modification on AVIII; J–P, *Neohydronomus pulchellus* Hustache: J, habitus, lateral; K, head, anterior; L, antenna; M, mandible; N, labrum, clypeus and epistoma, anterior; O, mala and labium, posterior; P, terminal segments, dorsal; Q, spiracles, showing alignment. After May (1994).

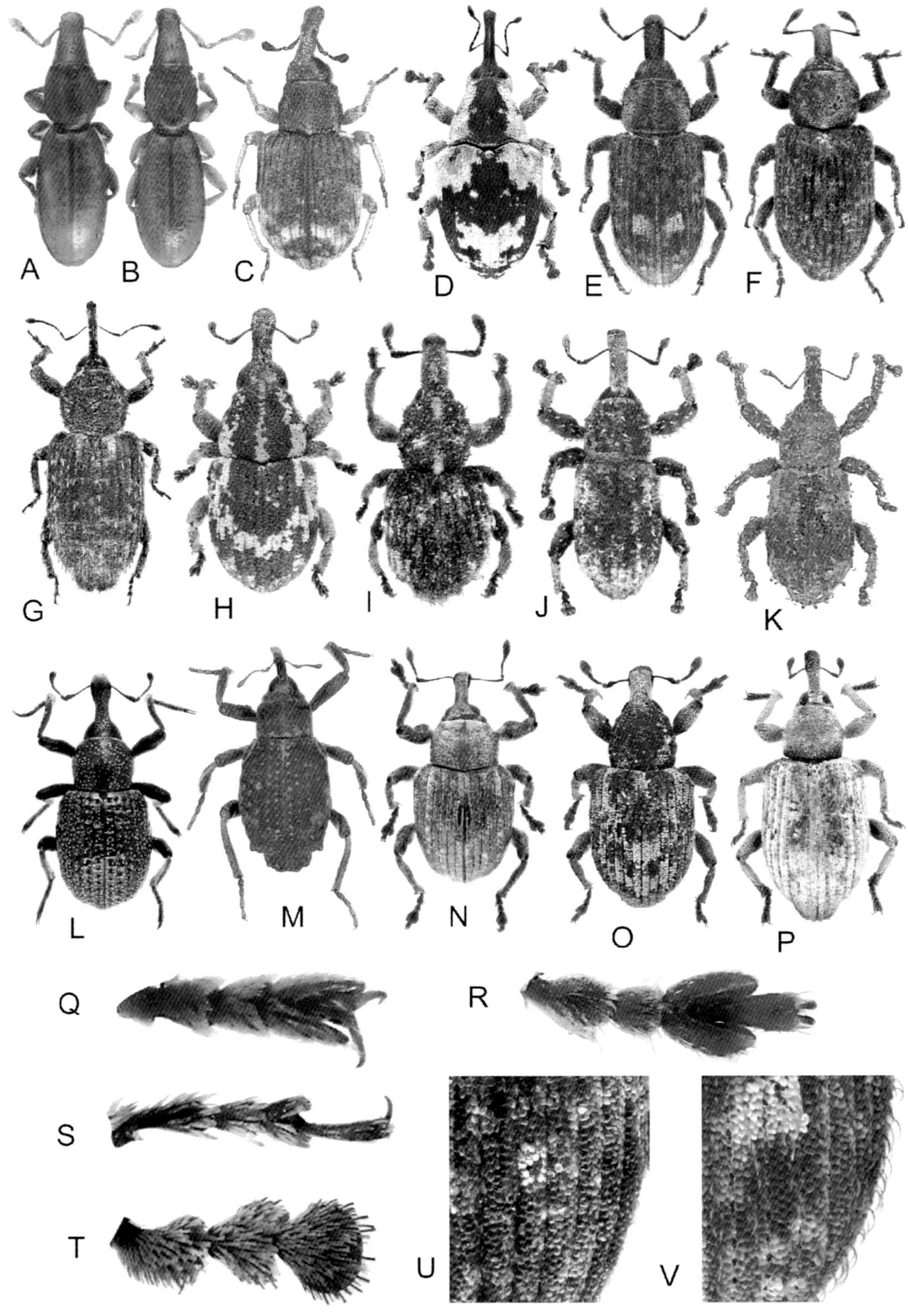

Fig. 44.11. Characters for key to genera of Bagoinae and Brachycerinae. A, *Hexonymus reginalis* Kuschel; B, *Myrtonymus moorei* Kuschel; C, *Azollaebagous clarenciensis* (Blackburn); D, *Aonychus hopei* Boheman; E, '*Bagoopsis*'; F, *Echinocnemus exsul* Faust; G, *Tadius barbatus* Lea; H, *Athor*; I, *Baeosomus uvidus* (Oke); J, *Cenchrena*; K, *Cenchrena alternata* Lea; L, *Cyrtobagous salviniae* Calder & Sands; M, *Hydrotimetes natans* Kolbe; N, *Neochetina eichhorniae* Warner; O, *Neohydronomus affinis* Hustache; P, *Niphobolus deceptor* Blackburn; Q–T, tarsi and claws: Q, *Niphobolus*; R, *Neochetina*; S, *Tadius*; T, *Aonychus*; U, *Echinocnemus*; V, '*Bagoopsis*'.

Fig. 44.12. Structural details of Brachycerinae (Tanysphyrini). A–K, *Athor* sp.: Blundell's Ck, ACT; A, head, lateral; B, antenna; C, mouthparts; D, prothorax, ventral; E, pterothorax, ventral; F, abdominal ventrites; G, elytron, lateral; H, scutellum, dorsal; I, tarsus, dorsal; J, leg; K, leg; L–T, undescr. genus, Yanac VIC, ♂: L, head, lateral; M, antenna; N, prothorax, ventral; O, pterothorax, ventral; P, scutellum, dorsal Q, elytron, lateral; R, abdominal ventrites; S, foreleg; T, hindleg.

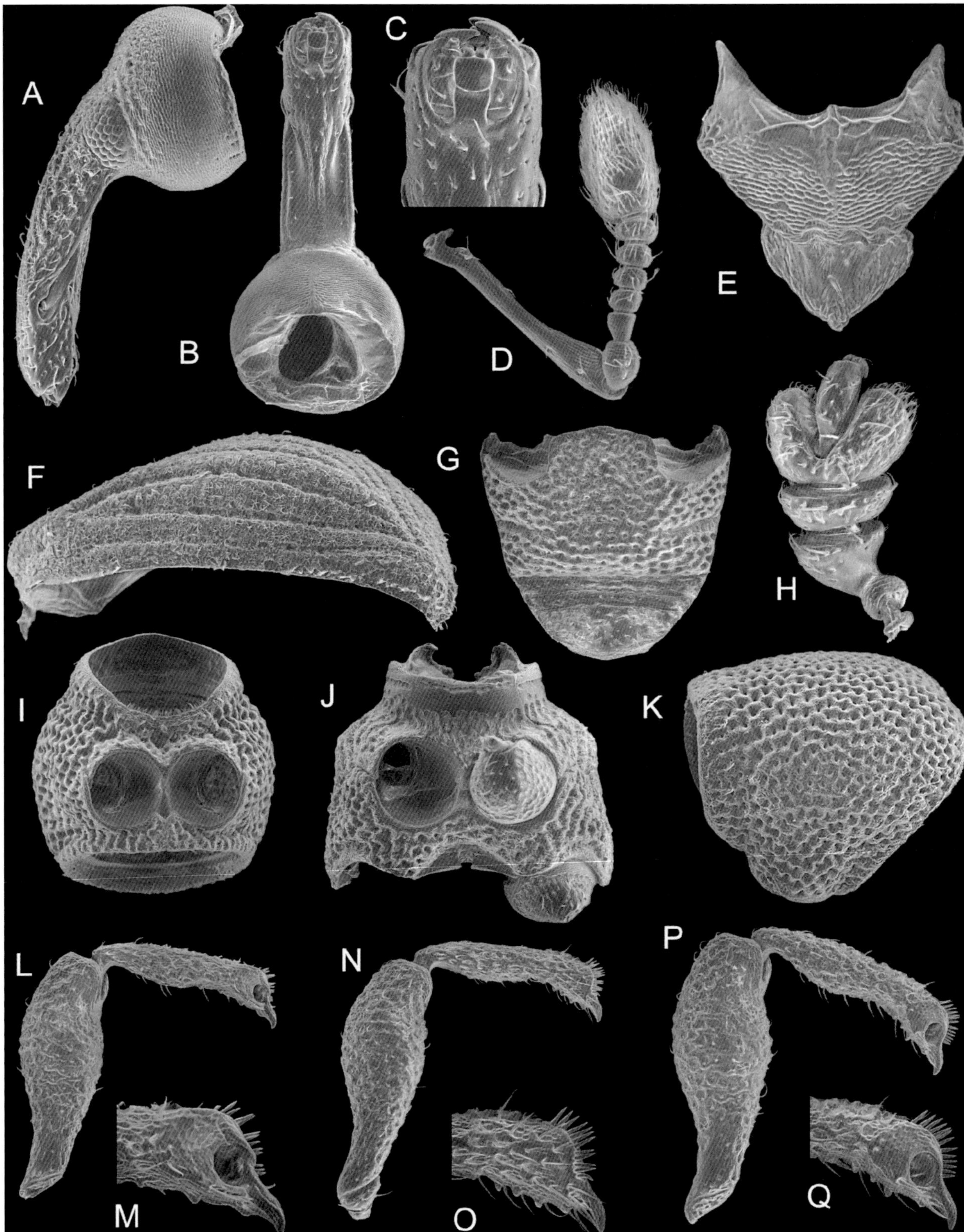

Fig. 44.13. Structural details of Brachycerinae, *Baeosomus* sp., ♂. A–B, head: A, lateral; B, ventral; C, mouthparts; D, antenna; E, scutellum, dorsal; F, elytron, lateral; G, abdominal ventrites; H, tarsus, dorsal; I, prothorax, ventral; J, pterothorax, ventral; K, prothorax, lateral; L, foreleg; M, protibial apex; N, middle leg; O, mesotibial apex; P, hindleg; Q, metatibial apex.

Key to the Australian genera of Brachycerinae

Modified after Zimmerman (1993).

1. Body subglabrous, with widely dispersed fine setae (Fig. 44.11A, B); eyes absent; body length ~1.0 mm 2
– Body with dense to scattered setae or squamae (scales) (Fig. 44.11C–P); eyes present; body length usually >1.0 mm 3
2(1). Funicles 6-segmented ..... ***Hexonymus*** **Kuschel**
– Funicles 7-segmented ..... ***Myrtonymus*** **Kuschel**
3(1). Prosternum with deep canal to receive rostrum in repose, ending in mesoventral receptacle 4
– Prosternum with or without canal, without mesoventral receptacle 5
4(3). Tarsi 3-segmented, tarsal claws absent (Fig. 44.11T) ..... ***Aonychus*** **Schoenherr**
– Tarsi 5-segmented, tarsal claws present (Fig. 44.11S) ..... ***Tadius*** **Pascoe**
5(3). Funicles 7-segmented (Fig. 44.12B, M) 6
– Funicles 6-segmented (Fig. 44.13D) 9
6(5). Tarsal claws absent (Fig. 44.11J) ..... ***Cenchrena*** **Pascoe**
– Tarsal claws present 7
7(6). Elytral interstriae with abundant erect, arcuate, sharp, needle-like setae (Fig. 44.11E, V) ..... **'*Bagoopsis*' Faust**
– Elytral interstriae with decumbent, inconspicuous, squamiform setae (Fig. 44.11F, U) 8
8(7). Body length approximately 2.0 mm (Fig. 44.11H); tarsal segments 5 about as long as segments 3 (e.g. Figs 44.11Q, R, 44.12I) ..... ***Athor*** **Broun**
– Body length between 3.5–4.6 mm (Fig. 44.11F); tarsal segments 5 about twice the length of segments 3 (e.g. Fig. 44.11S) ..... ***Echinocnemus*** **Schoenherr**
9(5). Body with aquifuge plastron; tarsi slender, segments subequal in width 10
– Body without aquifuge plastron; tarsal segments 2 and 3 broader than 1 11
10(9). Dorsum with scattered small squamae (Fig. 44.11L) ..... ***Cyrtobagous*** **Hustache**
– Dorsum with dense squamae (e.g. Fig. 44.11M) ..... ***Hydrotimetes*** **Kolbe**
11(9). Prothorax without ocular lobes; scutellar shield visible; tarsal segments 5 not extending beyond apex of 3 (Fig. 44.11Q) ..... ***Niphobolus*** **Blackburn**
– Prothorax with or without ocular lobes; scutellar shield concealed; tarsal segments 5 extending beyond apex of 3 (Fig. 44.11R) 12
12(11). Body length 3.0 mm or more; tibiae with hydrofuge setae; prothorax with large ocular lobes ..... ***Neochetina*** **Hustache**
– Body length ~2.0 mm; tibiae without hydrofuge setae; prothorax with feeble ocular lobes 13
13(12). Eyes small, narrow, vertical, in lateral view about 4–5 ommatidia wide (Fig. 44.13A) ..... ***Baeosomus*** **Broun**
– Eyes large, broad, in lateral view more than 5 ommatidia wide ..... ***Neohydronomus*** **Hustache**

Classification of the Australian genera of Brachycerinae

Tribe Erirhinini Schoenherr, 1825

Aonychus Schoenherr, 1844 (Pl. 47E, F)

Type species, by original designation: *Aonychus hopei* Boheman, 1844.

Australian species. *A. argus* Lea, 1905b, *A. barbatus* Lea, 1927e, *A. bicruciatus* Lea, 1905b, *A. hopei* Boheman, 1844 (*in* Schoenherr 1844), *A. lineatus* Pascoe, 1870b, *A. lituratus* Lea, 1927e, *A. luctuosus* Pascoe, 1872c, *A. pachypus* Lea, 1905b, *A. striatus* Lea, 1905b.

Distribution. Australia (all states and territories except TAS).

Notes. The genus is endemic to Australia and has no know close relatives elsewhere. Several undescribed species are known in addition to the nine described (Zimmerman 1993). The hosts and life histories of the genus are virtually unknown, except that adults are usually collected on species of Proteaceae such as *Grevillea* and *Hakea*.

'*Bagoopsis*' Faust, 1881 (Pl. 47G)

Type species, by subsequent designation (Zimmerman, 1993: 133): *Bagoopsis volgensis* Faust, 1881.

Australian species. '*Bagoopsis*' sp. n. (Zimmerman, 1993: 146).

Distribution. Australia (WA).

Notes. The undescribed species from the Kimberley area in WA that has been assigned to the Palaearctic genus *Bagoopsis* by Zimmerman (1993) does not, in fact, belong in this genus but, together with another similar species from further south, represents a different, undescribed Australian genus related to *Echinocnemus*.

Echinocnemus Schoenherr, 1843 (Pl. 47H, I; Fig. 44.9J–P)

Type species, by original designation: *Erirhinus squameus* Boheman, 1836 (*in* Schoenherr 1836).

Australian species. *E. exsul* Faust, 1887, *E. latus* (Lea, 1928), *E. malevolens* (Lea, 1899a), *E. rostralis* (Lea, 1928c), *E. vegrandis* (Lea, 1899a).

Distribution. Australia (QLD, NT, NSW, VIC, SA, WA), New Caledonia, Oriental, southern Palaearctic and Afrotropical regions.

Notes. The genus contains about 42 species (Oberprieler 2014b) worldwide but is poorly understood taxonomically, both in Australia and elsewhere. In addition to the five named species in Australia (Zimmerman 1993) there are at least a dozen undescribed ones known, mostly from the Northern Territory. Specimens are commonly collected at light near water. The larvae probably feed mainly on the roots of semi-aquatic grasses, but larvae of one species have been found feeding on the water fern *Marsilea* (May 1994).

Tadius Pascoe, 1885 (Pl. 47J)

Type species, by monotypy: *Tadius erirhinoides* Pascoe, 1885.
Australian species. *T. barbatus* Lea, 1928b, *T. erirhinoides* Pascoe, 1885.
Distribution. Australia (QLD, NT), New Guinea, Oriental region.
Notes. The genus only contains one or two species (Zimmerman 1993; Oberprieler 2014b). It appears to be associated with orchids (Oberprieler 2014b), but nothing specific is recorded about its life history.

Tribe Myrtonymini Kuschel, 1990

Hexonymus Kuschel, 2014 (Pl. 47K)

Type species, by original designation: *Hexonymus reginalis* Kuschel, 2014.
Australian species. *H. reginalis* Kuschel, 2014b.
Distribution. Australia (QLD).
Notes. The genus only contains a single described species, but it is likely that undescribed ones exist.

Myrtonymus Kuschel, 1990 (Pl. 47L; Fig. 44.12)

Type species, by original designation: *Myrtonymus zelandicus* Kuschel, 1990.
Australian species. *M. australicus* Kuschel, 2014b, *M. eucalypti* Kuschel, 2014b, *M. moorei* Kuschel, 2014b, *M. peckorum*, Kuschel, 2014b.
Distribution. Australia (QLD, NSW, WA), New Guinea, New Caledonia, New Zealand.
Notes. The genus currently contains six described species, but several unnamed ones are known, and the genus may be much more diverse and widespread than indicated by the few known species and specimens. The adults have been found in rotten bark, log litter and humus and soil around myrtaceaous trees, and the larvae are suspected to feed on fine rootlets.

Tribe Tanysphyrini Gistel, 1856

Athor Broun, 1909 (Pl. 48A, B; Fig. 44.12A–K)

Type species, by monotypy: *Athor arcifera* Broun, 1909a.
Australian species. *Athor* spp. (Zimmerman, 1993: 151–152).
Distribution. Australia (QLD, NSW, VIC, SA, TAS, WA), New Zealand.
Notes. Zimmerman (1993) assigned three unnamed species to *Athor* but additional species are also known.

Baeosomus Broun, 1904 (Pl. 48C; Fig. 44.13A–Q)

Type species, by monotypy: *Baeosomus tacitus* Broun, 1904.
Australian species. *B. uvidus* (Oke, 1931), *Baeosomus* spp. (Zimmerman, 1993: 153, as *Bryocatus*).
Distribution. Australia (VIC, TAS), New Zealand.
Notes. The genus *Baeosomus* contains approximately 25 species in New Zealand and surrounding islands. The assignment of *B. uvidus* and other Australian species to it (Zimmerman 1993) require reappraisal.

Cenchrena Pascoe, 1874 (Pl. 48D–F)

Type species, by subsequent designation (Setliff, 2007: 10): *Cenchrena fasciata* Pascoe, 1874c.
Australian species. *C. alternata* (Lea, 1911a).
Distribution. Australia (QLD, NSW), New Guinea and surrounding islands.
Notes. The genus contains four described and several undescribed species. It was only recognised as belonging in Brachycerinae by Pullen *et al.* (2014), owing to its pedotectal aedeagus. Nothing appears known about its hosts and life history.

Cyrtobagous Hustache, 1929 (Pl. 48G; Figs 44.8, 44.10A–B)

Type species, by original designation: *Cyrtobagous singularis* Hustache, 1929.
Australian species. *C. salviniae* Calder & Sands, 1985.
Distribution. Australia (QLD, NT, NSW), native to South America, also introduced to the USA, Botswana, Namibia, South Africa, Zambia, Fiji, New Guinea.
Notes. The single species in Australia was introduced from Brazil for the control of the aquatic fern *Salvinia molesta* (Salviniaceae) and is now widespread in Queensland, New South Wales and the Northern Territory (Zimmerman 1993; Pullen *et al.* 2014).

Hydrotimetes Kolbe, 1911 (Pl. 48H)

Type species, by monotypy: *Hydrotimetes natans* Kolbe, 1911.
Australian species. *H. natans* Kolbe, 1911.

Distribution. Australia (QLD), native to South America.

Notes. The single species in Australia was introduced from Argentina in 2023 for the control of the aquatic weed *Cabomba caroliniana* (Cabombaceae). Further releases are planned in northern New South Wales and the Northern Territory, but it remains to be seen whether and where the species establishes.

Neochetina Hustache, 1926 (Pl. 48I; Fig. 44.10C–I)

Type species, by subsequent designation (O'Brien, 1976: 165): *Neochetina bruchi* Hustache, 1926.

Australian species. *N. bruchi* Hustache, 1926, *N. eichhorniae* Warner, 1970.

Distribution. Australia (QLD, NSW, NT), native to South America.

Notes. The two species in Australia were introduced from Uruguay for the control of Water Hyacinth (*Eichhornia crassipes,* Pontederiaceae) and have established successfully (Zimmerman 1993; Pullen *et al.* 2014).

Neohydronomus Hustache, 1926 (Pl. 48J; Fig. 44.10J–P)

Type species, by monotypy: *Neohydronomus pulchellus* Hustache, 1926.

Australian species. *N. affinis* Hustache, 1926.

Distribution. Australia (QLD), native to South America.

Notes. The single species in Australia was introduced from South America for the control of Water Lettuce (*Pistia stratiotes,* Araceae) (Zimmerman 1993; Pullen *et al.* 2014).

Niphobolus Blackburn, 1893 (Pl. 48K, L)

Type species, by monotypy: *Niphobolus deceptor* Blackburn, 1893.

Australian species. *N. deceptor* Blackburn, 1893b.

Distribution. Australia (QLD, NSW, NT).

Notes. The genus is native to Australia and comprises only one described species, but a few undescribed ones are known. The weevils are sometimes collected in large numbers at light near water bodies, and the larvae probably develop on aquatic or semi-aquatic grasses.

Curculionidae: Cyclominae Schoenherr, 1826

Introduction. The Australian **Cyclominae** constitute a diverse group of four tribes (Amycterini, Aterpini, Listroderini and Rhythirrinini), previously placed in different subfamilies until reclassified by Oberprieler (2010) with some further modifications by Oberprieler *et al.* (2007). The small tribe Notiomimetini, also included in Cyclominae by Oberprieler (2010, 2014c), has since been shown to be a more basal lineage of Curculionidae related to Bagoinae (Gillett *et al.* 2018; Escalona & Oberprieler 2021) but it is listed here in Cyclominae as a matter of convenience. The subfamily also includes three non-Australian tribes, i.e. Cyclomini, Dichotrachelini and Hipporhinini (Oberprieler 2010, 2014c), and recent molecular analyses indicate that also the Gonipterini and several small Australian genera currently classified in Curculioninae: Storeini belong in Cyclominae (Baird *et al.* 2021; Haran *et al.* 2023). Its proper composition and monophyly as well as those of most of its tribes are therefore still unclear. The Cyclominae are one of the main lineages of the CEGH clade of 'higher' Curculionidae (Gunter *et al.* 2016) and appear to be sister-group of Entiminae (Gillet *et al.* 2014; Shin *et al.* 2018; Baird *et al.* 2021), but the relationships among all lineages of the CEGH clade remain to be properly resolved.

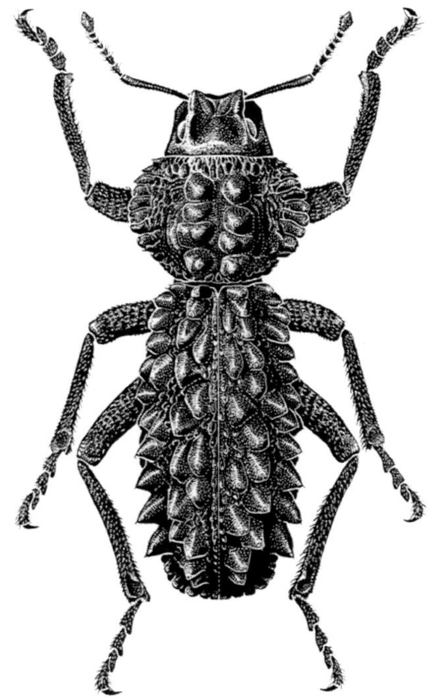

Fig. 44.14. *Gagatophorus draco* (Macleay).

A comprehensive account of the systematics and biology of the subfamily in its current concept was provided by Oberprieler (2014c), and the larvae of several genera were studied by May (1994), with some of her illustrations reproduced here (Figs 44.22–44.29). There are a number of undescribed eyeless genera and species (Pl. 57H–L) found in calcareous aquifers in north Western Australia and their placement could be within Cyclominae.

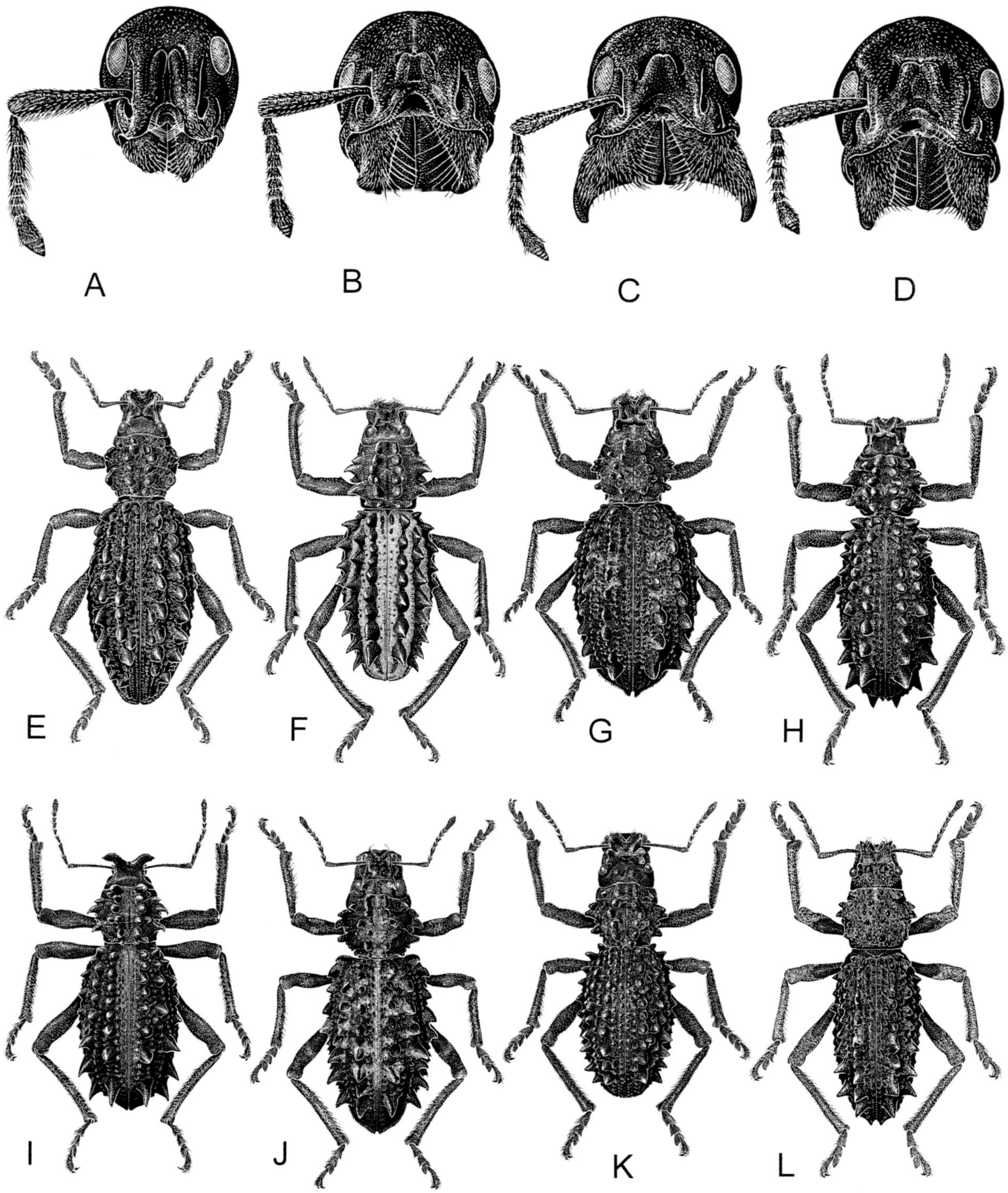

Fig. 44.15. Cyclominae, head structures and habitus, Amycterini. A–D, anterior view of some Amycterini heads showing mandibles: A, *Talaurinus carinatus* Ferguson; B, *Amycterus forficulatus* (Macleay); C, *Amycterus miracula* (Macleay); D, *Amycterus carteri* (Ferguson); E–L, dorsal habitus of *Acantholophus* species: E, *A. alpicola* Ferguson; F, *A. amycteroides* Macleay; G, *A. angusticollis* Ferguson; H, *A. apicalis* Macleay; I, *A. crassidens* Macleay; J, *A. dixoni* Ferguson; K, *A. echidna* Macleay; L, *A. eximius* (Macleay). After Zimmerman (1993).

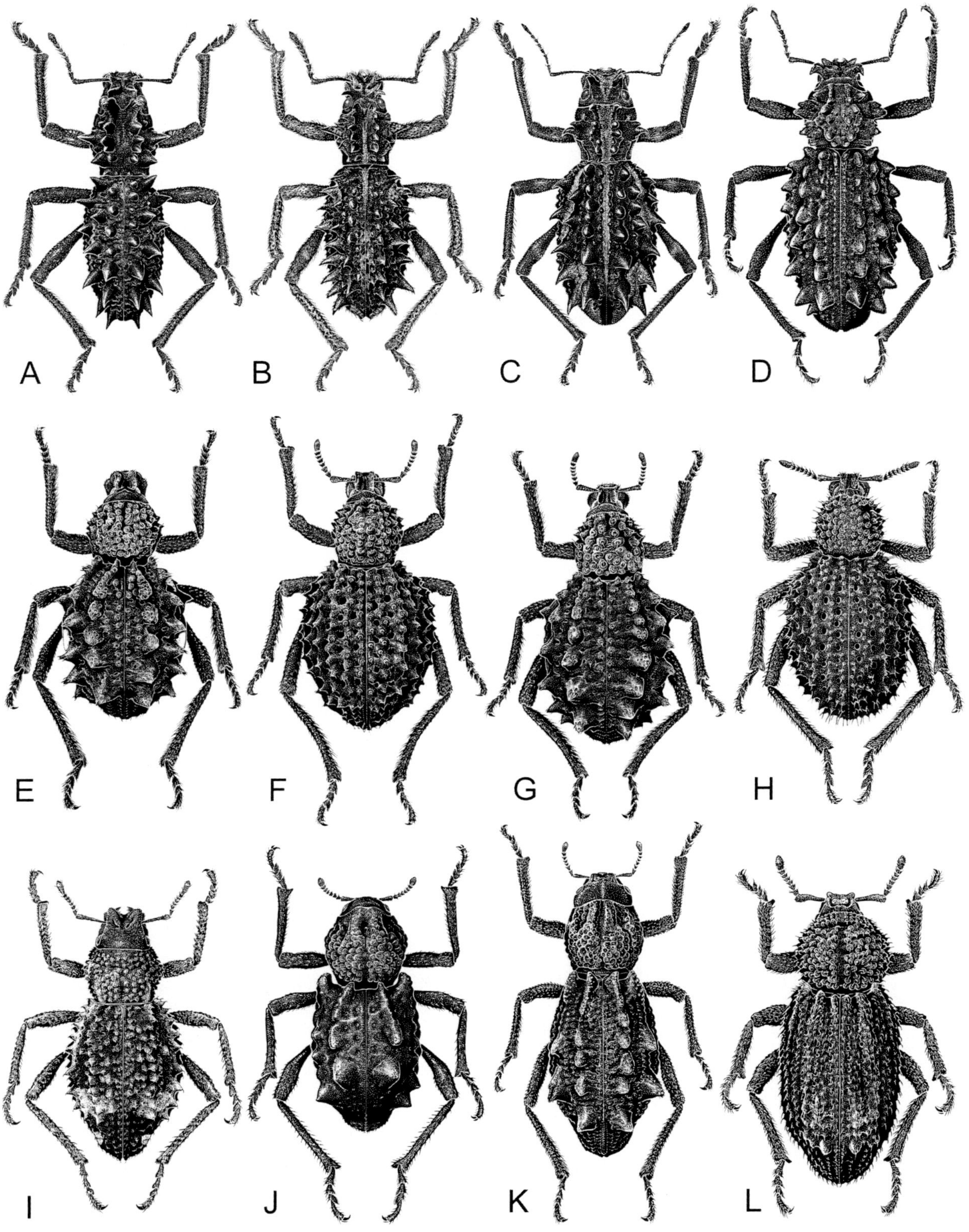

Fig. 44.16. Cyclominae, dorsal habitus, Amycterini. A, *Acantholophus gladiator* Pascoe; B, *Acantholophus krefftii* Macleay; C, *Acantholophus lateralis* (Boheman); D, *Acantholophus marshami* (Kirby); E, *Acherres globicollis* Lea; F, *Acherres granulatus* Ferguson; G, *Acherres mamillatus* Pascoe; H, *Acherres pilosus* Ferguson; I, *Achorostoma hystricosum* (Boheman); J, *Aedriodes fastigiatus* Pascoe; K, *Aedriodes nodipennis* (Boheman); L, *Alexirhea aurita* Pascoe. After Zimmerman (1993).

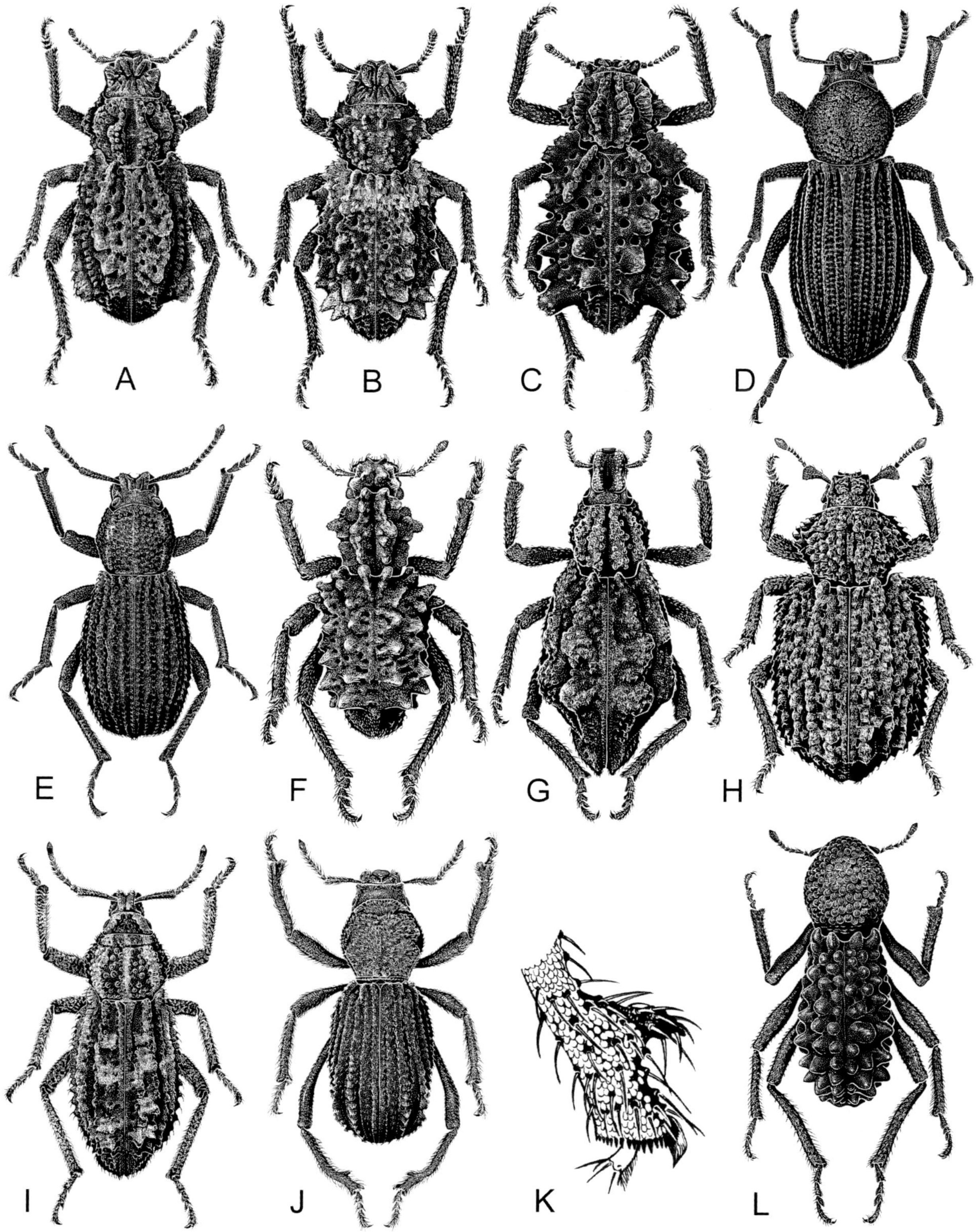

Fig. 44.17. Cyclominae, dorsal habitus, Amycterini. A, *Amorphorhinus obliteratus* Ferguson; B, *Amorphorhinus polyacanthus* Pascoe; C, *Amorphorhinus tuberculatus* Ferguson; D, *Amycterus mirabilis* (Kirby); E, *Amycterus riverinae* (Macleay); F, *Anascoptes fasciatus* Ferguson; G, *Atychoria funesta* Pascoe; H, *Brachyrothus auritus* (Ferguson); I, *Chriotyphus acromialis* Pascoe; J–K, *Cubicorhynchus calcaratus* Macleay: J, habitus; K, metatibial apex; L, *Cucullothorax horridus* Ferguson. After Zimmerman (1993).

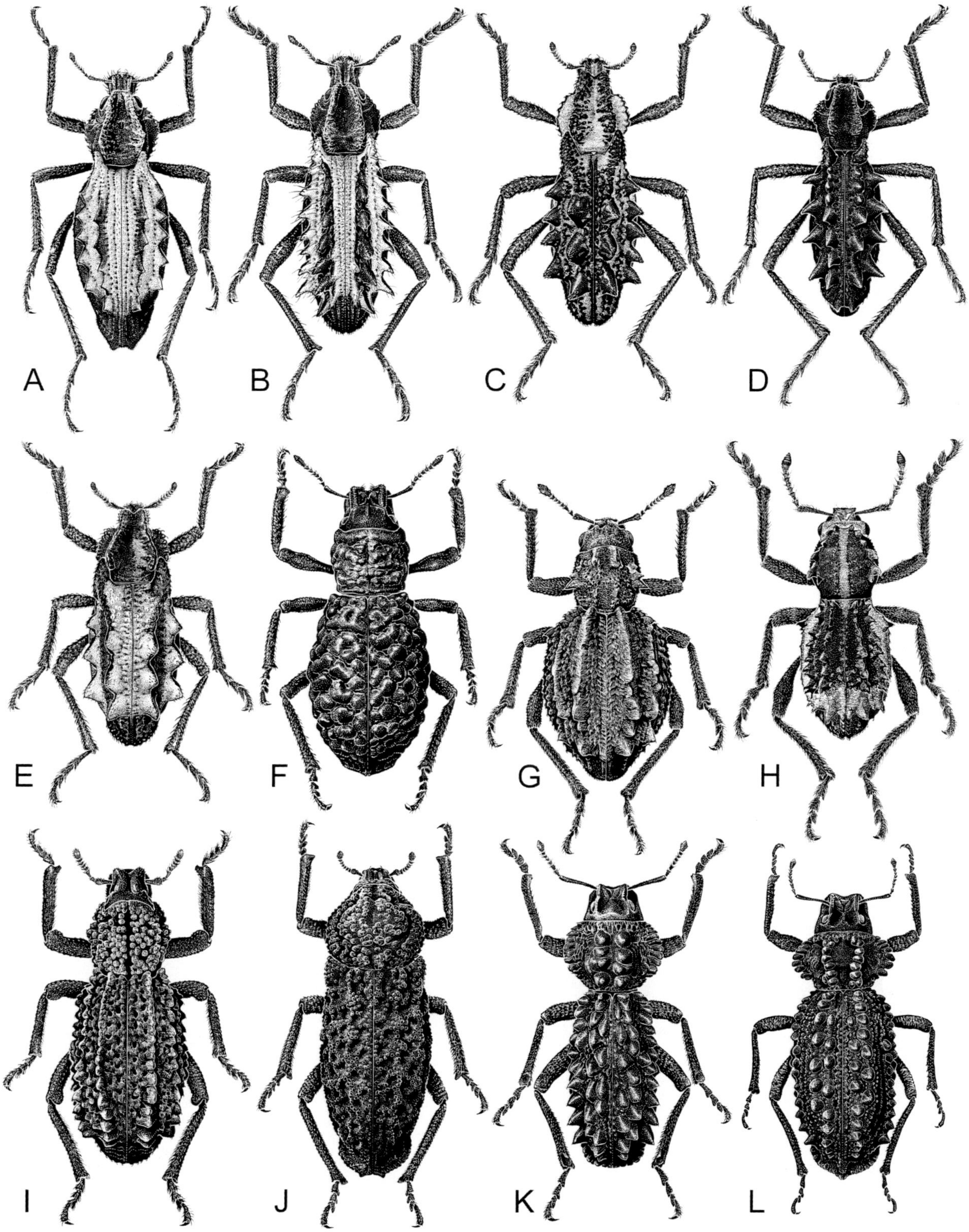

Fig. 44.18. Cyclominae, dorsal habitus, Amycterini. A, *Dialeptopus collaris* (Boheman); B, *Dialeptopus echinatus* Lea; C, *Dialeptopus lindensis* Blackburn; D, *Dialeptopus macilentus* Pascoe; E, *Dialeptopus validus* Blackburn; F, *Dicherotropis damelii* (Macleay); G, *Ennothus fallax* Pascoe; H, *Ennothus metricus* (Pascoe); I, *Euomella retusa* (Pascoe); J, *Euomus insculptus* (Boheman); K–L, *Gagatophorus draco* Macleay: K, ♂; L, ♀. After Zimmerman (1993).

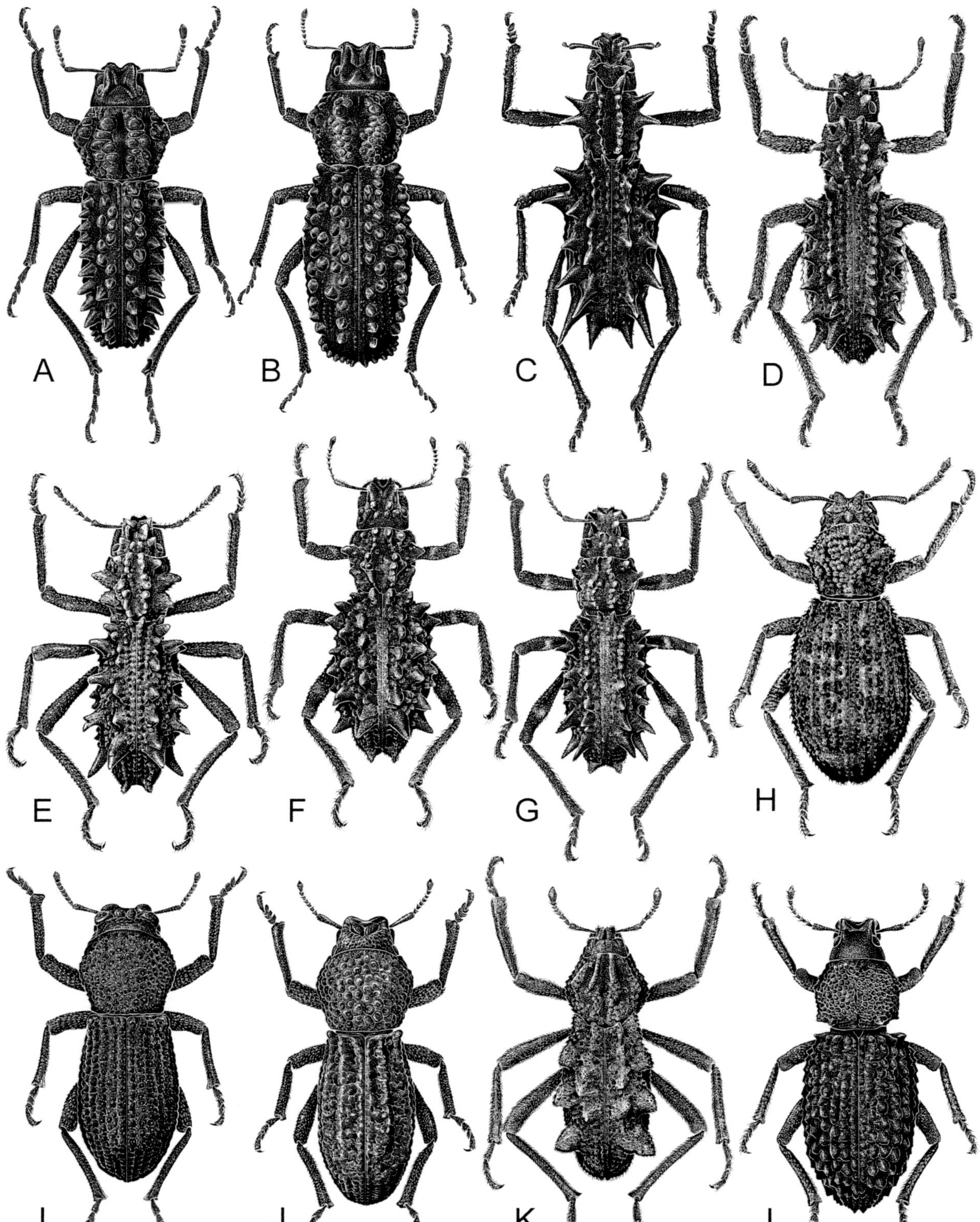

Fig. 44.19. Cyclominae, dorsal habitus, Amycterini. A–B, *Gagatophorus leichhardtii* (Macleay): A, ♂; B, ♀; C, *Hyborrhinus aculeatus* (Ferguson); D. *Hyborrhinus bicornutus* (Macleay); E, *Hyborrhinus coenosus* (Boheman); F, *Hyborrhinus furcatus* (Macleay); G. *Hyborrhinus prodigus* (Macleay); H, *Hypotomops angulicollis* (Lea); I, *Lataurinus macrocephalus* (Ferguson); J, *Lataurinus rugiceps* (Macleay); K, *Melanegis stygius* Pascoe; L, *Molochtus gagates* Pascoe. After Zimmerman (1993).

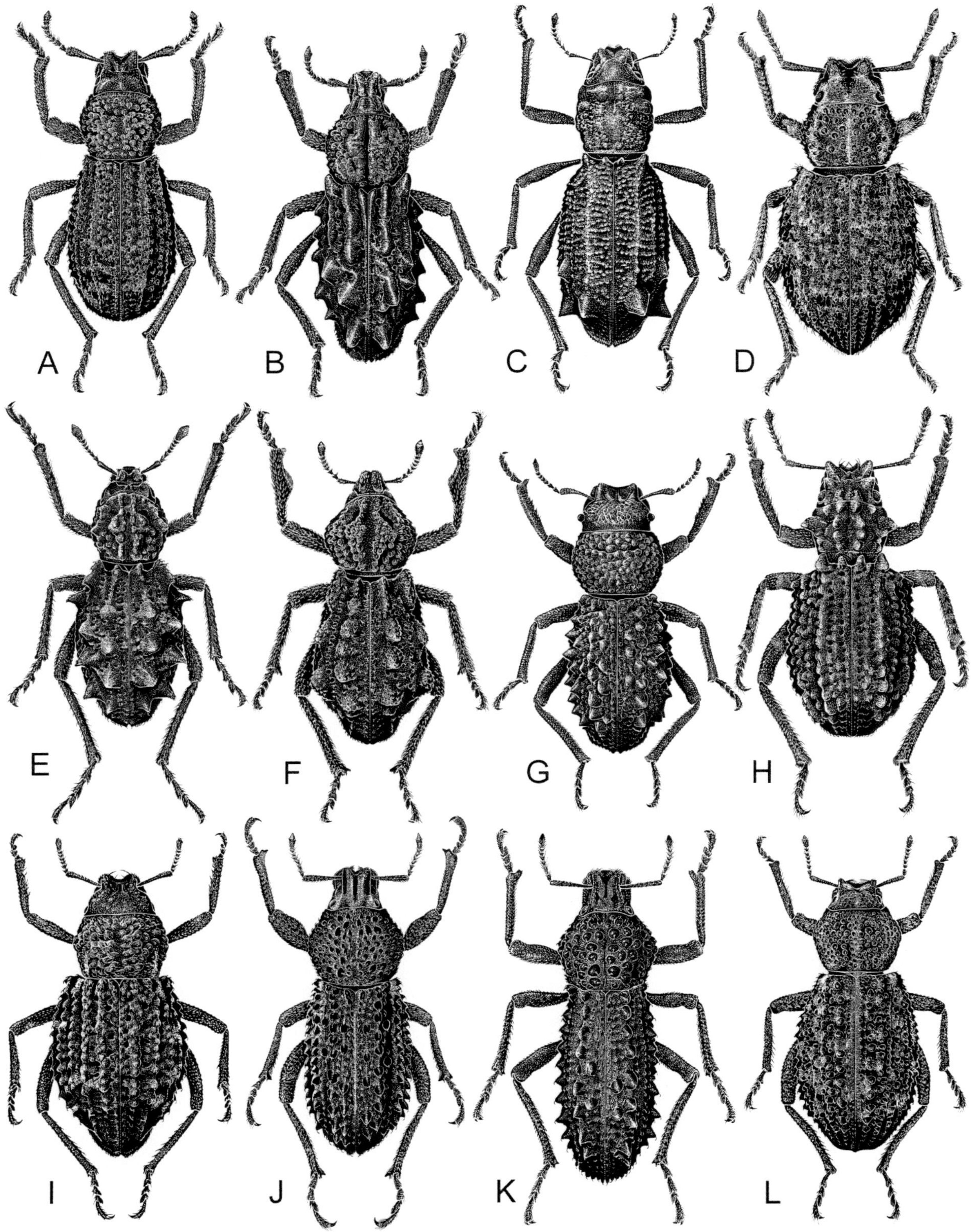

Fig. 44.20. Cyclominae, dorsal habitus, Amycterini. A, *Myotrotus obtusus* Pascoe; B, *Mythites tuberculatus* Schoenherr; C, *Neohyborrhynchus rugosus* (Macleay); D, *Notonophes cichlodes* (Pascoe); E, *Oditesus indutus* Pascoe; F, *Oditesus tibialis* Lea; G, *Ophthalamycterus laticeps* (Macleay); H, *Parahyborrhynchus convexiusculus* (Macleay); I, *Pseudonotonophes dumosus* (Macleay); J, *Sclerorinus adelaidae* Macleay; K, *Sclerorinus blackburni* Ferguson; L, *Sclerorrhinella manglesii* (Boheman). After Zimmerman (1993).

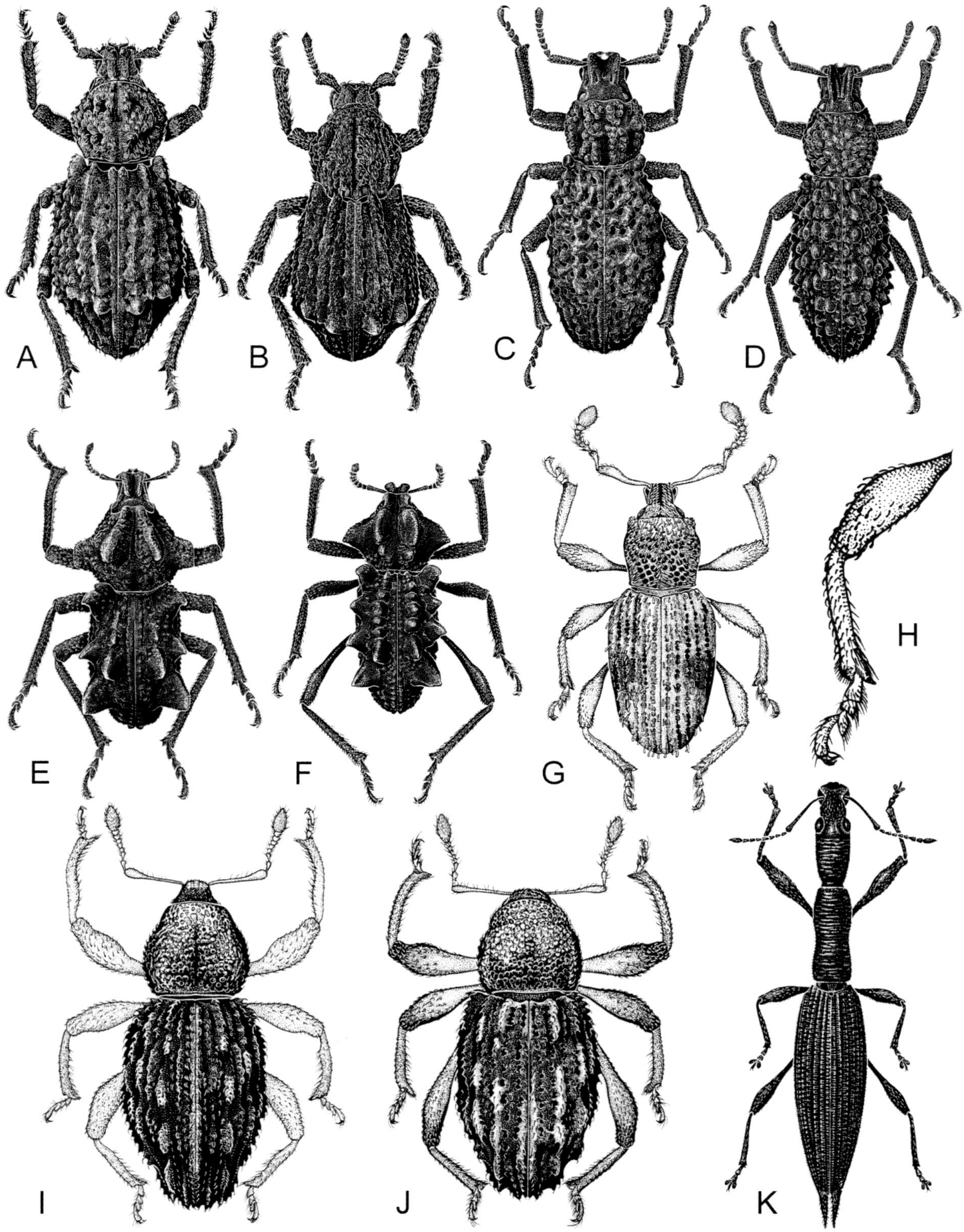

Fig. 44.21. Cyclominae, dorsal habitus, A–F Amycterini, G–J Listroderini and K Aterpini. A, *Sosytelus indemnis* (Pascoe); B, *Sosytelus lobatus* Pascoe; C, *Talaurinus foveipennis* Ferguson; D, *Talaurinus typicus* Macleay; E, *Tetralophus excursus* Pascoe; F, *Tetralophus sculpturatus* Waterhouse; G–H, *Mandalotina atranotata* Oke: G, habitus; H, middle leg; I, *Mandalotina bicolor* Oke; J, *Mandalotina varia* Oke; K, Aterpini, *Rhadinosomus lacordairei* Pascoe. A–F, after Zimmerman (1993); K, after Lawrence & Ślipiński (2013).

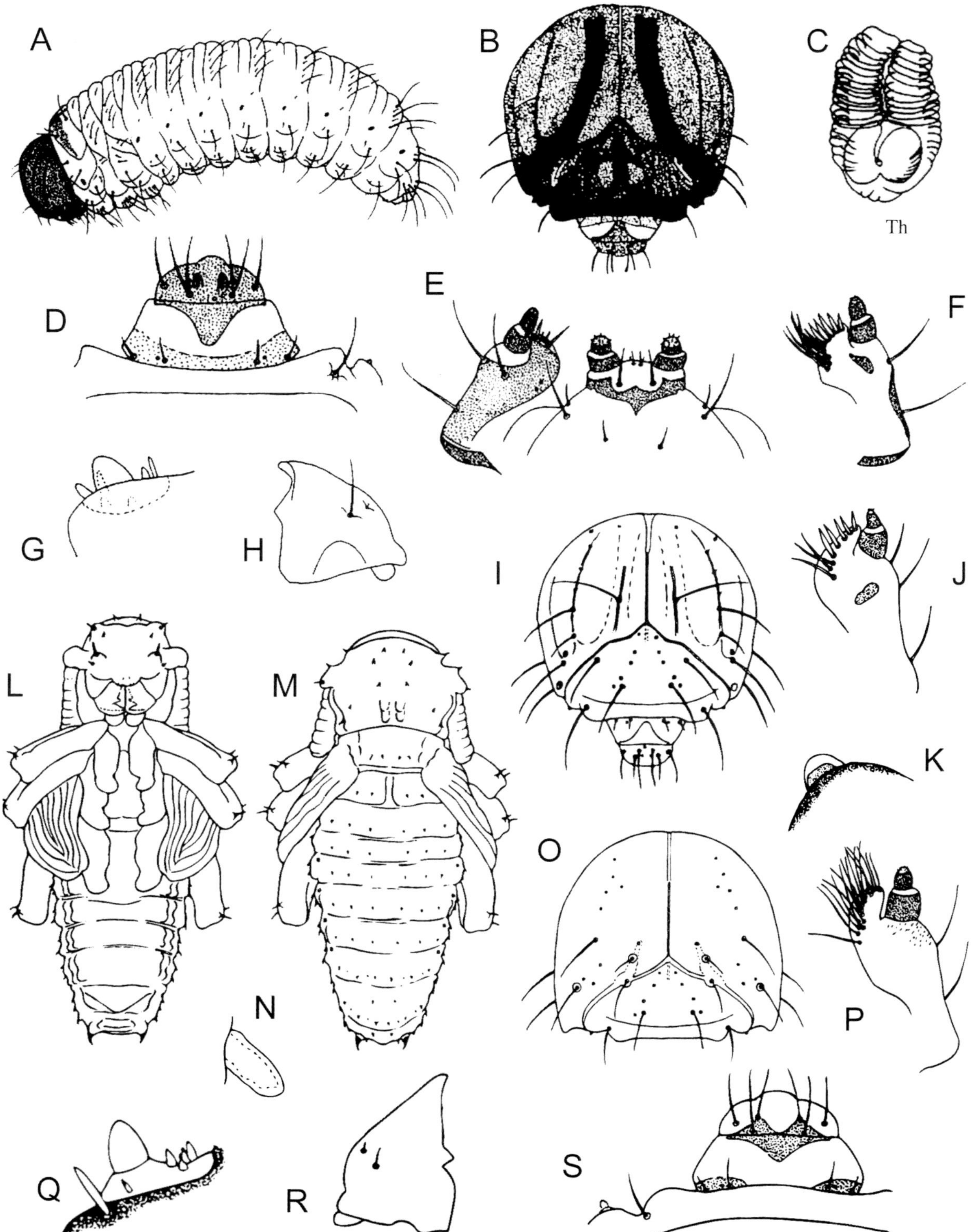

Fig. 44.22. Cyclominae, larvae and pupae, Amycterini. A–H, *Acantholophus marshami* (Kirby), larva: A, habitus, lateral; B, head, anterior; C, thoracic spiracle; D, labrum, clypeus, epistoma and antenna, anterior; E, mala and labium, posterior; F, mala, anterior; G, antenna; H, mandible; I–K, *Cubicorhynchus crenicollis* (Waterhouse), larva: I, head, anterior; J, mala, anterior; K, antenna; L–N, *Cubicorhynchus crenicollis* (Waterhouse), pupa: L, habitus, ventral; M, habitus, dorsal; N, secondary pterotheca with adult wing inside; O–S, *Dialeptopus echinatus* Lea, larva: O, head, anterior; P, mala, anterior; Q, antenna; R, mandible; S, labrum, clypeus, epistome and antenna, anterior. After May (1994).

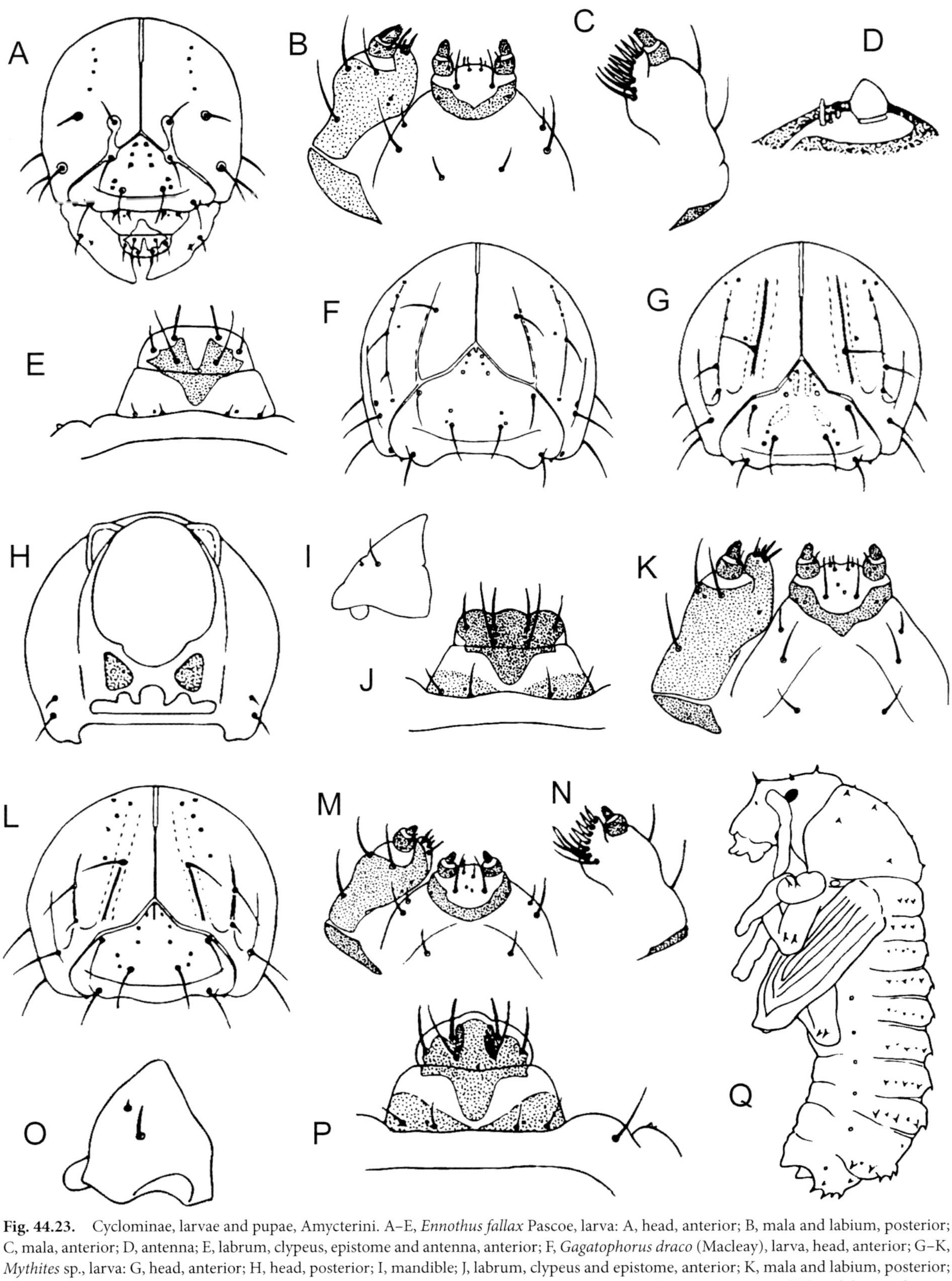

Fig. 44.23. Cyclominae, larvae and pupae, Amycterini. A–E, *Ennothus fallax* Pascoe, larva: A, head, anterior; B, mala and labium, posterior; C, mala, anterior; D, antenna; E, labrum, clypeus, epistome and antenna, anterior; F, *Gagatophorus draco* (Macleay), larva, head, anterior; G–K, *Mythites* sp., larva: G, head, anterior; H, head, posterior; I, mandible; J, labrum, clypeus and epistome, anterior; K, mala and labium, posterior; L–P, *Amycterus abnormis* (Macleay), larva: L, head, anterior; M, mala and labium, posterior; N, mala, anterior; O, mandible; P, labrum, clypeus, epistome and antenna, anterior; Q, *Amycterus abnormis* (Macleay), pupa, habitus, lateral. After May (1994).

Fig. 44.24. Cyclominae, larvae and pupae, Amycterini. A–E, *Sclerorinus* sp., larva: A, head, anterior; B, mala and labium, posterior; C, mala, anterior; D, antenna; E, labrum, clypeus and epistoma, anterior; F–J, *Sosytelus pubescens* (Lea), larva: F, head, anterior; G, mala and labium, posterior; H, antenna; I, labrum, clypeus, epistome and antenna, anterior; J, mandible; K–L, *Sosytelus pubescens* (Lea), pupa: K, head, dorsolateral; L, habitus, dorsal; M–O, *Talaurinus rugifer* (Boisduval), larva: M, head, anterior; N, mala and labium, posterior; O, labrum, mala, clypeus, epistome and antenna, anterior; P–Q, *Talaurinus rugifer* (Boisduval), pupa: P, pronotum and mesonotum; Q, habitus, lateral. After May (1994).

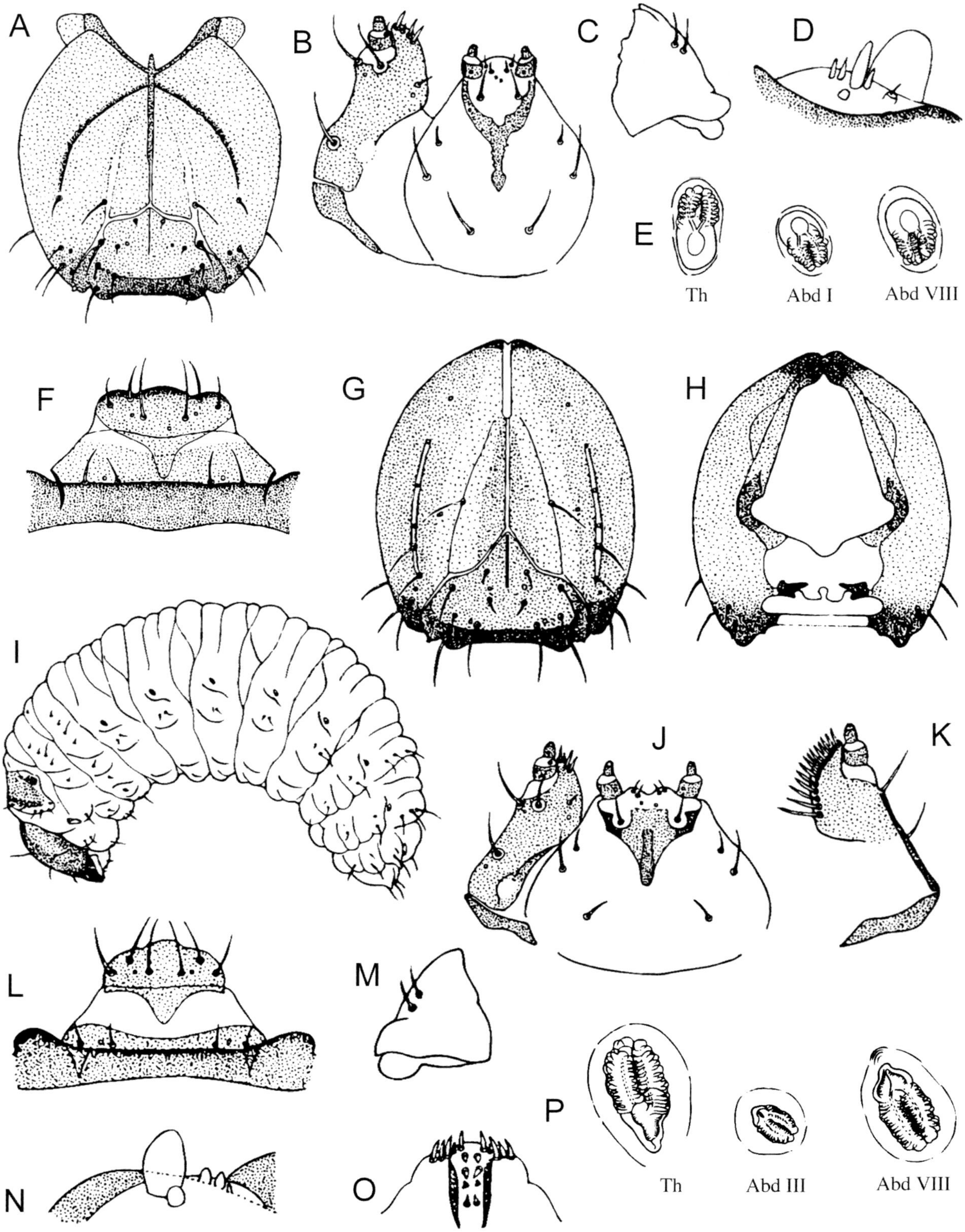

Fig. 44.25. Cyclominae, larvae, Aterpini. A–F, *Aesiotes leucurus* Pascoe, larva: A, head, anterior; B, mala and labium, posterior; C, mandible; D, antenna; E, spiracles, showing alignment; F, labrum, mala, clypeus and epistome, anterior; G–P, *Chrysolopus spectabilis* (Fabricius), larva: G, head, anterior; H, head, posterior; I, habitus, lateral; J, mala and labium, posterior; K, mala, anterior; L, labrum, mala, clypeus and epistome, anterior; M, mandible; N, antenna, posterior; O, epipharyngeal lining; P, spiracles, showing alignment. After May (1994).

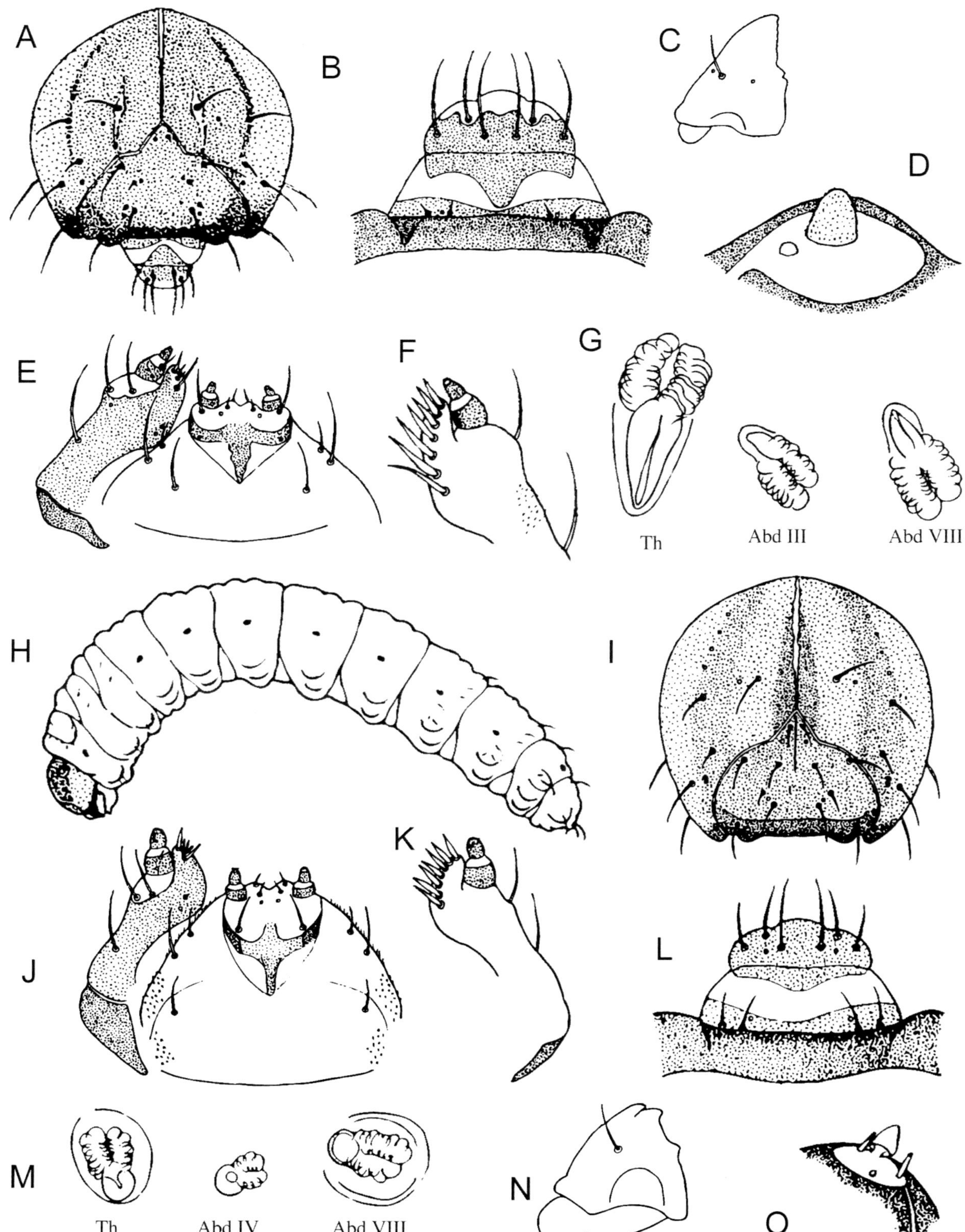

Fig. 44.26. Cyclominae, larvae, Aterpini. A–G, *Pelororhinus* sp., larva: A, head, anterior; B, labrum, mala, clypeus and epistome, anterior; C, mandible; D, antenna, posterior; E, mala and labium, posterior; F, mala, anterior; G, spiracles, showing alignment; H–O, *Rhadinosomus lacordarei* Pascoe, larva: H, habitus, lateral; I, head, anterior; J, mala and labium, posterior; K, mala, anterior; L, labrum, mala, clypeus and epistome, anterior; M, spiracles, showing alignment; N, mandible; O, antenna. After May (1994).

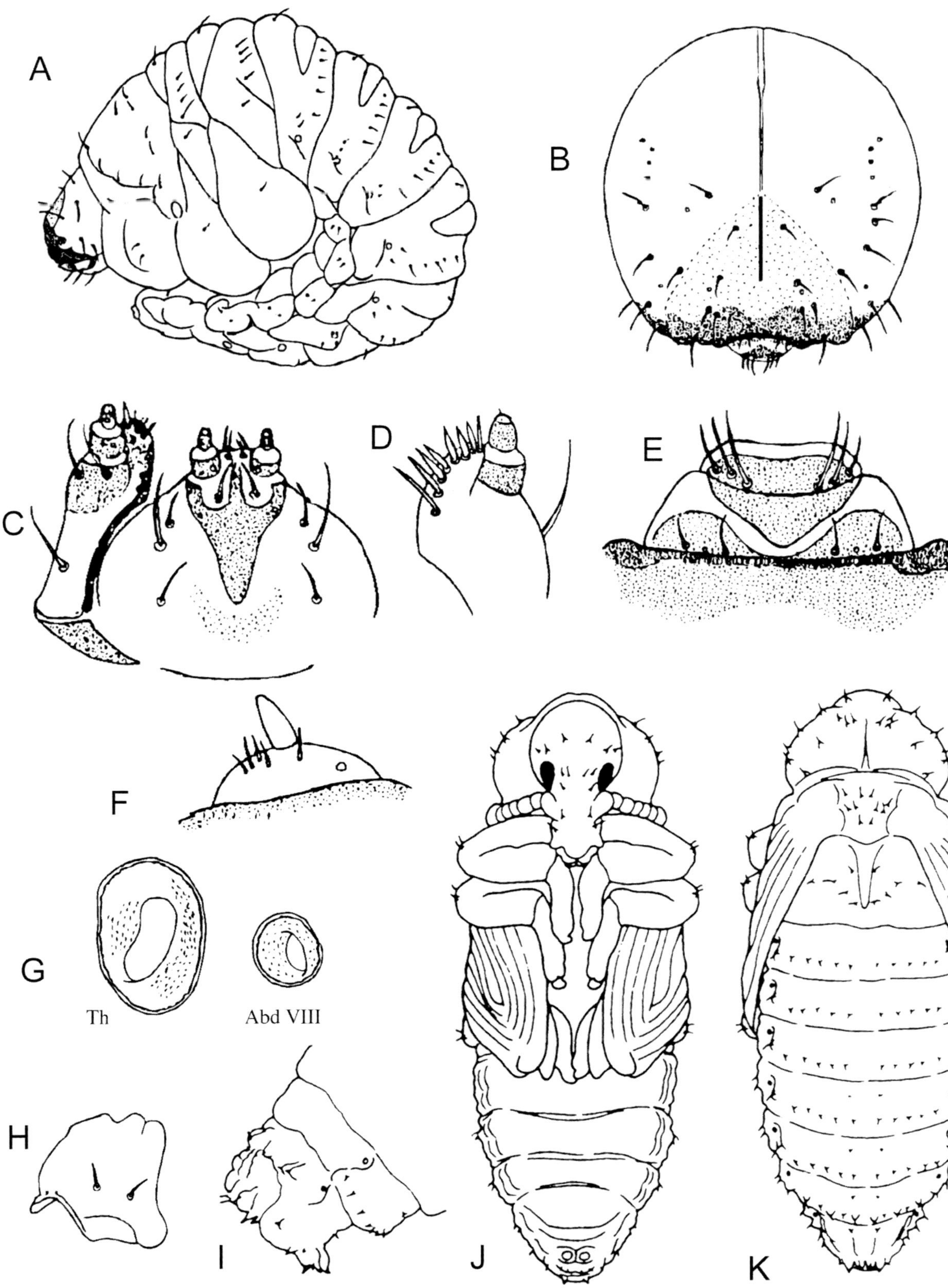

Fig. 44.27. Cyclominae, larvae and pupae, Aterpini. A–H, *Strongylorhinus clarki* Marshall, larva: A, habitus, lateral; B, head, anterior; C, mala and labium, posterior; D, mala, anterior; E, labrum, clypeus and epistome, anterior; F, antenna; G, spiracles, showing alignment; H, mandible; I–K, *Strongylorhinus clarki* Marshall, pupa: I, terminal segments, lateral; J, habitus, ventral; K, habitus, dorsal. After May (1994).

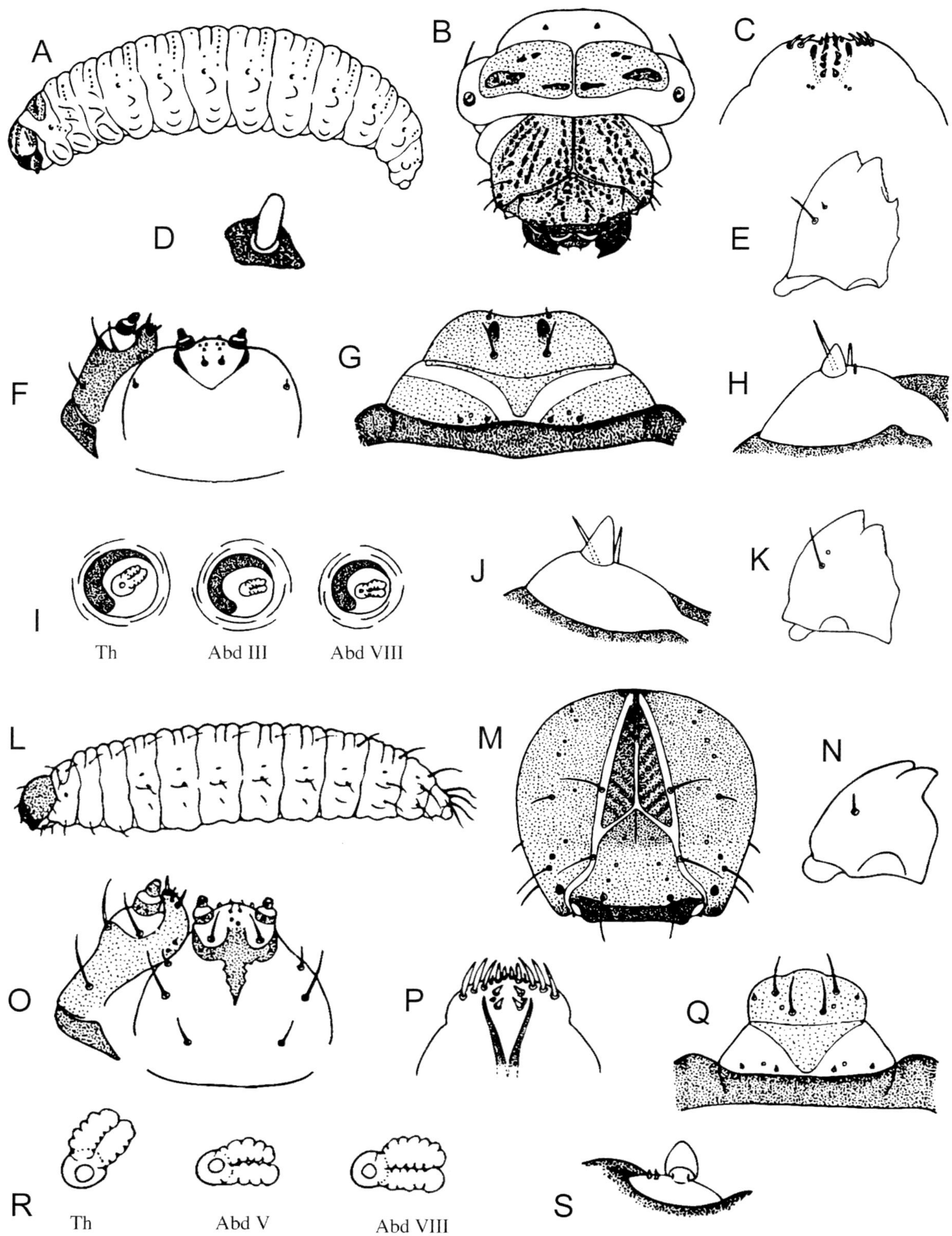

Fig. 44.28. Cyclominae, larvae, Listroderini. A–I, *Listroderes difficilis* Germain, larva (New Zealand): A, habitus, lateral; B, head and prothorax, anterior; C, epipharyngeal lining; D, antenna; E, mandible; F, mala and labium, posterior; G, labrum, clypeus and epistome, anterior; H, antenna; I, spiracles, showing alignment; J–K, *Listroderes delaiguei* Germain, larva (New Zealand); J, antenna; K, mandible; L–S, *Listronotus bonariensis* (Kuschel), larva (New Zealand): L, habitus, lateral; M, head, anterior; N, mandible; O, mala and labium, posterior; P, epipharyngeal lining; Q, labrum, clypeus and epistome, anterior; R, spiracles, showing alignment; S, antenna. After May (1994).

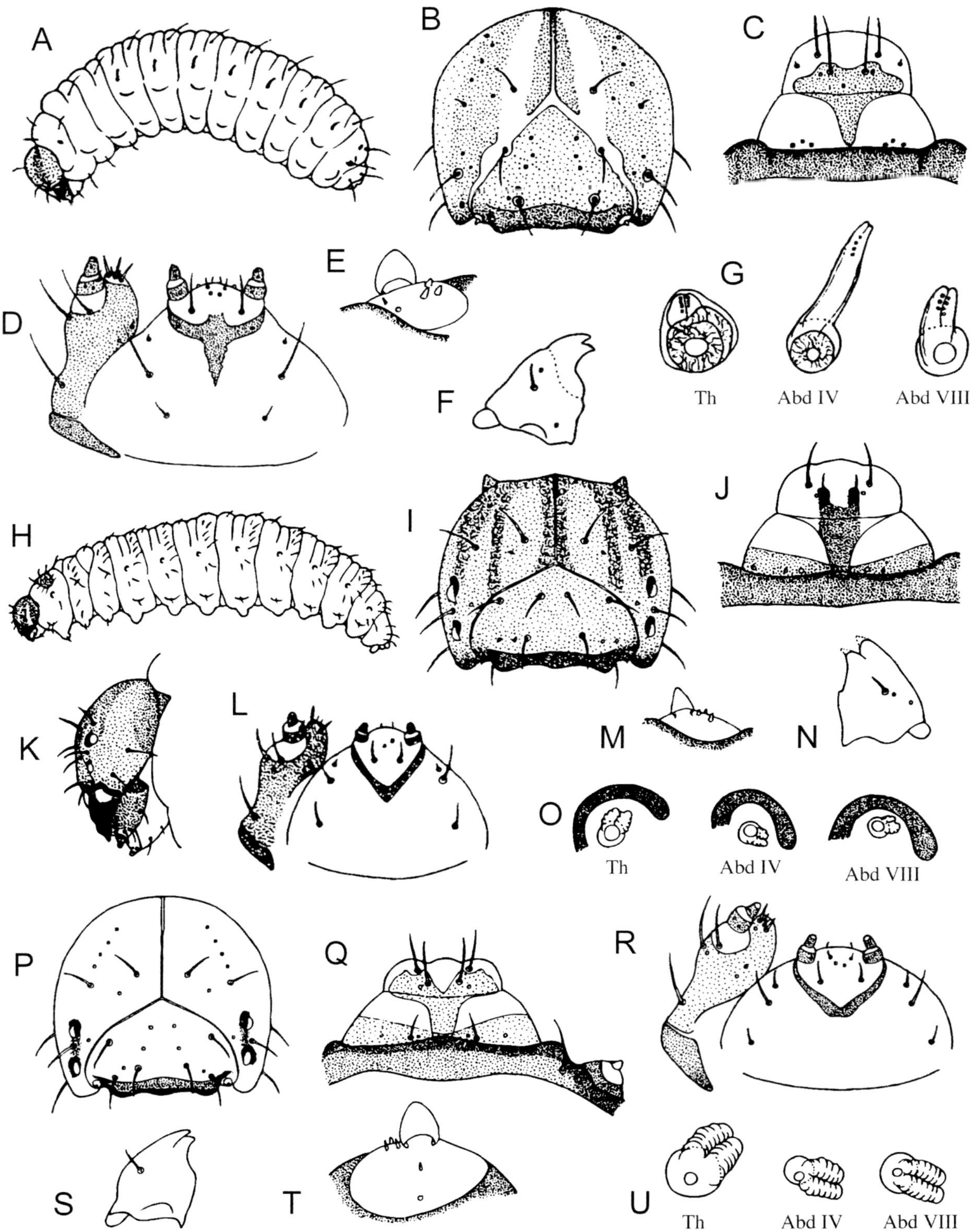

Fig. 44.29. Cyclominae, larvae, Listroderini and Rhythirrinini. A–G, *Steriphus diversipes* (Pascoe), larva (New Zealand): A, habitus, lateral; B, head, anterior; C, labrum, clypeus and epistome, anterior; D, mala and labium, posterior; E, antenna; F, mandible; G, spiracles, showing alignment; H–O, *Ethemaia sellata* Pascoe, larva: H, habitus, lateral; I, head, anterior; J, labrum, clypeus and epistome, anterior; K, head, lateral; L, mala and labium, posterior; M, antenna; N, mandible; O, spiracles, showing alignment; P–U, *Ophryota squamibunda* Pascoe, larva: P, head, anterior; Q, labrum, clypeus, epistome and antenna, anterior; R, mala and labium, posterior; S, mandible; T, antenna; U, spiracles, showing alignment. After May (1994).

Key to the Australian tribes and genera of Cyclominae

1. Body robust, usually with tubercles or carina (Fig. 44.30A); rostrum robust, short to very short; mandibles with more than 10 setae (plurisetose) (Figs 44.32E, 44.33C) 2
– Body robust or not; rostrum shape variable (Fig. 44.30B–M); mandibles with fewer than 10 long setae (paucisetose) 53
2(1). Antennal funicles 6-segmented; mandibles blade-like elongate (Fig. 44.33F) and usually with a single subapical tooth; maxillae coarsely sculptured, with palpifer covering short 3-segmented palpi; labial palps concealed in ventral view ..... **AMYCTERINI** ... 3
– Antennal funicles 7-segmented; mandibles broad with two subapical teeth; maxillae not coarsely sculptured, with palpifer not covering palp; labial palps exposed in ventral view ..... ***Aesiotes*** **Pascoe**
3(2). Prementum and maxillae deeply withdrawn and mostly concealed (Fig. 44.31J) 4
– Prementum and maxillae widely exposed (Fig. 44.32J) 23
4(3). Antennal scapes sinuous, asymmetrical, notably expanded and depressed apically 5
– Antennal scapes differently shaped 6
5(4). Meso- and metatibiae with two coarse subapical spines on outer side, surrounding a corbel-like plate; mandibles produced downward into triangular points; apex of sternite 8 slightly protruding at apical angles of ventrite 5 ..... ***Talaurinellus*** **Zimmerman**
– Not such species; apices of sternite 8 concealed ..... ***Antalaurinus*** **Zimmerman**
6(4). Tarsal segments 5 flattened (Fig. 44.30m) ..... ***Molochtus*** **Pascoe** (part)
– Tarsal segments 5 not flattened 7
7(6). Head with pointed tubercles above eyes (Fig. 44.30a) 8
– Head lacking such tubercles 10
8(7). Eyes with anterior margin emarginate ..... ***Notonophes*** **Sloane** (part)
– Eyes with anterior margin entire 9
9(8). Rostrum with epistomal plate below level of base of rostrum (Fig. 44.30h) ..... ***Hypotomops*** **Baker & Thompson** (part)
– Rostrum with epistomal plate about at same level as base of rostrum 31
10(7). Rostrum short; underside of rostrum slightly produced forming a chin extending only slightly below level of ventral surface of head (Fig. 44.32L) ..... ***Pseudonotonophes*** **Ferguson**
– Rostrum variable; underside of rostrum with strongly produced chin (Fig. 44.32E) 11
11(10). Antennal scapes long and slender, flattened and thin (Fig. 44.30k), except at apex; body length 20–30 mm; prothorax laterally explanate and tuberculate (Fig. 44.30k) ..... ***Gagatophorus*** **Jekel**
– Characters in different combination than above 12
12(11). Rostrum with a pair of large dorsal calli; antennal scapes very long, reaching to anterior margin of pronotum, cylindrical and strongly clavate apically ..... ***Xenommamycterus*** **Peterson**
– Rostrum without a pair of large dorsal calli; antennal scapes variable in length 13
13(12). Eyes strongly protuberant (Fig. 44.30i), subhemispherical ..... ***Ophthalamycterus*** **Ferguson**
– Eyes nearly flat or only moderately convex 14
14(13). Metafemora almost reaching apex of abdominal ventrite 5 ..... ***Achorostoma*** **Baker & Thompson**
– Metafemora not extending beyond apex of abdominal ventrite 4 15
15(14). Head with dorsal margin of scrobes at level of inner margin of eyes 16
– Head with dorsal margin of scrobes well above inner margin of eyes 18
16(15). Rostrum medially with a low callose area adjacent to epistome ..... ***Sclerorrhinella*** **Ferguson**
– Rostral medially without a callose area adjacent to epistome 17
17(16). Head and rostrum with surface notably uneven (Fig. 44.30g); scrobes directed ventrad in front of eye; setae on granules of pronotum and elytra minute ..... ***Lataurinus*** **Ferguson**
– Head and rostrum with surface more or less even; scrobes directed toward eye; setae of granules of pronotum and elytra comparatively long and conspicuous ..... ***Notonophes*** **Sloane** (part)
18(15). Eyes somewhat elongated vertically and overlapped by or very close to ocular lobes of prothorax ..... ***Myotrotus*** **Pascoe**
– Characters in different combination than above 19
19(18). Suprascrobal margins forming elevated carinae 20
– Suprascrobal margins not forming elevated carinae 22
20(19). Sternite VIII in male with anal forceps ..... ***Amycterus*** **Schoenherr**
– Sternite VIII in male without anal forceps 21
21(20). Rostrum in dorsal view with two prominent longitudinal carinae ..... ***Talaurinus*** **Macleay**
– Rostrum in dorsal without prominent longitudinal carinae ..... ***Sclerorinus*** **Macleay**
22(19). Mandibles finely multistriate longitudinally around cutting edges; scrobes expanded towards eyes ..... ***Hypotomops*** **Baker & Thompson** (part)
– Mandibles densely punctate around cutting edges, with large, coarse setiferous punctures scattered on background of dense, small punctures; scrobes narrow in front of eyes ..... ***Dicherotropis*** **Ferguson**
23(3). Tarsal segments 5 dorsoventrally flattened (Fig. 44.30m) ..... ***Molochtus*** **Pascoe** (part)
– Tarsal segments 5 not flattened 24
24(23). Antennal scapes expanded and apically asymmetrical 25
– Antennal scapes not expanded and apically asymmetrical 27
25(24). Granules and tubercles of prothorax and elytra bearing conspicuous, erect, long, sharp setae; rostrum with transverse sulcus ..... ***Alexirhea*** **Pascoe**
– Characters in different combination than above 26
26(25). Head and rostrum dorsally separated by transverse sulcus (Fig. 44.30l) ..... ***Sosytelus*** **Pascoe**
– Head and rostrum dorsally not separated by transverse sulcus ..... ***Brachyrothus*** **Marshall**

27(24). Antennal scapes extending to posterior margin of eye in repose ... **28**

– Antennal scape not reaching middle of length of eye in repose ... **40**

28(27). Pronotum greatly inflated to form a domelike, granulate structure ... ***Cucullothorax* Ferguson**

– Pronotum not inflated to form a domelike structure ... **29**

29(28). Antennal scapes not reaching beyond eye in repose ... **30**

– Antennal scapes reaching anterior margin of pronotum in repose ... **31**

30(29). Rostrum with transverse groove in front of eyes; body form elongate ... ***Mythites* Schoenherr** (part)

– Rostrum without transverse groove in front of eyes; body form rotund ... ***Acherres* Pascoe** (part)

31(29). Eyes unusually large, occupying half or more of sides of head; head without tuberculiform process above each eye ... ***Chriotyphus* Pascoe**

– Eyes smaller, head often with tuberculiform process above each eye (Fig. 44.30b) ... **32**

32(31). Scrobes directed beneath eye ... **33**

– Scrobes directed mostly toward eye ... **35**

33(32). Postmentum developed into prominent tongue-like process (Fig. 44.31I–J) ... ***Anascoptes* Pascoe** (part)

– Postmentum with only moderately protuberant process ... **34**

34(33). Prothorax about as long as wide, with distinct ocular lobes ... ***Neohyborrhynchus* Ferguson**

– Prothorax broader than long, without ocular lobes ... ***Parahyborrhynchus* Ferguson**

35(32). Antennal scapes reaching posterior margin of eye in repose ... ***Acherres* Pascoe** (part)

– Antennal scapes not reaching posterior margin of eye in repose ... **36**

36(35). Procoxal cavities separated ... ***Cubicorhynchus* Lacordaire** (part)

– Procoxal cavities contiguous ... **37**

37(36). Pronotum and elytra lacking any large, pointed tubercle ... **38**

– Pronotum and elytra with numerous pointed tubercles (Fig. 44.30A) ... **39**

38(37). Head and rostrum dorsally separated by transverse sulcus ... ***Cubicorhynchus* Lacordaire** (part)

– Head and rostrum dorsally not separated by transverse sulcus ... ***Notonophes* Sloane** (part)

39(37). Postmentum notably produced ... ***Hyborrhinus* Marshall**

– Postmentum not so produced ... ***Acantholophus* Boisduval** (part)

40(27). Prothorax with ocular lobes obsolete ... **41**

– Prothorax with moderate to large ocular lobes ... **43**

41(40). Postmentum with anterior edge not produced ... ***Acherres* Pascoe** (part)

– Postmentum with anterior edge produced ... **42**

42(41). Postmentum developed into prominent tongue-like process (Fig. 44.31I–J); eyes small to moderate ... ***Anascoptes* Pascoe** (part)

– Postmentum not so enlarged; eyes very large ... ***Ennothus* Pascoe**

43(40). Antennal scapes short, about as long as eye diameter ... ***Aedriodes* Pascoe**

– Antennal scapes longer than eye diameter ... **44**

44(43). Antennal scapes not reaching middle of eye in repose ... **45**

– Antennal scapes reaching beyond middle of eye in repose ... **52**

45(44). Postmentum forming low, broad convexity on rim around oral cavity (Fig. 44.32I) ... **46**

– Postmentum with strong protuberance conspicuously interrupting contour of lower rim of oral cavity ... **47**

46(45). Body large, heavy, about 13–23 mm long; eyes visible in dorsal view ... ***Euomus* Schoenherr** (part)

– Body smaller; eyes in dorsal view covered by supraocular processes (Fig. 44.30j) ... ***Amorphorhinus* Lacordaire**

47(45). Rostrum with long carinae extending to middle of eyes ... ***Atychoria* Pascoe**

– Rostrum with carinae reaching only base of rostrum ... **48**

48(47). Tarsal segments 1 not longer than apical width of tibiae ... ***Euomella* Ferguson**

– Tarsal segments 1 distinctly longer than apical width of tibiae ... **49**

49(48). Rostrum with enlarged, elevated carinae forming subtriangular processes above level of eyes ... ***Oditesus* Pascoe**

– Rostrum without such carinae ... **50**

50(49). Pronotum extending over head ... ***Dialeptopus* Pascoe**

– Pronotum not extending over head ... **51**

51(50). Metafemora reaching anterior part of ventrite 5 in repose ... ***Euomus* Schoenherr** (part)

– Metafemora reaching apex of elytra in repose ... ***Tetralophus* Waterhouse**

52(44). Antennal funicles with segment 2 similar to 3 ... ***Mythites* Schoenherr** (part)

– Antennal funicles with segment 2 about as long as segment 2 and 3 combined ... ***Melanegis* Pascoe**

53(1). Size small, body length 1.2–1.7 mm; eyes strongly reduced to a few ommatidia (e.g. Fig. 44.38A) or absent (*Tasmanica*); antennal funicles with 6 or fewer segments ... **54**

– Size small to large, body length 0.8–17.4 mm; eyes of normal size (e.g. Fig. 44.30M); antennal funicles usually with 7 segments, rarely fewer ... **56**

54(53). Eyes absent ... ***Tasmanica* Lea**

– Eyes present ... **55**

55(54). Antennal funicles 7-segmented; femora unarmed ... ***Wollastonicis* Lea**

– Antennal funicles 4-segmented; femora weakly dentate ... ***Notiomimetes* Wollaston**

56(53). Body sparsely setose, setae hair-like; scrobes open to eyes (Fig. 44.30M); tibiae usually mucronate ... **57**

– Body densely or sparsely setose, with setae scale-like and forming patterns (Fig. 44.30B–K); scrobes rarely open to eyes; tibiae mucronate or not ... **58**

57(56). Antennal funicles with segment 2 × as long as segment 1 ... ***Aphela* Pascoe**

– Antennal funicles with segments 2 and 1 subequal in length ... ***Psaldus* Pascoe**

58(56). Tarsi with segment 5 short, not or barely protruding beyond margin of 3 (Fig. 44.30o–p) ... 59

– Tarsi with segment 5 normal, clearly extending beyond 3 (Fig. 44.30q) ... 62

59(58). Tarsal claws absent (Fig. 44.30n) ... ***Atelicus* Waterhouse**

– Tarsal claws present ... 60

60(59). Tarsal claws divaricate ... 61

– Tarsal claws not divaricate, almost parallel ... ***Kershawcis* Lea**

61(60). Body surface with scale-like setae forming setose tufts on pronotum and elytra; scutellar shield projecting over elytra ... ***Aromagis* Pascoe**

– Body surface with uniform, appressed vestiture not forming tufts of setae on pronotum and elytra (Fig. 44.30I); scutellar shield at same level as elytra ... ***Strongylorhinus* Schoenherr**

62(58). Bases of prothorax and elytra about equally wide (Fig. 44.30F); eyes in lateral view separated from anterior margin of prothorax by ≥2 × eye width (Fig. 44.30F) ... 63

– Base of prothorax often narrower than base of elytra; eyes in lateral view separated from anterior margin of prothorax by <2 × eye length ... 65

63(62). Head behind eyes (temple) about 2 × as long as rostrum ... ***Rhadinosomus* Schoenherr**

– Head behind eyes <2 × as long as rostrum ... 64

64(63). Head behind eyes (temple) shorter than rostrum; epistome with posterior edge delimited by a ridge ... ***Ophthalmorychus* Blackburn**

– Head behind eyes about as long as rostrum (Fig. 44.30F); epistome with posterior edge not delimited by a ridge ... ***Euthyphasis* Pascoe**

65(62). Rostrum with epistome distinct, its posterior edge raised forming a ridge or sharp carina (e.g. Fig. 44.30f); prothorax without ocular lobes; tibiae with spurs (Fig. 44.30N) ... 66

– Rostrum with epistome and frons indistinct, without posterior carina; prothorax with or without ocular lobes absent or present; tibiae with or without out spurs ... 70

66(65). Elytra each with tubercle at apical third (Fig. 44.30K); bases of prothorax and elytra about equal in width (Fig. 44.30K); elytra with humeral angles inconspicuous ... ***Pocius* Zimmerman & Oberprieler**

– Elytra each with two or more tubercles; base of prothorax clearly narrower than base of elytra; elytra with humeral angles well defined ... 67

67(66). Rostrum in dorsal view narrowing towards apex (Fig. 44.30c) ... ***Ophryota* Pascoe**

– Rostrum in dorsal view parallel-sided (Fig. 44.30d–f) ... 68

68(67). Head in dorsal view with eyes not protuberant (Fig. 44.30d) ... ***Medicasta* Pascoe**

– Head in dorsal view with eyes protuberant (Fig. 44.30e–f) ... 69

69(68). Rostrum with epistome forming a glabrous plate, its posterior edge with well-defined carina extending down sides onto venter (Fig. 44.30f) ... ***Aparete* Pascoe**

– Rostrum with epistome not forming a plate (Fig. 44.30e), its posterior edge irregular not extending down sides to venter ... ***Ethemaia* Pascoe**

70(65). Tibial apices with mucro and one (Fig. 44.30N) or two spurs, sometimes spurs present only on metatibiae ... 71

– Tibial apices often with mucro but without spurs, if mucro absent then one spur usually present (*Mandalotina*) ... 77

71(70). Scutellar shield reduced and hardly visible, surface below level of elytra ... ***Anomocis* Lea**

– Scutellar shield not reduced, clearly visible, surface at same level as elytra ... 72

72(71). Pro- and mesotibiae uncinate; tibial spur formula 0-0-2 ... ***Cechides* Pascoe**

– Pro- and mesotibiae mucronate rather than uncinate; tibial spur formula variable ... 73

73(72). Rostrum slender, often as long as or longer than pronotum (Fig. 44.30J); scrobes well defined, reaching eyes ... 74

– Rostrum robust, shorter than pronotum; scrobes not well defined, not reaching eyes ... 76

74(73). Antennal funicles with segment 1 subequal to or shorter than 2 ... ***Listronotus* Jekel**

– Antennal funicles with segment 1 longer than 2 ... 75

75(74). Antennal scapes surpassing posterior margin of eye in repose ... ***Steriphus* Erichson**

– Antennal scapes not reaching anterior margin of eye in repose ... ***Anorthorhinus* Blackburn**

76(73). Prothorax longer than wide ... ***Methypora* Pascoe**

– Prothorax about as long as wide ... ***Listroderes* Schoenherr**

77(70). Pro- and mesotibiae uncinate ... 78

– Pro- and mesotibiae apex mucronate, truncate ... 83

78(77). Femora with small ventral tooth ... 79

– Femora without small ventral tooth ... 80

79(78). Protibiae with inner margin coarsely dentate ... ***Chrysolopus* Germar**

– Protibiae with inner margin simple, not dentate ... ***Dixoncis* Oke**

80(78). Eyes protuberant (Fig. 44.30E) ... ***Chrysophoracis* Lea**

– Eyes not protuberant ... 81

81(80). Pronotum and elytral disc tuberculate ... ***Cyllorhamphus* Erichson**

– Pronotum and elytral disc without tubercles ... 82

82(81). Rostrum shorter than prothorax; prothorax widest at base ... ***Alphitopis* Pascoe**

– Rostrum longer than prothorax; prothorax widest at middle ... ***Oenopus* Marshall**

83(77). Head and rostrum often with complex tubercles or protuberances (Fig. 44.30H), rarely without tubercles (*Nemestra*, Fig. 44.30G); mandibles bidentate (*Rhinoplethes*) or without teeth, snipper-like ... 84

– Head and rostrum without tubercles or protuberances; mandibles broad and bidentate ... 88

84(83). Head and/or rostrum with tubercles and protuberances ... 85

– Head and rostrum without tubercles or protuberances (*Nemestra*, Fig. 44.30G) ... 87

85(84). Scrobes broad and open ... ***Pelororhinus* Schoenherr**

– Scrobes usually sharp and narrow ... 86

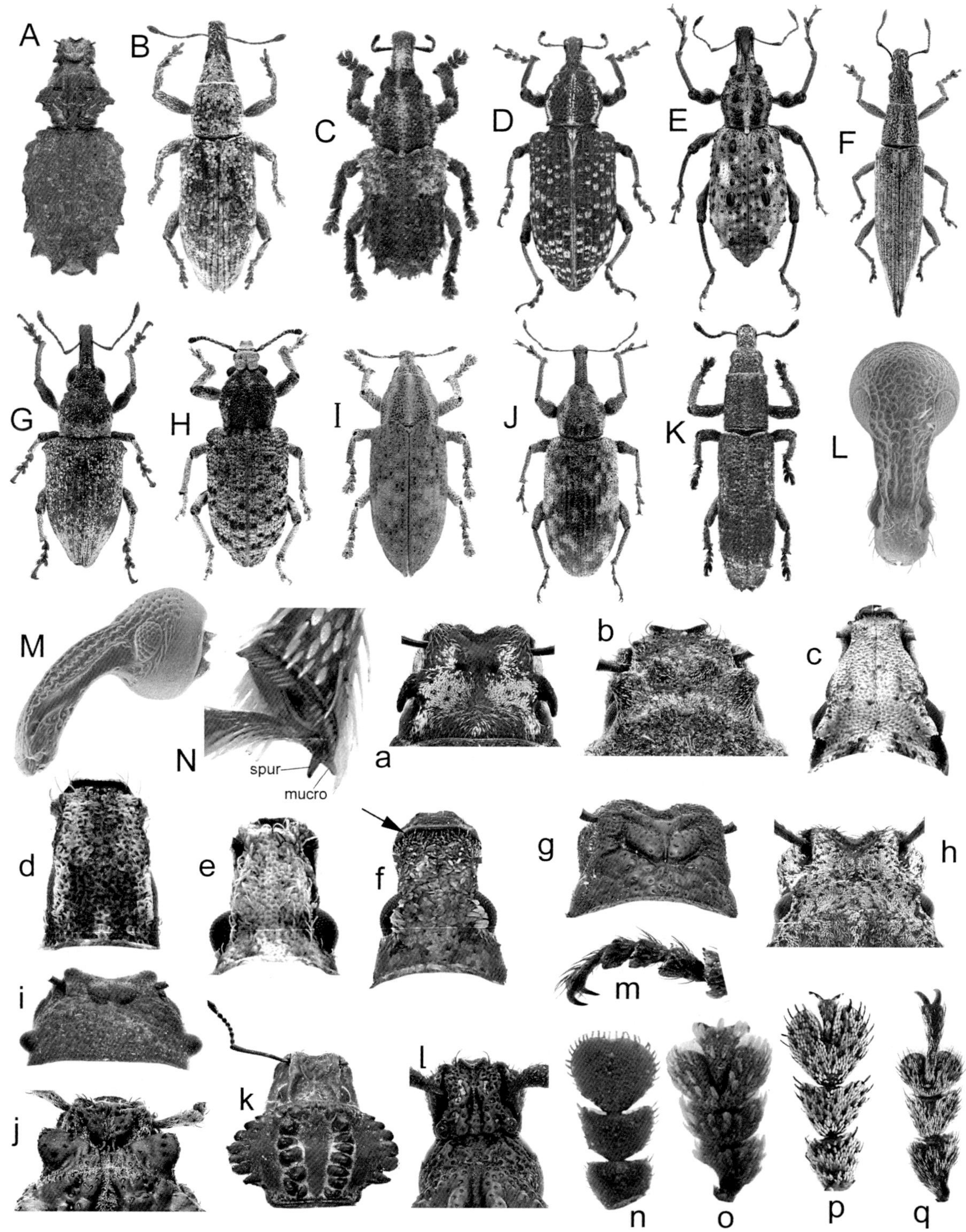

Fig. 44.30. Cyclominae, characters for key to genera. A, *Acantholophus denticollis* Macleay; B, *Acalonoma pusilla* Blackburn; C, *Aromagis echinata* Pascoe; D, *Chrysolopus spectabilis* (Fabricius); E, *Chrysophoracis pulcher* Lea; F, *Euthyphasis acuta* Pascoe; G, *Nemestra incerta* Pascoe; H, *Rhinaria cristata* Kirby; I, *Strongylorhinus clarki* Marshall; J, *Steriphus sericeus* (Blackburn); K, *Pocius parallelus* (Lea); L, *Neosyagrius,* head lateral view; M, *Mandalotina,* head lateral view; N, *Steriphus,* spur and mucro; a, *Notonophes*; b, *Anascoptes*; c, *Ophryota*; d, *Medicasta*; e, *Ethemaia*; f, *Aparete*; g, *Lataurinus*; h, *Hypotomops*; i, *Ophthalamycterus*; j, *Amorphorhinus*; k, *Gagatophorus*; l, *Sosytelus*; m–q, tarsi: m, *Molochtus*; n, *Atelicus atrophus*; o, *Aromagis*; p, *Strongylorhinus*; q, *Chrysolopus*.

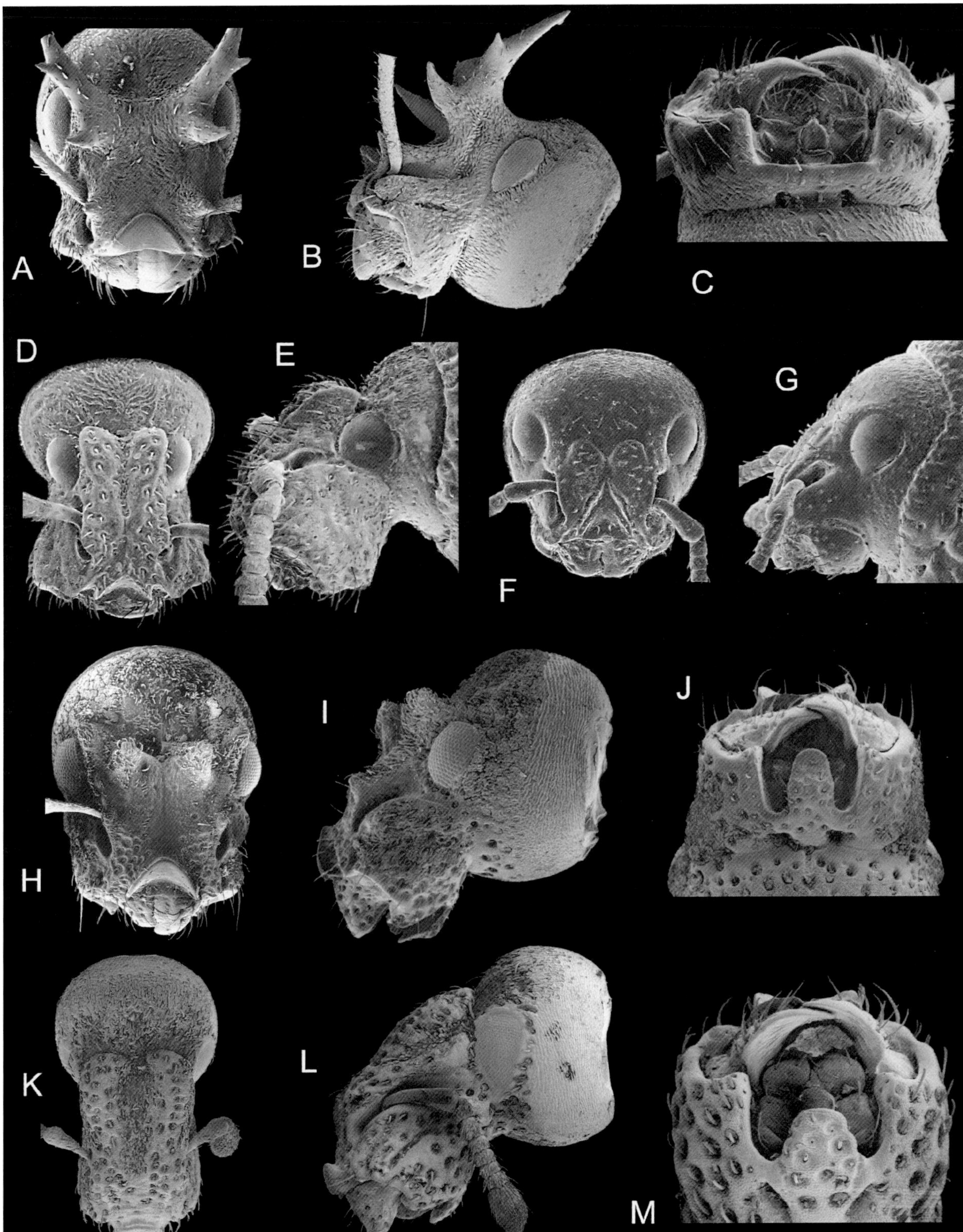

Fig. 44.31. Cyclominae, head structure, Amycterini. A–C, *Acantholophus niveovittatus* Blackburn: A, head, dorsal; B, head, lateral; C, mouthparts; D–E, *Acherres pilosus* Ferguson, head: D, dorsal; E, lateral; F–G, *Aedriodes fastigiatus* Pascoe, head: F, dorsal; G, lateral; H–J, *Anascoptes muricatus* Pascoe: H, head, dorsal; I, lateral; J, mouthparts; K–M, *Atychoria funesta* Pascoe: K, head, dorsal; L, lateral; M, mouthparts. After Zimmerman (1993).

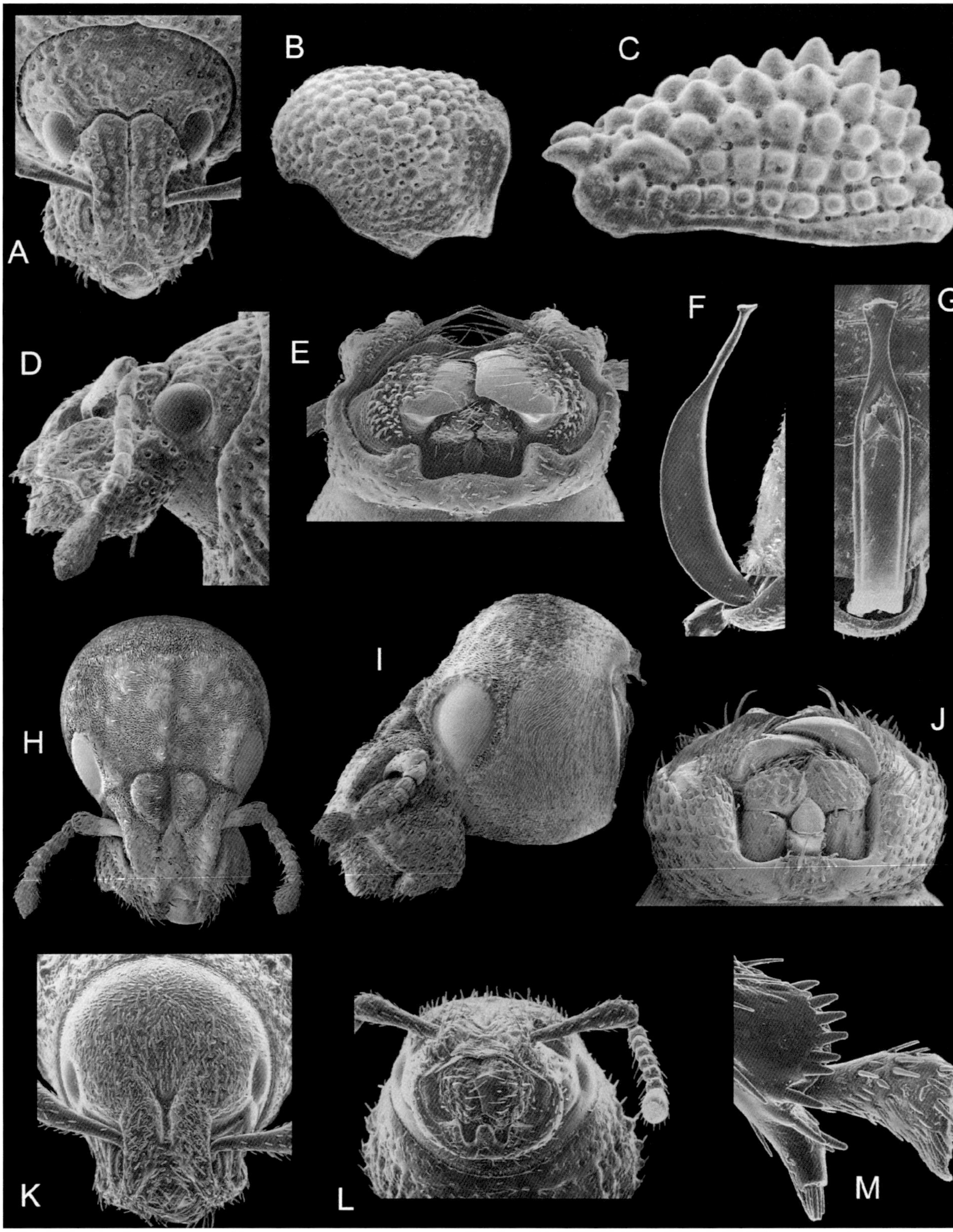

Fig. 44.32. Cyclominae, structural details, Amycterini. A–D, *Cucullothorax horridus* Ferguson: A, head, dorsal; B, prothorax, lateral; C, left elytron, lateral; D, head, lateral; E, *Talaurinus typicus* MacLeay, mouthparts; F–G, *Dialeptopus macilentus* Pascoe, penis: F, lateral; G, dorsal; H–J, *Euomus stephensii* (Gyllenhal): H, head, dorsal; I, lateral; J, mouthparts; K–M, *Pseudonotonophes lemmus* (Pascoe): K, head, dorsal; L, mouthparts; M, metatibial apex. After Zimmerman (1993).

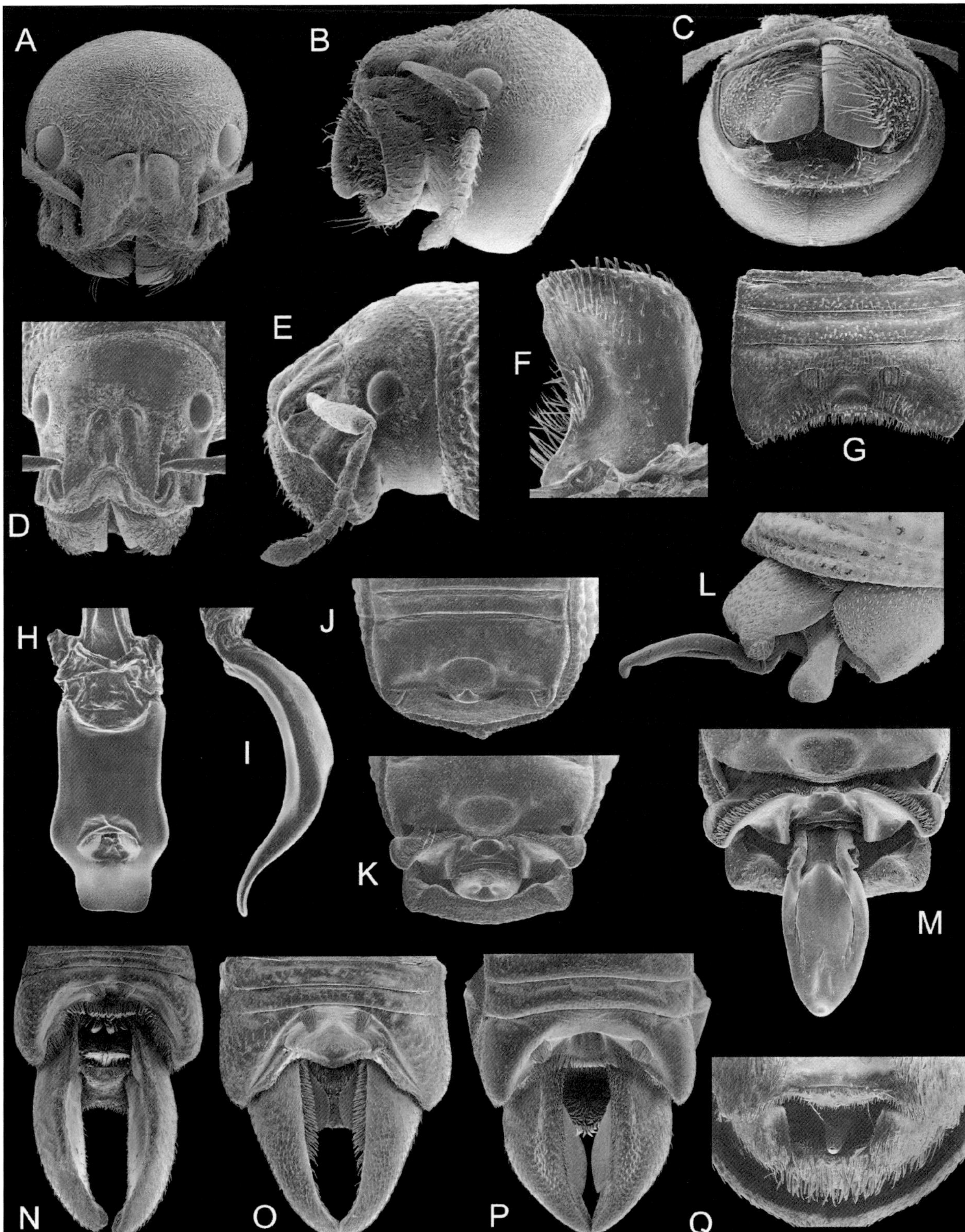

Fig. 44.33. Cyclominae, structural details, Amycterini, *Amycterus* sp. A–C, *Amycterus* sp.: A, head, dorsal; B, lateral; C, mouthparts; D–I, *A. fergusoni* (Sharp): D, head, dorsal; E, lateral; F, mandible; G, abdominal ventrites; H, penis, dorsal; I, lateral; J–M, *A. morbillosus* Boisduval: J, abdominal apex, closed; K, abdominal apex, open; L, open with extruded penis, lateral; M, open with extruded penis, dorsal; N–Q, abdominal apex: N, *A. caudatus* (Macleay); O, *A. elongatus* (Macleay); P, *A. talpa* Schoenherr; Q, *A. rayneri* (Macleay), showing tip of penis. After Zimmerman (1993).

86(85). Eyes protuberant (Fig. 44.30H); rostrum with dorsal surface smooth ..... ***Rhinaria*** **Kirby**
– Eyes flattened, not protuberant; rostrum with dorsal surface longitudinally carinate ..... ***Rhinoplethes*** **Pascoe**
87(84). Eyes entire, not emarginate; scutellar shield above elytral level ..... ***Nemestra*** **Pascoe**
– Eyes with anterior margin emarginate; scutellar shield below elytral level ..... ***Iphisaxus*** **Pascoe**
88(83). Prosternum broadly, shallowly excavate **89**
– Prosternum flat, not excavate **90**
89(88). Procoxae separated; pronotal and elytral tubercles with tufts of long scale-like setae ..... ***Aades*** **Schoenherr**
– Procoxae contiguous; pronotal and elytral tubercles without tufts of long scale-like setae ..... ***Aterpodes*** **Zimmerman**
90(88). Rostrum with lateral expansions around antennal insertions (pterygia) **91**
– Rostrum without lateral expansions around antennal insertions **92**
91(90). Prothorax longer than wide ..... ***Mandalotina*** **Oke**
– Prothorax wider than long ..... ***Neosyagrius*** **Lea**
92(90). Body length >9 mm ..... ***Aoplocnemis*** **Schoenherr**
– Body length <3.7 mm **93**
93(92). Rostrum longer than prothorax ..... ***Phrenozemia*** **Pascoe**
– Rostrum shorter than prothorax (Fig. 44.30B) ..... ***Acalonoma*** **Pascoe**

Classification of the Australian genera of Cyclominae

Tribe Amycterini Waterhouse, 1854

The Amycterini are the largest tribe of Cyclominae in Australia, an iconic endemic group of 39 genera and more than 400 species (Pullen *et al.* 2014) of mostly large, robust, terricolous (ground-dwelling) weevils (Pl. 6, Fig. 44.14) with a very short rostrum (resembling some Entiminae), unusual, large, chisel-like mandibles and often an upper surface covered with multiple large tubercles or spines. The tribe was revised by Zimmerman (1993) (treated as a subfamily), the treatment featuring keys to the genera and species, summaries of their life histories, diverse taxonomic and nomenclatural notes, many exceptional habitus illustrations (most of which reproduced here; Figs 44.14–44.20, 44.14A–F) and drawings of male and female terminalia. Oberprieler (2010) provided an updated diagnosis of the tribe and a comparison of the ovipositors of all genera. In their peculiar ovipositor, with the styli of the gonocoxites modified into claw-like organs apparently adapted for digging, the Amycterini resemble the African-Palaearctic tribe Hipporhinini (Oberprieler 2010), but molecular phylogenetic analyses to date indicate that they are most closely related to Aterpini (McKenna *et al.* 2009; Baird *et al.* 2021; Haran *et al.* 2023). Amycterines feed on a variety of monocotyledonous plants, especially Poaceae and Liliaceae, their larvae living in the soil and feeding on underground stems, crowns, tubers and rhizomes (Howden 1986).

Acantholophus Boisduval, 1835 (Pl. 49A–D; Figs 44.15E–L, 44.16A–D, 44.22A–H, 44.31A–C)

Type species, by subsequent designation (Schoenherr, 1847: 57): *Curculio marshami* Kirby, 1819.

Australian species. *A. adelaidae* Waterhouse, 1854, *A. alpicola* Ferguson, 1915b, *A. ambiguus* Zimmerman, 1993, *A. amycteroides* Macleay, 1865, *A. angusticollis* Ferguson, 1915a, *A. apicalis* Macleay, 1865, *A. aureolus* (Boheman, 1842) (*in* Schoenherr 1842b), *A. bivittatus* (Boheman, 1842) (*in* Schoenherr 1842b), *A. blackburni* Ferguson, 1915b, *A. browni* Ferguson, 1915b, *A. crassidens* Macleay, 1865, *A. crenaticollis* Macleay, 1865, *A. cupreomicans* Ferguson, 1921b, *A. denticollis* Macleay, 1865, *A. dixoni* Ferguson, 1915a, *A. doddi* Ferguson, 1921b, *A. dumosus* (Boheman, 1842) (*in* Schoenherr 1842b), *A. echidna* Macleay, 1865, *A. echinatus* Boisduval, 1835, *A. eximius* (Macleay, 1866, *A. foveirostris* Lea, 1911b, *A. franklinensis* Blackburn, 1890a, *A. gladiator* Pascoe, 1874c, *A. granulatus* Sloane, 1893, *A. gravicollis* Macleay, 1866, *A. halmaturinus* Ferguson, 1915b, *A. humeralis* Macleay, 1865, *A. hypoleucus* (Boheman, 1842) (*in* Schoenherr 1842b), *A. hystrix* (Boheman, 1842) (*in* Schoenherr 1842b), *A. krefftii* Macleay, 1865, *A. lateralis* (Boheman, 1842) (*in* Schoenherr 1842b), *A. marshami* (Kirby, 1819a), *A. mastersi* Macleay, 1866, *A. maximus* (Macleay, 1865), *A. nanus* Ferguson, 1921b, *A. niveovittatus* Blackburn, 1890c, *A. ocelliger* Ferguson, 1921b, *A. parvulus* Ferguson, 1921b, *A. planicollis* Waterhouse, 1854, *A. scabrosus* Macleay, 1865, *A. scaphirostris* Ferguson, 1915b, *A. sellatus* Ferguson, 1921b, *A. simplex* Pascoe, 1874c, *A. simulator* Ferguson, 1915b, *A. sordidus* Ferguson, 1921b, *A. spinifer* Macleay, 1865, *A. squalidus* Macleay, 1865, *A. squamosus* Macleay, 1865, *A. subtridentatus* Ferguson, 1921b, *A. suttoni* Carter, 1936, *A. suturalis* (Boheman, 1842) (*in* Schoenherr 1842b), *A. tasmaniensis* Lea,, 1911c, *A. tatei* Blackburn, 1896, *A. terraereginae* Ferguson, 1921b, *A. tragocephalus* Ferguson, 1921b, *A. transitus* Macleay, 1865, *A. tribulus* Macleay, 1866, *A. tridentatus* Macleay, 1865.

Distribution. Australia (QLD, NT, NSW, VIC, SA, TAS, WA).

Notes. This large genus was reviewed by Zimmerman (1993), and a list of its species was provided by Pullen *et al.* (2014). Its larvae are exceptional for Amycterini in that they feed on the pale green leaf bases in the crowns of grasstrees (*Xanthorrhoea*; Asphodelaceae), but associations with grasses (Poaceae) and sedges (Cyperaceae) have also been reported (Moore 1978; Howden 1986; May 1994).

Acherres Pascoe, 1872 (Pl. 49E; Figs 44.16E–H, 44.31D–E)

Type species, by monotypy: *Acherres mamillatus* Pascoe, 1872.

Australian species. *A. globicollis* Lea, 1904, *A. granulatus* Ferguson, 1915b, *A. mamillatus* Pascoe, 1872b, *A. pilosus* Ferguson, 1915b.

Distribution. Australia (SA, WA)

Notes. The life histories and larval hostplants of the genus remain unknown.

Achorostoma Baker & Thompson, 1980 (Pl. 49F; Fig. 44.16I)

Type species, by original designation: *Amycterus hystricosus* Boheman, 1842.

Australian species. *A. hystricosum* (Boheman, 1842) (*in* Schoenherr 1842b).

Distribution. Australia (WA).

Notes. The life history and larval hostplants of the single species are unknown.

Aedriodes Pascoe, 1872 (Pl. 49G; Figs 44.16J, K, 44.31F, G)

Type species, by subsequent designation (Ferguson, 1923: 404): *Aedriodes fastigiatus* Pascoe, 1872.

Australian species. *A. crawshawi* Ferguson, 1923, *A. fastigiatus* Pascoe, 1872b, *A. humeralis* Lea, 1904, *A. inuus* Pascoe, 1872b, *A. mendosus* Pascoe, 1872b, *A. mucronatus* Ferguson, 1923, *A. nodipennis* (Boheman, 1842) (*in* Schoenherr 1842b), *A. tibialis* Ferguson, 1923.

Distribution. Australia (WA).

Notes. The life histories and larval hostplants of the genus are unknown.

Alexirhea Pascoe, 1874 (Pl. 49H; Fig. 44.16L)

Type species, by subsequent designation (Ferguson, 1923: 411): *Alexirhea aurita* Pascoe, 1874.

Australian species. *A. aurita* Pascoe, 1874c, *A. falsifica* Pascoe, 1874c.

Distribution. Australia (WA).

Notes. The life histories and larval hostplants of the genus remain unknown.

Amorphorhinus Lacordaire, 1863 (Pl. 49I–K; Fig. 44.17A–C)

Type species, by monotypy: *Brachycerus australis* Germar, 1848.

Australian species. *A. arcanus* Pascoe, 1882, *A. australis* (Germar, 1848), *A. obliteratus* Ferguson, 1923, *A. polyacanthus* Pascoe, 1870e, *A. rugicollis* (Lea, 1895b), *A. tuberculatus* Ferguson, 1923.

Distribution. Australia (QLD, NSW, SA, WA).

Notes. The life histories and larval hostplants of the genus are unknown.

Amycterus Schoenherr, 1823 (Pl. 50A, B; Figs 44.15B–D, 44.17D–E, 44.23L–Q, 44.33A–Q)

Type species, by original designation: *Amycterus talpa* Schoenherr, 1823.

Australian species. *A. abnormis* (Macleay, 1865), *A. amplicollis* (Ferguson, 1909, *A. amplipennis* (Ferguson, 1909, *A. anthracoides* (Ferguson, 1921a), *A. brevicauda* (Ferguson, 1914b), *A. breviformis* (Ferguson, 1909), *A. carteri* (Ferguson, 1909), *A. caudatus* (Macleay, 1865), *A. cultratus* (Ferguson, 1909), *A. cuneicaudatus* (Ferguson, 1909), *A. decipiens* (Dohrn, 1872), *A. durvillei* Boheman, 1842 (*in* Schoenherr 1842b), *A. elongatus* (Macleay, 1865), *A. exasperatus* Erichson, 1842, *A. falciformis* (Macleay, 1865), *A. fergusoni* (Sharp, 1920), *A. flavosetosus* (Ferguson, 1909), *A. flavovarius* (Ferguson, 1909), *A. forficulatus* (Macleay, 1865), *A. foveatus* (Macleay, 1865), *A. frenchi* (Ferguson, 1909), *A. helmsi* (Ferguson, 1914c), *A. hopsoni* (Ferguson, 1921a), *A. kosciuskoanus* (Ferguson, 1909), *A. leai* (Ferguson, 1915a), *A. mastersii* (Macleay, 1865), *A. metasternalis* (Ferguson, 1909), *A. mirabilis* (Kirby, 1819b), *A. miracula* (Macleay, 1865), *A. mirus* Boheman, 1842 (*in* Schoenherr 1842), *A. monticola* (Ferguson, 1909), *A. morbillosus* Boisduval, 1835, *A. orthodoxus* (Lea, 1910b), *A. panduriformis* (Ferguson, 1912), *A. perlatus* (Ferguson, 1909), *A. pilbara* Peterson, 2013b, *A. posticus* Boisduval, 1835, *A. rayneri* (Macleay, 1865), *A. riverinae* (Macleay, 1865), *A. rufipes* (Blackburn, 1896), *A. sloanei* (Ferguson, 1909), *A. squalidus* Boheman, 1842 (*in* Schoenherr 1842b), *A. subcostatus* (Macleay, 1865), *A. sublaevigatus* (Ferguson, 1909), *A. sulcipennis* (Ferguson, 1909), *A. talpa* Schoenherr, 1823, *A. taylori* (Ferguson, 1914c), *A. tenebricosus* (Ferguson, 1912), *A. tessellatus* (Pascoe, 1874a), *A. tomentosus* Boisduval, 1835, *A. variegatus* (Macleay, 1865), *A. variolosus* (Ferguson, 1909), *A. wilcoxii* (Macleay, 1865).

Distribution. Australia (QLD, NT, NSW, VIC, SA, TAS, WA).

Notes. This large iconic genus was last revised by Zimmerman (1993), and a list of its species was provided by Pullen *et al.* (2014). The males are unusual among weevils by having a pair of large 'forceps' derived from sternite VIII. The larvae as known (*A. abnormis*, *A. elongatus*) feed on subterranean crowns and perhaps roots of *Nasella* and probably other grasses (Poaceae) (Howden 1986; May 1994).

***Anascoptes* Pascoe, 1874 (Pl. 49L; Figs 44.17F, 44.31H–J)**

Type species, by monotypy: *Anascoptes muricatus* Pascoe, 1874.

Australian species. *A. fasciatus* Ferguson, 1921c, *A. muricatus* Pascoe, 1874c, *A. obliteratus* Ferguson, 1921c.

Distribution. Australia (WA).

Notes. The life histories and larval hostplants are unknown.

***Antalaurinus* Zimmerman, 1993 (Pl. 50C)**

Type species, by original designation: *Talaurinus clavicornis* Ferguson, 1912.

Australian species. *A. clavicornis* (Ferguson, 1912), *A. pulverulentus* (Macleay, 1865).

Distribution. Australia (QLD, NSW).

Notes. The life histories and larval hostplants are unknown.

***Atychoria* Pascoe, 1865 (Pl. 50D; Figs 44.17G, 44.31K–M)**

Type species, by monotypy: *Atychoria funesta* Pascoe, 1865.

Australian species. *A. funesta* Pascoe, 1865.

Distribution. Australia (SA).

Notes. The life history and larval hostplants of the single species are unknown.

***Brachyrothus* Marshall, 1939 (Pl. 50E; Fig. 44.17H)**

Type species, by monotypy: *Brachymycterus auritus* Ferguson, 1915.

Australian species. *B. auritus* (Ferguson, 1915b).

Distribution. Australia (WA).

Notes. The life history and larval hostplants of the single species remain unknown. The genus was described by Marshall (1939b).

***Chriotyphus* Pascoe, 1874 (Pl. 50F; Fig. 44.17I)**

Type species, by monotypy: *Chriotyphus acromialis* Pascoe, 1874.

Australian species. *C. acromialis* Pascoe, 1874c, *C. tibialis* Ferguson, 1914.

Distribution. Australia (WA).

Notes. The larvae appear to feed in clumps of grasses (Howden 1986).

***Cubicorhynchus* Lacordaire, 1863 (Pl. 50G, H; Figs 44.17J, K, 44.22I–N)**

Type species, by subsequent designation (Zimmerman, 1993: 252): *Cubicorhynchus calcaratus* Macleay, 1865.

Australian species. *C. aureomaculatus* Ferguson, 1915b, *C. aurigena* (Blackburn, 1899), *C. bohemani* (Boheman, 1842) (*in* Schoenherr 1842b), *C. calcaratus* Macleay, 1865, *C. crenicollis* (Waterhouse, 1854), *C. curvipes* Ferguson, 1915b, *C. dohrnii* (Waterhouse, 1854), *C. globicollis* Lea, 1911b, *C. illidgei* Ferguson, 1916c, *C. maculatus* Macleay, 1865, *C. minor* Ferguson, 1916c, *C. modestus* Sloane, 1893, *C. mussoni* Blackburn, 1892b, *C. occultus* Sloane, 1893, *C. serratipes* Ferguson, 1916c, *C. setosus* Ferguson, 1916c, *C. sordidus* Ferguson, 1916c, *C. sparsus* Ferguson, 1916c, *C. spinicollis* Macleay, 1866, *C. sterilis* Pascoe, 1874c, *C. strigicollis* Ferguson, 1915b, *C. taurus* Blackburn, 1895b, *C. valgus* Lea, 1909b.

Distribution. Australia (QLD, NSW, VIC, SA, WA).

Notes. The larvae as known develop in clumps and crowns of grasses, feeding on underground stems and perhaps roots (Howden 1986; May 1994).

***Cucullothorax* Ferguson, 1915 (Pl. 50I; Figs 44.17L, 44.32A–D)**

Type species, by monotypy: *Cucullothorax horridus* Ferguson, 1915.

Australian species. *C. horridus* Ferguson, 1915b.

Distribution. Australia (SA, WA).

Notes. Adults of the single species have been found in association with *Stypandra glauca* (Asphodelaceae) (Howden 1986), but its life history is not known.

***Dialeptopus* Pascoe, 1870 (Pl. 50J; Figs 44.18A–E, 44.22O–S, 44.32F, G)**

Type species, by subsequent designation (Ferguson, 1923: 419): *Amycterus collaris* Boheman, 1842.

Australian species. *D. collaris* (Boheman, 1842) (*in* Schoenherr 1842b), *D. echinatus* Lea, 1896, *D. ferreus* Pascoe, 1870e, *D. lindensis* Blackburn, 1890c, *D. macilentus* Pascoe, 1870e, *D. monachus* Pascoe, 1870e, *D. plantaris* Pascoe, 1872c, *D. serricollis* Pascoe, 1872c, *D. validus* Blackburn, 1890c. **Distribution.** Australia (VIC, SA, WA).

Notes. The larva of *D. echinatus* apparently develops in the root tubers of *Dichopodon capillipes* (Asparagaceae) (Howden 1986; May 1994).

***Dicherotropis* Ferguson, 1912 (Pl. 50K; Fig. 44.18F)**

Type species, by subsequent designation (Ferguson, 1913: 388): *Talaurinus damelii* Macleay, 1865.

Australian species. *D. cavirostris* (Lea, 1911b), *D. damelii* (Macleay, 1865).

Distribution. Australia (WA).

Notes. The life history and larval hostplants of the genus remain unknown.

Ennothus Pascoe, 1872 (Pl. 50L; Figs 44.18G–H, 44.23A–E)

Type species, by monotypy: *Ennothus fallax* Pascoe, 1872.
Australian species. *E. fallax* Pascoe, 1872, *E. metricus* (Pascoe, 1874c)
Distribution. Australia (WA).
Notes. The larvae develop in the tubers of *Dichopodon capillipes* (Asparagaceae) (Howden 1986; May 1994).

Euomella Ferguson, 1923 (Pl. 51A; Fig. 44.18I)

Type species, by original designation: *Euomus retusus* Pascoe, 1872.
Australian species. *E. retusa* (Pascoe, 1872c).
Distribution. Australia (WA).
Notes. The life history and larval hostplants of the single species are unknown.

Euomus Schoenherr, 1847 (Pl. 51B; Figs 44.18J, 44.32H–J)

Type species, by original designation: *Amycterus insculptus* Boheman 1842.
Australian species. *E. insculptus* Boheman, 1842 (*in* Schoenherr 1842b), *E. scorpio* (Boisduval, 1835), *E. stephensii* (Gyllenhal, 1834) (*in* Schoenherr 1834b)
Distribution. Australia (WA).
Notes. The life history and larval hostplants of the genus are unknown.

Gagatophorus Jekel, 1865 (Pl. 51C; Figs 44.14, 44.18K–L, 44.19A, B, 44.23F)

Type species, by original designation: *Amycterus schoenherri* Hope, 1835.
Australian species. *G. boisduvalii* (Boisduval, 1835), *G. draco* (Macleay, 1826), *G. leichhardtii* (Macleay, 1865), *G. obsoletus* (Ferguson, 1914a), *G. schoenherri* (Hope, 1835), *G. tibialis* (Ferguson, 1914a).
Distribution. Australia (QLD, WA).
Notes. Adults of *G. draco* have been found feeding on stems and larvae at the roots of *Lepidobolus preissianus* (Restionaceae) (Howden 1986; May 1994).

Hyborrhinus Marshall, 1946 (Pl. 51D; Fig. 44.19C–G)

Type species, by subsequent designation (Ferguson, 1921c: 394): *Amycterus coenosus* Boheman, 1842.
Australian species. *H. aculeatus* (Ferguson, 1921c), *H. bicornutus* (Macleay, 1866), *H. coenosus* (Boheman, 1842) (*in* Schoenherr 1842b), *H. furcatus* (Macleay, 1865), *H. prodigus* (Macleay, 1866).
Distribution. Australia (SA, WA).
Notes. Adults of *H. prodigus* have been found in debris under *Daviesia teretifolia* (Fabaceae) plants (Howden 1986), but the larvae may rather feed on grasses like those of *Cubicorhynchus* and related genera do.

Hypotomops Baker & Thompson, 1978 (Pl. 51E; Fig. 44.19H)

Type species, by original designation: *Notonophes angulicollis* Lea, 1911.
Australian species. *H. angulicollis* (Lea, 1911c), *H. angustior* (Ferguson, 1914a), *H. leai* (Ferguson, 1914a).
Distribution. Australia (QLD, NT, WA).
Notes. The life history and larval hostplants are unknown.

Lataurinus Ferguson, 1912 (Pl. 51F; Fig. 44.19I, J)

Type species, by original designation: *Talaurinus rugiceps* Macleay, 1865.
Australian species. *L. macrocephalus* (Ferguson, 1912), *L. rugiceps* (Macleay, 1865).
Distribution. Australia (WA).
Notes. The life history and larval hostplants of the genus are unknown.

Melanegis Pascoe, 1870 (Pl. 51G; Fig. 44.19K)

Type species, by monotypy: *Melanegis stygius* Pascoe, 1870.
Australian species. *M. stygius* Pascoe, 1870e.
Distribution. Australia (WA).
Notes. Nothing is known about the life history and larval hostplants of the single species.

Molochtus Pascoe, 1874 (Pl. 51H; Fig. 44.19L)

Type species, by monotypy: *Molochtus gagates* Pascoe, 1874.
Australian species. *M. gagates* Pascoe, 1874c, *M. hercules* Ferguson, 1915b, *M. rotundicollis* Ferguson, 1916c, *M. tibialis* Sloane, 1893.
Distribution. Australia (SA, WA).
Notes. The larvae presumably develop in clumps and crowns of grasses like those of *Cubicorhynchus* do.

Myotrotus Pascoe, 1874 (Pl. 51I; Fig. 44.20A)

Type species, by monotypy: *Myotrotus* obtusus Pascoe, 1874.
Australian species. *M. apicihirtus* (Ferguson, 1912), *M. obtusus* Pascoe, 1874c.
Distribution. Australia (QLD).
Notes. The life history and larval hostplants of the genus are unknown.

Mythites **Schoenherr, 1847 (Pl. 51J; Figs 44.20B, 44.23G–K)**

Type species, by original designation: *Mythites tuberculatus* Schoenherr, 1847.

Australian species. *M. arboricola* Ferguson, 1923, *M. asperatus* Pascoe, 1872b, *M. basalis* (Boisduval, 1835), *M. coxi* Ferguson, 1923, *M. foveipennis* Lea, 1910a, *M. frater* Lea, 1910a, *M. granulatus* Lea, 1911b, *M. montanus* Ferguson, 1923, *M. perfossus* (Germar, 1848), *M. poropteroides* Lea, 1910a, *M. rugosus* Ferguson, 1923, *M. sulcicollis* (Germar, 1848), *M. tuberculatus* Schoenherr, 1847.

Distribution. Australia (QLD, NSW, VIC, TAS, SA, WA).

Notes. The larvae develop in underground stems and rhizomes of *Dianella revoluta* (Asphodelaceae) (Howden 1986; May 1994) and probably also of related taxa.

Neohyborrhynchus **Ferguson, 1921 (Pl. 51K; Fig. 44.20C)**

Type species, by original designation: *Hyborrhynchus rugosus* Macleay, 1865.

Australian species. *N. rugosus* (Macleay, 1865).

Distribution. Australia (WA).

Notes. Nothing is recorded about the life history and larval hostplants of the single species, but its larvae probably develop in clumps of grasses like those of *Cubicorhynchus* do. The genus was described by Ferguson (1921c).

Notonophes **Sloane, 1893 (Pl. 51L; Fig. 44.20D)**

Type species, by original designation: *Cubicorhynchus cichlodes* Pascoe, 1874.

Australian species. *N. auriger* Ferguson, 1914a, *N. cichlodes* (Pascoe, 1874c), *N. dilataticeps* (Blackburn, 1896), *N. gascoynensis* Baker, 1972, *N. hirsutinus* Baker, 1972, *N. taurus* Ferguson, 1914a.

Distribution. Australia (NT, WA).

Notes. Adults and a larva of *N. auriger* have been found at the base of tussocks of *Triodia basedowii* (Poaceae) (Howden 1986).

Oditesus **Pascoe, 1872 (Pl. 52A; Fig. 44.20E, F)**

Type species, by subsequent designation (Ferguson, 1923: 425): *Oditesus indutus* Pascoe, 1872.

Australian species. *O. buceros* Pascoe, 1872b, *O. indutus* Pascoe, 1872b, *O. sulcirostris* Pascoe, 1872b, *O. tibialis* Lea, 1904.

Distribution. Australia (WA).

Notes. Nothing is recorded about the life history and larval hostplants of the genus.

Ophthalamycterus **Ferguson, 1912 (Pl. 52B; Fig. 44.20G)**

Type species, by monotypy: *Talaurinus laticeps* Macleay, 1866.

Australian species. *O. laticeps* (Macleay, 1866).

Distribution. Australia (WA).

Notes. The life history and larval hostplants of the single species are unknown.

Parahyborrhynchus **Ferguson, 1921 (Pl. 52C; Fig. 44.20H)**

Type species, by original designation: *Acantholophus convexiusculus* Macleay, 1866.

Australian species. *P. convexiusculus* (Macleay, 1866), *P. crassiusculus* (Macleay, 1866).

Distribution. Australia (WA).

Notes. Nothing is recorded about the life history and larval hostplants of the genus, but the larvae probably develop in clumps of grasses like those of *Cubicorhynchus* do. The genus was described by Ferguson (1921c).

Pseudonotonophes **Ferguson, 1914 (Pl. 52D; Figs 44.20I, 44.32K–M)**

Type species, by original designation: *Talaurinus dumosus* Macleay, 1865.

Australian species. *P. dumosus* (Macleay, 1865), *P. gilesi* Ferguson, 1914a, *P. lemmus* (Pascoe, 1874c).

Distribution. Australia (WA).

Notes. The life history and larval hostplants of the genus are unknown.

Sclerorinus **Macleay, 1865 (Pl. 52F, G; Figs 44.20J, K, 44.24A–E)**

Type species, by subsequent designation (Zimmerman, 1993: 394): *Sclerorinus adelaidae* Macleay, 1865.

Australian species. *S. adelaidae* Macleay, 1865, *S. albovittatus* Ferguson, 1914b, *S. amycteroides* Ferguson, 1914c, *S. angustior* Ferguson, 1916b, *S. angustipennis* Sloane, 1893, *S. apicalis* Macleay, 1865, *S. aterrimus* Ferguson, 1916b, *S. besti* Ferguson, 1921a, *S. biordinatus* Macleay, 1866, *S. blackburni* Ferguson, 1914b, *S. browni* Ferguson, 1916b, *S. bubalus* (Olivier, 1807), *S. carteri* Ferguson, 1916a, *S. convexus* (Sloane, 1893), *S. daveyi* Ferguson, 1921a, *S. dilaticollis* Macleay, 1865, *S. dimidiatus* Macleay, 1866, *S. dixoni* Ferguson, 1915a, *S. dolens* (Boisduval, 1835), *S. elderi* Sloane, 1893, *S. elongatus* (Boheman, 1842) (*in* Schoenherr 1842b), *S. exilis* Macleay, 1865, *S. germari* Macleay, 1866, *S. goudiei* Ferguson, 1915a, *S. hopei* (Boheman, 1842) (*in* Schoenherr 1842b), *S. horridus* Macleay, 1865, *S. inconstans* Lea, 1911b, *S. inornatus* Ferguson,

1915a, *S. insignis* (Sloane, 1893), *S. kirbyi* (Boisduval, 1835), *S. laticollis* Macleay, 1866, *S. longus* Macleay, 1865, *S. mastersii* Macleay, 1866, *S. meliceps* Pascoe, 1874c, *S. molestus* Pascoe, 1874c, *S. molossus* (Pascoe, 1874c), *S. mucronipennis* Ferguson, 1914b, *S. neglectus* Ferguson, 1914b, *S. nigrospinosus* (Donovan, 1805), *S. noctis* (Sloane, 1893), *S. obliteratus* Macleay, 1865, *S. oblongatus* Ferguson, 1916a, *S. obscurus* (Sloane, 1893), *S. occidentalis* Sloane, 1893, *S. parvulus* Macleay, 1865, *S. pilularius* Macleay, 1866, *S. queenslandicus* Ferguson, 1916b, *S. regularis* Ferguson, 1914b, *S. riverinae* Macleay, 1865, *S. sabulosus* Macleay, 1866, *S. sloanei* Ferguson, 1921a, *S. sordidus* Macleay, 1865, *S. spencei* (Boheman, 1842) (*in* Schoenherr 1842b), *S. squalidus* Macleay, 1865, *S. stewartii* Macleay, 1865, *S. stutchburyi* Macleay, 1865, *S. subcarinatus* Ferguson, 1916b, *S. sublineatus* (Germar, 1848), *S. subsequens* Macleay, 1865, *S. tristis* (Boisduval, 1835), *S. tuberculosus* Macleay, 1865, *S. verrucosus* Macleay, 1865, *S. vestitus* Macleay, 1865.

Distribution. Australia (QLD, NT, NSW, VIC, TAS, SA, WA).

Notes. Larvae of an unidentified species of *Sclerorinus* have been found in stocks of *Stipa* (Poaceae) (May 1994).

Sclerorrhinella Ferguson, 1912 (Pl. 52E; Fig. 44.20L)

Type species, by original designation: *Amycterus manglesii* Boheman, 1842.

Australian species. *S. crawshawi* Ferguson, 1921a, *S. geniculata* (Pascoe, 1874c), *S. granuliceps* Ferguson, 1913, *S. manglesii* (Boheman, 1842) (*in* Schoenherr 1842b), *S. melanopsis* (Pascoe, 1874c).

Distribution. Australia (WA).

Notes. Nothing is recorded about the life history and larval hostplants of the genus.

Sosytelus Pascoe, 1872 (Pl. 52H; Figs 44.21A, B, 44.24F–L)

Type species, by monotypy: *Sosytelus lobatus* Pascoe, 1872.

Australian species. *S. carteri* (Ferguson, 1923), *S. hardcastlei* (Lea, 1910a), *S. indemnis* (Pascoe, 1882), *S. lobatus* Pascoe, 1872b, *S. pithecius* (Pascoe, 1872b), *S. pubescens* Lea, 1911b, *S. rudis* (Blackburn, 1896).

Distribution. Australia (QLD, NT, NSW, VIC, SA, WA).

Notes. The life history of *S. pubescens* has been recorded by Howden (1986); its larvae feed on the underground stems of *Bulbine alata* (Asphodelaceae) and pupate in earthen cells nearby.

Talaurinellus Zimmerman, 1993 (Pl. 52J)

Type species, by original designation: *Talaurinus sulciventris* Ferguson, 1912.

Australian species. *T. sulciventris* (Ferguson, 1912).

Distribution. Australia (QLD).

Notes. The life history and larval hostplants of the single species are unknown.

Talaurinus Macleay, 1865 (Pl. 52I; Figs 44.15A, 44.21C, D, 44.24M–Q, 44.32E)

Type species, by original designation: *Talaurinus typicus* Macleay, 1865.

Australian species. *T. acromialis* Ferguson, 1912, *T. acutipennis* Ferguson, 1912, *T. alternans* Macleay, 1865, *T. alternatus* Macleay, 1865, *T. angularis* Ferguson, 1912, *T. angustatus* Macleay, 1865, *T. angustus* Ferguson, 1914c, *T. brevior* Ferguson, 1914b, *T. bubaroides* Ferguson, 1914b, *T. bucephalus* (Olivier, 1807), *T. carbonarius* Pascoe, 1874c, *T. carinatifrons* Ferguson, 1914b, *T. carinatior* Ferguson, 1915a, *T. carinatus* Ferguson, 1914c, *T. carpentariae* Ferguson, 1912, *T. carteri* Ferguson, 1912, *T. catenulatus* Macleay, 1865, *T. caviceps* Macleay, 1866, *T. confusus* Ferguson, 1914c, *T. costipennis* Ferguson, 1912, *T. crassiceps* Sloane, 1893, *T. crenulatus* Ferguson, 1913, *T. euomoides* Macleay, 1865, *T. fergusoni* Carter, 1937, *T. flaveolus* Ferguson, 1914b, *T. fossulatus* Ferguson, 1912, *T. foveatus* Macleay, 1865, *T. foveipennis* Ferguson, 1912, *T. foveogranulatus* Ferguson, 1912, *T. gayndahensis* Ferguson, 1912, *T. griseus* Macleay, 1865, *T. halmaturinus* Ferguson, 1914b, *T. howittii* Macleay, 1865, *T. humeralis* Macleay, 1865, *T. hystrix* Ferguson, 1915a, *T. illidgei* Ferguson, 1912, *T. imitator* Blackburn, 1896, *T. impressicollis* Macleay, 1865, *T. inaequalis* Blackburn, 1896, *T. incanescens* Macleay, 1865, *T. inconspicuus* Ferguson, 1912, *T. irroratus* Ferguson, 1912, *T. kirbii* (Macleay, 1826), *T. lacunosus* Macleay, 1865, *T. laevicollis* Pascoe, 1874c, *T. leai* Ferguson, 1915a, *T. longipes* Ferguson, 1912, *T. maculipennis* Lea, 1911b, *T. megalongensis* Ferguson, 1912, *T. miliaris* Ferguson, 1912, *T. mitchellii* Macleay, 1865, *T. multigranulatus* (Lea, 1910b), *T. mythitoides* Ferguson, 1912, *T. niveovittatus* Ferguson, 1912, *T. pallidus* Macleay, 1865, *T. papulosus* Macleay, 1865, *T. parallelus* Macleay, 1865, *T. parvus* Ferguson, 1912, *T. perplexus* Ferguson, 1915a, *T. plagiatus* Ferguson, 1914b, *T. posticalis* Ferguson, 1915a, *T. prypnoides* Ferguson, 1912, *T. regularis* Sloane, 1893, *T. roei* (Boheman, 1842) (*in* Schoenherr 1842b), *T. rugicollis* Macleay, 1865, *T. rugifer* (Boisduval, 1835), *T. scaber* (Boisduval, 1835), *T. scabricollis* Ferguson, 1913, *T. scabrosus* Macleay, 1865, *T. scalpularis* Ferguson, 1912, *T. semispinosus* (Boheman, 1842) (*in* Schoenherr 1842b), *T. septentrionalis* Ferguson, 1912, *T. simplex* Ferguson, 1915a, *T. simplicipes* Lea, 1911b, *T. simulator* Pascoe, 1874c, *T. sobrinus* Ferguson, 1912, *T. sphaerulatus* Macleay, 1865, *T. spiniger* Ferguson, 1914b, *T. squamosus* Macleay, 1865, *T. strangulatus* Blackburn, 1895a, *T. subvittatus*

Ferguson, 1912, *T. suttoni* Carter, 1939, *T. tuberculosus* Zimmerman, 1993, *T. tumulosus* Ferguson, 1912, *T. typicus* Macleay, 1865, *T. vermicollis* Ferguson, 1912, *T. verrucosus* (Boisduval, 1835), *T. vitticollis* Ferguson, 1914c.

Distribution. Australia (QLD, NT, NSW, VIC, SA, WA).

Notes. The adults of several species have been observed feeding on the tough leaves of *Lomandra* (Asparagaceae), and the larvae of *T. rugifer* and *T. subvittatus* have been found among the roots of the same plant genus (Howden 1986; Zimmerman 1993; May 1994).

Tetralophus **Waterhouse, 1853 (Pl. 52K; Figs 44.21E, F)**

Type species, by monotypy: *Tetralophus sculpturatus* Waterhouse, 1853.

Australian species. *T. excursus* Pascoe, 1870e, *T. sculpturatus* Waterhouse, 1853.

Distribution. Australia (VIC, TAS, SA).

Notes. The adults of *T. sculpturatus* have been recorded feeding on leaves and stems of *Thelymitra pauciflora* (Orchidaceae) (Porch 2009), and the larvae probably feed on or in the tubers of these orchids.

Xenommamycterus **Peterson, 2013 (Pl. 52L)**

Type species, by original designation: *Talaurinus capito* Pascoe, 1874.

Australian species. *X. capito* (Pascoe, 1874c).

Distribution. Australia (WA).

Notes. Adults of the only described species have been found under clumps of *Triodia pungens* (Poaceae) during the day and ascending up the stems in the evening (R. Oberprieler, pers. obs.).

Tribe Aterpini Lacordaire, 1863

The Aterpini are an eclectic group of 25 genera and 114 species in Australia (Oberprieler 2014c; Pullen *et al.* 2014) but also occur in New Guinea, New Zealand, South America and South Africa (Oberprieler *et al.* 2007; Oberprieler 2010, 2014c). The tribe is poorly characterised morphologically and includes some unusual forms (Pls 8A–F, H, 9A–D, H) such as the iridescent green-mottled *Chrysolopus spectabilis* (Fabricius), named the Diamond Weevil or Botany Bay Weevil, the species of *Rhinaria* with their large and complex head tubercles, the elongate *Rhadinosomus* species (Fig. 44.21K) with their acutely tipped elytra and *Strongylorhinus*, whose larvae live gregariously in large stem galls on eucalypts. The exact composition and monophyly of the Aterpini remain elusive, relevant molecular phylogenetic analyses to date (McKenna *et al.* 2009; Gillett *et al.* 2014; Gunter *et al.* 2016; Shin *et al.* 2018; Baird *et al.* 2021; Haran *et al.* 2023) including too few and different genera and producing incongruent relationships among these. It nonetheless appears that a large number of the genera as currently included in Aterpini are more or less closely related to each other.

Aades **Schoenherr, 1823 (Pl. 53A)**

Type species, by subsequent designation (Schoenherr, 1833a: 313): *Curculio cultratus* Fabricius, 1775.

Australian species. *A. bifoveifrons* (Lea, 1916), *A. cultratus* (Fabricius, 1775), *A. foveipennis* (Lea, 1915c), *A. griseatus* (Pascoe, 1872a), *A. scoparius* (Erichson, 1842).

Distribution. Australia (QLD, NSW, TAS, VIC), New Guinea.

Notes. The genus is also represented in New Guinea by a single described species, *Aades franklini* (Heller). Adults of *A. cultratus* are usually found on different genera of Myrtaceae (Froggatt 1895).

Acalonoma **Pascoe, 1875 (Pl. 53B)**

Type species, by monotypy: *Acalonoma reductum* Pascoe, 1875.

Australian species. *A. pusillum* Blackburn, 1894c, *A. reductum* Pascoe, 1875.

Distribution. Australia.

Notes. Nothing is recorded about the life history and larval hostplants of the genus.

Aesiotes **Pascoe, 1865 (Pl. 53C, D; Fig. 44.25A–F)**

Type species, by monotypy: *Aesiotes notabilis* Pascoe, 1865.

Australian species. *A. leucurus* Pascoe, 1873c, *A. morosus* Pascoe, 1873c, *A. notabilis* Pascoe, 1865.

Distribution. Australia (QLD, NSW, VIC).

Notes. The genus is associated with conifers, *A. leucurus* with *Callitris* (Cupressaceae) and *A. notabilis* with *Agathis* and *Araucaria* (Araucariaceae). The larvae of *A. notabilis* tunnel in the phloem and cambium of Hoop Pine, *Araucaria cunninghamii*, and on maturity construct a dense cocoon of shredded wood for pupation (Brimblecombe 1945), and the habits of the larvae of *A. leucurus* are similar (Froggatt 1923). Both species also attack introduced species of *Pinus* (Pinaceae), specifically stressed trees, and sometimes in large numbers (Froggatt 1912). *Aesiotes notabilis* has been regarded as a pest in plantations of *A. cunninghamii* subjected to the practice of green-pruning (of lower green branches), which creates wounds that attract female weevils to lay their eggs on bark nearby. The species has also been introduced to the Philippines (Heller 1912) and was intercepted in China in logs imported from Australia (Oberprieler 2014c). The larva and pupa of *A. leucurus* were described by May (1994).

Alphitopis **Pascoe, 1870 (Pl. 53E)**

Type species, by monotypy: *Alphitopis nivea* Pascoe, 1870.

Australian species. *A. nivea* Pascoe, 1870d.

Distribution. Australia (QLD, NSW, WA).

Notes. The larvae of the single species develop in follicles and seeds of *Banksia* (Proteaceae), boring through the bracts and woody axis of the infructescences, and pupate in the follicles or the axis (Scott 1982). The adults have been observed feeding on *Banksia* pollen. Carnaby's Black Cockatoo (*Calyptorhynchus latirostris* Carnaby, Psittacidae) preys on the larvae, the birds extracting them from the infructescences (Scott & Black 1981).

Anomocis Lea, 1915 (Pl. 53F)

Type species, by monotypy: *Anomocis apicalis* Lea, 1915.
Australian species. *A. apicalis* Lea, 1915c.
Distribution. Australia.
Notes. Adults have been found feeding on leaves of *Lomandra* (Asparagaceae) at night (Oberprieler 2014c).

Aoplocnemis Schoenherr, 1845 (Pl. 53G, H)

Type species, by original designation: *Aoplocnemis rufipes* Boheman, 1845.
Australian species. *A. alboguttata* Chevrolat, 1879a, *A. armipennis* Lea, 1910b, *A. bifasciculata* Lea, 1915a, *A. clavipes* (Fabricius, 1801), *A. dorsonotata* Chevrolat, 1879a, *A. fulva* Chevrolat, 1879a, *A. guttigera* Pascoe, 1882, *A. lineata* Pascoe, 1872b, *A. loweri* Blackburn, 1894b, *A. maxima* Lea, 1915a, *A. phalerata* (Erichson, 1842), *A. rufipes* Boheman, 1845 (*in* Schoenherr 1845), *A. semicincta* Chevrolat, 1879a, *A. suturalis* Pascoe, 1882.
Distribution. Australia (NSW, NT, TAS, VIC, WA).
Notes. The adults are usually found feeding and mating on grass seed-heads, and the larvae bore in grass stems. Adults of *A. rufipes* have also been reported damaging hop vines (*Humulus lupulus*, Cannabaceae) in Tasmania (Oberprieler 2014c).

Aromagis Pascoe, 1865 (Pl. 54A)

Type species, by monotypy: *Aromagis echinata* Pascoe, 1865.
Australian species. *A. echinata* Pascoe, 1865, *A. horrens* Pascoe, 1883a, *A. saginata* Lea, 1898a.
Distribution. Australia (NSW, SA, VIC).
Notes. The life history and larval hostplants of the genus appear unknown.

Atelicus Waterhouse, 1862 (Pl. 53I, J)

Type species, by subsequent designation (Lea, 1927e: 146): *Atelicus inaequalis* Waterhouse.
Australian species. *A. abruptus* Pascoe, 1882, *A. atrophus* Pascoe, 1870e, *A. brevicornis* (Blanchard, 1853), *A. crassipes* Pascoe, 1882, *A. ferrugineus* Waterhouse, 1862, *A. fusiformis* Lea, 1927e, *A. guttatus* Pascoe, 1870e, *A. latericollis* Lea, 1927e, *A. miniatus* Pascoe, 1872a, *A. variabilis* Lea, 1898a.
Distribution. Australia (QLD, NSW, VIC, TAS, SA, WA).
Notes. Adults of *Atelicus* have been reared from hymenopteran galls on *Acacia* (Lea 1898a) but also from *Eucalyptus* spp. (Oberprieler 2014c).

Aterpodes Zimmerman, 1994 (Pl. 53K, L)

Type species, by original designation: *Aterpus kubus* Boheman, 1842.
Australian species. *A. abruptus* (Blackburn, 1892b), *A. horrens* (Gyllenhal, 1834) (*in* Schoenherr 1834a), *A. kubus* (Boheman, 1842) (*in* Schoenherr 1842a), *A. raucus* (Blackburn, 1892b), *A. seriatus* (Boisduval, 1835), *A. tuberculatus* Gyllenhal, 1834 (*in* Schoenherr 1834a).
Distribution. Australia (NSW, VIC, TAS, SA).
Notes. Adults of *A. kubus* have been reported rasping bark and cutting leaves from petioles of young eucalypt trees, and the larvae to tunnel in the taproot and other large roots near the base of the trunk, sometimes girdling and killing roots or entire plants, before pupating in earthen cells next to the tunnel entrance (Elliott *et al.* 1998).

Cechides Pascoe, 1872 (Pl. 54B)

Type species, by monotypy: *Cechides amoenus* Pascoe, 1872.
Australian species. *C. amoenus* Pascoe, 1872c.
Distribution. Australia (WA).
Notes. The life history of the single species is similar to that of *Alphitopis nivea*, the larvae also developing and pupating in follicles and seeds of *Banksia* (Proteaceae) and the adults feeding on pollen (Scott 1982).

Chrysolopus Germar, 1817 (Pl. 54C–E; Fig. 44.25G–P)

Type species, by subsequent designation (Schoenherr, 1823: column 1142): *Curculio spectabilis* Fabricius, 1775.
Australian species. *C. detritus* Chevrolat, 1879a, *C. spectabilis* (Fabricius, 1775).
Distribution. Australia (QLD, NSW, SA, VIC).
Notes. Adults of *Ch. spectabilis* are often found on shrubs of many *Acacia* species, feeding on leaves, and the larvae bore in the main roots, which they can hollow out completely (Froggatt 1893; Hawkeswood 1991; May 1994). The larva and pupa of *Ch. spectabilis* were described by May (1994).

Chrysophoracis Lea, 1915 (Pl. 54F)

Type species, by original designation: *Chrysophoracis pulcher* Lea, 1915.
Australian species. *C. amplipennis* Lea, 1915a, *C. pulcher* Lea, 1915a.

Distribution. Australia (TAS).
Notes. Adults have been collected from logs and roots of *Nothofagus* (Nothofagaceae) in Tasmania (Oberprieler 2014c), and the larvae probably bore in the roots.

Cyllorhamphus Erichson, 1842 (Pl. 54G, H)

Type species, by monotypy: *Cyllorhamphus tuberosus* Erichson, 1842.
Australian species. *C. angustus* Lea, 1914, *C. mimicus* Lea, 1914, *C. tuberosus* Erichson, 1842.
Distribution. Australia (QLD, NSW, VIC, TAS), Papua New Guinea.
Notes. The host include *Banksia* species.

Dixoncis Oke, 1931 (Pl. 54I)

Type species, by original designation: *Dixoncis pictus* Oke, 1931.
Australian species. *D. pictus* Oke, 1931.
Distribution. Australia (VIC).
Notes. Adults of the single species have been reared from mature fruits of *Hakea* (Proteaceae) and also been found on *Lomatia tinctoria*, of the same family (Oberprieler 2014c).

Euthyphasis Pascoe, 1875 (Pl. 54J, K)

Type species, by monotypy: *Euthyphasis acuta* Pascoe, 1875.
Australian species. *E. acuta* Pascoe, 1875, *E. funerea* Blackburn, 1901, *E. lineata* Lea, 1904, *E. parva* Blackburn, 1894c, *E. sordidata* Lea, 1904.
Distribution. Australia.
Notes. The larvae of *Eu. funerea* develop in stems of *Lordhowea amygdalifolia* (formerly in *Senecio*) (Asteraceae) (Oberprieler 2014c).

Iphisaxus Pascoe, 1870 (Pl. 54L)

Type species, by monotypy: *Iphisaxus asper* Pascoe, 1870.
Australian species. *I. aethiops* Pascoe, 1883a, *I. asper* Pascoe, 1870c.
Distribution. Australia (WA).
Notes. The genus appears associated with *Eucalyptus* (Oberprieler 2014c), but nothing is recorded about its life history.

Kershawcis Lea, 1908 (Pl. 55A)

Type species, by monotypy: *Kershawcis cylindricus* Lea, 1908.
Australian species. *K. cylindricus* Lea, 1908c.
Distribution. Australia.
Notes. Nothing is known about the life history and larval hostplants of this rare genus.

Nemestra Pascoe, 1872 (Pl. 55B)

Type species, by monotypy: *Nemestra incerta* Pascoe, 1872.
Australian species. *N. incerta* Pascoe, 1872c.
Distribution. Australia (WA).
Notes. The genus appears to be associated with *Allocasuarina* (Casuarinaceae) as well as *Hakea*, the larvae probably boring in twigs or stems (Oberprieler 2014c).

Oenopus Marshall, 1943 (Pl. 55C)

Type species, by monotypy: *Oenochroma rubeta* Pascoe, 1872b.
Australian species. *O. guttatus* (Boheman, 1834) (*in* Schoenherr 1834b), *O. triquetrus* (Lea, 1899a).
Distribution. Australia (NSW).
Notes. Adults of *O. guttatus* have been reared from mature fruits of *Hakea* (Proteaceae) and also been found on *Persoonia*, of the same family (Oberprieler 2014c).

Ophthalmorychus Blackburn, 1894 (Pl. 55D)

Type species, by monotypy: *Ophthalmorychus angustus* Blackburn, 1894.
Australian species. *O. angustus* Blackburn, 1894c.
Distribution. Australia.
Notes. The life history and larval hostplants of the rare single species are unknown.

Pelororhinus Schoenherr, 1834 (Pl. 55E–G; Fig. 44.26A–G)

Type species, by original designation: *Pelororhinus argentosus* Gyllenhal, 1834.
Australian species. *P. amplipennis* Lea, 1904, *P. angustatus* Fåhraeus, 1842 (*in* Schoenherr 1842a), *P. argentosus* Gyllenhal, 1834 (*in* Schoenherr 1834a), *P. bisulcatus* (Lea, 1904), *P. carinirostris* (Lea, 1929a), *P. caudatus* (Lea, 1904), *P. cavirostris* (Pascoe, 1883a), *P. crassus* Blackburn, 1892b, *P. debilis* (Blackburn, 1892b), *P. diversus* (Pascoe, 1883a), *P. elegans* (Lea, 1911c), *P. facetus* (Pascoe, 1865), *P. grandis* (Lea, 1909c), *P. interruptus* (Lea, 1915a), *P. interstitialis* Lea, 1911c, *P. longirostris* (Lea, 1915c), *P. maculiventris* (Lea, 1923), *P. margaritaceus* Erichson, 1842, *P. nigrivitta* (Lea, 1929a), *P. perdix* (Pascoe, 1873c), *P. proximus* Blackburn, 1892b, *P. pulicosus* (Lea, 1908b), *P. pusio* (Schoenherr, 1843), *P. sparsus* Germar, 1848, *P. stellio* (Pascoe, 1865), *P. sulcirostris* (Lea, 1904), *P. transversus* (Boisduval, 1835), *P. variegatus* (Boisduval, 1835).
Distribution. Australia (QLD, NSW, VIC, TAS, SA, WA).

Notes. Adults of *Pelororhinus* have been reared from roots of *Hakea sericea* (Proteaceae), while *P. perdix* (Pascoe) can damage strawberries, the larvae tunnelling in the crowns and petioles (French 1893).

Rhadinosomus Schoenherr, 1840 (Pl. 55H; Figs 44.21K, 44.26H–O)

Type species, by original designation: *Curculio acuminatus* Fabricius, 1775.

Australian species. *R. lacordairei* Pascoe, 1870b.

Distribution. Australia, New Zealand.

Notes. The larvae of the single species in Australia tunnel in stems of young eucalypts (Elliot *et al.* 1998) but also in woody brachyscelid galls (Froggatt 1907) as well as in the crowns of strawberry plants (May 1994). The genus was described by Schoenherr (1840b).

Rhinaria Kirby, 1819 (Pl. 55I, J)

Type species, by original designation: *Rhinaria cristata* Kirby, 1819.

Australian species. *R. aberrans* Lea, 1904, *R. caliginosa* Pascoe, 1872a, *R. concavirostris* Lea, 1904, *R. convexirostris* Lea, 1904, *R. cristata* Kirby, 1819a, *R. excavata* Boisduval, 1835, *R. fasciata* Pascoe, 1873c, *R. favosa* Lea, 1904, *R. foveipennis* Pascoe, 1872a, *R. granulosa* (Fåhraeus, 1842) (*in* Schoenherr 1842a), *R. grisea* Boisduval, 1835, *R. rugosa* Boisduval, 1835, *R. signifera* Pascoe, 1883a, *R. tessellata* Pascoe, 1883a, *R. tibialis* Blackburn, 1892b, *R. tragocephala* Lea, 1904.

Distribution. Australia (QLD, NSW, SA, VIC, WA).

Notes. The larvae generally appear to generally be stem-borers, mostly in *Eucalyptus* but also in several other Myrtaceae like *Chamaelaucium, Regelia* and *Thryptomene* (May 1994, as *Pelolorhinus*).

Rhinoplethes Pascoe, 1870 (Pl. 55K)

Type species, by monotypy: *Rhinoplethes foveatus* Pascoe, 1870.

Australian species. *R. foveatus* Pascoe, 1870c.

Distribution. Australia (WA).

Notes. Adults of the single species are generally found on foliage, flowers and fruits of *Banksia*, *Dryandra*, *Hakea* and also *Stirlingia* (Proteaceae) (Oberprieler 2014c), and the larvae presumably bore in the roots or trunks of these plants.

Strongylorhinus Schoenherr, 1847 (Pl. 55L; Fig. 44.27)

Type species, by original designation: *Strongylorhinus ochraceus* Schoenherr, 1847.

Australian species. *S. clarki* Marshall, 1930, *S. ochraceus* Schoenherr, 1847.

Distribution. Australia (NSW, VIC, TAS, SA, WA).

Notes. The larvae develop gregariously in stems and thick branches of eucalypts, inciting large woody galls of 4–5 times the stem diameter that contain up to about 200 larvae, each in its own cell with a hard black fecal cap, from which they appear to feed on the softer outer layers of wood (Lea 1898a; French 1909; Froggatt 1926; Oberprieler 2014c).

Tribe Listroderini LeConte, 1876

The Listroderini are largely a South American group but currently include eight genera with 41 described species as well as a handful of undescribed taxa (e.g. Pl. 9E) in Australia (two genera with five species introduced) (Pullen *et al.* 2014; Escalona & Oberprieler 2021) and a few also in New Zealand and South Africa (Oberprieler 2010, 2014c). *Mandalotina* and *Tasmanica* were previously classified in Notiomimetini (Oberprieler 2010, 2014c) but transferred to Listroderini by Escalona & Oberprieler (2021) as the relationship of the true (littoral) genera of the tribe (*Aphela*, *Notiomimetes*, *Psaldus* and *Wollastonicis*) to Bagoinae became evident (Gillett *et al.* 2018; Escalona & Oberprieler 2021) but that of *Mandalotina* to Aterpini (Baird *et al.* 2021). The apparent relationship of also *Steriphus* to Aterpini (Baird *et al.* 2021) makes it unlikely that any of the Australian and New Zealand genera currently placed in Listroderini actually belong in this tribe, but their true relationships and classification can only become clearer from more comprehensive phylogenetic analyses. The listroderine species introduced in Australia are mostly regarded as minor horticultural pests, in particular *Listroderes difficilis* Germain, the Vegetable Weevil, and *Listronotus bonariensis* (Kuschel), the Argentine Grass-stem Weevil, whereas *Listronotus setosipennis* (Hustache) is a biological control agent of the weed *Parthenium hysterophorus*.

Anorthorhinus Blackburn, 1890 (Pl. 56A)

Type species, by monotypy: *Anorthorhinus pictipes* Blackburn, 1890.

Australian species. *A. apicalis* Lea, 1899a, *A. brevicornis* Lea, 1899a, *A. pictipes* Blackburn, 1890b.

Distribution. Australia.

Notes. Nothing is recorded about the life history and larval hostplants of the genus.

Listroderes Schoenherr, 1826 (Pl. 56B; Fig. 44.28A–K)

Type species, by original designation: *Listroderes costirostris* Schoenherr, 1826.

Australian species. *L. delaiguei* Germain, 1895, *L. difficilis* Germain, 1895, *L. foveatus* (Lea, 1928c).

Distribution. Australia (QLD, NSW, Norfolk Is., VIC, TAS, SA, WA), native to South and central America, introduced in U.SA, Spain, South Africa, Japan, New Zealand.

Notes. The three species in Australia were accidentally introduced from South America in the late 19th and early 20th Centuries (Pullen *et al.* 2014). Their populations in Australia are parthenogenetic (comprising females only), and all are regarded as minor to serious agricultural and horticultural pests in the country, attacking a wide variety of crops. Their larvae feed at night on leaves, roots and tubers underground, and the adults also feed on leaves (May 1993, 1994; Oberprieler 2014c).

Listronotus Jekel, 1865 (Pl. 56C; Fig. 44.28L–S)

Type species, by subsequent designation (Henderson, 1940: 231): *Rhynchaenus caudatus* Say, 1824.

Australian species. *L. bonariensis* (Kuschel, 1955), *L. setosipennis* (Hustache, 1926).

Distribution. Australia (QLD, NSW, Norfolk Is., VIC, TAS, SA, WA), native to South and North America, introduced in the Pacific region, including New Zealand.

Notes: *Listronotus bonariensis*, known as the Argentine Grass-stem Weevil, was accidentally introduced to Australia from New Zealand by 1962 (Pullen *et al.* 2014) and is regarded as a serious pest of crop grasses including cereals, its larvae tunnelling in the stems. The larva and pupa were described by May (1993, 1994). By contrast, *L. setosipennis* was deliberately introduced from South America in the 1980s and 1990s for the biological control of the weed *Parthenium hysterophorus* (Asteraceae) (Pullen *et al.* 2014).

Mandalotina Oke, 1931 (Pl. 56H; Figs 44.21G–J, 44.34, 44.35)

Type species, by original designation: *Mandalotina atranotata* Oke, 1931.

Australian species. *M. atranotata* Oke, 1931, *M. bicolor* Oke, 1931, *M. varia* Oke, 1931.

Distribution. Australia (NSW, VIC).

Notes. Traditionally classified in Rhythirrinini (e.g. Alonzo-Zarazaga & Lyal 1999), this genus was tentatively included in Notiomimetini by Oberprieler (2010, 2014c) and Pullen *et al.* (2014) but placed in Listroderini by Escalona & Oberprieler (2021), although it appears to be more closely related to Aterpini (Baird *et al.* 2021). It is a composite genus and needs to be restricted to include only *M. atranotata* (and several undescribed species), whereas *M. bicolor* and *M. varia* apparently belonging in Molytinae (Pullen *et al.* 2014). The species occur in leaf litter, but nothing is known about their life histories.

Methypora Pascoe, 1865 (Pl. 56D)

Type species, by monotypy: *Methypora postica* Pascoe, 1865.

Australian species. *M. postica* Pascoe, 1865, *M. tibialis* Lea, 1911c.

Distribution. Australia (QLD, VIC).

Notes. The life histories and larval hostplants of the two species remain unknown.

Neosyagrius Lea, 1904 (Pl. 56I; Fig. 44.36)

Type species, by monotypy: *Neosyagrius cordipennis* Lea, 1904.

Australian species. *N. cordipennis* Lea, 1904 (*in* Froggatt 1904), *N. porosus* Marshall, 1922, *N. striatus* Marshall, 1922.

Distribution. Australia (NSW, QLD, VIC, WA), New Zealand.

Notes. Like *Mandalotina*, this genus was traditionally classified in Rhythirrinini (e.g. Alonzo-Zarazaga & Lyal 1999) but later included in Notiomimetini (Oberprieler 2010, 2014c; Pullen *et al.* 2014), but its similarity to *Mandalotina* indicates that it, too, should be placed in Listroderini for the time being (pending more comprehensive phylogenetic analyses of the Cyclominae). One species, *N. cordipennis*, has been introduced in New Zealand (Kuschel 1972). The species also occur in leaf litter, but nothing is known about their life histories.

Steriphus Erichson, 1842 (Pl. 56E, F; Fig. 44.29A–G)

Type species, by monotypy: *Steriphus solidus* Erichson, 1842.

Australian species. *S. abjectus* (Lea, 1899a), *S. albidosparsus* (Lea, 1928c), *S. alpinus* (Lea, 1928c), *S. angusticollis* (Pascoe, 1870d), *S. caudatus* (Pascoe, 1870d), *S. curvisetosus* (Lea, 1928c), *S. diversipes* (Pascoe, 1870d), *S. humeralis* (Lea, 1928c), *S. incontaminatus* (Lea, 1899a), *S. inermis* (Lea, 1928c), *S. irrasus* (Lea, 1899a), *S. longus* (Lea, 1928c), *S. major* (Blackburn, 1890b), *S. mecaspis* (Lea, 1899a), *S. metallicus* (Lea, 1928c), *S. mucronatus* (Lea, 1928c), *S. murinus* (Pascoe, 1870d), *S. parvicornis* (Lea, 1928c), *S. parvoniger* (Lea, 1928c), *S. parvus* (Blackburn, 1890b), *S. sericeus* (Blackburn, 1890b), *S. solidus* Erichson, 1842, *S. stenoderes* (Lea, 1928c), *S. vittatus* (Blackburn, 1893b).

Distribution. Australia (QLD, NSW, SA, VIC), New Zealand.

Notes. The taxonomy of the genus needs revision. The larvae often have piercing abdominal spiracles for tapping air from plant parts under water and may live aquatically or semi-aquatically, although those of *S. caudatus* and *S. diversipes* live under weeds in drier pastures and fallow fields and feed on germinating seeds, seedlings and roots of various grasses (Allen 1973; May 1993, 1994; Oberprieler 2014c).

Tasmanica Lea, 1900 (Pl. 56L; Fig. 44.40)

Type species, by monotypy: *Tasmanica myrmecophila* Lea, 1900.

Australian species. *T. myrmecophila* Lea, 1900a.
Distribution. Australia (NSW, TAS, WA).
Notes. Like *Mandalotina*, this genus was included in Notiomimetini by Oberprieler (2010, 2014c) and Pullen *et al.* (2014) but more recently transferred to Listroderini (Escalona & Oberprieler 2021). It currently includes only one described species in Tasmania, but several undescribed ones are in collections (Escalona & Oberprieler 2021). *Tasmanica myrmecophila* and other species are sometimes found in ant nests, but most specimens have been extracted from leaf litter. All species are anophthalmic, but similar species with rudimentary eyes may belong in the genus as well. It is in need of taxonomic study.

Tribe Notiomimetini Wollaston, 1873

As now restricted (Escalona & Oberprieler 2021), the Notiomimetini are a small group of only four genera with seven described species (and a few undescribed ones known) and almost restricted to Australia, except for one species of *Aphela* occurring in the eastern Palaearctic region. All species are small to minute (5.4–1.2 mm SL), often microphthalmic littoral weevils found under seagrass and drift litter along beaches. The tribe Notiomimetini was resurrected by Oberprieler (2010) from synonymy in Emphyastini and then included also three small, leaflitter-inhabiting Australian genera (*Mandalotina*, *Neosyagrius*, *Tasmanica*), but recent molecular phylogenetic analyses (Gillett *et al.* 2018; Baird *et al.* 2021) showed that *Aphela* and *Psaldus* are closely related to Bagoinae but *Mandalotina* to Aterpini, thus necessitating the restriction of Notiomimetini to include only *Aphela*, *Notiomimetes*, *Psaldus* and *Wollastonicis*. In addition, we consider that Notiomimetini may need to be classified as a tribe of Bagoinae. Although some putative notiomimetine larvae have been found, nothing is known about their feeding modes and life histories.

Aphela Pascoe, 1865 (Pl. 56G)

Type species, by monotypy: *Aphela helopoides* Pascoe, 1865.
Australian species. *A. algarum* Pascoe, 1870d, *A. helopoides* Pascoe, 1865, *A. phalerioides* Pascoe, 1870d.
Distribution. Australia (NSW, Lord Howe Is., VIC, TAS, SA, WA), New Zealand, Korea, Japan, Kuril Is., East Siberia.
Notes. Besides the species in Australia and New Zealand, the genus also includes *A. gotoi* (Chûjo & Voss, 1960) in the eastern Palaearctic region (sometimes placed in a separate genus, *Isonycholips* Chûjo & Voss). *Aphela algarum* is deemed to be adventive in New Zealand (Kuschel 1972), but in Australia it comprises a complex of several species, making it uncertain whether the species in New Zealand is *A. algarum* and whether it is indeed introduced. All species live on beaches under seagrass and drift litter and have broadened legs for burrowing, but their larvae and feeding substrates are still unknown.

Notiomimetes Wollaston, 1873 (Pl. 56J; Figs 44.37, 44.38)

Type species, by monotypy: *Notiomimetes pascoei* Wollaston, 1873.
Australian species. *N. pascoei* Wollaston, 1873b.
Distribution. Australia (NSW, VIC, SA, WA).
Notes. The genus comprises one described and one undescribed species as known. Both species live under mats of seagrass on beaches, and specimens have also been collected in rock holes along the beach. Their larvae and life histories are unknown.

Psaldus Pascoe, 1870 (Pl. 56K; Fig. 44.39)

Type species, by monotypy: *Psaldus liosomoides* Pascoe, 1870.
Australian species. *P. liosomoides* Pascoe, 1870d.
Distribution. Australia (NSW, VIC SA, TAS, WA).
Notes. The genus contains only a single but widespread species along the southern Australian coast. It can be common and numerous under mats of seagrass on the beach. Its putative larva is undescribed and its life history unknown.

Wollastonicis Lea, 1909 (Pl. 57A)

Type species, by monotypy: *Wollastonicis minutus* Lea, 1909.
Australian species. *W. minutus* Lea, 1909b.
Distribution. Australia (SA, WA).
Notes. The genus *Wollastonicis* also comprises one described and one undescribed species, but their identities are confused and require taxonomic attention. Both also live under mats of seagrass on beaches, but their larvae and life histories are unknown.

Tribe Rhythirrinini Lacordaire, 1863

The tribe Rhythirrinini is essentially a small African group but currently includes six Australian genera with 32 described species (Oberprieler 2010, 2014c; Pullen *et al.* 2014), although the placement of the Australian taxa in the tribe is tentative (Oberprieler 2010, 2014c), and based largely on similarities in their ovipositor structure, their true relationships are yet unclear. The molecular phylogenetic analysis of Baird *et al.* (2021) suggests that *Ethemaia* may be related to Gonipterini but *Rhythirrinus* instead to African members of Hipporhinini (Baird *et al.* unpubl.). Little is known about the hostplants and life histories of the Australian genera, except that the larvae of *Ethemaia* (Fig. 44.29H–O) and *Ophryota* are nocturnal ectophytic leaf-feeders on small herbaceous plants (May 1994; Oberprieler 2014c) and probably pupate in the soil. A similar lifestyle is likely to apply to the other genera as well.

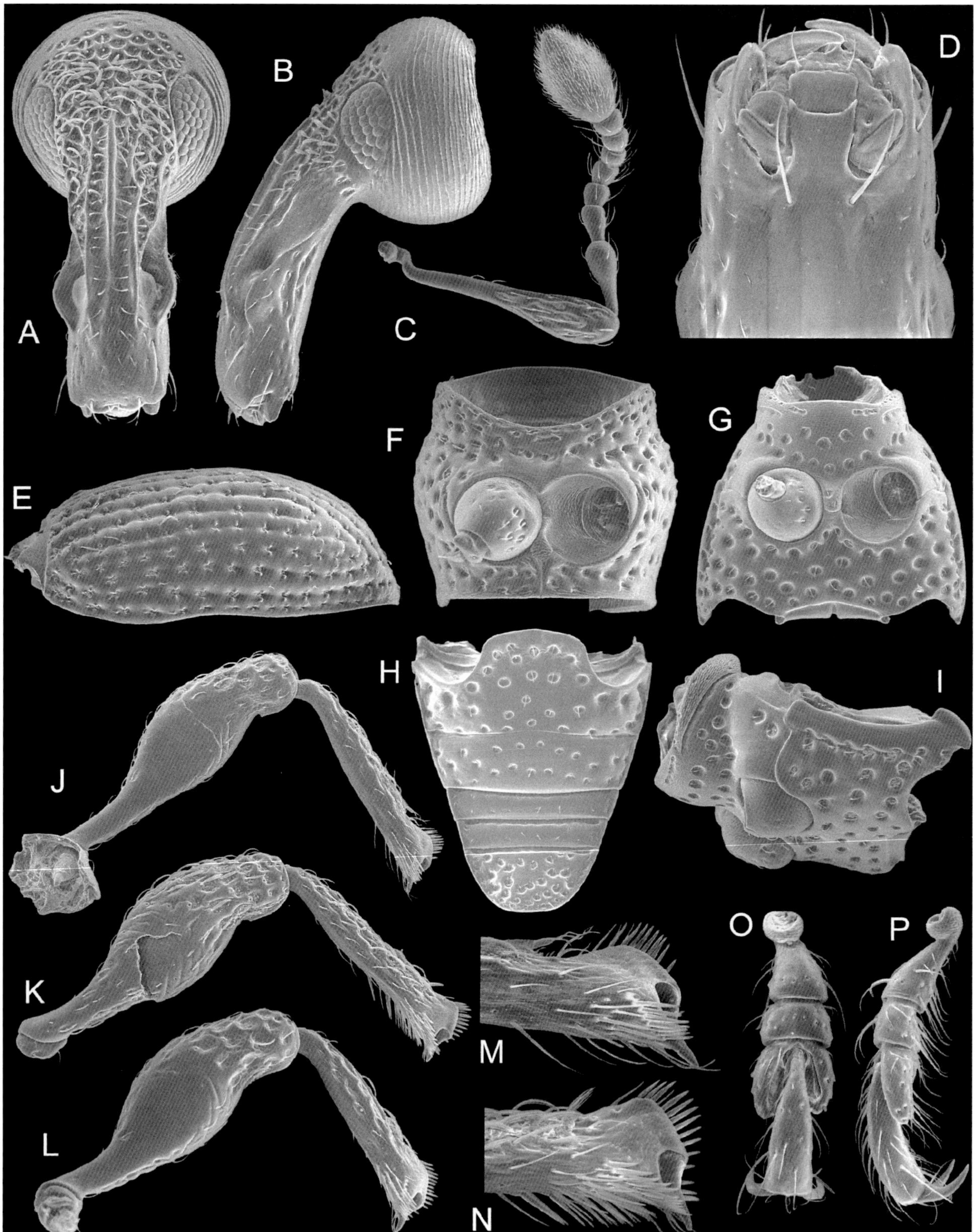

Fig. 44.34. Cyclominae, structural details, Listroderini, *Mandalotina* sp., Errinunda N.P., VIC, ♂. A–B, head: A, dorsal; B, lateral; C, right antenna, dorsal; D, mouthparts, ventral; E, left elytron, lateral; F, prothorax, ventral; G, pterothorax ventral; H, abdominal ventrites; I, pterothorax lateral; J, foreleg; K, middle leg; L, hindleg; M, protibial apex; N, mesotibial apex; O–P, tarsi: O, dorsal; P, lateral.

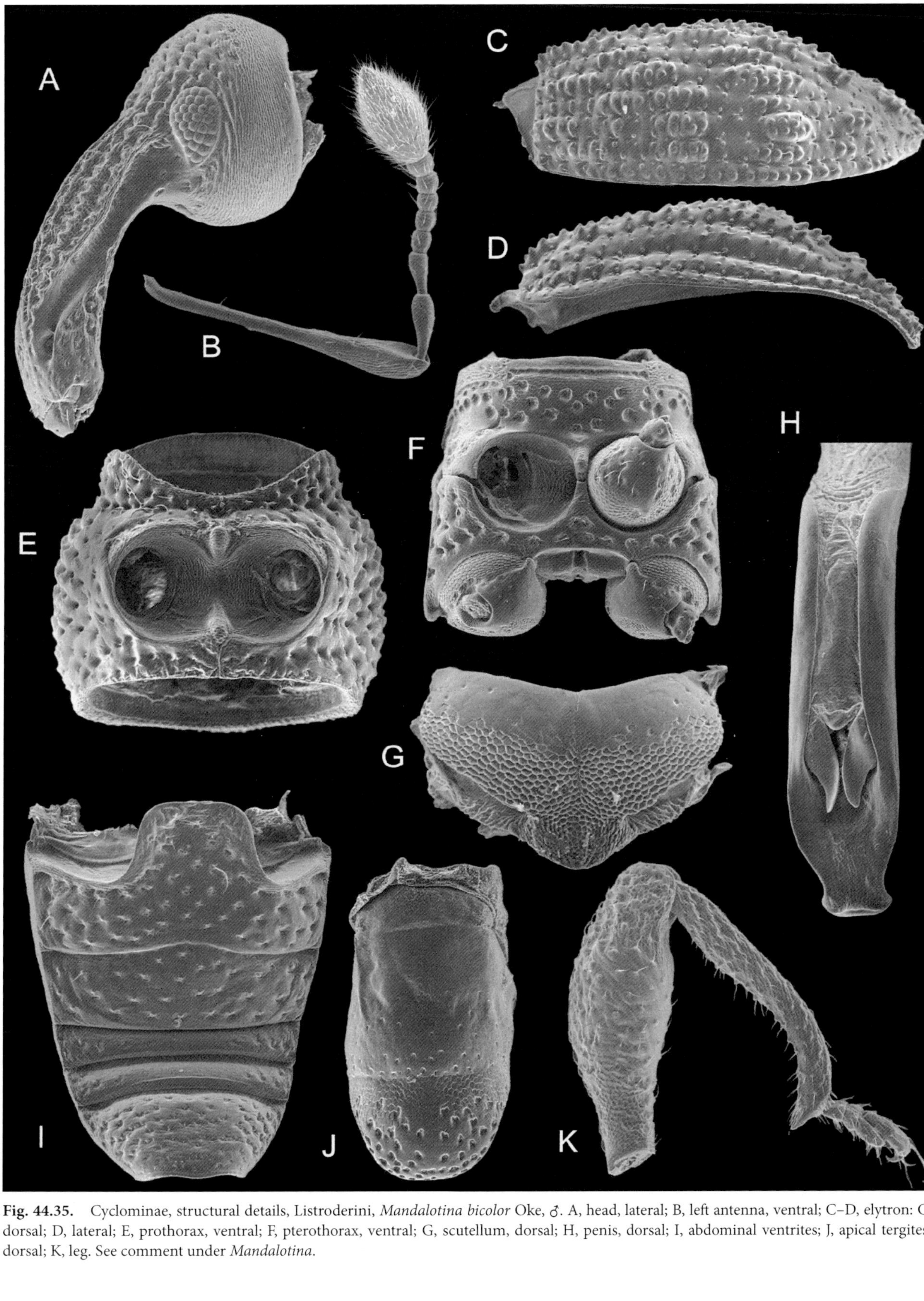

Fig. 44.35. Cyclominae, structural details, Listroderini, *Mandalotina bicolor* Oke, ♂. A, head, lateral; B, left antenna, ventral; C–D, elytron: C, dorsal; D, lateral; E, prothorax, ventral; F, pterothorax, ventral; G, scutellum, dorsal; H, penis, dorsal; I, abdominal ventrites; J, apical tergites, dorsal; K, leg. See comment under *Mandalotina*.

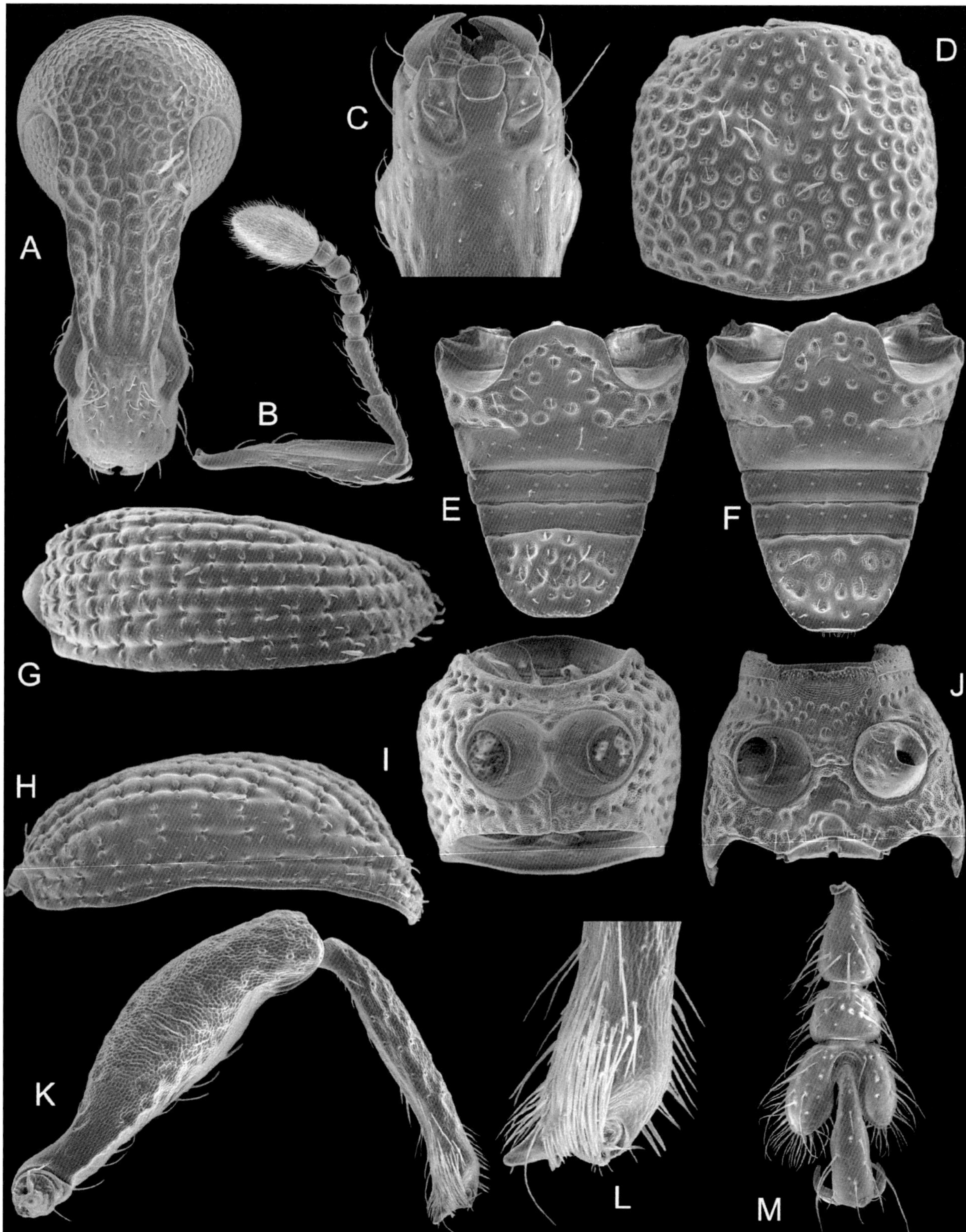

Fig. 44.36. Cyclominae, structural details, Listroderini, *Neosyagrius* sp., Barrington Tops N.P., NSW. A, head, dorsal; B, left antenna, ventral; C, mouthparts, ventral; D, prothorax, ventral; E–F, abdominal ventrites: E, ♂; F, ♀; G–H, elytron: G, dorsal; H, lateral; I, prothorax, ventral; J, pterothorax, ventral; K, leg; L, tibial apex; M, tarsus, dorsal.

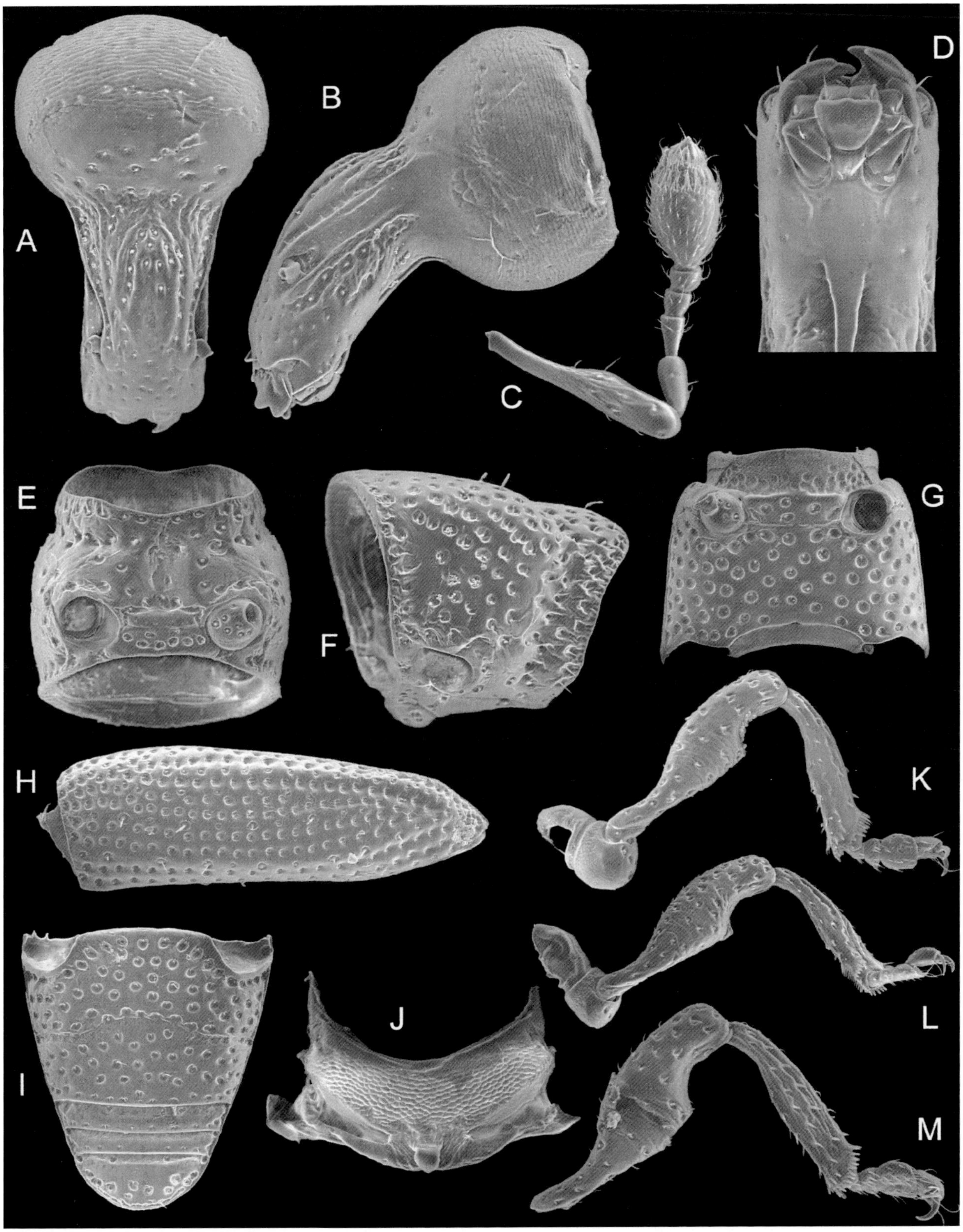

Fig. 44.37. Bagoinae, structural details, Notiomimetini, *Notiomimetes pascoei* Wollaston. A–B, head: A, dorsal; B, lateral; C, right antenna, dorsal; D, mouthparts, ventral; E–F, prothorax: E, ventral; F, lateral; G, pterothorax, ventral; H, left elytron, lateral; I, abdominal ventrites; J, scutellum, dorsal; K, middle leg; L, foreleg; M, hindleg.

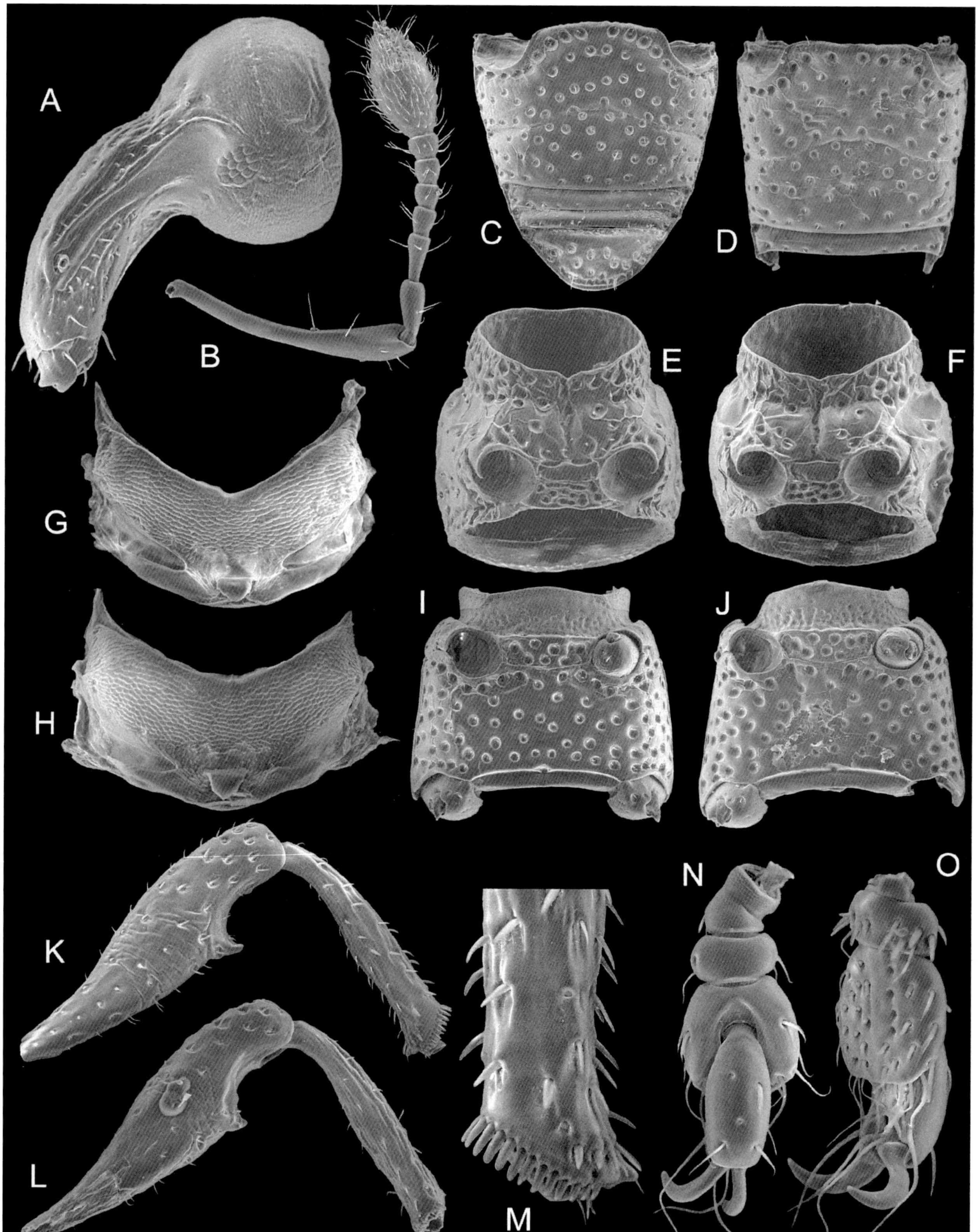

Fig. 44.38. Bagoinae, structural details, Notiomimetini, *Notiomimetes* sp., Albany, WA. A, head, lateral; B, left antenna, ventral; C–D, abdominal ventrites: C, ♂; D, ♀; E–F, prothorax, ventral: E, ♂; F, ♀; G–H, scutellum, dorsal; G, ♂; H, ♀; I–J, pterothorax, ventral: I, ♂; J, ♀; K, leg; L, leg; M, tibial apex; N–O, tarsus: N, dorsal; O, lateral.

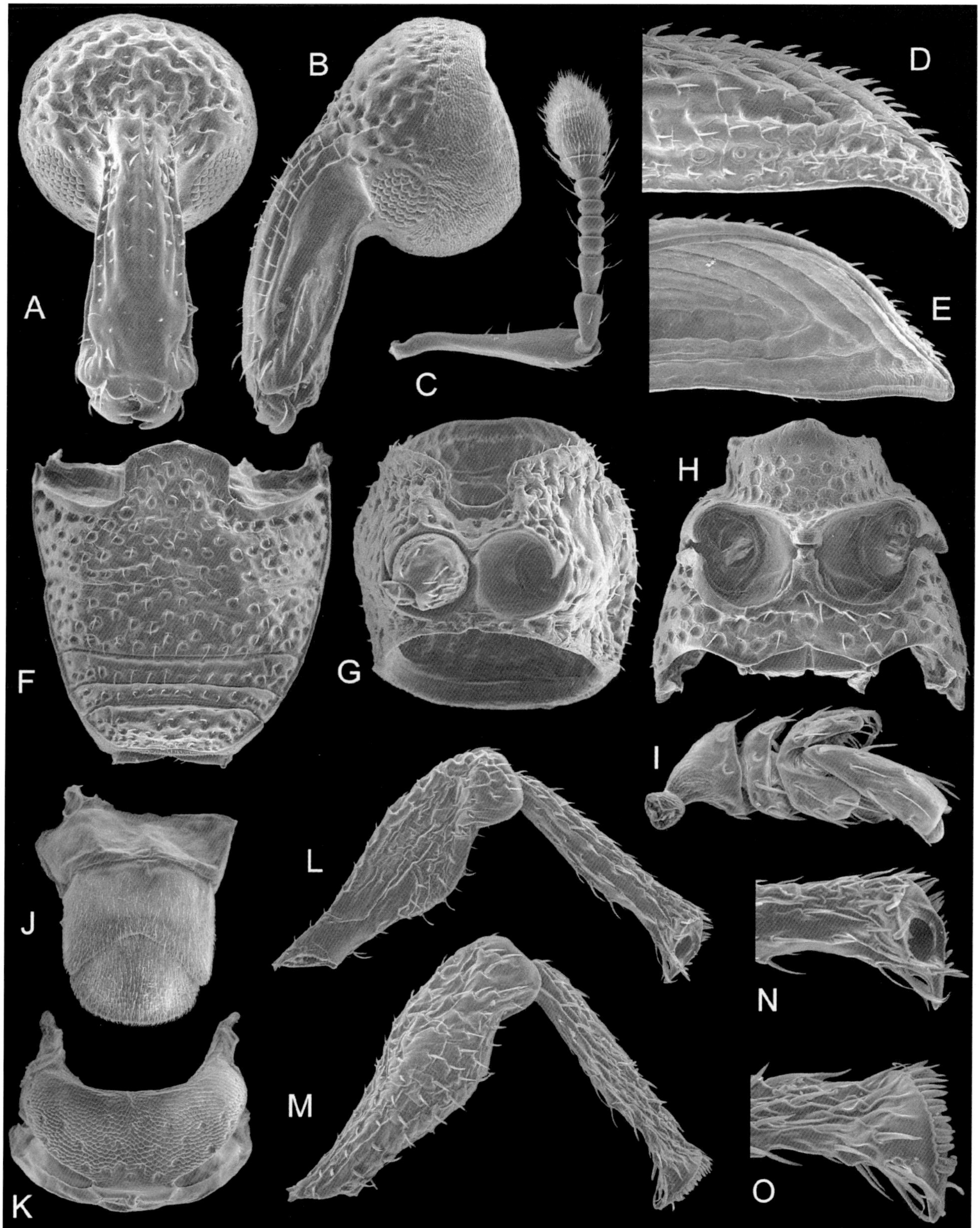

Fig. 44.39. Bagoinae, structural details, Notiomimetini, *Psaldus liosomoides* Pascoe. A–B, head: A, dorsal; B, lateral; C, left antenna, ventral; D–E, elytral apex: D, lateral; E, inner lateral; F, abdominal ventrites; G, prothorax, ventral; H, pterothorax, ventral; I, tarsus, dorsal; J, apical tergites, dorsal; K, scutellum, dorsal; L–M, legs: N, tibial apex, inner; O, tibial apex, outer.

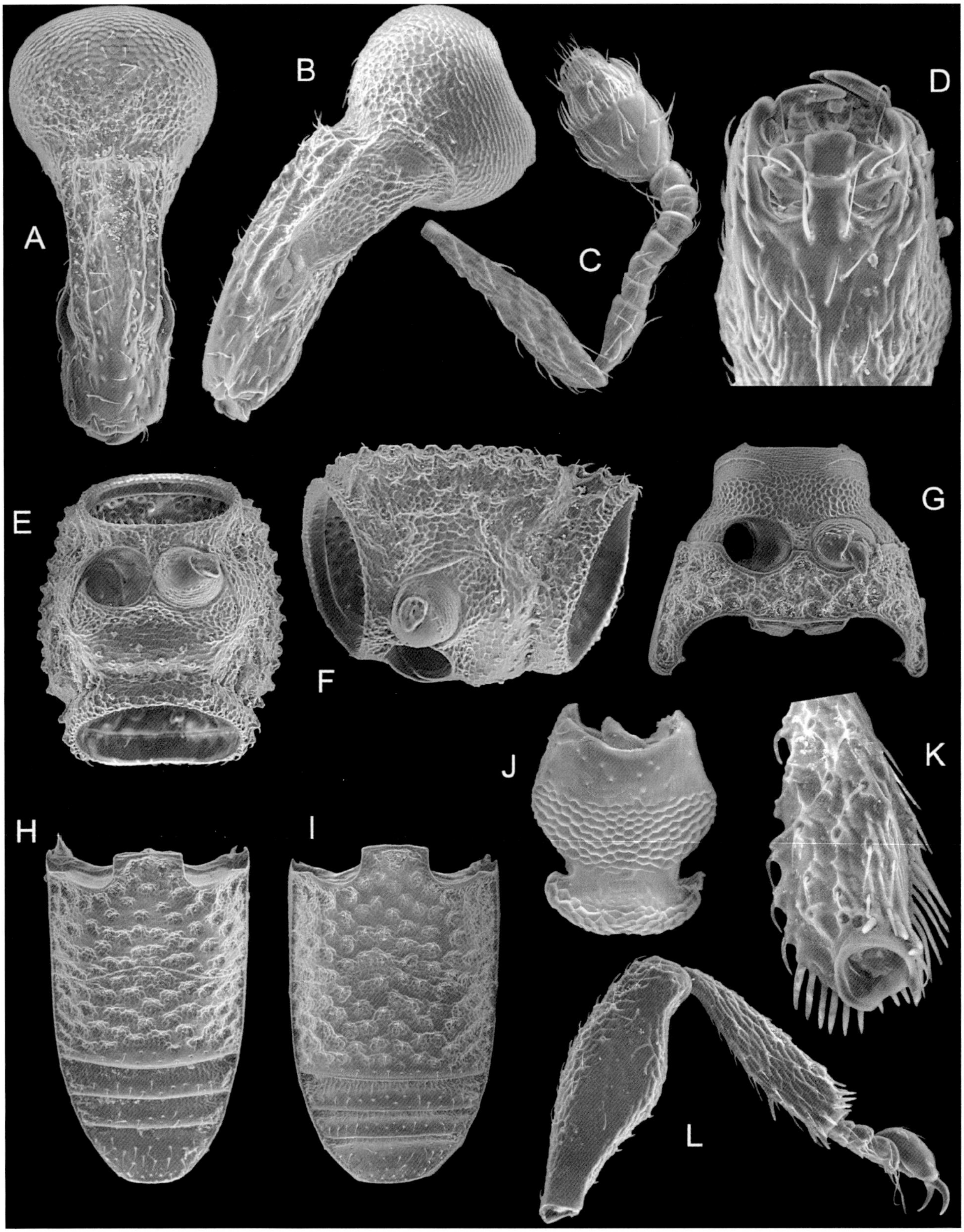

Fig. 44.40. Cyclominae, structural details, Listroderini, *Tasmanica* sp., Black Mt., ACT. A–B, head: A, dorsal; B, lateral; C, right antenna, dorsal; D, mouthparts, ventral; E–F, prothorax: E, ventral; F, lateral; G, pterothorax, ventral; H–I, abdominal ventrites: H, ♂; I, ♀; J, scutellum, dorsal; K, tibial apex; L, leg.

Aparete **Pascoe, 1871 (Pl. 57B)**

Type species, by monotypy: *Aparete palpebrosa* Pascoe, 1871.
Australian species. *A. longipes* Lea, 1914, *A. palpebrosa* Pascoe, 1871c.
Distribution. Australia.
Notes. The adults are usually found on saltbushes (*Atriplex, Maireana*, e.g. Pl. 8G) and similar Amaranthaceae, and it is likely that the larvae feed on the leaves of these plants at night in a similar fashion as those of *Ethemaia* do.

Ethemaia **Pascoe, 1865 (Pl. 57C; Fig. 44.29H–O)**

Type species, by subsequent designation (Lea, 1914: 325): *Ethemaia sellata* Pascoe, 1865.
Australian species. *E. adusta* Pascoe, 1865, *E. alternata* Lea, 1914, *E. angusticollis* Pascoe, 1883a, *E. apicalis* Lea, 1904, *E. curtula* Pascoe, 1883a, *E. emarginata* Lea, 1904, *E. ferruginea* Lea, 1915b, *E. funerea* Lea, 1904, *E. griffithi* Lea, 1909c, *E. mirabilis* Lea, 1914, *E. sellata* Pascoe, 1865, *E. vagans* Lea, 1904.
Distribution. Australia.
Notes. The genus is poorly delineated and requires taxonomic revision. The greenish larva of *E. sellata* climbs onto small plants of Malvaceae (*Alyogyne*, *Malva*) and Boraginaceae (*Echium*, *Heliotropium*) at night using its abdominal prolegs (pedal lobes expanded for adhesion) and feeds on the leaves (Froggatt 1900; Swan 1935; May 1994; Oberprieler 2010, 2014c). The larvae of the other species probably lead a similar ectophytic nocturnal life.

Medicasta **Pascoe, 1870 (Pl. 57D)**

Type species, by monotypy: *Medicasta leucura* Pascoe, 1870.
Australian species. *M. assimilis* (Pascoe, 1883a), *M. beltanensis* (Blackburn, 1893a), *M. ferrugata* (Pascoe, 1883a), *M. geometrica* (Lea, 1911c), *M. leptopsoides* Lea, 1904, *M. leucura* Pascoe, 1870b, *M. parallela* (Blackburn, 1893a), *M. personata* (Lea, 1904), *M. sordida* (Pascoe, 1870c, *M. variabilis* (Blackburn, 1893a).
Distribution. Australia.
Notes. Like *Ethemaia*, the concept of this genus is unclear and in need of taxonomic revision (Oberprieler 2010). As for that genus, adults of *M. geometrica* have been encountered on Amaranthaceae (*Salsola*) (Oberprieler 2010), and the larvae of all species may also be ectophytic nocturnal leaf feeders.

Ophryota **Pascoe, 1872 (Pl. 57E; Fig. 44.29P–U)**

Type species, by monotypy: *Ophryota squamibunda* Pascoe, 1872.
Australian species. *O. hystricosa* (Lea, 1914), *O. nodosa* (Blackburn, 1893a), *O. rapax* Blackburn, 1896, *O. squamibunda* Pascoe, 1872c, *O. subangulata* Lea, 1915b.
Distribution. Australia.
Notes. Adults have been found on *Lepidium* (Brassicaceae), and the larvae are also adapted for climbing onto plants (May 1994; Oberprieler 2010), suggesting an ectophytic life history like those of *Aparete*, *Ethemaia* and *Medicasta*.

Phrenozemia **Pascoe, 1872 (Pl. 57F)**

Type species, by monotypy: *Phrenozemia lyproides* Pascoe, 1872.
Australian species. *P. lunata* Pascoe, 1873a, *P. lyproides* Pascoe, 1872b.
Distribution. Australia (WA).
Notes. Adults have been collected on *Goodenia stobbsiana* (Goodeniaceae) (Oberprieler 2010, 2014c) but the larvae are unknown; they may also be ecotyphytic nocturnal leaf feeders.

Pocius **Zimmerman & Oberprieler, 2014 (Pl. 57G)**

Type species, by original designation: *Methypora parallela* Lea, 1895.
Australian species. *P. parallelus* (Lea, 1895a).
Distribution. Australia.
Notes. The genus was described in Pullen *et al.* (2014). The single species is poorly represented in collections, and nothing is known about its larval hosts and life history.

Curculionidae: Entiminae Schoenherr, 1823

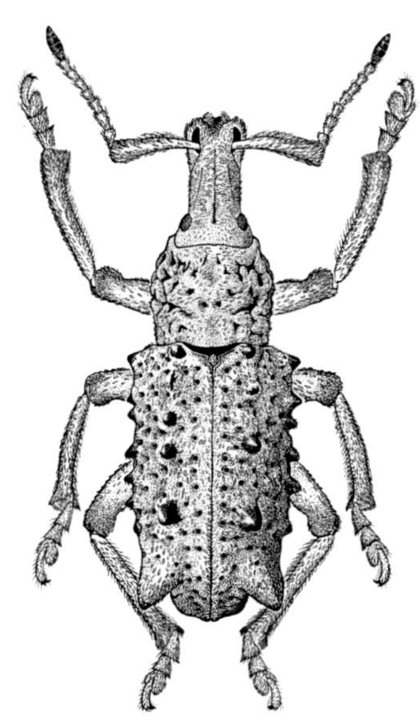

Fig. 44.41. *Leptopius setosus* (Lea).

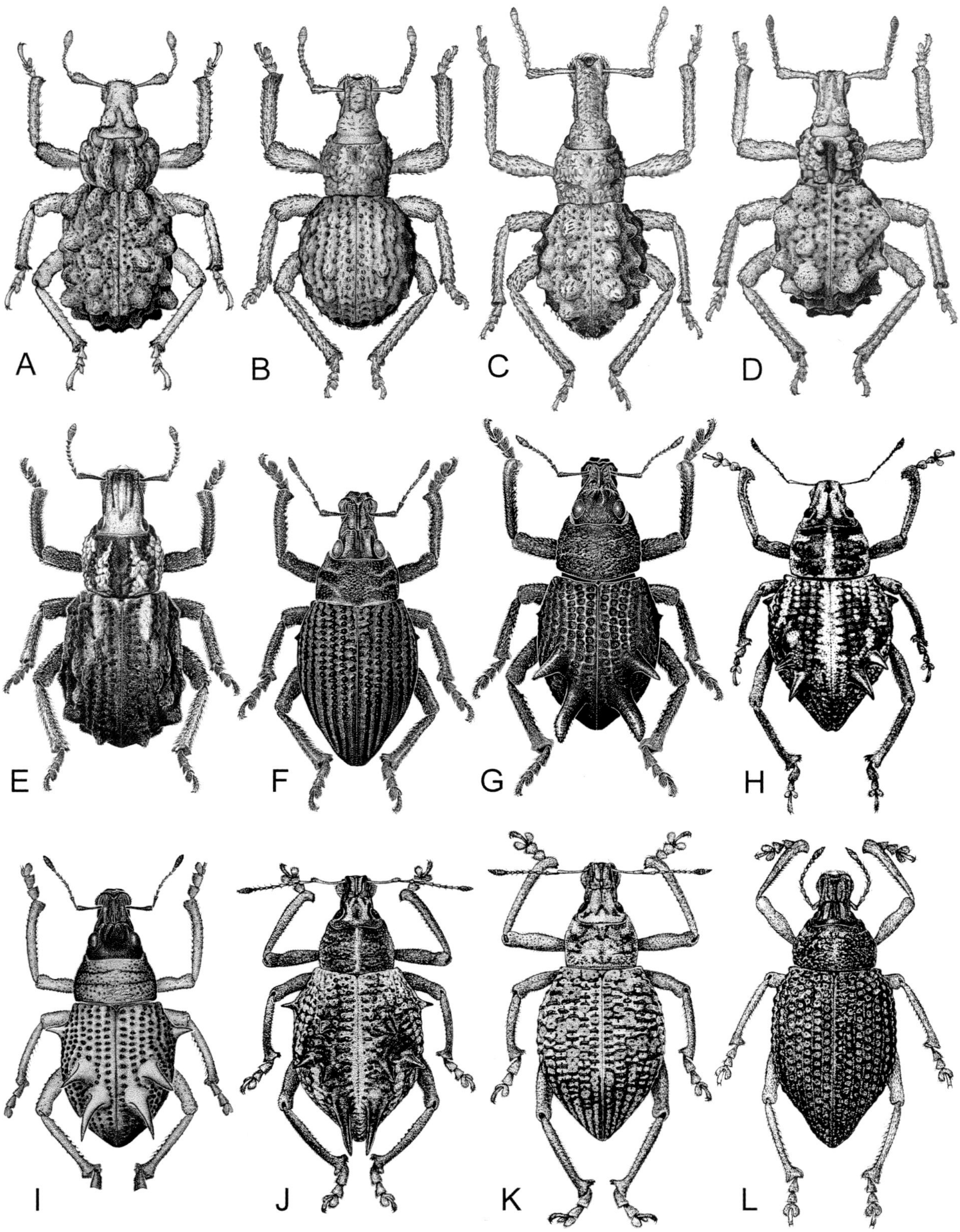

Fig. 44.42. Entiminae dorsal habitus, Leptopiini. A, *Amisallus basipennis* Lea; B, *Amisallus crassirostris* (Lea); C, *Amisallus phymatodis* (Lea); D, *Amisallus tuberculifrons Lea*; E, *Amnemus superciliaris* (Pascoe); F, *Catasarcus aerosus* Thompson; G, *Catasarcus albipectus* Thompson; H, *Catasarcus albisparsus* Pascoe; I, *Catasarcus albuminosus* Pascoe; J, *Catasarcus armatus* Blackburn; K, *Catasarcus aspergetus* Thompson; L, *Catasarcus asphaltinus* Thompson. A–G, I, by S. Monteith; H, J–L by C. A. O'Brien (*in* Thompson 1968).

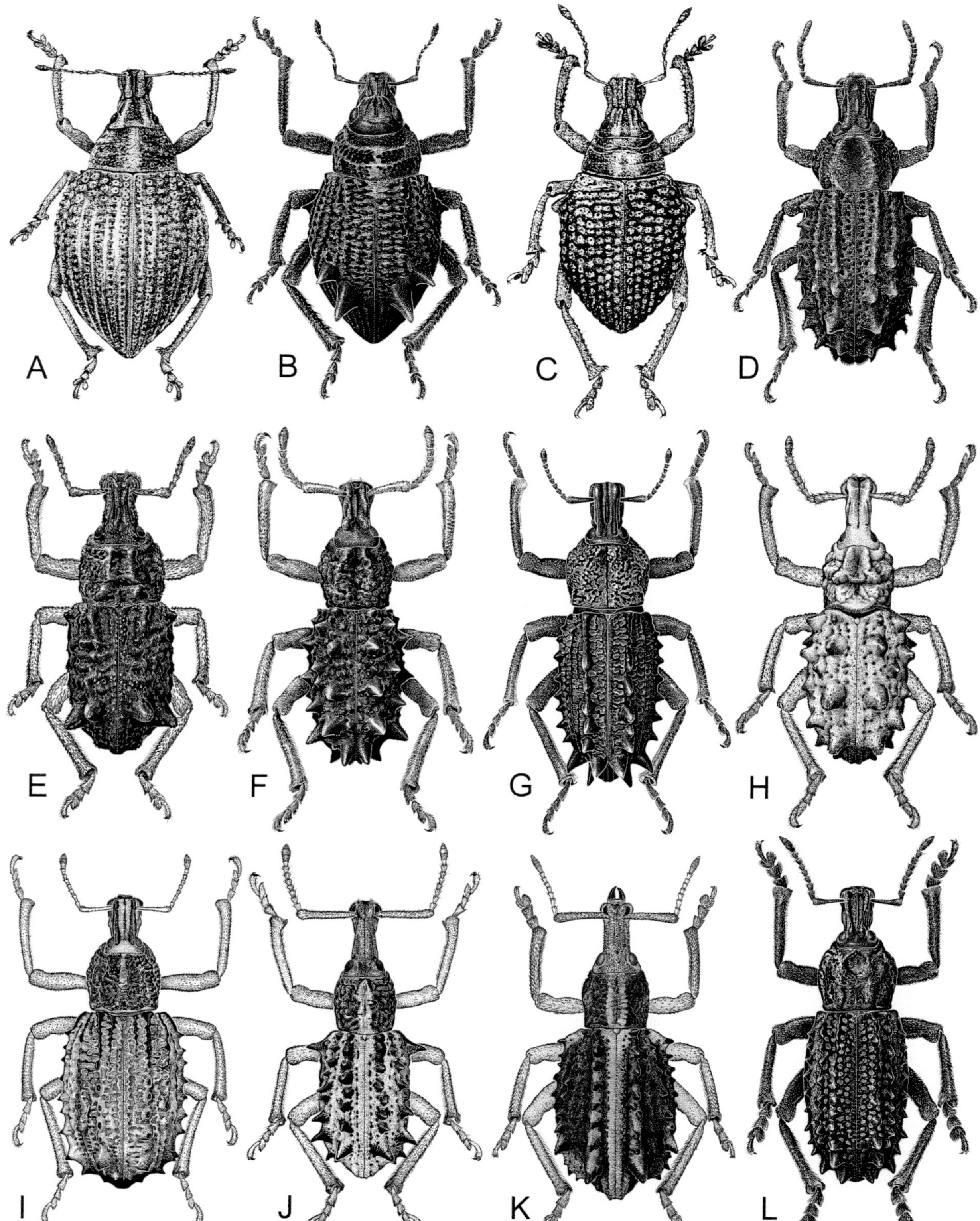

Fig. 44.43. Entiminae dorsal habitus, Leptopiini. A, *Catasarcus sericeus* Blackburn; B, *Catasarcus spinipennis* Fåhraeus; C, *Catasarcus transversalis* Germar; D, *Leptopius acerbus* (Pascoe); E, *Leptopius argillaceus* (Pascoe); F, *Leptopius baileyi* (Blackburn); G, *Leptopius biordinatus* (Blackburn); H, *Leptopius brachystylus* (Lea); I, *Leptopius cacozelus* (Lea); J, *Leptopius castelnaui* (Lea); K, *Leptopius clavus* (Fabricius); L, *Leptopius colossus* (Pascoe). A, C by C. A. O'Brien (*in* Thompson 1968).

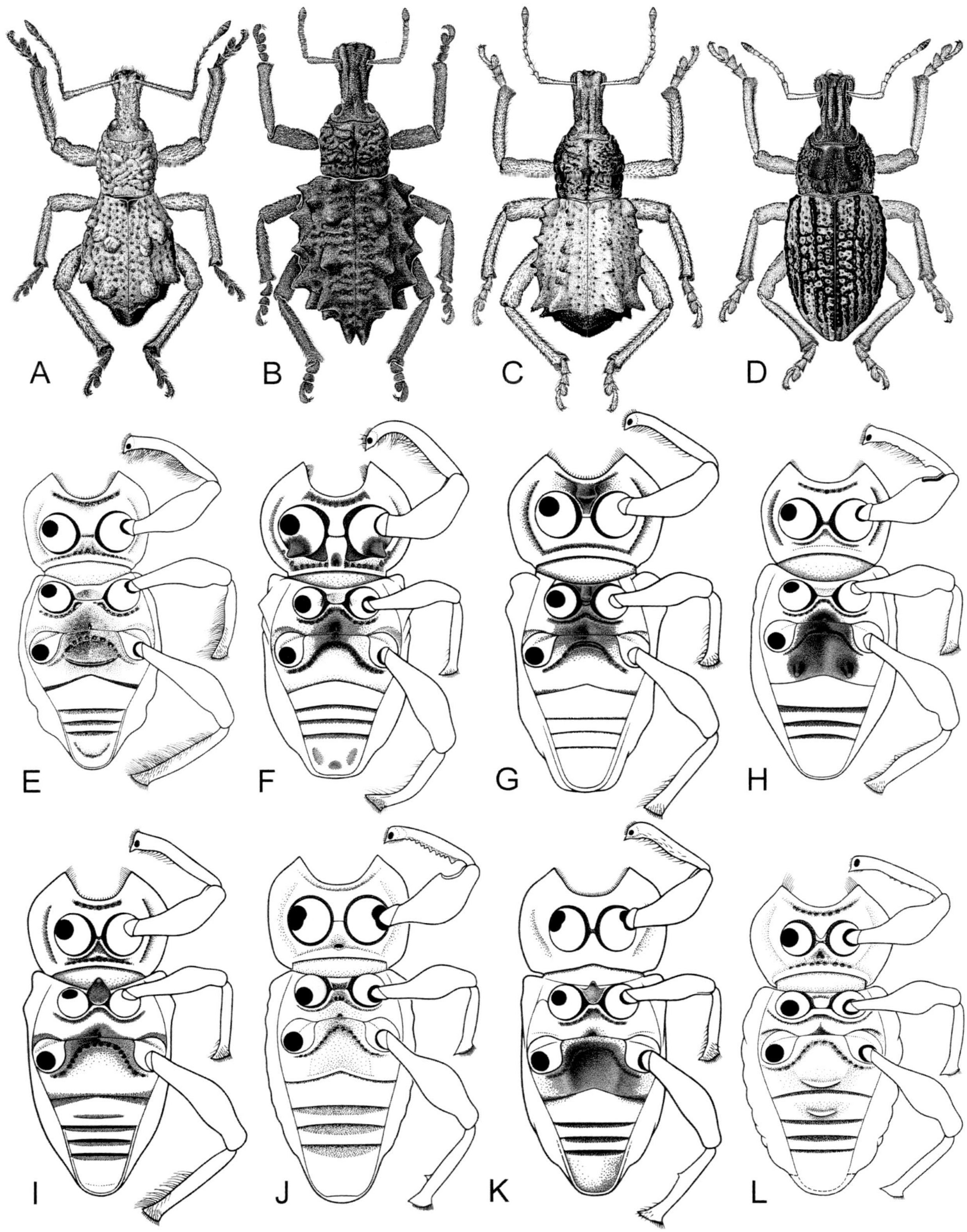

Fig. 44.44. Entiminae, dorsal and ventral habitus, Leptopiini. A, *Leptopius tetraphysodes* (Pascoe); B, *Leptopius tribulus* (Fabricius); C, *L. tuberculatus* (Macleay); D, *L. vacillans* (Lea); E, *Mandalotus arciferus* Lea; F, *M. armicoxis* Lea; G, *M. armipectus* Lea; H, *M. armivarius* Lea; I, *M. auchmeresthes* Lea; J, *M. bryophilus* Oke; K, *M. crudus* Erichson; L, *M. explanicollis* Oke.

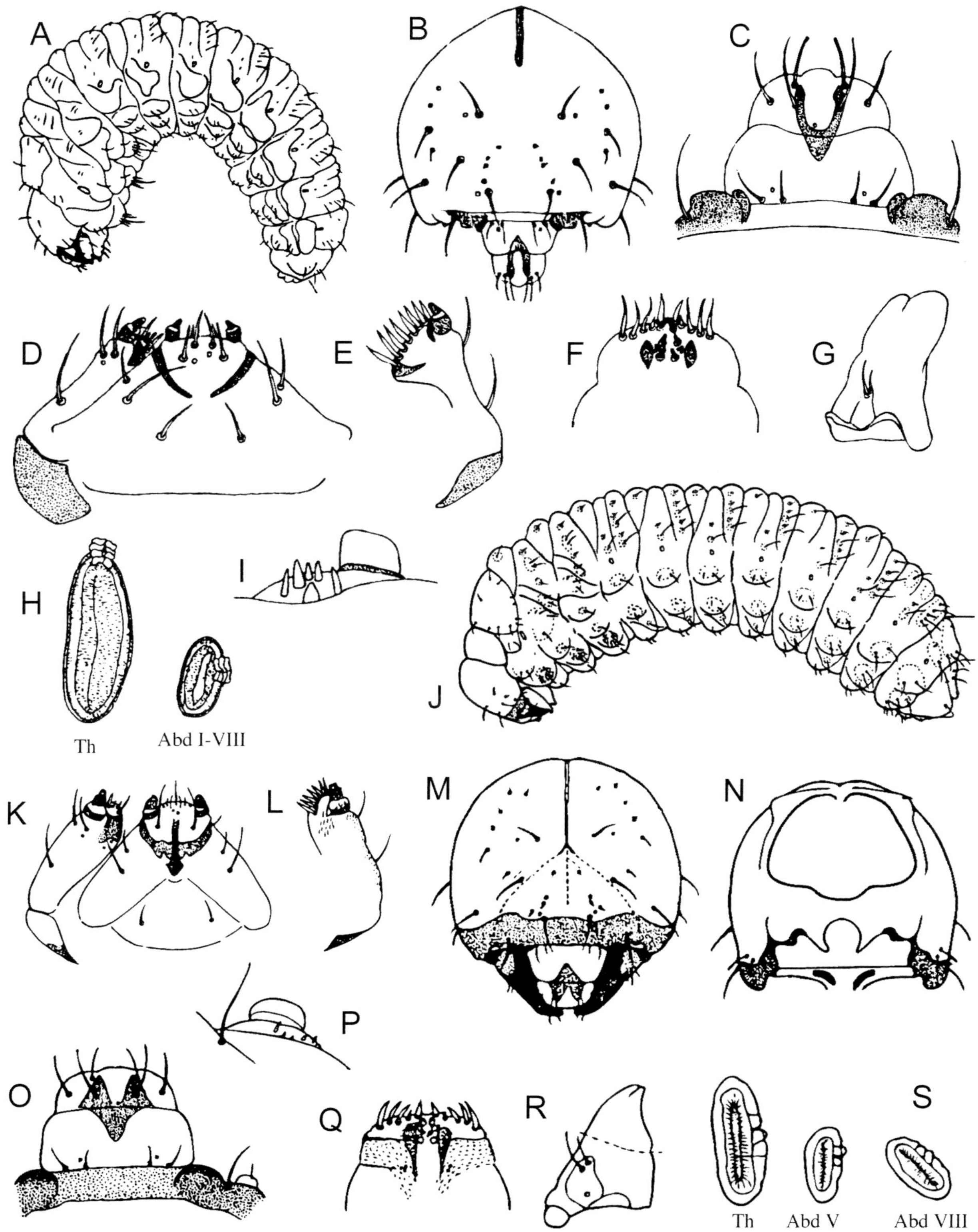

Fig. 44.45. Entiminae larvae, Leptopiini. A–I, *Leptopius robustus* (Olivier): A, habitus, lateral; B, head, anterior; C, labrum, clypeus and epistoma, anterior; D, mala and labium, posterior; E, mala, anterior; F, epipharyngeal lining; G, mandible; H, antenna; I, spiracles, showing alignment; J–S, *Perperus sp.*: J, habitus, lateral; K, mala and labium, posterior; L, mala, anterior; M, head, anterior; N, head, posterior; O, labrum, clypeus, epistoma and antenna, anterior; P, antenna; Q, epipharyngeal lining; R, mandible; S, spiracles, showing alignment. After May (1994).

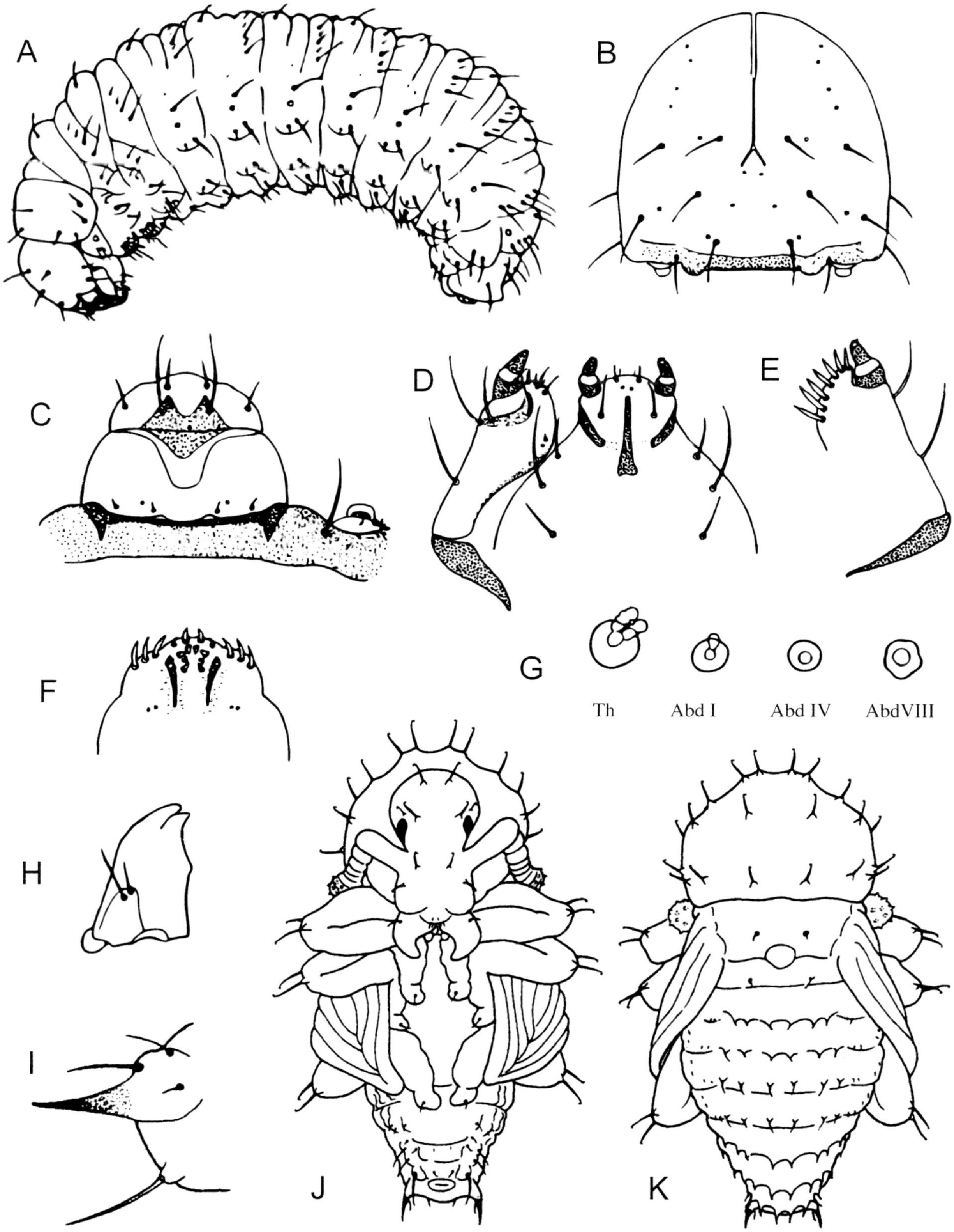

Fig. 44.46. Entiminae larva and pupa, Leptopiini, *Mandalotus miricollis* (Broun) (New Zealand). A–H, larva: A, habitus, lateral; B, head, anterior; C, labrum, clypeus, epistoma and antenna, anterior; D, mala and labium, posterior; E, mala, anterior; F, epipharyngeal lining; G, spiracles, showing alignment; H, mandible; I–K, pupa: I, pseudocercus; J, habitus, ventral; K, dorsal. After May (1994).

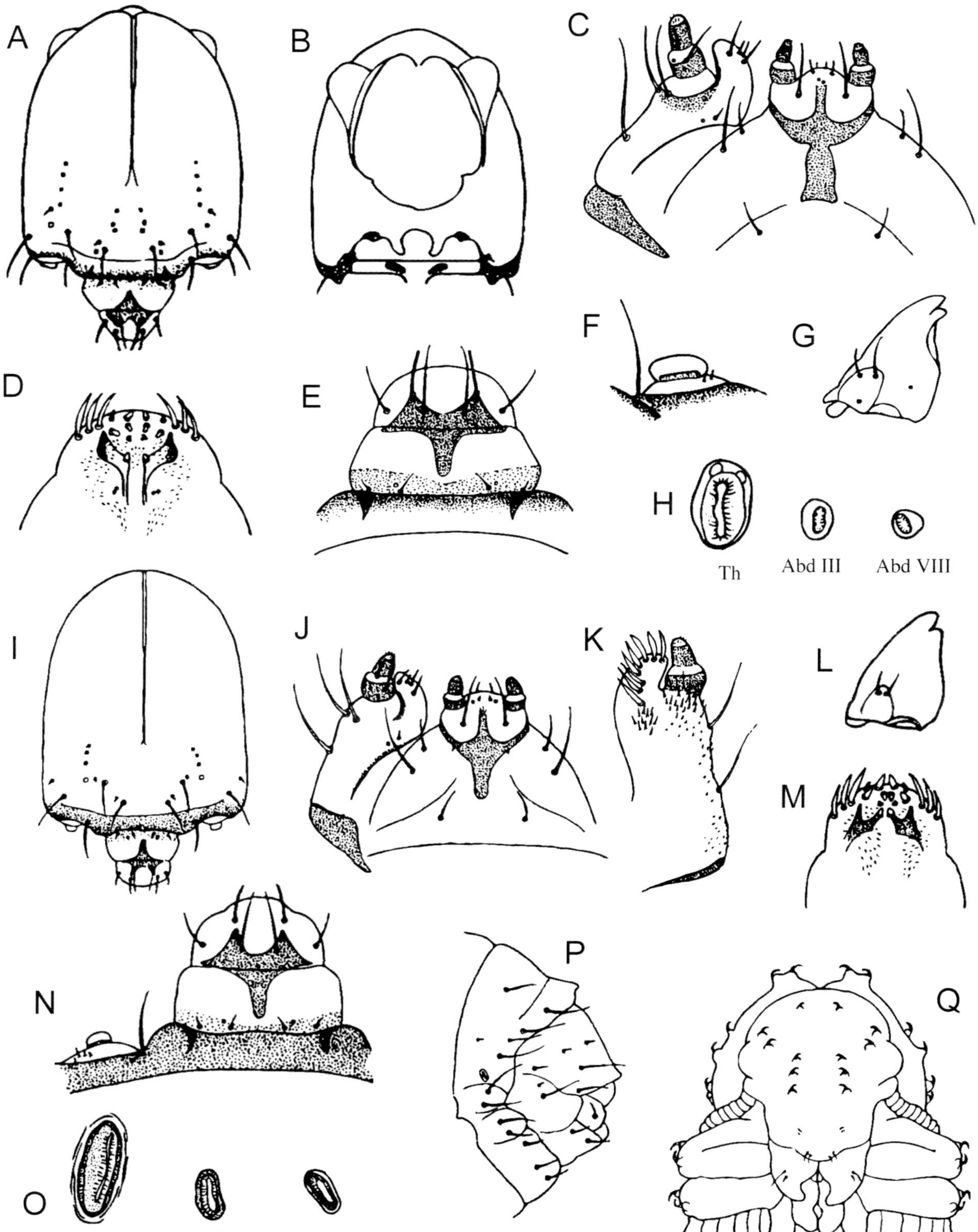

Fig. 44.47. Entiminae larvae and pupa, Naupactini. A–H, *Naupactus cervinus* Boheman, larva (New Zealand): A, head, anterior; B, head, posterior; C, mala and labium, posterior; D, epipharyngeal lining; E, labrum, clypeus and epistoma, anterior; F, antenna; G, mandible; H, spiracles, showing alignment; I–O, *Atrichonotus taeniatulus* (Berg), larva (New Zealand, USA): I, head, anterior; J, mala and labium, posterior; K, mala, anterior; L, mandible; M, epipharyngeal lining; N, labrum, clypeus, epistoma and antenna, anterior; O, spiracles, showing alignment; P–Q, *Atrichonotus taeniatulus* (Berg), pupa: P, terminal segments, lateral; Q, habitus, ventral, anterior part. After May (1994).

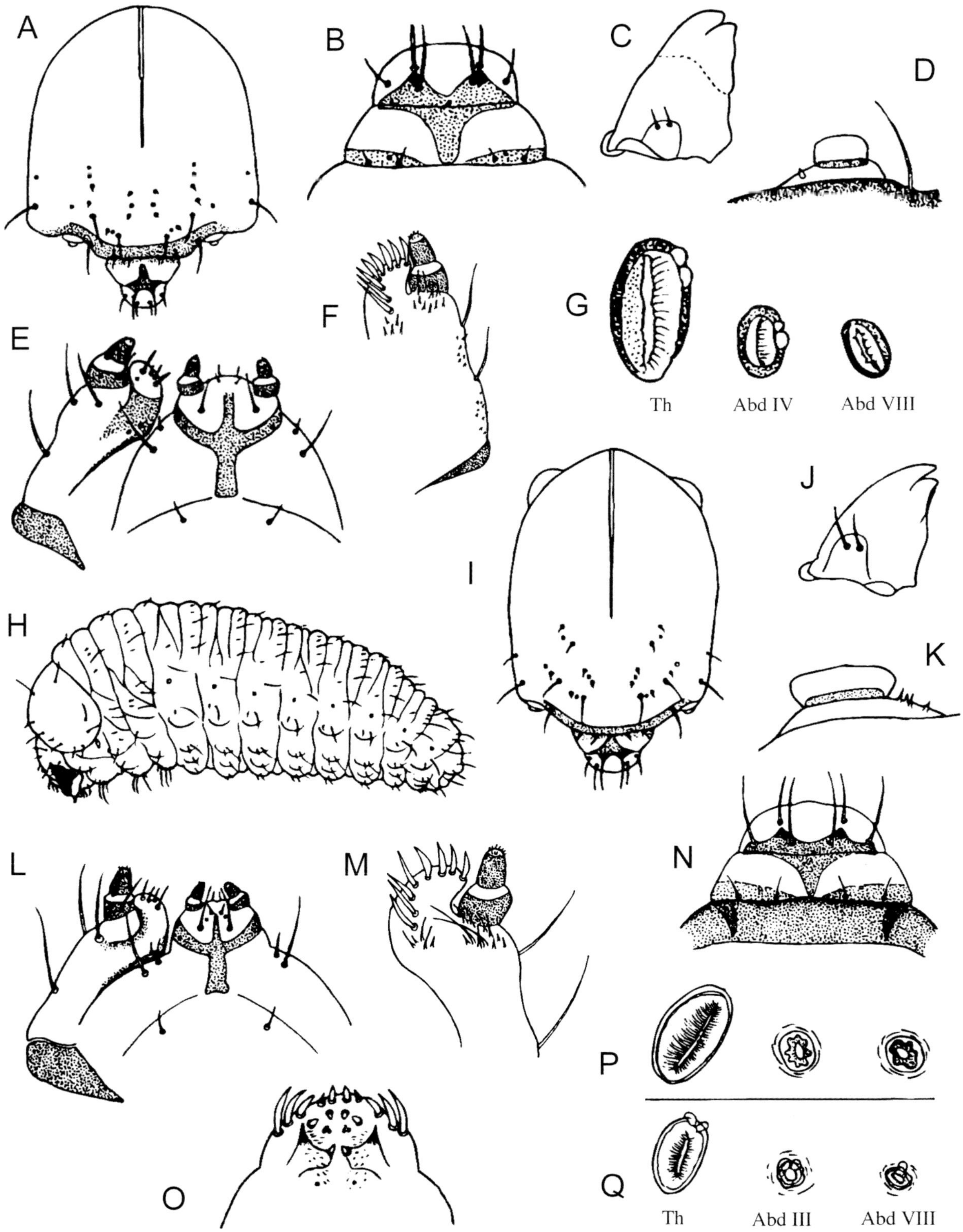

Fig. 44.48. Entiminae larvae, Naupactini. A–G, *Floresianus sordidus* Hustache (New Zealand): A, head, anterior view; B, labrum, clypeus and epistoma, anterior; C, mandible; D, antenna; E, mala and labium, posterior; F, mala, anterior; G, spiracles, showing alignment; H–Q, *Naupactus leucoloma* Boheman, (New Zealand): H, habitus, lateral; I, head, lateral; J, mandible; K, antenna; L, mala and labium, posterior; M, mala, anterior; N, labrum, clypeus and epistoma, anterior; O, epipharyngeal lining; P, spiracles, final instar, showing alignment; Q, spiracles, first instar, showing alignment. After May (1994).

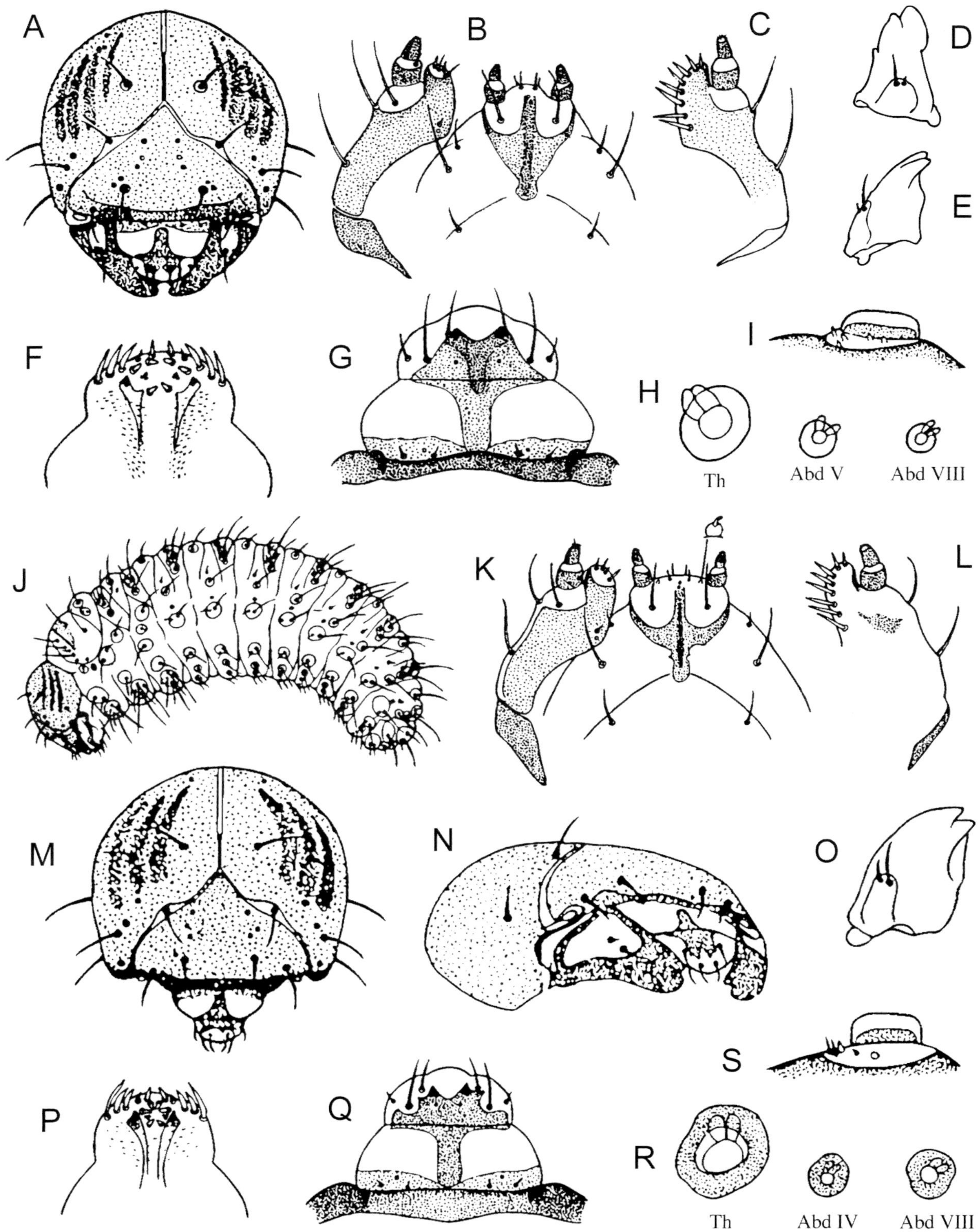

Fig. 44.49. Entiminae larvae, Otiorhynchini. A–I, *Otiorhynchus rugosostriatus* (Goeze) (New Zealand): A, head, anterior; B, mala and labium, posterior; C, mala, anterior; D, mandible, anterior; E, mandible, lateral; F, epipharyngeal lining; G, labrum, clypeus and epistoma, anterior; H, spiracles, showing alignment; I, antenna; J–S, *Otiorhynchus sulcatus* (Fabricius) (New Zealand): J, habitus, lateral; K, mala and labium, posterior; L, mala, anterior; M, head, anterior; N, head, lateral; O, mandible; P, epipharyngeal lining; Q, labrum, clypeus and epistoma, anterior; R, spiracles, showing alignment; S, antenna. After May (1994).

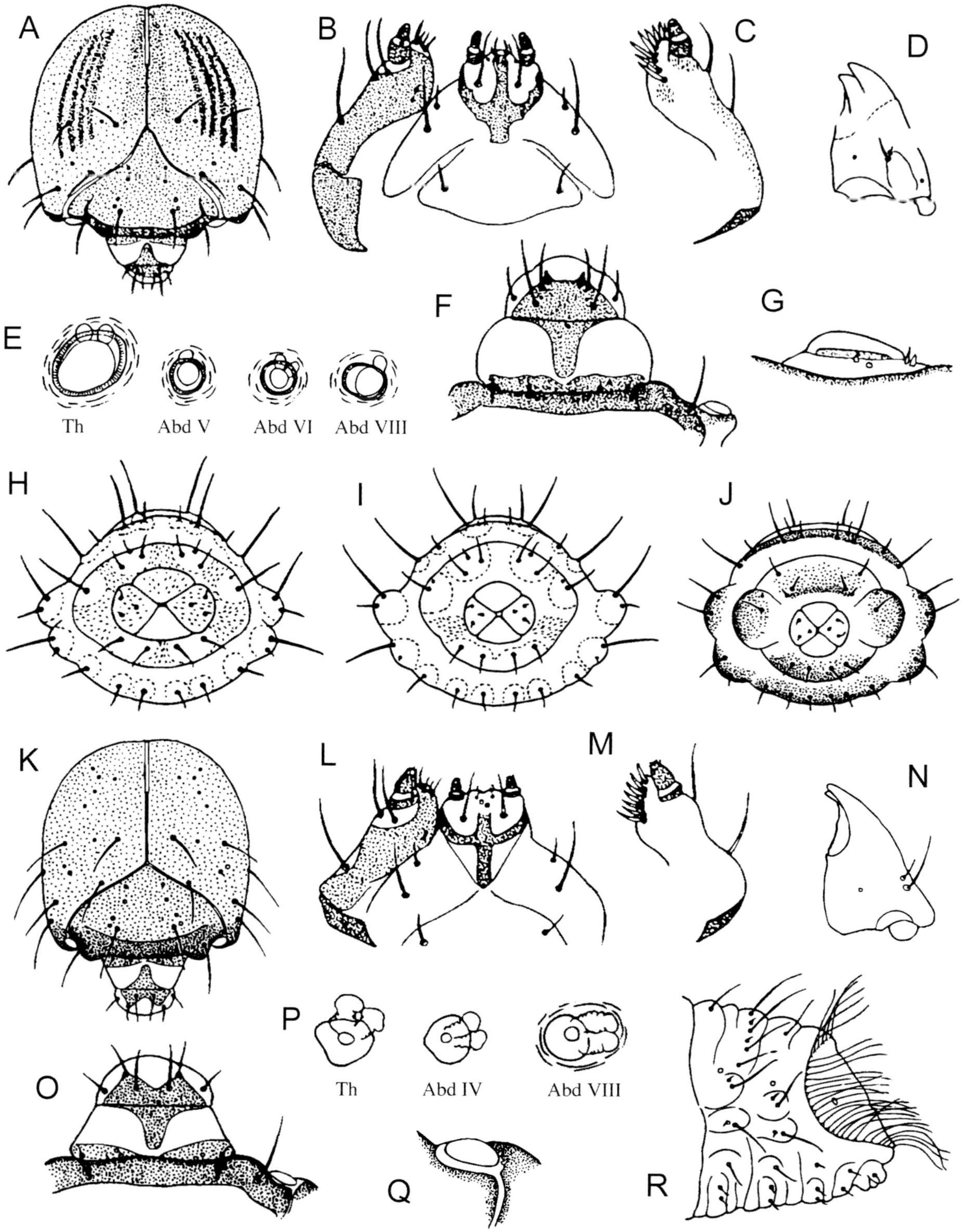

Fig. 44.50. Entiminae larvae, Oosomini, Otiorhynchini and Ottistirini. A–G, *Phlyctinus callosus* (Schoenherr), (New Zealand): A, head, anterior; B, mala and labium, posterior; C, mala, anterior; D, mandible; E, spiracles, showing alignment; F, labrum, clypeus and epistoma, anterior; G, antenna; H–J, terminal segments, caudal: H, *Otiorhynchus rugosostriatus* (Goeze) (New Zealand); I, *Otiorhynchus cribricollis* Gyllenhal; J, *Phlyctinus callosus* (Schoenherr); K–R, *Maleuterpes spinipes* Blackburn (New Zealand): K, head, anterior; L, mala and labium, posterior; M, mala, anterior; N, mandible; O, labrum, clypeus, epistoma and antenna, anterior; P, spiracles, showing alignment; Q, antenna; R, terminal segments, lateral. After May (1994).

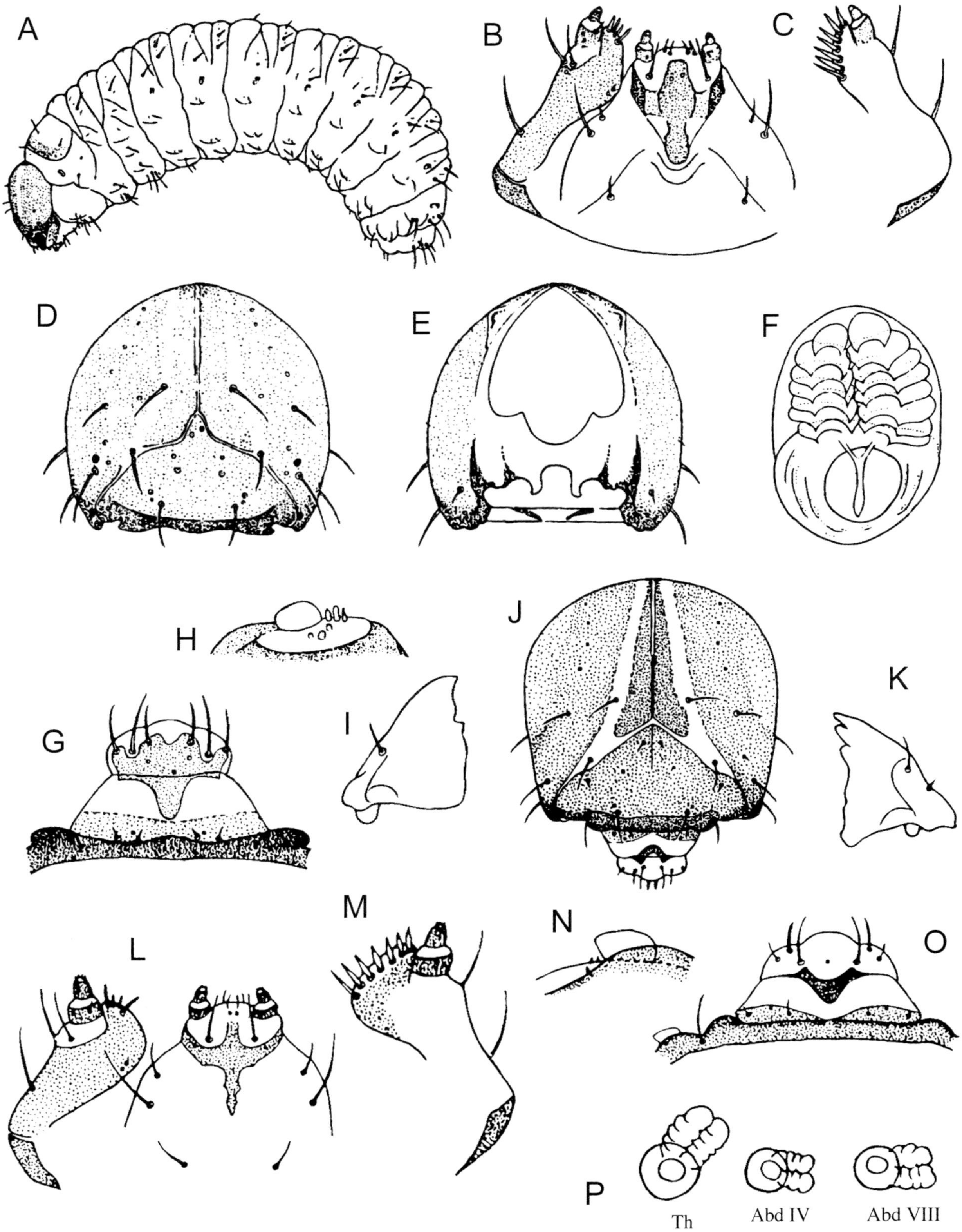

Fig. 44.51. Entiminae larvae, Pachyrhynchini and Sitonini. A–I, *Pantorhytes stanleyanus* (White) (Papua New Guinea): A, habitus, lateral; B, mala and labium, posterior; C, mala, anterior; D, head, anterior; E, head, posterior; F, spiracle; G, labrum, clypeus and epistoma, anterior; H, antenna; I, mandible; J–P, *Sitona discoideus* Gyllenhal (New Zealand): J, head, anterior; K, mandible; L, mala and labium, posterior; M, mala, anterior; N, antenna; O, labrum, clypeus, epistoma and antenna, anterior; P, spiracles, showing alignment. After May (1994).

Introduction. The Entiminae are a relatively well-defined subfamily of Curculionidae, also referred to as Adelognatha in previous classification systems. Members of this group are characterised by their mandibles having a deciduous process or cusp (Fig. 44.54F–I), which assists the adult weevil eclosing from its pupal case but is usually shed afterwards, leaving a distinct flat, circular to lenticular scar on the anterior side of the mandible (Figs 44.54A–C, H, 44.55A–C, J L), and by the concealment of all or most of the maxillae and labial palps by the enlarged prementum (Fig. 44.54A, F, H–I). The subfamily occurs world-wide and is very diverse, including in Australia, from where more than hundred genera and several hundreds of species have been described and numerous further, undescribed genera and species have accumulated in collections (Pullen *et al.* 2014). The classification of the Entiminae into tribes is still unsettled and in large parts incongruent, suffering from a lack of comprehensive phylogenetic analyses of the subfamily to be able to identify natural groups and relationships. Twelve tribes are currently recognised as occurring in Australia, namely Celeuthetini, Cyphicerini, Ectemnorhinini, Embrithini, Leptopiini, Naupactini, Oosomini, Otiorhynchini, Ottistirini, Pachyrhynchini, Sitonini and Tanymecini, of which the Embrithini, Naupactini, Oosomini, Otiorhynchini and Sitonini are introduced and have no native representatives in the country (Oberprieler & Zimmerman 2020).

In the style of his monograph series *Australian Weevils*, E. C. Zimmerman drafted a partial, unpublished manuscript on the group, including many superb drawings (e.g. Figs 44.42A–I, 44.43B, D–L, 44.44A–L), from which Oberprieler & Zimmerman (2020) published a foundational revision of the 11 smaller tribes (all but Leptopiini, e.g. Pls 11, 12) which provides a comprehensive source of information on the morphology, taxonomy, biology and classification of the Australian genera and species. May (1994) described and illustrated the larvae of several species, some of her illustrations reproduced here (Figs 44.45–44.51).

Several species of Entiminae are of agricultural and horticultural significance in Australia, occasionally reaching minor to major pest status, especially the introduced species of the genera *Naupactus*, *Otiorhynchus*, *Phlyctinus* and *Sitona*. Adult Entiminae are generally non-specific leaf feeders, whereas their larvae mostly feed ectophytically on roots in the soil.

Key to the Australian genera of Entiminae

Modified after Oberprieler & Zimmerman (2020), except for the Leptopiini. The genera *Chaodius* and *Telenica* (Pls 65F and 69L, respectively) are not included due to the unavailability of specimens. The genus *Rhyncholobus* (Fig. 44.60X), restricted to Christmas Is., also not included, is distinctive in its multisetose mandibles and large body size (12.4 mm). The taxonomic boundary between *Psapharus* and *Polyphrades* is not clear, and the differences between them given in the key pertain only to their type species. The concepts of many other genera of Leptopiini also require revision, and the numerous undescribed genera already known may key out incorrectly to described genera.

1. Prementum small, leaving maxillae and labial palps exposed (mouthparts phanerognathous) (Fig. 44.54E); mandibles without deciduous cusp; prothorax without ocular lobes; metatibiae without corbel (Fig. 44.58A) 2

– Prementum enlarged, covering maxillae and labial palps (mouthparts adelognathous) (Fig. 44.54F); mandibles usually with deciduous cusp or its scar (Fig. 44.55A); prothorax with or without ocular lobes; metatibiae with or without corbel 3

2(1). Vestiture consisting of sparse setae only; mandibles paucisetose, with 3–4 long and a few small setae (Fig. 44.54E); scrobes absent; tarsal claws with long, cylindrical ventrobasal seta; restricted to subantarctic Heard and McDonald Islands .... **ECTEMNORHININI**

– Vestiture consisting of setae and dense squamae (scales); mandibles plurisetose and squamose; scrobes narrow, sharply defined, curving down in front of eyes; tarsal claws with long, flattened seta on outside (Fig. 44.59E); not occurring on subantarctic islands .... **SITONINI ... *Sitona* Germar**

3(1). Antennal scrobes T-shaped or sharply angled posteriorly, with vertical stem (Fig. 44.53G); epistome usually separated from frons and genae by distinct epistomal groove extending down rostrum laterally .... **OTTISTIRINI**

– Antennal scrobes linear to curved, narrow to broadly open, without vertical stem; epistome not separated from frons and genae by a groove 4

4(3). Rostrum dorsally separated from head by curved or angled transverse groove (Fig. 44.52F); eyes protuberant; prothorax laterally without ocular lobes; metatibiae with false corbel .... **CELEUTHETINI**

– Rostrum dorsally not separated from head by a transverse groove; metatibiae without false corbel; eyes various; prothorax laterally with ocular lobes or without 5

5(4). Antennal scrobes open, visible in dorsal view; prothorax laterally without ocular lobes; metatibiae with two fused spurs (fused at sockets or together at base) (Fig. 44.57J) .... **OTIORHYNCHINI ... *Otiorhynchus* Germar**

– Antennal scrobes variously formed; prothorax laterally with or without ocular lobes; metatibiae without spurs 6

6(5). Mandibles without deciduous cusp or its scar; prothorax without ocular lobes; metatibiae without corbel .... **PACHYRHYNCHINI ... *Pantorhytes* Faust**

– Mandibles with deciduous cusp (Fig. 44.55A) or its scar; prothorax with or without ocular lobes; metatibiae with or without corbel 7

7(6). Antennal scrobes entirely visible in dorsal view, broadly open (Fig. 44.52G) 8

– Antennal scrobes lateral (Fig. 44.52A), not or only antennal insertions visible in dorsal view, narrow 10

8(7). Body covered with dense, imbricate squamae; metatibiae with squamose corbel (Fig. 44.58H) surrounded by thick, peg-like, sparse setae; tarsal claws connate at base (Fig. 44.59D) .... **EMBRITHINI ... *Afrophloeus* Borovec & Oberprieler**

– Body covered with sparse to tessellate squamae; metatibiae without corbel; tarsal claws free 9

9(8). Rostrum with large nasal plate (Fig. 44.52D); eyes strongly protruding; prothorax without ocular lobes; elytral declivity with 6 low, elongate tubercles .... **OOSOMINI ... *Phlyctinus* Schoenherr**

– Rostrum without nasal plate, apically usually flared and excised, with antennal insertions next to base of mandibles (Fig. 44.52J), except in tany-rhynchiform rostra, Fig. 44.52H; eyes flat or slightly rounded; prothorax with or without ocular lobes; elytral declivity without tubercles .... **CYPHICERINI**

10(7). Prothorax laterally with tuft of ocular vibrissae (Fig. 44.56B); metatibiae usually with narrow bare bevel (Fig. 44.58I) .... **TANYMECINI**

– Prothorax laterally without tuft of ocular vibrissae, with or without ocular lobe; metatibiae with or without bevel or corbel 11

11(10). Epistome-frons small, not forming nasal plate; rostrum short and about as broad as head; scrobes narrow, sharply defined, curved down in front of eyes; prothorax laterally without ocular lobes; tarsal claws free; parthenogenetic (only females present) .... **NAUPACTINI**
– Not with this combination of characters; prothorax with ocular lobes (Fig. 44.56A) or, if ocular lobes absent, with open scrobes and/or large nasal plate and/or narrow rostrum; bisexual (both sexes present) .... **LEPTOPIINI**

CELEUTHETINI

1. Elytra with 12 striae each, 12th (lateral one) abbreviated at level of metacoxa .... ***Celeuthetes* Schoenherr**
– Elytra with 10 striae each **2**
2(1). Pronotum and elytra with more or less uniformly spaced, non-coalescent squamae; funicle segments 1 and 2 longer than others (2 as long as 3 + 4) .... ***Celeuthetinus* Oberprieler & Zimmerman**
– Pronotum and elytra with patches of coalescent whitish squamae; funicle segments 1 and 2 subequal to others (2 shorter than 3 + 4) **3**
3(2). Elytra each with a large, flattened process on top of declivity; rostrum not declivous at apex .... ***Apirocalus* Pascoe**
– Elytra without process; rostrum declivous at apex **4**
4(3). Head constricted behind eyes, the latter strongly protruding .... ***Idiopsodes* Thompson**
– Head not constricted behind eyes, the latter less prominently protruding .... ***Oribius* Marshall**

CYPHICERINI

1. Rostrum elongate, laterally and dorsally continuous with head, laterally and ventrally usually partly glabrous (tanyrhynchiform), (Fig. 44.53B); eyes flat, depressed onto head **2**
– Rostrum short and broad, laterally and dorsally more or less set off from head, laterally and ventrally squamose; eyes slightly to distinctly protruding from head surface **9**
2(1). Posterior pronotal margin sinuate, median part shortly, broadly, roundly produced caudad to fit into shallow emargination of elytral bases **3**
– Posterior pronotal margin straight to slightly arcuate, medially not produced to fit into emargination of elytral bases **6**
3(2). Pronotum trapezoidal, sides straight; eyes dorsally almost contiguous; head behind eyes abruptly broadened to about twice width of rostrum (Fig. 44.52H) .... ***Matesia* Lea**
– Pronotum subrectangular, sides arcuate; eyes dorsally well separated; head behind eyes gradually broadened to only slightly more than width of rostrum **4**
4(3). Antennae inserted apically on rostrum; rostrum at antennal insertions broader than long part between insertions and eyes .... ***Tanymyllocerus* Oberprieler & Zimmerman**
– Antennae inserted medially on rostrum; rostrum evenly cylindrical, not broader across antennal insertions **5**
5(4). Antennal sockets closed anteriorly by confluent carinae of inner and outer socket margins, outer carina extending anteriad to corners of epistome; eyes dorsally separated by more than apical width of scape; rostrum and head densely squamose dorsally, sparsely squamose laterally and/or ventrally .... ***Ecmyllocerus* Zimmerman**
– Antennal sockets open anteriorly, inner margin of sockets not carinate, outer carina not extending to corners of epistome; eyes dorsally separated by less than apical width of scape; rostrum and head only sparsely squamose dorsally, laterally and ventrally nude, glabrous .... ***Euphalia* Pascoe**
6(2). Eyes small, round, slightly protruding, placed on sides of head, dorsally separated by twice their width and by nearly width of rostrum at base .... ***Apomyllocerus* Oberprieler & Zimmerman**
– Eyes larger, elongate, flat, flush with surface of head, placed dorsolaterally on head, dorsally separated by less than their width and less than width of rostrum at base **7**
7(6). Posterior corners of prothorax bluntly produced over elytral humeri; outer margin of antennal sockets sharply carinate, extending anteriad to rostral margin; apex of protibiae in female with outer angle produced into long, slightly curved, blunt spine .... ***Atmesia* Pascoe**
– Posterior corners of prothorax not extending over elytral humeri; apex of protibiae in female without outer spine **8**
8(7). Elytra with sparse, long, erect, pale setae; eyes dorsally separated by about width of scape; outer margin of antennal sockets sharply carinate, extending anteriad to rostral margin .... ***Tesiama* Oberprieler & Zimmerman**
– Elytra without erect setae; eyes dorsally separated by about twice width of scape; outer margin of antennal sockets sharp but not carinate, not extending anteriad to rostral margin (Fig. 44.61L) .... ***Sematia* Zimmerman**
9(1). Elytral humeri obsolete, elytra ovate (flightless species) .... ***Synomus* Pascoe**
– Elytral humeri developed, prominent (volant species) **10**
10(9). Anterior margin of prothorax laterally with ocular lobes and/or vibrissae **11**
– Anterior margin of prothorax laterally straight, without ocular lobes or vibrissae **13**
11(10). Anterior margin of prothorax laterally with prominent to faint ocular lobes bearing at least a few fine vibrissae; pronotum not much broader than long, anterior margin almost straight .... ***Hackeria* Lea**
– Anterior margin of prothorax laterally without ocular lobes, only vibrissae; pronotum transverse, about twice as wide as long, about as wide as elytra across humeri, anterior margin distinctly arcuate .... ***Syrphax* Marshall**
13(10). Anterior margin of antennal sockets at apex of rostrum (Fig. 44.61E); epistome-frons glabrous, triangular to crescentic, at least its apex caudally extending to behind a line drawn between apices of antennal sockets, thus deeply incising squamose epifrons and directing long postfrontal and postepistomal setae to project obliquely mesad, sides of frons often raised into C- or V-shaped carina; mandibles broad, more or less horizontal and most of inner margins touching each other to form cutting edges; habitus as Fig. 44.60A .... ***Myllocerus* Schoenherr**
– Anterior margin of antennal sockets behind apex of rostrum; epistome-frons not extending between antennal sockets, thus not incising squamose epifrons, postfrontal and postepistomal setae projecting anteriad; mandibles narrow and subvertical, not much wider than scar of deciduous appendages, touching each other only at tips **14**

14(13). Epistome-frons oblique, positioned in front of antennal sockets, pinched into slight to strong median carina extending onto epifrons between antennal sockets; sockets enclosed posteriorly by cariniform margin; bases of elytra and pronotum deeply sinuate; pronotum trapezoidal, much wider at base than at apex .... ***Paratitinia*** **Zimmerman & Oberprieler**

– Epistome-frons vertical or subvertical, positioned below anterior margins of antennal sockets in about right angle to epifrons, not pinched into median carina; epifrons elevated above antennal sockets, between them about as wide as scape; sockets open posteriorly; bases of elytra and pronotum straight to slightly sinuate; pronotum subrectangular, base and apex about equally wide .... ***Titinia*** **Pascoe**

ECTEMNORHININI

1. Pronotum with elongate median impression; head, pronotum and elytra densely covered with long, pale brown setae, elytra also with dark brown setae forming a pattern; interstriae 5 and 6 conspicuously confluent behind base of elytron .... ***Canonopsis*** **Waterhouse**

– Pronotum evenly convex, without median impression; head, pronotum and elytra sparsely covered with small iridescent setae or hair-scales; interstriae 5 and 6 not confluent behind base of elytron 2

2(1). Labial palps 2-segmented; epistome asymmetrical, right side projecting further forward; tibiae without mucro .... ***Ectemnorhinus*** **Waterhouse**

– Labial palps 3-segmented; epistome symmetrical; tibiae with or without mucro 3

3(2). Metanepisternal sutures distinct (Fig. 44.56E); tibiae with mucro; tarsal segment 5 longer than tarsites 1–3; tarsites 1–3 without ventral pads; habitus as Fig. 44.60B .... ***Palirhoeus*** **Kuschel**

– Metanepisternal sutures obsolete; tibiae without mucro; tarsal segment 5 shorter than tarsites 1–3; tarsites 1–3 with ventral pads .... ***Bothrometopus*** **Jeannel**

LEPTOPIINI

1. Eyes absent .... ***Undarobius*** **Escalona & Oberprieler**

– Eyes present 2

2(1). Rostrum long and narrow, in lateral view longer than prothorax, sides glabrous, in cross-section subtriangular (tanyrhynchiform) (Fig. 44.53B) 3

– Rostrum mostly short and broad, in lateral view shorter than prothorax, with sides setose, in cross section oval to subrectangular 6

3(2). Prothorax with ocular lobes; metatibiae with squamose corbel (Fig. 44.58G); tarsal claws free (Fig. 44.59B); habitus as Fig. 44.60L .... ***Lipothyrea*** **Pascoe**

– Prothorax with or without ocular lobes; metatibiae with corbel variable; tarsal claws connate (Fig. 44.59F–H) 4

4(3). Metatibiae with corbel poorly defined .... ***Xynaea*** **Pascoe**

– Metatibiae with corbel well defined, its surface setose or squamose 5

5(4). Prothorax without ocular lobes or vibrissae; metatibiae with squamose corbel; habitus as Fig. 44.60V .... ***Xyneella*** **Zimmerman**

– Prothorax without ocular lobes and vibrissae; metatibiae with setose corbel; habitus as Fig. 44.60T .... ***Synaptonyx*** **Waterhouse**

6(2). Rostrum dorsally separated from head by a fine groove (e.g. Fig. 44.53I); mandibles paucisetose (e.g. Fig. 44.55E); prothorax without ocular lobes; metatibiae without corbel 7

– Not with this combination of characters; rostrum dorsally not separated from head by a groove 11

7(6). Tarsal claws connate (Fig. 44.59D–J) 8

– Tarsal claws free (Fig. 44.59A–C) 9

8(7). Body surface covered with appressed squamae mixed with sparse, erect hair-scales or setae; eyes protuberant; rostrum dorsally separated from head by a straight transverse groove with a short median perpendicular groove (endocarina) towards vertex; mouthparts with maxillae and labial palps covered by prementum .... ***Merimnetes*** **Schoenherr**

– Body surface covered with imbricate squamae mixed with sparse, erect hair-scales or setae; eyes slightly protuberant; rostrum dorsally separated from head by a curved transverse groove, without a perpendicular median groove (endocarina); mouthparts with maxillae and labial palps partly exposed .... ***Rhyparophilus*** **Schoenherr**

9(7). Body length ~2.4 mm; eyes reduced; vestiture consisting of long, erect setae; habitus as Fig. 44.60J .... ***Howeocis*** **Lea**

– Body length 4.0 mm or more; eyes not reduced; vestiture variable 10

10(9). Body vestiture with dispersed black, erect, acute, long setae .... ***Agroicus*** **Jekel**

– Body vestiture with decumbent squamae; habitus as Fig. 44.60H .... ***Geosomus*** **Faust**

11(6). Prothorax with weak ocular lobes weak or without; tarsal claws connate at base (Fig. 44.59D, F–J) or single (Fig. 44.59K) 12

– Prothorax with or without ocular lobes; tarsal claws free (Fig. 44.59A–C) 24

12(11). Tarsal claws connate at base, equal 13

– Tarsal claw connate at base, strongly unequal (Fig. 44.59I, J) or single (Fig. 44.59K) 23

13(12). Rostrum dorsally with a median groove; metatibiae with setose corbel .... ***Zymaus*** **Pascoe**

– Rostrum dorsally without median groove; metatibiae with squamose or bare corbel 14

14(13). Body rounded in shape, about as long as wide (Fig. 44.60C); posterior angles of prothorax projecting posteriad (Fig. 44.60C); metatibiae with squamose corbel .... ***Basedowia*** Lea

– Body elongate; posterior angles of prothorax not projecting; metatibiae with or without corbel, if corbel present not squamose 15

15(14). Labium with palps exposed; prothorax with evident ocular lobes and vibrissae; metatibiae without corbel .... ***Mandalotus*** **Erichson** (part)

– Not with this combination of characters 16

16(15). Head capsule with median groove (endocarina); mouthparts with maxillae and labial palps exposed (e.g. Fig. 44.54F); metatibiae with narrow bare corbel .... ***Neomerimnetes*** **Blackburn**

– Head capsule without median groove (endocarina); mouthparts with maxillae and labial palps covered by prementum (e.g. Fig. 44.54H); metatibiae with broad, squamose or setose corbel 17

17(16). Body length ~2 mm; tarsal claws free, fused at base and divergent .... ***Amandalotus* Zimmerman & Oberprieler**
– Body more than 5 mm; tarsal claws connate at base, parallel or subparallel **18**
18(17). Rostrum with tubercles at base and apical third (Fig. 44.60O); prothorax without ocular lobes .... ***Platypterocis* Lea**
– Rostrum without tubercles; prothorax with weak to prominent ocular lobes **19**
19(18). Eyes protuberant in dorsal view; mouthparts with labial palps and part of the maxillae exposed .... ***Asceparnus* Schoenherr**
– Eyes not protuberant in dorsal view; mouthparts with labial palps and maxillae covered by prementum, rarely part of the maxillae exposed **20**
20(19). Distance between antennal insertions about half interocular distance (e.g. Fig. 44.53J) **21**
– Distance between antennal insertions about equal to interocular distance **22**
21(20). Eyes circular in outline; rostrum shorter than head capsule; habitus as Fig. 44.60I .... ***Hadrorhinus* Schoenherr**
– Eyes ovoid in outline; rostrum longer than head capsule .... ***Oophthalmus* Marshall**
22(20). Mandibles plurisetose (more than 10 long setae, including appressed squamose ones) .... ***Psapharus* Schoenherr**
– Mandibles paucisetose (3–4 long setae); habitus as Fig. 44.60P .... ***Polyphrades* Schoenherr**
23(12). Rostrum with median groove extending from basal third to between eyes; base of rostrum without transverse groove .... ***Essolithna* Pascoe**
– Rostrum without median groove in basal third; base of rostrum with transverse but medially interrupted groove .... ***Pascoellus* Oke**
24(11). Forehead with median groove and rostrum with transverse deep groove near base (Figs 44.42F–L, 44.43A–C); mesanepisterna and mesoventrite fused; elytra laterally near base often with a spine or tubercle; metatibiae with setose or glabrous corbel; habitus as Fig. 44.60E .... ***Catasarcus* Schoenherr**
– Not with this combination of characters; rostrum without transverse deep groove near base **25**
25(24). Mandibles paucisetose; prothorax without ocular lobes or vibrissae; metatibiae without corbel; tarsal claws unequal .... ***Esmelina* Pascoe**
– Mandibles pauci- or plurisetose; prothorax with or without ocular lobes and/or vibrissae; metatibiae with or without corbel; tarsal claws equal in length **26**
26(25). Eyes subcircular in outline (Fig. 44.53E); rostrum in lateral view with nasal plate vertical (Fig. 44.53E); mandibles blade-like, resembling those of Amycterini (Cyclominae), without apical tooth, densely setose and squamose; habitus as Fig. 44.60D .... ***Bothynorhynchus* Schoenherr**
– Not with this combination of characters **27**
27(26). Eyes strongly protuberant (Fig. 44.53K); distance between antennal insertions less than half interocular distance (Fig. 44.53K); mandibles densely setose and squamose, broad, with single apical tooth; elytra with lateral pointed tubercle behind humeral angle .... ***Ochrometa* Pascoe**
– Not with this combination of characters **28**
28(27). General body shape of *Leptopius* (e.g. Figs 44.42A–E, 44.43D–L, 44.44A–D, 44.60K); eyes often narrow and ovoid in outline (Figs 44.56A, 44.62D, H, but see 44.62F); rostrum robust (Figs 44.53D, 44.62D, G, H), as long as or longer than head capsule in lateral view, in ventral view with occipital grooves (Fig. 44.62H) and ventral longitudinal groove between them; mandibles plurisetose, rarely paucisetose (*Enchymus*, *Uroleptops*); maxillae and labium covered by prementum or maxillae partly exposed; prothorax with ocular lobes and/or vibrissae; metatibiae with or without corbel, when present often large and densely setose **29**
– Not with this combination of characters; mouthparts with labial palps exposed or partially visible (*Pseudotimareta*), if not visible (e.g. *Allotimareta*, *Mandalotus*, *Timareta*) then rostrum without occipital groove, metatibiae without well-defined corbel and tibiae expanded towards apex **44**
29(28). Metatibiae with large, well-defined corbel, often densely setose (Fig. 44.58F) but sometimes sparsely so (*Lysizone*) **30**
– Metatibiae with small or poorly defined corbel or without corbel **36**
30(29). Mouthparts with maxillae partly exposed; prothorax with or without ocular lobes, vibrissae present **31**
– Mouthparts with maxillae not exposed, completely covered by prementum; prothorax with well-defined ocular lobes, vibrissae present or absent **33**
31(30). Prothorax with prominent, well defined ocular lobes and vibrissae; elytra without humeral angle .... ***Centyres* Pascoe**
– Prothorax with ocular lobes weak or absent; elytra with well-defined humeral angle **32**
32(31). Antennal scrobes directed toward eyes .... ***Catastygnus* Pascoe**
– Antennal scrobes directed to below eyes .... ***Peripagis* Pascoe**
33(30). Antennal scrobes open, poorly defined and directed towards eyes; metatibiae with sparsely setose, mostly glabrous bevel .... ***Lysizone* Pascoe**
– Antennal scrobes well defined and directed to below eyes; metatibiae with densely setose bevel (Fig. 44.58J) **34**
34(33). Prementum with preapical transverse row of at least 6 setae .... ***Onesorus* Pascoe**
– Prementum without preapical transverse row of setae, with more than 10 setae spread over surface forming different patterns **35**
35(34). Mesotibiae with mucro entirely visible .... ***Rhinoscapha* Montrouzier**
– Mesotibiae with mucro covered by setae and only partially visible .... ***Leptopius* Oke**
36(29). Mandibles paucisetose, with ~4–5 long setae and a few squamae (Fig. 44.55E) **37**
– Mandibles plurisetose, with more than 5 long setae and more than 10 short setae or squamae **40**
37(36). Prothorax with prominent ocular lobes .... ***Amisallus* Schoenherr**
– Prothorax without ocular lobes **38**
38(37). Body length ~3.6 mm; antennal scrobes with margins poorly defined; metatibiae with well-defined narrow corbel .... ***Ecrizothis* Blackburn**
– Body length 6.5 mm or more; antennal scrobes well defined; metatibiae without corbel (Fig. 44.58B) or with bevel (Fig. 44.58B) **39**
39(38). Elytra in apical third with 4 pointed tubercles; metatibiae with bevel; habitus as Fig. 44.60U .... ***Uroleptops* Lea**
– Elytra in apical third without tubercles; metatibiae without bevel (Fig. 44.58B) .... ***Enchymus* Pascoe**
40(36). Prementum with approximately nine setae **41**
– Prementum with only two long setae **42**
41(40). Body covered with sparse vestiture or patches of scale-like setae; eyes flattened; mandibles without permanent cusp; prothorax with ocular lobes; scutellar shield not visible between elytra; protibiae cylindrical .... ***Stenocorynus* Schoenherr**
– Body covered with sparse fine setae; eyes protuberant; mandibles with permanent cusp (Fig. 44.55A); prothorax without ocular lobes; scutellar shield visible between elytra; protibiae flattened (Fig. 44.57E); habitus as Fig. 44.60Q .... ***Prostomus* Schoenherr**

42(40). Prothorax without ocular lobes, vibrissae present; habitus as Fig. 44.60R .... ***Prypnus*** **Schoenherr**
– Prothorax with ocular lobes, vibrissae absent **43**
43(42). Elytra laterally prominent behind humerus .... ***Baryopadus*** **Pascoe**
– Elytra laterally not prominent behind humerus .... ***Amnemus*** **Marshall**
44(28). Prothorax with distinct ocular lobes and vibrissae **46**
– Prothorax without ocular lobes or, if present, very weak and antennal scrobes without margins defined, vibrissae if present short **48**
46(45). Rostrum without dorsal median carina; metatibiae with setose corbel .... ***Perperus*** **Schoenherr**
– Rostrum with dorsal median carina; metatibiae with very narrow or not well-defined corbel or bevel **47**
47(46). Elytra without tubercles .... ***Scotasmus*** **Schoenherr**
– Elytral disc with weak tubercles and more prominent and pointed tubercles at apical third; habitus as Fig. 44.60G .... ***Gastrocis*** **Lea**
48(44). Rostrum with nasal plate vertical in lateral view (e.g. 44.53E); distance between antennal insertions less than half interocular distance; antennal scrobes weak and incipient; elytra apically extending into pointed tubercle .... ***Evadomorpha*** **Blackburn**
– Rostrum with nasal plate horizontal or oblique in lateral view; distance between antennal insertions slightly less than or equal to interocular distance; antennal scrobes well developed or with margins poorly defined; elytra without pointed apical tubercle **49**
49(48). Eyes not protuberant (e.g. 44.52J); protibiae usually without denticles along inner margin (e.g. Fig. 44.44E–I, L, but see Fig. 44.44J (*Mandalotus bryophilus* Oke)) **50**
– Eyes usually protuberant (Fig. 44.53H, except *Ocynoma*); protibiae with denticles along inner margin (e.g. Fig. 44.57A–B) **53**
50(49). Metatibiae with narrow, bare corbel fringed with blunt, peg-like setae **51**
– Metatibiae without corbel, not fringed with blunt, peg-like setae **52**
51(50). Prementum convex; dorsal vestiture consisting of round, mostly imbricate squamae and two or more irregular rows of needle-like setae on interstriae .... ***Pseudotimareta*** **Zimmerman & Oberprieler**
– Prementum flattened; dorsal vestiture consisting of separate, lanceolate squamae and single row of blunt setae on interstriae .... ***Allotimareta*** **Zimmerman & Oberprieler**
52(50). Mandibles with pointed ventral tooth, prominent in some species; habitus as Fig. 44.60S .... ***Timareta*** **Pascoe**
– Mandibles without pointed ventral tooth; habitus as Fig. 44.60M .... ***Mandalotus*** **Erichson** (part)
53(49). Eyes not protuberant; antennal scapes in repose reaching apical third of prothorax; metatibiae with corbel .... ***Ocynoma*** **Pascoe**
– Eyes protuberant; antennal scapes in repose not reaching apical third of prothorax; metatibiae without corbel **54**
54(53). Elytra with humeral angle slightly displaced posteriad **55**
– Elytra without humeral angle **56**
55(54). Elytra with apex rounded, not extending into tubercle .... ***Prosayleus*** **Schoenherr**
– Elytra with apex extending into pointed tubercle; habitus as Fig. 44.60F .... ***Evas*** **Pascoe**
56(54). Rostrum in lateral view with antennal insertions separated from eye by about length of eye .... ***Asceparnodes*** **Zimmerman & Oberprieler**
– Rostrum in lateral view with antennal insertions separated from eye by more than eye length .... ***Decienus*** **Pascoe**

NAUPACTINI

1. Elytra with small, recumbent, unobtrusive setae (not protruding from integument) 2
– Elytra with conspicuous suberect setae, at least on declivity 3
2(1). Elytra with median and dorsolateral pair of black vittae almost to apex, interstriae 3 as broad as 2; metatibiae with narrow squamose corbel and ascending posterior setal comb; antennae thin, sparsely setose, clubs elongate, 2.5 × as long as broad .... ***Atrichonotus*** **Buchanan**
– Elytral disc with a pair of short, broad, dark, truncate vittae, interstriae 3 broader than 2; metatibiae without corbel and ascending posterior setal comb; antennae thick, densely squamose .... ***Eurymetopus*** **Schoenherr**
3(1). Elytra with 2 shallows, broken, black V-shaped vittae; squamae dense, broad, imbricate; scapes reaching to middle of eye in repose; funicle segments 2 slightly shorter than 1 .... ***Floresianus*** **Hustache**
– Elytra without dark V-shaped vittae; squamae sparse (separate); scapes reaching to posterior margin of eye in repose; funicle segments 2 at least 1.5 × as long as 1; habitus as Fig. 44.60N .... ***Naupactus*** **Dejean**

OTTISTIRINI

1. Epistome-frons posteriorly well angled (triangular), not separated from epifrons by a groove; scrobes angled, dorsal horizontal groove not extending posteriad; antennal clubs elongate; mandibles with 3 long setae, mandibular cusps apical; prementum small; procoxae contiguous (Fig. 44.56D) .... ***Tregoleus*** **Oberprieler & Zimmerman**
– Epistome-frons posteriorly straight, sometimes minutely angled in middle, separated dorsally from epifrons by a groove sometimes extending onto sides of rostrum; scrobes T-shaped (e.g. Fig. 44.53E), dorsal horizontal groove extending posteriad; antennal clubs short; mandibles with 2 long setae, mandibular cusps ventral; prementum large; procoxae contiguous or separated (Fig. 44.56C, D) 2
2(1). Groove separating frons and epifrons weak, only dorsal, not extending onto sides of rostrum 3
– Groove separating frons and epifrons strong, deep, extending onto sides of rostrum 4
3(2). Procoxae contiguous, cavities confluent; odd elytral interstriae costate and at least interstriae 3 with declivital crest; eyes closer together than distance between eye and middle of anterior epistomal margin .... ***Reidius*** **Oberprieler & Zimmerman**
– Procoxae not contiguous, cavities separated by narrow septum; all interstriae flat; eyes as far apart as distance between eye and middle of anterior epistomal margin; habitus as Fig. 44.60W .... ***Eutinophaea*** **Pascoe**
4(2). Procoxae contiguous, cavities confluent .... ***Monteithus*** **Oberprieler & Zimmerman**
– Procoxae not contiguous, cavities separated by narrow septum 5

5(4). Profemora in male with large distal spine shearing over similar basal tibial tooth, both femora and tibiae adentate in female; procoxae diverging .... ***Maleuterpes*** Blackburn
– Profemora and -tibiae in both sexes adentate; procoxae more or less parallel **6**
6(5). Mandibles without deciduous cusp; prementum short, transverse .... ***Ottistirodes*** **Oberprieler & Zimmerman**
– Mandibles with ventral deciduous cusp; prementum quadratic **7**
7(6). Elytral interstriae 1, 3, 5 costate, 1 often with tubercle on top of declivity; protibiae inside with distal row of ~4 small teeth .... ***Weirius*** **Oberprieler & Zimmerman**
– Elytral interstriae flat; protibiae inside without distal row of teeth .... ***Ottistirellus*** **Oberprieler & Zimmerman**

TANYMECINI

1. Mouthparts with prementum large, fully covering maxillae (Fig. 44.54I) **2**
– Mouthparts with prementum smaller, largely exposing maxillae **3**
2(1). Pronotum evenly convex; scapes reaching anterior pronotal margin in repose; setae on pronotum and elytra small, broad, testaceous, recumbent; tarsal claws free, divergent .... ***Tanymecus*** **Germar**
– Pronotum on disc cross-like impressed; scapes reaching middle of eye in repose; setae on pronotum and elytra very fine, acute, long and erect in male but short and recumbent in female; tarsal claws basally connate, parallel .... ***Hypomeces*** **Schoenherr**
3(2). Tarsal claws free, divergent; aedeagus and ovipositor long and slender, endophallus with long flagellum **4**
– Tarsal claws basally connate, parallel; aedeagus and ovipositor short and broad, endophallus without flagellum .... ***Jenningsius*** **Oberprieler & Zimmerman**
4(3). Mandibles with deciduous cusp or its scar situated on pronounced pedestal, outside with dense long setae and smaller, narrow, white squamae .... ***Pullenius*** **Oberprieler & Zimmerman**
– Mandibles with deciduous cusp or its scar not on pedestal, outside with fewer long setae and no squamae .... ***Homoeotrachelus*** **Faust**

Classification of the Australian genera of Entiminae

Tribe Celeuthetini Lacordaire, 1863

Apirocalus Pascoe, 1881 (Pl. 62A)

Type species, by monotypy: *Apirocalus cornutus* Pascoe, 1881.
Australian species. *A. cornutus* Pascoe, 1881.
Distribution. Northern Australia (QLD), New Guinea.

Celeuthetes Schoenherr, 1842 (Pl. 62B; Fig. 44.52F)

Type species, by original designation: *Sphaeromus australis* (Gyllenhal, 1834)
Australian species. *C. australis* (Gyllenhal, 1834) (*in* Schoenherr 1834b).
Distribution. Australia (QLD), New Guinea, Solomon Is., Larat Is., Maluku Is., Vanuatu.

Celeuthetinus Oberprieler & Zimmerman, 2020 (Pl. 62C)

Type species, by original designation: *Celeuthetinus gnomus* Oberprieler & Zimmerman, 2020.
Australian species. *C. gnomus* Oberprieler & Zimmerman, 2020.
Distribution. Australia (NT).

Idiopsodes Thompson, 1977 (Pl. 62D)

Type species, by original designation: *Idiopsis grisea* Faust, 1897.
Australian species. *I. ocularis* (Lea, 1927b).
Distribution. Australia (QLD), New Guinea.

Oribius Marshall, 1956 (Pl. 62E; Figs 44.54C, 44.55E, 44.58K)

Type species, by original designation: *Coptorrhynchus generosus* Faust, 1897.
Australian species. *O. albivarius* (Lea, 1927b), *O. equinus* (Lea, 1927f), *O. gestroi* (Pascoe, 1885), *O. jansoni* (Pascoe, 1871a), *O. tessellatus* (Blanchard, 1853), *O. trivittatus* (Lea, 1927b).
Distribution. Australia (QLD, NT (Wessel Is.)), New Guinea.

Tribe Cyphicerini Lacordaire, 1863

Apomyllocerus Oberprieler & Zimmerman, 2020 (Pl. 62F)

Type species, by original designation: *Apomyllocerus ovatus* Oberprieler & Zimmerman, 2020.
Australian species. *A. ovatus* Oberprieler & Zimmerman, 2020.
Distribution. Australia (WA).

Atmesia Pascoe, 1870 (Pl. 62G)

Type species, by monotypy: *Atmesia marginata* Pascoe, 1870.
Australian species. *A. marginata* Pascoe, 1870c.
Distribution. Australia (SA).

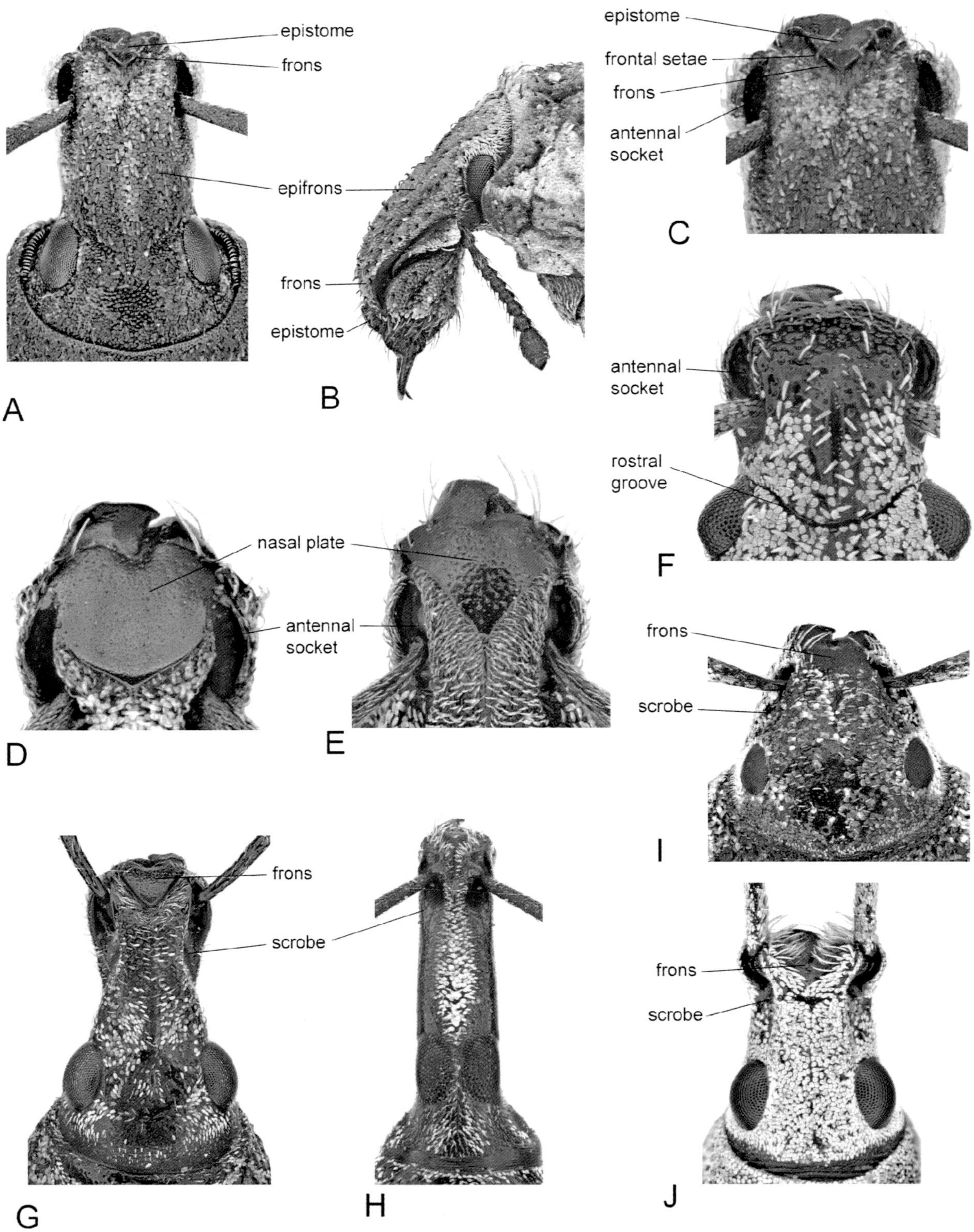

Fig. 44.52. Entiminae, structural details of rostrum. A, undescribed genus and species, dorsal; B, *Baryopadus corrugatus* Pascoe, lateral; C, undescribed genus and species, epistome, dorsal; D, *Polyphrades gibbipennis* Lea, nasal plate, dorsal; E, *Phlyctinus callosus* (Schoenherr), nasal plate, dorsal; F, *Celeuthetes australis* (Gyllenhal), rostral groove, dorsal; G, *Ecrizothis boviei* Lea, dorsal; H, *Matesia maculata* Lea, dorsal; I, *Basedowia basicollis* Lea, dorsal; J, *Myllocerus aphthosus* Pascoe, dorsal. After Oberprieler & Zimmerman (2020).

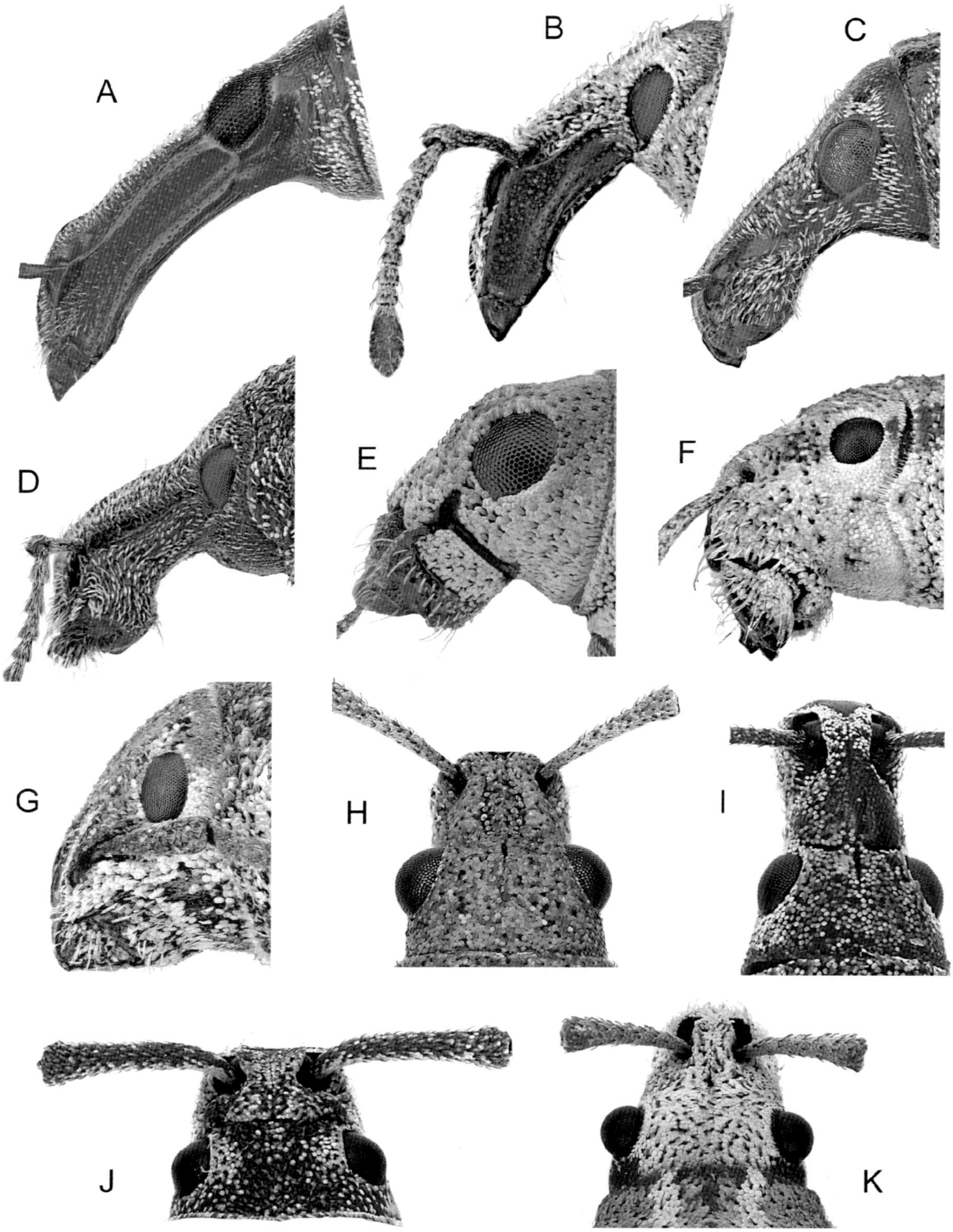

Fig. 44.53. Entiminae, shapes of head and rostrum. A, *Matesia maculata* Lea, lateral (Cyphicerini); B, *Lipothyrea arrowi* Lea, lateral (Leptopiini); C, *Ecrizothis boviei* Lea, lateral (Leptopiini); D, genus near *Onesorus* (Mt. Benstread Creek, NT), lateral; E, *Bothynorhynchus decorum* (Blackburn), lateral (Leptopiini); F, *Basedowia basicollis* Lea, lateral (Leptopiini); G, *Ottistirodes murinus* (Lea), lateral (Ottistirini); H, *Evas latipennis* Lea, dorsal (Leptopiini); I, *Merimnetes uniformis* Boheman, dorsal (Leptopiini); J, *Hadrorhinus lepidopterus* Gyllenhal, dorsal (Leptopiini); K, *Ochrometa amoena* Pascoe, dorsal (Leptopiini). After Oberprieler & Zimmerman (2020).

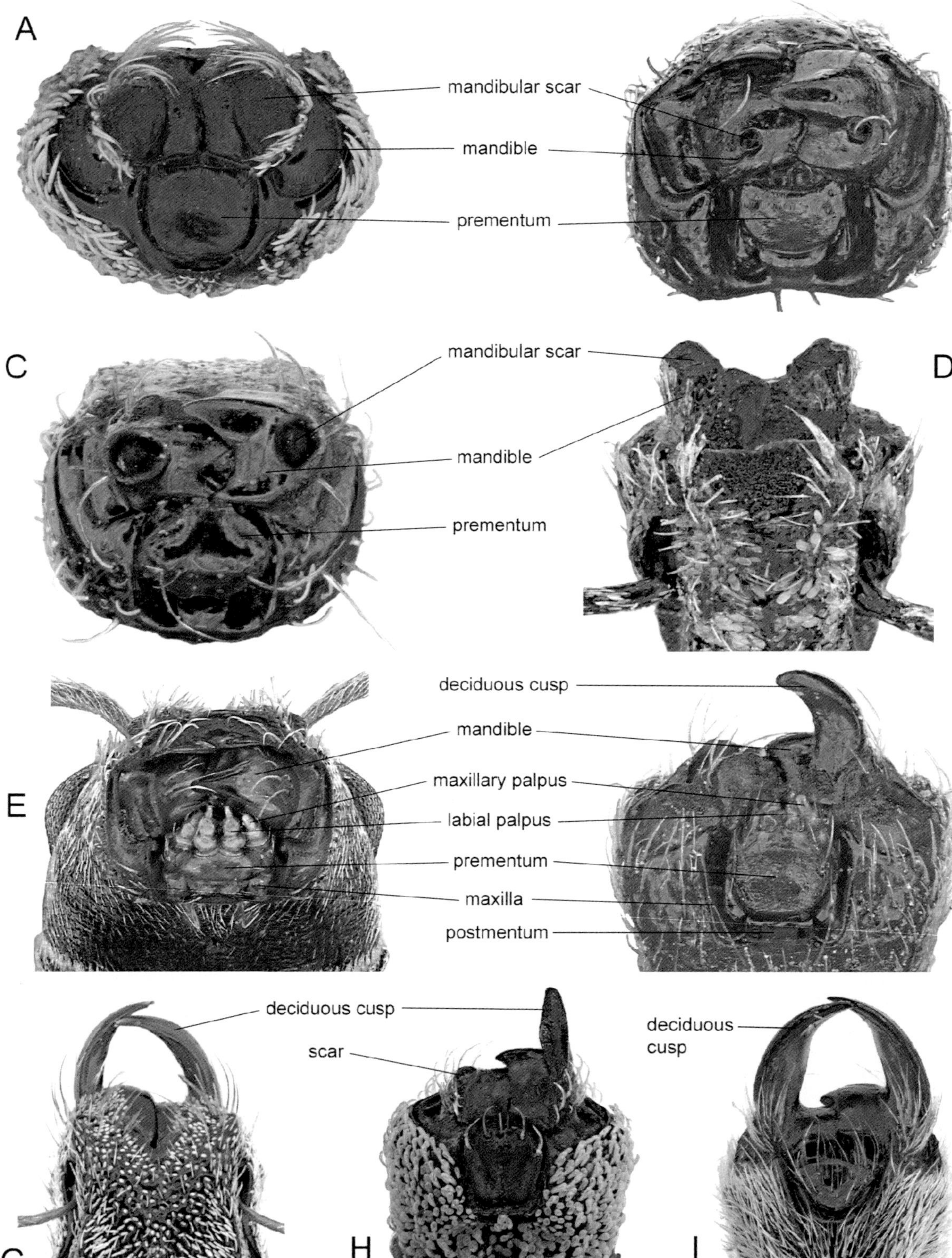

Fig. 44.54. Entiminae, structural details of mouthparts. A, *Leptopius setosus* (Lea), apicoventral; B, *Christensenia antarctica* Brinck, apicoventral; C, *Oribius* sp., apicoventral; D, *Onesorus* sp., dorsal; E, *Canonopsis sericea* Waterhouse, ventral; F, *Otiorhynchus rugosostriatus* (Goeze), ventral; G, *Naupactus leucoloma* Boheman, dorsal; H, *Onesorus armipennis* Lea, ventral; I, *Hypomeces obscurus* (Macleay), ventral. After Oberprieler & Zimmerman (2020).

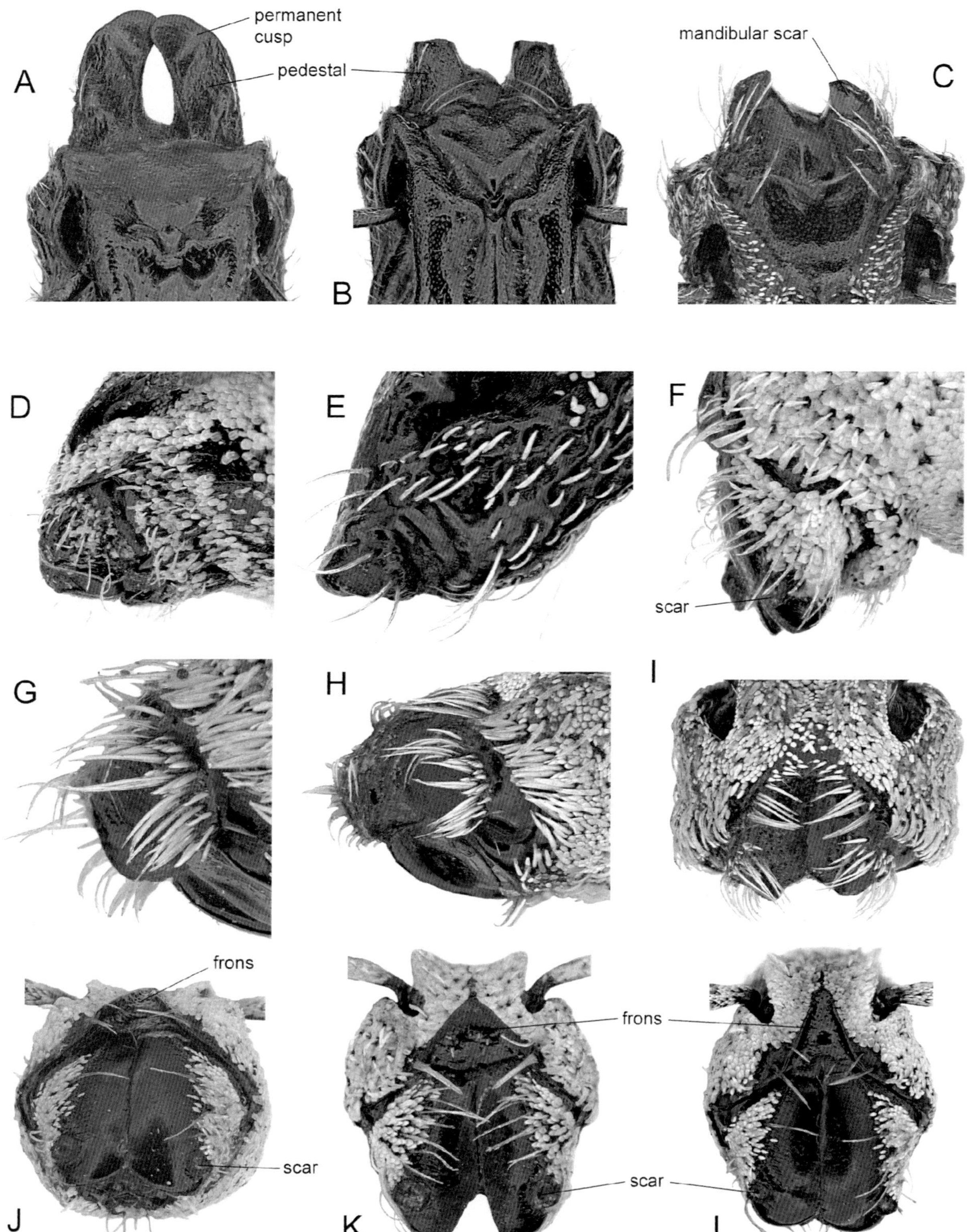

Fig. 44.55. Entiminae, shapes of mandibles. A–B, *Prostomus scutellatus* (Fabricius), dorsal: A, ♂; B, ♀; C, genus near *Onesorus*, dorsal; D, *Basedowia basicollis* Lea, lateral; E, *Oribius* sp., lateral; F, *Bothynorhynchus decorum* (Blackburn), lateral; G–I, *Leptopius setosus* (Lea), lateral and dorso-apical views: J, *Evas lineata* Pascoe, ventral; K, *Bothynorhynchus decorum*, apical view; L, *Evas lineata* Pascoe, dorso-apical view. After Oberprieler & Zimmerman (2020).

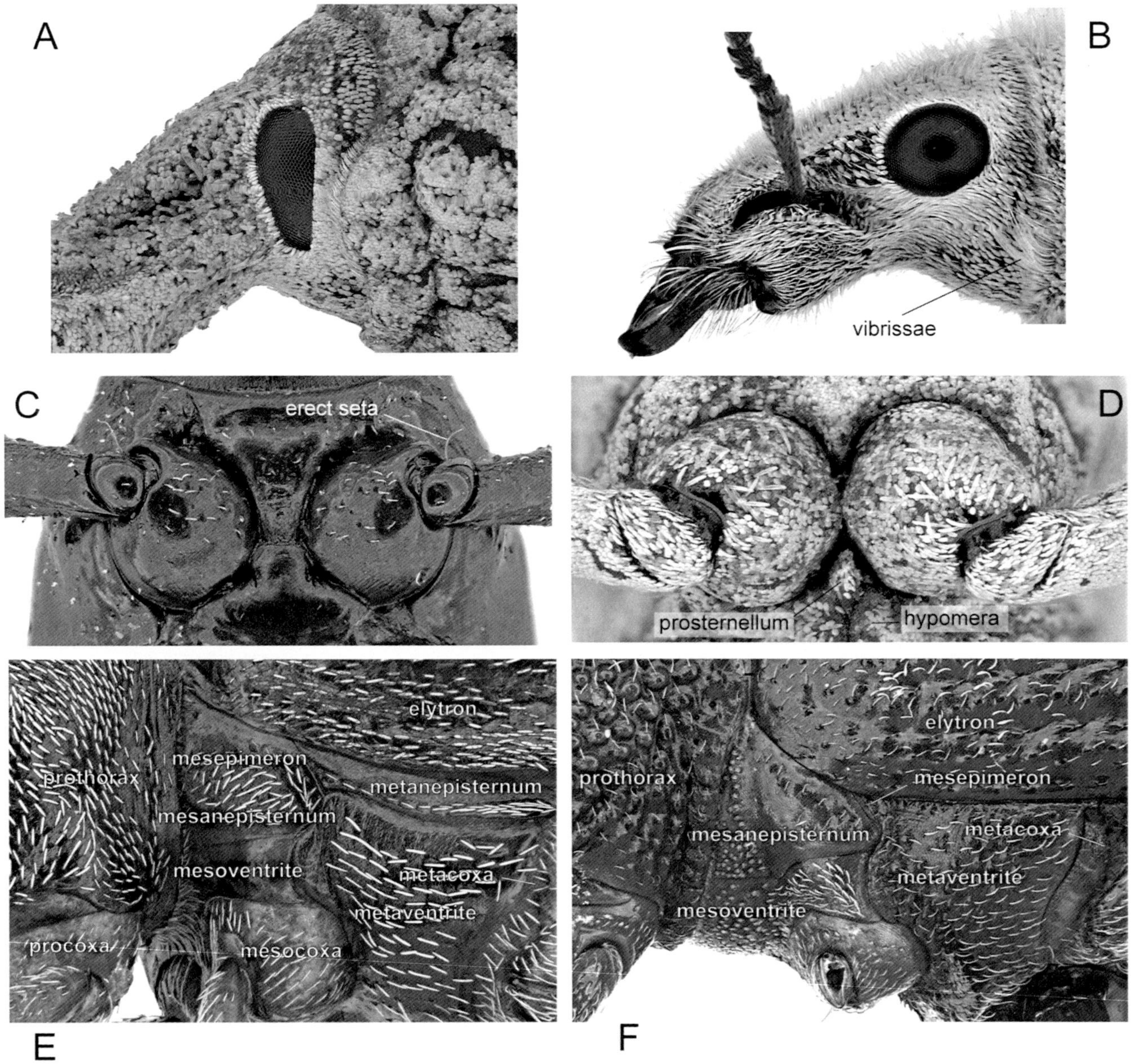

Fig. 44.56. Entiminae, structural details of thorax. A; D, *Leptopius setosus* (Lea); A, ocular lobe of prothorax; D, contiguous procoxae; B, *Hypomeces obscurus* (Macleay), lateral prothoracic margin; C, *Behrensiellus glabratus* (Pascoe), separate procoxae; E, *Christensenia antarctica* Brinck, lateral thoracic sclerites; F, *Otiorhynchus sulcatus* (Fabricius), lateral thoracic sclerites. After Oberprieler & Zimmerman (2020).

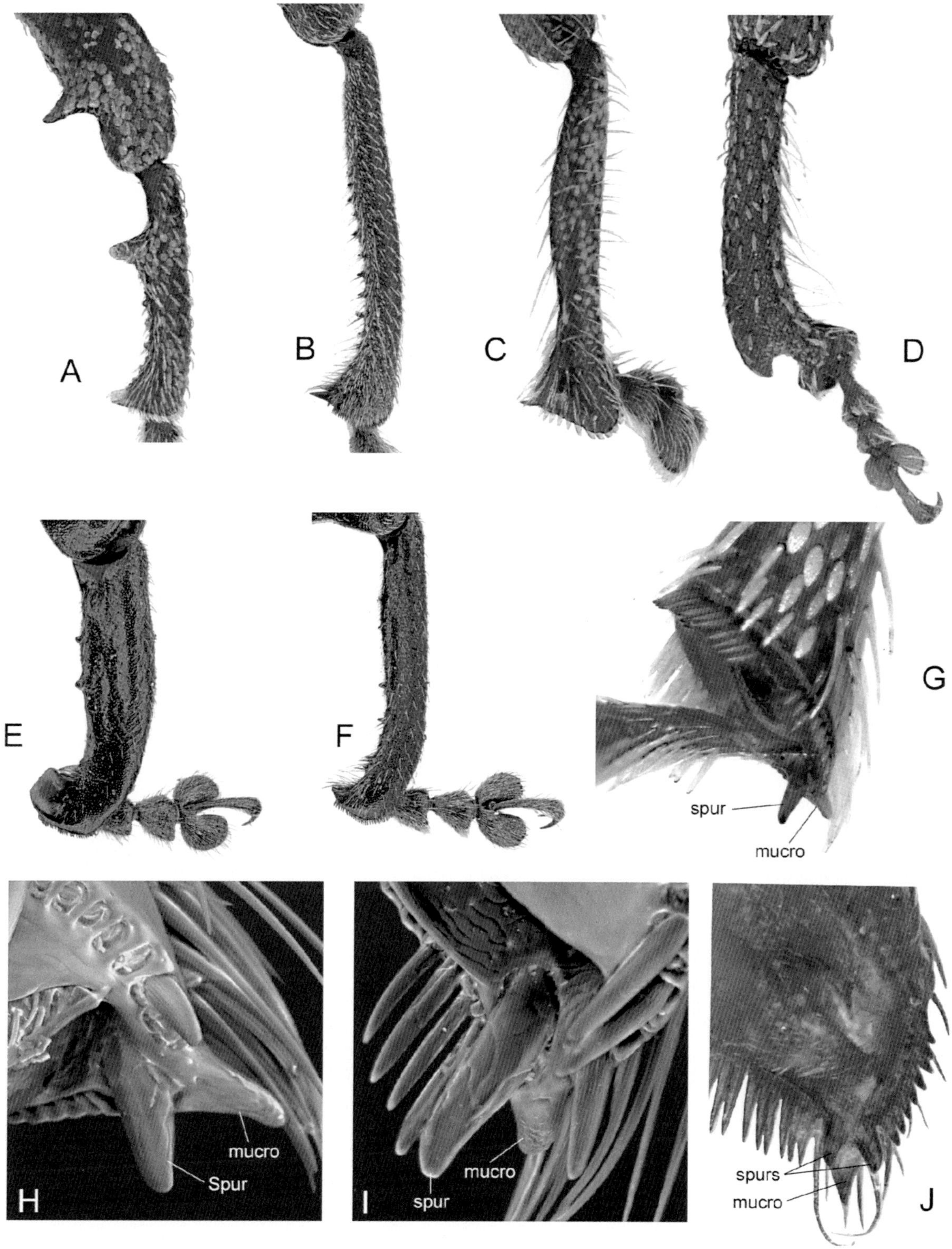

Fig. 44.57. Entiminae, shapes of protibiae and structural details of metatibial apex. A, *Maleuterpes spinipes* Blackburn, dentate protibia, ♂; B, *Prypnus trituberculatus* Germar, dentate protibia; C, *Timareta* sp., apically explanate protibia; D, *Mandalotus dentipes* Lea, apically bent and excised protibia, ♂; E–F, *Prostomus scutellatus* (Fabricius), apically curved protibia: E, ♂; F, ♀; G–I, *Steriphus sericeus* (Blackburn) (Cyclominae: Listroderini): metatibial spur and mucro; J, *Otiorhynchus sulcatus* (Fabricius), fused metatibial spurs and mucro. After Oberprieler & Zimmerman (2020).

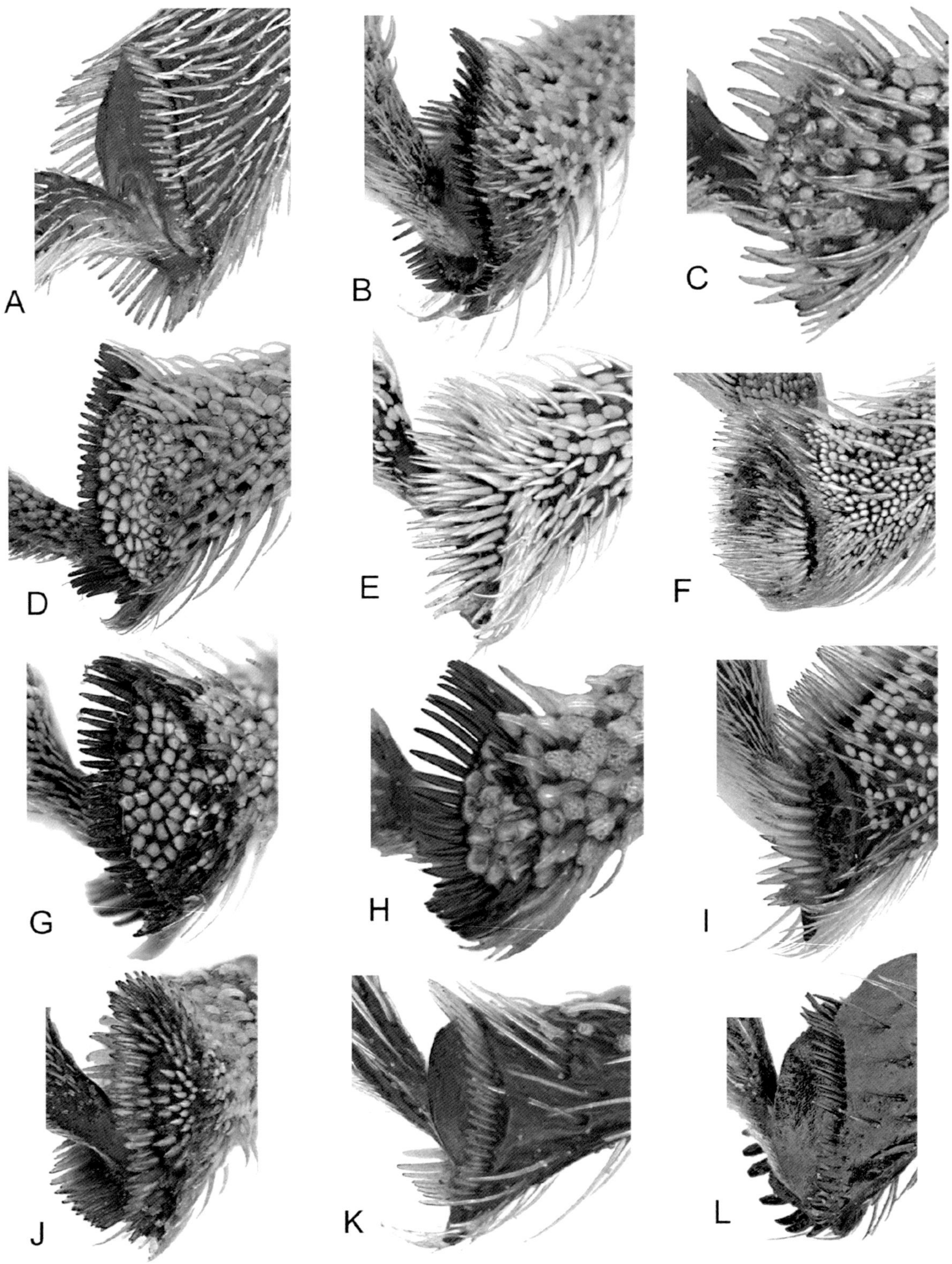

Fig. 44.58. Entiminae, structural details of metatibial apices. A, *Canonopsis sericea* Waterhouse, without corbel or bevel; B, *Enchymus humeralis* Pascoe, without corbel or bevel; C, *Xynaea saginata* Pascoe, with squamose bevel; D, *Basedowia basicollis* Lea, with squamose bevel; E, *Onesorus tarsalis* Lea, with setose corbel; F, *Leptopius setosus* Lea, with setose corbel; G, *Polyphrades laticollis* Fåhraeus, with squamose corbel; H, *Xyneella uniformis* (Lea), with squamose corbel; I, *Hypomeces obscurus* (Macleay), with bare rimmed bevel; J, *Onesorus tarsalis* Lea, with narrow bevel; K, *Oribius tessellatus* (Blanchard), with false corbel; L, *Behrensiellus glabratus* (Pascoe), with false corbel. After Oberprieler & Zimmerman (2020).

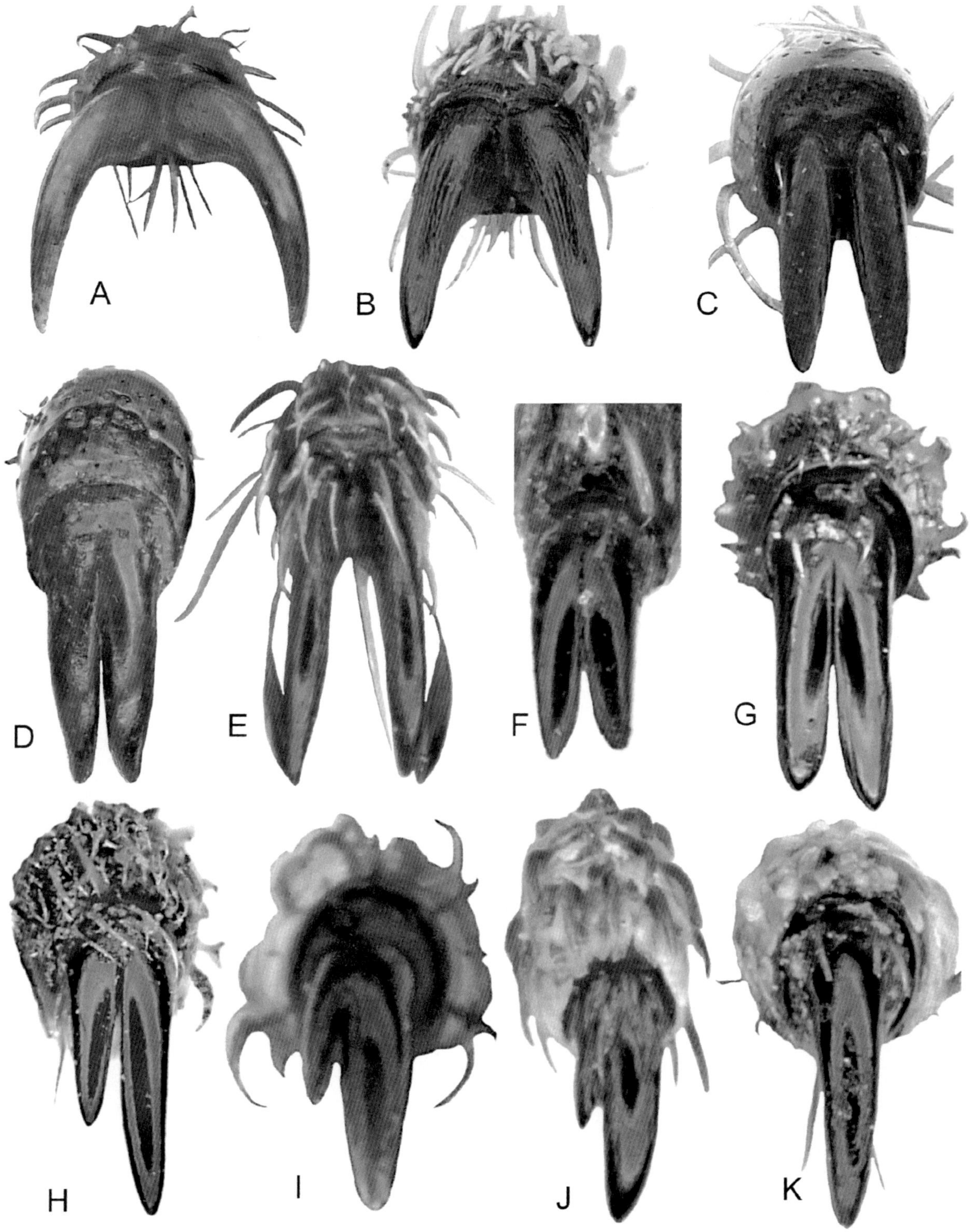

Fig. 44.59. Entiminae, shapes of tarsal claws. A, *Canonopsis sericea* Waterhouse, free, widely divergent; B, *Leptopius setosus* (Lea), free, narrowly divergent; C, *Zymaus angustus* Lea, free, parallel; D, genus near *Onesorus,* connate. equal; E, *Charagmus gressorius* (Fabricius), free, with long, distally flattened ventrobasal seta; F, *Basedowia basicollis* Lea, basally connate, equal; G, *Polyphrades* sp., basally connate, unequal; H, *Psapharus mistothes* (Herbst), basally connate, unequal; I, *Essolithna kingiae* Lea, basally connate, unequal; J, *Essolithna rhomba* Pascoe, basally connate, unequal; K, *Essolithna pluviata* Pascoe, single. After Oberprieler & Zimmerman (2020).

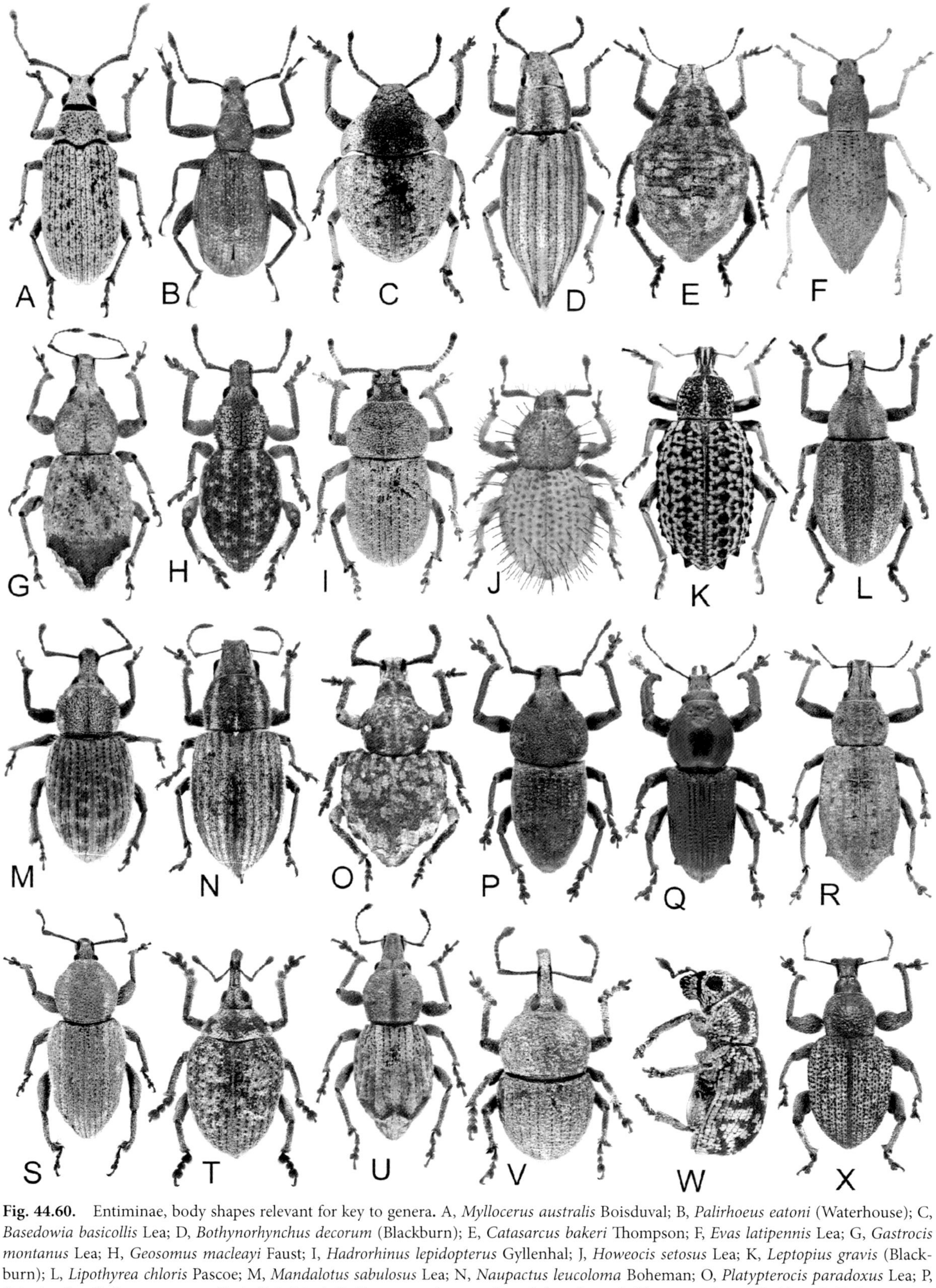

Fig. 44.60. Entiminae, body shapes relevant for key to genera. A, *Myllocerus australis* Boisduval; B, *Palirhoeus eatoni* (Waterhouse); C, *Basedowia basicollis* Lea; D, *Bothynorhynchus decorum* (Blackburn); E, *Catasarcus bakeri* Thompson; F, *Evas latipennis* Lea; G, *Gastrocis montanus* Lea; H, *Geosomus macleayi* Faust; I, *Hadrorhinus lepidopterus* Gyllenhal; J, *Howeocis setosus* Lea; K, *Leptopius gravis* (Blackburn); L, *Lipothyrea chloris* Pascoe; M, *Mandalotus sabulosus* Lea; N, *Naupactus leucoloma* Boheman; O, *Platypterocis paradoxus* Lea; P, *Polyphrades*; Q, *Prostomus scutellaris* (Fabricius); R, *Prypnus quinquenodosus* Schoenherr; S, *Timareta pilosa* (Blackburn); T, *Synaptonyx ovatus* Waterhouse; U, *Uroleptops impar* Lea; V, *Xyneella uniformis* (Lea); W, *Eutinophaea nana* Pascoe, ♂; X, *Rhyncholobus andrewsi* Gahan.

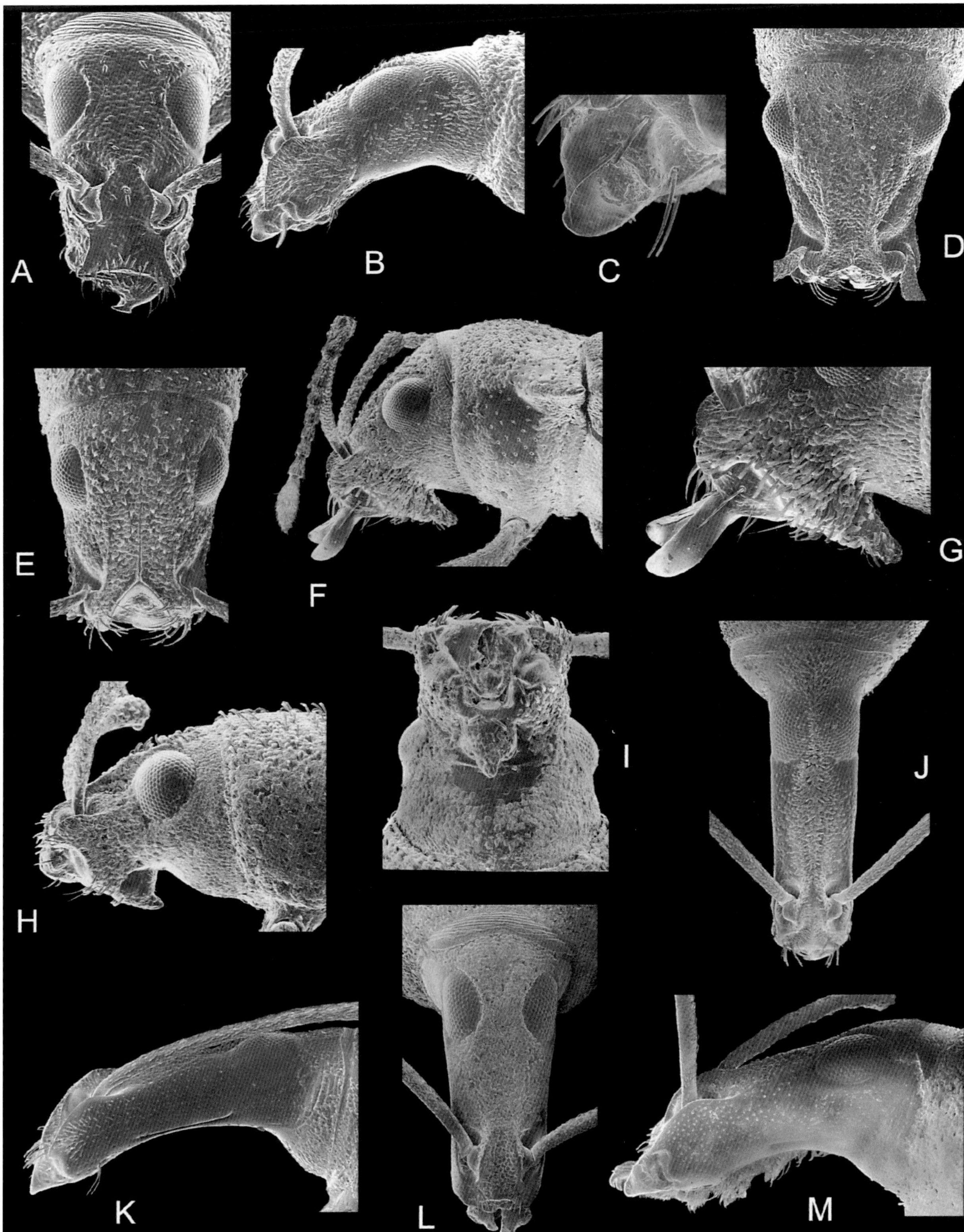

Fig. 44.61. Entiminae, shapes of head, Cyphicerini. A–C, *Ecmyllocerus trivitticollis* (Lea), head: A, dorsal; B, lateral views; C, mandibles, lateral view; D, *Myllocerus rusticus* Pascoe, head, dorsal view; E, *Myllocerus trepidus* Pascoe, head, lateral view; F–G, *Myllocerus laticollis* Lea: F, head; G, mandibles, lateral view; H–I, *Myllocerus mirabilis* Lea, head, lateral and ventral views; J–K, *Matesia maculata* Lea, head, dorsal and lateral views; L–M, *Sematia glaucina* (Pascoe), head, dorsal and lateral views. After Oberprieler & Zimmerman (2020).

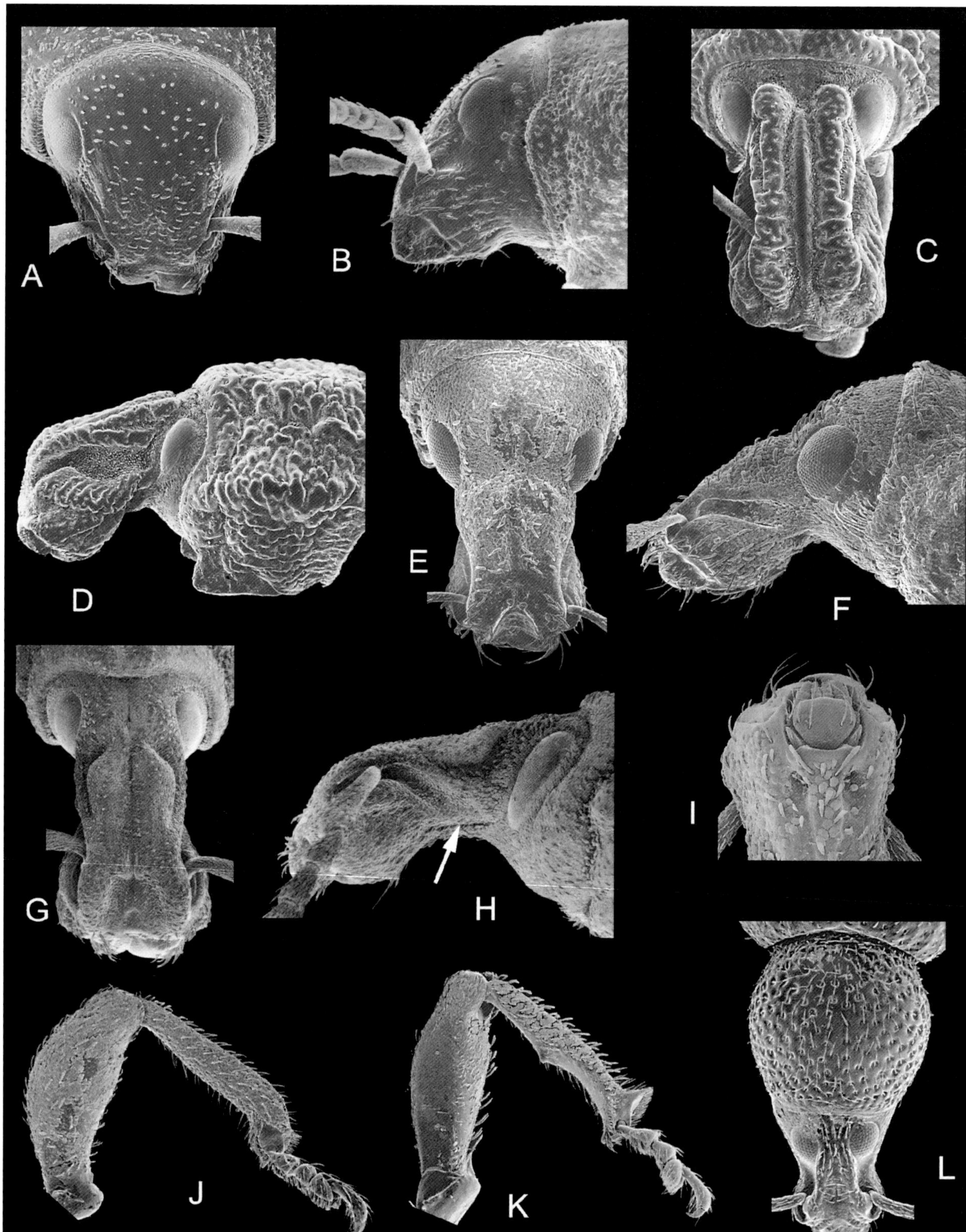

Fig. 44.62. Entiminae, shapes of head and legs, Leptopiini. A–B, *Basedowia basicollis* Lea, head, dorsal and lateral; C–D, *Leptopius echidna* (Macleay): head, dorsal and lateral; E–F, *Leptopius puellaris* (Pascoe), head, dorsal and lateral; G–H, *Leptopius t-squameus* (Lea): head, dorsal and lateral; I, *Merimnetes* sp., mouthparts, ventral; J–K, *Amandalotus cordipennis* (Lea): J, foreleg; K, hindleg; L, *Otiorhynchus cribricollis* Gyllenhal, head, dorsal.

Ecmyllocerus Zimmerman, 1994 (Pl. 62H; Fig. 44.61A–C)

Type species, by original designation: *Myllocerus trivitticollis* Lea, 1929.
Australian species. *E. trivitticollis* (Lea, 1929a).
Distribution. Australia (SA).

Euphalia Pascoe, 1870 (Pl. 62I)

Type species, by monotypy: *Euphalia pardalis* Pascoe, 1870.
Australian species. *E. pardalis* Pascoe, 1870c.
Distribution. Australia (WA).

Hackeria Lea, 1911 (Pl. 62J)

Type species, by monotypy: *Hackeria viridivaria* Lea, 1911.
Australian species. *H. aberrans* (Lea, 1917c), *H. chrysidea* (Pascoe, 1885), *H. gnopholota* (Lea, 1925), *H. hardcastlei* (Lea, 1911a), *H. nigrovaria* (Lea, 1914), *H. viridivaria* Lea, 1911b.
Distribution. Australia (QLD, NT, WA).

Matesia Lea, 1904 (Pl. 62K; Figs 44.52H, 44.53A, 44.61J, K)

Type species, by monotypy: *Matesia maculata* Lea, 1904.
Australian species. *M. maculata* Lea, 1904.
Distribution. Australia (WA).

Myllocerus Schoenherr, 1823 (Pl. 62L, 63A, B; Figs 44.52J, 44.61D–I)

Type species, by original designation: *Curculio curvicornis* Fabricius, 1792.
Australian species. *M. abstersus* (Pascoe, 1870b), *M. aeruginosus* Lea, 1925, *M. amblyrhinus* Lea, 1905a, *M. angustibasis* Lea, 1914, *M. anoplus* Lea, 1915a, *M. aphthosus* Pascoe, 1869, *M. armipectus* Lea, 1914, *M. armipes* Lea, 1925, *M. ashi* Lea, 1917c, *M. aurifex* Pascoe, 1869, *M. australis* Boisduval, 1835, *M. baeodontomerus* Lea, 1915a, *M. basicollis* Lea, 1911b, *M. bicolor* (Blackburn, 1892b), *M. bifasciatipennis* Lea, 1925, *M. bilineater* Lea, 1915c, *M. canalicornis* Lea, 1909c, *M. carinatus* Lea, 1904, *M. castor* Lea, 1909c, *M. chaunoderus* Lea, 1925, *M. cinerascens* Pascoe, 1869, *M. confinis* Lea, 1914, *M. constricticollis* Lea, 1914, *M. cyrtops* Lea, 1914, *M. darwini* Blackburn, 1889, *M. decipiens* Lea, 1925, *M. decretus* Pascoe, 1869, *M. doddi* Lea, 1914, *M. echinatus* Lea, 1905a, *M. elegans* Lea, 1905a, *M. eremitus* (Blackburn, 1892b), *M. exilis* Lea, 1905a, *M. foveiceps* Lea, 1909c, *M. foveifrons* Lea, 1911a, *M. fugitivus* Lea, 1914, *M. fuscomaculatus* Lea, 1915a, *M. herbaceus* Pascoe, 1869, *M. hilli* Lea, 1914, *M. incisicollis* (Lea, 1909b), *M. inermipes* Lea, 1929a, *M. intercoxalis* Lea, 1914, *M. latibasis* Lea, 1914, *M. laticollis* Lea, 1905a, *M. longicornis* (Pascoe, 1869), *M. longus* Lea, 1914, *M. maculatus* (Blackburn, 1892c), *M. mastersi* Lea, 1911b, *M. melvillensis* Lea, 1914, *M. microps* (Lea, 1911b), *M. minusculus* Lea, 1914, *M. mirabilis* Lea, 1911b, *M. multimaculatus* Lea, 1911a, *M. nasutus* Pascoe, 1869, *M. niveus* Lea, 1905a, *M. obliquifasciatus* Lea, 1917c, *M. obscurus* Lea, 1925, *M. perarmatus* Lea, 1929a, *M. pictus* Lea, 1925, *M. pollux* Lea, 1909c, *M. prosternalis* Lea, 1914, *M. pteroderes* Lea, 1929a, *M. quadricolor* Lea, 1917c, *M. rugicollis* Lea, 1905a, *M. rusticus* Pascoe, 1869, *M. scitulus* Lea, 1925, *M. setistriatus* Lea, 1914, *M. sordidus* Lea, 1905a, *M. speciosus* Blackburn, 1894c, *M. squamicornis* Lea, 1914, *M. subrostralis* Lea, 1914, *M. suturalis* Lea, 1905a, *M. tatei* Blackburn, 1869, *M. taylori* Lea, 1909c, *M. terreus* (Pascoe, 1869), *M. torridus* Blackburn, 1894c, *M. trepidus* Pascoe, 1885, *M. tricarinirostris* Lea, 1925, *M. trilineatus* Lea, 1905a, *M. trisinuatus* Lea, 1929a, *M. tristis* Lea, 1914, *M. usitatus* Lea, 1904, *M. varius* Lea, 1914, *M. viridimicans* Lea, 1917c, *M. viridipictus* (Lea, 1909c), *M. zopherus* Lea, 1917b.
Distribution. Australia (QLD, NSW, NT, SA, VIC, WA), Oriental, eastern Palaearctic, Afrotropical regions.

Paratitinia Zimmerman & Oberprieler, 2014 (Pl. 63C)

Type species, by original designation: *Myllocerus ceratorhinus* Lea, 1914.
Australian species. *P. ceratorhina* (Lea, 1914), *P. hoplosterna* (Lea, 1929a), *P. obliquicollis* (Lea, 1929a).
Distribution. Australia (QLD, WA).
Notes. The genus was described in Pullen *et al.* (2014).

Sematia Zimmerman, 1994 (Pl. 63D; Fig. 44.61L, M)

Type species, by original designation: *Atmesia glaucina* Pascoe, 1872.
Australian species. *S. glaucina* (Pascoe, 1872c).
Distribution. Australia (WA).

Synomus Pascoe, 1885 (Pl. 63E)

Type species, by monotypy: *Synomus cephalotes* Pascoe, 1885.
Australian species. *S. aeruginosus* Lea, 1905a, *S. angustipennis* (Lea, 1917c), *S. cephalotes* Pascoe, 1885, *S. chloris* (Pascoe, 1870b), *S. inconspicuus* Lea, 1914, *S. ornatus* (Faust, 1886), *S. ovipennis* Lea, 1914, *S. subapterus* (Lea, 1911b).
Distribution. Australia (QLD, NT, SA, WA).

Syrphax Marshall, 1944 (Pl. 63F)

Type species, by original designation: *Myllocerus incurvus* Lea, 1909c.

Australian species. *S. blackburni* (Lea, 1914), *S. bovilli* (Blackburn, 1894c), *S. fieldi* (Lea, 1914), *S. incurvus* (Lea, 1909c).
Distribution. Australia (QLD, NT, WA).

Tanymyllocerus Oberprieler & Zimmerman, 2020 (Pl. 63G)

Type species, by original designation: *Tanymyllocerus tenuirostris* Oberprieler & Zimmerman, 2020.
Australian species. *T. tenuirostris* Oberprieler & Zimmerman, 2020.
Distribution. Australia (WA).

Tesiama Oberprieler & Zimmerman, 2020 (Pl. 63H)

Type species, by original designation: *Tesiama hystricosa* Oberprieler & Zimmerman, 2020.
Australian species. *T. hystricosa* Oberprieler & Zimmerman, 2020
Distribution. Australia (WA).

Titinia Pascoe, 1869 (Pl. 63I)

Type species, by monotypy: *Titinia ignara* Pascoe, 1869.
Australian species. *T. brevicollis* Blackburn, 1892b, *T. grata* (Pascoe, 1885), *T. laeta* Blackburn, 1892b, *T. tenuis* (Germar, 1848), *T. villosipennis* (Lea, 1917c).
Distribution. Australia (QLD, NSW, SA, VIC, WA).

Tribe Ectemnorhinini Lacordaire, 1863

Bothrometopus Jeannel, 1940 (Pl. 63J, K)

Type species, by original designation: *Bothrometopus fasciatus* Jeannel, 1940.
Australian species. *B. brevis* (Waterhouse, 1875), *B. gracilipes* (Waterhouse, 1875).
Distribution. Australia (Heard Is., McDonald Is.), Crozet Is., Kerguelen Is., Marion Is., Prince Edward Is.

Canonopsis Waterhouse, 1875 (Pl. 63L; Figs 44.54E, 44.58A, 44.59A)

Type species, by monotypy: *Canonopsis sericea* Waterhouse, 1875.
Australian species. *C. sericea* Waterhouse, 1875.
Distribution. Australia (Heard Is., McDonald Is.), Crozet Is., Kerguelen Is.

Ectemnorhinus Waterhouse, 1853 (Pl. 64A)

Type species, by monotypy: *Ectemnorhinus viridis* Waterhouse, 1853.
Australian species. *E. viridis* Waterhouse, 1853.
Distribution. Australia (Heard Is., McDonald Is.), Crozet Is., Kerguelen Is., Marion Is., Prince Edward Is.

Palirhoeus Kuschel, 1971 (Pl. 64B)

Type species, by original designation: *Ectemnorrhinus eatoni* Waterhouse, 1876
Australian species. *P. eatoni* (Waterhouse, 1876).
Distribution. Australia (Heard Is.), Crozet Is., Kerguelen Is., Marion Is., Prince Edward Is.

Tribe Embrithini Marshall, 1942

Afrophloeus Borovec & Oberprieler, 2013 (Pl. 64C)

Type species, by original designation: *Trachyphloeus spathulatus* Boheman, 1843.
Australian species. *A. squamifer* (Boheman, 1843) (*in* Schoenherr 1843).
Distribution. Australia (SA), southern Africa.
Notes. The species was accidentally introduced from South Africa in 1992 (Borovec & Oberprieler 2013; Oberprieler & Zimmerman 2020).

Tribe Leptopiini Oke, 1951

Agroicus Jekel, 1875 (Pl. 64D)

Type species, by monotypy: *Prosayleus comosus* Germar, 1848.
Australian species. *A. ateropterus* (Boheman, 1840) (*in* Schoenherr 1840a), *A. concolor* (Blackburn, 1894c), *A. intermedius* (Blackburn, 1894c), *A. lineatus* (Blackburn, 1894c), *A. subfasciatus* (Blackburn, 1894c).
Distribution. Australia (QLD, NSW, Lord Howe Is., VIC, TAS, SA, NT, WA).

Allotimareta Zimmerman & Oberprieler, 2014 (Pl. 64E)

Type species, by original designation: *Timareta setistriata* Lea, 1915.
Australian species. *A. setistriata* (Lea, 1915c).
Distribution. Australia (WA).
Notes. The genus was described in Pullen *et al.* (2014).

Amandalotus Zimmerman & Oberprieler, 2014 (Pl. 64F; Fig. 44.62J, K)

Type species, by original designation: *Mandalotus cordipennis* Lea, 1926d.
Australian species. *A. cordipennis* (Lea, 1926).
Distribution. Australia.
Notes. The genus was described in Pullen *et al.* (2014).

Amisallus Schoenherr, 1840 (Pl. 64G; Fig. 44.42A–D)

Type species, by original designation: *Amisallus tuberosus* Boheman, 1840.

Australian species. *A. basipennis* Lea, 1917a, *A. crassirostris* (Lea, 1906b), *A. nodosus* Erichson, 1842, *A. phymatodis* (Lea, 1906b), *A. tuberculifrons* Lea, 1917a, *A. tuberosus* Boheman, 1840 (*in* Schoenherr 1840a), *A. whitei* (Waterhouse, 1853).

Distribution. Australia (QLD, NSW, VIC, TAS, WA).

Notes. Specimens have been found on *Ceratopetalum apetalum* (Cunoniaceae).

Amnemus Marshall, 1952 (Pl. 64H; Fig. 44.42E)

Type species, by monotypy: *Leptops quadrituberculatus* Boheman, 1834.

Australian species. *A. quadrituberculatus* (Boheman, 1834) (*in* Schoenherr 1834a), *A. superciliaris* (Pascoe, 1870b).

Distribution. Australia (QLD, NSW).

Hosts. The larva develops in the roots of several legumes, whereas the adults feed on several plants.

Asceparnodes Zimmerman & Oberprieler, 2014 (Pl. 64I)

Type species, by original designation: *Timareta duplicata* Lea, 1909.

Australian species. *A. duplicatus* (Lea, 1909c).

Distribution. Australia.

Notes. The genus was described in Pullen *et al.* (2014).

Asceparnus Schoenherr, 1847 (Pl. 64J)

Type species, by original designation: *Asceparnus nodipennis* Schoenherr, 1847.

Australian species. *A. nodipennis* Schoenherr, 1847, *A. subfasciatus* (Blackburn, 1894c), *A. xanthorrhoeae* (Lea, 1909c).

Distribution. Australia (SA, WA).

Baryopadus Pascoe, 1870 (Pl. 64K; Fig. 44.52A–C)

Type species, by monotypy: *Baryopadus corrugatus* Pascoe, 1870.

Australian species. *B. corrugatus* Pascoe, 1870d.

Distribution. Australia (QLD, NSW, SA).

Notes. Larvae of the single described species have been recorded feeding on roots and the adults on the foliage of several plants, such as cowpea, peanut, sunflower and beans.

Basedowia Lea, 1930 (Pl. 64L; Figs 44.52I, 44.53F, 44.55D, 44.58D, 44.59F; 44.62A, B)

Type species, by monotypy: *Basedowia basicollis* Lea, 1930.

Australian species. *B. basicollis* Lea, 1930c.

Distribution. Australia (QLD, SA, NT).

Notes. The host plants includes *Acacia* species (Fabaceae).

Bothynorhynchus Schoenherr, 1842 (Pl. 65A; Figs 44.53E, 44.55F, K)

Type species, by original designation: *Bothynorhynchus lascivus* Boheman, 1842.

Australian species. *B. decorum* (Blackburn, 1896), *B. frigidus* (Blackburn, 1892b), *B. lascivus* Boheman, 1842 (*in* Schoenherr 1842b), *B. rugiceps* (Lea, 1904).

Distribution. Australia (NSW, VIC, SA).

Notes. The head structure of the genus resembles that of several genera of Amycterini, and for many years it was associated with this tribe.

Catasarcus Schoenherr, 1840 (Pl. 65B, C; Figs 44.42F–L, 44.43A–C)

Type species, by original designation: *Catasarcus bilineatus* Fåhraeus, 1840.

Australian species. *C. aerosus* Thompson, 1968, *C. albipectus* Thompson, 1968, *C. albisparsus* Pascoe, 1870a, *C. albuminosus* Pascoe, 1870a, *C. armatus* Blackburn, 1894c, *C. aspergetus* Thompson, 1968, *C. asphaltinus* Thompson, 1968, *C. azureipes* Thompson, 1968, *C. bakeri* Thompson, 1968, *C. bicolor* Thompson, 1968, *C. bilineatus* Fåhraeus, 1840 (*in* Schoenherr 1840a), *C. carbo* Pascoe, 1870a, *C. carinaticeps* Lea, 1909c, *C. cicatricosus* Pascoe, 1870a, *C. concretus* Pascoe, 1870a, *C. coruscus* Thompson, 1968, *C. cygnensis* Thompson, 1968, *C. echidna* Pascoe, 1870a, *C. frontalis* Thompson, 1968, *C. griseus* Pascoe, 1870a, *C. hopei* Fåhraeus, 1840 (*in* Schoenherr 1840a), *C. impressipennis* (Boisduval, 1835), *C. inaequalis* Thompson, 1968, *C. intermedius* Pascoe, 1870a, *C. laevior* Thompson, 1968, *C. latheticus* Thompson, 1968, *C. lepidus* Pascoe, 1870a, *C. longicornis* Pascoe, 1870a, *C. marginispinis* Pascoe, 1870a, *C. memnonius* Pascoe, 1870a, *C. militaris* Peterson, 1995, *C. murex* Thompson, 1968, *C. nephelodes* Thompson, 1968, *C. obesus* Thompson, 1968, *C. opimus* Pascoe, 1870a, *C. pallidiventris* Thompson, 1968, *C. rugulosus* Boheman, 1845 (*in* Schoenherr 1845), *C. sericeus* Blackburn, 1894c, *C. spinipennis* Fåhraeus, 1840 (*in* Schoenherr 1840a), *C. transversalis* Germar, 1848, *C. ustulatus* Thompson, 1968, *C. varus* Thompson, 1968.

Distribution. Australia (western VIC, SA, NT, WA).

Notes. The genus was revised by Thompson (1968). The species have been associated with several plants, including *Leucadendron*, *Hakea rostrata* (Proteaceae), *Leptospermum*, *Eucalyptus diversicolor*, *Melaleuca* (Myrtaceae), *Acacia*, *Jacksonia*, *Kennedia* (Fabaceae), *Casuarina* (Casuarinaceae) and

Xanthorrhoea (Asphodelaceae), but both adults and larvae of most species are probably not host-specific. The genus is common in WA, and specimens are often found by beating vegetation during the day.

Catastygnus Pascoe, 1871 (Pl. 65D)

Type species, by subsequent designation (Pullen *et al.* 2014: 148): *Catastygnus scutellaris* Pascoe, 1871.
Australian species. *C. densus* Lea, 1908a, *C. elegans* Lea, 1908a, *C. limbatus* Pascoe, 1871a, *C. limbifer* Faust, 1900 (*in* Heller, 1900), *C. ochreipennis* Lea, 1908a, *C. rivulosus* Pascoe, 1871a, *C. robustus* (Lea, 1910b), *C. scutellaris* Pascoe, 1871a, *C. stigma* Pascoe, 1871a, *C. textilis* Pascoe, 1871a, *C. v-albus* (Lea, 1910b), *C. variabilis* Lea, 1908b.
Distribution. Australia (QLD, NSW, including Lizard Is.).

Centyres Pascoe, 1871 (Pl. 65E)

Type species, by monotypy: *Centyres turgidus* Pascoe, 1871.
Australian species. *C. ovis* Pascoe, 1872b, *C. turgidus* Pascoe, 1871a.
Distribution. Australia (QLD).

Chaodius Pascoe, 1870 (Pl. 65F)

Type species, by monotypy: *Chaodius nigrescens* Pascoe, 1870.
Australian species. *C. nigrescens* Pascoe, 1870d.
Distribution. Australia (WA).

Decienus Pascoe, 1873 (Pl. 65G)

Type species, by monotypy: *Decienus sphasodes* Pascoe, 1873.
Australian species. *D. sphasodes* Pascoe, 1873b.
Distribution. Australia (SA, WA).

Ecrizothis Blackburn, 1899 (Pl. 65H; Figs 44.52G, 44.53C)

Type species, by monotypy: *Ecrizothis inaequalis* Blackburn, 1899.
Australian species. *E. blackburni* Oke, 1931, *E. boviei* Lea, 1911b, *E. inaequalis* Blackburn, 1899, *E. terminalis* Oke, 1931.
Distribution. Australia (VIC).
Notes. Specimens have been recorded from apple, cherry and gooseberry plants.

Enchymus Pascoe, 1871 (Pl. 65I; Fig. 44.58B)

Type species, by monotypy: *Enchymus punctonotatus* Pascoe, 1871.
Australian species. *E. humeralis* Pascoe, 1872b, *E. punctonotatus* Pascoe, 1871a.
Distribution. Australia (NSW, VIC, SA, WA).
Notes. The host plants include *Eucalyptus pilularis* (Myrtaceae) and *Hakea* species (Proteaceae).

Esmelina Pascoe, 1870 (Pl. 65J)

Type species, by monotypy: *Esmelina flavovittata* Pascoe, 1870.
Australian species. *E. flavovittata* Pascoe, 1870c, *E. geonomoides* (Pascoe, 1883a), *E. stenocera* Lea, 1904.
Distribution. Australia (NSW).

Essolithna Pascoe, 1870 (Pl. 65K, L; Fig. 44.59I–K)

Type species, by subsequent designation (Pullen *et al.* 2014: 150): *Essolithna pluviata* Pascoe, 1870.
Australian species. *E. cordipennis* Lea, 1904, *E. jonesi* Lea, 1917c, *E. kingiae* Lea, 1904, *E. maculata* Lea, 1904, *E. mediofusca* Lea, 1914, *E. pluviata* Pascoe, 1870b, *E. rhombus* Pascoe, 1870b, *E. seriata* Blackburn, 1892c, *E. terrena* Lea, 1904.
Distribution. Australia (WA).

Evadomorpha Blackburn, 1896 (Pl. 66A)

Type species, by monotypy: *Evas everardensis* Blackburn, 1892.
Australian species. *E. everardensis* (Blackburn, 1892c).
Distribution. Australia (SA).

Evas Pascoe, 1870 (Pl. 66B, C; Fig. 44.55J, L)

Types species, by subsequent designation (Pullen *et al.* 2014: 150): *Evas crassirostris* Pascoe, 1870.
Australian species. *E. argenteiventris* Pascoe, 1870d, *E. crassirostris* Pascoe, 1870d, *E. elliptica* Lea, 1904, *E. latipennis* Lea, 1904, *E. lineata* Pascoe, 1873a.
Distribution. Australia (SA, WA).
Notes. Apparent host plants are *Banksia* and *Grevillea* species (Proteaceae).

Gastrocis Lea, 1915 (Pl. 66D)

Type species, by monotypy: *Gastrocis montanus* Lea, 1915.
Australian species. *G. montanus* Lea, 1915a.
Distribution. Australia (TAS).
Notes. Specimens have been found in moss.

Geosomus Faust, 1892 (Pl. 66E)

Type species, by monotypy: *Geosomus macleayi* Faust, 1892.

Australian species. *G. dispar* (Germar, 1848), *G. macleayi* Faust, 1892b, *G. sublineatus* (Lea, 1911a).
Distribution. Australia (VIC, SA).

Hadrorhinus Schoenherr, 1834 (Pl. 66F)

Type species, by original designation: *Hadrorhinus lepidopterus* Gyllenhal, 1834.
Australian species. *H. basirostris* (Lea, 1914), *H. collaris* (Lea, 1914), *H. concinnus* (Lea, 1908a), *H. lepidopterus* Gyllenhal, 1834 (*in* Schoenherr 1834b).
Distribution. Australia (NT, WA).
Notes. The host plants seem to be grasses.

Howeocis Lea, 1926 (Pl. 66G)

Type species, by monotypy: *Howeocis setosus* Lea, 1926.
Australian species. *H. microps* (Lea, 1926d), *H. nodipennis* (Lea, 1926d), *H. setosus* Lea, 1926a.
Distribution. Australia (NSW, Lord Howe Is.).

Leptopius Oke, 1951 (Pl. 66H–J; Figs 44.41, 44.43D–L, 44.44A–D, 44.45A–I, 44.54A, 44.55G–I, 44.56A, D, 44.58F, 44.59B, 44.62C–H)

Type species, by original designation: *Curculio robustus* Olivier, 1807.
Australian species. *L. acerbus* (Pascoe, 1870b), *L. acutispinis* (Pascoe, 1882), *L. areolatus* (Blackburn, 1892c), *L. argillaceus* (Pascoe, 1873b), *L. baileyi* (Blackburn, 1889), *L. biordinatus* (Blackburn, 1892c), *L. brachystylus* (Lea, 1904), *L. cacozelus* (Lea, 1906b), *L. castelnaui* (Lea, 1911b), *L. clavus* (Fabricius, 1775), *L. colossus* (Pascoe, 1870b), *L. concaviceps* (Lea, 1906b), *L. concinnus* (Lea, 1917a), *L. contrarius* (Blackburn, 1892c), *L. crenatus* (Boheman, 1842) (*in* Schoenherr 1842a), *L. duboulayi* (Pascoe, 1870b), *L. duponti* (Boisduval, 1835), *L. ebeninus* (Pascoe, 1870b), *L. echidna* (Macleay, 1826), *L. elongatus* (Lea, 1906b), *L. etheridgei* (Olliff, 1889), *L. fasciculatus* (Lea, 1906b), *L. ferox* (Lea, 1916), *L. ferus* (Pascoe, 1870d), *L. fraterculus* (Lea, 1906b), *L. frontalis* (Blackburn, 1889), *L. fumatus* (Lea, 1917a), *L. furfuraceus* (Pascoe, 1882), *L. gladiator* (Lea, 1906b), *L. glaucus* (Pascoe, 1882), *L. globicollis* (Lea, 1906b), *L. graniventris* (Lea, 1917a), *L. gravis* (Blackburn, 1892c), *L. hercules* (Lea, 1906b), *L. hirsutus* (Lea, 1906b), *L. horni* (Blackburn, 1896), *L. horridus* (Lea, 1904), *L. humeralis* (Germar, 1848), *L. hypocritus* (Pascoe, 1871a), *L. iliacus* (Pascoe, 1871a), *L. incomptus* (Pascoe, 1883a), *L. inermis* (Lea, 1916), *L. insignis* (Blackburn, 1889), *L. insularis* (Lea, 1916), *L. interruptus* (Lea, 1906b), *L. intricatus* (Lea, 1917a), *L. laticollis* (Lea, 1906b), *L. latipennis* (Lea, 1916), *L. longipes* (Lea, 1916), *L. maleficus* (Lea, 1904), *L. minor* (Lea, 1917a), *L. mirabilis* (Lea, 1926a), *L. mucidus* (Lea, 1917a), *L. multinodosus* (Lea, 1906b), *L. muricatus* (Pascoe, 1873b), *L. murinus* (Lea, 1917a), *L. musimon* (Pascoe, 1873b), *L. nigropunctatus* (Lea, 1906b), *L. nitidicollis* (Lea, 1916), *L. nitidiventris* (Lea, 1916), *L. niveus* (Lea, 1906b), *L. nodicollis* (Lea, 1904), *L. obesus* (Lea, 1910b), *L. orthodoxus* (Lea, 1906b), *L. palmensis* (Blackburn, 1896), *L. parvicornis* (Lea, 1906b), *L. polyacanthus* (Pascoe, 1870b), *L. puellaris* (Pascoe, 1882), *L. punctiger* (Pascoe, 1883a), *L. quadridens* (Fabricius, 1775), *L. recurvus* (Lea, 1914), *L. retusus* (Pascoe, 1870b), *L. rhizophagus* (Lea, 1914), *L. robustus* (Olivier, 1807), *L. scaber* (Lea, 1917a), *L. setosus* (Lea, 1904), *L. spencei* (Boheman, 1834) (*in* Schoenherr 1834a), *L. spinipennis* (Fåhraeus, 1842) (*in* Schoenherr 1842a), *L. spinosus* (Fåhraeus, 1842) (*in* Schoenherr 1842), *L. sulcicollis* (Blackburn, 1892c), *L. suturalis* (Lea, 1916), *L. t-squameus* (Lea, 1910b), *L. tempeensis* (Blackburn, 1896), *L. tetraphysodes* (Pascoe, 1871a), *L. tribulus* (Fabricius, 1775), *L. tuberculatus* (Macleay, 1826), *L. vacillans* (Lea, 1910b), *L. variegatus* (Lea, 1916), *L. vermicosus* (Pascoe, 1883a), *L. ziczac* (Lea, 1915b).
Distribution. Australia (QLD, NSW, Lord Howe Is. (two species), VIC, TAS, SA, NT, WA), New Zealand.
Notes. The genus is only recorded from Tasmania by one species, *L. duponti*, which is associated with calcified pupal cocoons commonly found in SA (Tilley *et al.* 1997), similar records from bauxite deposits in north Queensland correspond with beetles species that remain undetermined (Eggleton & Taylor 2008). *Leptopius robustus* has been recorded from New Zealand (Leschen *et al.* 2023). The host plants recorded include species of *Acacia* and *Cassia* (Fabaceae) and several non-native plants such as sugarcane, grapes, apricot and apple trees.

Lipothyrea Pascoe, 1882 (Pl. 66K, L; Fig. 44.53B)

Type species, by monotypy: *Lipothyrea chloris* Pascoe, 1882.
Australian species. *L. arrowi* Lea, 1911b, *L. chloris* Pascoe, 1882.
Distribution. Australia (QLD).
Notes. Host plants recorded are grasses (Poaceae) such as *Alloteropsis semialata, Coelorachis rottboellioides, Megathyrsus maximus, Melinis repens* and *Themeda triandra* (A. Howden, unpublished notes).

Lysizone Pascoe, 1870 (Pl. 67A)

Type species, by monotypy: *Lysizone alternata* Pascoe, 1870.
Australian species. *L. alternata* Pascoe, 1870c, *L. conifer* (Lea, 1910b).
Distribution. Australia (QLD, NT, WA).
Notes. Specimens have been recorded from *Cyamopsis tetragonoloba* (Fabaceae).

Mandalotus Erichson, 1842 (Pl. 67C, D; Figs 44.44E–L, 44.46, 44.57D)

Type species, by subsequent designation (May, 1993: 54): *Mandalotus crudus* Erichson, 1842.

Australian species. *M. abdominalis* Lea, 1914, *M. acutangulus* Lea, 1911a, *M. advenus* (Blackburn, 1892b), *M. albonotatus* Lea, 1907a, *M. alpinus* Lea, 1926d, *M. amplicollis* Lea, 1904, *M. angustipictus* Lea, 1911a, *M. angustus* Lea, 1914, *M. arciferus* Lea, 1907a, *M. arcuatus* Lea, 1907a, *M. armicoxis* Lea, 1927b, *M. armipectus* Lea, 1909c, *M. armivarius* Lea, 1926d, *M. auchmeresthes* Lea, 1926d, *M. avenaceus* Lea, 1907a, *M. bicarinatus* Lea, 1912, *M. bilobicollis* Lea, 1926d, *M. bimaculatus* Lea, 1926d, *M. bivitticollis* Lea, 1926d, *M. blackburni* Lea, 1907a, *M. blackmorei* Lea, 1907a, *M. brevicarinatus* Lea, 1926d, *M. bryophagus* Lea, 1907a, *M. bryophilus* Oke, 1931, *M. campylocnemis* Lea, 1904, *M. canalicornis* Lea, 1931b, *M. carinatipes* Lea, 1911a, *M. carteri* Lea, 1907a, *M. cellaris* (Pascoe, 1873b), *M. ciliatus* Lea, 1914, *M. cinereus* Lea, 1931b, *M. coatesi* Lea, 1907a, *M. collaris* Lea, 1926d, *M. contortus* Lea, 1931b, *M. corrugicollis* Lea, 1929a, *M. coxalis* Lea, 1909c, *M. crawfordi* (Blackburn, 1890b), *M. crudus* Erichson, 1842, *M. decipiens* Lea, 1914, *M. denticulatus* Lea, 1926d, *M. dentipes* Lea, 1911a, *M. dolens* Lea, 1929a, *M. egenus* Oke, 1931, *M. excavatus* Lea, 1904, *M. exilis* Oke, 1934, *M. explanicollis* Oke, 1931, *M. femoralis* Lea, 1929a, *M. fergusoni* Lea, 1907a, *M. ferrugineus* Lea, 1914, *M. fimbriatus* Lea, 1929a, *M. foveatus* Lea, 1912, *M. fuligineus* (Pascoe, 1870e), *M. funereus* Lea, 1926d, *M. geminatus* Lea, 1907a, *M. glaber* (Blackburn, 1892b), *M. graminicola* Oke, 1931, *M. granicollis* Lea, 1931b, *M. granulatus* Lea, 1907a, *M. gymnogaster* Lea, 1926d, *M. hoplocnemus* Lea, 1926d, *M. hoplosternus* Lea, 1916, *M. hoplostethus* (Pascoe, 1870e), *M. humeralis* Lea, 1907a, *M. hypulus* Lea, 1926d, *M. hystricosus* Lea, 1926d, *M. imitator* Lea, 1907a, *M. imponderosus* Lea, 1926d, *M. impressicollis* Oke, 1931, *M. incisipes* Lea, 1931b, *M. incisus* Lea, 1909c, *M. inconspicuus* Lea, 1916, *M. insignipes* Lea, 1926d, *M. insignis* Lea, 1931b, *M. insularis* Lea, 1926d, *M. intercoxalis* Lea, 1907a, *M. interocularis* Lea, 1911a, *M. inusitatus* Lea, 1907a, *M. irrasus* Lea, 1911a, *M. laminatipes* Lea, 1915a, *M. laminipectus* Lea, 1916, *M. latebricola* Lea, 1926d, *M. latens* Lea, 1914, *M. latus* Lea, 1912, *M. leai* Oke, 1931, *M. litoralis* Lea, 1907a, *M. longicollis* Lea, 1907a, *M. lucaris* Oke, 1934, *M. luciphilus* Oke, 1934, *M. lutosus* Lea, 1914, *M. macrops* Lea, 1926d, *M. maculatus* Lea, 1907a, *M. magnicollis* Lea, 1917a, *M. medcoxalis* Lea, 1926d, *M. medianus* Lea, 1931b, *M. melancholicus* Lea, 1931b, *M. mesosternalis* Lea, 1909c, *M. metasternalis* Lea, 1909c, *M. microscopicus* Lea, 1917a, *M. minusculus* Oke, 1931, *M. minutus* Lea, 1914, *M. miricollis* (Broun, 1917), *M. modicus* Lea, 1931b, *M. multicarinatus* Lea, 1926d, *M. murrayi* Lea, 1923, *M. muscivorus* Lea, 1909c, *M. niger* Lea, 1907a, *M. nodicollis* Lea, 1907a, *M. norfolcensis* Lea, 1926d, *M. obliquus* Oke, 1934, *M. ochreonotatus* Lea, 1907a, *M. octagonalis* Oke, 1931, *M. oculivorus* Lea, 1931b, *M. oxyomus* Lea, 1926d, *M. pallidus* Lea, 1904, *M. parentheticus* Lea, 1931b, *M. pentagonalis* Lea, 1926d, *M. pentagonoderes* Lea, 1929a, *M. piliventris* Lea, 1904, *M. postcoxalis* Lea, 1926d, *M. prosternalis* Lea, 1909c, *M. puncticollis* Lea, 1926d, *M. punctiventris* (Blackburn, 1890b), *M. pyrifer* Lea, 1926d, *M. raui* Lea, 1916, *M. recticarinatus* Lea, 1926d, *M. reticulatus* Lea, 1904, *M. rufimanus* Lea, 1914, *M. rufipes* Lea, 1912, *M. sabulosus* Lea, 1907a, *M. scaber* Lea, 1904, *M. seticollis* Lea, 1907a, *M. setistriatus* Lea, 1926d, *M. setosus* Lea, 1907a, *M. severini* Lea, 1911b, *M. similis* Lea, 1907a, *M. simulator* Lea, 1907a, *M. spurcus* Lea, 1904, *M. squalidus* Lea, 1914, *M. squamibundus* Lea, 1911a, *M. squamosus* Lea, 1926d, *M. sterilis* Erichson, 1842, *M. sternocerus* Lea, 1927b, *M. striatus* Lea, 1926d, *M. subglaber* Lea, 1904, *M. subhumeralis* Lea, 1916, *M. suturalis* Lea, 1904, *M. sydneyensis* Lea, 1907a, *M. taylori* Lea, 1907a, *M. tenuicornis* Lea, 1912, *M. tenuis* Lea, 1926d, *M. tibialis* Lea, 1916, *M. transversus* Lea, 1914, *M. trisinuatus* Lea, 1907a, *M. tuberculiventris* Lea, 1907a, *M. tuberipennis* Lea, 1929a, *M. uniformis* Lea, 1926d, *M. vacillans* Lea, 1907a, *M. valgus* (Pascoe, 1870c), *M. variabilis* Lea, 1907a, *M. vigilans* Lea, 1926d, *M. villosipes* Lea, 1929a.

Distribution. Australia (QLD, NSW, Lord Howe Is., VIC, TAS, SA, Norfolk Is.), New Zealand (introduced).

Notes. *Mandalotus* is a genus of ground weevils quite common in leaf litter but also in grass tussocks, moss and lichen, under stones and among roots of beach plants. The adults appear to be mainly nocturnal. A teneral adult has been found in a fruiting body of *Mesophellia* (Mesophelliaceae), a genus of subterranean truffle-like fungi. *Mandalotus* is particularly diverse in NSW and VIC, and the species number seems to be much smaller in QLD, TAS and SA, and one species, *M. miricollis*, has been recorded from New Zealand, its larva there described by May (1977, 1994). The adults of some species display remarkable sexual dimorphism, particularly on the ventral side and hindlegs. A few species have been indicated to be minor agricultural pests (e.g. *M. spurcus*).

Merimnetes Schoenherr, 1842 (Pl. 67E, F; Fig. 44.62I)

Type species, by original designation: *Merimnetes uniformis* Boheman, 1842.

Australian species. *M. australis* (Boisduval, 1835), *M. celmisiae* Lea, 1910a, *M. decipiens* Lea, 1910a, *M. montanus* Lea, 1910a, *M. oblongus* (Blanchard, 1853), *M. simplicipennis* Lea, 1910a, *M. uniformis* Boheman, 1842 (*in* Schoenherr 1842b), *M. viridis* Lea, 1910a.

Distribution. Australia (NSW, VIC, TAS).

Notes. Host plants recorded are *Celmisia* and *Olearia* (Asteraceae), *Leptospermum* (Myrtaceae), *Nothofagus cunninghamii* (Nothofagaceae) and *Pomaderris eliptica* (Rhamnaceae).

Neomerimnetes Blackburn, 1900 (Pl. 67G)

Type species, by original designation: *Neomerimnetes destructor* Blackburn, 1900.

Australian species. *N. destructor* Blackburn, 1900, *N. flindersiae* Marshall, 1938a, *N. inflatus* Lea, 1910a, *N. sobrinus* Lea, 1911b.

Distribution. Australia (QLD, NSW, SA).

Notes. *Flindersia australis* (Rutaceae) has been recorded as a host plant of the genus.

Ochrometa Pascoe, 1870 (Pl. 67H)

Type species, by monotypy: *Ochrometa amoena* Pascoe, 1870.

Australian species. *O. amoena* Pascoe, 1870b.

Distribution. Australia (WA).

Notes. The Kangaroo Paw *Anigozanthos pulcherrimus* (Haemodoraceae) has been recorded as a host plant of the genus.

Ocynoma Pascoe, 1873 (Pl. 67I)

Type species, by subsequent designation (Pullen *et al.* 2014: 166): *Ocynoma antennata* Pascoe, 1873.

Australian species. *O. antennatum* Pascoe, 1873b, *O. cordipenne* Pascoe, 1873b.

Distribution. Australia (WA).

Onesorus Pascoe, 1870 (Pl. 67J, K; Figs 44.54D, H, 44.58E, J, 44.59D)

Type species, by subsequent designation (Pullen *et al.* 2014:166): *Onesorus maculosus* Pascoe, 1870.

Australian species. *O. armipennis* Lea, 1916, *O. candidus* Pascoe, 1870c, *O. farinosus* (Blackburn, 1896), *O. hoplocnemus* Lea, 1917a, *O. maculosus* Pascoe, 1870c, *O. obesus* Pascoe, 1870c, *O. ocularis* Lea, 1916, *O. pullatus* Lea, 1916, *O. tarsalis* Lea, 1916, *O. tigrinus* Pascoe, 1870c, *O. vermicollis* Lea, 1916.

Notes. The genus is composite and probably needs to be restricted to only the type and a few closely related species. Adults have been recorded from *Eucalyptus* species (Myrtaceae).

Distribution. Australia (QLD).

Oophthalmus Marshall, 1943 (Pl. 67L)

Type species, by monotypy: *Oops pistor* Germar, 1848.

Australian species. *O. pistor* (Germar, 1848), *O. rostralis* (Blackburn, 1894c), *O. viridis* (Oke, 1931).

Distribution. Australia (VIC, NSW, SA).

Notes. Specimens have been recorded from *Eucalyptus* species (Myrtaceae).

Pascoellus Oke, 1951 (Pl. 68A)

Type species, by monotypy: *Pephricus echimys* Pascoe, 1870.

Australian species. *P. echimys* (Pascoe, 1870d), *P. fissiceps* (Lea, 1904), *P. militaris* (Lea, 1904), *P. rattulus* (Pascoe, 1882), *P. squalidus* (Blackburn, 1892a), *P. vittaticeps* (Blackburn, 1894c).

Distribution. Australia (SA, WA).

Peripagis Pascoe, 1870 (Pl. 68B)

Type species, by monotypy: *Peripagis rufipes* Pascoe, 1870.

Australian species. *P. rufipes* Pascoe, 1870d.

Distribution. Australia (NSW).

Perperus Schoenherr, 1842 (Pl. 68C, D; Fig. 44.45J–S)

Type species, by original designation: *Perperus innocuus* Boheman, 1842.

Australian species. *P. angustibasis* Lea, 1911c, *P. conloni* Lea, 1908c, *P. convexipennis* Blackburn, 1901, *P. innocuus* Boheman, 1842 (*in* Schoenherr 1842a), *P. insularis* Boheman, 1842 (*in* Schoenherr 1842a), *P. languidus* (Erichson, 1842), *P. lateralis* (Boisduval, 1835), *P. obscurus* Boheman, 1842 (*in* Schoenherr 1842a), *P. sinuatus* (Blackburn, 1892b), *P. tuberculatus* Lea, 1910b, *P. urticarum* Pascoe, 1873b, *P. variegatus* Pascoe, 1873b.

Distribution. Australia (QLD, NSW, VIC, Bass Strait Is.).

Notes. Specimens have been recorded feeding on leaves of apple and grape as well as of *Hakea* (Proteaceae).

Platypterocis Lea, 1914 (Pl. 68E)

Type species, by monotypy: *Platypterocis paradoxus* Lea, 1914.

Australian species. *P. paradoxus* Lea, 1914.

Distribution. Australia (SA).

Polyphrades Schoenherr, 1840 (Pl. 68F–H; Figs 44.52D, 44.58G, 44.59G)

Type species, by original designation: *Polyphrades cinereus* Fåhraeus, 1840 (*in* Schoenherr 1840a).

Australian species. *P. apicalis* Lea, 1917a, *P. basalis* Lea, 1916, *P. biplagiatus* Pascoe, 1870e, *P. cordatus* Lea, 1917a, *P. cordipennis* Lea, 1908a, *P. crassicornis* Lea, 1914, *P. duriusculus* Lea, 1916, *P. emblematicus* Lea, 1908a,

P. exoletus Lea, 1904, *P. extenuatus* Lea, 1904, *P. farinosus* Lea, 1930c, *P. fortis* Blackburn, 1894c, *P. gibbipennis* Lea, 1930c, *P. granicollis* Lea, 1908a, *P. halmaturinus* Lea, 1916, *P. hartmeyeri* Lea, 1909b, *P. inconspicuus* Blackburn, 1894c, *P. insignipennis* Lea, 1914, *P. laetus* Blackburn, 1894c, *P. laminatus* Blackburn, 1894c, *P. laticollis* Fåhraeus, 1840 (*in* Schoenherr 1840a), *P. latipennis* Pascoe, 1870e, *P. latus* Lea, 1908a, *P. longipennis* Pascoe, 1872a, *P. macrops* Lea, 1916, *P. maestus* Lea, 1916, *P. marmoratus* Lea, 1914, *P. maurulus* (Blackburn, 1892c), *P. modestus* Blackburn, 1894c, *P. murinus* Fåhraeus, 1840 (*in* Schoenherr 1840a), *P. nanus* (Gyllenhal, 1833) (*in* Schoenherr 1833b), *P. obscurus* (Germar, 1848), *P. ortyx* Pascoe, 1870e, *P. paganus* Boheman, 1845 (*in* Schoenherr 1845), *P. pardalotus* Pascoe, 1870e, *P. parvus* Lea, 1908a, *P. perplexus* Blackburn, 1894c, *P. pictus* Blackburn, 1894c, *P. planipennis* Lea, 1916, *P. pusillus* Pascoe, 1870e, *P. raui* Lea, 1917c, *P. rugulosus* Blackburn, 1892c, *P. satelles* Blackburn, 1890c, *P. tumidulus* Blackburn, 1892c, *P. uniformis* Lea, 1909b.

Distribution. Australia (QLD, NSW, VIC, SA, NT, WA).

Notes. The concept of the genus is unclear, and it may have to be restricted to a smaller number of species. Specimens have been recorded from *Cassia* (Fabaceae).

Prosayleus Schoenherr, 1840 (Pl. 68I)

Type species, by original designation: *Prosayleus hopei* Boheman, 1840.

Australian species. *P. hopei* Boheman, 1840 (*in* Schoenherr 1840a), *P. latipennis* Lea, 1910b.

Distribution. Australia (NSW, SA).

Notes. Specimens have been recorded from *Hakea* species (Proteaceae).

Prostomus Schoenherr, 1823 (Pl. 68J; Figs 44.55A, B, 44.57E, F)

Type species, by subsequent designation (Schoenherr, 1826: 142): *Curculio scutellaris* Fabricius, 1787.

Australian species. *P. murinus* (Lea, 1911a), *P. scutellaris* (Fabricius, 1787).

Distribution. Australia (TAS).

Prypnus Schoenherr, 1823 (Pl. 68K, L; Fig. 44.57B)

Type species, by original designation: *Prypnus quinquenodosus* Schoenherr, 1823.

Australian species. *P. angustus* Lea, 1911b, *P. australis* (Boisduval, 1835), *P. canaliculatus* Gyllenhal, 1842 (*in* Schoenherr 1842a), *P. fallax* Gyllenhal, 1842 (*in* Schoenherr 1842a), *P. fausti* Chadwick, 1965, *P. glaber* Lea, 1911b, *P. quinquenodosus* Schoenherr, 1823, *P. squalidus* (Gyllenhal, 1833) (*in* Schoenherr 1833b), *P. tenebriosus* (Gyllenhal, 1833) (*in* Schoenherr 1833b), *P. tibialis* Lea, 1911b, *P. trituberculatus* Germar, 1848.

Distribution. Australia (NSW, SA, WA).

Psapharus Schoenherr, 1823 (Pl. 69A–C; Fig. 44.59H)

Type species, by original designation: *Curculio infaustus* Olivier, 1807.

Australian species. *P. aureolus* (Pascoe, 1872a), *P. infaustus* (Olivier, 1807), *P. mastersii* (Pascoe, 1871c), *P. mistothes* (Herbst, 1797), *P. nitidilabris* (Germar, 1848), *P. oesalon* (Pascoe, 1870e), *P. punctipennis* (Pascoe, 1871c), *P. ruficornis* (Blackburn, 1893b), *P. strigiceps* (Lea, 1911c), *P. vestitus* (Pascoe, 1870f).

Distribution. Australia (QLD, NSW, VIC, TAS, SA, WA).

Notes. A teneral specimen was found in a fruiting body of *Mesophellia* (Mesophelliaceae), a genus of subterranean truffle-like fungi.

Pseudotimareta Zimmerman & Oberprieler, 2014 (Pl. 69D)

Type species, by original designation: *Timareta subterranea* Lea, 1908.

Australian species. *P. inconstans* (Lea, 1909c), *P. intermixta* (Lea, 1909c), *P. puncticollis* (Lea, 1909c), *P. subterranea* (Lea, 1908c), *P. swanseaensis* (Lea, 1909c).

Distribution. Australia (TAS).

Notes. The genus was described in Pullen *et al.* (2014).

Rhinoscapha Montrouzier, 1855 (Pl. 69E)

Type species, by monotypy: *Rhinoscapha bicincta* Montrouzier, 1855.

Australian species. *R. cobaltinata* Heller, 1914, *R. doriae* Pascoe, 1885.

Distribution. Australia (QLD), New Guinea (incl. Trobriand Is., Woodlark Is.), Solomon Is., Indonesia (Aru Is., Java, Maluku Is., Sunda Is.), Vanuatu, Fiji, New Caledonia.

Rhyparophilus Schoenherr, 1845 (Pl. 69F, G)

Type species, by original designation: *Rhyparophilus alternans* Boheman, 1845.

Australian species. *R. alternans* Boheman, 1845 (*in* Schoenherr 1845), *R. ammophilus* (Lea, 1909c), *R. goudiei* (Lea, 1931b), *R. herbivorus* (Lea, 1909c), *R. howensis* (Lea, 1926d), *R. pondericornis* (Lea, 1911a).

Distribution. Australia (NSW, Lord Howe Is., VIC, TAS).

Scotasmus Schoenherr, 1842 (Pl. 69H, I)

Type species, by original designation: *Scotasmus carinirostris* Boheman, 1842.

Australian species. *S. carinirostris* Boheman, 1842 (*in* Schoenherr 1842a), *S. costirostris* (Lea, 1908), *S. litoralis* (Lea, 1911), *S. parvicornis* (Lea, 1911).
Distribution. Australia (TAS, King Is.).
Notes. The larvae have been found under pasture and strawberry plants. Also *Leptospermum* (Myrtaceae) seems to be a host.

Stenocorynus Schoenherr, 1823 (Pl. 69J)

Type species, by original designation: *Curculio crenulatus* Fabricius, 1775.
Australian species. *S. albolineatus* (Macleay, 1826), *S. alleni* Lea, 1908b, *S. apicipennis* Lea, 1910b, *S. aridus* Pascoe, 1872a, *S. crenulatus* (Fabricius, 1775), *S. cretatus* Lea, 1908b, *S. lateralis* Boheman, 1834 (*in* Schoenherr 1834a), *S. lepidurus* Lea, 1916, *S. marginatus* Pascoe, 1885, *S. minor* Lea, 1908b, *S. ovalipennis* (Pascoe, 1871a), *S. subfasciatus* (Pascoe, 1870d), *S. variabilis* (Blackburn, 1889), *S. vittatus* Pascoe, 1872a.
Distribution. Australia (QLD, NT).

Synaptonyx Waterhouse, 1853 (Pl. 69K)

Type species, by monotypy: *Synaptonyx ovatus* Waterhouse, 1853.
Australian species. *S. ovatus* Waterhouse, 1853.
Distribution. Australia (QLD, WA).

Telenica Pascoe, 1872 (Pl. 69L)

Type species, by subsequent designation (Pullen *et al.* 2014: 175): *Telenica sublimbata* Pascoe, 1872.
Australian species. *T. nebulosa* Pascoe, 1872c, *T. sublimbata* Pascoe, 1872c.
Distribution. Australia (WA).

Timareta Pascoe, 1872 (Pl. 70A, B; Fig. 44.57C)

Type species, by subsequent designation (Pullen *et al.* 2014: 175): *Timareta figurata* Pascoe, 1872.
Australian species. *T. crinita* Pascoe, 1874a, *T. figurata* Pascoe, 1872c, *T. hamata* Lea, 1922, *T. incisipes* Lea, 1922, *T. infortunata* Lea, 1931b, *T. pilipes* (Pascoe, 1870e), *T. pilosa* (Blackburn, 1890b), *T. pusilla* Blackburn, 1894c.
Distribution. Australia (SA, WA).
Notes. The species occur in sandy areas and on offshore islands. The adults are nocturnal and the larvae feed among on roots.

Undarobius Escalona & Oberprieler, 2021 (Pl. 70C)

Type species, by original designation: *Undarobius howarthi* Escalona & Oberprieler, 2021.
Australian species. *U. howarthi* Escalona & Oberprieler, 2021, *U. irvini* Escalona & Oberprieler, 2021
Distribution. Australia (QLD).
Notes. Both species appear to only occur deep in lava caves in Undara Volcanic N.P., in which the adults and larvae probably feed on tree roots that penetrate the roof of the caves.

Uroleptops Lea, 1915 (Pl. 70D)

Type species, by original designation: *Uroleptops impendens* Lea, 1915.
Australian species. *U. impar* Lea, 1915a, *U. impendens* Lea, 1915a.
Distribution. Australia (NSW, VIC).

Xynaea Pascoe, 1865 (Pl. 70E; Fig. 44.58C)

Type species, by monotypy: *Xynaea saginata* Pascoe, 1865.
Australian species. *X. saginata* Pascoe, 1865.
Distribution. Australia (SA).

Xyneella Zimmerman, 1994 (Pl. 70F; Fig. 44.58H)

Type species, by monotypy: *Xynaea uniformis* Lea, 1904.
Australian species. *X. uniformis* (Lea, 1904).
Distribution. Australia (WA).

Zymaus Pascoe, 1870 (Pl. 70G; Fig. 44.59C)

Type species, by monotypy: *Zymaus binodosus* Pascoe, 1870.
Australian species. *Z. angustus* Lea, 1910b, *Z. binodosus* Pascoe, 1870e.
Distribution. Australia (QLD).

Tribe Naupactini Gistel, 1848

Atrichonotus Buchanan, 1939 (Pl. 70H; Fig. 44.47I–Q)

Type species, by original designation: *Naupactus taeniatulus* Berg, 1881.
Australian species. *A. taeniatulus* Berg, 1881.
Distribution. Australia (NSW, WA), Argentina, Brazil, Chile, Uruguay, southern USA, New Zealand.
Notes. The single species in Australia was accidentally introduced from Argentina in about 1938 (Pullen *et al.* 2014; Oberprieler & Zimmerman 2020).

Eurymetopus Schoenherr, 1840 (Pl. 70I)

Type species, by original designation: *Eurymetopus fallax* Boheman, 1840 (*in* Schoenherr 1840b).
Australian species. *E. birabeni* Kuschel, 1945.
Distribution. Australia (NSW, VIC), Argentina, Brazil, Uruguay.

Notes. The single species in Australia was accidentally introduced from South America in about 1962 (Pullen *et al.* 2014; Oberprieler & Zimmerman 2020).

Floresianus Hustache, 1939 (Pl. 70J; Fig. 44.48A–G)

Type species, by original designation: *Floresianus sordidus* Hustache, 1939.
Australian species. *F. sordidus* Hustache, 1939.
Distribution. Australia (NSW), Argentina, Brazil, Uruguay, New Zealand.
Notes. The single species in Australia was accidentally introduced from Argentina in about 1941 (Pullen *et al.* 2014; Oberprieler & Zimmerman 2020).

Naupactus Dejean, 1821 (Pl. 70K; Fig. 44.54G)

Type species, by subsequent designation (Champion, 1911: 232): *Curculio rivulosus* Olivier, 1790.
Australian species. *N. cervinus* Boheman, 1840 (*in* Schoenherr 1840b), *N. leucoloma* Boheman, 1840 (*in* Schoenherr 1840b), *N. minor* (Buchanan, 1942), *N. peregrinus* (Buchanan, 1939).
Distribution. Australia (QLD, NSW, SA, TAS, VIC, WA), America, St. Helena, South Africa, Kermadec Is., New Zealand.
Notes. All the species in Australia have been accidentally introduced from South America (Pullen *et al.* 2014; Oberprieler & Zimmerman 2020). Whether *N. minor* is established in Australia is uncertain (Oberprieler & Zimmerman 2020).

Tribe Oosomini Lacordaire, 1863

Phlyctinus Schoenherr, 1826 (Pl. 70L; Figs 44.50A–G, J, 44.52E)

Type species, by original designation: *Peritelus callosus* Schoenherr, 1826.
Australian species. *P. callosus* (Schoenherr, 1826).
Distribution. Australia (NSW, SA, TAS, VIC, WA, Norfolk Is.), Africa, St. Helena, Kermadec Is., New Zealand.
Notes. The single species in Australia was accidentally introduced from South Africa in about 1888 (Pullen *et al.* 2014; Oberprieler & Zimmerman 2020).

Tribe Otiorhynchini Schoenherr, 1826

Otiorhynchus Germar, 1822 (Pl. 71A; Figs 44.49A–S, 44.50H, I, 44.54F, 44.56F, 44.57J, 44.62L)

Type species, by original designation: *Otiorhynchus carcelloides* Stierlin, 1888.
Australian species. *O. cribricollis* Gyllenhal, 1834 (*in* Schoenherr 1834b), *O. rugosostriatus* (Goeze, 1777), *O. sulcatus* (Fabricius, 1775).
Distribution. Australia (ACT, NSW, SA, TAS, VIC, WA), Palaearctic region, introduced in North America, Chile, Juan Fernandez Is., St. Helena, Hawaii, Madagascar, New Zealand.
Notes. The three species in Australia are all accidental introductions from Europe (Pullen *et al.* 2014; Oberprieler & Zimmerman 2020).

Tribe Ottistirini Heller, 1825

Eutinophaea Pascoe, 1870 (Pl. 71B, C)

Type species, by monotypy: *Eutinophaea nana* Pascoe, 1870.
Australian species. *E. nana* Pascoe, 1870d.
Distribution. Australia, New Guinea?

Maleuterpes Blackburn, 1894 (Pl. 71D; Fig. 44.50K–R)

Type species, by monotypy: *Maleuterpes spinipes* Blackburn, 1894.
Australian species. *M. spinipes* Blackburn, 1894c.
Distribution. Australia (QLD, NSW, NT, SA, VIC), New Guinea, New Caledonia, Indonesia, New Zealand (introduced).

Monteithus Oberprieler & Zimmerman, 2020 (Pl. 71E)

Type species, by original designation: *Eutinophaea falcata* Lea, 1904.
Australian species. *M. falcatus* (Lea, 1904), *M. subviridis* (Lea, 1930a).
Distribution. Australia (QLD).

Ottistirellus Oberprieler & Zimmerman, 2020 (Pl. 71F)

Type species, by original designation: *Eutinophaea variegata* Lea, 1914.
Australian species. *O. variegatus* (Lea, 1914).
Distribution. Australia (QLD).

Ottistirodes Oberprieler & Zimmerman, 2020 (Pl. 71G, H; Fig. 44.53G)

Type species, by original designation: *Eutinophaea murina* Lea, 1930.
Australian species. *O. murinus* (Lea, 1930a).
Distribution. Australia (QLD, NT).

Reidius Oberprieler & Zimmerman, 2020 (Pl. 71I)

Type species, by original designation: *Eutinophaea bicristata* Lea, 1929.
Australian species. *O. bicristatus* (Lea, 1929b), *O. fasciculatus* (Lea, 1930a), *O. setistriatus* (Lea, 1930a).
Distribution. Australia (QLD, NSW).

Tregoleus Oberprieler & Zimmerman, 2020 (Pl. 71J)

Type species, by original designation: *Tregoleus griseus* Oberprieler & Zimmerman, 2020.
Australian species. *T. griseus* Oberprieler & Zimmerman, 2020.
Distribution. Australia (QLD).

Weirius Oberprieler & Zimmerman, 2020 (Pl. 71K)

Type species, by original designation (Oberprieler & Zimmerman, 2020: 430): *Eutinophaea suturalis* Lea, 1930.
Australian species. *O. suturalis* (Lea, 1930a).
Distribution. Australia (QLD, NSW).

Tribe Pachyrhynchini Schoenherr, 1826

Pantorhytes Faust, 1892 (Pl. 71L; Fig. 44.51A–I)

Type species, by original designation: *Pachyrhynchus chrysomelas* Montrouzier, 1855.
Australian species. *P. stanleyanus* (White, 1852).
Distribution. Australia (QLD), New Guinea (including Trobriand Is.), Solomon Is., Fiji.
Reference. Faust, 1892c.

Tribe Sitonini Gistel, 1848

Sitona Germar, 1817 (Pl. 72A; Fig. 44.51J–P)

Type species, by subsequent designation (Schoenherr, 1823: column 1141): *Curculio lineatus* Linnaeus, 1758.
Australian species. *S. discoideus* Gyllenhal, 1834 (*in* Schoenherr 1834a).
Distribution. Australia (ACT, QLD, NSW, SA, TAS, VIC, Norfolk Is.), Nearctic and Palaearctic regions, introduced in New Zealand, southern Africa and South America.
Notes. The single species in Australia was accidentally introduced from the Mediterranean region in the 1950s (Pullen *et al.* 2014; Oberprieler & Zimmerman 2020).

Tribe Tanymecini Lacordaire, 1863

Homoeotrachelus Faust, 1886 (Pl. 72B, C)

Type species, by subsequent designation (Pullen *et al.* 2014: 182): *Homoeotrachelus australasiae* Faust, 1886.
Australian species. *H. australasiae* Faust, 1886, *H. hadromerus* Lea, 1904, *H. tricarinatus* Lea, 1905a.
Distribution. Australia (QLD, SA, NT, VIC, WA).

Hypomeces Schoenherr, 1823 (Pl. 72D; Figs 44.54I, 44.56B, 44.58I)

Type species, by original designation: *Curculio squamosus* Fabricius, 1792.
Australian species. *H. obscurus* (Macleay, 1826).
Distribution. Australia (NT), New Guinea, Indonesia, India, Burma, China, Philippines, Thailand.

Jenningsius Oberprieler & Zimmerman, 2020 (Pl. 72E)

Type species, by original designation: *Jenningsius debbiae* Oberprieler & Zimmerman, 2020.
Australian species. *J. debbiae* Oberprieler & Zimmerman, 2020, *J. williamsi* Oberprieler & Zimmerman, 2020.
Distribution. Australia (QLD, NSW).

Pullenius Oberprieler & Zimmerman, 2020 (Pl. 72F)

Type species, by original designation: *Pullenius kimberii* Oberprieler & Zimmerman, 2020.
Australian species. *P. kimberii* Oberprieler & Zimmerman, 2020.
Distribution. Australia (NSW).

Tanymecus Germar, 1817 (Pl. 72G)

Type species, by subsequent designation (Schoenherr, 1823: column 1144): *Curculio palliatus* Fabricius, 1787.
Australian species. *Tanymecus* near *T. boettcheri* Voss, 1922.
Distribution. Australia (NT), Oriental, Palaearctic, Nearctic (introduced) and Afrotropical regions.
Notes. The single species in Australia was accidentally introduced in about 1984 (Pullen *et al.* 2014; Oberprieler & Zimmerman, 2020).

Unplaced to tribe

Rhyncholobus Gahan, 1900 (Pl. 72H; Fig. 44.60X)

Type species, by subsequent designation (Pullen *et al.* 2014: 183): *Rhyncholobus rossi* Gahan, 1900.
Australian species. *R. andrewsi* Gahan, 1900, *R. discoidalis* Gahan, 1900, *R. rossi* Gahan, 1900, *R. vittatus* Gahan, 1900.
Distribution. Australia (Christmas Is.).

Curculionidae: Curculioninae Latreille, 1802

Introduction. The large subfamily Curculioninae, comprising approximately 350 genera and 4500 species world-wide, is morphologically and taxonomically poorly defined, as are most of the 34 or so tribes currently recognised in it (Caldara *et al.* 2014). Recent molecular-phylogenetic analyses including reasonable representations of the subfamily (Shin *et al.* 2018; Leschen *et al.* 2022; Haran *et al.* 2023) failed to recover Curculioninae as a distinct, monophyletic group, several tribes clustering with representatives of other subfamilies of the CCCMS clade of higher Curculionidae (Gunter *et al.* 2016) and some (Styphlini, Viticiini) even falling in the other major

Fig. 44.63. *Curculio simulator* Zimmerman & Oberprieler.

clade, CEGH. Furthermore, several Australian genera currently classified in Storeini *sensu lato* (Pullen *et al.* 2014) also belong in the CEGH clade (Haran *et al.* 2023). This indicates that the composition and classification of the Curculioninae both globally and in Australia is still largely unresolved and prone to undergo significant changes in the future. Nonetheless, nine of the ten tribes (except Storeini) currently recognised as occurring in Australia are reasonably well defined morphologically or can be differentiated by a combination of characters. However, the Australian faunas of none of them have been comprehensively revised, only the Cranopoeini (Kuschel 2009) and Cryptoplini (Jennings & Oberprieler 2018) having received any taxonomic attention in recent times.

The host plants and life histories of the Australian Curculioninae are also poorly known; the available information was summarised by Caldara *et al.* (2014) and the larvae of only eight species are described (May 1994). The Acalyptini, comprising only three described genera but several undescribed ones, are flower visitors as adults and mainly associated with palms, but their larvae remain unknown. The four species of two of the five genera of Ceutorhynchini, *Mogulones* (Pl. 17E) and *Trichosirocalus*, have been introduced for the biological of Paterson's Curse (*Echium*) and thistles, respectively. Cranopoeini, represented by four described genera, are small weevils characterised by waxy outgrowths on the pronotum and sometimes the elytra and often found on Myrtaceae such as eucalypts, but nothing is specifically recorded about their hosts and habits. Cryptoplini, containing six genera and about 100 species (Jennings & Oberprieler 2018), are mostly associated with Myrtaceae (*Eucalyptus*, *Syzygium*, e.g. Pl. 15F–H), the larvae developing in flower buds, fruits and galls, but those of *Kuschelorhynchus* (Pl. 17C–D) develop in fruits and seeds of cultivated *Macadamia* (Proteaceae) and are considered a pest in plantations of this tree species. Only two genera of Curculionini (Pl. 20B) occur in Australia; the larvae of *Curculio* as known live in fruits of *Ficus* (Moraceae) and *Syzygium* (Myrtaceae) and those of the single species of *Pycnochirus* in fruits of *Alectryon* (Sapindaceae). Eugnomini comprise eight described genera and 41 species in Australia (Pullen *et al.* 2014) as well as several undescribed taxa; the adults are mainly flower visitors and the larvae develop under bark of dead branches and logs or in conflorescences and infructescenses of *Banksia* and probably other Proteaceae. The tribe Ochyromerini is not yet properly recorded from Australia but represented by several undescribed taxa and probably by '*Eudela*' *castanea* Lea (the species not belonging in this genus); the larvae often have the spiracles of segment VIII located on elongated protuberances, apparently an adaptation for breathing in moist environments of fleshy fruits or rotting plant tissues. Members of the two genera of Rhamphini in Australia have greatly enlarged metafemora and readily jump to escape; their larvae are mostly leaf-miners but in some species also develop in seed pods. Smicronychini are represented in Australia by only three species of *Smicronyx*, of which one, *S. lutulentus* Dietz, has been introduced as a biological control agent of the weed *Parthenium hysterophorus* (Asteraceae).

The large tribe Storeini, divided by Pullen *et al.* (2014) into Storeini *sensu stricto* (six genera) and *sensu lato* (34 genera), is a composite group, several genera of the latter cluster recently shown not to belong in Curculioninae but in the CEGH clade of Curculionidae instead (Haran *et al.* 2023). The larvae of Storeini as known develop in flowers, fruits and seed pods of various plants, with some species of *Elleschodes* involved in the pollination of *Eupomatia* (Eupomatiaceae). The larvae of *Cydmaea*, *Dicomada* and *Erytenna* develop in fruits of *Hakea* and *Grevillea* (Proteaceae), and some of their species have been employed in the biological control of *Hakea sericea* in South Africa. Although these last genera and some others are now known not to belong in Curculioninae, they are still treated under this subfamily here.

Key to the Australian tribes and genera of Curculioninae

The following genera are not included in the key: *Diapelmus*, *Meripherellus*, *Mesembrinocis*, *Minia* and *Phaeodica*.

1. Body length approximately 1.1–2.6 mm; maxillary palps 1- to 2-segmented; antennae inserted in basal half of rostrum; metafemora swollen and modified for jumping (Fig. 44.67J–K), larger than pro- and mesofemora and with internal apodeme; metatibial apices without uncus or mucro; tarsal claws widely divergent .... **RHAMPHINI**

– Not with this combination of characters; metafemora not swollen or modified for jumping 2

2(1). Rostrum stout and straight, sometimes apically flattened (Fig. 44.75A); maxillary and labial palps 3-segmented; femora dentate with one or more unequal teeth; tibiae with large uncus and smaller premucro surrounded by setae (Fig. 44.68f); tarsal segment 5 narrow or reduced to absent (e.g. Fig. 44.68o), barely projecting beyond tarsal segment 3; tarsal claws simple and either connate, single or absent; penis with tectal plate fused with pedon at base; habitus as Fig. 44.67E–F .... **CRYPTOPLINI**

– Not with this combination of characters 3

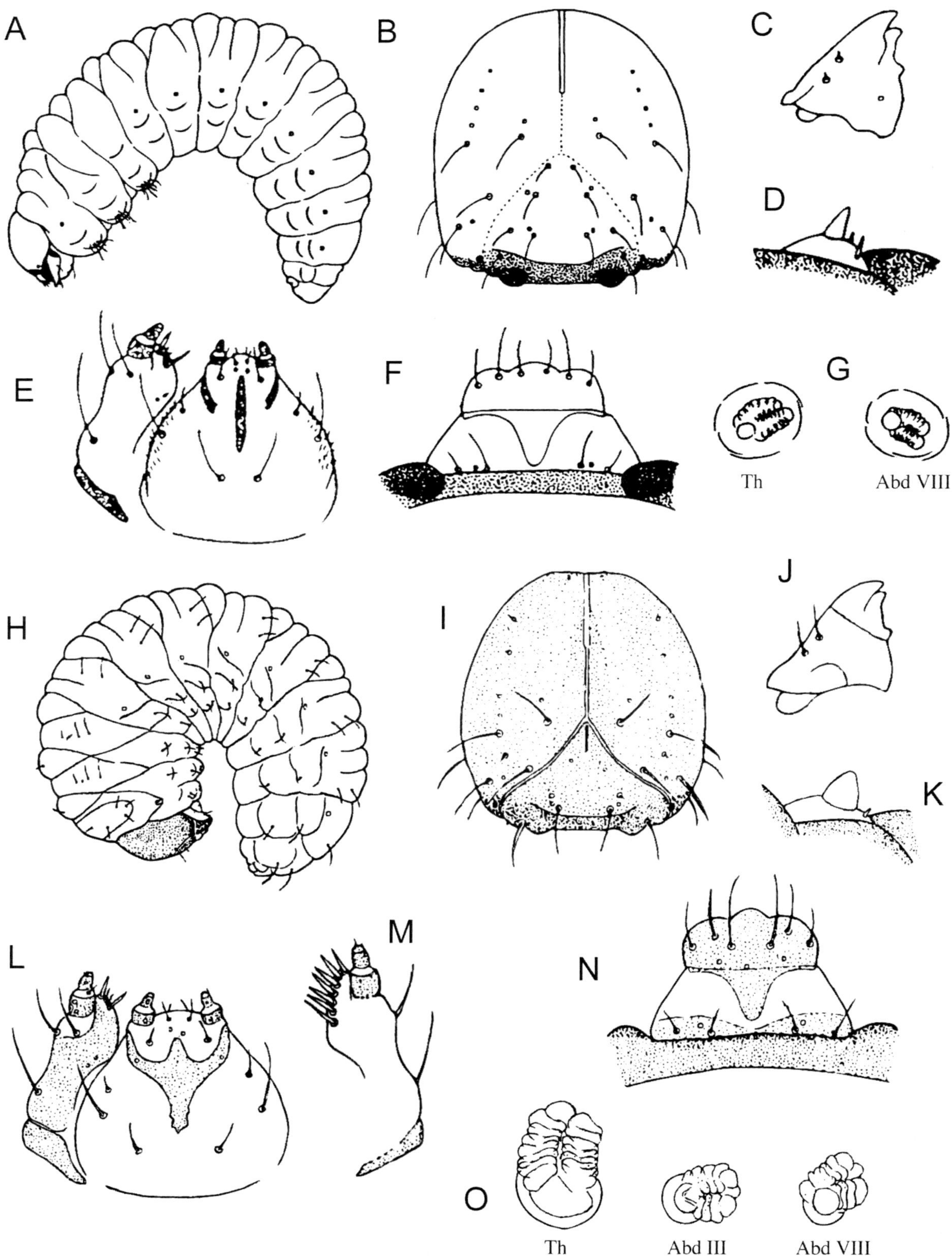

Fig. 44.64. Curculioninae, larvae, Cryptoplini and Curculionini. A–G, *Cryptoplus* sp.: A, habitus, lateral; B, head, anterior; C, mandible; D, antenna; E, mala and labium, posterior; F, labrum, clypeus and epistoma, anterior; G, spiracles, showing alignment; H–N, *Curculio simulator* Zimmerman & Oberprieler: H, habitus, lateral; I, head, anterior; J, mandible; K, antenna; L, mala and labium, posterior; M, mala, anterior; N, labrum, clypeus and epistoma, anterior; O, spiracles, showing alignment. After May (1994).

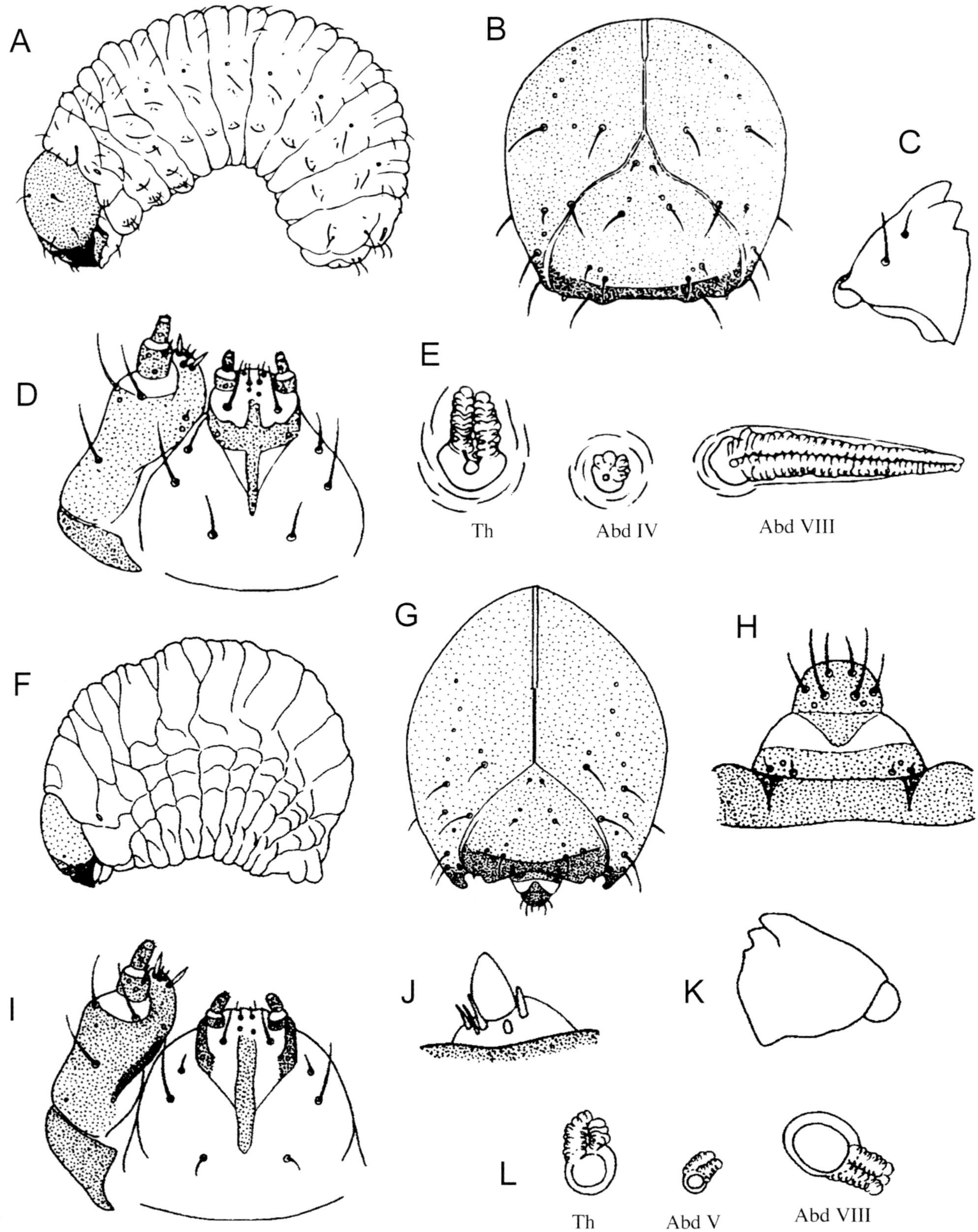

Fig. 44.65. Curculioninae, larvae, Eugnomini. A–E, *Myossita banksiae* Lea: A, habitus, lateral; B, head, anterior; C, mandible; D, labium and mala, posterior; E, spiracles, showing alignment; F–L, *Myossita cirrifera* Pascoe: F, habitus, lateral; G, head, anterior; H, labrum, clypeus and epistoma, anterior; I, mala and labium, posterior; J, antenna; K, mandible; L, spiracles, showing alignment. After May (1994).

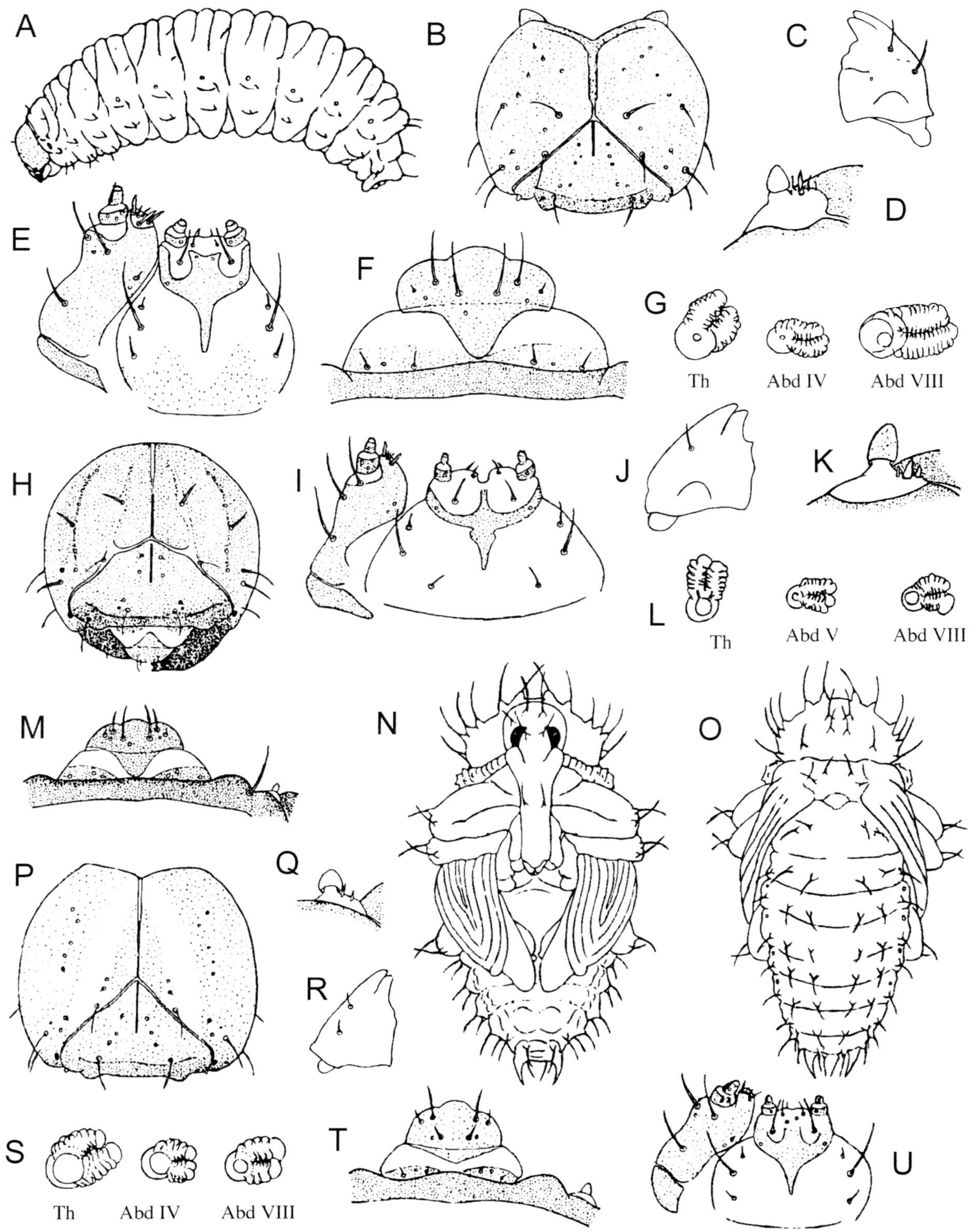

Fig. 44.66. Curculioninae, larvae and pupa, Storeini. A–G, *Elleschodes* sp., larva: A, habitus, lateral; B, head, anterior; C, mandible; D, antenna; E, mala and labium, posterior; F, labrum, clypeus and epistoma, anterior; G, spiracles, showing alignment; H–M, *Erytenna consputa* Pascoe, larva: H, head, anterior; I, mala and labium, posterior; J, mandible; K, antenna; L, spiracles, showing alignment; M, labrum, clypeus, epistoma and antenna, anterior; N–O, *Storeus albosignatus* (Blackburn), pupa (New Zealand): N, habitus, ventral; O, habitus, dorsal; P–U, *S. albosignatus*, larva (New Zealand): P, head, anterior; Q, antenna; R, mandible; S, spiracles, showing alignment; T, labrum, clypeus, epistoma and antenna, anterior; U, mala and labium, posterior. After May (1994).

3(2). Body covered with waxy secretions (Fig. 44.67D) sometimes forming long projections on pronotum; rostrum slightly flattened; rostrum in females (except *Cranopoeus*) with two deep foveae in antennal scrobe; prothorax slightly constricted around apical third, forming a broad collar (Fig. 44.74J), prosternum broadly emarginate (Fig. 44.68C) .... **CRANOPOEINI**

– Not with this combination of characters; body not covered with waxy secretions **4**

4(3). Body length approximately 3.0 mm; head with rostrum in lateral view distinctly set off from head (Fig. 44.67T); eyes large (Fig. 44.67T) and subcontiguous ventrally, separated by about width of one ommatidium; prosternum emarginate (e.g. Fig. 44.68F); elytra with patches of scales (Fig. 44.68X); tibial apices with mucro; tarsal claws connate .... **SMICRONYCHINI** ... ***Smicronyx* Schoenherr**

– Not with this combination of characters; eyes not subcontiguous ventrally, separated by width of 5 or more ommatidia width; tarsal claws often free **5**

5(4). Prosternum with canal (e.g. Fig. 44.68B); procoxae separate; mesepimera large (Fig. 44.68N), often ascending between pronotum and elytral base and visible dorsally; pygidium exposed with transverse carina in basal half, locking elytral apex (Fig. 44.68e) .... **CEUTORHYNCHINI**

– Not with this combination of characters; pygidium without a transverse carina in basal half (e.g. Fig. 44.68c–d) **6**

6(5). Mandibles long, subtriangular, articulating obliquely, or broad and bidentate; rostrum very long and slender (Fig. 44.67G), longer in females; elytra narrowly impressed around scutellar shield; pygidium often exposed; procoxae contiguous; tarsi broad and flat .... **CURCULIONINI**

– Not with this combination of characters **7**

7(6). Prothorax without ocular lobes; prosternum not emarginate (Fig. 44.68G); procoxae contiguous; tibial apices uncinate (e.g. Fig. 44.68f) but uncus often reduced to absent in metatibiae; femora with large flat tooth (e.g. Fig. 44.68i); tarsal claws appendiculate (e.g. Fig. 44.68n); tegmen without parameres; habitus as Fig. 44.67P .... **OCHYROMERINI**

– Not with this combination of characters; femora without tooth or, if present, on all femora, small (e.g. Fig. 44.68k) or large and flat and exodont mandibles (Fig. 44.67Z) **8**

8(7). Eyes large, protuberant (Fig. 44.67H–I); mandibles exodont (Fig. 44.67Z) or not; procoxae contiguous; femora with large tooth (Fig. 44.68i), on metafemora often larger than on others; tarsal segment 2 dorsally often (not in *Stephanorhynchus* and *Rhopalomerus*; Fig. 44.67H–I) projecting medially over base of tarsal segment 3 (Fig. 44.67l) .... **EUGNOMINI**

– Not with this combination of characters; mandibles not exodont **9**

9(8). Body vestiture consisting mainly of setae (e.g. Fig. 44.68P); prosternum not emarginate (Fig. 44.68A), distance in front of procoxae larger than length of procoxal cavity; procoxae narrowly separate; femora rarely with small ventral teeth (some *Tithene*); tibiae rarely with apical spines (uncus, mucro); tergite VII exposed as pygidium in both sexes; habitus as Fig. 44.67A .... **ACALYPTINI**

– Not with this combination of characters; body vestiture consisting of squamae or subglabrous (scattered setae only); tergite VIII exposed as pygidium in males only; habitus as Fig. 44.67L–O .... **STOREINI**

ACALYPTINI

1. Head and rostrum in lateral view with outline slightly downcurved (Fig. 44.67R); rostrum slim and long, in lateral view more than twice as long as head capsule (Fig. 44.67R); mesocoxal process slightly prominent (Fig. 44.68M); femora without ventral prominence or spine; habitus as Fig. 44.67A .... ***Tithene* Pascoe**

– Head and rostrum in lateral view with broadly curved outline (Fig. 44.67U); mesocoxal process not prominent; rostrum robust and shorter, in lateral view about twice as long as head; femora with a small ventral prominence or spine (Fig. 44.68j) **2**

2(1). Dorsal vestiture of pronotum and elytral uniform (Fig. 44.68Q), not forming patterns .... ***Eudelodes* Zimmerman**

– Dorsal vestiture of pronotum and elytral dorsal consisting of pale yellow to pale brown setae, forming irregular patterns .... ***Epamoebus* Blackburn**

EUGNOMINI

1. Tarsal segment 2 dorsally projecting medially over segment 3 (Fig. 44.68l); mandibles exodont (outer margin with at least one tooth, Fig. 44.67Z) and/or curved outwards or not exodont (outer margin without tooth) and curved inwards **2**

– Tarsal segment 2 dorsally not projecting medially over segment 3 (Fig. 44.68m); mandibles without outer tooth and curved inwards **9**

2(1). Funicles 6-segmented (Fig. 44.67W) **3**

– Funicles 7-segmented (Fig. 44.67X–Y) **4**

3(2). Body colour brown, legs conspicuously paler, antennal clubs with apical 2 segments dark; mandibles not exodont .... ***Bothrophasis* Zimmerman & Oberprieler**

– Body colour black; surface smooth to finely reticulate; mandibles long and narrow with long curved inner tooth and shorter outer one (exodont) .... ***Aptilonotus* Zimmerman & Oberprieler**

4(2). Mandibles exodont, outer margin with at least one tooth and/or curved outwards **5**

– Mandibles not exodont, outer margin without tooth, curved inward; habitus as Fig. 44.67H .... ***Ancyttalia* Zimmerman**

5(4). Epistome produced into triangular process extending over mandibles **6**

– Epistome not produced over mandibles .... ***Meripherinus* Lea**

6(5). Body dorsally smooth, black or dark brown, finely setose, pronotal base with patches of dense squamae (Fig. 44.67I); disk of metaventrite (Fig. 44.68O) and abdominal ventrite 1 (Fig. 44.68b) in male conspicuously elevated and flattened .... ***Orpha* Pascoe**

– Body dorsally brown with dense uniform vestiture; metaventrite and abdominal ventrite 1 in male not elevated and flattened (e.g. Fig. 44.68a) **7**

7(6). Mandibles with two external teeth .... ***Meriphus* Erichson**

– Mandibles with a single apical tooth .... ***Myossita* Pascoe**

9(1). Head in lateral view with eyes separated from anterior margin of prothorax by more than twice eye length; habitus as Fig. 44.67C .... ***Stephanorhynchus* White**

– Head in lateral view with eyes separated from anterior margin of prothorax by less than eye length .... ***Rhopalomerus* Blanchard**

CEUTORHYNCHINI

Modified after Colonnelli (2004).

1. Base of pronotum extending into spine between elytra (Fig. 44.68S) and over scutellar shield; habitus as Fig. 44.67B ..... ***Mecysmoderes*** **Schoenherr**
– Base of pronotum not extending into spine between elytra and over scutellar shield (Fig. 44.68R) 2
2(1). Metafemora notably thicker than pro- and mesofemora 3
– Metafemora similar in shape to pro- and mesofemora 4
3(2). Rostrum short, robust, almost straight, bare and shining in front of antennal insertions; scutellar shield reduced to a carina (Fig. 44.68T) ..... ***Hypurus*** **Rey**
– Rostrum long, broadly curved (Fig. 44.67S), punctate and sparsely setose; scutellar shield not reduced to a carina ..... ***Hypohypurus*** **Hustache**
4(2). Funicles 6-segmented (e.g. Fig. 44.67W) ..... ***Trichosirocalus*** **Colonnelli**
– Funicles 7-segmented 5
5(4). Femora with ventral tooth ..... ***Mogulones*** **Reitter**
– Femora without ventral tooth ..... ***Rhinoncus*** **Schoenherr**

CURCULIONINI

1. Mandibles bidentate, curved; femora simple; tarsal claws simple and very close together, appearing connate ..... ***Pycnochirus*** **Berg**
– Mandibles simple, without tooth; femora with ventral tooth; tarsal claws divaricate and dentate; habitus as Fig. 44.67G ..... ***Curculio*** **Linnaeus**

CRANOPOEINI

1. Funicles 7-segmented 2
– Funicles 5- or 6-segmented 4
2(1). Antennal scrobes in females with deep fovea covered with ceriferous vestiture 3
– Antennal scrobes in females without deep fovea ..... ***Cranopoeus*** **Marshall**
3(2). Tarsal claws paired, connate ..... ***Cranoides*** **Kuschel**
– Tarsal claw single ..... ***Enneaeus*** **Kuschel**
4(1). Antennal scrobes in females with deep fovea covered with ceriferous vestiture; tarsal claw single ..... ***Fergusoniella*** **Marshall**
– Antennal scrobes in females without deep fovea; tarsal claws absent (e.g. Fig. 44.68o) ..... ***Cratoscelocis*** **Lea**

CRYPTOPLINI

1. Funicles 6-segmented, last segment appressed to club, this appearing 5-segmented; tarsal segment 4 minute, hidden in apical cleft of tarsal segments 3 (Fig. 44.68o), usually clawless 2
– Funicles 7-segmented, clubs 4-segmented; tarsal segments 4 distinct and exposed, bearing one or two claws 3
2(1). Rostrum short, straight to slightly curved; penis with flat, symmetrical endophallic sclerite; spermathecal gland bulbous ..... ***Cryptoplus*** **Erichson**
– Rostrum long and thin, distinctly curved; penis with bent, twisted, double endophallic sclerite; spermathecal gland mushroom-shaped; habitus as Fig. 44.67F ..... ***Zeopus*** **Pascoe**
3(1). Tarsal claws single, rarely apically bifid (fusion incomplete) ..... ***Haplonyx*** **Schoenherr**
– Tarsal claws paired, divergent or parallel 4
4(3). Rostrum apically broadened in dorsal view only, narrowed or straight in lateral view; pronotum with two pairs of small, low admedian tubercles; elytra with three to four pairs of small, low tubercles ..... ***Menechirus*** **Hartmann**
– Rostrum apically broadened in dorsal and lateral view; pronotum with two pairs of large, fasciculate admedian tubercles and one to two pairs of smaller lateral ones; elytra with seven or eight pairs of larger, sometimes fasciculate tubercles 5
5(4). Funicle segments 7 closely appressed to club, funicles appearing 6-segmented; body sparsely covered with black hair-scales; elytra with eight pairs of large fasciculate tubercles on interstriae 3 and 5, apically rounded; claws closely associated and parallel ..... ***Sigastus*** **Pascoe**
– Funicle segments 7 separate from club, funicles distinctly 7-segmented; body densely covered with greenish-grey scales, reddish-brown on pronotal tubercles; elytra with one pair of huge, compressed, squamose tubercles on interstriae 3 and with four smaller tubercles anterolaterally of these, apically truncate; claws separate and slightly divergent; habitus as Fig.44.67E ..... ***Kuschelorhynchus*** **Jennings & Oberprieler**

RHAMPHINI

1. Rostrum flattened and glabrous; antennae inserted at base of rostrum, funicles 7-segmented; habitus as Fig. 44.67J ..... ***Rhamphus*** **Clairville**
– Rostrum cylindrical with scattered setae; antennae inserted at basal third of rostrum, funicles 6-segmented; habitus as Fig. 44.67K ..... ***Orchestes*** **Illiger**

STOREINI

1. Mandibles exodont, outer margin with at least one tooth (e.g. Fig. 44.67Z) 2
– Mandibles not exodont, outer margin without tooth 4
2(1). Body length approximately 4.6 mm; funicles 7-segmented; tarsal claws present ..... ***Plaesiorhinus*** **Blackburn**
– Body length approximately 1.2–2.1 mm; funicles 6-segmented; tarsal claws absent (e.g. Fig. 44.68o) 3

3(2). Body vestiture consisting of dense, imbricate squamae, integument of pronotum and elytra not or barely visible; prothorax about as wide as elytra . ... ***Micraonychus* Lea**
– Body vestiture consisting of sparse, separate squamae, integument of pronotum and elytra cuticle visible; prothorax narrower than elytra .... ***Misophrice* Pascoe**
4(1). Prosternum with well-defined canal, bordered by distinct carinae (Fig. 44.68D–E) **5**
– Prosternum without canal (Fig. 44.68G) or anteriorly broadly emarginate and concave or canal not defined by carinae (Fig. 44.68K) **12**
5(4). Prosternum with walls of canal with deep hole or window (Storeini s. str., e.g. Fig. 44.68E) **6**
– Prosternum with walls of canal without deep pit (e.g. Fig. 44.68J) **11**
6(5). Body integument black, smooth with sparse, minute setae (Fig. 44.67L); femora with ventral tooth .... ***Leucomelacis* Lea**
– Body integument brown, covered with dense, fine hair-scales; femora with or without ventral tooth **7**
7(6). Body length approximately 1.7 mm; vestiture consisting of closely appressed hair-scales about as long as wide; rostral canal extending from prosternum to metaventrite .... ***Microberosiris* Lea**
– Body length 2.4–7.5 mm; vestiture mixed, consisting of decumbent squamae and erect hair-scales clearly longer than wide; rostral canal restricted to prosternum **8**
8(7). Forehead with median groove; rostrum with ventral side (partly corresponding to ventral margin of scrobes) explanate (Fig. 44.67V) .... ***Placorrhinus* Marshall**
– Forehead without a median groove; rostrum with ventral side not explanate **9**
9(8). Prothorax without ocular lobes .... ***Storeus* Schoenherr**
– Prothorax with ocular lobes **10**
10(9). Scutellar shield with dense scales (Fig. 44.68U); meso- and metafemora with ventral tooth (Fig. 44.68h) .... ***Pseudostoreus* Lea**
– Scutellar shield with sparse setae; meso- and metafemora without ventral tooth .... ***Emplesis* Pascoe**
11(5). Antennal funicles 7-segmented (Fig. 44.88B); scutellar shield about as long as wide (Fig. 44.68V); elytra lateral striae very fine or not evident (Fig. 44.88E); mid femur without ventral spine (Fig. 44.68g) .... ***Rhinidotasia* Lea**
– Antennal funicles 6-segmented (Fig. 44.85A); scutellar shield longer than wide (Fig. 44.85H); elytra lateral striae evident (Fig. 44.68W) ; mid femur with ventral spine (Fig. 44.87K) .... ***Imathia* Pascoe**
12(4). Body length approximately 2.7 mm; integument brown with sparse, elongate hair-scales; pro- and mesotibiae in male with deep preapical emargination .... ***Sellechus* Lea**
– Not with this combination of characters **13**
13(12). Funicles 5-segmented .... ***Anarciarthrum* Blackburn**
– Funicles 6-segmented .... ***Dyschoenium* Blackburn**
– Funicles 7-segmented **14**
14(13). Pro- and mesotibiae with long, pointed uncus and premucro **15**
– Pro and mesotibiae with either uncus or mucro **16**
15(14). Femora without ventral tooth; tarsal claws simple (e.g. Fig. 44.68l) .... ***Ixamine* Pascoe**
– Femora with ventral tooth; tarsal claws dentate (e.g. Fig. 44.68n) .... ***Hibberticola* Lea**
16(14). Vestiture of pronotum and elytra consisting of dense, imbricate hair-scales; funicles 7-segmented; prosternum medially excavate for reception of rostrum (Fig. 44.68I); procoxae and their cavities separate; tibial apices with mucro; tarsal claws divaricate .... ***Cisowhitea* Lea**
– Not with this combination of characters; prosternum not excavate for reception of rostrum **17**
17(16). Dorsal vestiture of body consisting of short decumbent setae mixed with sparse long erect setae; tibial apices with prominent uncus; tarsal segments 5 longer than 1–3 combined, tarsal claws simple, large .... ***Olbiodorus* Blackburn**
– Not with this combination of characters; vestiture consisting only of long, erect setae, tibial apices without uncus **18**
18(17). Body length approximately 1.2 mm; body vestiture consisting of appressed, mostly non- overlapping hair-scales; rostrum mostly glabrous except for few scattered hair-scales near base; scapes curved at about apical third; tarsal segments 5 slightly projecting beyond segment 3 .... ***Misophricoides* Rheinheimer**
– Not with this combination of characters; body length greater than 2.0 mm, rostrum often setose; tarsal segments 5 clearly projecting beyond segment 3 .... **19**
19(18). Tarsal claws dentate or laminate (Fig. 44.68n) **20**
– Tarsal claws simple **23**
20(19). Prosternum with anterior border broadly emarginate and concave (Fig. 44.68K) .... ***Elleschodes* Blackburn**
– Prosternum with anterior border not emarginate, flat or slightly convex **21**
21(20). Pro- and mesotibiae with small uncus (e.g. Fig. 44.94J), habitus as Pl. 79E .... **'*Tychius*' *clavivarius* Lea**
– Pro- and mesotibiae with mucro **22**
22(21). Femora without ventral tooth .... ***Ochrophoebe* Pascoe**
– Femora with small, pointed tooth .... ***Agestra* Pascoe**
23(19). Scrobes with large, setiferous punctures (Fig. 44.67Q, abraded); procoxae contiguous .... ***Cydmaea* Pascoe**
– Scrobes without setiferous punctures; procoxae contiguous or not **24**
24(23). Rostrum with lateral grooves directed towards eyes; scrobes directed ventrad **25**
– Rostrum without lateral grooves directed towards eyes; scrobes often directed towards eyes **26**
25(24). Body length approximately 2.5 mm; surface mostly glabrous or with few patches of hair-scales .... ***Phaunaeus* Lea**
– Body length approximate 7.0 mm or more; surface covered with setae (Fig. 44.67M) .... ***Myllorhinus* Boisduval**
26(24). Prosternum not emarginate anteriorly (Fig. 44.68H); prothorax without ocular lobes **27**
– Prosternum broadly emarginate anteriorly; prothorax mostly with ocular lobes **29**
27(26). Antennae inserted in basal quarter of rostrum (Fig. 44.67O); prothorax about as wide as elytra .... ***Orichora* Pascoe**
– Antennae inserted near middle of rostrum; prothorax narrower than elytra **28**

28(27). Rostrum with three dorsal carinae; tarsal claws divaricate ..... *Epacticus* Blackburn
– Rostrum without dorsal carinae; tarsal claws parallel or subparallel ..... *Paryzeta* Pascoe
29(26). Tarsal claws parallel or subparallel ..... *Abethas* Zimmerman
– Tarsal claws divaricate, forming a 180° angle (Fig. 44.68m) 30
30(29). Procoxae separate (Fig. 44.83H) ..... *Dicomada* Pascoe
– Procoxae contiguous (Fig. 44.68L) 31
31(30). Prothorax 0.7 × as long as wide ..... *Erytenna* Pascoe
– Prothorax more than 0.8 × as long as wide 32
32(31). Vestiture with iridescent squamae ..... *Themelia* Blackburn
– Vestiture without iridescent squamae 33
33(32). Vestiture with white hair-scales on basal half of pronotum and elytral humeri; elytra in apical third with yellow hair-scales ..... *Empira* Pascoe
– Vestiture not as above 34
34(33). Vestiture consisting of appressed, more or less concolorous hair-scales ..... *Empolis* Blackburn
– Vestiture consisting of decumbent and erect setae and appressed hair-scales with contrasting coloration ..... *Encosmia* Blackburn

Classification of the Australian genera of Curculioninae

Tribe Acalyptini Thomson, 1859

Epamoebus Blackburn, 1893 (Pl. 73A)

Type species, by monotypy: *Epamoebus scutellaris* Blackburn, 1893.
Australian species. *E. flavipennis* (Lea, 1915a), *E. insularis* Lea, 1909b, *E. scutellaris* Blackburn, 1893b, *E. ziczac* Lea, 1909b.
Distribution. Australia (NSW, VIC, TAS, SA, WA).
Notes. Specimens are often found on *Acacia*.

Eudelodes Zimmerman, 1994 (Pl. 73B; Fig. 44.69)

Type species, by original designation: *Amorphoidea bicolor* Faust, 1899.
Australian species. *E. ater* (Lea, 1915a), *E. bicolor* (Faust, 1899).
Distribution. Australia, New Guinea.
Notes. The larvae develop in flowers of *Hibiscus tiliaceus* (Malvaceae) (Caldara *et al.* 2014).

Tithene Pascoe, 1874 (Pl. 73C, D; Fig. 44.70)

Type species, by monotypy: *Tithene microcephala* Pascoe, 1874c.
Australian species. *T. linospadicis* (Rheinheimer, 1991), *T.* sp. (*Parimera* new species, Zimmerman, 1992: 394, pl. 501).
Distribution. Australia, Indonesia (Borneo, Java).
Notes. The species are associated with palms (Rheinheimer 1991; Caldara *et al.* 2014).

Tribe Ceutorhynchini Gistel, 1848

Hypohypurus Hustache, 1920 (Pl. 73E; Fig. 44.71)

Type species, by subsequent designation (Colonnelli, 1997: 387): *Hypohypurus perrieri* Hustache, 1920.
Australian species. *H. medvedevi* (Korotyaev, 1981).
Distribution. Australia (NT), Oriental and Afrotropical regions.
Note. This species is probably introduced.

Hypurus Rey, 1882 (Pl. 73F; Fig. 44.72)

Type species, by monotypy: *Ceutorhynchus bertrandi* Perris, 1852.
Australian species. *H. bertrandi* (Perris, 1852).
Distribution. Australia (QLD), Europe, North Africa, central Asia, introduced in Pacific, Oriental, Afrotropical, Nearctic and Neotropical regions.
Notes. *H. bertrandi* is native to the Mediterranean region and was first reported from Australia in 1993, reared from leaves of *Portulaca oleracea* in Brisbane (McFadyen 1994); in the ANIC there are also specimens collected in 1983 at Tinaroo Creek near Mareeba, Queensland.

Mecysmoderes Schoenherr, 1837 (Pl. 73G; Fig. 44.73)

Type species, by original designation: *Mecysmoderes euglyptus* Gyllenhal, 1837 (*in* Schoenherr 1837).
Australian species. *Mecysmoderes* sp. (Zimmerman, 1992: 162, pl. 385).
Distribution. Australia, Oriental and southern Palaearctic regions.

Mogulones Reitter, 1916 (Pl. 73H, I)

Type species, by subsequent designation (Wagner, 1927: 203): *Curculio geographicus* Goeze, 1777.
Australian species. *M. geographicus* (Goeze, 1777), *M. larvatus* (Schultze, 1897).
Distribution. Australia, Palaearctic region.
Notes. The two species present in Australia were introduced for the control of *Echium plantagineum* (Boraginaceae) (Sheppard *et al.* 1999).

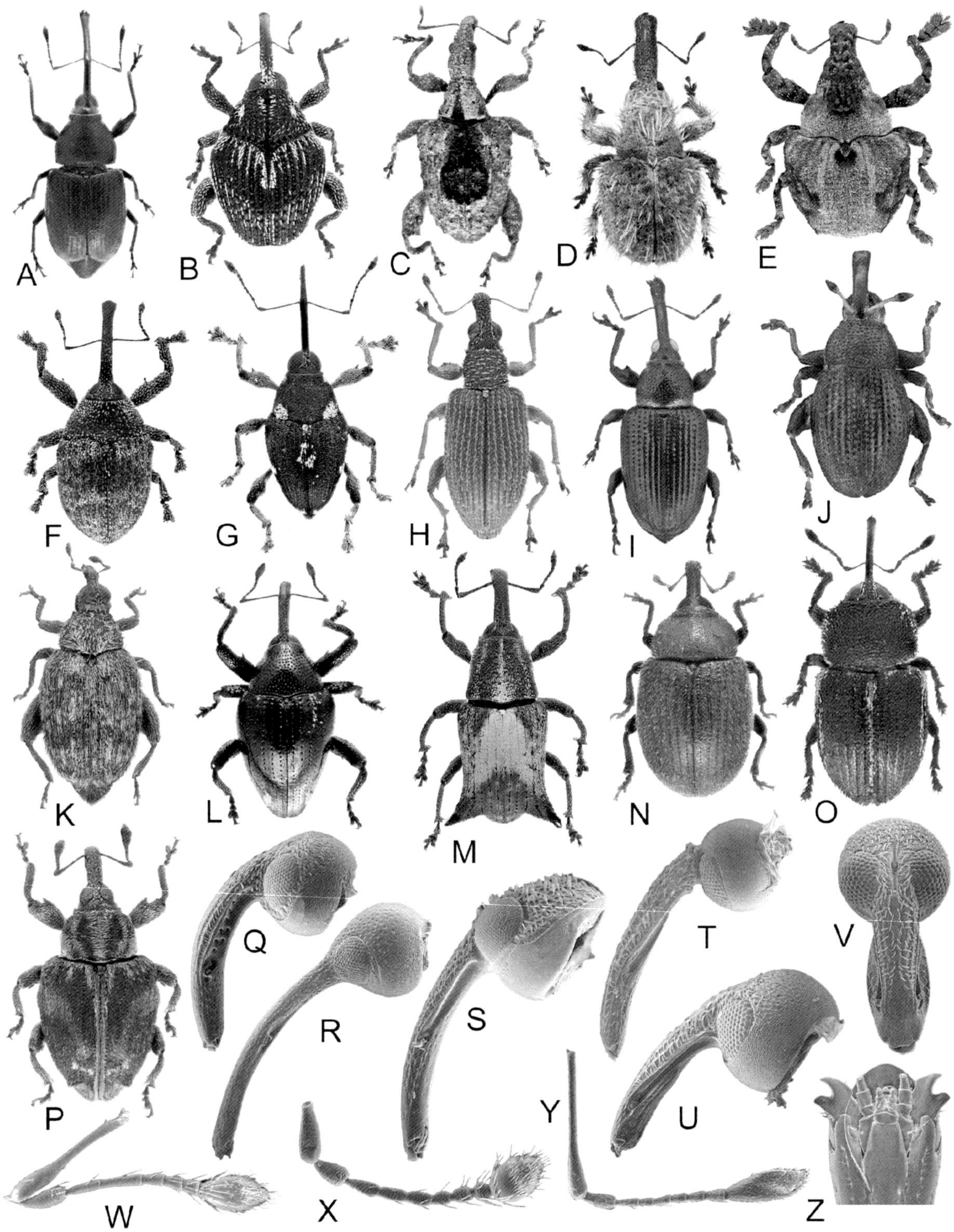

Fig. 44.67. Curculioninae, characters for key to genera. A–P, habitus, dorsal: A, *Tithene*; B, *Mecysmoderes*; C, *Stephanorhynchus crassus* Broun; D, *Cranoides*; E, *Kuschelorhynchus macadamiae* Jennings & Oberprieler; F, *Zeopus*; G, *Curculio simulator* Zimmerman & Oberprieler; H, *Ancyttalia*; I, *Orpha flavicornis* Pascoe; J, *Rhamphus*; K, *Orchestes australiae* Lea; L, *Leucomelacis quadrinotatus* Lea; M, *Myllorhinus insignis* (Lea); N, *Elleschodes hamiltoni* Blackburn; O, *Orichora trivirgata* Pascoe; P, Ochyromerini undet. genus; Q–U, head, lateral: Q, *Cydmaea*; R, *Tithene*; S, *Hypohypurus*; T, *Smicronyx*; U, *Pseudostoreus*; V, *Placorrhinus*, head, dorsal; W–Y, antenna: W, *Mecysmoderes*; X, *Hypurus*; Y, *Tithene*; Z, mouthparts, *Orpha*.

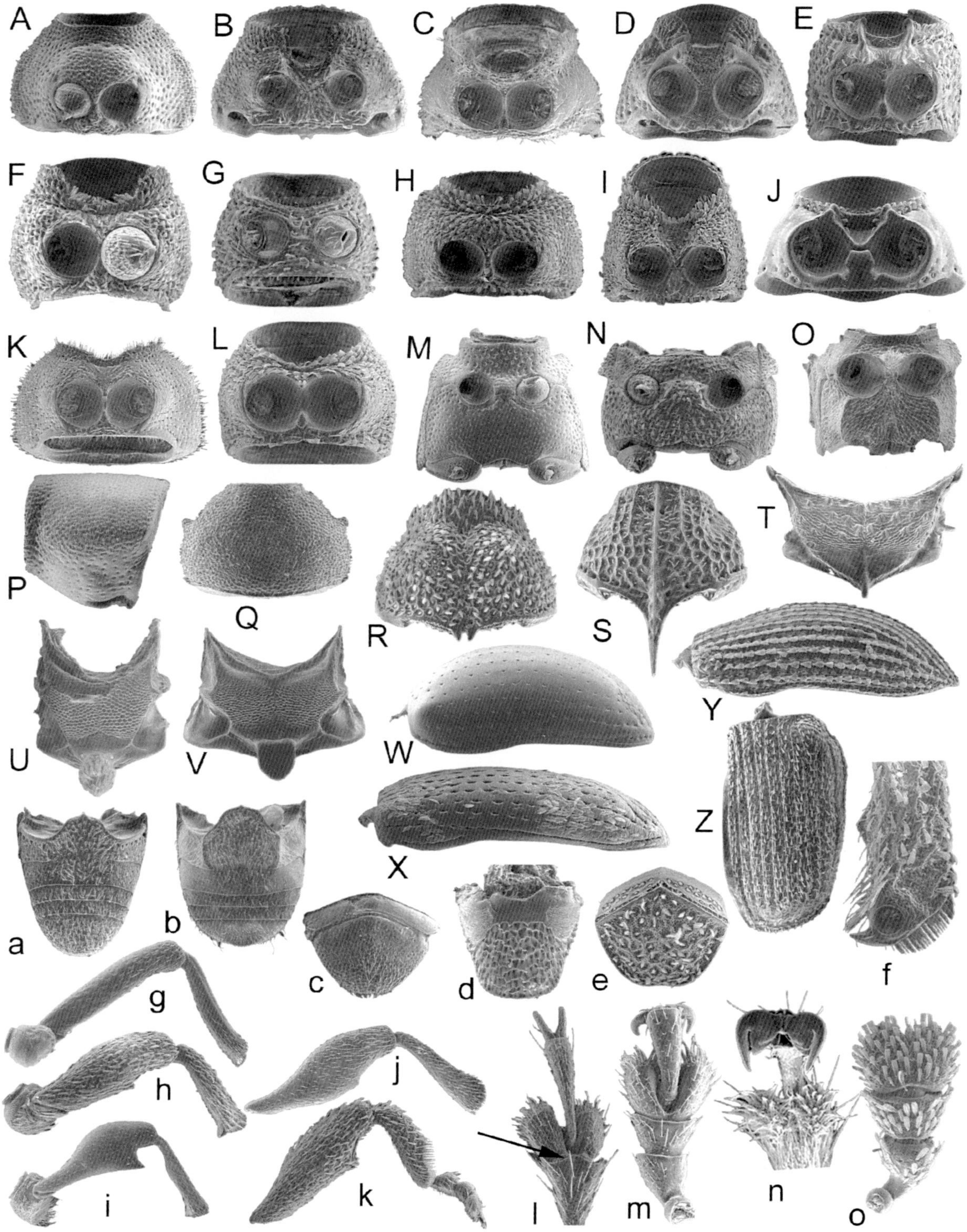

Fig. 44.68. Curculioninae, characters for key to genera. A–L, prothorax, ventral: A, *Eudelodes*; B, *Hypurus*; C, *Enneaeus*; D, *Leucomelacis*; E, *Placorrhinus*; F, *Encosmia*; G, *Misophricoides*; H, *Orichora*; I, *Cisowhitea*; J, *Imathia*; K, *Elleschodes*; L, *Paryzeta*; M–O, pterothorax, ventral: M, *Tithene*; N, *Hypurus*; O, *Orpha*; P, *Tithene*, pronotum, lateral; Q–S, pronotum, dorsal: Q, *Eudelodes*; R, *Hypohypurus*; S, *Mecysmoderes*; T–V, scutellum, dorsal: T, *Hypurus*; U, *Pseudostoreus*; V, *Rhinidotasia*; W–Y, elytron, lateral: W, *Imathia*; X, *Smicronyx*; Y, *Misophricoides*; Z, elytron, dorsal, *Hypurus*; a–b, abdominal ventrites: a, *Pseudostoreus*; b, *Orpha*; c–e, pygidium: c, *Eudelodes*; d, *Cydmaea*; e, *Hypohypurus*; f, tibial apex, *Cryptoplus*; g–k, legs: g, *Rhinidotasia*; h, *Pseudostoreus*; i, *Orpha*; j, *Eudelodes*; k, *Hypohypurus*; l–o, tarsi: l, *Meriphus*; m, *Eudelodes*; n, *Meriphus*; o, *Cryptoplus*.

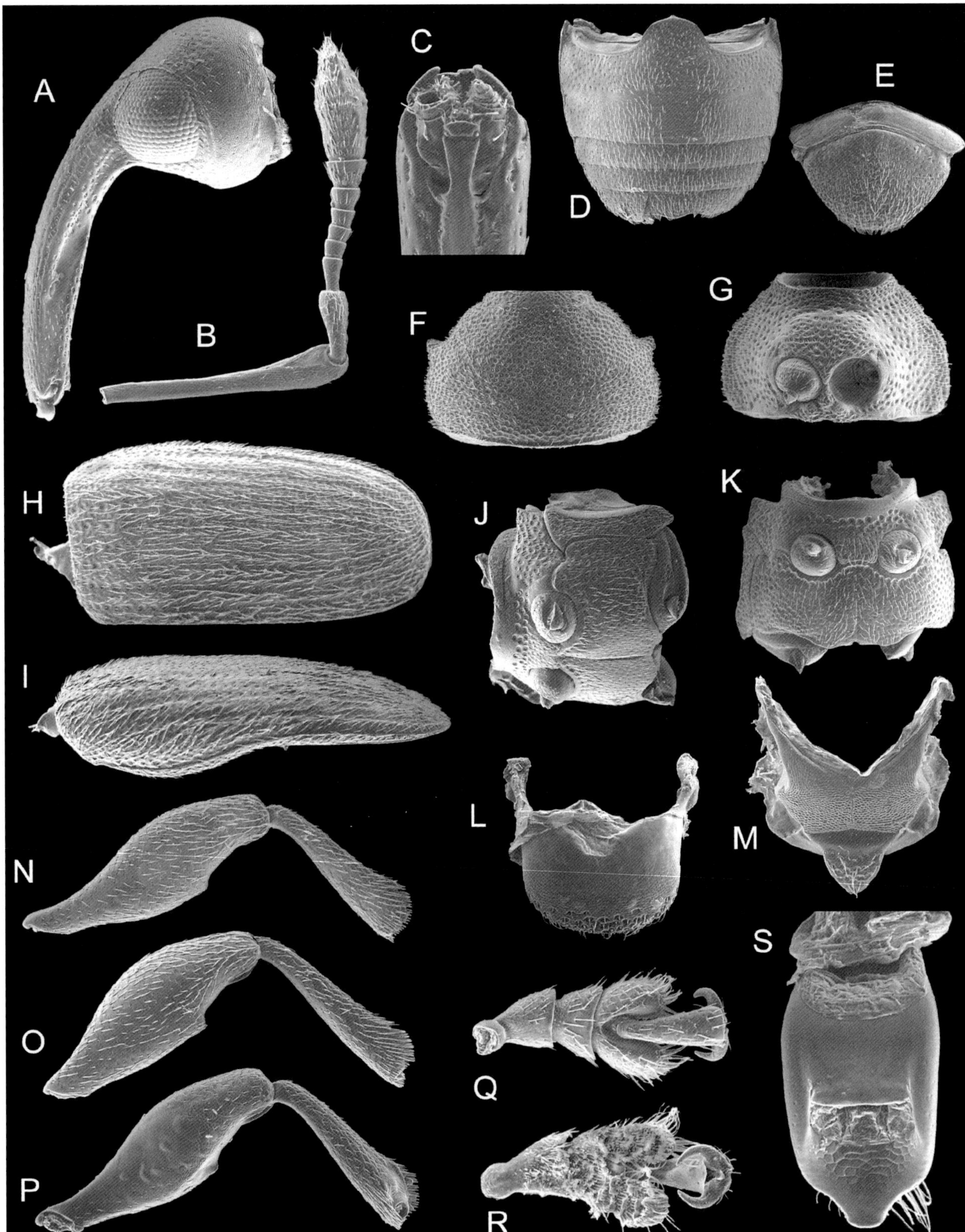

Fig. 44.69. Curculioninae, structural details, Acalyptini, *Eudelodes bicolor* (Faust). A, head, lateral; B, antenna; C, mouthparts, ventral; D, abdominal ventrites; E, pygidium, dorsal; F–G, prothorax: F, dorsal, ♂; G, ventral, ♀; H–I, elytron: H, dorsal; I, lateral; J–K, pterothorax: J, lateral; K, ventral; L, pygidium, dorsal; M, scutellum; N, foreleg; O, middle leg; P, hindleg; Q–R, tarsus: Q, dorsal; R, ventral; S, penis, dorsal.

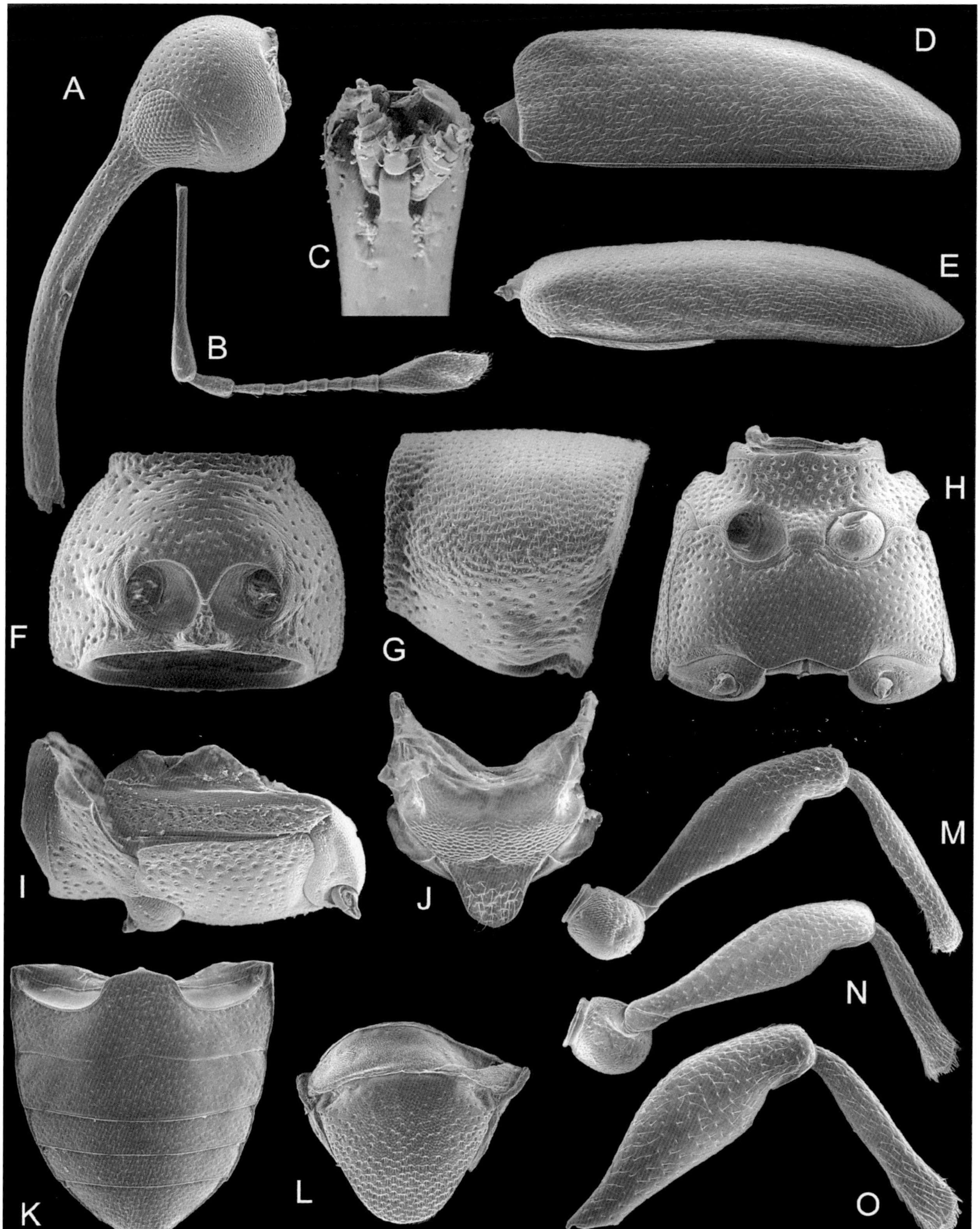

Fig. 44.70. Curculioninae, structural details, Acalyptini, *Tithene linospadicis* (Rheinheimer). A, head, lateral; B, antenna; C, mouthparts, ventral; D–E, elytron: D, dorsal; E, lateral; F–G, prothorax: F, ventral; G, lateral; H–I, pterothorax: H, ventral; I, lateral; J, scutellum, dorsal; K, abdominal ventrites; L, pygidium, dorsal; M–N, legs; O, hindleg.

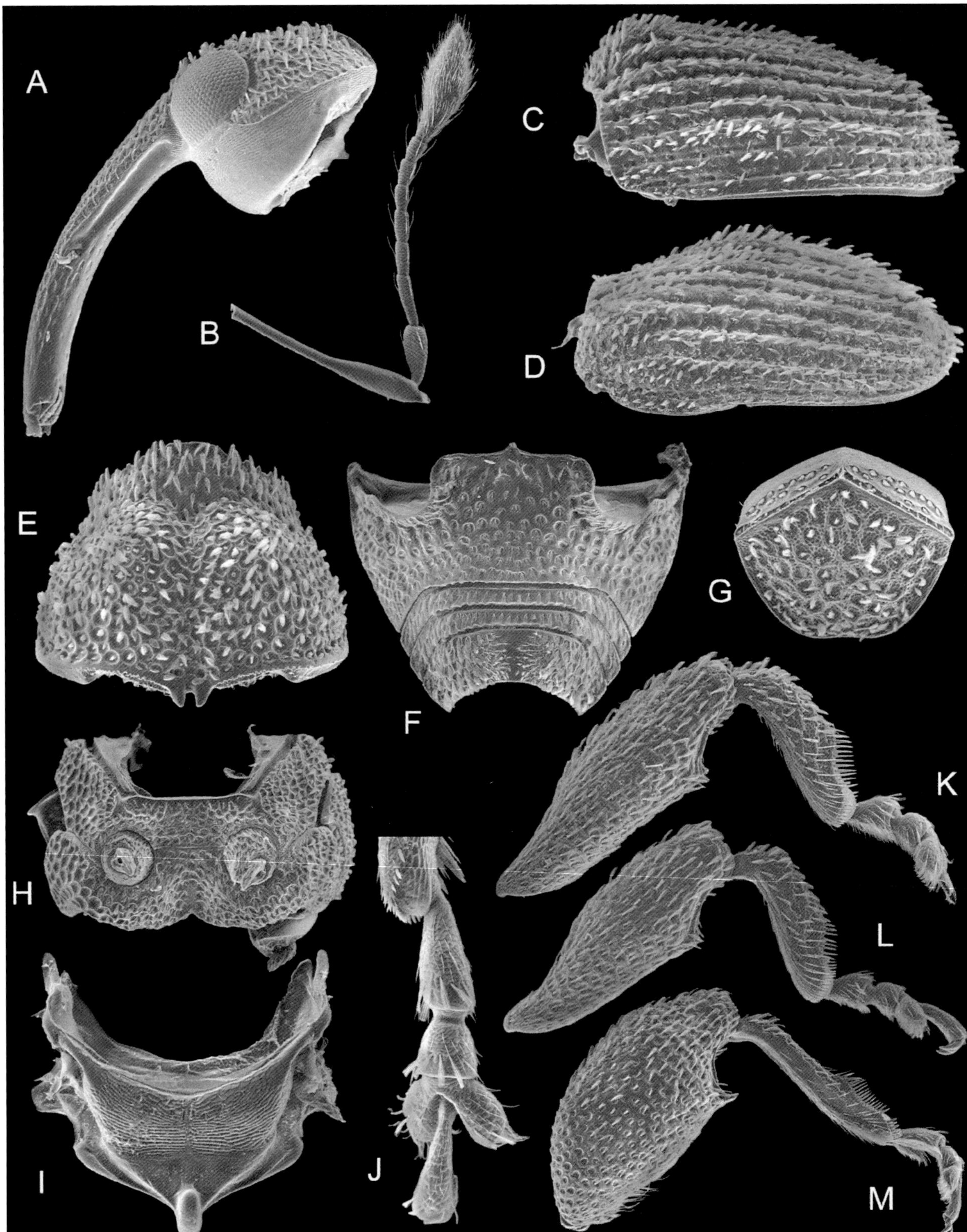

Fig. 44.71. Curculioninae, structural details, Ceutorhynchini, *Hypohypurus medvedevi* (Korotyaev). A, head, lateral; B, antenna; C–D, elytron: C, dorsal; D, lateral; E, prothorax, dorsal; F, abdominal ventrites; G, pygidium, dorsal; H, pterothorax, ventral; I, scutellum, dorsal; J, tarsus, dorsal; K, foreleg; L, middle leg; M, hindleg.

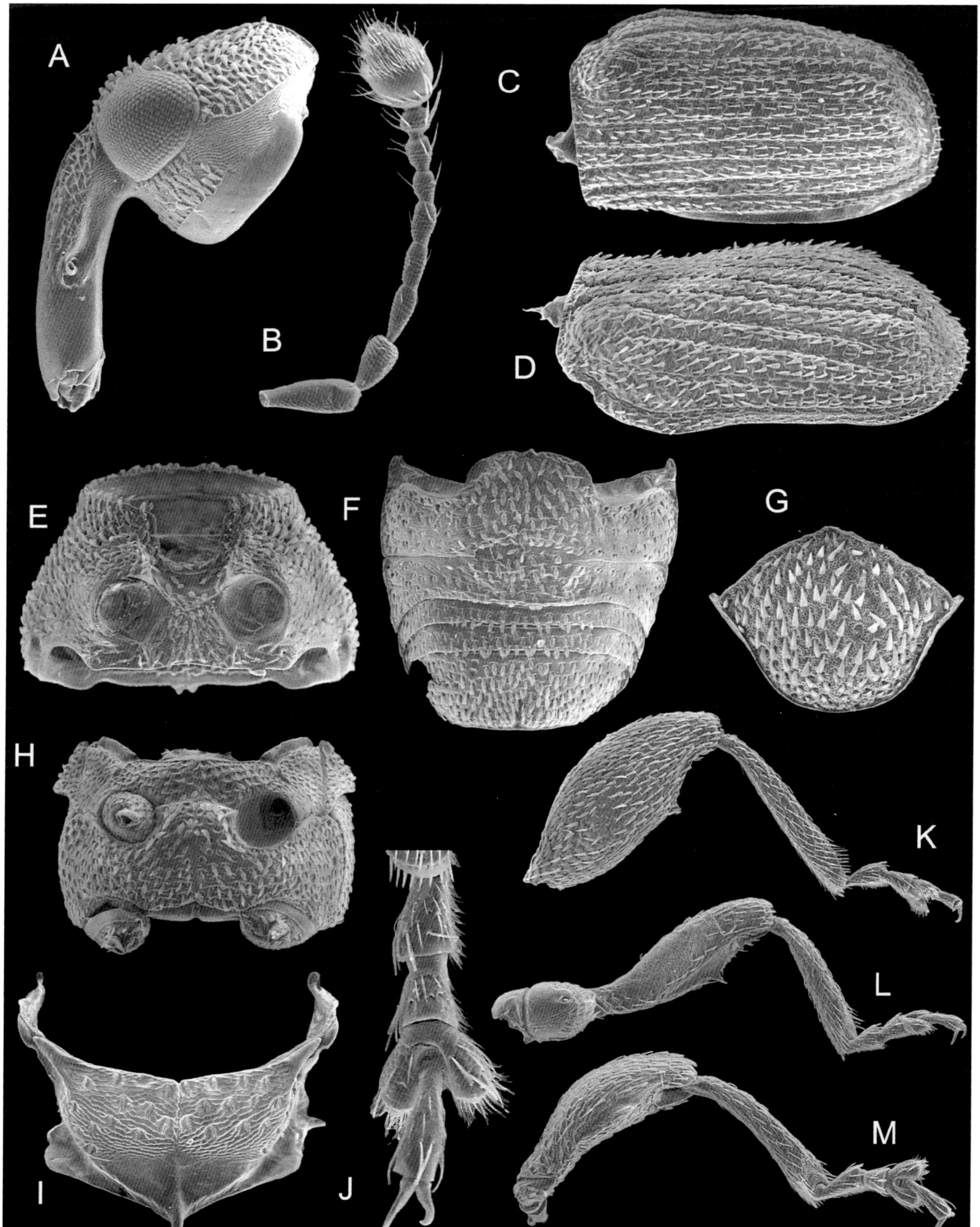

Fig. 44.72. Curculioninae, structural details, Ceutorhynchini, *Hypurus bertrandi* (Perris). A, head, lateral; B, antenna; C–D, elytron: C, dorsal; D, lateral; E, prothorax, ventral; F, abdominal ventrites; G, pygidium; H, pterothorax, ventral; I, scutellum; J, tarsus, dorsal; K, hindleg; L–M, legs.

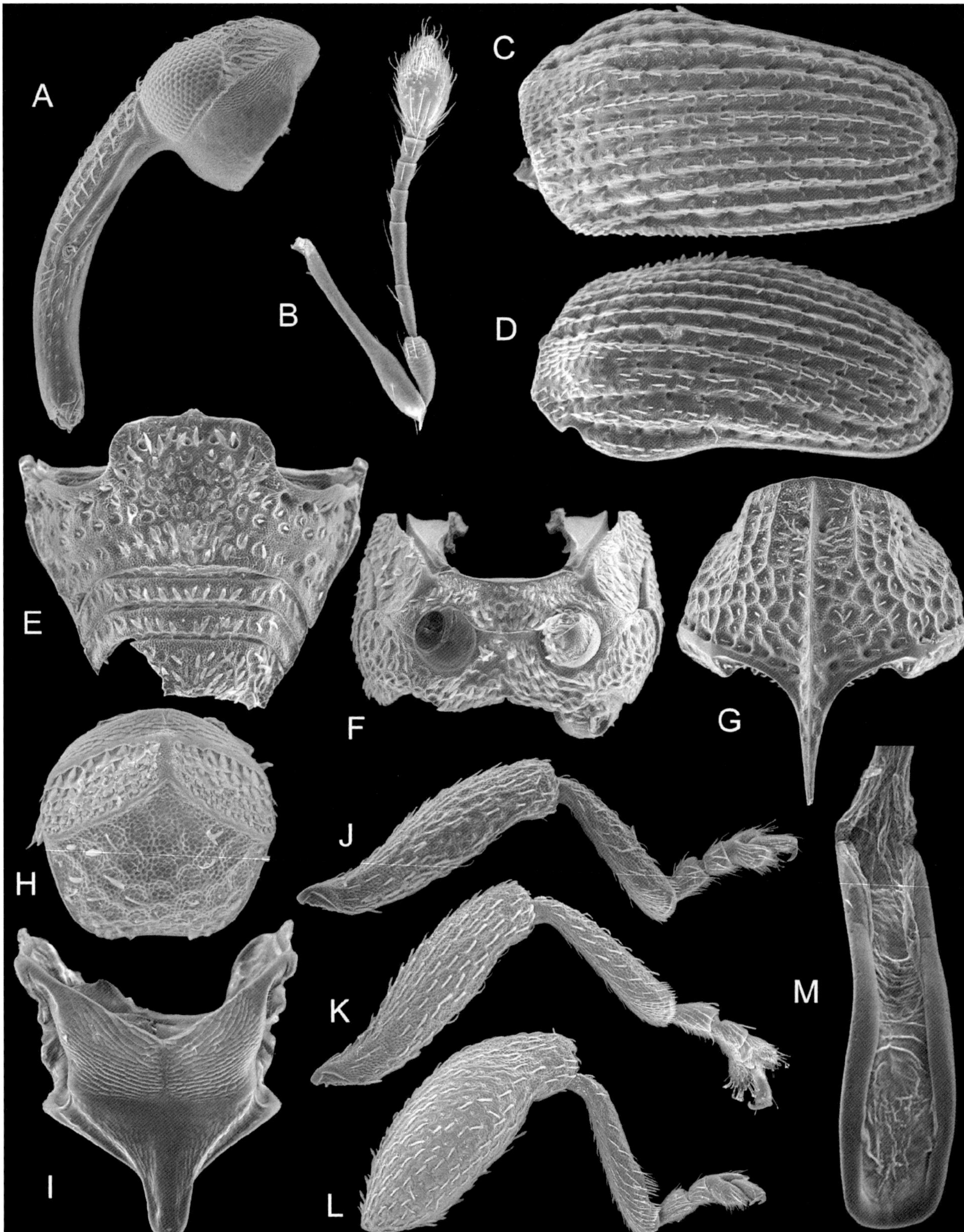

Fig. 44.73. Curculioninae, structural details, Ceutorhynchini, *Mecysmoderes* sp. Mt Webb N.P., QLD. A, head, lateral; B, antenna; C–D, elytron: C, dorsal; D, lateral; E, abdominal ventrites; F, pterothorax, ventral; G, prothorax, dorsal; H, pygidium; I, scutellum; J, foreleg; K, middle leg; L, hindleg; M, penis, dorsal.

Rhinoncus Schoenherr, 1825 (Pl. 73J)

Type species, by subsequent designation (Westwood, 1838: 38): *Curculio pericarpius* Linnaeus, 1758.
Australian species. *R. australis* Oke, 1931, *R. nigriventris* Pascoe, 1873a.
Distribution. Australia, New Zealand, Oriental, Palaearctic, Nearctic and Afrotropical regions.
Notes. The larvae of the single species present in Australia develop in stems of *Rumex australis* (Polygonaceae) (Julien & Matthews 1980).

Trichosirocalus Colonnelli, 1979 (Pl. 73K)

Type species, by original designation: *Curculio troglodytes* Fabricius, 1787.
Australian species. *T. briesei* Alonso-Zarazaga & Sánchez-Ruiz, 2002, *T. horridus* (Panzer, 1801).
Distribution. Australia, Palaearctic region, introduced in USA and New Zealand.
Notes. The two species in Australia were introduced for the control of *Carduus* and *Onopordum* thistles (Asteraceae) (Woodburn 1997; Briese *et al.* 2002).

Tribe Cranopoeini Kuschel, 2009

Kuschel (2009) compared this tribe with and enumerated its differences from Cryptoplini, and he also provided a key to all genera, a diagnosis of each genus and illustrations of several morphological characters.

Cranoides Kuschel, 2009 (Pl. 74A)

Type species, by original designation: *Cranoides avitus* Kuschel, 2009.
Australian species. *Cranoides* sp.
Distribution. Australia (QLD), New Caledonia.
Notes. This undescribed species, so far known from a single female, was recorded from Australia by Pullen *et al.* (2014).

Cranopoeus Marshall, 1931 (Pl. 74B)

Type species, by original designation: *Cranopoeus turritus* Marshall, 1931.
Australian species. *Cranopeous* sp.
Distribution. Australia (NSW), New Guinea, New Caledonia, Pacific region.
Notes. A single species has been recorded from Australia (Kuschel 2009) but remains undescribed. The genus currently contains six described species, but at least 30 have been identified, many of them occurring in New Caledonia (Kuschel 2009). The host plants include Euphorbiaceae, Myrtaceae and Sonneratiaceae.

Cratoscelocis Lea, 1927 (Pl. 74C)

Type species, by monotypy: *Cratoscelocis foveicollis* Lea, 1927.
Australian species. *C. foveicollis* Lea, 1927e.
Distribution. Australia (NSW).

Enneaeus Kuschel, 2009 (Pl. 74D; Fig. 44.74)

Type species, by original designation: *Fergusonia bituberculata* Lea, 1911.
Australian species. *E. bituberculatus* (Lea, 1911b).
Distribution. Australia (SA, WA).
Notes. Specimens have been collected from *Eucalyptus wandoo* (Myrtaceae) in the Stirling Ranges N.P., Western Australia.

Fergusoniella Alonso-Zarazaga & Lyal, 1999 (Pl. 74E, F)

Type species, by subsequent designation (Alonso-Zarazaga & Lyal, 1999: 76): *Fergusonia cristata* Lea, 1911.
Australian species. *F. cristata* (Lea, 1911b).
Distribution. Australia (QLD, NSW).

Tribe Cryptoplini Lacordaire, 1863

Cryptoplus Erichson, 1842 (Pl. 74G; Figs 44.64A–G, 44.75)

Type species, by monotypy: *Cryptoplus perdix* Erichson, 1842.
Australian species. *C. albus* (Lea, 1927e), *C. basalis* (Lea, 1927e), *C. basipennis* (Lea, 1927e), *C. fasciatus* (Lea, 1927e), *C. ferrugineus* (Lea, 1927e), *C. inconspicuus* (Lea, 1927e), *C. intermedius* (Lea, 1927e), *C. intermixtus* (Lea, 1927e), *C. latirostris* (Lea, 1927e), *C. longirostris* (Lea, 1910b), *C. maculipennis* (Lea, 1927e), *C. maestus* (Lea, 1910a), *C. marmoratus* (Lea, 1927e), *C. minimus* (Lea, 1910a), *C. multimaculatus* (Lea, 1927e), *C. nigrirostris* (Chevrolat, 1879b), *C. nuceus* (Pascoe, 1870b), *C. orbiculatus* (Lea, 1910b), *C. ornatipennis* (Blackburn, 1894a), *C. parvus* (Lea, 1927e),*C. perdix* Erichson, 1842, *C. pictus* (Lea, 1927e), *C. puncticollis* (Lea, 1910a), *C. quinquecarinatus* (Lea, 1927e), *C. rostralis* (Lea, 1927e), *C. rubiginosus* (Pascoe, 1870b), *C. rufirostris* (Lea, 1927e), *C. sobrius* (Lea, 1915d), *C. tibialis* (Lea, 1898a), *C. trifasciatus* (Lea, 1909a), *C. uniformis* (Lea, 1898a), *C. variegatus* (Lea, 1910a), *C. vertebralis* (Lea, 1927e).
Distribution. Australia (QLD, NSW, VIC, TAS, SA, WA).
Notes. The genus comprises 33 described species but needs revision. It is largely associated with *Eucalyptus*, the larvae developing in flower buds (Jennings & Oberprieler 2018).

Haplonyx Schoenherr, 1836 (Pl. 74H)

Type species, by original designation: *Haplonyx spencei* Gyllenhal, 1836.

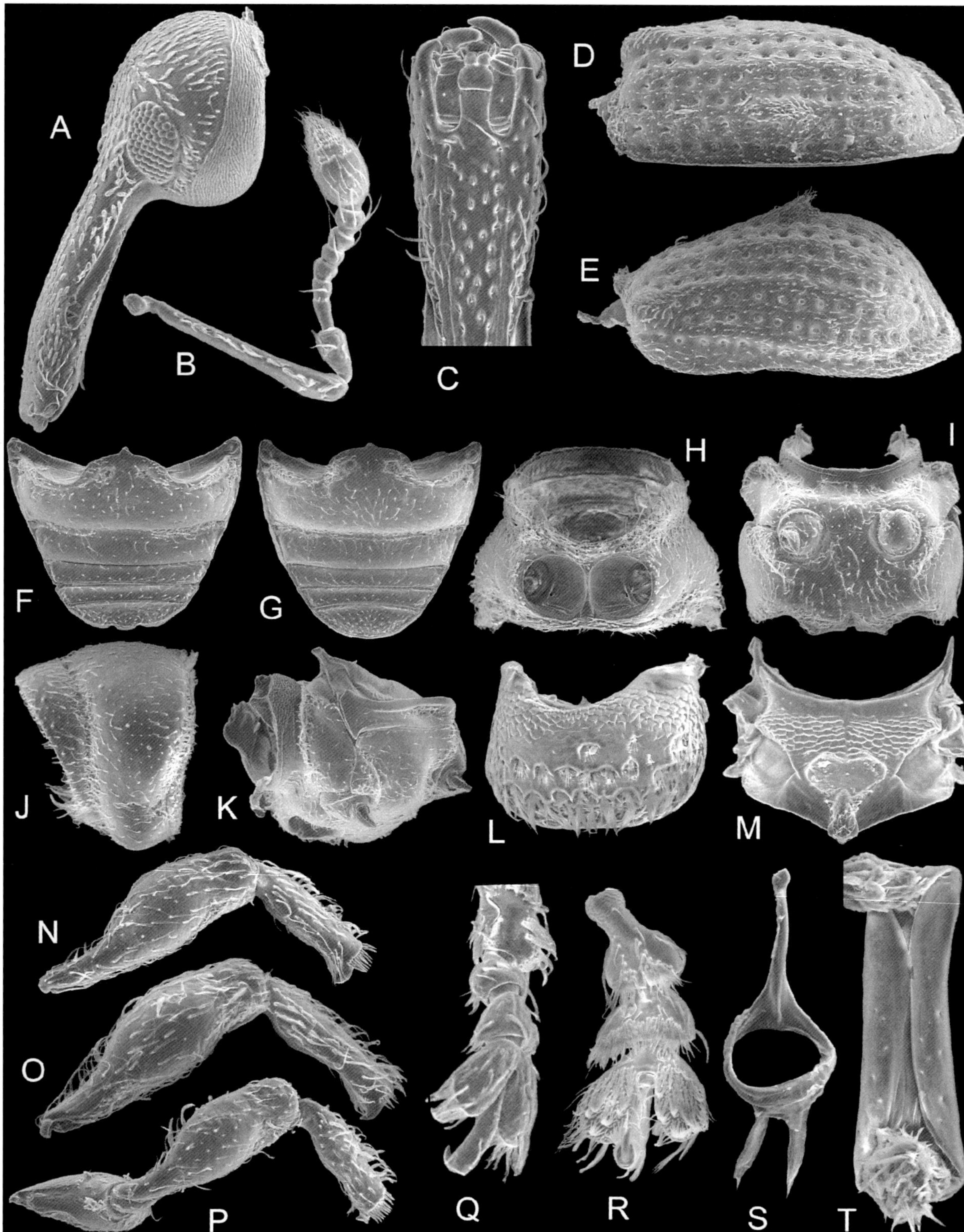

Fig. 44.74. Curculioninae, structural details, Cranopoeini, *Enneaeus bituberculatus* (Lea). A, head, lateral; B, antenna; C, mouthparts, ventral; D–E, elytron: D, dorsal; E, lateral; F–G, abdominal ventrites: F, ♂; G, ♀; H; J, prothorax; H, ventral and J, lateral; I; K, pterothorax; I, ventral; K, lateral; L, pygidium, dorsal; M, scutellum, dorsal; N, foreleg; O, middle leg; P, hindleg; Q–R, tarsus: Q, dorsal; R, ventral; S, tegmen, dorsal; T, penis, dorsal.

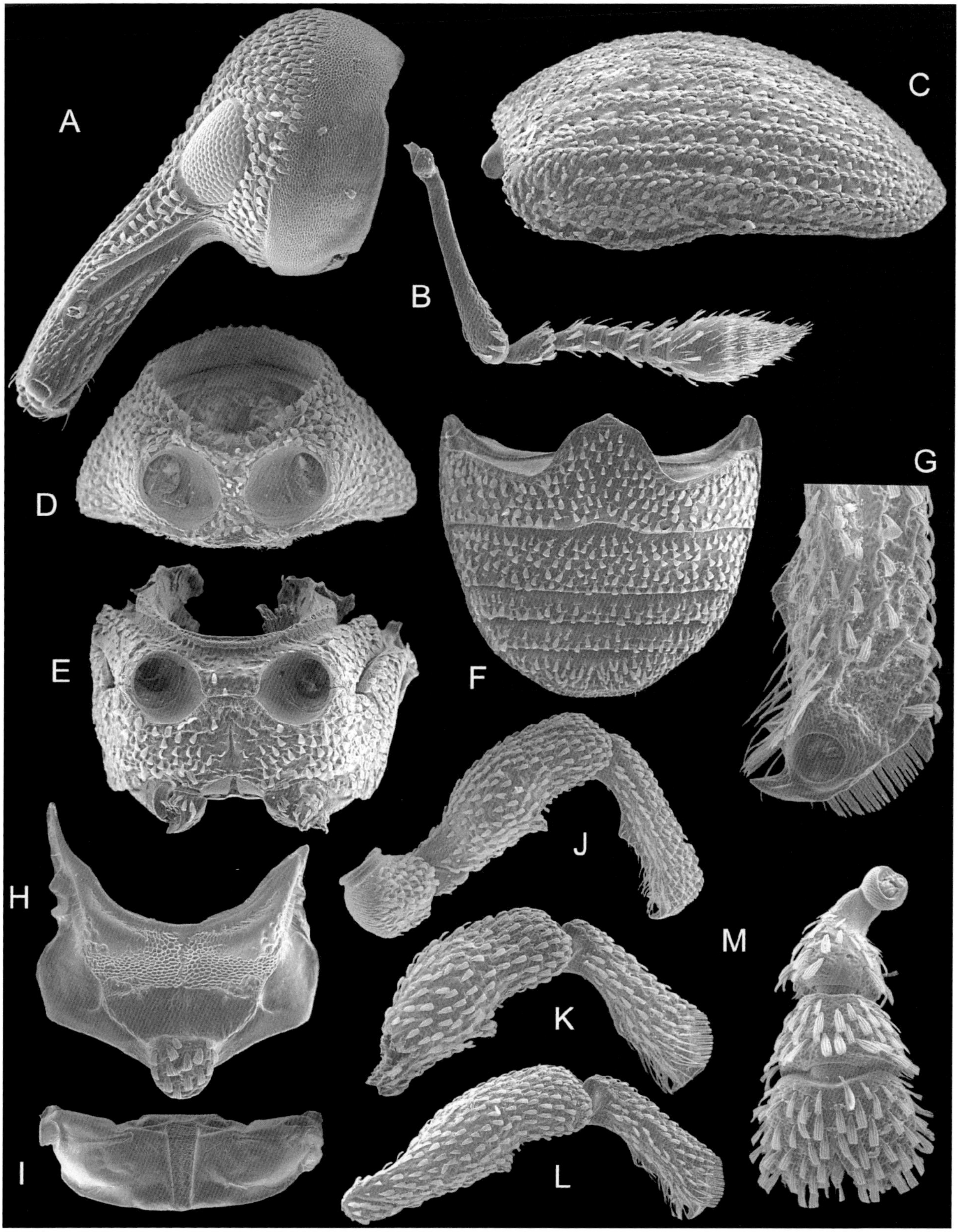

Fig. 44.75. Curculioninae, structural details, Cryptoplini, *Cryptoplus nigrirostris* (Chevrolat). A, head, lateral; B, antenna; C, elytron, lateral; D, prothorax, ventral; E, pterothorax, ventral; F, abdominal ventrites; G, tibial apex showing uncus and premucro; H, scutellum; I, metanotum; J–L, legs; M, tarsus, dorsal.

Australian species. *H. abnormis* Chevrolat, 1879b, *H. albofasciatus* Chevrolat, 1879b, *H. alboguttatus* Chevrolat, 1879b, *H. albosparsus* Lea, 1898a, *H. annularis* Lea, 1928a, *H. ater* Lea, 1898a, *H. bifasciatus* Lea, 1928a, *H. brevirostris* Lea, 1928a, *H. casuarinae* (Lea, 1909a), *H. centralis* Pascoe, 1870c, *H. cioniformis* Chevrolat, 1879b, *H. cionoides* Pascoe, 1870c, *H. circularis* Lea, 1928a, *H. elongatus* Lea, 1928a, *H. ericeus* Pascoe, 1870c, *H. fallaciosus* Pascoe, 1870c, *H. fasciculatus* Boheman, 1844 (*in* Schoenherr 1844), *H. foveipennis* Lea, 1928a, *H. frontalis* Chevrolat, 1879b, *H. haemorrhoidalis* (Fabricius, 1775), *H. hopei* Boheman, 1844 (*in* Schoenherr 1844), *H. kirbyi* Fåhraeus, 1844 (*in* Schoenherr 1844), *H. latus* Lea, 1910a, *H. leucopholus* Lea, 1928a, *H. longipilosus* Lea, 1898a, *H. lucius* Pascoe, 1870c, *H. magniceps* Lea, 1898a, *H. maialis* Pascoe, 1870c, *H. maximus* Lea, 1928a, *H. mediochreatus* Lea, 1915d, *H. mediocinctus* Chevrolat, 1879b, *H. melaspis* Chevrolat, 1879b, *H. modicus* Lea, 1910a, *H. montanus* Lea, 1911b, *H. mucidus* Lea, 1910a, *H. multicolor* Lea, 1915a, *H. nasutus* Lea, 1910b, *H. nigrolineatus* Lea, 1915a, *H. niveodispersus* Lea, 1910b, *H. obliquatus* Lea, 1928a, *H. pectoralis* Chevrolat, 1879b, *H. posticalis* Chevrolat, 1879b, *H. punctipennis* Lea, 1898a, *H. punctum* Chevrolat, 1879b, *H. rubiginosus* Chevrolat, 1879b, *H. rufobrunneus* Lea, 1928a, *H. rufulus* Chevrolat, 1879b, *H. schoenherri* Boheman, 1859, *H. scoparius* Lea, 1928a, *H. seminudus* Lea, 1911b, *H. serratipennis* Lea, 1928a, *H. sexvittatus* Chevrolat, 1879b, *H. sordidus* Lea, 1898a, *H. spencei* Gyllenhal, 1836 (*in* Schoenherr 1836), *H. submaculatus* Lea, 1928a, *H. suturalis* Chevrolat, 1879b, *H. suturellus* Schenkling & Marshall, 1936, *H. tasmanicus* Lea, 1928a, *H. tubicen* Chevrolat, 1879b, *H. unidentatus* Lea, 1910b, *H. vicinus* Chevrolat, 1879b.

Distribution. Australia (QLD, NSW, VIC, TAS, SA, NT, WA).

Notes. The genus currently contains approximately 61 described species, but numerous synonymies are indicated and several undescribed species are known (Jennings & Oberprieler 2018). The larvae develop in flower and fruit buds of *Eucalyptus* and *Melaleuca*, and adults have also been reared from galls on *Eucalyptus* and *Leptospermum*.

Kuschelorhynchus Jennings & Oberprieler, 2018 (Pl. 74I)

Type species, by monotypy: *Kuschelorhynchus macadamiae* Jennings & Oberprieler, 2018.

Australian species. *K. macadamiae* Jennings & Oberprieler, 2018.

Distribution. Australia (QLD, NSW).

Notes. The larvae of the single species develop in the fruits and seeds of cultivated *Macadamia* trees, and the species is regarded as a pest in the horticultural industry of macadamia nuts in Australia (Jennings & Oberprieler 2018).

Menechirus Hartmann, 1901 (Pl. 74J)

Type species, by original designation: *Menechirus oculatus* Hartmann, 1901.

Australian species. *M. fuscodorsalis* Heller, 1922, *M. howdenae* Jennings & Oberprieler, 2018, *M. mundus* Jennings & Oberprieler, 2018, *M. parryi* Jennings & Oberprieler, 2018.

Distribution. Australia (QLD, NSW, NT), New Guinea, Philippines (introduced).

Notes. The genus was revised by Jennings & Oberprieler (2018). In addition to the four species in Australia, it contains another described one in New Guinea, and one, *M. fuscodorsalis*, has been introduced to the Philippines. The larvae develop in flowers and fruit buds of *Syzygium* and probably *Rhodomyrtus* (Myrtaceae) (Juniper & Britton 2010; Jennings & Oberprieler 2018).

Sigastus Pascoe, 1865 (Pl. 74K)

Type species, by monotypy: *Sigastus fascicularis* Pascoe, 1865.

Australian species. *S. fascicularis* Pascoe, 1865.

Distribution. Australia (NSW, SA, WA).

Notes. The genus now contains a single species (Jennings & Oberprieler 2018). Its larva develops in woody homopteran galls on *Eucalyptus* (Myrtaceae) (Jennings & Oberprieler 2018).

Zeopus Pascoe, 1872 (Pl. 74L)

Type species, by monotypy: *Zeopus storeoides* Pascoe, 1872.

Australian species. *Z. storeoides* Pascoe, 1872c.

Distribution. Australia (NSW, SA, VIC, WA).

Notes. The single species appears to be associated with *Eucalyptus* (Myrtaceae), but its larva and life history are unknown (Jennings & Oberprieler 2018).

Tribe Curculionini Latreille, 1802

Curculio Linnaeus, 1758 (Pl. 75A; Figs 44.63, 44.64H–N)

Type species, by subsequent designation (Latreille, 1810: 430): *Curculio nucum* Linnaeus, 1758.

Australian species. *C. aequalis* (Lea, 1904), *C. amoenus* Fabricius, 1775, *C. appendiculatus* (Lea, 1911b), *C. bicruciatus* (Motschulsky, 1866), *C. delicatulus* (Lea, 1904), *C. intricatus* (Lea, 1904), *C. mastersii* (Pascoe, 1872a), *C. simulator* Zimmerman & Oberprieler, 2014 (*in* Pullen *et al.* 2014), *C. submaculatus* (Lea, 1904), *C. sulfureosignatus* (Heller, 1927).

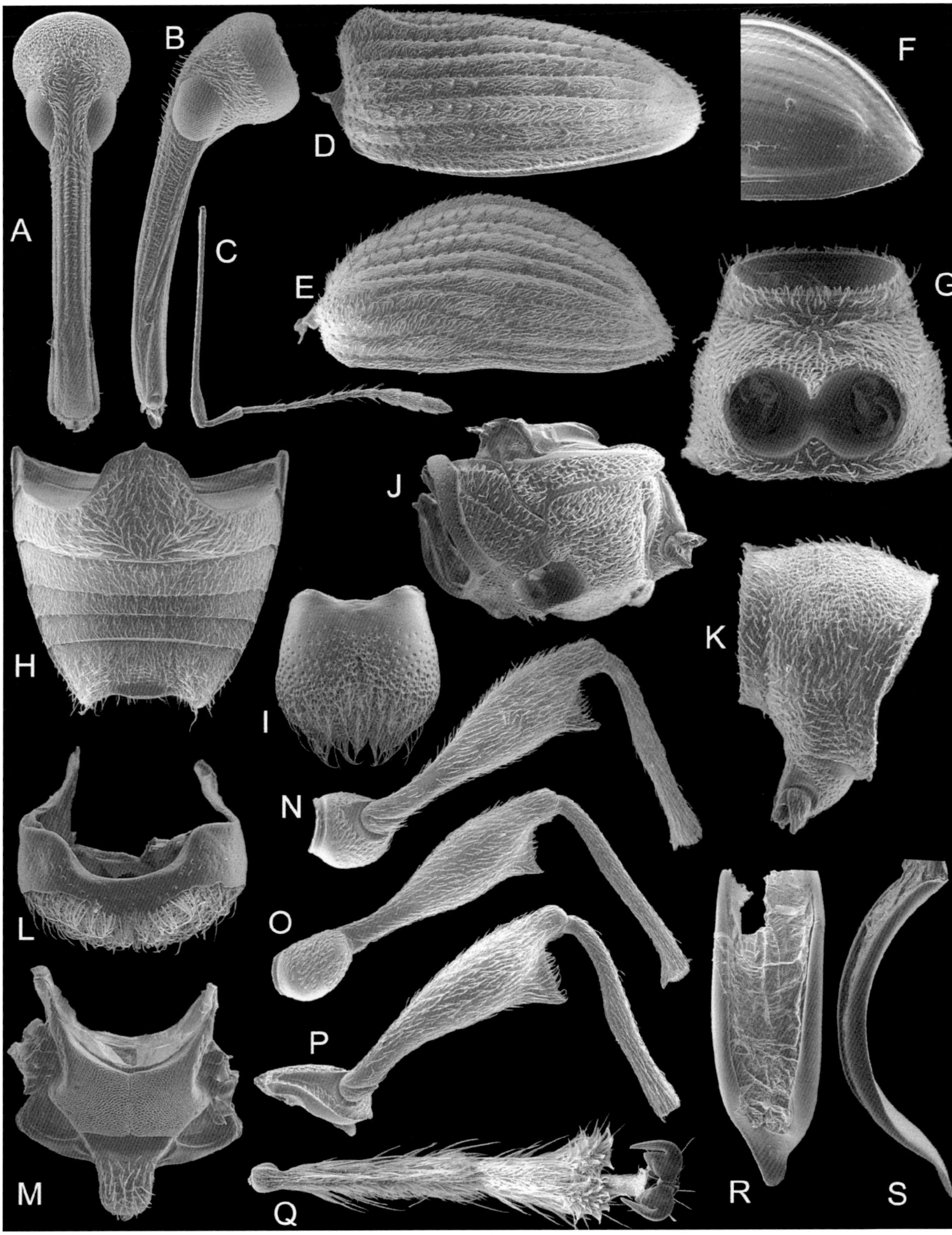

Fig. 44.76. Curculioninae, structural details, Eugnomini, *Meriphus fimbriatus* Lea, ♂. A–B, head: A, dorsal; B, lateral; C, antenna; D–F, elytron: D, dorsal; E, lateral; F, apex, inner lateral; G; K, prothorax; G, ventral; K, lateral; H, abdominal ventrites; I, pygidium; J, pterothorax, lateral; L, pygidium, dorsal; M, scutellum, dorsal; N, middle leg; O, foreleg; P, hindleg; Q, tarsus, ventral; R–S, penis: R, dorsal; S, lateral.

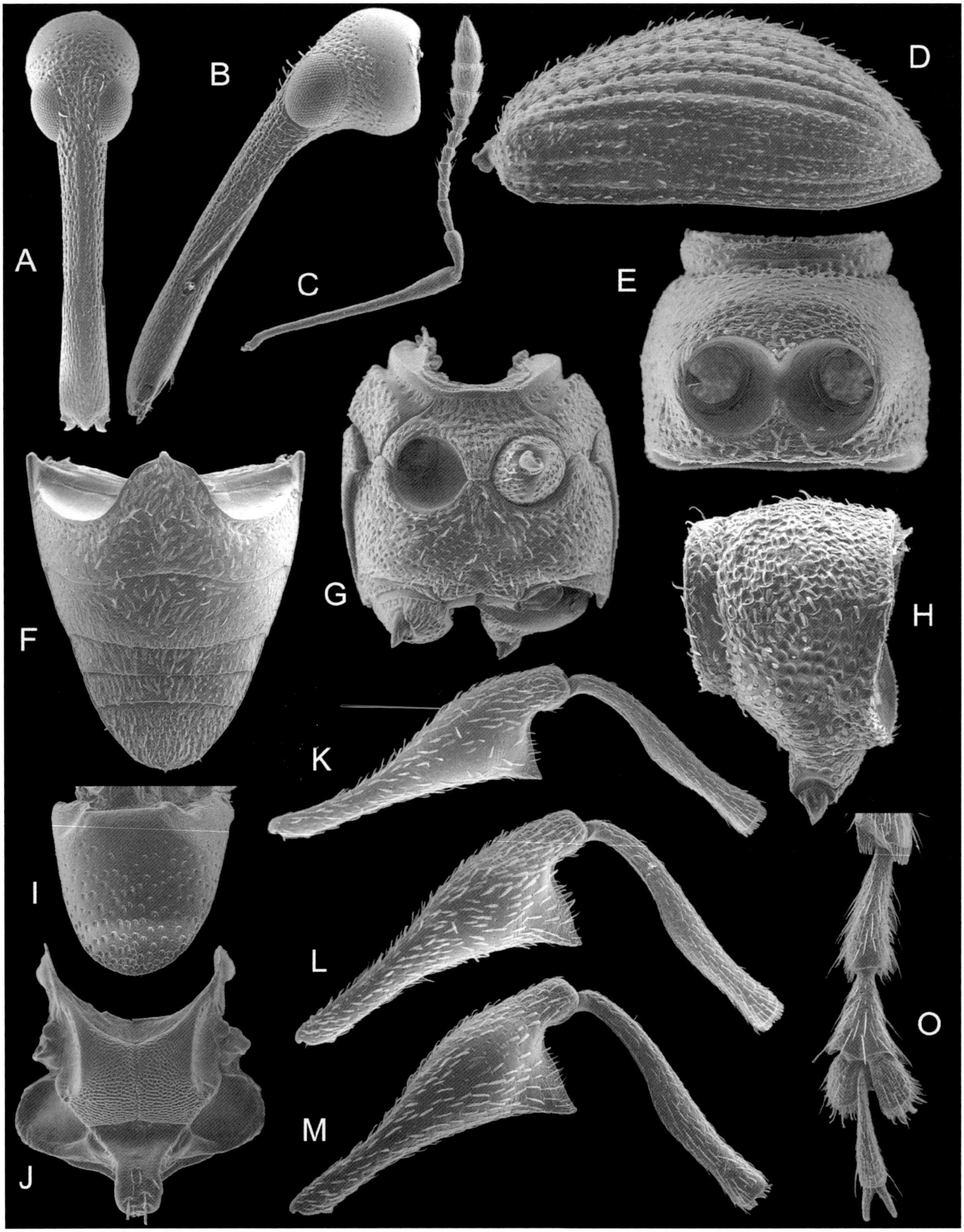

Fig. 44.77. Curculioninae, structural details, Eugnomini, *Meriphus longirostris* Pascoe, ♀. A–B, head: A, dorsal; B, lateral; C, antenna; D, elytron, lateral; E; H, prothorax; E, ventral; H, lateral; I, pygidium; J, scutellum; K, foreleg; L, middle leg; M, hindleg; O, tarsus, dorsal.

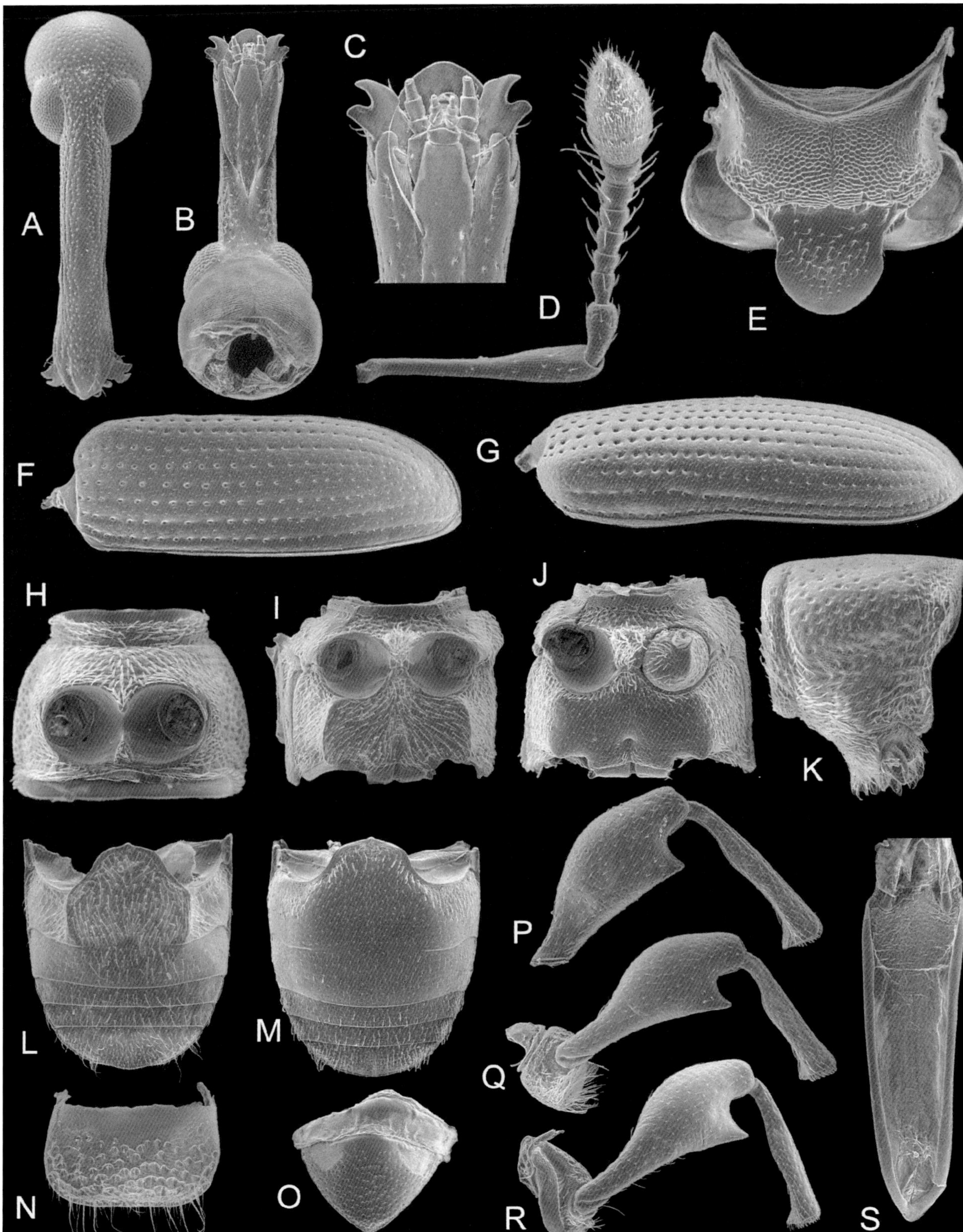

Fig. 44.78. Curculioninae, structural details, Eugnomini, *Orpha flavicornis* Pascoe. A–B, head: A, dorsal; B, ventral; C, mouthparts; D, antenna; E, scutellum; F–G, elytron: F, dorsal; G, lateral; H; K, prothorax; H, ventral; K, lateral; I–J, pterothorax, ventral: I, ♂; J, ♀; L–M, abdominal ventrites: L, ♂; M, ♀; N–O, terminalia; P, foreleg; Q, middle leg; R, hindleg; S, penis, dorsal.

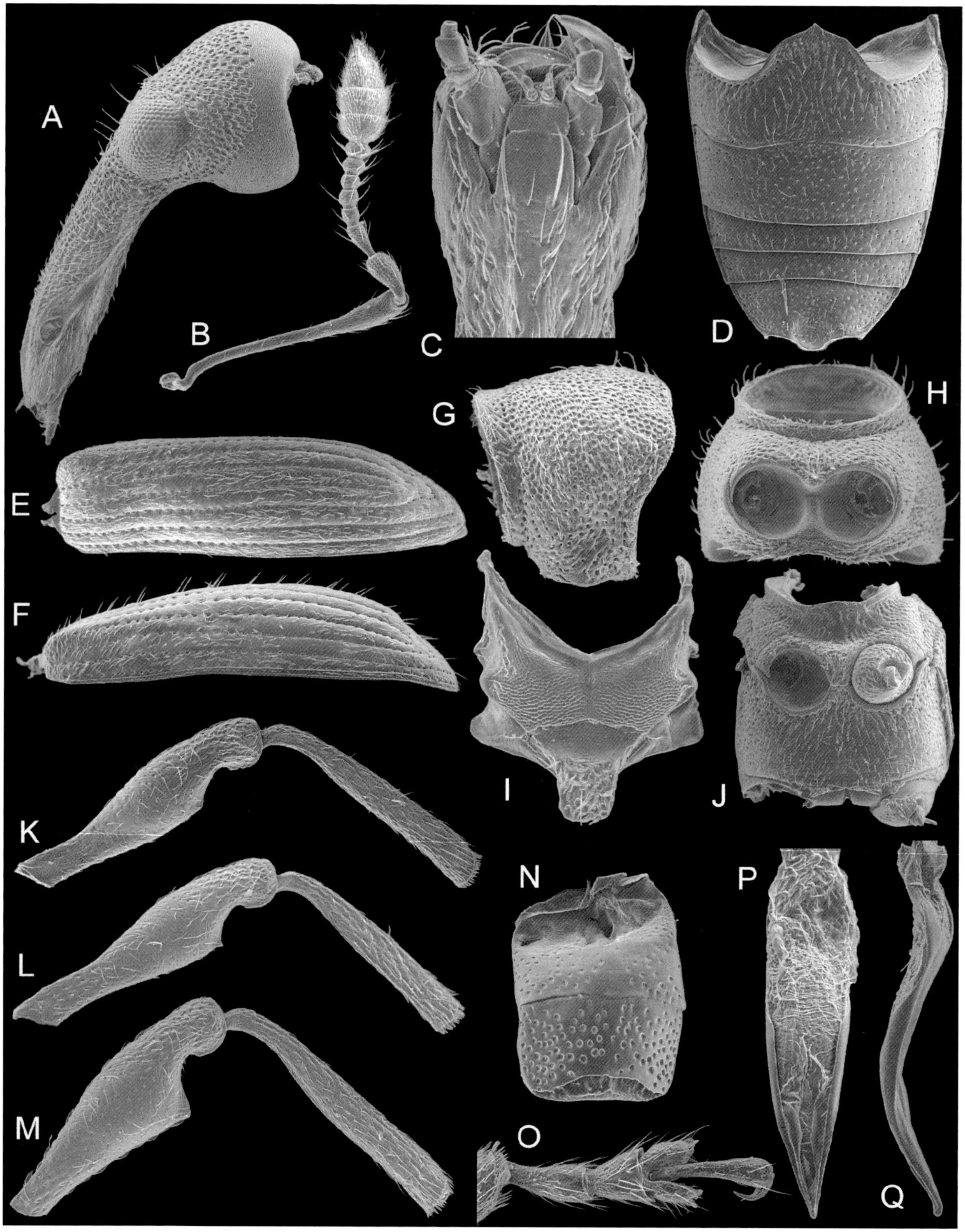

Fig. 44.79. Curculioninae, structural details, Eugnomini, *Rhopalomerus piceosetosus* (Lea). A, head, lateral; B, antenna; C, mouthparts; D, abdominal ventrites; E–F, elytron: E, dorsal; F, lateral; G–H, prothorax: G, lateral; H, ventral; I, scutellum; J, pterothorax, ventral; K, foreleg; L, middle leg; M, hindleg; N, apical tergites, dorsal; O, tarsus, dorsal; P–Q, penis: P, dorsal; Q, lateral.

Distribution. Australia (QLD, NSW, NT, WA), New Guinea, Oriental, Holarctic, Afrotropical and Nearctic regions.

Notes. The larvae of the Australian species seem to develop in figs (*Ficus*, Moraceae) and *Syzygium* (Myrtaceae) (Zimmerman 1994; Howden 1995), as they do in Africa.

Pycnochirus Berg, 1898 (Pl. 75B)

Type species, by monotypy: *Megachirus fuscovarius* Faust, 1895.

Australian species. *P. fuscovarius* (Faust, 1895).

Distribution. Australia (NSW, QLD), New Guinea.

Notes. The single species in Australia has been collected from fruits of *Alectryon coriaceus* (Sapindaceae) (Rheinheimer 1989).

Tribe Eugnomini Lacordaire, 1863

Ancyttalia Zimmerman, 1994 (Pl. 75C)

Type species, by original designation: *Cyttalia sydneyensis* Blackburn, 1894.

Australian species. *A. acaciae* (Lea, 1899a), *A. apicalis* (Lea, 1906a), *A. erichsoni* (Pascoe, 1870d), *A. impura* (Lea, 1899a), *A. inornata* (Lea, 1899a), *A. maculata* (Lea, 1899a), *A. munda* (Blackburn, 1892b), *A. nigra* (Lea, 1899a), *A. nigriclava* (Lea, 1899a), *A. oleariae* (Lea, 1906a), *A. parva* (Lea, 1899a), *A. rufipes* (Lea, 1899a), *A. sydneyensis* (Blackburn, 1894a), *A. tarsalis* (Blackburn, 1894a), *A. ventralis* (Pascoe, 1870d).

Distribution. Australia (NSW, Lord Howe Is., VIC, TAS, SA, WA, Norfolk Is.), New Guinea.

Notes. Adults are generally found on flowers of various plant species, including species of *Acacia*, *Callistemon*, *Dendrobium*, *Olearia* and many others. The larvae develop under the bark of dead branches and logs.

Aptilonotus Zimmerman & Oberprieler, 2014 (Pl. 75D)

Type species, by original designation: *Meriphus ater* Lea, 1915c.

Australian species. *A. ater* (Lea, 1915c).

Distribution. Australia (QLD).

Notes. The genus was described in Pullen *et al.* (2014).

Bothrophasis Zimmerman & Oberprieler, 2014 (Pl. 75E)

Type species, by original designation: *Meripherellus nigriclavus* Lea, 1916.

Australian species. *B. nigriclavus* (Lea, 1916).

Distribution. Australia (QLD).

Notes. The genus was described in Pullen *et al.* (2014).

Meripherinus Lea, 1915 (Pl. 75F)

Type species, by monotypy: *Meripherinus fimbriatus* Lea, 1915.

Australian species. *M. fimbriatus* Lea, 1915a.

Distribution. Australia (NSW).

Meriphus Erichson, 1842 (Pl. 75G; Figs 44.76, 44.77)

Type species, by monotypy: *Meriphus fullo* Erichson, 1842.

Australian species. *M. australis* (Boisduval, 1835), *M. coronatus* Pascoe, 1873a, *M. fullo* Erichson, 1842, *M. granulatus* Lea, 1911c, *M. guttatus* Pascoe, 1870d, *M. humeralis* Blackburn, 1892b, *M. lateroalbus* Lea, 1915a, *M. lineatus* Blackburn, 1890c, *M. longirostris* Pascoe, 1871a, *M. raucus* Blackburn, 1890c, *M. tuberculatus* Lea, 1909c.

Distribution. Australia (QLD, NSW, VIC, TAS, SA, WA).

Notes. The adults are generally found on flowers (e.g. *Aotus lanigera*, *Hibertia scandens*, *Banksia*, *Isopogon*), but the life history of the larvae is not recorded.

Myossita Pascoe, 1865 (Pl. 75H; Fig. 44.65A–L)

Type species, by monotypy: *Myossita rufula* Pascoe, 1865.

Australian species. *M. banksiae* Lea, 1899a, *M. cirrifera* Pascoe, 1870d, *M. crucigera* Blackburn, 1892b, *M. melanocephala* Pascoe, 1870d, *M. melanosoma* Lea, 1915b, *M. persimilis* (Pascoe, 1872b), *M. rufula* Pascoe, 1865, *M. sublineata* Pascoe, 1875, *M. tabida* Pascoe, 1871a, *M. tessellata* Lea, 1915b.

Distribution. Australia (QLD, NSW, VIC, TAS, SA, WA).

Notes. The larvae as known develop in conflorescences and infructescences of *Banksia* (Proteaceae) (Scott 1982).

Orpha Pascoe, 1870 (Pl. 75I; Fig. 44.78)

Type species, by monotypy: *Orpha flavicornis* Pascoe, 1870.

Australian species. *O. flavicornis* Pascoe, 1870d.

Distribution. Australia.

Rhopalomerus Blanchard, 1849 (Pl. 75J; Fig. 44.79)

Type species, by monotypy: *Rhopalomerus tenuirostris* Blanchard, 1849 (*in* Gay 1849).

Australian species. *R. piceosetosus* (Lea, 1906a).

Distribution. Australia (TAS), New Zealand, Argentina, Chile.

Stephanorhynchus White, 1846 (Pl. 73L)

Type species, by monotypy: *Stephanorhynchus curvipes* White, 1846.

Australian species. *Stephanorhynchus crassus Broun, 1880.*

Distribution. Australia (TAS), New Zealand.

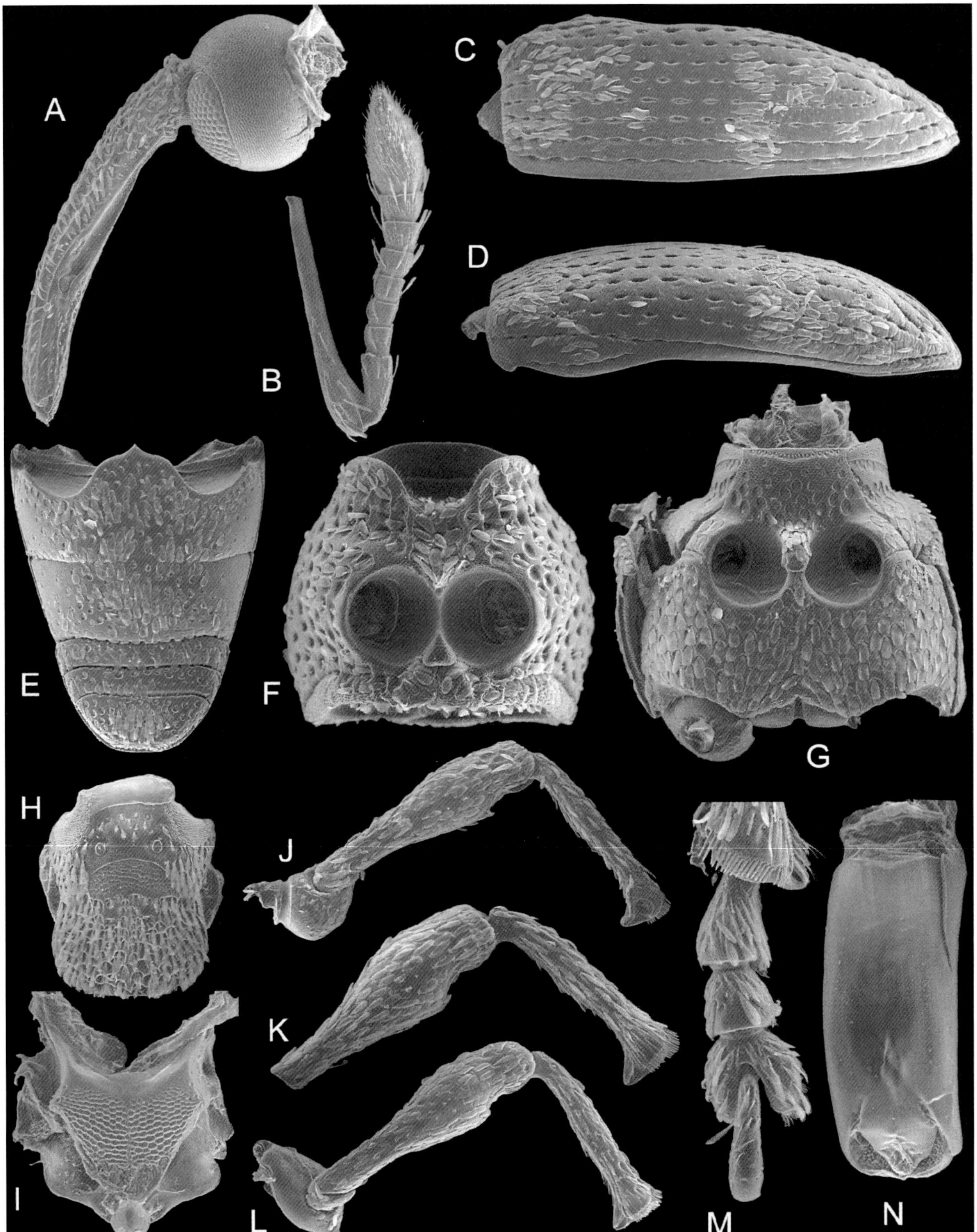

Fig. 44.80 Curculioninae, structural details, Smicronychini, *Smicronyx longirostris* (Lea), ♂. A, head, lateral; B, antenna; C–D, elytron: C, dorsal; D, lateral; E, abdominal ventrites; F, prothorax, ventral; G, pterothorax, ventral; H, pygidium; I, scutellum; J, foreleg; K, middle leg; L, hindleg; M, tarsus, dorsal; N, penis, dorsal.

Notes. The single species in Australia appears to represent a recent accidental introduction from New Zealand to Tasmania. The larva develops in seed capsules of *Pittosporum* species (Pittosporaceae). The genus was revised by Mazur & Brown (2024).

Tribe Ochyromerini Voss, 1935

The tribe is represented in Australia by several undescribed taxa (e.g. Pl. 79J–L) in Queensland and the Northern Territory as well as probably by '*Eudela*' *castanea* Lea, 1915a (Pl. 79I), a species requiring placement in a different genus (Pullen *et al.* 2014).

Tribe Rhamphini Rafinesque, 1815

Orchestes Illiger, 1798 (Pl. 75K)

Type species, by subsequent monotypy (Creutzer, 1799: 125): *Orchestes signifer* Creutzer, 1799.
Australian species. *O. australiae* Lea, 1925.
Distribution. Australia (NSW, QLD), Holarctic and Afrotropical regions.
Notes. Adults of *O. australiae* Lea have been reared from seed pods of *Brachychiton populneus* (Malvaceae) (Caldara *et al.* 2014), but the larvae of undescribed species mine in leaves of *Argyrodendron* (Malvaceae) (Maunsell *et al.* 2017).

Rhamphus Clairville, 1798 (Pl. 75L)

Type species, by monotypy: *Rhamphus flavicornis* Clairville, 1798.
Australian species. *R. acaciae* Lea, 1895a, *R. amplipennis* Lea, 1925, *R. australis* Blackburn, 1890a, *R. distinguendus* Blackburn, 1890a, *R. megalops* Lea, 1925, *R. microscopicus* Lea, 1925, *R. perpusillus* (Pascoe, 1875), *R. setistriatus* Lea, 1926b.
Distribution. Australia (QLD, NSW, VIC, TAS, SA, WA), Palaearctic region.
Notes. The adults of the Australian species are often less than 1 mm in length. They are usually found on *Acacia* (Fabaceae), and the larvae probably mine in the leaves of the plants.

Tribe Smicronychini Seidlitz, 1891

Smicronyx Schoenherr, 1843 (Pl. 76A; Fig. 44.80)

Type species, by original designation: *Micronyx reichii* Gyllenhal, 1835 (*in* Schoenherr 1835).
Australian species. *S. longirostris* (Lea, 1926b), *S. lutulentus* Dietz, 1894, *S. zherichini* Karasyov & Okrajko, 1998.
Distribution. Australia, Holarctic, Afrotropical and Neotropical regions.
Notes. *S. lutulentus* was introduced into Australia from Mexico between 1981 and 1983 for the control of *Parthenium hysterophorus* (Asteraceae) (Dhileepan & McFadyen 2012).

Tribe Storeini Lacordaire, 1863

Storeini *sensu stricto*

The genera of this group are identifiable by their possession of a distinct cleft or 'peep hole' in the walls of the prosternal canal in front of the procoxae (pectoral canal), e.g. Fig. 44.68E (Caldara *et al.* 2014; Pullen *et al.* 2014).

Emplesis Pascoe, 1870 (Pl. 76B)

Type species, by subsequent designation (Lea, 1927d: 35): *Emplesis scolopax* Pascoe, 1870.
Australian species. *E. aenigmatica* Blackburn, 1896, *E. albifasciata* Lea, 1927c, *E. albifrons* Lea, 1927c, *E. alphabetica* Lea, 1927c, *E. alternata* Lea, 1927c, *E. angusta* Lea, 1927c, *E. apiciventris* Lea, 1927c, *E. assimilis* Blackburn, 1890b, *E. basipennis* Lea, 1927c, *E. bellula* (Lea, 1899a), *E. bifoveata* Lea,1927c, *E. bituberculata* Lea, 1927c, *E. brachyderes* (Lea, 1899a), *E. brevimana* Lea, 1927c, *E. canaliculata* (Lea, 1899a), *E. composita* Lea, 1927c, *E. consueta* (Lea, 1899a), *E. cordipennis* Lea, 1927c, *E. costirostris* Lea, 1927c, *E. cryptorhyncha* (Lea, 1899a), *E. curvirostris* Lea, 1927c, *E. cylindrirostris* Lea, 1927c, *E. cyphirrhina* (Lea, 1899a), *E. dispar* (Lea, 1899a), *E. dorsalis* (Lea, 1899a), *E. elliptica* (Lea, 1899a), *E. ephippiger* (Lea, 1899a), *E. femoralis* (Lea, 1899a), *E. ferruginea* Lea, 1927c, *E. filirostris* Pascoe, 1873a, *E. grata* Lea, 1927c, *E. gravis* Blackburn, 1890b, *E. grisea* Lea, 1927c, *E. ignobilis* (Lea, 1899a), *E. illota* Lea, 1927c, *E. impotens* (Lea, 1899a), *E. inamoena* (Lea, 1899a), *E. indistincta* (Lea, 1899a), *E. inscripta* Lea, 1927c, *E. interioris* Blackburn, 1894a, *E. intermixta* Lea, 1927c, *E. interocularis* Lea, 1927c, *E. interrupta* Lea, 1927c, *E. intricata* Lea, 1927c, *E. invenusta* Lea, 1927c, *E. invidiosa* (Lea, 1899a), *E. juvenca* (Lea, 1899a), *E. lata* Lea, 1927c, *E. leucomela* Lea, 1927c, *E. leucophaea* Lea, 1927c, *E. lilliputana* Lea, 1927c, *E. lineigera* Pascoe, 1870d, *E. lithostrota* (Lea, 1899a), *E. longicollis* Lea, 1927c, *E. longirostris* (Boheman, 1859), *E. macrosticta* Lea, 1927c, *E. macrostyla* (Lea, 1899a), *E. masculina* Lea, 1927c, *E. medfasciata* Lea, 1927c, *E. mediocris* (Lea, 1899a), *E. metasternalis* (Lea, 1926b), *E. microsticta* Lea, 1927c, *E. miscella* Lea, 1927c, *E. monticola* Blackburn, 1892b, *E. multiarticulata* (Lea, 1899a), *E. munda* Blackburn, 1890b, *E. nana* Lea, 1927c, *E. nigriclava* Lea, 1927c, *E. nigrirostris* Lea, 1927c, *E. nigrofasciata* (Lea, 1899a), *E. niveiceps* (Lea, 1899a), *E. notata* Blackburn, 1890b, *E. obliqua* Lea, 1927c, *E. occidentalis* (Lea, 1899a), *E. ocellata* Blackburn, 1892b, *E. ovalisticta* Lea, 1927c, *E. pallida* Lea, 1927c, *E. parilis* Lea, 1927c, *E. parvidens* Lea, 1927c, *E. parvula* (Lea, 1899a),

E. paupercula (Lea, 1899a), *E. picta* Lea, 1927c, *E. pictipennis* Lea, 1927c, *E. pulicosa* Lea, 1927c, *E. rectirostris* Lea, 1927c, *E. remissa* Faust, 1888, *E. scolopax* Pascoe, 1870d, *E. setipennis* Lea, 1927c, *E. simplex* Pascoe, 1870d, *E. sordida* Lea, 1927c, *E. squamirostris* Lea, 1927c, *E. squamivaria* Lea, 1927c, *E. stenoderes* Lea, 1927c, *E. storeoides* Pascoe, 1873a, *E. sublecta* Lea, 1927c, *E. submunda* Lea, 1927c, *E. subtibialis* Lea, 1927c, *E. subuniformis* Lea, 1927c, *E. suturalis* Lea, 1927c, *E. tarsalis* Lea, 1927c, *E. tessellata* (Lea, 1899a), *E. tibialis* Lea, 1927c, *E. trisinuata* Lea, 1927c, *E. tuberculifrons* (Lea, 1899a), *E. umbrosa* Blackburn, 1890b, *E. vitticollis* Lea, 1927c.

Distribution. Australia, New Zealand.

Notes. The genus is poorly defined and difficult to separate taxonomically from *Storeus* (Pullen *et al.* 2014); both genera require taxonomic revision. Adults are generally found on *Eucalyptus* but those of *E. bifoveata* mostly on *Syzygium* (Rheinheimer 1989, 1996; Caldara *et al.* 2014), in whose fruits its larvae develop, also in New Zealand where the species is adventive.

Leucomelacis Lea, 1928 (Pl. 76C; Fig. 44.93)

Type species, by original designation: *Leucomelacis quadrinotatus* Lea, 1928.

Australian species. *L. albohumeralis* Lea, 1928a, *L. quadrinotatus* Lea, 1928a.

Distribution. Australia.

Notes. Adults have been reared from fruits of *Syzygium* (Myrtaceae) (Caldara *et al.* 2014).

Microberosiris Lea, 1907 (Pl. 76D)

Type species, by monotypy: *Microberosiris exilis* Lea, 1907.

Australian species. *M. albus* Lea, 1913b, *M. exilis* Lea, 1907b.

Distribution. Australia (SA, WA).

Placorrhinus Marshall, 1948 (Pl. 76E; Fig. 44.94)

Type species, by original designation: *Placorrhinus proteus* Marshall, 1948.

Australian species. *P. proteus* Marshall, 1948.

Distribution. Australia.

Notes. Adults have been reared collected on *Doryphora sassafras* (Atherospermataceae) (Caldara *et al.* 2014).

Pseudostoreus Lea, 1899 (Pl. 76F; Fig. 44.95)

Type species, by monotypy: *Pseudostoreus placitus* Lea, 1899.

Australian species. *P. amoenus* (Lea, 1899a), *P. placitus* Lea, 1899c.

Distribution. Australia.

Storeus Schoenherr, 1843 (Pl. 76G; Fig. 44.66N–U)

Type species, by original designation: *Storeus variegatus* Boheman, 1843.

Australian species. *S. acutidens* Lea, 1927d, *S. albosignatus* (Blackburn, 1890b), *S. amplipennis* Lea, 1927d, *S. apicalis* Lea, 1927d, *S. arcuatus* Lea, 1899a, *S. armipennis* Lea, 1927d, *S. baeodontus* Lea, 1927d, *S. captiosus* Lea, 1899a, *S. carinirostris* Lea, 1927d, *S. cognatus* Lea, 1927d, *S. contortus* Lea, 1899a, *S. eurypterus* Lea, 1927d, *S. falsus* Lea, 1899a, *S. fasciculatus* Lea, 1927d, *S. fimbripes* Lea, 1927d, *S. hoplocnemus* Lea, 1927d, *S. humeralis* Lea, 1927d, *S. hystricosus* Lea, 1927d, *S. inconspicuus* Lea, 1927d, *S. inconstans* Lea, 1927d, *S. insularis* Lea, 1927d, *S. inustus* Lea, 1927d, *S. laetus* Lea, 1927d, *S. longirostris* (Lea, 1915d), *S. magnus* Lea, 1929a, *S. majusculus* (Blackburn, 1893b), *S. maximus* Lea, 1927d, *S. modestus* (Pascoe, 1875), *S. preapicalis* Lea, 1929a, *S. pulchricollis* Lea, 1911b, *S. scutellaris* Lea, 1927d, *S. seticollis* Lea, 1927d, *S. setosus* Lea, 1899a, *S. signatus* Boheman, 1843 (*in* Schoenherr 1843), *S. simplicipennis* (Lea, 1928a), *S. specularis* Lea, 1927d, *S. squamibundus* (Lea, 1915d), *S. squamipictus* Lea, 1929a, *S. tenuirostris* Lea, 1927d, *S. variegatus* Boheman, 1843 (*in* Schoenherr 1843), *S. ventralis* Lea, 1927d, *S. ziczac* (Lea, 1931a).

Distribution. Australia, New Zealand (introduced), New Caledonia, Fiji, Philippines.

Notes. The genus is poorly defined and difficult to separate taxonomically from *Emplesis* (Pullen *et al.* 2014); some species likely belong in that genus. Whether the *Storeus* species described from New Caledonia (*S. australis* (Montrouzier)), Fiji (*S. aurifer* Lea, *S. inermis* Lea, *S. minimus* Lea) and the Philippines (*S. buchanani* Zimmerman, *S. luzonae* Zimmerman) truly belong in the genus also needs to be ascertained. In Australia, several species occur on *Ficus* (Moraceae) but *S. albosignatus* on *Acacia* (Fabaceae) and *S. baeodontus* on *Dodonea cuneata* (Sapindaceae) (Caldara *et al.* 2014), although this range may merely reflect the poor composition of *Storeus*. *Storeus albosignatus* has been introduced to New Zealand.

Storeini *sensu lato*

The genera of this group (e.g. Pl. 15B, E, G–H) were classified in Tychiini by Zimmerman (1992) but in Curculioninae without tribal assignment by Zimmerman (1994) and Alonso-Zarazaga & Lyal (1999). Pullen *et al.* (2014) later placed them in the tribe Storeini in the wide sense of Kuschel (1990), though transferring some (*Byrsodes*, *Eniopea*, *Glaucopela*, *Lexithia*, *Nedyleda* and *Olanaea*) to Hyperini, which is now treated as a subfamily in the CEGH clade of Curculionidae (Gillet *et al.* 2014; Gunter *et al.* 2016; Shin *et al.* 2018; Baird *et al.* 2021; Haran *et al.* 2023). Haran *et al.* (2023) recently showed that several other genera (*Cydmaea*, *Empolis*, *Encosmia*, *Epacticus*, *Erytenna*, *Misophrice*, *Myllorhinus*, *Ochrophoebe*, *Orichora*, *Phaunaeus*) also belong in the CEGH clade (of uncertain placement for now), and some others will undoubtedly have to be transferred from Curculioninae to the CEGH clade as well. Given the flux of the situation, we here follow the composition of Storeini *sensu lato* as in Pullen *et al.* (2014).

Abethas Zimmerman, 1999 (Pl. 76H)

Type species, by subsequent designation (Zimmerman, 1994: 662): *Ethas varians* Blackburn, 1894.

Australian species. *A. eruditis* (Blackburn, 1894a), *A. varians* (Blackburn, 1894a).

Distribution. Australia (NSW, VIC).

Notes. The genus was renamed *Abethas* by Zimmerman in Alonso-Zarazaga & Lyal (1999) to replace the junior homonym *Anethas* Zimmerman, 1994. It probably belongs in the CEGH clade of Curculionidae rather than in Curculioninae and Storeini.

Agestra Pascoe, 1873 (Pl. 76I)

Type species, by monotypy: *Agestra suturalis* Pascoe, 1873.

Australian species. *A. daviesae* Rheinheimer, 1995, *A. dilatatae* Rheinheimer, 1995, *A. horridae* Rheinheimer, 1995, *A. rubiginea* Pascoe, 1875, *A. suturalis* Pascoe, 1873a.

Distribution. Australia (ACT, WA).

Notes. Adults have been collected from species of *Bossiaea*, *Daviesia* and *Mirbelia* (Fabaceae) (Rheinheimer 1995; Caldara *et al.* 2014).

Anarciarthrum Blackburn, 1890 (Pl. 76J)

Type species, by monotypy: *Anarciarthrum viride* Blackburn, 1890.

Australian species. *A. viride* Blackburn, 1890b.

Distribution. Australia (SA, WA).

Notes. This monotypic genus is closely related to *Misophrice* and also belongs in the CEGH clade of Curculionidae rather than in Curculioninae and Storeini.

Cisowhitea Lea, 1915 (Pl. 76K; Fig. 44.81)

Type species, by monotypy: *Cisowhitea longicollis* Lea, 1915.

Australian species. *C. intermixta* Lea, 1928a, *C. longicollis* Lea, 1915e, *C. marmorata* Lea, 1928a.

Distribution. Australia (QLD, SA, NT, VIC).

Cydmaea Pascoe, 1872 (Pl. 76L; Fig. 44.82)

Type species, by subsequent designation (Pullen *et al.* 2014: 205): *Cydmaea bimaculata* Pascoe, 1872.

Australian species. *C. aemula* Lea, 1928c, *C. basalis* Lea, 1928a, *C. bimaculata* Pascoe, 1872a, *C. binotata* Lea, 1899a, *C. brevicornis* Lea, 1899a, *C. cara* Lea, 1899a, *C. cordipennis* Lea, 1928c, *C. crassirostris* Blackburn, 1894b, *C. diversa* Blackburn, 1890b, *C. dorsalis* Lea, 1899a, *C. eucalypti* Lea, 1899a, *C. exilis* Lea, 1928c, *C. fasciata* Lea, 1899a, *C. filirostris* Lea, 1899a, *C. fumosa* Blackburn, 1896, *C. gemmea* Lea, 1928c, *C. grisea* Lea, 1899a, *C. inconspicua* Lea, 1899a, *C. inconstans* Lea, 1928c, *C. indistincta* Lea, 1928c, *C. intermixta* Lea, 1928c, *C. interocularis* Lea, 1928c, *C. invalida* Blackburn, 1890b, *C. latirostris* Lea, 1928c, *C. leucomela* Lea, 1928c, *C. lineata* Blackburn, 1890c, *C. luctuosa* Pascoe, 1872a, *C. metasternalis* Lea, 1928c, *C. mixta* Blackburn, 1894a, *C. modesta* Lea, 1899a, *C. moerens* Lea, 1899a, *C. monobia* Lea, 1928c, *C. multimaculata* Lea, 1928c, *C. murina* Lea, 1928c, *C. nasalis* Lea, 1928c, *C. notaticollis* Pascoe, 1875, *C. nymphoides* Lea, 1899a, *C. obscura* Blackburn, 1890a, *C. pusilla* Pascoe, 1872a, *C. rostralis* Lea, 1899a, *C. ruficornis* Lea, 1928c, *C. rufipes* Lea, 1899a, *C. scutellaris* Lea, 1928c, *C. selligera* Pascoe, 1875, *C. setipennis* Lea, 1928c, *C. sordida* Lea, 1928c, *C. soror* Lea, 1928c, *C. subuniformis* Lea, 1928c, *C. suturalis* Lea, 1915c, *C. teramocera* Lea, 1899a, *C. tibialis* Lea, 1915c, *C. uniformis* Lea, 1899a, *C. viridis* Lea, 1928c, *C. viridula* Pascoe, 1872a, *C. vitticollis* Lea, 1928c.

Distribution. Australia (QLD, NSW, VIC, TAS, SA, NT, WA).

Notes. This genus belongs in the CEGH clade of Curculionidae (Haran *et al.* 2023). Its known hosts include *Hakea* and *Grevillea* (Proteaceae). The larva of *C. binotata* bores in shoots of *Hakea*, and the species has been introduced in South Africa for the biocontrol of the invasive *Hakea sericea* but seemingly died out there (Caldara *et al.* 2014).

Diapelmus Erichson, 1842

Type species, by monotypy: *Diapelmus mendax* Erichson, 1842.

Australian species. *D. bryophagus* (Lea, 1908a), *D. mendax* Erichson, 1842, *D. similis* (Lea, 1908a).

Distribution. Australia (TAS).

Dicomada Pascoe, 1873 (Pl. 77A; Fig. 44.83)

Type species, by subsequent designation (Pullen *et al.* 2014: 207): *Dicomada litigiosa* Pascoe, 1873.

Australian species. *D. litigiosa* Pascoe, 1873a, *D. murina* Pascoe, 1875, *D. ovalis* Pascoe, 1873a, *D. rufa* Blackburn, 1890c, *D. terrea* Pascoe, 1873a.

Distribution. Australia (NSW, SA, WA), South Africa (introduced).

Notes. The genus also belongs in the CEGH clade of Curculionidae. The larva of *D. rufa* develops in flower buds of *Hakea* (Proteaceae), and the species has been introduced to South Africa for the control of *Hakea sericea* (Gordon 1999; Caldara *et al.* 2014).

Dyschoenium Blackburn, 1890 (Pl. 77B)

Type species, by monotypy: *Dyschoenium flavum* Blackburn, 1890.

Australian species. *D. flavum* Blackburn, 1890b, *D. minutissimum* (Boheman, 1859).

Distribution. Australia (QLD, NSW).

Elleschodes Blackburn, 1897 (Pl. 77C, D; Figs 44.66A–G, 44.92)

Type species, by monotypy: *Elleschodes hamiltoni* Blackburn, 1897.

Australian species. *E. basipennis* Lea, 1909c, *E. castelnaui* (Lea, 1908b), *E. compactus* Lea, 1908a, *E. concinnus* (Lea, 1908a), *E. decipiens* (Lea, 1908b), *E. ellipticus* Lea, 1908b, *E. eucalypti* Lea, 1908c, *E. hamiltoni* Blackburn, 1897, *E. hystricosus* Lea, 1915d, *E. inconstans* Lea, 1909b, *E. macrops* Lea, 1915d, *E. modicus* Lea, 1908a, *E. nigrirostris* Lea, 1908a, *E. pallidus* Lea, 1908a, *E. pictus* Lea, 1908a, *E. placidus* Lea, 1908a, *E. rufimanus* Lea, 1908b, *E. rufulus* Lea, 1908a, *E. scutellaris* Lea, 1908a, *E. suturalis* Lea, 1908a, *E. tenuirostris* Lea, 1911c, *E. uniformis* Lea, 1908a, *E. v-albus* Lea, 1915d, *E. varipes* (Lea, 1908b), *E. wellingtoniensis* (Lea, 1908a).

Distribution. Australia (QLD, NSW, VIC, TAS, SA, WA).

Notes. The genus is an artificial group and needs division (Pullen *et al.* 2014). Its known host plants include *Eucalyptus*, *Eupomatia* and *Pultenaea*. *Elleschodes hamiltoni* and relatives are closely associated with *Eupomatia*, specifically as their only known pollinator (Hamilton 1897; Armstrong & Irvine 1990; Williams 2021).

Empira Pascoe, 1874 (Pl. 77E)

Type species, by monotypy: *Empira variegata* Pascoe, 1874.

Australian species. *E. variegata* Pascoe, 1874a.

Distribution. Australia (WA).

Empolis Blackburn, 1890 (Pl. 77F)

Type species, by monotypy: *Empolis angustatus* Blackburn, 1890.

Australian species. *E. abacetus* Lea, 1915a, *E. angustatus* Blackburn, 1890b, *E. granulatus* Lea, 1915c, *E. leai* Blackburn, 1893b, *E. longipes* Blackburn, 1893b, *E. niveodispersus* Lea, 1915c, *E. squamosus* Lea, 1915a.

Distribution. Australia (NSW, VIC, SA).

Notes. This genus belongs in the CEGH clade of Curculionidae (Haran *et al.* 2023) rather than in Curculioninae and Storeini.

Encosmia Blackburn, 1893 (Pl. 77G; Fig. 44.84)

Type species, by subsequent designation (Pullen *et al.* 2014: 210): *Encosmia adelaidae* Blackburn, 1893.

Australian species. *E. adelaidae* Blackburn, 1893b, *E. alba* Lea, 1926c, *E. albifascia* Lea, 1926c, *E. apicalis* Lea, 1926c, *E. basalis* Lea, 1926c, *E. bivittata* Lea, 1926c, *E. blackburni* Lea, 1926c, *E. cornuta* Blackburn, 1894a, *E. cryptoderma* Lea, 1926c, *E. curvirostris* Lea, 1926c, *E. fasciata* Lea, 1915c, *E. fasciculata* Lea, 1926c, *E. infuscata* Blackburn, 1893b, *E. interrupta* Lea, 1926c, *E. medioglabra* Lea, 1926c, *E. melanostetha* Lea, 1926c, *E. minor* Lea, 1926c, *E. montana* Lea, 1926c, *E. nigriclava* Lea, 1926c, *E. pulicosa* Lea, 1926c, *E. ruficornis* Lea, 1926c, *E. suturalis* Lea, 1926c, *E. variegata* Lea, 1926c, *E. ventralis* Lea, 1926c.

Distribution. Australia (QLD, NSW, VIC, TAS, SA, WA).

Notes. This genus also belongs in the CEGH clade of Curculionidae (Haran *et al.* 2023) rather than in Curculioninae and Storeini. Specimens have been collected from *Eucalyptus*.

Epacticus Blackburn, 1893 (Pl. 77H)

Type species, by subsequent designation (Pullen *et al.* 2014: 211): *Epacticus ruber* Blackburn, 1893.

Australian species. *E. aspratilis* Lea, 1917b, *E. bimaculatus* Lea, 1917c, *E. nigrirostris* Blackburn, 1893b, *E. occidentalis* Lea, 1917c, *E. ruber* Blackburn, 1893b, *E. suturalis* Lea, 1917b, *E. whitei* Lea, 1917c.

Distribution. Australia (SA).

Notes. The genus also belongs in the CEGH clade of Curculionidae (Haran *et al.* 2023) but is composite and needs restricting to just a few of the current species (Pullen *et al.* 2014). It is associated with Myrtaceae (Caldara *et al.* 2014) but the larvae and life histories are unknown.

Erytenna Pascoe, 1870 (Pl. 77I; Fig. 44.66H–M)

Type species, by subsequent designation (Pullen *et al.* 2014: 211): *Erytenna consputa* Pascoe, 1870.

Australian species. *E. consputa* Pascoe, 1870d.

Distribution. Australia (NSW, VIC, SA, WA), South Africa (introduced).

Notes. This genus belongs in the CEGH clade of Curculionidae (Haran *et al.* 2023) rather than in Curculioninae and Storeini. The larva of the single species develops in the fruits of *Hakea* species, feeding on the seeds, and the species has been introduced in South Africa for the biocontrol of *Hakea sericea* and is one of the main agents reducing the impacts of the weed (Esler *et al.* 2010; Caldara *et al.* 2014).

Hibberticola Lea, 1908 (Pl. 77J)

Type species, by monotypy: *Hibberticola echinatus* Lea, 1908.
Australian species. *H. echinatus* Lea, 1908a, *H. punctulatus* (Blackburn, 1890c).
Distribution. Australia (SA).
Notes. Adults of *H. echinata* have been found in galls on *Hibbertia sericea* (Dilleniaceae) (Lea 1908a), which may be incited by the larvae.

Imathia Pascoe, 1885 (Pl. 77K; Figs 44.85, 44.87)

Type species, by subsequent designation (Zimmerman, 1967: 192): *Imathia bella* Pascoe, 1885.
Australian species. *I. amplipennis* (Lea, 1928a), *I. carinirostris* (Lea, 1928a), *I. imbellis* (Lea, 1928a), *I. parvicollis* (Lea, 1928a).
Distribution. Australia, New Guinea, Pacific Is., Oriental region to southern Japan (Ryukyu Is.).
Notes. The genus appears to be composite (Pullen *et al.* 2014) and requires revision and proper delimitation.

Ixamine Pascoe, 1870 (Pl. 77L)

Type species, by monotypy: *Ixamine atomaria* Pascoe, 1870.
Australian species. *I. alba* (Lea, 1926c), *I. atomaria* Pascoe, 1870d, *I. banksiae* (Lea, 1926c), *I. leucomelas* (Lea, 1926c), *I. selligera* (Pascoe, 1873a).
Distribution. Australia (SA, WA).

Meripherellus Lea, 1915 (Pl. 79F)

Type species, by monotypy: *Meripherellus apicalis* Lea, 1915.
Australian species. *M. apicalis* Lea, 1915c.
Distribution. Australia (QLD).

Mesembrinocis Lea, 1915 (Pl. 79G)

Type species, by monotypy: *Mesembrinocis variegatus* Lea, 1915.
Australian species. *M. variegatus* Lea, 1915d.
Distribution. Australia (NSW).

Micraonychus Lea, 1915 (Pl. 78A)

Type species, by subsequent designation (Pullen *et al.* 2014: 213). *Micraonychus decipiens* Lea, 1915.
Australian species. *M. casuarinae* Lea, 1915d, *M. cinerascens* Lea, 1916, *M. coelosternus* Lea, 1927e, *M. decipiens* Lea, 1915d, *M. ilotus* Lea, 1927e, *M. interruptus* Lea, 1927e, *M. maculatus* Lea, 1915d, *M. nigrirostris* Lea, 1915d, *M. rufimanus* Lea, 1916, *M. sordidus* Lea, 1915d.
Distribution. Australia (QLD, NSW, VIC, TAS, SA).
Notes. This genus appears related to *Misophrice* and also belongs in the CEGH clade of Curculionidae rather than in Storeini. It likewise lives on *Allocasuarina* (Casuarinaceae) (Caldara *et al.* 2014) but the larvae are unknown.

Minia Pascoe, 1883 (Pl. 79H)

Type species, by monotypy: *Minia opalescens* Pascoe, 1883.
Australian species. *M. opalescens* Pascoe, 1883a.
Distribution. Australia.
Notes. The genus belongs in the CEGH clade of Curculionidae, not in Curculioninae and Storeini. The single species appears to be known only from the holotype, and its type locality is uncertain; whereas the description states 'Clarence River', alluding to the river near Grafton in New South Wales, the label of the holotype gives Champion Bay (Geraldton, WA) as origin of the specimen. It is possible that 'Clarence' River is a misprint of Chapman River.

Misophrice Pascoe, 1872 (Pl. 78B)

Type species, by monotypy: *Misophrice hispida* Pascoe, 1872.
Australian species. *M. albolineata* Lea, 1927e, *M. alternata* Lea, 1899a, *M. amplicollis Lea, 1906a*, *M. amplipennis* Lea, 1915a, *M. apionoides* Lea, 1906a, *M. argentata* Blackburn, 1890b, *M. arida* Lea, 1917c, *M. barretti* Lea, 1927e, *M. blackburni* Lea, 1914, *M. brevisetosa* Lea, 1914, *M. carteri* Lea, 1911a, *M. clathrata* Lea, 1899a, *M. cristatifrons* Lea, 1911a, *M. cylindrica* Lea, 1899a, *M. dispar* Blackburn, 1894a, *M. dissentanea* Lea, 1899a, *M. dubia* Lea, 1914, *M. evanida* Lea, 1915a, *M. fenestrata* Lea, 1899a, *M. gloriosa* Lea, 1906a, *M. griffithi* Lea, 1915a, *M. hispida* Pascoe, 1872b, *M. hoblerae* Lea, 1911a, *M. inconstans* Lea, 1914, *M. inflata* Lea, 1906a, *M. insularis* Lea, 1914, *M. lata* Lea, 1927e, *M. minima* Lea, 1915a, *M. munda* Blackburn, 1890b, *M. nigriceps* Lea, 1915a, *M. nigripes* Lea, 1906a, *M. nigriventris* Lea, 1899a, *M. oblonga* Blackburn, 1890a, *M. orthorrhina* Lea, 1911a, *M. parallela* Blackburn, 1890b, *M. quadraticollis* Blackburn, 1894a, *M. rufiventris* Lea, 1914, *M. setosa* Lea, 1915a, *M. setulosa* Blackburn, 1890b, *M. soror* Lea, 1914, *M. spilota* Blackburn, 1894a, *M. squamibunda* Lea, 1899a, *M. squamiventris* Lea, 1906a, *M. squamosa* Blackburn, 1890b, *M. submetallica* Blackburn, 1890b, *M. subvariabilis* Lea, 1927e, *M. tuberculata* Lea, 1914, *M. ursa* Lea, 1927e, *M. v-alba* Lea, 1915a, *M. variabilis* Blackburn, 1890b, *M. vicina* Lea, 1906a, *M. viridisquama* Lea, 1899a, *M. vitiata* Lea, 1899a, *M. vittata* Lea, 1927e, *M. wardi* Lea, 1927e.
Notes. Specimens can be found on *Allocasuarina*.
Distribution. Australia (QLD, NSW, VIC, TAS, SA, NT, WA).
Notes. This genus has recently been shown not to belong in Curculioninae but in the CEGH clade of Curculionidae

(Haran *et al.* 2023). The species are associated with *Allocasuarina* and *Casuarina* (Casuarinaceae) (Lea 1906a; Caldara *et al.* 2014) and deemed to have potential as biological control agents of invasive *Casuarina* species in Florida, USA (Wheeler *et al.* 2011). The adults feed on pollen, but the larvae and their sites of development are unknown.

Misophricoides Rheinheimer, 1990 (Pl. 78C; Fig. 44.86)

Type species, by original designation: *Misophricoides callitriphilus* Rheinheimer, 1990.
Australian species. *M. callitriphilus* Rheinheimer, 1990.
Distribution. Australia.
Notes. The single species is associated with *Callitris* (Cupressaceae) (Rheinheimer 1990).

Myllorhinus Boisduval, 1835 (Pl. 78D, E)

Type species, by monotypy: *Myllorhinus bicaudatus* Boisduval, 1835.
Australian species. *M. aurifer* (Lea, 1915a), *M. bicaudatus* Boisduval, 1835, *M. dentiferus* (Boheman, 1843) (*in* Schoenherr 1843), *M. insignis* (Lea, 1915c), *M. posticus* (Chevrolat, 1879a), *M. simplex* (Blackburn, 1894c), *M. strenuus* (Blackburn, 1894a).
Distribution. Australia (QLD, NSW, VIC, TAS, SA, WA).
Notes. The genus belongs in the CEGH clade of Curculionidae (Haran *et al.* 2023) rather than in Curculioninae and Storeini. Its hosts include several species of *Eucalyptus* and *Syzygium* (Myrtaceae), the larvae boring in shoots (Jones & Potts 2000).

Ochrophoebe Pascoe, 1872 (Pl. 78F)

Type species, by monotypy: *Ochrophoebe uniformis* Pascoe, 1872.
Australian species. *O. uniformis* Pascoe, 1872a.
Distribution. Australia (WA).
Notes. The genus belongs in the CEGH clade of Curculionidae (Haran *et al.* 2023) rather than in Curculioninae and Storeini.

Olbiodorus Blackburn, 1893 (Pl. 78G)

Type species, by monotypy: *Olbiodorus hirsutus* Blackburn, 1893.
Australian species. *O. hirsutus* Blackburn, 1893b.
Distribution. Australia (SA).

Orichora Pascoe, 1870 (Pl. 78I, J; Fig. 44.89)

Type species, by monotypy: *Orichora trivirgata* Pascoe, 1870.
Australian species. *O. trivirgata* Pascoe, 1870c.
Notes. Specimens have been collected from inflorescences of *Banksia* species.
Distribution. Australia (WA).
Notes. This genus belongs in the CEGH clade of Curculionidae (Haran *et al.* 2023) rather than in Curculioninae and Storeini. Specimens have been collected from inflorescences of *Banksia* species (Proteaceae).

Paryzeta Pascoe, 1873 (Pl. 78H; Fig. 44.90)

Type species, by monotypy: *Paryzeta musiva* Pascoe, 1873.
Australian species. *P. coxii* (Lea, 1915c), *P. musiva* Pascoe, 1873a, *P. nigrovaria* (Lea, 1909b), *P. seriata* (Blackburn, 1894a), *P. vittata* Blackburn, 1894a.
Distribution. Australia (NSW, SA, WA).
Notes. Specimens have been collected on *Melaleuca* plants (Myrtaceae).

Phaeodica Pascoe, 1874

Type species, by subsequent designation (Zimmerman, 1994: 669): *Phaeodica fulvicornis* Pascoe, 1874.
Australian species. *P. fulvicornis* Pascoe, 1874a.
Notes. Specimens of this genus were not available to us.
Distribution. Australia (WA).
Notes. The single species appears not to have been collected in recent times.

Phaunaeus Lea, 1910 (Pl. 78K, L)

Type species, by monotypy: *Phaunaeus longirostris* Lea, 1910.
Australian species. *P. longirostris* Lea, 1910a, *P. medioalbus* Lea, 1915c, *P. trilinealbus* Lea, 1915c.
Distribution. Australia (QLD, NSW).
Notes. This genus belongs in the CEGH clade of Curculionidae (Haran *et al.* 2023) rather than in Curculioninae and Storeini, but it is a composite one, the three included species not being congeneric (Pullen *et al.* 2014).

Plaesiorhinus Blackburn, 1893 (Pl. 79A)

Type species, by monotypy: *Plaesiorhinus notatus* Blackburn, 1893.
Australian species. *P. buxifoliae* Rheinheimer, 1994, *P. notatus* Blackburn, 1893b.
Distribution. Australia (SA).
Notes. The species appears associated with *Bossiaea* and *Daviesia* (Fabaceae) species (Rheinheimer 1994; Caldara *et al.* 2014), but the larvae and their breeding sites are unknown.

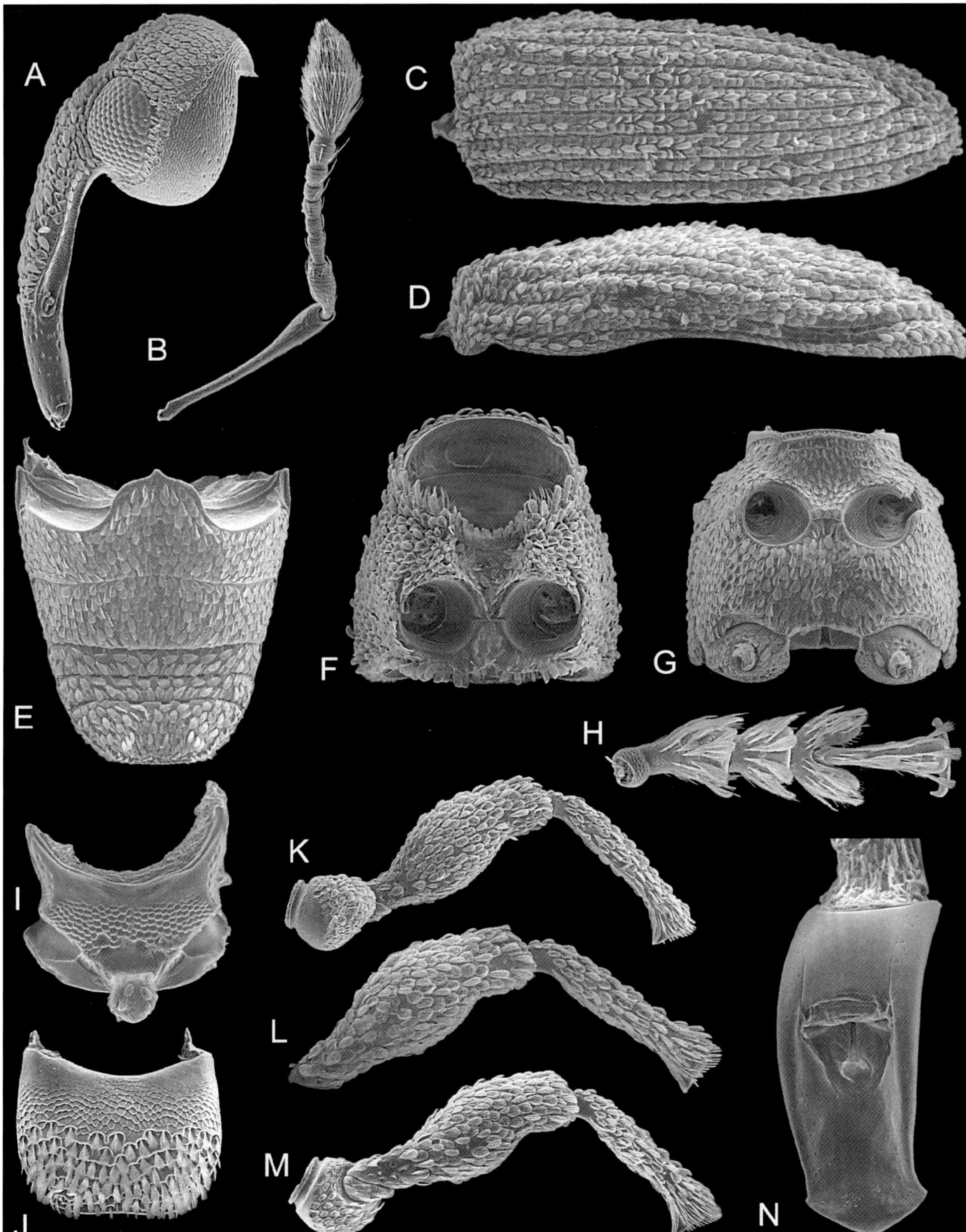

Fig. 44.81. Curculioninae, structural details, Storeini, *Cisowhitea longicollis* Lea. A, head, lateral; B, antenna, C–D, elytron: C, dorsal; D, lateral; E, abdominal ventrites; F, prothorax, ventral; G, pterothorax, ventral; H, tarsus, dorsal; I, scutellum; J, pygidium; K–M, legs; N, penis, dorsal.

Fig. 44.82. Curculioninae, structural details, Storeini, *Cydmaea binotata* Lea. A–B, head, lateral: A, ♂; B, ♀; C, antenna; D, mouthparts; E, prothorax, ventral; F, pterothorax; G–H, elytron: G, dorsal; H, lateral; I–J, abdominal ventrites: I, ♂; J, ♀; K–L, pygidium: K, ♂; L, ♀; M, scutellum; N–O, midleg: P, hindleg; Q, penis, dorsal.

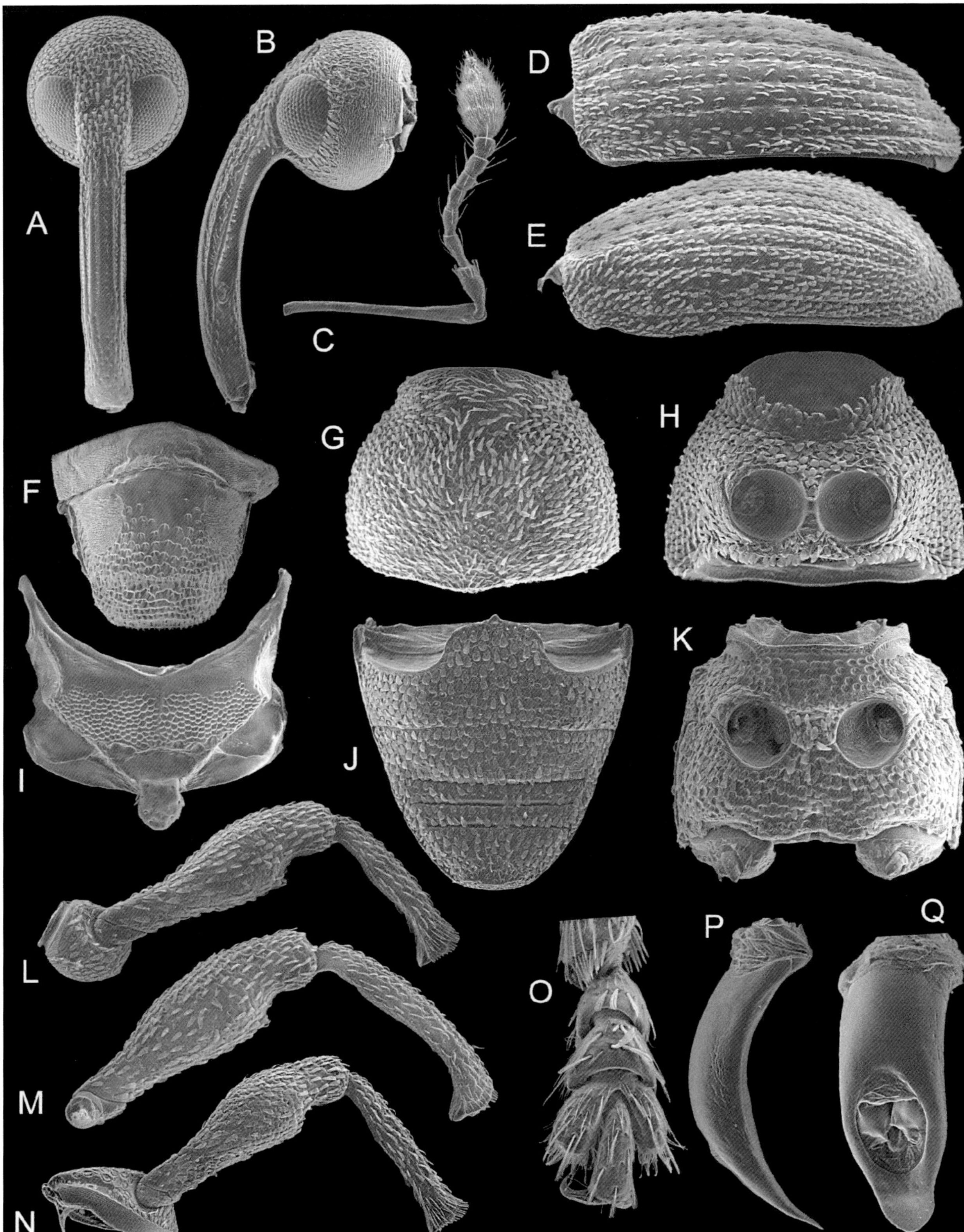

Fig. 44.83. Curculioninae, structural details, Storeini, *Dicomada rufa* Blackburn, ♂. A–B, head: A, dorsal; B, lateral; C, antenna; D–E, elytron: D, dorsal; E, lateral; F, pygidium; G–H, prothorax: G, dorsal; H, ventral; I, scutellum; J, abdominal ventrites; K, pterothorax, ventral; L, middle leg; M, foreleg; N, hindleg; O, tarsus, dorsal; P–Q, penis: P, lateral; Q, dorsal.

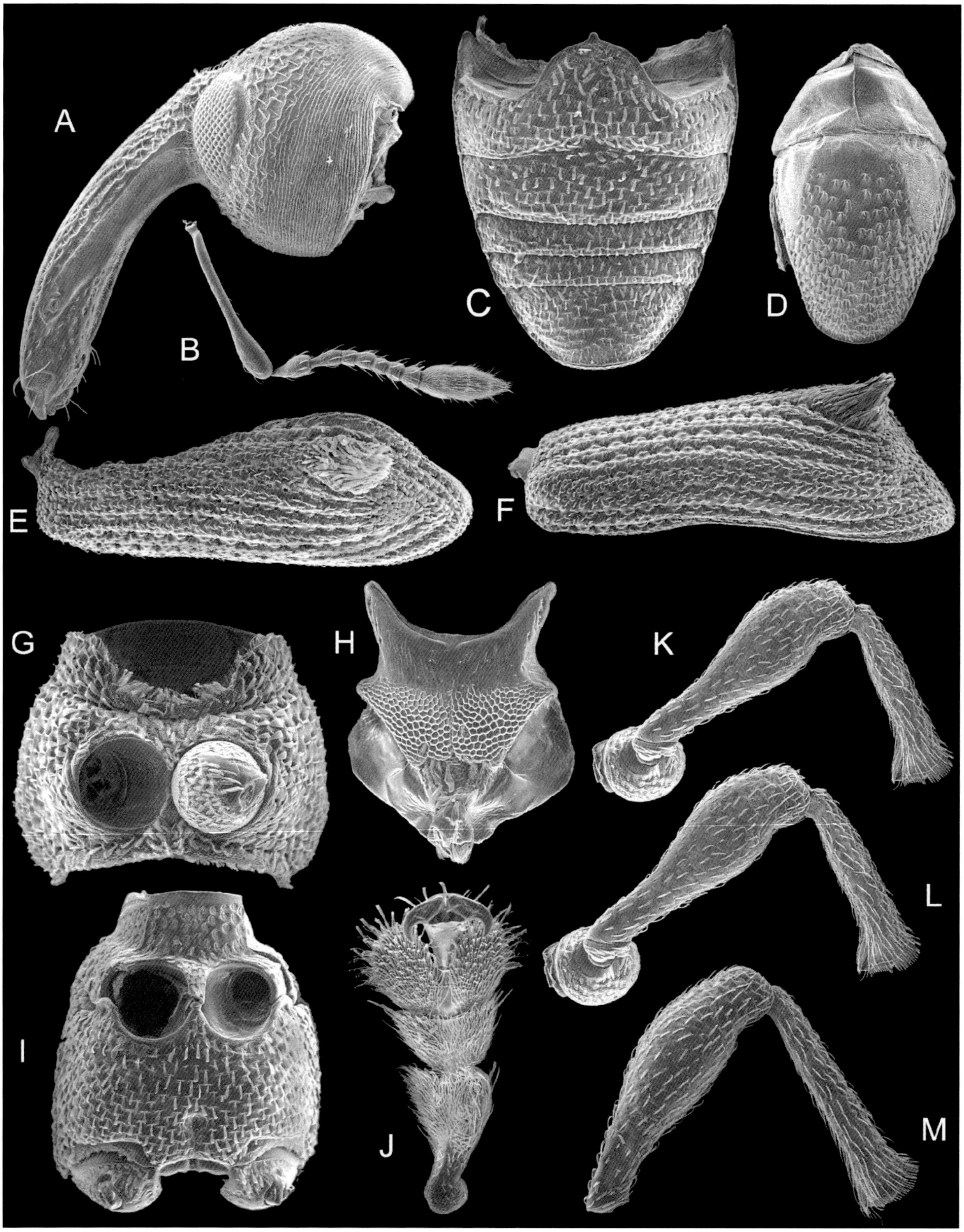

Fig. 44.84. Curculioninae, structural details, Storeini, *Encosmia cornuta* Blackburn, ♀. A, head, lateral; B, antenna; C, abdominal ventrites; D, pygidium; E–F, elytron: E, dorsal; F, lateral; G, prothorax, ventral; H, scutellum; I, pterothorax, ventral; J, tarsus, ventral; K, foreleg; L, middle leg; M, hindleg.

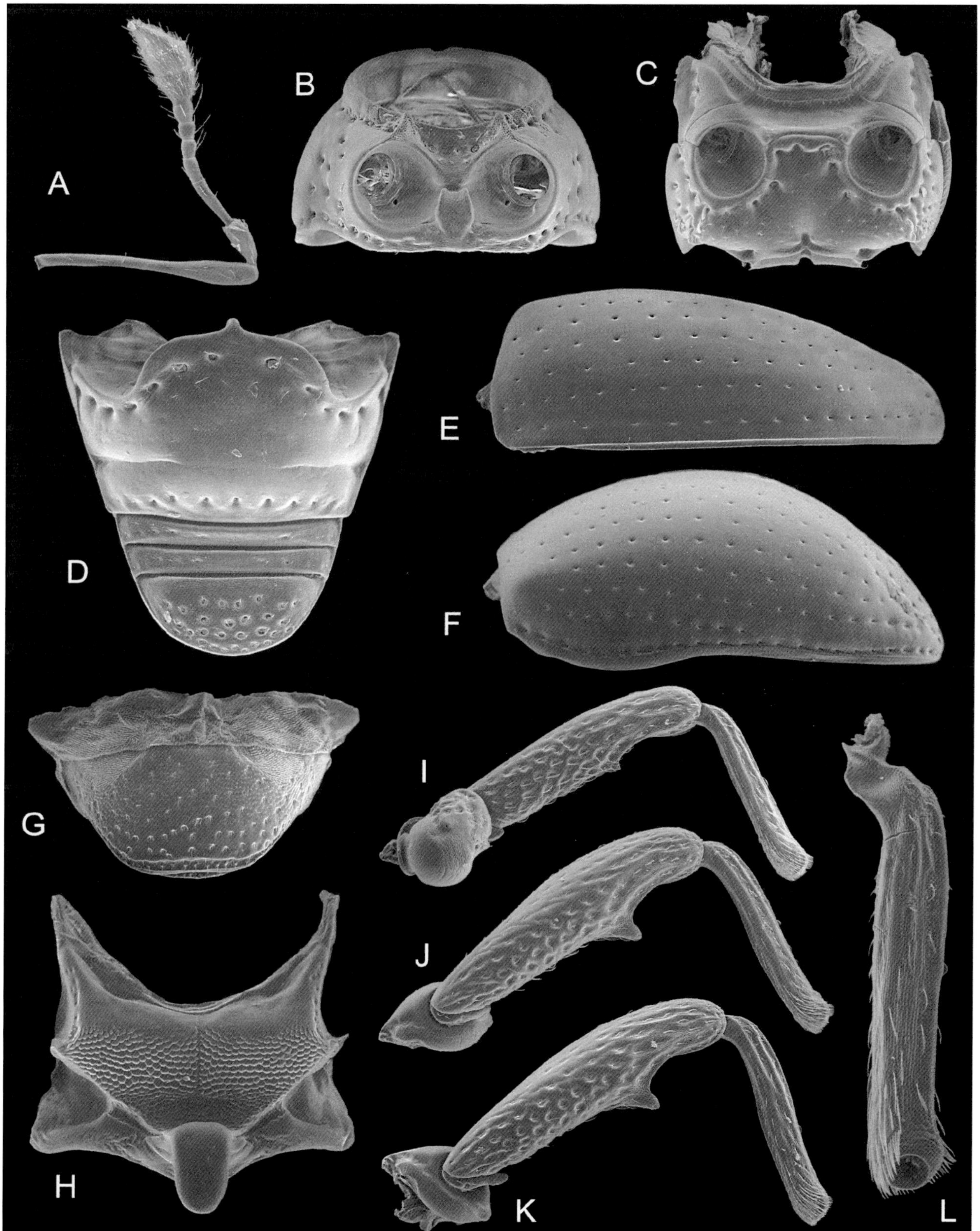

Fig. 44.85. Curculioninae, structural details, Storeini, *Imathia amplipennis* (Lea). A, antenna; B, prothorax, ventral; C, pterothorax, ventral; D, abdominal ventrites; E–F, elytron: E, dorsal; F, lateral; G, pygidium; H, scutellum; I, foreleg; J, hindleg; K, hindleg; L, tibia, inner view.

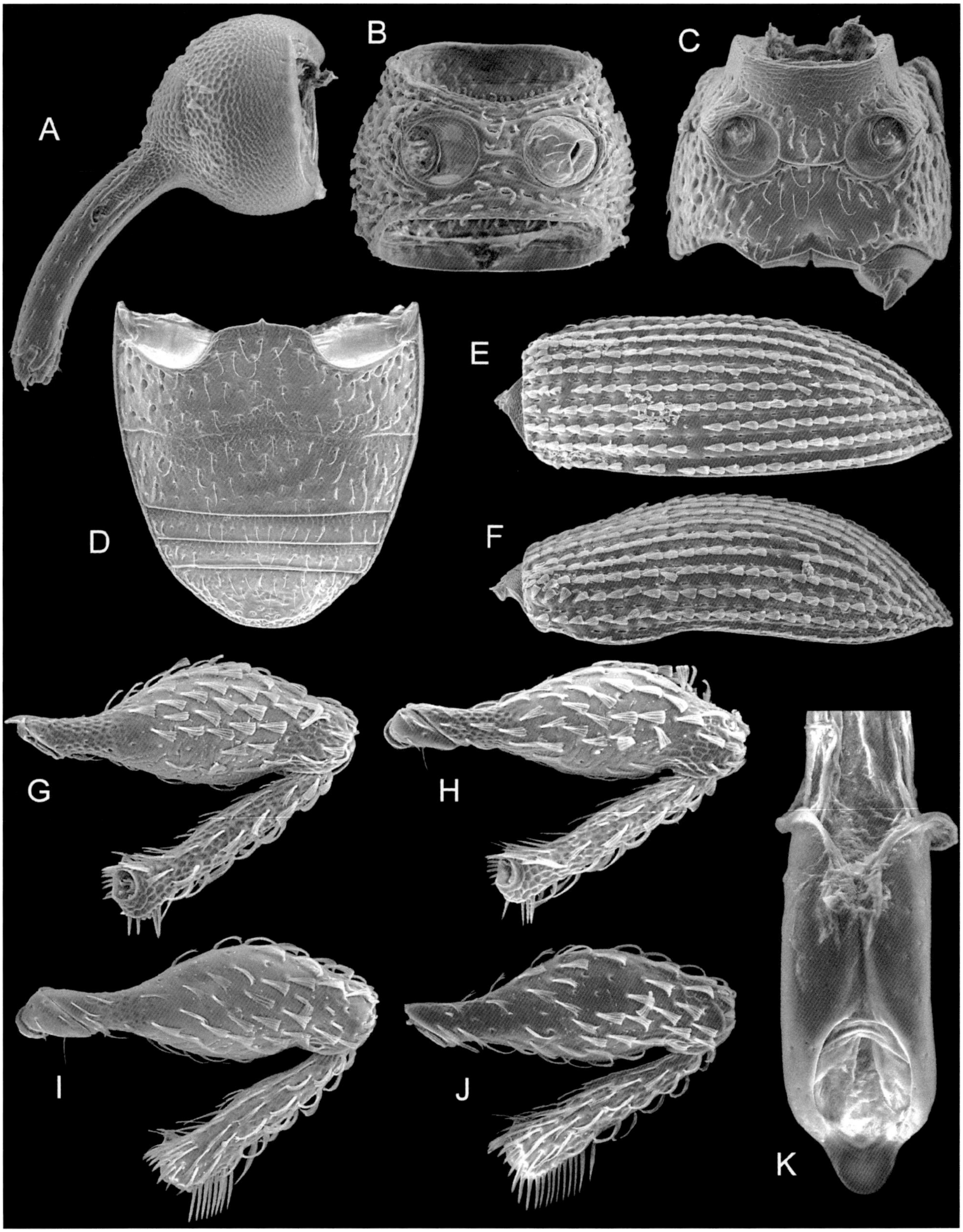

Fig. 44.86. Curculioninae, structural details, Storeini, *Misophricoides callitriphilus* Rheinheimer, ♂. A, head, lateral; B, prothorax, ventral; C, pterothorax, ventral; D, abdominal ventrites; E–F, elytron: E, dorsal; F, lateral; G–J, legs; K, penis, dorsal.

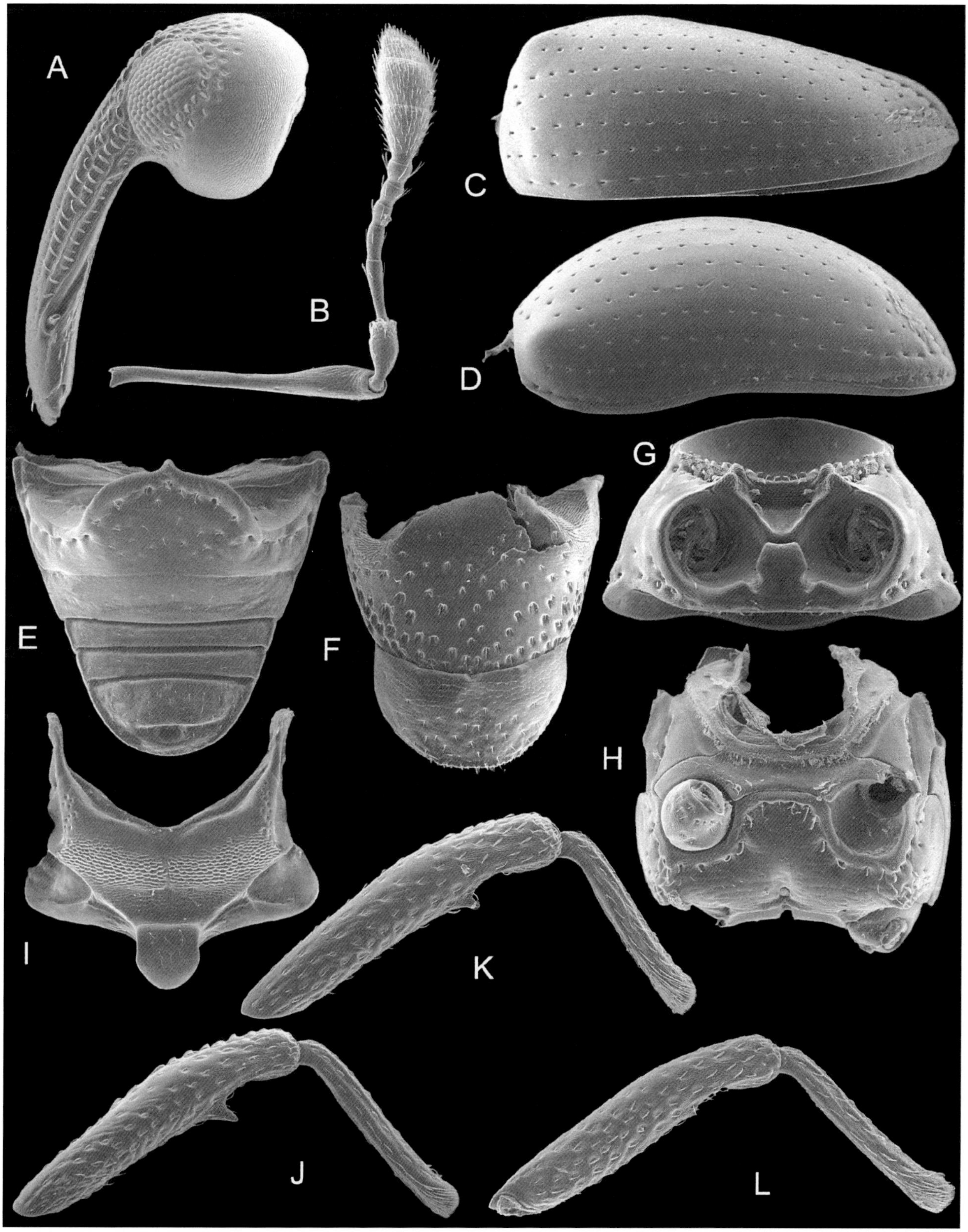

Fig. 44.87. Curculioninae, structural details, Storeini, *Imathia* sp., Millaa Millaa Falls, QLD. A, head, lateral; B, antenna; C–D, elytron: C, dorsal; D, lateral; E, abdominal ventrites; F, apical tergites, dorsal; G, prothorax, ventral; H, pterothorax, ventral; I, scutellum; J, foreleg; K, middle leg; L, hindleg.

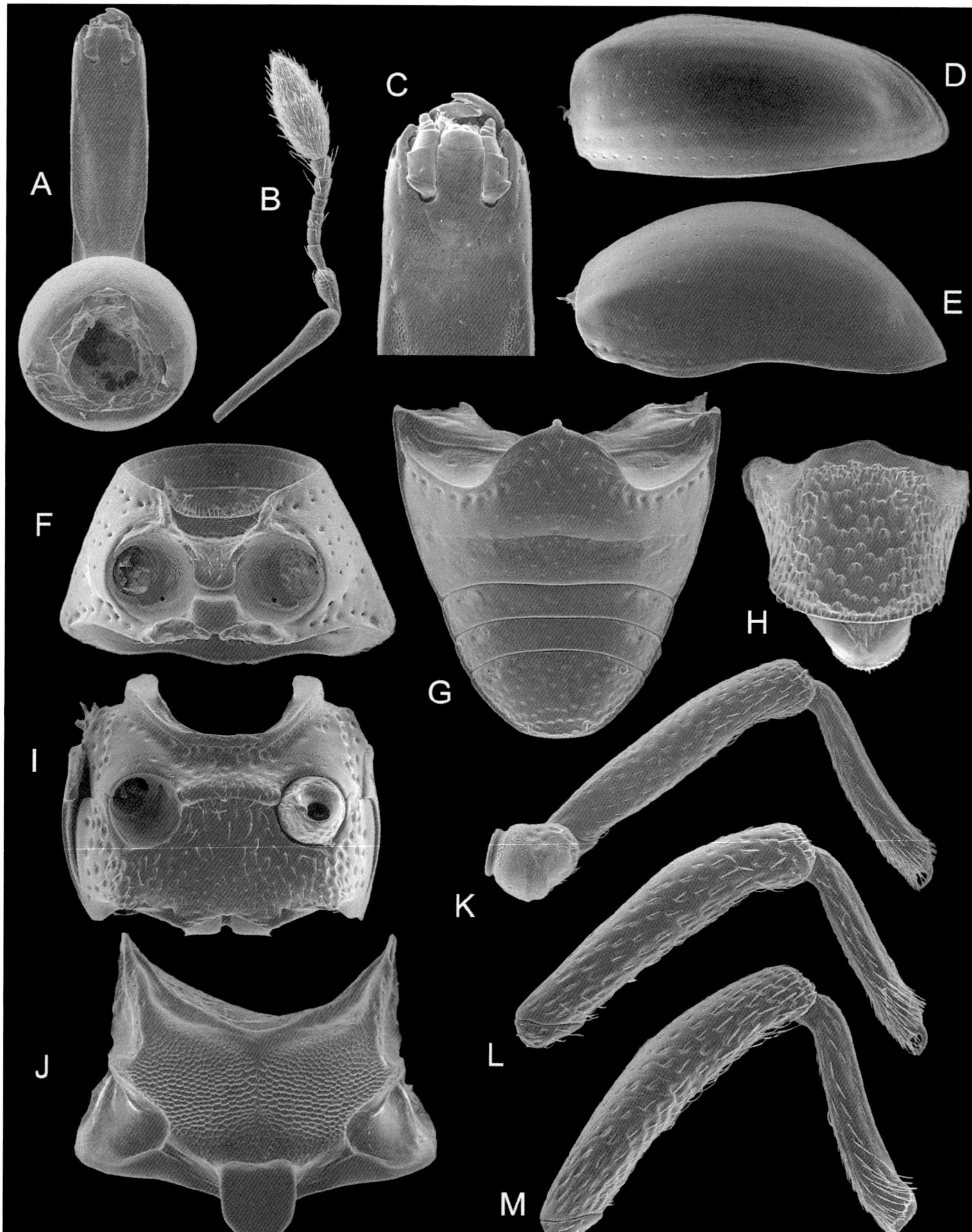

Fig. 44.88. Curculioninae, structural details, Storeini, *Rhinidotasia* sp., Mossman, QLD, ♀. A, head, ventral; B, antenna; C, mouthparts; D–E, elytron: D, dorsal; E, lateral; F, prothorax, ventral; G, abdominal ventrites; H, pygidium; I, pterothorax, ventral; J, scutellum; K, middle leg; L, foreleg; M, hindleg.

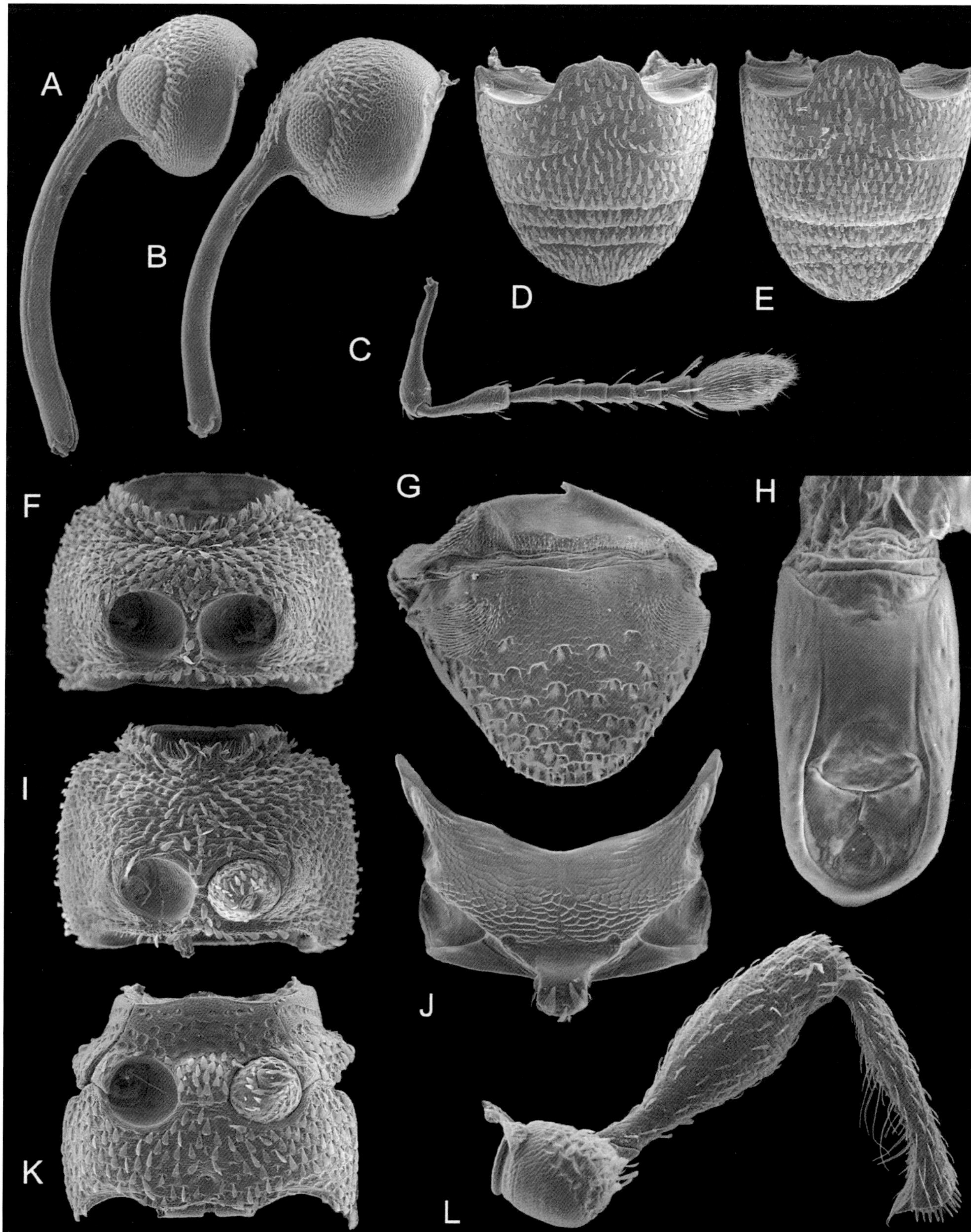

Fig. 44.89. Curculioninae, structural details, Storeini, *Orichora trivirgata* Pascoe. A–B, head, lateral: A, ♂; B, ♀; C, antenna; D–E, abdominal ventrites: D, ♂; E, ♀; F; I, prothorax, ventral; F, ♀; I, ♂; G, pygidium; H, penis, dorsal; J, scutellum; K, pterothorax, ventral; L, leg.

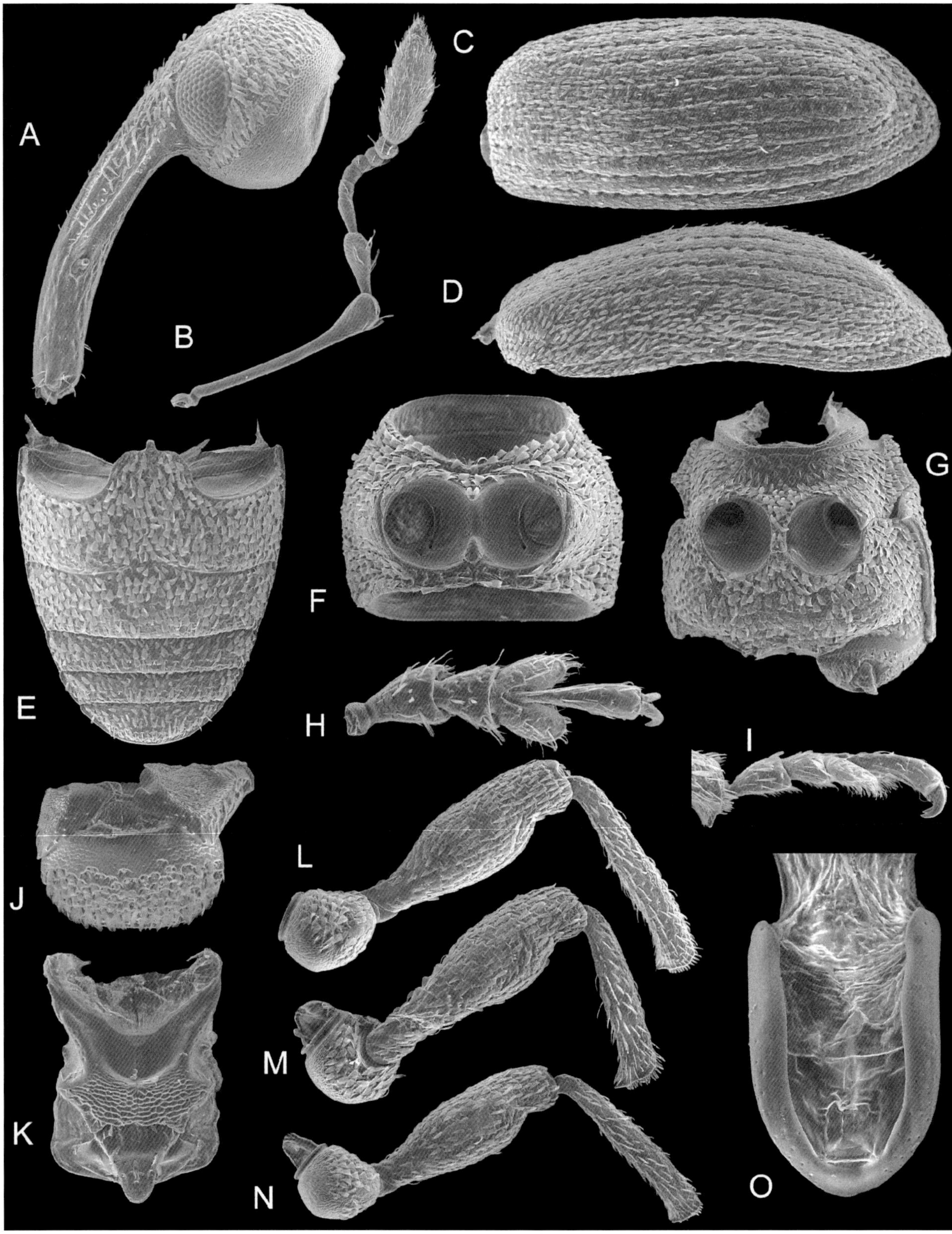

Fig. 44.90. Curculioninae, structural details, Storeini, *Paryzeta musiva* Pascoe. A, head, lateral; B, antenna; C–D, elytron: C, dorsal; D, lateral; E, abdominal ventrites; F, prothorax, ventral; G, pterothorax, ventral; H–I, tarsi and claws: H, dorsal; I, lateral; J, pygidium; K, scutellum; L–N, legs: O, penis, dorsal.

Fig. 44.91 Curculioninae, structural details, Storeini, undetermined genus, Mt. Barker, WA, ♂. A, head, lateral; B, antenna; C–D, elytron: C, dorsal; D, lateral; E, abdominal ventrites; F, prothorax, ventral; G, pterothorax, ventral; H, pygidium; I, scutellum; J–K, legs: L, hindleg; M, penis, dorsal.

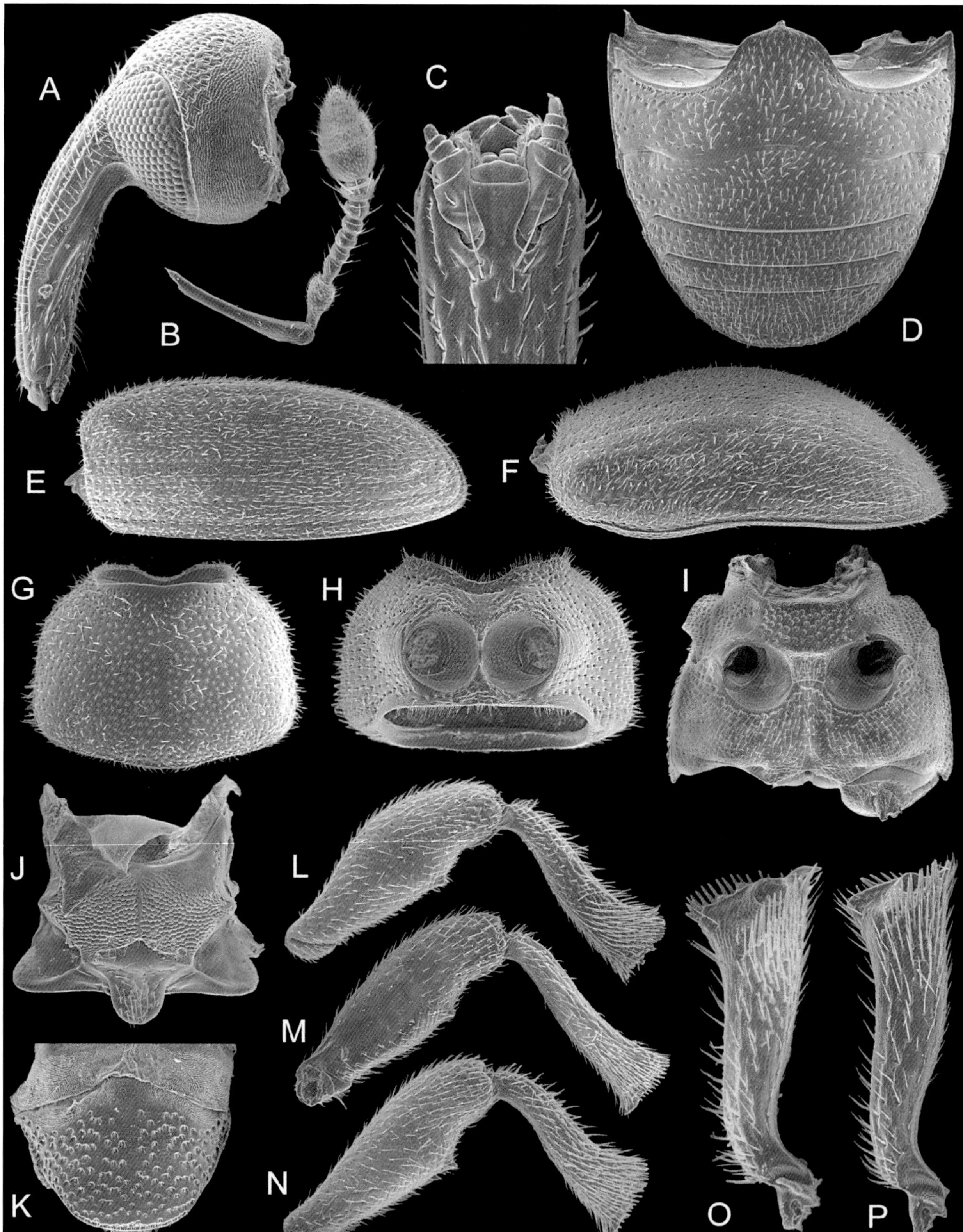

Fig. 44.92. Curculioninae, structural details, Storeini, *Elleschodes hamiltoni* Blackburn, ♀. A, head, lateral; B, antenna; C, mouthparts; D, abdominal ventrites; E–F, elytron: E, dorsal; F, ventral; G–H, prothorax: G, dorsal; H, ventral; I, pterothorax, ventral; J, scutellum; K, pygidium; L, foreleg; M, middle leg; N, hindleg; O–P, tibiae.

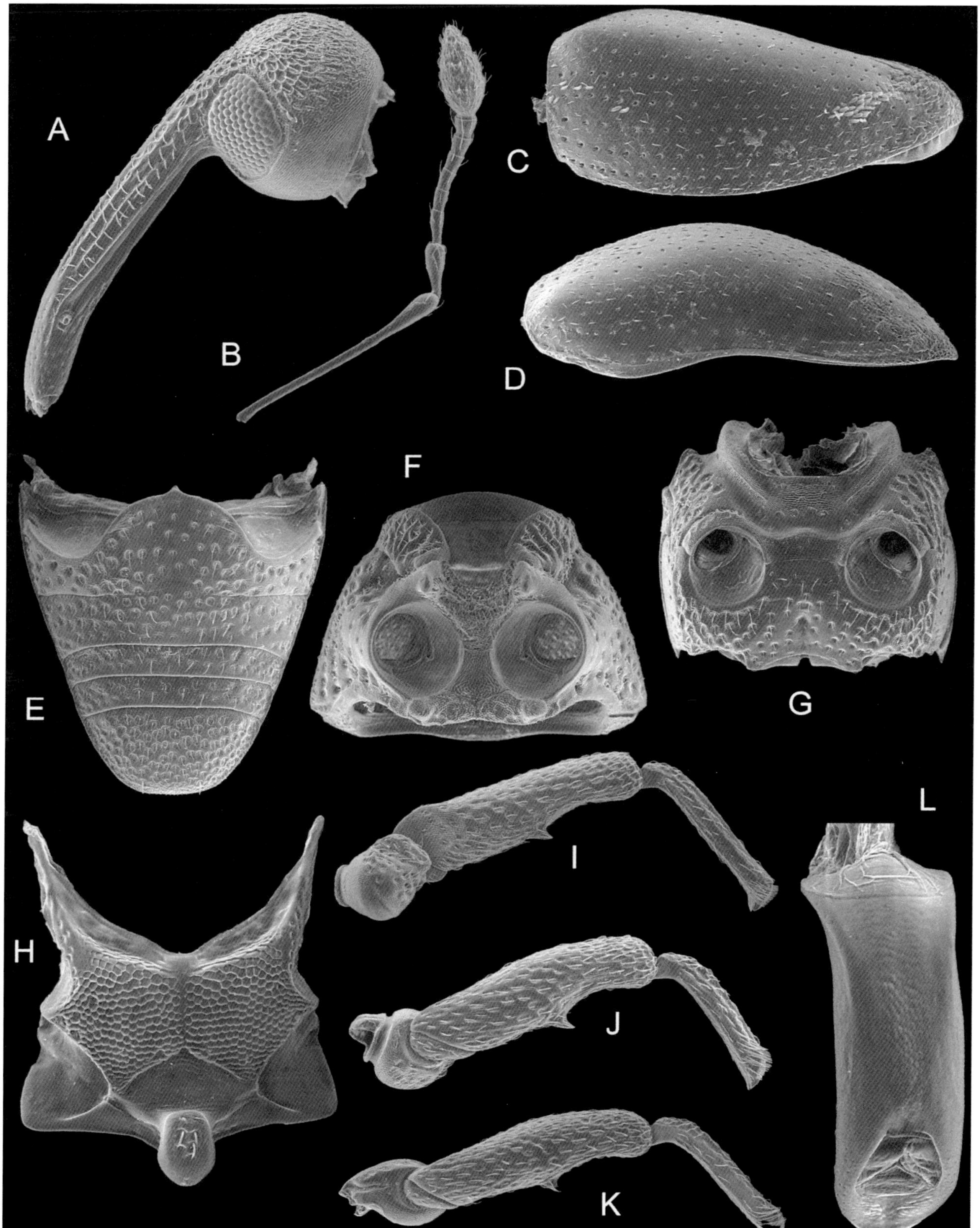

Fig. 44.93. Curculioninae, structural details, Storeini, *Leucomelacis quadrinotatus* Lea, ♂. A, head, lateral; B, antenna; C–D, elytron: C, dorsal; D, lateral; E, abdominal ventrites; F, prothorax, ventral; G, pterothorax, ventral; H, scutellum; I, foreleg; J, middle leg; K, hindleg; L, penis, dorsal.

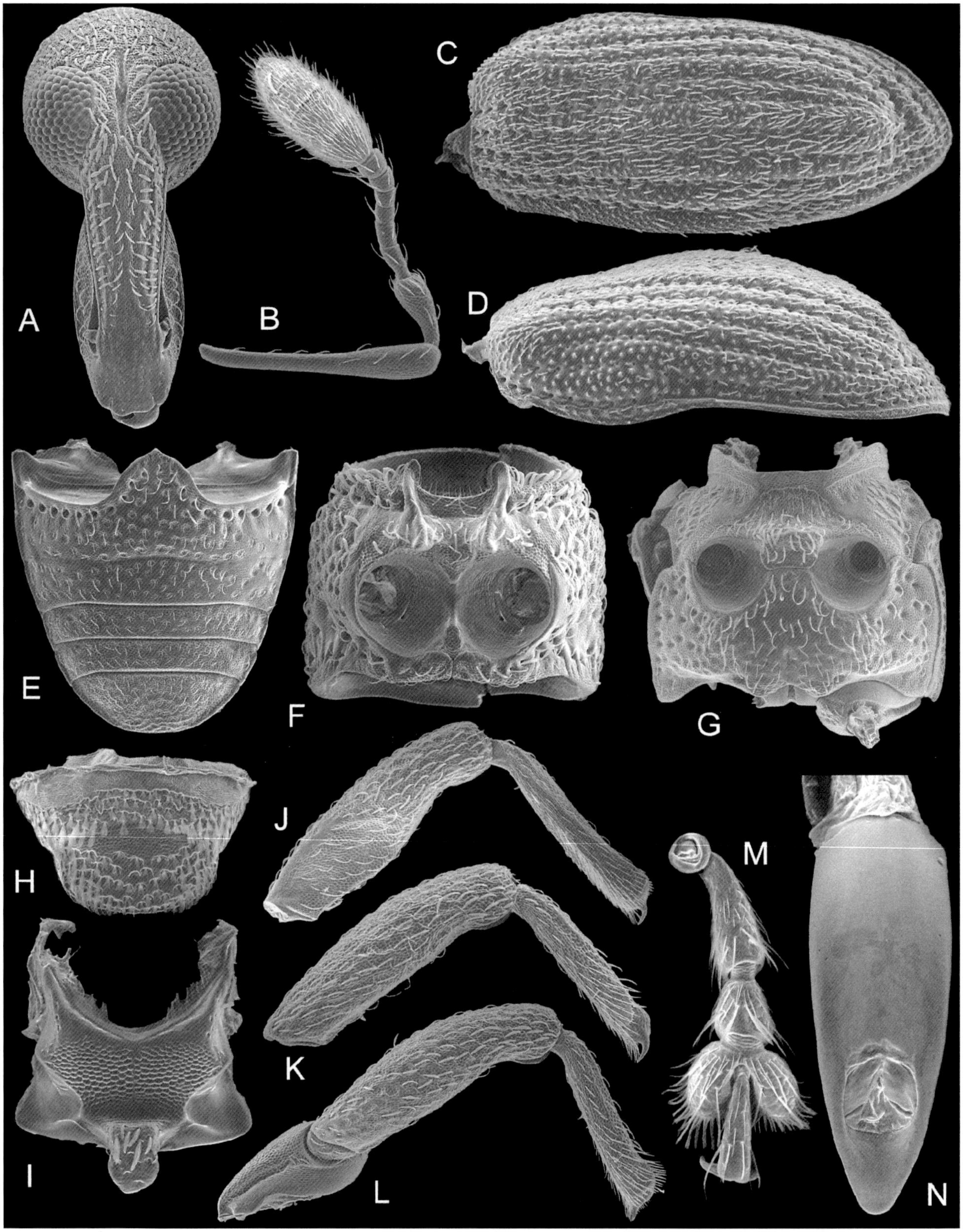

Fig. 44.94. Curculioninae, structural details, Storeini, *Placorrhinus proteus* Marshall. A, head, dorsal; B, antenna; C–D, elytron: C, dorsal; D, lateral; E, abdominal ventrites; F, prothorax, ventral; G, pterothorax, ventral; H, pygidium; I, scutellum; J, foreleg; K, middle leg; L, hindleg; M, tarsus, dorsal; N, penis, dorsal.

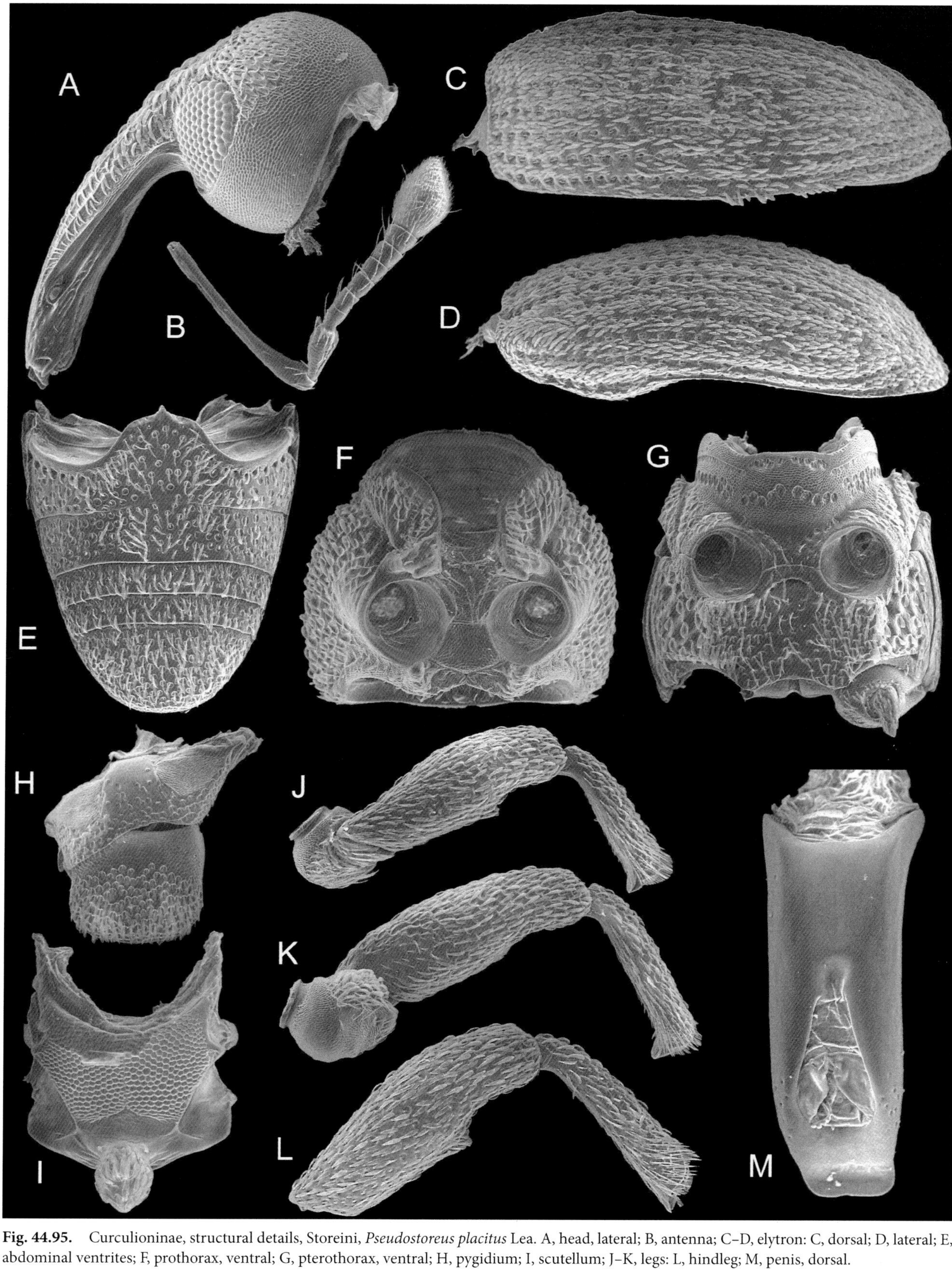

Fig. 44.95. Curculioninae, structural details, Storeini, *Pseudostoreus placitus* Lea. A, head, lateral; B, antenna; C–D, elytron: C, dorsal; D, lateral; E, abdominal ventrites; F, prothorax, ventral; G, pterothorax, ventral; H, pygidium; I, scutellum; J–K, legs: L, hindleg; M, penis, dorsal.

Rhinidotasia Lea, 1928 (Pl. 79B; Fig. 44.88)

Type species, by original designation: *Rhinidotasia suturalis* Lea, 1928.
Australian species. *R. cribrosa* Lea, 1928a, *R. edentata* Lea, 1928a, *R. suturalis* Lea, 1928a.
Distribution. Australia.

Sellechus Lea, 1908 (Pl. 79C)

Type species, by monotypy: *Sellechus tibialis* Lea, 1908.
Australian species. *S. tibialis* Lea, 1908a.
Distribution. Australia (NSW).

Themelia Blackburn, 1894 (Pl. 79D)

Type species, by monotypy: *Themelia inconspicua* Blackburn, 1894.
Australian species. *T. inconspicua* Blackburn, 1894a.
Distribution. Australia (NSW).
Notes. This genus also belongs in the CEGH clade of Curculionidae rather than in Curculioninae and Storeini. Adults have been collected on flowers of *Aotus ericoides* (Fabaceae) and *Leptospermum flavescens* (Myrtaceae). Undetermined taxa from WA (Fig. 44.91) seem to be related to this genus.

Curculionidae: Molytinae Schoenherr, 1823

Introduction. The **Molytinae** are by far the most speciose weevil subfamily in Australia and include several different forms (Pls 19, 20A, E, G–H, 22) but typically they are long-legged weevils with a long rostrum, which is often directed backwards and fits into a deep canal formed between the procoxae and terminating in a receptacle on the mesoventrite, as in the large tribe Cryptorhynchini (Fig. 44.99C). The life histories of most genera are poorly known, but many, in particular the Cryptorhynchini, develop in dead wood and leaf litter. The genus *Tentegia* is unusual in that its larvae, as far as known, develop in dung pellets of marsupials, which the adult weevils collect on the ground and roll and cache under logs (Wassell 1966; Escalona *et al.* 2023). Some species of Molytinae in Australia are deemed to be agricultural pests, such as the Mango Seed Weevil, *Sternochetus mangiferae* (Fabricius), whereas other are beneficial to human endeavours by functioning as biological control agents of weeds. The composition and classification of the Molytinae is poorly resolved, with the concepts and constitutions of many included tribes uncertain and sometimes controversial and their relationships unclear.

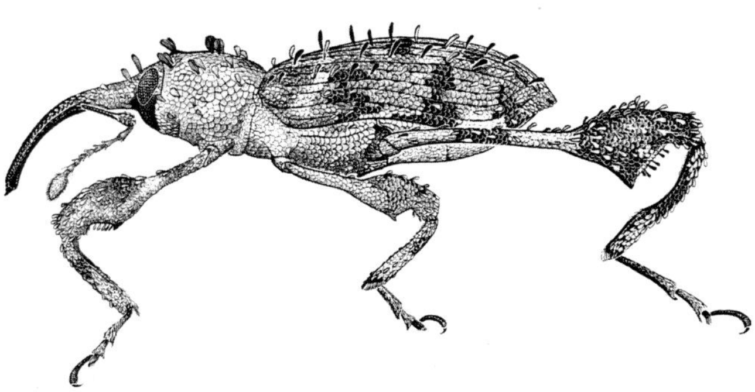

Fig. 44.96. *Acicnemis meriones* Pascoe.

The tribal classification of Pullen *et al.* (2014) is followed here. The tribe Aedemonini comprises eleven genera of species resembling Cryptorhynchini but with their pectoral canal not ending in a mesosternal receptacle; Zimmerman (1994) revised their taxonomy at generic level (as Mechistocerinae) but did not attend to that of the species. The Cleogonini include the large genera *Lybaeba* and *Melanterius* (Pl. 15D, F) whose larvae develop in seeds of *Acacia*, and several species of the latter genus are important biocontrol agents of invasive *Acacia* species in South Africa. The Cryptorhynchini (Pl. 21) are by far the most speciose molytine tribe in Australia, including many very small (monotypic) genera juxtaposed with some very large and varied ones (e.g. *Decilaus*, *Poropterus*) as well as numerous undescribed genera and species, especially inhabiting leaf litter but also specialised habitats such as subterranean aquifers (Pl. 98E), and the Australian fauna is in need of comprehensive taxonomic revision. The Mesoptiliini are a smaller group of wood-boring weevils with truncate elytra exposing a horizontal pygidium; their taxonomy at generic level was revised by Zimmerman (1994). Orthorhinini in their current concept are widespread in Australia, especially the so-called Elephant Weevil (*Orthorhinus cylindrirostris* (Fabricius)), and also include the species of largest weevil in the country, *Eurhamphus fasciculatus* Shuckard, whose larvae develop in wood of Hoop Pine (*Araucaria cunninghamii*, Araucariaceae). Psepholacini are closely related to Cryptorhynchini but include species with a much shorter rostrum and resembling bark beetles (Scolytinae) in general appearance. The *Tranes* group of genera includes mainly taxa associated with cycads, several species constituting the sole or main pollinators of these plants, but the species of *Paratranes* breed in the flower stalks of *Xanthorrhoea* species (Asphodelaceae); the group was recently revised in several papers by Y. Hsiao and R. Oberprieler. Tribes with much smaller representations in Australia are the Ithyporini, Lixini, Molytini, Paipalesomini, Sophrorhinini and Trachodini. A full list of the molytine taxa occurring in Australia is presented by Pullen *et al.* (2014), including also some incorrectly or doubtfully recorded species and species introduced for biological control of weeds but not established.
A comprehensive overview of the morphology, biology and distribution of the subfamily in a narrower sense (excluding Lixini and Mesoptiliini) was provided by Lyal (2014a), with Lixini and Mesoptiliini similarly covered by Meregalli (2014) and Lyal (2014b), respectively. May (1994) described and illustrated the immature stages of some molytine genera present in Australia; some of her illustrations are reproduced here (Figs 44.100, 44.104–44.107, 44.150, 44.151, 44.155, 44.162).

Key to the Australian tribes and genera of Molytinae

The genera of Orthorhinini are included in the first section of the key.

1. Antennal funicles 6-segmented; prosternum without prosternal canal; base of elytra forming convex lobe over base of pronotum; protibiae with large ventral tooth bordered by small teeth; tarsal claws parallel, basally connate, strongly bifid with inner teeth large and close together .... **MECYSOLOBINI**
– Not with this combination of characters; if tarsal claws appendiculate then divaricate and prosternum with canal (Cleogonini) **2**
2(1). Mouthparts without labial palps; tarsal segments 5 with pointed ventral projection on each side; habitus as Fig. 44.97e .... **PAIPALESOMINI** ... ***Peribleptus* Schoenherr**
– Mouthparts with labial palps; tarsal segments 5 without ventral projections **3**
3(2). Prosternum with canal continued posteriad to between hypomeral lobes, its walls there extending into laminate flanges or tubercles (e.g. *Dystropicus*) **4**
– Prosternum without canal or with canal only in front of procoxae (not between hypomeral lobes or on mesoventrite) **5**
4(3). Mandibles with two or more teeth; prosternal canal with flanges of hypomeral lobes meeting posteriorly to form closed receptacle, not extending onto mesoventrite .... **AEDEMONINI**
– Mandibles without teeth; prosternal canal with flanges of hypomeral lobes posteriorly open, continued onto mesoventrite; habitus as Fig. 44.97Z .... **ITHYPORINI** ... ***Camptorhinus* Schoenherr**
5(3). Labial palps 1-segmented; prosternum without canal (e.g. Fig. 44.98e); tarsal claws connate .... **LIXINI**
– Not with this combination of characters; labium often with palpiger and 2-segmented palps (Fig. 44.110D) **6**
6(5). Body vestiture consisting of dense scales and large, erect hair-scales on base of rostrum, between eyes (Fig. 44.98G) and on pronotum and elytra; prothorax with ocular lobes; prosternum without canal (Fig. 44.98f); mesoventral process narrower than width of mesocoxa; metafemora strongly clavate (Fig. 44.99R), with large triangular tooth, about twice as long than on meso- and profemora (Fig. 44.97i); tarsal segments 1 longer than 2+3; metatarsi about as long as tibia .... **TRACHODINI** ... ***Acicnemis* Fairmaire**
– Not with this combination of characters **7**
7(6). Prosternum in front of procoxae often a narrow strip (Fig. 44.99B); elytral apices truncate (Fig. 44.99J, K); abdominal ventrite 2 with lateral posterior angles projecting over ventrite 3 (Fig. 44.98w, except *Notomagdalis*, Fig. 44.153G); pygidium exposed and transverse in *Notomagdalis* (Fig. 44.153J); habitus as Fig. 44.97c .... **MESOPTILIINI**
– Not with this combination of characters **8**
8(7). Body robust and ovoid (Fig. 44.97f), integument brown (e.g. Pl. 101); rostrum variable, in lateral view shorter than head capsule (*Protohylastes*, Pl. 101H) or longer than head (Fig. 44.159A), densely to sparsely setose dorsally; prosternum with canal (e.g. Figs 44.98c, 44.159F), if canal weakly impressed then mesotibiae with external tooth; mesoventral receptacle sometimes absent (*Catocalephe*, *Wiburdia*), when present often incomplete, i.e. posterior wall absent (*Neozeneudes*, *Therebiosoma*, e.g. Fig. 44.98l), when complete or cavernous, i.e. enclosing apex of rostrum (*Calpurnius*, *Oreda*, e.g. Fig. 44.98p), then mesotibiae with tooth on external margin; metatibiae with false corbels (pseudocorbels of Zimmerman 1994, Fig. 44.58K) but oblique and poorly defined in *Calpurnius* and *Oreda* .... **PSEPHOLACINI**
– Not with this combination of characters; metatibiae rarely with false corbels, if present prosternum often without canal (e.g. *Demyrsus*, *Siraton*) **9**
9(8). Prosternum with a well-defined canal (Fig. 44.98W, X, Z, a, b) **10**
– Prosternum without canal (Fig. 44.98e, f) or with a weak impression without defined borders **12**
10(9). Prosternum and mesoventrite with a canal (e.g. Fig. 44.98b, k); mesoventral receptacle present (e.g. Fig. 44.98k–p, r, s) ..... **CRYPTORHYNCHINI**
– Prosternum with or without a canal, if present then rarely extending onto mesoventrite (*Eprias*, *Deretiosus*); mesoventral receptacle absent or mesoventrite depressed between mesocoxae to accommodate rostrum (*Deretiosus*, Fig. 44.160H; *Eprias*) **11**
11(10). Mesoventrite flat or meso- and metaventrite depressed between mesocoxae to accommodate rostrum (*Eprias*, *Deretiosus*); internal margin of tibiae with shallow sulcus at base and carina towards apex; tarsal claws simple .... **SOPHRORHININI**
– Meso- and metaventrite not depressed to accommodate rostrum; internal margin of tibiae without sulcus at base, rarely with carina towards apex; tarsal claws simple or appendiculate .... **CLEOGONINI** (part)
12(9). Body length approximately 3.0 mm, shape rounded (Fig. 44.97E), about 1.17 × as long as wide; scutellar shield not visible; abdominal process wider than a metacoxa (Fig. 44.99H); tarsal segments 4 and 5 absent (Fig. 44.99f) .... ***Achelocis* Lea**
– Body length variable, often > 1.17 × as long as wide; scutellar shield visible or not; abdominal process variable; tarsal segments 4 and 5 present (Fig. 44.99g) **13**
13(12). Labial palps not exposed (Fig. 44.98F); procoxae contiguous (Fig. 44.99A); mesocoxae separated by less than mesocoxal width; metanepisternal sutures with sclerolepidia **14**
– Labial palps visible; procoxae contiguous or not; mesocoxae separated by variable distance; metanepisternal sutures with or without sclerolepidia **15**
14(13). Body length approximately 1.6 mm; rostrum flattened (Fig. 44.98E); eyes separated by less than width of rostrum; mesosternal process 0.9 × a transverse coxal diameter .... ***Eristinus* Lea**
– Body length approximately 3.5 mm; rostrum subcylindrical (Fig. 44.98U); eyes separated by distance similar to width of rostrum; mesosternal process 0.5 × a transverse coxal diameter (Fig. 44.99D) .... ***Thaumastophasis* Wollaston**
15(13). Body length < 1.5 mm; eyes reduced to 1 or 2 ommatidia; antennal funicles 6-segmented; Norfolk Is. .... ***Reyesiella* Alonso-Zarazaga & Lyal**
– Not with this combination of characters **16**
16(15). Body length approximately 5.4 mm; tarsal claws toothed .... ***Eurymelanterius* Zimmerman & Oberprieler**
– Body length variable; tarsal claws simple **17**
17(16). Body length > 20.0 mm; cuticle black, densely covered with white to pale brown, appressed, elongate, hair-like scales, some very long and erect in two dorsolateral clumps on pronotal disc and in variously sized and positioned clumps on elytra; elytra with sutural interstriae slightly swollen towards base and at base of interstria 3, otherwise interstriae flat .... **ORTHORHININI** ... ***Eurhamphus* Shuckard**
– Not with this combination of characters; body length < 15.0 mm **18**

18(17). Body length > 7.0 mm; pronotum with median ridge or carina; elytra with odd interstriae variously raised, usually granulate and sparsely setose to partially glabrous .... **ORTHORHININI** ... ***Ilacuris* Pascoe**
– Not with this combination of characters; pronotum without median ridge or carina **19**
19(18). Rostrum straight and cylindrical (Fig. 44.157B); femora with ventral tooth; tibiae and tarsi flattened **20**
– Not with this combination of characters **22**
20(19). Metafemora with larger tooth than on pro- and mesofemora; metatibiae broad at base and then narrowed towards apex .... **ORTHORHININI** ... ***Parorthorhinus* Kuschel**
– Femora with similar-sized teeth; metatibiae of similar width at base and apex **21**
21(20). Prosternum reduced to narrow strip (e.g. Fig. 44.99B), shorter than antennal scape .... **ORTHORHININI** ... ***Euthebus* Pascoe**
– Prosternum about as long as antennal scape (Fig. 44.98d); habitus as Fig. 44.97d .... **ORTHORHININI** ... ***Orthorhinus* Schoenherr**
22(19). Body length < 5 mm; elytral disc with interstriae 3 raised, interstriae 5 raised from anterior one-fifth of length to elytral declivity, there abruptly ending .... **ORTHORHININI** ... ***Imbilius* Marshall**
– Not with this combination of characters **23**
23(22). Scutellar shield not visible **24**
– Scutellar shield visible **25**
24(23). Eyes narrowly separated dorsally, distance between them smaller than width of rostrum at base .... ***Opsittis* Pascoe**
– Eyes widely separated dorsally, distance between them larger than width of rostrum at base; prosternum and mesoventrite as in Fig. 44.98i, j respectively .... ***Syagrius* Pascoe**
25(23). Rostrum in lateral view almost straight (Fig. 44.98V); prosternal canal absent, procoxae not or barely projecting below prosternum; protibiae gradually narrowing towards apex .... **ORTHORHININI** ... ***Notopissodes* Zimmerman & Oberprieler**
– Rostrum in lateral view broadly curved; prosternal canal present or absent; procoxae well projecting below prosternum; protibiae parallel side or slightly expanded at the apex **26**
26(25). Body length approximately 4.6 mm; prosternum deeply emarginate anteriorly and with canal (Fig. 44.98g); mesocoxae well separated (Fig. 44.98t) .... ***Exeiratus* Broun**
– Not with this combination of characters; if prosternum with weak canal (*Demyrsus*) body length > 10.8 mm **27**
27(26). Procoxal cavities contiguous; protibiae with tooth or projection in apical quarter .... **MOLYTINI** ... ***Aclees* Schoenherr**
– Procoxal cavities narrowly to broadly separate (e.g. Fig. 44.98Y); protibiae without tooth or projection in apical quarter **28**
28(27). Body length approximately 4.8 mm; habitus as Fig. 44.97a; Norfolk Is. .... **JUANORHININI** ... ***Arecophaga* Broun**
– Body length < 6.1 mm; mainland Australia and Lord Howe Is. .... ***Tranes* group**

AEDEMONINI

1. Profemora strongly clavate (e.g. Fig. 44.99S) **2**
– Profemora subparallel or sublinear (e.g. Fig. 44.99V) **4**
2(1). Scutellar shield densely squamose (e.g. Fig. 44.113D); metatibiae with external margin slightly sinuous .... ***Cincius* Zimmerman**
– Scutellar shield mostly glabrous (e.g. Fig. 44.102H); metatibiae with external margin strongly sinuous **3**
3(2). Body ventrally with dense, mostly imbricate vestiture; elytral intervals with prostrate setae; habitus as Fig. 44.97B .... ***Mechistocerus* Fauvel**
– Body ventrally with cuticle exposed, with few sparse setae; elytral intervals with slanting-erect setae .... ***Magius* Zimmerman**
4(1). Rostral canal with small tubercles behind procoxae, not extending onto mesoventrite .... ***Dystropicus* Pascoe**
– Rostral canal with walls forming flanges behind procoxae and extending onto mesoventrite **5**
5(4). Metatibiae robust, expanding towards apex, with strong external tooth in apical quarter and false corbels at apex .... ***Derbyiella* Lea**
– Metatibiae equally wide throughout length, without external tooth in apical quarter and without false corbels **6**
6(5). Rostral canal extending onto metaventrite, mesoventral receptacle distinctly open .... ***Damippus* Zimmerman**
– Rostral canal barely passing anterior margin of mesocoxae, mesoventral receptacle deeply cavernous **7**
7(6). Metafemora extending to elytral apices (Fig. 44.97A), ventral tooth opposite of or closer to elytral apices .... ***Diatassa* Pascoe**
– Metafemora either not reaching elytral apices or slightly exceeding them but ventral tooth always distant from elytral apices **8**
8(7). Metaventrite between meso- and metacoxae about as long as metacoxa .... ***Rhadinomerus* Faust**
– Metaventrite between meso- and metacoxae longer than metacoxa **9**
9(8). Tarsal segments 5 extending beyond apex of segment 3 for by less than length of 3; elytral humeri slightly produced and moderately embracing prothorax .... ***Epirus* Zimmerman**
– Tarsal segments 5 extending beyond apex of segment 3 by more than length of 3; elytral humeri not embracing base of prothorax **10**
10(9). Pronotum strongly convex; elytra striae with inconspicuous, small setae in punctures; abdominal ventrite 2 about as long as 3 + 4 in both sexes; antennae in both sexes inserted much further from apex of rostrum than length of club; protibiae of male not fimbriate; habitus as Fig. 44.97D .... ***Roecus* Zimmerman**
– Pronotum almost flat; elytral striae with conspicuous, large, broad squamae in punctures; female with abdominal ventrite 2 almost as long as ventrites 3–5, male with ventrite 2 only as long as 3 + 4; antennae of male inserted at about length of club from apex of rostrum; protibiae of male conspicuously fimbriate on inside; habitus as Fig. 44.97C .... ***Parendymia* Kirsch**

CLEOGONINI

1. Prosternum anteriorly deeply emarginate (Fig. 44.98W); mesoventral receptacle absent, mesoventral process with slight emargination; scutellar shield not visible; elytral bases with interstriae forming tubercles (Fig. 44.97F); femora without ventral tooth .... ***Cycloporopterus* Lea**
– Not with this combination of characters; elytral bases with interstriae not forming tubercles **2**

2(1). Prosternal canal with lateral walls forming ridges (Fig. 44.98X); mesoventrite slightly concave without lateral flanges; elytral striae carinate (Fig. 44.99Q); abdominal ventrites 1 and 2 medially fused, suture not visible (Fig. 44.99G) .... ***Arthriticosoma*** **Lea**
– Not with this combination of characters; abdominal ventrites 1 and 2 not fused medially 3
3(2). Body with metallic cuticle; pronotum strongly plicate; mesoventrite slightly convex, not excavate; elytra smooth, interstriae flat; tarsal claws connate .... ***Chalcodermus*** **Dejean**
– Not with this combination of characters; tarsal claws free 4
4(3). Tarsal claws appendiculate and often divaricate 5
– Tarsal claws simple and more or less parallel 7
5(4). Body vestiture consisting of dense squamae, cuticle brown; profemora without tooth .... ***Conotrachelus*** **Dejean**
– Body vestiture consisting of dispersed hair-like setae, cuticle black; profemora with small, ventral pointed tooth 6
6(5). Mesoventrite concave, with lateral flanges to accommodate rostrum in repose .... ***Moechius*** **Pascoe**
– Mesoventrite flat, without lateral flanges .... ***Neomelanterius*** **Lea**
7(4). Scutellar shield not visible; femora without ventral tooth .... ***Teutheria*** **Pascoe**
– Scutellar shield visible; femora with ventral tooth 8
8(7). Elytra irregularly tuberculate (Fig. 44.97G), striae and interstriae not distinguishable .... ***Hybophorus*** **Waterhouse**
– Elytra not tuberculate (Fig. 44.97H), striae and interstriae clearly distinguishable 9
9(9). Body vestiture consisting of sparse, pale, unicoloured setae; rostrum in lateral view equally thick; abdominal ventrites 1–5 at same level .... ***Melanterius*** **Erichson**
– Body vestiture consisting of dense, testaceous, bi- or tricoloured scales; rostrum in lateral view gradually flattened towards apex; abdominal ventrites 1 and 2 inflated, at lower level that 3–5, 2 shorter than 1 and in oblique plane .... ***Lybaeba*** **Pascoe**

CRYPTORHYNCHINI

The following 30 genera are not included in the key because of unavailability of authentic specimens for detailed examination, although color photographs of most of them are presented: *Albius, Amorphocis, Ampagiosoma, Ancocis, Anepigraphocis, Anilaus, Anoplocis, Athyreocis, Austrectopsis, Coptocelis, Diaphorocis, Didorus, Eucalyptocis, Euoropis, Eurycis, Eurymetopocis, Exithioides, Gymnocis, Hoplodecilaus, Leucothyreocis, Poropterculus, Poropterinus, Pseudotepperia, Rhaebocnemocis, Scleropoides, Scotinocis, Sophronocis, Sympediosoma, Tepalicus, Triptocis*. The identity of '*Cryptorhynchus*' *teter* Boisduval has not been established, and its name currently is a *nomen dubium* (Pullen *et. al* 2014). The genus *Glyptoporopterus* is also not included as its presence in Australia is doubtful, given that the single species described from Australia, *G. asper* Lea, appears only known in the country from its holotype (recorded from New South Wales without specific locality) but occurs on Misima Is. of Papua New Guinea, suggesting that the holotype is mislabeled and *G. asper* does not occur in Australia. The genus *Homorthorrhinus* was also not included in the key, the only known Australian specimen (id. by Zimmerman, Pl. 99D) is similar to *H. perversus* (Pascoe, 1885) but lack a main characteristic of the genus (hind tibia without pointed tooth), and its diagnosis is pending further work.

1. Body length approximately 12.8 mm; antennal clubs 3-segmented, basal segment more than 2 × as long terminal two; pronotum with median carina, projecting over head; scutellar shield not visible; abdominal ventrites 1 and 2 fused; tibiae without apical comb of setae .... ***Plagiocorynus*** **Waterhouse**
– Not with this combination of characters 2
2(1). Body length approximately 15.4 mm; cuticle black; scutellar shield not visible; epipleura about as wide as mesofemur; Lord Howe Is. .... ***Hybomorphus*** **Saunders & Jekel**
– Not with this combination of characters 3
3(2). Scutellar shield often visible; metanepisterna separated by metanepisternal suture from metaventrite (e.g. Fig. 44.110J); metanepisterna longer than wide, reaching metacoxa 4
– Scutellar shield often absent or strongly reduced but well developed and prominent in some genera, e.g. *Ophrythyreocis*; metanepisterna partly or completely fused to metaventrite (metanepisternal sutures reduced or absent) (e.g. Figs 44.103K, 44.117K); metanepisterna, when present, reduced to small triangular sclerite next to mesepimeron 106
4(3). Rostrum straight (Fig. 44.98P); elytral bases projecting anteriad, overlapping base of pronotum; elytral apices produced into pointed tubercle; tibiae flattened .... ***Eutyrhinus*** **Dejean**
– Not with this combination of characters 5
5(4). Body length approximately 8.5 mm; cuticle smooth and shining, pronotum and elytra mostly glabrous, setae restricted to apex and base of pronotum and forming at least two transverse lines on elytra; mesoventral receptacle U-shaped (Fig. 44.98I); protibiae without apical comb or reduced to less than 10 setae .... ***Enteles*** **Schoenherr**
– Not with this combination of characters 6
6(5). Body length often > 14.0 mm, rarely less; pronotum with well-defined median carina; femora linear 7
– Not with this combination of characters 9
7(6). Antennal clubs with oblique sutures; protibiae with small ventral tooth covered by setae; tibiae with internal edge carinate .... ***Dysopirhinus*** **Roelofs**
– Not with this combination of characters; antennal clubs with transverse sutures 8

8(7). Scutellar shield glabrous; mesoventral receptacle without posterior wall; habitus as Fig. 44.97X ... ***Hypsophorus*** **Dejean**

– Scutellar shield setose; mesoventral receptacle with posterior wall ... ***Orphanistes*** **Pascoe**

9(6). Mesoventral receptacle without posterior wall, or posterior wall barely raised above meso- and metaventrites (e.g. Fig. 44.98k), leaving rostral apex exposed when in repose; femora robust with ventral spine ... **10**

– Mesoventral receptacle with posterior wall (e.g. Fig. 44.98l–p), cavernous, covering rostral apex (except *Tychreus*); femora variable ... **20**

10(9). Body length < 3.1 mm ... **11**

– Body length > 6.0 mm ... **13**

11(10). Dorsal vestiture comprised of long erect setae (Fig. 44.97L) ... ***Andracalles*** **Kuschel**

– Dorsal vestiture comprised of appressed scale-like setae ... **12**

12(11). Elytra with sides subparallel (Pl. 86B) ... ***Ephrycus*** **Pascoe**

– Elytra with sides broadly rounded (Pl. 96C) ... ***Cardiopterocis*** **Lea**

13(10). Tibial apex with surface of false corbel concave ... ***Isax*** **Pascoe**

– Tibial apex without false corbel, if present surface forming a flange ... **14**

14(13). Rostrum with about about as wide at base and apex (Fig. 44.98D); pronotum transverse, about 0.6 × as long as wide, with median prominence or longitudinal carina; mesoventral receptacle not extending beyond middle of mesocoxae (Fig. 44.98o); scutellar shield not visible ... ***Tamphilus*** **Zimmerman**

– Not with this combination of characters; pronotum longer than wide ... **15**

15(14). Body length approximately 8.5 mm; pronotum with posterior angles projecting posteriad; elytra with humeral angles clipping base of prothorax; meso- and metatibiae with carina on internal margin ... ***Nechyrus*** **Pascoe**

– Not with this combination of characters ... **16**

16(15). Body length approximately 7.2 mm; rostrum short and broad, slightly expanded towards apex; base of pronotum with median pointed tubercle; legs long, metafemora extending beyond abdominal apex (Fig. 44.97Q); tibiae flattened ... ***Eufaustia*** **Lea**

– Not with this combination of characters; rostrum often subparallel ... **17**

17(16) Prosternal canal without well-defined walls; mesoventral receptacle not reaching metaventrite ... ***Athanasius*** **Zimmerman**

– Prosternal canal with well-defined walls; mesoventral receptacle reaching metaventrite ... **18**

18(17). Prosternal canal divided by transverse carina; elytra without humeral angles ... ***Pseudonidistus*** **Lea**

– Prosternal canal not interrupted by transverse carina; elytra with humeral angles ... **19**

19(18). Head capsule setose behind eyes; pronotum with median carina; elytra with anterior angles not projecting anteriad, disc with tubercles ... ***Neomystocis*** **Lea**

– Head capsule glabrous behind eyes (Fig. 44.98H); pronotum without median carina; elytra with anterior angles slightly projecting anteriad, disc without tubercles ... ***Onidistus*** **Pascoe**

20(9). Body length approximately 9.0 mm; antennae 9-segmented, clubs 1-segmented; pronotum transverse; femora linear ... ***Blepiarda*** **Pascoe**

– Not with this combination of characters; antennal clubs 3-segmented ... **21**

21(20). Body length approximately 12.0 mm; funicles with segment 7 2 × as long as segment 6 and longer than club, widened towards apex; rostrum in males with curved, pointed tubercles; femora linear; metafemora without ventral tooth ... ***Glochinorhinus*** **Waterhouse**

– Not with this combination of characters; funicle with segment 7 about as long as segment 6 and shorter than club ... **22**

22(21). Femora without ventral tooth; at least one pair of tibiae with basal spine at setal comb (Fig. 44.108K); rostrum often sexually dimorph, with antennae inserted near base and rest of rostrum glabrous in female but antenna inserted in about middle (Fig. 44.98B) and rostrum with uniform vestiture in male ... **23**

– Femora with or without ventral tooth; tibiae without basal spine in setal comb ... **29**

23(22). Body mostly glabrous (Fig. 44.97M) with few scattered setae on apical third of elytra; head capsule glabrous (Fig. 44.98K); abdominal ventrites 1 and 2 laterally fused to elytra ... ***Bepharus*** **Pascoe**

– Body densely covered with appressed and imbricate scale-like setae, cuticle not visible unless abraded; head capsule with dense to sparse vestiture; abdominal ventrites not fused to elytra ... **24**

24(23). Body length approximately 4.0 mm; body vestiture consisting of scale-like appressed setae and erect flattened setae on pronotum and elytral interstriae ... ***Ephrycinus*** **Lea**

– Body length > 4.0 mm; body vestiture without erect flattened setae over pronotum and elytral interstriae ... **25**

25(24). Elytra with fine irregular punctures not forming striae, with longitudinal lateral carina and declivity at level of humeral angles; Lord Howe Is. ... ***Idiopterocis*** **Lea**

– Not with this combination of characters; elytra without lateral carina and declivity ... **26**

26(25). Pronotal disc flattened ... **27**

– Pronotal disc slightly convex ... **28**

27(26). Body length approximately 7.5 mm; Western Australia ... ***Metacymia*** **Pascoe**

– Body length approximately 11.0 mm; habitus as Fig. 44.97K; Lord Howe Is. ... ***Aethreus*** **Pascoe**

28(26). Pronotal base straight; femora linear; tarsal segments 3 about as wide as segment 2; habitus as Fig. 44.97I ... ***Achopera*** **Pascoe**

– Pronotal base projecting slightly posteriad medially; femora enlarged medially; tarsal segments 3 wider than segment 2 ... ***Mitrastethus*** **Redtenbacher**

29(22). Body length approximately 6.3 mm; antennae inserted near basal third of rostrum; rostrum straight, subcylindrical, glabrous beyond antennal insertions; scapes shorter than clubs; metanepisterna large, with anterodorsal margin produced and elytra broadly emarginate; femora with ventral tooth; habitus as Fig. 44.97U ... ***Platytenes*** **Pascoe**

– Not with this combination of characters; metanepisterna with anterodorsal margin not produced and elytra not broadly emarginate ... **30**

30(29). Dorsal vestiture composed of erect setae with multifurcate apex and one appressed, scale-like seta near base; pronotum with lateral carina, disc flattened; epipleura as broad as mesofemora ... ***Emydica*** **Pascoe**

– Not with this combination of characters ... **31**

31(30). Protibiae with apical third bent inwards; mesoventral receptacle deep (Fig. 44.99C) **32**
– Protibiae with apical third straight; mesoventral receptacle variable **33**
32(31). Pronotum with disc flattened; femora without ventral tooth; abdominal segment 1 with deep basal groove on each side .... ***Paleticus* Pascoe**
– Pronotum with disc convex; mesoventral receptacle deep (Fig. 44.99C); femora with ventral tooth; abdominal segment 1 without basal grooves .... ***Pachyporopterus* Lea**
33(31). Body length approximately 7.5 mm; vestiture composed of rounded appressed scales; elytra with interstriae between striae 4 and 5 forming a raised carina marking elytral declivity; metafemora with ventral tooth; tarsal claws divaricate; habitus as Fig. 44.97P; Lord Howe Is. .... ***Embaphiodes* Pascoe**
– Not with this combination of characters; elytra without such declivity **34**
34(33). Body length approximately 7.0 mm; prothorax transverse, about 0.6 × as long as wide; abdominal ventrite 1 with stridulatory files on lateral posterior angles .... ***Meniomorpha* Lea**
– Body length and prothorax variable; abdominal ventrite 1 without stridulatory files on lateral posterior angles **35**
35(34). Body length approximately 4.6 mm, shape oval, 1.41 × as long as wide (Fig. 44.97Y); cuticle black; scutellar shield not visible; epipleura broad and emarginate to accommodate pro- and metafemora; femora sulcate to accommodate tarsi; Western Australia .... ***Tetengia* Lea**
– Not with this combination of characters **36**
36(35). Body approximately 2.7 × as long as wide; pronotum with or without median carina; posterior angles of pronotum projecting and overlapping base of elytra; elytra with humeral angles projecting anteriad; femora without ventral tooth **37**
– Not with this combination of characters; posterior angles of pronotum not projecting **40**
37(36). Pronotum without median carina; pronotal base not projecting medially .... ***Niconotus* Pascoe**
– Pronotum with median carina; pronotal base projecting medially **38**
38(37). Elytra gradually narrowing from middle towards apex .... ***Pseudomydaus* Lea**
– Elytra parallel-sided, narrowing from apical third **39**
39(38). Body length approximately 9.0 mm .... ***Omydaus* Pascoe**
– Body length approximately 5.0 mm .... ***Agenopus* Pascoe**
40(36). Eyes dorsally surrounded by a groove; pronotum often with median carina and dense coarse punctures; elytra often with interstriae carinate; femora with ventral tooth; metatibia widened and flattened towards apex, with apical external angle broadly rounded, setal comb horizontal, not ascending, false corbels with prominent flange .... ***Tyrtaeosus* Pascoe**
– Not with this combination of characters, metatibia rarely widened or flattened towards apex **41**
41(40). Body length approximately 10.0 mm; eyes narrowly separated dorsally; scutellar shield prominent, elongate and glabrous, separated from elytra by deep sulcus; femora with ventral tooth; meso- and metatibia flattened, external border forming a sharp carina, apex forming an angle .... ***Metraniomorpha* Lea**
– Not with this combination of characters; meso- and metatibia not forming an angle **42**
42(41). Body length approximately 3.5 mm; body densely setose; scutellar shield setose; mesoventral receptacle almost reaching posterior border of metaventrite; femora without ventral tooth .... ***Tituacus* Marshall**
– Not with this combination of characters **43**
43(42). Femora with ventral tooth or minute spine or prominence **44**
– Femora without ventral tooth **86**
44(43). Body length approximately 4.4 mm; eyes protuberant (Fig. 44.98I); pronotum with four tufts of long setae, two anteriorly and two on disc; scutellar shield setose; mesoventral receptacle with narrow canal (Fig. 44.98m); elytral disc with tubercles; metanepisternal sutures with sclerolepidia; femora slightly clavate (Fig. 44.99S) .... ***Acrotychreus* Lea**
– Not with this combination of characters **45**
45(44). Body length approximately 4.0 mm or less; elytra mostly glabrous or with few sparse setae or with patches of scales, cuticle shiny **46**
– Body length variable and elytra setose **48**
46(45). Mandibles narrow, with single tooth .... ***Leptidotasia* Zimmerman**
– Mandibles bidentate **47**
47(46). Pronotum widest at base and with minute setae .... ***Alatidotasia* Lea**
– Pronotum widest at midline with scale-like setae, their colour dark brown except for three white patches, on disc and at sides (Fig. 44.97R) .... ***Hoplidotasia* Lea**
48(45). Body length 5.5–7.2 mm; profemora with two ventral teeth; metanepisternal sutures with sclerolepidia; tibia sulcate internally **49**
– Not with this combination of characters; profemora with single ventral tooth **50**
49(48). Body length approximately 5.5 mm; protibiae with external border broadly curved .... ***Sybulus* Pascoe**
– Body length about 7.2 mm; protibiae with external border angled .... ***Critomerus* Lea**
50(48). Body length >6.2 mm; scutellar shield prominent as a tubercle, glabrous or with sides setose; profemora with tooth or spine near apical third **51**
– Not with this combination of characters; scutellar shield not prominent **53**
51(50). Pronotum with median carina; scutellar shield rounded and glabrous; elytral apex acuminate .... ***Bleptocis* Lea**
– Pronotum without median carina; scutellar shield partly setose; elytral apex broadly rounded **52**
52(51). Scutellar shield prominent and pointed .... ***Odosyllis* Pascoe**
– Scutellar shield prominent and dull .... ***Lophocheirus* Marshall**
53(50). Body length approximately 4.3 mm, body elongate and dorsally flattened; vestiture consisting of appressed scale-like setae; tibiae flattened, internal margin emarginate at base; tarsal segments 3 not bilobed .... ***Griffithia* Lea**
– Not with this combination of characters **54**

54(53). Body length approximately 6.4 mm; prothorax transverse, 0.6 × as long as wide (Fig. 44.97N), disc concave, base projecting posteriad; elytral disc convex .... ***Brachyphyes* Marshall**
– Not with this combination of characters; pronotum flattened or convex 55
55(54). Body length approximately 9.2 mm; pronotum with median carina; femora linear with minute ventral tooth; tibial apex with surface blunt, without flange .... ***Mecryptorhynchus* Zimmerman & Oberprieler**
– Not with this combination of characters 56
56(55). Body length 10–16.0 mm; pronotal disc with median carina; scutellar shield glabrous; elytra with humeral angle slightly projecting anteriad; femora linear, with small ventral tooth; metatibia with surface of false corbel with flange and coarse microsculpture or striae; protibiae with setal comb as a single row, meso- and metatibiae with multisetose comb (forming a double row); mesotibiae externally with basal tooth .... ***Axionicus* Pascoe**
– Not with this combination of characters; metatibiae with surface of false corbel smooth 57
57(56). Scutellar shield not visible 58
– Scutellar shield visible, at level of elytra or slightly above 60
58(57). Body in dorsal view subparallel; elytra without humeral angles; legs slender and long; habitus as Fig. 44.97V .... ***Nototragopus* Heller**
– Body widened toward apex, sides convergent from apical third; elytra with humeral angles; legs robust 59
59(58). Femora with ventral tooth exposed, not covered by setae; protibiae with internal border broadly sinuate .... ***Paletonidistus* Lea**
– Femora with ventral tooth covered by setae; protibiae with internal border mostly straight .... ***Euryporopterus* Lea**
60(57). Scutellar shield very small, glabrous and narrow; mesoventral receptacle with lateral walls diagonally truncate to accommodate procoxae; meso- and metatibiae setose above setal comb .... ***Perissops* Pascoe**
– Not with this combination of characters 61
61(60). Body length approximately 8.3 mm; eyes with anterior border deeply sulcate; pronotum with median carina interrupted in middle by a tubercle; tibiae flattened, external border forming carina, anterior surface with two (protibiae) or one (meso- and metatibiae) longitudinal carinae .... ***Cyphoderocis* Lea**
– Not with this combination of characters; pronotum without median tubercle 62
62(61). Head capsule with three carinae, median one extending onto rostrum and interrupted between eyes; rostrum longer than prothorax, mostly glabrous, antennae inserted in basal third; metafemora extending beyond abdominal apex; femora with prominent ventral tooth, larger on profemora 63
– Not with this combination of characters 64
63(62). Body length approximately 11.2 mm; pronotum widest near base .... ***Hyparinus* Pascoe**
– Body length approximately 6.4 mm; pronotum widest at anterior third .... ***Methidrysis* Pascoe**
64(62). Body length approximately 11.2 mm; eyes with dorsal border deeply sulcate; antennal clubs elongate, with oblique sutures; pronotum with median carina; scutellar shield glabrous and rounded; legs long (Fig. 44.97S); metafemora extending beyond abdominal apex; meso- and metatibiae with apex blunt, without a flange .... ***Pezichus* Waterhouse**
– Not with this combination of characters 65
65(64). Body length approximately 5.3 mm; antennal clubs elongate, with oblique sutures; abdominal ventrite 5 concave .... ***Bothynacrum* Lea**
– Not with this combination of characters; abdominal ventrite five flat or slightly convex 66
66(65). Elytra covered with dense scale-like setae except for glabrous epipleura; metanepisternal sutures lined with sclerolepidia; abdominal ventrites at two different levels, ventrite 2 steeply inclined .... ***Mallus* Marshall**
– Not with this combination of characters; abdominal ventrites at same level 67
67(66). Body length approximately 7.3 mm; pronotum slightly concave anteriorly and projecting over head capsule; mesoventral receptacle extending below level of mesocoxae; elytra with humeri projecting anteriad, clipping base of prothorax, with preapically projecting tubercles .... ***Ouroporopterus* Lea**
– Not with this combination of characters 68
68(67). Body length approximately 3.8 mm; prothorax with walls of prosternal canal with deep pit; metanepisternal sutures lined with sclerolepidia; tibiae with fine carina on outer side; metatibiae with uncus apex bifurcate .... ***Tyrtaeosellus* Lea**
– Not with this combination of characters; metatibiae with uncus apex not bifurcate (Fig. 44.99X–Y) 69
69(68). Body length approximately 4.3 mm; body densely setose; pronotum with deep irregular grooves or punctures, posterior margin projecting medially; meso- and metaventrites and abdominal ventrites 1 and 2 with large punctures (Fig. 44.141F, H, J); pro- and mesofemora without ventral tooth .... ***Pseudapries* Lea**
– Not with this combination of characters 70
70(69). Pronotum with posterior border projecting posteriad; meso- and metatibial apices with double setal comb, forming two parallel rows; metanepisternal sutures lined with sclerolepidia .... ***Sternochetus* Pierce**
– Not with this combination of characters; meso- and metatibial apices with one setal comb 71
71(70). Pronotum with median carina 72
– Pronotum without median carina 75
72(71). Pronotal median carina complete, extending from anterior to posterior border 73
– Pronotal median carina incomplete 74
73(72). Profemora without ventral tooth or fine carina .... ***Notocryptorhynchus* Lea**
– Profemora with ventral tooth .... ***Hexymus* Pascoe**
74(72). Body length approximately 11.2 mm; pronotal carina present in apical half .... ***Gygaeus* Pascoe**
– Body length approximately 7.8 mm; pronotal carina present in basal half .... ***Metyrus* Pascoe**
75(71). Metafemora with large tooth, several times larger than tooth on pro- and mesofemora, about as wide as femur 76
– Metafemora with minute tooth or spine, subequal in size to those on pro- and mesofemora 77

76(75). Elytra with uniform brown setae .... ***Cratomerocis* Lea**
– Elytra bicoloured, setae brown with white sides .... ***Hyperiosoma* Lea**
77(75). Body length approximately 5.7 mm; pronotum with white scales forming an X on each side; elytra in apical third with patches of white setae; mesoventral receptacle prominent; metanepisternal sutures without sclerolepidia .... ***Amleanus* Marshall**
– Not with this combination of characters **78**
78(77). Body length approximately 4.9 mm; vestiture with dense, imbricate, appressed setae; elytral suture prominent behind scutellar shield .... ***Menios* Pascoe**
– Not with this combination of characters **79**
79(78). Body about 2.0 × as long as wide, without tubercles; vestiture uniform, consisting of decumbent setae, without long erect setae; scutellar shield elongate and glabrous .... ***Metyrculus* Lea**
– Not with this combination of characters **80**
80(79). Tibial apices with false corbel, bordered by a flange **81**
– Tibial apex without false corbel or flange **82**
81(80). Rostrum slightly flattened; scutellar shield glabrous; tibiae margin angulate at base .... ***Tepperia* Lea**
– Rostrum subcylindrical (e.g. Fig. 44.98J); scutellar shield setose; tibiae margin broadly curved at base .... ***Episodiocis* Lea**
82(80). Body length approximately 8.2 mm; head capsule anteriorly glabrous, smooth and shiny; eyes with dorsal border deeply sulcate; rostrum tricarinate at base; scutellar shield glabrous; femora with small teeth .... ***Notocalviceps* Lea**
– Not with this combination of characters **83**
83(82). Prothorax widest at middle; pronotum convex .... ***Pseudometyrus* Lea**
– Prothorax widest at base; pronotum flattened **84**
84(83). Pronotum and elytra (Fig. 44.99O) without tubercles .... ***Phloeoglymma* Pascoe**
– Pronotum apex with tubercles or tufts of erect setae, elytra with tubercles in apical third **85**
85(84). Vestiture consisting of dense and imbricate brown setae with simple apex .... ***Achoperinus* Lea**
– Vestiture consisting of dense and imbricate white to pale-brown setae and palmate apex .... ***Chimades* Pascoe**
86(43). Scutellar shield not exposed **87**
– Scutellar shield visible **90**
87(86). Mesoventral receptacle extending below mesocoxae; metanepisternal sutures without sclerolepidia; abdominal ventrite 1 with deep groove on each side of abdominal process .... ***Scolyphrus* Pascoe**
– Mesoventral receptacle not extending below mesocoxae; metanepisternal sutures lined with sclerolepidia; abdominal ventrite 1 without grooves on abdominal process **88**
88(87). Body length often < 3.0 mm; eyes reduced, coarsely facetted (Fig. 44.98N); habitus as Fig. 44.97O .... ***Decilaus* Pascoe** (part)
– Body length > 6.0 mm; eyes large, finely facetted **89**
89(88). Elytra approximately 0.8 × as long as wide; metatibial apices with two setal combs .... ***Tragopus* Schoenherr**
– Elytra approximately 1.9 × as long as wide; metatibial apex with one setal comb .... ***Pseudoporopterus* Lea**
90(86). Elytral apices slightly projecting; mesoventral receptacle not covering apex of rostrum, cavernous, slightly overlapping metaventrite .... ***Tychreus* Pascoe**
– Mesoventral receptacle covering apex of rostrum, rarely reaching metaventrite **91**
91(90). Rostrum wider at base than apex (Fig. 44.98C); scutellar shield glabrous; elytra prominent and tuberculate behind scutellar shield; femora grooved for reception of tibiae; tibiae on inside with two rows of setae .... ***Orochlesis* Pascoe**
– Not with this combination of characters **92**
92(91). Tibiae with apical comb not evident, covered by dense vestiture **93**
– Tibiae with apical comb distinct, not covered by tibial vestiture **94**
93(92). Body approximately 1.6 × as long as wide; rostrum longer than prothorax; Western Australia .... ***Paratituacia* Lea**
– Body approximately 2.0 × as long as wide; rostrum shorter than prothorax; New South Wales, Tasmania .... ***Ethocis* Lea**
94(92). Body length approximately 6.9 mm; pronotum with median sulcus .... ***Stenoporopterus* Lea**
– Body length variable; pronotum surface flat or slightly convex without median sulcus **95**
95(94). Body length approximately 6.2 mm; body densely setose with tufts of setae on pronotum and elytra; pronotal base projecting posteriad; base of elytra overlapping base of pronotum .... ***Tylocis* Lea**
– Not with this combination of characters **96**
96(95). Body length approximately 1.5 mm .... ***Microcryptorhynchus* Lea**
– Body length >3.4 mm **97**
97(96). Pronotum uniformly convex with dense coarse punctures; tibiae with external border carinate, apex formed by flange and diagonally truncate .... ***Osaces* Pascoe**
– Not with this combination of characters **98**
98(97). Scutellar shield mostly glabrous **99**
– Scutellar shield densely setose **100**
99(98). Body length approximately 6.0 mm; pronotum with appressed and imbricate scale-like setae .... ***Evaniocis* Lea**
– Body length approximately 3.4 mm; pronotum with erect setae, surface deeply and densely punctate .... ***Eutinobothrus* Faust**
100(98). Body length approximately 10.0 mm; eyes with dorsal margin prominent; anterior area of pronotum with median bituberculate prominence .... ***Oemethylus* Pascoe**
– Not with this combination of characters **101**

101(100). Pronotum with weak median carina; abdominal process narrower than mesocoxa, triangular .... ***Orthoporopterus*** **Lea**
– Pronotum without median carina; abdominal process as wide as or wider than mesocoxa, often transverse **102**
102(101). Pronotum without defined lateral margin .... ***Pteroporopterus*** **Lea**
– Pronotum with lateral margin well defined and densely setose **103**
103(102). Pronotal disc raised .... ***Agathicis*** **Lea**
– Pronotal disc flattened **104**
104(103). Pronotum and elytral disc bordered by dense setose and prominent thick crest .... ***Imalithus*** **Pascoe**
– Pronotum and elytral disc not bordered by a crest **105**
105(104). Scutellar shield with erect setae .... ***Miotus*** **Pascoe**
– Scutellar shield with appressed setae .... ***Chaetectetorus*** **Schoenherr**
106(3). Body length approximately 8.6 mm, dorsal side without prominent tubercles; head with longitudinal sulcus between eyes; rostrum with three dorsal carinae, one median and two lateral; eyes with deep groove along dorsal margin; abdominal ventrite 1 with postcoxal grooves; profemora with ventral minute spine or carina, meso- and metafemora without ventral tooth .... ***Platyporopterus*** **Lea**
– Not with this combination of characters **107**
107(106). Body length approximately 3.7 mm, dorsum without tubercles or prominences; vestiture consisting of dense and imbricate white or black scale like setae, cuticle not visible .... ***Neodecilaus*** **Lea**
– Not with this combination of characters **108**
108(107). Scutellar shield visible, often at level of elytra or highly reduced (e.g. some species of *Poropterus*); elytra clearly emarginate around scutellar shield **109**
– Scutellar shield not visible; elytra not or barely emarginated around scutellar shield **116**
109(108). Head capsule often depressed behind eyes (Fig. 44.98M); eyes protuberant; prothorax subquadrate; mesoventral receptacle slightly projecting below level of mesocoxae; femora with ventral tooth .... ***Exithius*** **Pascoe**
– Not with this combination of characters; femora often without ventral tooth **110**
110(109). Rostrum straight (Fig. 44.98Q); metafemora flattened and expanded (Fig. 44.99U) with dorsal border carinate, fitting into grooves on abdominal ventrite 1 .... ***Ampagia*** **Pascoe**
– Not with this combination of characters **111**
111(110). Scutellar shield prominent, projecting above elytra level **112**
– Scutellar shield flattened or not projecting above elytral level **113**
112(111). Scutellar shield rounded, setose; mesoventral receptacle not projecting below level of mesocoxae; elytra apices without tubercles .... ***Ophrythyreocis*** **Lea**
– Scutellar shield pointed; elytra apices with tubercles .... ***Baeodontocis*** **Lea**
113(111). Mesoventral receptacle with posterior wall reduced (Fig. 44.98k) or barely raised below meso- and metaventrites, leaving rostrum apex exposed in repose .... ***Tapinocis*** **Lea**
– Mesoventral receptacle with well-developed posterior wall (Fig. 44.142F), cavernous, projecting below level of mesocoxae **114**
114(113). Body length approximately 10.0 mm; pronotum and elytral disc with prominent tubercles (Fig. 44.97J); base of elytra overlapping base of pronotum .... ***Acidinus*** **Zimmerman**
– Not with this combination of characters **115**
115(114). Body length approximately 4.6 mm; antennal insertions near basal third of rostrum .... ***Roptoperus*** **Lea**
– Body length > 10.0 mm; antennal insertion near apical third of rostrum .... ***Poropterus*** **Schoenherr**
116(108). Body oval; femora with ventral tooth at least on one pair; meso- and metatibial apices with uncus, premucro and suprauncal projection (minute in *Mesalcus*) **117**
– Not with this combination of characters **119**
117(116). Metanepisterna with dense to imbricate cluster of appressed scales at anterior angle between mesepimeron and elytron **118**
– Metanepisterna without dense cluster of appressed scales at anterior angle between mesepimeron and elytron .... ***Ecildaus*** **Lea**
118(117). Body length > 5.0 mm, body approximately 1.5 × as long as wide; abdominal ventrites 1 and 2 separated by suture; femora with tooth on all legs or absent (*T. favosa*) .... ***Tentegia*** **Pascoe**
– Body length approximately 3.3 mm body approximately 1.1 × as long as wide (Fig. 44.97W); abdominal ventrites 1 and 2 fused; profemora with small ventral tooth, absent on meso- and metafemora .... ***Mesalcus*** **Zimmerman**
119(116). Tarsal segments 3 not bilobed, as wide as segments 1 and 2 (Fig. 44.99i) **120**
– Tarsal segments 3 bilobed, wider than segments 1 and 2 (Fig. 44.99e) **123**
120(119). Body length > 15.0 mm .... ***Mormosintes*** **Pascoe**
– Body length < 8.0 mm **121**
121(120). Body length approximately 3.0 mm; vestiture composed of dispersed, erect setae; prosternal canal with sparse setae; elytra laterally glabrous, smooth, shiny; tarsal segments 5 with ventral apex not projecting between claws .... ***Zenoporopterus*** **Lea**
– Not with this combination of characters; body length > 6.0 mm **122**
122(121). Body length approximately 6.8 mm; vestiture composed of dense, appressed setae; scrobes about half the length of the rostrum (Fig. 44.98A); prosternal canal densely setose (Fig. 44.122I); elytra with uniform vestiture; abdominal ventrite 1 overlapping ventrite 2 (Fig. 44.99F); tarsal segments 5 with ventral apex projecting between claws .... ***Elaeagna*** **Pascoe**
– Not with this combination of characters .... ***Hyposcolyphrus*** **Zimmerman & Oberprieler**
123(119). Body length < 6.0 mm, cuticle often subglabrous and shining; mesoventral receptacle cavernous (Fig. 44.98s); metathoracic spiracle displaced and visible (Fig. 44.146H); elytral disc laterally emarginate (Fig. 44.99L); tibial uncus close to external angle (Fig. 44.99d); tarsal claws highly reduced (Fig. 44.146N) .... ***Trigonopterus*** **Fauvel**
– Not with this combination of characters **124**

124(123). Body length < 2.0 mm, cuticle often subglabrous and shining; eyes displaced to front of head; rostrum with scrobes extended beneath eyes; mesoventral receptacle prominent (Fig. 44.126K); metanepisternal sutures lined with sclerolepidia; abdominal ventrites 1 and 2 partly fused and forming plate projecting posteriad over ventrite 3; femora without ventral tooth .... ***Genuacalles* Rheinheimer**
– Not with this combination of characters **125**
125(124). Pronotum with median carina **126**
– Pronotum without median carina **127**
126(125). Body length approximately 3.1 mm; rostrum dorsally carinate .... ***Tropidotasia* Lea**
– Body length approximately 10.8 mm; rostrum dorsally without carina .... ***Neoporopterus* Faust**
127(125). Femora with ventral tooth or spine (Fig. 44.99W) at least on one pair **128**
– Femora without ventral tooth or spine (Fig. 44.99V) **134**
128(127). Body length approximately 5.2 mm; mesoventral receptacle well projecting below mesocoxae; elytra without humeral angles; metafemora without ventral tooth .... ***Drassicus* Pascoe**
– Not with this combination of characters; femora with ventral tooth or spine on all legs **129**
129(128). Body length approximately 6.8 mm, body oval and strongly convex, approximately 1.4 × as long as wide; mesoventral receptacle flattened .... ***Salcus* Pascoe**
– Not with this combination of characters; body elongate approximately 1.9 × as long as wide **130**
130(129). Body length approximately 13.2 mm; elytra with humeral angles projecting anteriad; apex of elytra with prominent lateral tubercles; abdominal ventrites 1 and 2 prominent and fused, with deep groove between them .... ***Tapinosomus* Faust**
– Not with this combination of characters; body length approximately 5.6 mm **131**
131(130). Body length approximately 6.5 mm, body 1.6 × as long as wide; meso- and metatibiae with carina on inside .... ***Imaliodes* Pascoe**
– Not with this combination of characters **132**
132(131). Elytral disc without tubercles; abdominal ventrites 1 and 2 fused, suture not visible .... ***Poropterellus* Lea**
– Elytral disc with tubercles (Fig. 44.99M); abdominal ventrites 1 and 2 connate, separated by suture (Fig. 44.99E) **133**
133(132) Mesoventral receptacle well projecting below level of mesocoxae (Fig. 44.98r) .... ***Terporopus* Lea**
– Mesoventral receptacle not projecting below level of mesocoxae .... ***Cairnsicis* Lea**
134(127). Body length approximately 5.8 mm; rostrum longer than pronotum; pronotum with median carina; abdominal ventrites 1 and 2 fused .... ***Myrtesis* Pascoe**
– Not with this combination of characters; pronotum without median carina **135**
135(134). Body length approximately 3.0 mm; elytra with flat, imbricate scales, interstriae carinate; tibiae without scale-like setae .... ***Tomweirius* Zimmerman & Oberprieler**
– Not with this combination of characters; elytral interstriae not carinate **136**
136(135). Body length approximately 2.5 mm; body mostly glabrous, cuticle smooth and shiny .... ***Gymnoporopterus* Lea**
– Not with this combination of characters **137**
137(136). Body length approximately 2.5 mm, body oval, about as long as wide, highly convex (Fig. 44.97T); outline of head and rostrum in profile evenly curved (Fig. 44.98L); prosternum shorter than procoxa (Fig. 44.98Z); elytra convex (Fig. 44.99N); abdominal ventrite 1 often with anteromedian deep groove (Fig. 44.99I) .... ***Parasalcus* Zimmerman**
– Body length variable, body elongate, > 1.4 × as long as wide; outline of head and rostrum in profile variable; prosternum variable in length; elytra convex or not (Fig. 44.99M, O); abdominal ventrite 1 without deep anteromedian groove **138**
138(137). Body length 4.0–6.0 mm; tibial apices with double comb of setae .... ***Atragopus* Zimmerman & Oberprieler**
– Not with this combination of characters; tibial apices with one comb of setae (Fig. 44.99Z) **139**
139(138). Body length often < 4.0 mm; pronotum and elytra without tubercles or tuberculate tufts of setae .... ***Decilaus* Pascoe (part)**
– Body length > 6.4 mm; pronotum and elytra with tubercles or tuberculate tufts of setae **140**
140(139). Pronotum approximately 0.7 × as long as elytra, apical third tapered .... ***Poropteroides* Lea**
– Pronotum approximately 0.5 × as long as elytra, elytra apical third variable **141**
141(140). Head flattened behind eyes (Fig. 44.98O); metaventrite with deep groove on either side .... ***Microporopterus* Lea**
– Head slightly convex behind eyes (Fig. 44.114B); metaventrite without deep lateral grooves .... ***Brachyporopterus* Lea**

LIXINI

1. Vestiture consisting of long, dense setae (Fig. 44.97b); rostrum in lateral view shorter than head .... ***Rhinocyllus* Germar**
– Vestiture consisting of shorter, dense setae; rostrum in lateral view longer than head **2**
2(1). Pronotum approximately 0.7 × as long as wide, posterior angles slightly projecting posteriad, ocular lobes weak .... ***Larinus* Dejean**
– Pronotum approximately 0.9 × as long as wide, posterior angles not projecting posteriad, ocular lobes prominent .... ***Lixus* Fabricius**

MECYSOLOBINI

1. Tarsal claws weakly bifid, inner teeth small .... ***Cylindralcides* Heller**
– Tarsal claws strongly bifid, inner teeth large and close together, giving a tridentate appearance .... ***Sternuchopsis* Heller**

MESOPTILIINI

Modified after Zimmerman (1994).

1. Antennal insertions at base of rostrum, next to eyes and foveiform (Fig. 44.98R); rostrum narrow, nearly straight and subcylindrical .... ***Aethemagdalis* Zimmerman**
– Not with this combination of characters; antennal insertions well separated from eyes (e.g. Fig. 44.98S) **2**

2(1). Eyes anteriorly deeply emarginate .... ***Atopomagdalis* Zimmerman**
– Eyes anteriorly not deeply emarginate **3**
3(2). Basal margin of elytra almost straight or broadly V-shaped; scutellar shield located in elytral depression; abdominal ventrite 2 with posterior angles not curved over ventrite 3; ventrite 3 not narrower at sides; pygidium usually narrow and transverse . ***Notomagdalis* Zimmerman**
– Not with this combination of characters **4**
4(3). Tarsi with paired claws of equal length **5**
– Metatarsi with single claws or one claw of each pair rudimentary **7**
5(4). Prothorax with posterior angles wide and low, overlapped by elytra; prosternum short .... ***Allolaemosaccus* Zimmerman**
– Not with this combination of characters **6**
6(5). Antennal scapes reaching or surpassing anterior margin of eye; eye separation variable; posterolateral corners of prothorax usually acute or subacute; habitus as Fig. 44.97c .... ***Neolaemosaccus* Hustache**
– Antennal scapes not reaching eye; eyes subcontiguous, separated by about width of two or three ommatidia; posterolateral corners of prothorax broad and obtuse .... ***Melaemosaccus* Zimmerman**
7(4). Antennal scapes not reaching eye; tarsi either with single claw or with one long, normal claw and a rudimentary second claw; tarsal segments 5 extending far beyond apical fringe setae of tarsal segment 3 .... ***Zimmermanianthus* Alonso-Zarazaga & Lyal**
– Antennal scapes capable reaching anterior margin of eye; pro- and mesotarsi with very small single or paired claws; metatarsi with single claws; tarsal segments 5 reduced and their apices not or only slightly extending beyond tarsal segment 3 .... ***Laemosaccellus* Zimmerman**

PSEPHOLACINI

Modified after Zimmerman (1994).

1. Rostrum very short, shorter than eye; antennal scapes reaching eye; mesoventrite without concave receptacle .... ***Protohylastes* Wood**
– Not with this combination of characters; rostrum longer than eye **2**
2(1). Mesotibiae on outside with broad, convex, flange-like expansion or strong dentiform process in basal half **3**
– Mesotibiae on outside without such expansion or process but sometimes with large tooth near middle or outer margin carinate or multidenticulate **6**
3(2). Head and rostrum in lateral view with dorsal contours discontinuous; prosternum deeply emarginate anteriorly, lateral walls of prosternal canal dentiform; mesoventrite with anterior margin of intercoxal process concave .... ***Oreda* White**
– Not with this combination of characters; prosternum with anterior margin not or barely emarginate **4**
4(3). Head with interocular area about half as wide as base of rostrum; mesoventral receptacle cavernous; femora without tooth or sulcus .... ***Psepholacipus* Lea**
– Head with interocular area almost as wide as base of dorsum of rostrum; mesoventral receptacle variable; meso- and metafemora on ventral side sulcate and dentate **5**
5(4). Dorsal vestiture dense; basal elytral margin deeply sinuous; rostrum with dorsal margin of scrobes expanded; antennal segments 2 unusual, laterally compressed and conspicuously enlarged; mesoventral receptacle open, its medial line only slightly tumid behind middle .... ***Neozeneudes* Lea**
– Dorsal vestiture sparse; basal elytral margin almost straight; dorsal edge of scrobes not so expanded; antennal segments 2 not laterally compressed or conspicuously enlarged; mesoventral receptacle cavernous .... ***Calpurnius* Zimmerman**
6(2). Rostrum with sides below scrobe often densely setose (Fig. 44.98T); mesotibiae on outside with large, broad tooth near middle (Fig. 44.99T); habitus as Fig.44.97f .... ***Psepholax* White**
– Rostrum with sides glabrous or sparsely setose; mesotibiae on outside carinate or multidenticulate or serrate before any apical or subapical armature if present **7**
7(6). Vestiture sparse, leaving much of shiny cuticle clearly exposed **8**
– Vestiture mostly dense, cuticle not exposed (e.g. Fig. 44.99P) **10**
8(7). Mesoventrite without receptacle for rostrum .... ***Catocalephe* Blackburn**
– Mesoventrite with receptacle for rostrum **9**
9(8). Antennae inserted antemedially on rostrum; scrobes in dorsal view with ventral margin narrowly visible; mesoventral receptacle a simple open cavity; femora notably broad; metatibiae strongly flattened; tibial margins with strong, terminal, spinelike process overhanging false corbel; habitus as Fig. 44.97g .... ***Sympiezoscelus* Waterhouse**
– Antennae inserted postmedially on rostrum; scrobes in dorsal view with ventral margin concealed; mesoventral receptacle fully developed, deep and cavernous; femora and tibiae at most moderately wide; tibial apices variable .... ***Celidaus Lea***
10(7). Antennae unusual with scape short and not reaching the eye, funicular segments 2 flattened, expanded, subcircular and other segments strongly transverse; mesoventral receptacle open; dorsal vestiture squamose .... ***Zeneudes* Pascoe**
– Not with this combination of characters; scapes extending close to eye **11**
11(10). Scrobes with distal parts visible in dorsal view; head with interocular distance shorter than width of rostrum at apex **12**
– Scrobes mostly or entirely hidden in dorsal view; head with interocular distance about as long as width of rostrum at apex **13**
12(11). Meso- and metatibiae on outside cariniform; mesoventrite with intercoxal process broad, concave or almost flat .... ***Psepholasoma* Lea**
– Meso- and metatibiae on outside multidentate from base to superuncus; mesoventrite with intercoxal process concave and each side strongly produced into large, long, sharp process .... ***Wiburdia* Lea**
13(11). Mesoventrite with intercoxal process concave, forming open, cuplike receptacle; procoxae apically produced into obtuse, angulate, setose process .... ***Therebiosoma* Lea**
– Not with this combination of characters; procoxae rounded **14**
14(13). Rostral canal appearing deep and narrow between procoxae; body robust; WA .... ***Therebus* Pascoe**
– Rostral canal wide, as wide or wider between coxae as apical width of rostrum; body narrow; QLD **15**

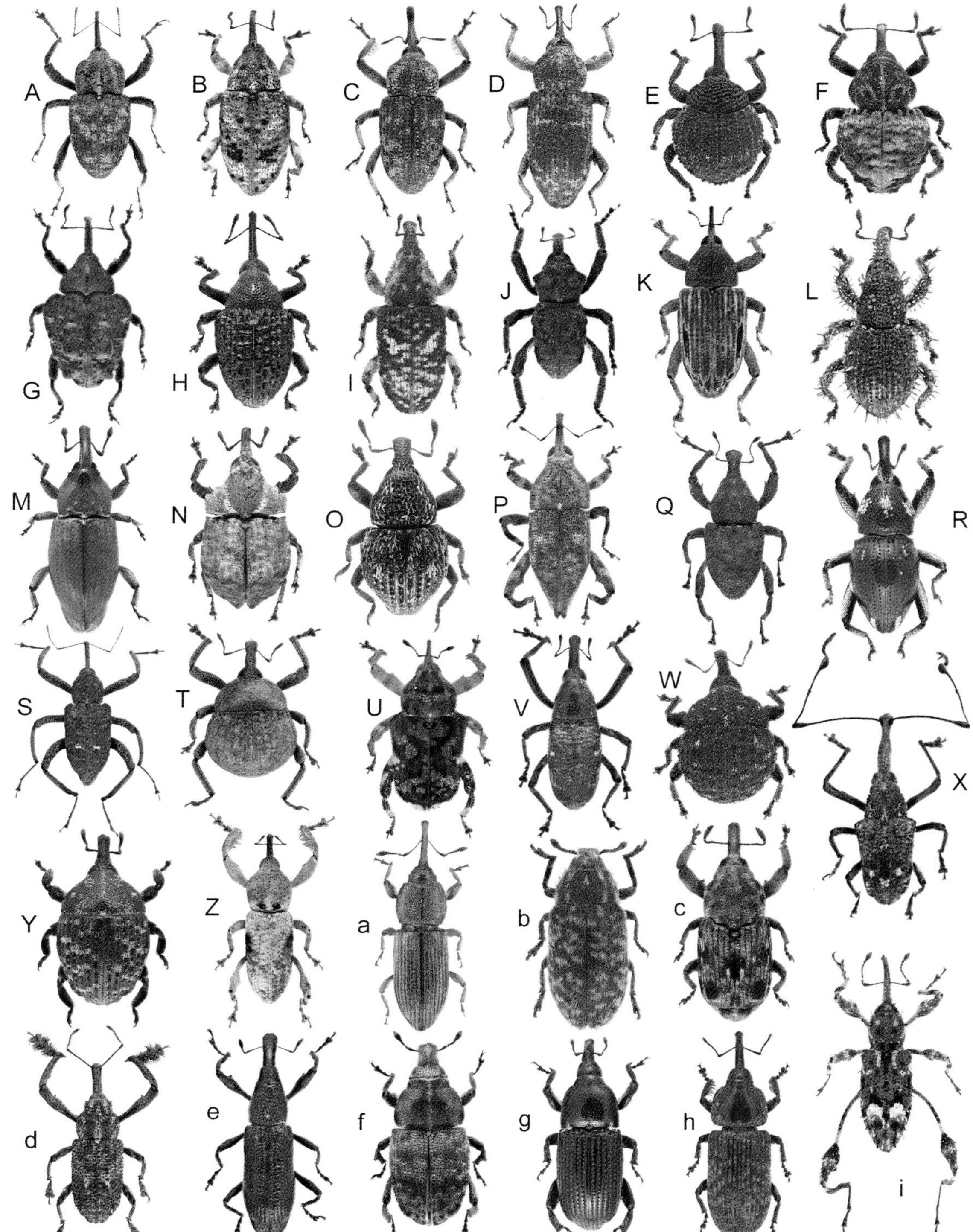

Fig. 44.97. Molytinae, characters for key to genera, habitus. A, *Diatassa denticulata* (Pascoe); B, *Mechistocerus calidris* (Pascoe); C, *Parendymia pilipes* Kirsch; D, *Roecus vulneratus* (Lea); E, *Achelocis rudis* Lea; F, *Cycloporopterus mysticus* Lea; G, *Hybophorus rufotuberosus* Waterhouse; H, *Melanterius porcatus* Erichson; I, *Achopera*; J, *Acidinus sedecimtuberculatus* (Lea); K, *Aethreus cicatricosus* Pascoe; L, *Andracalles fasciculatus* (Lea); M, *Bepharus ellipticus* Pascoe; N, *Brachyphyes thoracicus* (Lea); O, *Decilaus suturaelevata* (Rheinheimer); P, *Embaphiodes phyxidatus* Pascoe; Q, *Eufaustia mirabilis* Lea; R, *Hoplidotasia torresensis* Lea; S, *Pezichus binotatus* Lea; T, *Parasalcus latissimus* (Pascoe); U, *Platytenes varius* Pascoe; V, *Nototragopus plagiatus* (Pascoe); W, *Mesalcus*; X, *Hypsophorus dromedarius* (Boisduval); Y, *Tetengia solenopa* Lea; Z, *Camptorhinus inornatus* Lea; a, *Arecophaga*; b, *Rhinocyllus conicus* (Frölich); c, *Neolaemosaccus chadwicki* (Janczyk); d, *Orthorhinus cylindrirostris* (Fabricius); e, *Peribleptus dealbatus* (Boisduval); f, *Psepholax egerius* Pascoe; g, *Sympiezoscelus spencei* Waterhouse; h, *Tranes vigorsii* Boheman; i, *Acicnemis meriones* Pascoe.

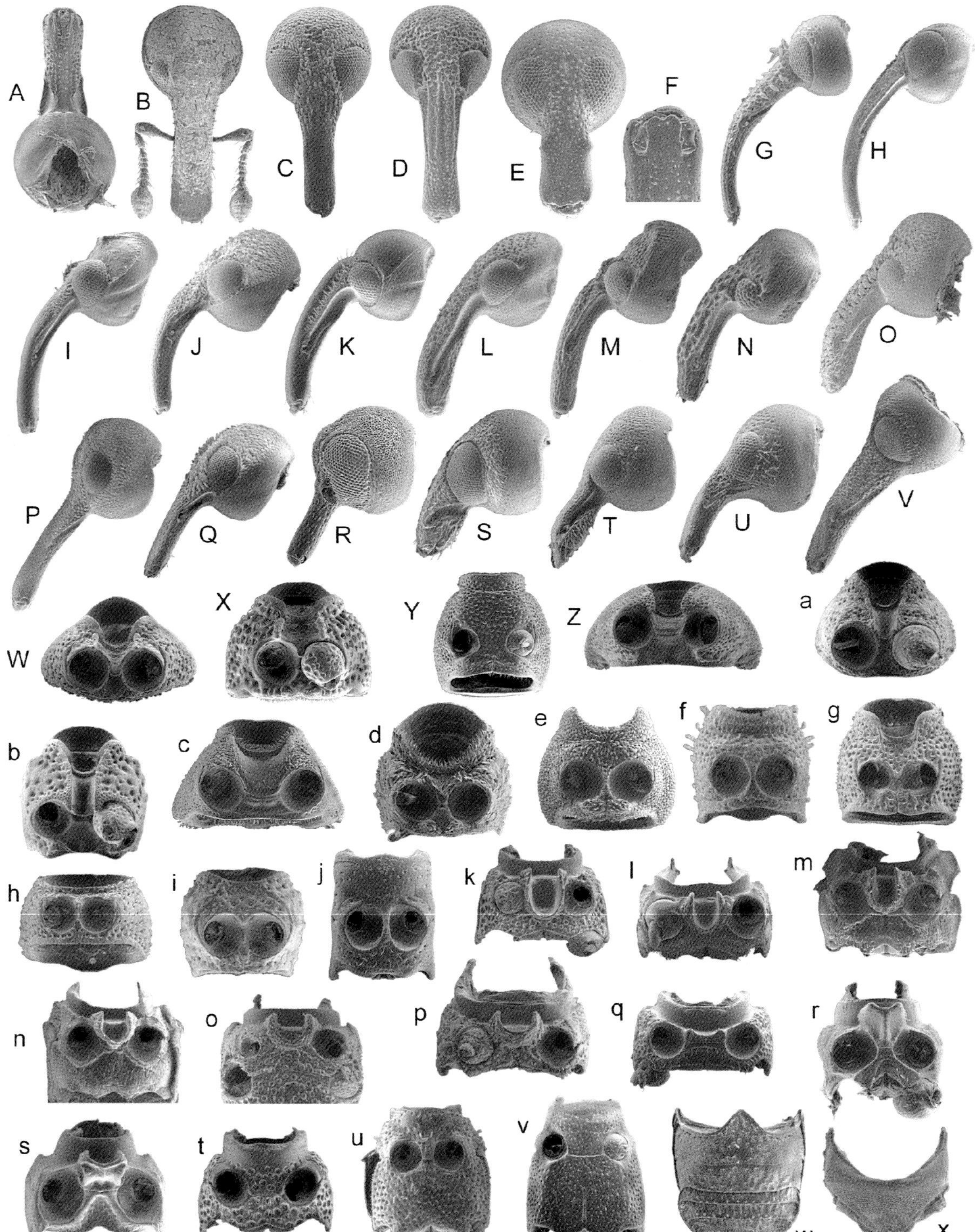

Fig. 44.98. Molytinae, characters for key to genera. A, *Elaeagna*, head, ventral; B–D, head, dorsal: B, *Achopera*; C, *Orochlesis*; D, *Tamphilus*; E–F, *Eristinus*: E, head, dorsal; F, mouthparts; G–V, head, lateral: G, *Acicnemis*; H, *Onidistus*; I, *Acrotychreus*; J, *Phloeoglymma*; K, *Bepharus*; L, *Parasalcus*; M, *Exithius*; N, *Decilaus*; O, *Microporopterus*; P, *Eutyrhinus*; Q, *Ampagia*; R, *Aethemagdalis*; S, *Notomagdalis*; T, *Psepholax*; U, *Thaumastophasis*; V, *Notopissodes*; W–i, prothorax, ventral: W, *Cycloporopterus*; X, *Arthriticosoma*; Y, *Arecophaga*: Z, *Parasalcus*; a, *Onidistus*; b, *Terporopus*; c, *Psepholax*; d, *Orthorhinus*; e, *Notopissodes*; f, *Acicnemis*; g, *Exeiratus*; h, *Eristinus*; i, *Syagrius*; j–v, pterothorax, ventral: j, *Syagrius*; k, *Tapinocis*; l, *Enteles*; m, *Acrotychreus*; n, *Chimades*; o, *Tamphilus*; p, *P. latissimus*; q, *C. mysticus*; r, *Terporopus*; s, *Trigonopterus*; t, *Exeiratus*; u, *Acicnemis*; v, *Arecophaga*; w, *Aethemagdalis*, abdominal ventrites; x, *Tamphilus*, scutellum.

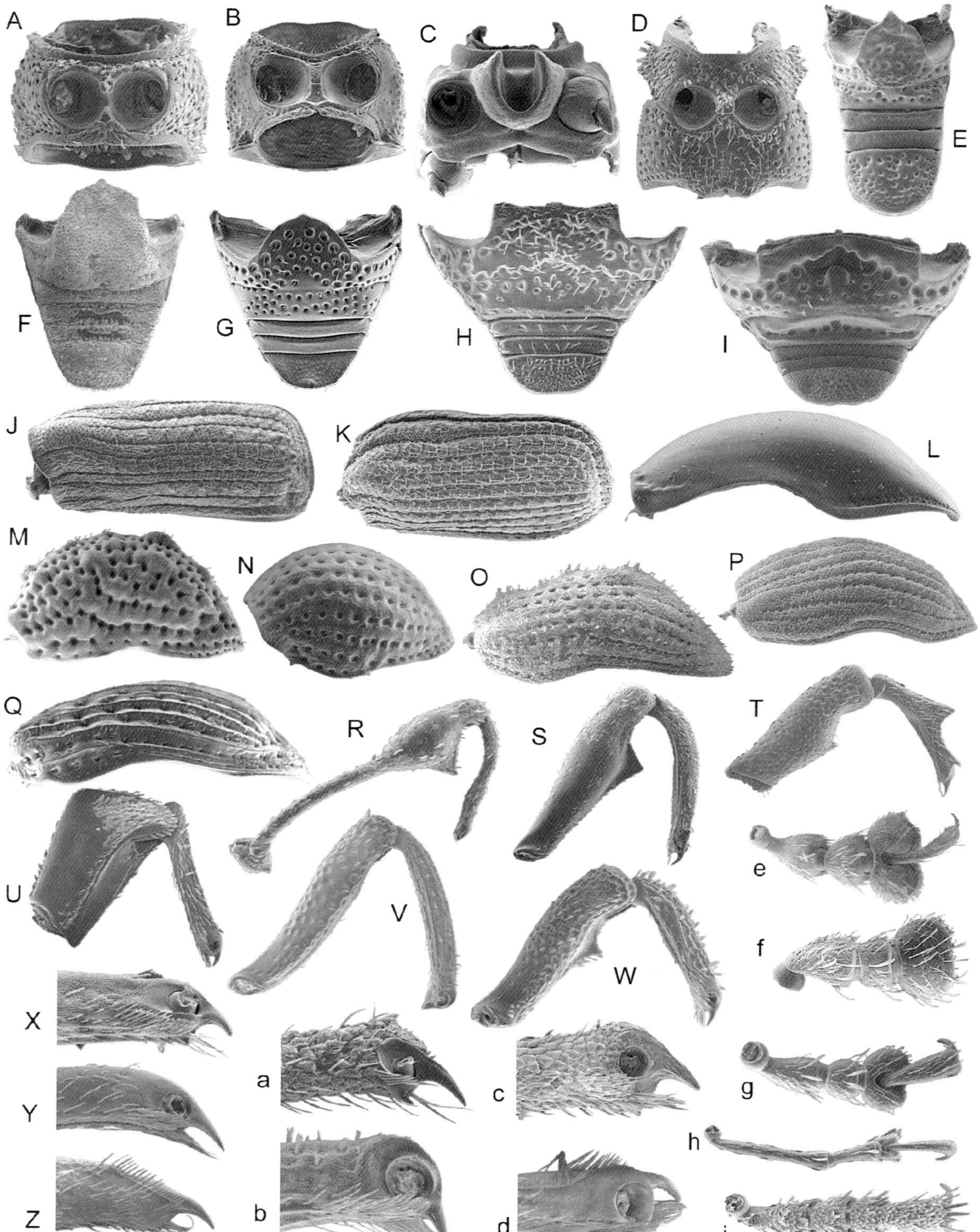

Fig. 44.99. Molytinae, characters for key to genera. A, *Thaumastophasis*, prothorax, ventral; B, *Notomagdalis*, prothorax, ventral; C, *Paleticus*, pterothorax, ventral; D, *Thaumastophasis*, pterothorax, ventral; E–I, abdominal ventrites: E, *Terporopus*; F, *Elaeagna*; G, *Arthriticosoma*; H, *Achelocis*; I, *Parasalcus*; J–K, elytron, dorsal: J, *Zimmermanianthus*; K, *Aethemagdalis*; L–Q, elytron, lateral: L, *Trigonopterus*; M, *Terporopus*; N, *Parasalcus*; O, *Phloeoglymma*; P, *Psepholax*; Q, *Arthriticosoma*; R–W, legs: R, *Acicnemis*; S, *Acrotychreus*; T, *Psepholax*; U, *Ampagia*; V, *Parasalcus*; W, *Tamphilus*; X–d, tibial apex: X, *Terporopus*; Y, *Enteles*; Z, *Onidistus*; a, *Achopera*; b, *Parasalcus*; c, *Elaeagna*; d, *Trigonopterus*; e–i, tarsi, dorsal: e, *Parasalcus*; f, *Achelocis*; g, *Tapinocis*; h, *Acicnemis*; i, *Elaeagna*.

15(14). Mesoventrite with anterior margin of intercoxal process reaching beyond anterior margins of mesocoxae; elytral striae 10 joining or nearly joining 9 above metacoxae .... ***Pseudotherebus*** **Lea**
– Mesoventrite with anterior margin not reaching beyond anterior margins of mesocoxae; elytral striae 9 and 10 distinctly separated from base to apex .... ***Glechinus*** **Pascoe**

SOPHRORHININI

1. Prosternum with canal but not extending to mesoventrite .... ***Colobodes*** **Schoenherr**
– Prosternum and mesoventrite with rostral canal 2
2(1). Prothorax about 0.9 × as long as wide .... ***Perrhaebius*** **Pascoe**
– Prothorax transverse, around 0.6 × as long as wide 3
3(2). Prothorax with sides explanate near apical third (Fig. 44.160E) .... ***Deretiosus*** **Pascoe**
– Prothorax with sides not explanate .... ***Eprias*** **Heller**

TRANES GROUP

Modified after Zimmerman (1994), Hsiao & Oberprieler (2020a, b, 2022).

1. Metanepisternal sutures lined with sclerolepidia (e.g. Fig. 44.110J) 2
– Metanepisternal suture not lined with sclerolepidia 3
2(1). Dorsum of body and legs densely covered with brownish to whitish, coarse setae; antennal clubs as long as or longer than funicles; surface of pronotum punctorugulose .... ***Demyrsus*** **Pascoe**
– Dorsum of body and legs lacking setae, at most sparsely and finely pubescent; antennal clubs distinctly shorter than funicles; surface of pronotum sparsely punctate .... ***Siraton*** **Hustache**
3(1). Body and legs completely black; procoxae contiguous 4
– Body and legs orange to dark brown; procoxae separate 5
4(3). Femora not sulcate beneath; elytra densely covered with coarse yellowish setae in apical third; Lord Howe Is. .... ***Howeotranes*** **Zimmerman**
– Femora distinctly sulcate beneath; elytral derm nearly nude, sparsely covered with short setae; mainland Australia .... ***Paratranes*** **Zimmerman**
5(3). Forehead narrower than width of rostrum at base; pronotum completely punctorugulose; meso- and metatibiae with distal setal comb restricted to apical margin .... ***Miltotranes*** **Zimmerman**
– Forehead as wide as rostrum at base; pronotum mostly flat, only punctorugulose laterally; meso- and metatibiae with distal setal comb extending to apical quarter of tibia; habitus as Fig. 44.97h .... ***Tranes*** **Schoenherr**

Classification of the Australian genera of Molytinae

Tribe Aedemonini Faust, 1898

Cincius Zimmerman, 1994 (Pl. 80A, B)

Type species, by original designation: *Mechistocerus triangulifer* Lea, 1931.
Australian species. *C. triangulifer* (Lea, 1931a).
Distribution. Australia.

Damippus Zimmerman, 1994 (Pl. 80C, D)

Type species, by original designation: *Mecistocerus tenuirostris* Lea, 1907.
Australian species. *D. tenuirostris* (Lea, 1907b).
Distribution. Australia.

Derbyiella Lea, 1907 (Pl. 80E)

Type species, by monotypy: *Derbyia laminata* Lea, 1900.
Australian species. *D. laminata* (Lea, 1900b).
Distribution. Australia (WA).
Reference. Lea (1907b).

Diatassa Pascoe, 1871 (Pl. 80F)

Type species, by monotypy: *Diatassa phalerata* Pascoe, 1871c.
Australian species. *D. denticulata* (Pascoe, 1875).
Distribution. Australia, Philippines, Maluku Is., New Guinea.

Dystropicus Pascoe, 1885 (Pl. 80G)

Type species, by monotypy: *Dystropicus squalidus* Pascoe, 1885.
Australian species. *D. hystricosus* (Lea, 1909d), *D. sanctus* (Lea, 1931a), *D. squalidus* Pascoe, 1885, *D. tanyrhynchus* (Lea, 1907b).
Distribution. Australia (QLD), India, New Guinea.

Epirus Zimmerman, 1994 (Pl. 80H)

Type species, by original designation: *Mecistocerus maerens* Lea, 1907.
Australian species. *E. maerens* (Lea, 1907b).
Distribution. Australia.

Magius Zimmerman, 1994 (Pl. 80I)

Type species, by original designation: *Mechistocerus magnidens* Lea, 1928.
Australian species. *M. cancellatus* (Lea, 1909c), *M. magnidens* (Lea, 1928b).
Distribution. Australia.

Mechistocerus Fauvel, 1862 (Pl. 80J)

Type species, by monotypy: *Coelosternus impressus* Montrouzier, 1861.
Australian species. *M. calidris* (Pascoe, 1885), *M. multimaculatus* Lea, 1928b.
Distribution. Australia, Melanesia, Japan, China, Mexico (?), Afrotropical, Oriental.

Parendymia Kirsch, 1877 (Pl. 80K)

Type species, by monotypy: *Parendymia pilipes* Kirsch, 1877.
Australian species. *P. pilipes* Kirsch, 1877.
Distribution. Australia to Philippines and New Guinea, Micronesia and Melanesia to Samoa.

Rhadinomerus Faust, 1892 (Pl. 80L)

Type species, by monotypy: *Mecistocerus mastersi* Pascoe, 1870.
Australian species. *R. basalis* (Lea, 1913b), *R. dentivarius* (Lea, 1931a), *R. duplicatus* (Lea, 1911b), *R. egens* (Lea, 1907b), *R. fimbritarsis* (Lea, 1931a), *R. interocularis* (Lea, 1928b), *R. leai* (Hustache, 1936), *R. marmoreus* (Lea, 1913b), *R. mastersi* (Pascoe, 1870e), *R. metasternalis* (Lea, 1913b), *R. punctiventris* (Lea, 1911b), *R. setiventris* (Lea, 1928b).
Distribution. Australia (QLD, NSW, NT, WA), Oriental, Kuril Is., China, Japan, Afrotropical.
Reference. Faust (1892a).

Roecus Zimmerman, 1999 (Pl. 80M)

Type species, by original designation: *Mechistocerus cylindricus* Lea, 1913.
Australian species. *R. cylindricus* (Lea, 1913b), *R. subglobicollis* (Lea, 1928b), *R. vulneratus* (Lea, 1907b).
Distribution. Australia.
Notes. The genus was described in Alonso-Zarazaga & Lyal (1999).

Tribe Cleogonini Gistel, 1848

Achelocis Lea, 1927 (Pl. 81A; Fig. 44.101)

Type species, by monotypy: *Achelocis rudis* Lea, 1927.
Australian species. *A. rudis* Lea, 1927e.
Distribution. Australia (QLD).

Arthriticosoma Lea, 1899 (Pl. 81B; Fig. 44.102)

Type species, by monotypy: *Arthriticosoma vigilans* Lea, 1899.
Australian species. *A. vigilans* Lea, 1899c.
Distribution. Australia.

Chalcodermus Dejean, 1835 (Pl. 81C)

Type species, by subsequent designation (Schoenherr, 1837: 378): *Rhynchaenus calidus* Fabricius, 1801.
Australian species. *C. serripes* Fåhraeus, 1837 (*in* Schoenherr 1837).
Distribution. Australia, America.
Notes. The single species in Australia was introduced from Brazil, Mexico and Venezuela between 1996 and 2000 for the control of the weed *Mimosa pigra* (Fabaceae) (Heard *et al.* 1999; Pullen *et al.* 2014).

Conotrachelus Dejean, 1835 (Pl. 81D)

Type species, by subsequent designation (Schoenherr, 1837: 392): *Balaninus diaconitus Klug,* 1829.
Australian species. *C. albocinerus* Fiedler, 1940.
Distribution. Australia, America.
Notes. The single species in Australia was introduced from South America for the control of the weed *Parthenium hysterophorus* (Asteraceae) (McFadyen 2000; Pullen *et al.* 2014).

Cycloporopterus Lea, 1908 (Pl. 81E; Fig. 44.103)

Type species, by monotypy: *Cycloporopterus mysticus* Lea, 1908.
Australian species. *C. mysticus* Lea, 1908b.
Distribution. Australia.

Eurymelanterius Zimmerman & Oberprieler, 2014 (Pl. 81F)

Type species, by original designation: *Eurymelanterius monteithi* Zimmerman & Oberprieler, 2014.
Australian species. *E. monteithi* Zimmerman & Oberprieler, 2014.
Distribution. Australia.
Notes. The genus was described in Pullen *et al.* (2014).

Hybophorus Waterhouse, 1853 (Pl. 81G)

Type species, by monotypy: *Hybophorus rufotuberosus* Waterhouse, 1853.
Australian species. *H. rufotuberosus* Waterhouse, 1853.
Distribution. Australia.

Lybaeba Pascoe, 1873 (Pl. 81H, I)

Type species, by subsequent designation (Pullen *et al.* 2014: 222): *Lybaeba subfasciata* Pascoe, 1873.
Australian species. *L. acutidens* Lea, 1899c, *L. aestuans* (Pascoe, 1873a), *L. affluens* (Pascoe, 1874b), *L. albomaculata* (Lea, 1928a), *L. alternata* (Lea, 1928a), *L. amplicornis* Lea, 1909c, *L. amplipennis* (Lea, 1899c), *L. apicalis* (Lea,

1913a), *L. apicispina* (Lea, 1928a), *L. aulica* (Lea, 1928a), *L. basipennis* (Lea, 1928a), *L. bimaculiceps* (Lea, 1928a), *L. blackburni* Lea, 1909c, *L. cinnamomea* (Pascoe, 1872a), *L. cognata* (Lea, 1928a), *L. concinna* Lea, 1899c, *L. congrua* (Lea, 1899c), *L. consanguinea* Lea, 1899c, *L. falcata* (Lea, 1928a), *L. famelica* Lea, 1899c, *L. ferruginea* (Lea, 1928a), *L. fervida* (Pascoe, 1873a), *L. filirostris* (Lea, 1928a), *L. florida* (Pascoe, 1875), *L. fugitiva* (Pascoe, 1875), *L. funerea* (Lea, 1913b), *L. graniventris* (Lea, 1928a), *L. heterodoxa* (Lea, 1928a), *L. hybrida* (Lea, 1913b), *L. hypoleuca* (Lea, 1928a), *L. hypolissa* (Lea, 1928a), *L. impolita* (Lea, 1899c), *L. inaequalis* Lea, 1899c, *L. incisipes* (Lea, 1931a), *L. inconstans* (Lea, 1928a), *L. inermis* (Lea, 1913b), *L. insignita* (Elston, 1919), *L. insuavis* (Lea, 1928a), *L. majorina* Lea, 1899c, *L. metasternalis* (Lea, 1915d), *L. minuscula* (Lea, 1928a), *L. mollis* Lea, 1899c, *L. nigriclava* (Lea, 1931a), *L. nigrirostris* (Lea, 1928a), *L. nigrosuturalis* (Lea, 1928a), *L. nigrovaria* Lea, 1899c, *L. niveodispersa* (Lea, 1915d), *L. nodipennis* (Lea, 1913b), *L. orthodoxa* (Lea, 1928a), *L. pallida* (Lea, 1928a), *L. pallidicollis* (Lea, 1913b), *L. parvicollis* Lea, 1899c, *L. picta* Lea, 1899c, *L. pictipennis* (Lea, 1931a), *L. porphyrea* (Pascoe, 1873a), *L. potens* (Lea, 1928a), *L. pretiosa* Lea, 1911b, *L. remota* (Blackburn, 1893a), *L. repanda* Pascoe, 1873a, *L. setirostris* (Lea, 1928a), *L. setosa* (Oke, 1931), *L. silacea* (Lea, 1913b), *L. squamivaria* Lea, 1909a, *L. subaurifera* (Lea, 1928b), *L. subfasciata* Pascoe, 1873a, *L. subglobosa* (Lea, 1928a), *L. subsulfurea* (Lea, 1928a), *L. sulfurea* (Lea, 1913a), *L. suturalis* (Lea, 1913b), *L. tantilla* Lea, 1899c, *L. tenuirostris* (Lea, 1915d), *L. tricolor* Lea, 1911b, *L. trifasciata* Lea, 1911b, *L. truncatidens* (Lea, 1931a), *L. tuberculata* (Lea, 1915d), *L. venusta* (Oke, 1931).

Distribution. Australia (QLD, NSW, VIC, SA, WA).

Notes. The host plants include *Acacia* species (Fabaceae) and *Geijera parviflora* (Rutaceae).

Melanterius Erichson, 1842 (Pl. 81J–L; Fig. 44.100)

Type species, by subsequent designation (Lea, 1899c: 206): *Melanterius porcatus* Erichson, 1842.

Australian species. *M. abbreviatus* Pinzón-Navarro *et al.* 2017, *M. aberrans* Lea, 1899c, *M. acaciae* Lea, 1899c, *M. antennalis* Lea, 1899c, *M. aratus* Pascoe, 1885, *M. arenaceus* Lea, 1928a, *M. atronitens* Lea, 1931a, *M. baridioides* Lea, 1913b, *M. bicalcaratus* Lea, 1913b, *M. bidentatus* Lea, 1899a, *M. biseriatus* Lea, 1913b, *M. cardiopterus* Lea, 1913b, *M. castaneus* Lea, 1899c, *M. cicatricosus* Lea, 1928a, *M. compositus* Lea, 1909c, *M. confusus* Lea, 1913a, *M. conspiciendus* Lea, 1909c, *M. cordipennis* Lea, 1899a, *M. corosus* (Boisduval, 1835), *M. costatus* (Lea, 1899c), *M. costipennis* Lea, 1905a, *M. curvirostris* Lea, 1931a, *M. curvistriatus*, Pinzón-Navarro *et al.* 2017, *M. ellipticus* Lea, 1913a, *M. elusus* (Casey, 1892), *M. fasciculatus* Lea, 1913a, *M. femoralis* Lea, 1928a, *M. humeralis* Lea, 1928a, *M. imitator* Lea, 1913a, *M. incisus* Lea, 1928a, *M. incomptus* Lea, 1899c, *M. inconspicuus* (Lea, 1899c), *M. insularis* Lea, 1928a, *M. interstitialis* Lea, 1899c, *M. labeculosus* Lea, 1928a, *M. lamellatus* Lea, 1913a, *M. lateralis* Lea, 1928a, *M. laticornis* Lea, 1913a, *M. latipennis* Lea, 1928a, *M. legitimus* Lea, 1909c, *M. leptorrhynchus* Lea, 1911c, *M. leucophaeus* Lea, 1913a, *M. maculatus* Lea, 1899c, *M. maestus* Lea, 1913a, *M. mediocris* Lea, 1913a, *M. microtrichius* Lea, 1928a, *M. minor* Lea, 1913a, *M. modestus* Lea, 1928a, *M. modicus* Lea, 1928a, *M. multimaculatus* Lea, 1928a, *M. nemorhinus* Lea, 1913a, *M. oleosus* Lea, 1928a, *M. parvidens* Lea, 1899c, *M. persimilis* Lea, 1909c, *M. porcatus* Erichson, 1842, *M. porosus* Lea, 1899c, *M. psittacoides* Pinzón-Navarro *et al.* 2017, *M. pungalinae* Pinzón-Navarro *et al.* 2017, *M. rufirostris* Lea, 1928a, *M. rufus* Lea, 1928a, *M. semiporcatus* Erichson, 1842, *M. semiporosus* Lea, 1908b, *M. servulus* Pascoe, 1872a, *M. setistriatus* Lea, 1931a, *M. solitus* Lea, 1899c, *M. squamipennis* Lea, 1928a, *M. stenocnemis* Lea, 1913b, *M. strabonus* Lea, 1899c, *M. submaculatus* Lea, 1928a, *M. tenuis* Lea, 1899c, *M. tesseymanni* Pinzón-Navarro *et al.* 2017, *M. tibialis* Lea, 1928a, *M. tropicus* Lea, 1928a, *M. unidentatus* Lea, 1899c, *M. ventralis* Lea, 1899c, *M. villosipes* Lea, 1931a, *M. vinosus* Pascoe, 1872a, *M. vulgivagus* Lea, 1899c.

Distribution. Australia (NSW, NT, QLD, SA, TAS, VIC, WA).

Notes. *Melanterius* is tightly associated with *Acacia* (Fabaceae), the larvae developing in the seeds and several species employed as biological control agents of invasive wattles in South Africa. A summary of the known host associations, together with the description of some new species, was published by Pinzón-Navarro *et al.* (2017).

Moechius Pascoe, 1872 (Pl. 82A)

Type species, by monotypy: *Moechius anaglyptus* Pascoe, 1872.

Australian species. *M. anaglyptus* Pascoe, 1872.

Distribution. Australia (QLD).

Neomelanterius Lea, 1899 (Pl. 82B, C)

Type species, by subsequent designation (Setliff, 2007: 10): *Melanterius carinicollis* Pascoe, 1875.

Australian species. *N. carinicollis* (Pascoe, 1875), *N. interruptus* Lea, 1913b, *N. longirostris* Lea, 1899c, *N. setipennis* (Lea, 1928a), *N. subtuberculatus* Lea, 1910b.

Distribution. Australia (QLD, NSW), New Guinea.

Notes. Specimens have been collected from blossoms of *Syzygium smithii* (Myrtaceae).

Teutheria Pascoe, 1875 (Pl. 82D)

Type species, by monotypy: *Teutheria insculpta* Pascoe, 1875.

Australian species. *T. insculpta* Pascoe, 1875.

Distribution. Australia.

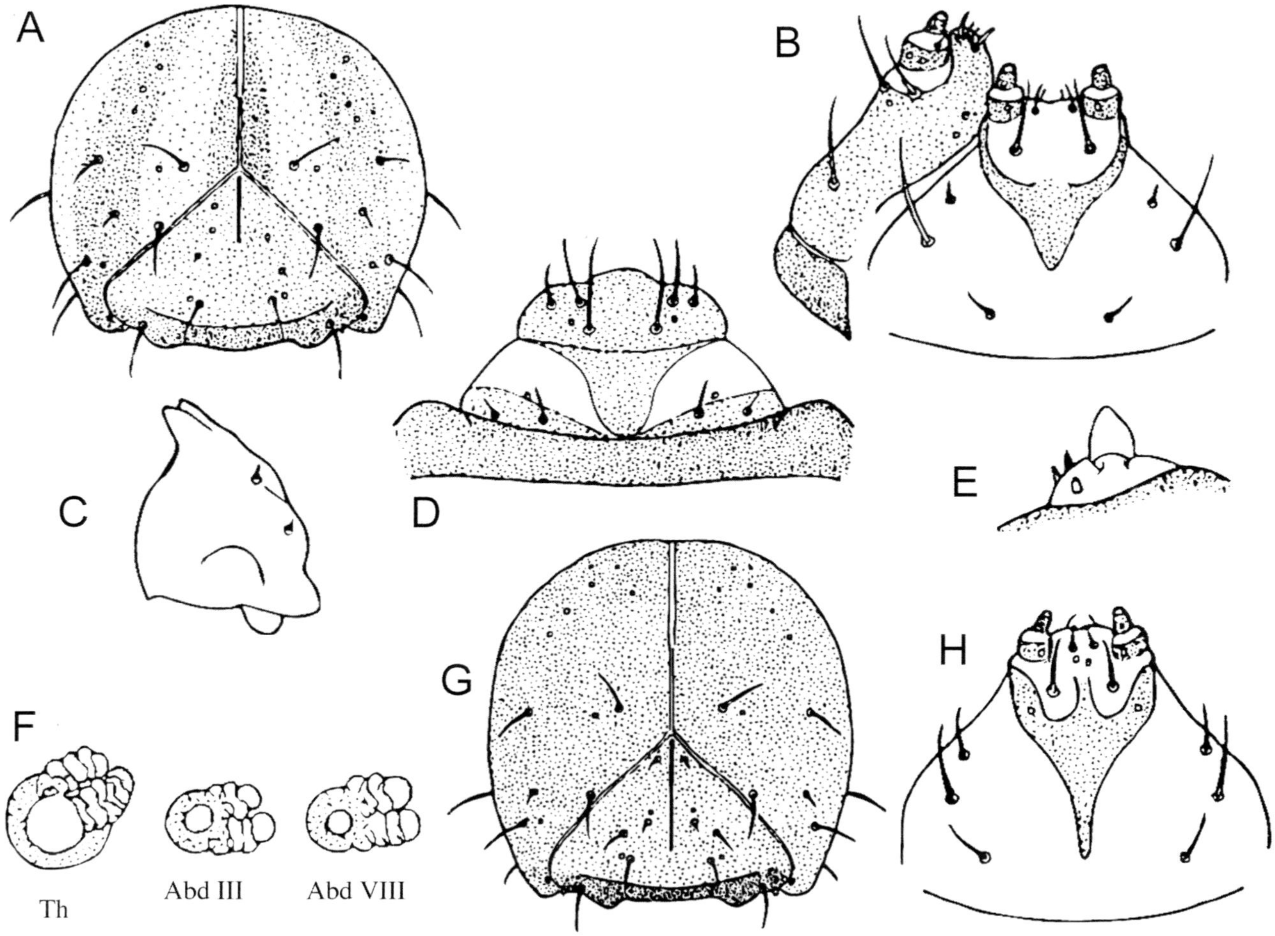

Fig. 44.100. Cleogonini, structural details of larvae. A–F, *Melanterius maculatus* Lea, larva: A, head, dorsal; B, mala and labium, ventral; C, mandible; D, labrum, clypeus and epistoma; E, antenna; F, spiracles, showing alignment; G–H, *Melanterius costipennis* Lea, larva: G, head, dorsal; H, labium. After May (1994).

Tribe Cryptorhynchini Schoenherr, 1825

Achopera Pascoe, 1870 (Pl. 83A; Fig. 44.108)

Type species, by subsequent designation (Pullen *et al.* 2014: 230): *Achopera lachrymosa* Pascoe, 1870.

Australian species. *A. alba* Lea, 1931b, *A. alternata* Lea, 1910c, *A. araucariae* Marshall, 1948, *A. bifasciata* Lea, 1913b, *A. isabellina* Lea, 1913b, *A. lachrymosa* Pascoe, 1870e, *A. longiventris* Lea, 1931b, *A. maculata* Pascoe, 1870e, *A. microps* Lea, 1931b, *A. multimaculata* Lea, 1931b, *A. parva* Lea, 1913b, *A. pictiventris* Lea, 1931b, *A. subalba* Lea, 1931b, *A. subcylindrica* Lea, 1931b, *A. subulosa* Lea, 1908c, *A. uniformis* Pascoe, 1870e, *A. xanthorrhoeae* Lea, 1908b.

Distribution. Australia, Chile, UK.

Notes. Murray *et al.* (2014) documented the presence of *A. alternata* Lea in the UK.

Achoperinus Lea, 1909 (Pl. 83B)

Type species, by monotypy: *Cryptorhynchus infulatus* Erichson, 1842.

Australian species. *A. infulatus* (Erichson, 1842).

Distribution. Australia.

Reference. Lea (1909d).

Acidinus Zimmerman, 1994 (Pl. 83C)

Type species, by monotypy: *Illidgea sedecimtuberculata* Lea, 1912.

Australian species. *A. sedecimtuberculatus* (Lea, 1912)

Distribution. Australia.

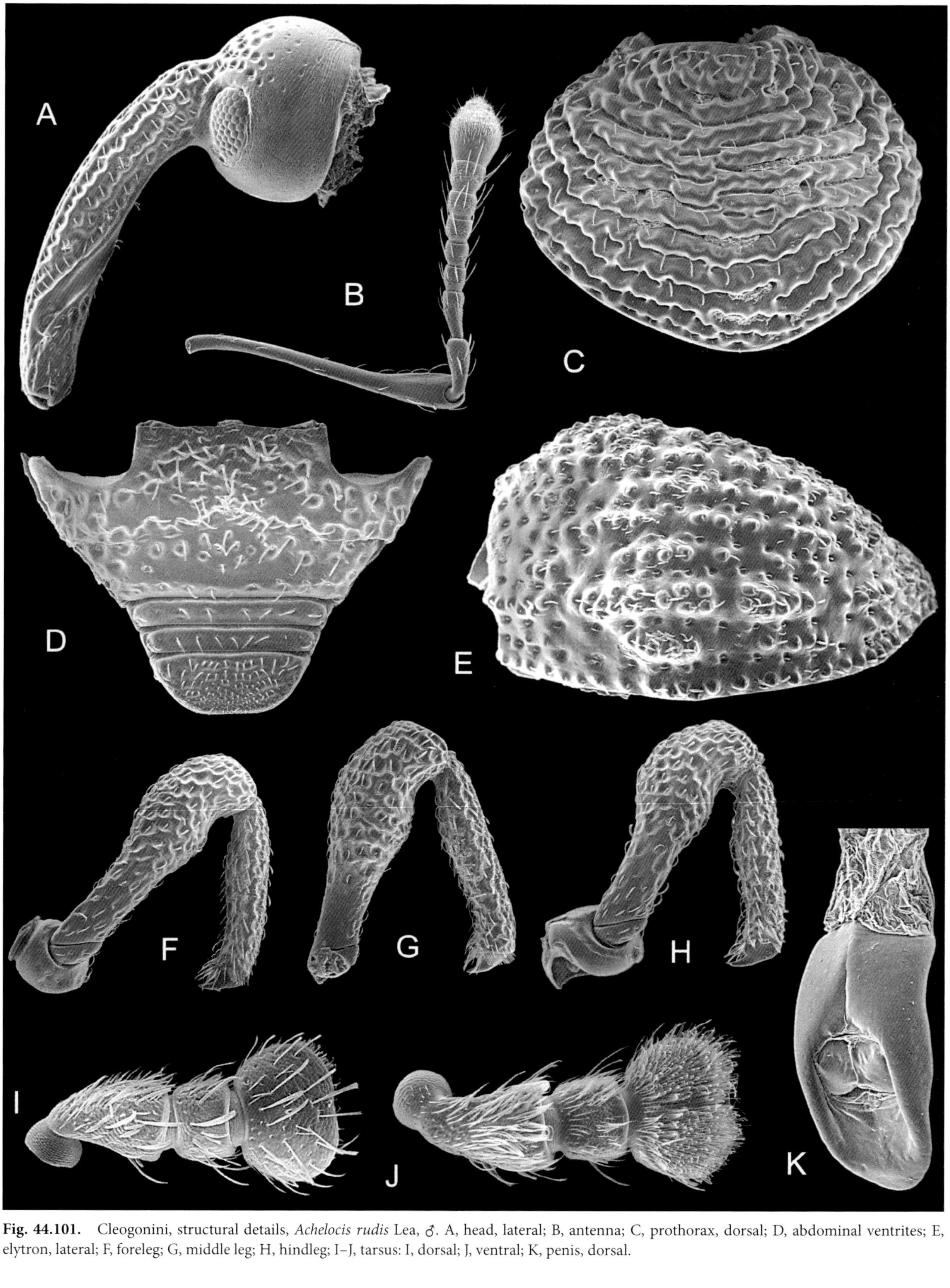

Fig. 44.101. Cleogonini, structural details, *Achelocis rudis* Lea, ♂. A, head, lateral; B, antenna; C, prothorax, dorsal; D, abdominal ventrites; E, elytron, lateral; F, foreleg; G, middle leg; H, hindleg; I–J, tarsus: I, dorsal; J, ventral; K, penis, dorsal.

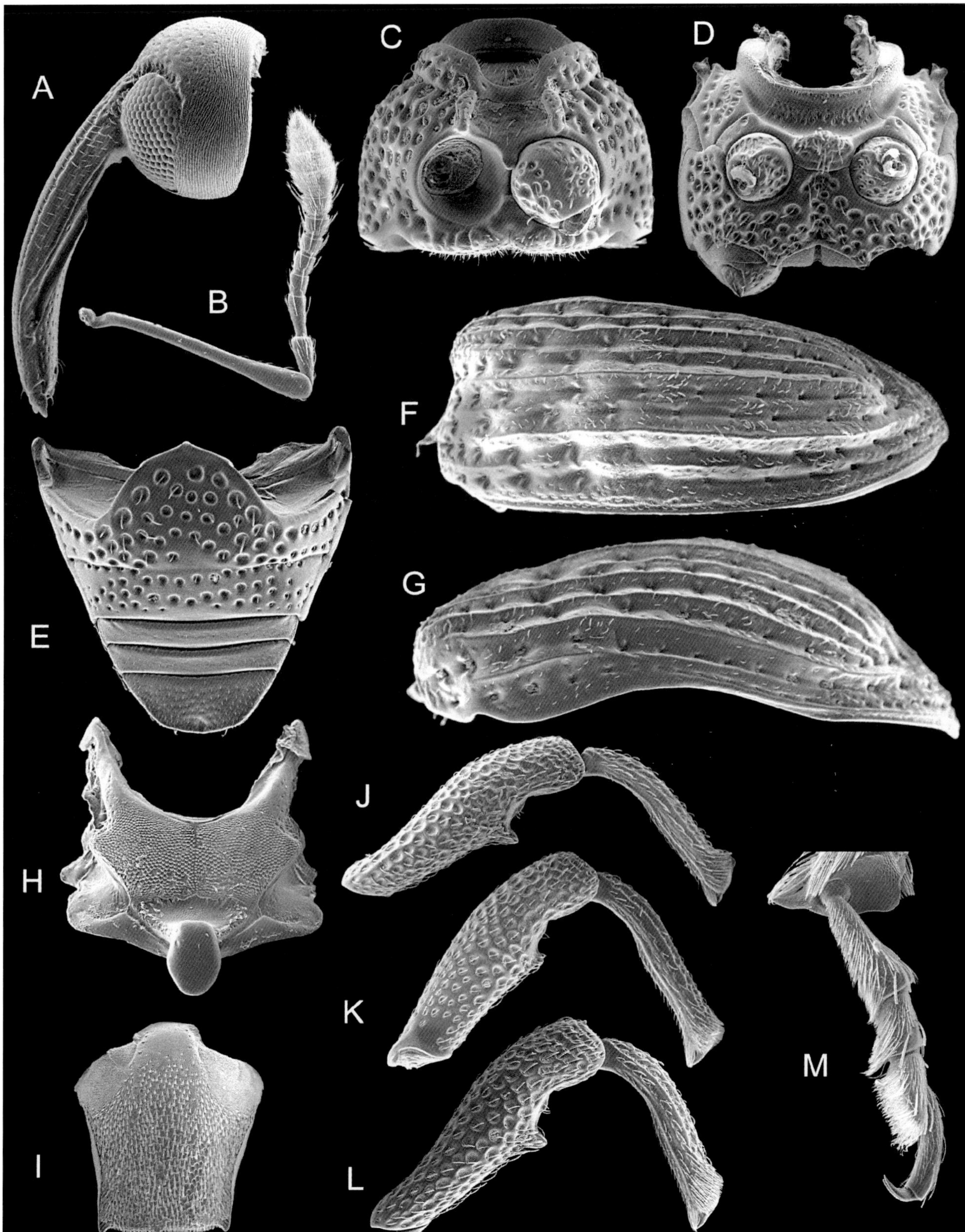

Fig. 44.102. Cleogonini, structural details, *Arthriticosoma vigilans* Lea. A, head, lateral; B, antenna; C, prothorax, ventral; D, pterothorax, ventral; E, abdominal ventrites; F–G, elytron: F, dorsal; G, lateral; H, scutellum; I, pygidium; J, foreleg; K, middle leg; L, hindleg; M, tarsus, lateral.

Figs 44.103. Cleogonini, structural details, *Cycloporopterus mysticus* Lea, ♂. A–B, head: A, dorsal; B, lateral; C, antenna; D–F, elytron: D, dorsal; E, lateral; F, inner lateral; G, abdominal ventrites; H–I, prothorax: H, lateral; I, ventral; J–K, pterothorax: J, ventral; K, lateral; L, pygidium; M, scutellum; N, foreleg; O, middle leg; P, hindleg; Q–R, penis: Q, dorsal; R, lateral.

Acrotychreus Lea, 1909 (Pl. 83D; Fig. 44.109)

Type species, by monotypy: *Tychreus fasciculatus* Lea, 1895.
Australian species. *A. fasciculatus* (Lea, 1895b).
Distribution. Australia (NSW, QLD).
Reference. Lea (1909d).

Aethreus Pascoe, 1875 (Pl. 83E)

Type species, by monotypy: *Aethreus cicatricosus* Pascoe, 1875.
Australian species. *A. cicatricosus* Pascoe, 1875.
Distribution. Australia (Lord Howe Is.).
Notes. Specimens have been collected on Kentia Palm (*Howea forsteriana*, Arecaceae).

Agathicis Lea, 1913 (Pl. 83F)

Type species, by monotypy: *Agathicis distinctus* Lea, 1913.
Australian species. *A. distinctus* Lea, 1913b.
Distribution. Australia.

Agenopus Pascoe, 1872 (Pl. 83G; Fig. 44.110)

Type species, by monotypy: *Agenopus agricola* Pascoe, 1872.
Australian species. *A. agricola* Pascoe, 1872b.
Distribution. Australia.

Alatidotasia Lea, 1910 (Pl. 83H)

Type species, by monotypy: *Alatidotasia rubriventris* Lea, 1910.
Australian species. *A. elliptica* Lea, 1913b, *A. exclamationis* (Fabricius, 1775), *A. rubriventris* Lea, 1910b.
Distribution. Australia (QLD).

Albius Zimmerman, 1994 (Pl. 95A, B)

Type species, by original designation: *Coelosternus humerifer* Boheman, 1844.
Australian species. *A. humerifer* (Boheman, 1844) (*in* Schoenherr 1844).
Distribution. Australia.

Amleanus Marshall, 1936 (Pl. 83I)

Type species, by original designation: *Tepperia bicrucicollis* Lea, 1928.
Australian species. *A. bicrucicollis* (Lea, 1928a).
Distribution. Australia.

Amorphocis Lea, 1913 (Pl. 95C)

Type species, by original designation: *Amorphocis mirus* Lea, 1913.
Australian species. *A. mirus* Lea, 1913a.
Distribution. Australia.

Ampagia Pascoe, 1870 (Pl. 83J, K; Fig. 44.111A–H)

Type species, by monotypy: *Ampagia erinacea* Pascoe, 1870.
Australian species. *A. abdominalis* (Pascoe, 1871c), *A. alata* Lea, 1913c, *A. cognata* Lea, 1913c, *A. erinacea* Pascoe, 1870d, *A. femoralis* (Erichson, 1842), *A. hieroglyphica* Lea, 1929c, *A. hystricosa* Lea, 1913b, *A. leucomela* Lea, 1929c, *A. montivaga* (Olliff, 1889), *A. nigrinasus* (Chevrolat, 1881a), *A. obscura* Lea, 1929c, *A. squamibunda* Lea, 1929c, *A. squamigera* (Olliff, 1889), *A. tarsalis* (Lea, 1913a), *A. v-album* Lea, 1929c.
Distribution. Australia (including Lord Howe Is.), Malay Archipelago, Melanesia, New Zealand.

Ampagiosoma Lea, 1913 (Pl. 95D)

Type species, by monotypy: *Ampagiosoma convexum* Lea, 1913.
Australian species. *A. convexum* Lea, 1913c.
Distribution. Australia.

Ancocis Lea, 1913 (Pl. 95E, F)

Type species, by monotypy: *Ancocis collaris* Lea, 1913.
Australian species. *A. collaris* Lea, 1913a.
Distribution. Australia.

Andracalles Kuschel, 1982 (Pl. 83L)

Type species, by original designation: *Acalles spurcus* Broun, 1881.
Australian species. *A. fasciculatus* (Lea, 1928b), *A. norfolcensis* (Lea, 1928b).
Distribution. Australia (including Norfolk Is., Lord Howe Is.), Pacific region, New Zealand.

Anepigraphocis Lea, 1905 (Pl. 95G)

Type species, by monotypy: *Anepigraphocis basiventris* Lea, 1905.
Australian species. *A. basiventris* Lea, 1905b.
Distribution. Australia (WA).

Anilaus Pascoe, 1870 (Pl. 95H)

Type species, by monotypy: *Anilaus sordidus* Pascoe, 1870.
Australian species. *A. sordidus* Pascoe, 1870e.
Distribution. Australia (QLD), New Caledonia, Pacific Islands.

Anoplocis Lea, 1913 (Pl. 95I, J)

Type species, 1913 by monotypy: *Anoplocis ferrugineus* Lea, 1913.
Australian species. *A. ferrugineus* Lea, 1913a.
Distribution. Australia (NSW).

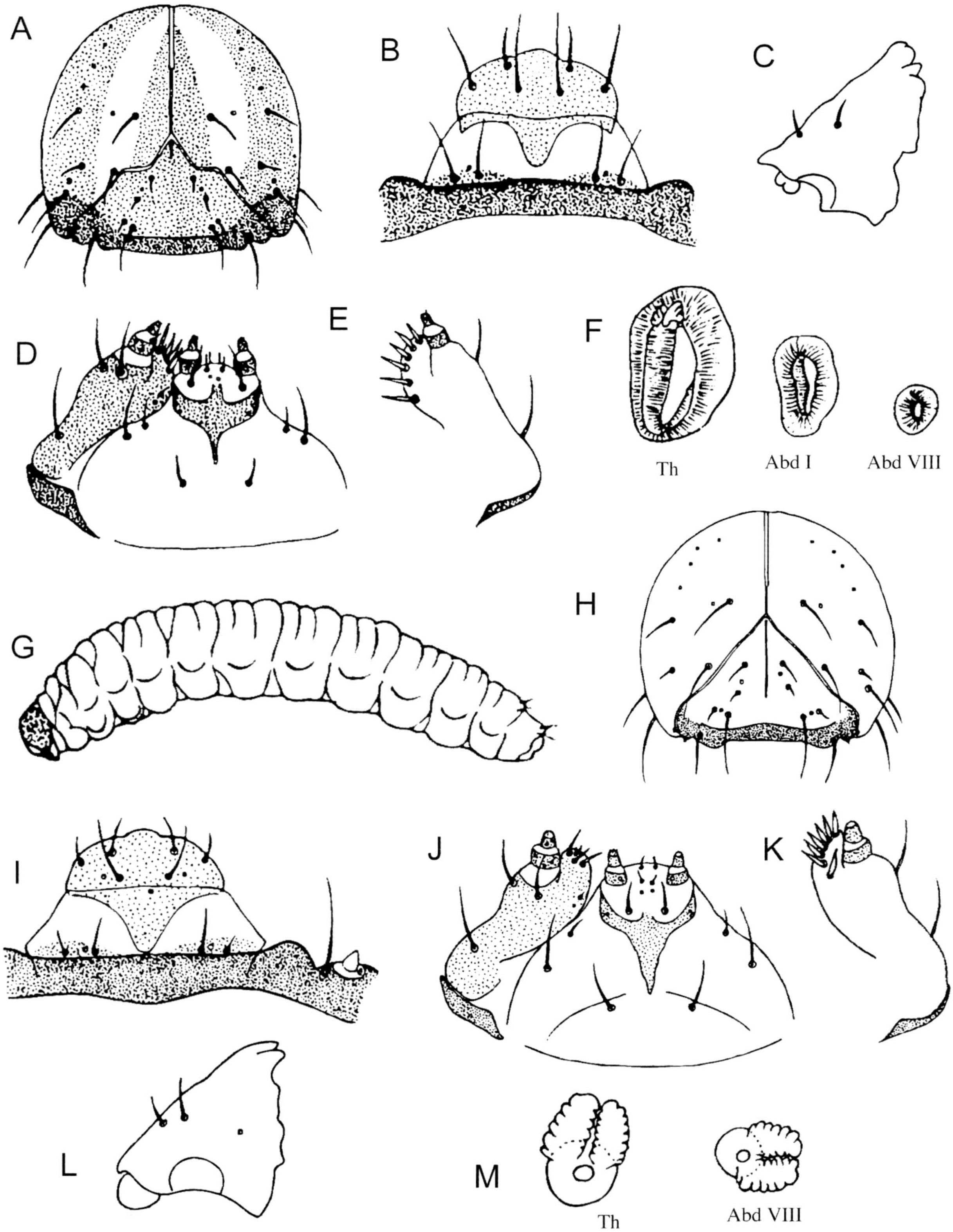

Fig. 44.104. Cryptorhynchini, structural details of immature stages. A–F, *Enteles vigorsii* Gyllenhal, larva: A, head, dorsal; B, labrum, clypeus and epistoma; C, mandible; D, mala and labium, ventral; E, mala, dorsal; F, spiracles, showing alignment; G–M, *Euscepes batatae* (Waterhouse), larva (Fiji): G, habitus, lateral; H, head, dorsal; I, labrum, clypeus and epistoma; J, mala and labium, ventral; K, mala, dorsal; L, mandible; M, spiracles, showing alignment. After May (1994).

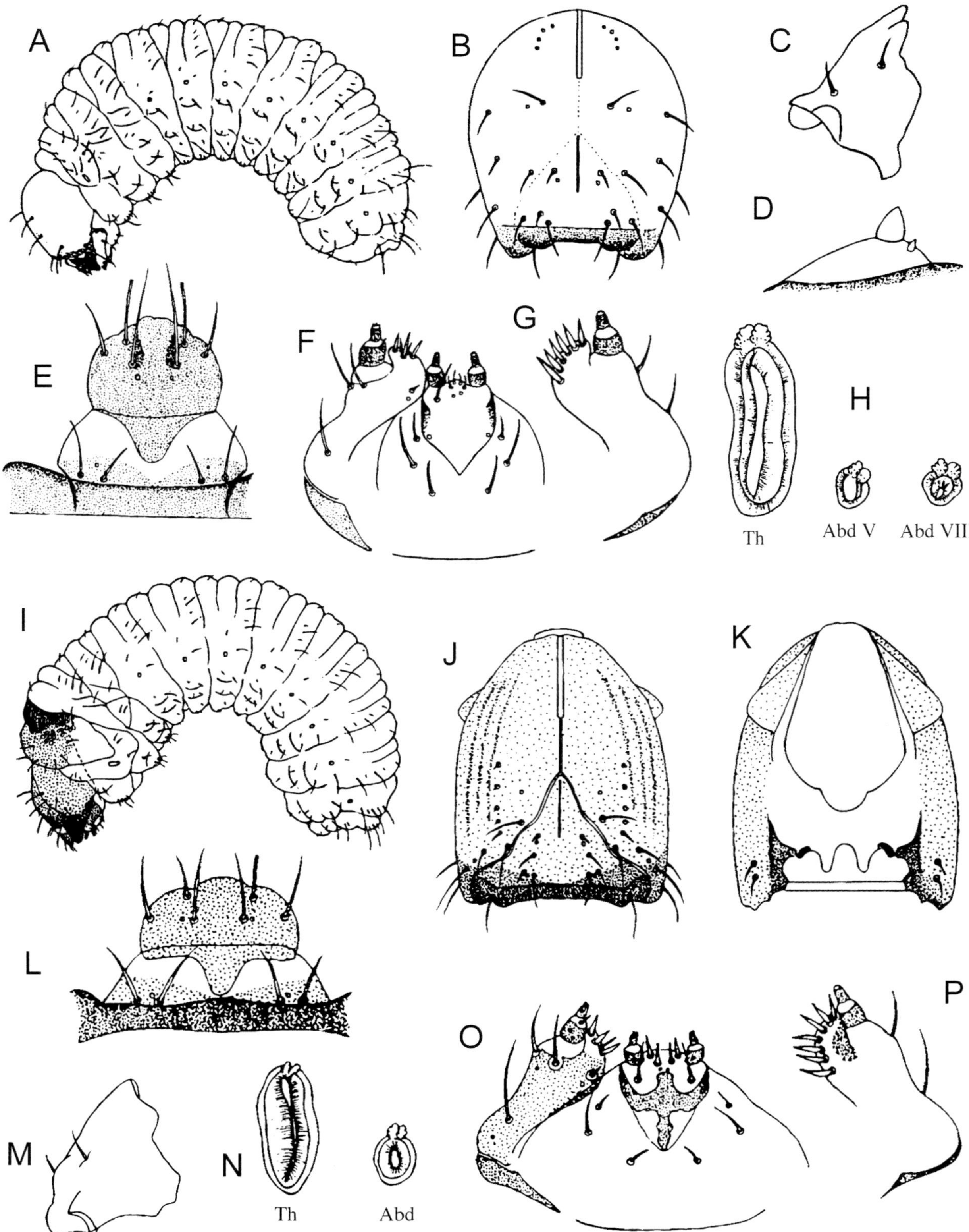

Fig. 44.105. Cryptorhynchini, structural details of immature stages. A–H, *Mitrastethus australiae* Lea, larva (Papua New Guinea): A, habitus, lateral; B, head, dorsal; C, mandible; E, labrum, clypeus and epistoma; F, mala and labium, ventral; G, mala, dorsal; H, spiracles, showing alignment; I–P, *Perissops variegatus* Lea, larva: I, habitus, lateral; J, head, dorsal; K, head, ventral; L, labrum, clypeus and epistoma; M, mandible; N, spiracles, showing alignment; O, mala and labium, ventral; P, mala, dorsal. After May (1994).

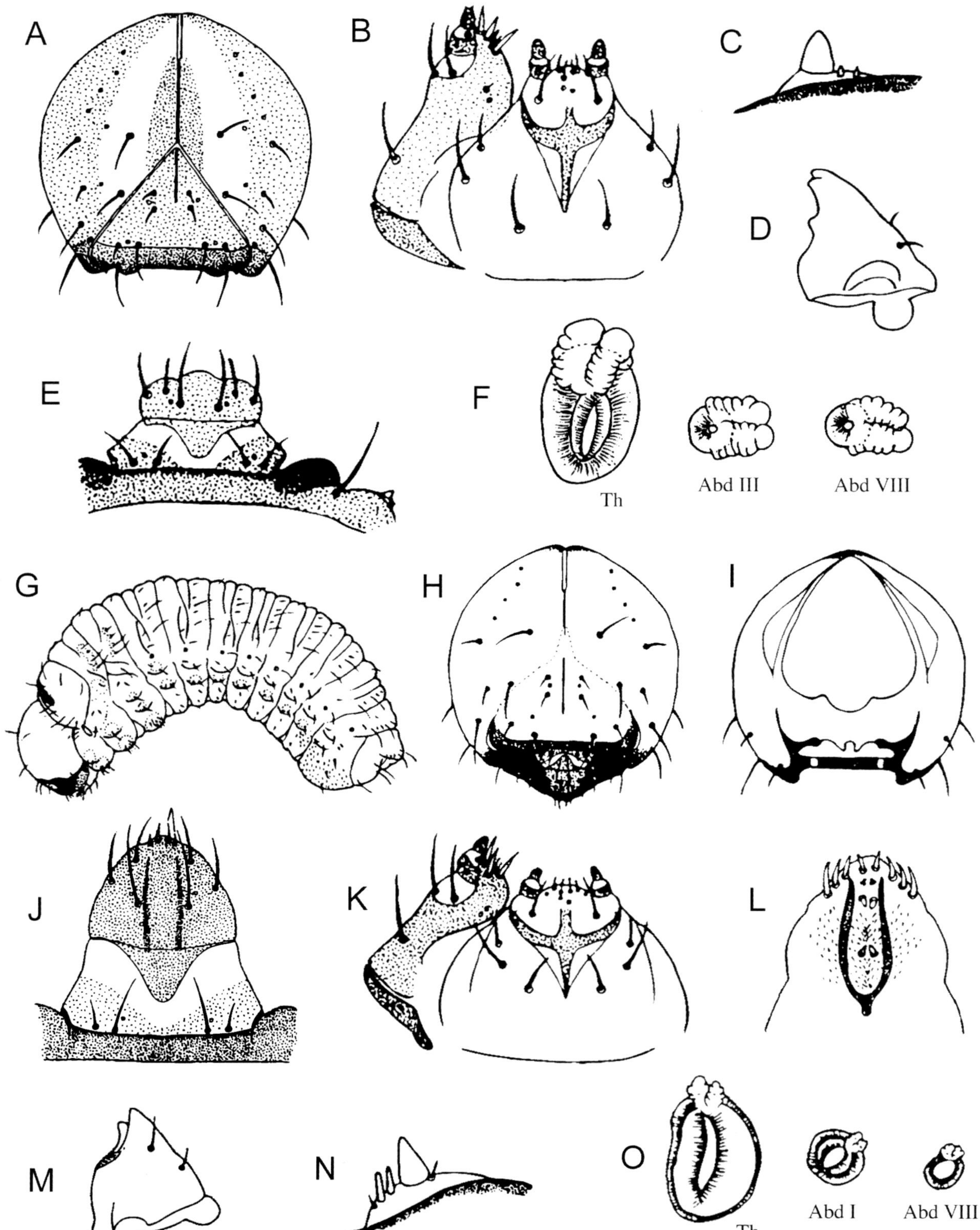

Fig. 44.106. Cryptorhynchini, structural details of immature stages. A–F, *Phloeoglymma dorsale* (Pascoe), larva: A, head, dorsal; B, mala and labium, ventral; C, antenna; D, mandible; E, labrum, clypeus, epistoma and antenna; F, spiracles, showing alignment; G–O, *Psepholax sulcatus* White, larva (New Zealand): G, habitus, lateral; H, head, dorsal; I, head, ventral; J, labrum, clypeus and epistoma; K, mala and labium, ventral; L, epipharyngeal lining; M, mandible; N, antenna; O, spiracles, showing alignment. After May (1994).

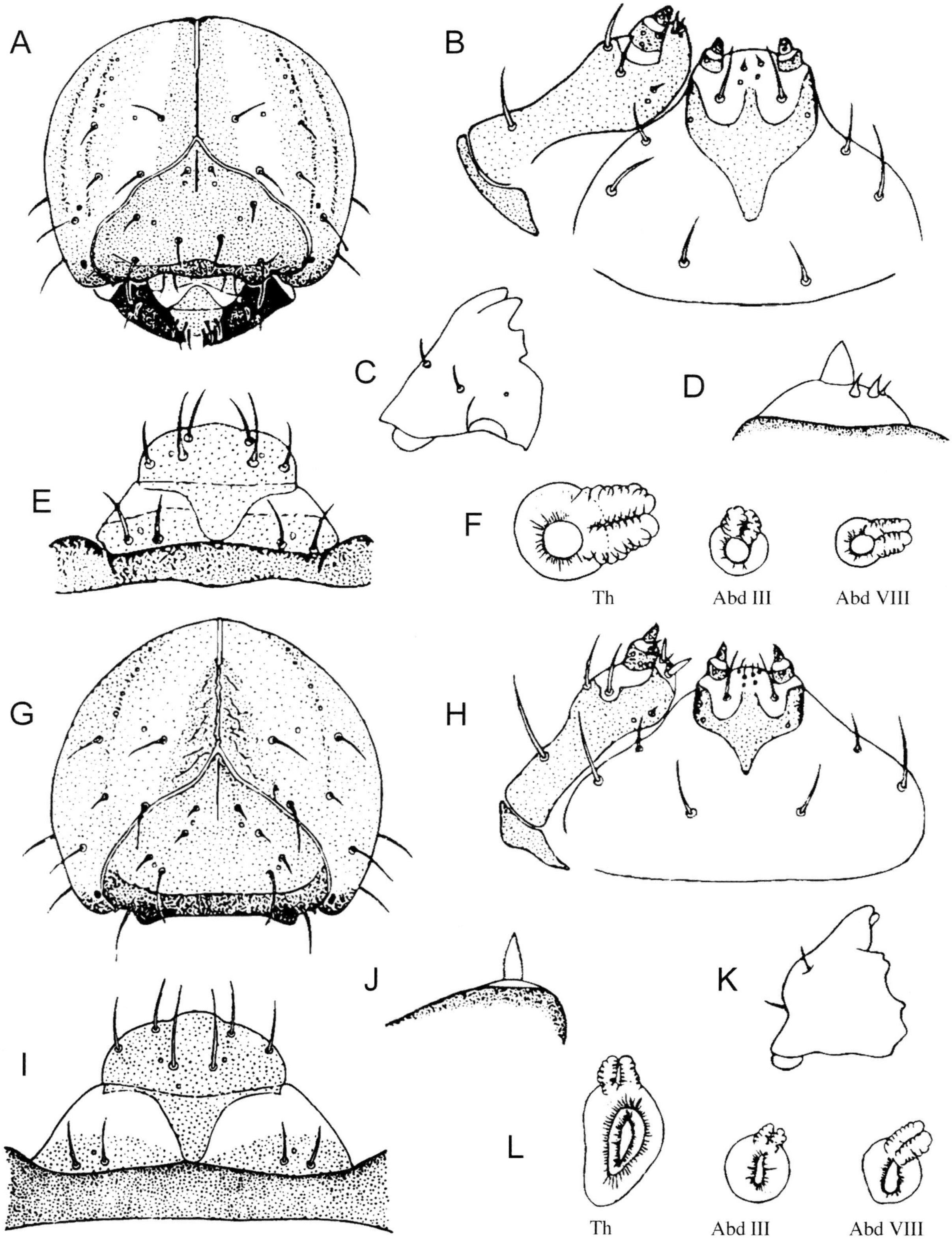

Fig. 44.107. Cryptorhynchini, structural details of immature stages. A–F, *Sternochetus mangiferae* (Fabricius), larva: A, head, dorsal; B, mala and labium, ventral; C, mandible; D, antenna; E, labrum, clypeus and epistoma; F, spiracles, showing alignment; G–L, *Tepperia sterculiae* Lea, larva: G, head, dorsal; H, mala and labium, ventral; I, labrum, clypeus and epistoma; J, antenna; K, mandible; L, spiracles, showing alignment. After May (1994).

Athanasius Zimmerman, 1994 (Pl. 84A)

Type species, by original designation: *Neomystocis latipennis* Lea, 1915.
Australian species. *A. latipennis* (Lea, 1915d).
Distribution. Australia (QLD), New Guinea.

Athyreocis Lea, 1913 (Pl. 95K, L)

Type species, by subsequent designation (Pullen *et al.* 2014: 234): *Athyreocis tarsalis* Lea, 1913b.
Australian species. *A. albonotatus* Lea, 1913b, *A. echinatus* Lea, 1913b, *A. nigronotatus* Lea, 1913b, *A. semicircularis* Lea, 1929a, *A. tarsalis* Lea, 1913b, *A. tibialis* Lea, 1913b, *A. triangulifer* Lea, 1929a.
Distribution. Australia (QLD).

Atragopus Zimmerman & Oberprieler, 2014 (Pl. 84B)

Type species, by original designation: *Anchithyrus caliginosus* Lea, 1912.
Australian species. *A. caliginosus* (Lea, 1912), *A. reticulatus* (Lea, 1912).
Distribution. Australia (QLD).
Notes. The genus was described in Pullen *et al.* (2014).

Austrectopsis Lea, 1912 (Pl. 96A, B)

Type species, by monotypy: *Austrectopsis oblonga* Lea, 1912.
Australian species. *A. oblonga* Lea, 1912.
Distribution. Australia (QLD).

Axionicus Pascoe, 1870 (Pl. 84C)

Type species, by monotypy: *Axionicus insignis* Pascoe, 1870.
Australian species. *A. insignis* Pascoe, 1870b.
Distribution. Australia (QLD, NSW, VIC).
Notes. The single species is associated with *Brachychiton* species (Malvaceae), which has given it its common name of Kurrajong Weevil.

Baeodontocis Lea, 1913 (Pl. 84D; Fig. 44.111I–O)

Type species, by monotypy: *Baeodontocis megapholus* Lea, 1913.
Australian species. *B. megapholus* Lea, 1913a.
Distribution. Australia (NSW).

Bepharus Pascoe, 1870 (Pl. 84E; Fig. 44.112)

Type species, by monotypy: *Bepharus ellipticus* Pascoe, 1870.
Australian species. *B. ellipticus* Pascoe, 1870d.
Distribution. Australia (QLD).
Notes. The single species has been recorded from *Araucaria cunninghamii* (Araucariaceae).

Blepiarda Pascoe, 1865 (Pl. 84F)

Type species, by subsequent designation (Zimmerman 1939: 325): *Blepiarda undulata* Pascoe, 1865.
Australian species. *B. undulata* Pascoe, 1865, *B. vitiata* Pascoe, 1871c.
Distribution. Australia (QLD, NSW, VIC), Malay Archipelago, Fiji, New Guinea, Solomon Is.
Notes. A recorded host plant is *Polyscias elegans* (Araliaceae).

Bleptocis Lea, 1905 (Pl. 84G)

Type species, by monotypy: *Bleptocis collaceratus* Lea, 1905.
Australian species. *B. collaceratus* Lea, 1905b.
Distribution. Australia (QLD, NSW, VIC).

Bothynacrum Lea, 1902 (Pl. 92F)

Type species, by monotypy: *Bothynacrum storeoides* Lea, 1902.
Australian species. *B. storeoides* Lea, 1902.
Distribution. Australia (QLD, NSW).

Brachyphyes Marshall, 1939 (Pl. 84H; Fig. 44.113)

Type species, by monotypy: *Brachycis thoracicus* Lea, 1913.
Australian species. *B. thoracicus* (Lea, 1913b).
Distribution. Australia (QLD, NSW).
Reference. Marshall (1939a).

Brachyporopterus Lea, 1908 (Pl. 84I; Fig. 44.114)

Type species, by monotypy: *Brachyporopterus apicigriseus* Lea, 1908.
Australian species. *B. apicigriseus* Lea, 1908c, *B. montanus* Lea, 1913a, *B. vermiculatus* Lea, 1912.
Distribution. Australia (QLD, NSW).

Cairnsicis Lea, 1912 (Pl. 84J)

Type species, by monotypy: *Cairnsicis opalescens* Lea, 1912.
Australian species. *C. opalescens* Lea, 1912.
Distribution. Australia (QLD).

Cardiopterocis Lea, 1913 (Pl. 96C, D)

Type species, by original designation: *Cardiopterocis variegatus* Lea, 1913.
Australian species. *C. variegatus* Lea, 1913b.
Distribution. Australia (QLD).

Chaetectetorus Schoenherr, 1844 (Pl. 84K; Fig. 44.115)

Type species, by original designation: *Gasterocercus bifasciatus* Boheman, 1837.

Australian species. *C. bifasciatus* (Boheman, 1837) (*in* Schoenherr 1837), *C. cinerascens* Lea, 1913a, *C. clitellae* Pascoe, 1870e, *C. egenus* Lea, 1909d, *C. haedulus* Pascoe, 1870e, *C. latus* Pascoe, 1870e, *C. setosus* (Boheman, 1837) (*in* Schoenherr 1837).
Distribution. Australia (QLD, SA, VIC, WA), New Guinea, Pacific region.

Chimades Pascoe, 1870 (Pl. 84L; Fig. 44.116)

Type species, by monotypy: *Chimades lanosus* Pascoe, 1870.
Australian species. *C. lanosus* Pascoe, 1870e.
Distribution. Australia (NSW, VIC).

Coptocelis Chevrolat, 1881

Type species, by monotypy: *Gasterocercus nigroaeneus* Chevrolat, 1881.
Australian species. *C. nigroaenea* (Chevrolat, 1881b).
Distribution. Australia.
Notes. The identity of the single Australian species and the occurrence of the genus in the country are unclear (Pullen *et al.* 2014).

Cratomerocis Lea, 1913 (Pl. 85A)

Type species, by monotypy: *Cratomerocis flavonotatus* Lea, 1913.
Australian species. *C. flavonotatus* Lea, 1913b.
Distribution. Australia (QLD).

Critomerus Lea, 1903 (Pl. 85B)

Type species, by monotypy: *Critomerus emblematicus* Lea, 1903.
Australian species. *C. iliacus* (Pascoe, 1871c).
Distribution. Australia (QLD), Malay Archipelago.

Cyphoderocis Lea, 1913 (Pl. 85C)

Type species, by monotypy: *Cyphoderocis dorsalis* Lea, 1913.
Australian species. *C. dorsalis* Lea, 1913a.
Distribution. Australia (QLD).

Decilaus Pascoe, 1870 (Pl. 85D, E; Figs 44.117–44.121)

Type species, by monotypy: *Decilaus squamosus* Pascoe, 1870.
Australian species. *D. abdominalis* Lea, 1930b, *D. acerosus* (Erichson, 1842), *D. albohumeralis* Lea, 1930b, *D. albonotatus* Lea, 1913a, *D. ambiguus* (Lea, 1912), *D. apicatus* Lea, 1905a, *D. auricomus* Lea, 1905a, *D. bifurcatus* Lea, 1913b, *D. bryophilus* Lea, 1913a, *D. calviceps* Lea, 1913b, *D. cribricollis* (Pascoe, 1874b), *D. cuniculosus* Lea, 1905a, *D. curvipes* Lea, 1913b, *D. curvirostris* Lea, 1930b, *D. distans* (Pascoe, 1874b), *D. episternalis* Lea, 1930b, *D. erythromelas* Lea, 1930b, *D. erythropholus* Lea, 1930b, *D. expletus* (Pascoe, 1874b), *D. foraminosus* (Pascoe, 1874b), *D. foveiventris* Lea, 1898c, *D. hispidus* Lea, 1898c, *D. humeralis* Lea, 1913b, *D. hystricosus* Lea, 1913a, *D. imparipunctatus* (Rheinheimer, 1993), *D. infaustus* (Pascoe, 1874b), *D. irrasus* Lea, 1905a, *D. lateralis* Lea, 1913a, *D. lineifer* Lea, 1930b, *D. litoralis* Lea, 1898c, *D. longirostris* Lea, 1913a, *D. major* Lea, 1908c, *D. medioalbus* Lea, 1930b, *D. megapholus* Lea, 1930b, *D. memnonius* (Pascoe, 1874b), *D. mirabilis* Lea, 1913b, *D. mixtus* Lea, 1908c, *D. mollis* Lea, 1908c, *D. moluris* Lea, 1898c, *D. nigriclavus* Lea, 1913a, *D. nigrohumeralis* Lea, 1930b, *D. nigronotatus* Lea, 1913a, *D. nitidirostris* Lea, 1930b, *D. noctivagus* Lea, 1905a, *D. nucleatus* (Pascoe, 1874b), *D. ordinarius* Lea, 1913b, *D. ovatus* Lea, 1905a, *D. parvidens* Lea, 1930b, *D. perditus* (Pascoe, 1874b), *D. ruficornis* Lea, 1930b, *D. semicalviceps* Lea, 1930b, *D. seriatopunctatus* Lea, 1909c, *D. sobrinus Lea, 1908c*, *D. spissus* Lea, 1905a, *D. squamosus* Pascoe, 1870d, *D. striatus* Lea, 1913a, *D. subfasciculatus* Lea, 1930b, *D. subterraneus* Lea, 1898c, *D. suturaelevata* (Rheinheimer, 1993), *D. suturalis* Lea, 1913a, *D. triangulifer* Lea, 1930b, *D. variegatus* Lea, 1913a, *D. wilsoni* Lea, 1930b, *D. xanthorrhoeae* Lea, 1898c.
Distribution. Australia (NSW, QLD, VIC, SA, TAS, WA).
Notes. The genus is not demonstrably monophyletic but likely an artificial conglomerate of various small wingless Australian cryptorhynchines (Pullen *et al.* 2014). Specimens are commonly collected from leaf litter as well as moss.

Diaphorocis Lea, 1913 (Pl. 96E)

Type species, by monotypy: *Diaphorocis variegatus* Lea, 1913a.
Australian species. *D. variegatus* Lea, 1913.
Distribution. Australia (QLD).

Didorus Zimmerman & Oberprieler, 2014 (Pl. 96F)

Type species, by original designation: *Cryptorhynchus pictifrons* Lea, 1911.
Australian species. *D. pictifrons* (Lea, 1911b).
Distribution. Australia.
Notes. The genus was described in Pullen *et al.* (2014).

Drassicus Pascoe, 1872 (Pl. 85F)

Type species, by subsequent designation (Heller, 1916: 324): *Drassicus nigricornis* Pascoe, 1872.
Australian species. *D. illotus* Pascoe, 1872b, *D. nigricornis* Pascoe, 1872b.
Distribution. Australia (QLD).

Dysopirhinus **Roelofs, 1880 (Pl. 85G)**

Type species, by monotypy: *Dysopirhinus gestroi* Roelofs, 1880.
Australian species. *D. grandis* Lea, 1903, *D. quadrinotatus* Lea, 1913a.
Distribution. Australia (QLD), New Guinea.

Ecildaus **Lea, 1912 (Pl. 85H)**

Type species, by subsequent designation (Pullen *et al.* 2014: 240): *Ecildaus personatus* Lea, 1912.
Australian species. *E. bassiae* (Marshall, 1936), *E. glabricornis* Lea, 1912, *E. inconstans* (Lea, 1913a), *E. melancholicus* Lea, 1912, *E. personatus* Lea, 1912.
Distribution. Australia (NSW, QLD).

Elaeagna **Pascoe, 1870 (Pl. 85I; Fig. 44.122)**

Type species, by monotypy: *Elaeagna squamibunda* Pascoe, 1870.
Australian species. *E. bassiae* Marshall, 1934, *E. doddi* Marshall, 1934, *E. humeralis* Marshall, 1934, *E. inepta* Lea, 1898c, *E. nodipennis* Lea, 1913a, *E. squamibunda* Pascoe, 1870e, *E. variolaris* Lea, 1898c.
Distribution. Australia (NSW, NT, QLD, VIC, WA).
Notes. Recorded hostplants of the genus are *Bassia* and *Atriplex* species (Amaranthaceae).

Embaphiodes **Pascoe, 1874 (Pl. 85J)**

Type species, by monotypy: *Embaphiodes pyxidatus* Pascoe, 1874.
Australian species. *E. pyxidatus* Pascoe, 1874b.
Distribution. Australia (Lord Howe Is.).
Notes. Specimens have been reared from wood and the Kentia Palm (*Howea forsteriana*, Arecaceae) has been recorded as a host plant.

Emydica **Pascoe, 1885 (Pl. 85K)**

Type species, by monotypy: *Emydica platynota* Pascoe, 1885.
Australian species. *E. brevisetosa* Lea, 1913b.
Distribution. Australia (QLD), New Guinea.

Enteles **Schoenherr, 1837 (Pl. 85L; Figs 44.104A–F, 44.123A–J)**

Type species, by original designation: *Enteles vigorsii* Gyllenhal, 1837.
Australian species. *E. vicinus* Faust, 1888, *E. vigorsii* Gyllenhal, 1837 (*in* Schoenherr 1837).
Distribution. Australia (NSW, QLD).
Notes. Specimens of *E. vigorsii* have been reared from branches of *Flindersia* (Rutaceae) and several other plants as well as from cycad cones.

Ephrycinus **Lea, 1909 (Pl. 86A)**

Type species, by monotypy: *Ephrycinus pilistriatus* Lea, 1909.
Australian species. *E. pilistriatus* Lea, 1909d.
Distribution. Australia (Lord Howe Is.).
Notes. Kentia Palm (*Howea forsteriana,* Arecaceae) is recorded as a host plant of the single species.

Ephrycus **Pascoe, 1870 (Pl. 86B)**

Type species, by monotypy: *Ephrycus obliquus* Pascoe, 1870.
Australian species. *E. brachystylus* Lea, 1909d, *E. erythraeus* Lea, 1909d, *E. minor* Lea, 1913b, *E. obliquus* Pascoe, 1870e, *E. parvus* Lea, 1908c, *E. variabilis* Lea, 1913a.
Distribution. Australia (QLD, NSW, TAS, VIC, WA).
Notes. Host plants of the genus include *Eucalyptus* species (Myrtaceae) such as White Gum (*E. viminalis*) and stringybarks (Tepper 1887).

Episodiocis **Lea, 1905 (Pl. 86C)**

Type species, by original designation: *Episodiocis microderes* Lea, 1905.
Australian species. *E. inconstans* Lea, 1913b, *E. microderes* Lea, 1905b.
Distribution. Australia (NSW, QLD).

Ethocis **Lea, 1913 (Pl. 86D)**

Type species, by subsequent designation (Pullen *et al.* 2014: 242): *Ethocis discicollis* Lea, 1913.
Australian species. *E. bifasciatus* Lea, 1913a, *E. discicollis* Lea, 1913a.
Distribution. Australia (NSW, TAS).

Eucalyptocis **Lea, 1913 (Pl. 96G)**

Type species, by monotypy: *Eucalyptocis fasciculatus* Lea, 1913.
Australian species. *E. fasciculatus* Lea, 1913a.
Distribution. Australia (TAS).
Notes. Specimens have been collected under bark of *Eucalyptus* species (Myrtaceae).

Eufaustia **Lea, 1912 (Pl. 86E)**

Type species, by monotypy: *Eufaustia mirabilis* Lea, 1912.
Australian species. *E. mirabilis* Lea, 1912.
Distribution. Australia (NSW).

Euoropis Pascoe, 1874 (Pl. 96H)

Type species, by monotypy: *Euoropis castanea* Pascoe, 1874.
Australian species. *E. castanea* Pascoe, 1874b.
Distribution. Australia (WA).

Eurycis Lea, 1913 (Pl. 96I, J)

Type species, by monotypy: *Eurycis multinodosus* Lea, 1913.
Australian species. *E. multinodosus* Lea, 1913b.
Distribution. Australia.

Eurymetopocis Lea, 1913 (Pl. 96K, L)

Type species, by monotypy: *Eurymetopocis bilobus* Lea, 1913.
Australian species. *E. bilobus* Lea, 1913b.
Distribution. Australia (QLD).

Euryporopterus Lea, 1908 (Pl. 86F; Fig. 44.123K–Q)

Type species, by subsequent designation (Pullen *et al.* 2014: 243): *Euryporopterus funereus* Lea, 1908.
Australian species. *E. annulipes* (Pascoe, 1874b), *E. cryptodermus* Lea, 1913b, *E. funereus* Lea, 1908b, *E. invidus* (Pascoe, 1870e), *E. tenuifasciatus* Lea, 1912.
Distribution. Australia (NSW, QLD, SA, VIC).

Eutinobothrus Faust, 1896 (Pl. 86H)

Type species, by original designation: *Cryptorhynchus pilosellus* Boheman, 1844.
Australian species. *E. pilosellus* (Boheman, 1844) (*in* Schoenherr 1844).
Distribution. Australia, Mexico, Central and South America.
Notes. The single species in Australia was accidentally introduced from Mexico between 1997 and 1999 for the control of the weed *Sida* species (Malvaceae), but its establishment in the country is in need of confirmation (Palmer *et al.* 2010; Pullen *et al.* 2014).

Eutyrhinus Dejean, 1835 (Pl. 86G; Fig. 44.124)

Type species, by original designation: *Curculio meditabundus* Fabricius, 1775.
Australian species. *E. meditabundus* (Fabricius, 1775), *E. spinipennis* (Waterhouse, 1853).
Distribution. Australia (QLD, NSW, NT, VIC, Lord Howe and Norfolk Is.), Malay Archipelago, New Guinea, Solomon Is., New Zealand.
Notes. Twenty-nine species in 18 families have been recorded as host plants of the variable and widespread *E. meditabundus*, including *Araucaria cunninghamii* (Araucariaceae), *Ficus macrophylla* (Moraceae), *Grevillea* and *Banksia* (Proteaceae) and *Sterculia diversifolia* (Malvaceae) (Hawkeswood 1991). The genus was re-described by Lyal (1993) with notes on the biology of *E. squamiger* White.

Evaniocis Lea, 1913 (Pl. 86J)

Type species, by subsequent designation (Pullen *et al.* 2014: 243): *Evaniocis ellipticus* Lea, 1913.
Australian species. *E. albicans* Lea, 1913a, *E. ellipticus* Lea, 1913a.
Distribution. Australia (QLD).

Exithioides Lea, 1912 (Pl. 97A, B)

Type species, by monotypy: *Exithioides punctatus* Lea, 1912.
Australian species. *E. punctatus* Lea, 1912.
Distribution. Australia (NSW).

Exithius Pascoe, 1870 (Pl. 86I; Fig. 44.125)

Type species, by monotypy: *Exithius capucinus* Pascoe, 1870d.
Australian species. *E. athyreus* Lea, 1913a, *E. auchmeresthes* Lea, 1913b, *E. basipennis* Lea, 1913a, *E. bidentatus* Lea, 1930b, *E. brevis* Lea, 1912, *E. carinicollis* Lea, 1930b, *E. cariosus* (Erichson, 1842), *E. conjunctus* Lea, 1913a, *E. conspiciendus* Lea, 1912, *E. edentatus* Lea, 1930b, *E. ephippiatus* Lea, 1913b, *E. episternalis* Lea, 1930b, *E. ferrugineus* Lea, 1912, *E. fumatus* Lea, 1909a, *E. inamabilis* Lea, 1912, *E. insularis* Lea, 1930b, *E. intermixtus* Lea, 1913a, *E. loculiferus* Lea, 1913a, *E. loculosus* Lea, 1912, *E. megapholus* Lea, 1913a, *E. microps* Lea, 1930b, *E. murinus* Lea, 1930b, *E. musculus* (Pascoe, 1872c), *E. obliquus* Lea, 1913b, *E. obscurus* Lea, 1913b, *E. occidentalis* Lea, 1913a, *E. parvidens* Lea, 1930b, *E. pullatus* Lea, 1930b, *E. sculptilis* Lea, 1912, *E. semicalviceps* Lea, 1930b, *E. simulator* Lea, 1909a, *E. squamosis* Lea, 1930b, *E. stenocerus* Lea, 1913a, *E. tenebrosus* Lea, 1913a, *E. tricarinatus* Lea, 1913a, *E. trisinuatus* Lea, 1913a, *E. tropidopterus* Lea, 1913b.
Distribution. Australia (NSW, VIC, TAS, WA).

Genuacalles Rheinheimer, 1993 (Pl. 86K; Fig. 44.126)

Type species, by original designation: *Genuacalles erectosetosus* Rheinheimer, 1993.
Australian species. *G. erectosetosus* Rheinheimer, 1993, *G. parvoniger* (Lea, 1913a), *G. trivirgatus* (Lea, 1913b).
Distribution. Australia (NSW, QLD).
Notes. Specimens are often collected in leaf litter.

Glochinorhinus Waterhouse, 1853 (Pl. 86L)

Type species, by monotypy: *Glochinorhinus doubledayi* Waterhouse, 1853.
Australian species. *G. doubledayi* Waterhouse, 1853, *G. evanidus* Lea, 1895b.
Distribution. Australia (QLD).
Notes. *Ficus macrophylla* (Moraceae) has been recorded as a host plant of the genus.

Griffithia Lea, 1911 (Pl. 87A)

Type species, by monotypy: *Griffithia anomala* Lea, 1911.
Australian species. *G. anomala* Lea, 1911b.
Distribution. Australia (NSW, TAS).

Gygaeus Pascoe, 1885 (Pl. 87B)

Type species, by monotypy: *Gygaeus prodigus* Pascoe, 1885.
Australian species. *G. prodigus* Pascoe, 1885, *G. squamiventris* Lea, 1913b.
Distribution. Australia (QLD), New Guinea.

Gymnocis Lea, 1931 (Pl. 97C, D)

Type species, by monotypy: *Gymnocis impunctatus* Lea, 1931.
Australian species. *G. impunctatus* Lea, 1931b.
Distribution. Australia (QLD).
Notes. Specimens of the single species have been found under bark of *Araucaria cunninghamii* (Araucariaceae).

Gymnoporopterus Lea, 1912 (Pl. 87C; Fig. 44.127)

Type species, by monotypy: *Gymnoporopterus pictipes* Lea, 1912.
Australian species. *G. pictipes* Lea, 1912.
Distribution. Australia (QLD).

Hexymus Pascoe, 1871 (Pl. 87D)

Type species, by monotypy: *Hexymus tuberosus* Pascoe, 1871.
Australian species. *H. alatus* Lea, 1913a, *H. australis* (Boisduval, 1835), *H. monachus* Pascoe, 1872c, *H. tuberosus* Pascoe, 1871c.
Distribution. Australia (QLD, NSW).

Hoplidotasia Lea, 1928 (Pl. 87E; Fig. 44.97R)

Type species, by monotypy: *Hoplidotasia torresensis* Lea, 1928.
Australian species. *H. torresensis* Lea, 1928a.
Distribution. Australia (QLD).

Hoplodecilaus Lea, 1912 (Pl. 97E, F)

Type species, by monotypy: *Hoplodecilaus marmoratus* Lea, 1912.
Australian species. *H. marmoratus* Lea, 1912.
Distribution. Australia (WA).

Hybomorphus Saunders & Jekel, 1855 (Pl. 87F)

Type species, by original designation: *Hybomorphus melanosomus* Saunders & Jekel, 1855.
Australian species. *H. melanosomus* Saunders & Jekel, 1855.
Distribution. Australia (Lord Howe Is.).
Notes. The single species of the genus is listed as a Species Presumed Extinct in Schedule 1 of the NSW Threatened Species Conservation Act, not having been seen on Lord Howe Is. for the last 150 years (Pullen *et al.* 2014).

Hyparinus Pascoe, 1885 (Pl. 87G)

Type species, by monotypy: *Hyparinus dispar* Pascoe, 1885.
Australian species. *H. brevipes* Lea, 1915d, *H. dispar* Pascoe, 1885, *H. tenuirostris* Lea, 1913a.
Distribution. Australia (QLD, NSW), New Guinea.

Hyperiosoma Lea, 1902 (Pl. 97I)

Type species, by monotypy: *Hyperiosoma falcatum* Lea, 1902.
Australian species. *H. falcatum* Lea, 1902.
Distribution. Australia (QLD).

Hyposcolyphrus Zimmerman & Oberprieler, 2014 (Pl. 87H)

Type species, by original designation: *Scolyphrus semipunctatus* Lea, 1913.
Australian species. *H. semipunctatus* (Lea, 1913d).
Distribution. Australia.
Notes. The genus was described in Pullen *et al.* (2014).

Hypsophorus Dejean, 1835 (Pl. 87K)

Type species, by monotypy: *Cryptorhynchus dromedarius* Boisduval, 1835.
Australian species. *H. cristatus* (Pascoe, 1870b), *H. dromedarius* (Boisduval, 1835), *H. hirticornis* (Lea, 1913b), *H. schoenherri* (Waterhouse, 1853), *H. tectus* (Lea, 1903).
Distribution. Australia (QLD, NSW), New Guinea.

Idiopterocis Lea, 1931 (Pl. 87I; Fig. 44.128)

Type species, by monotypy: *Idiopterocis trilinealbus* Lea, 1931.
Australian species. *I. trilinealbus* Lea, 1931b.

Distribution. Australia (Lord Howe Is.).
Notes. Specimens have been collected from leaf litter.

Imaliodes Pascoe, 1870 (Pl. 87J)

Type species, by subsequent designation (Pullen *et al.* 2014: 248): *Imaliodes subfasciatus* Pascoe, 1870.
Australian species. *I. binodosus* Lea, 1913b, *I. edentatus* Lea, 1910b, *I. frater* Lea, 1912, *I. latipennis* Lea, 1913b, *I. longipes* Lea, 1913b, *I. nodulosus* Pascoe, 1872b, *I. oopterus* Lea, 1913a, *I. ovipennis* Lea, 1912, *I. scitulus* Lea, 1912, *I. scrofa* Pascoe, 1874b, *I. squamirostris* Lea, 1913a, *I. subfasciatus* Pascoe, 1870e, *I. terreus* Pascoe, 1870e, *I. tibialis* (Lea, 1905a), *I. ventralis* Lea, 1913a.
Distribution. Australia (QLD, NSW, WA), Marshall Is., Maluku Is.

Imalithus Pascoe, 1870 (Pl. 87L)

Type species, by monotypy: *Imalithus patella* Pascoe, 1870.
Australian species. *I. patella* Pascoe, 1870c.
Distribution. Australia (QLD, NSW).
Notes. Specimens of the single species have been reared from *Araucaria* (Araucariaceae) in New Guinea.

Isax Pascoe, 1865 (Pl. 88A)

Type species, by monotypy: *Isax gallinago* Pascoe, 1865.
Australian species. *I. gallinago* Pascoe, 1865, *I. planipennis* Lea, 1911b, *I. tricostirostris* Lea, 1931b.
Distribution. Australia (QLD, NSW, Lord Howe Is.).

Leptidotasia Zimmerman, 1994 (Pl. 88B)

Type species, by original designation: *Alatidotasia macrops* Lea, 1913.
Australian species. *L. macrops* (Lea, 1913).
Distribution. Australia (QLD).

Leucothyreocis Lea, 1913 (Pl. 97G, H)

Type species, by original designation: *Leucothyreocis convexus* Lea, 1913.
Australian species. *L. convexus* Lea, 1913b.
Distribution. Australia (QLD).

Lophocheirus Marshall, 1915 (Pl. 88C)

Type species, by original designation: *Odosyllis gemmata* Pascoe, 1885.
Australian species. *L. cruciger* (Pascoe, 1885), *L. laminatus* (Lea, 1913a).
Distribution. Australia (QLD), Malay Archipelago.

Mallus Marshall, 1948 (Pl. 88D)

Type species, by original designation: *Mallus costatus* Marshall, 1948.
Australian species. *M. costatus* Marshall, 1948.
Distribution. Australia (QLD).
Notes. The host plant of the single species is *Araucaria cunninghamii* (Araucariaceae).

Mecryptorhynchus Zimmerman & Oberprieler, 2014 (Pl. 88E)

Type species, by original designation: *Cryptorhynchus stigmaticus* Pascoe, 1870.
Australian species. *M. stigmaticus* (Pascoe, 1870), *M. verus* (Lea, 1902).
Distribution. Australia (QLD).
Notes. The genus was described in Pullen *et al.* (2014).

Meniomorpha Lea, 1908 (Pl. 88F)

Type species, by monotypy: *Meniomorpha inconstans* Lea, 1908.
Australian species. *M. inconstans* Lea, 1908b.
Distribution. Australia (QLD, NT).

Menios Pascoe, 1870 (Pl. 88G)

Type species, by monotypy: *Menios internatus* Pascoe, 1870.
Australian species. *M. albifasciatus* Lea, 1910c, *M. alternatus* Lea, 1913a, *M. ferrugineus* Lea, 1931b, *M. internatus* Pascoe, 1870e, *M. nebulosus* Lea, 1910c, *M. poecilopterus* Lea, 1931b, *M. sinuatus* Lea, 1913b, *M. sordidatus* Lea, 1908c, *M. spurcus* Lea, 1931b.
Distribution. Australia (QLD, NSW, NT, SA, VIC, WA).
Notes. Specimens of *Menios internatus* have been reared from stems of the New South Wales Waratah, *Telopea speciosissima* (Proteaceae).

Mesalcus Zimmerman, 1994 (Pl. 88H)

Type species, by original designation: *Salcus circularis* Lea, 1928.
Australian species. *M. circularis* (Lea, 1928b).
Distribution. Australia (QLD).

Metacymia Pascoe, 1870 (Pl. 88I; Fig. 44.129A–H)

Type species, by monotypy: *Metacymia marmorea* Pascoe, 1870.
Australian species. *M. marmorea* Pascoe, 1870e.
Distribution. Australia (WA).
Notes. The larvae of the single species develop in flowering spikes of *Xanthorrhoea* species (Asphodelaceae).

Methidrysis **Pascoe, 1870 (Pl. 88J)**

Type species, by monotypy: *Methidrysis afflicta* Pascoe, 1870.
Australian species. *M. afflicta* Pascoe, 1870e.
Distribution. Australia (QLD, NSW).

Metraniomorpha **Lea, 1903 (Pl. 88K)**

Type species, by monotypy: *Metraniomorpha entima* Lea, 1903.
Australian species. *M. entima* Lea, 1903.
Distribution. Australia (NSW).
Notes. Specimens of the single species have been reared from seeds of *Austrocallerya australis* (Fabaceae).

Metyrculus **Lea, 1909 (Pl. 89A, B)**

Type species, by monotypy: *Metyrculus bimaculatus* Lea, 1909.
Australian species. *M. bimaculatus* Lea, 1909d, *M. cinerascens* Lea, 1913b, *M. mediofasciatus* Lea, 1913b, *M. mediomaculatus* Lea, 1913b, *M. postscutellaris* Lea, 1931b, *M. semicircularis* Lea, 1931b, *M. sinuatus* Lea, 1913b, *M. sordidus* Lea, 1931b.
Distribution. Australia (QLD, NSW, WA, Norfolk Is.).

Metyrus **Pascoe, 1872 (Pl. 88L; Fig. 44.130)**

Type species, by monotypy: *Metyrus collaris* Pascoe, 1872c.
Australian species. *M. albicollis* (Germar, 1848).
Distribution. Australia (NSW, QLD, VIC, SA, WA).
Notes. The single species is associated with mistletoes of the genus *Amyema* (Loranthaceae), the adults feeding on the leaves and resting on twigs, where they mimic the droppings of the Mistletoebird (*Dicaeum hirundinaceum*) and are sometimes attended by ants, and the larvae developing in the stems (McMillan 1987).

Microcryptorhynchus **Lea, 1908 (Pl. 89C; Fig. 44.131)**

Type species, by monotypy: *Microcryptorhynchus pygmaeus* Lea, 1908.
Australian species. *M. cylindricollis* Lea, 1912, *M. echinatus* Lea, 1912, *M. evanescens* Lea, 1928b, *M. howensis* Lea, 1928b, *M. interruptus* Lea, 1928b, *M. oreas* Lea, 1928b, *M. pygmaeus* Lea, 1908c, *M. rufimanus* Lea, 1928b, *M. rufirostris* Lea, 1928b, *M. setosus* Lea, 1928b.
Distribution. Australia (including Lord Howe and Norfolk Is.), China, Japan, Taiwan, Pacific region, New Zealand.
Notes. The genus was re-described by Lyal (1993) with several notes on the constitution of the genus.

Microporopterus **Lea, 1898 (Pl. 89D; Fig. 44.129J–N)**

Type species, by subsequent designation (Setliff, 2007: 10): *Poropterus tumulosus* Pascoe, 1873a.
Australian species. *M. curvirostris* Lea, 1898c, *M. regularis* Lea, 1898c, *M. tumulosus* (Pascoe, 1873a).
Distribution. Australia (NSW, VIC), New Guinea.

Miotus **Pascoe, 1885 (Pl. 89E, F)**

Type species, by monotypy: *Miotus styphloides* Pascoe, 1885.
Australian species. *M. crux* (Marshall, 1938a), *M. elumbis* (Lea, 1910c), *M. foveicollis* (Lea, 1910c), *M. histrio* (Pascoe, 1885), *M. humeralis* (Lea, 1913a), *M. jubatus* (Lea, 1910c), *M. pediculosus* (Lea, 1909a).
Distribution. Australia (QLD, NSW, WA), New Guinea.

Mitrastethus **Redtenbacher, 1868 (Pl. 89G; Figs 44.105A–H, 44.132)**

Type species, by monotypy: *Mitrastethus baridioides* Redtenbacher, 1868.
Australian species. *M. baridioides* Redtenbacher, 1868, *M. lateralis* Lea, 1931b.
Distribution. Australia (QLD, NSW, Norfolk Is.), New Caledonia, New Guinea, New Zealand.
Notes. Hostplants of *M. baridioides* are various conifers, both native (*Agathis*, *Araucaria*) and, in New Zealand, introduced (*Juniperus*, *Pinus*, *Pseudotsuga*), the larvae tunnelling in the sapwood. *Mitrastethus lateralis* may not belong in this genus, appearing closely related to *Aethreus cicatricosus*. The genus was re-described by Lyal (1993).

Mormosintes **Pascoe, 1865 (Pl. 89H)**

Type species, by monotypy: *Mormosintes rubus* Pascoe, 1865.
Australian species. *M. rubus* Pascoe, 1865.
Distribution. Australia (QLD), New Guinea.

Myrtesis **Pascoe, 1865 (Pl. 89I)**

Type species, by monotypy: *Myrtesis caligata* Pascoe, 1865.
Australian species. *M. caligata* Pascoe, 1865, *M. nasuta* Lea, 1912, *M. pullata* Lea, 1912.
Distribution. Australia (NSW, QLD).

Nechyrus **Pascoe, 1871 (Pl. 89J)**

Type species, by subsequent designation (Heller, 1916: 325): *Nechyrus lemur* Pascoe, 1871c.
Australian species. *N. cribratus* Lea, 1928b, *N. incomptus* Pascoe, 1872b, *N. latipennis* Lea, 1907b, *N. legitimus* Lea, 1907b, *N. mollipes* Lea, 1907b.

Distribution. Australia (NSW, QLD, Lord Howe Is.), Malay Archipelago, New Caledonia.

Neodecilaus Lea, 1912 (Pl. 89K)

Type species, by subsequent designation (Pullen *et al.* 2014: 253): *Neodecilaus picus* Lea, 1912.
Australian species. *N. gratus* Lea, 1912, *N. picus* Lea, 1912.
Distribution. Australia (QLD).

Neomystocis Lea, 1905 (Pl. 89L)

Type species, by monotypy: *Neomystocis squamiventris* Lea, 1905.
Australian species. *N. squamiventris* Lea, 1905b, *N. viridis* Lea, 1913a.
Distribution. Australia (QLD, NSW), New Guinea.

Neoporopterus Faust, 1899 (Pl. 90A; Fig. 44.133A–J)

Type species, by original designation: *Poropterus varicosus* Pascoe, 1873a.
Australian species. *N. lithodermus* (Boisduval, 1835).
Distribution. Australia (NSW).

Niconotus Pascoe, 1870 (Pl. 90B; Fig. 44.133K–T)

Type species, by monotypy: *Niconotus tarphioides* Pascoe, 1870.
Australian species. *N. stenotarsus* Lea, 1913d, *N. tarphioides* Pascoe, 1870e.
Distribution. Australia (QLD, NSW).
Notes. In the type species, *N. tarphioides*, the posterior angles of the pronotum project over the elytral bases, as they do in *Agenopus, Omydaus* and *Pseudomydaus*, whereas in *N. stenotarsus* the basis of the pronotum is not extended in this way (Pl. 90B). In the key to genera above, the couplet keying to *Niconotus* only applies to *N. tarphioides*. Fig. 44.133K–T depicts an undescribed species close to *N. tarphioides*.

Notocalviceps Lea, 1912 (Pl. 90C)

Type species, by subsequent designation (Pullen *et al.* 2014: 254): *Notocalviceps punctipennis* Lea, 1912.
Australian species. *N. punctipennis* Lea, 1912, *N. rarus* Lea, 1912.
Distribution. Australia (QLD, NSW).

Notocryptorhynchus Lea, 1903 (Pl. 90D)

Type species, by monotypy: *Notocryptorhynchus sinuatus* Lea, 1903.
Australian species. *N. sinuatus* Lea, 1903.
Distribution. Australia (QLD), New Guinea.

Nototragopus Heller, 1929 (Pl. 90E, F)

Type species, by monotypy: *Tragopus plagiatus* Pascoe, 1872.
Australian species. *N. plagiatus* (Pascoe, 1872b).
Distribution. Australia (QLD).

Odosyllis Pascoe, 1874 (Pl. 97J)

Type species, by subsequent designation (Pascoe, 1885: 277): *Odosyllis congesta* Pascoe, 1874c.
Australian species. *O. scutellaris* Lea, 1913b.
Distribution. Australia (QLD), New Guinea, Cambodia, Malay Archipelago.

Oemethylus Pascoe, 1870 (Pl. 90G)

Type species, by monotypy: *Oemethylus lumbaris* Pascoe, 1870.
Australian species. *O. lumbaris* Pascoe, 1870e, *O. triangularis* Lea, 1911c.
Distribution. Australia (NSW, QLD).
Notes. Specimens of *O. triangularis* have been reared from *Passiflora herbertiana* (Passifloraceae).

Omydaus Pascoe, 1871 (Pl. 90H)

Type species, by monotypy: *Omydaus plinthoides* Pascoe, 1871c.
Australian species. *O. confusus* Lea, 1912, *O. contractus* Lea, 1912, *O. impressicollis* Lea, 1912, *O. longus* Lea, 1913a, *O. luridus* (Fabricius, 1775), *O. nigrofasciculatus* Lea, 1913a, *O. oblongopunctatus* Lea, 1911c, *O. parviceps* Lea, 1913b, *O. rostralis* Lea, 1913a, *O. sculptilis* Lea, 1913b, *O. subfasciculatus* Lea, 1912, *O. tibialis* Lea, 1913b.
Distribution. Australia (QLD, NSW).

Onidistus Pascoe, 1870 (Pl. 90I; Fig. 44.134)

Type species, by subsequent designation (Heller, 1916: 325): *Onidistus nodipennis* Pascoe, 1870.
Australian species. *O. araneus* Pascoe, 1870e, *O. nodipennis* Pascoe, 1870e, *O. subfornicatus* Lea, 1912.
Distribution. Australia (QLD), New Caledonia.
Notes. The species *O. araneus* has been reared from *Argyrodendron actinophyllum* (Malvaceae).

Ophrythyreocis Lea, 1913 (Pl. 90J)

Type species, by subsequent designation (Pullen *et al.* 2014: 256): *Ophrythyreocis exophthalmus* Lea, 1913.

Australian species. *O. alternatus* Lea, 1930b, *O. cyclothyreus* (Lea, 1913a), *O. exophthalmus* Lea, 1913b, *O. fasciculatus* Lea, 1930b, *O. ferrugineus* Lea, 1913a, *O. microps* Lea, 1913a, *O. niger* Lea, 1930b, *O. tibialis* Lea, 1930b, *O. valgus* (Lea, 1909c), *O. vigilans* Lea, 1930b.
Distribution. Australia (NSW, QLD, SA, TAS, VIC, WA).

Orochlesis Pascoe, 1871 (Pl. 90K; Fig. 44.135)

Type species, by subsequent designation (Lea, 1913a: 327, in footnote): *Orochlesis annularis* Pascoe, 1871.
Australian species. *O. annularis* Pascoe, 1871c, *O. cornuta* Zimmerman, 1936, *O. delta* Lea, *O. inermis* (Lea, 1902), *O. personata* (Pascoe, 1883b).
Distribution. Australia (QLD), Japan, Taiwan, Malay Archipelago, Melanesia.
Notes. The genus was revised by Zimmerman (1936).

Orphanistes Pascoe, 1870 (Pl. 90L)

Type species, by monotypy: *Orphanistes eustictus* Pascoe, 1870b.
Australian species. *O. eustictus* Pascoe, 1870.
Distribution. Australia (QLD).

Orthoporopterus Lea, 1910 (Pl. 91A; Fig. 44.136A–H)

Type species, by monotypy: *Orthoporopterus elongatus* Lea, 1910b.
Australian species. *O. elongatus* Lea, 1910.
Distribution. Australia (QLD).

Osaces Pascoe, 1883 (Pl. 91B)

Type species, by monotypy: *Osaces naso* Pascoe, 1883.
Australian species. *O. naso* Pascoe, 1883b.
Distribution. Australia (QLD).
Notes. The host plant includes *Citrus* (Rutaceae).

Ouroporopterus Lea, 1912 (Pl. 91C)

Type species, by monotypy: *Ouroporopterus diurus* Lea, 1912.
Australian species. *O. diurus* Lea, 1912, *O. squamiventris* Lea, 1913a.
Distribution. Australia (QLD, NSW).

Pachyporopterus Lea, 1913 (Pl. 91D)

Type species, by monotypy: *Poropterus satyrus* Pascoe, 1873.
Australian species. *P. humeralis* Lea, 1913a, *P. satyrus* (Pascoe, 1873a).
Distribution. Australia (TAS).
Reference. Lea (1913d).

Paleticus Pascoe, 1870 (Pl. 91E; Fig. 44.136I–R)

Type species, by subsequent designation (Lea, 1898c: 212): *Paleticus pedestris* Pascoe, 1870.
Australian species. *P. arciferus* Lea, 1898c, *P. basalis* Lea, 1913a, *P. confinis* Pascoe, 1870e, *P. convexicollis* Lea, 1913a, *P. cordipennis* (Pascoe, 1872c), *P. frontalis* Pascoe, 1870e, *P. inflatus* Lea, 1913b, *P. laticollis* Pascoe, 1870e, *P. lutulentus* Lea, 1913b, *P. pedestris* Pascoe, 1870e, *P. quadraticollis* Lea, 1898c, *P. subereus* (Pascoe, 1870e), *P. subparallelus* Lea, 1898c, *P. tetricus* (Pascoe, 1874b).
Distribution. Australia (NSW, QLD), New Guinea.

Paletonidistus Lea, 1912 (Pl. 91F; Fig. 44.137)

Type species, by monotypy: *Paletonidistus trisinuatus* Lea, 1912.
Australian species. *P. foveicollis* Lea, 1913b, *P. trisinuatus* Lea, 1912.
Distribution. Australia (QLD, NSW).

Parasalcus Zimmerman, 1994 (Pl. 91G; Fig. 44.138)

Type species, by original designation: *Salcus latissimus* Pascoe, 1875.
Australian species. *P. latissimus* (Pascoe, 1875).
Distribution. Australia (QLD).

Paratituacia Lea, 1907 (Pl. 98A)

Type species, by monotypy: *Paratituacia dorsosignata* Lea, 1907.
Australian species. *P. dorsosignata* Lea, 1907b.
Distribution. Australia (WA).
Notes. Species of *Acacia* (Fabaceae) have been indicated as host plants of the single species.

Perissops Pascoe, 1871 (Pl. 91H; Fig. 44.105I–P)

Type species, by original designation: *Enteles ocellatus* Redtenbacher, 1868.
Australian species. *P. abacetus* Lea, 1929a, *P. albonotatus* Lea, 1903, *P. brevicollis* Lea, 1903, *P. carus* Lea, 1903, *P. funiculatus* Lea, 1928a, *P. granicollis* Lea, 1913a, *P. granulatus* Lea, 1903, *P. intricatior* Lea, 1903, *P. intricatus* Lea, 1903, *P. medionotatus* Lea, 1913a, *P. mucidus* Pascoe, 1871c, *P. multimaculatus* Lea, 1903, *P. ocellatus* (Redtenbacher, 1868), *P. ochreonotatus* Lea, 1913b, *P. parvus* Lea, 1913b, *P. piscicorpus* Lea, 1913a, *P. robiginosus* Lea, 1903, *P. semicalvus* Lea, 1903, *P. squamivarius* Lea, 1929a, *P. tarsalis* Lea, 1913a, *P. variegatus* Lea, 1903.

Distribution. Australia (QLD, NSW, Lord Howe Is.), New Caledonia, New Guinea.
Notes. *Ficus* species (Moraceae) are indicated as host plants of *P. granulatus*.

Pezichus Waterhouse, 1853 (Pl. 91K)

Type species, by monotypy: *Pezichus binotatus* Waterhouse, 1853.
Australian species. *P. binotatus* Waterhouse, 1853, *P. decipiens* Lea, 1913b, *P. gracilis* Lea, 1905b, *P. parallelirostris* Lea, 1913a.
Distribution. Australia (QLD, NSW).
Notes. *Ficus* species (Moraceae) are indicated as host plants of *P. binotatus*.

Phloeoglymma Pascoe, 1870 (Pl. 91I, J; Figs 44.106A–F, 44.139)

Type species, by monotypy: *Phloeoglymma alternans* Pascoe, 1870.
Australian species. *P. alternans* Pascoe, 1870e, *P. dorsale* (Pascoe, 1873c), *P. longirostre* Lea, 1913a, *P. mixtum* Lea, 1908c, *P. pallidum* Lea, 1915d.
Distribution. Australia (NSW, VIC, SA, WA).
Notes. Adults of *P. alternans* have been found in flowers of *Ozothamnus diosmifolius* (Asteraceae), and *Ricinocarpus pinifolius* (Euphorbiaceae) is recorded as a host plant of *P. dorsale*.

Plagiocorynus Waterhouse, 1853 (Pl. 91L)

Type species, by monotypy: *Plagiocorynus quadrituberculatus* Waterhouse, 1853.
Australian species. *P. quadrituberculatus* Waterhouse, 1853.
Distribution. Australia (QLD).

Platyporopterus Lea, 1898 (Pl. 92A)

Type species, by original designation: *Poropterus porrigineus* Pascoe, 1872.
Australian species. *P. fetus* Lea, 1913a, *P. porrigineus* (Pascoe, 1872c), *P. salebrosus* Lea, 1908b.
Distribution. Australia (NSW, VIC).
Reference. Lea (1898c).

Platytenes Pascoe, 1870 (Pl. 92B)

Type species, by monotypy: *Platytenes varius* Pascoe, 1870.
Australian species. *P. varius* Pascoe, 1870c.
Distribution. Australia (QLD), Malay Archipelago.
Notes. Setliff & Larson (2009) revised the genus indicating its association with palms.

Poropterculus Lea, 1912 (Pl. 97K, L)

Type species, by monotypy: *Poropterculus subnitidus* Lea, 1912.
Australian species. *P. subnitidus* Lea, 1912
Distribution. Australia (WA).

Poropterellus Lea, 1912 (Pl. 98B)

Type species, by monotypy: *Poropterellus intercoxalis* Lea, 1912.
Australian species. *P. abdominalis* Lea, 1913a, *P. intercoxalis* Lea, 1912.
Distribution. Australia (QLD).

Poropterinus Lea, 1912 (Pl. 98C)

Type species, by monotypy: *Poropterinus trilobus* Lea, 1912.
Australian species. *P. trilobus* Lea, 1912.
Distribution. Australia (NSW).

Poropteroides Lea, 1898 (Pl. 92C)

Type species, by monotypy: *Poropteroides dichotomus* Lea, 1898.
Australian species. *P. dichotomus* Lea, 1898c.
Distribution. Australia (NSW).

Poropterus Schoenherr, 1844 (Pl. 92D, E; Fig. 44.140A–K)

Type species, by original designation: *Poropterus antiquus* Boheman, 1844 (*in* Schoenherr 1844).
Australian species. *P. abstersus* Boheman, 1844 (*in* Schoenherr 1844), *P. alboscutellaris* Lea, 1911b, *P. angustatus* Lea, 1898b, *P. angustus* Lea, 1928a, *P. antiquus* (Erichson, 1842), *P. astheniatus* Lea, 1898b, *P. basalis* Lea, 1928a, *P. basipennis* Lea, 1913a, *P. bisignatus* Pascoe, 1872c, *P. bituberculatus* Lea, 1898b, *P. carinicollis* Lea, 1909c, *P. cavernosus* Lea, 1905a, *P. cavirostris* Lea, 1898b, *P. chevrolatii* Waterhouse, 1853, *P. communis* Lea, 1898b, *P. conifer* (Erichson, 1842), *P. constrictifrons* Lea, 1913b, *P. convexus* Lea, 1913b, *P. corvus* Lea, 1898b, *P. crassicornis* Lea, 1898b, *P. crassipes* Lea, 1913a, *P. cribratus* Lea, 1928a, *P. cryptodermus* Lea, 1928a, *P. difficilis* Lea, 1898b, *P. ellipticus* Pascoe, 1871c, *P. fasciculatus* Lea, 1898b, *P. ferrugineus* Lea, 1928a, *P. flexuosus* Pascoe, 1871c, *P. foveatus* Lea, 1905a, *P. foveipennis* Pascoe, 1872c, *P. griseus* Lea, 1928a, *P. harpagus* Lea, 1898b, *P. humeralis* Lea, 1909c, *P. idolus* Lea, 1898b, *P. impendens* Lea, 1928a, *P. intermedius* Lea, 1898b, *P. inusitatus* Lea, 1905a, *P. jekelii* Waterhouse, 1853, *P. latipennis* Lea, 1913a, *P. lissorhinus* Lea, 1905a, *P. longipes* Lea, 1905a, *P. lutulentus* Lea, 1898b, *P. magnus* Lea,

1913a, *P. melancholicus* Lea, 1911b, *P. mollis* Lea, 1913a, *P. montanus* Lea, 1909c, *P. morbillosus* Pascoe, 1871c, *P. multicolor* Lea, 1913a, *P. obesus* Lea, 1928a, *P. oniscus* Pascoe, 1873a, *P. ornaticollis* Lea, 1913b, *P. orthodoxus* Lea, 1898b, *P. parallelus* Lea, 1898b, *P. parryi* Waterhouse, 1853, *P. parvidens* Lea, 1913b, *P. pictus* Lea, 1928a, *P. platyderes* Lea, 1913a, *P. posterior* Lea, 1928a, *P. posticalis* Lea, 1913a, *P. punctipennis* Lea, 1928a, *P. python* Pascoe, 1881c, *P. rhyticephalus* Lea, 1905a, *P. rubetra* (Erichson, 1842), *P. setipes* Lea, 1928a, *P. simsoni* Lea, 1913a, *P. sphacelatus* Pascoe, 1871c, *P. stenogaster* Lea, 1913a, *P. submaculatus* Lea, 1928a, *P. succisus* (Erichson, 1842), *P. sulciventris* Lea, 1909c, *P. sylvicola* Lea, 1928a, *P. trifoveiventris* Lea, 1913a, *P. undulatus* Lea, 1911b, *P. variabilis* Lea, 1898b, *P. verres* Pascoe, 1871c, *P. waterhousii* Pascoe, 1871c, *P. westwoodii* Waterhouse, 1853, *P. zopherus* Lea, 1898b.

Distribution. Australia (QLD, NSW, VIC, TAS, SA, Lord Howe Is.), New Guinea.

Notes. *Poropterus* is a composite genus that needs to be restricted to a much smaller group of species around *P. antiquus*, the others requiring placement in other existing or in new genera. The name of the type species (*Poropterus antiquus* Boheman) requires further clarification. E. C. Zimmerman synonymised it with the older *Cryptorhynchus antiquus* Erichson, 1842 in his draft text on the Australian Cryptorhynchini, but the synonymy was only published by Pullen *et al.* (2014), who did not, however, indicate it as such. Although it appears to be correct, it requires confirmation from comparison of the type specimens.

Pseudapries Lea, 1908 (Pl. 92G; Fig. 44.141)

Type species, by monotypy: *Pseudapries corticalis* Lea, 1908.

Australian species. *P. corticalis* Lea, 1908b, *P. gronopoides* (Pascoe, 1885), *P. nebulosus* Lea, 1910c, *P. ptychoderes* Lea, 1910c, *P. squamiseriatus* Lea, 1913a.

Distribution. Australia (QLD, WA).

Pseudometyrus Lea, 1910 (Pl. 92H; Fig. 44.140L–S)

Type species, by subsequent designation (Alonso-Zarazaga & Lyal, 1999): *Pseudometyrus laminatus* Lea, 1910.

Australian species. *P. antares* (Erichson, 1842), *P. bicaudatus* Lea, 1913a, *P. cylindricus* Lea, 1910c, *P. laminatus* Lea, 1910c, *P. persimilis* Lea, 1910c, *P. placidus* Lea, 1910c, *P. sirius* (Erichson, 1842), *P. vicarius* Lea, 1910c.

Distribution. Australia (QLD, NSW, VIC, TAS).

Pseudomydaus Lea, 1912 (Pl. 98D)

Type species, by monotypy: *Pseudomydaus tenuis* Lea, 1912.

Australian species. *P. tenuis* Lea, 1912.

Distribution. Australia (NSW).

Pseudonidistus Lea, 1912 (Pl. 92I)

Type species, by monotypy: *Pseudonidistus cordatus* Lea, 1912.

Australian species. *P. calviceps* Lea, 1913b, *P. cordatus* Lea, 1912.

Distribution. Australia (QLD, NSW).

Pseudoporopterus Lea, 1898 (Pl. 92J)

Type species, by monotypy: *Poropterus lemur* Pascoe, 1881.

Australian species. *P. cruciatus* (Fabricius, 1775), *P. cyclopterus* (Lea, 1913a), *P. delirus* (Pascoe, 1874b), *P. lemur* (Pascoe, 1881), *P. rugosus* (Pascoe, 1885), *P. sulcifrons* Lea, 1913b.

Distribution. Australia (QLD), New Guinea, Sulawesi.

Reference. Lea (1898c).

Pseudotepperia Lea, 1903 (Pl. 98F, G)

Type species, by monotypy: *Pseudotepperia compta* Lea, 1903.

Australian species. *P. compta* Lea, 1903.

Distribution. Australia (WA).

Pteroporopterus Lea, 1912 (Pl. 92K)

Type species, by monotypy: *Pteroporopterus lacunosus* Lea, 1912.

Australian species. *P. lacunosus* Lea, 1912.

Distribution. Australia (QLD).

Rhaebocnemocis Lea, 1913 (Pl. 98H)

Type species, by monotypy: *Rhaebocnemocis posterus* Lea, 1913.

Australian species. *R. posterus* Lea, 1913b.

Distribution. Australia (QLD).

Roptoperus Lea, 1908 (Pl. 93A, B; Fig. 44.142A–K)

Type species, by monotypy: *Roptoperus tasmaniensis* Lea, 1908.

Australian species. *R. basalis* Lea, 1913a, *R. bryophilus* Lea, 1913a, *R. calviceps* Lea, 1913b, *R. carinatus* Lea, 1930b, *R. fuligineus* Lea, 1913a, *R. lissorhinus* Lea, 1930b, *R. longus* Lea, 1913a, *R. occidentalis* Lea, 1912, *R. pallidicornis* Lea, 1913a, *R. scutellaris* Lea, 1913b, *R. tasmaniensis* Lea, 1908c, *R. terraereginae* Lea, 1912.

Distribution. Australia (QLD, NSW, TAS, VIC, WA).

Notes. Specimens are commonly found in leaf litter and moss.

Salcus Pascoe, 1870 (Pl. 92L; Fig. 44.142L–T)

Type species, by monotypy: *Salcus globosus* Pascoe, 1870.

Australian species. *S. elevatus* Pascoe, 1875, *S. globosus* Pascoe, 1870b.
Distribution. Australia (QLD), New Caledonia, New Guinea.

Scleropoides Lea, 1902 (Pl. 98I)

Type species, by subsequent designation (Pullen *et al.* 2014: 267): *Scleropoides squamicollis* Lea, 1902.
Australian species. *S. squamicollis* Lea, 1902, *S. trianguliferus* Lea, 1902.
Distribution. Australia (NSW).

Scolyphrus Pascoe, 1874 (Pl. 93C)

Type species, by original designation: *Scolyphrus obesus* Pascoe, 1874.
Australian species. *S. obesus* Pascoe, 1874b.
Distribution. Australia (QLD, NSW).

Scotinocis Lea, 1913 (Pl. 98J, K)

Type species, by original designation: *Scotinocis sagittifer* Lea, 1913.
Australian species. *S. sagittifer* Lea, 1913a.
Distribution. Australia (NSW).

Sophronocis Lea, 1913 (Pl. 99A, B)

Type species, by monotypy: *Sophronocis albonotatus* Lea, 1913.
Australian species. *S. albonotatus* Lea, 1913b.
Distribution. Australia (QLD).

Stenoporopterus Lea, 1908 (Pl. 93D)

Type species, by monotypy: *Stenoporopterus canaliculatus* Lea, 1908.
Australian species. *S. canaliculatus* Lea, 1908b.
Distribution. Australia (QLD, NSW).

Sternochetus Pierce, 1917 (Pl. 93E; Fig. 44.107A–F)

Type species, by subsequent designation (Buchanan, 1939: 82): *Curculio mangiferae* Fabricius, 1775.
Australian species. *S. mangiferae* (Fabricius, 1775).
Distribution. Australia (QLD), southern USA, Hawaii, China, Japan, Solomon Is.
Notes. *Sternochetus mangiferae* is a widespread pest species of cultivated mango (*Mangifera indica*) that was accidentally introduced from India (Pullen *et al.* 2014). The other two pest species, *S. gravis* (Fabricius) and *S. olivieri* (Faust), are not known to be present in Australia.

Sybulus Pascoe, 1871 (Pl. 93F)

Type species, by subsequent designation (Alonso-Zarazaga, 2013: 61): *Sybulus peccuarius* Pascoe, 1871c.
Australian species. *S. yorkensis* Lea, 1907b.
Distribution. Australia (QLD), Japan, Malay Archipelago.

Sympediosoma Lea, 1907 (Pl. 98L)

Type species, by subsequent designation (Pullen *et al.* 2014: 268): *Sympediosoma albifrons* Lea, 1907.
Australian species. *S. albifrons* Lea, 1907b, *S. obliquifasciatum* Lea, 1907b.
Distribution. Australia (QLD).

Tamphilus Zimmerman, 1994 (Pl. 93G; Fig. 44.143A–J)

Type species, by original designation: *Acalles amplicollis* Fairmaire, 1849.
Australian species. *T. amplicollis* (Fairmaire), 1849.
Distribution. Australia (QLD), Pacific region, New Guinea, New Caledonia.

Tapinocis Lea, 1913 (Pl. 93H; Fig. 44.143K–U)

Type species, by subsequent designation (Pullen *et al.* 2014: 270): *Tapinocis corticalis* Lea, 1913.
Australian species. *T. abundans* Lea, 1928a, *T. constrictus* Lea, 1928a, *T. corticalis* Lea, 1913a, *T. humeralis* Lea, 1928a, *T. scutellaris* Lea, 1913a, *T. setosus* Lea, 1928a, *T. subapicalis* Lea, 1913a.
Distribution. Australia (TAS, WA, Lord Howe and Norfolk Is.).

Tapinosomus Faust, 1899 (Pl. 93I)

Type species, by original designation: *Poropterus exitiosus* Pascoe, 1871.
Australian species. *T. exitiosus* (Pascoe, 1871c).
Distribution. Australia (QLD, NSW).

Tentegia Pascoe, 1873 (Pl. 93J, K)

Type species, by monotypy: *Tentegia favosa* Pascoe, 1873.
Australian species. *T. amplipennis* Lea, 1930c, *T. bisignata* (Pascoe, 1874b), *T. favosa* Pascoe, 1873c, *T. grossbechleri* Escalona *et al.* 2023, *T. parva* Blackburn, 1896, *T. quadriseriata* Lea, 1912, *T. sana* Faust, 1892b, *T. spenceri* Blackburn, 1896, *T. stupida* (Fabricius, 1775), *T. tompsetti* Escalona *et al.* 2023, *T. tortipes* Lea, 1912, *T. weiri* Escalona *et al.* 2023.
Distribution. Australia (QLD, NSW, NT, SA, WA).
Notes. The genus was revised by Escalona *et al.* (2023), who also provided notes and video recordings of the dung-rolling behaviour of *T. stupida*. *Tentegia* is the only known

weevil genus with coprophagous species, the larvae developing in dung pellets of marsupials collected and cached by the adults (Wassell 1966). Adults of *T. amplipennis* and *T. stupida* and have been observed rolling kangaroo or wallaby dung pellets, and *T. bisignata* has been reared from possum pellets. The habits and life histories of the other species are unknown, but the larvae of at least some others likely also breed in marsupial dung collected by the adults.

Tepalicus Lea, 1912 (Pl. 99E)

Type species, by monotypy: *Tepalicus semicalvus* Lea, 1912.
Australian species. *T. semicalvus* Lea, 1912.
Distribution. Australia (QLD).

Tepperia Lea, 1903 (Pl. 93L; Figs 44.107G–L, 44.144)

Type species, by subsequent designation (Marshall, 1936: 195): *Tepperia sterculiae* Lea, 1903.
Australian species. *T. brachychitonis* Marshall, 1936, *T. bryanti* Marshall, 1936, *T. major* Lea, 1903, *T. sterculiae* Lea, 1903.
Distribution. Australia (QLD, NSW, WA).
Notes. Specimens of *T. brachychitonis* have been collected from trunks of *Brachychiton* (Malvaceae) and *T. sterculiae* from those of *B. populneus.*

Terporopus Lea, 1912 (Pl. 94A; Fig. 44.145)

Type species, by monotypy: *Terporopus tenuicornis* Lea, 1912.
Australian species. *T. tenuicornis* Lea, 1912.
Distribution. Australia (QLD).

Tetengia Lea, 1912 (Pl. 94B)

Type species, by monotypy: *Tetengia solenopa* Lea, 1912.
Australian species. *Tetengia solenopa* Lea, 1912.
Distribution. Australia (WA).
Notes. Label data indicate that the single species may be associated with ants.

Tituacus Marshall, 1943 (Pl. 94C)

Type species, by monotypy: *Tituacia ostracion* Pascoe, 1870.
Australian species. *T. ostracion* (Pascoe, 1870e).
Distribution. Australia (WA).

Tomweirius Zimmerman & Oberprieler, 2014 (Pl. 94D)

Type species, by original designation: *Tomweirius mirus* Zimmerman & Oberprieler, 2014.
Australian species. *T. mirus* Zimmerman & Oberprieler, 2014.
Distribution. Australia (QLD).
Notes. The genus was described in Pullen *et al.* (2014).

Tragopus Schoenherr, 1837 (Pl. 94E)

Type species, by monotypy: *Tragopus asper* Boheman, 1837 (*in* Schoenherr 1837).
Australian species. *T. muticus* (Lea, 1908b).
Distribution. Australia (QLD, NSW), Solomon Is., Malay Archipelago.

Trigonopterus Fauvel, 1862 (Pl. 94F; Fig. 44.146)

Type species, by monotypy: *Trigonopterus insignis* Fauvel, 1862.
Australian species. *T. aequalis* (Pascoe, 1872b), *T. albidosparsus* (Lea, 1913c), *T. allaetus* Riedel, 2016, *T. athertonensis* Riedel, 2016, *T. australinasutus* Riedel, 2016, *T. australis* Riedel, 2016, *T. bisignatus* Riedel, 2016, *T. bisinuatus* Riedel, 2016, *T. boolbunensis* Riedel, 2016, *T. cooktownensis* Riedel, 2016, *T. daintreensis* Riedel, 2016, *T. deplanatus* Riedel, 2016, *T. evanidus* (Pascoe, 1872b), *T. finniganensis* Riedel, 2016, *T. fraterculus* Riedel, 2016, *T. garradungensis* Riedel, 2016, *T. hasenpuschi* Riedel, 2016, *T. hartleyensis* Riedel, 2016, *T. kurandensis* Riedel, 2016, *T. laetus* (Lea, 1913c), *T. lewisensis* Riedel, 2016, *T. montanus* Riedel, 2016, *T. monteithi* Riedel, 2016, *T. mossmanensis* Riedel, 2016, *T. oberprieleri* Riedel, 2016, *T. robertsi* Riedel, 2016, *T. rostralis* (Lea, 1928a), *T. sculptirostris* (Lea, 1928a), *T. squamosus* (Lea, 1928a), *T. striatipennis* (Lea, 1928a), *T. terraereginae* Riedel, 2016, *T. yorkensis* Riedel, 2016.
Distribution. Australia (QLD), Malay Archipelago, Melanesia.
Notes. The Australian species of the genus were revised by Riedel & Taenzler (2016).

Triptocis Lea, 1913 (Pl. 99C)

Type species, by monotypy: *Triptocis puncticollis* Lea, 1913.
Australian species. *Triptocis puncticollis* Lea, 1913b.
Distribution. Australia (QLD).

Tropidotasia Lea, 1928 (Pl. 94G; Fig. 44.147)

Type species, by monotypy: *Tropidotasia femoralis* Lea, 1928.
Australian species. *T. femoralis* Lea, 1928a.
Distribution. Australia (Lord Howe Is.).

Fig. 44.108. Cryptorhynchini, structural details, *Achopera araucariae* Marshall, ♀. A–B, head: A, dorsal; B, lateral; C, antenna; D, scutellum; E–F, elytron: E, lateral; F, inner lateral; G, prothorax, ventral; H, pterothorax, ventral; I–L, middle leg: I, anterior; J, posterior; K, tibial apex, anterior; L, tibial apex, posterior M, tarsus and claws, dorsal.

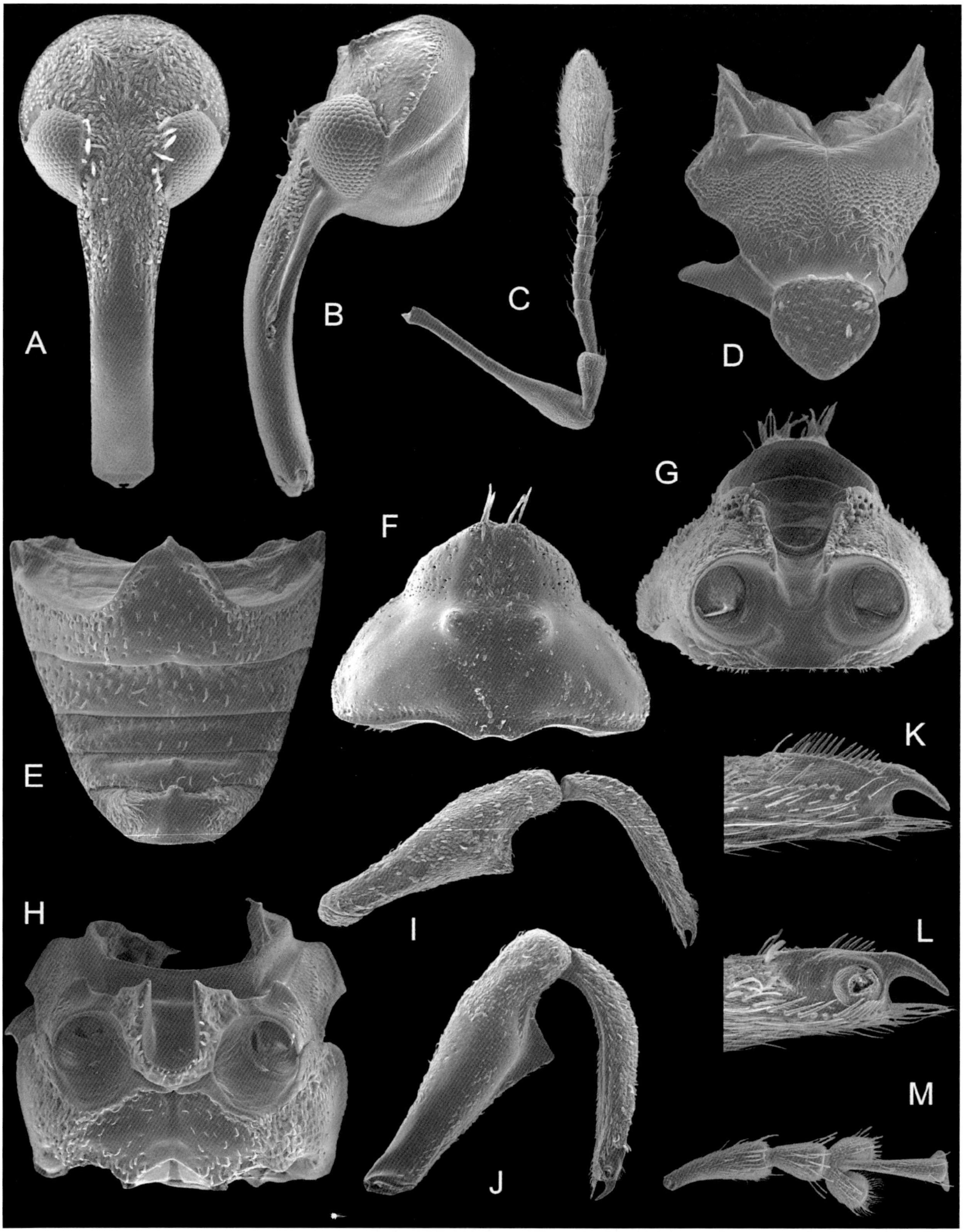

Fig. 44.109. Cryptorhynchini, structural details, *Acrotychreus fasciculatus* (Lea), ♀. A–B, head: A, dorsal; B, lateral; C, antenna; D, scutellum; E, abdominal ventrites; F–G, prothorax; F, dorsal; G, ventral; H, pterothorax, ventral; I, midleg; J, hindleg; K, mesotibial apex; L, metatibial apex; M, tarsus, dorsal.

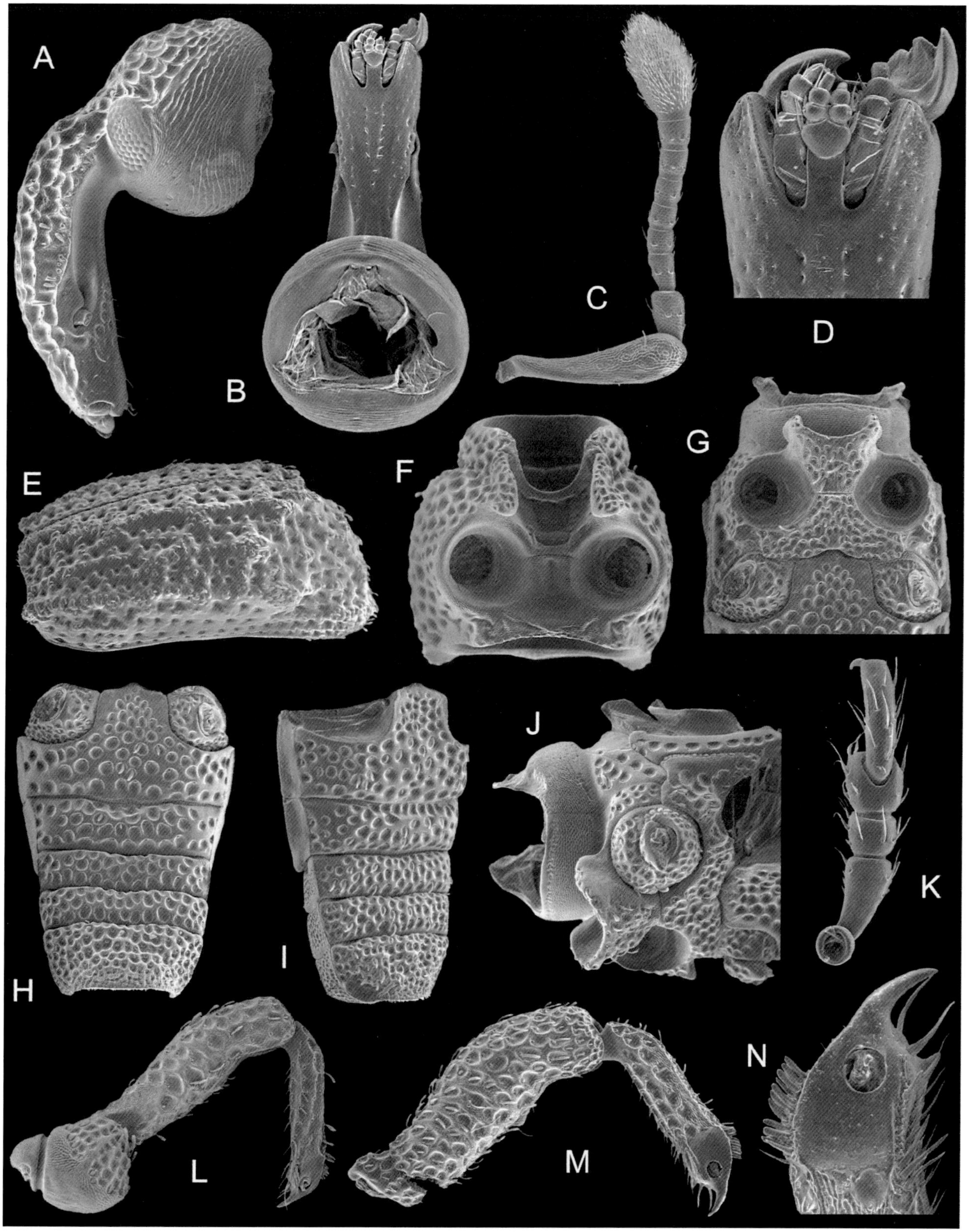

Fig. 44.110. Cryptorhynchini, structural details, *Agenopus* sp., ♂. A–B, head: A, lateral; B, ventral; C, antenna; D, mandibles; E, elytron, lateral; F, prothorax, ventral; G; J, pterothorax; G, ventral; J, lateral; H–I, abdominal ventrites: H, ventral; I, oblique lateral; K, tarsus, dorsal; L, foreleg; M, hindleg; N, metatibial apex.

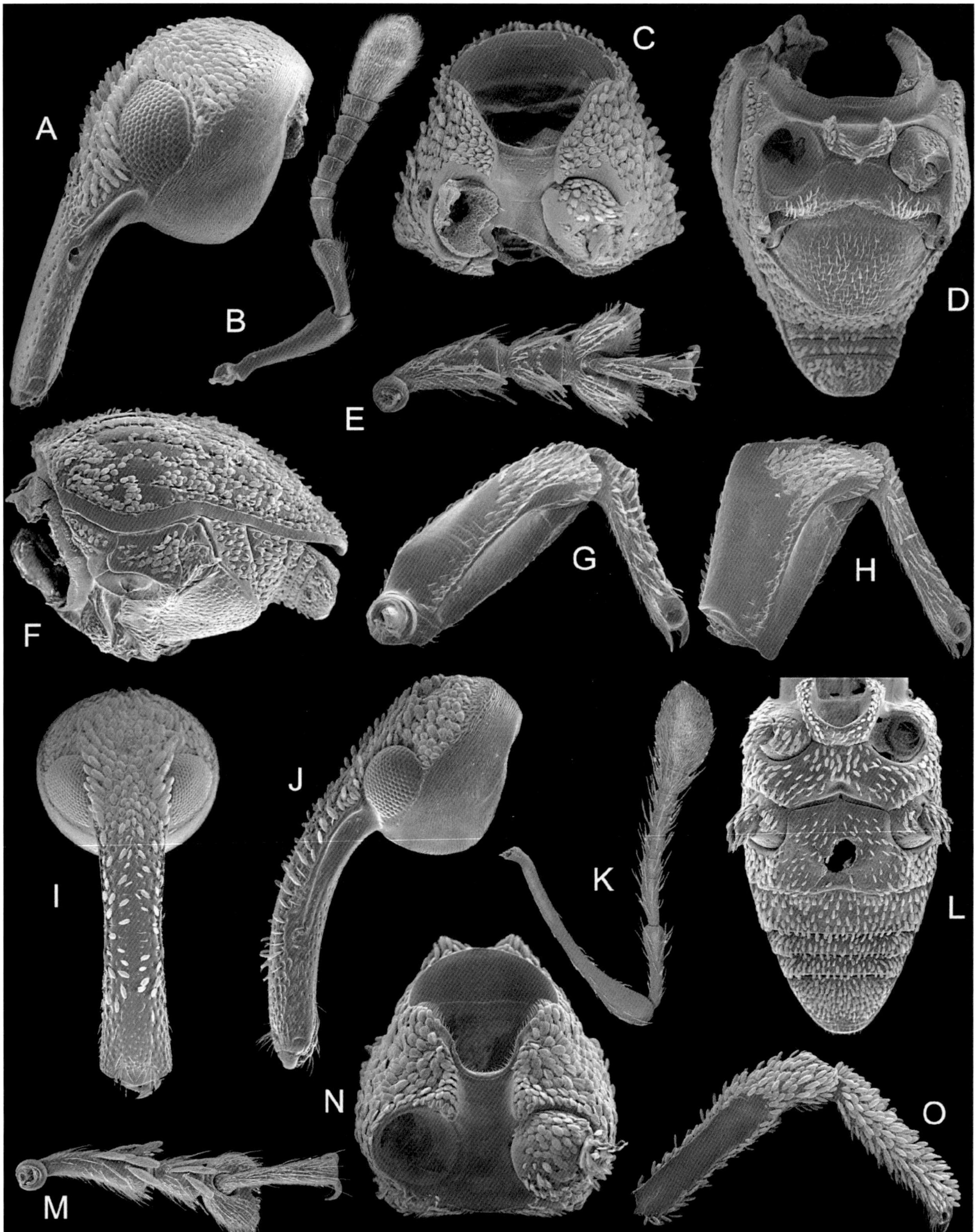

Fig. 44.111. Cryptorhynchini, structural details, *Ampagia* and *Baeodontocis*. A–H, *Ampagia* sp.: A, head, lateral; B, antenna; C, prothorax, ventral; D, pterothorax and abdominal ventrites E, tarsus, dorsal; F, pterothorax and elytron, lateral; G, midleg; H, hindleg; I–O, *Baeodontocis* sp.: I, head, dorsal; J, head, lateral; K, antenna; L, pterothorax and abdominal ventrites; M, tarsus, dorsal; N, prothorax, ventral; O, middle leg.

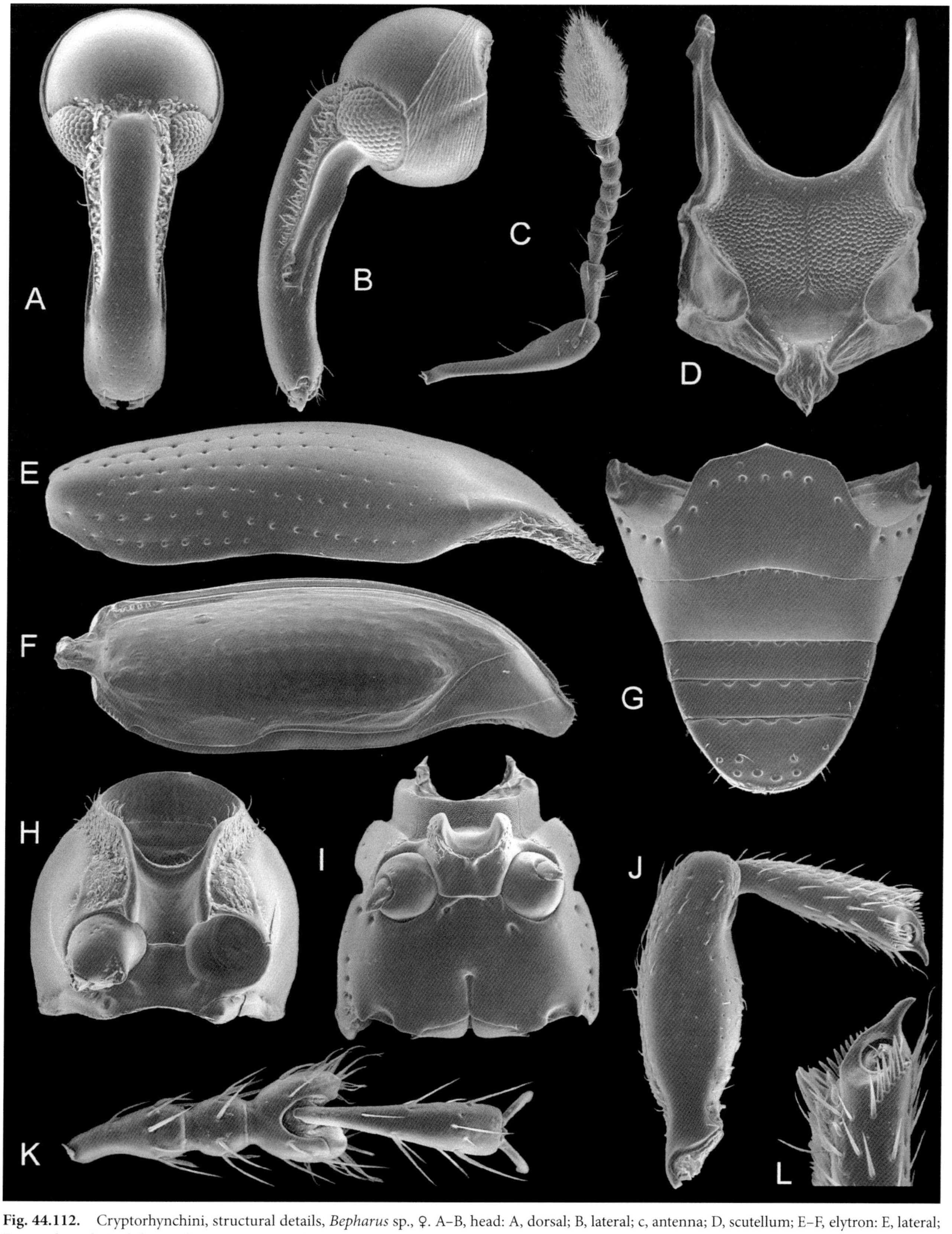

Fig. 44.112. Cryptorhynchini, structural details, *Bepharus* sp., ♀. A–B, head: A, dorsal; B, lateral; c, antenna; D, scutellum; E–F, elytron: E, lateral; F, inner lateral; G, abdominal ventrites; H, prothorax, ventral; I, pterothorax, ventral; J, leg; K, tarsus, dorsal; L, tibial apex.

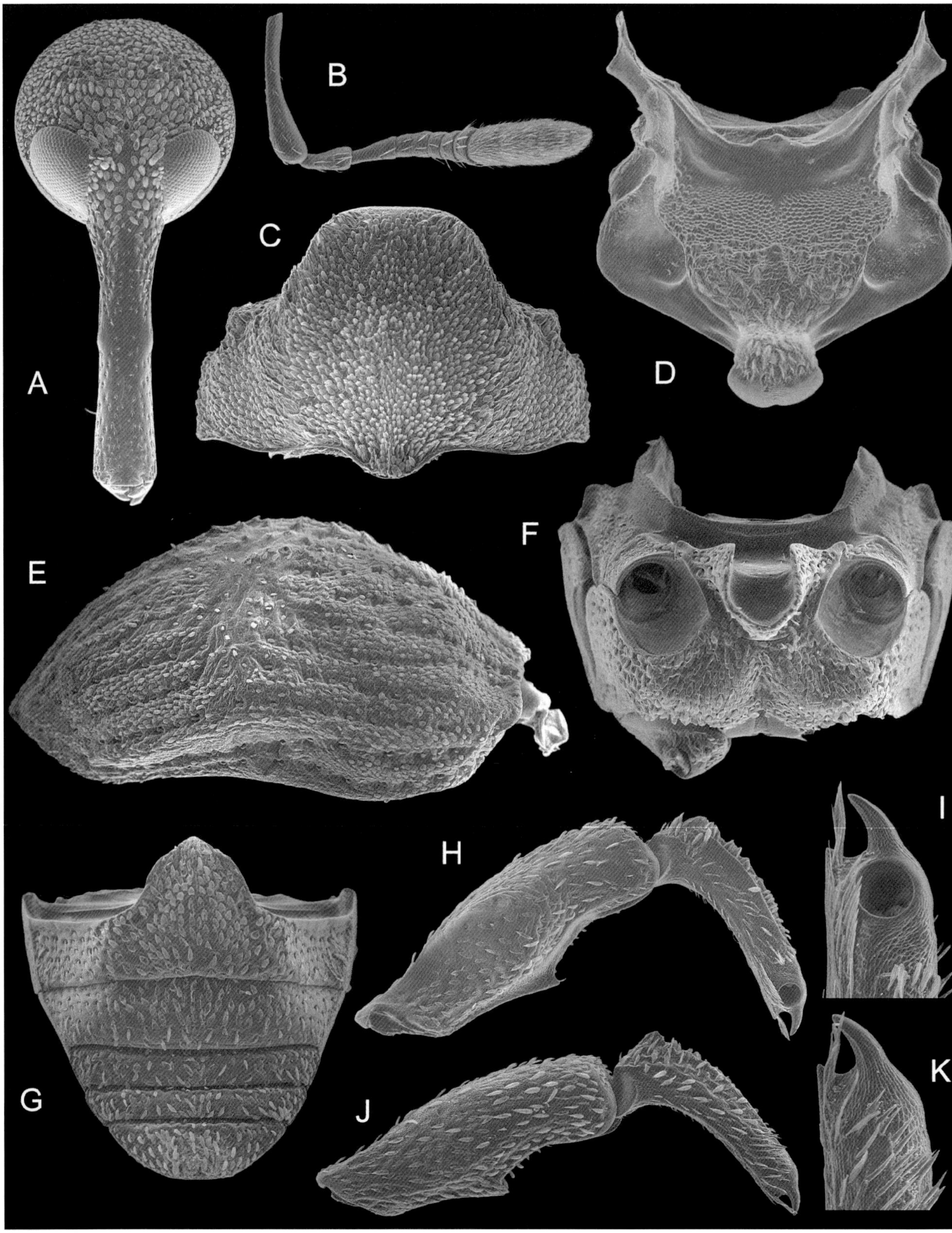

Fig. 44.113. Cryptorhynchini, structural details, *Brachyphyes thoracicus* (Lea). A, head, dorsal; B, antenna; C, prothorax, dorsal; D, scutellum; E, elytron, lateral; F, pterothorax, ventral; G, abdominal ventrites; H, leg; I, tibial apex; J, leg; K, tibial apex.

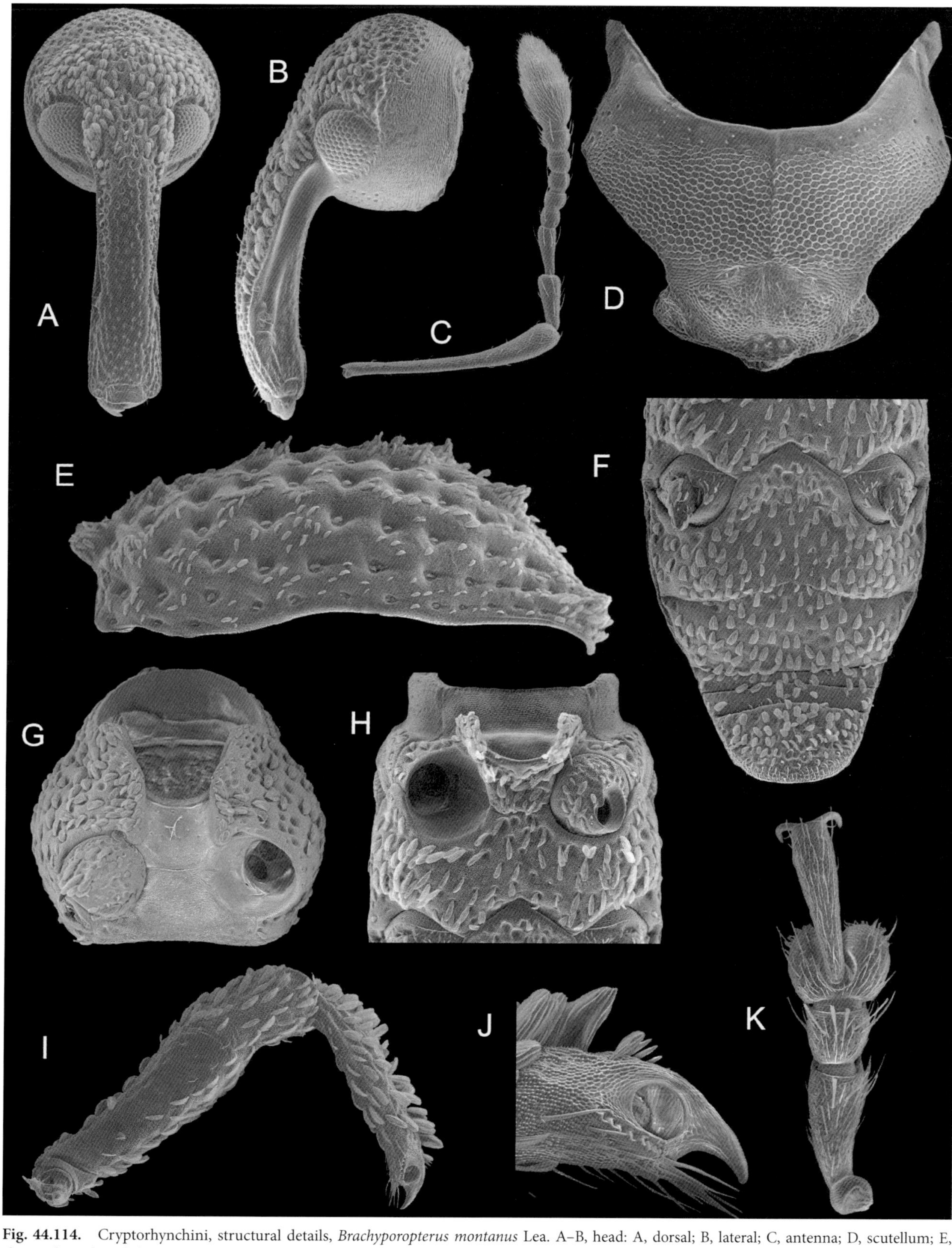

Fig. 44.114. Cryptorhynchini, structural details, *Brachyporopterus montanus* Lea. A–B, head: A, dorsal; B, lateral; C, antenna; D, scutellum; E, elytron, lateral; F, abdominal ventrites; G, prothorax, ventral; H, pterothorax, ventral; I, middle leg; J, tibial apex; K, tarsus and claws.

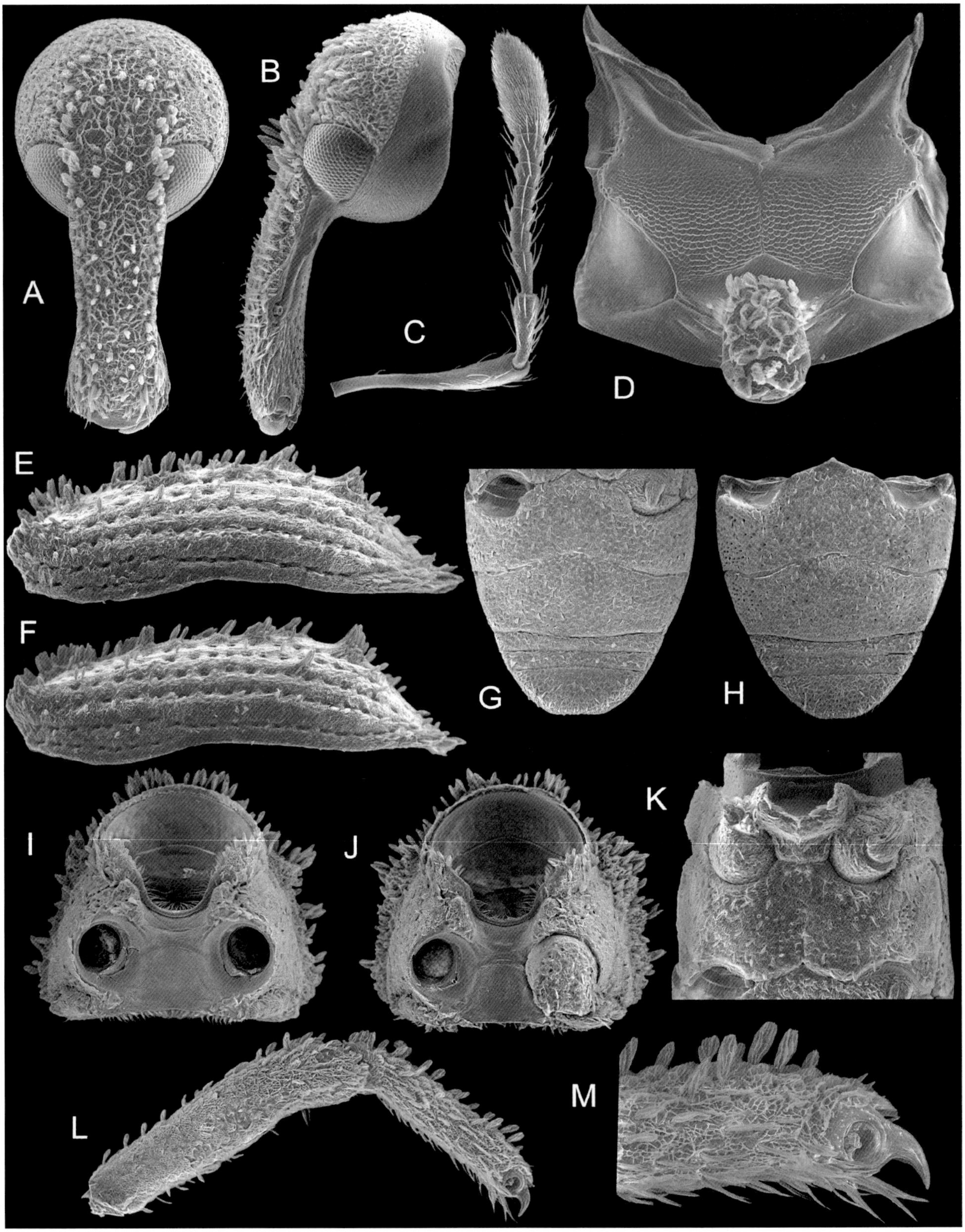

Fig. 44.115. Cryptorhynchini, structural details, *Chaetectetorus bifasciatus* (Boheman). A–B, head: A, dorsal; B, lateral; C, antenna; D, scutellum; D, scutellum; E–F, elytron, lateral: E, ♂; F, ♀; G–H, abdominal ventrites: G, ♂; H, ♀; I–J, prothorax, ventral: I, ♂; J, ♀; K, pterothorax, ventral; L, middle leg; M, tibial apex.

Fig. 44.116. Cryptorhynchini, structural details, *Chimades* sp., ♀. A, head, dorsal; B, antenna; C, scutellum; D, elytral apex, lateral; E, abdominal ventrites; F, prothorax, ventral; G, pterothorax, ventral; H, middle leg; I, tarsus and claws, dorsal; J, tibial apex.

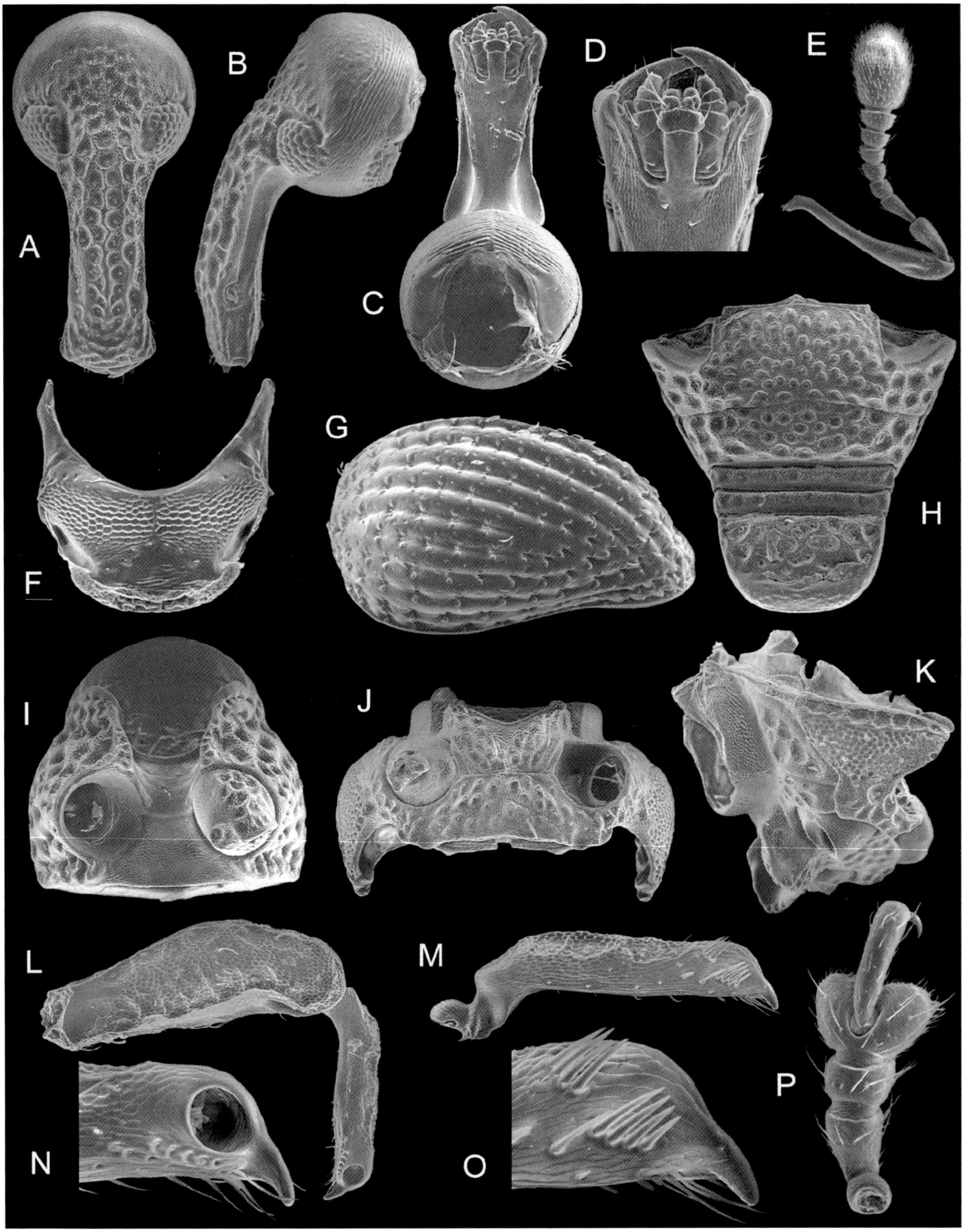

Fig. 44.117. Cryptorhynchini, structural details, *Decilaus* sp., ♂. A–C, head: A, dorsal; B, lateral; C, ventral; D, mouthparts; E, antenna; F, scutellum; G, elytron, lateral; H, abdominal ventrites; I, prothorax, ventral; J–K, pterothorax: J, ventral; K, lateral; L, middle leg; M, mesotibia; N, tibial apex, posterior; O, tibial apex, anterior; P, tarsus, dorsal.

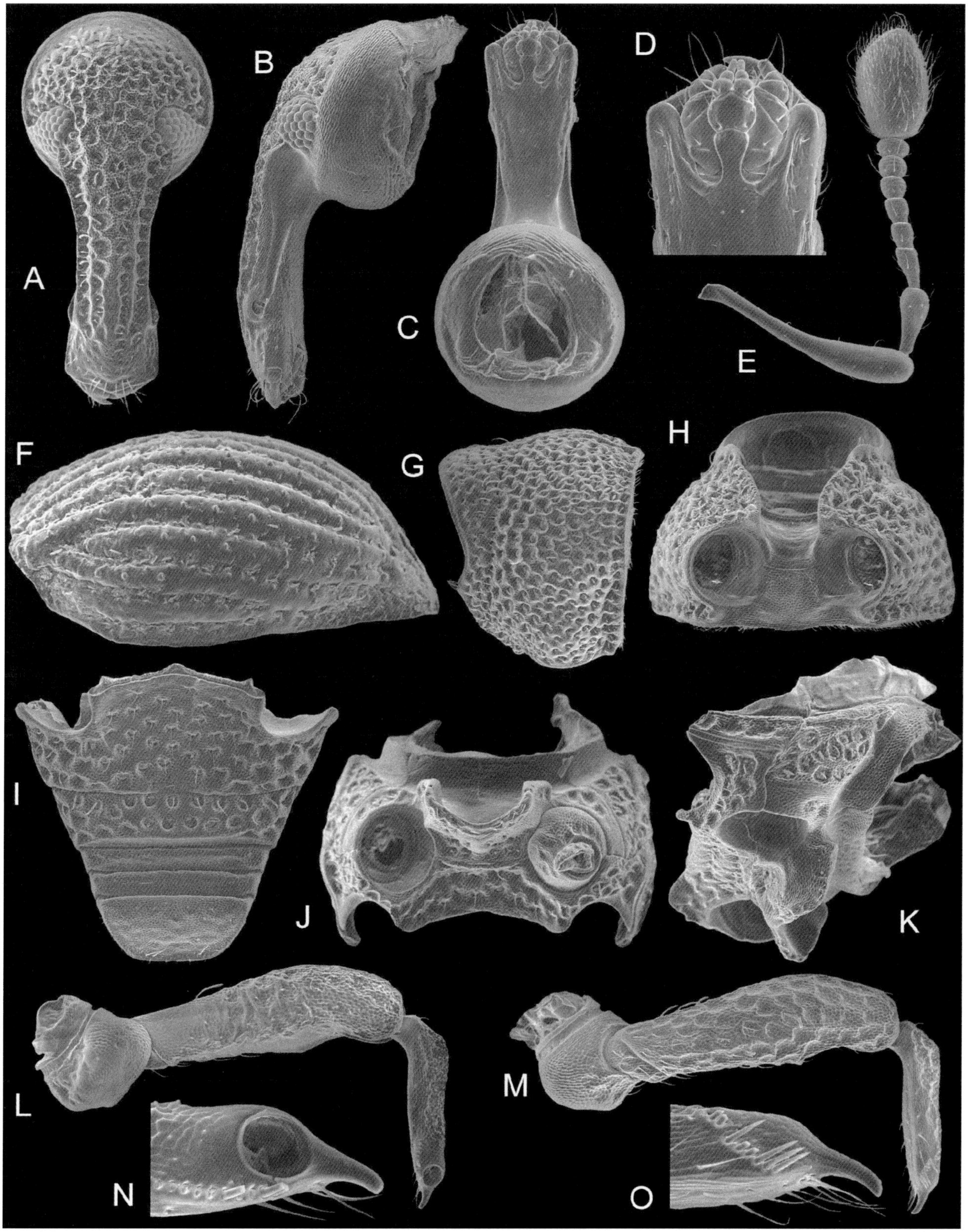

Fig. 44.118. Cryptorhynchini, structural details, *Decilaus nigronotatus* Lea, ♂. A–C, head: A, dorsal; B, lateral; C, ventral; D, mouthparts; E, antenna; F, elytron, lateral; G–H, prothorax: G, lateral; H, ventral; I, abdominal ventrites; J–K, pterothorax: J, ventral; K, lateral; L, middle leg; M, foreleg; N, mesotibial apex; O, protibial apex.

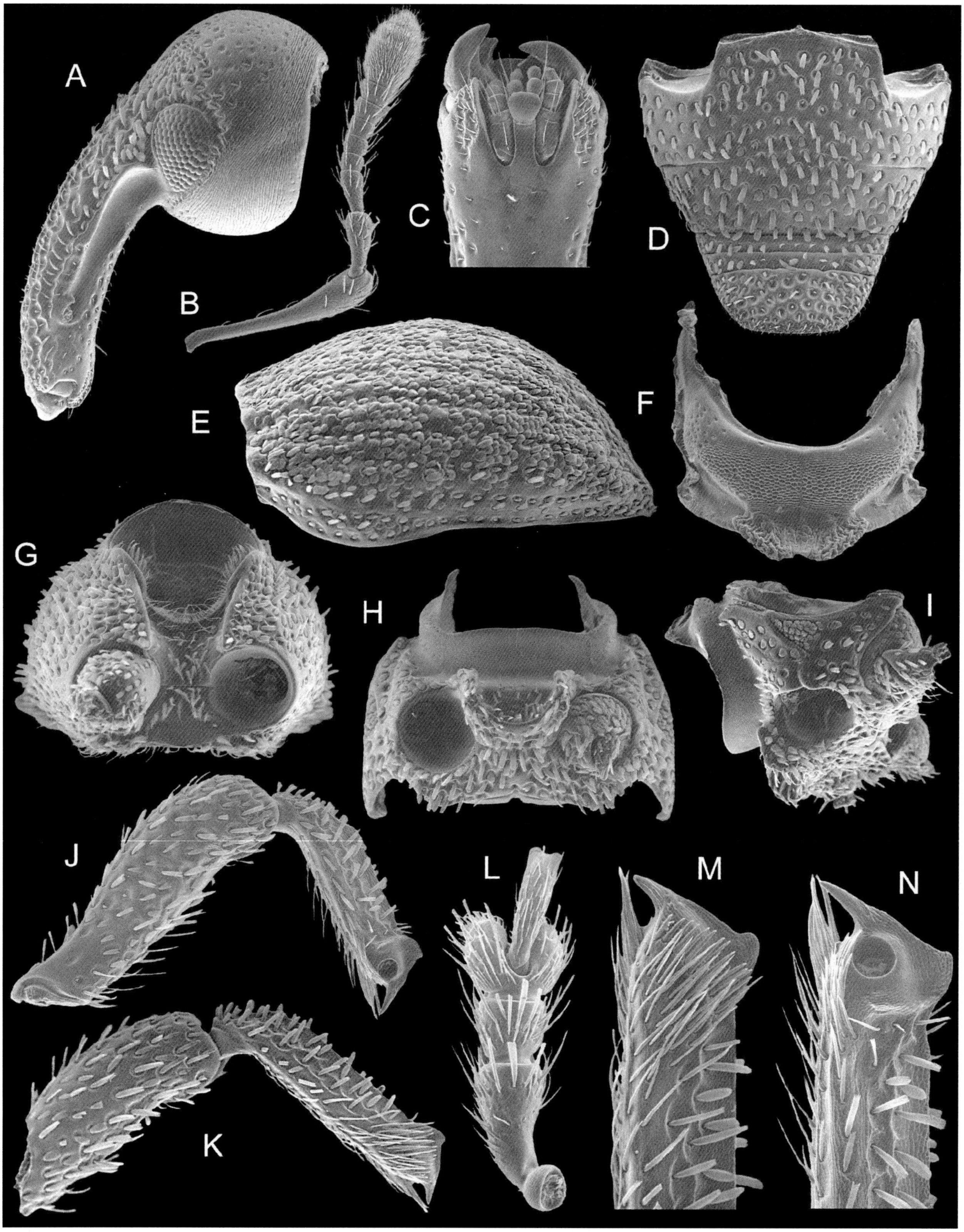

Fig. 44.119. Cryptorhynchini, structural details, *Decilaus* nr *squamosus*, Dongara, WA. A, head, lateral; B, antenna; C, mouthparts; D, abdominal ventrites; E, elytron, lateral; F, scutellum; G, prothorax, ventral; H–I, pterothorax: H, ventral; I, lateral; J, hindleg; K, middle leg; L, tarsus, dorsal; M, tibial apex, anterior; N, tibial apex, posterior.

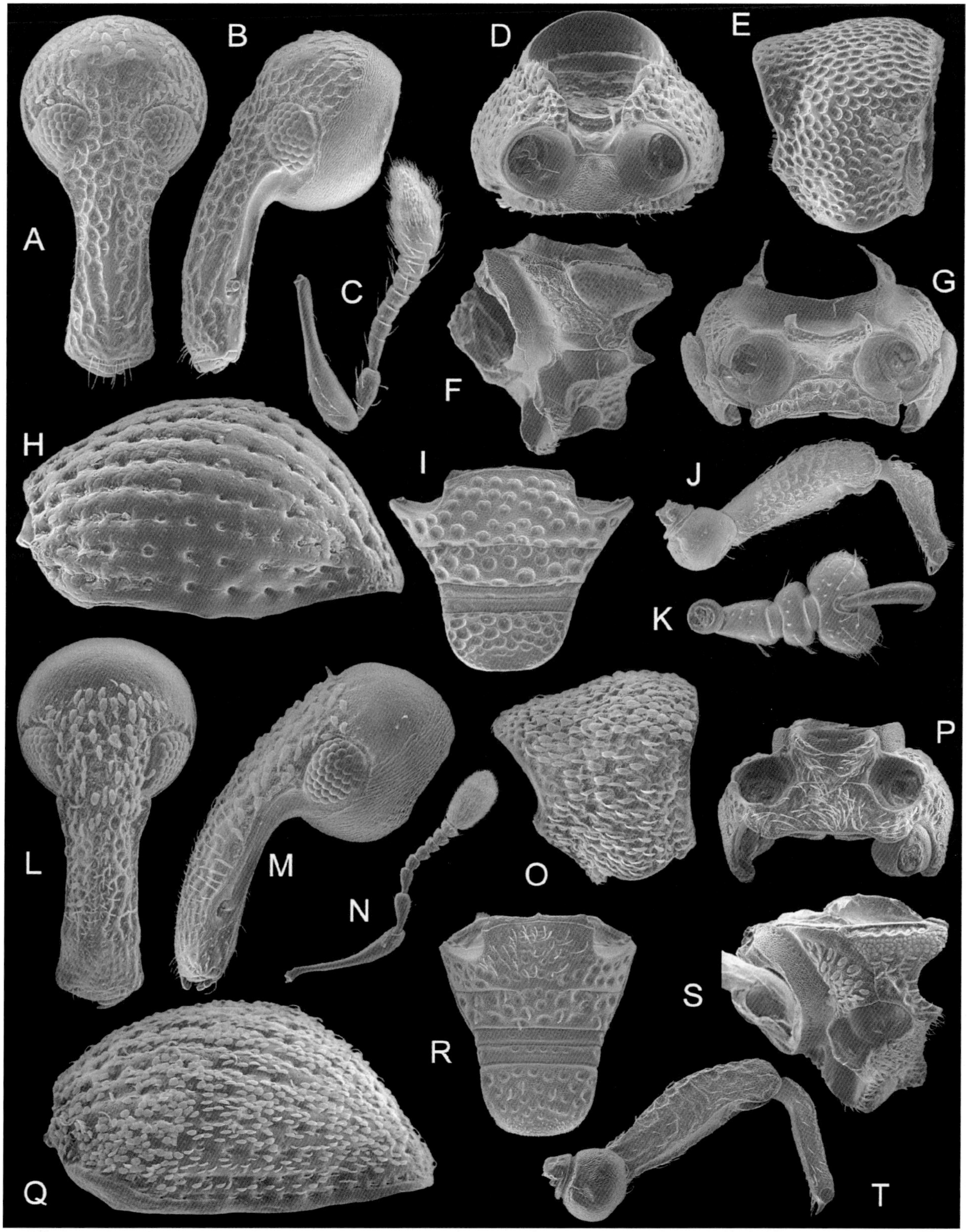

Fig. 44.120. Cryptorhynchini, structural details, *Decilaus* spp. A–K, *Decilaus* sp.: ♂; A, head, dorsal; B, head, lateral; C, antenna; D, prothorax, ventral; E, prothorax, lateral; F, pterothorax, lateral; G, pterothorax, ventral; H, elytron, lateral; I, abdominal ventrites; J, foreleg; K, tarsus, dorsal; L–T, *Decilaus* sp., ♂: L, head, dorsal; M, head, lateral; N, antenna; O, prothorax, lateral; P, pterothorax, ventral; Q, elytron, lateral; R, abdominal ventrites; S, pterothorax, lateral; T, midleg.

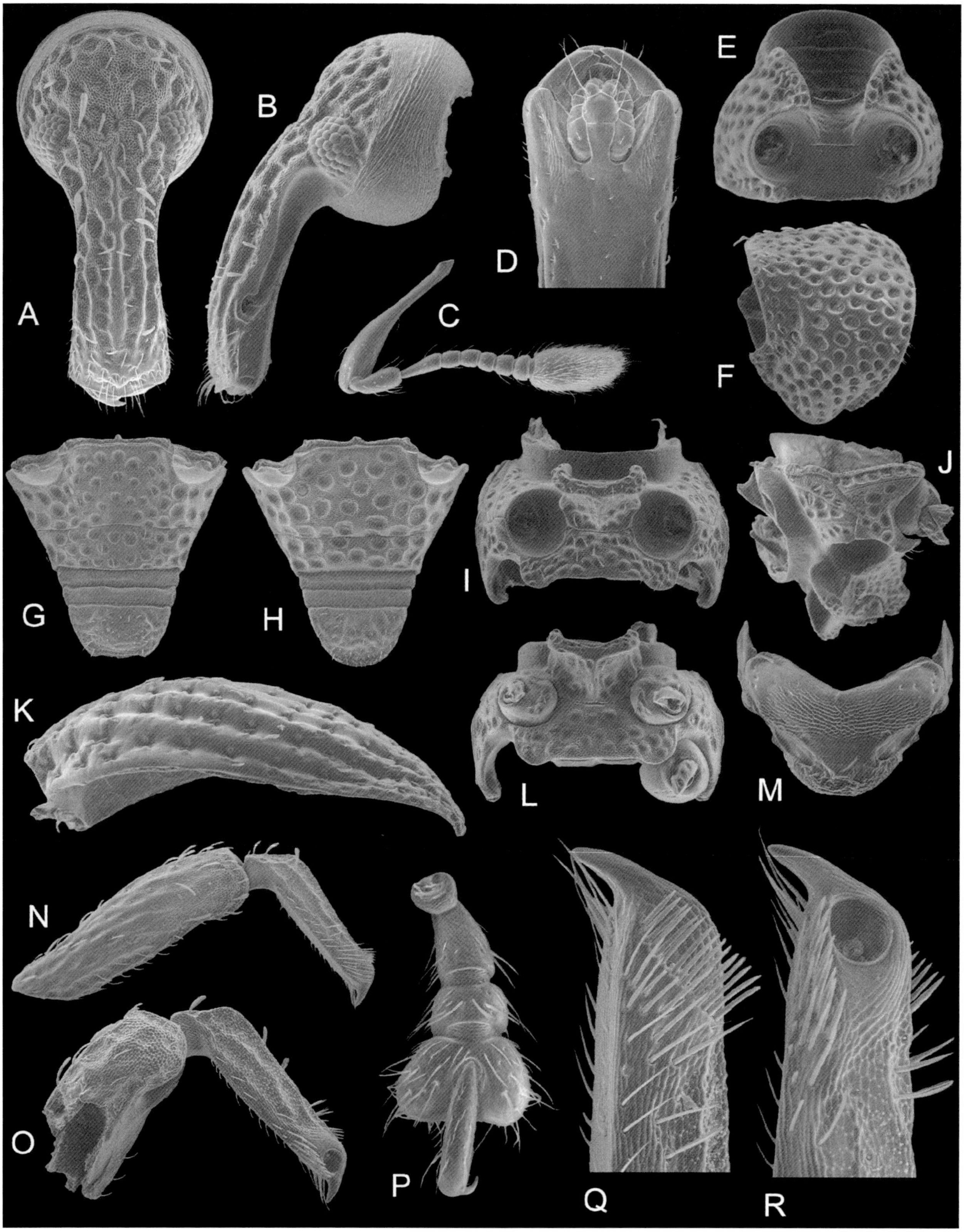

Fig. 44.121. Cryptorhynchini, structural details, *Decilaus* sp. A–B, head: A, dorsal; B, lateral; C, antenna; D, mouthparts; E–F, prothorax; E, ventral; F, lateral; G–H, abdominal ventrites: G, ♂; H, ♀; I–J, pterothorax, ♂; I, ventral; J, lateral; K, elytron, lateral; L, pterothorax, ventral, ♀; M, scutellum; N, middle leg; O, foreleg; P, tarsus, dorsal; Q, tibial apex, anterior; R, tibial apex, posterior.

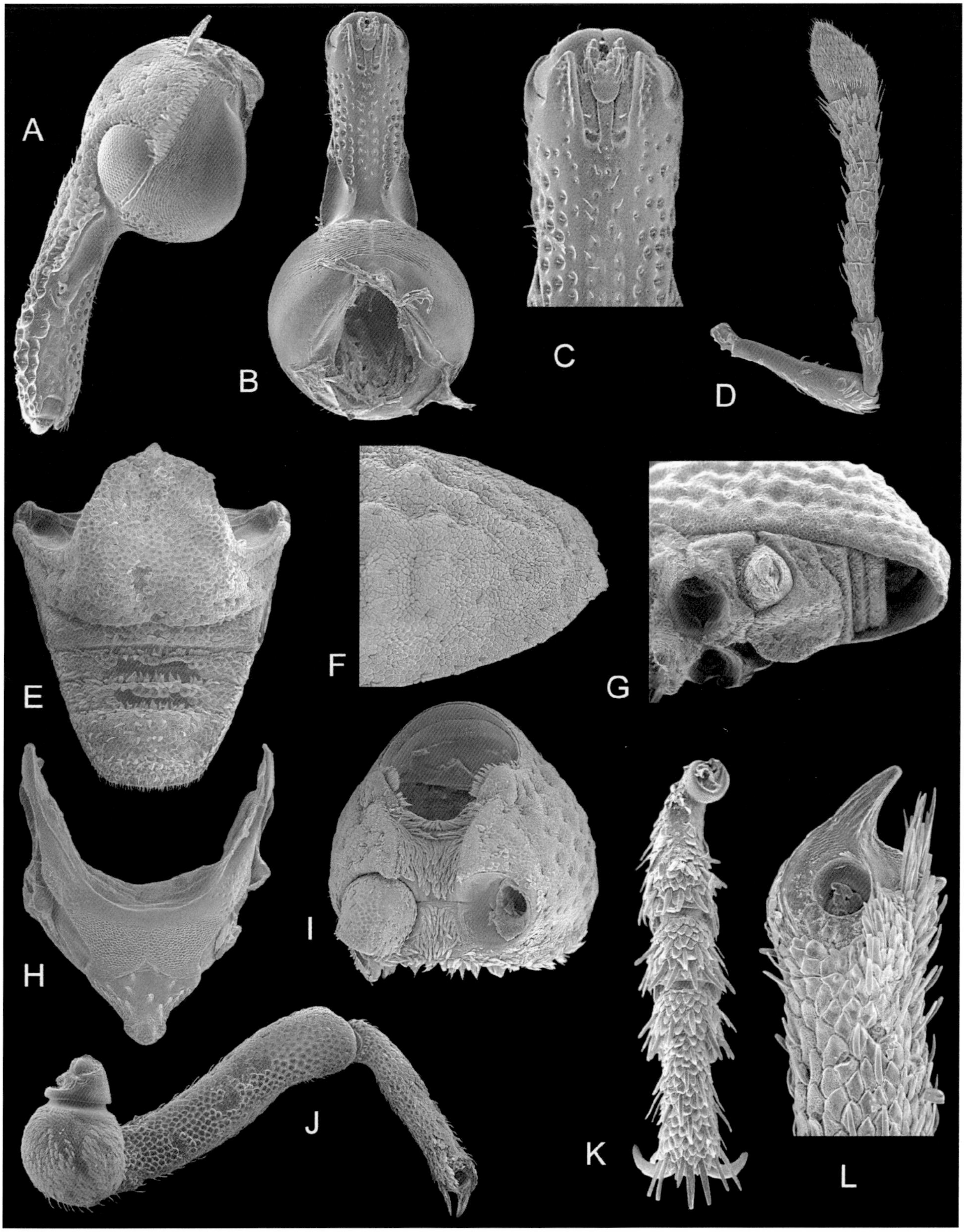

Fig. 44.122. Cryptorhynchini, structural details, *Elaeagna* sp., ♂. A–B, head: A, lateral; B, ventral; C, mouthparts; D, antenna; E, abdominal ventrites; F, elytral apex, lateral; G, pterothorax and abdominal ventrites; H, scutellum; I, prothorax, ventral; J, foreleg; K, tarsus, dorsal; L, protibial apex.

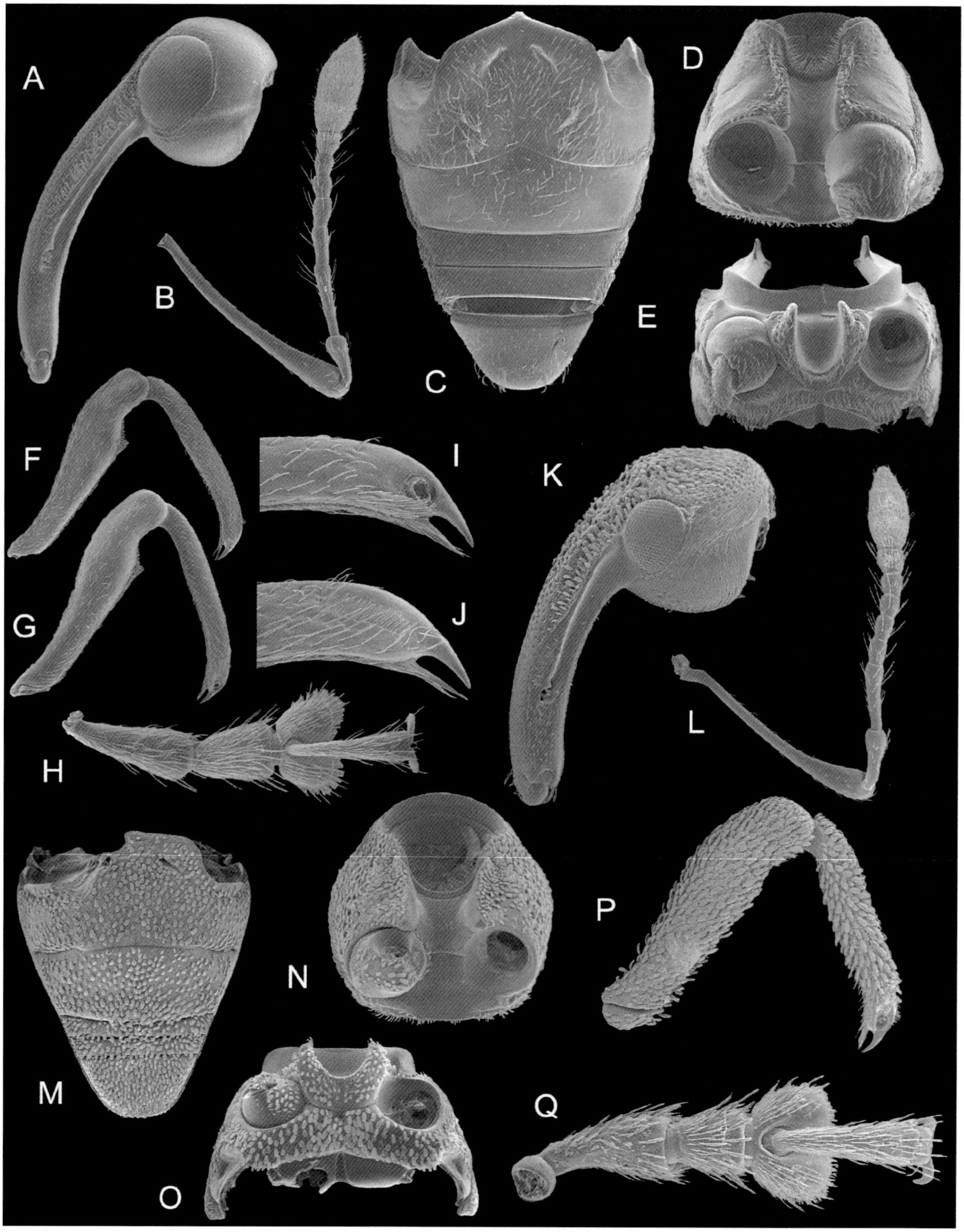

Fig. 44.123. Cryptorhynchini, structural details, *Enteles* and *Euryporopterus*. A–J, *Enteles vigorsii* Gyllenhal, ♂: A, head, lateral; B, antenna; C, abdominal ventrites; D, prothorax, ventral; E, pterothorax, ventral; F, middle leg, anterior; G, middle leg, posterior; I, tibial apex, posterior; J, tibial apex, anterior; K–Q, *Euryporopterus invidus* (Pascoe), ♀: K, head, lateral; L, antenna; M, abdominal ventrites; N, prothorax, ventral; O, pterothorax, ventral; P, middle leg; Q, tarsus, dorsal.

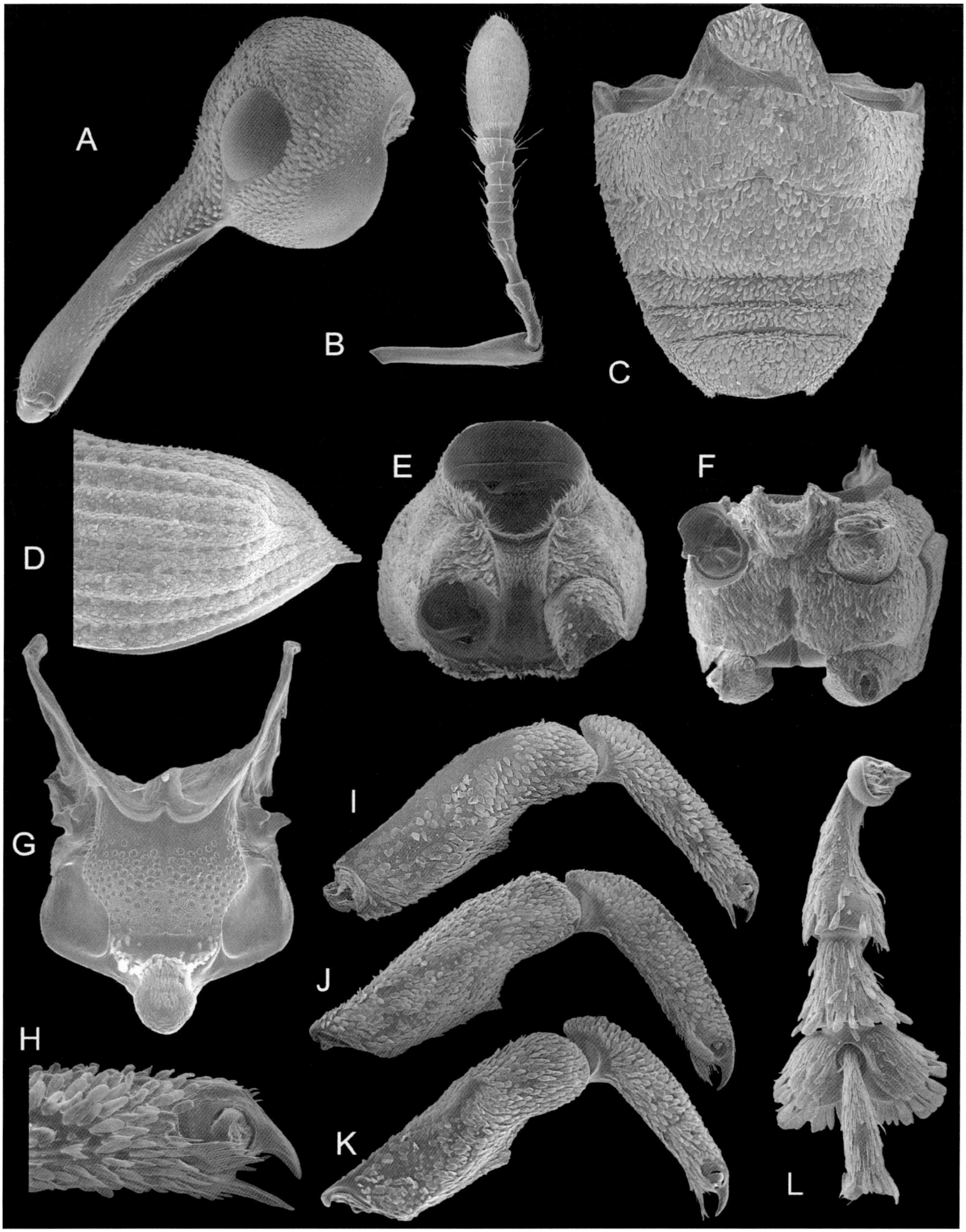

Fig. 44.124. Cryptorhynchini, structural details, *Eutyrhinus* sp., ♀. A, head, lateral; B, antenna; C, abdominal ventrites; D, elytral apex, lateral; E, prothorax, ventral; F, pterothorax, ventral; G, scutellum; H, mesotibial apex; I, foreleg; J, middle leg; K, hindleg; L, tarsus, dorsal.

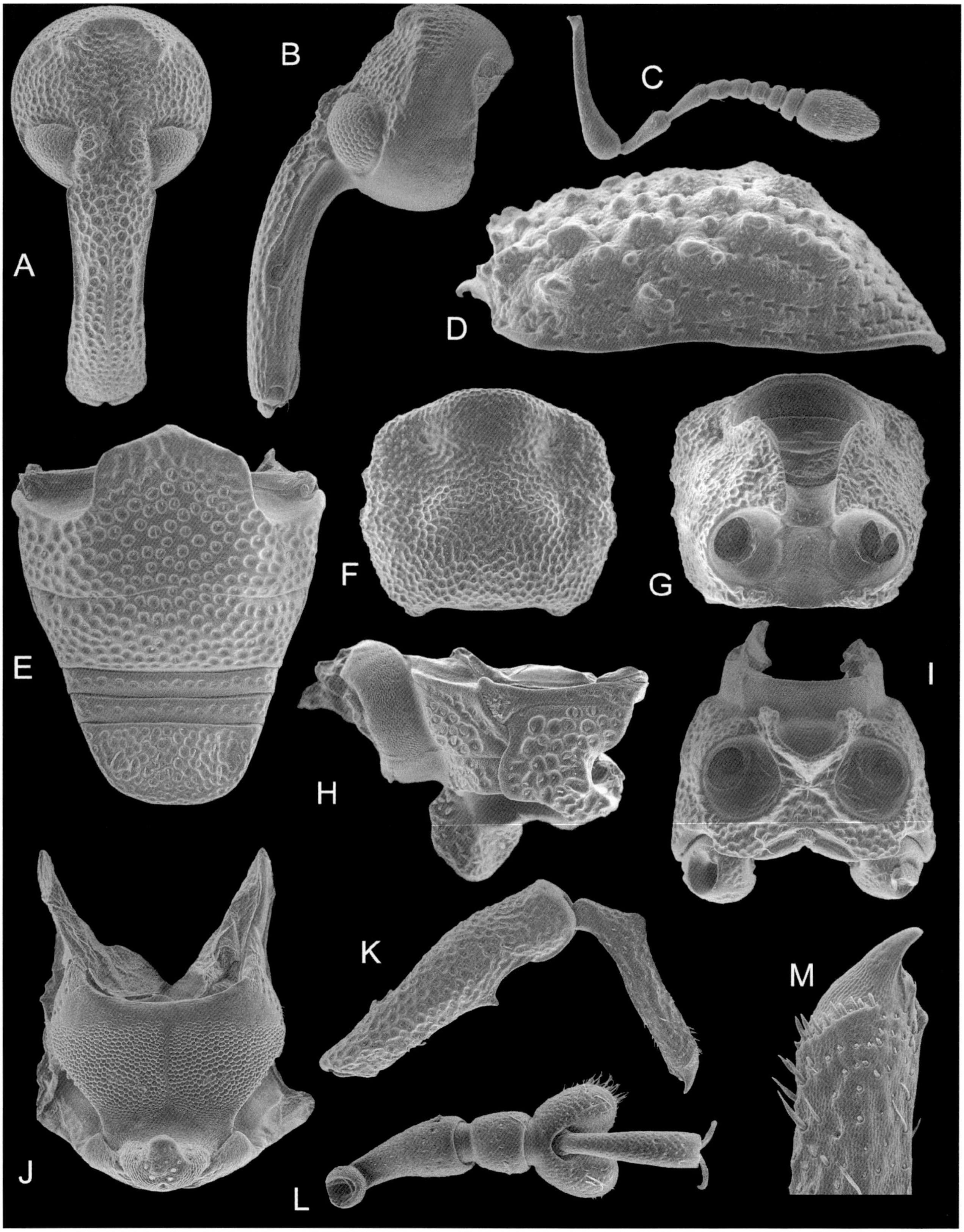

Fig. 44.125. Cryptorhynchini, structural details, *Exithius cariosus* (Erichson). A–B, head: A, dorsal; B, lateral; C, antenna; D, elytron, lateral; F–G, prothorax: F, dorsal; G, ventral; H–I, pterothorax: H, lateral; I, ventral; J, scutellum; K, middle leg; L, tarsus and claws, dorsal; M, tibial apex.

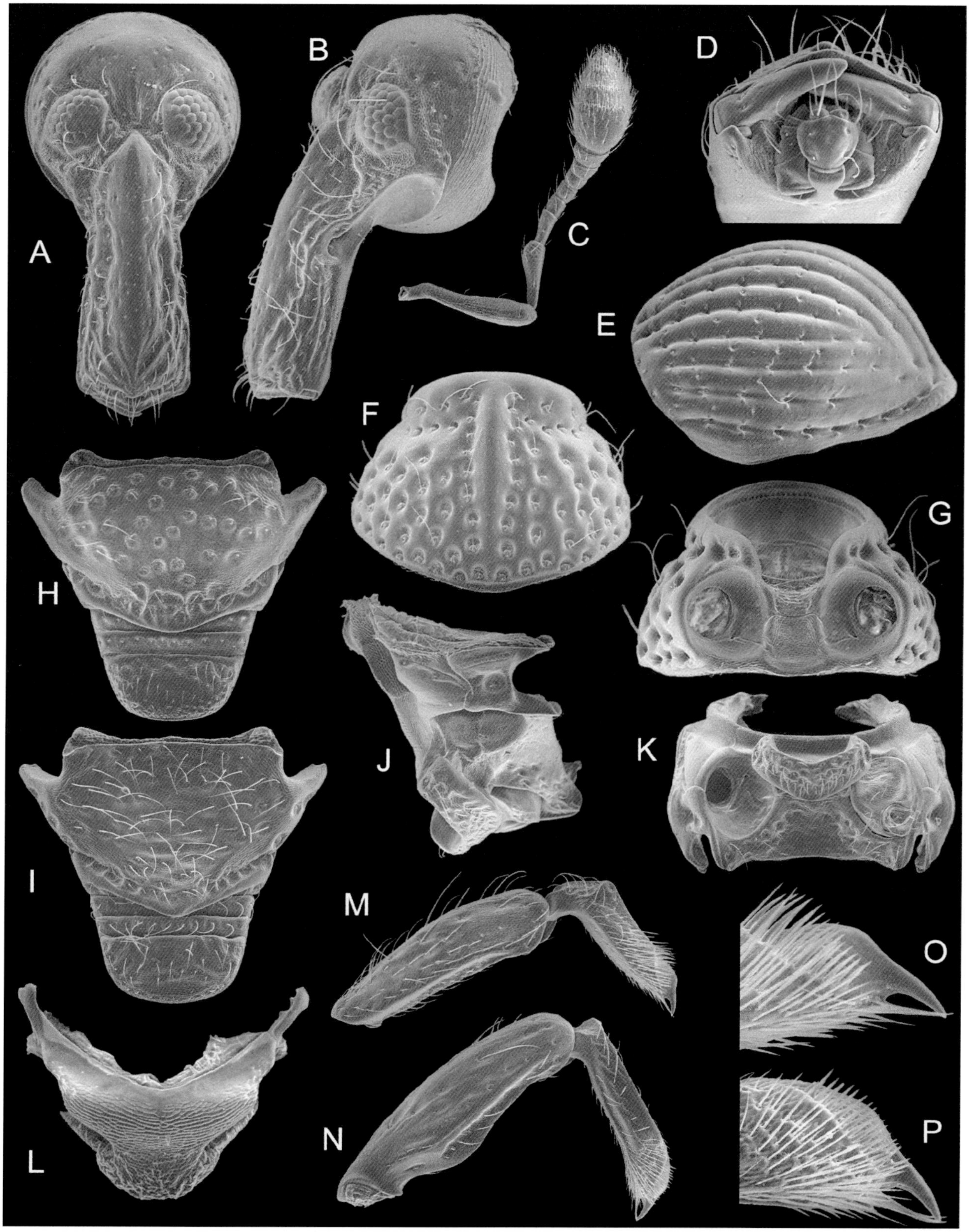

Fig. 44.126. Cryptorhynchini, structural details, *Genuacalles erectosetosus* Rheinheimer. A–B, head: A, dorsal; B, lateral; C, antenna; D, mouthparts; E, elytron, lateral; F–G, prothorax: F, dorsal; G, ventral; H–I, abdominal ventrites: H, ♀; I, ♂; J–K, pterothorax: J, lateral; K, ventral; L, scutellum; M, middle leg; N, hindleg; O, mesotibial apex; P, metatibial apex.

Fig. 44.127. Cryptorhynchini, structural details, *Gymnoporopterus pictipes*, Mt. Tozer, ♀. A–B, head: A, dorsal; B, lateral; C, antenna; D–E, elytron: D, lateral; E, inner lateral; F, abdominal ventrites; G–H, prothorax: G, lateral; H, ventral; I, pterothorax and abdominal ventrites, lateral; J, scutellum; K, middle leg; L, mesotibial apex.

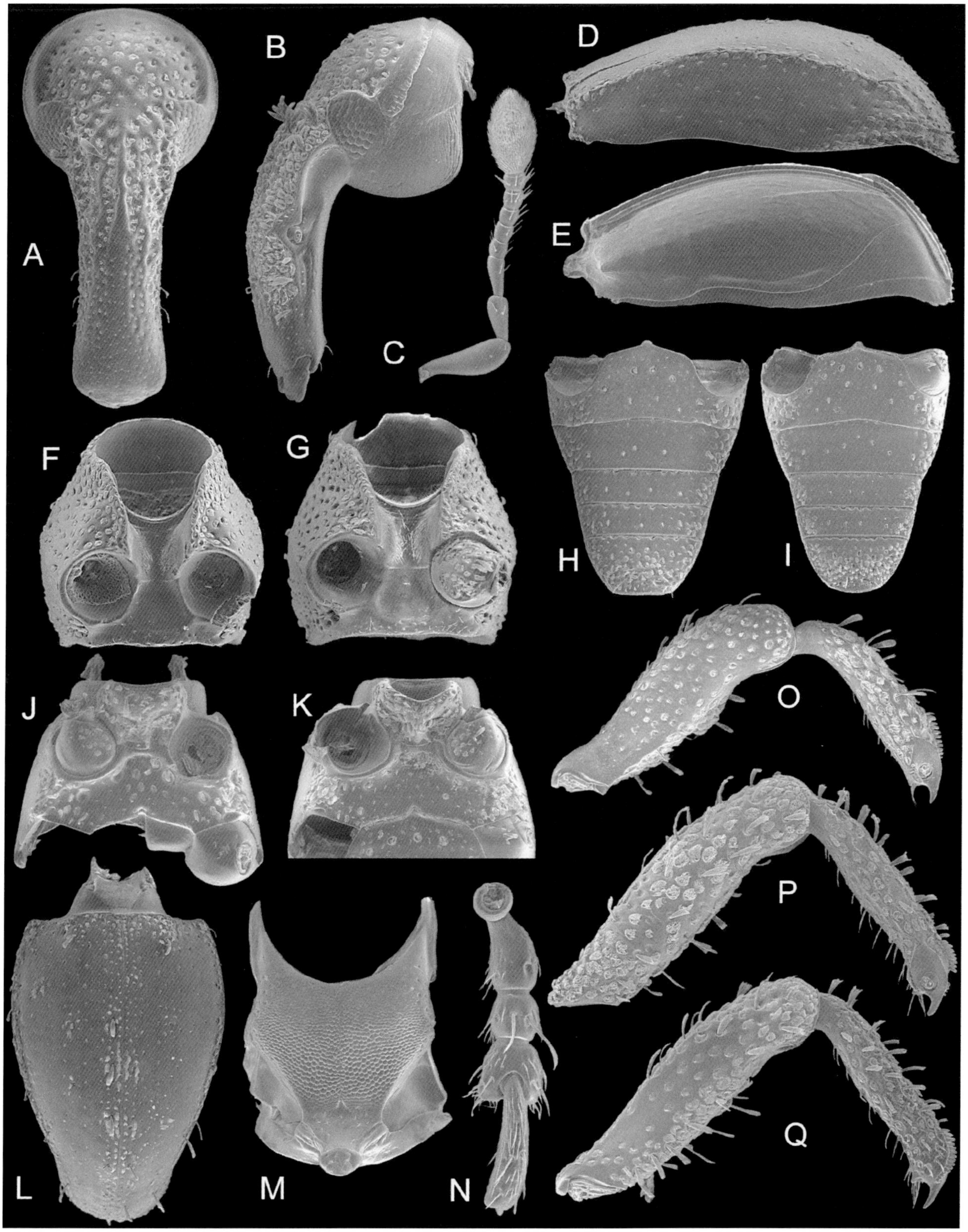

Fig. 44.128. Cryptorhynchini, structural details, *Idiopterocis* sp. A–B, head: A, dorsal; B, lateral; C, antenna; D–E, elytron: D, lateral; E, ventral; F–G, prothorax, ventral: F, ♂; G, ♀; H–I, abdominal ventrites: H, ♂; I, ♀; J–K, pterothorax, ventral: J, ♂; K, ♀; L, elytron, dorsal; M, scutellum; N, tarsus, dorsal; O, foreleg; P, middle leg; Q, hindleg.

Fig. 44.129. Cryptorhynchini, structural details, *Metacymia* and *Microporopterus*. A–H, *Metacymia* sp.: A, head, dorsal B, head, lateral; C, antenna; D, abdominal ventrites; E, prothorax; F, prothorax, ventral; G, pterothorax and abdominal ventrites, lateral; H, tarsus and claws, dorsal; J–N, *Microporopterus regularis* Lea: J, head, lateral; J, prothorax, ventral; K, pterothorax and abdominal ventrites, ventral; L, tarsus, dorsal; M, antenna; N, middle leg.

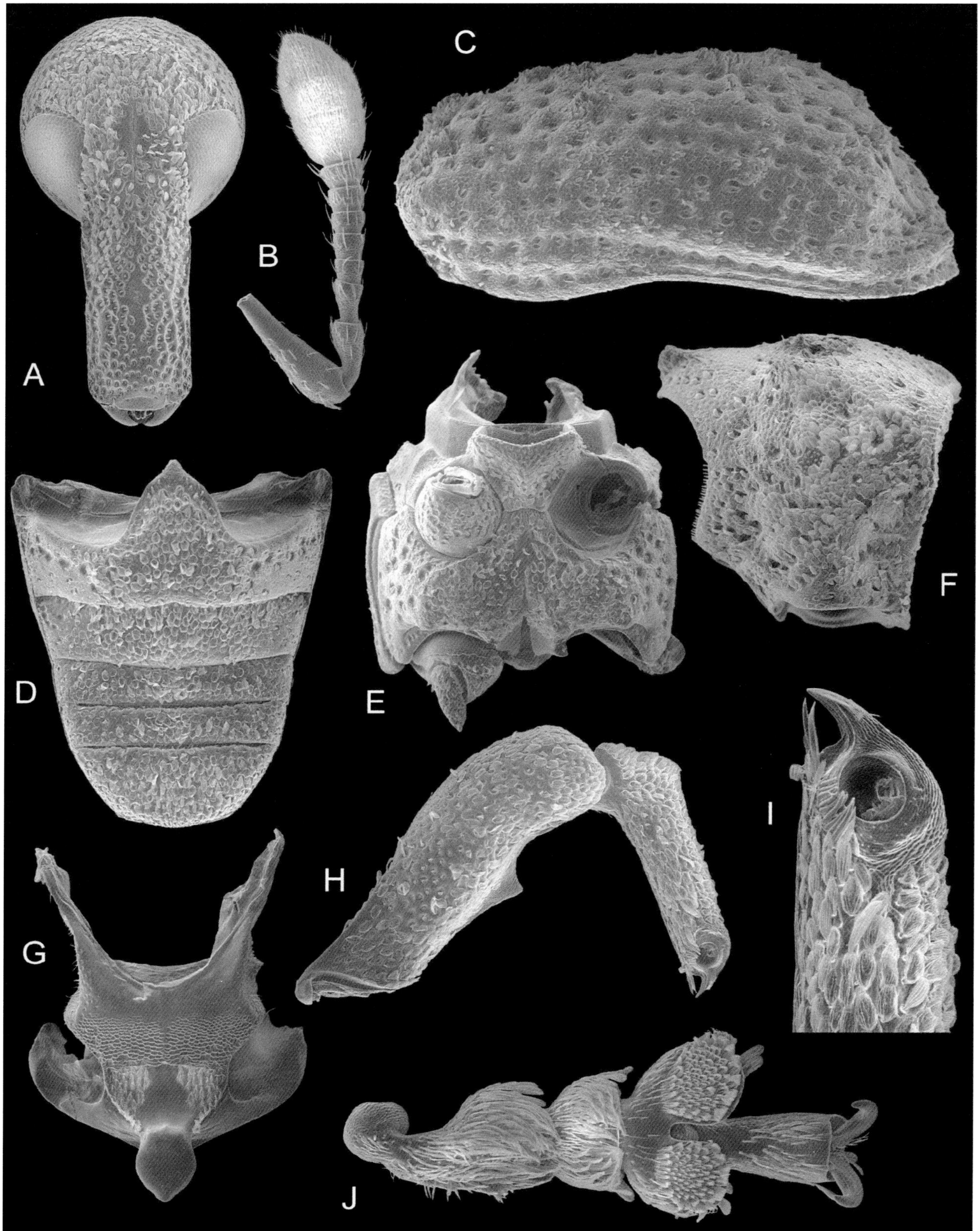

Fig. 44.130. Cryptorhynchini, structural details, *Metyrus albicollis* (Germar), ♂. A, head, dorsal; B, antenna; C, elytron, lateral; D, abdominal ventrites; E, pterothorax, ventral; F, prothorax, lateral; G, scutellum; H, middle leg; I, mesotibial apex; J, tarsus and claws, ventral.

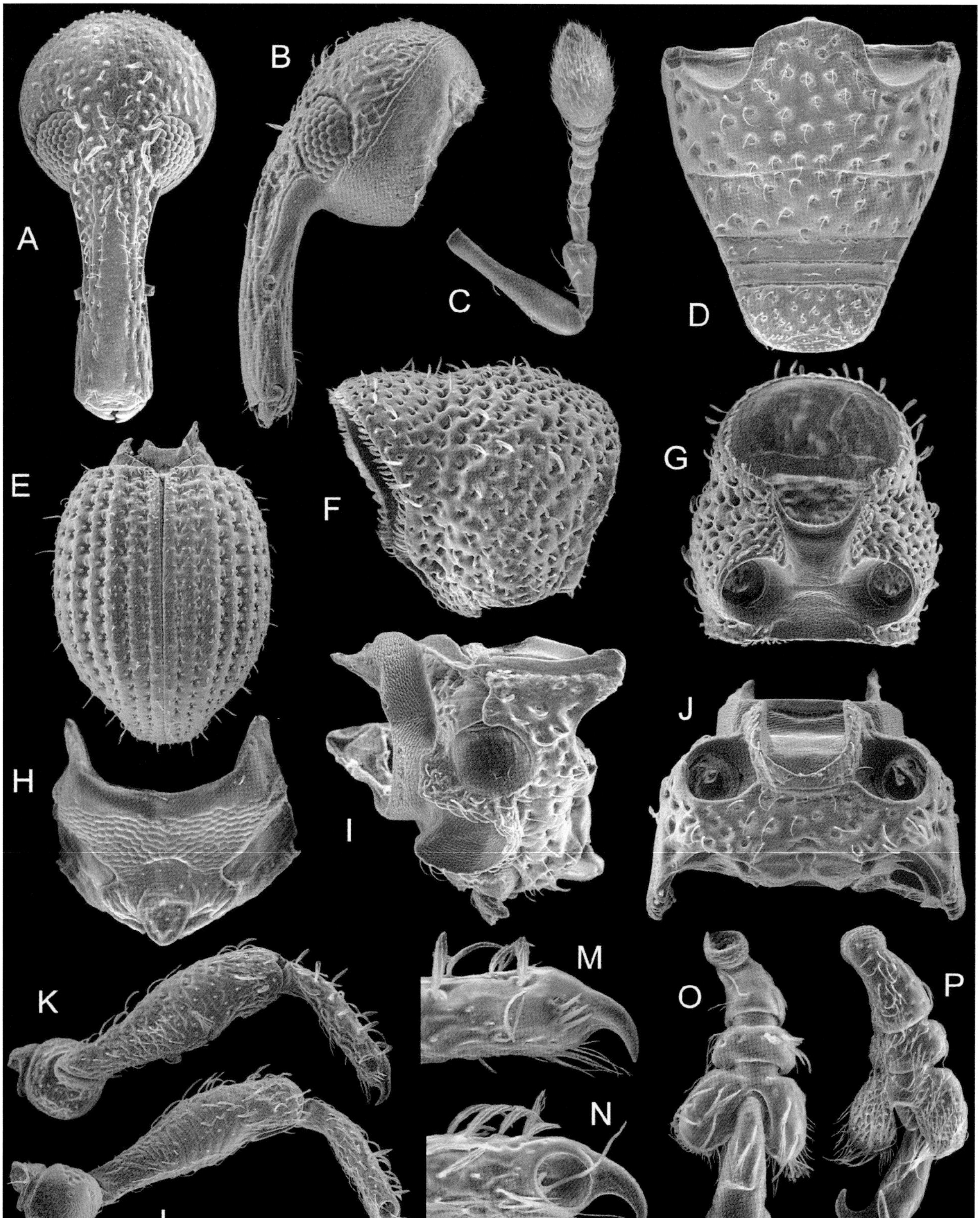

Fig. 44.131. Cryptorhynchini, structural details, *Microcryptorhynchus* sp., ♀. A–B, head: A, dorsal; B, lateral; C, antenna; D, abdominal ventrites; E, elytron, dorsal; F–G, prothorax: F, lateral and G, ventral; H, scutellum; I–J, pterothorax: I, lateral; J, ventral; K, foreleg, anterior; L, foreleg, posterior; M, tibial apex; N, tibial apex; O–P, tarsus and claws: O, dorsal; P, ventral.

Fig. 44.132. Cryptorhynchini, structural details, *Mitrastethus* sp., ♀. A–B, head: A, dorsal; B, lateral; C, antenna; D, abdominal ventrites; E, elytron, lateral; F, prothorax, ventral; G, pterothorax, ventral; H, scutellum; I, middle leg; J, mesotibial apex; K, tarsus, dorsal.

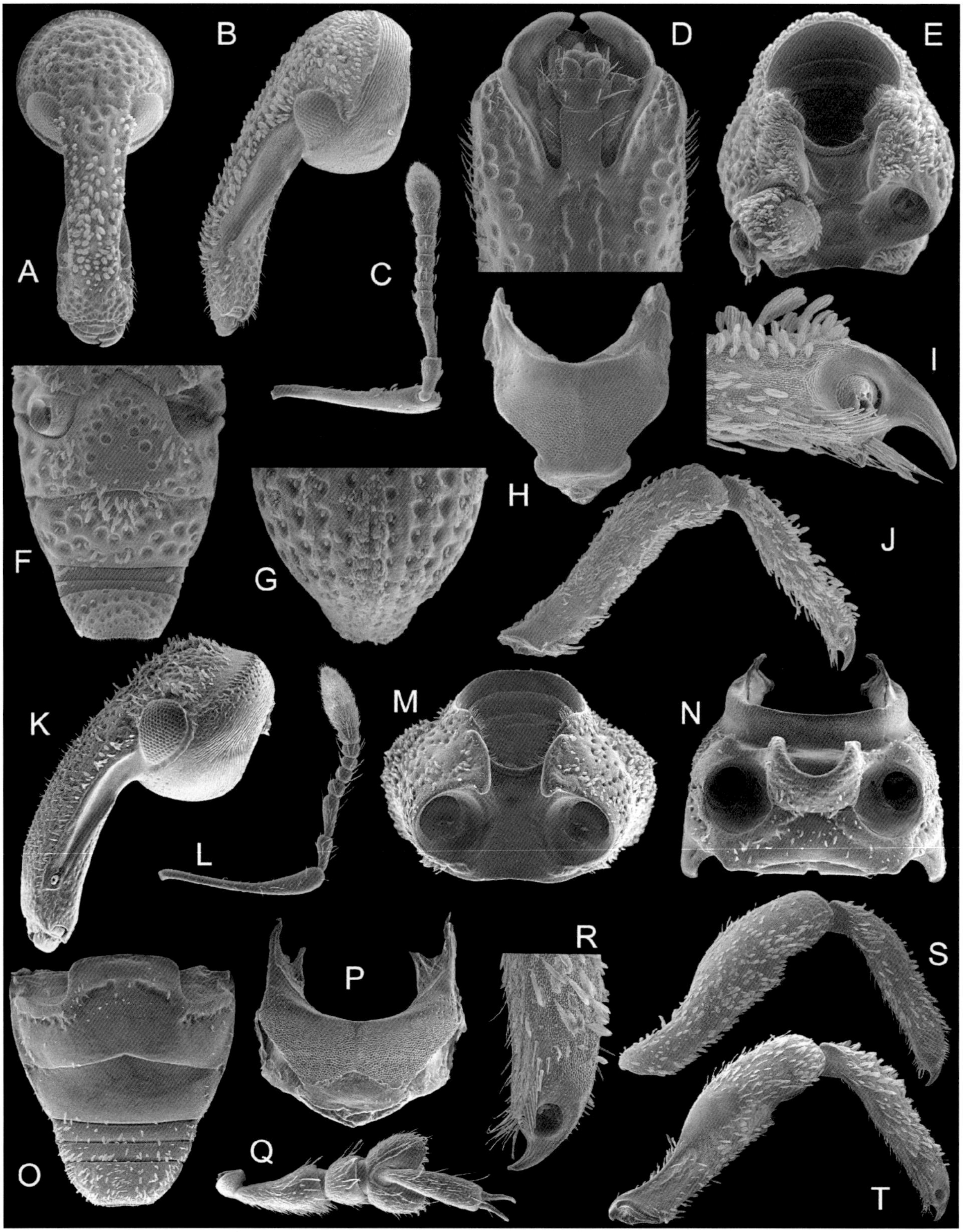

Fig. 44.133. Cryptorhynchini, structural details, *Neoporopterus* and *Niconotus*. A–J, *Neoporopterus lithodermus* (Boisduval), ♂: A, head, dorsal; B, head, lateral; C, antenna; D, mouthparts; E, prothorax, ventral; F, abdominal ventrites; G, elytral apex, dorsal; H, scutellum; I, mesotibial apex; J, middle leg; K–T, *Niconotus* sp.: K, head, lateral; L, antenna; M, prothorax, ventral; N, pterothorax, ventral; O, abdominal ventrites; P, scutellum; Q, tarsus and claws, dorsal; R, mesotibial apex; S, middle leg, anterior; T, middle leg, posterior.

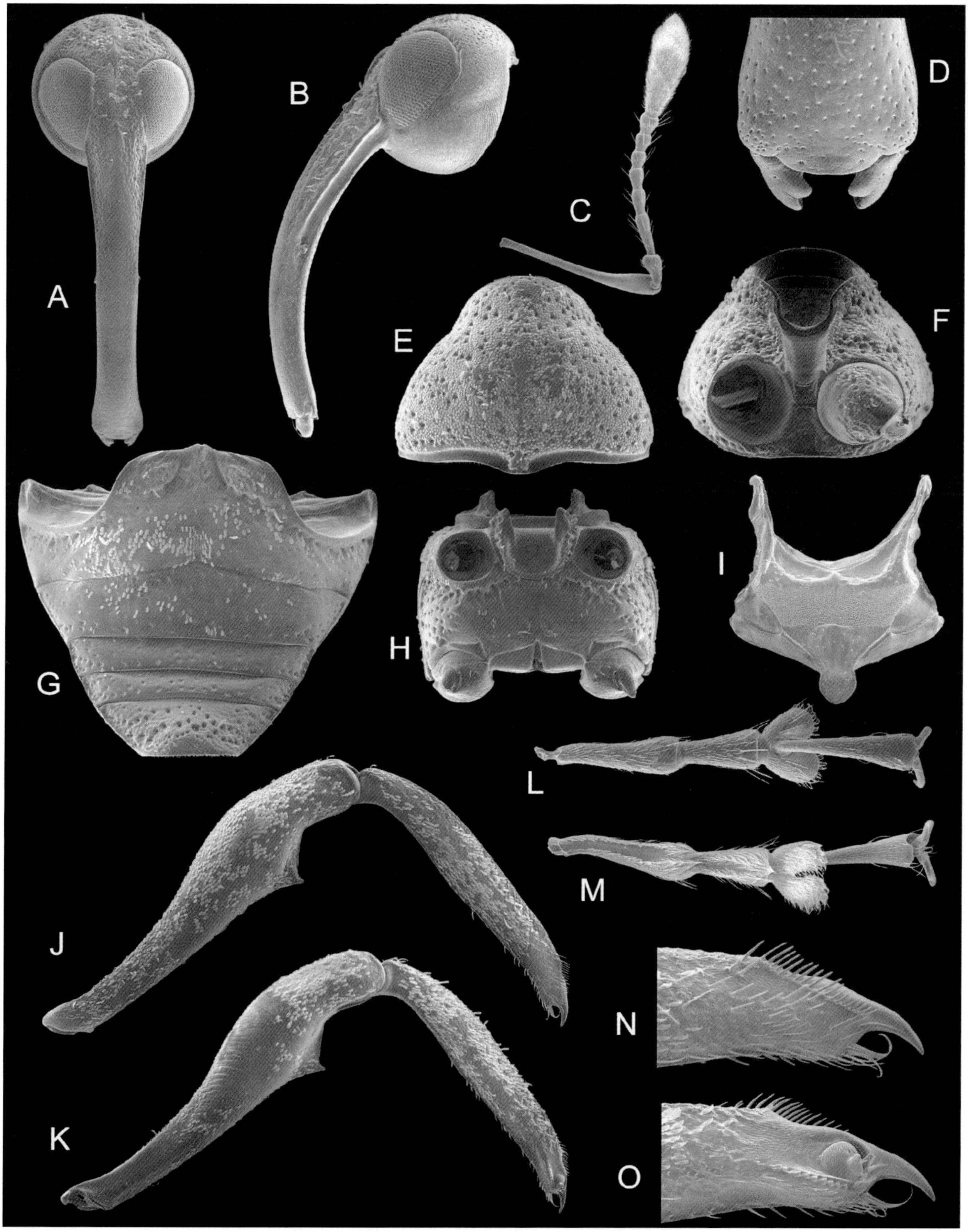

Fig. 44.134. Cryptorhynchini, structural details, *Onidistus* sp. A–B, head: A, dorsal; B, lateral; C, antenna; D, rostrum with mandibles, dorsal; E–F, prothorax: E, dorsal; F, ventral; G, abdominal ventrites; H, pterothorax, ventral; I, scutellum; J, middle leg, anterior; K, middle leg, posterior; L–M, tarsus and claws: L, dorsal; M, ventral; N–O, tibial apex: N, outer; O, inner.

Fig. 44.135. Cryptorhynchini, structural details, *Orochlesis* sp. A–B, head: A, dorsal; B, lateral; C, antenna; D, elytron, lateral; E–F, abdominal ventrites: E, ♂; F, ♀; G, prothorax, ventral; H, scutellum; I, tarsus and claws, dorsal; L, pterothorax, ventral; K, middle leg; L, hindleg.

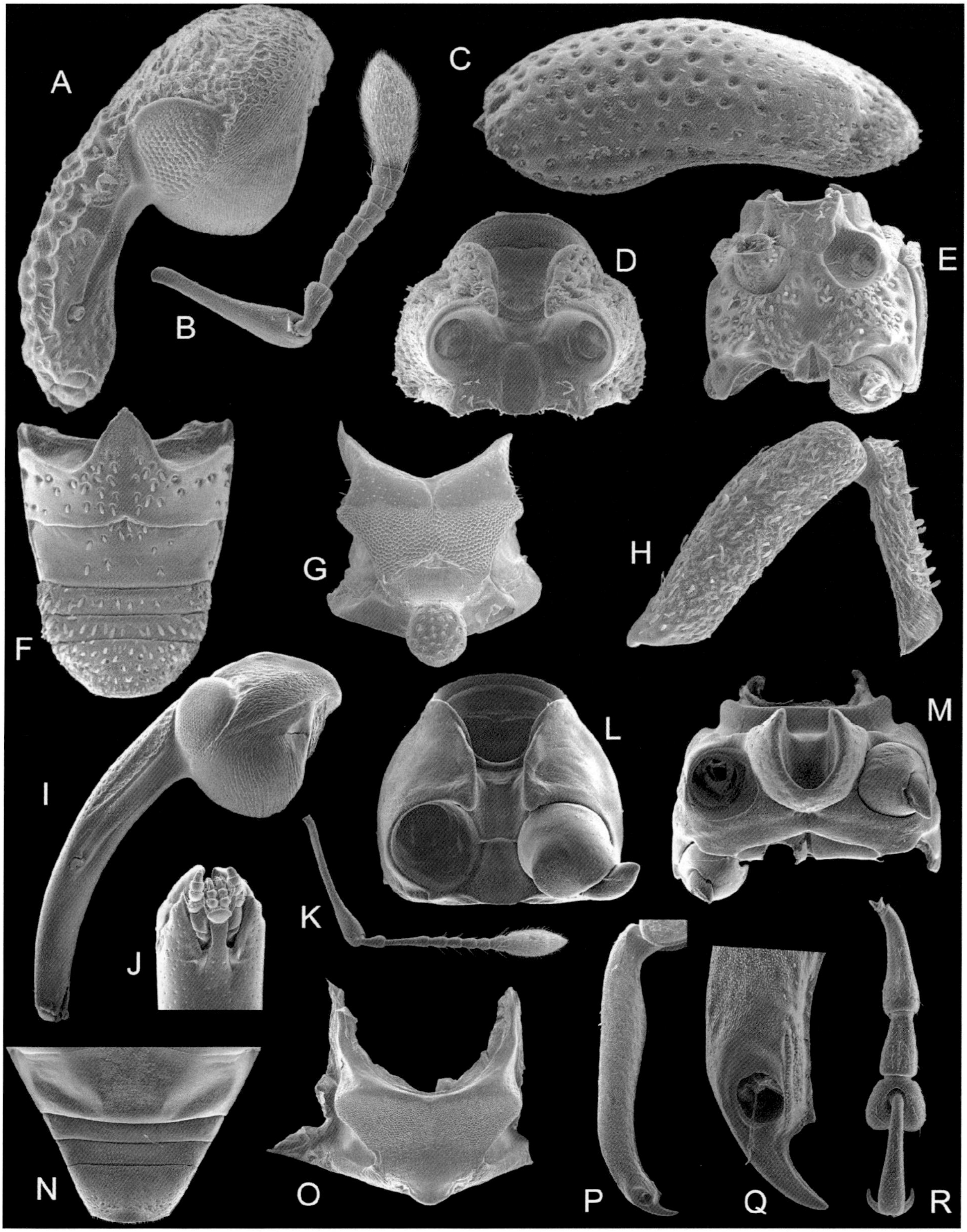

Fig. 44.136. Cryptorhynchini, structural details, *Orthoporopterus* and *Paleticus*. A–H, *Orthoporopterus elongatus* Lea, ♂: A, head, lateral; B, antenna; C, elytron, lateral; D, prothorax, ventral; E, pterothorax, ventral; F, abdominal ventrites; G, scutellum; H, middle leg; I–R, *Paleticus subereus* Pascoe: I, head, lateral; J, mouthparts; K, antenna; L, prothorax, ventral; M, pterothorax, ventral; N, abdominal ventrites; O, scutellum; P, protibia; Q, protibial apex; R, tarsus, dorsal.

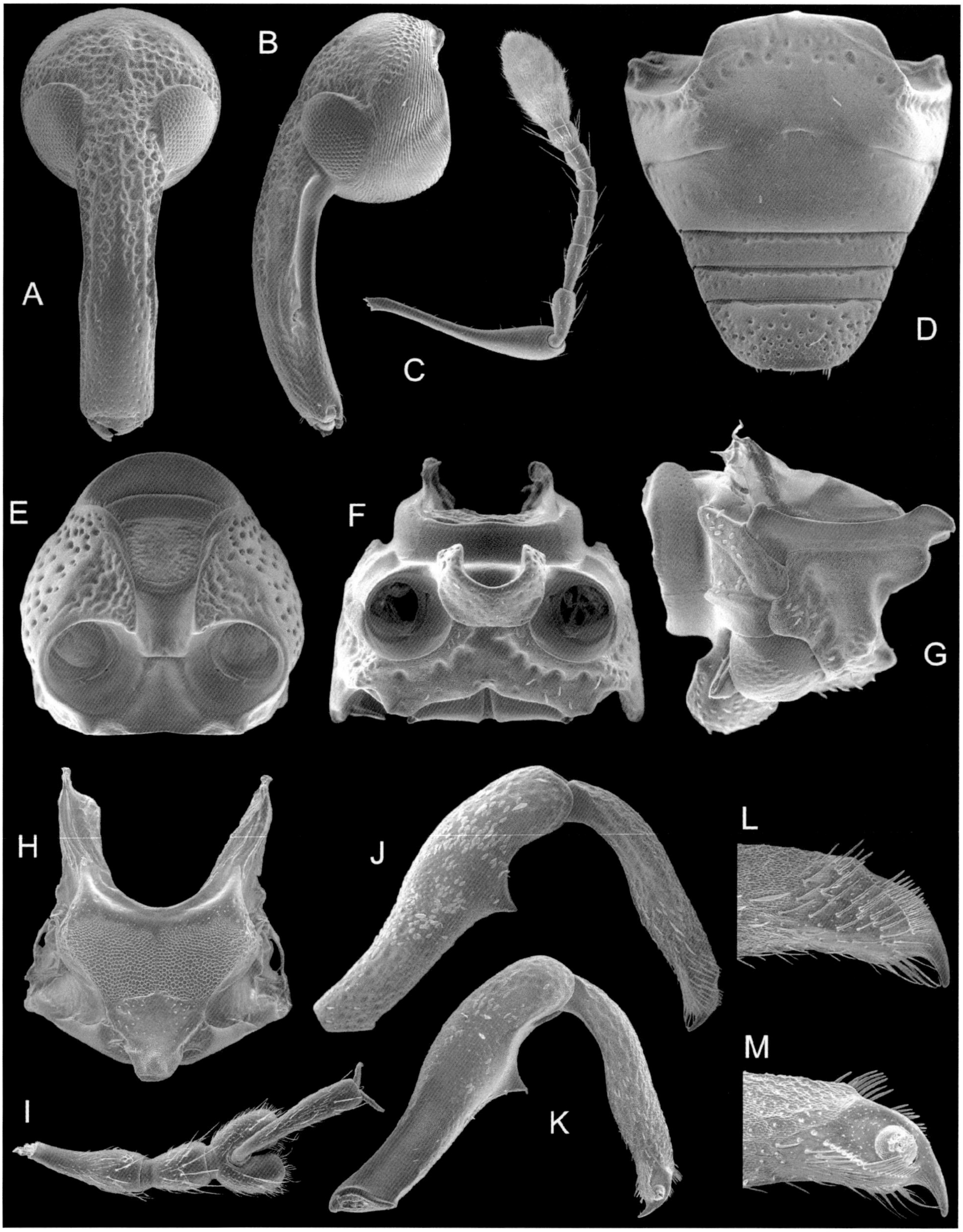

Fig. 44.137. Cryptorhynchini, structural details, *Paletonidistus trisinuatus* Lea. A–B, head: A, dorsal; B, lateral; C, antenna; D, abdominal ventrites; E, prothorax, ventral; F–G, pterothorax: F, ventral; G, lateral; H, scutellum; I, tarsus and claws, dorsal; J–M, legs: middle leg; J, anterior; K, posterior; L, tibial apex, anterior; M, tibial apex, posterior.

Fig. 44.138. Cryptorhynchini, structural details, *Parasalcus latissimus* (Pascoe). A, head, lateral; B, antenna; C, abdominal ventrites; D, elytron, lateral; E, prothorax, ventral; F, scutellum; G, pterothorax, ventral; H, middle leg; I, mesotibial apex; J, tarsus, dorsal.

Fig. 44.139. Cryptorhynchini, structural details, *Phloeoglymma dorsale* (Pascoe), ♀. A, head, lateral; B, antenna; C, abdominal ventrites; D, prothorax, ventral; E, pterothorax, ventral; F, scutellum; G, elytron, lateral; H, tarsus and claws, dorsal; I, middle leg; J, mesotibial apex.

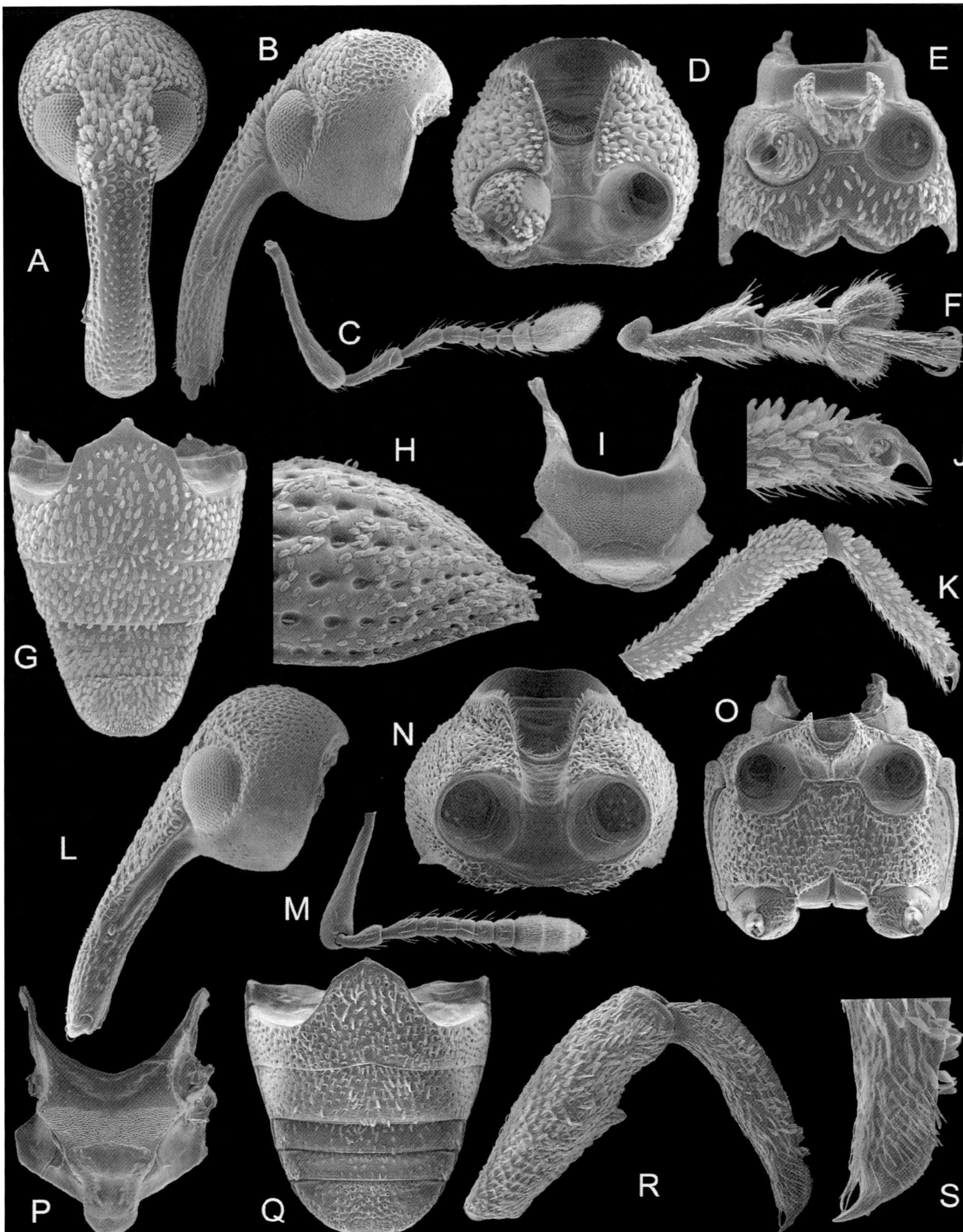

Fig. 44.140. Cryptorhynchini, structural details, *Poropterus* and *Pseudometyrus*. A–K, *Poropterus bituberculatus* Lea, ♀: A, head, dorsal; B, head, lateral; C, antenna; D, prothorax, ventral; E, pterothorax, ventral; F, tarsus and claws, dorsal; G, abdominal ventrites; H, elytral apex, lateral; J, mesotibial apex; K, middle leg; L–S, *Pseudometyrus antares* (Erichson), ♂: L, head, lateral; M, antenna; N, prothorax, ventral; O, pterothorax, ventral; P, scutellum; Q, abdominal ventrites; R, middle leg; S, mesotibial apex.

Fig. 44.141. Cryptorhynchini, structural details, *Pseudapries corticalis* Lea. A–B, head: A, dorsal; B, lateral; C, mouthparts; D, scutellum; E, antenna; F–G, abdominal ventrites: F, ♂; G, ♀; H–I, prothorax; H, dorsal; I, ventral; J, pterothorax, ventral; K, elytron, lateral; L, tarsus and laws, dorsal; M, middle leg; N, hindleg.

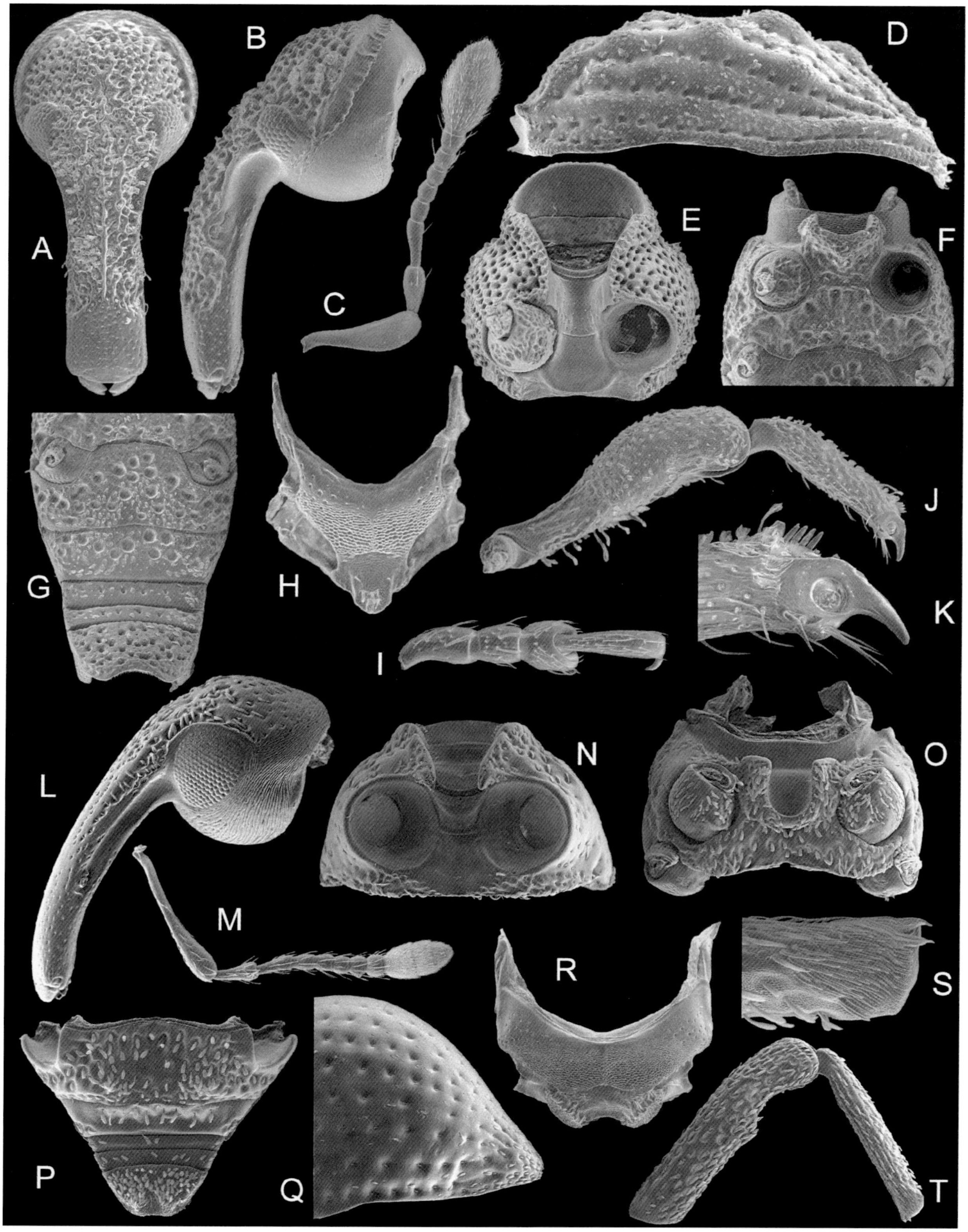

Fig. 44.142. Cryptorhynchini, structural details, *Roptoperus* and *Salcus*. A–K, *Roptoperus basalis* Lea, ♂: A, head, dorsal; B, head, lateral; C, antenna; D, elytron, lateral; E, prothorax, ventral; F, pterothorax, ventral; G, abdominal ventrites; H, scutellum; I, tarsus, dorsal; J, middle leg; K, mesotibial apex; L–T, *Salcus elevatus* Pascoe, ♂: L, head, lateral; M, antenna; N, prothorax, ventral; O, pterothorax, ventral; P, abdominal ventrites; Q, elytral apex, lateral; R, scutellum; S, mesotibial apex; T, middle leg.

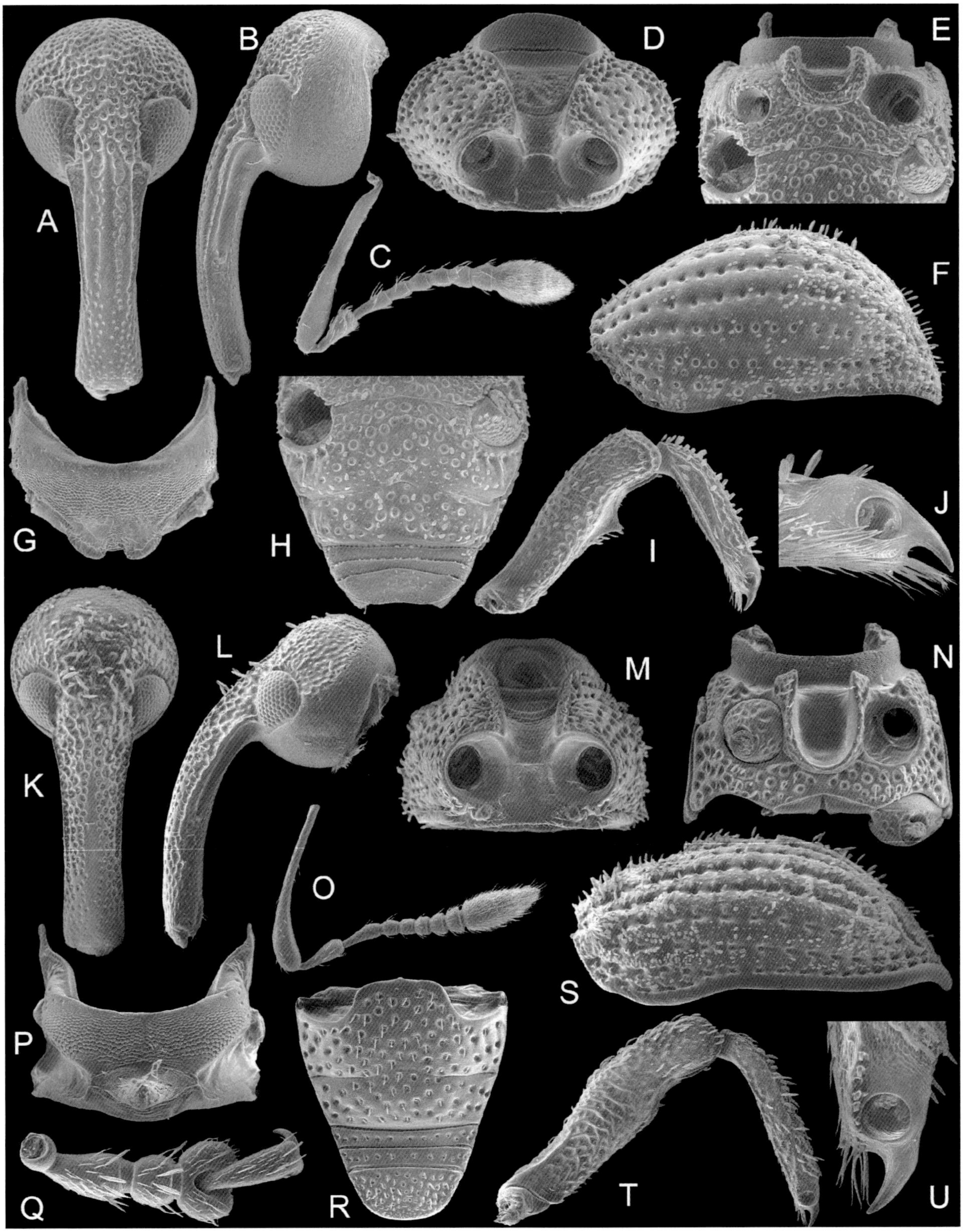

Fig. 44.143. Cryptorhynchini, structural details, *Tamphilus* and *Tapinocis*. A–J, *Tamphilus* sp., ♀: A, head, dorsal; B, head, lateral; C, antenna; D, prothorax, ventral; E, pterothorax, ventral; F, elytron, lateral; G, scutellum; H, abdominal ventrites; I, middle leg; J, mesotibial apex; K–U, *Tapinocis* nr *corticalis,* Mt. Oakleigh TAS, ♀: K, head, dorsal; L, head, lateral; M, prothorax, ventral; N, pterothorax, ventral; O, antenna; P, scutellum; Q, tarsus, dorsal; R, abdominal ventrites; S, elytron, lateral; T, middle leg; U, mesotibial apex.

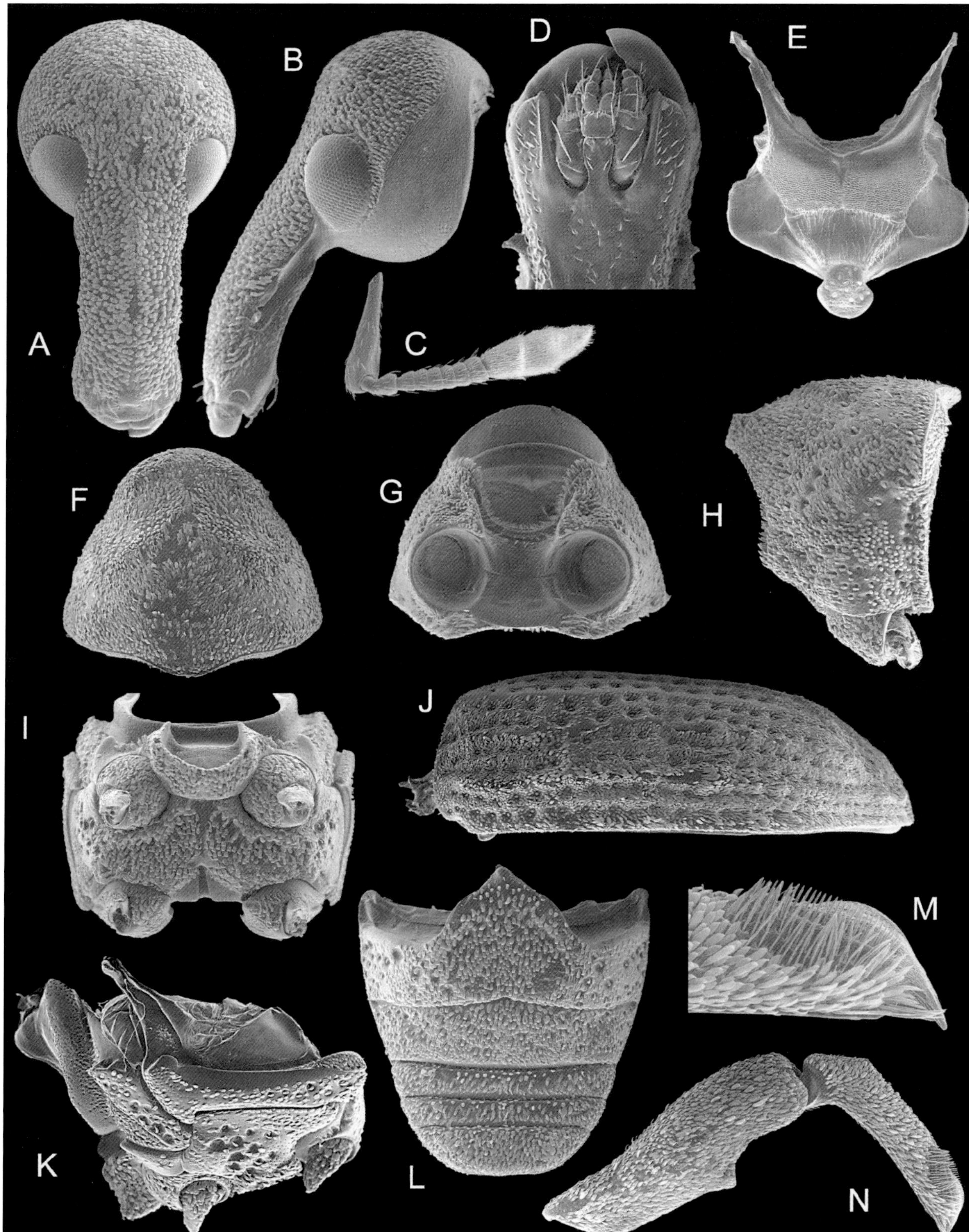

Fig. 44.144. Cryptorhynchini, structural details, *Tepperia sterculiae* (Lea). A–B, head: A, dorsal; B, lateral; C, antenna; D, mouthparts; E, scutellum; F–H, prothorax: F, dorsal; G, ventral; H, lateral; I; K, pterothorax; I, ventral; K, lateral; J, elytron, lateral; L, abdominal ventrites; M, mesotibial apex; N, middle leg.

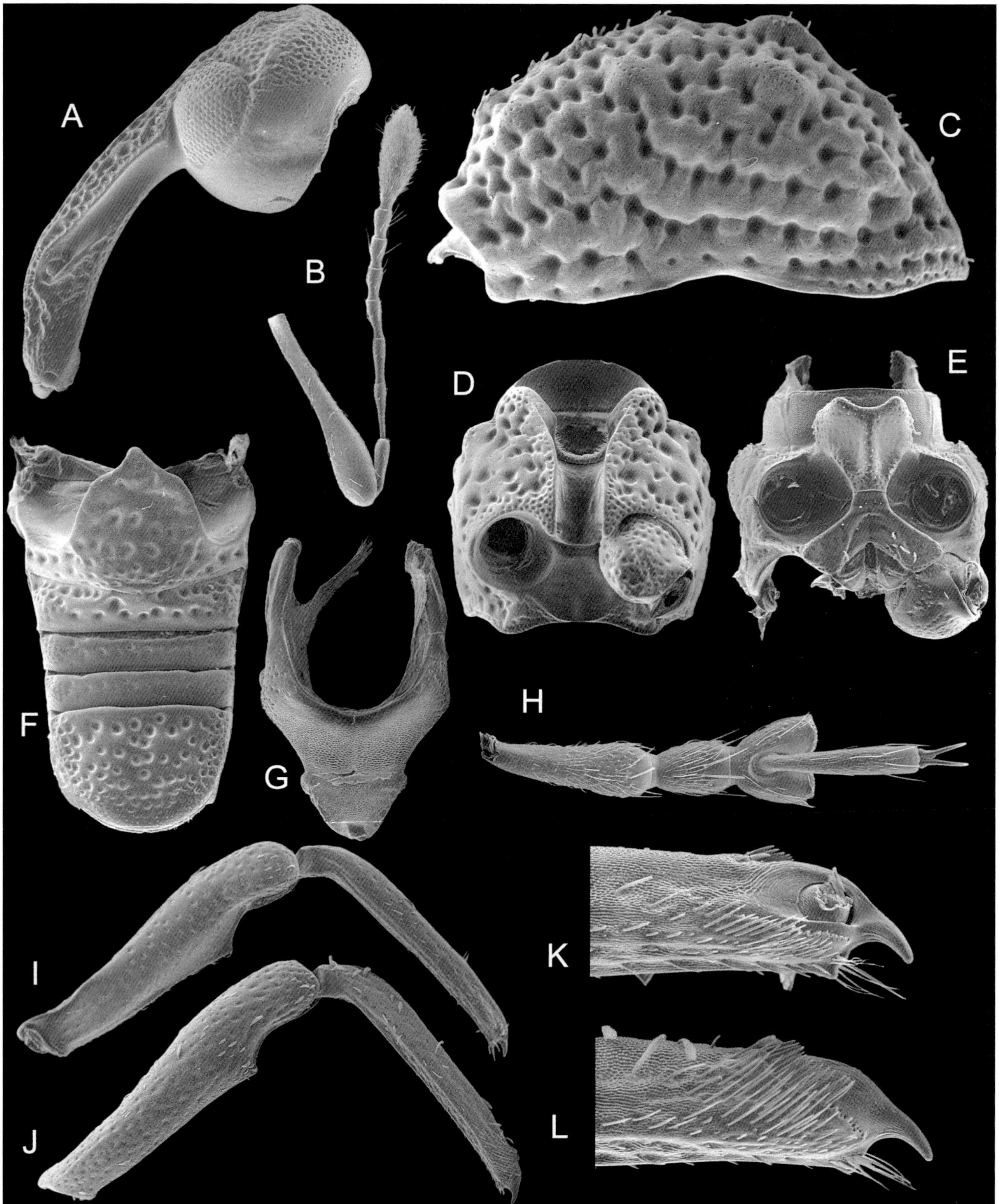

Fig. 44.145. Cryptorhynchini, structural details, *Terporopus tenuicornis* Lea ♂. A, head, lateral; B, antenna; C, elytron, lateral; D, prothorax, ventral; E, pterothorax, ventral; F, abdominal ventrites; G, scutellum; H, tarsus and claws, dorsal; I–L, middle leg: I, posterior; J, anterior; K, tibial apex, anterior; L, tibial apex, posterior.

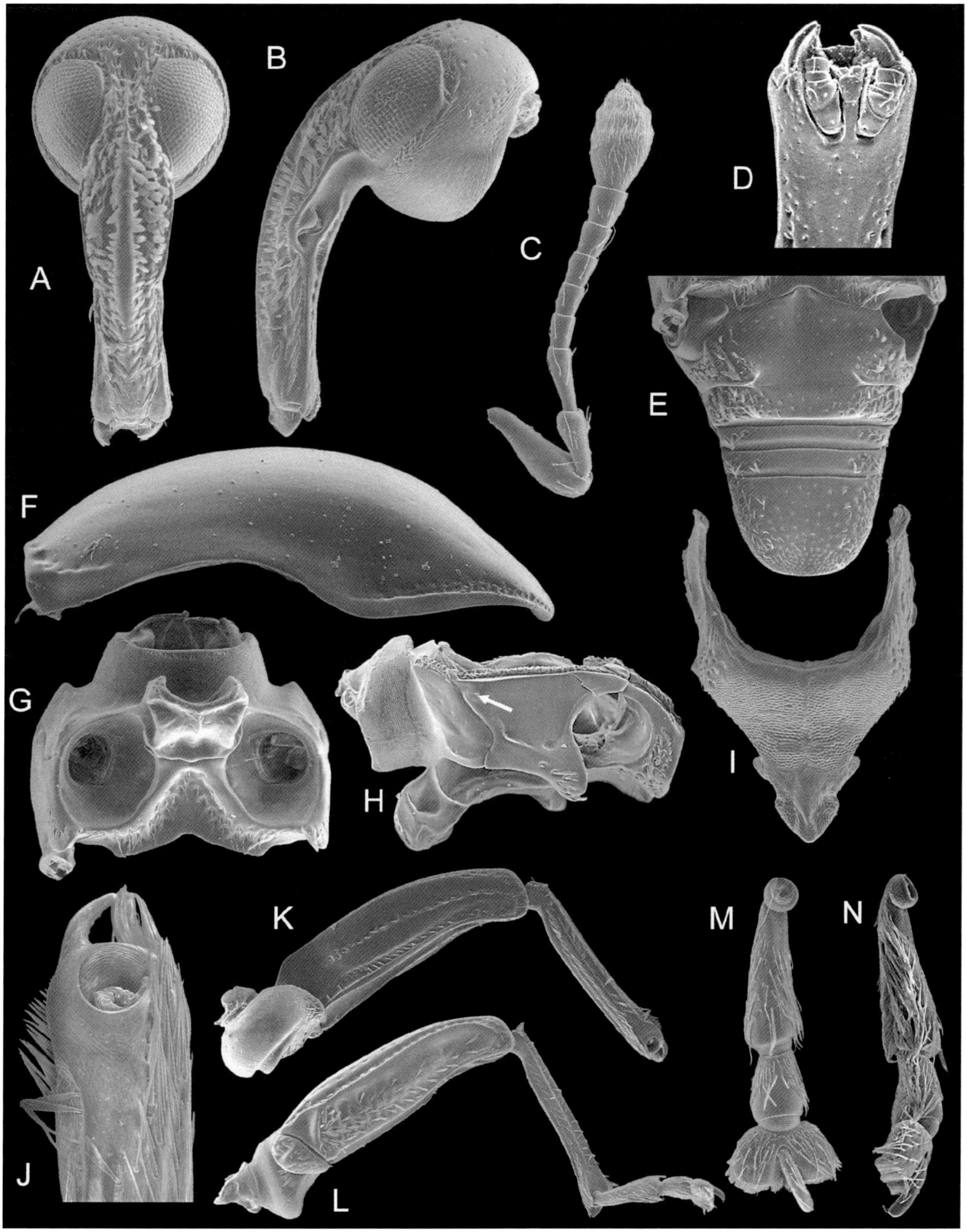

Fig. 44.146. Cryptorhynchini, structural details, *Trigonopterus rostralis* (Lea). A–B, head: A, dorsal; B, lateral; C, antenna; D, antenna; E, abdominal ventrites; F, elytron, lateral; G–H, pterothorax: G, ventral; H, lateral; I, scutellum; J, protibial apex; K, hindleg; L, foreleg; M–N, tarsus: M, dorsal; N, lateral.

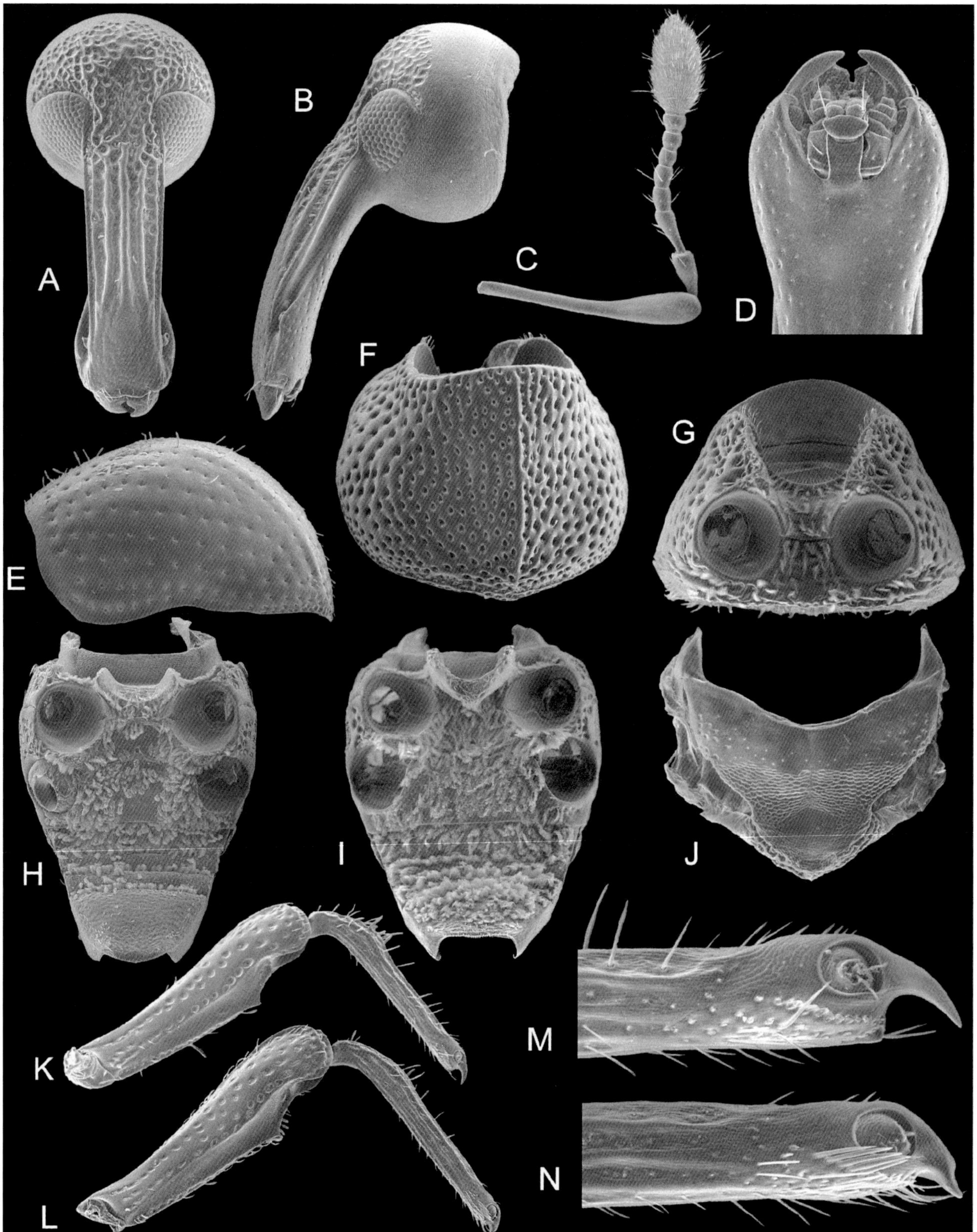

Fig. 44.147. Cryptorhynchini, structural details, *Tropidotasia* sp. A–B, head: A, dorsal; B, lateral; C, antenna; D, mouthparts; E, elytron, lateral; F–G, prothorax: F, dorsal; G, ventral; H–I, abdominal ventrites: H, ♂; I, ♀; J, scutellum; K, middle leg; L, foreleg; M, mesotibial apex; N, protibial apex.

Fig. 44.148. Cryptorhynchini, structural details, *Tyrtaeosus alternatus* Lea. A–B, head: A, dorsal; B, lateral; C, antenna; D, elytron, lateral; E, abdominal ventrites; F–G, prothorax: F, dorsal; G, ventral; H–I, pterothorax: H, lateral; I, ventral; J, scutellum; K, hindleg; L, foreleg; M, metatibial apex; N, protibial apex; O, tarsus and claws, dorsal.

Tychreus Pascoe, 1870 (Pl. 94H)

Type species, by monotypy: *Tychreus camelus* Pascoe, 1870.
Australian species. *T. aberrans* Lea, 1913a, *T. camelus* Pascoe, 1870e, *T. coryssomerus* Lea, 1908b, *T. dilaticollis* Lea, 1908b, *T. discicollis* Lea, 1913b, *T. fumosus* Lea, 1910c, *T. furvus* Lea, 1910c, *T. incanus* Lea, 1913b, *T. insularis* Lea, 1913a, *T. lanifer* Lea, 1913b, *T. latifrons* Lea, 1913b, *T. longicornis* Lea, 1913b, *T. nigronotatus* Lea, 1913a, *T. reversus* Lea, 1910c, *T. sellatus* Pascoe, 1873c.
Distribution. Australia (NSW, QLD, TAS, VIC, WA, Lord Howe Is.).
Notes. *Atherosperma moschatum* (Atherospermataceae) has been recorded as a host plant of the genus.

Tylocis Lea, 1913 (Pl. 99F)

Type species, by original designation: *Tylocis squamibundus* Lea, 1913.
Australian species. *T. squamibundus* Lea, 1913a.
Distribution. Australia (NSW).

Tyrtaeosellus Lea, 1913 (Pl. 94I)

Type species, by subsequent designation (Pullen *et al.* 2014: 272): *Tyrtaeosellus coxalis* Lea, 1913.
Australian species. *T. alternatus* Lea, 1913a, *T. attenuatus* Lea, 1913b, *T. coxalis* Lea, 1913b, *T. nigrofasciatus* Lea, 1913b.
Distribution. Australia (QLD, NSW).

Tyrtaeosus Pascoe, 1870 (Pl. 94J, K; Fig. 44.148)

Type species, by subsequent designation (Faust, 1899: 75): *Tyrtaeosus microthorax* Pascoe, 1870.
Australian species. *T. aberrans* Lea, 1913b, *T. aemulus* Lea, 1902, *T. aequus* Lea, 1902, *T. albolineatus* Lea, 1915d, *T. alternatus* Lea, 1913b, *T. apicicollis* Lea, 1913b, *T. assimilis* Lea, 1902, *T. basiventris* Lea, 1913b, *T. bicolor* Pascoe, 1885, *T. bifasciatus* Lea, 1913a, *T. bifoveiceps* Lea, 1913b, *T. biseriatus* Lea, 1902, *T. bivulneratus* Lea, 1913b, *T. brevirostris* Lea, 1913b, *T. carinatus* Lea, 1902, *T. castor* Lea, 1902, *T. cinerascens* Lea, 1902, *T. concinnus* Lea, 1913b, *T. concretus* Pascoe, 1885, *T. corpulentus* Lea, 1902, *T. crassirostris* Lea, 1902, *T. dolosus* Lea, 1902, *T. flavonotatus* Lea, 1913b, *T. foveipennis* Lea, 1913b, *T. foveiventris* Lea, 1913b, *T. ichthyosomus* Lea, 1902, *T. imitator* Lea, 1902, *T. incallidus* Pascoe, 1870e, *T. interstitialis* Lea, 1911c, *T. lateralis* Pascoe, 1870e, *T. laterarius* Lea, 1902, *T. macrops* Lea, 1913a, *T. majorinus* Lea, 1913b, *T. melanterioides* Lea, 1915d, *T. microthorax* Pascoe, 1870e, *T. mixtus* Lea, 1913a, *T. modicus* Lea, 1913b, *T. nigromaculatus* Lea, 1913b, *T. pardalis* Pascoe, 1885, *T. pollux* Lea, 1902, *T. pulcher* Lea, 1913a, *T. punctirostris* Lea, 1913b, *T. quadratolineatus* Lea, 1902, *T. religiosus* Lea, 1902, *T. squamiceps* Lea, 1913b, *T. squamivarius* Lea, 1913a, *T. subopacus* Lea, 1913b, *T. trianguliferus* Lea, 1913a, *T. trilinealbus* Lea, 1913b, *T. urens* Lea, 1902, *T. ustulatus* Pascoe, 1870e, *T. vetustus* Pascoe, 1870e.
Distribution. Australia (QLD, NSW, SA, TAS, VIC, WA, Norfolk Is.), New Guinea.
Notes. Specimens have been collected on *Eucalyptus* and on flowers of Livistona palms.

Zenoporopterus Lea, 1912 (Pl. 94L)

Type species, by monotypy: *Zenoporopterus mirus* Lea, 1912.
Australian species. *Z. mirus* Lea, 1912.
Distribution. Australia (NSW).

Tribe Ithyporini Lacordaire, 1865

Camptorhinus Schoenherr, 1825 (Pl. 82E, F)

Type species, by original designation: *Curculio statua* Fabricius, 1792.
Australian species. *C. crinipes Gahan*, *C. dorsalis* (Boisduval, 1835), *C. inornatus* Lea, 1913c, *C. interstitialis* Lea, 1913a.
Distribution. Australia, Oriental, Afrotropical, Pacific, southern and eastern Palaearctic regions.

Tribe Juanorhinini Aurivillius, 1926

Arecophaga Broun, 1880 (Pl. 82G, H; Fig. 44.149)

Type species, by monotypy: *Arecophaga varia* Broun, 1880.
Australian species. *Arecophaga* sp.
Distribution. Australia, New Zealand.

Tribe Lixini Schoenherr, 1823

Larinus Dejean, 1821 (Pl. 82I)

Type species, by subsequent designation (Westwood, 1840: 36): *Curculio planus* Fabricius, 1792.
Australian species. *L. latus* (Herbst, 1783).
Distribution. Australia, NE USA, Palaearctic, Afrotropical, Madagascar, India, Sumatra, Taiwan.
Notes. The single species in Australia was introduced from Greece around 1992 (Pullen *et al.* 2014).

Lixus Fabricius, 1801 (Pl. 82J; Fig. 44.150A–G)

Type species, by subsequent designation (Latreille, 1810: 430): *Curculio paraplecticus* Linnaeus, 1758.
Australian species. *L. albolineatus* Lea, 1899b, *L. breweri* Pascoe, 1874a, *L. cardui* Olivier, 1807, *L. immundus*

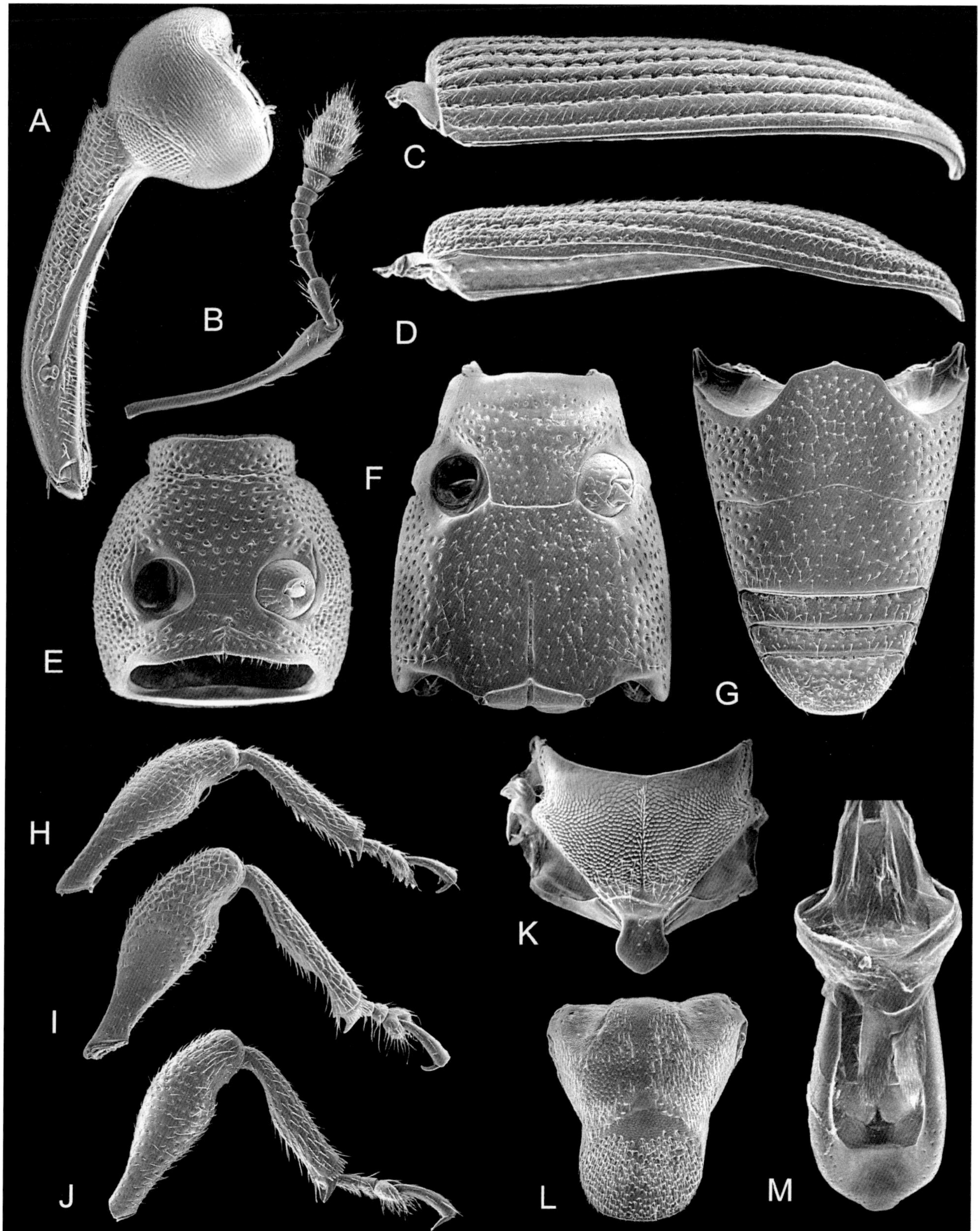

Fig. 44.149. Juanorhinini, structural details, *Arecophaga* sp. A, head, lateral; B, antenna; C–D, elytron: C, dorsal; D, lateral; E, prothorax, ventral; F, pterothorax, ventral; G, abdominal ventrites; H, foreleg; I, middle leg; J, hindleg; K, scutellum; L, pygidium; M, penis, dorsal.

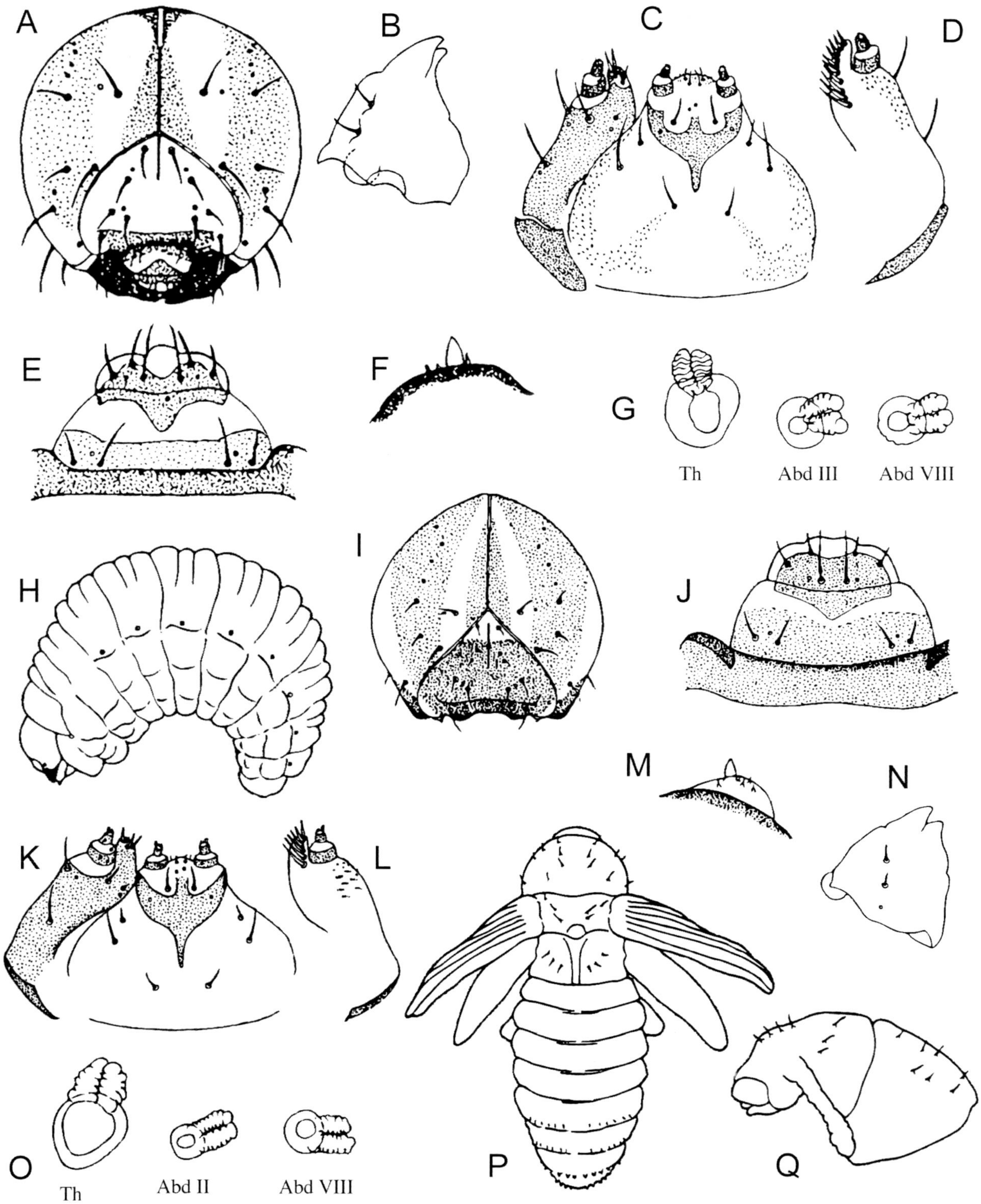

Fig. 44.150. Lixini, structural details of immature stages. A–G, *Lixus mastersii* Pascoe, larva (Fiji and Australia): A, head, dorsal; B, mandible; C, mala and labium, ventral; D, mala, dorsal; E, labrum, clypeus and epistoma; F, antenna; G, spiracles, showing alignment; H–O, *Rhinocyllus conicus* (Frölich), larva (New Zealand): H, habitus, lateral; I, head, dorsal; J, labrum, clypeus and epistoma; K, mala and labium, ventral; L, mala, dorsal; M, antenna; N, mandible; O, spiracles, showing alignment; P–Q, *Rhinocyllus conicus* (Frölich), pupa (New Zealand): P, habitus, dorsal; Q, head and prothorax, lateral. After May (1994).

Boheman, 1859, *L. imponderosus* Lea, 1911a, *L. mastersii* Pascoe, 1874a, *L. tasmanicus* Germar, 1848.

Distribution. Australia (NSW, QLD, NT, VIC, SA, TAS, WA), cosmopolitan.

Notes. *Lixus cardui* was introduced from the Mediterranean region in 1993 for the biological control of *Onopordium* thistles (Pullen *et al.* 2014). Another species introduced as a biocontrol agent, *L. linearis* Olivier, 1807 from Morocco to control *Rumex* species (Polygonaceae), has not established (Pullen *et al.* 2014). The taxonomy of the native Australian species is in need of revision, and none probably belong in *Lixus*. Their recorded hosts are Amaranthaceae (*Amaranthus*, *Alternanthera*) and Asteraceae (*Chrysanthemun*, *Senecio*, *Xanthium*).

Rhinocyllus Germar, 1817 (Pl. 82K; Fig. 44.150H–Q)

Type species, by monotypy: *Curculio thaumaturgus* Rossi, 1794.

Australian species. *R. conicus* (Frölich, 1792).

Distribution. Australia, Africa, central and eastern Canada, USA, Europe, Central Asia, Syria, Turkey, New Zealand.

Notes. The single species in Australia was introduced from Europe between 1988 and 1999 for the biocontrol of *Carduus*, *Cirsium* and *Silybum* thistles (Pullen *et al.* 2014).

Tribe Mecysolobini Reitter, 1913

Cylindralcides Heller, 1918 (Pl. 100A)

Type species, by subsequent designation (Alonso-Zarazaga, 2013: 115): *Alcides longirostris* Heller, 1918.

Australian species. *C. bubo* (Fabricius, 1801).

Distribution. Australia (QLD, NT), Oriental.

Sternuchopsis Heller, 1918 (Pl. 100B)

Type species, by subsequent designation (Voss, 1956: 1153): *Alcides pectoralis* Boheman, 1836 (*in* Schoenherr 1836).

Australian species. *S. gallus* (Pascoe, 1887b), *S. pentasticta* (Ancey, 1881), *S. pusilla* (Pascoe, 1885).

Distribution. Australia, Siberia, Japan, Central Asia, Africa, Philippines, Solomon Is., Sri Lanka, New Guinea.

Notes. The genus name *Sternuchopsis* is feminine in gender (Alonso-Zarazaga & Lyal 1999), therefore the adjectival species names *pentasticta* and *pusilla* have to agree in gender and take a feminine suffix.

Tribe Mesoptiliini Lacordaire, 1863

Aethemagdalis Zimmerman, 1994 (Pl. 100C; Fig. 44.152)

Type species, by original designation: *Laemosaccus instabilis* Lea, 1896.

Australian species. *A. fulvirostris* (Pascoe, 1873c), *A. instabilis* (Lea, 1896).

Distribution. Australia.

Allolaemosaccus Zimmerman, 1994 (Pl. 100D)

Type species, by original designation: *Laemosaccus latirostris* Lea, 1926.

Australian species. *A. latirostris* (Lea, 1926a).

Distribution. Australia.

Atopomagdalis Zimmerman, 1994 (Pl. 100E)

Type species, by original designation: *Atopomagdalis renoculus* Zimmerman, 1994.

Australian species. *A. renoculus* Zimmerman, 1994.

Distribution. Australia.

Laemosaccellus Zimmerman, 1994 (Pl. 100F)

Type species, by original designation: *Laemosaccus cryptonyx* Pascoe, 1872.

Australian species. *L. cryptonyx* (Pascoe, 1872a), *L. festivus* (Lea, 1896), *L. magdaloides* (Pascoe, 1873c), *L. semiustus* (Pascoe, 1873c).

Distribution. Australia.

Melaemosaccus Zimmerman, 1994 (Pl. 100G)

Type species, by original designation: *Laemosaccus ocularis* Pascoe, 1873.

Australian species. *M. angulatus* (Janczyk, 1959), *M. bidentatus* (Lea, 1926a), *M. longiceps* (Pascoe, 1873c), *M. ocularis* (Pascoe, 1873c).

Distribution. Australia.

Neolaemosaccus Hustache, 1937 (Pl. 100H, I; Fig. 44.151)

Type species, by original designation: *Laemosaccus subsignatus* Boheman, 1844.

Australian species. *N. argenteus* (Lea, 1896), *N. ater* (Lea, 1896), *N. bilobus* (Lea, 1899b), *N. brevipennis* (Pascoe, 1870b), *N. brevis* (Lea, 1899b), *N. calotrichus* (Lea, 1926a), *N. carinicollis* (Lea, 1896), *N. catenatus* (Pascoe, 1871c), *N. chadwicki* (Janczyk, 1966), *N. compactus* (Lea, 1896), *N. crucicollis* (Lea, 1896), *N. cylindricus* (Lea, 1926a), *N.*

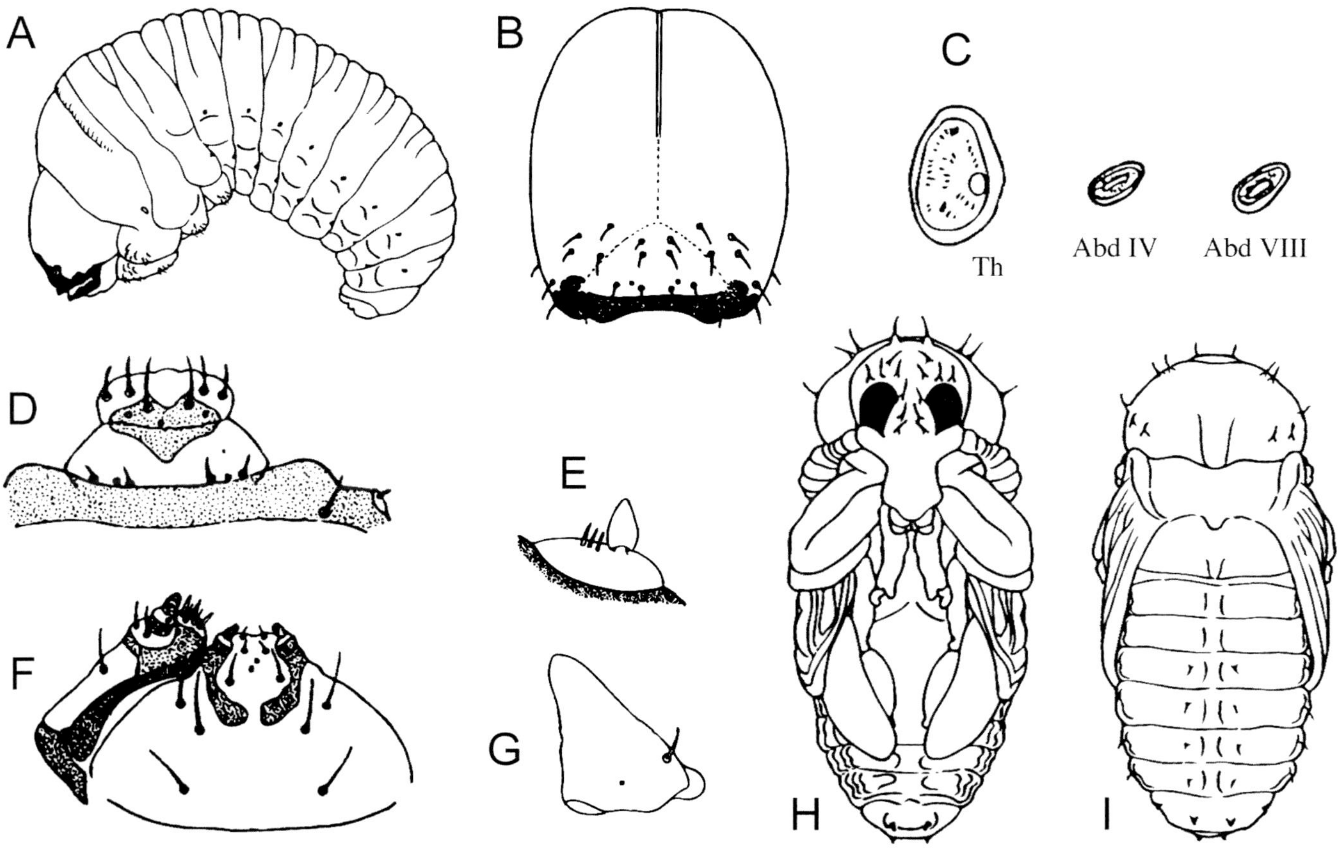

Fig. 44.151. Mesoptiliini, structural details of immature stages. A–I, *Neolaemosaccus narinus* (Pascoe), larva (New Zealand, present in Australia): A, habitus, lateral; B, head, dorsal; C, spiracles, showing alignment; D, labrum, clypeus, epistoma and antenna; E, antenna; F, mala and labium, ventral; G, mandible; H–I, *Neolaemosaccus narinus* (Pascoe), pupa (New Zealand, present in Australia): H, habitus, ventral; I, habitus, dorsal. After May (1994).

dapsilis (Pascoe, 1872a), *N. dubius* (Lea, 1896), *N. electilis* (Pascoe, 1871c), *N. funereus* (Pascoe, 1873c), *N. gibbosus* (Pascoe, 1873c), *N. hieroglyphicus* (Lea, 1911c), *N. longimanus* (Pascoe, 1872a), *N. marmoratus* (Lea, 1926a), *N. melanocephalus* (Boheman, 1843) (*in* Schoenherr 1843), *N. melanoceps* Zimmerman, 1994, *N. microps* (Lea, 1926a), *N. narinus* (Pascoe, 1872a), *N. nigriceps* (Lea, 1911b), *N. nigrirostris* (Lea, 1926a), *N. niveonotatus* (Lea, 1926a), *N. notatus* (Pascoe, 1871c), *N. obscurus* (Lea, 1896), *N. pascoei* (Lea, 1896), *N. peccuarius* (Pascoe, 1871c), *N. pubicollis* (Lea, 1926a), *N. quadriseriatus* (Lea, 1926a), *N. querulus* (Pascoe, 1873c), *N. rivularis* (Lea, 1899b), *N. rufipennis* (Lea, 1896), *N. rufipes* (Lea, 1896), *N. scutellaris* (Lea, 1926a), *N. semicrudus* (Lea, 1926a), *N. subsignatus* (Boheman, 1844) (*in* Schoenherr 1844), *N. synopticus* (Pascoe, 1870b), *N. tantulus* (Pascoe, 1870b), *N. tarsalis* (Pascoe, 1873c), *N. tenuirostris* (Lea, 1926a), *N. thoreyi* (Janczyk, 1957), *N. triangulatus* (Janczyk, 1957), *N. triseriatus* (Lea, 1926a), *N. tropicus* (Lea, 1911b), *N. ustulus* (Pascoe, 1871c), *N. variabilis* (Lea, 1896), *N. varius* (Bovie, 1909) (*in* Lea & Bovie, 1909), *N. ventralis* (Lea, 1896).

Distribution. Australia (QLD, NSW, VIC, TAS, SA, NT, WA), New Zealand.

Notes. *Neolaemosaccus narinus* has been introduced to New Zealand and it is associated with acacias (May 1993, 1994).

Notomagdalis Zimmerman, 1994 (Pl. 100J; Fig. 44.153)

Type species, by original designation: *Laemosaccus biseriatus* Lea, 1926.

Australian species. *N. biseriata* (Lea, 1926a), *N. cossonoides* (Lea, 1896), *N. imitator* (Lea, 1926a), *N. inermis* (Lea, 1909c), *N. mamillata* (Lea, 1909c), *N. rufimanus* (Lea, 1908c), *N. stenotarsus* (Lea, 1909c), *N. subcylindrica* (Lea, 1926a).

Distribution. Australia.

Zimmermanianthus Alonso-Zarazaga & Lyal, 1999 (Pl. 100K; Fig. 44.154)

Type species, by original designation: *Laemosaccus frater* Lea, 1899.

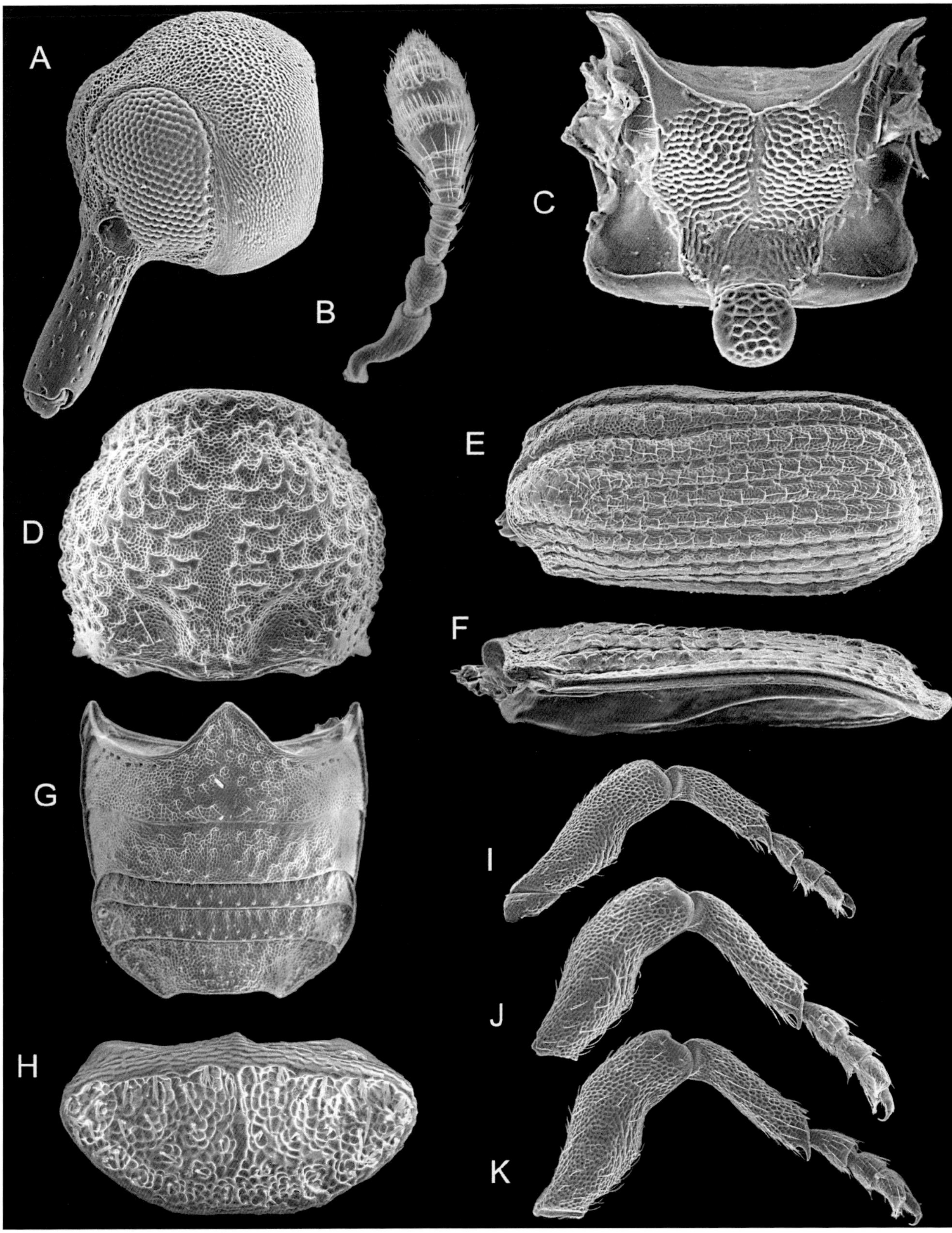

Fig. 44.152. Mesoptiliini, structural details, *Aethemagdalis instabilis* (Lea). A, head, lateral; B, antenna; C, scutellum; D, prothorax; dorsal; E–F, elytron: E, dorsal; F, lateral; G, abdominal ventrites; H, pygidium; I, foreleg; J, middle leg; K, hindleg.

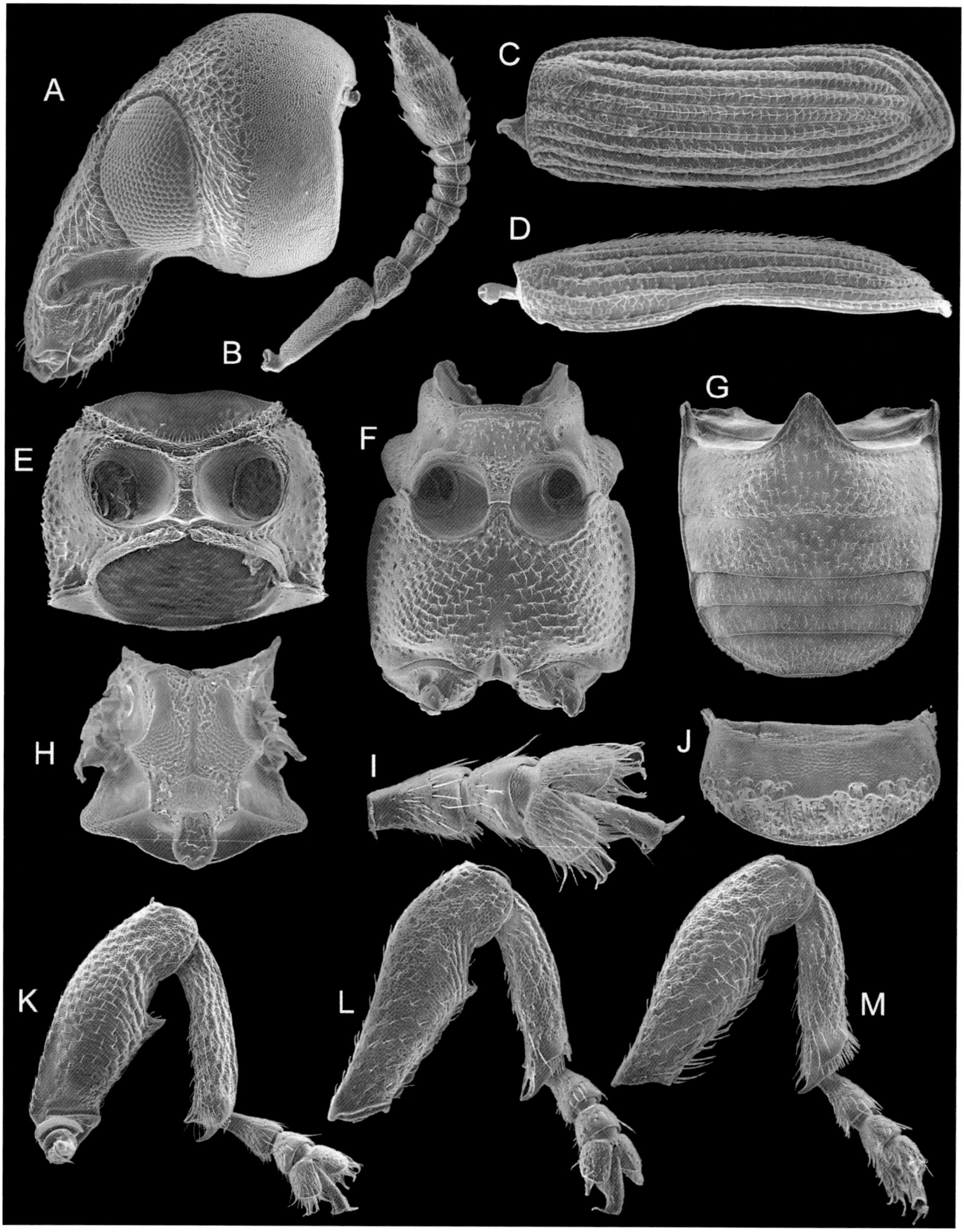

Fig. 44.153. Mesoptiliini, structural details, *Notomagdalis biseriata* (Lea). A, head, lateral; B, antenna; C–D, elytron: C, dorsal; D, lateral; E, prothorax, ventral; F, pterothorax, ventral; G, abdominal ventrites; H, scutellum; I, tarsus and claws, dorsal; J, pygidium; K, foreleg; L, middle leg; M, hindleg.

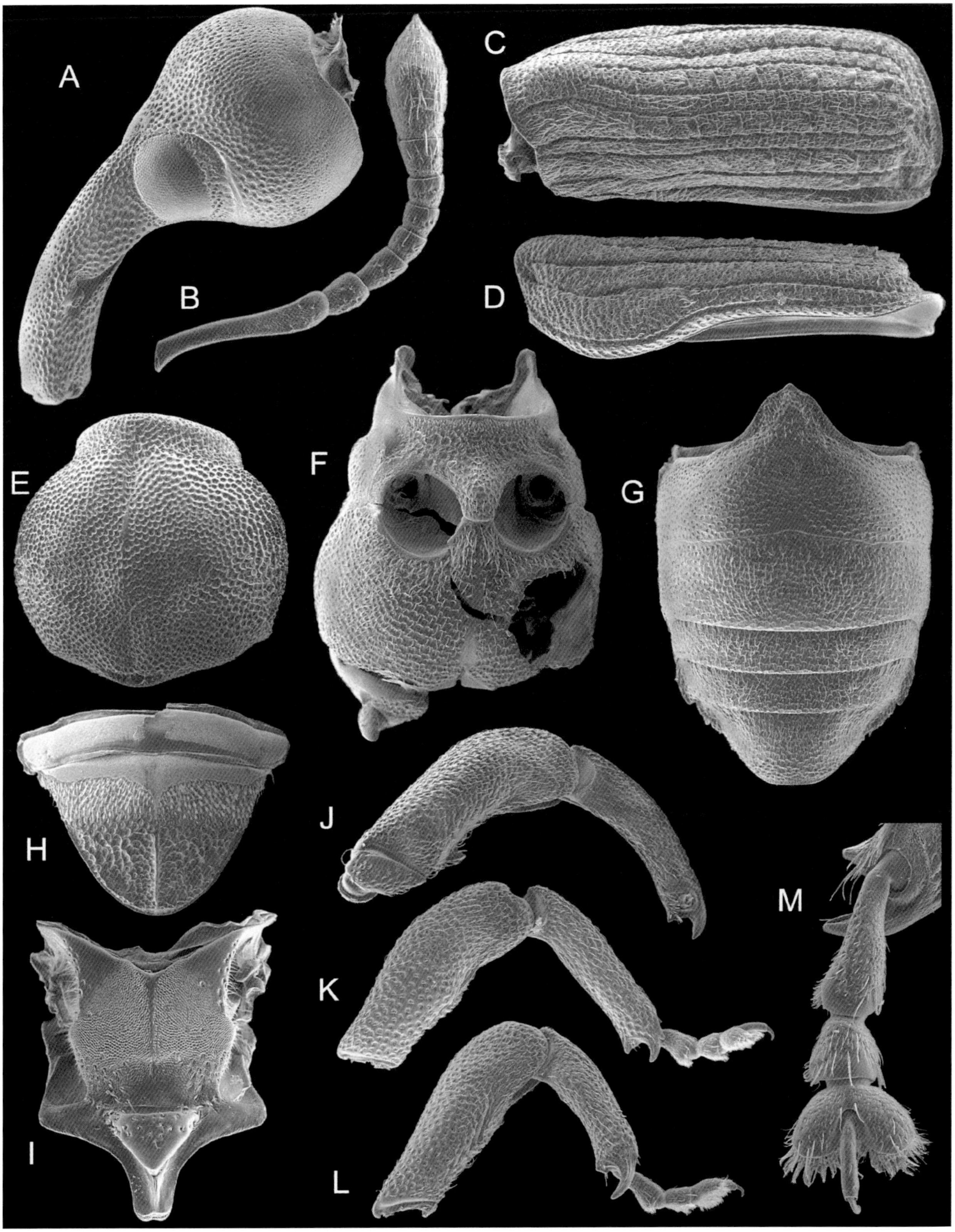

Fig. 44.154. Mesoptiliini, structural details, *Zimmermanianthus frater* (Lea), ♀. A, head, lateral; B, antenna; C–D, elytron: C, dorsal; D, lateral; E, prothorax, dorsal; F, pterothorax, ventral; G, abdominal ventrites; H, pygidium; I, scutellum; J, foreleg; K, middle leg; L, hindleg; M, tarsus and claw, dorsal.

Australian species. *Z. frater* (Lea, 1899b), *Z. judaicus* (Lea, 1899b).
Distribution. Australia.

Tribe Molytini Kirby, 1837

Aclees Schoenherr, 1835 (Pl. 100L)

Type species, by original designation: *Aclees cribratus* Gyllenhal, 1835 (*in* Schoenherr 1835).
Australian species. *Aclees* sp.
Distribution. Australia (QLD, NSW), Afrotropical, Oriental, Micronesia, Solomon Is.
Notes. The single species in Australia has generally been referred to as *A. porosus* Pascoe, 1871 (e.g. Pullen *et al.* 2014), but this species occurs further north, around the Maluku Islands of Indonesia (Meregalli *et al.* 2020), and the species in Australia is undescribed (M. Meregalli, pers. comm. 2020). The larvae develop in the stems of fig trees (Ficus spp.) and perhaps other Moraceae.

Tribe Orthorhinini Jekel, 1865

Eurhamphus Shuckard, 1838 (Pl. 99G)

Type species, by monotypy: *Eurhamphus fasciculatus* Shuckard, 1838.
Australian species. *E. fasciculatus* Shuckard, 1838.
Distribution. Australia (NSW, QLD).
Notes. The larvae of the single species bore in trunks of *Araucaria cunninghamii* and *A. bidwillii* (Araucariaceae), and the adults are often found on the trunks of dying trees (Froggatt 1927; Monteith & Baldwin 2018). The genus was revised by Anderson *et al.* (2018).

Euthebus Pascoe, 1870 (Pl. 72J)

Type species, by monotypy: *Euthebus troglodytes* Pascoe, 1870.
Australian species. *E. troglodytes* Pascoe, 1870e.
Distribution. Australia.
Notes. The straight rostrum and flattened tarsi of the single species indicate an affinity of this genus with *Orthorhinus* and *Parorthorhinus*, and the genus is here classified in Orthorhinini.

Homorthorrhinus Voss, 1960 (Pl. 99D)

Type species, by original designation: *Orthorhinus brachypus* Pascoe, 1885.
Australian species. *Homorthorrhinus* sp.
Distribution. Australia (QLD).
Notes. The only known Australian specimen of *Homorthorrhinus* (id. by Zimmerman) is similar to *H. perversus* (Pascoe, 1885) but lack the pointed tooth in the hind tibia observed in specimens from overseas and mentioned by Kuschel (2008) in the diagnosis of the genus and key to the *Orthorhinus* genus group. The identity of the Australian specimen requires further verification.

Ilacuris Pascoe, 1865 (Pl. 99H)

Type species, by monotypy: *Ilacuris laticollis* Pascoe, 1865.
Australian species. *I. laticollis* Pascoe, 1865, *I. suttoni* Anderson & Setliff, 2018.
Distribution. Australia (NSW, QLD), Papua New Guinea.
Notes. The genus was revised by Anderson *et al.* (2018). In addition to the two species in Australia, another, *I. papuana* Anderson & Setliff, 2018, occurs in New Guinea. The larvae as known develop in trunks and limbs of *Araucaria* species (Araucariaceae).

Imbilius Marshall, 1938 (Pl. 99I)

Type species, by original designation: *Imbilius araucariae* Marshall, 1938.
Australian species. *I. araucariae* Marshall, 1938a.
Distribution. Australia (QLD).
Notes. The genus was reviewed by Anderson *et al.* (2018), but a second species in New Guinea remains undescribed. Both species have been reared from branches of *Araucaria cunninghamii* (Araucariaceae), the larvae tunnelling under the bark.

Notopissodes Zimmerman & Oberprieler, 2014 (Pl. 99J; Fig. 44.156)

Type species, by original designation: *Notopissodes pictus* Zimmerman & Oberprieler, 2014.
Australian species. *N. pictus* Zimmerman & Oberprieler, 2014, *N. variegatus* Oberprieler, 2018.
Distribution. Australia (NSW, QLD).
Notes. The genus was described in Pullen *et al.* (2014) and revised by Anderson *et al.* (2018). A third, undescribed species occurs in New Guinea. Both this and *N. pictus* have been reared from branches and trunks of *Araucaria cunninghamii* (Araucariaceae), the larvae apparently developing under the bark.

Orthorhinus Schoenherr, 1826 (Pl. 99K; Figs 44.155, 44.157, 44.158)

Type species, by original designation: *Curculio cylindrirostris* Fabricius, 1775.
Australian species. *O. aspredo* Pascoe, 1882a, *O. carbonarius* Pascoe, 1882a, *O. cylindrirostris* (Fabricius, 1775), *O. klugii*

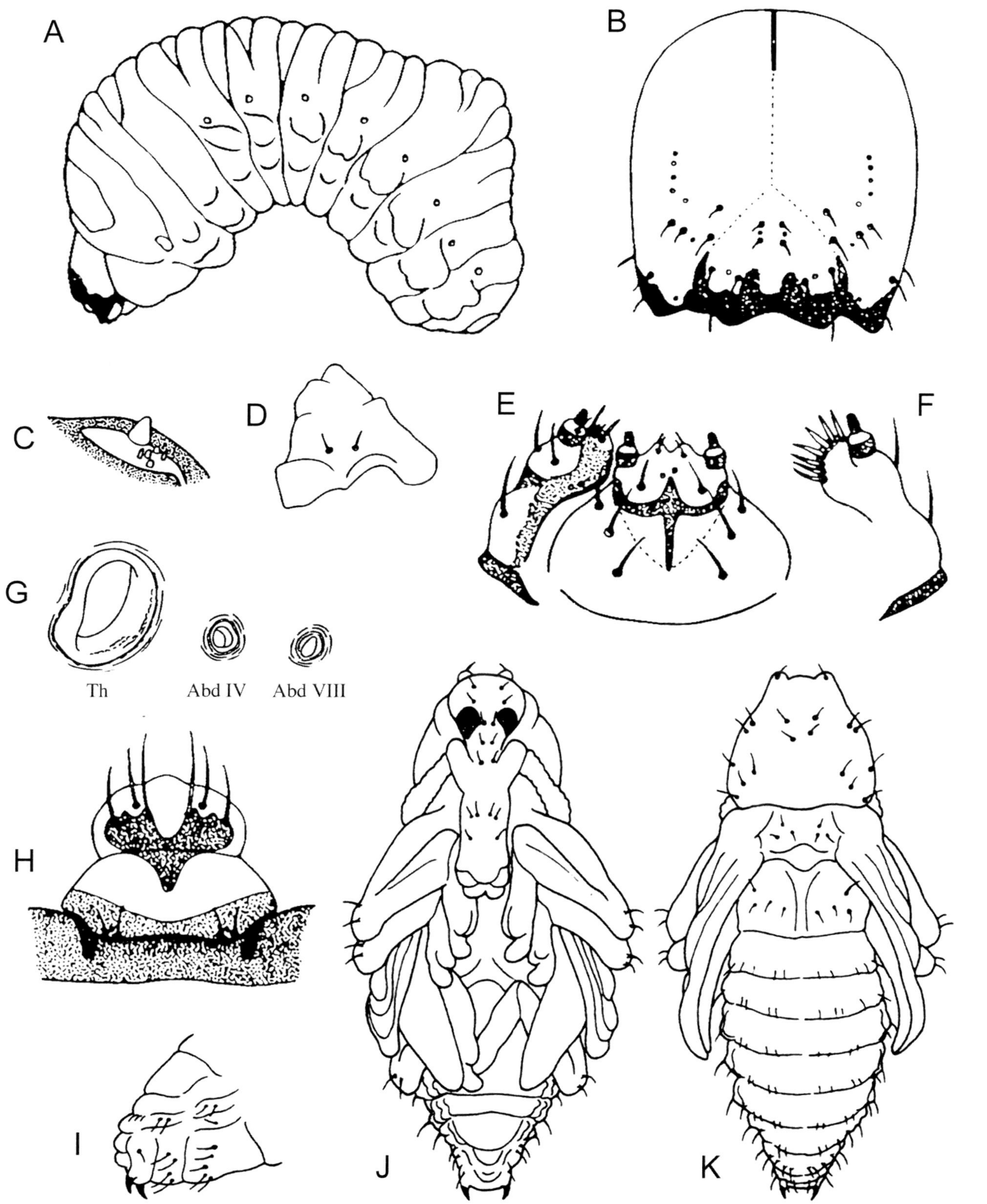

Fig. 44.155. Orthorhinini, structural details of immature stages. A–H, *Orthorhinus cylindrirostris* (Fabricius), larva: A, habitus, lateral; B, head, dorsal; C, antenna, ventral; D, mandible; E, mala and labium, ventral; F, mala, ventral; G, spiracles, showing alignment; H, labrum, clypeus and epistoma; I–K, *Orthorhinus cylindrirostris* (Fabricius), pupa: I, terminal segments, lateral; J, habitus, ventral; K, habitus, dorsal. After May (1994).

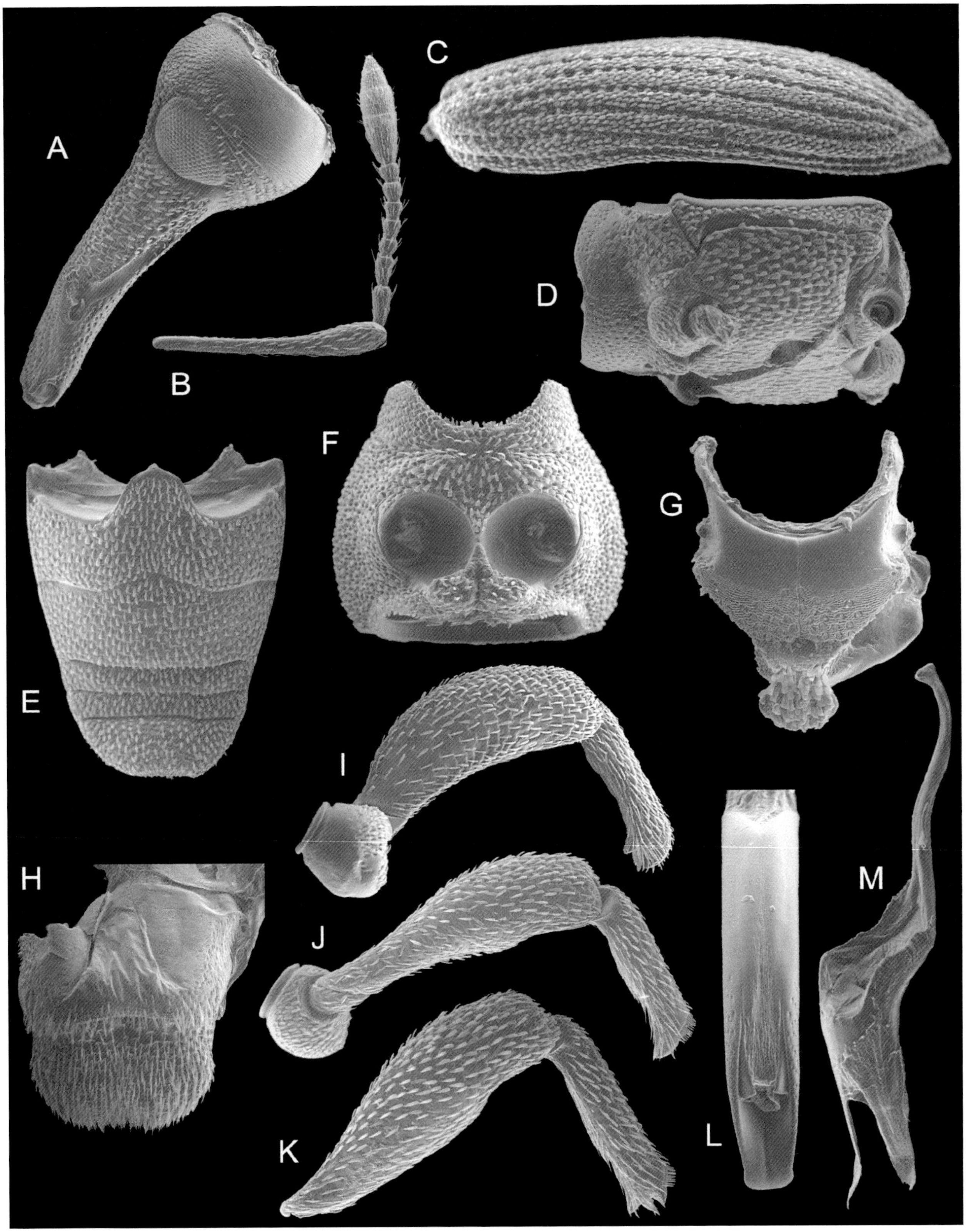

Fig. 44.156. Orthorhinini, structural details of immature stages. *Notopissodes pictus* Zimmerman & Oberprieler, ♂. A, head, lateral; B, antenna; C, elytron, lateral; D, pterothorax, lateral; E, abdominal ventrites; F, prothorax, ventral; G, scutellum; H, pygidium; I, foreleg; J, middle leg; K, hindleg; L–M, penis: L, dorsal; M, lateral.

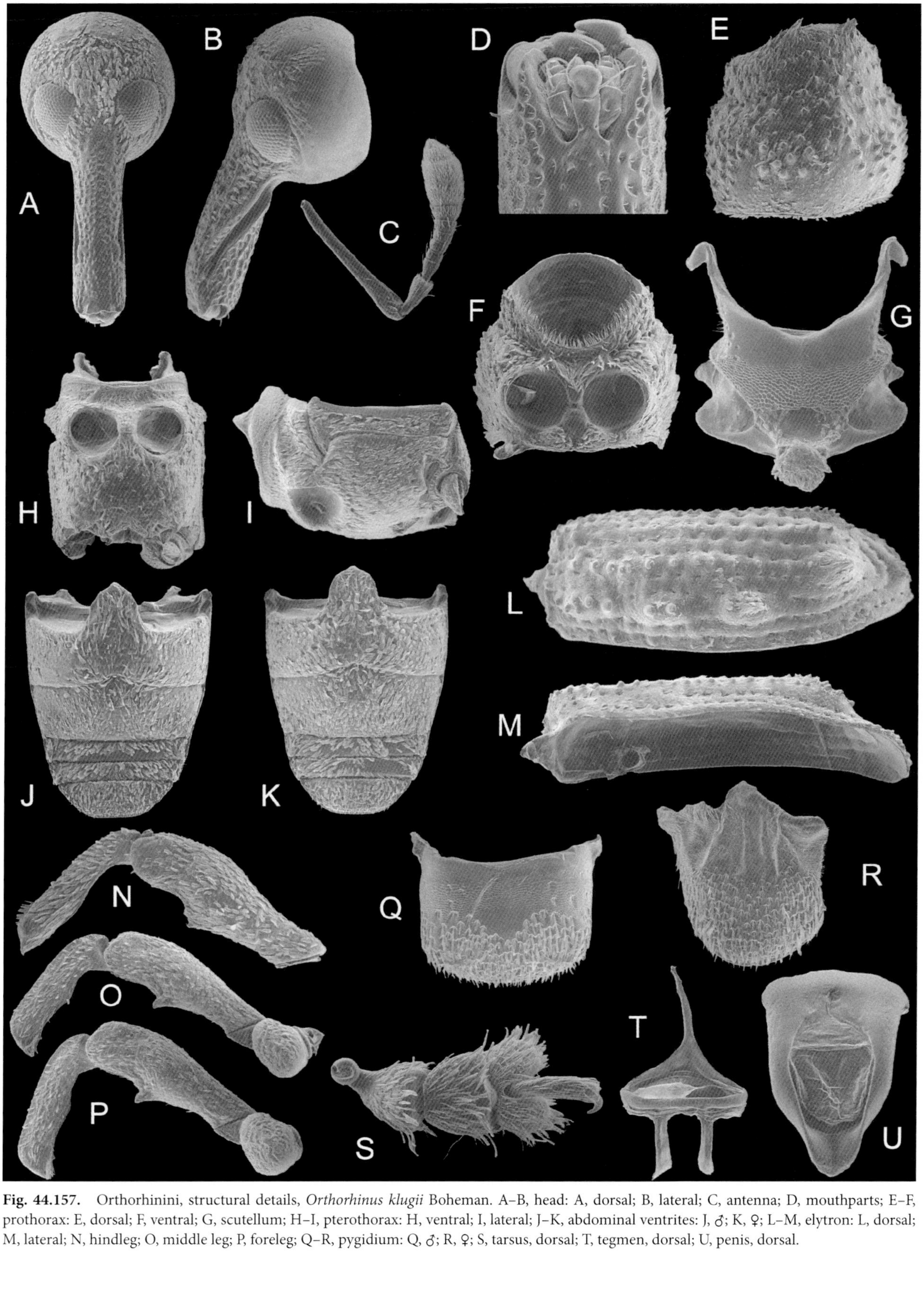

Fig. 44.157. Orthorhinini, structural details, *Orthorhinus klugii* Boheman. A–B, head: A, dorsal; B, lateral; C, antenna; D, mouthparts; E–F, prothorax: E, dorsal; F, ventral; G, scutellum; H–I, pterothorax: H, ventral; I, lateral; J–K, abdominal ventrites: J, ♂; K, ♀; L–M, elytron: L, dorsal; M, lateral; N, hindleg; O, middle leg; P, foreleg; Q–R, pygidium: Q, ♂; R, ♀; S, tarsus, dorsal; T, tegmen, dorsal; U, penis, dorsal.

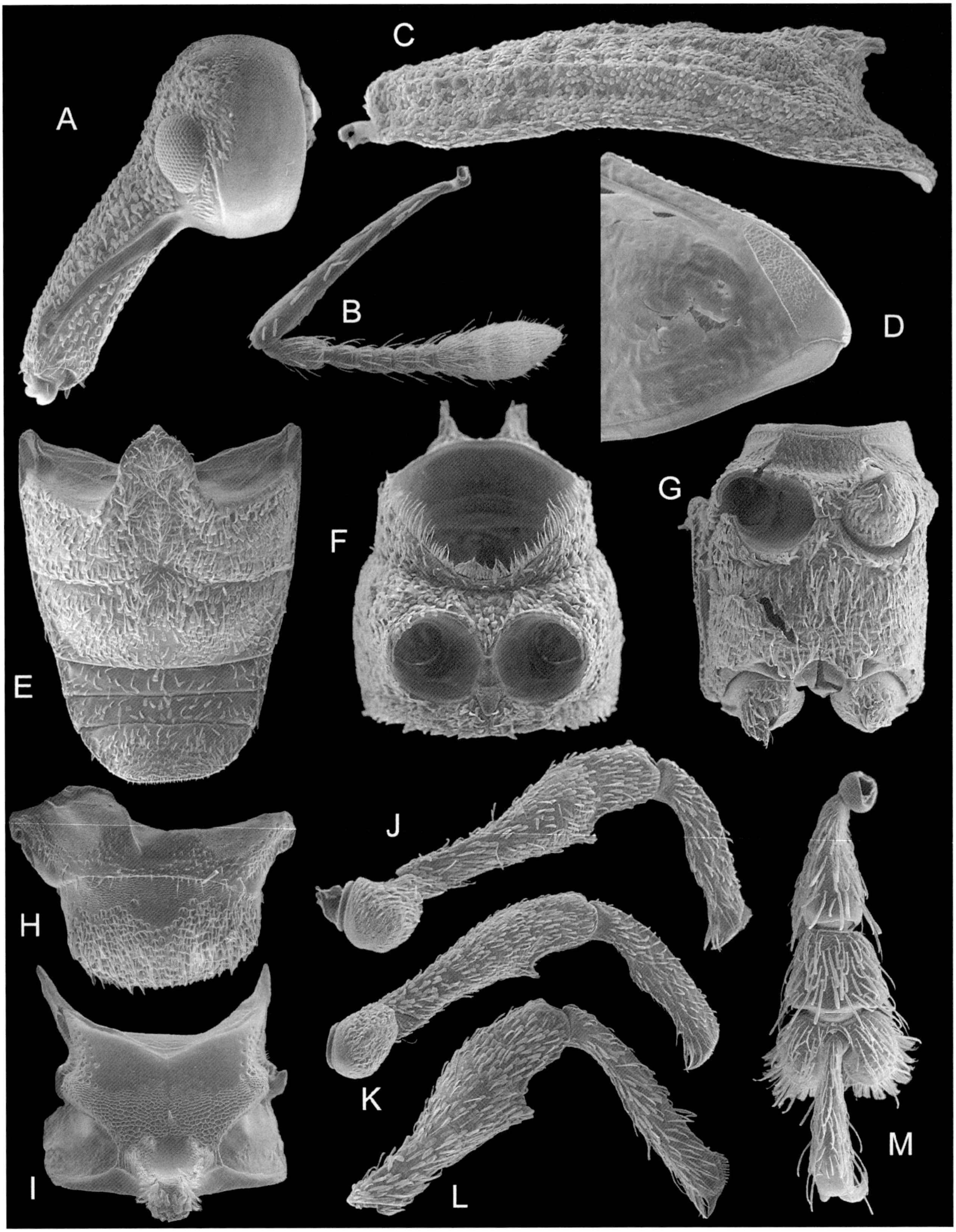

Fig. 44.158. Orthorhinini, structural details, *Orthorhinus* sp. A, head, lateral; B, antenna; C, elytron, lateral; D, elytral apex, inner lateral; E, abdominal ventrites; I, scutellum; J, foreleg; K, middle leg; L, hindleg; M, tarsus, dorsal.

Boheman, 1835 (*in* Schoenherr 1835), *O. laetus* Saunders & Jekel, 1855, *O. lateralis* Pascoe, 1882a.

Distribution. Australia, New Guinea, New Caledonia, Fiji, New Hebrides.

Notes. The common and widespread *O. cylindrirostris* and some other species are in need of revision and proper delimitation (Pullen *et al.* 2014).

Parorthorhinus Kuschel, 2008 (Pl. 99L)

Type species, by original designation: *Orthorhinus aethiops* Boisduval, 1835.

Australian species. *P. aethiops* (Boisduval, 1835), *P. bicolor* (Blackburn, 1892b), *P. carinatus* (Pascoe, 1873a), *P. heilipoides* (Pascoe, 1870f), *P. infidus* (Pascoe, 1873a), *P. lepidotus* (Erichson, 1842), *P. meleagris* (Pascoe, 1870d), *P. variegatus* (Saunders & Jekel, 1855).

Distribution. Australia.

Tribe Paipalesomini Marshall, 1932

Peribleptus Schoenherr, 1843 (Pl. 82L)

Type species, by original designation: *Peribleptus scalptus* Boheman, 1843 (*in* Schoenherr 1843).

Australian species. *P. dealbatus* (Boisduval, 1835).

Distribution. Australia, China, Japan, Taiwan, Myanmar, India, Pakistan, Malay Archipelago, New Guinea.

Notes. The single species recorded from Australia appears to represent an accidental introduction from New Guinea not established here (Pullen *et al.* 2014).

Tribe Psepholacini Lacordaire, 1865

Calpurnius Zimmerman, 1994 (Pl. 101A)

Type species, by subsequent designation (Zimmerman, 1994: 648): *Xestocis niger* Lea, 1913.

Australian species. *C. castaneus* (Lea, 1913a), *C. niger* (Lea, 1913a).

Distribution. Australia (QLD, Norfolk Is.).

Catocalephe Blackburn, 1895 (Pl. 101B)

Type species, by monotypy: *Catocalephe minans* Blackburn, 1895.

Australian species. *C. minans* Blackburn, 1895b.

Distribution. Australia.

Celidaus Lea, 1909 (Pl. 101C)

Type species, by monotypy: *Celidaus michaelseni* Lea, 1909.

Australian species. *C. michaelseni* Lea, 1909b.

Distribution. Australia.

Glechinus Pascoe, 1871 (Pl. 101D, E)

Type species, by monotypy: *Glechinus talpa* Pascoe, 1871c.

Australian species. *G. trichocerus* (Montrouzier, 1861).

Distribution. Australia, New Caledonia.

Neozeneudes Lea, 1900 (Pl. 101F)

Type species, by monotypy: *Neozeneudes dives* Lea, 1900.

Australian species. *N. dives* Lea, 1900b.

Distribution. Australia.

Oreda White, 1846 (Pl. 101G)

Type species, by monotypy: *Oreda notata* White, 1846.

Australian species. *O. notata* White, 1846.

Distribution. Australia, Marianas Is., New Caledonia, New Zealand.

Protohylastes Wood, 1973 (Pl. 101H)

Type species, by original designation: *Protohylastes annosus* Wood, 1973.

Australian species. *P. latirostris* (Pascoe, 1873a), *P. marmoratus* (Lea, 1913a).

Distribution. Australia.

Psepholacipus Lea, 1900 (Pl. 101I)

Type species, by subsequent designation (Pullen *et al.* 2014: 286): *Psepholacipus fossilis* Lea, 1900.

Australian species. *P. fossilis* Lea, 1900b, *P. minor* Lea, 1900b.

Distribution. Australia (QLD).

Psepholasoma Lea, 1913 (Pls 101J, 102A)

Type species, by monotypy: *Psepholasoma rostrale* Lea, 1913.

Australian species. *P. cylindricum* (Lea, 1913a), *P. rostrale* Lea, 1913a.

Distribution. Australia (TAS).

Psepholax White, 1843 (Pl. 101K; Figs 44.106G–O, 44.159)

Type species, by monotypy: *Psepholax sulcatus* White, 1843 (*in* White & Doubleday, 1843).

Australian species. *P. basalis* Lea, 1913a, *P. egerius* Pascoe, 1873a, *P. humeralis* Lea, 1900b, *P. lateripennis* Lea, 1913a, *P. leoninus* Lea, 1900b, *P. mastersii* Pascoe, 1873a, *P. pascoei* Olliff, 1888, *P. subconicollis* Lea, 1913a.

Distribution. Australia (NSW, QLD, Norfolk Is.), Chile, New Caledonia, Solomon Is., New Zealand.

Notes. The adults of *Psepholax* are mostly found in wood of dead or dying trees (Lyal 2014a). The species *P. dentipes* (Boheman) from Chile can be found in large numbers in dying or recently dead trees of *Nothofagus* (Nothofagaceae) as well as other evergreen trees (Fierro *et al.* 2023).

Pseudotherebus Lea, 1900 (Pl. 102B)

Type species, by monotypy: *Pseudotherebus sculptipennis* Lea, 1900.
Australian species. *P. sculptipennis* Lea, 1900b.
Distribution. Australia.

Sympiezoscelus Waterhouse, 1853 (Pl. 102C)

Type species, by monotypy: *Sympiezoscelus spencei* Waterhouse, 1853.
Australian species. *S. foveiventris* Lea, 1931b, *S. norfolcensis* Lea, 1913a, *S. spencei* Waterhouse, 1853.
Distribution. Australia (including Norfolk Is.), New Guinea.
Notes. Species of *Sympiezoscelus* have been considered as minor pests of *Araucaria cunninghamii* (Araucariaceae) trees in Bulolo and Wau, New Guinea (Schneider 1999).

Therebiosoma Lea, 1900 (Pl. 102D)

Type species, by monotypy: *Therebiosoma rhinarioides* Lea, 1900.
Australian species. *T. rhinarioides* Lea, 1900b, *T. variegatum* Lea, 1913b.
Distribution. Australia (NSW, QLD).

Therebus Pascoe, 1872 (Pl. 102E)

Type species, by original designation: *Therebus cepuroides* Pascoe, 1872.
Australian species. *T. cepuroides* Pascoe, 1872c.
Distribution. Australia (WA), Philippines, Taiwan.

Wiburdia Lea, 1908 (Pl. 102F)

Type species, by monotypy: *Wiburdia scrobiculata* Lea, 1908.
Australian species. *W. dentipes* Lea, 1913a, *W. scrobiculata* Lea, 1908c.
Distribution. Australia (NSW, VIC).

Zeneudes Pascoe, 1874 (Pl. 102G)

Type species, by monotypy: *Zeneudes sterculiae* Pascoe, 1874.
Australian species. *Z. sterculiae* Pascoe, 1874c.
Distribution. Australia (QLD).

Tribe Sophrorhinini Lacordaire, 1865

Colobodes Schoenherr, 1837 (Pl. 102H)

Type species, by original designation: *Colobodes billbergii* Boheman, 1837 (*in* Schoenherr 1837).
Australian species. *Colobodes* sp.
Distribution. Australia, Russia, China, Japan, Taiwan, Bangladesh, Myanmar, Cambodia, Malay Archipelago, New Guinea.

Deretiosus Pascoe, 1871 (Pl. 102L; Fig. 44.160)

Type species, by monotypy: *Deretiosus aridus* Pascoe, 1871.
Australian species. *D. alphabeticus* Lea, 1931a, *D. aridus* Pascoe, 1871c, *D. verrucifer* Faust, 1899.
Distribution. Australia (QLD), China, Malay Archipelago, Pacific region, New Guinea, Solomon Is.

Eprias Heller, 1922 (Pl. 102I)

Type species, by original designation: *Ocoblodes vanus* Heller, 1922.
Australian species. *E. aspratilis* (Lea, 1909d).
Distribution. Australia, Philippines.

Perrhaebius Pascoe, 1874 (Pl. 102J, K)

Type species, by monotypy: *Perrhaebius ephippiger* Pascoe, 1874c.
Australian species. *P. tibialis* (Lea, 1928b).
Distribution. Australia (QLD), Java, Philippines, Sumatra, New Guinea, Maluku Is.

Tribe Trachodini Gistel, 1848

Acicnemis Fairmaire, 1849 (Pl. 103A; Figs 44.96, 44.161)

Type species, by subsequent designation (Fairmaire, 1881: 299): *Acicnemis variegata* Fairmaire, 1849.
Australian species. *A. andrewsi Gahan*, *A. arachnopus* Hubenthal, 1917, *A. meriones* Pascoe, 1872c, *A. personata* Hubenthal, 1917, *A. sannio* (Pascoe, 1872c), *A. sororia* Pascoe, 1885, *A. spilonota* Pascoe, 1885.
Distribution. Australia (QLD), Oriental Region, New Guinea, Melanesia, Pacific region.
Notes. Morimoto & Miyakawa (1995) redescribed the genus with emphasis on the species occurring in Japan.

Tranes group

Demyrsus Pascoe, 1872 (Pl. 103B, C; Fig. 44.162A–H)

Type species, by monotypy: *Demyrsus meleoides* Pascoe, 1872.

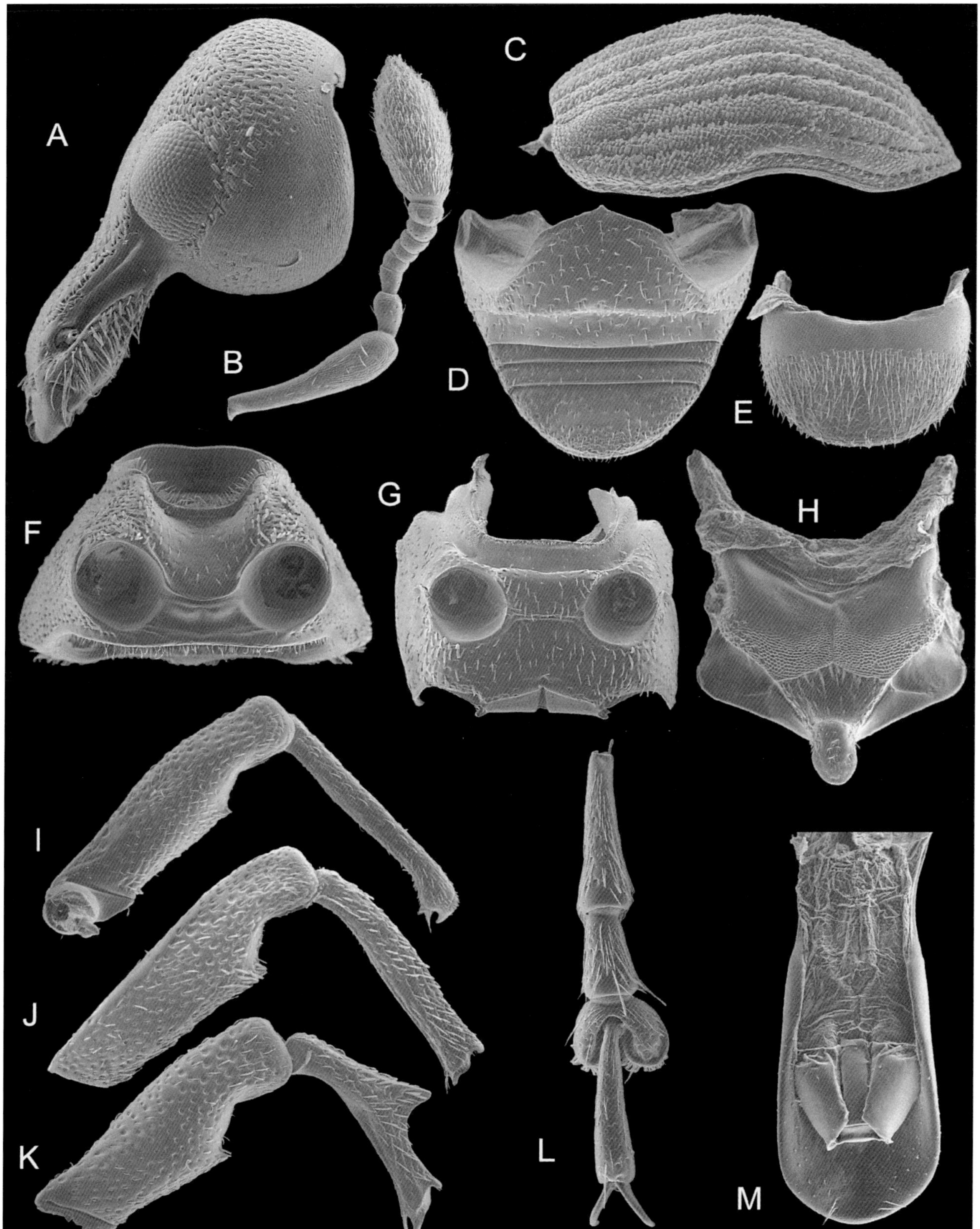

Fig. 44.159. Psepholacini, structural details, *Psepholax* sp., ♂. A, head, lateral; B, antenna; C, elytron, lateral; D, abdominal ventrites; E, pygidium; F, prothorax, ventral; G, pterothorax, ventral; H, scutellum; I, foreleg; J, middle leg; K, hindleg; L, tarsus and claws, dorsal; M, penis, dorsal.

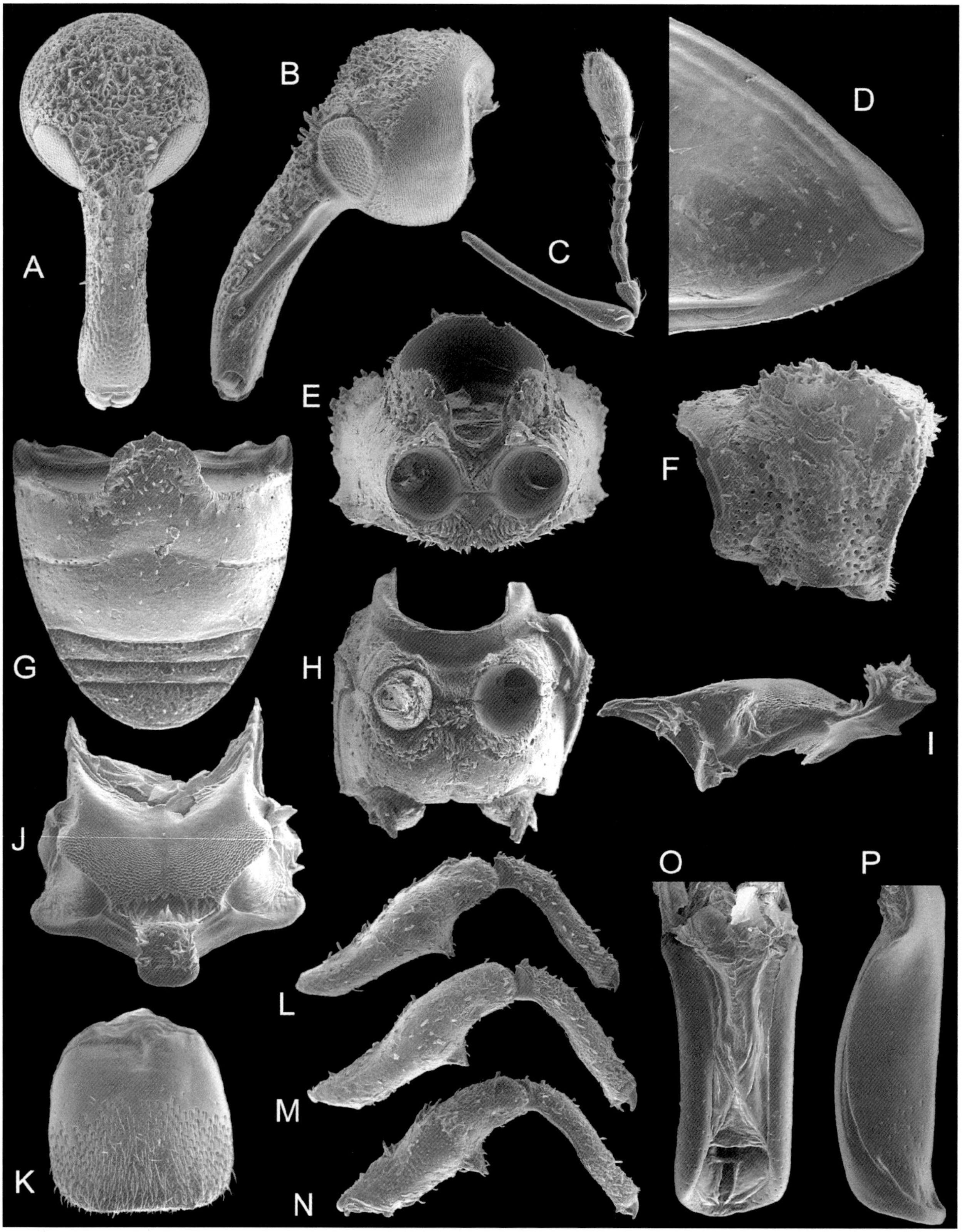

Fig. 44.160. Sophrorhinini, structural details, *Deretiosus aridus* Pascoe, ♂. A–B, head: A, dorsal; B, lateral; C, antenna; D, elytral apex, ventral; E–F, prothorax: E, ventral; F, lateral; G, abdominal ventrites; H, pterothorax, ventral; I–J, scutellum: I, lateral; J, dorsal; K, pygidium; L, foreleg; M, middle leg; N, hindleg; O–P, penis: O, dorsal; P, lateral.

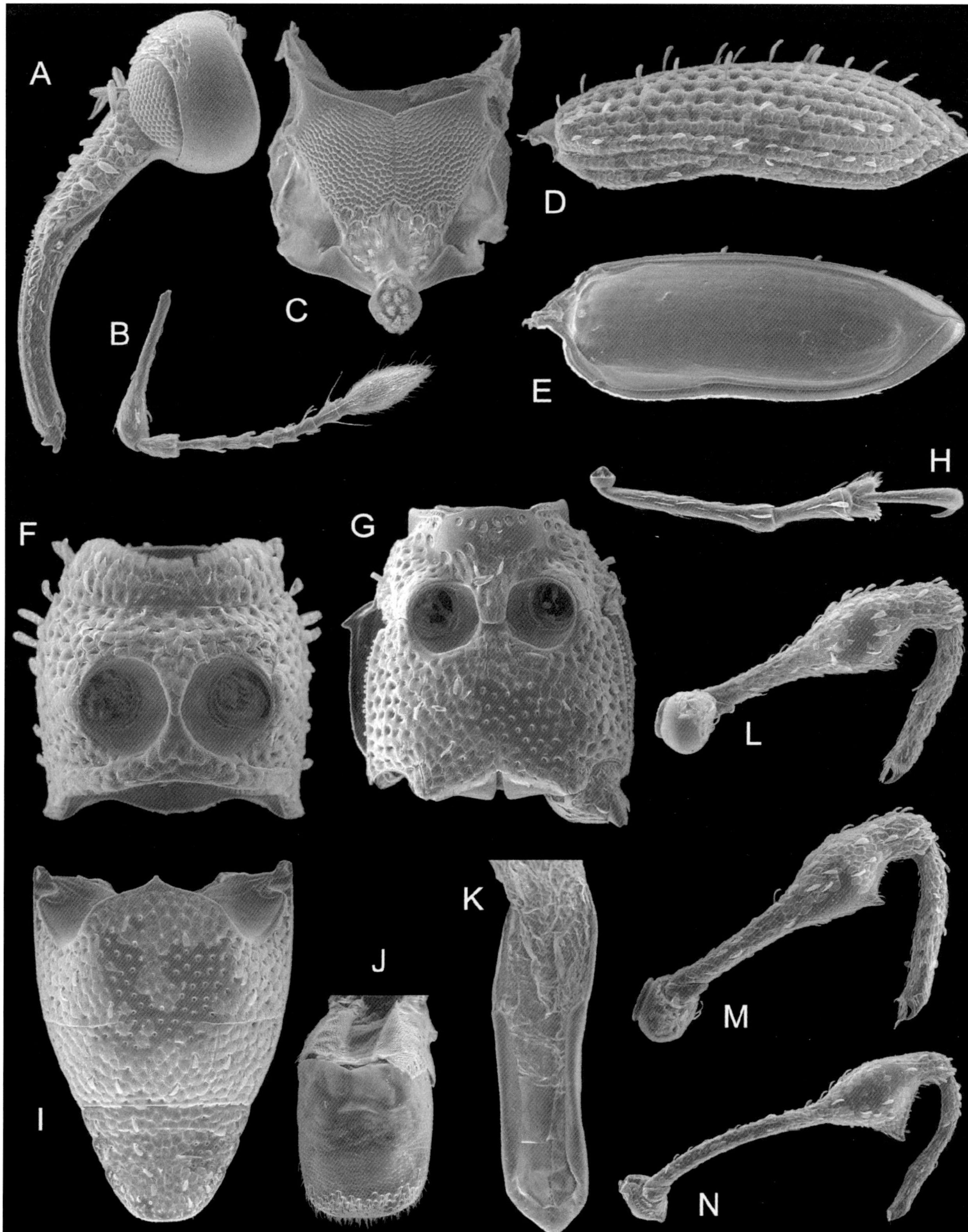

Fig. 44.161. Trachodini, structural details, *Acicnemis* sp., ♂. A, head, lateral; B, antenna; C, scutellum; D–E, elytron: D, lateral; E, ventral; F, prothorax, ventral; G, pterothorax, ventral; H, tarsus, dorsal; I, abdominal ventrites; J, pygidium; K, penis, dorsal; L, foreleg; M, middle leg; N, hindleg.

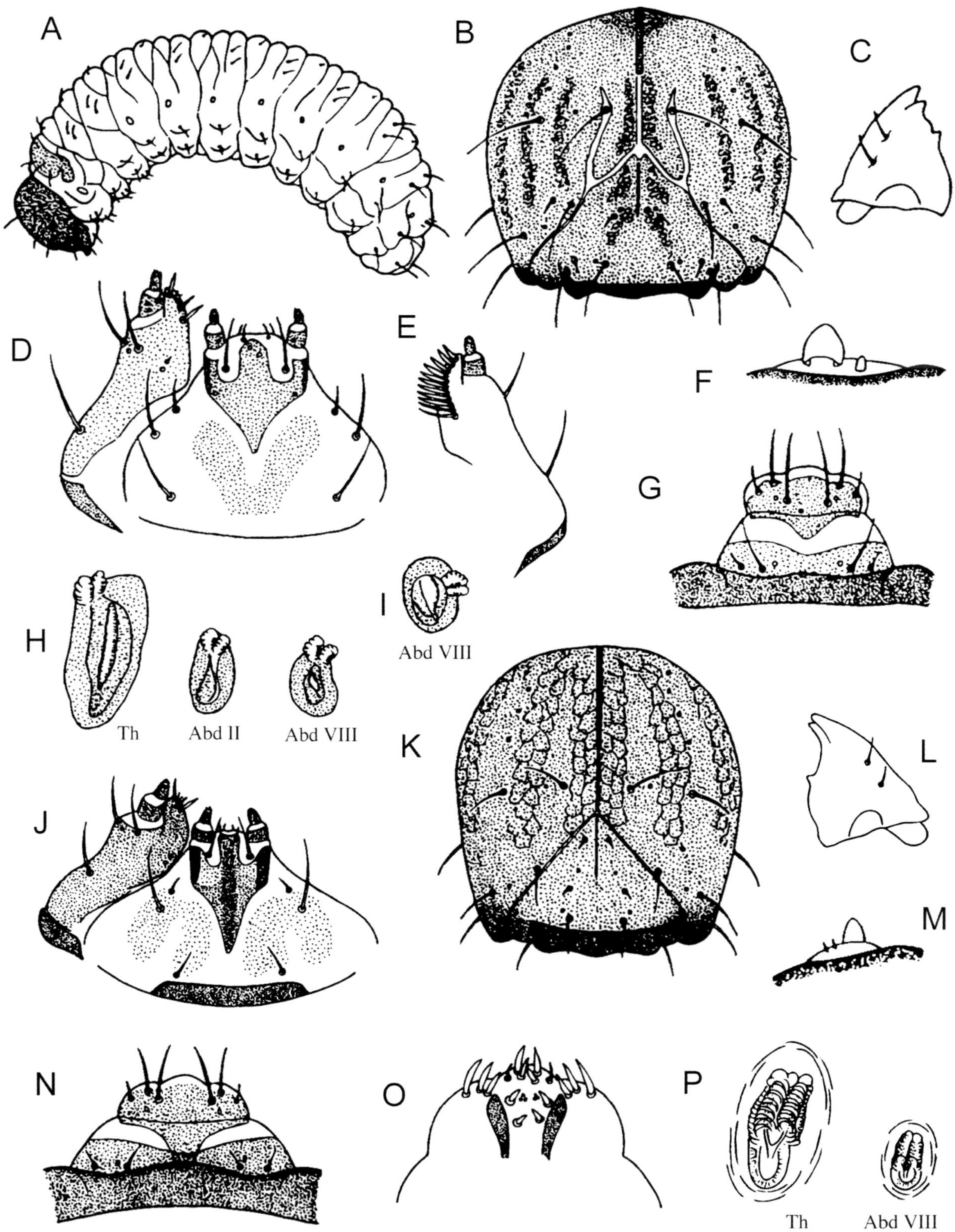

Fig. 44.162. *Tranes* group, structural details of immature stages. A–H, *Demyrsus meleoides* Pascoe, larva: A, habitus, lateral; B, head, lateral; C, mandible; D, mala and labium, ventral; E, mala, dorsal; F, antenna; G, labrum, clypeus and epistoma; H, spiracles, showing alignment; I, *Siraton roei* (Boheman), larva, Abd VIII spiracle; J, mala and labium, ventral; K–P, *Tranes lyterioides* (Pascoe), larva: K, head, dorsal; L, mandible; M, antenna; N, labrum, clypeus and epistoma; O, epipharyngeal lining; P, spiracles, showing alignment. After May (1994).

Australian species. *D. digmon* Hsiao & Oberprieler, 2020b, *D. meleoides* Pascoe, 1872a.
Distribution. Australia (QLD, NSW).
Notes. The genus was revised by Hsiao & Oberprieler (2020b). The larvae of both species develop in cycads trunks, and *D. meleoides* has been introduced with cycads in Italy and South Africa but did not establish there (Hsiao & Oberprieler 2020b).

Howeotranes Zimmerman, 1994 (Pl. 103D)

Type species, by original designation: *Tranes insularis* Pascoe, 1874.
Australian species. *H. insularis* (Pascoe, 1874a).
Distribution. Australia (Lord Howe Is.).
Notes. The single species had not been collected for many decades and was surmised to be extinct, but a few specimens have been found on Lord Howe Is. recently by the naturalist Ian Hutton. It appears to be associated with palms.

Miltotranes Zimmerman, 1994 (Pl. 103E, F)

Type species, by original designation *Tranes prosternalis* Lea, 1929.
Australian species. *M. prosternalis* (Lea, 1929a), *M. subopacus* (Lea, 1929a), *M. wilsoni* Hsiao & Oberprieler, 2022.
Distribution. Australia (QLD).
Notes. Hsiao & Oberprieler (2022) revised the genus. The larvae develop in sporophylls and axes of male cones of *Bowenia* cycads (Zamiaceae), and the adults act as pollinators of their hosts (Hsiao & Oberprieler 2020a).

Paratranes Zimmerman, 1994 (Pl. 103G)

Type species, by original designation: *Tranes monopticus* Pascoe, 1870.
Australian species. *P. monopticus* (Pascoe, 1870d), *P. zimmermani* Hsiao & Oberprieler, 2021.
Distribution. Australia (QLD, NSW, SA, WA).
Notes. The genus was revised by Hsiao & Oberprieler (2021). Both species are associated with grasstrees, *Xanthorrhoea* (Asphodelaceae), the adults mostly occurring at the base of flower stalks and between young leaves (Froggatt 1896, 1907; Mulder 1964; Hawkeswood 1985).

Siraton Hustache, 1934 (Pl. 103H–J; Fig. 44.162I)

Type species, by monotypy: *Tranes internatus* Pascoe, 1870.
Australian species. *S. internatus* (Pascoe, 1870d), *S. roei* (Boheman, 1843) (*in* Schoenherr 1843).
Distribution. Australia (QLD, NSW, WA).
Notes. *Siraton internatus* has been considered as a pest of ornamental cycads in Europe and USA (Oberprieler & Caldara 2012; Hsiao & Liao 2024).

Tranes Schoenherr, 1843 (Pl. 103K, L; Fig. 44.162K–P)

Type species, by original designation: *Tranes vigorsii* Boheman, 1843.
Australian species. *T. insignipes* Lea, 1929a, *T. lyterioides* (Pascoe, 1875), *T. sparsus* Boheman, 1843 (*in* Schoenherr 1843), *T. vigorsii* Boheman, 1843 (*in* Schoenherr 1843).
Distribution. Australia (NSW).
Notes. In addition to the four described species, another six undescribed ones occur in Australia (Hsiao & Oberprieler, in prep.). The larvae of all species develop in sporophylls and axes of male cones of *Macrozamia* and *Lepidozamia* cycads (Zamiaceae), and the adults of several species act as the sole or main pollinators of their hosts (Chadwick 1993; Hall *et al.* 2004; Terry *et al.* 2005, Hsiao & Oberprieler, in prep.). Hsiao *et al.* (2023) inferred the phylogenetic relationships between the species based on molecular data.

Unplaced to Tribe

Eristinus Lea, 1915 (Pl. 72I; Fig. 44.163)

Type species, by subsequent designation (Pullen *et al.* 2014: 289): *Eristinus eucalypti* Lea, 1915.
Australian species. *E. eucalypti* Lea, 1915c, *E. flavipes* Lea, 1915c, *E. sobrinus* Lea, 1915c.
Distribution. Australia (QLD, NSW).

Exeiratus Broun, 1914 (Pl. 72K; Figs 44.164, 44.165)

Type species, by monotypy: *Exeiratus setarius* Broun, 1914.
Australian species. *E. bigranulatus* (Lea, 1928a), *E. carinatus* (Lea, 1928a), *E. insignipes* (Lea, 1928a), *E. sordidus* (Lea, 1928a).
Distribution. Australia (NSW, VIC, TAS, SA, southwestern WA), New Zealand.
Notes. The weevils are usually found in leaf litter, moss and under bark.

Opsittis Pascoe, 1870 (Pl. 72L)

Type species, by monotypy: *Opsittis atomaria* Pascoe, 1870.
Australian species. *O. atomaria* Pascoe, 1870e.
Distribution. Australia.
Notes. The genus is currently monotypic, but several undescribed species and similar undescribed genera are in collections such as the ANIC.

Reyesiella Alonso-Zarazaga & Lyal, 1999

Type species, by monotypy: *Idus caecus* Broun, 1893.

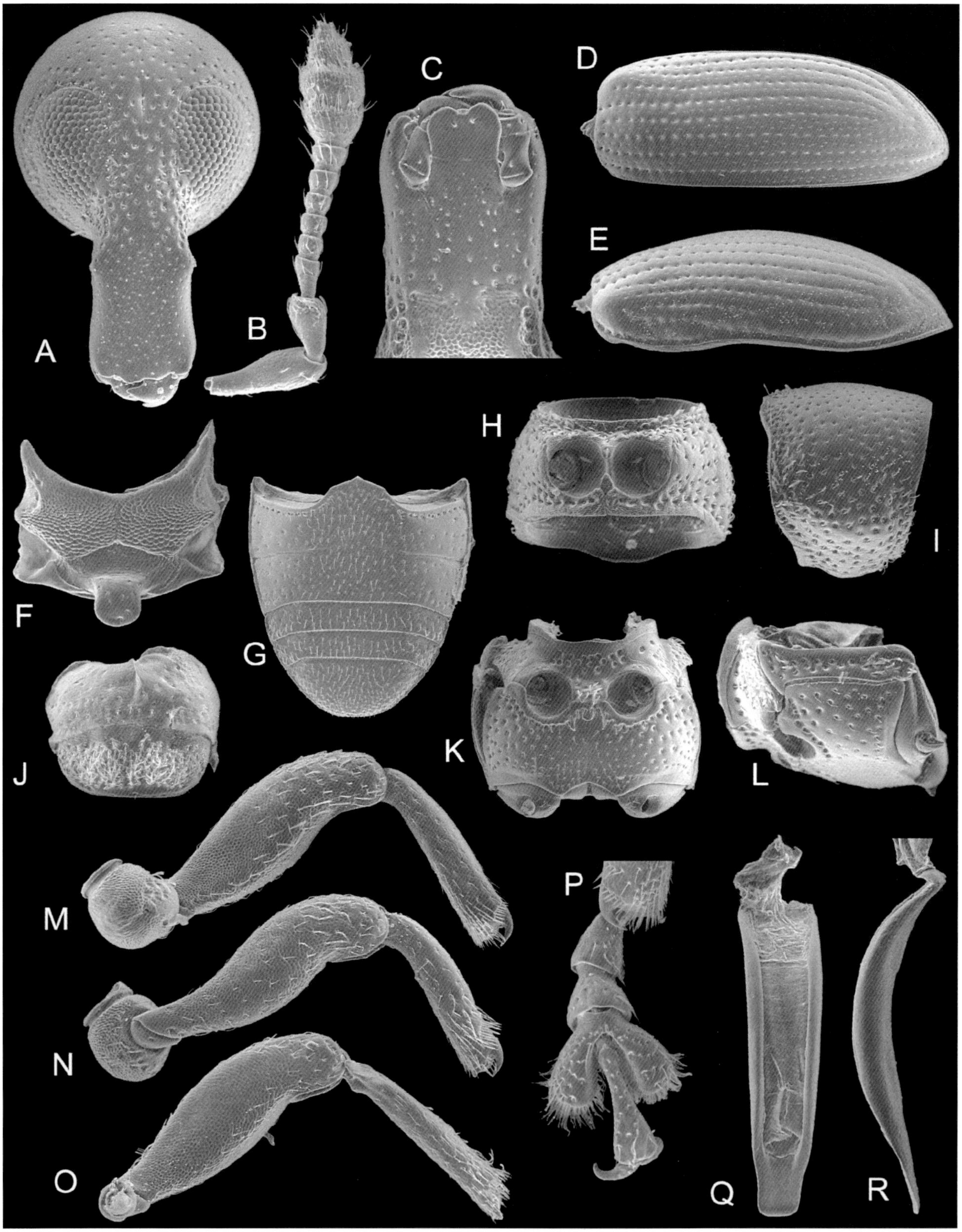

Fig. 44.163. Molytinae tribe unplaced, structural details, *Eristinus* sp., ♂. A, head, dorsal; B, antenna; C, mouthparts; D–E, elytron: D, dorsal; E, lateral; F, scutellum; G, abdominal ventrites; H–I, prothorax: H, ventral; I, lateral; J, pygidium; K–L, pterothorax: K, ventral; L, lateral; M, foreleg; N, middle leg; O, hindleg; P, tarsus, dorsal; Q–R, penis: Q, dorsal; R, lateral.

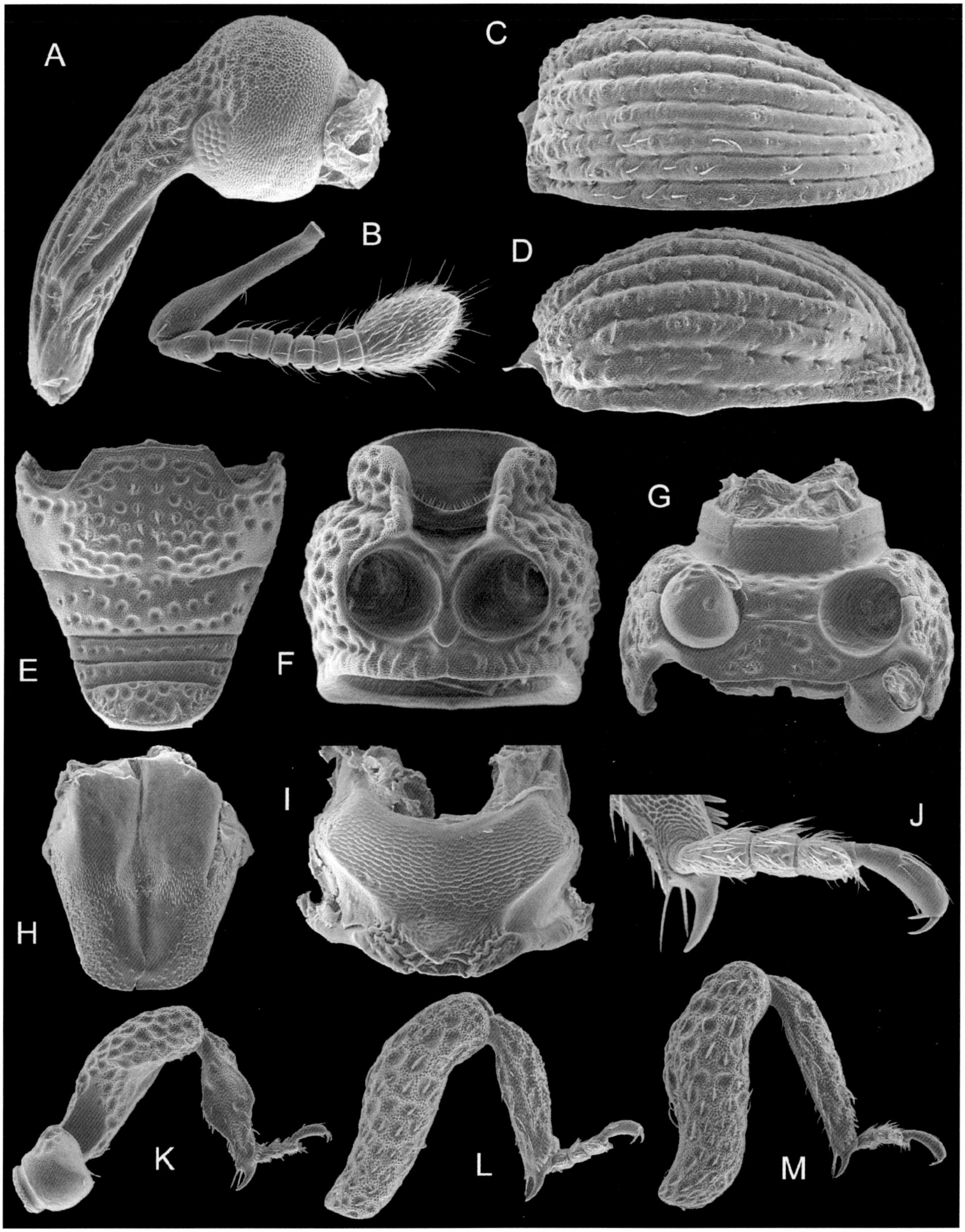

Fig. 44.164. Molytinae tribe unplaced, structural details, *Exeiratus insignipes* (Lea). A, head, lateral; B, antenna; C–D, elytron: C, dorsal; D, lateral; E, abdominal ventrites; F, prothorax, ventral; G, pterothorax, ventral; H, pygidium; I, scutellum; J, tarsus and claws, lateral; K, foreleg; L, middle leg; M, hindleg.

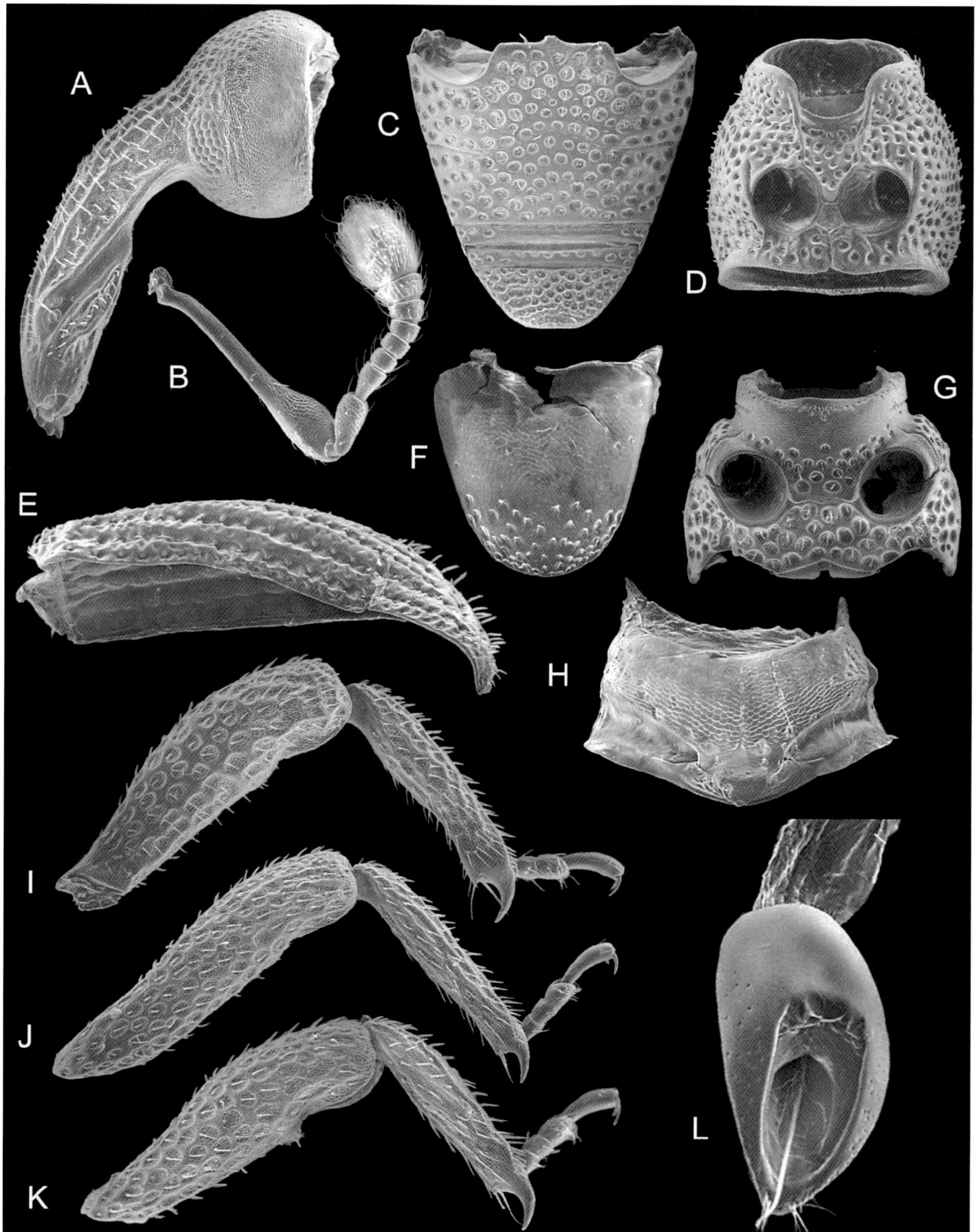

Fig. 44.165. Molytinae tribe unplaced, structural details, *Exeiratus sordidus* (Lea), ♂. A, head, lateral; B, antenna; C, abdominal ventrites; D, prothorax, ventral; E, elytron, lateral; F, pygidium; G, pterothorax, ventral; H, scutellum; I, foreleg; J, middle leg; K, hindleg; L, penis, dorsal.

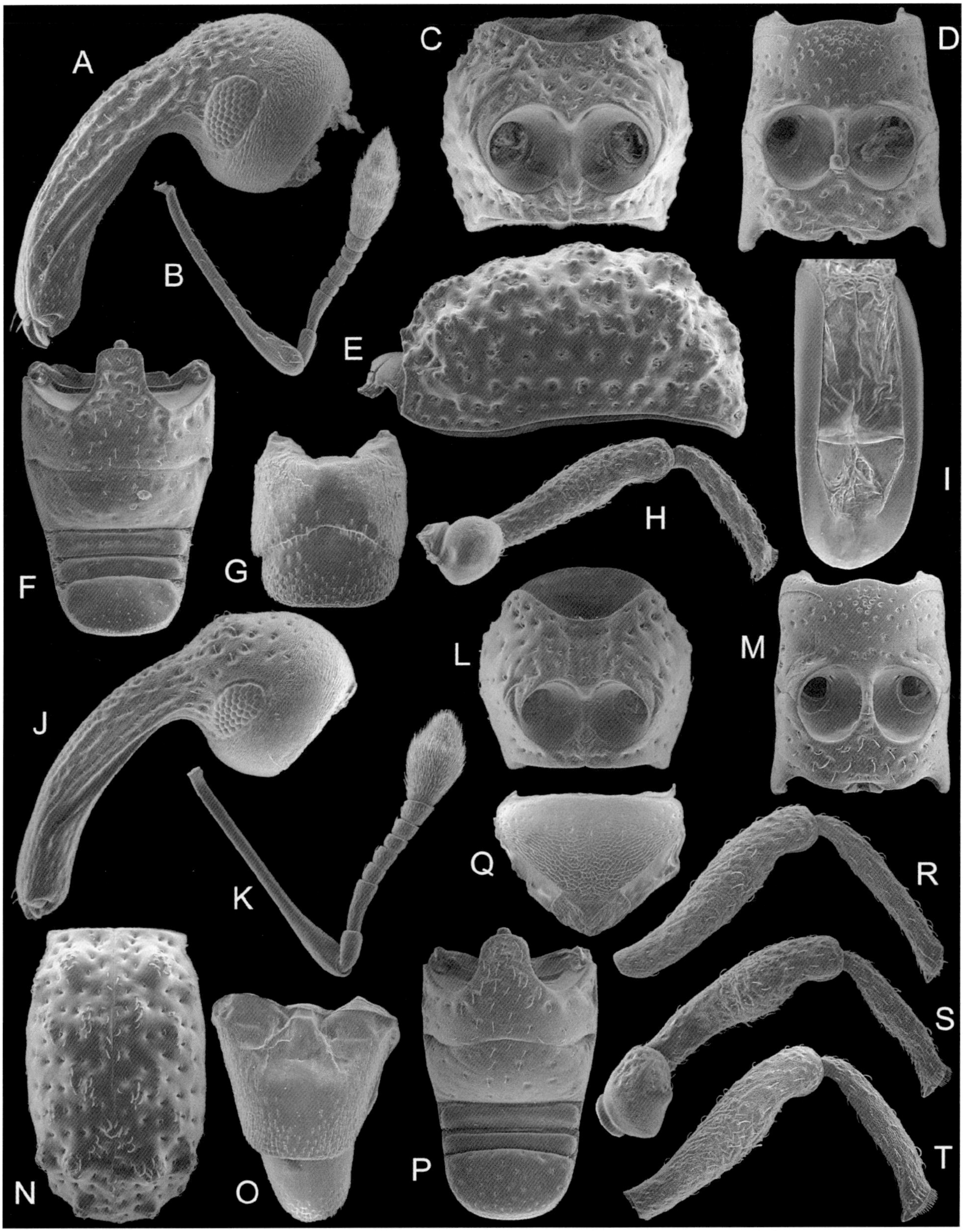

Fig. 44.166. Molytinae tribe unplaced, structural details, *Syagrius* spp. A–I, *Syagrius* sp.: ♂; A, head, lateral; B, antenna; C, prothorax, ventral; D, pterothorax, ventral; E, elytron, lateral; F, abdominal ventrites; G, pygidium; H, foreleg; I, penis, dorsal; J–T, *Syagrius* sp.: J, head, lateral; K, antenna; L, prothorax, ventral; M, pterothorax, ventral; N, elytron, dorsal; O, pygidium; P, abdominal ventrites; R, foreleg; S, middle leg; T, hindleg.

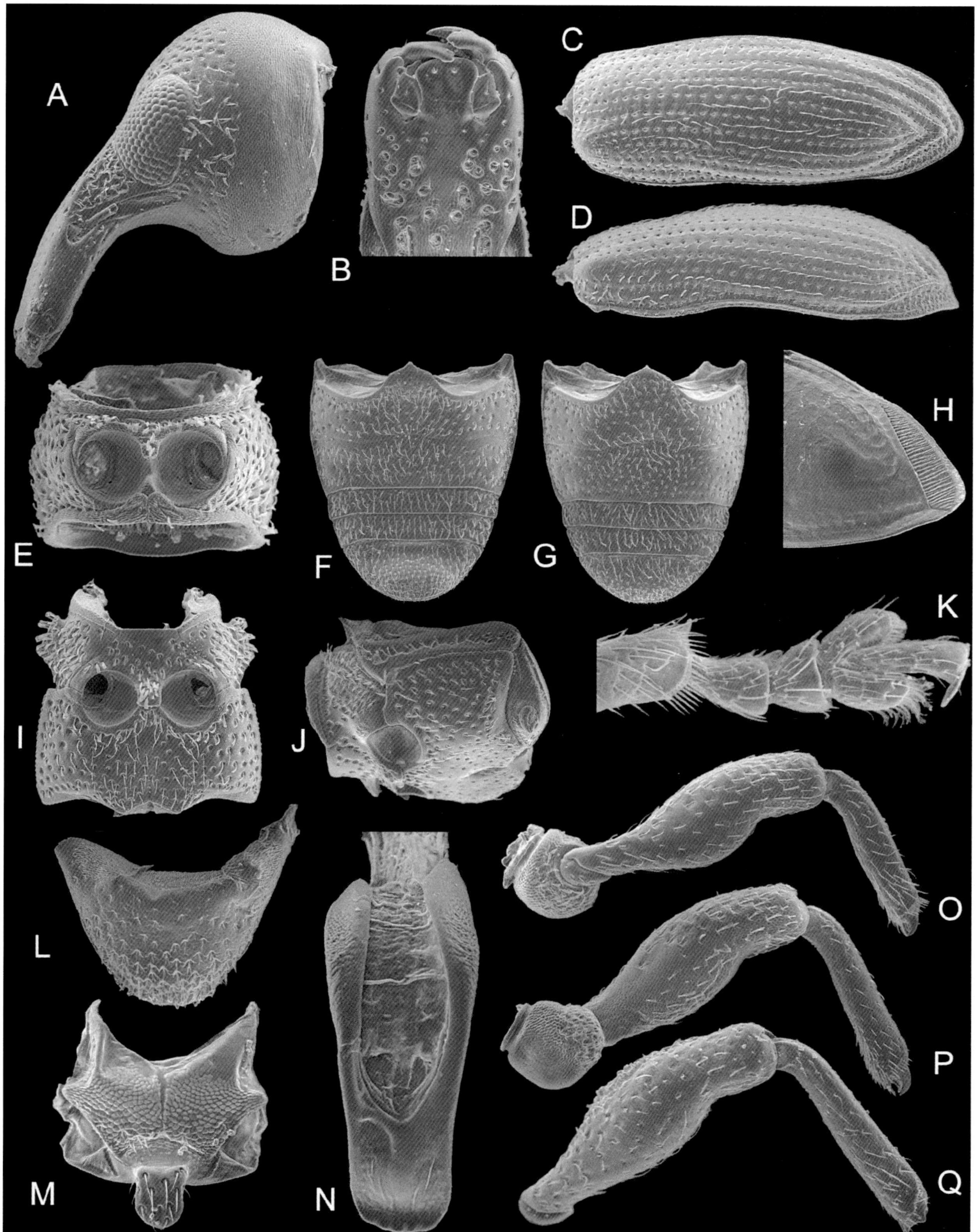

Fig. 44.167. Molytinae tribe unplaced, structural details, *Thaumastophasis* sp. A, head, lateral; B, mouthparts; C–D, elytron: C, dorsal; D, lateral; E, prothorax, ventral; F–G, abdominal ventrites: F, ♂; G, ♀; H, elytral apex, inner lateral; I–J, pterothorax: I, ventral; J, lateral; K, tarsus, lateral; L, pygidium; M, scutellum; N, penis, dorsal; O, foreleg; P, middle leg; Q, hindleg.

Australian species. *Reyesiella* sp.
Distribution. Australia (Norfolk Is.), New Zealand.
Notes. The genus was described as *Idus* from New Zealand but renamed because of its name being a junior homonym. Only a single specimen appears known from Australia to date, collected from forest litter on Norfolk Is. (Kuschel 1972). The New Zealand species *Reyesiella caecus* (Broun) was illustrated by Grebennikov (2010).

Syagrius Pascoe, 1875 (Pl. 61K; Fig. 44.166)

Type species, by monotypy: *Syagrius fulvitarsis* Pascoe, 1875.
Australian species. *S. costicollis* Marshall, 1922, *S. fulvitarsis* Pascoe, 1875, *S. intrudens* Waterhouse, 1903, *S. pembertoni* Marshall, 1922, *S. squamipes* Marshall, 1922.
Distribution. Australia (NSW, QLD), Hawaii, Ireland.
Notes. The species are associated with ferns, occasionally causing damage to these plants in greenhouses. *Syagrius intrudens* and *S. fulvitarsis* have been introduced in England and Hawaii, where their larvae can cause damage to their fern hosts (Marshall 1922; Thompson & Howell 2001).

Thaumastophasis Wollaston, 1873 (Pl. 61L; Fig. 44.167)

Type species, by monotypy: *Thaumastophasis oculata* Wollaston, 1873.
Australian species. *T. bicolor* (Blackburn, 1892b), *T. blackburni* (Lea, 1911b), *T. inconstans* (Lea, 1911b), *T. oculata* Wollaston, 1873b, *T. orbitalis* (Boheman, 1843) (*in* Schoenherr 1843), *T. pallida* (Lea, 1908c), *T. pallidicornis* (Lea, 1911b), *T. setosa* (Blackburn, 1892b), *T. uniformis* (Lea, 1915c).
Distribution. Australia (QLD, NSW, VIC, TAS, SA, WA).

Curculionidae: Conoderinae Schoenherr, 1833

Introduction. The wide concept of Conoderinae as used here is that of Zherikhin & Egorov (1990) and Zherikhin & Gratshev (1995), under the name Baridinae, as adopted by, e.g. Oberprieler *et al.* (2007) and Pullen *et al.* (2014), although it may not be a natural one (which also applies to most other current subfamilies of Curculionidae). Prena *et al.* (2014) presented an overview of the subfamily, albeit in a slightly different composition. Conoderinae are mainly characterised by their large ascending mesanepisterna, visible dorsally between prothorax and elytra, and a small, exposed pygidium. They are poorly represented in Australia with only three tribes, Baridini, Campyloscelini and Conoderini, and 21 genera (Pullen *et al.* 2014). The Baridini are a large and very diverse tribe in tropical regions of the world but in Australia represented by only eleven genera, four of them introduced, and most species falling into the large but composite genus *Baris*. Both adults and larvae of the two genera of Campyloscelini in Australia, *Phaenomerus* and *Tomicoproctus*, inhabit the tunnels of Platypodinae, in which they feed on dead adults and larvae but may be cleptobionts or scavengers rather than predators (Thompson 1996; May 1997). The adults of Conoderini have large, sometimes dorsally contiguous eyes and long legs (Pl. 20F) and are active runners and fliers on tree trunks, and their larvae are mainly wood-borers, but almost nothing is known about the life histories of the Australian species. Apart from the campylosceline genus *Phaenomerus*, which was revised by Thompson (1996), no other Australian conoderines have received taxonomic attention in recent times. May (1984, 1994, 1997) described the larvae of some genera, and some of her illustrations are reproduced here (Fig. 44.169).

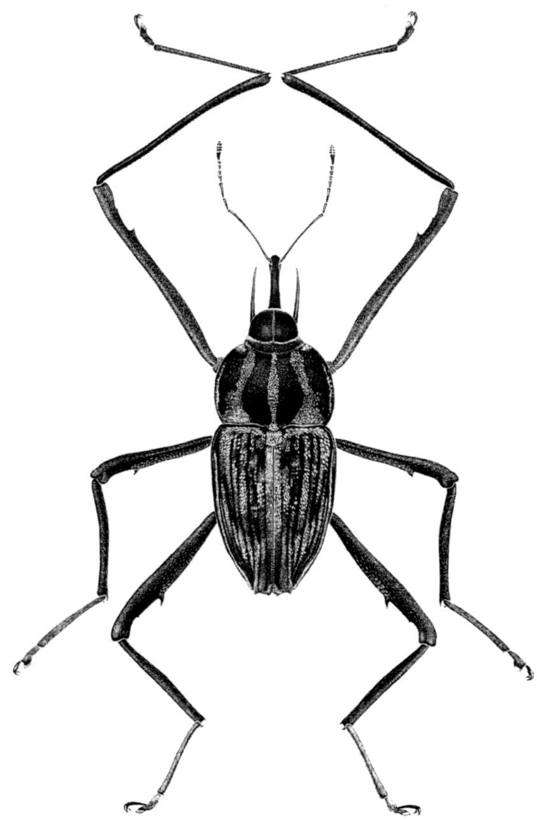

Fig. 44.168. *Mecopus doryphorus* Quoy & Gaimard.

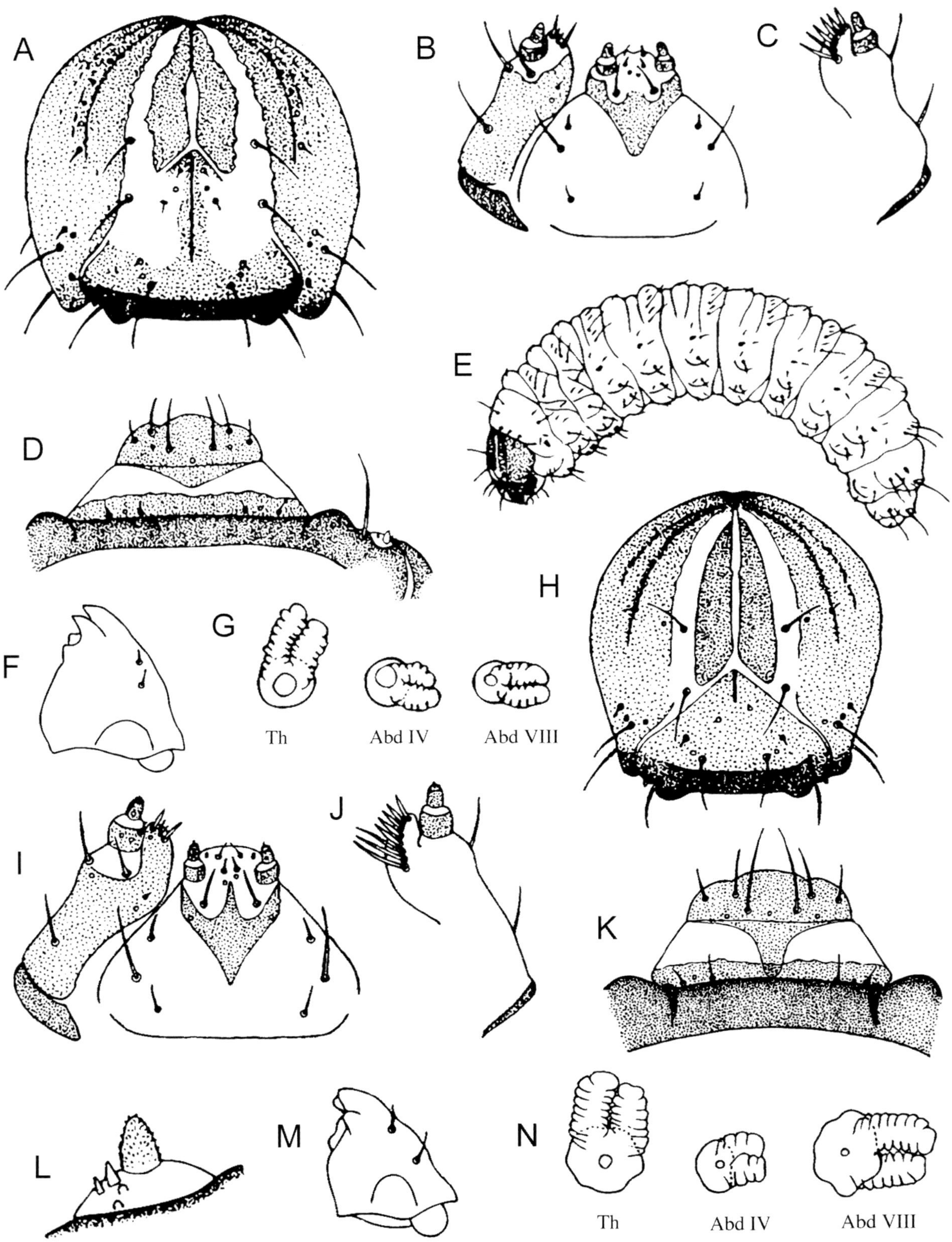

Fig. 44.169. Baridini, structural details of immature stages. A–G, *Linogeraeus urbanus* (Boheman) larva (New Zealand): A, head, dorsal; B, mala and labium, ventral; C, mala, dorsal; D, labrum, clypeus and epistoma; E, habitus, lateral; F, mandible; G, spiracles, showing alignment; H–N, *Orchidophilus aterrimus* (Waterhouse), larva: H, head, dorsal; I, mala and labium, ventral; J, mala, dorsal; K, labrum, clypeus and epistoma; L, antenna; M, mandible; N, spiracles, showing alignment. After May (1994).

Key to the Australian genera of Conoderinae

The genera *Brimoides* and *Othippia* (Conoderini) are not included in the key.

1. Body cylindrical (Fig. 44.170H, J); femora with large ventral tooth, longer on metafemora; metatibiae with two strong apical hooks .... CAMPYLOSCELINI ... 2
– Not with this combination of characters; metatibiae with single apical hook (uncus) (Fig. 44.177M) 3
2(1). Head and rostrum convex; rostrum longer than head capsule, narrow and cylindrical (Fig. 44.170J) .... ***Phaenomerus*** **Schoenherr**
– Head and rostrum flattened; rostrum about as long as head capsule, broad and flattened (Fig. 44.170H) .... ***Tomicoproctus*** **Faust**
3(1). Eyes dorsally subcontiguous (Fig. 44.170I, L, M, T) and large (Fig. 44.170S), interocular distance less than width of antennal clubs .... CONODERINI ... 4
– Eyes dorsally well separated (Fig. 44.170F) and often smaller (Fig. 44.170Q, R), interocular distance larger than width of antennal clubs .... BARIDINI ... 9
4(3). Prosternal canal widening apicad (Fig. 44.170b); elytra emarginate by abdominal ventrite 1 (Fig. 44.177H) .... ***Agametis*** **Pascoe**
– Prosternal canal parallel-sided (Fig. 44.170d); elytra not emarginate by abdominal ventrite 1 5
5(4). Profemora narrowly subcylindrical, with small ventral tooth (Fig. 44.170L) 6
– Profemora robust, often with large ventral tooth (Fig. 44.170M) 7
6(5). Pronotum with basal margin projecting medially over scutellar shield (Fig. 44.178H); metatibiae with single ventral tooth; habitus as Fig. 44.170K .... ***Lobotrachelus*** **Schoenherr**
– Pronotum with basal margin not projecting medially over scutellar shield; metatibiae with two ventral teeth .... ***Mecopus*** **Schoenherr**
7(5). Pronotum with basal margin projecting medially over scutellar shield; elytral bases not projecting behind scutellar shield; habitus as Fig. 44.170M .... ***Metialma*** **Pascoe**
– Pronotum basal margin not projecting medially over scutellar shield; elytral bases projecting behind scutellar shield 8
8(7). Pronotum and elytral disc with yellow setae .... ***Lamitema*** **Lea**
– Pronotum and elytral disc without yellow setae .... ***Temialma*** **Lea**
9(3). Body densely setose (Fig. 44.170C, O) or sparsely setose with tubercles and clusters of setae (Fig. 44.170D) 10
– Body glabrous or subglabrous (Fig. 44.170A, E, F, N) 12
10(9). Rostrum glabrous and smooth; mandibles exodont .... ***Linogeraeus*** **Casey**
– Rostrum setose with microsculpture; mandibles not exodont 11
11(10). Femora with ventral tooth .... ***Nyella*** **Oke**
– Femora without ventral tooth .... ***Cosmobaris*** **Casey**
12(9). Rostrum with deep dorsal groove at the base (Fig. 44.170G) .... ***Acythopeus*** **Pascoe**
– Rostrum without deep dorsal groove at the base 13
13(12). Body surface dull, with coarse microsculpture (Fig. 44.170E); mesotibiae in males with a tubercle near apical third; abdomen as Fig. 44.170P .... ***Orchidophilus*** **Buchanan**
– Body surface smooth and shining, with or without fine microsculpture; mesotibiae in males without tubercle near apical third 14
14(13). Prosternal canal present (e.g. Fig. 44.170b–d), though sometimes weakly 15
– Prosternal canal absent 16
15(14). Pronotum with fine punctures, distance between the punctures larger than a puncture diameter; prosternal canal well defined; pygidium not exposed .... ***Gymnobaris*** **Lea**
– Pronotum with coarse punctures, distance between the punctures smaller than a puncture diameter; prosternal canal poorly defined or weak; pygidium exposed; habitus as Fig. 44.170B .... ***Baris*** **Germar**
16(14). Rostrum, pronotum and elytra with scattered flattened setae, several times longer than a puncture .... ***Apinocis*** **Lea**
– Rostrum, pronotum and elytra without long flattened setae, setae when present minute, as long as a puncture 17
17(16). Pronotum with each puncture bearing minute seta (Fig 44.170a); pygidium mostly exposed .... ***Myctides*** **Pascoe**
– Pronotum with minute weak glabrous punctures; pygidium barely exposed .... ***Ipsichora*** **Pascoe**

Classification of the Australian genera of Conoderinae

Tribe Baridini Schoenherr, 1836

Acythopeus Pascoe, 1874 (Pl. 105A, B)

Type species, by subsequent designation (Heller, 1940: 106): *Acythopeus tristis* Pascoe, 1874.

Australian species. *A. bigeminatus* Pascoe, 1874c.

Distribution. Australia, Africa and Oriental Regions to New Guinea.

Note. The generic assignment of *A. bigeminatus* needs to be re-assessed; it is likely that *Acythopeus* does not occur in Australia (J. Prena pers. comm.).

Apinocis Lea, 1927 (Pl. 104A)

Type species, by monotypy: *Apinocis variipennis* Lea, 1927.

Australian species. *A. variipennis* Lea, 1927b.

Distribution. Australia (NSW, VIC), America.

Notes. The single species in Australia represents an accidental introduction from Argentina before 1927 (Pullen *et al.* 2014). Its larvae mine in grass stems.

Baris Germar, 1817 (Pl. 104B, C; Fig. 44.171)

Type species, by subsequent designation (Curtis, 1839: nr. 766): *Curculio artemisiae* Herbst, 1795.

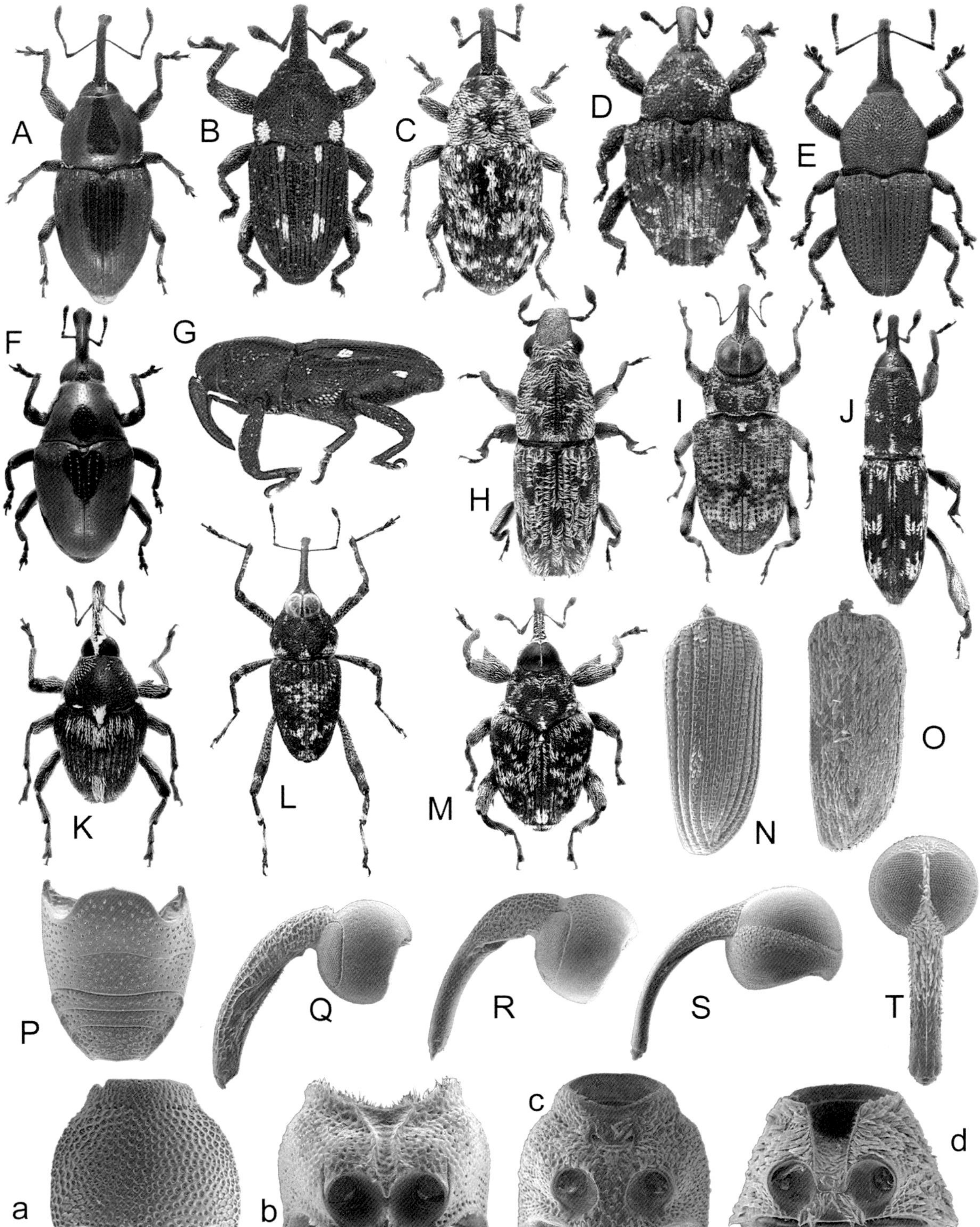

Fig. 44.170. Conoderinae, characters for key to genera. A–M, habitus dorsal: A, *Apinocis*; B, *Baris*; C, *Cosmobaris discolor* (Boheman); D, *Nyella tuberculata* Oke; E, *Orchidophilus aterrimus* (Waterhouse); F, *Solenobaris decipiens* Lea; G, *Acythopeus bigeminatus* Pascoe; H, *Tomicoproctus*; I, *Agametis bifasciata* Lea; J, *Phaenomerus auriceps* Thompson; K, *Lobotrachelus stigma* Pascoe; L, *Mecopus tipularius* Pascoe; M, *Metialma australiae* Lea; N–O, elytron, dorsal: N, *Baris*; O, *Cosmobaris*; P, *Orchidophilus*, abdominal ventrites; Q–T, heads, lateral and dorsal: Q, *Cosmobaris*; R, *Baris*; S, *Agametis*; T, *Lobotrachelus*; a–d, prothorax: a, *Baris*, dorsal; b, *Agametis*, ventral; c, *Cosmobaris*, ventral; d, *Lobotrachelus*, ventral.

Australian species. *B. albigutta* Lea, 1906a, *B. albopicta* Lea, 1906a, *B. amoenula* (Boheman, 1836) (*in* Schoenherr 1836), *B. angophorae* Lea, 1906a, *B. apicinivea* Lea, 1915d, *B. australiae* Lea, 1906a, *B. australis* (Boisduval, 1835), *B. barronensis* Lea, 1927b, *B. basirostris* Lea, 1906a, *B. bituberculata* Lea, 1927b, *B. bryanti* Lea, 1915d, *B. cairnsensis* Lea, 1927b, *B. cyaneotincta* Lea, 1927b, *B. devia* Lea, 1906a, *B. dilatatifrons* Lea, 1927b, *B. elliptica* Lea, 1906a, *B. episternalis* Lea, 1927b, *B. foveata* Lea, 1931c, *B. foveifrons* Zimmerman, 1992, *B. geraldtonensis* Lea, 1927b, *B. hoplocnemis* Lea, 1927b, *B. illepida* Lea, 1927b, *B. latericollis* Lea, 1927b, *B. livida* Zimmerman, 1992, *B. longicollidis* Hustache, 1938, *B. melanochroa* Lea, 1927b, *B. melanostetha* Lea, 1927b, *B. microscopica* Lea, 1906a, *B. monobia* Lea, 1931c, *B. niveodispersa* Lea, 1927b, *B. niveonotata* Lea, 1906a, *B. orthodoxa* Lea, 1927b, *B. parvonigra* Lea, 1927b, *B. porosa* Lea, 1906a, *B. pulchriparva* Lea, 1927b, *B. pulchripennis* Lea, 1927b, *B. quadrisignata* (Boheman, 1836), *B. setipennis* Lea, 1927b, *B. setistriata* Lea, 1927b, *B. sororia* Lea, 1906a, *B. sublaminata* Lea, 1906a, *B. subopaca* Lea, 1906a, *B. tenuistriata* Lea, 1906a, *B. transversicollis* Lea, 1915a, *B. trisinuata* Lea, 1927b, *B. vagans* Lea, 1906a.

Distribution. Australia (NSW, QLD, NT, WA), cosmopolitan.

Notes. The Australian species classified in *Baris* represent different genera (Pullen *et al.* 2014; E. C. Zimmerman unpublished notes; J. Prena pers. comm.). *Baris leucospila* Pascoe, 1885, earlier recorded from Australia (Lea 1908b), does not occur in the country (Pullen *et al.* 2014).

Cosmobaris Casey, 1920 (Pl. 104D; Fig. 44.172)

Type species, by monotypy: *Cosmobaris americana* Casey, 1920.

Australian species. *C. discolor* (Boheman, 1836) (*in* Schoenherr 1836).

Distribution. Australia, Holarctic region and Africa.

Notes. The single species in Australia was accidentally introduced by about 1927 (Pullen *et al.* 2014).

Gymnobaris Lea, 1906 (Pl. 104E)

Type species, by monotypy: *Gymnobaris polita* Lea, 1906.

Australian species. *G. polita* Lea1906a.

Distribution. Australia (QLD).

Ipsichora Pascoe, 1874 (Pl. 104F; Fig. 44.173)

Type species, by subsequent designation (Lea, 1931c: 167): *Ipsichora cupido* Pascoe, 1874c.

Australian species. *I. bimaculibasis* Lea, 1927b, *I. desiderabilis* Lea, 1906a, *I. duplicata* Lea, 1906a, *I. macleayi* Lea, 1906a, *I. mesosternalis* Lea, 1906a, *I. tibialis* Lea, 1927b.

Distribution. Australia (QLD, NSW), Maluku Is., New Guinea.

Linogeraeus Casey, 1920 (Pl. 104G; Figs 44.169A–G, 44.174)

Type species, by original designation: *Centrinus lineellus* LeConte, 1859.

Australian species. *L. urbanus* (Boheman, 1859).

Distribution. Australia (QLD, NSW, VIC), New Zealand, America.

Notes. The single species in Australia was accidentally introduced from South America by 1927 (Pullen *et al.* 2014). Its larva, described by May (1984), mines in stolons of the turf grass *Paspalum distichum* (Poaceae).

Myctides Pascoe, 1874 (Pl. 104H; Fig. 44.175)

Type species, by monotypy: *Myctides barbatus* Pascoe, 1874c.

Australian species. *M. balaninirostris* Lea, 1906a, *M. familiaris* Pascoe, 1885, *M. imberbis* Lea, 1906a.

Distribution. Australia (QLD), Malay Archipelago, New Guinea, SE USA (introduced).

Notes. The larvae develop in fruits of *Syzygium* species (Myrtaceae).

Nyella Oke, 1931 (Pl. 104I)

Type species, by original designation: *Nyella tuberculata* Oke, 1931.

Australian species. *N. tuberculata* Oke, 1931.

Distribution. Australia (VIC).

Notes. The type specimens were collected on *Myrsine variabilis* (Primulaceae).

Orchidophilus Buchanan, 1935 (Pl. 104J; Figs 44.169H–N, 44.176)

Type species, by original designation: *Orchidophilus peregrinator* Buchanan, 1935.

Australian species. *O. aterrimus* (Waterhouse, 1874).

Distribution. Australia (NT, QLD), American Samoa, Hawaii, New Caledonia and Papua New Guinea.

Notes. A taxonomic synopsis of the genus was published by Prena (2008), together with accounts of its habits and life history. *Orchidophilus aterrimus* was accidentally introduced to Australia from Southeast Asia by 1904 and appears well established in the Northern Territory and northern Queensland (Pullen *et al.* 2014). Its larvae develop in stems of various orchids and can cause serious damage to these plants in gardens and nurseries (Pullen *et al.* 2014). The species *O. epidendri* (Murray) and *O. insidiosus* Prena have been recorded from Australia but do not seem to be established here (Pullen *et al.* 2014).

Solenobaris Lea, 1906 (Pl. 104K)

Type species, by subsequent designation (Setliff, 2007: 10): *Solenobaris decipiens* Lea, 1906.
Australian species. *S. cryptodon* Lea, 1931c, *S. cryptorhynchoides* Lea, 1927b, *S. decipiens* Lea, 1906a, *S. edentata* Lea, 1906a, *S. parvipunctata* Lea, 1931c.
Distribution. Australia, Fiji, New Guinea.

Tribe Campyloscelini Schoenherr, 1845

Phaenomerus Schoenherr, 1836 (Pl. 105C)

Type species, by original designation: *Phaenomerus sundewalli* Boheman, 1836 (*in* Schoenherr 1836).
Australian species. *P. auriceps* Thompson, 1996, *P. exilis* Pascoe, 1872c, *P. peregrinator* Thompson, 1996.
Distribution. Australia (NSW, QLD), Melanesia, Oriental, Afrotropical, Madagascar, Japan.
Notes. Thompson (1996) revised the species of the Australian Region and summarised the available information about their habits and life histories. Both adults and larvae inhabit the tunnels of ambrosia beetles (Platypodinae), in which they feed on the fungi and dead beetles but appear to be scavengers rather than predators (Thompson 1996; May 1997). Recorded host trees include *Alstonia* (Apocynaceae), *Corymbia* and *Eucalyptus* (Myrtaceae) and *Ceratopetalum succirubrum* (Cunoniaceae).

Tomicoproctus Faust, 1898 (Pl. 105D)

Type species, by original designation: *Tomicoproctus eichhoffi* Faust, 1898.
Australian species. *T. maculatus* (Schedl, 1936).
Distribution. Australia (QLD, NSW), Africa, New Guinea.
Notes. Only one species of the genus is recorded from Australia, but May (1997) reported the larvae of two species collected from tunnels of *Crossotarsus armipennis* Lea, in which they appear to live as cleptobionts living off the ambrosia fungi.

Tribe Conoderini Schoenherr, 1833

Agametis Pascoe, 1870 (Pl. 105G; Fig. 44.177)

Type species, by monotypy: *Agametis festiva* Pascoe, 1870c.
Australian species. *A. bifasciata* Lea, 1910a.
Distribution. Australia, Malay Archipelago, New Guinea.

Brimoides Kojima & Lyal, 2002 (Pl. 105E)

Type species, by original designation: *Brimoides morimotoi* Kojima & Lyal, 2002.
Australian species. *B. exilis* (Pascoe, 1885).
Distribution. Australia, Sri Lanka, Taiwan, Philippines, Malaysia.

Lamitema Lea, 1910 (Pl. 105H)

Type species, by monotypy: *Lamitema decipiens* Lea, 1910.
Australian species. *L. decipiens* Lea, 1910a.
Distribution. Australia.

Lobotrachelus Schoenherr, 1838 (Pl. 105I; Fig. 44.178)

Type species, by original designation: *Lobotrachelus vestitus* Rosenschoeld, 1838 (*in* Schoenherr 1838).
Australian species. *L. stigma* Pascoe, 1874c.
Distribution. Australia, Afrotropical and Oriental regions.

Mecopus Schoenherr, 1825 (Pl. 105J, K; Fig. 44.168)

Type species, by original designation: *Rhynchaenus bispinosus* Fabricius, 1801.
Australian species. *M. australasiae* Heller, 1893, *M. bispinosus* (Weber, 1801), *M. doryphorus* Quoy & Gaimard, 1824, *M. macleayi* Lea, 1910a, *M. phthisicus* Lea, 1899b, *M. pictus* Lea, 1910a, *M. rufipes* Heller, 1893, *M. sobrinus* Lea, 1910b, *M. terraereginae* Heller, 1893, *M. tipularius* Pascoe, 1870f.
Distribution. Australia (NSW, QLD), Afrotropical and Oriental regions.

Metialma Pascoe, 1871 (Pl. 105L)

Type species, by subsequent designation (Marshall, 1939b: 10): *Metialma scenica* Pascoe, 1871b.
Australian species. *M. australiae* Lea, 1910a.
Distribution. Australia, Afrotropical and Oriental regions.

Othippia Pascoe, 1874 (Pl. 105F)

Type species, by subsequent designation (Setliff, 2007: 10): *Othippia distigma* Pascoe, 1874c.
Australian species. *O. guttula* Pascoe, 1885.
Distribution. Australia, Malay Archipelago, Madagascar.

Temialma Lea, 1910 (Pl. 105M)

Type species, by monotypy: *Temialma suturalis* Lea, 1910.
Australian species. *T. suturalis* Lea, 1910a.
Distribution. Australia.

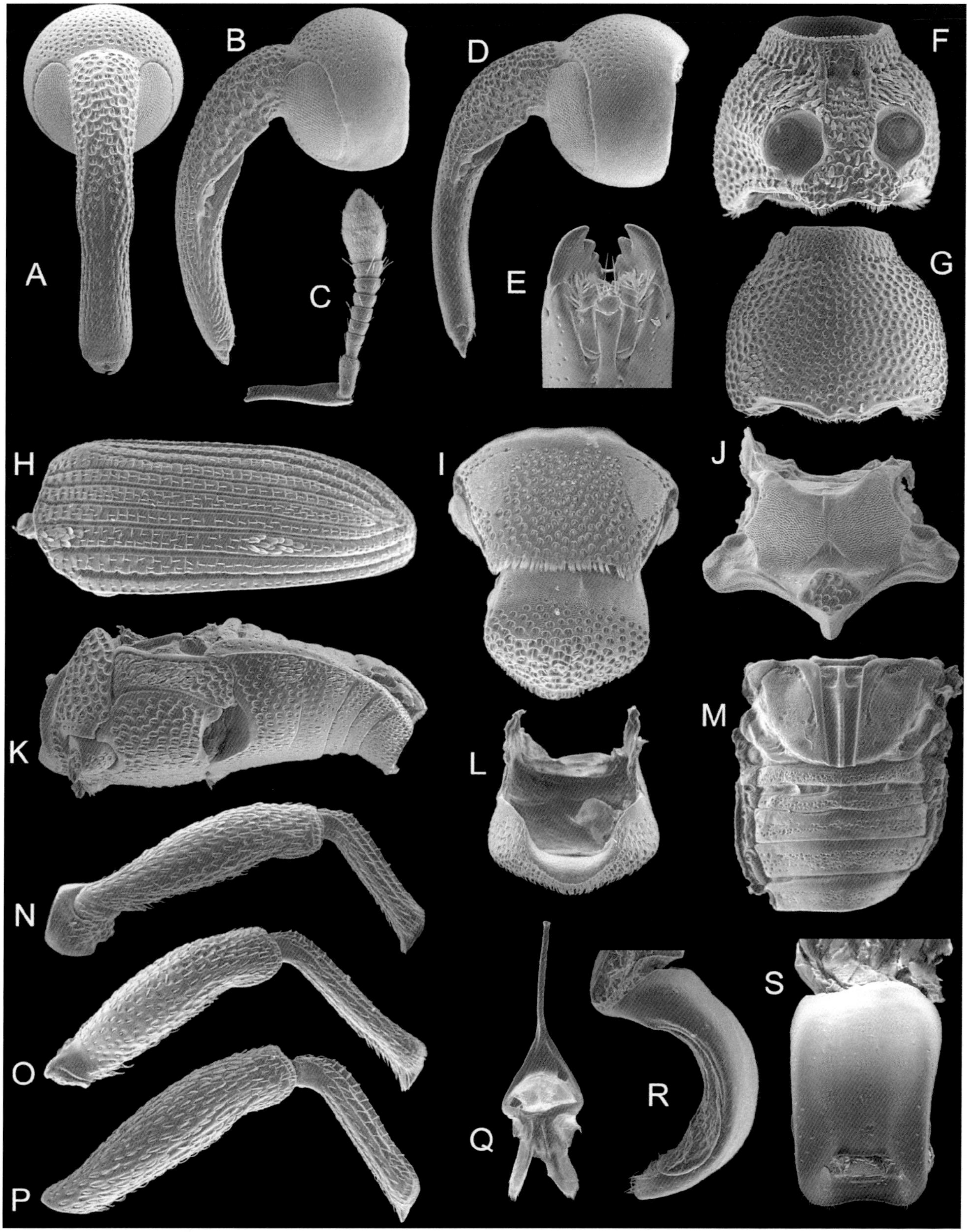

Fig. 44.171. Baridini, structural details, *Baris* sp., Cairns QLD, ♂ (except D and E, ♀). A–B, head: A, dorsal; B, lateral; C, antenna; D, head, lateral; E, mouthparts; G, pronotum; H, elytron, dorsal; I, terminalia, dorsal; J, scutellum; L, pygidium, ventral; M, metanotum and abdominal tergites; N, foreleg; O, middle leg; P, hindleg; Q, tegmen; R–S, penis: R, lateral; S, dorsal.

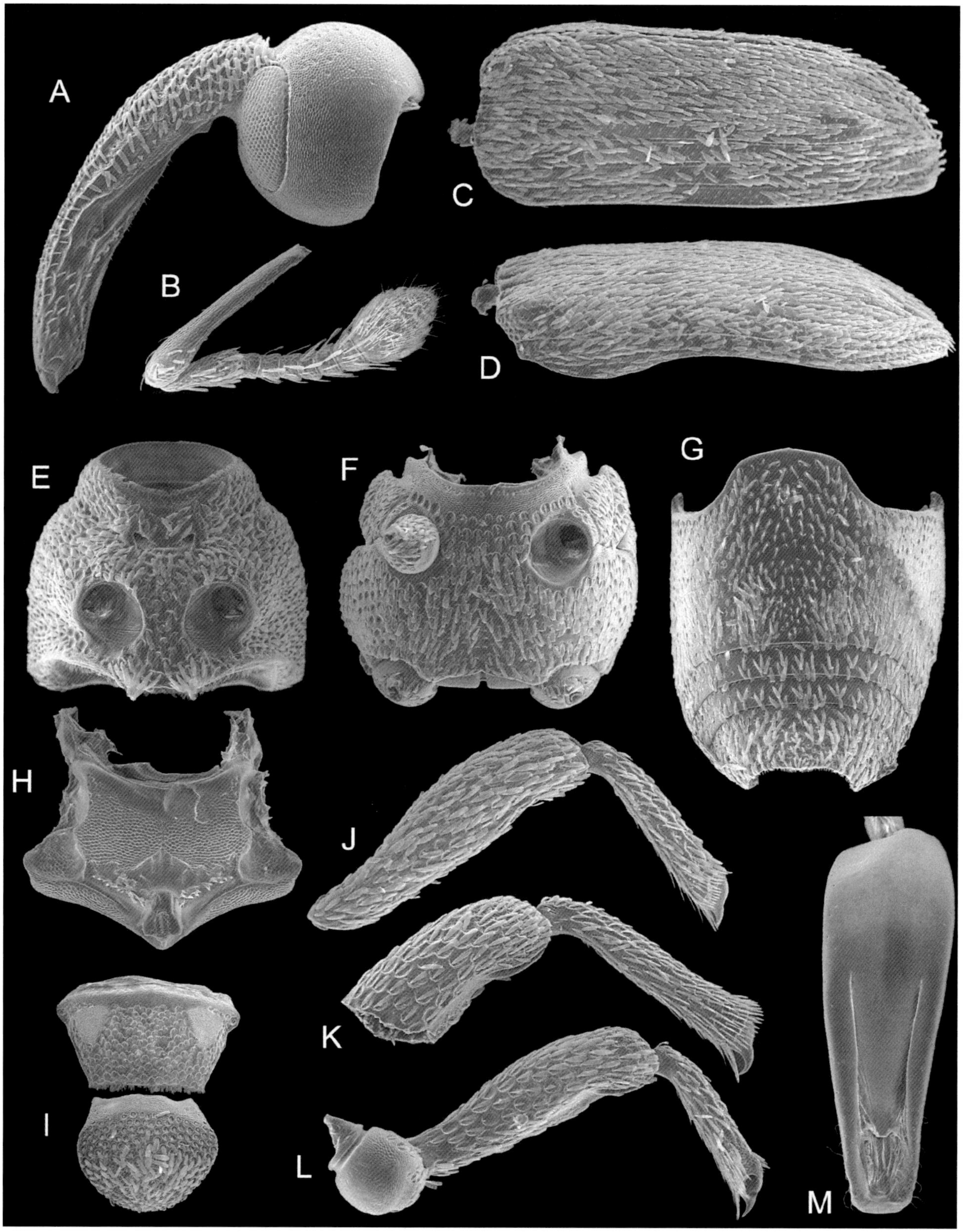

Fig. 44.172. Baridini, structural details, *Cosmobaris discolor* (Boheman), ♂. A, head, lateral; B, antenna; C–D, elytron: C, dorsal; D, lateral; E, prothorax, ventral; F pterothorax, ventral; G, abdominal ventrites; H, scutellum; I, terminalia; J, hindleg; K, middle leg; L, foreleg; M, penis, dorsal.

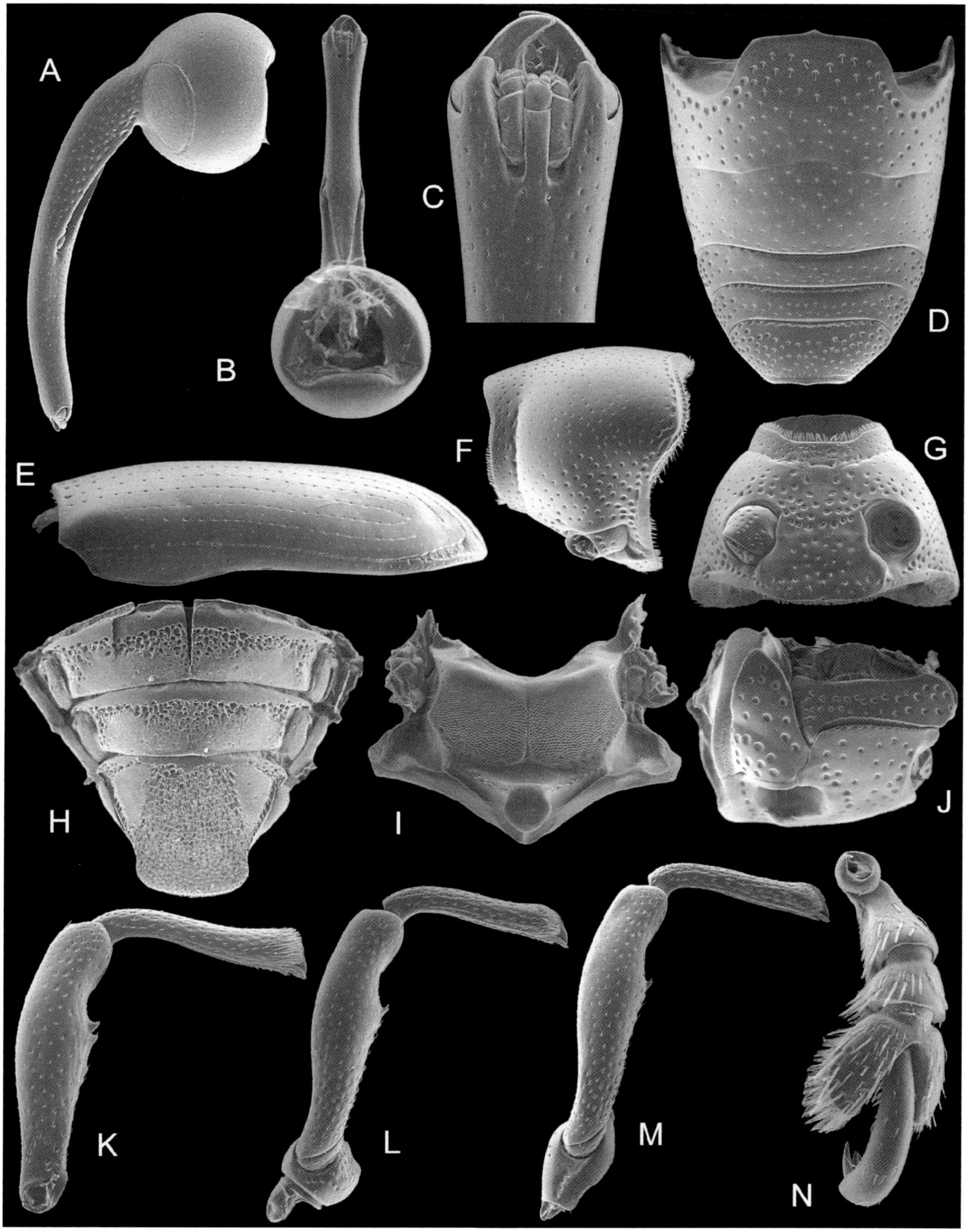

Fig. 44.173. Baridini, structural details, *Ipsichora desiderabilis* Lea. A–B, head: A, lateral; B, ventral; C, mouthparts; D, abdominal ventrites; E, elytron, lateral; F–G, prothorax: F, lateral; G, ventral; H, abdominal tergites; I, scutellum; J, pterothorax, lateral; K, middle leg; L, foreleg; M, hindleg; N, tarsus, dorsal.

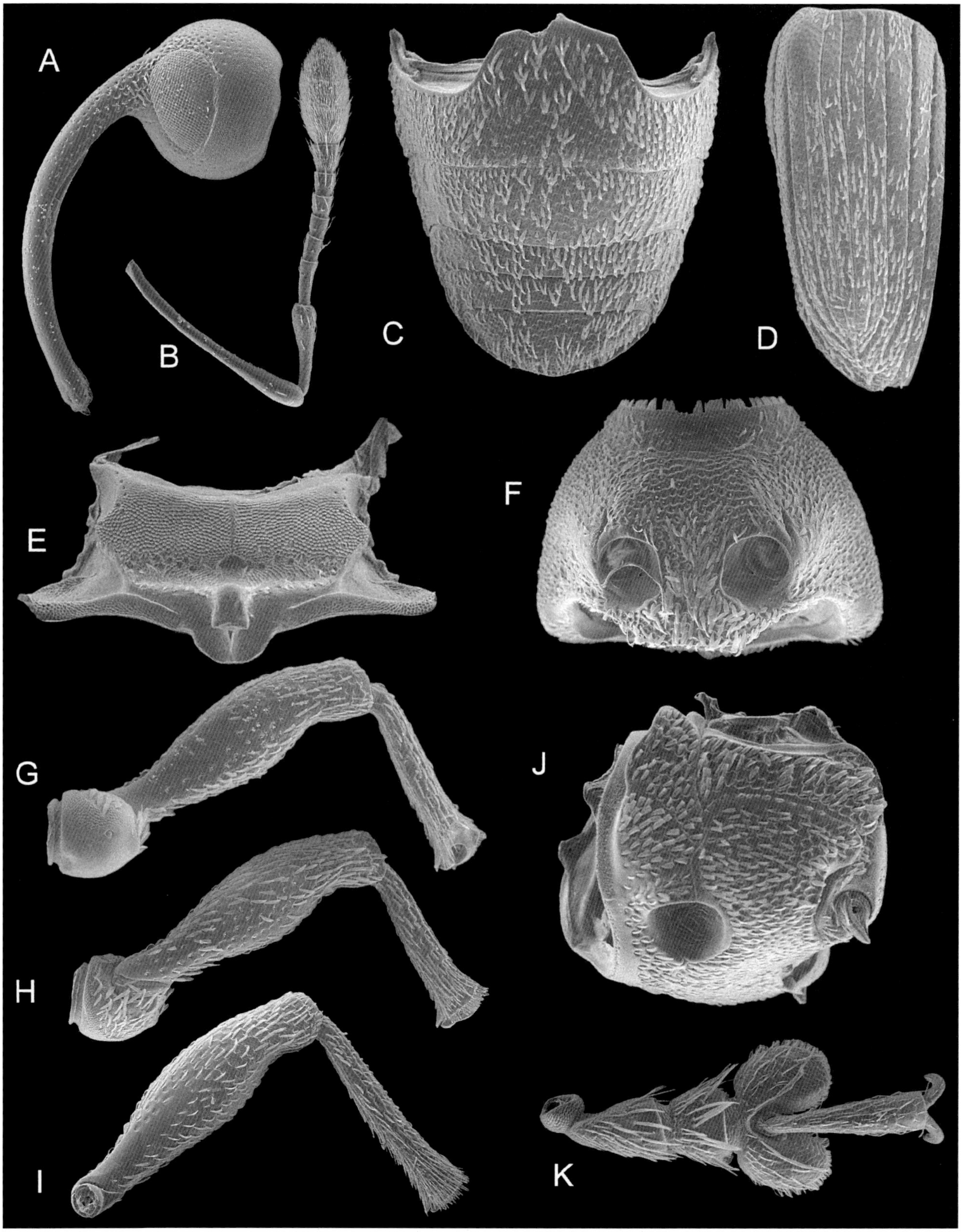

Fig. 44.174. Baridini, structural details, *Linogeraeus urbanus* (Boheman). A, head, lateral; B, antenna; C, abdominal ventrites; D, elytron, dorsal; E, scutellum; F, prothorax, ventral; G; H, middle legs; I, hindleg; J, pterothorax, lateral; K, tarsus, dorsal.

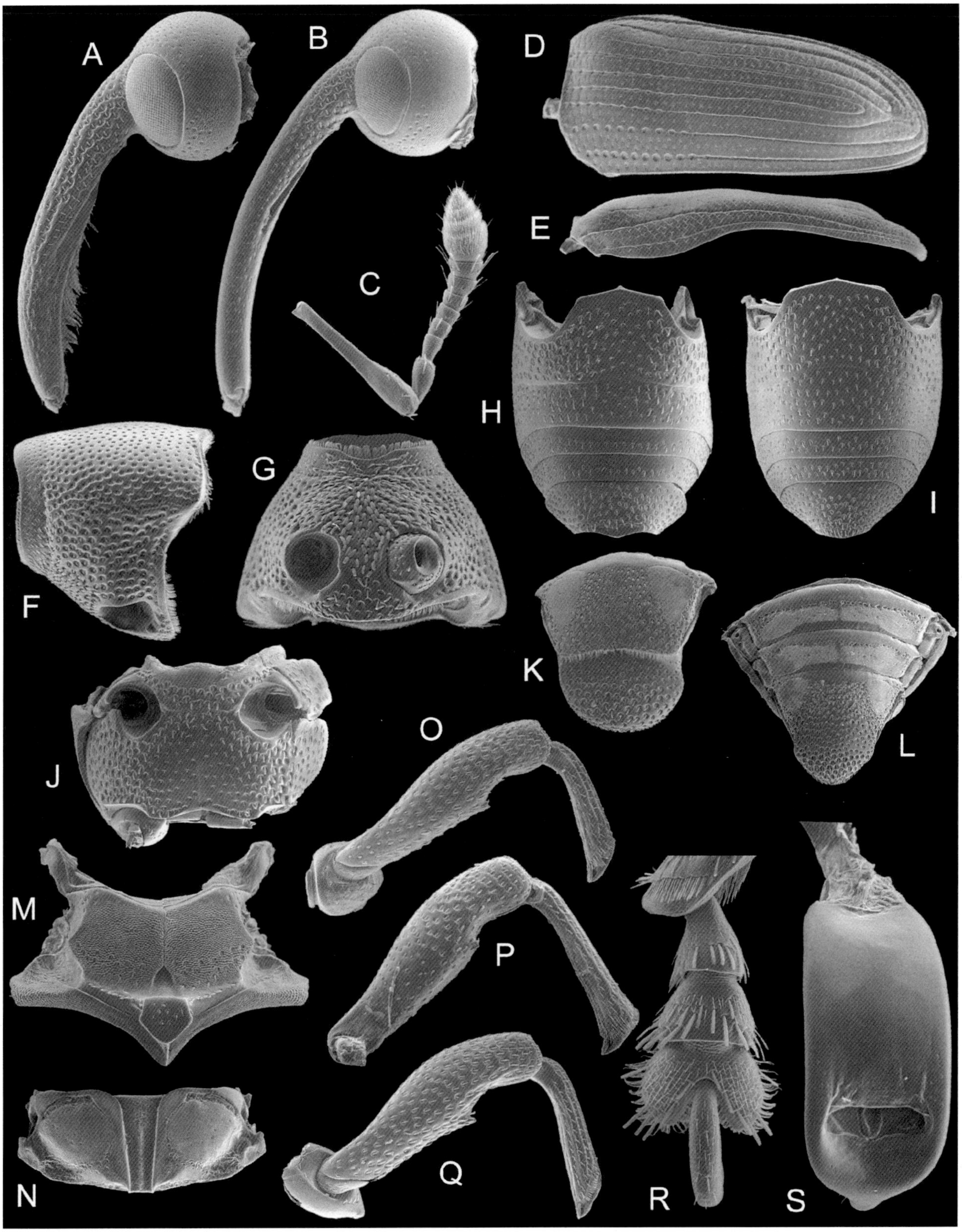

Fig. 44.175. Baridini, structural details, *Myctides imberbis* Lea. A–B, head, lateral: A, ♂; B, ♀; C, antenna; D–E, elytron: D, dorsal; E, lateral; F–G, prothorax: F, lateral; G, ventral; H–I, abdominal ventrites: H, ♂; I, ♀; J, pterothorax, ventral; K–L, terminalia, dorsal: K, ♂; L, ♀; M, scutellum; N, metanotum; O, foreleg; P, middle leg; Q, hindleg; R, tarsus, dorsal; S, penis, dorsal.

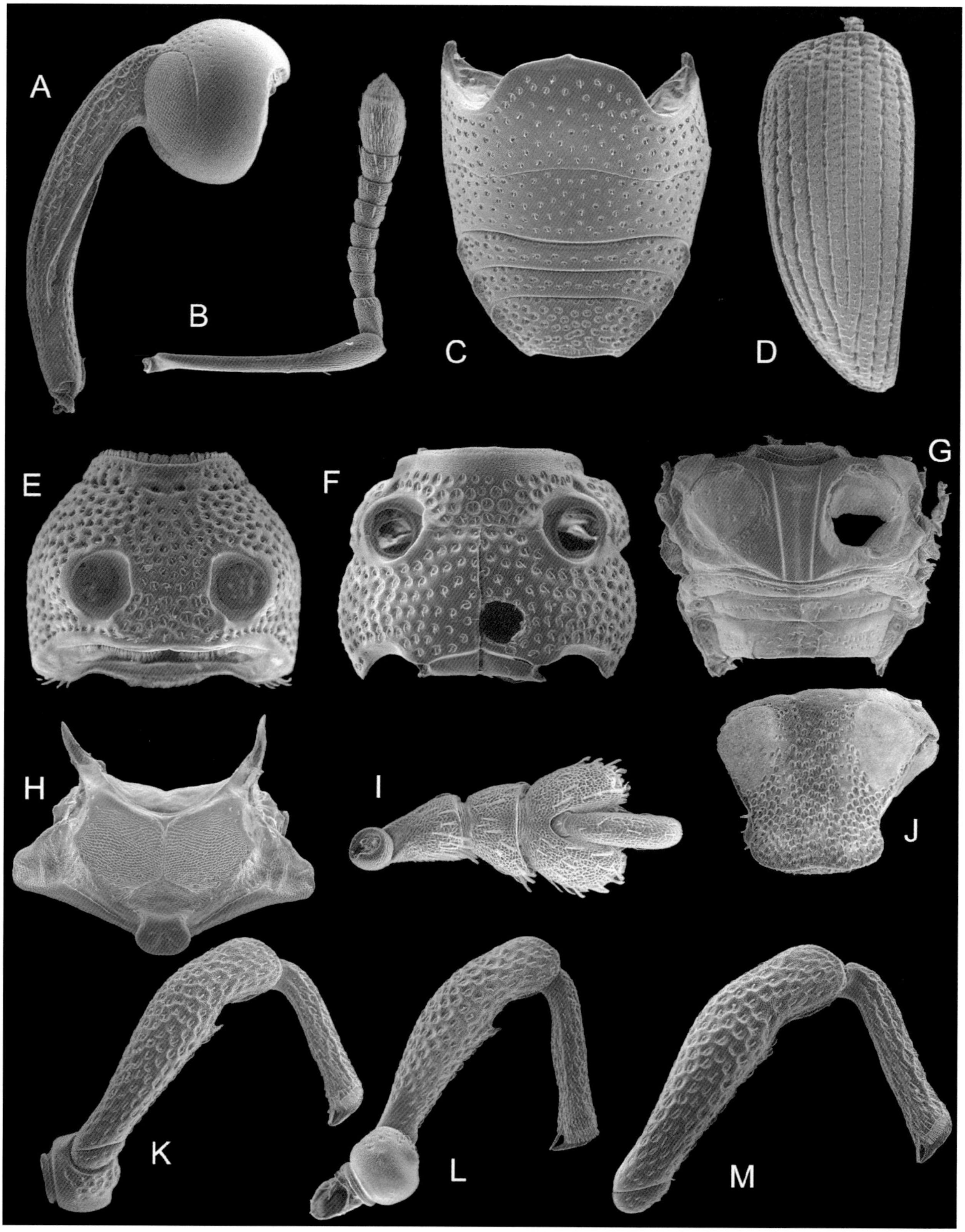

Fig. 44.176. Baridini, structural details, *Orchidophilus aterrimus* (Waterhouse), ♀. A, head, lateral; B, antenna; C, abdominal ventrites; D, elytron, dorsal; E, prothorax, ventral; F, pterothorax, ventral; G, metanotum and tergites; H, scutellum; I, tarsus, dorsal; J, pygidium; K, middle leg; L, foreleg; M, hindleg.

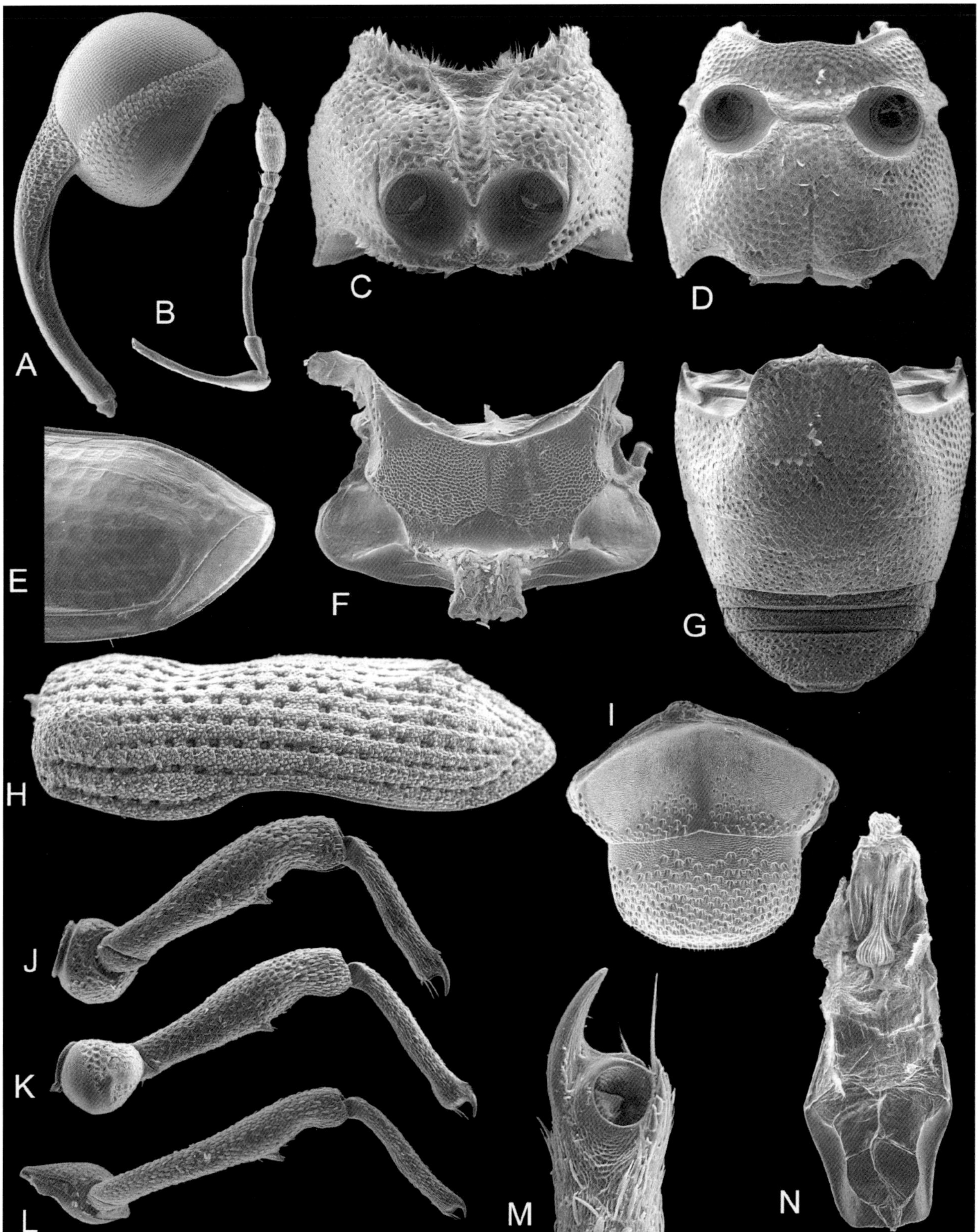

Fig. 44.177. Conoderini, structural details, *Agametis bifasciata* Lea. A, head, lateral; B, antenna; C, prothorax, ventral; D, pterothorax, ventral; E, elytral apex, inner lateral; F, scutellum; G, abdominal ventrites H, elytron, lateral; I, terminalia, dorsal; J, middle leg; K, foreleg; L, hindleg; M, tibia apex; N, penis, dorsal.

Fig. 44.178. Conoderini, structural details, *Lobotrachelus stigma* Pascoe, ♀. A–B, head: A, dorsal; B, lateral; C, antenna; D–F, elytron: D, lateral; E, inner lateral; F, dorsal; G–H, prothorax: G, ventral; H, dorsal; I, abdomen; J, pterothorax, ventral; K, terminalia, dorsal; L, ventrites, posterior; M, scutellum; N, foreleg; O, middle leg; P, hindleg.

Curculionidae: Cossoninae Schoenherr, 1825

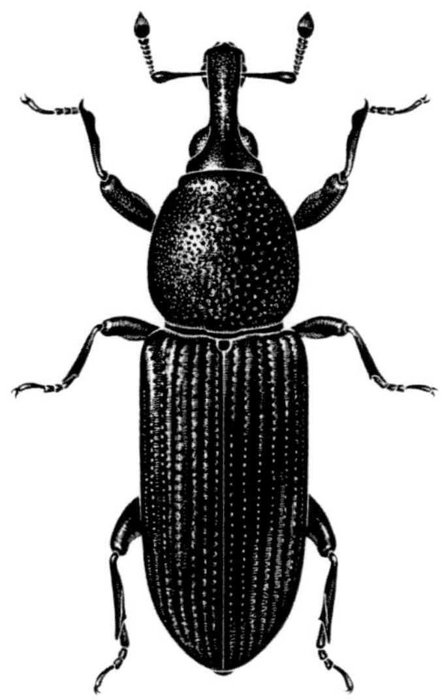

Fig. 44.179. *Cossonus simsoni* Lea.

Introduction. The Cossoninae are mostly small, black to brown weevils with an elongate body and sparse, minute setae (e.g. Pl. 16). They occur worldwide, and many species are widespread due to human trade of wood and wood products. Their larvae develop mostly in dead, old and moist wood, but some tunnel under bark or in palm and fern fronds, and the recently described genus *Zimmiodes* (Fig. 44.180) inhabits male cones of *Cycas* (Cycadaceae) and pollinates its hosts (Hsiao & Oberprieler 2024). Many species have the eyes reduced or lost, such as *Conlonia litoralis* Lea, which occurs in wracks of seagrass along beaches of the southern coast, and *Halorhynchus caecus* Wollaston, which lives on roots of plants in coastal environments. The taxonomy of the subfamily has received very little attention globally in recent times, including its division into natural (monophyletic) tribes. A study of the Australian fauna was commenced by G. Kuschel in New Zealand and later continued by E. C. Zimmerman, but neither completed this work and published much of the information they had gathered, and the taxonomy of the taxa in Australia remains very poorly resolved. As a result, we found it impossible to draw up a workable and reliable key to the genera and therefore limit our coverage to illustrating the habitus of almost all genera and the diagnostic features of the most salient taxa, using also Zimmerman's unpublished notes and several of his photographs. May (1993) described the larvae of some New Zealand genera that also occur in Australia or may have affinities to the Australia taxa, and some of her illustrations are reproduced here (Figs 44.181, 44.182, Fig. 44.182A–G). An overview of the taxonomy and biology of the subfamily globally was provided by Jordal (2014).

Classification of the Australian genera of Cossoninae

Agrilochilus Broun, 1880 (Pl. 106A)

Type species, by monotypy: *Agrilochilus prolixus* Broun, 1880.
Australian species. *Agrilochilus* sp.
Distribution. Australia (Lord Howe Is.), New Zealand.

Allopentarthrum Kuschel, 1986 (Pl. 106B; Fig. 44.183A–E)

Type species, by original designation: *Pentarthrum castaneum* Voss, 1934.
Australian species. *A. elumbe* (Boheman, 1838) (*in* Schoenherr 1838).
Distribution. Australia (Lord Howe Is.), South and Central America, Africa, Malaysia, Japan, New Guinea, Samoa, Hawaii.
Notes. The single species in Australia was accidentally introduced from the Neotropical region by 1916 (Pullen *et al.* 2014). The genus was described in Wibmer & O'Brien (1986).

Aphanocorynes Wollaston, 1873 (Pl. 106C; Fig. 44.189J–O)

Type species, by monotypy: *Aphanocorynes depressus* Wollaston, 1873.
Australian species. *A. depressus* Wollaston, 1873b, *A. procerus* Olliff, 1889, *A. rugosipennis* (Lea, 1915d).
Distribution. Australia (SA, Lord Howe Is.), Samoa, New Guinea.

Catolethrobius Voss, 1939 (Pl. 106D; Fig. 44.183F–J)

Type species, by monotypy: *Catolethrus silvestris* Kolbe, 1910.
Australian species. *C. silvestris* (Kolbe, 1910).

Fig. 44.180. *Zimmiodes australis* Hsiao & Oberprieler.

Distribution. Australia (Qld, Lord Howe Is.), Seychelles, Reunion, Siberia, New Zealand.

Notes. The single species in Australia seems to represent an accidental introduction from the Seychelles (Pullen *et al.* 2014).

Choerorhinus Fairmaire, 1858 (Pl. 106E)

Type species, by monotypy: *Choerorhinus squalidus* Fairmaire, 1858.

Australian species. *Choerorhinus* sp.

Distribution. Australia (NT), Mediterranean region, Afghanistan, Japan, Africa.

Notes. A single undescribed species is recorded from Australia (Pullen *et al.* 2014).

Conlonia Lea, 1908 (Pl. 106F; Fig. 44.184A–F)

Type species, by monotypy: *Conlonia litoralis* Lea, 1908.

Australian species. *C. litoralis* Lea, 1908c.

Distribution. Australia (NSW, SA, VIC, WA).

Notes. Specimens occur under mats of seagrass along the south Australian coast.

Coptocorynus Marshall, 1948 (Pl. 106G; Fig. 44.185)

Type species, by original designation: *Coptocorynus araucariae* Marshall, 1948.

Australian species. *C. araucariae* Marshall, 1948.

Distribution. Australia (QLD).

Cossonideus Wollaston, 1873 (Pl. 106H)

Type species, by monotypy: *Cossonideus pascoei* Wollaston, 1873.

Australian species. *C. lineola* (Fabricius, 1775), *C. pascoei* Wollaston, 1873b.

Distribution. Australia (QLD).

Cossonus Clairville, 1798 (Pl. 106I; Figs 44.179, 44.186, 44.187)

Type species, by subsequent designation (Latreille, 1810: 431): *Curculio linearis* Fabricius, 1775.

Australian species. *C. arctatus* (Pascoe, 1885), *C. bilineatus* (Pascoe, 1885), *C. castelnaui* (Lea, 1911b), *C. coptorhinus* Lea, 1915a, *C. excavatus* Pascoe, 1885, *C. frenchi* Lea, 1915a, *C. hackeri* Lea, 1909c, *C. incisus Pascoe, 1885*, *C. indigens* Pascoe, 1885, *C. integricollis* Lea, 1896, *C. irregularis* Lea, 1927b, *C. macilentus* Lea, 1915a, *C. nigroapicalis* Lea, 1909c, *C. nitidirostris* Lea, 1915d, *C. platynotus* Lea, 1911b, *C. porosternus* Lea, 1915a, *C. praeustus* Redtenbacher, 1868, *C. simsoni* Lea, 1911c, *C. subpubescens* (Wollaston, 1873b), *C. variipennis* Gahan, 1900, *C. vicarius* Lea, 1909c.

Distribution. Cosmopolitan.

Dryotribus Horn, 1873 (Pl. 106J; Fig. 44.184F–J)

Type species, by monotypy: *Dryotribus mimeticus* Horn, 1873.

Australian species. *D. mimeticus* Horn, 1873.

Distribution. Australia (WA, QLD), Galapagos Is., The Caribbean, Mexico, SE USA, Hawaii, Wake Is., Sunda Is., China, Japan.

Eiratus Pascoe, 1877 (Pl. 106K)

Type species, by monotypy: *Eiratus parvulus* Pascoe, 1877.

Australian species. *Eiratus* sp.

Distribution. Australia (QLD), New Zealand.

Entium Sharp, 1878 (Pl. 106L)

Type species, by monotypy: *Entium aberrans* Sharp, 1878.

Australian species. *Entium* sp.

Distribution. Australia (QLD, Lord Howe Is.), New Zealand.

Exomesites Broun, 1886 (Pl. 107A)

Type species, by monotypy: *Exomesites optimus* Broun, 1886.

Australian species. *Exomesites* sp.

Distribution. Australia (Lord Howe Is.), New Zealand.

Halorhynchus Wollaston, 1873 (Pl. 107B; Fig. 44.188)

Type species, by monotypy: *Halorhynchus caecus* Wollaston, 1873.

Australian species. *H. caecus* Wollaston, 1873b, *H. geniculatus* Lea, 1900a.

Distribution. Australia (WA, SA, VIC), Madagascar (?).

Notes. The status of *H. geniculatus* requires assessment; it is probably conspecific with *H. caecus*. The recently described *H. remii* Chamorro & Steiner from Madagascar (Chamorro & Steiner 2022) evidently belongs in a different genus, and *Halorhynchus* is apparently endemic to Australia. The weevils occur in coastal habitats along the southern coasts among the roots of spinifex grasses and saltbushes (*Atriplex*, Amaranthaceae) (Lea 1900a).

Hexarthroides Lea, 1896 (Pl. 109J; Fig. 44.189A–D)

Type species, by monotypy: *Hexarthroides punctulatus* Lea, 1896.

Australian species. *H. punctulatus* Lea, 1896.

Distribution. Australia (NSW).

Himatium Wollaston, 1873 (Pl. 107C, D)

Type species, by monotypy: *Himatium pubescens* Wollaston, 1873b.

Australian species. *Himatium* sp.

Distribution. Australia (QLD, Lord Howe Is.), Central and North America, Siberia, Japan, Malay Archipelago, India, Marianas, Sri Lanka, Africa.

Hoplocossonus Lea, 1911 (Pl. 107E)

Type species, by subsequent designation (Pullen *et al.* 2014: 301): *Cossonus lethargicus* Olliff, 1889.
Australian species. *H. lethargicus* (Olliff, 1889).
Distribution. Australia (Lord Howe Is.).
Reference. Lea (1911b).

Macrorhyncolus Wollaston, 1873 (Pl. 107F, G)

Type species, by monotypy: *Macrorhyncolus crassiusculus* Wollaston, 1873a.
Australian species. *M. littoralis* (Broun, 1880).
Distribution. Australia, Chile, Northern America, Southern Africa, Japan, New Zealand, Sri Lanka.

Mastersinella Lea, 1896 (Pl. 107H; Fig. 44.189E–I)

Type species, by monotypy: *Mastersinella octoarticulata* Lea, 1896.
Australian species. *M. octoarticulata* Lea, 1896.
Distribution. Australia (QLD).

Microcossonus Wollaston, 1873

Type species, by monotypy: *Microcossonus wallacei* Wollaston, 1873a.
Australian species. *M. pandani* Lea, 1896.
Distribution. Australia (NSW).

Microtribus Wollaston, 1873 (Pl. 107I)

Type species, by monotypy: *Microtribus huttoni* Wollaston, 1873b.
Australian species. *Microtribus* sp. (Pullen *et al.* 2014).
Distribution. Australia (QLD, Norfolk Is.), New Zealand, Samoa, St. Paul Is.

Morronella Alonso-Zarazaga & Lyal, 1999 (Pl. 107J)

Type species, by monotypy: *Heteropsis lawsoni* Wollaston, 1873b.
Australian species. *Morronella* sp.
Distribution. Australia (NSW), New Zealand.

Ochronanus Pascoe, 1885 (Pl. 107K)

Type species, by monotypy: *Ochronanus pygmaeus* Pascoe, 1885.
Australian species. *Ochronanus* sp.

Distribution. Australia (QLD, Lord Howe Is., TAS), Kuril Is., Japan, China, Malay Archipelago, Samoa, New Guinea, Comoros, Reunion.

Omeretes Marshall, 1938 (Pl. 107L, Fig. 44.190)

Type species, by original designation: *Omeretes podocarpi* Marshall, 1938.
Australian species. *O. onychialis* Marshall, 1938a, *O. podocarpi* Marshall, 1938a.
Distribution. Australia (QLD, VIC), India.

Orthotemnus Wollaston, 1873 (Pl. 108A; Fig. 44.191A–E)

Type species, by original designation: *Orthotemnus reflexus* Wollaston, 1873b.
Australian species. *O. disparilis* Pascoe, 1885, *O. polixus* (Erichson), 1842.
Distribution. Australia, Maluku Is., New Guinea, Madagascar, Reunion, Seychelles, Siberia.

Oxydema Wollaston, 1873 (Pl. 108B; Fig. 44.191F–O)

Type species, by subsequent designation (Zimmerman, 1940: 286): *Oxydema fusiformis* Wollaston, 1873b.
Australian species. *O. australis* (Wollaston, 1873b), *O. major* (Wollaston, 1873b), *O. maxima* (Lea, 1915a), *O. procer* (Boheman, 1838) (*in* Schoenherr 1838).
Distribution. Australia (QLD), Kuril Is., Philippines, Sri Lanka, Sumatra, Hawaii, Marianas, Maluku Is., New Guinea, Samoa, Seychelles.
Notes. *Oxydema procer* is probably an accidental introduction in Australia (Pullen *et al.* 2014).

Pachyops Wollaston, 1873 (Pl. 108C, D)

Type species, by monotypy: *Pachyops cylindricus* Wollaston, 1873b.
Australian species. *P. incertus* Gahan, 1900, *P. insularis* (Lea, 1927b).
Distribution. Australia (Christmas Is.), Malay Archipelago, New Zealand.

Pentamimus Wollaston, 1873 (Pl. 108E)

Type species, by subsequent designation (Pullen *et al.* 2014: 303): *Pentamimus rhyncoliformis* Wollaston, 1873b.
Australian species. *P. australis* (Erichson, 1842), *P. nepeanianus* (Olliff, 1888).
Distribution. Australia (mostly TAS).

Pentarthrocis Lea, 1922 (Pl. 108F; Fig. 44.192A–E)

Type species, by monotypy: *Pentarthrocis ammophilus* Lea, 1922.

Australian species. *P. ammophilus* Lea, 1922.
Distribution. Australia (SA, WA).

Pentarthrum Wollaston, 1854 (Pl. 108G; Figs 44.181I–R, 44.193F)

Type species, by monotypy: *Pentarthrum huttoni* Wollaston, 1854.
Australian species. *P. foveiceps* Lea, 1927b, *P. foveiventre* Lea, 1927b, *P. interoculare* Lea, 1927b, *P. millingtoni* Olliff, 1888, *P. nigrum* Wollaston, 1873b, *P. orthodoxum* Lea, 1927b.
Distribution. Australia (SA, TAS), Argentina, Chile, Juan Fernandez Is., Canada, Europe, Japan, Africa, Oriental and Pacific regions, New Guinea, New Zealand, Chatham Is., Hawaii.

Pselactus Broun, 1886 (Pl. 108H)

Type species, by monotypy: *Pselactus punctatus* Broun, 1886.
Australian species. *P. spadix* (Herbst, 1795).
Distribution. Australia, cosmopolitan.
Notes. The single species in Australia is of Palaearctic origin but widely distributed by human trade and now seemingly naturalised in Australia (Pullen *et al.* 2014).

Rhyncolosoma Champion, 1914 (Pl. 108I, J)

Type species, by original designation: *Phloeophagosoma dubium* Gahan, 1900.
Australian species. *R. dubium* (Gahan, 1900).
Distribution. Australia (Christmas Is.), Samoa and Seychelles.

Rhyncolus Germar, 1817 (Pl. 108K)

Type species, by subsequent designation (ICZN, 1991: 268): *Curculio ater* Linnaeus, 1758.
Australian species. *Rhyncolus* sp.
Distribution. Australia, cosmopolitan.
Notes. The occurrence of the genus in Australia is doubtful (Pullen *et al.* 2014).

Sericotrogus Wollaston, 1873 (Pl. 108L)

Type species, by monotypy: *Sericotrogus subaenescens* Wollaston, 1873b.
Australian species. *S. subaenescens* Wollaston
Distribution. Australia (Lord Howe and Norfolk Is.), New Zealand.

Stenotrupis Wollaston, 1873 (Pl. 109A, B)

Type species, by subsequent designation (Champion, 1914: 465): *Stenotrupis crassifrons* Wollaston, 1873b.
Australian species. *Stenotrupis* sp.
Distribution. Australia (Lord Howe Is.), The Caribbean and Central America, Pacific region, New Guinea, New Zealand, Malay Archipelago, Madagascar, Reunion, Seychelles.

Stereoborus Wollaston, 1873 (Pl. 109C; Fig. 44.192F–O)

Type species, by subsequent designation (Voss, 1956: 133): *Stereoborus robustus* Wollaston, 1873b.
Australian species. *S. brevirostris* Lea, 1915d, *S. inductus* Pascoe, 1885, *S. interstitialis* Lea, 1896, *S. laporteae* Lea, 1896.
Distribution. Australia, Maluku Is., New Guinea.

Stereoderus Wollaston, 1873 (Pl. 109L)

Type species, by subsequent designation (Setliff, 2007: 10): *Stereoderus barbatus* Wollaston, 1873b.
Australian species. *S. lucens* Marshall, 1938b, *S. macleayi* Lea, 1896.
Distribution. Australia (QLD), India, Fiji, Maluku Is., New Caledonia, New Guinea, Samoa, Solomon Is.

Stereomimetes Wollaston, 1873 (Pl. 109K; Fig. 44.193G–K)

Type species, by original designation: *Stereomimetes crassicornis* Wollaston, 1873.
Australian species. *S. crassicornis* Wollaston, 1873b.
Distribution. Australia (WA).

Stilbocara Broun, 1893 (Pl. 109D)

Type species, not yet designated: *Stilbocara* Broun, 1893: 387.
Australian species. *Stilbocara* sp.
Distribution. Australia (QLD, Lord Howe Is.), New Zealand.

Temnorrhamphus Champion, 1914 (Pl. 109E)

Type species, by original designation: *Temnorrhamphus latirostris* Champion, 1914.
Australian species. *Temnorrhamphus* sp.
Distribution. Australia (Christmas Is.), Seychelles.

Toura Broun, 1909 (Pl. 109F; Fig. 44.193A–E)

Type species, by original designation: *Pentarthrum longirostre* Wollaston, 1873b.
Australian species. *Toura* sp.
Distribution. Australia (Lord Howe Is.), New Zealand.
Reference. Broun (1909b).

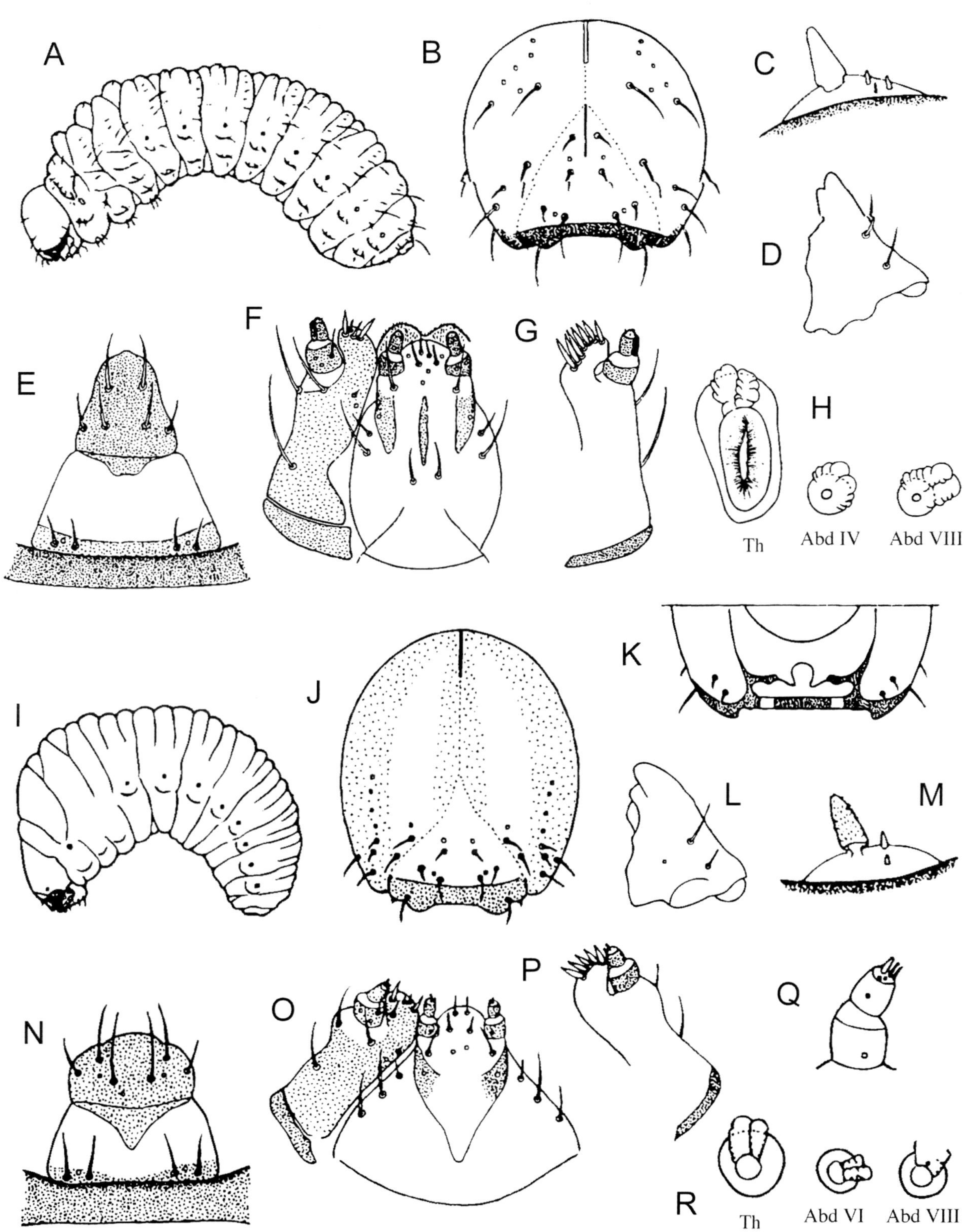

Fig. 44.181. Cossoninae, structural details of immature stages. A–H, *Mesites pallidipennis* Boheman, larva (New Zealand): A, habitus, lateral; B, head, dorsal; C, antenna; D, mandible; E, labrum, clypeus and epistoma; F, mala and labium, ventral; G, mala, dorsal; H, spiracles, showing alignment; I–R, *Pentarthrum zealandicum* Wollaston, larva (New Zealand): I, habitus, lateral; J, head, dorsal; K, head, ventral, anterior portion; L, mandible; M, antenna; N, labrum, clypeus and epistoma; O, mala and labium, ventral; P, mala, dorsal; Q, labial palp; R, spiracles, showing alignment. After May (1994).

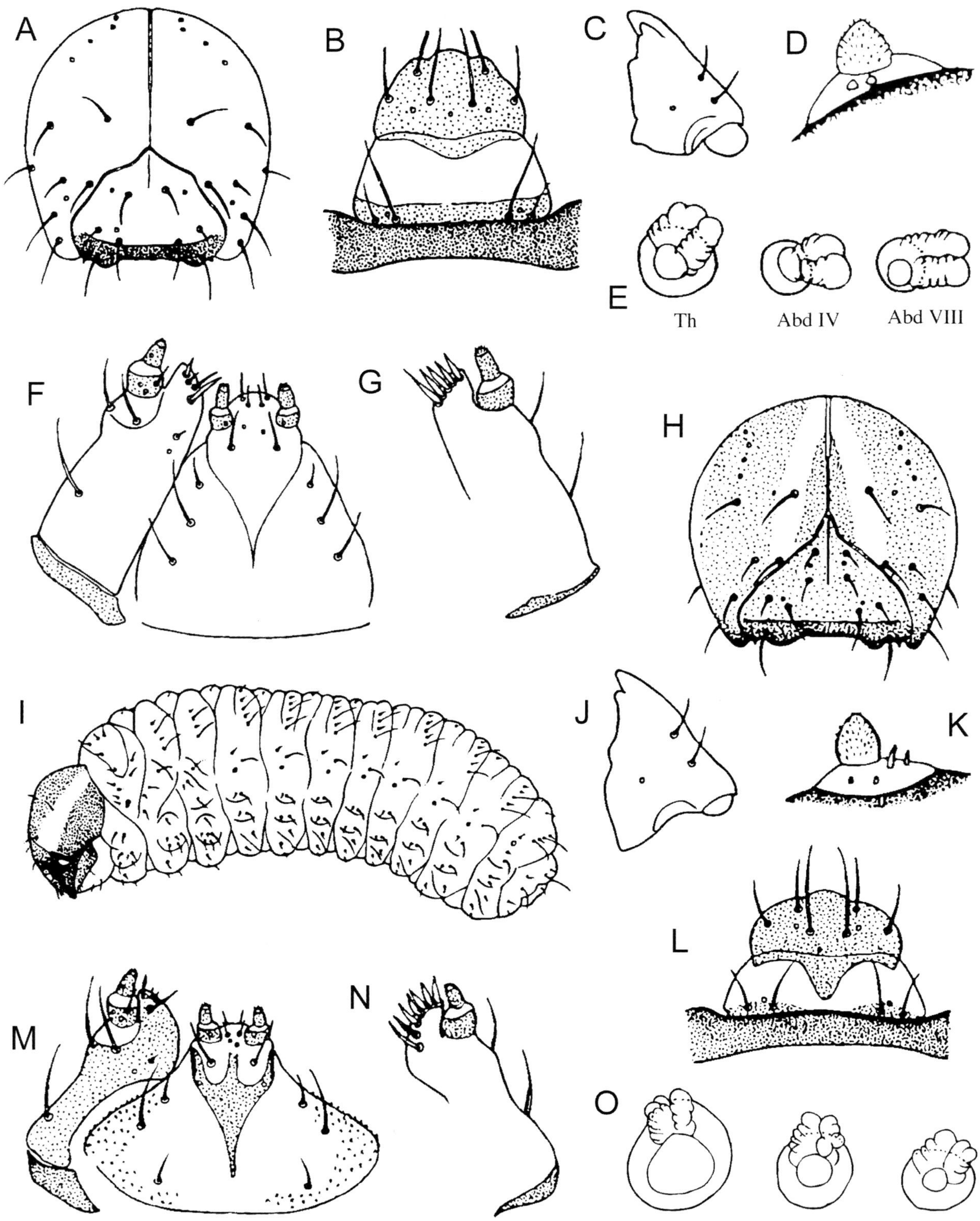

Fig. 44.182. Cossoninae, structural details of immature stages, *Phloeophagosoma* and *Xenocnema*. A–G, *Phloeophagosoma thoracicum* Wollaston, larva (New Zealand): A, head, dorsal; B, labrum, clypeus and epistoma; C, mandible; D, antenna; E, spiracles, showing alignment; F, mala and labium, ventral; G, mala, dorsal; H–O, *Xenocnema spinipes* Wollaston, larva (New Zealand): H, head, dorsal; I, habitus, lateral; J, mandible; K, antenna; L, labrum, clypeus and epistoma; M, mala and labium, ventral; N, mala, dorsal; O, spiracles, showing alignment. After May (1994).

Fig. 44.183. Cossoninae, structural details, *Allopentarthrum* and *Catolethrobius*. A–E, *Allopentarthrum elumbe* (Boheman): A, habitus, dorsal; B, habitus, ventral; C, habitus, lateral; D, head, dorsal; E, head, lateral; F–J, *Catolethrobius silvestris* (Kolbe): F, head, lateral; G, habitus, lateral; H, habitus, dorsa; I, tarsus, ventral; J, habitus, ventral.

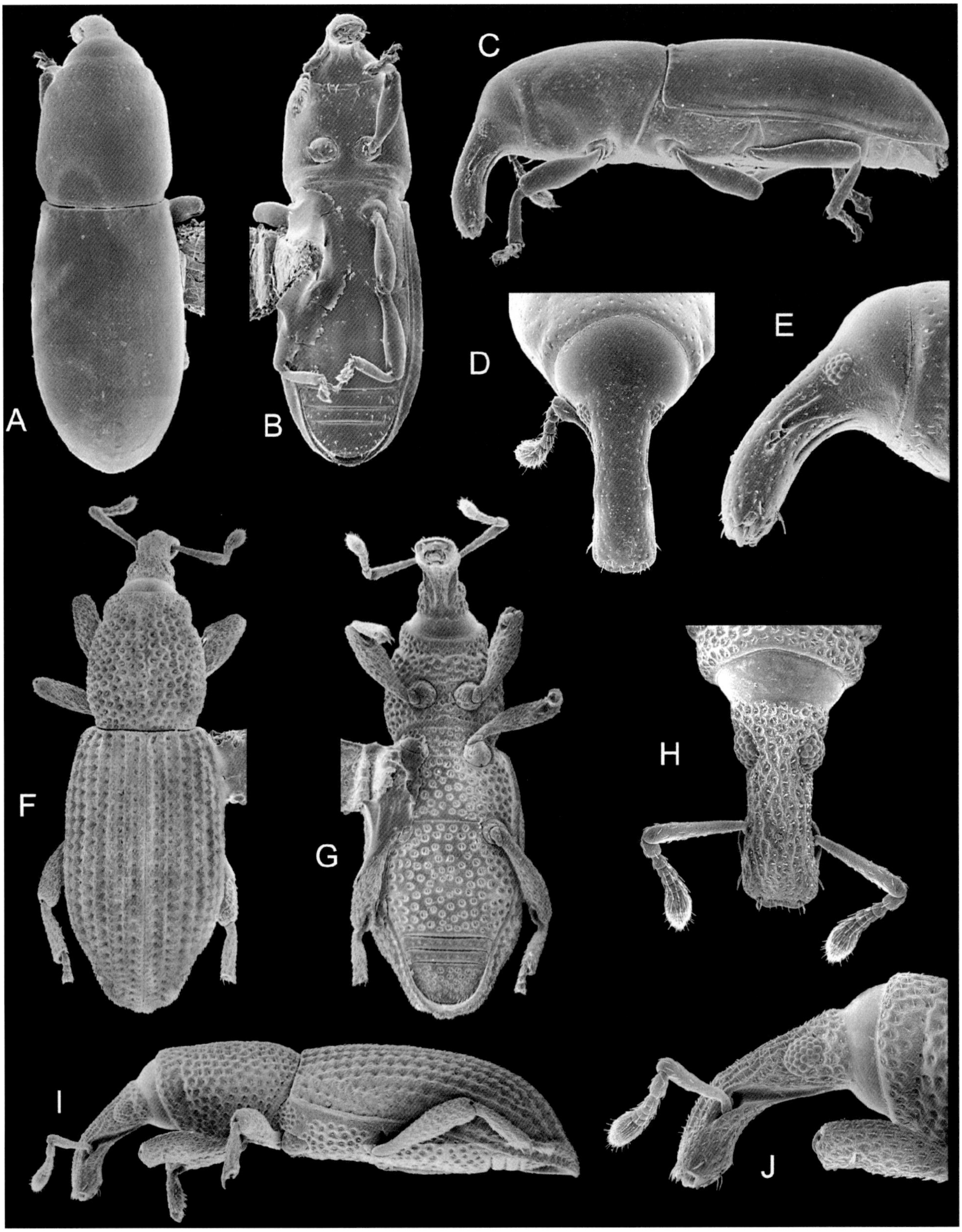

Fig. 44.184. Cossoninae structural details, *Conlonia* and *Dryotribus*. A–F, *Conlonia* sp.: A, habitus dorsal; B, habitus, ventral; C, habitus, lateral; D, head, dorsal; E, head, lateral; F–J, *Dryotribus mimeticus* Horn: A, habitus, dorsal; B, habitus, ventral; H, head, dorsal; I, habitus, lateral; J, head, lateral.

Fig. 44.185. Cossoninae, structural details, *Coptocorynus* spp. A–F, *Coptocorynus araucariae* Marshall: A, habitus, dorsal; B, habitus, ventral; C, habitus, lateral; D, head, lateral; E, head, dorsal; F, antenna; G–L, *Coptocorynus* sp.: G, habitus, lateral; H, habitus, dorsal; I, habitus, ventral; J, head, dorsal; K, head, lateral; L, antenna.

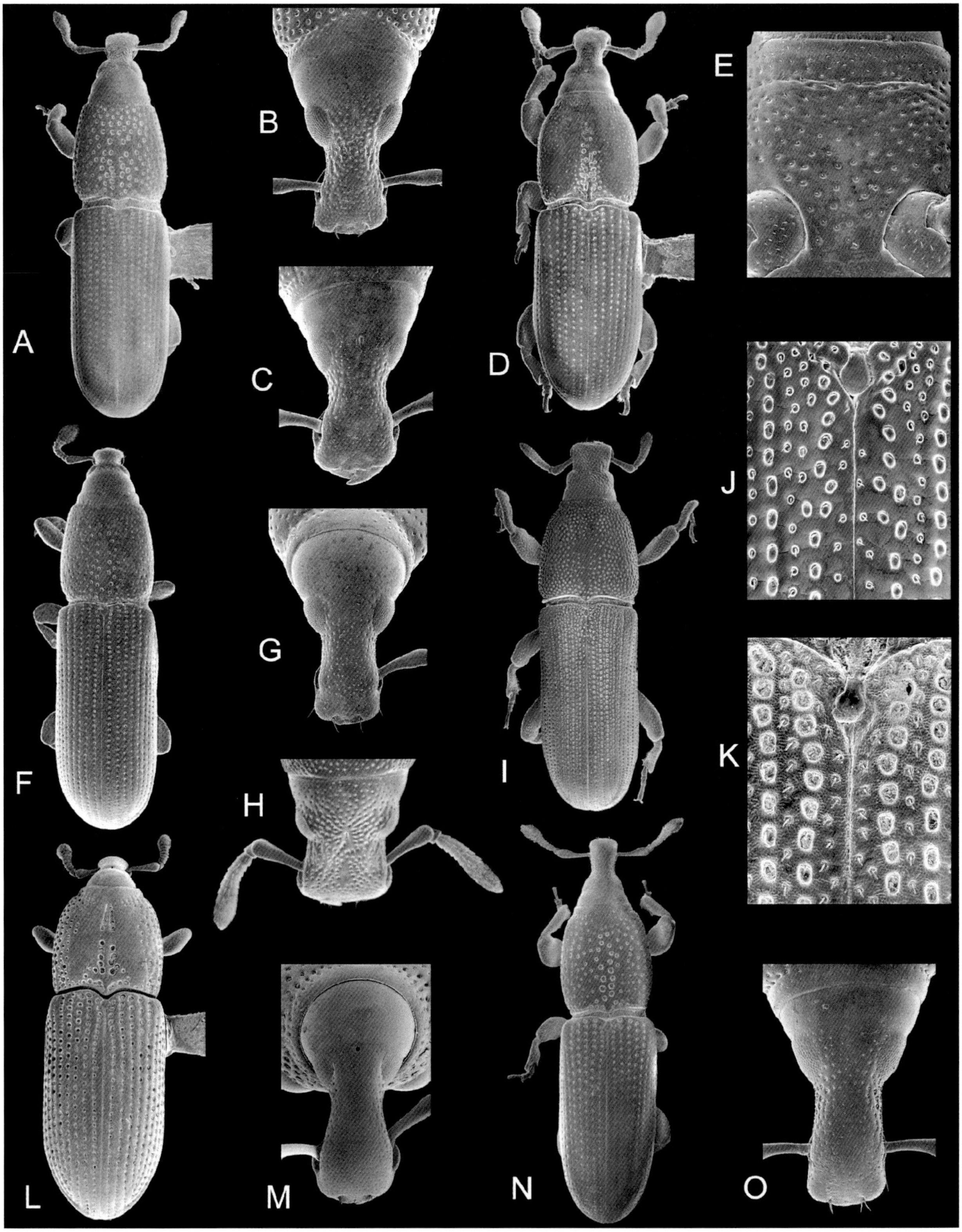

Fig. 44.186. Cossoninae, structural details, *Cossonus* spp. A–B, *Cossonus arctatus* (Pascoe): A, habitus, dorsal; B, head, dorsal; C–E, *Cossonus excavatus* Pascoe: C, head, dorsal; D, habitus, dorsal; E, prothorax, ventral; F–G, *Cossonus frenchi* Lea: F, habitus, dorsal; G, head, dorsal; H–J, *Cossonus hackeri* Lea: H, head, dorsal; I, habitus, dorsal; J, elytron; K, *Cossonus subpubescens* Wollaston, elytron; L–M, *Cossonus incisus* Pascoe: L, habitus, dorsal; M, head, dorsal; N–O, *Cossonus indigens* Pascoe: N, habitus, dorsal; O, head, dorsal.

Fig. 44.187. Cossoninae, structural details, *Cossonus* spp. A–D, *Cossonus integricollis* Lea: A, habitus, dorsal; B, head, dorsal; C, head and prothorax, dorsal; D, head and prothorax, ventral; E–F, *Cossonus macilentus* Lea; E, habitus, dorsal; F, head, dorsal; G–I, *Cossonus porosternus* Lea: G, head, dorsal; H, habitus, dorsal; I, prothorax, ventral; J–K, *Cossonus simsoni* Lea: J, habitus, dorsal; K, head and prothorax, ventral; L–M, *Cossonus subpubescens* (Wollaston): L, habitus, dorsal; M, head and prothorax, dorsal.

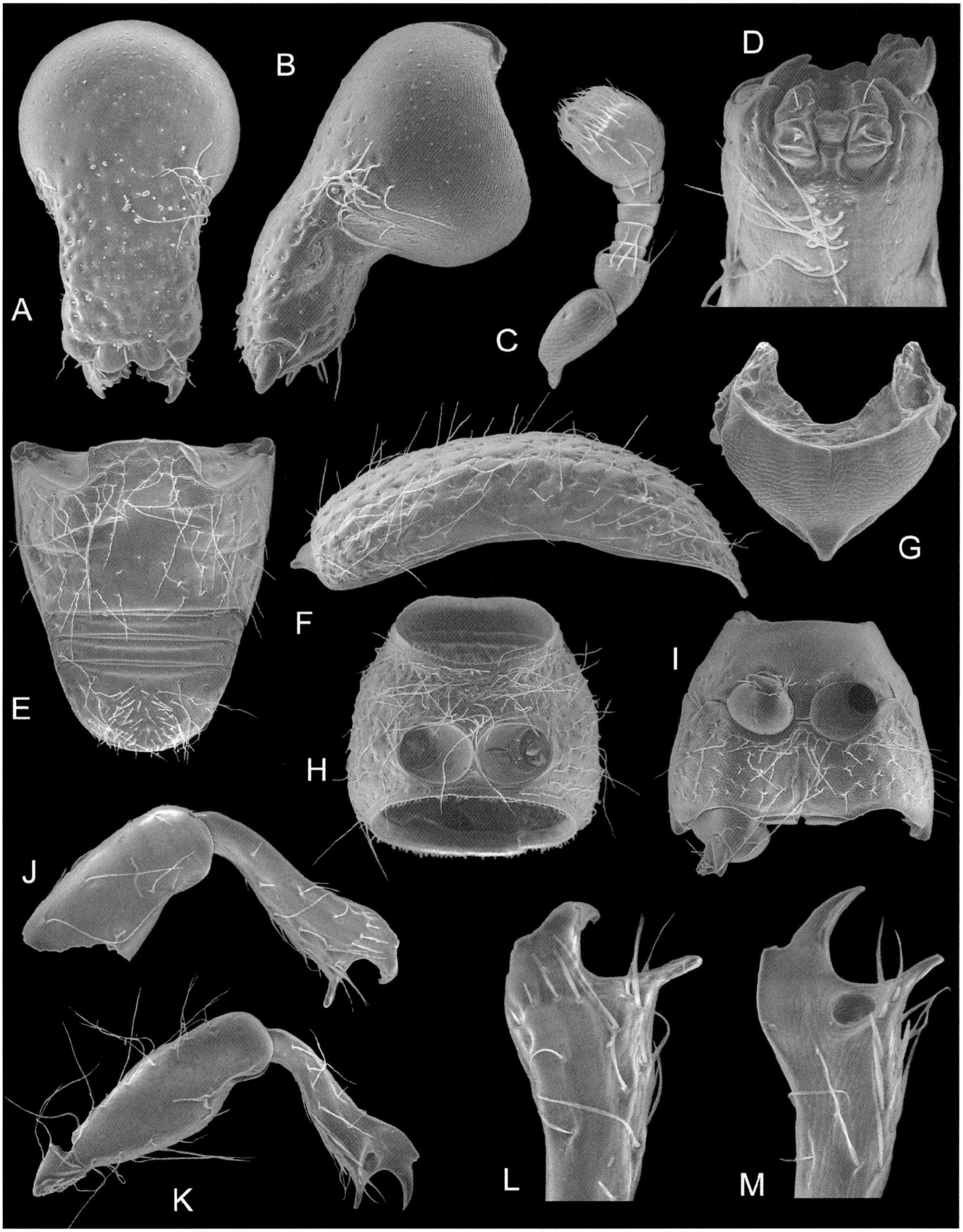

Fig. 44.188. Cossoninae, structural details, *Halorhynchus caecus* Wollaston. A–B, head: A, dorsal; B, lateral; C, antenna; D, mouthparts; E, abdominal ventrites; F, elytron, lateral; G, scutellum; H, prothorax, ventral; I, pterothorax, ventral; J–K, legs; L–M, tibiae apex.

Fig. 44.189. Cossoninae, structural details, *Hexarthroides, Mastersinella* and *Aphanocorynes*. A–D, *Hexarthroides punctulatus* Lea; A, habitus, dorsal; B, habitus, ventral; C, habitus, lateral; D, head, dorsal; E–I, *Mastersinella octoarticulata* Lea: E, head, dorsal; F, habitus, dorsal; G, habitus, ventral; H, habitus, lateral; I, head, lateral; J–O, *Aphanocorynes rugosipennis* (Lea): J, head, dorsal; K, habitus, dorsal; L, habitus, ventral; M, habitus, lateral; N, head, lateral; O, tarsus, dorsal.

Fig. 44.190. Cossoninae, structural details, *Omeretes* spp. A–G, *Omeretes onychialis* Marshall: A, habitus, dorsal; B, habitus, ventral; C, habitus, lateral; D, head, dorsal; E, head, ventral; F, mesotarsus lateral; G, mesotarsus, dorsal; H–M, *Omeretes podocarpi* Marshall: H, habitus, dorsal; I, habitus, ventral; J, habitus, lateral; K, elytron, posterior; L, head, lateral; M, head, dorsal.

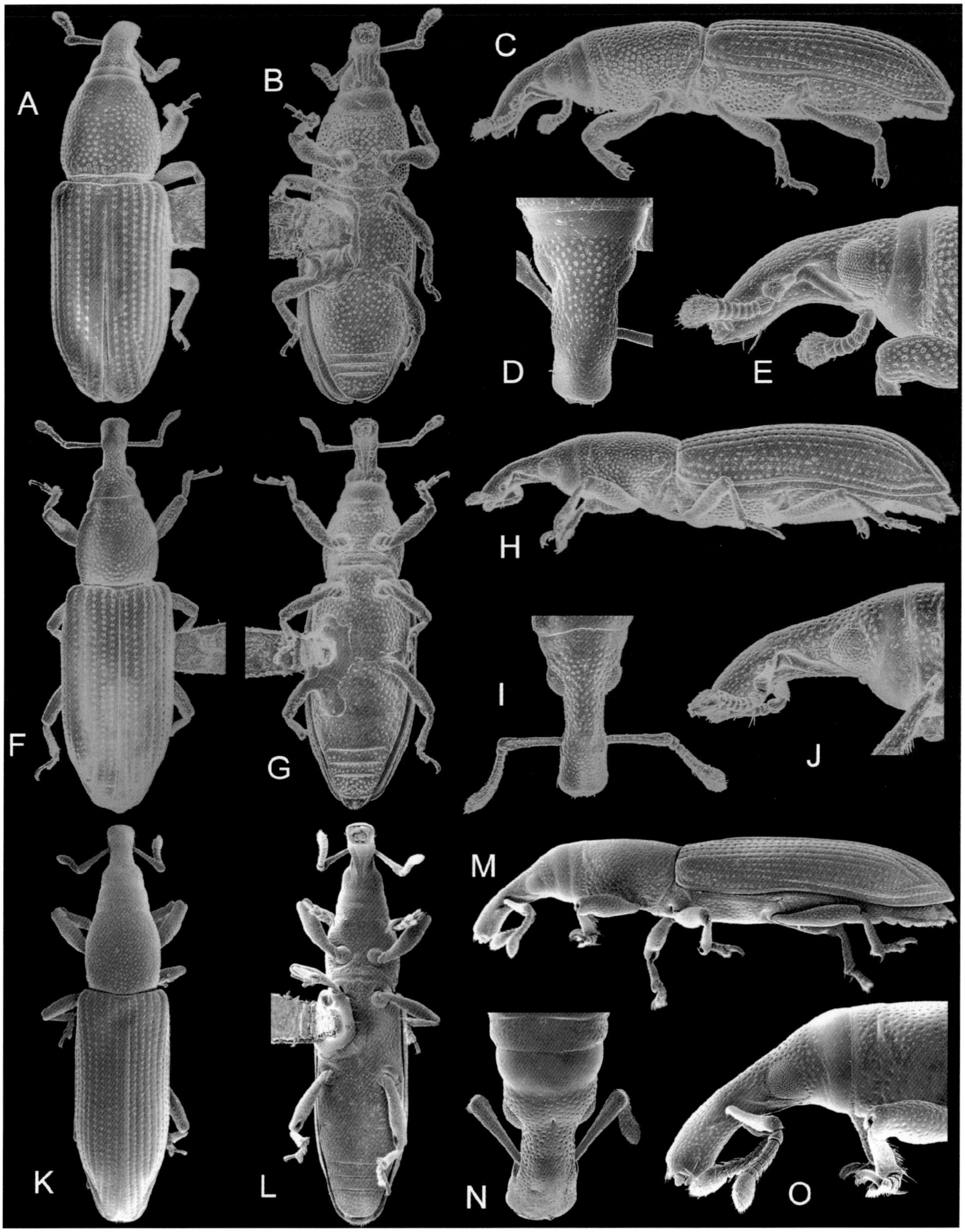

Fig. 44.191. Cossoninae, structural details, *Orthotemnus* and *Oxydema*. A–E, *Orthotemnus disparilis* Pascoe: A, habitus, dorsal; B, habitus, ventral; C, habitus, lateral; D, head, dorsal; E, head, lateral; F–J, *Oxydema australis* (Wollaston): F, habitus, dorsal; G, habitus, ventral; H, habitus, lateral; I, head, dorsal; J, head, lateral; K–O, *Oxydema major* (Wollaston): K, habitus, dorsal; L, habitus, ventral; M, habitus, lateral; N, head, dorsal; O, head, lateral.

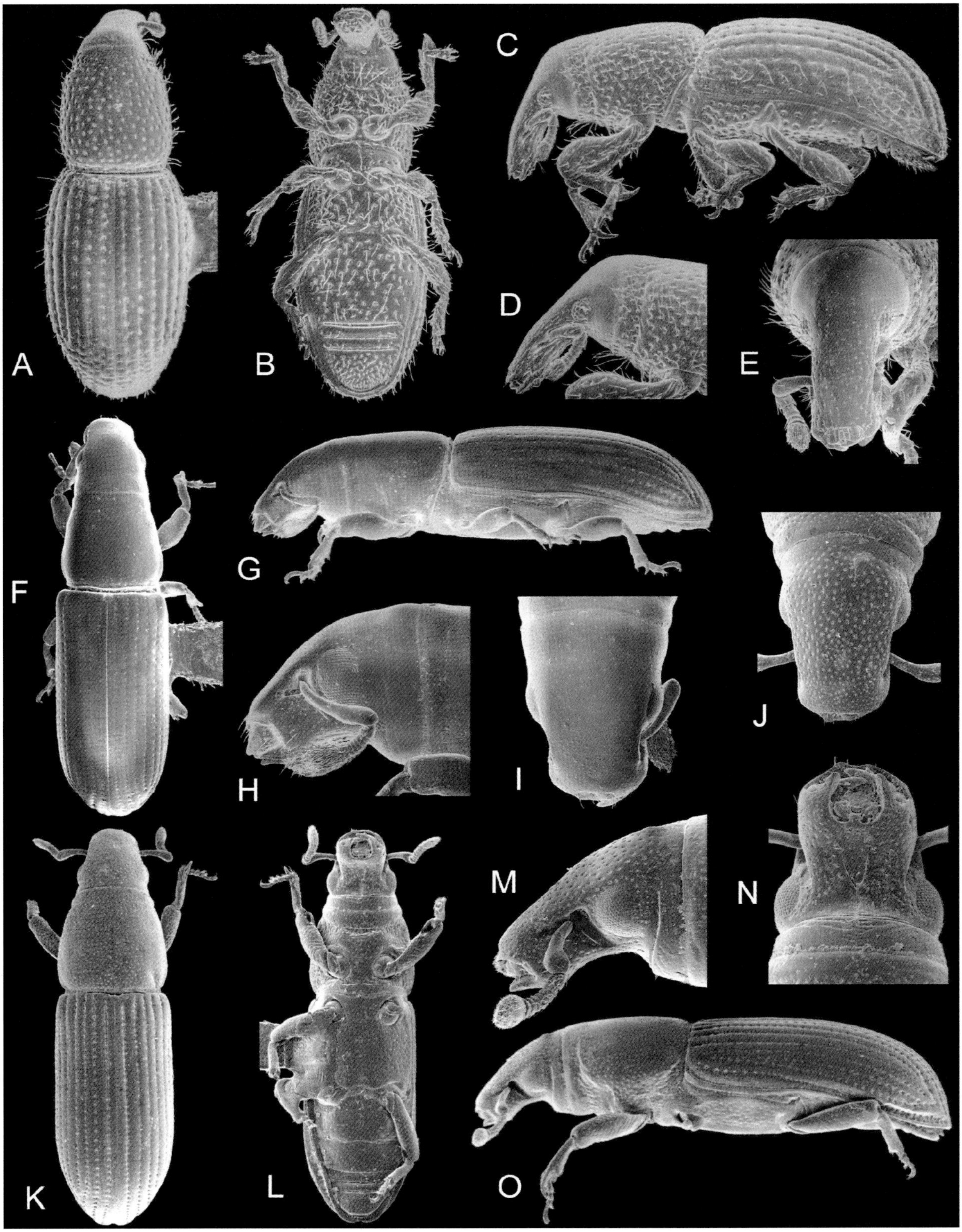

Fig. 44.192. Cossoninae, structural details, *Pentarthrocis* and *Stereoborus*. A–E, *Pentarthrocis ammophilus* Lea: A, habitus, dorsal; B, habitus, ventral; C, habitus, lateral; D, head, lateral; E, head, dorsal; F–I, *Stereoborus brevirostris* Lea: F, habitus, dorsal; G, habitus, lateral; H, head, lateral; I, head, dorsal; J–O, *Stereoborus laporteae* Lea: J, head, dorsal; K, habitus, dorsal; L, habitus, ventral; M, head, lateral; N, head, ventral; O, habitus, lateral.

Fig. 44.193. Cossoninae, structural details, *Toura*, *Pentarthrum* and *Stereomimetes*. A–E, *Toura* sp.: A, habitus, dorsal; B, habitus, ventral; C, head, dorsal; D, head, lateral; E, habitus, lateral; F, *Pentarthrum millingtoni* Olliff, head, lateral; G–K, *Stereomimetes crassicornis* Wollaston: G, rostrum, ventral; H, head, dorsal; I, habitus, dorsal; J, head, lateral; K, tarsus, dorsal.

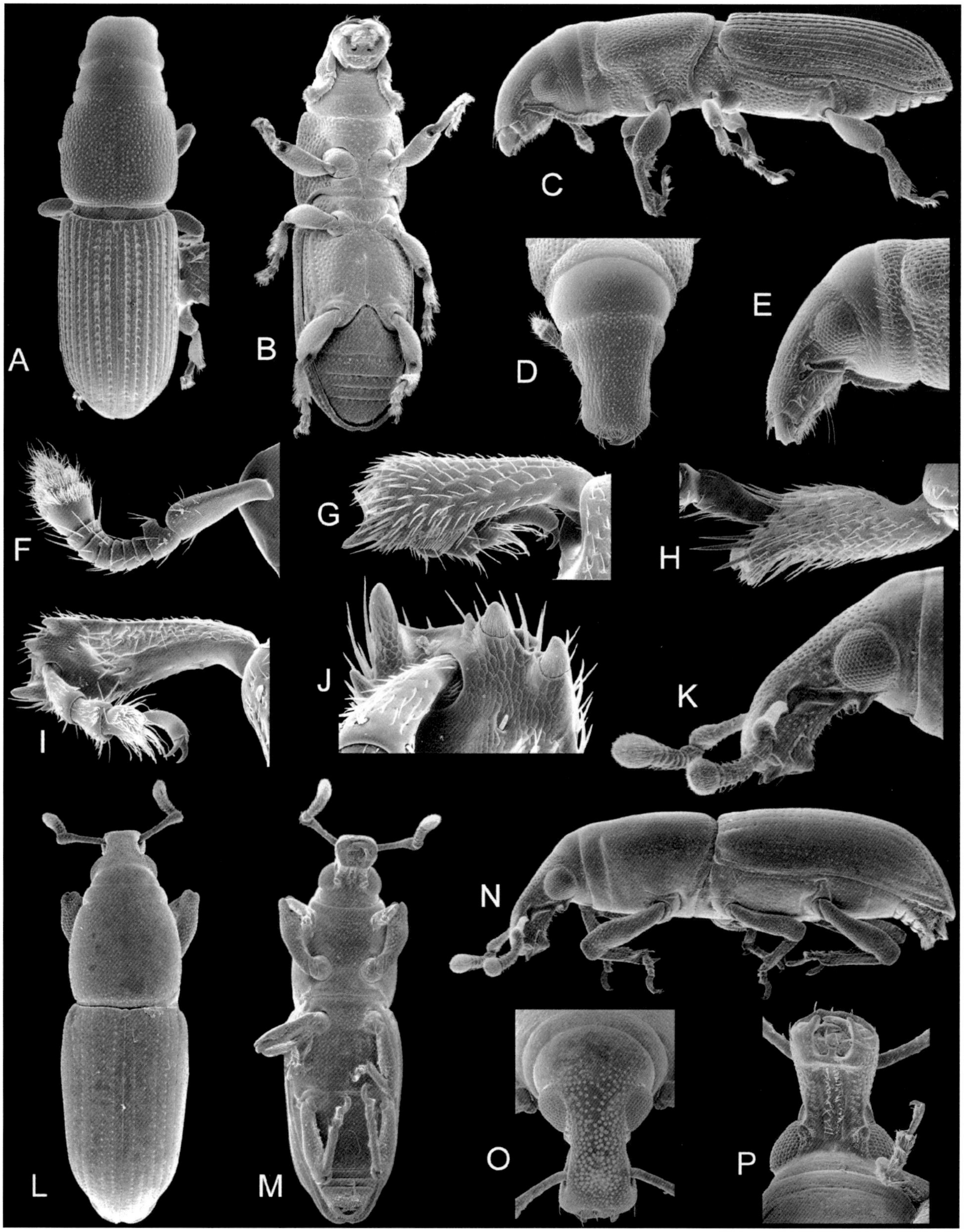

Fig. 44.194. Cossoninae, structural details, *Xenocnema* and *Xenotrupis*. A–J, *Xenocnema spinipes* Wollaston: A, habitus, dorsal; B, habitus, ventral; C, habitus, lateral; D, head, dorsal; E, head, lateral; F, antenna; G, right foreleg; H, left foreleg; I, tibia and tarsus, lateral; J, tibial spurs and spines; K–P, *Xenotrupis* sp.; K, head, lateral; L, habitus, dorsal; M, habitus, ventral; N, haabitus, lateral; O, head, dorsal; P, head, ventral.

Unas Broun, 1909 (Pl. 109G)

Type species, by monotypy: *Pentarthrum piceum* Broun, 1880.

Australian species. *U. longicollis* (Montrouzier, 1861).

Distribution. Australia (Norfolk and Lord Howe Islands), New Zealand.

Xenocnema Wollaston, 1873 (Pl. 109H; Figs 44.182H–O, 44.194A–J)

Type species, by monotypy: *Xenocnema spinipes* Wollaston, 1873.

Australian species. *X. spinipes* Wollaston, 1873b.

Distribution. Australia, New Caledonia, New Zealand.

Xenotrupis Wollaston, 1873 (Pl. 109I; Fig. 44.194K–P)

Type species, by monotypy: *Xenotrupis fusiformis* Wollaston, 1873b.

Australian species. *Xenotrupis* sp.

Distribution. Australia, Maluku Is., New Guinea.

Zimmiodes Hsiao & Oberprieler, 2024 (Fig. 44.180)

Type species, by original designation: *Zimmiodes australis* Hsiao & Oberprieler, 2024.

Australian species. *Zimmiodes australis* Hsiao & Oberprieler, 2024.

Distribution. Australia (QLD, NT), New Guinea.

Notes. The two known species are associated with the cycad genus *Cycas* (Cycadaceae), the larvae developing in the sporophylls and axes of male cones and the adults being the main or sole pollinators of their hosts (Hall & Walter 2018; Hsiao & Oberprieler 2024). *Zimmiodes australis* has been recorded from several apparent species of *Cycas* (*C. armstrongii, C. arnhemica, C. canalis, C. maconochiei, C. media banksii, C. media ensata, C. megacarpa, C. ophiolitica, C. orientis, C. platyphylla, C. semota*), suggesting that these are closely related if not conspecific (Hsiao & Oberprieler 2024).

Unplaced Curculionidae

Introduction. The placements of tribal groups (Gonipterini, Hyperinae, Phrynixini, Viticiini) and four genera (*Ethadomorpha, Hypsomus, Symbothynus, Thechia*) belonging in the CEGH clade are as yet uncertain, and, following Pullen *et al.* (2014) and Oberprieler *et al.* (2014b), they are here treated separately. The **Gonipterini** have since been shown to fall in the subfamily Cyclominae as currently composed (Shin *et al.* 2018; Baird *et al.* 2021; Haran *et al.* 2023) and are seemingly close to the Australian genera currently placed in Rhythirrinini, such as *Ethemaia* (Baird *et al.* 2021), but their precise relationships and status remain to be resolved. The **Hyperinae** evidently constitute the sister-group of the subfamilies Cyclominae and Entiminae (though with the position of Sitonini still unresolved) (Gillett *et al.* 2014; Shin *et al.* 2018; Baird *et al.* 2021; Haran *et al.* 2023) and were consequently re-elevated to subfamily level by Shin *et al.* (2018). The **Phrynixini** appear to be the basal-most lineage in the CEGH clade (Gunter *et al.* 2016; Shin *et al.* 2018; Baird *et al.* 2021; Haran *et al.* 2023) and will likely have to be raised to subfamily level as well when their precise composition and relationships become clear. The **Viticiini** appear to be fall in the current concept of Cyclominae as well, related to similarly short-snouted species of the genera *Acalonoma* and *Rhadinosomus* (Gunter *et al.* 2016), currently placed in Aterpini (see above), and to include the aberrant genus *Geochus* (Leschen *et al.* 2022), but their status and proper composition also remain unclear. *Ethadomorpha* seems to be a fairly basal genus in CEGH as well but may also be related to *Acalonoma* according to Gunter *et al.* (2016), but its relationships and taxonomic status, along with those of the seemingly related *Symbothynus* and *Thechia*, remain unresolved. The African genus *Hypsomus*, represented in Australia by one introduced and undescribed species, has recently been shown to belong in the otherwise Palaearctic tribe Styphlini, whose status in the CEGH clade also remains to be elucidated from a larger phylogenetic analysis of this clade.

Tribe Gonipterini Lacordaire, 1863

Introduction. The Gonipterini are a small group distributed in the Australo-Pacific region, with seven genera in Australia and one, *Pterogonius* Kuschel, in New Caledonia (10 species) and Fiji

Fig. 44.195. *Oxyops fasciculatus* Redtenbacher.

(2 species). Their host plants all belong to the family Myrtaceae, with species of *Corymbia*, *Eucalyptus*, *Kunzea*, *Leptospermum*, *Melaleuca* and *Verticordia* recorded in Australia (Oberprieler *et al.* 2014b) and *Syzygium* in New Caledonia (Kuschel 2014a). Three species of *Gonipterus* have been accidentally introduced to several countries in Africa, America, Europe and New Zealand, where both adults and larvae can cause severe defoliation of eucalypt trees in plantations and cause great economic losses (Mapondera *et al.* 2012; Oberprieler *et al.* 2014b). In contrast, *Oxyops vitiosa* has been used as a biological control agent of the invasive *Melaleuca quinquenervia* in Florida, USA (Center *et al.* 2000). The larvae and adults of most genera are ectophytic leaf feeders (Pl. 13A–F, H), but those of *Syarbis* are leaf-miners and those of *Pantoreites* and *Prophaesia* unknown (Oberprieler *et al.* 2014b). Pupation occurs in the soil. The larvae of a few species have been described by May (1994) (Fig. 44.196). An overview of the distribution, biology, morphology, taxonomy and phylogeny of the tribe was provided by Oberprieler *et al.* (2014b).

Key to the Australian genera of Gonipterini

1.	Tarsal segments 4 and 5 absent (e.g. Fig. 44.197a)	***Syarbis*** **Pascoe**
–	Tarsal segments 4 and 5 present (e.g. Fig. 44.197b)	2
2(1).	Rostrum about as long as pronotum (Fig. 44.197E); antennal insertions near middle of rostrum	***Prophaesia*** **Pascoe**
–	Rostrum shorter than pronotum (Fig. 44.197B–D); antennal insertions near apical third of rostrum	3
3(2).	Head abruptly constricted behind the eyes (Fig. 44.197C)	***Iptergonus*** **Lea**
–	Head not abruptly constricted behind the eyes	4
4(3).	Eyes flattened, not prominent (Fig. 44.197D)	***Pantoreites*** **Pascoe**
–	Eyes convex, and prominent above the level of the head (Fig. 44.197B)	5
5(4).	Antennal clubs poorly defined, antennae incrassate (Fig. 44.197A)	***Bryachus*** **Pascoe**
–	Antennal clubs clearly differentiated from the funicle (Fig. 44.197B); funicles with segments subequal in size, narrower than the club	6
6(5).	Body dorsal vestiture uniform; elytra with well defined humeral angles	***Gonipterus*** **Schoenherr**
–	Body dorsal vestiture including clusters of erect setae or sparse and dispersed; elytra with or without humeral angles	***Oxyops*** **Schoenherr**

Classification of the Australian genera of Gonipterini

Bryachus Pascoe, 1870 (Pl. 58A; Figs 44.196A–H, 44.197A)

Type species, by monotypy: *Bryachus squamicollis* Pascoe, 1870.

Australian species. *B. squamicollis* Pascoe, 1870c.

Distribution. Australia (QLD, NSW, VIC, SA, NT, WA).

Notes. The larvae of the single species feed on *Eucalyptus* leaves.

Gonipterus Schoenherr, 1833 (Pls 13A, C, D, 58B, C; Figs 44.196I–N, 44.197B)

Type species, by original designation: *Gonipterus lepidotus* Gyllenhal, 1833.

Australian species. *G. balteatus* Pascoe, 1870c, *G. cinnamomeus* Pascoe, 1870c, *G. citrophagus* Lea, 1898a, *G. conicollis* Lea, 1927a, *G. crassipes* Lea, 1898a, *G. exaratus* Fåhraeus, 1840 (*in* Schoenherr 1840b), *G. excavifrons* Lea, 1898a, *G. ferrugatus* Pascoe, 1870c, *G. geminatus* Lea, 1898a, *G. gibberus* Boisduval, 1835, *G. inconspicuus* Lea, 1927a, *G. intermedius* Lea, 1927a, *G. lateritius* (Blackburn, 1889), *G. lepidotus* Gyllenhal, 1833 (*in* Schoenherr 1833b), *G. notographus* Boisduval, 1835, *G. parallelicollis* Lea, 1927a, *G. platensis* (Marelli, 1926), *G. pulverulentus* Lea, 1898a, *G. scutellatus* Gyllenhal, 1833 (*in* Schoenherr 1833b), *G. suturalis* Gyllenhal, 1833 (*in* Schoenherr 1833b), *G. xanthorrhoeae* Lea, 1898a.

Distribution. Australia (QLD, NSW, VIC, TAS, SA, NT, WA), introduced in New Zealand, southern to eastern Africa, western and eastern South America, eastern North America, western Europe.

Notes. Three species of the genus have been accidentally introduced outside of their native Australian range: *G. platensis*, native to Tasmania, in WA, New Zealand, western Europe (Canary Is., Portugal, Spain), eastern and western South America (Argentina, Brazil, Chile, Colombia, Uruguay) and USA (California, Hawaii); *G. pulverulentus*, native to VIC and TAS, in Argentina, Brazil and Uruguay; *G.* sp. (undescribed), native to south-eastern Australia, in WA, southern France, Italy and the eastern side of Africa (South Africa, Lesotho, Eswatini, Mozambique, Zimbabwe, Malawi, Rwanda, Uganda, Kenya) including Mauritius and Madagascar as well as St. Helena in the Atlantic Ocean (Mapondera *et al.* 2012; Oberprieler *et al.* 2014b; Barratt *et al.* 2018). Being considered pests of eucalypt plantations in these regions, the weevils are being biologically controlled by egg-parasitic wasps of the genus *Anaphes* (Mymaridae) (e.g. Garcia *et al.* 2019). The known distribution of the main species in Australia and their host and parasitoid associations were explored by Garcia *et al.* (2019) and Schröder *et al.* (2021), but the genus requires taxonomic revision to properly assess the number of species and their taxonomic delineations. The genus was described in Schoenherr (1833a).

Iptergonus Lea, 1908 (Pl. 58D, Fig. 44.197C)

Type species, by original designation: *Gonipterus cionoides* Pascoe, 1870.

Australian species. *I. aberrans* (Lea, 1898a), *I. bifurcatus* Lea, 1908a, *I. cionoides* (Pascoe, 1870c), *I. niveopictus* Lea, 1908a.

Distribution. Australia (QLD, NSW, SA, WA).

Notes. The constricted head behind the eyes of the genus resembles the condition of *Syarbis*. The recorded host plants are *Kunzea phylicoides* and *Verticordia*.

Oxyops Schoenherr, 1826 (Pl. 58E–G; Figs 44.195, 44.196O–P)

Type species, by original designation: *Oxyops clathratus* Schoenherr, 1826.

Australian species. *O. alphabeticus* Lea, 1917c, *O. amplipennis* Lea, 1910b, *O. arciferus* Pascoe, 1870c, *O. areolicollis* Lea, 1927a, *O. aulicus* Pascoe, 1870c, *O. bilunaris* Pascoe, 1870c, *O. calidus* Pascoe, 1873b, *O. carinirostris* Lea, 1927a, *O. clathratus* Schoenherr, 1826, *O. concretus* Pascoe, 1870c, *O. convexus* (Olivier, 1791), *O. crassirostris* Pascoe, 1870c, *O. decipiens* Lea, 1908a, *O. excavatus* Boisduval, 1835, *O. farinosus* Pascoe, 1871a, *O. fasciatus* (Boisduval, 1835), *O. fasciculatus* Redtenbacher, 1868, *O. floreus* Pascoe, 1873b, *O. frenchi* Lea, 1908a, *O. gemellus* Pascoe, 1870c, *O. gibbus* (Fabricius, 1792), *O. griffithi* Lea, 1908a, *O. griseus* Lea, 1915c, *O. hopei* Boheman, 1835, *O. hyperoides* (Pascoe, 1871a), *O. insignis* Lea, 1927a, *O. irrasus* Pascoe, 1870c, *O. leucopholus* Lea, 1927a, *O. lugubris* (Blackburn, 1892b), *O. marginalis* Pascoe, 1870c, *O. mastersii* Pascoe, 1873b, *O. meles* Pascoe, 1873b, *O. memnonius* Pascoe, 1873b, *O. microlepis* Lea, 1927a, *O. modestus* Lea, 1898a, *O. modicus* Blackburn, 1889, *O. mucronatus* Lea, 1908a, *O. multiarmatus* Lea, 1927a, *O. multidentatus* Lea, 1898a, *O. niveosparsus* Pascoe, 1882, *O. nodicollis* Lea, 1910b, *O. pallidus* Lea, 1908a, *O. parvicollis* Lea, 1915c, *O. parvoscabrus* Lea, 1927a, *O. pictipennis* Blackburn, 1894a, *O. placidus* Blackburn, 1894a, *O. platyodontus* Lea, 1927a, *O. posticalis* Lea, 1910b, *O. pruinosus* Pascoe, 1873b, *O. reticulatus* (Boisduval, 1835), *O. rufus* Lea, 1908a, *O. rutilus* Pascoe, 1873b, *O. scaber* Lea, 1908a, *O. scabrosus* (Boisduval, 1835), *O. scoparius* Lea, 1908b, *O. semicircularis* Lea, 1927a, *O. sepulchralis* (Pascoe, 1870c), *O. siccus* Blackburn, 1896, *O. soror* Lea, 1910b, *O. sparsutus* Pascoe, 1873b, *O. spenceri* Blackburn, 1896, *O. tessellatus* Lea, 1927a, *O. tuberculatus* Perroud, 1865 (*in* Perroud & Montrouzier 1865), *O. turbidus* (Pascoe, 1871a), *O. vacillans* Lea, 1917c, *O. vitiosus* Pascoe, 1870c.

Distribution. Australia (QLD, NSW, VIC, TAS, SA, NT, WA).

Notes. *Oxyops vitiosa* has been employed as a biological control agent of the invasive *Melaleuca quinquenervia* (Myrtaceae) in Florida, USA. Other species of the genus have been reported to feed on leaves of *Corymbia variegata*, *Eucalyptus longifolia, Eucalyptus leucoxylon and Melaleuca quinquenervia*.

Pantoreites Pascoe, 1870 (Pls 13B, 58H, I; Fig. 44.197D)

Type species, by subsequent designation (Pullen *et al.* 2014: 126): *Pantoreites virgatus* Pascoe, 1870.

Australian species. *P. arctatus* (Pascoe, 1870c), *P. clarki* Marshall, 1936, *P. cretatus* Pascoe, 1873b, *P. fusiformis* Lea, 1927a, *P. illuminatus* Lea, 1898a, *P. major* Lea, 1908a, *P. micans* Lea, 1898a, *P. scenicus* Pascoe, 1870c, *P. trilinealbus* Lea, 1908a, *P. trivirgatus* Lea, 1914, *P. virgatus* Pascoe, 1870c, *P. vittatus* Pascoe, 1872c.

Distribution. Australia (NSW, VIC, TAS, SA, NT, WA).

Notes. Host plants recorded for the genus are *Leptospermum scoparium* and *Melaleuca argentea*, but the larvae have not been found and possibly feed endophytically.

Prophaesia Pascoe, 1870 (Pl. 58J; Fig. 44.197E)

Type species, by subsequent designation (Pullen *et al.* 2014: 126): *Prophaesia albilatera* Pascoe, 1870.

Australian species. *P. alba* Lea, 1927a, *P. albilatera* Pascoe, 1870d, *P. confusa* Pascoe, 1873a, *P. cretata* Pascoe, 1870d, *P. florea* Pascoe, 1883a, *P. tenuirostris* Lea, 1927a.

Distribution. Australia (NSW, VIC, TAS, SA, WA).

Notes. The hosts plants and larvae of the genus appear unknown.

Syarbis Pascoe, 1865 (Pls 13E, 58K, L; Fig. 44.197F)

Type species, by monotypy: *Syarbis pachypus* Pascoe, 1865.

Australian species. *S. albivittis* Lea, 1927a, *S. alcyone* Lea, 1898a, *S. deyrollei* (Roelofs, 1867), *S. emarginata* (Roelofs, 1866), *S. eucalypti* Lea, 1914, *S. fasciculatissima* Lea, 1898a, *S. gonipteroides* Pascoe, 1870b, *S. goudiei* Lea, 1910b, *S. haagii* (Roelofs, 1866), *S. nervosa* Pascoe, 1873b, *S. nigra* (Roelofs, 1867), *S. nubila* (Roelofs, 1866), *S. pachypus* Pascoe, 1865, *S. porcata* Lea, 1908b, *S. posthumeralis* Lea, 1927a, *S. pulchella* Lea, 1914, *S. pulchripennis* Lea, 1898a, *S. punctipennis* (Roelofs, 1867), *S. sciurus* Pascoe, 1870b, *S. semilineata* Pascoe, 1873b, *S. simulans* Lea, 1898a, *S. subnitida* (Roelofs, 1866).

Distribution. Australia (QLD, NSW, VIC, TAS, SA, NT, WA).

Notes. The larvae as known mine in leaves of *Corymbia* and *Eucalyptus*, but nothing specifically is known about the life histories of the various species.

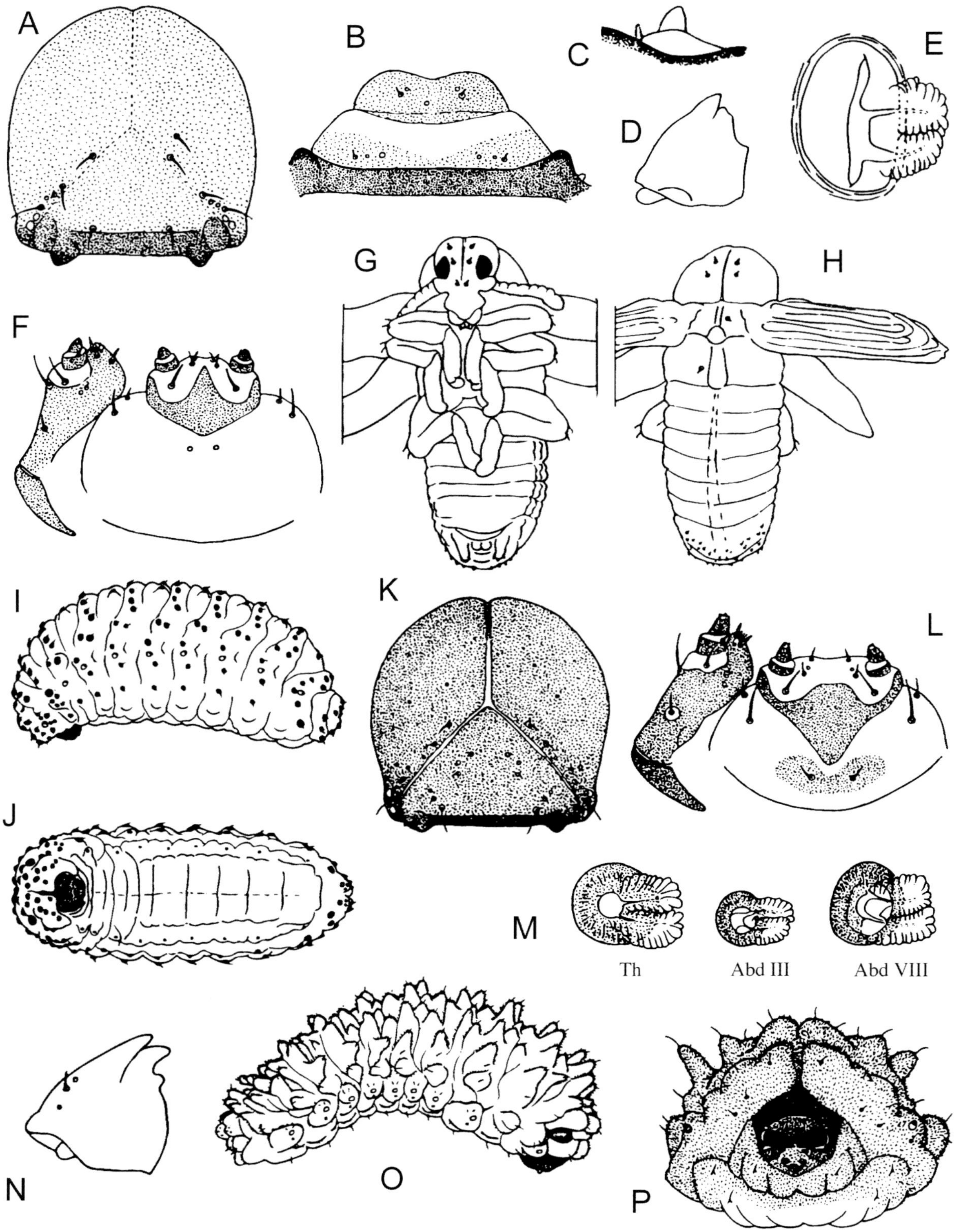

Fig. 44.196. Gonipterini, structural details of immature stages. A–F, *Bryachus squamicollis* Pascoe, larva: A, head, dorsal; B, labrum, clypeus and epistoma; C, antenna; D, mandible; E, spiracle; F, mala and labium, ventral; G–H, *Bryachus squamicollis* Pascoe, pupa: G, habitus ventral; H, habitus, dorsal, I–N, *Gonipterus scutellatus* Gyllenhall, larva (New Zealand): I, habitus, lateral; J, habitus, ventral; K, head, dorsal; L, mala and labium, ventral; M, spiracles, showing alignment; N, mandible; O–P, *Oxyops concretus* Pascoe, larva: O, habitus, lateral; P, anterior. After May (1994).

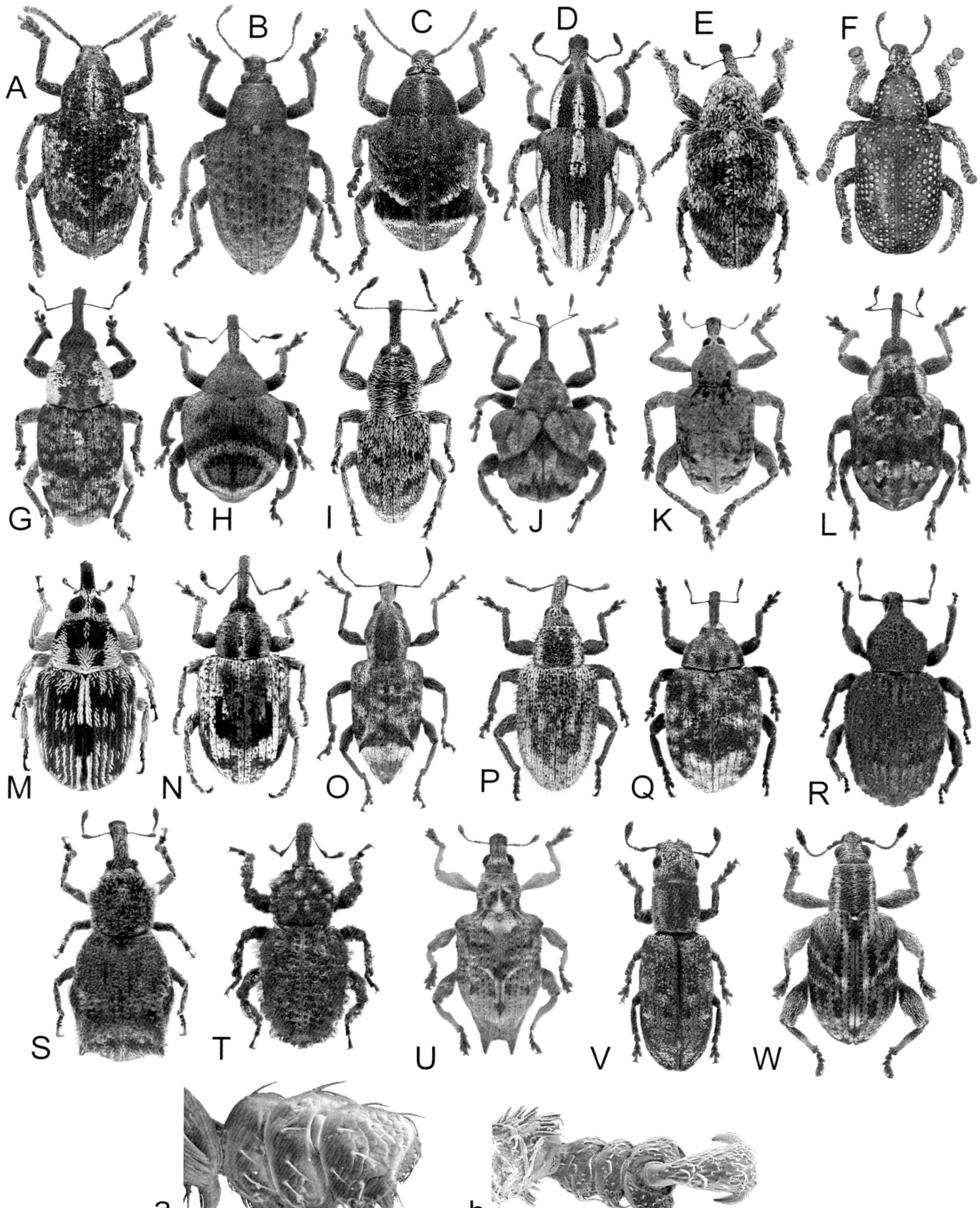

Fig. 44.197. Characters for key to unplaced genera. A, *Bryachus squamicollis* Pascoe; B, *Gonipterus notographus* Boisduval; C, *Iptergonus cionoides* (Pascoe); D, *Pantoreites trivirgatus* Lea; E, *Prophaesia tenuirostris* Lea; F, *Syarbis pachypus* Pascoe; G, *Byrsodes binodipennis* (Lea); H, *Cassythicola*; I, *Eniopea viridisquama* Lea; J, *Euhackeria insignis* Lea; K, *Eurychirus bituberculatus* Waterhouse; L, *Gerynassa nodulosa* Pascoe; M, *Glaucopela unicolor* Pascoe; N, *Lexithia rufipennis* Pascoe; O, *Lycosura bispinosa* Pascoe; P, *Olanaea nigricollis* Pascoe; Q, *Xeda notabilis* Blackburn; R, *Cisolea umbratilis* Oke; S, *Crossixus punctatus* Kuschel; T, *Sclerixus victoriae* (Oke); U, *Austrocis bicaudatus* Zimmerman & Oberprieler; V, *Peliocis subcylindricus* Lea; W, *Platynotocis pyriformis* Lea; a–b, tarsi: a, *Geochus*; b, *Crossixus*.

Subfamily Hyperinae Lacordaire, 1863

Introduction. This subfamily is distributed worldwide but most diverse and best studied in the Holarctic region. The Australian fauna is smaller and relatively poorly known (Pl. 15A, C), with several genera transferred to it from Curculioninae only recently and possibly more to follow (Oberprieler *et al.* 2014b; Pullen *et al.* 2014). Several further genera and species remain to be described. The subfamily is poorly defined morphologically, and the relationships of the Australian taxa with those of the Northern Hemisphere are in need of investigation. The most salient feature of the subfamily is that the larvae are generally mobile ectophytic feeders of leaves, shoots and inflorescences and spin a meshed aerial cocoon for pupation, although the larvae of *Gerynassa* and possibly other Australasian genera mine in leaves of *Eucalyptus* and likely other Myrtaceae (Oberprieler *et al.* 2014b). An overview of the distribution, biology, morphology, taxonomy and phylogeny of the subfamily (treated as a tribe *incertae sedis*) was provided by Oberprieler *et al.* (2014b).

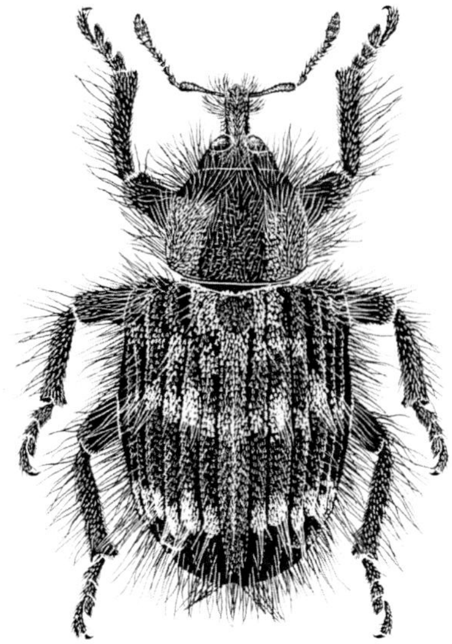

Fig. 44.198. *Agriochaeta crinita* Pascoe.

Key to the Australian genera of Hyperinae

1.	Prosternum with canal or at least with weak impression	2
–	Prosternum without canal or weak impression	5
2(1).	Body vestiture including dense long erect setae; procoxae separated by prosternal process	***Agriochaeta*** **Pascoe**
–	Body vestiture without dense long setae; procoxae contiguous	3
3(1).	Head with interocular distance smaller than width of rostrum	***Nothyperus*** **Marshall**
–	Head with interocular distance larger than width of rostrum	4
4(3).	Elytra with tubercles or prominences near apical third; Fig. 44.197G	***Byrsodes*** **Marshall**
–	Elytra without tubercles or prominences near apical third	***Glaucopela*** **Pascoe**
5(1).	Scutellar shield protuberant above level of elytra; habitus as Fig. 44.197O	***Lycosura*** **Pascoe**
–	Scutellar shield not or barely protruding above level of elytra	6
6(5).	Elytral disc with two median pointed tubercles; habitus as Fig. 44.197K	***Eurychirus*** **Waterhouse**
	Elytral disc without median tubercles	7
7(6).	Body oval (Fig. 44.197H); scutellar shield lower than level of elytra; Fig. 44.197H	***Cassythicola*** **Lea**
–	Body elongate (e.g. Fig. 44.197I); scutellar shield at same height as elytra	8
8(7).	Antennal funicles 6-segmented; Fig. 44.197N	***Lexithia*** **Pascoe**
–	Antennal funicles 7-segmented	9
9(8).	Antennal funicles with segment 2 as long as or longer than 1	10
–	Antennal funicles with segment 2 shorter than 1	11
10(9).	Elytra wider at base (Fig. 44.197J)	***Euhackeria*** **Lea**
–	Elytra subparallel between base and apical third (Fig. 44.197L)	***Gerynassa*** **Pascoe**
11(9).	Tarsal claws divaricate; Fig. 44.197I	***Eniopea*** **Pascoe**
–	Tarsal claws subparallel	12
12(11).	Body vestiture consisting of appressed and erect setae (Fig. 44.197P)	***Olanaea*** **Pascoe**
–	Body vestiture consisting of appressed setae only (Fig. 44.197Q)	***Nedyleda*** **Pascoe –** ***Xeda*** **Pascoe**

Classification of the Australian genera of Hyperinae

Agriochaeta Pascoe, 1872 (Pl. 59A, B; Fig. 44.198)

Type species, by monotypy: *Agriochaeta crinita* Pascoe, 1872.

Australian species. *A. crinita* Pascoe, 1872b.

Distribution. Australia (QLD).

Notes. The adults live on *Acacia pendula* (Fabaceae); the larvae are unknown but probably feed on the leaves of the same and similar plant species.

Byrsodes Marshall, 1939 (Pl. 59C, D; Fig. 44.197G)

Type species, by monotypy: *Byrsia cerata* Pascoe, 1883.

Australian species. *B. binodipennis* (Lea, 1931a), *B. ceratus* (Pascoe, 1883b).

Distribution. Australia (QLD, WA).

Notes. The two species in the genus are not congeneric, *B. ceratus* being related to *Glaucopela* but *B. binodipennis* to *Erytenna* (Oberprieleret *al.* 2014b; Pullen *et al.* 2014). Specimens have been found on mistletoes growing on

Grevillea and *Hakea* (Proteaceae). The genus was described by Marshall (1939a).

Cassythicola Lea, 1910 (Pl. 59E; Fig. 44.197H)

Type species, by subsequent designation (Pullen *et al.* 2014: 128): *Cassythicola rotundatus* Lea, 1910.
Australian species. *C. longirostris* Lea, 1910b, *C. medius* Lea, 1915c, *C. rotundatus* Lea, 1910b.
Distribution. Australia (NSW, WA).
Notes. Specimens have been found on *Eucalyptus*, but *Cassythicola rotundatus* appears to be associated with the parasitic vine *Cassytha* (Lauraceae). Several undescribed species are known.

Eniopea Pascoe, 1873 (Pl. 59F; Fig. 44.197I)

Type species, by monotypy: *Eniopea amoena* Pascoe, 1873.
Australian species. *E. amoena* Pascoe, 1873a, *E. bivittata* Lea, 1909a, *E. ferrugata* (Boheman, 1843) (*in* Schoenherr 1843), *E. nebulosa* Lea, 1911b, *E. posticalis* Blackburn, 1893b, *E. subcaerulea* Lea, 1908c, *E. tenebricosa* Blackburn, 1893b, *E. tenuis* Lea, 1911b, *E. viridisquama* Lea, 1911b.
Distribution. Australia (NSW, VIC, TAS, WA).
Notes. Specimens have been found on *Leptospermum* species (Myrtaceae).

Euhackeria Lea, 1910 (Pl. 59G; Fig. 44.197J)

Type species, by monotypy: *Euhackeria insignis* Lea, 1910.
Australian species. *E. insignis* Lea, 1910b.
Distribution. Australia (NSW, QLD).

Eurychirus Waterhouse, 1853 (Pl. 59H; Fig. 44.197K)

Type species, by monotypy: *Eurychirus bituberculatus* Waterhouse, 1853.
Australian species. *E. alleni* Lea, 1910a, *E. bituberculatus* Waterhouse, 1853, *E. obtusatus* Lea, 1929a.
Distribution. Australia (QLD, NSW).

Gerynassa Pascoe, 1873 (Pl. 59I; Fig. 44.197L)

Type species, by subsequent designation (Pullen *et al.* 2014: 129): *Gerynassa nodulosa* Pascoe, 1873.
Australian species. *G. affinis* Blackburn, 1892b, *G. andersoni* Blackburn, 1893b, *G. basalis* Pascoe, 1873a, *G. minor* Blackburn, 1892b, *G. nodulosa* Pascoe, 1873a, *G. picticornis* Blackburn, 1893b, *G. simplex* Blackburn, 1893b.
Distribution. Australia (NSW, VIC, TAS, SA, WA).
Notes. The taxonomy of the genus requires revision, and most species are difficult to delineate and identify. The larvae of an undescribed species mine in tips of *Eucalyptus* leaves but spin meshed cocoons typical of the subfamily (Oberprieler *et al.* 2014b), and it is likely that the other species have similar life styles.

Glaucopela Pascoe, 1874 (Pl. 59J, K; Fig. 44.197M)

Type species, by monotypy: *Glaucopela unicolor* Pascoe, 1874.
Australian species. *G. acaciae* Marshall, 1938a, *G. distincta* Blackburn, 1894a, *G. fuscomarmorea* Blackburn, 1894a, *G. hakeae* (Lea, 1899a), *G. instabilis* Lea, 1899a, *G. interioris* Lea, 1915e, *G. nidicola* Lea, 1928c, *G. rufa* Lea, 1899a, *G. unicolor* Pascoe, 1874a, *G. varipes* Blackburn, 1890c.
Distribution. Australia (NSW, NT, SA, WA).
Notes. The genus is an artificial assemblage of several unrelated species and probably needs to be restricted to only the type species, which appears related to *Byrsodes* (Pullen *et al.* 2014). In the key above, it is only represented by *G. unicolor*, not by different species such as *G. acaciae* (Pl. 59J). The latter species has been reared from galls on *Acacia*, and *G. varipes* also appears associated with this genus (Oberprieler *et al.* 2014b), their larvae probably developing in galls or twigs.

Lexithia Pascoe, 1870 (Pl. 59L; Fig. 44.197N)

Type species, by monotypy: *Lexithia rufipennis* Pascoe, 1870.
Australian species. *L. alternata* (Blackburn, 1890b), *L. aurulenta* (Pascoe, 1873a), *L. griseola* (Pascoe, 1873a), *L. latipennis* (Lea, 1926b), *L. rufipennis* Pascoe, 1870d, *L. setosa* (Pascoe, 1873a).
Distribution. Australia (NSW, SA, WA).
Notes. This is also an artificial genus that needs restricting to only the type species and perhaps a few others (Pullen et al. 2014).

Lycosura Pascoe, 1875 (Pl. 60A; Fig. 44.197O)

Type species, by monotypy: *Lycosura bispinosa* Pascoe, 1875.
Australian species. *L. bispinosa* Pascoe, 1875, *L. breweri* (Pascoe, 1873b).
Distribution. Australia (WA).

Nedyleda Pascoe, 1872 (Pl. 60B)

Type species, by monotypy: *Nedyleda semiusta* Pascoe, 1872.
Australian species. *N. obsti* (Lea, 1911c), *N. semiusta* Pascoe, 1872c.
Distribution. Australia (WA).
Notes. The difference between *Nedyleda* and *Xeda* is unclear and requires investigation.

Nothyperus Marshall, 1948 (Pl. 60C)

Type species, by original designation: *Nothyperus cionodes* Marshall, 1948.
Australian species. *N. cionodes* Marshall, 1948
Distribution. Australia (QLD, NSW).
Notes. The host plant of the single species appears to be *Citrus glauca* (Rutaceae) (Oberprieler *et al.* 2014b).

Olanaea Pascoe, 1873 (Pl. 60D; Fig. 44.197P)

Type species, by monotypy: *Olanaea nigricollis* Pascoe, 1873.
Australian species. *O. laeta* Blackburn, 1890b, *O. maculata* Blackburn, 1890b, *O. mentitrix* Blackburn, 1894a, *O. metropolitana* Blackburn, 1894a, *O. nigricollis* Pascoe, 1873a, *O. ornata* (Pascoe, 1875), *O. variabilis* (Lea, 1915c).
Distribution. Australia (NSW, VIC, SA, WA).
Notes. *Hakea* (Proteaceae) has been recorded as a possible host plant for the genus.

Xeda Pascoe, 1873 (Pls 14A, 60E; Fig. 44.197Q)

Type species, by subsequent designation (Pullen *et al.* 2014: 131): *Xeda amplipennis* Pascoe, 1873.
Australian species. *X. acaciae* (Pascoe, 1873a), *X. amplipennis* Pascoe, 1873a, *X. bilineata* Pascoe, 1873a, *X. fasciata* Lea, 1915e, *X. magistra* Blackburn, 1894a, *X. notabilis* Blackburn, 1894a.
Distribution. Australia (NSW, SA, WA).
Notes. The species are associated with *Acacia* (Fabaceae), their larvae probably feeding on the leaves but possibly also inside the seed pods (Oberprieler *et al.* 2014b).

Fig. 44.199. *Sclerixus victoriae* (Oke).

Tribe Phrynixini Kuschel, 1964

Introduction. The Phrynixini are a small group of peculiar small-eyed, brachypterous to apterous weevils (Pl. 20C–D) apparently restricted to Australia and New Zealand and some of its islands. The adults live in leaf litter and among mosses and lichens, and the larvae seemingly develop in decaying wood (Lyal 2014a) or may bore in roots or stems. The group has traditionally included a number of genera associated with ferns (*Syagrius* in Australia and relatives *Megacolabus* Broun and *Rystheus* Broun in New Zealand), but Gunter *et al.* (2016) showed that *Syagrius* belongs in the CCCMS clade near *Psepholax*, whereas the true Phrynixini are a basal lineage of the CEGH clade, and Leschen *et al.* (2022) also found *Rystheus* falling outside of Phrynixini. The genus *Cisolea*, placed in Phrynixini by Pullen *et al.* (2014), has since been shown to belong in Cyclominae, possibly in Aterpini (Gunter *et al.* 2016, Baird *et al.* 2021, Haran *et al.* 2023), but is still treated in Phrynixini here. The Australian Phrynixini were revised by Kuschel (1972), and Lyal (2014a) provided an overview of the tribe, although in both cases including the fern-associated *Syagrius* group.

Key to Australian genera of Phrynixini

Modified after Kuschel (1972).

1. Tarsal segments 4 and 5 absent (Fig. 44.200I); habitus as Fig. 44.197R ..... ***Cisolea* Oke**
– Tarsal segments 4 and 5 present 2
2(1). Basal margin of pronotum and elytra with a highly raised, sharp edge; areas around scutellar shield depressed; elytra with oblique postmedian impression 3
– Basal margin of pronotum and elytra without highly raised sharp edge; area around scutellar shield area not depressed; elytra without oblique postmedian impression 4
3(2). Body with ventral surface nearly impunctate; forehead between eyes about as wide as rostrum near base (Fig. 44.203A); surface of pronotum uneven, with raised and deeply impressed areas ..... ***Mallixus* Kuschel**
– Body with ventral surface densely punctate; forehead between eyes distinctly wider than rostrum near base (Fig. 44.201A); surface of pronotum even, without raised or deeply impressed areas; habitus as Fig. 44.197S ..... ***Crossixus* Kuschel**

4(2). Scrobes in dorsal view narrowly but distinctly exposed anteriorly; sides of prothorax and some or all striae of elytra with distinct granules; trochanters with one long erect seta .... ***Dinichus*** **Pascoe**
– Scrobes in dorsal view not visible; sides of prothorax and striae of elytra with distinct granules; trochanters without erect seta 5
5(4). Elytra with base of interstriae 3 flat, without a tuft .... ***Lophixus*** **Kuschel**
– Elytra with base of interstriae 3 distinctly convex, with elongate tuft; habitus as Fig. 44.197T .... ***Sclerixus*** **Kuschel**

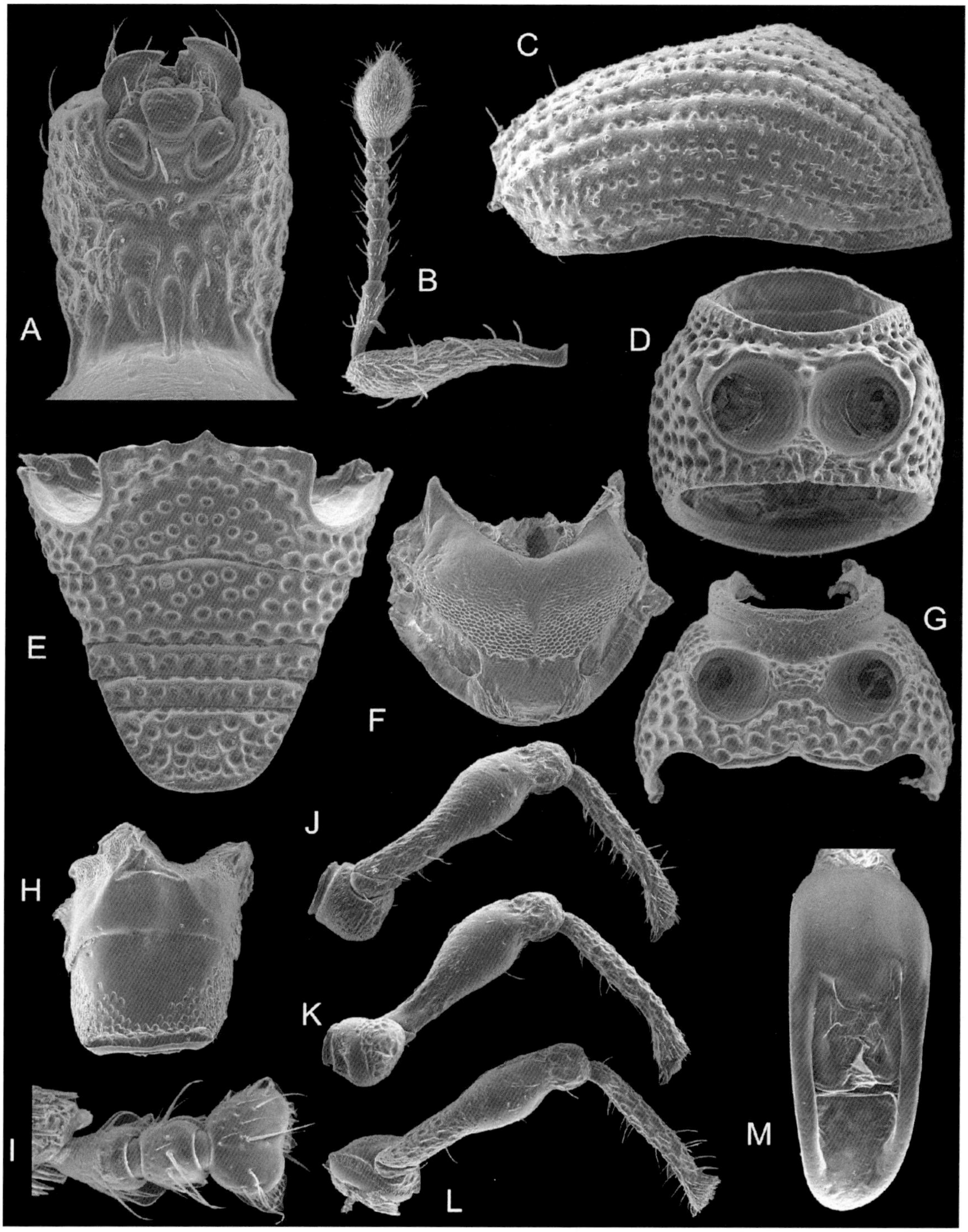

Fig. 44.200. Phrynixini, structural details, *Cisolea* sp., ♂. A, mouthparts; B, antenna; C, elytron, lateral; D, prothorax, ventral; E, abdominal ventrites; F, scutellum; G, pterothorax; H, pygidium, dorsal; I, tarsus, dorsal; J, foreleg; K, middle leg; L, hindleg; M, penis, dorsal.

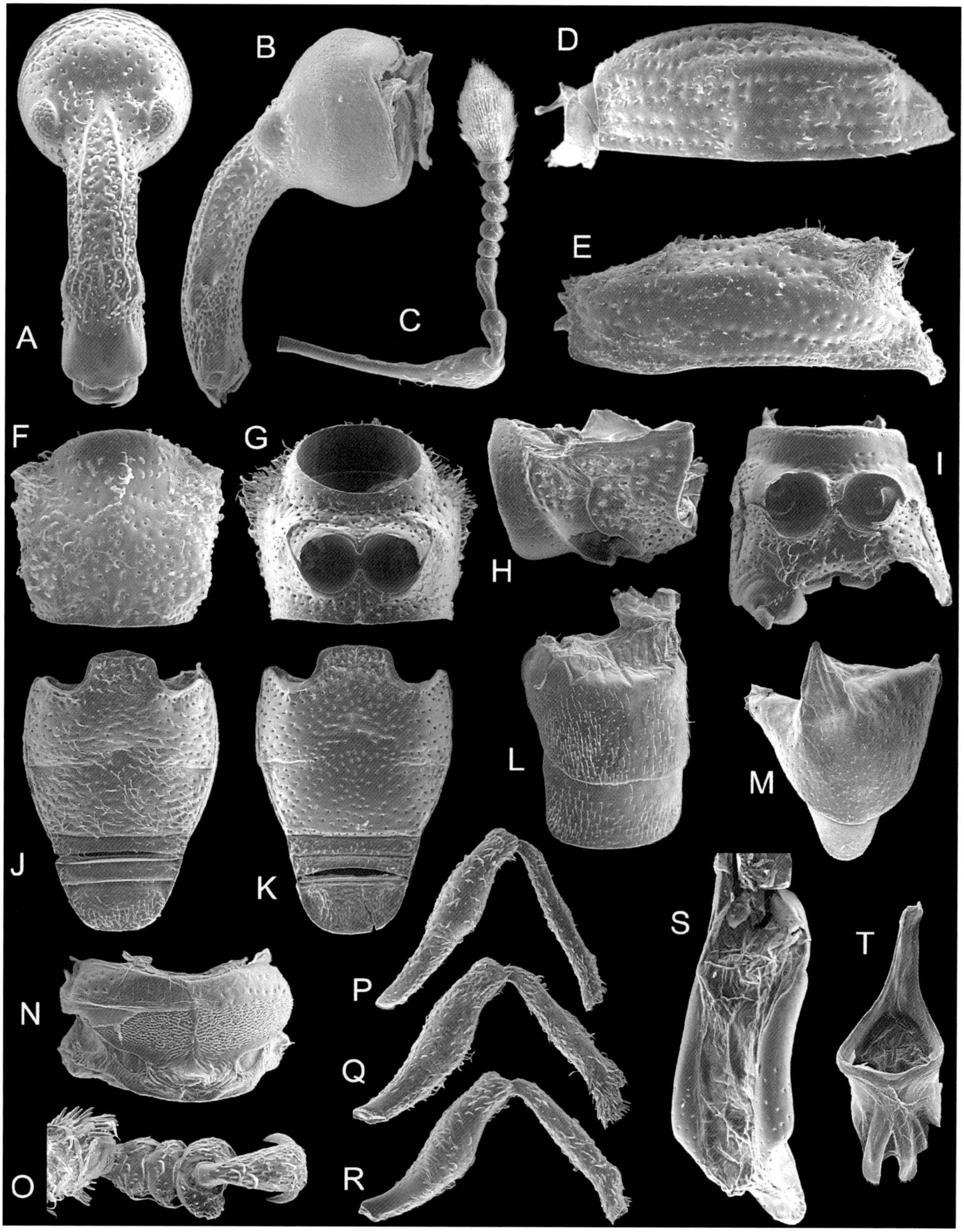

Fig. 44.201. Phrynixini, structural details, *Crossixus punctatus* Kuschel. A–B, head: A, dorsal; B, lateral; C, antenna; D–E, elytron: D, dorsal; E, lateral; F–G, prothorax: F, dorsal; G, ventral; H–I, pterothorax: H, lateral; I, ventral; J–K, abdominal ventrites: J, ♂; K, ♀; L–M, pygidium, dorsal: L, ♂; M, ♀; N, scutellum; O, tarsus, dorsal; P, foreleg; Q, middle leg; R, hindleg; S, penis, dorsal; T, tegmen, dorsal.

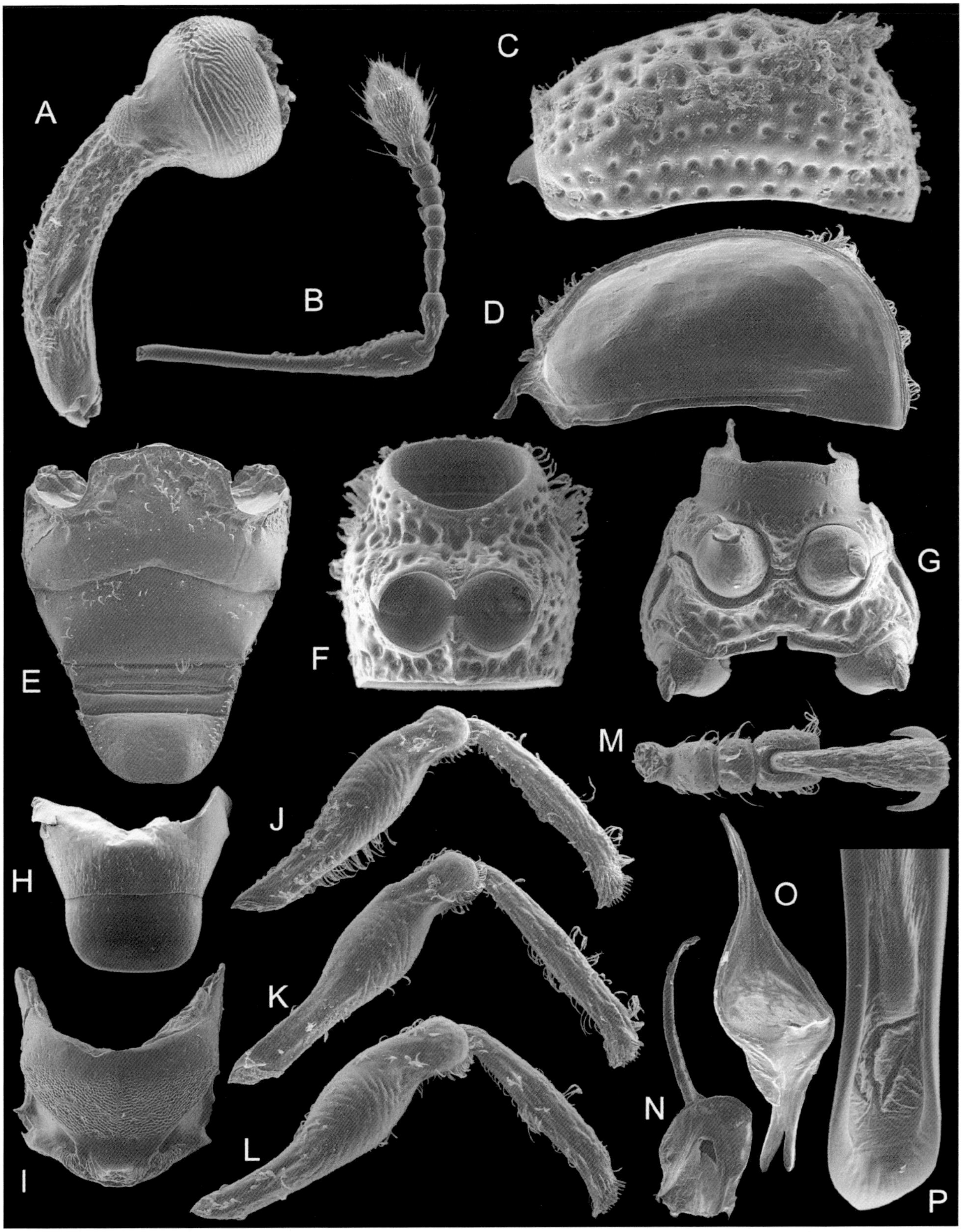

Fig. 44.202. Phrynixini, structural details, *Sclerixus victoriae* (Oke), ♂. A, head, lateral; B, antenna; C–D, elytron: C, lateral; D, inner lateral; E, abdominal ventrites; F, prothorax, ventral; G, pterothorax, ventral; H, pygidium, dorsal; I, scutellum; J, foreleg; K, middle leg; L, hindleg; M, tarsus, dorsal; N, terminalia; O, tegmen; P, penis, dorsal.

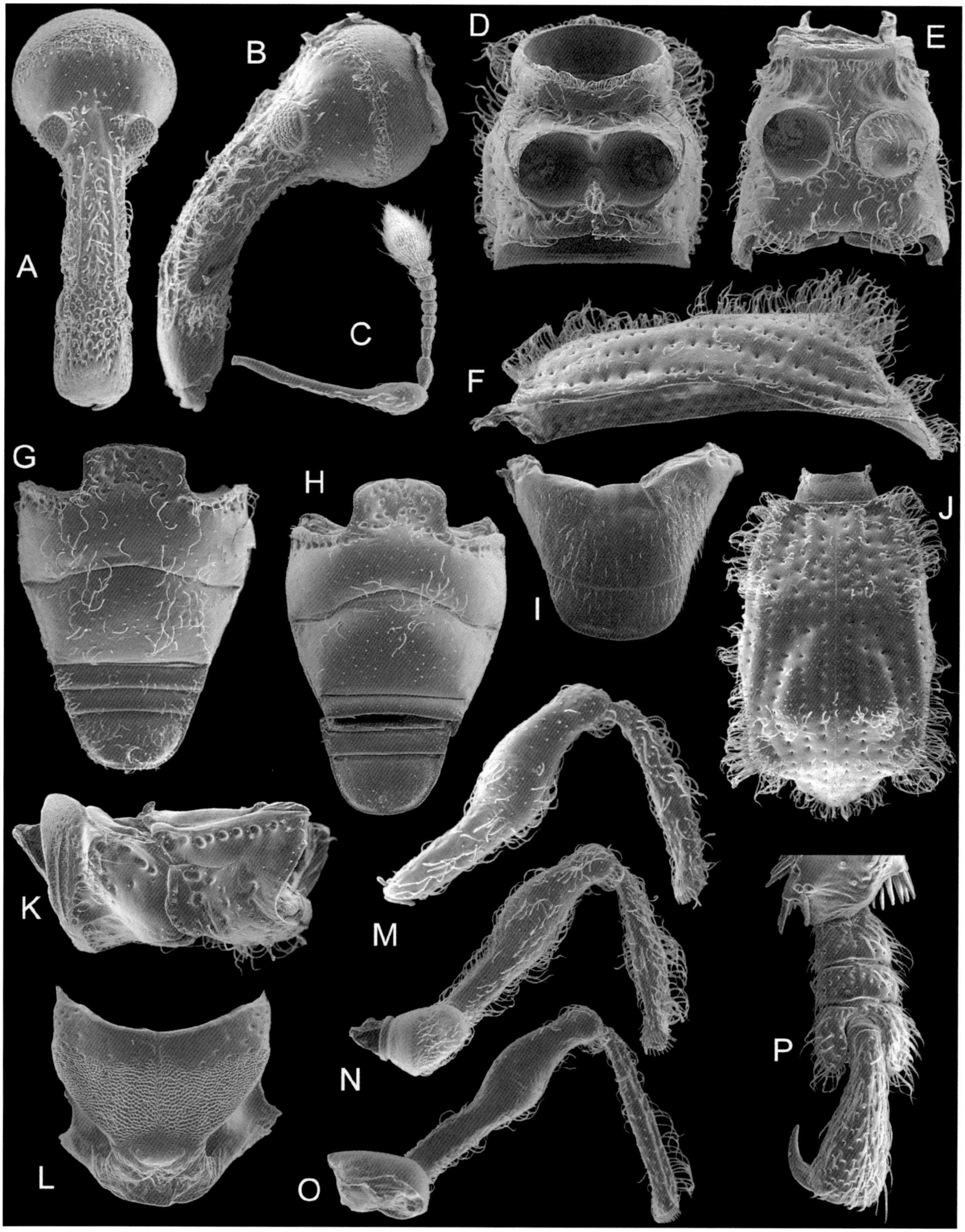

Fig. 44.203. Phrynixini, structural details, *Mallixus sylvicola* (Oke). A–B, head: A, dorsal; B, lateral; C, antenna; D–E, prothorax, ventral: D, ♂; E, ♀; F, elytron, lateral; G–H, abdominal ventrites: G, ♂; H, ♀; I, pygidium, dorsal; J, elytra, dorsal; K, pterothorax, lateral; L, scutellum; M, foreleg; N, middle leg; O, hindleg; P, tarsus, dorsal.

Classification of the Australian genera of Phrynixini

Cisolea Oke, 1934 (Pl. 60F; Figs 44.195R, 44.200)

Type species, by original designation: *Cisolea umbratilis* Oke, 1934.
Australian species. *C. umbratilis* Oke, 1934.
Distribution. Australia (QLD, NSW).
Notes. The genus has been shown not to belong in Phrynixini (see above).

Crossixus Kuschel, 1972 (Pl. 60G; Figs 44.195S, b, 44.201)

Type species, by original designation: *Crossixus punctatus* Kuschel, 1972.
Australian species. *C. punctatus* Kuschel, 1972.
Distribution. Australia (TAS).
Notes. Specimens have been collected from moss and lichens.

Dinichus Pascoe, 1887 (Pl. 60H)

Type species, by monotypy: *Dinichus terreus* Pascoe, 1887.
Australian species. *D. mucronatus* Kuschel, 1972, *D. terreus* Pascoe, 1887a.
Distribution. Australia (TAS).
Notes. Specimens have been collected from moss and lichens.

Lophixus Kuschel, 1972 (Pl. 60I)

Type species, by original designation: *Phrynixus major* Oke, 1931.
Australian species. *L. major* (Oke, 1931).
Distribution. Australia (TAS).

Mallixus Kuschel, 1972 (Pl. 60J; Fig. 44.203)

Type species, by original designation: *Phrynixus sylvicola* Oke, 1931.
Australian species. *M. sylvicola* (Oke, 1931).
Distribution. Australia (VIC).

Sclerixus Kuschel, 1972 (Pls 20C, D, 60K; Figs 44.199, 44.195T, 44.202)

Type species, by original designation: *Phrynixus victoriae* Oke, 1931.
Australian species. *S. lasius* Kuschel, 1972, *S. victoriae* (Oke, 1931).
Distribution. Australia (VIC).

Tribe Viticiini Morimoto, 1983

Introduction. The Viticiini are a small tribe of short-snouted weevils distributed in the Australo-Pacific region, Japan and Indonesia. The peculiar genus *Geochus* has recently been shown to belong in this tribe as well (Leschen *et al.* 2022). The larvae as known mine in both living and dead leaves (May 1992, 1993; Oberprieler 2010; Caldara *et al.* 2014; Maunsell *et al.* 2017). An overview of the tribe was provided by Caldara *et al.* (2014).

Fig. 44.204. *Austrocis* sp.

Key to Australian genera of Viticiini

1.	Tarsal segments 4 and 5 absent (e.g. Fig. 44.197a)	***Geochus*** **Broun**
–	Tarsal segments 4 and 5 present (e.g. Fig. 44.197b)	**2**
2(1).	Elytral apices with prominent pointed projection (Fig. 44.197U)	***Austrocis*** **Zimmerman & Oberprieler**
–	Elytral apices without prominent pointed projection	**3**
3(2).	Metafemora larger than pro- and mesofemora (Fig. 44.197W) and with ventral tooth	***Platynotocis*** **Lea**
–	Femora of similar size (Fig. 44.197U), metafemora without ventral tooth	**4**

4(3).	Cuticle dark brown and mostly glabrous and smooth with patches of white setae	*Tivicis* Morimoto
–	Cuticle pale brown and uniformly setose	5
5(4).	Vestiture consisting of long suberect flattened setae	*Eupholocis* Lea
–	Vestiture consisting of appressed setae (Fig. 44.197V)	*Peliocis* Lea

Classification of the Australian genera of Viticiini

Austrocis Zimmerman & Oberprieler, 2014 (Pl. 61A, Figs 44.195U, 44.204, 44.207)

Type species, by original designation: *Austrocis bicaudatus* Zimmerman & Oberprieler, 2014.
Australian species. *A. bicaudatus* Zimmerman & Oberprieler, 2014.
Distribution. Australia (NSW).
Notes. The genus and single species were described in Pullen *et al.* (2014). The larva mines in leaves of *Elaeocarpus reticulatus* (Elaeocarpaceae) (Pullen *et al.* 2014).

Eupholocis Lea, 1930 (Pl. 61B)

Type species, by original designation: *Eupholocis dentipes* Lea, 1930.
Australian species. *E. dentipes* Lea, 1930a.
Distribution. Australia (QLD), New Guinea.

Geochus Broun, 1882 (Pl. 60L; Figs 44.195a, 44.205, 44.206)

Type species, by original designation: *Geophilus inaequalis* Broun, 1880.
Australian species. *G. howensis* Lea, 1927e.
Distribution. Australia (Lord Howe Is.), New Caledonia, New Zealand.
Notes. The taxonomic position and phylogenetic relationships of the genus were uncertain for a long time (Lyal 2014a; Pullen *et al.* 2014), until Leschen *et al.* (2022) showed from a molecular-phylogenetic analysis that it belongs in Viticiini. It is most diverse in New Zealand and only sparsely represented in Australia, with several species on Lord Howe Is., and apparently related taxa occur in New Caledonia. The adults are found commonly in leaf litter, and the larvae mine dead leaves on the ground (May 1992, 1993).

Peliocis Lea, 1930 (Pl. 61C; Figs 44.195V, 44.208)

Type species, by original designation: *Peliocis subcylindricus* Lea, 1930.
Australian species. *P. subcylindricus* Lea, 1930a.
Distribution. Australia (NSW, QLD).

Platynotocis Lea, 1930 (Pl. 61D; Figs 44.195W, 44.209)

Type species, by original designation: *Platynotocis pyriformis* Lea, 1930.
Australian species. *P. albomaculatus* Zimmerman & Oberprieler, 2014, *P. angulipennis* Zimmerman & Oberprieler, 2014, *P. pyriformis* Lea, 1930a.
Distribution. Australia (QLD, NSW).
Notes. The genus contains additional undescribed species in Australia. The larva of *P. angulipennis* mines in leaves of *Polyosma cunninghamii* (Escallioniaceae) (Maunsell *et al.* 2017).

Tivicis Morimoto, 1983 (Pl. 61E)

Type species, by original designation: *Tivicis maculatus* Morimoto, 1983.
Australian species. *Tivicis* sp.
Distribution. Australia (north QLD), New Guinea, Japan.
Notes. Undescribed species are known from New Guinea (Pullen *et al.* 2014) and probably also occur in Australia.

Genera unplaced to subfamily and tribe

Introduction. Four genera belonging in the CEGH clade but of uncertain affinity were treated separately by Pullen *et al.* (2014) and are so here as well. *Hypsomus*, represented by an introduced but undescribed species in Australia, has recently been found to belong in the tribe Styphlini (Haran *et al.* 2023), but the relationships of the three Australian genera remain unknown.

Classification of the Australian Genera

Ethadomorpha Blackburn, 1901 (Pl. 61F; Fig. 44.210)

Type species, by monotypy: *Ethadomorpha clauda* Blackburn, 1901.
Australian species. *E. australis* (Erichson, 1842), *E. clauda* Blackburn, 1901, *E. puer* (Blackburn, 1893b).
Distribution. Australia (NSW, VIC, WA).

Hypsomus Schoenherr, 1836 (Pl. 61G)

Type species, by original designation: *Hypsomus scapha* Boheman, 1936 (*in* Schoenherr 1836).
Australian species. *Hypsomus* sp.
Distribution. Australia (WA, SA?), South Africa.
Notes. The single species in Australia was introduced in WA from South Africa in the late 1960s, probably with imported grass, and attacks cultivated grains and some other pasture crops (Pullen *et al.* 2014). The genus has recently been shown to belong in the tribe Styphlini, which falls into the CEGH clade (Haran *et al.* 2023).

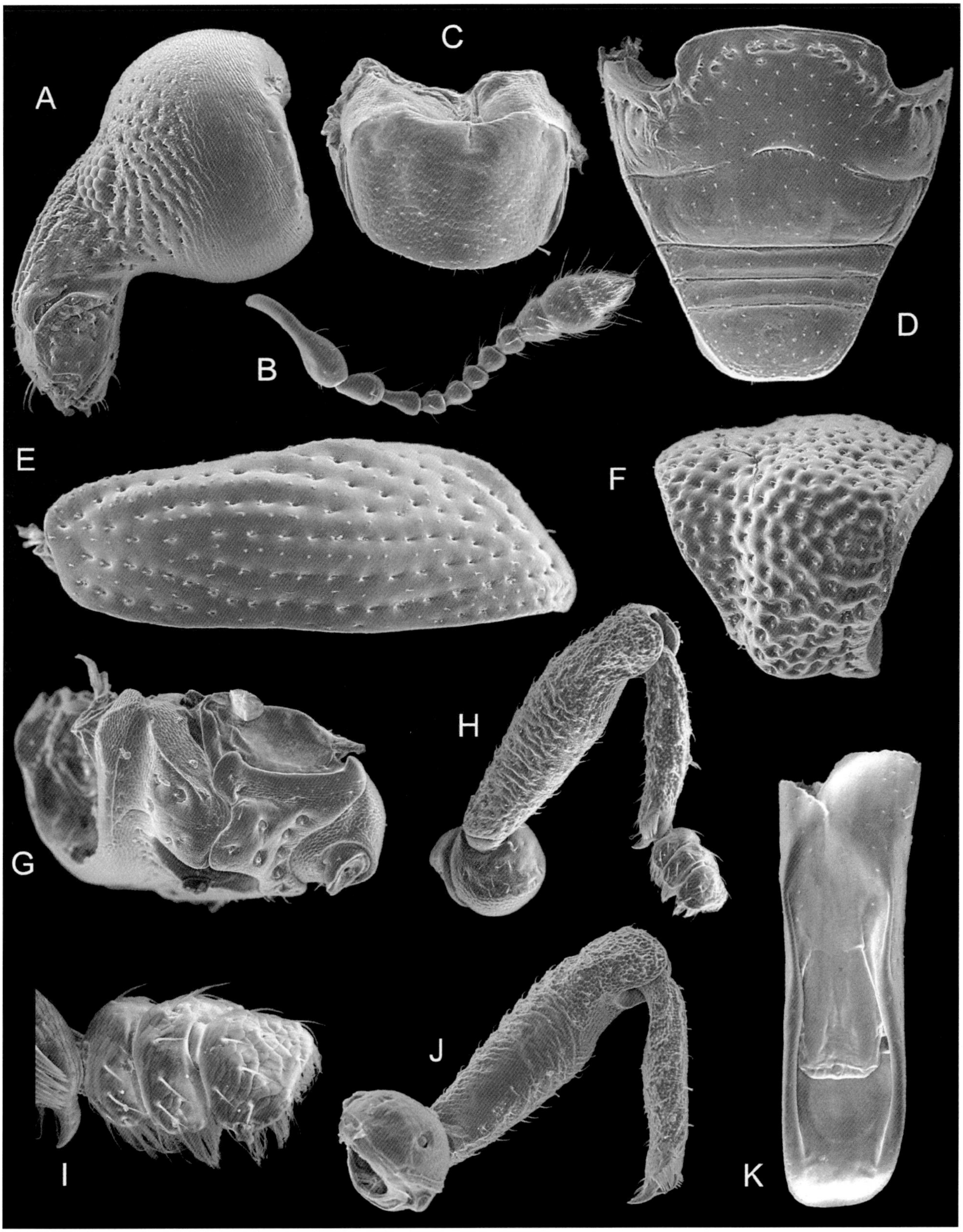

Fig. 44.205. Viticiini, structural details, *Geochus* sp., ♂. A, head, lateral; B, antenna; C, pygidium, dorsal; D, abdominal ventrites; E, elytron, lateral; F, prothorax, lateral; G, pterothorax, lateral; H, foreleg; I, tarsus, lateral; J, hindleg; K, penis, dorsal.

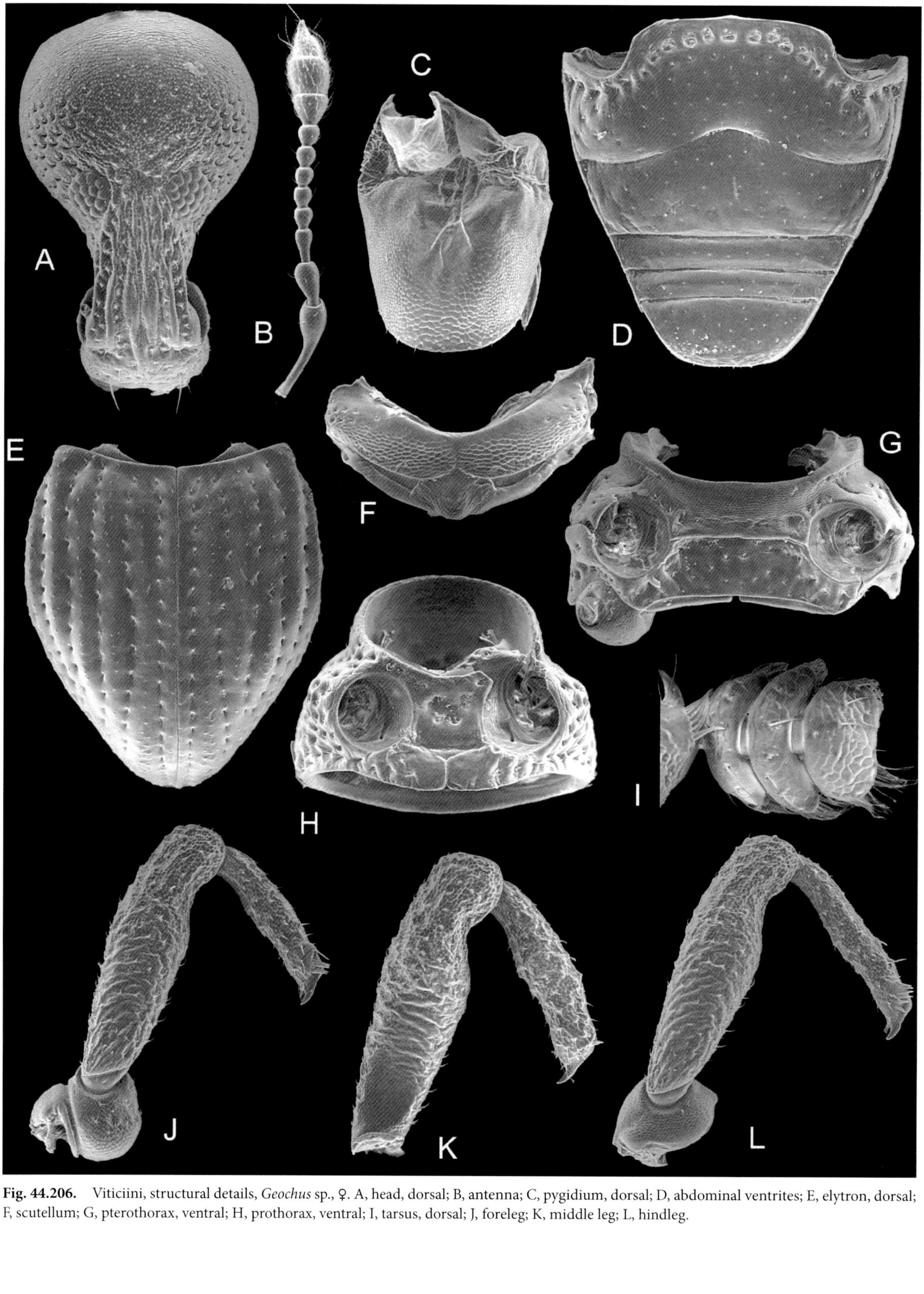

Fig. 44.206. Viticiini, structural details, *Geochus* sp., ♀. A, head, dorsal; B, antenna; C, pygidium, dorsal; D, abdominal ventrites; E, elytron, dorsal; F, scutellum; G, pterothorax, ventral; H, prothorax, ventral; I, tarsus, dorsal; J, foreleg; K, middle leg; L, hindleg.

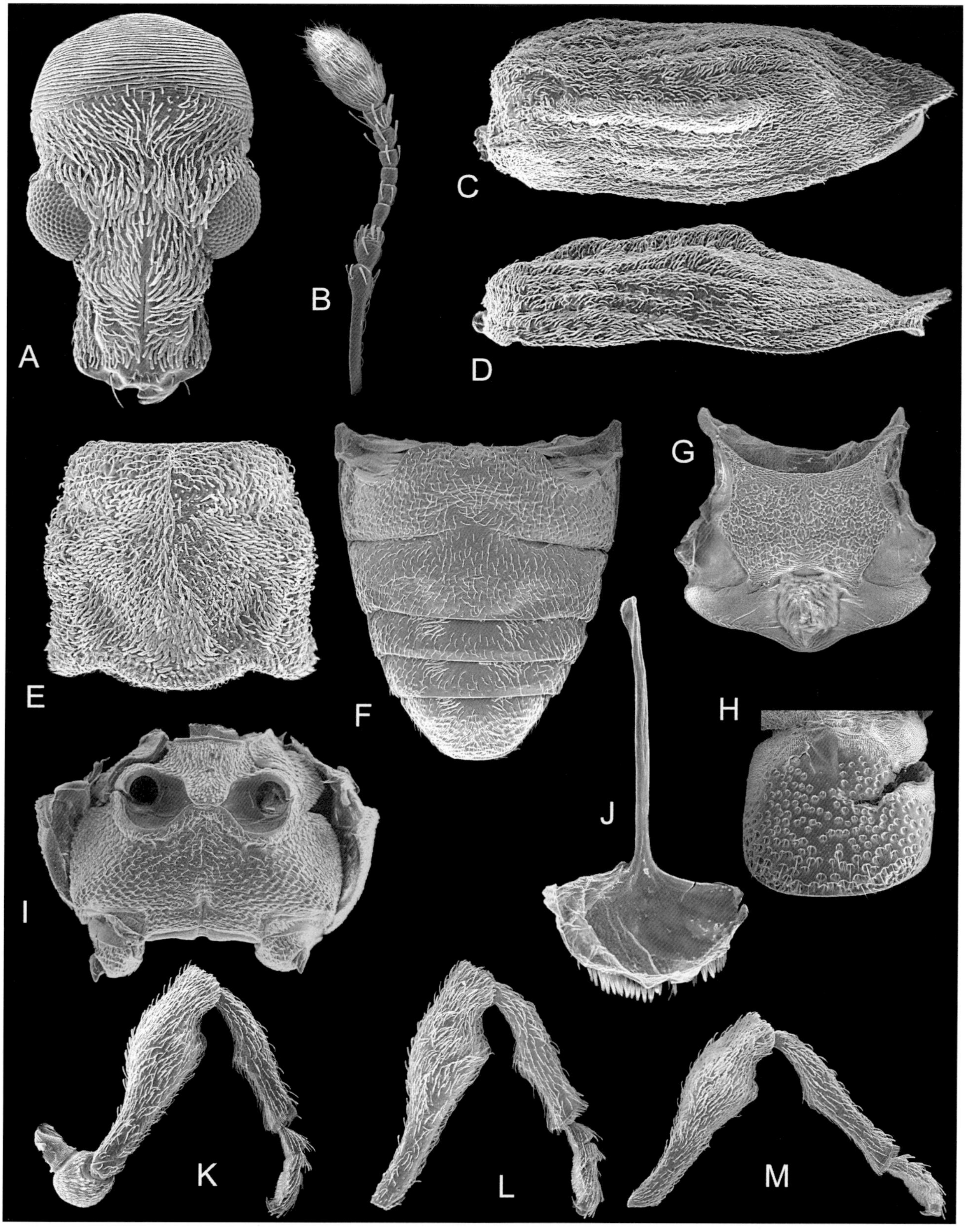

Fig. 44.207. Viticiini, structural details, *Austrocis bicaudatus* Zimmerman & Oberprieler, ♀. A, head, dorsal; B, antenna; C–D, elytron: C, dorsal; D, lateral; E, prothorax, dorsal; F, abdominal ventrites; G, scutellum; H, pygidium, dorsal; I, pterothorax, ventral; J, sternite VIII; K, foreleg; L, middle leg; M, hindleg.

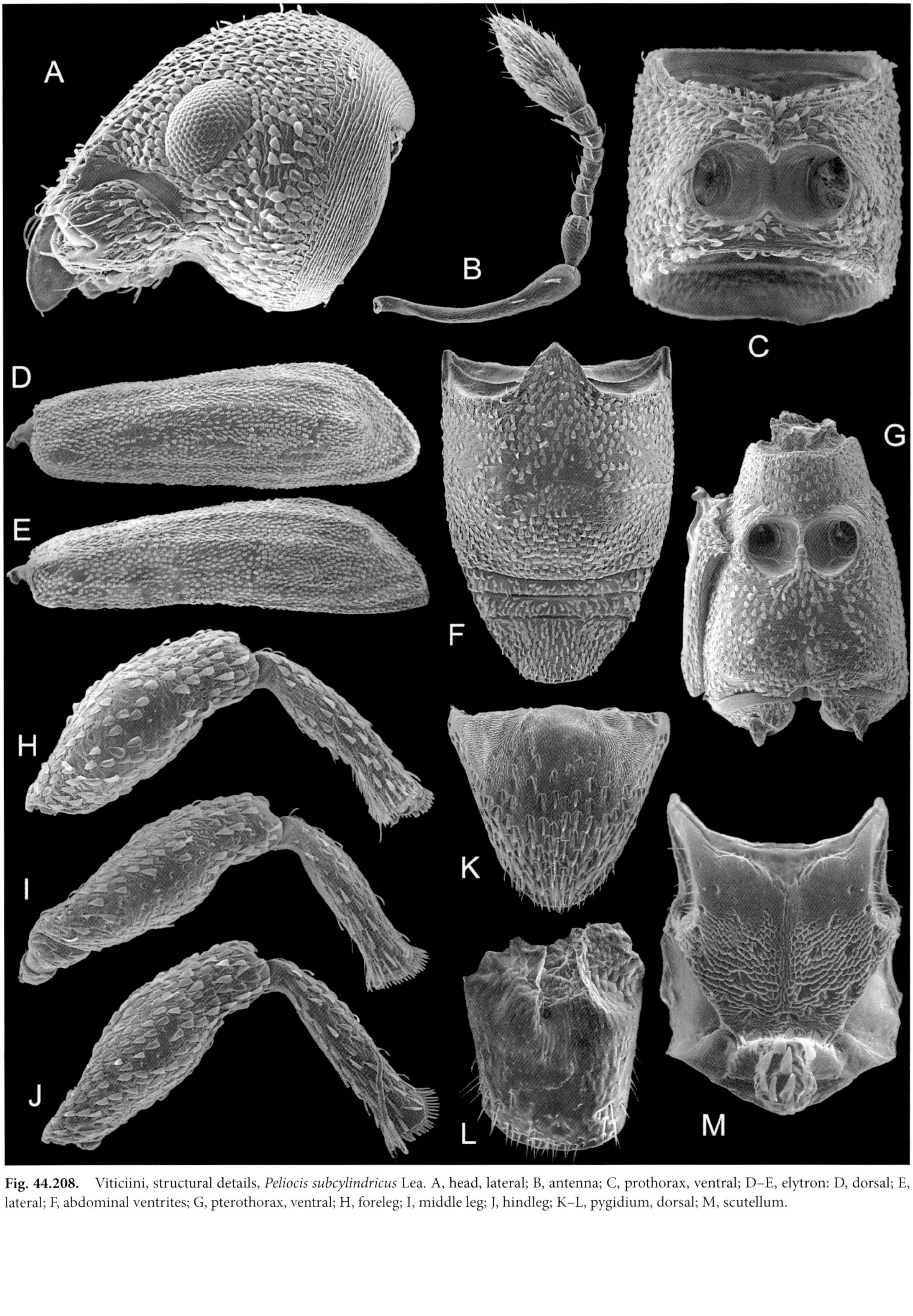

Fig. 44.208. Viticiini, structural details, *Peliocis subcylindricus* Lea. A, head, lateral; B, antenna; C, prothorax, ventral; D–E, elytron: D, dorsal; E, lateral; F, abdominal ventrites; G, pterothorax, ventral; H, foreleg; I, middle leg; J, hindleg; K–L, pygidium, dorsal; M, scutellum.

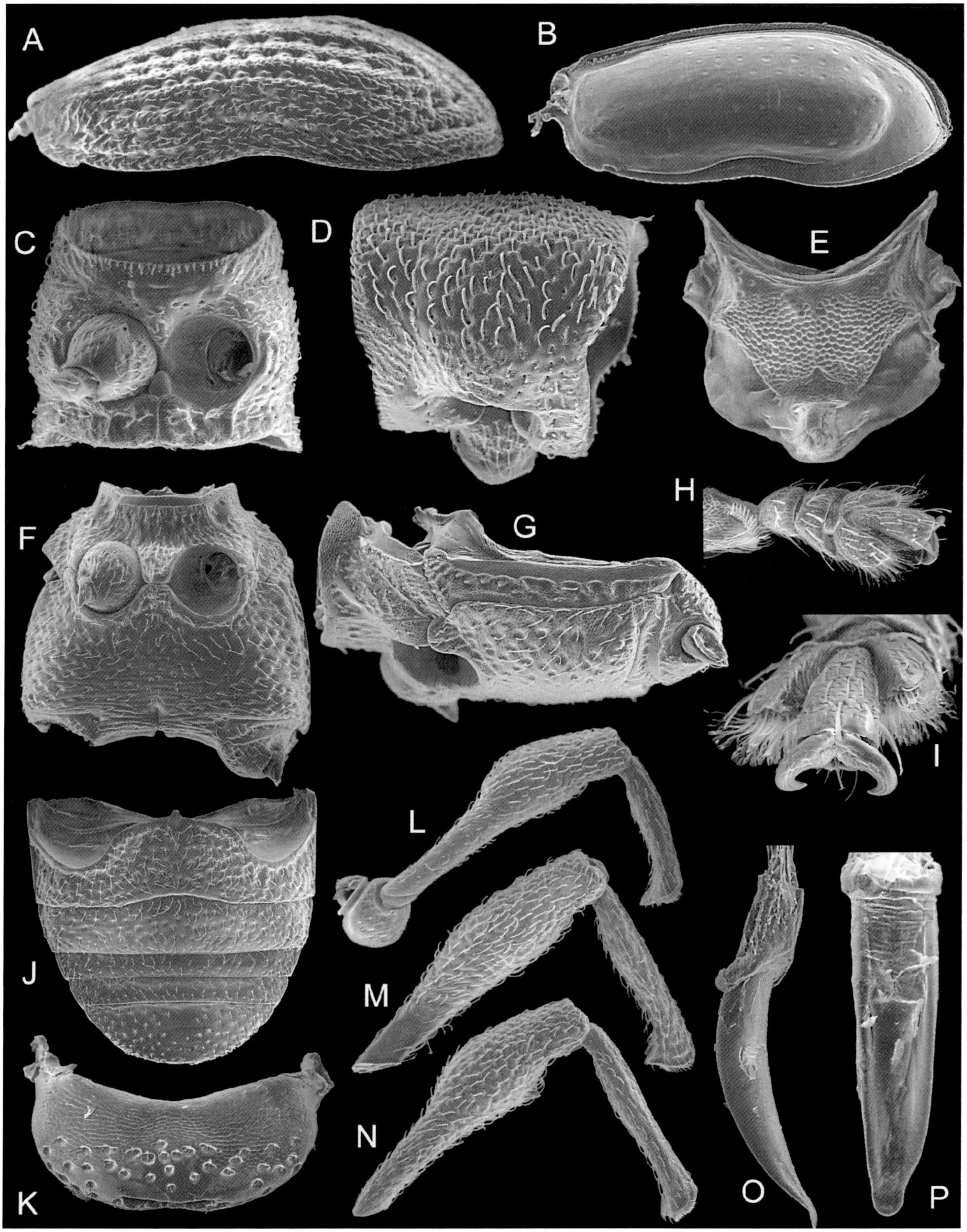

Fig. 44.209. Viticiini, structural details, *Platynotocis pyriformis* Lea. A–B, elytron: A, lateral; B, inner lateral; C–D, prothorax: C, ventral; D, lateral; E, scutellum; F–G, pterothorax: F, ventral; G, lateral; H, tarsus, dorsal; I, claws, anterior; J, abdominal ventrites; K, scutellum; L, foreleg; M, middle leg; N, hindleg; O–P, penis: O, lateral; P, dorsal.

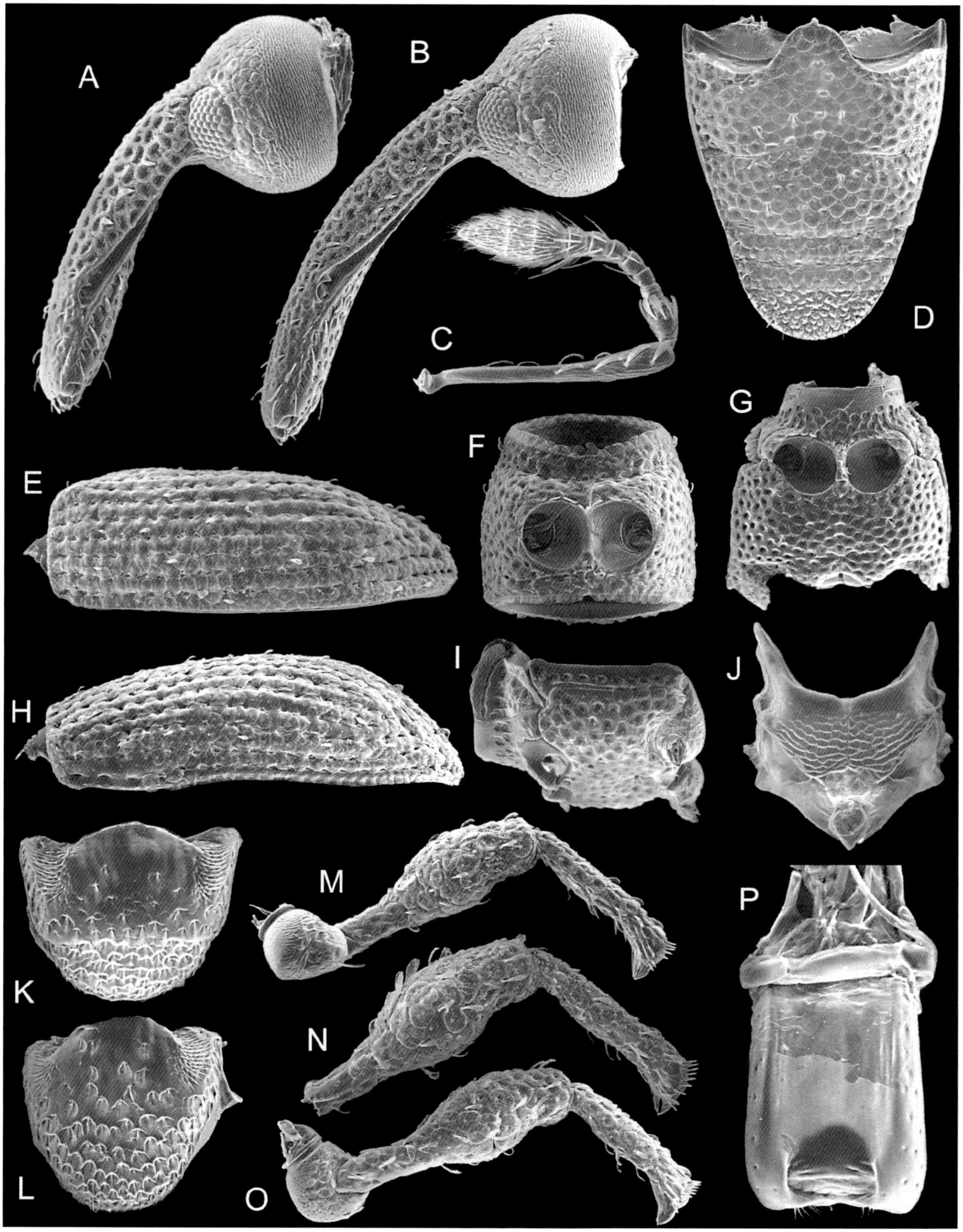

Fig. 44.210. Unplaced, structural details, *Ethadomorpha australis* (Erichson). A–B, head, lateral: A, ♂; B, ♀; C, antenna; D, abdominal ventrites, ♀; E; H, elytron; E, dorsal; H, lateral; F, prothorax, ventral; G; I, pterothorax; G, ventral; I, lateral; J, scutellum; K–L, pygidium, dorsal: K, ♂; L, ♀; M, foreleg; N, middle leg; O, hindleg; P, penis, dorsal.

Symbothynus Blackburn, 1890 (Pl. 61H, I)

Type species, by monotypy: *Symbothynus squalidus* Blackburn, 1890.

Australian species. *S. nasutus* Blackburn, 1894a, *S. squalidus* Blackburn, 1890c.

Distribution. Australia (NSW, SA, WA)

Notes. The two species currently included in the genus are not congeneric (Pullen *et al.* 2014). Specimens of the type species have been found on *Gahnia clarkei* (Cyperaceae) (Caldara *et al.* 2014).

Thechia Pascoe, 1873 (Pl. 61J)

Type species, by subsequent monotypy (Pascoe, 1874c: 25): *Thechia pygmaea* Pascoe, 1874.

Australian species. *T. brevirostris* Lea, 1926b, *T. cinerascens* Lea, 1911a, *T. latipennis* Lea, 1915a, *T. longirostris* Lea, 1926b, *T. mollis* Lea, 1927e, *T. pygmaea* Pascoe, 1874c.

Distribution. Australia.

Notes. The genus in its current concept is artificial and appears to need restriction to only comprise the type species (Pullen *et al.* 2014).

Acknowledgements

We would like to thank Bruno Bell, Jiri Lochman, Nick Monaghan, Dash Huang, Nick Volpe, Jean and Fred Hort, Nicholas Fisher, Jessa Thurman, Bernhard Jacobi and Nick Porch for their photographs of weevils in their natural habitat; Christine Lambkin and Karin Koch (Queensland Museum, Brisbane), Nik Tatarnic (Western Australian Museum, Perth), Ken Walker and Simon Hinkley (Museum Victoria, MV, Melbourne), and Ben Parslow (South Australia Museum, Adelaide) for access to specimens used for imaging. We also would also like to thank Otto Bell, Kim Pullen, Jens Prena and Chris Reid for providing feedback and suggestions. Thanks to Nicole Fisher (ANIC-CSIRO) for her assistance with photographic equipment.

References

Allen PG (1973) Biology of *Desiantha caudata* Pascoe (Coleoptera: Curculionidae) in South Australia. *Australian Journal of Entomology* **12**(3), 201–206. doi:10.1111/j.1440-6055.1973.tb01660.x

Alonso-Zarazaga MA (2013) New nomenclatural and taxonomic acts, comments – Curculionidae: Cryptorhynchinae. In *Catalogue of Palearctic Coleoptera. Vol. 7. Curculionoidea I.* (Eds I Lobl and A Smetana) pp. 66–77. Apollo Books, Stenstrup.

Alonso-Zarazaga MA, Lyal CHC (1999) *A World Catalogue of Families and Genera of Curculionoidea (Insecta: Coleoptera) (Excepting Scolytidae and Platypodidae).* Entomopraxis, Barcelona, 315 pp.

Alonso-Zarazaga MA, Sánchez-Ruiz M (2002) Revision of the *Trichosirocalus horridus* (Panzer) species complex, with description of two new species infesting thistles (Coleoptera: Curculionidae, Ceutorhynchinae). *Australian Journal of Entomology* **41**(3), 199–208. doi:10.1046/j.1440-6055.2002.00291.x

Ancey CF (1881) Diagnoses de curculionides nouveaux. *Le Naturaliste* **1**, 372.

Anderson RS (2002) The Dryophthoridae of Costa Rica and Panama: checklist with keys, new synonymy and descriptions of new species of *Cactophagus, Mesocordylus, Metamasius* and *Rhodobaenus. Zootaxa* **80**(1), 1–94. doi:10.11646/zootaxa.80.1.1

Anderson RS, Marvaldi AE (2014) 3.7.3 Dryophthorinae Schoenherr, 1825. In *Handbook of Zoology, Coleoptera, Beetles Volume 3: Morphology and Systematics (Phytophaga).* (Eds RAB Leschen and RG Beutel) pp. 226–230. De Gruyter, Berlin/Boston, xii + 675 pp.

Anderson RS, Oberprieler RG, Setliff GP (2018) A review of the *Araucaria*-Associated Weevils of the tribe Orthorhinini (Coleoptera: Curculionidae: Molytinae), with description of new species of *Ilacuris* Pascoe, 1865 and *Notopissodes* Zimmerman & Oberprieler, 2014 and a new genus, *Kuschelorhinus* Anderson & Setliff. *Diversity* **10**(3), 54pp. doi:10.3390/d10030054

Armstrong JE, Irvine AK (1990) Functions of staminodia in the beetle-pollinated flowers of *Eupomatia laurina. Biotropica*, **22**(4), 429–431.

Baird HP, Shin S, Oberprieler RG, Hullé M, Vernon P, *et al.* (2021) Fifty million years of beetle evolution along the Antarctic Polar Front. *Proceedings of the National Academy of Sciences* **118**(24), e2017384118. doi:10.1073/pnas.2017384118

Baker FHU (1972) A revision of the genus *Notonophes* Sloane (Coleoptera: Curculionidae: Amycterinae). *Journal of the Australian Entomological Society* **11**, 114–128. doi:10.1111/j.1440-6055.1972.tb01612.x

Baker FHU, Thompson RT (1978) A new genus of the tribe Acantholophini (Coleoptera: Curculionidae: Amycterinae). *Journal of the Australian Entomological Society* **17**(3), 247–255. doi:10.1111/j.1440-6055.1978.tb00153.x

Baker FHU, Thompson RT (1980) A further new genus of the tribe Acantholophini (Coleoptera: Curculionidae: Amycterinae). *Journal of the Australian Entomological Society* **19**(1), 63–67. doi:10.1111/j.1440-6055.1980.tb00963.x

Barratt BIP, Cock MJW, Oberprieler RG (2018) Weevils as targets for biological control, and the importance of taxonomy and phylogeny for efficacy and biosafety. *Diversity* **10**(3), 73, 1–19. doi:10.3390/d10030073

Bedoya CL, Hofstetter RW, Nelson XJ, Hayes M, Miller DR, *et al.* (2021) Sound production in bark and ambrosia beetles. *Bioacoustics* **30**(1), 58–73. doi:10.1080/09524622.2019.1686424

Berg C (1881) Entomologisches aus dem Indianergebiet der Pampa. *Stettiner Entomologische Zeitung* **42**, 36–71.

Berg C (1898) Sustitución de nombres genéricos. *Comunicaciones del Museo Nacional de Buenos Aires* **1**(1), 16–19.

Blackburn T (1889) Further notes on Australian Coleoptera, with descriptions of new genera and species. *Proceedings of the Linnean Society of New South Wales (Series 2)* **3**, 1387–1506.

Blackburn T (1890a) Further notes on Australian Coleoptera, with descriptions of new genera and species. Part VII. *Transactions of the Royal Society of South Australia* **13**, 82–93.

Blackburn T (1890b) Notes on Australian Coleoptera, with descriptions of new species. Part VII. *Proceedings of the Linnean Society of New South Wales (Series 2)* **5**, 303–366.

Blackburn T (1890c) Notes on Australian Coleoptera, with descriptions of new species. Part VIII. *Proceedings of the Linnean Society of New South Wales (Series 2)* **5**, 553–592.

Blackburn T (1892a) Further notes on Australian Coleoptera, with descriptions of new genera and species. Part XII. *Transactions of the Royal Society of South Australia* **15**(2), 207–261.

Blackburn T (1892b) Notes on Australian Coleoptera with descriptions of new species. Part XI. *Proceedings of the Linnean Society of New South Wales (Series 2)* **7**, 65–151.

Blackburn T (1892c) Coleoptera [Scientific results of the Elder Exploring Expedition]. *Transactions of the Royal Society of South Australia* **16**, 16–61.

Blackburn T (1893a) Coleoptera (continued) [Scientific results of the Elder Exploring Expedition]. *Transactions of the Royal Society of South Australia* **16**(2), 177–202.

Blackburn T (1893b) Further notes on Australian Coleoptera, with descriptions of new genera and species. XIV. *Transactions of the Royal Society of South Australia* **17**, 294–315.

Blackburn T (1894a) Further notes on Australian Coleoptera with descriptions of new genera and species. Part XV. *Transactions of the Royal Society of South Australia* **18**, 139–168.

Blackburn T (1894b) Notes on Australian Coleoptera with descriptions of new species. Part XIII. *Proceedings of the Linnean Society of New South Wales (Series 2)* **8**, 185–208.

Blackburn T (1894c) Notes on Australian Coleoptera, with descriptions of new species. Part XIV. *Proceedings of the Linnean Society of New South Wales (Series 2)* **8**, 245–286.

Blackburn T (1895a) Further notes on Australian Coleoptera, with descriptions of new genera and species. XVII. *Transactions of the Royal Society of South Australia* **19**, 27–60.

Blackburn T (1895b) Further notes on Australian Coleoptera, with descriptions of new genera and species. Part XVIII. *Transactions of the Royal Society of South Australia* **19**, 201–258.

Blackburn T (1896) Coleoptera (exclusive of the Carabidae). In *Report on the Work of the Horn Scientific Expedition to Central Australia. Part ii. Zoology*. (Ed. B Spencer) pp. 254–308. Dulau, London.

Blackburn T (1897) Further notes on Australian Coleoptera, with descriptions of new genera and species. Part XXI. *Transactions of the Royal Society of South Australia* **21**, 28–39.

Blackburn T (1899) Further notes on Australian Coleoptera, with descriptions of new genera and species. Part XXV. *Transactions of the Royal Society of South Australia* **23**, 22–101.

Blackburn T (1900) On some new genera and species of Australian Coleoptera. *Proceedings of the Royal Society of Victoria* **12**, 206–233.

Blackburn T (1901) Further notes on Australian Coleoptera with descriptions of new genera and species. Part XXVIII. *Transactions of the Royal Society of South Australia* **25**, 15–44.

Blanchard E (1853) Description des Insectes. In *Voyage au Pole Sud et dans l'Océanie sur les corvettes l'Astrolabe et la Zélée; exécuté par ordre du Roi pendant les années 1837–1828–1839–1840, sous le commandement de M. J. Dumont d'Urville, capitaine de vaisseau; publié par ordre du government, sous la direction supérieure de M. Jacquinot, capitaine de vasseau, commandant de la Zélée. Zoologie. Tome quatrième.* (Eds JB Hombron and H Jacquinot) pp. 1–422. Gide & Baudry, Paris.

Boheman CH (1859) Coleoptera. In *Kongliga Svenska Fregatten Eugenies Resa omkring Jorden under befäl af C. A. Virgin Åren 1851–1853. Vetenskapliga iakttagelser på H. M. Konung Oscar Den Förstes befallning.* (Ed. CA Virgin) pp. 1–217. K. Svenska Vetenskaps Akademien. P. A. Norstedt & Söner, Stockholm. doi:10.5962/bhl.title.2467

Boisduval JA (1835) *Voyage de découvertes de l'Astrolabe, Exécuté par ordre du Roi, pendant les années 1826–1827–1828–1829, sous le commandement de M. J. Dumont d'Urville. Faune entomologique de l'Océan Pacifique, avec l'illustration des insectes nouveaux recueillis pendant le voyage. Deuxième Partie. Coléoptères et autres Ordres.* J. Tastu, Paris, vii + 716 pp.

Borovec R, Oberprieler RG (2013) *Afrophloeus*, a new genus of African weevils of the tribe Embrithini (Coleoptera: Curculionidae: Entiminae), with description of a new species and notes on the composition of Embrithini. *Zootaxa* **3693**(3), 365–378. doi:10.11646/zootaxa.3693.3.7

Briese DT, Thomann T, Vitou J (2002) Impact of the rosette crown weevil *Trichosirocalus briesei* on the growth and reproduction of *Onopordum* thistles. *Journal of Applied Ecology* **39**(4), 688–698.

Broun T (1880) *Manual of the New Zealand Coleoptera*. Part 1. Colonial Museum and Geological Survey Department, Wellington, pp. 1–651.

Broun T (1881) *Manual of the New Zealand Coleoptera. Part 2.* Colonial Museum and Geological Survey Department, Wellington, pp. i–viii, 653–744.

Broun T (1882) Change of nomenclature of N.Z. beetles. *New Zealand Journal of Science* **1**(3), 128.

Broun T (1886) *Manual of the New Zealand Coleoptera*. Parts 3–4. Colonial Museum and Geological Survey Department, Wellington, pp. i–xvi, 745–973.

Broun T (1893) Descriptions of new Coleoptera from New Zealand (continued). *Annals and Magazine of Natural History* Series 6, **12**(71), 374–392. doi:10.1080/00222939308677637

Broun T (1904) Descriptions of new genera and species of New Zealand Coleoptera (Concluded). *Annals and Magazine of Natural History (Series 7)* **14**(80), 105–127. doi:10.1080/03745480409442977

Broun T (1909a) Descriptions of new genera and species of New Zealand Coleoptera (Continued). *Annals and Magazine of Natural History (Series 8)* **4**(19), 51–71. doi:10.1080/00222930908692639

Broun T (1909b) Revision of the New Zealand Cossonidae, with descriptions of new genera and species. *Transactions of the New Zealand Institute* **41**, 151–215.

Broun T (1914) Descriptions of new genera and species of Coleoptera (Part II). *Bulletin of the New Zealand Institute* **1**, 79–142.

Broun T (1917) Descriptions of new genera and species of Coleoptera (Part V). *Bulletin of the New Zealand Institute* **1**, 347–474. doi:10.5962/bhl.title.17491

Browne FG (1972) Larvae of the principal old world genera of the Platypodinae (Coleoptera: Platypodidae). *Transactions of the Royal Society of London* **124**(2), 167–190. doi:10.1111/j.1365-2311.1972.tb00362.x

Buchanan LL (1935) A new genus and species of orchid weevils (Coleoptera, Curculionidae, Barinae). *Proceedings of the Hawaiian Entomological Society* **9**(1), 45–48.

Buchanan LL (1939) The species of *Pantomorus* of America north of Mexico. *United States Department of Agriculture, Miscellaneous Publication* **341**, 1–39.

Buchanan LL (1942) Four new species of white-fringed beetles (sub-genus *Graphognathus*) from the south-eastern part of the United States (Coleoptera: Curculionidae). *Bulletin of the Brooklyn Entomological Society* **37**, 107–110.

Caldara R, Franz NM, Oberprieler RO (2014) 3.7.10 Curculioninae Latreille, 1802. In *Handbook of Zoology. Volume 3. Morphology and Systematics (Phytophaga).* (Eds RG Beutel and RAB Leschen) pp. 589–592. De Gruyter, Berlin/Boston, xii + 675 pp.

Caldara R, O'Brien CW, Meregalli M (2017) A phylogenetic analysis of the aquatic weevil tribe Bagoini (Coleoptera: Curculionidae) based on morphological characters of adults. *Zootaxa* **4287**(1), 1–63. doi:10.11646/zootaxa.4287.1.1

Calder AA, Sands DPA (1985) A new Brazilian *Cyrtobagous* Hustache (Coleoptera: Curculionidae) introduced into Australia to control *Salvinia*. *Journal of the Australian Entomological Society* **24**, 57–64. doi:10.1111/j.1440-6055.1985.tb00185.x

Carter HJ (1936) Australian Coleoptera. Notes and new species. No. 10. *Proceedings of the Linnean Society of New South Wales* **61**, 98–110.

Carter HJ (1937) Some new Tenebrionidae in the South Australian Museum, together with notes and descriptions of other Australian Coleoptera. *Transactions of the Royal Society of South Australia* **61**, 121–144.

Carter HJ (1939) Australian Coleoptera. Notes and new species. No. XI (mostly Elateridae). *Proceedings of the Linnean Society of New South Wales* **64**, 297–330.

Casey TL (1892) Studies in the rhynchophorous subfamily Barinae of the Brazilian fauna. *Memoirs on the Coleoptera* **10**, 1–520.

Casey TL (1920) Some descriptive studies among the American Barinae. *Memoirs on the Coleoptera* **9**, 300–529.

Chadwick CE (1965) A check list of the Brachyderinae (Col., Curculionidae) occurring in Australia. *Journal of the Entomological Society of Australia (N.S.W.)* **2**, 21–34.

Chadwick CE (1993) The roles of *Tranes lyterioides* and *T. sparsus* Boh (col., Curculionidae) in the pollination of Macrozamia communis (Zamiaceae). In The *Biology, Structure and Systematics of the Cycadales. Proc. of Cycad*

90, the Second Int. Conf. on Cycad Biology. (Eds DW Stevenson and KJ Norstog) pp. 77–88. Palm & Cycad Societies of Australia, Milton, Australia.

Chamorro ML (2019) An illustrated synoptic key and comparative morphology of the larvae of Dryophthorinae (Coleoptera, Curculionidae) genera with emphasis on the mouthparts. *Diversity* **11**(4), 96pp. doi:10.3390/d11010004

Chamorro ML, Steiner W (2022) A new species of *Halorhynchus* from Madagascar (Coleoptera, Curculionidae, Cossoninae, Onycholipini). *ZooKeys* **1100**, 103–116. doi:10.11646/zootaxa.4287.1.1

Chamorro ML, de Medieiros BAS, Farrell BD (2021) First phylogenetic analysis of Dryophthorinae (Coleoptera, Curculionidae) based on structural alignment of ribosomal DNA reveals Cenozoic diversification. *Ecology and Evolution* **11**(5), 1984–1998. doi:10.1002/ece3.7131

Champion GC (1911) Rhynchophora. Curculionidae. Otiorhynchinae [part, 'Alatae'; and supplement to the Thecesterninae and Otiorhynchinae]. *Biologia Centrali-Americana, Insecta, Coleoptera* **4**(3), 178–240.

Champion GC (1914) Coleoptera, Curculionidae. [Reports of the Percy Sladen Trust Expedition to the Indian Ocean in 1905, under the Leadership of Mr. J. Stanley Gardner, M. A. Vol. 5. XIX]. *Transactions of the Linnean Society of London Zoology (Series 2)* **16**(4), 393–497.

Chevrolat LAA (1879a) Diagnoses de Curculionides Australiens. *Petites Nouvelles Entomologiques (2)* **305**, 309–310.

Chevrolat LAA (1879b) Diagnoses d'espèces nouvelles d'*Haplonyx*. *Le Naturaliste* **1**, 30–31, 38, 54–55, 60.

Chevrolat LAA (1881a) [Descriptions d'espèces de Curculionides de la Nouvelle-Guinée et d'Australie.]. *Annales de la Société Entomologique de France, Bulletin (Series 6)* **1**(2), LXVIII–LXIX.

Chevrolat LAA (1881b) Diagnoses de Curculionides exotiques. *Le Naturaliste* **1**(62), 494–495.

Chevrolat LAA (1885) Calandrides. Nouveaux genres et nouvelles espèces, observations, synonymies, doubles emplois de noms de genres et d'espèces, etc. 3e partie. *Annales de la Société Entomologique de France (Series 6)* **5**, 275–292.

Chûjo M, Voss E (1960) Neue Curculioniden-Subfamilie, -Gattungen und -Arten von Japan (Coleoptera, Curculionidae). *Memoirs of the Faculty of Liberal Arts and Education, Kagawa University* **2**(94), 1–17.

Clairville JP (1798) *Entomologie helvétique, ou catalogue des insectes de la Suisse rangés d'après une nouvelle méthode./Helvetische Entomologie, oder Verzeichniss der Schweizerischen Insekten, nach einer neuen Methode geordnet, mit Beschreibungen und Abbildungen.* Orell, Füssli & Compagnie, Zürich, 149 pp. doi:10.5962/bhl.title.15822

Colonnelli E (1979) Osservazioni sulla nomenclatura e sulla posizione sistematica di alcune entita Olartiche di Ceutorhynchinae. *Fragmenta Entomologica* **15**(1), 209–232.

Colonnelli E (1997) New Holarctic species of Ceutorhynchinae (Coleoptera: Curculionidae). *Beiträge zur Entomologie* **47**(2), 377–397.

Colonnelli E (2004) *Catalogue of Ceutorhynchinae of the world, with a key to genera (Insecta: Coleoptera: Curculionidae).* Argania editio, Barcelona, 124 pp.

Creutzer C (1799) *Entomologische Versuche.* Karl Schaumburg und Comp., Wien, 149 pp, 3 pls.

Curtis J (1839) *British Entomology; being Illustrations and Descriptions of the Genera of Insects found in Great Britain and Ireland: Containing Coloured Figures from Nature of the Most Rare and Beautiful Species, and in Many Instances of the Plants Upon Which they are found. Vol. 16.* The Author, London, 3 parts, 48 pls. [parts 181–192, pls. 722–769].

Dejean PFMA (1821) *Catalogue de la collection de Coléoptères de M. le Baron Dejean.* Crevot, Paris, i–viii, 1–138 pp. doi:10.5962/bhl.title.11259

Dejean PFMA (1835) Catalogue des coléoptères de la collection de M. le Comte Dejean. Deuxième edition. Livraison 4. Méquignon-Marvis Père et Fils, Paris, pp. 257–360.

Dhileepan K, McFadyen REC (2012) *Parthenium hysterophorus L. – parthenium.* In *Biological Control of Weeds in Australia.* (Eds MH Julien, REC McFadyen and JM Cullen) pp. 448–462. CSIRO Publishing, Collingwood, Victoria.

Dietz WG (1894) Revision of the genera and species of Desmori of North America. *Transactions of the American Entomological Society* **21**(2), 113–178.

Dohrn CA (1872) Exotisches. 4. Zur Rüsselkäfergruppe der Amycteriden [continued]. *Entomologische Zeitung* **33**(4–6), 143–165.

Donovan E (1805) *An Epitome of the Natural History of the Insects of New Holland, New Zealand, New Guinea, Otaheite, and other Islands in the Indian, Southern, and Pacific Oceans: including Figures and Descriptions of One Hundred and Fifty-three Species of the more Splendid, Beautiful, and Interesting Insects, hitherto Discovered in those Countries, and which for the most Part have not appeared in the Works of any Preceding Author.* Rivington, London, iv + [167] pp, [41] pls.

Eggleton RA, Taylor G (2008) Effects of some macrobiota on the Weipa Bauxite, northern Australia. *Australian Journal of Earth Sciences* **1**(55) (S1), S71–82.

Elliott HJ, Wylie FR, Ohmart CP (1998) *Insect Pests of Australian Forests: Ecology and Management.* Inkata Press, Melbourne, 214 pp.

Elston AH (1919) Australian Coleoptera. Part l. *Transactions of the Royal Society of South Australia* **42**, 343–348.

Erichson WF (1842) Beitrag zur Insecten-Fauna von Vandiemensland, mit besonderer Berücksichtigung der geographischen Verbreitung der Insecten. *Archiv für Naturgeschichte* **8**(1), 83–287. doi:10.5962/bhl.part.21656

Escalona HE, Oberprieler RG (2021) *Undarobius*, a new genus of cavernicolous weevils (Curculionidae: Entiminae: Leptopiini) from the Undara Lava Caves in north-eastern Australia, with an overview of anophthalmic and microphthalmic Australian Curculionidae. *Zootaxa* **5023**(2), 207–222. doi:10.11646/zootaxa.5023.2.2

Escalona HE, Jennings D, Oberprieler RO (2023) A dung-beetle impostor: revision of the Australian weevil genus *Tentegia* Pascoe and the dung-rolling behaviour of *Tentegia stupida* (Fabricius) (Curculionidae: Molytinae: Cryptorhynchini). *Annales Zoologici* **73**(1), 11–151. doi:10.3161/00034541ANZ2023.73.1.010

Esler KJ, van Wilgen BW, te Roller KS, Wood AR, van der Merwe JH (2010) A landscape-scale assessment of the long-term integrated control of an invasive shrub in South Africa. *Biological Invasions* **12**(1), 211–218. doi:10.1007/s10530-009-9443-2

Fabricius JC (1775) *Systema entomologiae, sistens insectorum classes, ordines, genera, species, adiectis synonymis, locis, descriptionibus, observationibus.* Officina Libraria Kortii, Flensburgi et Lipsiae, xxxii + 832 pp. doi:10.5962/bhl.title.36510

Fabricius JC (1787) *Mantissa insectorum sistens eorum species nuper detectas adiectis characteribus genericus, differentiis specificis, emendationibus, observationibus. Tom. I.* C. G. Proft, Hafniae, xx + 348 pp. doi:10.5962/bhl.title.11657

Fabricius JC (1792) *Entomologia systematica emendata et aucta. Secundum classes, ordines, genera, species adjectis synonymis, locis, observationibus, descriptionibus. Tomus I. Pars II.* C. G. Proft, Hafniae, xx + 538 pp.

Fabricius JC (1801) *Systema eleutheratorum secundum ordines, genera, species: adiectis synonymis, locis, observationibus, descriptionibus. Tomus II.* Bibliopolii Academici Novi, Kiliae, 687 pp.

Fairmaire L (1849) Essai sur les Coléoptères de la Polynésie. *Revue et Magasin de Zoologie Pure et Appliquée, (Series 2)* **1**, 277–291, 352–365, 410–422, 445–460, 504–516, 550–559.

Fairmaire L (1858) Miscellanea entomologica. Deuxième partie. *Annales de la Société Entomologique de France (Series 3)* **5**(4), 725–745.

Fairmaire L (1881) Essai sur les Coléoptères des îles Viti (Fidgi). *Annales de la Société Entomologique de France (Series 6)* **1**, 243–318.

Faust J (1881) Beiträge zur Kenntniss der Käfer des Europäischen und Asiatischen Russlands mit Einschluss der Küsten des kaspischen Meeres (3. Fortsetzung). *Horae Societatis Entomologicae Rossicae* **16**(3–4), 285–333.

Faust J (1886) Neue exotische Rüsselkäfer. *Deutsche Entomologische Zeitschrift* **30**(2), 337–372.

Faust J (1887) Zur Gattung *Echinocnemus* Sch. *Stettiner Entomologische Zeitung* **48**, 187–192.

Faust J (1888) Neue Rüsselkäfer aller Länder. *Stettiner Entomologische Zeitung* **49**, 284–311.

Faust J (1892a) Notizen über Rüsselkäfer. *Stettiner Entomologische Zeitung* **53**(1–3), 44–52.

Faust J (1892b) Fünf neue Curculioniden von Australien. *Stettiner Entomologische Zeitung* **53**, 179–184.

Faust J (1892c) Curculioniden aus dem Malayischen Archipel. *Stettiner Entomologische Zeitung* **53**, 184–228.

Faust J (1894) Viaggio di Leonardo Fea in Birmania e regioni vicine. LX. Curculionidae. *Annali del Museo Civico di Storia Naturale Genova* **34**, 153–370. doi:10.5962/bhl.title.48832

Faust J (1895) Sechs neue Curculioniden-Gattungen und ein neuer *Glochinorhinus*. *Stettiner Entomologische Zeitung* **56**(7–9), 212–228.

Faust J (1896) Reise von E. Simon in Venezuela. Curculionidae. Pars tertia. *Stettiner Entomologische Zeitung* **57**, 33–136.

Faust J (1897) Neue Gattungen und Arten in der Celeuthetiden-Gruppe. *Stettiner Entomologische Zeitung* **58**, 229–298.

Faust J (1898) Beitrag zur Kenntniss der Fauna von Kamerun, mit besonderer Berücksichtigung der afrikanischen Menemachiden, Isorhynchiden und Campylosceliden. *Deutsche Entomologische Zeitschrift* **1898**(1), 17–91.

Faust J (1899) Viaggio di Lamberto Loria nella Papuasia Orientale. XXIII. Curculionidae (Curculionidae von Neu Guinea). *Annali del Museo Civico di Storia Naturale Genova* **40**, 1–130.

Fauvel A (1862) Coléoptères de la Nouvelle-Calédonie, recueillis par M.\E. Déplanche, chirurgien de la marine impériale, 1858–59–60. *Notices Entomologiques* **1**, 1–68.

Ferguson EW (1909) Revision of the Amycteridae. *Proceedings of the Linnean Society of New South Wales* **34**, 524–585.

Ferguson EW (1912) Revision of the Amycterides. Part ii. *Talaurinus*. *Proceedings of the Linnean Society of New South Wales* **37**(1), 83–135.

Ferguson EW (1913) Revision of the Amycterides. Part ii. *Talaurinus* (continued). *Proceedings of the Linnean Society of New South Wales* **38**(2), 340–394.

Ferguson EW (1914a) Revision of the Amycterides. Part iii. *Notonophes*, *Macramycterus*, and genera allied to *Talaurinus*. *Proceedings of the Linnean Society of New South Wales* **39**(2), 217–252.

Ferguson EW (1914b) Notes on the Amycterides in the South Australian Museum, with descriptions of new species. Part I. *Transactions of the Royal Society of South Australia* **38**, 11–40.

Ferguson EW (1914c) Notes on Amycterides, with descriptions of new species. Part I. *Proceedings of the Royal Society of Victoria (New Series)* **26**, 243–255.

Ferguson EW (1915a) Notes on Amycterides, with descriptions of new species. Part II. *Proceedings of the Royal Society of Victoria (New Series)* **27**, 232–260.

Ferguson EW (1915b) Notes on the Amycterides in the South Australian Museum, with descriptions of new species. Part II. *Transactions of the Royal Society of South Australia*, **39** 57–90.

Ferguson EW (1916a) Revision of the Amycterides. Part IV. *Sclerorinus* (Section I). *Proceedings of the Linnean Society of New South Wales* **40**, 685–718.

Ferguson EW (1916b) Revision of the Amycterides. Part IV. *Sclerorinus* (Section II). *Proceedings of the Linnean Society of New South Wales* **40**, 759–805.

Ferguson EW (1916c) Revision of the Amycterides. Part V. *Molochtus* and *Cubicorrhynchus*. *Proceedings of the Linnean Society of New South Wales* **41**, 422–452.

Ferguson EW (1921a) Notes on Amycterides, with descriptions of new species, Part III. *Proceedings of the Royal Society of Victoria (New Series)* **34**(1), 17–32.

Ferguson EW (1921b) Revision of the Amycterides. Part VI. *Acantholophus*. *Proceedings of the Linnean Society of New South Wales* **46**(1), 19–75.

Ferguson EW (1921c) Revision of the Amycterides. Part VII. *Hyborrhynchus* and allied genera. *Proceedings of the Linnean Society of New South Wales* **46**(3), 393–406.

Ferguson EW (1923) Revision of the Amycterides (Coleoptera). Part VIII. The Euomides. *Proceedings of the Linnean Society of New South Wales* **48**(3), 381–435.

Fiedler C (1940) Monograph of the South American Weevils of the Genus *Conotrachelus*. British Museum (Natural History), London, v + 365 pp.

Fierro A, Vergara PM, Elgueta M, Carvajal MA, Alaniz AJ (2023) A saproxylic weevil acts as an ecosystem engineer: Impacts across multiple trophic levels. *Forest Ecology and Management* **527**, 120603. doi:10.1016/j.foreco.2022.120603

French C (1893) *A Handbook of the Destructive Insects of Victoria, Part II. With Notes on the Methods to be Adopted to Check and Extirpate them.* Robt. S. Brain, Government Printer, Melbourne, 222 pp.

French C (1909) *A Handbook of the Destructive Insects of Victoria, Part IV. With Notes on the Methods to be Adopted to Check and Extirpate them.* Osboldstone & Co. Printers, Melbourne, 195 pp.

Froggatt WW (1893) On the life-histories ol Australian Coleoptera. Part. 1. *Proceedings of the Linnaean Society, New South Wales* **8**, 27–42

Froggatt WW (1895) Life histories of Australian Coleoptera. Part III. *Proceedings of the Linnaean Society of New South Wales (2)* **10**, 325–336.

Froggatt WW (1896) The entomology of the grass-trees (*Xanthorrhoea*). *Proceedings of the Linnean Society of New South Wales* **21**, 74–87.

Froggatt WW (1900) Entomological notes on specimens received during 1899. *Agricultural Gazette of New South Wales* **11**, 639–651.

Froggatt WW (1904) Some fern and orchid pests. *Agricultural Gazette of New South Wales* **15**, 514–518.

Froggatt WW (1907) *Australian Insects*. William Brooks & Company Ltd, Sydney, 449 pp.

Froggatt WW (1912) A weevil (*Aesiotes leucurus*) destructive to Pine trees (*Pinus halepensis*). *Agricultural Gazette of New South Wales* **23**, 55–56.

Froggatt WW (1923) *Forest Insects of Australia*. Alfred James Kent, Government Printer, Sydney, 171 pp.

Froggatt WW (1926) Forest insects. No. 25. Shot hole borers (ambrosia beetles) (belonging to the genus *Platypus*). *Australian Forestry Journal* **9**, 256–260.

Froggatt WE (1927) *Forest Insects and Timber Borers*. Alfred James Kent, Government Printer, Sydney, 107 pp.

Frölich JA (1792) Bemerkungen über einige seltene Käfer aus der Insectensammlung des Herrn Hofr. und Prof. Rudolph in Erlangen. *Der Naturforscher* **26**, 68–165.

Gahan CJ (1900) Order 4. Coleoptera. In *A Monograph of Christmas Is. (Indian Ocean): Physical Features and Geology; with Descriptions of the Fauna and Flora by Numerous Contributors*. (Ed. CW Andrews) pp. 89–127. British Museum (Natural History), London. doi:10.5962/bhl.title.25555

Garcia A, Allen GR, Oberprieler RG, Ramos AP, Valente C, *et al.* (2019) Biological control of *Gonipterus*: uncovering the associations between eucalypts, weevils and parasitoids in their native range. *Forest Ecology and Management* **443**, 106–116. doi:10.1016/j.foreco.2019.04.004

Gardner JCM (1932) Immature stages of Indian Coleoptera (10) (Anthribidae), (ii) (Platypodidae). *Indian Forest Records Calcutta* **16**, 327–334.

Gardner JCM (1934) Immature stages of Indian Coleoptera (14) (Curculionidae). *Indian Forest Records Delhi* **20**(2), 1–48.

Gardner JCM (1938) Immature stages of Indian Coleoptera (24) (Curculionidae cont.). *Indian Forest Records (NS) Entomology* **3**, 227–261.

Gay C (1849) *Historia Fisica y Politica de Chile. Zoologia. Tomo cuarto.* C. Gay, Paris & Santiago, 511pp, 32pls.

Germain P (1895) Apuntes sobre los insectos de Chile. I. Estudio i descripción de los Listroderitos de Chile i tierras magallánicas de la colección del Museo Nacional i de la de Don Fernando Paulsen (Continuación). *Anales de la Universidad de Chile* **91**, 53–104.

Germar EF (1817) Miscellen und Correspondenz-Nachrichten. *Magazin der Entomologie* **2**, 339–341.

Germar EF (1822) *Fauna Insectorum Europae. Fasciculus 7*. Car. Aug. Kümmelii, Halae [Halle], 1 + 25 pp, 25 pls.

Germar EF (1824) *Insectorum species novae aut minus cognitae, descriptionibus illustratae: Coleoptera. Volumen Primum*. J. C. Hendel et Filii, Halae (Halle), xxiv + 624 pp.

Germar EF (1848) Beiträge zur Insektenfauna von Adelaide. *Linnaea Entomologica* **3**, 153–247.

Gillett C, Lyal C, Vogler A, Emerson B (2018) Statistical evaluation of monophyly in the 'broad-nosed weevils' through molecular phylogenetic analysis combining mitochondrial genome and single-locus sequences (Curculionidae: Entiminae, Cyclominae, and Hyperinae). *Diversity* **10**(2), 1–21 doi:10.3390/d10020021

Gillett C, Crampton-Platt A, Timmermans M, Jordal B, Brent CE *et al.* (2014) Bulk de novo mitogenome assembly from pooled total DNA elucidates the phylogeny of weevils (Coleoptera: Curculionoidea). *Molecular Biology and Evolution* **31**(8), 2223–2237. doi:10.1093/molbev/msu154

Goeze JAE (1777) *Entomologische Beyträge zu des Ritter Linné zwölften Ausgabe des Natursystems. Erster Theil*. Weidmanns Erben und Reich, Leipzig, xvi + 736 pp. doi:10.5962/bhl.title.45974

Gordon AJ (1999) A review of established and new insect agents for the biological control of *Hakea sericea* Schrader (Proteaceae) in South Africa. *African Entomology Memoir* **1**, 35–43.

Grebennikov V (2010) First *Alaocybites* weevil (Insecta: Coleoptera: Curculionoidea) from the Eastern Palaearctic: a new microphthalmic species and generic relationships. *Arthropod Systematics & Phylogeny* **68**(3), 331–365. doi:10.3897/asp.68.e31734

Guérin-Méneville FE (1833) *Iconographie du règne animal de G. Cuvier, ou représentation d'après nature de l'une des espèces les plus remarquables, et souvent non figurées, de chaque genre d'animaux, avec un texte descriptif mis au courant de la science. Ouvrage pouvant servir d'atlas a tous les traités de zoologie. Tome VII. Insectes*. Baillière, Paris, 110 pls. [Plates 36–39 bis published in 1833, plate 40 in 1835, after Cowan, 1971; the text only completed in 1844].

Guérin-Méneville FE (1834) *Iconographie du règne animal de G. Cuvier, ou représentation d'après nature de l'une des espèces les plus remarquables et souvent non figurées de chaque genre d'animaux. Tome II: Planches des Animaux invertébrés*. Baillière, Paris, 227 pl.

Gunter N, Oberprieler RG, Cameron SL (2016) Molecular phylogenetics of Australian weevils (Coleoptera: Curculionoidea): exploring relationships in a hyperdiverse lineage through comparison of independent analyses. *Austral Entomology* **55**(2), 217–233. doi:10.1111/aen.12173

Hall JA, Walter GH (2018) Pollination of the Australian cycad *Cycas ophiolitic* (Cycadaceae): the limited role of wind pollination in a cycad with beetle pollinator mutualists, and its ecological significance. *Journal of Tropical Ecology* **34**(2), 121–134. doi:10.1017/s0266467418000111

Hall JA, Walter GH, Bergstrom DM, Machin P (2004) Pollination ecology of the Australian cycad *Lepidozamia peroffskyana* (Zamiaceae). *Australian Journal of Botany* **52**(3), 333–343. doi:10.1071/BT03159

Hamilton AG (1897) On the fertilisation of *Eupomatia laurina* R. *Proceedings of the Linnean Society of New South Wales* **22**, 48–55. doi:10.5962/bhl.part.12702

Haran J, Li X, Shin S, Benoit L, Oberprieler RG *et al.* (2023) Phylogenomics illuminates the phylogeny of flower weevils (Curculioninae) and reveals ten independent origins of brood-site pollination mutualism in true weevils. *Proceedings of the Royal Society Biological Sciences, Series B* **290**(2008). doi:10.1098/rspb.2023.0889

Hartmann F (1901) Neue Rüsselkäfer aus Kaiser-Wilhelms-Land. *Deutsche Entomologische Zeitschrift* **2**, 273–295. doi:10.1002/MMND.48019000215

Hawkeswood TJ (1985) Notes on some beetles (Coleoptera) associated with *Xanthorrhoea johnsonii* (Xanthorrhoeaceae) in the Brisbane area, southeast Queensland. *Victorian Naturalist* **102**(5), 162–166.

Hawkeswood TJ (1991) Review of the history, biology and host plants of the Australian weevil *Chrysolopus spectabilis* (Fabricius) (Coleoptera: Curculionidae: Aterpinae). *Spixiana* **14**(1), 17–25.

Heard TA, Burcher JA, Forno IW (1999) *Chalcodermus serripes* (Coleoptera: Curculionidae) for biological control of *Mimosa pigra*: host relations and life cycle. *Biological Control* **15**(1), 1–9.

Heller KM (1893) Zygopiden-Studien mit besonderer Berücksichtigung der Gattung *Mecopus*. *Abhandlungen und Berichte des Königlichen Zoologischen und Anthropologisch-Ethnographischen Museums zu Dresden* **4**(2), 1–48.

Heller KM (1900) Systematische Aufzählung der Coleopteren. Mit Neubeschreibungen von Arten von E. Brenke und J. Faust. *Denkschriften der Medicinisch-Naturwissenschaftlichen Gesellschaft zu Jena* **8**, 615–626 (also paginated 99–110). [*in* Semon, R. (1894–1903) Zoologische Forschungsreisen in Australien und dem Malayischen Archipel mit Unterstützung des Herrn Paul von Ritter, ausgeführt in den Jahren 1891–1893 von Dr. Richard Semon. Issued in 16 parts in *Denkschriften der Medicinisch-Naturwissenschaftlichen Gesellschaft zu Jena*, 4–8.]

Heller KM (1912) Philippinische Rüsselkäfer. *Philippine Journal of Science D* 7(6), 347–415.

Heller KM (1914) Neue papuanische Käfer. *Deutsche Entomologische Zeitschrift* **1914**(3), 305–318.

Heller KM (1916) Die Käfer von Neu-Caledonien und den benachbarten Inselgruppen. In *Nova Caledonia. Forschungen in Neu-Caledonien und auf den Loyalty-Inseln. A. Zoologie. Vol. II, L. III*. (Eds F Sarasin and J Roux) pp. 229–365. C. W. Kreidel, Wiesbaden, pls. X–XI. doi:10.5962/bhl.title.6790

Heller KM (1918) Die philippinischen Arten der Rüsselkäfergattung *Alcides* Schönh. *Stettiner Entomologische Zeitung* **78**(2), 209–245.

Heller KM (1922) New Philippine Coleoptera. *Philippine Journal of Science* **19**(5), 523–627.

Heller KM (1927) Studien zur Systematik altweltlicher Balaninini II. *Stettiner Entomologische Zeitung* **88**, 175–288.

Heller KM (1929) Fauna Buruana. Coleoptera, Fam. Curculionidae. *Treubia* 7(Supplement), 491–524 [105–138 in reprint].

Heller KM (1940) Indomalayische Rüsselkäfer II (Coleoptera: Curculionidae). *Arbeiten über morphologische und taxonomische Entomologie* **7**(2), 92–113.

Henderson LS (1940) A revision of the genus *Listronotus* (Curculionidae, Coleoptera). *University of Kansas Science Bulletin* **26**(4), 215–337.

Herbst JFW (1783) Kritisches Verzeichniss meiner Insektensammlung. *Archiv der Insectengeschichte* **4**, 1–72.

Herbst JFW (1795) *Natursystem aller bekannten in- und ausländischen Insekten, als eine Fortsetzung der von Büffonschen Naturgeschichte. Der Käfer sechster Theil. Mit 38 illuminirten Kupfertafeln*. Pauli, Berlin, xxiv + 520 pp, pl. LX–XCV.

Herbst JFW (1797) *Natursystem aller bekannten in- und ausländischen Insekten; als eine Fortsezzung der von Büffonschen Naturgeschichte. Der Käfer siebenter Theil. Mit 26 illuminirten Kupfertafeln*. Pauli, Berlin, xi + 346 pp, pl. XCVI–CXVI.

Hope FW (1835) Description of a new curculionideous beetle from the Swan River (*Amycterus Schoenherri*). *Transactions of the Entomological Society of London* **2**(1), 68–69. doi:10.1111/j.1365-2311.1837.tb03183.x

Horn GH (1873) Contribution to a knowledge of the Curculionidae of the United States. *Proceedings of the American Philosophical Society* **13**, 407–469.

Howden AT (1986) Notes on the biology of adult and immature Amycterinae (Coleoptera, Curculionidae). *Proceedings of the Linnaean Society of NSW* **109**(2), 91–105.

Howden AT (1995) Structures related to oviposition in Curculionoidea. *Memoirs of the Entomological Society of Washington* **14**, 53–102.

Hsaio Y, Liao JR (2024) Predicting habitat suitability for the Australian cycad-attacking weevil (*Siraton internatus*) under climate change. *Biological Invasions* **8**, 1–6. doi:10.1007/s10530-024-03330-5

Hsiao Y, Oberprieler RG (2020a) Bionomics and rearing of *Miltotranes prosternalis* (Lea, 1929) (Coleoptera: Curculionidae), a mutualistic cycad pollinator in Australia. *Entomological Science* **23**(4), 369–73. doi:10.1111/ens.12434

Hsiao Y, Oberprieler RG (2020b) A review of the trunk-boring cycad weevils in Australia, with description of a second species of *Demyrsus* Pascoe, 1872 (Coleoptera: Curculionidae). *Austral Entomology* **59**(4), 677–700. doi:10.1111/aen.12498

Hsiao Y, Oberprieler RG (2021) A review of *Paratranes* Zimmerman, 1994, *Xanthorrhoea*-associated weevils of the *Tranes* group (Coleoptera, Curculionidae, Molytinae), with description of a new species. *European Journal of Taxonomy*, **767**(1), 117–141. doi:10.5852/ejt.2021.767.1493

Hsiao Y, Oberprieler RG (2022) Taxonomic revision of the genus *Miltotranes* Zimmerman, 1994 (Coleoptera: Curculionidae: Molytinae), the *Bowenia*-pollinating cycad weevils in Australia, with description of a new species and implications for the systematics of *Bowenia*. *Insects* 2022 May 12; **13**(5), 456. doi:10.3390/insects13050456

Hsiao Y, Oberprieler RG (2024) An integrative taxonomic and phylogenetic approach reveals a new genus of Australasian *Cycas*-pollinating weevils (Coleoptera: Curculionidae: Cossoninae). *Zoological Journal of the Linnean Society* **2024**, zlad190. doi:10.1093/zoolinnean/zlad190

Hsiao Y, Oberprieler RG, Zwick A, Zhou L, Slipinski SA (2023) Museomics unveil systematics, diversity and evolution of Australian cycad-pollinating weevils. *Proceedings of the Royal Society Biological Sciences, Series B* **290**. doi:10.1098/rspb.2023.1385

Hubenthal W (1917) Beiträge zur Kenntnis der Curculioniden-Gattung *Acicnemis* Lacordaire (Col.). *Archiv für Naturgeschichte* **83A**(8), 92–160.

Hustache A (1920) Contribution à l'étude des Ceuthorrhynchini (Col. Curculionidae). *Annales de la Société Entomologique de France* **88**(3–4), 329–344.

Hustache A (1926) Contribution à l'étude des Curculionides de la Republique Argentine (première note). *Anales del Museo Nacional de Historia Natural 'Bernardino Rivadavia'* **34**, 155–261.

Hustache A (1929) Nouveaux Curculionides de l'Amérique du sud. *Revista de la Sociedad Entomologica Argentina* **10**, 227–232.

Hustache A (1934) Un curculionide (Col.) nouveau de l'Île d'Elbe. *Bulletin de la Société Entomologique de France* **39**(17), 250–253.

Hustache A (1936) Curculionidae: Cryptorrhynchinae. In *Coleopterorum Catalogus. Pars 151.* (Eds W Junk and S Schenkling). W. Junk, 's-Gravenhage, 317 pp.

Hustache A (1937) Magdalinae (Coleoptera, Curculionidae) de l'Amérique meridionale. *Annals and Magazine of Natural History (Series 10)* **19**(110), 198–248. doi:10.1080/00222933708655256

Hustache A (1938) Curculionidae: Barinae. In *Coleopterorum Catalogus. Pars 163.* (Eds W Junk and S Schenkling) W. Junk, 's-Gravenhage, 219 pp.

Hustache A (1939) Curculionides nouveaux de l'Argentine et autres régions sud-americaines. *Anales de la Sociedad Científica Argentina* **128**, 38–124.

International Commission on Zoological Nomenclature (1991) Opinion 1655. *Curculio viridicollis* Fabricius, 1792 (currently *Phyllobius viridicollis*; Insecta, Coleoptera): specific name conserved, and *Rhyncolus Germar*, 1817: Curculio ater Linnaeus, 1758 designated as the type species. *Bulletin of Zoological Nomenclature* **48**(3), 268–269.

Illiger JCW (1798) *Verzeichniss der Käfer Preussens. Entworfen von Johann Gottlieb Kugelann.* Halle, Gerbauer, 510 pp.

Janczyk F (1957) Neue Curculioniden der zoologischen Sammlung des Naturhistorischen Museums. 1. Beitrag zur Kenntnis der Curculionidae. *Annalen des Naturhistorischen Museums in Wien* **61**, 241–248.

Janczyk F (1959) Neue Curculioniden der zoologischen Abteilung des Naturhistorischen Museums. 2. Beitrag zur Kenntnis der Curculionidae. *Annalen des Naturhistorischen Museums in Wien* **63**, 430–435.

Janczyk F (1966) Ein neuer *Laemosaccus* aus Neu-Südwales (Australien). 7. Beitrag zur Kenntnis der Curculionidae. *Annalen des Naturhistorischen Museums in Wien* **63**, 213–214.

Jeannel R (1940) Croisière du Bougainville aux îles australes françaises. III. Coléoptères. *Mémoires du Muséum National d'Histoire Naturelle (Nouvelle Series)* **14**, 63–201.

Jekel H (1865) Recherches sur la classification naturelle des Curculionides. 1e partie. *Annales de la Société Entomologique de France (Series 4)* **4**(3), 537–566.

Jekel H (1875) *Coleoptera Jekeliana, adjecta Eleutheratorum Bibliotheca. Enumération systématique & synonymique des Coléoptères européens & exotiques composant la collection d'Henri Jekel. Observations critiques. Descriptions d'espèces nouvelles. Reproduction et traduction de genres et espèces publiés dans des ouvrages rares français et etrangers de manière à former insensiblement la bibliothèque du coléoptériste. Livraison 2.* Autographie Typomorphe de l'Auteur, Paris, pp. 101–196.

Jennings D, Oberprieler RO (2018) A review of the tribe Cryptoplini (Coleoptera: Curculioninae), with revision of the genus *Menechirus* Hartmann, 1901 and description of a new genus associated with *Macadamia*. *Diversity* **10**(71), 1–32. doi:10.3390/d10030071

Jones T, Potts BM (2000) Genetic variation in susceptibility of *Eucalyptus globulus* coppice regrowth to shoot-feeding weevils (*Myllorhinus* spp.) *Tasforests* **12**, 147–153.

Jordal BH (2014) 3.7.11 Cossoninae Schoenherr, 1825. In *Handbook of Zoology, Coleoptera, Beetles Volume 3: Morphology and Systematics (Phytophaga).* (Eds RG Beutel and RAB Leschen) pp. 629–633. De Gruyter, Berlin/Boston, xii + 675 pp.

Julien MH, Matthews NC (1980) The biology of *Rhinoncus australis* Oke (Coleoptera: Curculionidae), a weevil attacking the weed *Emex australis* Steinheil in eastern Australia. *The Australian Entomologist*, **6**(5), 87–89.

Juniper PA, Britton DR (2010) Insects associated with the fruit of *Syzygium paniculatum* (Magenta Lillypilly) and *Syzygium australe* (Brush Cherry). *Australian Journal of Entomology* **49**(4), 296–303. http://onlinelibrary.wiley.com/journal/10.1111

Karasyov VP, Okrajko TV (1998) Contribution to knowledge of the Oriental and Australian species of *Smicronyx* Schoenherr (Insecta: Coleoptera: Curculionidae: Erirrhininae). *Reichenbachia* **32**, 277–285.

Kirby W (1819a) A century of insects, including several new genera described from his cabinet. *Transactions of the Linnean Society of London* **12**, 375–453. doi:10.1111/j.1095-8339.1817.tb00239.x

Kirby W (1819b) A description of several new species of insects collected in New Holland by Robert Brown, Esq. F.R.S. Lib. Linn. Soc. *Transactions of the Linnean Society of London* **12**, 454–482. doi:10.1111/j.1095-8339.1817.tb00240.x

Kirsch T (1877) Beitrag zur Kenntniss der Coleopteren-Fauna von Neu Guinea. *Mitteilungen aus dem K. Zoologischen Museum zu Dresden* **2**, 135–161.

Klug JCF (1829) *Preis-Verzeichniss vorräthiger Insectendoubletten des Königl. zoologischen Museums der Universität.* Berlin-Dahlen, 18 pp.

Kojima H, Lyal CHC (2002) New Oriental and Australian Conoderinae, with taxonomic notes on the tribe Othippiini (Coleoptera: Curculionidae). *Esakia* **42**, 161–174. doi:10.5109/2666

Kolbe HJ (1910) Beitrage zur Kenntniss der Brenthiden. I. Die Gruppe Cyphagoginae. *Stettiner Entomologische Zeitung* **53**, 162–175.

Kolbe HJ (1911) Ein neuer hygrophiler Rüsselkäfer aus Argentinien. (Col.). *Deutsche entomologische Zeitschrift* **3**, 315–317.

Korotyaev BA (1981) Novye i maloizvestnye dolgonosiki podsem. Ceutorhynchinae (Coleoptera, Curculionidae) iz palaearkticheskoj, indo-malajskoj i avstralijskoj oblastej. *Entomologicheskoe Obozrenie* **60**(1), 126–159. [*Also published as*: New and little-known weevils of the subfamily Ceutorhynchinae (Coleoptera, Curculionidae) from the Palearctic, Indo-Malayan and Australian regions. *Entomological Review* **60**(1), 112–146.]

Kuschel G (1945) Aportes entomológicos I (Curculionidae). *Anales de la Sociedad Científica Argentina* **139**, 120–136.

Kuschel G (1955) Nuevas sinonimias y anataciones sobre Curculionoidea. *Revista Chilena de Entomologia* **4**, 261–312.

Kuschel G (1964) Insects of Campbell Island. Coleoptera: Curculionidae of the Subantarctic Islands of New Zealand. *Pacific Insects Monograph* **7**, 416–493.

Kuschel G (1971) Curculionidae. In *Marion and Prince Edward Islands. Report on the South African Biological and Geological Expedition 1965–1966*. (Eds EM Van Zinderen Bakker, JM Winterbottom and RA Dyer) pp. 355–359. A.A. Balkema, Cape Town.

Kuschel G (1972) The Australian Phrynixinae. *New Zealand Journal of Science* **15**(2), 209–231.

Kuschel G (1982) Apionidae and Curculionidae (Coleoptera) from the Poor Knights Islands, New Zealand. *Journal of the Royal Society of New Zealand* **12**(3), 273–282. doi:10.1080/03036758.1982.10415349

Kuschel G (1990) Beetles in a suburban environment: a New Zealand case study. The identity and status of Coleoptera in the natural and modified habitats of Lynfield, Auckland (1974–1989). *DSIR Plant Protection Report* **3**, 118 pp.

Kuschel G (1995) A phylogenetic classification of Curculionoidea to families and subfamilies. *Memoirs of the Entomological Society of Washington* **14**, 5–33.

Kuschel G (2008) Curculionoidea (weevils) of New Caledonia and Vanuatu: basal families and some Curculionidae. In *Zoologia Neocaledonica 6. Biodiversity Studies in New Caledonia* (Ed. P Grandcolas) pp. 99–249. *Mémoires du Muséum National d'Histoire Naturelle* **197**, 326 pp.

Kuschel G (2009) [2008] New tribe, new genus and species for an Australasian weevil group with notes and keys (Coleoptera, Curculionidae). *Revue Française d'Entomologie (Nouvelle Series)* **30**(2–4), 41–66.

Kuschel G (2014a) The New Caledonian and Fijian species of Aterpini and Gonipterini (Coleoptera: Curculionidae). In *Zoologia Neocaledonica 8. Biodiversity Studies in New Caledonia* (Eds É Guilbert, T Robillard, H Jourdan, P Grandcolas) pp. 133–163. *Mémoires du Muséum National d'Histoire Naturelle* **206**, 315 pp.

Kuschel G (2014b) The blind weevils of Myrtonymina in New Caledonia and Australia (Curculionidae: Curculioninae: Erirhinini: Myrtonymina). In *Zoologia Neocaledonica 8. Biodiversity Studies in New Caledonia* (Eds É Guilbert, T Robillard, H Jourdan, P Grandcolas) pp. 165–180. *Mémoires du Muséum National d'Histoire Naturelle* **206**, 315 pp.

Lacordaire T (1863) *Histoire naturelle des Insectes. Genera des Coléoptères ou exposé méthodique et critique de tous les genres proposés jusqu'ici dans cet ordre d'insectes. Tome sixième.* Roret, Paris, 637 pp. doi:10.5962/bhl.title.67686

Latreille PA (1810) *Considérations générales sur l'ordre naturel des animaux, composant les classes des Crustacés, des Arachnides, et des Insectes; avec un tableau méthodique de leurs genres, disposés en familles.* F. Schoell, Paris, 444 pp. doi:10.5962/bhl.title.39620

Lawrence JF, Ślipiński SA (2013) *Australian Beetles. Vol. 1. Morphology, Classification and Keys.* CSIRO Publishing, Collingwood, Victoria, viii + 561 pp.

Lea AM (1895a) Descriptions of new species of Australian Coleoptera. *Proceedings of the Linnean Society of New South Wales (Series 2)* **9**, 589–634.

Lea AM (1895b) Descriptions of new species of Australian Coleoptera. Part II. *Proceedings of the Linnean Society of New South Wales (Series 2)* **10**, 224–319.

Lea AM (1896) Descriptions of new species of Australian Coleoptera. Part III. *Proceedings of the Linnean Society of New South Wales* **21**(3), 284–319.

Lea AM (1898a) Descriptions of new species of Australian Coleoptera. Part IV. *Proceedings of the Linnean Society of New South Wales* **22**(3), 584–638.

Lea AM (1898b) Revision of the Australian Curculionidae belonging to the subfamily Cryptorhynchides. Part I. *Proceedings of the Linnean Society of New South Wales* **22**(3), 449–513.

Lea AM (1898c) Revision of the Australian Curculionidae belonging to the subfamily Cryptorhynchides. Part II. *Proceedings of the Linnean Society of New South Wales* **23**(2), 178–217.

Lea AM (1899a) Descriptions of Australian Curculionidae with notes on previously described species. Part I. *Transactions of the Royal Society of South Australia* **23**, 137–197.

Lea AM (1899b) Descriptions of new species of Australian Coleoptera. Part V. *Proceedings of the Linnean Society of New South Wales* **23**(4), 521–645.

Lea AM (1899c) Revision of the Australian Curculionidae belonging to the subfamily Cryptorhynchides. Part III. *Proceedings of the Linnean Society of New South Wales* **24**(2), 200–270.

Lea AM (1900a) Descriptions of two new blind weevils from Western Australia and Tasmania. *Proceedings of the Linnean Society of New South Wales* **25**, 391–393.

Lea AM (1900b) Revision of the Australian Curculionidae belonging to the subfamily Cryptorhynchides. Part IV. *Proceedings of the Linnean Society of New South Wales* **24**(4), 522–546.

Lea AM (1902) Revision of the Australian Curculionidae belonging to the subfamily Cryptorhynchides. Part V. *Proceedings of the Linnean Society of New South Wales* **27**(3), 408–442.

Lea AM (1903) Revision of the Australian Curculionidae belonging to the Subfamily Cryptorhynchides. Part VI. *Proceedings of the Linnean Society of New South Wales* **28**(3), 643–679.

Lea AM (1904) Descriptions of Australian Curculionidae, with notes on previously described species. Part II. *Transactions of the Royal Society of South Australia* **28**, 77–134.

Lea AM (1905a) Descriptions of Australian Curculionidae, with notes on previously described species. Part III. *Transactions of the Royal Society of South Australia* **29**, 209–236.

Lea AM (1905b) Revision of the Australian Curculionidae belonging to the subfamily Cryptorhynchides. Part VII. *Proceedings of the Linnean Society of New South Wales* **30**(2), 235–258.

Lea AM (1906a) Descriptions of Australian Curculionidae, with notes on previously described species. Part IV. *Transactions of the Royal Society of South Australia* **30**, 71–103.

Lea AM (1906b) Notes on the genus *Leptops* with descriptions of new species. *Annales de la Société Entomologique de Belgique (Comptes-rendus)* **50**, 313–354.

Lea AM (1907a) Descriptions of Australian Curculionidae, with notes on previously described species. Part V. *Transactions of the Royal Society of South Australia* **31**, 129–168.

Lea AM (1907b) Revision of the Australian Curculionidae belonging to the subfamily Cryptorhynchides. Part VIII. *Proceedings of the Linnean Society of New South Wales* **32**(2), 400–430.

Lea AM (1908a) Descriptions of Australian Curculionidae, with notes on previously described species. Part VI. *Transactions of the Royal Society of South Australia* **32**, 203–251.

Lea AM (1908b) Notes on the Australian Curculionidae in the Belgian Museum with descriptions of new species. Part I. *Mémoires de la Société Entomologique de Belgique* **16**, 127–186.

Lea AM (1908c) The Coleoptera of King Island, Bass Strait. *Proceedings of the Royal Society of Victoria (New Series)* **20**(2), 143–207.

Lea AM (1909a) Curculionidae from various parts of Australia. *Mitteilungen aus dem Naturhistorischen Museum in Hamburg* **26** (2. Beiheft), 193–203.

Lea AM (1909b) Curculionidae. In *Die Fauna Südwest-Australiens. Ergebnisse der Hamburger südwest-australischen Forschungsreise 1905. Band II.* (Eds W Michaelsen and R Hartmeyer) pp. 215–232. Gustav Fischer, Jena. doi:10.5962/bhl.title.7416

Lea AM (1909c) Descriptions of Australian Curculionidae, with notes on previously described species. Part VII. *Transactions of the Royal Society of South Australia* **33**, 145–196.

Lea AM (1909d) Revision of the Australian Curculionidae belonging to the subfamily Cryptorhynchides. Part IX. *Proceedings of the Linnean Society of New South Wales* **33**(4), 701–732.

Lea AM (1910a) Descriptions of Australian Curculionidae, with notes on previously described species. Part VIII. *Transactions of the Royal Society of South Australia* **34**, 13–58.

Lea AM (1910b) Notes on Australian Curculionidae in the collection of the German Entomological National Museum of Berlin (Col.). *Deutsche Entomologische Zeitschrift* **1910**, 153–172, 505–526.

Lea AM (1910c) Revision of the Australian Curculionidae belonging to the subfamily Cryptorhynchides. Part X. *Proceedings of the Linnean Society of New South Wales* **34**(4), 593–635.

Lea AM (1911a) Descriptions of Australian Curculionidae, with notes on previously described species. Part IX. *Transactions of the Royal Society of South Australia* **35**, 62–88.

Lea AM (1911b) Notes on Australian Curculionidae in the Belgian Museum, with descriptions of new species. Part II. *Mémoires de la Société Entomologique de Belgique* **18**, 61–127.

Lea AM (1911c) Notes on Australian Curculionidae in the Berlin Museum with descriptions of new species. *Mitteilungen aus dem Zoologischen Museum in Berlin* **5**(2), 177–201.

Lea AM (1912) Descriptions of Australian Curculionidae, with notes on previously described species. Part X. *Transactions of the Royal Society of South Australia* **36**, 76–139.

Lea AM (1913a) Descriptions of Australian Curculionidae, with notes on previously described species. Part XI. *Transactions of the Royal Society of South Australia* **37**, 301–445.

Lea AM (1913b) Notes on Cryptorhynchides (Coleoptera, Curculionidae) in the South Australian Museum, with descriptions of new species. *Transactions of the Royal Society of South Australia* **37**, 182–300.

Lea AM (1913c) Revision of the Australian Curculionidae belonging to the subfamily Cryptorhynchides. Part XI. *Proceedings of the Linnean Society of New South Wales* **37**(4), 602–616.

Lea AM (1913d) Revision of the Australian Curculionidae belonging to the subfamily Cryptorhynchides. Part XII. *Proceedings of the Linnean Society of New South Wales* **38**(3), 451–489.

Lea AM (1914) Notes on some miscellaneous Coleoptera, with descriptions of new species. *Transactions of the Royal Society of South Australia* **38**, 249–344.

Lea AM (1915a) Descriptions of new species of Australian Coleoptera. Part X. *Proceedings of the Linnean Society of New South Wales* **39**(4), 650–694.

Lea AM (1915b) Descriptions of new species of Australian Coleoptera. Part XI. *Proceedings of the Linnean Society of New South Wales* **40**(3), 490–521.

Lea AM (1915c) On some Australian Malacodermidae and Curculionidae collected by Mr. G. E. Bryant. *Annals and Magazine of Natural History (Series 8)* **15**(88), 389–421. doi:10.1080/00222931508693655

Lea AM (1915d) On some Australian Malacodermidae and Curculionidae collected by Mr. G. E. Bryant [concluded]. *Annals and Magazine of Natural History (Series 8)* **15**(89), 452–481. doi:10.1080/00222931508693660

Lea AM (1915e) Insecta. [Scientific notes on an expedition into the north-western regions of South Australia, (i).] *Transactions of the Royal Society of South Australia* **39**, 793–801.

Lea AM (1916) Notes on some miscellaneous Coleoptera with descriptions of new species. Part II. *Transactions of the Royal Society of South Australia* **40**, 272–436.

Lea AM (1917a) Descriptions of new species of Australian Coleoptera. Part XII. *Proceedings of the Linnean Society of New South Wales* **41**, 720–745.

Lea AM (1917b) Descriptions of new species of Australian Coleoptera. Part XIII. *Proceedings of the Linnean Society of New South Wales* **42**, 545–582, pl. 29.

Lea AM (1917c) Insecta. [Results of the South Australian Museum expedition to Strzelecki and Cooper Creeks, September and October, 1916, (n).] *Transactions of the Royal Society of South Australia* **41**, 489–630, pls. 33–37.

Lea AM (1922) The flora and fauna of Nuyt's Archipelago and the Investigator Group. No. 4. Coleoptera. *Transactions of the Royal Society of South Australia* **46**, 295–303.

Lea AM (1923) The flora and fauna of Nuyt's Archipelago and the Investigator Group. No. 11. The Coleoptera of Pearson Island. *Transactions of the Royal Society of South Australia* **47**, 355–360, pl. 31.

Lea AM (1925) Descriptions of new species of Australian Coleoptera. Part XVIII. *Proceedings of the Linnean Society of New South Wales* **50**(4), 414–431.

Lea AM (1926a) Notes on some miscellaneous Coleoptera, with descriptions of new species. Part VI. *Transactions of the Royal Society of South Australia* **50**, 45–84.

Lea AM (1926b) On some Australian Coleoptera collected by Charles Darwin during the voyage of the Beagle. *Transactions of the Royal Entomological Society of London* **74**(2), 279–288. doi:10.1111/j.1365-2311.1926.tb02239.x

Lea AM (1926c) On some Australian Curculionidae. *Proceedings of the Linnean Society of New South Wales* **51**(3), 327–362.

Lea AM (1926d) On the genus *Mandalotus* (Coleoptera, Curculionidae). *Records of the South Australian Museum* **3**(2), 147–193.

Lea AM (1927a) Australian Curculionidae of the subfamily Gonipterides. *Proceedings of the Royal Society of Victoria (New Series)* **39**(1), 76–112.

Lea AM (1927b) Descriptions of new species of Australian Coleoptera. Part XIX. *Proceedings of the Linnean Society of New South Wales* **52**, 354–377.

Lea AM (1927c) On new species of *Emplesis* (Curculionidae). *Transactions of the Royal Society of South Australia* **51**, 93–122.

Lea AM (1927d) On *Storeus*, *Emplesis*, and other Curculionidae, mostly from Queensland. *Memoirs of the Queensland Museum* **9**(1), 35–62.

Lea AM (1927e) The clawless and apparently clawless Curculionidae of Australia. *Transactions of the Royal Society of South Australia* **51**, 144–168.

Lea AM (1927f) Two new species of Otiorhynchini from Australia. *Entomologische Mitteilungen* **16**(2), 129–130.

Lea AM (1928a) Australian Curculionidae of the subfamilies Haplonycides and Cryptorhynchides. *Transactions of the Royal Society of South Australia* **52**, 95–164, pls. 7–9.

Lea AM (1928b) Cryptorhynchides (Curculionidae) mostly from Australia. *Records of the South Australian Museum* **4**(1), 49–90.

Lea AM (1928c) New species of Erirhinides (Curculionidae). *Proceedings of the Linnean Society of New South Wales* **53**(4), 375–396.

Lea AM (1929a) Descriptions of new species of Australian Coleoptera. XX. *Proceedings of the Linnean Society of New South Wales* **54**(5), 519–549.

Lea AM (1929b) Notes on some miscellaneous Coleoptera, with descriptions of new species. Part VII. *Transactions of the Royal Society of South Australia* **53**, 203–244.

Lea AM (1929c) Notes on the coleopterous genus *Ampagia* (Curculionidae), with descriptions of new species. *Transactions of the Entomological Society of London* **77**(2), 185–194, pl. 16. doi:10.1111/j.1365-2311.1929.tb00687.x

Lea AM (1930a) Descriptions of new species of Australian Coleoptera. Part XXI. *Proceedings of the Linnean Society of New South Wales* **55**, 451–467.

Lea AM (1930b) On Australian Coleoptera. Part VI. *Records of the South Australian Museum* **4**(2), 243–273.

Lea AM (1930c) On some Coleoptera from northern Australia, collected by Dr. H. Basedow. *Transactions of the Royal Society of South Australia* **54**, 148–154.

Lea AM (1931a) Australasian Coleoptera. *Records of the South Australian Museum* **4**(3), 365–408.

Lea AM (1931b) Notes on some miscellaneous Coleoptera, with descriptions of new species. Part VIII. *Transactions of the Royal Society of South Australia* **55**, 39–64.

Lea AM (1931c) On Baridiinae (Curculionidae), mostly from New Guinea. *Proceedings of the Linnean Society of New South Wales* **56**(3), 139–171.

Lea AM, Bovie A (1909) Coleoptera. Fam. Curculionidae. Subfam. Belinae. In *Genera Insectorum. 91*me *Fascicule*. (Ed. P Wytsman) pp. 1–13. P. Wytsman, Bruxelles.

LeConte JL (1859) Catalogue of the Coleoptera of Fort Tejon, California. *Proceedings of the Academy Natural Sciences of Phildelphia* **11**, 69–90.

Leschen RA, Davis S, Brown S, Brav-Cubitt T, Buckley TR (2022) The enigmatic dead-leaf miner *Geochus* Broun (Coleoptera: Curculionidae): phylogenetic placement, a new species, and lectotype designations. *Coleopterists Bulletin* **76**(1), 1–35.

Leschen RA, Lawrence JF, Kuschel G, Thorpe S, Wang Q (2003) Coleoptera genera of New Zealand. *New Zealand Entomologist* **26**(1), 15–28.

Linnaeus C (1758) *Systema naturae per regna tria naturae, secundem classes, ordines, genera, species, cum characteribus, differentis, synonymis, locis. Editio decimal, reformata. Tom. I.* Laurentii Salvii, Holmiae, 824 pp. doi:10.5962/bhl.title.542

Linnaeus C (1763) Centuria Insectorum. *Amoenitates Academicae* **6**, 384–415.

Lyal CHC (1993) Cryptorhynchinae (Insecta: Coleoptera: Curculionidae). *Fauna of New Zealand* **29**, 1–305.

Lyal CHC (2014a) 3.7.7 Molytinae Schoenherr, 1823. In *Handbook of Zoology, Coleoptera, Beetles Volume 3: Morphology and Systematics (Phytophaga).* (Eds RG Beutel and RAB Leschen) pp. 529–570. De Gruyter, Berlin/Boston, xii + 675 pp.

Lyal CHC (2014b) 3.7.8 Mesoptiliinae Lacordaire, 1863. In *Handbook of Zoology, Coleoptera, Beetles Volume 3: Morphology and Systematics (Phytophaga).* (Eds RG Beutel and RAB Leschen) pp. 570–576. De Gruyter, Berlin/Boston, xii + 675 pp.

Lyal CHC, King T (1996) Elytro-tergal stridulation in weevils (Insecta: Coleoptera: Curculionoidea). *Journal of Natural History* **30**(5), 703–773. doi:10.1080/00222939600770391

Lyal CHC, Douglas DA, Hine SJ (2006) Morphology and systematic significance of sclerolepidia in the weevils (Coleoptera: Curculionoidea). *Systermatics and Biodiversity* **4**(2), 203–241. doi:10.1017/S1477200005001714

Macleay WJ (1865) The genera and species of Amycteridae. *Transactions of the Entomological Society of New South Wales* **1**(4), 199–298.

Macleay WJ (1866) New species of Amycteridae. *Proceedings of the Entomological Society of New South Wales* **1**(5), 318–340.

Macleay WS (1826) Annulosa. Catalogue of Insects, collected by Captain King, R. N. Appendix B. Containing a list and description of the subjects of natural history collected during Captain King's survey of the intertropical and western coasts of Australia. In *Narrative of a Survey of the Intertropical and Western Coasts of Australia, Performed between the Years 1818 and 1822. Vol. II.* (Ed. PP King) pp. 438–469. John Murray, London.

Mapondera T, Burgess T, Matsuki M, Oberprieler RG (2012) Identification and molecular phylogenetics of the cryptic species of the *Gonipterus scutellatus* complex (Coleoptera: Curculionidae: Gonipterini). *Australian Journal of Entomology* **51**(3), 175–188. doi:10.1111/j.1440-6055.2011.00853.x

Marelli CA (1926) La plaga de gorgojos de los eucaliptos, debida a *Dacnirótatus Bruchi* Mar. y experiencias sobre la desinsectación de los árboles atacados. *Memorias del Ministerio de Obras Públicas, Buenos Aires* **1924/25**, 597–646.

Marshall GAK (1915) Part II. Curculionidae. In *Report on the Coleoptera collected by the British Ornithologists' Union Expedition and the Wollaston Expedition in Dutch New Guinea*. (Eds GJ Arrow, GAK Marshall, CJ Gahan and KG Blair). *Transactions of the Zoological Society of London* **20**(17), 509–533. doi:10.1111/j.1469-7998.1912.tb07845.x

Marshall GAK (1922) On the Australian fern weevils. *Bulletin of Entomological Research* **13**(2), 169–174, pls. 6–8. doi:10.1017/s0007485300028042

Marshall GAK (1930) New Curculionidae with notes on synonymy. *Annals and Magazine of Natural History (Series 10)* **6**(35), 551–577. doi:10.1080/00222933008673252

Marshall GAK (1931) Curculionidae. *Insects of Samoa and Other Samoan Terrestrial Arthropoda, Part IV Coleoptera*. Fasc. **5**, 249–346.

Marshall GAK (1934) On the genus *Elaeagna*, Pasc. (Col. Curcul.). *Stylops* **3**(10), 222–225. doi:10.1111/j.1365-3113.1934.tb01525.x

Marshall GAK (1936) New Australian Curculionidae (Col.). *Proceedings of the Royal Entomological Society of London (B)* **5**(10), 189–195. doi:10.1111/j.1365-3113.1936.tb01309.x

Marshall GAK (1938a) New Australian Curculionidae (Col.). *Proceedings of the Royal Entomological Society of London (B)* **7**(1), 8–14. doi:10.1111/j.1365-3113.1938.tb01225.x

Marshall GAK (1938b) New Curculionidae (Col.) from southern Africa. *Annals and Magazine of Natural History (Series 11)* **1**(2), 178–195. doi:10.1080/00222933808526755

Marshall GAK (1939a) New tropical African Curculionidae (Col.). *Annals and Magazine of Natural History (Series 11)* **3**(18), 561–583. doi:10.1080/03745481.1939.9723637

Marshall GAK (1939b) On the curculionid tribe Coryssomerini (Coleopt.). *Annals and Magazine of Natural History (Series 11)* **4**(19), 1–32.

Marshall GAK (1943) New Indian Curculionidae (Col.). *Annals and Magazine of Natural History (Series 11)* **10**(62), 105–119. doi:10.1080/03745481.1943.9727999

Marshall GAK (1944) On the genera of the tribe Cyphicerini (Col., Curc.) – II. *Annals and Magazine of Natural History (Series 11)* **11**(79), 433–462. doi:10.1080/00222934408527446

Marshall GAK (1946) Taxonomic notes on Curculionidae (Col.). *Annals and Magazine of History (Series 11)* **13**(98), 93–98. doi:10.1080/00222934608654526

Marshall GAK (1948) Five new weevils from Queensland (Coleoptera, Curculionidae). *Annals and Magazine of Natural History (Series 11)* **14**(115), 485–493. doi:10.1080/00222934708654657

Marshall GAK (1952) Taxonomic notes on Curculionidae (Col.). *Annals and Magazine of Natural History (Series 12)* **5**(51), 261–270. doi:10.1080/00222935208654289

Marshall GAK (1956) *The otiorrhynchine Curculionidae of the tribe Celeuthetini (Col.).* Trustees of the British Museum, London, 134 pp.

Marvaldi AE (1997) Higher level phylogeny of Curculionidae (Coleoptera: Curculionoidea) based mainly on larval characters, with special reference to broad-nosed weevils. *Cladistics* **13**(4), 285–312. doi:10.1111/j.1096-0031.1997.tb00321.x

Marvaldi AE, Morrone JJ (2000) Phylogenetic systematics of weevils (Coleoptera: Curculionoidea): a reappraisal based on larval and adult morphology. *Insect Systematics & Evolution* **31**(1), 43–58.

Marvaldi AE, Sequeira AS, O'Brien CW, Farrell BD (2002) Molecular and morphological phylogenetics of weevils (Coleoptera, Curculionoidea): do niche shifts accompany diversification? *Systematic Biology* **51**(5), 761–785. doi:10.1080/10635150290102465

Maunsell SC, Burwell CJ, Morris RJ, McDonald WJ, Edwards E, *et al.* (2017) Host-plants of leaf-miners in Australian subtropical rainforest. *Austral Entomology* **56**(4), 403–411. doi:10.1111/aen.12252

May BM (1977) Immature stages of Curculionidae: larvae of the soil- dwelling weevils of New Zealand. *Journal of the Royal Society of New Zealand* **7**(2), 189–228. doi:10.1080/03036758.1977.10427160

May BM (1984) The immature stages of *Linogeraeus urbanus* (Coleoptera: Curculionidae: Baridinae) and its biology in Auckland, New Zealand. *New Zealand Entomologist* **8**(1), 29–35. doi:10.1080/00779962.1984.9722458

May BM (1992) *Geochus* Broun, 1882: an enigmatic genus now known to be a leafminer (Curculionidae: Curculionini). *New Zealand Entomologist* **15**(1), 22–25. doi:10.1080/00779962.1992.9722623

May BM (1993) Larvae of Curculionoidea (Insecta: Coleoptera): a systematic overview. *Fauna of New Zealand* **28**, 1–223. doi:10.7931/J2/FNZ.28

May BM (1994) An introduction to the immature stages of Australian Curculionoidea. In *Australian Weevils (Coleoptera: Curculionoidea). Volume II. Brentidae, Eurhynchidae, Apionidae and a Chapter on Immature Stages by Brenda May*. (Ed. EC Zimmerman) pp. 365–728. CSIRO Australia, Melbourne, x + 755 pp.

May BM (1997) Larvae of two weevil genera (Coleoptera: Curculionidae) associated with *Crossotarsus armipennis* Lea (Platypodidae) from New South Wales. *New Zealand Entomologist* **20**(1), 7–13. doi:10.1080/00779962.1997.9722662

Mazur MA, Brown SD (2024) A review of the genus *Stephanorhynchus* White, 1846 (Coleoptera: Curculionidae: Eugnomini), with a description of three new genera and one new species from New Zealand. *The European Zoological Journal* **91**(1), 457–549. doi:10.1080/24750263.2024.2319789

McFadyen REC (1994) Two New Insects from *Portulaca oleracea* L. in Australia. *Australian Journal of Entomology* **33**(1), 43–44. doi:10.1111/j.1440-6055.1994.tb00915.x

McFadyen REC (2000) Biology and host specificity of the stem galling weevil *Conotrachelus albocinereus* Fiedler (Col.: Curculionidae), a biocontrol agent for parthenium weed *Parthenium hysterophorus* L. (Asteraceae) in Queensland, Australia. *Biocontrol Science and Technology* **10**(2), 195–200. doi:10.1080/09583150029323

McKenna DD, Sequeira AS, Marvaldi AE, Farrell BD (2009) Temporal lags and overlap in the diversification of weevils and flowering plants. *Proceedings of the National Academy of Sciences of the USA* **106**(17), 7083–7088. doi:10.1073/pnas.0810618106

McKenna DD, Shin S, Ahrens D, Balke M, Beza-Beza C, *et al.* (2019) The evolution and genomic basis of beetle diversity. *Proceedings of the National Academy of Sciences* **116**(49), 24729–24737. doi:10.1073/pnas.1909655116

McMillan RP (1987) Notes on a mistletoe weevil *Metyrus albicollis* Germ. *Western Australian Naturalist* **17**, 20–21.

Meregalli M (2014) 3.7.6. Lixinae Schoenherr, 1823. In *Handbook of Zoology, Coleoptera, Beetles Volume 3: Morphology and Systematics (Phytophaga)*. (Eds RAB Leschen and RG Beutel) pp. 523–451529. De Gruyter, Berlin/Boston, xii + 675 pp.

Meregalli M, Boriani M, Taddei A, Hsu Ch-F, Tseng W-Zh, *et al.* (2020) A new species of *Aclees* from Taiwan with notes on other species of the genus (Coleoptera: Curculionidae: Molytinae). *Zootaxa* **4868**(1), 1–26. doi:10.11646/zootaxa.4868.1.1

Monteith G, Baldwin N (2018) David Behrens and his studies on the giant hoop pine weevil, *Eurhamphus fasciculatus*. *Entomological Society of Queensland News Bulletin* **45**(9), 177–181.

Montrouzier X (1855) Essai sur la faune l'île de Woodlark ou Moiou. *Annales des Sciences Physiques et Naturelles, d'Agriculture et d'Industrie, Société impériale d'Agriculture (etc.) de Lyon (Series 2)* **7**(1), 1–114.

Montrouzier X (1861)[1860] Essai sur la faune entomologique de la Nouvelle-Calédonie (Balade) et des îles des Pins, Art, Lifu, etc. (Suite). *Annales de la Société Entomologique de France (Series 3)* **8**, 867–916.

Moore BP (1978) *Life on Forty Acres: A Saga of Australian Rural Life*. EW Classey Ltd, Faringdon.

Morimoto K (1983) The family Curculionidae of Japan. II. Viticiinae, subfamily nov. *Esakia* **20**, 55–62.

Morimoto K, Miyakawa S (1995) The family Curculionidae of Japan. VIII. Subfamily Acicnemidinae. *Esakia* **35**, 17–62.

Motschulsky V de (1855) Voyages. Lettres de M. de Motschulsky à M. Ménétriés. Études Entomologiques **4**, 1–84. doi:10.5962/bhl.title.8176

Motschulsky V de (1866) Essai d'un catalogue des insectes de l'île de Ceylan. Supplément. *Bulletin de la Société des Naturalistes de Moscou* **39**(1), 393–446.

Mugu S, Piston D, Jordal B (2018) New molecular markers resolve the phylogenetic position of the enigmatic wood-boring weevils Platypodinae (Coleoptera: Curculionidae). *Arthropod Systematics and Phylogeny* **76**(1), 45–58. doi:10.3897/asp.76.e31938

Mulder RH (1964) Insects associated with '*Xanthorrhoea* sp.' *Journal of the Entomological Society of Australia (NSW)* **1**, 12.

Murray A (1869) Orchid-eating insects. *The Gardeners' Chronicle and Agricultural Gazette* **1869**, 1279.

Murray D, Murray C, Barclay MV, Thomson RT, Morris MG (2014) *Achopera alternata* Lea, 1910 (Curculionidae, Cryptorhynchinae), an Australian weevil new to the Northern Hemisphere, breeding out of doors in Britain. *The Coleopterist* **23**(2), 45–48.

Oberprieler RG (2010) A reclassification of the weevil subfamily Cyclominae (Coleoptera: Curculionidae). *Zootaxa* **2515**(1), 1–35. doi:10.11646/zootaxa.2515.1.1

Oberprieler RG (2014a) 3.7 Curculionidae Latreille, 1802. In *Handbook of Zoology, Coleoptera, Beetles Volume 3: Morphology and Systematics (Phytophaga)*. (Eds RAB Leschen and RG Beutel) pp. 423–424. De Gruyter, Berlin/Boston, xii + 675 pp.

Oberprieler RG (2014b) 3.7.1 Brachycerinae Billberg, 1820. In *Handbook of Zoology, Coleoptera, Beetles Volume 3: Morphology and Systematics (Phytophaga)*. (Eds RAB Leschen and RG Beutel) pp. 424–451. De Gruyter, Berlin/Boston, xii + 675 pp.

Oberprieler RG (2014c) 3.7.4 Cyclominae Schoenherr, 1826. In *Handbook of Zoology, Coleoptera, Beetles Volume 3: Morphology and Systematics (Phytophaga)*. (Eds RAB Leschen and RG Beutel) pp. 483–502. De Gruyter, Berlin/Boston, xii + 675 pp.

Oberprieler RG, Caldara R (2012) *Siraton devillei* Hustache (Coleoptera: Curculionidae), the mysterious weevil from the Isle of Elba: exiled no longer. *Zootaxa* **3573**, 55–58. doi:10.11646/zootaxa. 3573.1.6

Oberprieler RG, Jennings D (2016) Specialist Diagnostic Guide to Pest Weevils – *Rhynchophorus* Palm Weevils. *Reprint CSIRO National Research Collections Australia* **2016**, 1–18.

Oberprieler RG, Zimmerman EC (2020) *Australian Weevils (Coleoptera: Curculionoidea). Volume IV. Obituary: Elwood Curtin Zimmerman, Curculionidae: Entiminae Part I, Literature Consulted II*. CSIRO Australia, Melbourne, 649 pp.

Oberprieler RG, Anderson RS, Marvaldi AM (2014a) 3 Curculionoidea Latreille, 1802: Introduction, Phylogeny. In *Handbook of Zoology, Coleoptera, Beetles Volume 3: Morphology and Systematics (Phytophaga)*. (Eds RAB Leschen and RG Beutel) pp. 285–300. De Gruyter, Berlin/Boston, xii + 675 pp.

Oberprieler RG, Caldara R, Skuhrovec J (2014b) 3.7.2 Bagoini Thomson, 1859; Gonipterini Lacordaire, 1863; Hyperini Marseul, 1863. In *Handbook of Zoology, Coleoptera, Beetles Volume 3: Morphology and Systematics (Phytophaga)*. pp. 452–476. De Gruyter, Berlin/Boston, xii + 675 pp.

Oberprieler RG, Marvaldi AE, Anderson RS (2007) Weevils, weevils, weevils everywhere. *Zootaxa* **520**(1668), 491–520. doi:10.11646/zootaxa.1668.1.24

O'Brien CW (1976) Taxonomic revision of the New World subaquatic genus *Neochetina* (Coleoptera-Curculionidae-Bagoini). *Annals of the Entomological Society of America* **69**(2), 165–174. doi:10.1093/aesa/69.2.165

O'Brien CW, Askevold IS (1992) Systematics and evolution of weevils of the genus *Bagous* Germar (Coleoptera: Curculionidae), I. Species of Australia. *Transactions of the American Entomological Society* **118**(4), 331–452.

Oke C (1931) On some Australian Curculionoidea. *Proceedings of the Royal Society of Victoria (New Series)* **43**(2), 177–201.

Oke C (1934) On some Australian Curculionoidea. Part 2. *Proceedings of the Royal Society of Victoria (New Series)* **46**, 250–263.

Oke C (1951) The Coleoptera of the Russell Grimwade Expedition. *Memoirs of the National Museum of Victoria* **17**, 19–25.

Olivier AG (1790) *Entomologie, ou histoire naturelle des insectes, avec leurs caracteres génériques et spécifiques, leur description, leur synonymie, et leur figure enluminée. Coléoptères Vol. 2, Nos. 9–34*. Baudouin, Paris, 485 pp.

Olivier AG (1791) *Encyclopédie Méthodique. Histoire Naturelle. Insectes. Tome cinquième*. Panckoucke, Paris, 793 pp.

Olivier AG (1807) *Entomologie, ou histoire naturelle des insectes, avec leurs caractères génériques et spécifiques, leur description, leur synonymie, et leur figure enluminée. Coléoptères. Tome cinquième*. Desray, Paris, 612 pp. doi:10.5962/bhl.title.49479

Olliff AS (1888) IV. Insecta. In *Report on a Small Zoological Collection from Norfolk Island*. Millington, JAM. *Proceedings of the Linnean Society of New South Wales (Series 2)* **2**, 1001–1014.

Olliff AS (1889) The insect fauna of Lord Howe Island. *Memoirs of the Australian Museum* **2**(4), 75–98. doi:10.3853/j.0067-1967.2.1889.482

Palmer WA, Heard TA, Sheppard AW (2010) A review of Australian classical biological control of weeds programs and research activities over the past 12 years. *Biological Control* **52**(3), 271–287. doi:10.1016/j.biocontrol.2009.07.011

Panzer GWF (1801) *Faunae Insectorum Germanicae initia, oder Deutschlands Insecten. 84. Heft*. Felssecker, Nürnberg, 24 pp, 24 pls.

Pascoe FP (1865) On some new genera of Curculionidae. Part 1. *Journal of Entomology* **2**(13), 413–432.

Pascoe FP (1869) On some new Australian genera and species of Curculionidae belonging to the Otiorhynchinae. *Entomologist's Monthly Magazine* **6**, 99–105.

Pascoe FP (1870a) A revision of the genus *Catasarcus*. *Transactions of the Entomological Society of London* **1870**(1), 13–40. doi:10.1111/j.1365-2311.1870.tb01863.x

Pascoe FP (1870b) Contributions towards a knowledge of the Curculionidae [part]. *Journal of the Linnean Society of London, Zoology* **10**(47), 434–458. doi:10.1111/j.1096-3642.1870.tb00671.x

Pascoe FP (1870c) [Contributions towards a knowledge of the Curculionidae] [concl.]. *Journal of the Linnean Society of London, Zoology* **10**(48), 459–493. doi:10.1111/j.1096-3642.1870.tb00672.x

Pascoe FP (1870d) Descriptions of some genera and species of Australian Curculionidae. *Transactions of the Entomological Society of London* **1870**(2), 181–209. doi:10.1111/j.1365-2311.1870.tb01873.x

Pascoe FP (1870e) Descriptions of some genera and species of Australian Curculionidae. *Transactions of the Entomological Society of London* **1870**(4), 445–484. doi:10.1111/j.1365-2311.1870.tb01880.x

Pascoe FP (1870f) Further descriptions of Australian Curculionidae. *Transactions of the Entomological Society of London* **1870**(2), 209–212. doi:10.1111/j.1365-2311.1870.tb01874.x

Pascoe FP (1871a) Additions to the Australian Curculionidae. Part I. *Annals and Magazine of Natural History, (Series 4)* **8**(44), 89–99. doi:10.1080/00222937108696441

Pascoe FP (1871b) Catalogue of Zygopinae, a subfamily of Curculionidae, found by Mr Wallace in the Eastern Archipelago. *Annals and Magazine of Natural History (Series 4)* **7**(39), 198–266. doi:10.1080/00222937108696365

Pascoe FP (1871c) Contributions towards a knowledge of the Curculionidae. [Part II.] *Journal of the Linnean Society of London, Zoology* **11**(51), 154–218. doi:10.1111/j.1096-3642.1871.tb01661.x

Pascoe FP (1872a) Additions to the Australian Curculionidae. Part II. *Annals and Magazine of History (Series 4)* **9**(50), 132–142. doi:10.1080/00222937208696553

Pascoe FP (1872b) Additions to the Australian Curculionidae. Part III. *Annals and Magazine of Natural History (Series 4)* **10**(56), 84–101. doi:10.1080/00222937208696651

Pascoe FP (1872c) Contributions towards a knowledge of the Curculionidae. Part III. *Journal of the Linnean Society of London, Zoology* **11**(55), 440–492. doi:10.1111/j.1096-3642.1872.tb01668.x

Pascoe FP (1873a) Additions to the Australian Curculionidae. Part IV. *Annals and Magazine of Natural History (Series 4)* **11**(63), 178–199. doi:10.1080/00222937308696794

Pascoe FP (1873b) Additions to the Australian Curculionidae. Part V. *Annals and Magazine of Natural History (Series 4)* **12**(69), 230–239. doi:10.1080/00222937308680748

Pascoe FP (1873c) Additions to the Australian Curculionidae. Part V [continued]. *Annals and Magazine of Natural History (Series 4)* **12**(70), 278–286. doi:10.1080/00222937308680768

Pascoe FP (1874a) Additions to the Australian Curculionidae. Part VI. *Annals and Magazine of Natural History (Series 4)* **12**(77), 383–389. doi:10.1080/00222937408680885

Pascoe FP (1874b) Additions to the Australian Curculionidae. Part VII. *Annals and Magazine of Natural History (Series 4)* **12**(78), 412–419. doi:10.1080/00222937408680896

Pascoe FP (1874c) Contributions towards a knowledge of the Curculionidae. Part IV. *Journal of the Linnean Society of London, Zoology* **12**, 1–99. doi:10.1111/j.1096-3642.1875.tb02572.x

Pascoe FP (1875) Additions to the Australian Curculionidae. Part VIII. *Annals and Magazine of Natural History (Series 4)* **16**(91), 55–67. doi:10.1080/00222937508681122

Pascoe FP (1877) Descriptions of new genera and species of New-Zealand Coleoptera. Part IV. *Annals and Magazine of Natural History (Series 4)* **19**(110), 140–147. doi:10.1080/00222937708682109

Pascoe FP (1881) Descriptions of Curculionidae. Part 1. *Cistula Entomologica* **2**(24), 587–601.

Pascoe FP (1882) Additions to the Australian Curculionidae. Part IX. *Annals and Magazine of Natural History (Series 5)* **9**(53), 374–383. doi:10.1080/00222938209459054

Pascoe FP (1883a) Additions to the Australian Curculionidae. Part X. *Annals and Magazine of Natural History (Series 5)* **12**(72), 412–421. doi:10.1080/00222938309459652

Pascoe FP (1883b) Descriptions of some new genera and species of Curculionidae, mostly Asiatic. Part II. *Annals and Magazine of Natural History (Series 5)* **12**(68), 88–101. doi:10.1080/00222938309459584

Pascoe FP (1885) List of the Curculionidae of the Malay Archipelago collected by Dr. Odoardo Beccari, L. M. d'Albertis, and others. *Annali del Museo Civico di Storia Naturale Genova* **22**, 201–332.

Pascoe FP (1887a) Descriptions of some new genera and species of Curculionidae, mostly Asiatic. Part III. *Annals and Magazine of Natural History (Series 5)* **19**(113), 370–380. doi:10.1080/00222938709460259

Pascoe FP (1887b) Descriptions of some new genera and species of Curculionidae, mostly Asiatic. Part IV. *Annals and Magazine of Natural History (Series 5)* **20**(109), 348–361. doi:10.1080/00222938709460076

Perris E (1852) Seconde excursion dans les Grandes-Landes. Lettre adressée à M. Mulsant. *Annales de la Société Linnéenne de Lyon* **2**, 145–216.

Perroud BP, Montrouzier X (1865) Essai sur la faune entomologique de Kanala (Nouvelle Calédonie) et description de quelques espèces nouvelles ou peu connues. *Annales de la Société Linnéenne de Lyon (Series 2)* **11**, 46–257, 1 pl.

Peterson M (1995) A new *Catasarcus* species (Coleoptera: Curculionidae: Entiminae) from the Shark Bay-Kalbarri region of Western Australia. *Records of the Western Australian Museum* **17**, 163–167.

Peterson M (2013) A new genus of Amycterini from the Pilbara region of Western Australia (Coleoptera: Curculionidae: Cyclominae). *Journal of Insect Biodiversity* **1**(4), 1–12. doi:10.12976/jib/2013.1.4

Pierce WD (1917) *A Manual of Dangerous Insects Likely to be Introduced in the United States through Importations*. Government Printing Office, Washington, D.C., 256 pp.

Pinzón-Navarro SV, Jennings D, Oberprieler RG (2017) Host associations of *Melanterius* Erichson (Coleoptera: Curculionidae: Cleogonini), with a diagnosis and delimitation of the genus and description of five new species. *Zootaxa* **4298**(1), 1–77. doi:10.11646/zootaxa.4298.1.1

Prena J (2008) A synopsis of the orchid weevil genus *Orchidophilus* Buchanan (Coleoptera, Curculionidae, Baridinae), with taxonomic rectifications and description of one new species. *Zootaxa* **1783**(1), 18–30. doi:10.11646/zootaxa.1783.1.2

Prena J, Colonnelli E, Hespenheide HA (2014) 3.7.9 Conoderinae Schoenherr, 1833. In *Handbook of Zoology, Coleoptera Volume 3: Morphology and Systematics (Phytophaga)*. (Eds RAB Leschen and RG Beutel) pp. 577–589. Walter de Gruyter, Berlin, Germany.

Porch N (2009) The first record of an amycterine weevil (Coleoptera: Curculionidae: Amycterinae) feeding on Orchidaceae. *Australian Entomologist* **36**(1), 29–32.

Pullen K, Jennings D, Oberprieler RO (2014) Annotated catalogue of Australian weevils (Coleoptera: Curculionoidea). *Zootaxa* **3896**(1), 1–481. doi:10.11646/zootaxa.3896.1.1

Quoy JRC, Gaimard JP (1824) Zoologie. In *Voyage autour du Monde, entrepris par Ordre du Roi, sous le ministère et conformément aux instructions de S. Exc. M. le Vicomte du Bouchage, Secrétaire d'État au Departemente de la Marine, exécuté sur les corvettes de S. M. l'Uranie et la Physicienne, pendant les années 1817, 1818, 1819 et 1820*. (Ed. LCD Freycinet). Pillet Aîné, Paris, 712 pp. doi:10.5962/bhl.title.63985

Redtenbacher L (1868) *Reise der österreichischen Fregatte Novara um die Erde in den Jahren 1857, 1858, 1859 unter den Befehlen des Commodore B. von Wüllersdorf-Urbair. Zoologischer Theil, zweiter Band, 1. Abtheilung, A. 1: Coleoptera*. Carl Gerold's Sohn, Wien, 249 pp.

Reitter E (1916) *Fauna germanica. Die Kafer des deutschen Reiches. Nach der analytischen Methode, Bd V*. K. G. Lutz, Stuttgart, 343 pp.

Rheinheimer J (1989) Notes on the host plants of some adult Australian weevils (Coleoptera: Curculionidae). Australian *Entomological Magazine* **16**(1), 27–31.

Rheinheimer J (1990) *Misophricoides callitriphilus* gen. et sp. n., eine neue Rüsselkäfergattung und -art aus Australien (Insecta, Coleoptera, Curculionidae: Erirhininae). *Reichenbachia* **28**(7), 43–45.

Rheinheimer J (1991) Bemerkungen über die australischen Arten der Gattung *Amorphoidea* Motschulsky sowie Beschreibung einer neuen Art (Coleoptera: Curculionidae). *Koleopterologische Rundschau* **61**, 189–192.

Rheinheimer L (1993) Neue Cryptorhynchinae aus Australien (Coleoptera: Curculionidae). *Koleopterologische Rundschau* **63**, 317–324.

Rheinheimer J (1994) Ein neuer *Plaesiorhinus* Blackburn und Bemerkungen über australische Rüsselkäfer. (Coleoptera: Curculionidae). *Koleopterologische Rundschau* **64**, 275–277.

Rheinheimer J (1995) Die australische Rüsselkäfergattung *Agestra* Pascoe (Coleoptera: Curculionidae). *Koleopterologische Rundschau* **65**, 221–227.

Rheinheimer J (1996) Ein neuer *Storeus* aus Australien. (Coleoptera: Curculionidae). *Koleopterologische Rundschau* **66**, 223–226.

Riedel A, Taenzler R (2016) Revision of the Australian species of the weevil genus *Trigonopterus* Fauvel. *Zookeys* **556**, 97–162. doi:10.3897/zookeys.556.6126

Riedel A, Taenzler R, Pons J, Sudhardjono Y, Balke M (2016) Large-scale molecular phylogeny of Cryptorhynchinae (Coleoptera, Curculionidae) from multiple genes suggests American origin and later Australian radiation. *Systematic Entomology* **41**(2), 492–503. doi:10.1111/syen.12170

Roelofs W (1866) Notice sur un nouveau genre de curculionides d'Australie. *Annales de la Société Entomologique de Belgique (Comptes-rendus)* **10**, 243–250.

Roelofs W (1867) Notice sur le genre *Acroteriasis* – Suite. *Annales de la Société Entomologique de Belgique (Comptes-rendus)* **11**, 75–77, pl. 2.

Roelofs W (1879) Diagnoses de nouvelles espèces de Curculionides, Brenthides, Anthribides et Bruchides du Japon. *Annales de la Société Entomologique de Belgique (Comptes-rendus)* **22**, 53–55.

Roelofs W (1880) Descriptions de deux nouvelles espèces de Cholides et de deux nouvelles espèces de Cryptorhynchides. *Annales de la Société Entomologique de Belgique (Comptes-rendus)* **23**, 39–45.

Roelofs W (1892) Deux nouveaux genres et deux nouvelles especes du groupe des Rhynchophorides. *Notes from the Leyden Museum* **XIV**, 207–212.

Rossi P (1790) *Fauna Etrusca sistens insecta quae in povinciis Florentina et Pisana praesertim collegit Petrus Rossius. Tomus primus*. Typis Thomae Masi & Sociorum, Liburni, 348 pp.

Rossi P (1794) *Mantissa Insectorum, exhibens species nuper in Etruria collectas, adjectis Faunae Etruscae illustrationibus ac emendationibus. Tomus secundus*. Polloni, Pisa, 154 pp.

Saunders WW, Jekel H (1855) Descriptions de quelques Curculionites. *Annales de la Société Entomologique de France (Series 3)* **3**, 289–306.

Say T (1824) Descriptions of Coleopterous insects collected in the late expedition to the Rocky Mountains, performed by order of Mr. Calhoun, Secretary of War, under the command of Mayor Long. *Journal of the Academy of Natural Sciences* **3**(2), 238–282.

Schedl KE (1936) Scolytidae and Platypodidae. Contribution 35. The collection of the South Australian Museum. *Records of the South Australian Museum* **5**(4), 513–535.

Schenkling S, Marshall GAK (1936) Haplonychinae *in* Curculionidae: Prionomerinae, Aterpinae, Amalactinae, Haplonychinae, Omophorinae. In *Coleopterorum Catalogus. Pars 150*. (Eds W Junk and S Schenkling) pp. 2, 3, 8, 9, 11. W. Junk, 's-Gravenhage.

Schneider MF (1999) *Entomology - A Textbook for Students, Agriculturalists and Foresters in Papua New Guinea*. Bulolo University College; Training Manual No 19; 312 pp., 11 colour plates.

Schoenherr CJ (1823) Curculionides [Tabula synoptica Familiae Curculionidum]. *Isis von Oken* **1823**(10), columns 1132–1146.

Schoenherr CJ (1825) Continuatio Tabulae synopticae Familiae Curculionidum. *Isis von Oken* **1825**(5), columns 581–588.

Schoenherr CJ (1826) *Curculionidum dispositio methodica, cum generum characteribus, descriptionibus atque observationibus variis, seu Prodromus ad Synonymiae Insectorum. Partem IV*. Fredericus Fleischer, Lipsia, x + 338 pp. doi:10.5962/bhl.title.9327

Schoenherr CJ (1833a) *Genera et species Curculionidum, cum synonymia hujus familiae. Species novae aut hactenus minus cognitae, descriptionibus a Dom. Leonardo Gyllenhal, C. H. Boheman, et entomologis aliis illustratae. Tomus primus. – Pars prima*. Roret, Parisii, pp. i–xv, 1–381 + 3 pp. Corrigenda.

Schoenherr CJ (1833b) *Genera et species Curculionidum, cum synonymia hujus familiae. Species novae aut hactenus minus cognitae, descriptionibus a Dom. Leonardo Gyllenhal, C. H. Boheman, et entomologis aliis illustratae. Tomus primus. – Pars secunda*. Roret, Parisii/Fred. Fleischer, Leipzig, pp. 383–681. doi:10.5962/bhl.title.8952

Schoenherr CJ (1834a) *Genera et species curculionidum, cum synonymia hujus familiae. Species novae aut hactenus minus cognitae, descriptionibus a Dom. Leonardo Gyllenhal, C. H. Boheman, et entomologis aliis illustratae. Tomus secundus. – Pars prima*. Roret, Parisii/Fred. Fleischer, Lipsia, pp. 1–326. doi:10.5962/bhl.title.8952

Schoenherr CJ (1834b) *Genera et species curculionidum, cum synonymia hujus familiae. Species novae aut hactenus minus cognitae, descriptionibus a Dom. Leonardo Gyllenhal, C. H. Boheman, et entomologis aliis illustratae. Tomus secundus. – Pars secunda*. Roret, Parisii/Fred. Fleischer, Lipsia, pp. 329–669. doi:10.5962/bhl.title.8952

Schoenherr CJ (1835) *Genera et species curculionidum, cum synonymia hujus familiae. Species novae aut hactenus minus cognitae, descriptionibus a Dom. Leonardo Gyllenhal, C. H. Boheman, et entomologis aliis illustratae. Tomus tertius. – Pars prima*. Roret, Parisii/Fred. Fleischer, Lipsia, pp. 1–505. doi:10.5962/bhl.title.8952

Schoenherr CJ (1836) *Genera et species curculionidum, cum synonymia hujus familiae. Species novae aut hactenus minus cognitae, descriptionibus a Dom. Leonardo Gyllenhal, C. H. Boheman, et entomologis aliis illustratae. Tomus tertius. – Pars secunda.* Roret, Parisii/Fred. Fleischer, Lipsia, pp. 506–858 + 3 pp. Corrigenda. doi:10.5962/bhl.title.8952

Schoenherr CJ (1837) *Genera et species curculionidum, cum synonymia hujus familiae. Species novae aut hactenus minus cognitae, descriptionibus a Dom. Leonardo Gyllenhal, C. H. Boheman, et entomologis aliis illustratae. Tomus quartus. – Pars prima.* Roret, Parisii/Fred. Fleischer, Lipsia, pp. 1–600.

Schoenherr CJ (1838) *Genera et species curculionidum, cum synonymia hujus familiae. Species novae aut hactenus minus cognitae, descriptionibus a Dom. Leonardo Gyllenhal, C. H. Boheman et entomologis aliis illustratae. Tomus quartus. – Pars secunda.* Roret, Parisii/Fred. Fleischer, Lipsia, pp. 601–1121 + 3 pp. Corrigenda.

Schoenherr CJ (1840a) *Genera et species curculionidum, cum synonymia hujus familiae. Species novae aut hactenus minus cognitae, descriptionibus a Dom. L. Gyllenhal, C. H. Boheman, O. J. Fahraeus et entomologis aliis illustratae. Tomus quintus. – Pars secunda. Supplementum continens.* Roret, Parisii/Fred. Fleischer, Lipsia, pp. 465–970 (+ 4 pp. Corrigenda).

Schoenherr CJ (1840b) *Genera et species curculionidum, cum synonymia hujus familiae. Species novae aut hactenus minus cognitae, descriptionibus a Dom. L. Gyllenhal, C. H. Boheman, O. J. Fahraeus et entomologis aliis illustratae. Tomus sextus. – Pars prima. Supplementum continens.* Roret, Parisii/Fred. Fleischer, Lipsia, pp. 1–474. doi:10.5962/bhl.title.8952

Schoenherr CJ (1842a) *Genera et species curculionidum, cum synonymia hujus familiae. Species novae aut hactenus minus cognitae, descriptionibus a Dom. L. Gyllenhal, C. H. Boheman, O. J. Fahraeus et entomologis aliis illustratae. Tomus sextus. – Pars secunda. Supplementum continens.* Roret, Parisii/Fred. Fleischer, Lipsia, pp. 1–495. doi:10.5962/bhl.title.8952

Schoenherr CJ (1842b) *Genera et species curculionidum, cum synonymia hujus familiae. Species novae aut hactenus minus cognitae, descriptionibus a Dom. L. Gyllenhal, C. H. Boheman, O. J. Fahraeus et entomologis aliis illustratae. Tomus septimus. – Pars prima. Supplementum continens.* Roret, Parisii/Fred. Fleischer, Lipsia, pp. 1–479.

Schoenherr CJ (1843) *Genera et species curculionidum, cum synonymia hujus familiae. Species novae aut hactenus minus cognitae, descriptionibus a Dom. L. Gyllenhal, C. H. Boheman, O. J. Fahraeus et entomologis aliis illustratae. Tomus septimus. – Pars secunda. Supplementum continens.* Roret, Parisii/Fred. Fleischer, Lipsia, pp. 1–461.

Schoenherr CJ (1844) *Genera et species curculionidum, cum synonymia hujus familiae. Species novae aut hactenus minus cognitae, descriptionibus a Dom. L. Gyllenhal, C. H. Boheman, O. J. Fahraeus et entomologis aliis illustratae. Tomus octavus. – Pars prima. Supplementum continens.* Roret, Parisii/Fred. Fleischer, Lipsia, pp. 1–442.

Schoenherr CJ (1845) *Genera et species curculionidum, cum synonymia hujus familiae. Species novae aut hactenus minus cognitae, descriptionibus a Dom. L. Gyllenhal, C. H. Boheman, O. J. Fahraeus et entomologis aliis illustratae. Tomus octavus. – Pars secunda. Supplementum continens.* Roret, Parisii/Fred. Fleischer, Lipsia, pp. 1–504.

Schoenherr CJ (1847) *Mantissa secunda familiae Curculionidum. Seu descriptiones novorum quorundam generum Curculionidum.* Norstedt et filii, Holmia, 86 pp. doi:10.5962/bhl.title.5794

Schultze A (1897) Beschreibung neuer Ceuthorrhynchinen. *Deutsche Entomologische Zeitschrift* **1896**(2), 261–288.

Schröder ML, Nahrung HF, de Souza NM, Lawson SA, Slippers B, *et al.* (2021) Distribution of Gonipterus species and their egg parasitoids in Australia: implications for biological control. *Forests* **12**, 969. doi:10.3390/f12080969

Scott JK (1982) The impact of destructive insects on reproduction in six species of *Banksia* L. f. (Proteaceae). *Australian Journal of Zoology* **30**(6), 901–921.

Scott JK, Black R (1981) Selective predation by white-tailed black cockatoos on fruit of *Banksia attenuata* containing the seed-eating weevil *Alphitopis nivea*. *Australian Wildlife Research* **8**(2), 421–430.

Setliff GP (2007) Annotated checklist of weevils from the Papuan region (Coleoptera, Curculionoidea). *Zootaxa* **1536**(1536), 1–296. doi:10.11646/zootaxa.1536.1.1

Setliff GP, Larson PA (2009) The Indo-Australian weevil genus *Platytenes* Pascoe, 1870 (Coleoptera: Curculionidae: Cryptorhynchinae). *Insecta Mundi* **79**, 1–14.

Sharp D (1878) Descriptions of eight new species, and a new genus of Cossonides from New Zealand. *Transactions of the Entomological Society of London* **1878**(1), 9–14. doi:10.1111/j.1365-2311.1878.tb01939.x

Sharp D (1920) Studies in Rhynchophora. VII. On *Phalidura*. Amycteridae. *Entomologist's Monthly Magazine* **56**, 1–7.

Sheppard AW, Smyth MJ, Swirepik A (1999) Impact of the root-crown weevil (*Mogulones larvatus*) and other biocontrol agents on Paterson's curse in Australia: an update. In *Proceeding of the 12th Australian Weeds Conference* (Eds AC Bishop, M. Boersma and CD Barnes) pp 343–346. Tasmanian Weed Society, Hobart, Tasmania.

Shin S, Clarke DJ, Lemmon AR, Moriarty Lemmon E, Aitken AL, *et al.* (2018) Phylogenomic data yield new and robust insights into the phylogeny and evolution of weevils. *Molecular Biology and Evolution* **35**(4), 823–836. doi:10.1093/molbev/msx324

Shuckard WE (1838) Descriptions of some new genera of Coleoptera in the author's collection. *The Entomological Magazine* **5**(5), 505–513.

Sloane TG (1893) Amycterini [of the Elder Exploring Expedition]. *Transactions of the Royal Society of South Australia* **16**(2), 213–236.

Stierlin G (1888) Ueber Otiorrhynchus bicostatus und gemellatus. *Deutsche Entomologische Zeitschrift* **32**, 79–80.

Swan DC (1935) A weevil attacking mallow *(Malva parviflora* L. and *M. nicaeensis* All.) in South Australia. *Journal of Agriculture of South Australia* **38**, 1125–1128.

Tepper JGO (1887) *Common Native Insects of South Australia: a Popular Guide to South Australian Entomology. Part 1. Coleoptera or Beetles.* E. S. Wigg & Son, Adelaide, 46 pp.

Terry LI, Walter GH, Donaldson JS, Snow E, Forster PI, *et al.* (2005) Pollination of Australian *Macrozamia* cycads (Zamiaceae): effectiveness and behavior of specialist vectors in a dependent mutualism. *American Journal of Botany* **92**(6), 931–940. doi:10.3732/ajb.92.6.931

Thompson RT (1968) Revision of the genus *Catasarcus* Schönherr. *Bulletin of the British Museum (Natural History), Entomology* **22**(8), 359–454. doi:10.5962/bhl.part.9955

Thompson RT (1977) A revision of the New Guinea weevil genus *Apirocalus* Pascoe (Coleoptera: Curculionidae). *Bulletin of the British Museum (Natural History), Entomology* **36**(5), 193–280.

Thompson RT (1992) Observations on the morphology and classification of weevils (Coleoptera, Curculionoidea) with a key to major groups. *Journal of Natural History* **26**(4), 835–891. doi:10.1080/00222939200770511

Thompson RT (1996) The species of *Phaenomerus* Schönherr (Coleoptera: Curculionidae: Zygopinae) of the Australian Region. *Invertebrate Taxonomy* **10**(5), 937–993. doi:10.1071/it9960937

Thompson RT, Howell A (2001) The Australian fern weevil Syagrius intrudens Waterhouse (Curculionidae) confirmed on Guernsey. *The Coleopterist* **10**(2), 48–48.

Tilley DB, Barrows TT, Zimmerman EC (1997) Bauxitic insect pupal cases from northern Australia. *Alcheringa* **21**(2):157–160. doi:10.1080/03115519708619182

van Emden F (1938) On the taxonomy of Rhynchophora larvae (Coleoptera). *Transactions of the Royal Entomological Society of London* **87**, 1–37.

van Emden F (1952) On the taxonomy of Rhynchophora larvae: Adelognatha and Alophinae (Insecta: Coleoptera). *Proceedings of the Zoological Society of London* **122**, 651–795. doi:10.1111/j.1096-3642.1952.tb00248.x

Vaurie P (1971) Weevils of the tribe Sipalini (Coleoptera, Curculionidae, Rhynchophorinae). Part 3, The genus *Sipalinus*. *American Museum novitates* **2463**, 40–43.

Voss E (1922) Neue Rüsselkafer aus verschiedenen Erdteilcn. *Deutsche Entomologische Zeitschrift* **1922**(1), 166–174. doi:10.1002/mmnd.48019220116

Voss E (1934) Die Cossoninen Afrikas und Madagaskars des Deutschen Entomologischen Instituts der Kaiser-Wilhelm-Gesellschaft. *Arbeiten ueber Morphologische und Taxonomische Entomologie Berlin* **1**, 118–123, 186–194.

Voss E (1935) Fünf Rüssler aus den Tribus Prionomerini und Anthonomimi von Java (Col. Curc.). *Entomologische Blaetter Krefeld* **31**, 228–233.

Voss E (1939) Rhynchiten, Attelabinen und Cossoninen aus dem Kongo-Gebiet (Col., Curc.). *Revue de Zoologie et de Botanique Africaines* **32**(1), 42–82.

Voss E (1956) Die von Biro auf Neu-Guinea aufgefundenen Russelkafer (Col.). I. *Annales Historico-Naturales Musei Hungarici NS* **7**, 121–142.

Voss E (1960) Die von Biró auf Neu Guinea aufgefundenen Rüsselkäfer, III. (Coleoptera, Curculionidae.). *Annales Historico-Naturales Musei Nationalis Hungarici (Zoologica)* **52**, 313–346.

Wagner H (1927) Revision des Subgen. *Mogulones* Rttr. der Gattung *Ceutorrhynchus* Germ. (Curcul.). *Coleopterologisches Centralblatt* **2**(3–4), 201–217.

Warner RE (1970) *Neochetina eichhorniae*, a new species of weevil from waterhyacinth, and biological notes on it and *N. bruchi* (Coleoptera: Curculionidae: Bagoini). *Proceedings of the Entomological Society of Washington* **72**(4), 487–496.

Wassell JLH (1966) Coprophagous weevils (Coleoptera: Curculionidae). *Australian Journal of Entomology* **5**(1), 73–74. doi:10.1111/j.1440-6055.1966.tb00688.x

Waterhouse CO (1874) Description of a new species of *Baridius* (Coleoptera: Rhynchophora) from Singapore, which destroys orchids. *Entomologist's Monthly Magazine* **10**, 226–227.

Waterhouse CO (1875) On the Coleoptera of Kerguelen's Island. *Entomologist's Monthly Magazine* **12**, 54–57.

Waterhouse CO (1876) Description of a new species of *Ectemnorrhinus* from Kerguelen Island. *Entomologist's Monthly Magazine* **13**, 51–52.

Waterhouse CO (1903) Description of a new coleopterous insect belonging to the Curculionidae. *Annals and Magazine of Natural History (Series 7)* **11**(62), 229–230. doi:10.1080/00222930308678755

Waterhouse GR (1853) Abstract: 'Descriptions of new genera and species of Curculionidae'. *Proceedings of the Entomological Society of London (Series 2)* **2**(5), 99–105.

Waterhouse GR (1854) Notes on the species of *Amycterus* and allied genera, with descriptions of some new species. *Transactions of the Entomological Society of London (Series 2)* **3**(2), 75–80. doi:10.1111/j.1365-2311.1854.tb02662.x

Waterhouse GR (1862) Note on the *Strongylorhinus ochraceus* of Schönherr, and descriptions of two nearly allied species of Curculionidae, from Australia. *Transactions of the Entomological Society of London (Series 3)* **1**(3), 227–231. doi:10.1111/j.1365-2311.1862.tb00601.x

Weber F (1801) *Observationes entomologicae, continentes novorum quae condidit generum characteres, et nuper detectarum specierum descriptiones*. Bibliopolii Academici Novi, Kiliae, xii + 116 pp. doi:10.5962/bhl.title.8639

Westwood JO (1838) Synopsis of the genera of British insects. In *An Introduction to the Modern Classification of Insects, Founded on the Natural Habits and Corresponding Organisation of the Different Families.* (Ed. JO Westwood) pp. 1–48. Longman, Orme, Brown, Green & Longmans, London.

Wheeler GS, Taylor GS, Gaskin JF, Purcell MF (2011) Ecology and management of sheoak (*Casuarina* spp.), an invader of coastal Florida. *Journal of Coastal Research* **27**(3), 485–492.

White A (1846) Insects. In *The Zoology of the Voyage of H.M.S. Erebus & Terror, under the Command of Captain Sir James Clark Ross, R.N., F.R.S., during the Years 1839–43. Vol. II. Reptiles, Fishes, Crustacea, Insects, Mollusca.* (Eds J Richardson and JE Gray) pp. 1–24. E. W. Janson, London.

White A (1852) Appendix. Descriptions of some new species of Annulosa collected by Mr. John MacGillivray during the voyage of H.M.S. Rattlesnake. In *Narrative of the Voyage of H.M.S. Rattlesnake, Commanded by the late Captain Owen Stanley, R.N., F.R.S. etc, during the Years 1846–1850. Including Discoveries and Surveys in New Guinea, the Louisiade Archipelago, etc., to which is added the Account of Mr. E B. Kennedy's Expedition for the Exploration of the Cape York Pensinsula. In two volumes. Vol. II.* (Ed. J MacGillivray) pp. 387–395. T. & W. Boone, London. doi:10.5962/bhl.title.4847

White A, Doubleday E (1843) List of the annulose animals hitherto recorded as found in New Zealand, with the descriptions of some new species. In *Travels in New Zealand; With Contributions to the Geography, Geology, Botany and Natural History of that Country. In two volumes – Vol. II.* (Ed. E Dieffenbach) pp. 1–396. John Murray, London.

Wibmer GJ, O'Brien CW (1986) Annotated checklist of the weevils (Curculionidae *sensu lato*) of South America (Coleoptera: Curculionoidea). *Memoirs of the American Entomological Institute* **39**, i–xvi + 1–563.

Williams G (2021) *The Flowering of Australia's Rainforests: Pollination Ecology and Plant Evolution.* CSIRO Publsihing, Melbourne, Victoria, 288 pp.

Wollaston TV (1854) A revision of North American Cryphalini (Scolytidae, Coleoptera). *University of Kansas Science Bulletin* **36**(2), 959–1089.

Wollaston TV (1873a) On the Cossonidae of Japan. *Transactions of the Entomological Society of London* **1873**(1), 5–43. doi:10.1111/j.1365-2311.1873.tb00635.x

Wollaston TV (1873b) On the genera of the Cossonidae. *Transactions of the Entomological Society of London* **1873**(4), 427–657. doi:10.1111/j.1365-2311.1873.tb00645.x

Wood SL (1973) On the taxonomic status of Platypodidae and Scolytidae (Coleoptera). *Great Basin Naturalist* **33**(2), 77–90.

Woodburn TL (1997) Establishment in Australia of *Trichosirocalus horridus* a biological control agent for *Carduus nutans*, and preliminary assessment of its impact on plant growth and reproductive potential. *Biocontrol Science and Technology* **7**(4), 645–656. doi:10.1080/09583159730686

Zimmerman EC (1936) Review of the genus *Orochlesis*. *Occasional Papers of the Bernice P. Bishop Museum* **12**(1), 1–19.

Zimmerman EC (1939) *Blepiarda* of Fiji (Coleoptera: Curculionidae). *Proceedings of the Hawaiian Entomological Society* **10**(2), 325–328.

Zimmerman EC (1940) Synopsis of the genera of Hawaiian Cossoninae with notes on their origin and distribution (Coleoptera, Curculionidae). *Occasional Papers of the Bishop Museum Honolulu* **15**, 271–293.

Zimmerman EC (1967) *Imathia* and *Amblycnemus* (Coleoptera: Curculionidae: Cryptorhynchinae). *Pacific Insects* **9**(2), 187–196.

Zimmerman EC (1991) *Australian Weevils (Coleoptera: Curculionoidea). Volume V. Colour Plates 1–304.* CSIRO Australia, Melbourne, x + 633 pp, 303 pls.

Zimmerman EC (1992) *Australian Weevils (Coleoptera: Curculionoidea). Volume VI. Colour Plates 305–632.* CSIRO Australia, Melbourne, viii + 707 pp, 328 pls.

Zimmerman EC (1993) *Australian Weevils (Coleoptera: Curculionoidea). Volume III. Nanophyidae, Rhynchophoridae, Erirhinidae, Curculionidae: Amycterinae, Literature Consulted.* CSIRO Australia, Melbourne, x + 854 pp.

Zimmerman EC (1994) *Australian Weevils (Coleoptera: Curculionoidea). Volume 1. Orthoceri. Anthribidae to Attelabidae. The Primitive Weevils.* CSIRO Australia, Melbourne, xxxii + 741 pp.

45. CURCULIONIDAE: PLATYPODINAE SCHUCKARD, 1840

James R. M. Bickerstaff, Deborah S. Kent, Rolf G. Oberprieler, Debbie Jennings and Hermes E. Escalona

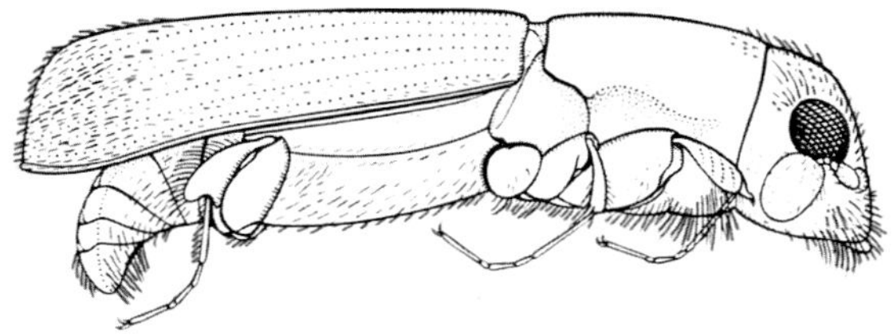

Fig. 45.1. *Platypus subgranosus* Schedl.

Introduction. The subfamily Platypodinae comprises more than 1400 species in 34 genera worldwide and is divided into two tribes, Tesserocerini and Platypodini (Jordal 2015; Kirkendall *et al.* 2015). All species occur in forested ecosystems and most in subtropical and tropical environments. The subfamily is characterised by high levels of endemism (Jordal 2015), but several species have been spread across continents by humans and are considered as pests in managed agricultural plantations and native forest ecosystems (Kirkendall & Faccoli 2010; Beaver 2013). Platypodines complete their entire life cycle in woody tissues of host trees, constructing tunnels or galleries in fallen logs and moribund or healthy trees, in which their larvae develop (Beaver 1989; Jordal 2014).

The monophyly of Platypodinae has now been firmly established and is rarely questioned, except for the inclusion of the genus *Mecopelmus* Blackman. The taxonomic status of the subfamily and its relationship to the other subfamily of wood-boring ambrosia beetles, Scolytinae, have been contentious (Jordal 2014, 2015). Previously often treated as a distinct family (e.g. Wood 1973; Thompson 1992; Alonso-Zarazaga 1999), Platypodinae are now universally recognised as a subfamily of Curculionidae (e.g. Crowson 1955; May 1993; Oberprieler *et al.* 2007; McKenna *et al.* 2009; Jordal 2014; Pullen *et al.* 2014; Shin *et al.* 2018). Due to their similar life styles and morphological adaptations and also to phylogenetic analyses based on morphological characters of the adults, they were previously thought to be closely related to Scolytinae (e.g. Kuschel 1995; Marvaldi *et al.* 2002; Jordal *et al.* 2011), but phylogenetic analyses based on characters of the larvae (Marvaldi 1997) and various molecular data (e.g. McKenna *et al.* 2009; Gillett *et al.* 2014; Mugu *et al.* 2018; Shin *et al.* 2018; Song *et al.* 2020; Smit *et al.* 2022) have consistently demonstrated a sister-group relationship between Platypodinae and Dryophthorinae. Fuller accounts of the historical treatment of the subfamily were provided by Wood (1993), Kuschel *et al.* (2000) and Jordal (2014).

The fossil record of Platypodinae is sparse, with most specimens identified and described belonging to the tribe Tesserocerini. Most fossils have been recorded from Baltic (Peris *et al.* 2017), Mexican and Dominican (Bright & Poinar 1994; Peris *et al.* 2015) and Ethiopian and Zhanpu (Solórzano-Kraemer *et al.* 2021) amber deposits of Eocene and Miocene origins. Fossil-calibrated phylogenetic analyses suggest an Upper Cretaceous origin of the Platypodinae, with the divergence between Platypodinae and Dryophthorinae occurring 85 Ma ago and the divergence between Tesserocerini and Platypodini about 75 Ma ago (Shin *et al.* 2018).

Biology. Most Platypodinae are xylomycetophagous, boring in wood but having a symbiont association with ambrosia fungi. Both the adult and larval life stages obtain nutrition from their fungal symbionts, which has given them the common name 'ambrosia beetles', but this name also applies to various lineages of Scolytinae with similar associations (Hulcr & Stelinski 2017). This life style is, however, absent in the American genera *Schedlarius* Wood and *Mecopelmus* Blackman, which feed on dead wood rather than on fungi (Jordal 2014). Platypodinae fungal symbionts typically belong to the ascomycete genus *Raffaelea*, but several species are also recorded to be associated with *Ambrosiozyma* yeasts (Hulcr & Stelinski 2017), whereas adults of the widespread *Crossotarsus imitratrix* (Schedl) are known to invade ambrosia beetle galleries and directly consume mycelia from the host species (Hulcr & Cognato 2010), a life style termed mycetocleptism. Platypodine ambrosia beetles transport fungal symbionts in specialised structures broadly known as mycangia (although the correct term is mycetangia, see Vega & Biedermann 2020). These structures are variable in size and shape, but in most species the mycangia are cuticular invaginations of various sizes and shapes on the pronotal disc, surrounded by glandular cells to reduce desiccation of spores or fungal masses (Mayers *et al.* 2022). In some other cases, like *Crossotarsus*, the mycangium is located inside the head capsule (Nakashima 1971; Mayers *et al.* 2022).

Platypodines typically live and reproduce in recently dead or dying trees, constructing branching tunnels, referred to as galleries, in their trunks. The galleries are typically bored through the sapwood and heartwood of the host, and fungal symbionts are cultivated on the walls (Hulcr & Stelinski 2017). Most species that construct galleries in dead wood are generalists and highly polyphagous (Wood & Bright 1992). Several Australian species, including *Austroplatypus incompertus* Schedl, *Platypus tuberculosus* Strohmeyer and *Notoplatypus elongatus* Lea, construct galleries in live *Eucalyptus* trees but apparently do not negatively impact the tree health (Browne 1971; Kent 2008a; Bickerstaff *et al.* 2020). In Tasmania, *Platypus subgranosus* Schedl is regarded as a pest of *Nothofagus cunninghamii* trees that are already infected with the fungus *Davidsonella australis* (Elliott *et al.* 1987; Kile & Hall 1988; Candy 1990). In other regions of the world, *Megaplatypus mutatus* (Chapuis), *Trachyostus ghanaensis* (Schedl) and *Dendroplatypus impar* (Schedl) attack live trees and can cause physiological stress, which can lead to host tree mortality (Roberts 1961; Neumann & Harris 1974; Alfaro *et al.* 2007; Zanuncio *et al.* 2010).

All Platypodinae species except *A. incompertus* are thought to be social to some degree, engaging in parental investment

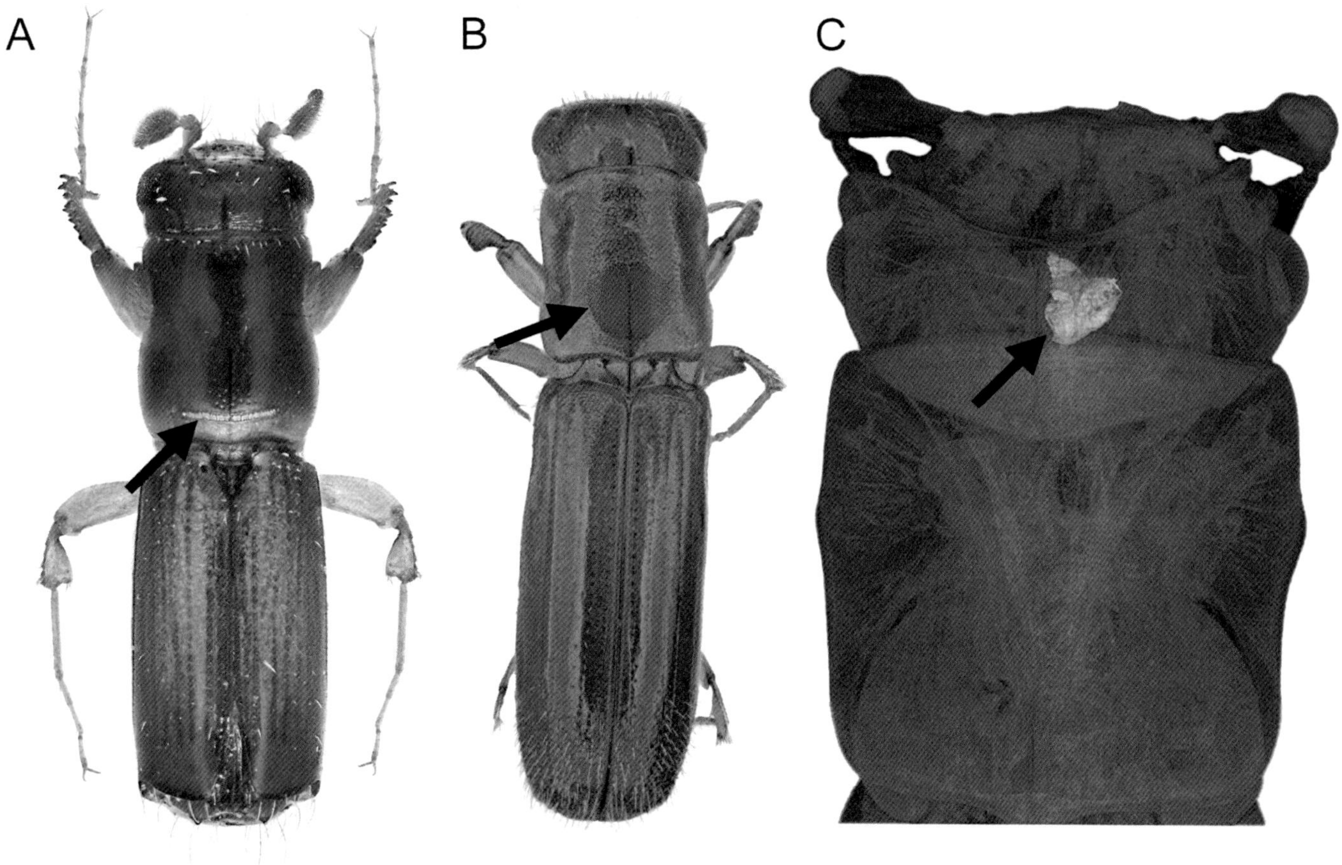

Fig. 45.2. Presence and location of mycangia in some Platypodinae. A, mycangial bar of *Diapus pusillimus* Chapuis; B, mycangial pronotal pits of *Treptoplatypus crenatus* (Chapuis); C, mycangium inside head capsule of *Crossotarsus mniszechi* Chapuis.

into offspring and alloparental care. All platypodine species are monogynous, and typically both parents remain in the gallery to contribute to fungal cultivation, nest hygiene and care for the offspring (Kirkendall *et al.* 1997). Males typically guard the entrance of the gallery to deter the entry of intruders using their specially adapted elytral declivity to plug the entrance (e.g. *Dinoplatypus*) and also to expel frass and debris from the gallery entrance (Kirkendall *et al.* 1997). Recently eclosed adults have been recorded to delay their dispersal and reproduction in order to contribute to alloparental care and assist in gallery maintenance (Kirkendall *et al.* 2015).

Austroplatypus incompertus is unique among Coleoptera in having evolved eusociality (Kent & Simpson 1992; Smith *et al.* 2018). The species forms a monotypic lineage of Platypodini and is the single transition to eusociality in Platypodinae (Jordal 2015). In contrast to most eusocial insect lineages (such as those in Hymenoptera), it has a diplodiploid sex determination system (Smith *et al.* 2009) and is considered to be primitively eusocial, as there are no morphological differences between queens and workers (Kent 2001; Smith *et al.* 2018). Colonies of *A. incompertus* construct galleries in live stringybark trees (*Eucalyptus* species). The colonies consist of an inseminated foundress queen, unmated daughter workers that remain in the gallery permanently, larvae and pupae and dispersing adult daughters and sons (Smith *et al.* 2018). Foundress queens copulate only once, as mature males have never been found in the galleries and are presumed to die outside (Smith *et al.* 2018), and *A. incompertus* is thus the only known platypodine species in which the gallery is not established by a founding pair. Furthermore, no inbreeding events have been detected in galleries, and daughter workers remain unmated (Smith *et al.* 2018), so that all reproductive output in the gallery is the result of a single mating of the foundress. Galleries of *A. incompertus* are long-lived and have been recorded to be maintained for up to 36 years (Harris *et al.* 1976). There is a higher proportion of daughter workers retained in younger colonies to assist in gallery maintenance, and as colonies age the proportion of dispersing daughters increases (Smith *et al.* 2018).

Several species of Platypodinae have been introduced across multiple continents, including *Crossotarsus externedentatus* (Fairmaire), *Euplatypus parallelus* (Fabricius) and *Megaplatypus mutatus*. Native to the Oriental region, *C. externedentatus* has been introduced to several countries around the world, including Australia, tropical eastern and

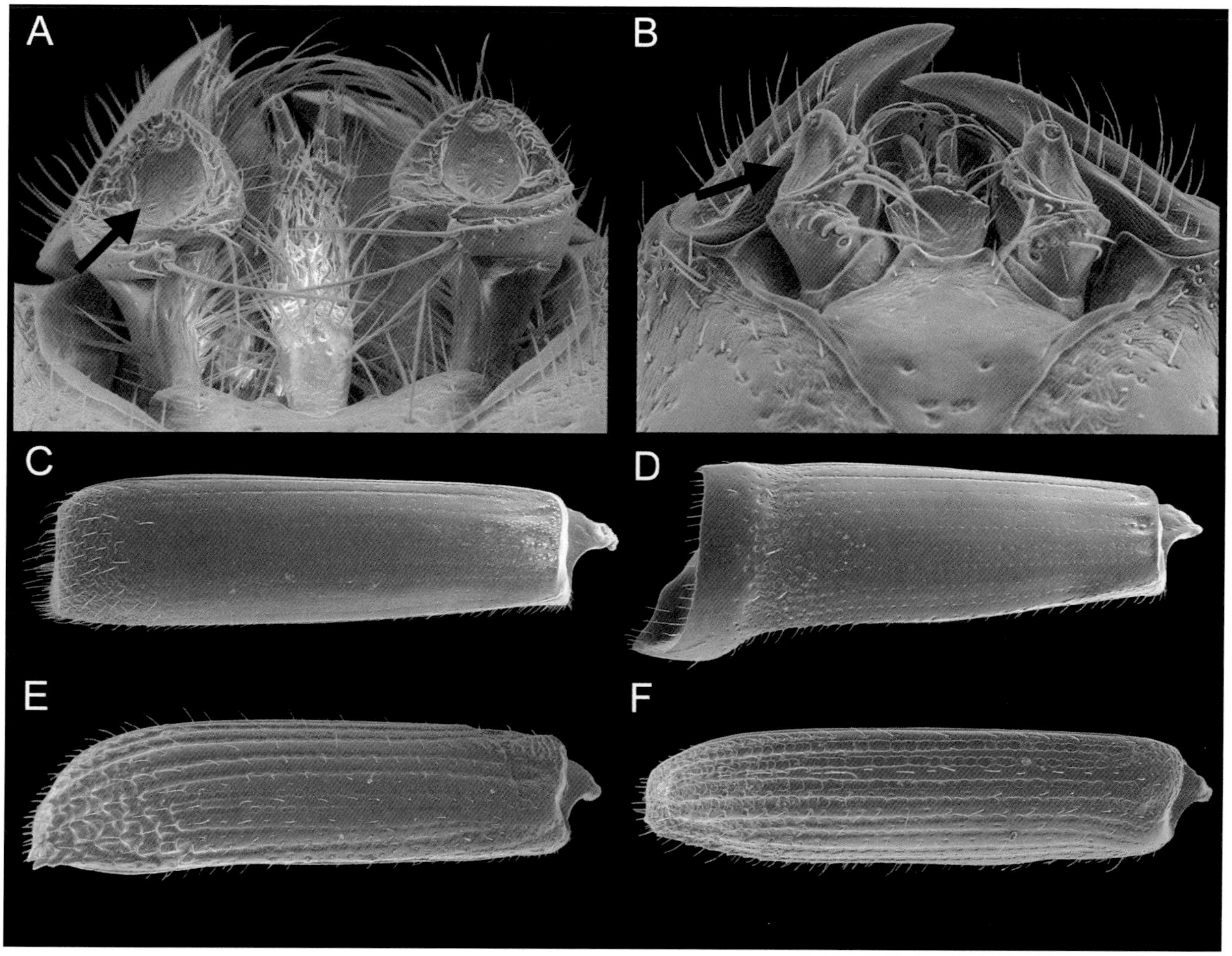

Fig. 45.3. Sexually dimorphic characters in Platypodinae. A–B, maxillary palps: A, ♀; B, ♂; C–D, *Dinoplatypus* sp., elytron: C, ♀; D, ♂; E–F, *Austroplatypus incompertus* (Schedl), elytron: E, apically spinose, ♀: F, apically unarmed, ♂.

southern Africa and Fiji and Samoa, and it has also been intercepted in Japan (Beaver & Liu 2013). This species is highly polyphagous and of economic concern as it attacks rubberwood logs in Thailand (Sittichaya & Beaver 2009) as well as chestnut (Puranasakul 2006), mango (Sittichaya 2012) and mahogany trees (Beaver 2000a). In Australia, individuals have been recorded near ports of entry in Sydney, but establishment of the species has not yet been confirmed (Bickerstaff *et al.* 2020). Another widely distributed species is *Euplatypus parallelus*, which has been introduced from its native range in Central and South America to Asia, Southeast Asia, Africa, Australia and Papua New Guinea. In Australia its distribution ranges from North Queensland to Barrington Tops in New South Wales. It is highly polyphagous (Beaver 2013; Silva *et al.* 2013) and known to attack rubber trees in Thailand (Sittichaya & Beaver 2009), and it also transmits the phytopathogenic wilt fungus *Fusarium oxysporum* (Bumrungsri *et al.* 2008; Ploetz *et al.* 2013). Of these three widely distributed platypodines, *M. mutatus* is the only species that is not present in Australia, but it has been introduced from South America to Italy (Kirkendall & Faccoli 2010) and is recorded to attack poplars, willows and fruit trees throughout subtropical and tropical South America (Alfaro *et al.* 2007) and *Eucalyptus* trees in plantations in Brazil (Zanuncio *et al.* 2010). The fungus *Raffaelea santoroi* has been identified to be its primary symbiont, but it poses no damage when inoculated into host trees, host mortality instead being caused by mechanical damage through the boring activity of the beetles (Alfaro *et al.* 2007; Ploetz *et al.* 2013).

Several species of Platypodinae are associated with phytopathogenic fungi. *Platypus quercivorus* Murayama and its symbiont, *Raffaelea quercivora*, are responsible for Japanese Oak Wilt (JOW) in Fagaceae host trees in Japan (Kuroda 2001; Kubono & Ito 2002). The symbiont fungus alone does not cause the death of the host, but mass attacks of *P. quercivorus* and inoculation with the phytopathogen result in tree mortality (Kinuura & Kobayashi 2006). Similar to JOW, Korean Oak Wilt (KOW) is caused by *P. koryoensis* Murayama inoculating susceptible

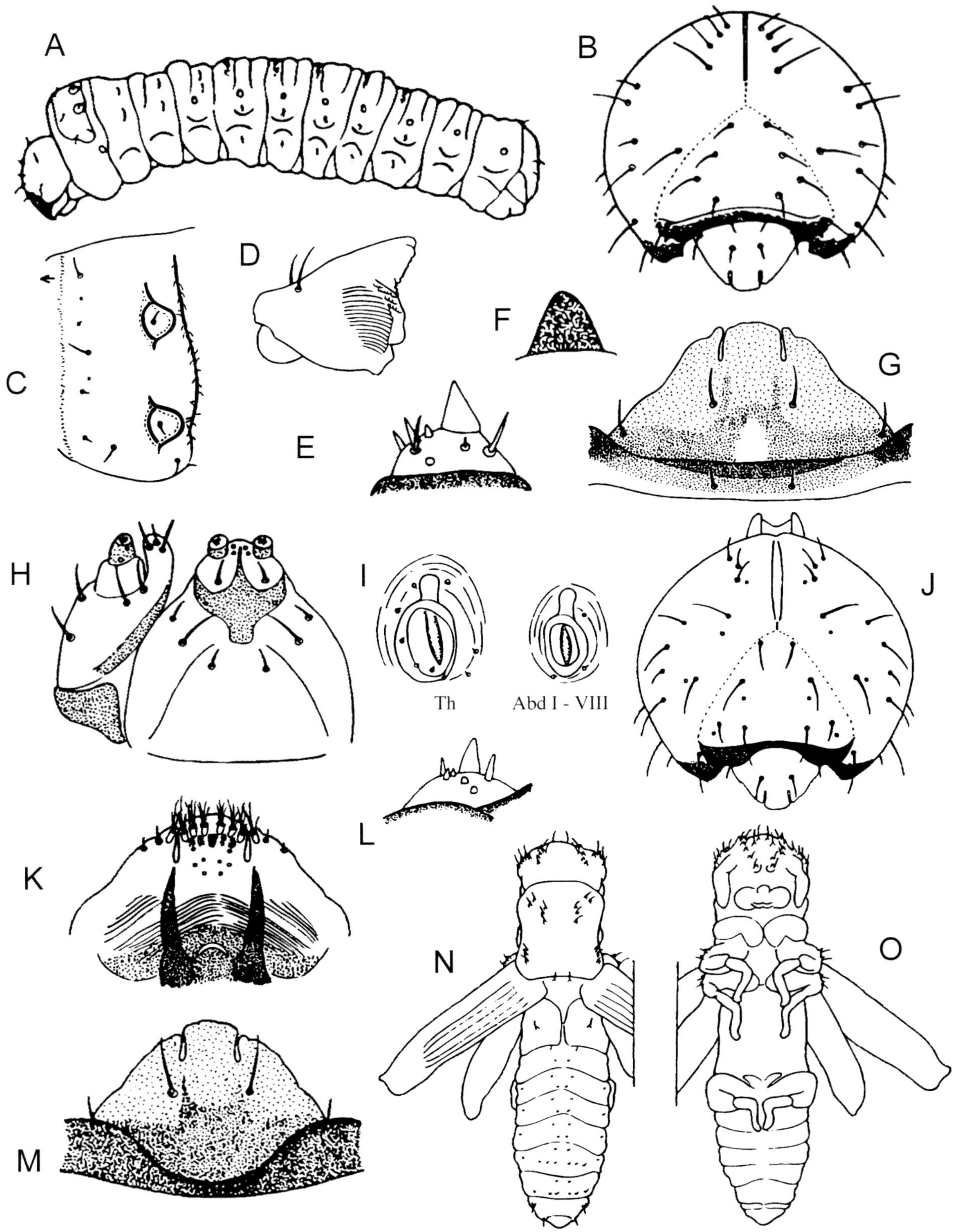

Fig. 45.4. Platypodini, structural details of immature stages. A–I, *Treptoplatypus crenatus* (Chapuis), larva: A, habitus, lateral; B, head, dorsal; C, pronotum, lateral; D, mandible; E, antenna; F, caudal horn of AIX; G, labrum and epistome, dorsal; H, mala and labium, ventral; I, spiracles, showing alignment; J–M, *Euplatypus parallelus* (Fabricius), larva: J, head, dorsal; K, epipharyngeal lining, ventral; L, antenna; M, labrum and epistome, dorsal; N–O, *E. parallelus*, pupa, habitus: N, dorsal; O, ventral. After May (1994).

Quercus trees with the phytopathogen *R. quercus-mongolicae* (Ploetz *et al.* 2013), with mortality of the hosts also caused by the fungus only when accompanied by boring damage by the beetles (Lee *et al.* 2011). In New Zealand, *Nothofagus* trees are stressed and potentially killed by the phytopathogen *Sporothrix nothofagi*, which is transmitted by *P. apicalis* (White), *P. gracilis* Broun and *Treptoplatypus caviceps* Broun (Ploetz *et al.* 2013). In Australia, no platypodines are known to be solely responsible for transmitting phytopathogenic fungi that lead to host tree mortality. *Platypus subgranosus* is associated with the pathogenic fungus *Davidsonella australis*, which can cause mortality of *N. cunninghamii*; however, the beetle does not inoculate the trees with the phytopathogen but only constructs galleries in already infected trees (Elliott *et al.* 1987; Kile & Hall 1988; Candy 1990).

Australian fauna. There are 43 species in nine genera and two tribes of Platypodinae known in Australia, although recent phylogenetic analyses indicate that some genera, particularly of Platypodini, are polyphyletic (Jordal 2015). Twenty-nine of these species are endemic or native to Australia and one, *Euplatypus parallelus*, is introduced. The remaining 13 species are all probable intercepts reported from Australian ports (Bickerstaff *et al.* 2020). Most Australian species belong in the tribe Platypodini, with *Platypus* being the most specious genus, containing 15 established species and seven probable intercepts. Wood (1993), in his revision of the Platypodinae, recognised 121 species of *Platypus* globally, most of them occurring in southeast Asia and the Pacific Islands. The second largest genus in Australia is *Crossotarsus*, containing five established species and two probable intercepts, and globally it comprises 90 species. The remaining Australian genera have fewer species, *Dinoplatypus* three established and three probable intercepts, *Diapus* one established and one probable intercept, *Treptoplatypus* two established and *Baiocis* one established species. Australia is also home to two endemic monotypic genera, *Notoplatypus* and *Austroplatypus*, which are the basal (early-diverging) lineages in their respective tribes and both constructing galleries in *Eucalyptus* trees (Jordal 2015; Bickerstaff *et al.* 2020).

In Australia, most platypodine species are found in wet mesic forest ecosystems along the east coast. North Queensland in particular hosts a significant portion of this diversity, and one species, *Treptoplatypus crenatus* (Chapuis), also occurs in tropical areas of the Northern Territory (Bickerstaff *et al.* 2020). Only a few species inhabit islands off the mainland of Australia, *P. subgranosus, P. tasmanicus* Schedl and *P. tuberculosus* Strohmeyer occurring in Tasmania, *P. omnivorus* Lea on Lord Howe Island and *P. norfolkensis* Schedl endemically on Norfolk Island.

Characteristics. Adults. Body length 1.85–7.8 mm, body typically elongate and cylindrical. Sexual dimorphism present in most species but varying; in Tesserocerini typically apparent by presence or absence of mycangia and armature of elytral declivity, in Platypodini typically also evident by presence of mycangia (except *Crossotarsus*), elytral modifications and armature in males (except *Austroplatypus*, in which female elytra armed) and inflated maxillary palps in females (except *Austroplatypus* and *Crossotarsus*). Forehead and vertex typically smooth in most species to densely punctate, usually sparsely setose. Eyes rounded. Antennal funicles 4-segmented in most species, 3-segmented in *Notoplatypus elongatus*; antennal clubs typically large and longer than scape in most species but small and shorter than scape in *Austroplatypus*, either partially or completely setose but a testaceous glabrous strip present in Tesserocerini. Submentum with lateral margins reaching genae in Platypodini but not in Tesserocerini. Mandibles with deciduous appendages in females of some *Diapus* species. Maxillary and labial palps typically 3-segmented, but labial palps 2-segmented in *Crossotarsus*; maxillary palps with segment 3 inflated in females of *Dinoplatypus, Platypus* and *Treptoplatypus*. Pronotum typically quadrate to constricted medially in dorsal view, typically smooth and shining but densely punctate in *Notoplatypus*; with mycangial pits in some genera, mostly females but also males of some *Platypus* species. Scutellum impressed below elytra. Elytral declivity highly variable, either truncate, rounded, forming vertical face surrounded by elevated carina, attenuated or ridged, often also armed with spines, teeth, tubercles or granules; setae on the declivity dense, sparse and variable in length to absent. Procoxae contiguous to widely separate. (Jordal 2014; Bickerstaff *et al.* 2020).

Larvae (final instar). Body elongate, cylindrical, 1–9 mm long; head 1–1.3 mm wide. Body soft and white, with few dark, coarse and short setae. Head free, subglobose, hypognathous, partially covered by pronotum, usually pale yellow but oral margins, mandibles and labrum reddish brown. Stemmata usually absent. Antennae minute, 1-segmented, conical, finely setose. Clypeus obsolete, reduced to a narrow membrane connecting labrum to epistome; labrum broadly trapezoidal; tormae subparallel, widened at base and fused with labrum. Mandibles triangulate with chisel-like cutting edge. Thorax in some genera with slightly raised patterns of brown bands varying in structure; spiracles broadly ovate and surrounded by 9–10 min setae. Abdomen in some genera armed with granules or tubercles. (Browne 1972; May 1994; Kent 2010).

Key to the Australian tribes and genera of Platypodinae

The key is more likely to work for males from couplet 7 onwards, as the remaining genera are defined on the shape of the elytral declivity in males and there are few characters suited to identify both sexes.

1. Submentum with lateral margins reaching genae (Fig. 45.5B); maxillary palps truncate (Fig. 45.5B) .... **PLATYPODINI** ... 3

– Submentum with lateral margins not reaching genae (Fig. 45.5A); maxillary palps conical (Fig. 45.5A) .... **TESSEROCERINI** ... 2

2(1). Antennal funicles 4-segmented (Fig. 45.5D, E); elytral declivity truncate in lateral view and armed with spines in males, unarmed in females; body length 1.85–2.85 mm .... ***Diapus* Lea**

– Antennal funicles 3-segmented (Fig. 45.5C); elytral declivity rounded in lateral view (Pl. 46N) and armed with spines in both sexes, larger in males; body length 6–6.5 mm .... ***Notoplatypus* Lea**

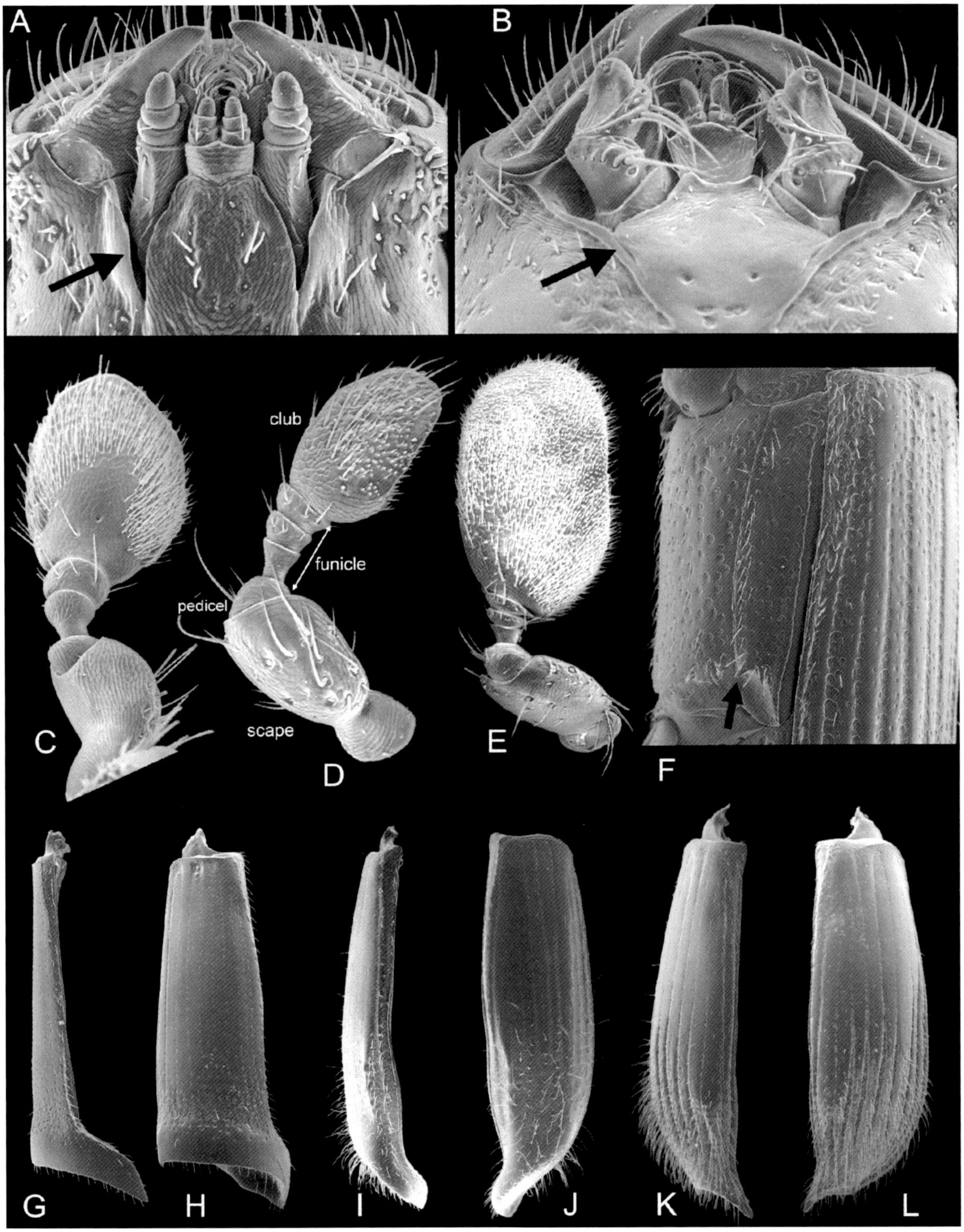

Fig. 45.5. Diagnostic characters for the key to the genera of Australian Platypodinae. A, *Notoplatypus elongatus* Lea, submentum, lateral margins not reaching genae, ♂; B, *Platypus omnivorus* Lea, submentum, lateral margins reaching genae, ♂; C–E, antennae: C, *N. elongatus* Lea, 3-segmented antennal funicle; D, *Austroplatypus incompertus* Schedl, 4-segmented antennal funicle and small club; E, *Treptoplatypus crenatus* (Chapuis), 4-segmented antennal funicle and large, densely setose club; F, *Euplatypus parallelus* (Fabricius), metanepisternum with posterior spines; G–H, *Dinoplatypus pseudocupulatus* (Schedl), keeled elytral declivity: G, lateral; H, dorsal; I–J, *T. crenatus* (Chapuis), attenuated elytral declivity: I, lateral; J, dorsal; K–L, *P. omnivorus* Lea, spined and furrowed elytral declivity: K, lateral; L, dorsal.

3(1). Labial palps 2-segmented; body length 3–7.8 mm .... ***Crossotarsus*** **Chapuis**
– Labial palps 3-segmented (Fig. 45.5A) 4
4(3). Body 4 × as long as wide; prothorax emarginate behind midline in dorsal view; elytral declivity truncate in lateral view; body length 2.3–2.5 mm .... ***Baiocis*** **Browne**
– Body 3.4–3.6 × as long as wide; prothorax quadrate to slightly emarginate in dorsal view; elytral declivity variable in lateral view 5
5(4). Antennal clubs relatively small, as long as scapes, basal section glabrous (Fig. 45.5D); body length 5–6 mm .... ***Austroplatypus*** **Browne**
– Antennal clubs relatively large, longer than scapes, surface setose (Fig. 45.5E) 6
6(5). Elytral declivity truncate and broad at apex, forming vertical surface surrounded by elevated carina (Fig. 45.5G, H); body length 2.8–5.5 mm .... ***Dinoplatypus*** **Wood**
– Elytral declivity not as above 7
7(6). Metanepisterna and metaventrite in males with posterior spines (Fig. 45.5F); elytral striae deeply punctate; elytral apex emarginate in dorsal view (Pl. 46I); body length 3.5–5.2 mm .... ***Euplatypus*** **Wood**
– Metanepisterna and metaventrite in males and females without spines; elytral striae faint; elytral apex emarginate or not in dorsal view 8
8(7). Elytral declivity in males attenuate (Fig. 45.5I), elytra with apex weakly emarginate or not in dorsal view (Fig. 45.5J); body length 4–5 mm .... ***Treptoplatypus*** **Schedl**
– Elytral declivity in males variable, attenuate with elytral apex wide and strongly emarginate or not, armed with spines, tubercles or striae and excavated (e.g. Fig. 45.5K, L); body length 2.6–7 mm .... ***Platypus*** **Herbst**

Classification of the Australian genera of Platypodinae

The generic classification used herein follows that of Pullen *et al.* (2014), with subsequent taxonomic changes by Beaver & Sanguansub (2015, 2020). The biogeographical origins of species (i.e., native, introduced, intercepted) are taken from Bickerstaff *et al.* (2020), with probable intercepts referring to species that have only been collected at or near ports of entry to Australia. As such, these represent species that may have been recorded in Australia but lack established populations.

Tribe Tesserocerini Strohmeyer, 1914

Diapus Chapuis, 1865 (Pl. 46M, N)

Type species, by subsequent designation (Hopkins, 1914: 121): *Diapus quadrispinatus* Chapuis, 1865.

Characteristics. Body length 1.85–2.35 mm. Frons and vertex mostly smooth with some punctures, sparsely setose. Antennal funicles 4-segmented, clubs setose and with a testaceous glabrous strip, either along posterior margin of club or medially. Submentum with lateral margins not reaching genae; labial and maxillary palps 3-segmented. Mandibles with deciduous appendages in females of some species. Pronotum constricted medially and without punctures. Posterior area of pronotal disc with mycangial bar, in *D. quinquespinatus* accompanied by pits. Scutellum impressed underneath elytra. Elytral declivity truncate in lateral view, unarmed in females but sometimes armed with spines in males, sparsely setose. Procoxae widely separate.

Australian species. Native: *D. pusillimus* Chapuis, 1865. Intercepted: *D. quinquespinatus* Chapuis, 1865.

Distribution. Australia (QLD), Africa (tropical countries), India, Madagascar, Micronesia, Papua New Guinea, Samoan Islands, Southeast Asia.

Notes. Highly polyphagous, constructs galleries in recently dead or dying trees (Bickerstaff *et al.* 2020). Species are monogamous with male-initiated galleries. Females use deciduous mandibular appendages to pull males out of the galleries they have initiated and complete gallery construction after mating (Roberts 1993; Beaver 2000b).

Notoplatypus Lea, 1910 (Pl. 46O; Fig. 45.6)

Type species, by monotypy: *Notoplatypus elongatus* Lea, 1910.

Characteristics. Body length 6.25–6.5 mm. Forehead densely punctate, sparsely setose. Antennal funicles 3-segmented, clubs setose and with testaceous, with glabrous strip in centre of posterior face. Submentum with lateral margins not reaching genae; labial and maxillary palps 3-segmented. Pronotum slightly constricted medially, densely punctate with mycangial pits present in both sexes but fewer in males. Scutellum impressed underneath elytra. Elytral declivity rounded in lateral view, unarmed in females and armed with spines in males, densely setose. Procoxae narrowly separate.

Australian species. Native: *N. elongatus* Lea, 1910.

Distribution. Australia (NSW, QLD, VIC).

Notes. Colonises living *Eucalyptus* trees by entering damaged trunks with exposed heartwoods and excavates both horizontal and vertical galleries through trunk (Bickerstaff *et al.* 2020; D. S. Kent & J. R. M. Bickerstaff, pers. obs.).

Tribe Platypodini Shuckard, 1840

Austroplatypus Browne, 1971 (Pl. 46A; Figs 45.7, 45.8)

Type species, by original designation: *Platypus incompertus* Schedl, 1968.

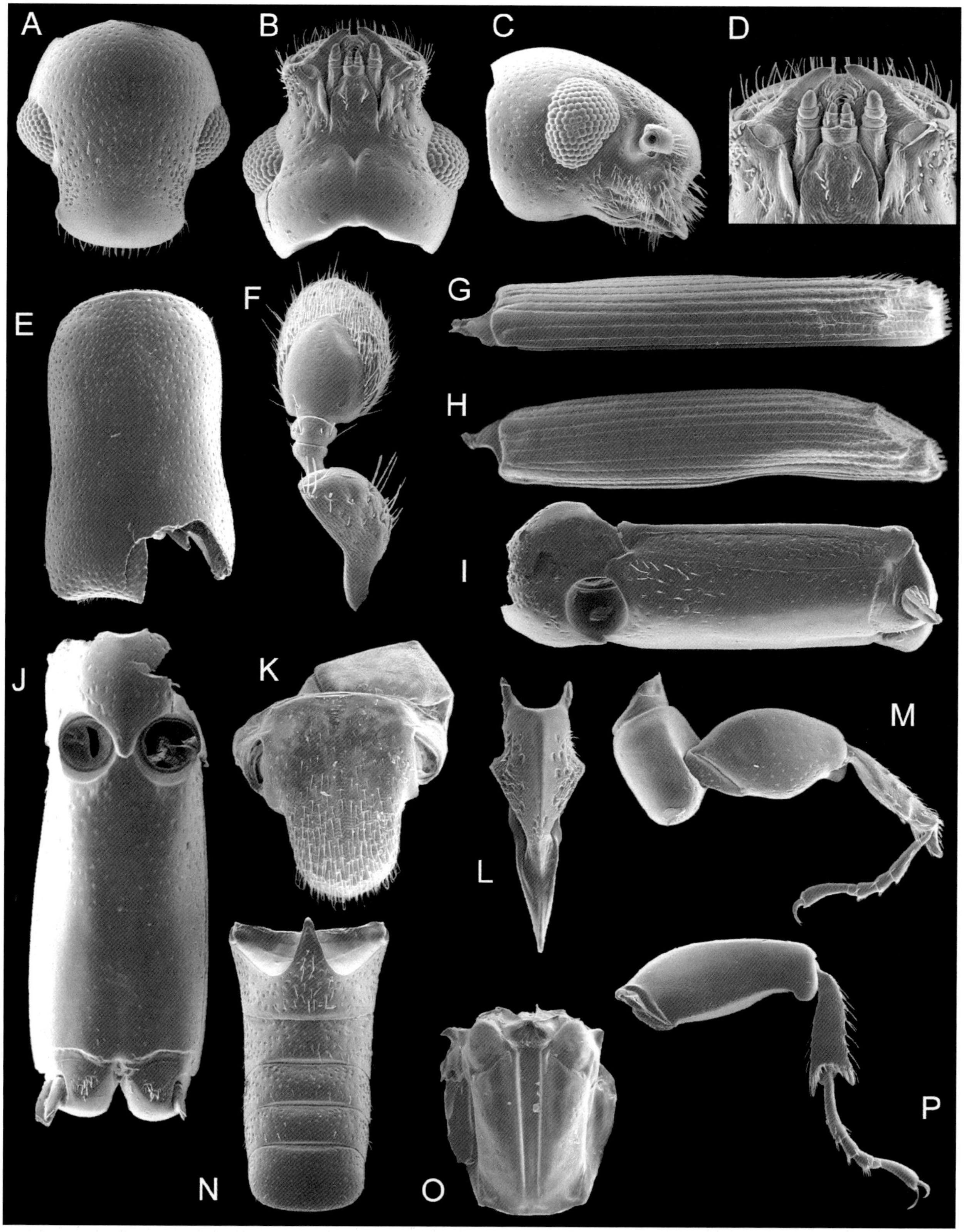

Fig. 45.6. Tesserocerini, structural details, *Notoplatypus elongatus* Lea, ♂. A–C, head: A, dorsal; B, ventral; C, lateral; D, mouthparts; E, prothorax, dorsal; F, antenna; G–H, elytron: G, dorsal; H, lateral; I–J, pterothorax: I, lateral; J, ventral; K, tergite VIII, dorsal; L, scutellum; M, foreleg; N, abdominal ventrites; O, metanotum; P, middle leg.

Fig. 45.7. Platypodini, structural details, *Austroplatypus incompertus* (Schedl), ♂. A, head, dorsal; B, antenna; C, head, lateral; D, mouthparts; E–F, prothorax: E, dorsal; F, ventral; G–H, elytron: G, dorsal; H, lateral; I–J, pterothorax: I, ventral; F, lateral; K, abdominal ventrites; L, scutellum; M, tergite VII, dorsal; N, aedeagus, dorsal; O, middle leg; P, foreleg; Q, hindleg.

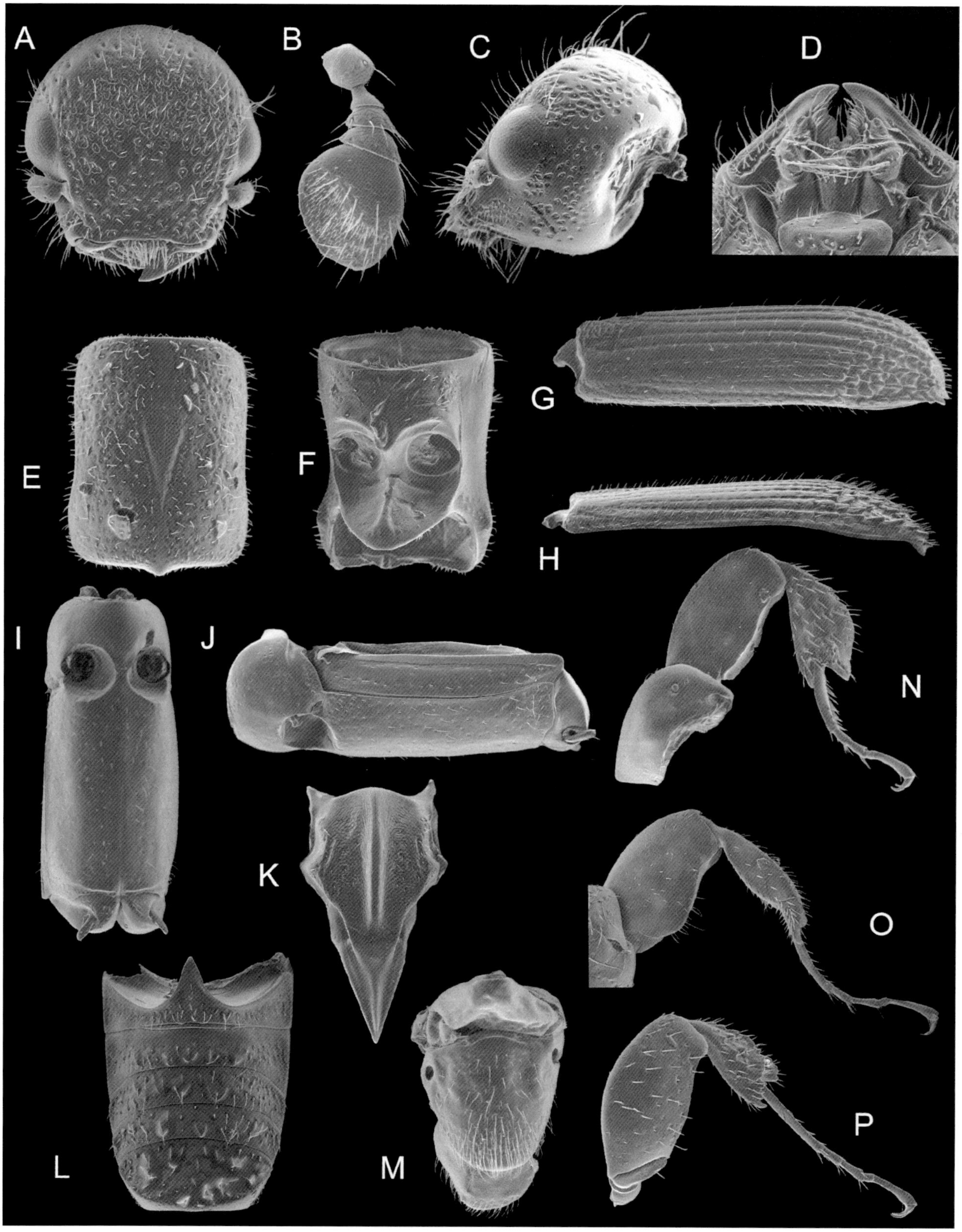

Fig. 45.8. Platypodini, structural details, *Austroplatypus incompertus* (Schedl), ♀. A, head, dorsal; B, antenna; C, head, lateral; D, mouthparts; E–F, prothorax: E, dorsal; F, ventral; G–H, elytron: G, dorsal; H, lateral; I–J, pterothorax: I, ventral; F, lateral; K, scutellum; L, abdominal ventrites; M, tergites VII–VIII, dorsal; N, middle leg; O, foreleg; P, hindleg.

Characteristics. Body length 5–6 mm. Forehead and vertex densely punctate, sparsely setose. Antennal funicles 4-segmented, antennal clubs setose at posterior end, small and shorter than scapes. Submentum with lateral margins reaching genae, labial and maxillary palps 3-segmented. Pronotum subquadrate, densely punctate and sparsely setose. Mycangial pits only present in females in centre of pronotal disc and caudate in shape. Scutellum impressed below elytra. Elytral declivity rounded, armed by spines in females and unarmed in males, sparsely setose. Interstriae deeply impressed. Procoxae contiguous.

Australian species. Native: *A. incompertus* (Schedl, 1968).

Distribution. Australia (NSW, VIC).

Notes. Colonises living *Eucalyptus* trees but apparently without damaging the hosts (Kent 2008a). Represents the only known eusocial beetle, colonies consisting of a foundress female, daughter workers and male and female offspring. Both queen and workers engage in fungal cultivation and care of the offspring, but how workers are chosen to stay in the gallery is unclear. Queen and daughter workers lack tarsi, thus unable to survive outside galleries (Smith *et al.* 2018). (Kent & Simpson 1992; Kent 2001, 2008a, b, 2010).

Baiocis Browne, 1962 (Pl. 46B, C)

Type species, by original designation: *Crossotarsus pernanulus* Schedl, 1935.

Characteristics. Body length 2.3–2.5 mm, 4 × as long as wide. Forehead and vertex mostly smooth, sparsely setose. Antennal funicles 4-segmented, clubs setose, large, longer than scapes. Submentum with lateral margins reaching genae; labial and maxillary palps 3-segmented. Pronotum greatly constricted, mostly smooth with some punctures, mycangial pits absent. Scutellum impressed below elytra. Elytral declivity truncate and unarmed, sparsely setose. Procoxae widely separate.

Australian species. Native: *B. pernanulus* (Schedl, 1935).

Distribution. Australia (QLD), Papua New Guinea, Southeast Asia, China.

Notes. Polyphagous, constructing galleries in dead and dying trees; no records of Australian host tree families (Bickerstaff *et al.* 2020).

Crossotarsus Chapuis, 1865 (Pl. 46D, E; Fig. 45.9)

Type species, by subsequent designation (Hopkins, 1914: 119): *Platypus wallacei* Thomson, 1858.

Characteristics. Body length 3–7.8 mm. Forehead and vertex mostly smooth, typically sparsely setose in males, in females occasionally densely covered with long golden setae. Antennal funicles 4-segmented, clubs setose, large, usually longer than scapes but scapes elongate in *Crossotarsus mniszechi*. Submentum with lateral margins reaching genae; labial palps 2-segmented, maxillary palps 3-segmented. Pronotum subquadrate with minor medial constriction, smooth, mycangial pits absent. Scutellum impressed below elytra. Elytral declivity truncate, unarmed in females and armed with spines in males, sparsely setose. Procoxae narrowly to widely separate.

Australian species. Native: *C. armipennis* Lea, 1910, *C. chalcographus* Schedl 1972, *C. lacordairei* Chapuis, 1865, *C. mniszechi* Chapuis, 1865, *C. subpellucidus* Lea, 1910. Intercepted: *C. externedentatus* (Fairmaire, 1850), *C. indomitus* Chapuis, 1865.

Distribution. Australia (NSW, QLD), Papua New Guinea, Bismarck Is., Pacific Islands, Indonesia, Malaysia, Japan, China; *Crossotarsus externedentatus* a widespread pest endemic to Asia and established in South Africa, Madagascar and tropical countries in Africa.

Notes. Highly polyphagous, most species constructing galleries in dead and dying trees (Bickerstaff *et al.* 2020); *C. externedentatus* recorded to attack live hosts, including *Eucalyptus* (Browne 1961; Beaver 2000a) and mango (Sittichaya 2012). The Conoderinae genera *Phaenomerus* Schoenherr and *Tomicoproctus* Faust have been reported from galleries of *C. armipennis* in New South Wales (May 1997). The names *Platypus quadricintus* Schedl, 1962 and *Crossotarsus nitescens* Schedl, 1979 were synonymised with *C. armipennis* by Beaver & Sanguansub (2020).

Dinoplatypus Wood, 1993 (Pl. 46F, G; Figs 45.10, 45.11)

Type species, by original designation: *Platypus cupulatus* Chapuis, 1865.

Characteristics. Body length 2.8–5.5 mm. Forehead and vertex mostly smooth, sparsely setose. Antennal funicles 4-segmented, clubs setose, large, longer than scapes. Submentum with lateral margins reaching genae; labial and maxillary palps 3-segmented. Pronotum subquadrate with minor medial constriction, smooth, mycangial pits sometimes present in both sexes. Scutellum impressed below elytra. Elytral declivity truncate, broad at apex, forming vertical face surrounded by elevated carina, face concave and shining, in males typically emarginate, forming V-shaped cleft, and setae absent or sparsely covering surface, in females similar but reduced compared to shape in males. Procoxae contiguous.

Australian species. Native: *D. luniger* (Motschulsky, 1863), *D. pallidus* (Chapuis, 1865), *D. pseudocupulatus* (Schedl, 1935). Intercepted: *D. cupulatus* (Chapuis, 1865), *D. forficula* (Chapuis, 1865).

Distribution. Australia (QLD), Papua New Guinea, India, Sri Lanka, Southeast Asia, Seychelle Is.

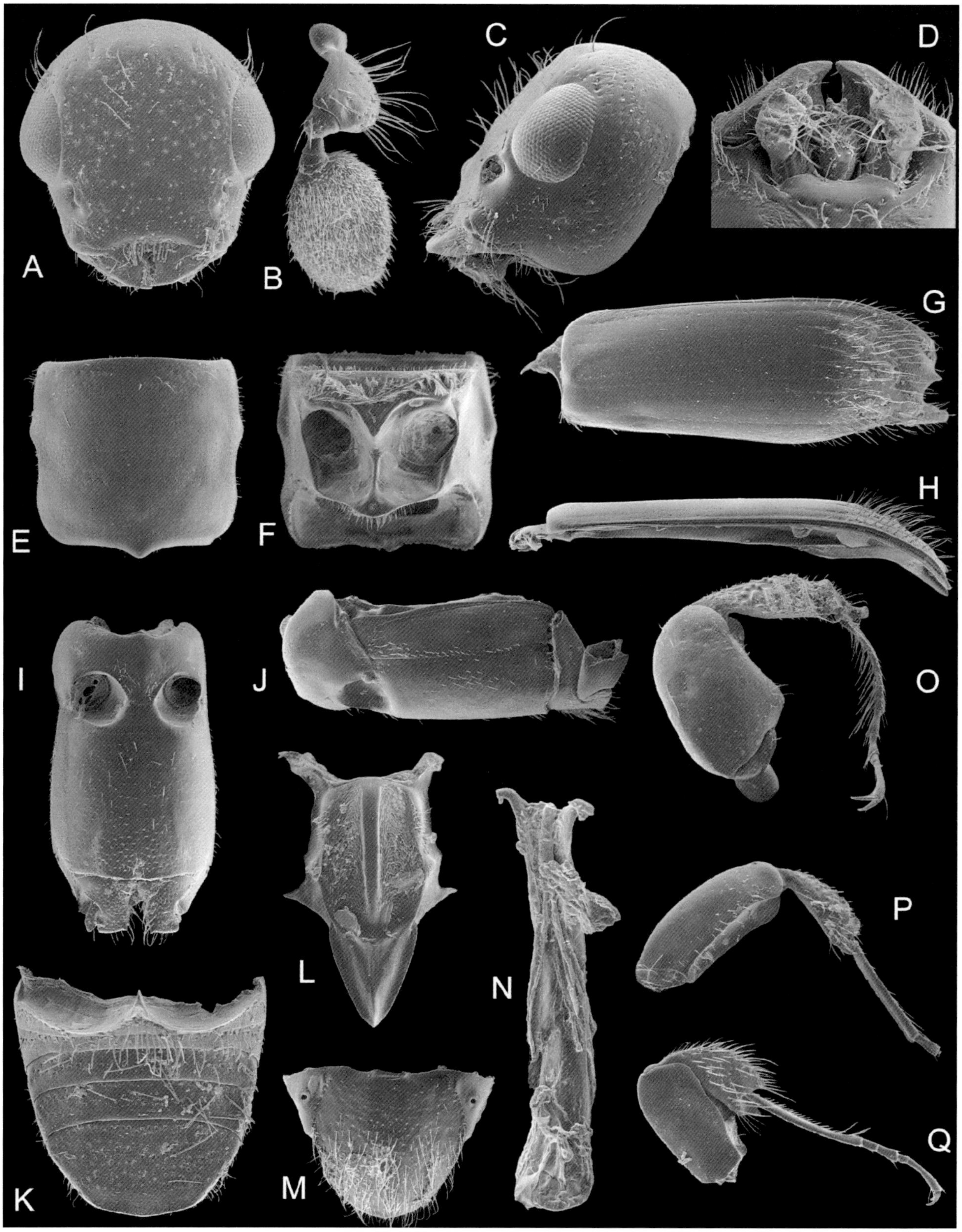

Fig. 45.9. Platypodini, structural details, *Crossotarsus armipennis* Lea, ♂. A, head, dorsal; B, antenna; C, head, lateral; D, mouthparts; E–F, prothorax: E, dorsal; F, ventral; G–H, elytron: G, dorsal; H, lateral; I–J, pterothorax: I, ventral; F, lateral; K, abdominal ventrites; L, scutellum; M, tergite VII, dorsal; N, aedeagus, dorsal; O, foreleg; P, middle leg; Q, foreleg.

Notes. Highly polyphagous, with numerous host families recorded (Bickerstaff *et al.* 2020); *D cupulatus* and *D. pseudocupulatus* recorded to infest mango trees in Thailand (Sittichaya 2012). The presence of *D. lepidus* (Chapuis, 1865) in Australia (Pullen *et al.* 2014) needs confirmation, as the current record may be based on a misidentification by Schedl (1936) (Bickerstaff *et al.* 2020).

Euplatypus Wood, 1993 (Pl. 46.I, J, Figs 45.4J–M, 45.5F)

Type species, by original designation: *Bostrichus parallelus* Fabricius, 1801.

Characteristics. Body length 3.5–5.2 mm. Forehead and vertex punctate, sparsely setose. Antennal funicles 4-segmented, clubs setose, large, longer than scapes. Submentum with lateral margins reaching genae; labial and maxillary palps 3-segmented. Pronotum subquadrate, smooth with some punctures, mycangia absent in both sexes. Scutellum impressed below elytra. Elytra in males truncate, slightly attenuate, apex deeply emarginate and sparsely setose, in females rounded and setose at apex; interstriae deeply grooved along elytral. Metanepisterna and metaventrite in males with posterior spines. Procoxae narrowly separate.

Australian species. Introduced: *E. parallelus* (Fabricius, 1801).

Distribution. Native to northern, central and southern Americas, introduced to Australia (NSW, QLD), Afrotropical countries, Indian Ocean islands and Madagascar, Asian countries including Sri Lanka, India, Indonesia and Papua New Guinea.

Notes. Single species in Australia, highly polyphagous, attacking trees of many families (Bickerstaff *et al.* 2020). *E. parallelus* is a widespread pest that can transmit wilt fungus *Fusarium oxysporum* (Beaver 2013), which is thought to lead to death of *Pterocarpus indicus* trees in southern Thailand (Bumrungsri *et al.* 2008).

Platypus Herbst, 1793 (Pl. 46.H, Figs 45.12, 45.13)

Type species, by monotypy: *Bostrichus cylindrus* Fabricius, 1793.

Characteristics. Body length 2.6–7 mm. Forehead and vertex smooth to highly punctate, typically sparsely setose. Antennal funicles 4-segmented, clubs setose, large, longer than scapes. Submentum with lateral margins reaching genae; labial and maxillary palps 3-segmented. Pronotum subquadrate with minor constriction, smooth with some punctures, mycangia present in females and sometimes in males. Scutellum impressed below elytra. Elytra with declivity in males highly variable in shape, attenuate, with striae, rounded or truncate, sometimes armed with tubercles and spines, in females often rounded, in both sexes often setose. Procoxae contiguous to narrowly separate.

Australian species. Native: *P. carbonescens* (Beeson, 1937), *P. dentipennis* (Schedl, 1964), *P. froggatti* Sampson, 1926, *P. hastulifer* Schedl, 1959, *P. lineellus* Schedl, 1972, *P. norfolkensis* Schedl, *P. omnivorus* Lea, 1904, *P. pilidens* Schedl, 1955, *P. queenslandi* Schedl, 1936, *P. scalaris* Schedl, 1980, *P. semigranosus* (Sampson, 1925), *P. subgranosus* Schedl, 1936, *P. tasmanicus* Schedl, 1942, *P. transversecarinatus* Schedl, 1942, *P. tuberculosus* Strohmeyer, 1910. Intercepted: *P. apicalis* (White, 1846), *P. bifurcus* (Schedl, 1938), *P. curtus* Chapuis, 1865, *P. geminatus* Chapuis, 1865, *P. gerstaeckeri* Chapuis, 1865, *P. hospes* Schedl, 1964, *P. jansoni* Chapuis, 1865.

Distribution. Present on every continent except Antarctica, including Australia (Lord Howe Is., NSW, Norfolk Is., QLD, TAS, VIC), Papua New Guinea, New Zealand, Oceanic Island countries, Southeast Asia.

Notes. Most species highly polyphagous, constructing galleries in dying or dead trees (Bickerstaff *et al.* 2020) but several species present in Australia associated with living trees, *P. tuberculosus* with preference for live, healthy *Eucalyptus* trees (Kent 2001) and *P. subgranosus* a pest of commercial *Nothofagus cunninghamii* plantations in Tasmania and secondarily associated with phytopathogen *Davidsonella australis* (Elliott *et al.* 1987; Kile & Hall 1988; Candy 1990). The name *Platypus quadrincinctus* Schedl, 1962 was synonymised with *Crossotarsus armipennis* Lea, 1910 by Beaver & Sanguansub (2020). *Platypus hybridus* Schedl, 1935 was listed as occurring in Australia by Pullen *et al.* (2014), as reported by Beaver & Liu (2013), but this is an erroneous record.

Treptoplatypus Schedl, 1972 (Pl. 46.K, L, Figs 45.4A–I, 45.14, 45.15)

Type species, by original designation: *Crossotarsus trepanatus* Chapuis, 1865.

Characteristics. Body length 4–5 mm. Forehead and vertex smooth to punctate, setose. Antennal funicles 4-segmented, clubs setose, large, longer than scapes. Cleft between submentum and genae absent; labial and maxillary palps 3-segmented. Pronotum subquadrate with minor constriction, smooth with some punctures, mycangia present in females. Elytra with declivity in males attenuate, acuminate, in females rounded, in both sexes sometimes setose. Proxocae contiguous.

Australian species. Native: *T. crenatus* (Chapuis, 1865), *T. solidus* (Walker, 1859).

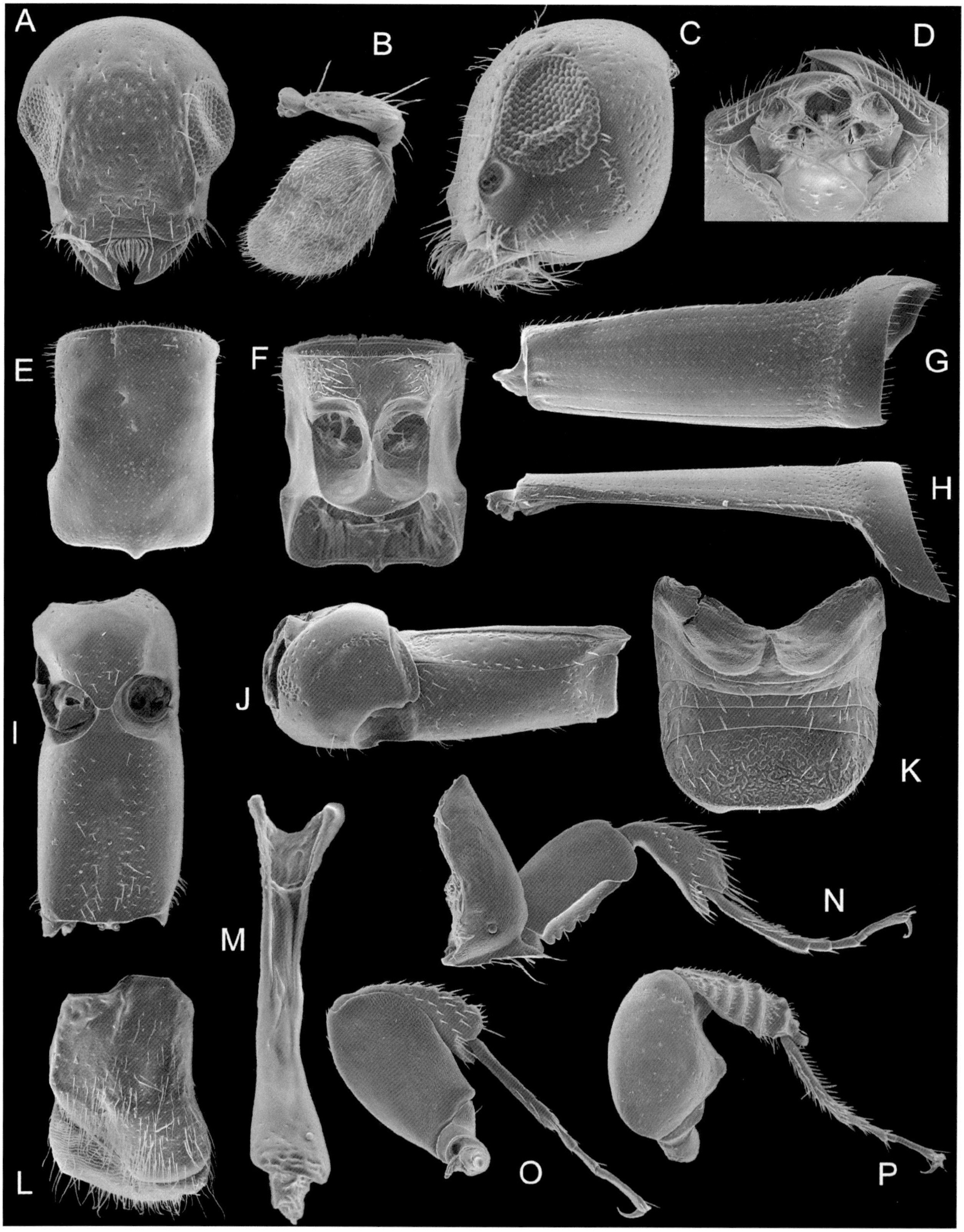

Fig. 45.10. Platypodini, structural details, *Dinoplatypus pseudocupulatus* (Schedl), ♂. A, head, dorsal; B, antenna; C, head, lateral; D, mouthparts; E–F, prothorax: E, dorsal; F, ventral; G–H, elytron: G, dorsal; H, lateral; I–J, pterothorax: I, ventral; J, lateral; K, abdominal ventrites; L, tergite VII, dorsal; M, aedeagus, dorsal; N, foreleg; O, hindleg; P, middle leg.

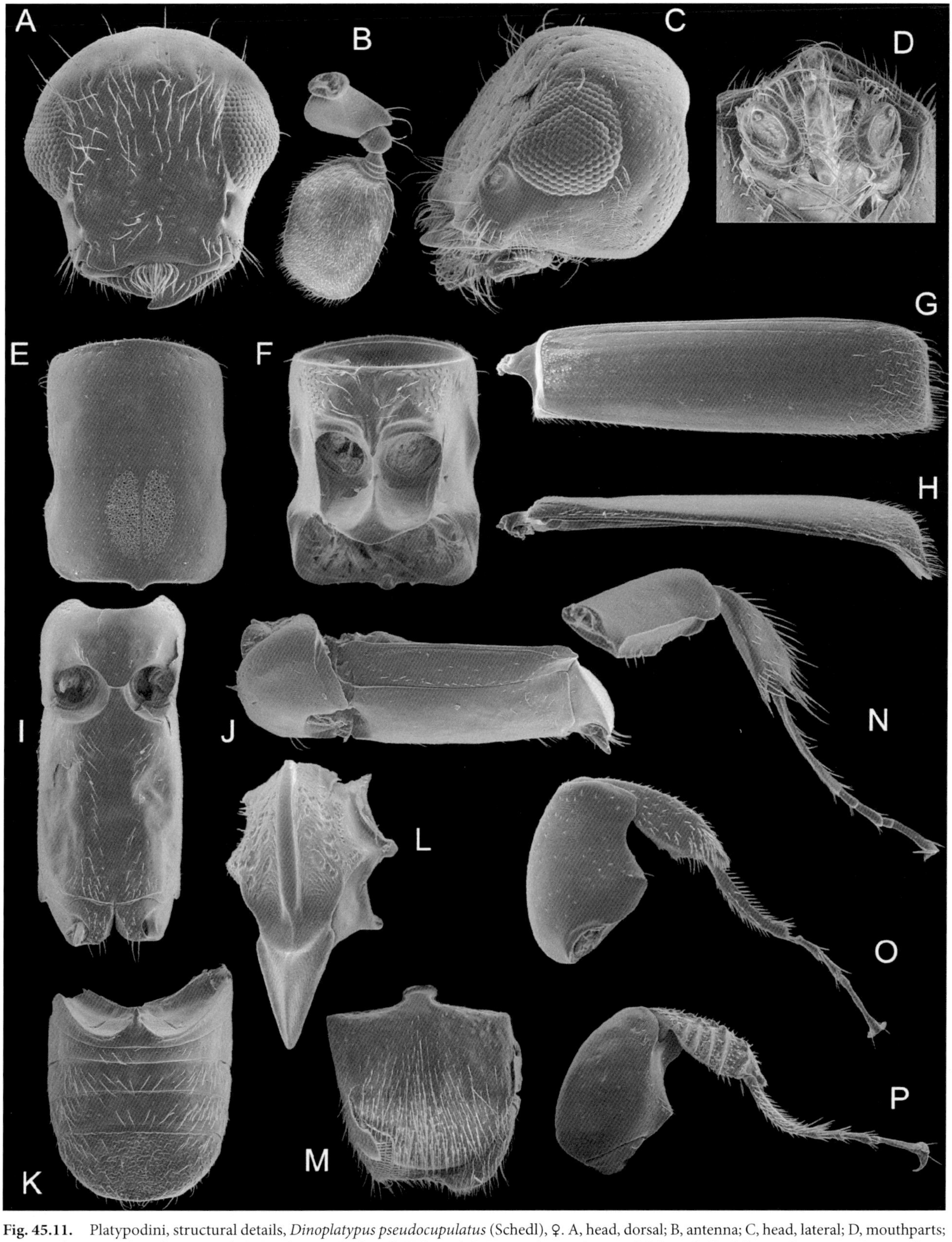

Fig. 45.11. Platypodini, structural details, *Dinoplatypus pseudocupulatus* (Schedl), ♀. A, head, dorsal; B, antenna; C, head, lateral; D, mouthparts; E–F, prothorax: E, dorsal; F, ventral; G–H, elytron: G, dorsal; H, lateral; I–J, pterothorax: I, ventral; J, lateral; K, abdominal ventrites; L, scutellum; M, tergite VII, dorsal; N, hindleg; O, middle leg; P, foreleg.

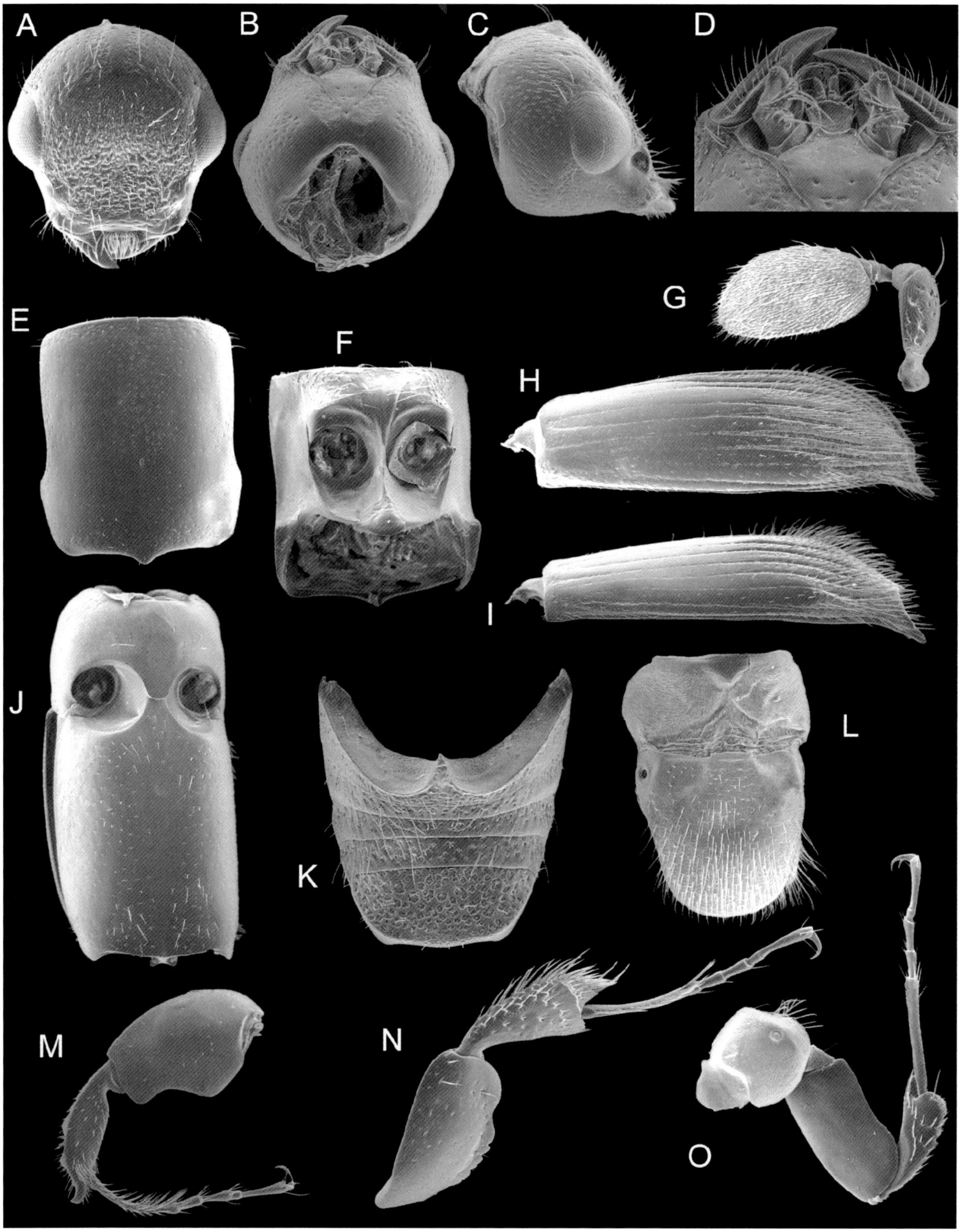

Fig. 45.12. Platypodini, structural details, *Platypus omnivorus* Lea, ♂. A–C, head: A, dorsal; B, ventral; C, lateral; D, mouthparts; E–F, prothorax: E, dorsal, F; ventral; G, antenna; H–I, elytron: H, dorsal; I, lateral; J, pterothorax, ventral; K, abdominal ventrites; L, tergites VII & VIII, dorsal; M, middle leg; N, foreleg; O, hindleg.

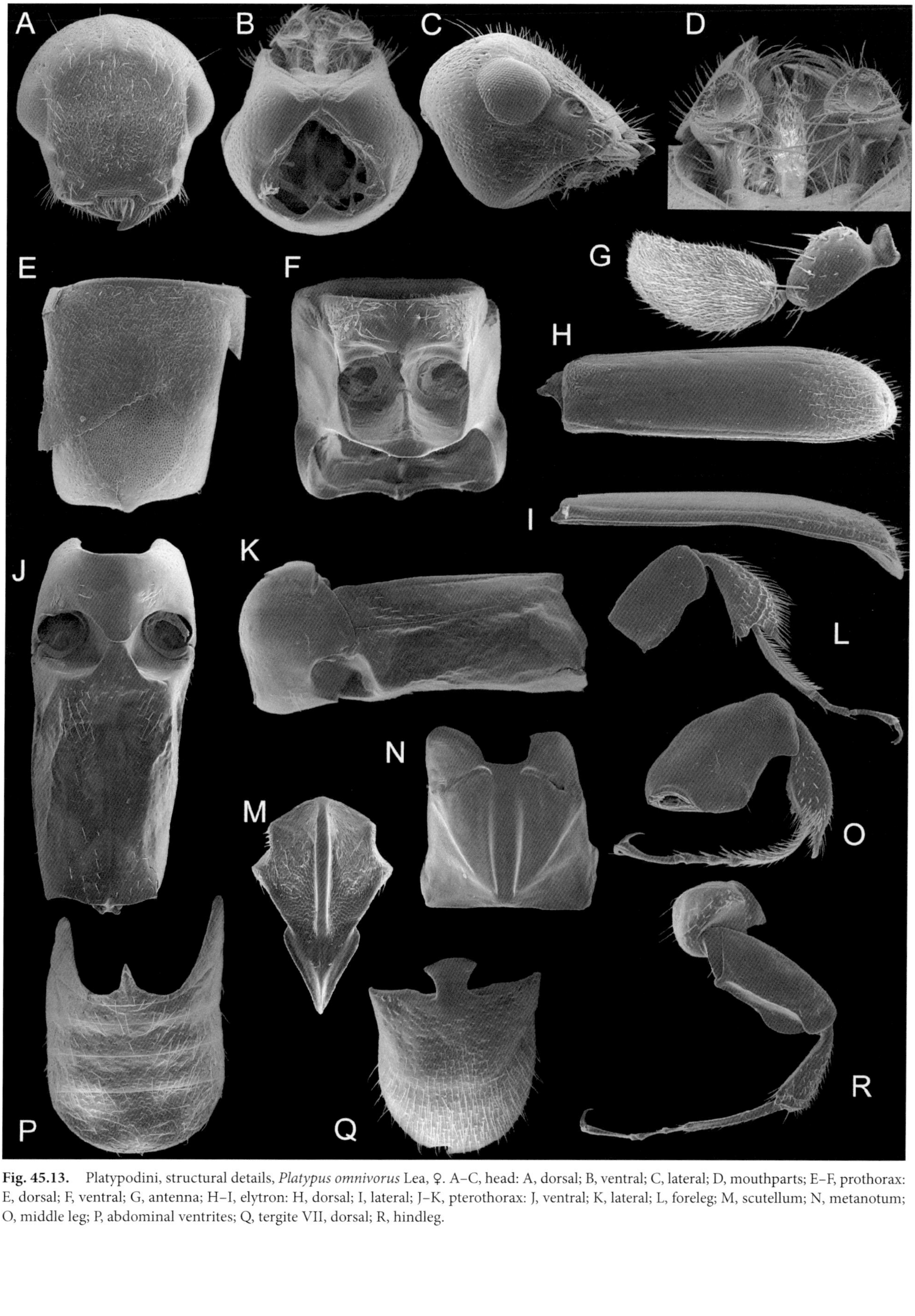

Fig. 45.13. Platypodini, structural details, *Platypus omnivorus* Lea, ♀. A–C, head: A, dorsal; B, ventral; C, lateral; D, mouthparts; E–F, prothorax: E, dorsal; F, ventral; G, antenna; H–I, elytron: H, dorsal; I, lateral; J–K, pterothorax: J, ventral; K, lateral; L, foreleg; M, scutellum; N, metanotum; O, middle leg; P, abdominal ventrites; Q, tergite VII, dorsal; R, hindleg.

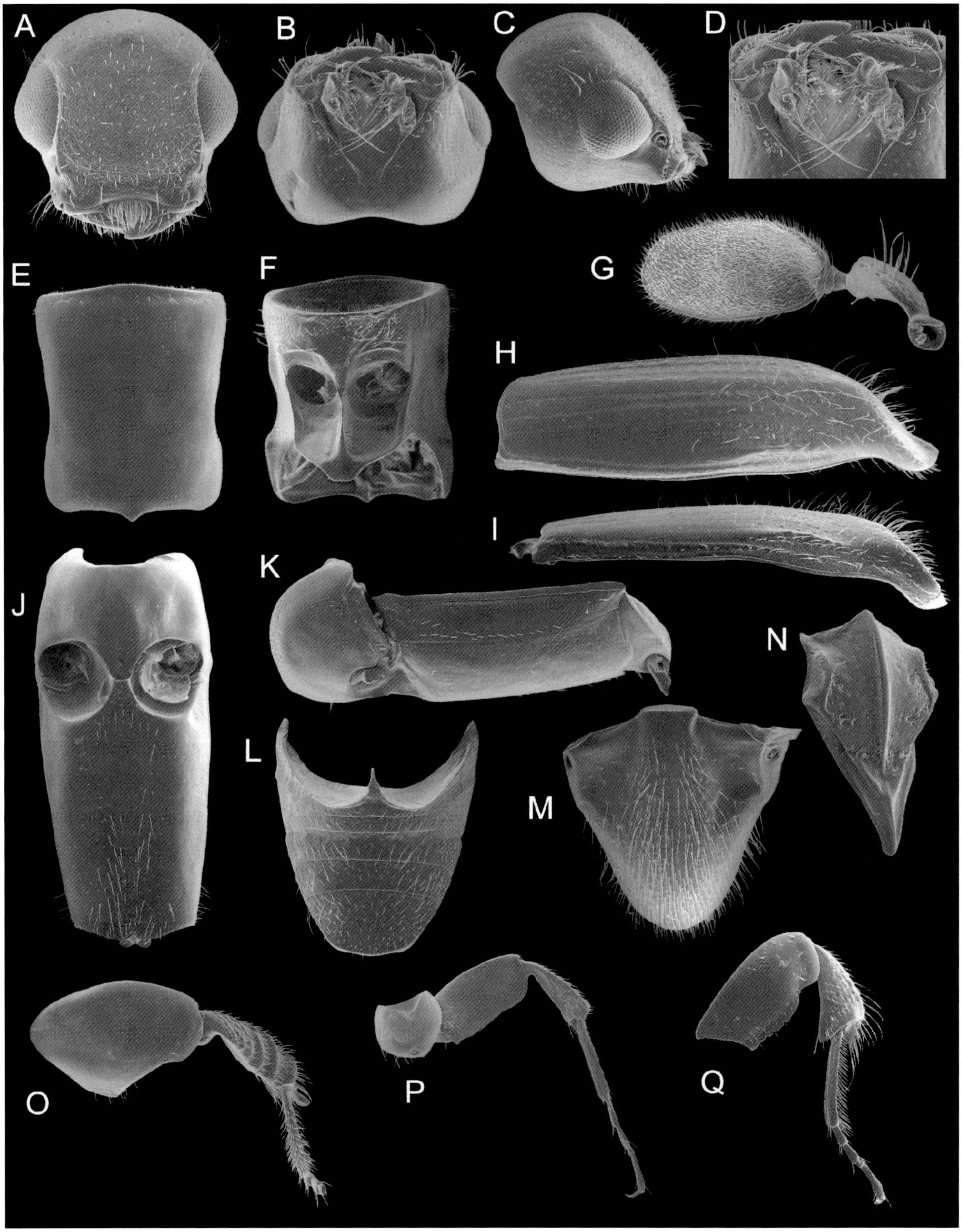

Fig. 45.14. Platypodini, structural details, *Treptoplatypus crenatus* (Chapuis), ♂. A–C, head: A, dorsal; B, ventral; C, lateral; D, mouthparts; E–F, prothorax: E, dorsal; F, ventral; G, antenna; H–I, elytron: H, dorsal; I, lateral; J–K, pterothorax: J, ventral; K, lateral; L, abdominal ventrites; M, tergite VII, dorsal; N, scutellum; O, foreleg; P, hindleg; Q, middle leg.

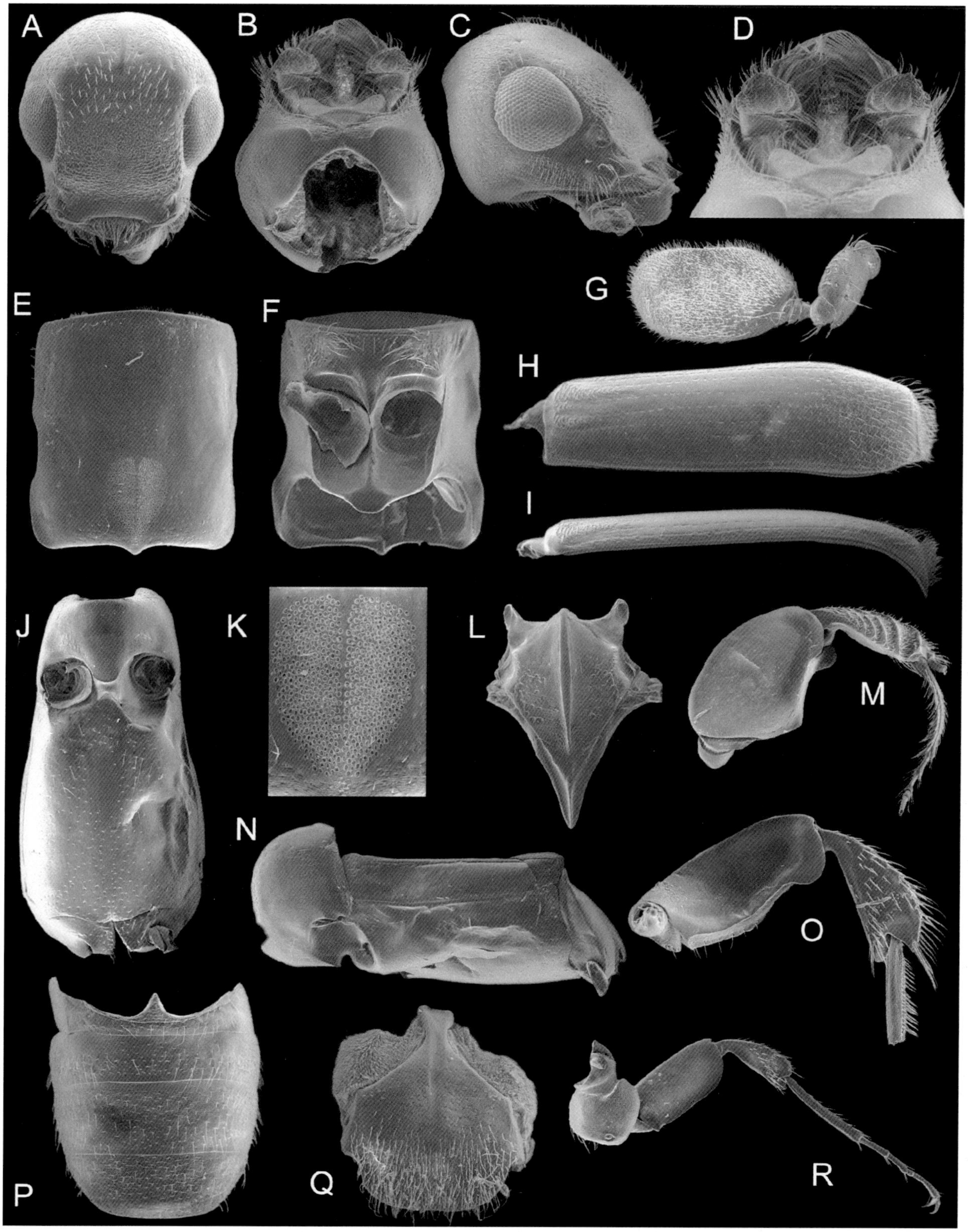

Fig. 45.15. Platypodini, structural details, *Treptoplatypus crenatus* (Chapuis), ♀. A–C, head: A, dorsal; B, ventral; C, lateral; D, mouthparts; E–F, prothorax: E, dorsal; F, ventral; G, antenna; H–I, elytron: H, dorsal; I, lateral; J, pterothorax, ventral; K, pronotal mycangia L, scutellum; M, foreleg; N, pterothorax, lateral; O, middle leg; P, abdominal ventrites; Q, tergite VII, dorsal; R, hindleg.

Distribution. Australia (NSW, NT, QLD), *T. solidus* also present in Southeast Asia (India, Japan, Korea, Nepal, Pacific Islands, Pakistan, Sri Lanka).

Notes. Polyphagous, recorded from numerous host tree families (Bickerstaff *et al.* 2020).

Acknowledgements

We thank Jiri Lochman for his photograph of *Treptoplatypus crenatus* in its natural habitat. JRMB would like to thank Markus Riegler (Western Sydney University) and Bjarte Jordal (University of Bergen, Norway) for their support during his PhD research on bark beetles.

References

Alfaro RI, Humble LM, Gonzalez P, Villaverde R, Allegro G (2007) The threat of the ambrosia beetle *Megaplatypus mutatus* (Chapuis) (=*Platypus mutatus* Chapuis) to world poplar resources. *Forestry* **80**(4), 471–479. doi:10.1093/forestry/cpm029

Alonso-Zarazaga MA (1999) *A World Catalogue of Families and Genera of Curculionoidea (Insecta: Coleoptera)(Excepting Scolytidae and Platypodidae).* Entomopraxis, Barcelona.

Beaver RA (1989) Insect–fungus relationships in the bark and ambrosia beetles. In *Insect-fungus Interactions.* (Eds N Wilding, NM Collins, PM Hammond and JF Weber) pp. 121–143. Academic Press, London. doi:10.1016/B978-0-12-751800-8.50011-2

Beaver RA (2000a) Ambrosia beetles (Coleoptera: Platypodidae) of the South Pacific. *The Canadian Entomologist* **132**(6), 755–763. doi:10.4039/Ent132755-6

Beaver RA (2000b) Studies on the genus *Diapus* Chapuis (Coleoptera: Platypodidae): new species and new synonymy. *Serangga* **5**, 247–260.

Beaver RA (2013) The invasive neotropical ambrosia beetle *Euplatypus parallelus* (Fabricius, 1801) in the Oriental region and its pest status (Coleoptera: Curculionidae, Platypodinae). *Entomologist's Monthly Magazine* **149**, 143–154.

Beaver RA, Liu LY (2013) A synopsis of the pin-hole borers of Thailand (Coleoptera: Curculionidae: Platypodinae). *Zootaxa* **3646**(4), 447–486. doi:10.11646/zootaxa.3646.4.7

Beaver RA, Liu LY (2018) A review of the genus *Baiocis* Browne, 1962 (Coleoptera: Curculionidae: Platypodinae), with new species, new synonymy and a key to males. *Zootaxa* **4434**(3), 481–501. doi:10.11646/zootaxa.4434.3.5

Beaver RA, Sanguansub S (2015) A review of the genus *Carchesiopygus* Schedl (Coleoptera: Curculionidae: Platypodinae), with keys to species. *Zootaxa* **3931**(3), 401–412. doi:10.11646/zootaxa.3931.3.4

Beaver RA, Sanguansub S (2020) New synonymy and taxonomic changes in Australian and Oriental pin-hole borers (Coleoptera: Curculionidae, Platypodinae). *Entomologist's Monthly Magazine* **156**(2), 79–86. doi:10.31184/M00138908.1562.4024

Bickerstaff JRM, Smith S, Kent DS, Beaver RA, Seago AE, *et al.* (2020) A review of the distribution and host plant associations of the platypodine ambrosia beetles (Coleoptera: Curculionidae: Platypodinae) of Australia, with an electronic species identification key. *Zootaxa* **4894**(1), 69–80. doi:10.11646/zootaxa.4894.1.3

Bright DE, Poinar GO (1994) Scolytidae and Platypodidae (Coleoptera) from Dominican Republic amber. *Annals of the Entomological Society of America* **87**(2), 170–194. doi:10.1093/aesa/87.2.170

Browne FG (1961) Taxonomic notes on Platypodidae (Coleoptera). *Annals and Magazine of Natural History* **4**(47), 641–656. doi:10.1080/00222936108651189

Browne FG (1971) *Austroplatypus,* a new genus of the Platypodidae (Coleoptera), infesting living *Eucalyptus* trees in Australia. *The Commonwealth Forestry Review* **50**(1), 49–50.

Browne FG (1972) Larvae of the principal old world genera of the Platypodinae (Coleoptera: Platypodidae). *Transactions of the Royal Entomological Society of London* **124**(2), 167–190. doi:10.1111/j.1365-2311.1972.tb00362.x

Bumrungsri S, Beaver RA, Phongpaichit S, Sittichaya W (2008) The infestation by an exotic ambrosia beetle, *Euplatypus parallelus* (F.) (Coleoptera: Curculionidae: Platypodinae) of Angsana trees (*Pterocarpus indicus* Willd.) in southern Thailand. *Songklanakarin Journal of Science and Technology* **30**(5), 570–582.

Candy SG (1990) *Biology of the Mountain Pinhole Borer, Platypus subgranosus Schedl, in Tasmania.* University of Tasmania.

Crowson R (1955) *The Natural Classification of the Families of Coleoptera.* Nathaniel Lloyd & Co., London.

Elliott HJ, Kile GA, Candy SG, Ratkowsky DA (1987) The incidence and spatial pattern of *Nothofagus cunninghamii* (Hook.) Oerst. attacked by *Platypus subgranosus* Schedl in Tasmania's cool temperate rainforest. *Australian Journal of Ecology* **12**(2), 125–138. doi:10.1111/j.1442-9993.1987.tb00934.x

Gillett C, Crampton-Platt A, Timmermans MJTN, Jordal BH, Emerson BC, *et al.* (2014) Bulk *de novo* mitogenome assembly from pooled total DNA elucidates the phylogeny of weevils (Coleoptera: Curculionoidea). *Molecular Biology and Evolution* **31**(8), 2223–2237. doi:10.1093/molbev/msu154

Harris J, Campbell K, Wright GM (1976) Ecological studies on the horizontal borer *Austroplatypus incompertus* (Schedl)(Coleoptera: Platypodidae). *Journal of the Entomological Society of Australia (NSW)* **9**, 11–21.

Hopkins AD (1914) List of generic names and their type-species in the coleopterous superfamily Scolytoidea. *Proceedings of the United States National Museum* **48**(2066), 115–136. doi:10.5479/si.00963801.2066.115

Hulcr J, Cognato AI (2010) Repeated evolution of crop theft in fungus-farming ambrosia beetles: mycocleptism in ambrosia beetles. *Evolution* **64**(11), 3205–3212. doi:10.1111/j.1558-5646.2010.01055.x

Hulcr J, Stelinski LL (2017) The ambrosia symbiosis: from evolutionary ecology to practical management. *Annual Review of Entomology* **62**(1), 285–303. doi:10.1146/annurev-ento-031616-035105

Jordal BH (2014) 3.7. 13. Platypodinae Shuckard, 1840. In *Handbook of Zoology. Arthropoda: Insecta. Coleoptera, Beetles. Volume 3: Morphology and Systematics (Phytophaga)* (Eds NP Kristensen and RG Beutel) pp. 642–648. Walter de Gruyter, Berlin, Germany.

Jordal BH (2015) Molecular phylogeny and biogeography of the weevil subfamily Platypodinae reveals evolutionarily conserved range patterns. *Molecular Phylogenetics and Evolution* **92**, 294–307. doi:10.1016/j.ympev.2015.05.028

Jordal BH, Sequeira AS, Cognato AI (2011) The age and phylogeny of wood boring weevils and the origin of subsociality. *Molecular Phylogenetics and Evolution* **59**(3), 708–724. doi:10.1016/j.ympev.2011.03.016

Kent DS (2001) *The Biology of the Ambrosia Beetle Austroplatypus incompertus (Schedl).* University of Sydney.

Kent DS (2008a) Distribution and host plant records of *Austroplatypus incompertus* (Scheld) (Coleoptera: Curculionidae: Platypodinae). *The Australian Entomologist* **35**(1), 1–6.

Kent DS (2008b) Mycangia of the ambrosia beetle *Austroplatypus incompertus* (Schedl) (Coleoptera: Curculionidae: Platypodinae). *Australian Journal of Entomology* **47**(1), 9–12. doi:10.1111/j.1440-6055.2007.00612.x

Kent DS (2010) The external morphology of *Austroplatypus incompertus* (Schedl) (Coleoptera, Curculionidae, Platypodinae). *ZooKeys* **56**, 141–140. doi:10.3897/zookeys.56.521

Kent DS, Simpson JA (1992) Eusociality in the beetle *Austroplatypus incompertus* (Coleoptera: Curculionidae). *Naturwissenschaften* **79**, 87–89. doi:10.1007/BF01131811

Kile GA, Hall MF (1988) Assessment of *Platypus subgranosus* as a vector of *Chalara australis*, causal agent of a vascular disease of *Nothofagus cunninghamii. New Zealand Journal of Forestry Science* **18**(2), 166–186.

Kinuura H, Kobayashi M (2006) Death of *Quercus crispula* by inoculation with adult *Platypus quercivorus* (Coleoptera: Platypodidae). *Applied Entomology and Zoology* **41**(1), 123–128. doi:10.1303/aez.2006.123

Kirkendall L, Faccoli M (2010) Bark beetles and pinhole borers (Curculionidae, Scolytinae, Platypodinae) alien to Europe. *ZooKeys* **56**, 227–251. doi:10.3897/zookeys.56.529

Kirkendall LR, Biedermann PHW, Jordal BH (2015) Evolution and diversity of bark and ambrosia beetles. In *Bark Beetles*. (Eds FE Vega and RW Hofstetter) pp. 85–156. Elsevier. doi:10.1016/B978-0-12-417156-5.00003-4

Kirkendall LR, Kent DS, Raffa KF (1997) Interactions among males, females and offspring in bark and ambrosia beetles: the significance of living in tunnels for the evolution of social behavior. In *The Evolution of Social Behavior in Insects and Arachnids*. (Eds JC Choe and BJ Crespi) pp. 181–215. Cambridge University Press. doi:10.1017/CBO9780511721953.010

Kubono T, Ito S (2002) *Raffaelea quercivora* sp. nov. associated with mass mortality of Japanese oak, and the ambrosia beetle (*Platypus quercivorus*). *Mycoscience* **43**(3), 255–260. doi:10.1007/S102670200037

Kuroda K (2001) Responses of *Quercus* sapwood to infection with the pathogenic fungus of a new wilt disease vectored by the ambrosia beetle *Platypus quercivorus*. *Journal of Wood Science* **47**(6), 425–429. doi:10.1007/BF00767893

Kuschel G (1995) A phylogenetic classification of Curculionoidea to families and subfamilies. *Memoirs of the Entomological Society of Washington* **14**, 5–33.

Kuschel G, Leschen RAB, Zimmerman EC (2000) Platypodidae under scrutiny. *Invertebrate Taxonomy* **14**(6), 771–808. doi:10.1071/IT00024

Lee JS, Haack RA, Choi WI (2011) Attack pattern of *Platypus koryoensis* (Coleoptera: Curculionidae: Platypodinae) in relation to crown dieback of Mongolian oak in Korea. *Environmental Entomology* **40**(6), 1363–1369. doi:10.1603/EN11138

Marvaldi AE (1997) Higher level phylogeny of Curculionidae (Coleoptera: Curculionoidea) based mainly on larval characters, with special reference to broad-nosed weevils. *Cladistics* **13**(4), 285–312. doi:10.1111/j.1096-0031.1997.tb00321.x

Marvaldi AE, Sequeira AS, O'Brien CW, Farrell, BD (2002) Molecular and morphological phylogenetics of weevils (Coleoptera, Curculionoidea): do niche shifts accompany diversification? *Systematic Biology* **51**(5), 761–785. doi:10.1080/10635150290102465

May BM (1993) *Larvae of Curculionoidea: (Insecta: Coleoptera); a systematic overview*. Manaaki Whenua Press.

May BM (1994) An introduction to the immature stages of Australian Curculionoidea. In *Australian Weevils 2*. (Ed. EC Zimmerman) pp. 365–728. CSIRO, Australia.

May BM (1997) Larvae of two weevil genera (Coleoptera: Curculionidae) associated with *Crossotarsus armipennis* Lea (Platypodidae) from New South Wales. *New Zealand Entomologist* **20**(1), 7–13. doi:10.1080/00779962.1997.9722662

Mayers CG, Harrington TC, Biedermann PHW (2022) Mycangia define the diverse ambrosia beetle–fungus symbioses. In *The Convergent Evolution of Agriculture in Humans and Insects*. (Eds TR Schultz, R Gawne and PN Peregrine) pp. 105–142. The MIT Press, Cambridge, MA. doi:10.7551/mitpress/13600.003.0013

McKenna DD, Sequeira AS, Marvaldi AE, Farrell BD (2009) Temporal lags and overlap in the diversification of weevils and flowering plants. *Proceedings of the National Academy of Sciences* **106**(17), 7083–7088. doi:10.1073/pnas.0810618106

Mugu S, Pistone D, Jordal BH (2018) New molecular markers resolve phylogenetic position of the enigmatic wood-boring weevils Platypodinae (Coleoptera: Curculionidae). *Arthropod Systematics & Phylogeny* **76**(1), 45–58. doi:10.3897/asp.76.e31938

Nakashima T (1971) Notes on the associated fungi and the mycetangia of the ambrosia beetle *Crossotarsus niponicus* Blandford: Coleoptera: Platypodidae. *Applied Entomology and Zoology* **6**(3), 131–137. doi:10.1303/aez.6.131

Neumann FG, Harris JA (1974) Pinhole borers in green timber. *Australian Forestry* **37**(2), 132–141. doi:10.1080/00049158.1974.10675605

Oberprieler RG, Marvaldi AE, Anderson RS (2007) Weevils, weevils, weevils everywhere. *Zootaxa* **1668**(1), 491–520. doi:10.11646/zootaxa.1668.1.24

Peris D, Kraemer MMS, Peñalver E, Delclòs X (2015) New ambrosia beetles (Coleoptera: Curculionidae: Platypodinae) from Miocene Mexican and Dominican ambers and their paleobiogeographical implications. *Organisms Diversity & Evolution* **15**(3), 527–542. doi:10.1007/s13127-015-0213-y

Peris D, Kraemer MMS, Smith SM, Cognato AI (2017) *Eoplatypus jordali* gen.n. *et* sp.n., the first described Platypodinae (Coleoptera: Curculionidae) from Baltic amber. *Arthropod Systematics & Phylogeny* **75**(2), 185–194. doi:10.3897/asp.75.e31898

Ploetz RC, Hulcr J, Wingfield MJ, de Beer ZW (2013) Destructive tree diseases associates with ambrosia and bark beetles: black swan events in tree pathology? *Plant Disease* **97**(7), 856–872. doi:10.1094/PDIS-01-13-0056-FE

Pullen KR, Jennings D, Oberprieler RG (2014) Annotated catalogue of Australian weevils (Coleoptera: Curculionoidea). *Zootaxa* **3896**(1), 1–481. doi:10.11646/zootaxa.3896.1.1

Puranasakul W (2006) *Diversity of Bark and Ambrosia Beetles (Coleoptera: Scolytidae and Platypodidae) in Mixed Evergreen and Deciduous Dipterocarp Forests in Northern Thailand*. Chiang Mai University, Chiang Mai.

Roberts H (1961) A description of the developmental stages of *Trachyostus aterrimus* (Schauf.), a West African Platypodid, and some remarks on its biology. 5th Report West African Timber Borer Research Unit 1961–1962, 29–46.

Roberts H (1993) The tribe Diapodini (Coleoptera: Platypodidae) of Papua New Guinea. *Bishop Museum Occassional Papers* **35**, 1–39.

Shin S, Clarke DJ, Lemmon AR, Moriarty Lemmon E, Aitken AL, *et al.* (2018) Phylogenomic data yield new and robust insights into the phylogeny and evolution of weevils. *Molecular Biology and Evolution* **35**(4), 823–836. doi:10.1093/molbev/msx324

Silva JCPD, Putz P, Silveira EDC, Flechtmann CAH (2013) Biological aspects of *Euplatypus parallelus* (F.) (Coleoptera, Curculionidae, Platypodinae) attacking *Hevea brasiliensis* (Willd. Ex A. Juss.) in São Paulo northwest, Brazil. *III Congresso Brasileiro de Heveicultura*. doi:10.13140/2.1.4830.5601

Sittichaya W (2012) Bark and ambrosia beetles (Coleoptera: Curculionidae: Scolytinae and Platypodinae) infesting mango trees (*Mangifera indica* L.) in Southern Thailand, with two new species recorded for Thailand. *Songklanakarin Journal of Science and Technology* **32**(2), 153–155.

Sittichaya W, Beaver RA (2009) Rubberwood-destroying beetles in the eastern and gulf areas of Thailand (Coleoptera: Bostrichidae, Curculionidae: Scolytinae and Platypodinae). *Songklanakarin Journal of Science and Technology* **31**(4), 381–387.

Smit SJ, Allsopp E, Nethavhani Z, Caleca V, Oberprieler RG *et al* (2022) Mitogenomics of the Olive Seed Weevil, *Anchonochranus oleae* Marshall, and implications for its phylogenetic position in Curculionidae. *Insects* **13**(7), 607. doi:10.3390/insects13070607

Smith SM, Beattie AJ, Kent DS, Stow AJ (2009) Ploidy of the eusocial beetle *Austroplatypus incompertus* (Schedl) (Coleoptera, Curculionidae) and implications for the evolution of eusociality. *Insectes Sociaux* **56**(3), 285–288. doi:10.1007/s00040-009-0022-4

Smith SM, Kent DS, Boomsma JJ, Stow AJ (2018) Monogamous sperm storage and permanent worker sterility in a long-lived ambrosia beetle. *Nature Ecology & Evolution* **2**(6), 1009–1018. doi:10.1038/s41559-018-0533-3

Solórzano-Kraemer MM, Hammel JU, Kun R, Xu CP, Cognato AI (2021) Miocene pinhole borer ambrosia beetles: new species of *Diapus*

(Coleoptera: Curculionidae: Platypodinae). *Palaeoworld* **32**(4), 658–668. doi:10.1016/j.palwor.2021.10.001

Song N, Li X, Yin X, Li X, Yin S, *et al.* (2020) The mitochondrial genome of *Apion squamigerum* (Coleoptera, Curculionoidea, Brentidae) and the phylogenetic implications. *PeerJ*, **8**, e8386. doi:10.7717/peerj.8386

Thompson RT (1992) Observations on the morphology and classification of weevils (Coleoptera, Curculionoidea) with a key to major groups. *Journal of Natural History* **26**(4), 835–891. doi:10.1080/00222939200770511

Vega FE, Biedermann PH (2020) On interactions, associations, mycetangia, mutualists and symbiotes in insect-fungus symbioses. *Fungal Ecology* **44**, 100909. doi:10.1016/j.funeco.2019.100909

Wood SL (1973) On the taxonomic status of Platypodidae and Scolytidae (Coleoptera). *The Great Basin Naturalist* **33**(2), 77–90. doi:10.5962/bhl.part.28150

Wood SL (1993) Revision of the genera of Platypodidae (Coleoptera). *The Great Basin Naturalist* **53**(3), 259–281. doi:10.5962/bhl.part.16605

Wood SL, Bright DE (1992) A catalog of Scolytidae and Platypodidae (Coleoptera). Part 2: Taxonomic index. *Great Basin Naturalist Memoirs* **13**, 1–1553

Zanuncio AJV, Pastori PL, Kirkendall LR, Lino-Neto J, Serrão JE, *et al.* (2010) *Megaplatypus mutatus* (Chapuis) (Coleoptera: Curculionidae: Platypodinae) attacks hybrid *Eucalyptus* L'Héritier De Brutelle clones in southern Espírito Santo, Brazil. *The Coleopterists Bulletin* **64**(1), 81–83. doi:10.1649/0010-065X-64.1.81

Plate 1. Anthribidae, Attelabidae. A, *Xynotropis* sp., Mt. Field N.P., TAS (Anthribidae: Anthribinae) © Nick Porch; B, *Phaulimia* sp., Bunya Mt., QLD (Anthribidae: Anthribinae) © Nick Monaghan; C, *Telala* sp., Kondalilla N.P., QLD (Anthribidae: Anthribinae) © Nick Monaghan; D, *Euops* sp., Blackburn Lake, VIC (Attelabidae: Attelabinae) © Nick Porch; E, *Metopum* sp., Wellington, TAS (Attelabidae: Rhynchitinae) © Bruno Bell; F, *Euops falcatus* (Guérin-Méneville) (Attelabidae: Attelabinae) © Nick Porch; G, *Euops falcatus* (Guérin-Méneville) (Attelabidae: Attelabinae) © Nick Monaghan.

Plate 2. Belidae. A, *Agnesiotis pilosula* Pascoe (Belinae: Agnesiotidini) © Nick Monaghan; B, *Sphinctobelus cinereus* (Blanchard) (Belinae: Pachyurini) © Nick Monaghan; C, *Isacanthodes* sp., Fitzgerald River N.P., WA (Belinae: Belini) © Jiri Lochman; D, *Pachyura australis* Hope (Belinae: Pachyurini) © Nick Monaghan; E, *Cyrotyphus vestitus* (Pascoe) (Belinae: Agnesiotidini) © Rolf Oberprieler; F, *Rhinotia simplicipennis* Lea (Belinae: Belini) © Nick Monaghan; G, *Rhinotia suturalis* (Macleay) (Belinae: Belini) © Nick Monaghan; H, *Rhinotia ruficornis* (Lea) (Belinae: Belini) © Nick Porch.

Plate 3. Belidae, Caridae, Brentidae. A, *Rhinotiodes spinipennis* (Lacordaire) (Belidae: Belinae) © Nick Monaghan; B, *Car* sp., Canberra, ACT (Caridae) © Rolf Oberprieler; C, *Car condensatus* Blackburn (Caridae) © Bruno Bell; D, *Car pini* Lea (Caridae) © Bruno Bell; E, *Ctenaphides porcellus* Pascoe, ♀ (Brentidae: Eurhynchinae) © Nick Porch; F, *Aporhina pulchra* Oberprieler (Brentidae: Eurhynchinae) © Nick Monaghan; G, *Schizoeupsalis promissa* (Pascoe), ♀ (Brentidae: Brentinae) © Nick Volpe; H, *Cordus* sp., Tremont, VIC (Brentidae: Brentinae) © Nick Porch.

Plate 4. Brentidae. A, *Ankleineella sulcicollis* (Pascoe) (Brentinae: Brentini) © Jiri Lochman; B, *Ithystenus hollandiae* (Boisduval) (Brentinae: Trachelizini) © Nick Monaghan; C, *Cerobates ophthalmicus* (Pascoe), ♂ (Brentinae: Cyphagogini) © Nick Monaghan; D, *Austronanodes alleni* (Lea) (Nanophyinae: Nanophyini) © Nick Monaghan; E, *Araiorrhinus howittii* (Pascoe) (Brentinae: Cyphagogini) © Nick Porch; F, *Apiotherium* sp., Kallista, VIC (Apioninae: Apionini) © Nick Porch; G, *Rhynolaccus exsertus* (Pascoe) (Apioninae: Myrmacicelini) © Nick Monaghan; H, *Alissapion* sp. (Apioninae: Rhadinocybini) © Bruno Bell.

Plate 5. Curculionidae. A, *Aromagis horrens* Pascoe (Cyclominae: Aterpini) © Dash Huang; B, *Aonychus lineatus* Pascoe (Brachycerinae, Erirhinini) © Bernhard Jacobi; C, *Treptoplatypus crenatus* (Chapuis) (Platypodinae: Platypodini) © Jiri Lochman; D, *Trigonotarsus calandroides* Guérin-Méneville (Dryophthorinae: Rhynchophorini) © Rolf Oberprieler; E, *Dryophthorus* sp., Fernshaw, VIC (Dryophthorinae: Dryophthorini) © Nick Porch; F, *Sipalinus gigas* (Fabricius) (Dryophthorinae: Orthognathini) © Nick Volpe; G, *Sipalinus gigas* (Fabricius), larva (Dryophthorinae: Orthognathini) © Rolf Oberprieler; H, *Trigonotarsus calandroides* Guérin-Méneville, larva (Dryophthorinae: Rhynchophorini) © Rolf Oberprieler.

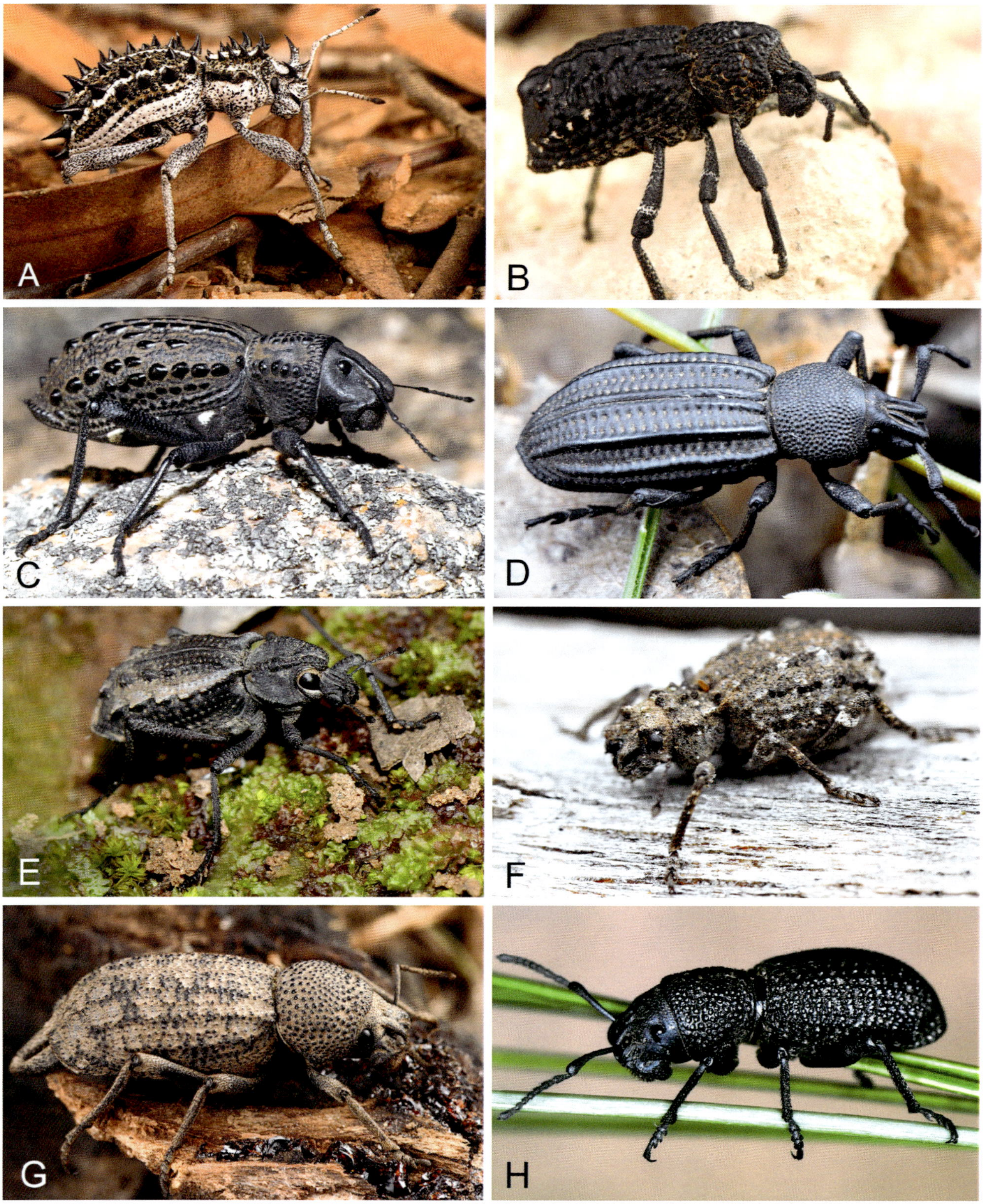

Plate 6. Curculionidae (Cyclominae: Amycterini). A, *Acantholophus niveovittatus* Blackburn © Dash Huang; B, *Mythites* sp., QLD © Dash Huang; C, *Gagatophorus obsoletus* (Ferguson), ♀ © Jiri Lochman; D, *Talaurinus impressicollis* Macleay © Nick Monaghan; E, *Dialeptopus echinatus* Lea, ♀ © Jean and Fred Hort; F, *Parahyborrhynchus convexiusculus* (Macleay) © Nick Porch; G, *Amycterus caudatus* (Macleay) © Dash Huang; H, *Xenommamycterus capito* (Pascoe) © Rolf Oberprieler.

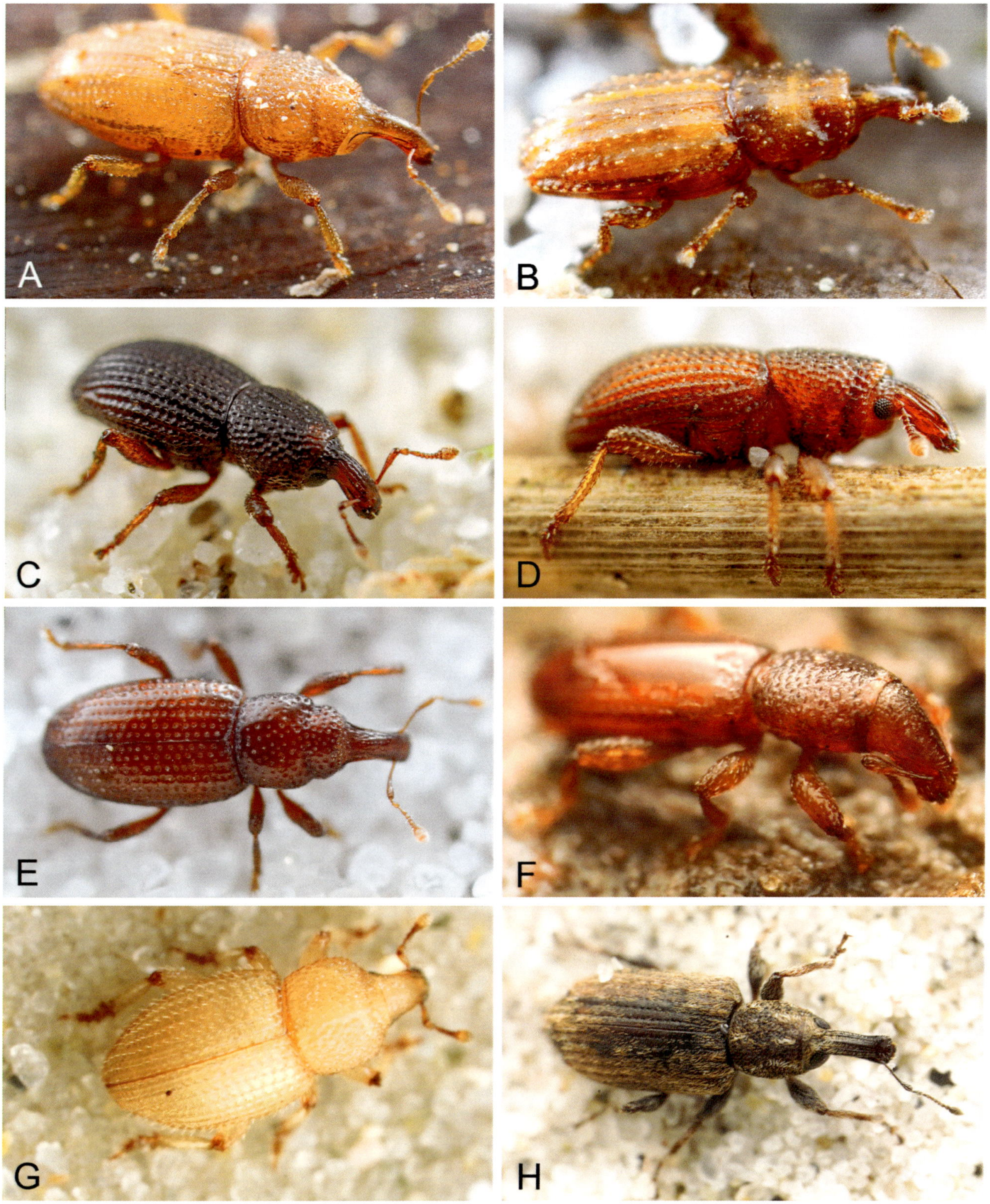

Plate 7. Curculionidae (Cyclominae, Notiomimetini). A, *Wollastonicis* sp., WA (Notiomimetini) © Nick Porch; B, *Wollastonicis* sp., Jervis Bay, NSW (Notiomimetini) © Nick Porch; C, D, *Psaldus liosomoides* Pascoe (Notiomimetini) © Nick Porch; E, *Notiomimetes* sp., WA (Notiomimetini) © Nick Porch; F, *Tasmanica* sp., Echo Sugarloaf, TAS (Listroderini) © Bruno Bell; G, *Aphela* sp., Schouten Island, TAS (Notiomimetini) © Bruno Bell; H, *Steriphus diversipes* (Pascoe), TAS (Listroderini) © Bruno Bell.

Plate 8. Curculionidae (Cyclominae). A, B, *Chrysolopus spectabilis* (Fabricius) (Aterpini) © Nick Monaghan; C, Undet. genus, Grampians, VIC (Aterpini) © Nick Porch; D, *Aades cultratus* (Fabricius) (Aterpini) © Nick Monaghan; E, *Aades griseatus* (Pascoe) (Aterpini) © Nick Monaghan; F, *Aoplocnemis* sp., Churchill N.P., VIC (Aterpini) © Nick Monaghan; G, *Aparete palpebrosa* Pascoe (Rhythirrinini) © Jean and Fred Hort; H, *Chrysophoracis pulcher* Lea (Aterpini) © Nick Porch.

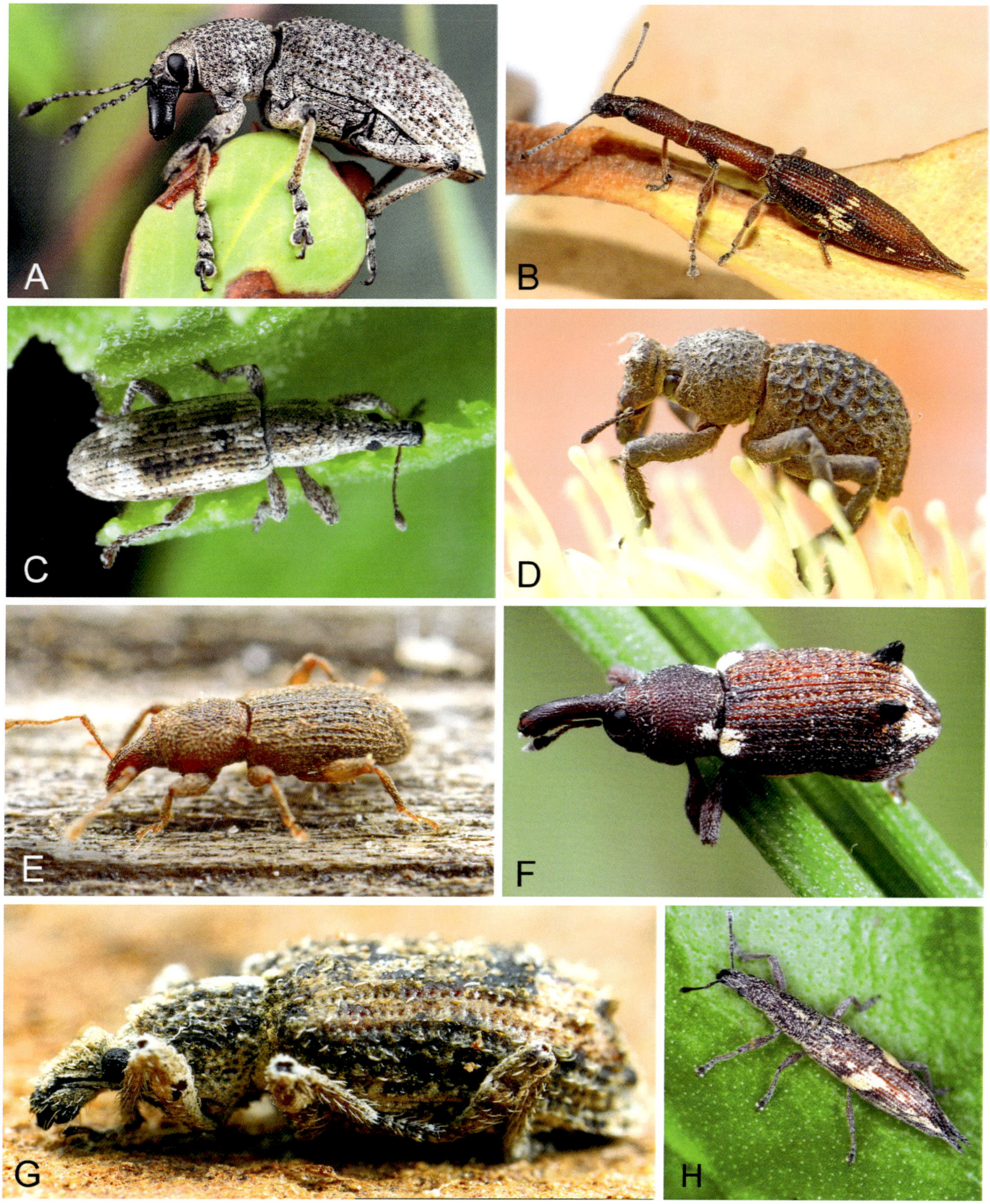

Plate 9. Curculionidae (Cyclominae). A, *Rhinaria granulosa* (Fåhraeus) (Aterpini) © Nick Monaghan; B, *Rhadinosomus lacordairei* Pascoe (Aterpini) © Jean and Fred Hort; C, *Acalonoma pusillum* Blackburn (Aterpini) © Nick Monaghan; D, *Rhinoplethes foveatus* Pascoe (Aterpini) © Nick Porch; E, Undet. genus, Porongurup Range N.P., WA (Listroderini) © Nick Porch; F, *Encosmia cornuta* Blackburn (Aterpini) © Nick Monaghan; G, *Ethemaia sellata* Pascoe (Rhythirrinini) © Nick Porch; H, *Euthyphasis acuta* Pascoe (Aterpini) © Nick Monaghan.

Plate 10. Curculionidae (Entiminae). A, *Catasarcus spinipennis* Fåhraeus (Leptopiini) © Jiri Lochman; B, *Titinia villosipennis* (Lea) (Cyphicerini) © Nick Monaghan; C, *Myllocerus longus* Lea (Cyphicerini) © Jiri Lochman; D, *Pseudotimareta intermixta* (Lea) (Leptopiini) © Nick Porch; E, *Titinia villosipennis* (Lea) (Cyphicerini) © Nick Porch; F, *Catasarcus pallidiventris* Thompson (Leptopiini) © Nick Monaghan; G, *Prypnus* sp., Kallista, VIC (Leptopiini) © Nick Porch; H, *Polyphrades* sp., Hattah-Kulkyne N.P., VIC (Leptopiini) © Nick Monaghan.

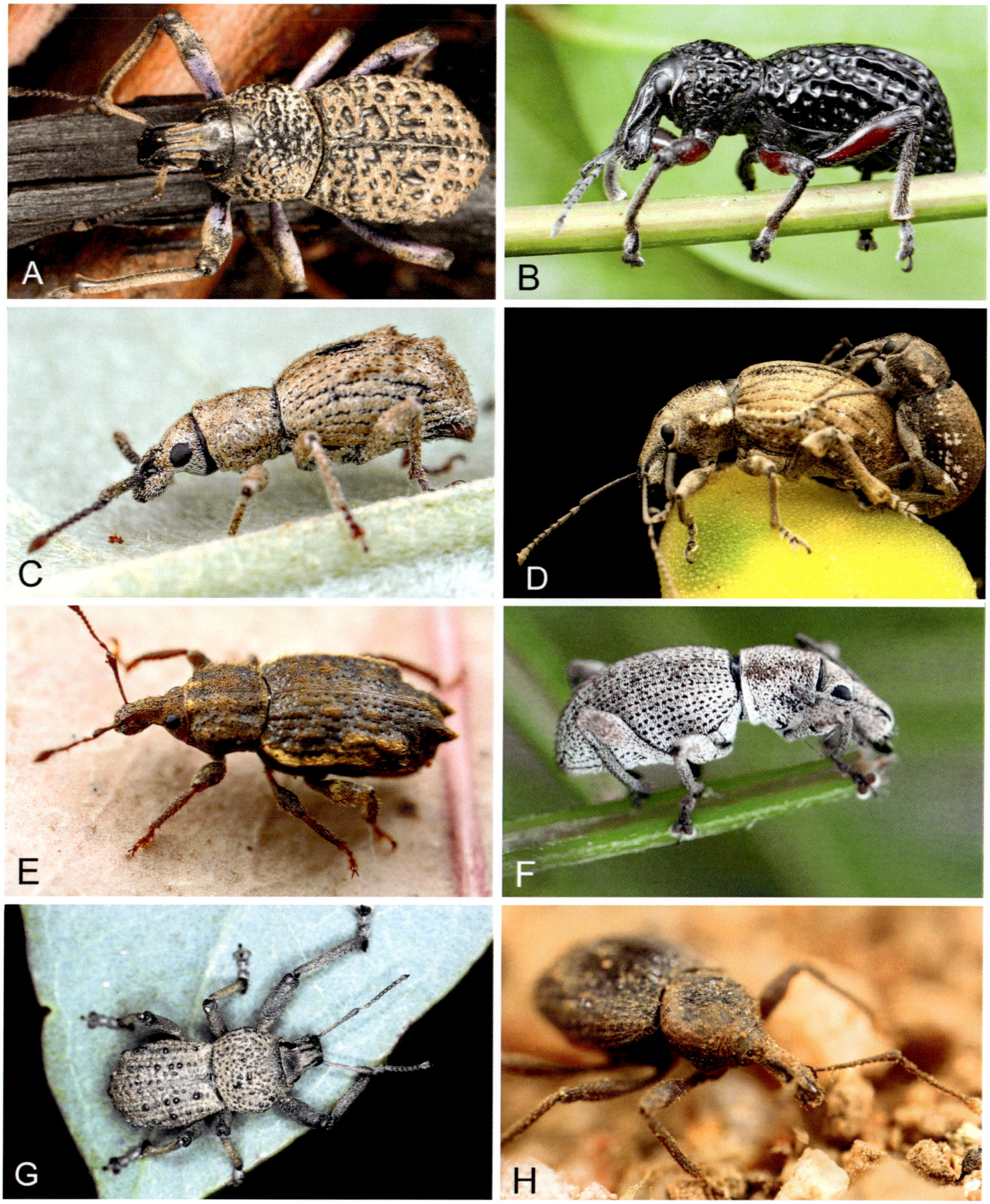

Plate 11. Curculionidae (Entiminae: Leptopiini). A, *Psapharus infaustus* (Olivier) © Dash Huang; B, *Zymaus angustus* Lea © Nick Monaghan; C, *Ecrizothis inaequalis* Blackburn (Leptopiini) © Nick Porch; D, *Perperus lateralis* (Boisduval) © Bruno Bell; E, *Mandalotus* sp., Weldborough, TAS © Nick Porch; F, *Merimnetes* sp., Canberra, ACT © Rolf Oberprieler; G, *Psapharus infaustus* (Olivier) © Rolf Oberprieler; H, *Undarobius howarthi* Escalona & Oberprieler © Hermes E. Escalona.

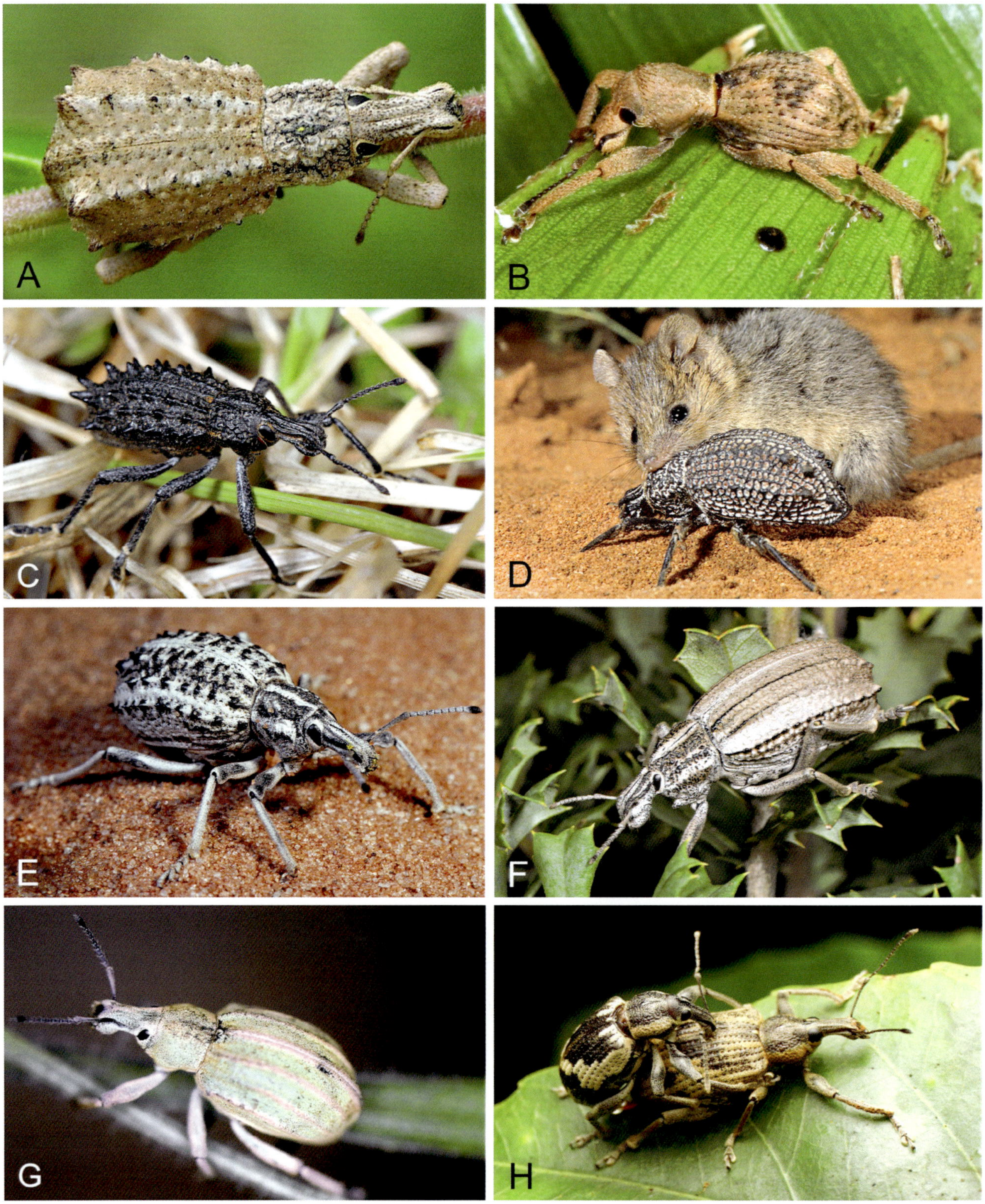

Plate 12. Curculionidae (Entiminae: Leptopiini). A, *Leptopius nodicollis* (Lea), ♀ © Nick Monaghan; B, *Leptopius puellaris* (Pascoe) © Nick Monaghan; C, *Leptopius* sp., Mornington N.P., VIC © Nick Monaghan; D, *Leptopius areolatus* (Blackburn) © Jiri Lochman; E, *Leptopius* sp., Rudall River N.P., WA © Jiri Lochman; F, *Leptopius colossus* (Pascoe) © Jiri Lochman; G, *Leptopius glaucus* (Pascoe) © Rolf Oberprieler; H, *Catastygnus* sp., Mary Cairncross Scenic Reserve, Maleny, QLD © Jessa Thurman.

Plate 13. Curculionidae (unplaced to subfamily). A, *Gonipterus* sp., Brisbane Ranges, QLD (Gonipterini) © Nick Porch; B, *Pantoreites fusiformis* Lea (Gonipterini) © Rolf Oberprieler; C, *Gonipterus* sp., larva, Dandenong Ranges N.P., VIC (Gonipterini) © Nick Monaghan; D, *Gonipterus* sp., Mapleton Falls N.P., QLD (Gonipterini) © Nick Monaghan; E, *Syarbis* sp., Noosa N.P., QLD (Gonipterini) © Nick Monaghan; F, *Oxyops* sp., Cathedral Range State Park, VIC (Gonipterini) © Nick Monaghan; G, *Atelicus guttatus* Pascoe (Cyclominae: Aterpini) © Bruno Bell; H, *Oxyops* sp., Karijini, WA (Gonipterini) © Rolf Oberprieler.

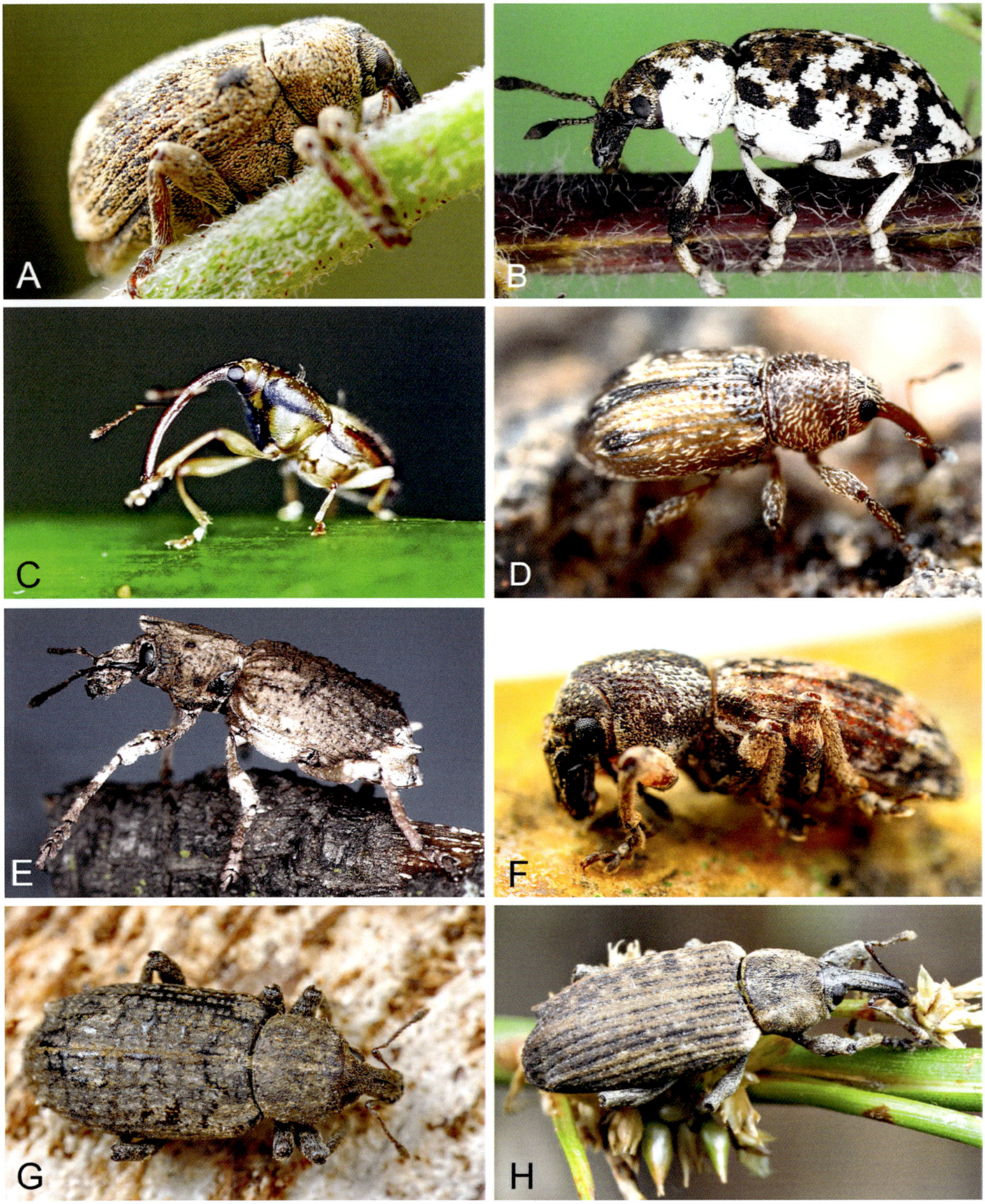

Plate 14. Curculionidae. A, *Xeda* sp., Ashwood, VIC (Hyperinae) © Nick Porch; B, *Pelororhinus perdix* (Pascoe) (Cyclominae: Aterpini) © Rolf Oberprieler; C, Undet. genus, Gove Peninsula, NT (Curculioninae: Acalyptini) © Rolf Oberprieler; D, *Misophrice amplicollis* Lea (Curculioninae: Storeini s.l.) © Bruno Bell; E, *Aesiotes leucurus* Pascoe (Cyclominae: Aterpini) © Rolf Oberprieler; F, *Pelororhinus* sp., Orford, TAS (Cyclominae: Aterpini) © Nick Porch; G, *Listroderes foveatus* (Lea) (Cyclominae: Listroderini) © Nick Porch; H, *Steriphus* sp., Warramate Hills, VIC (Cyclominae: Listroderini) © Nick Monaghan.

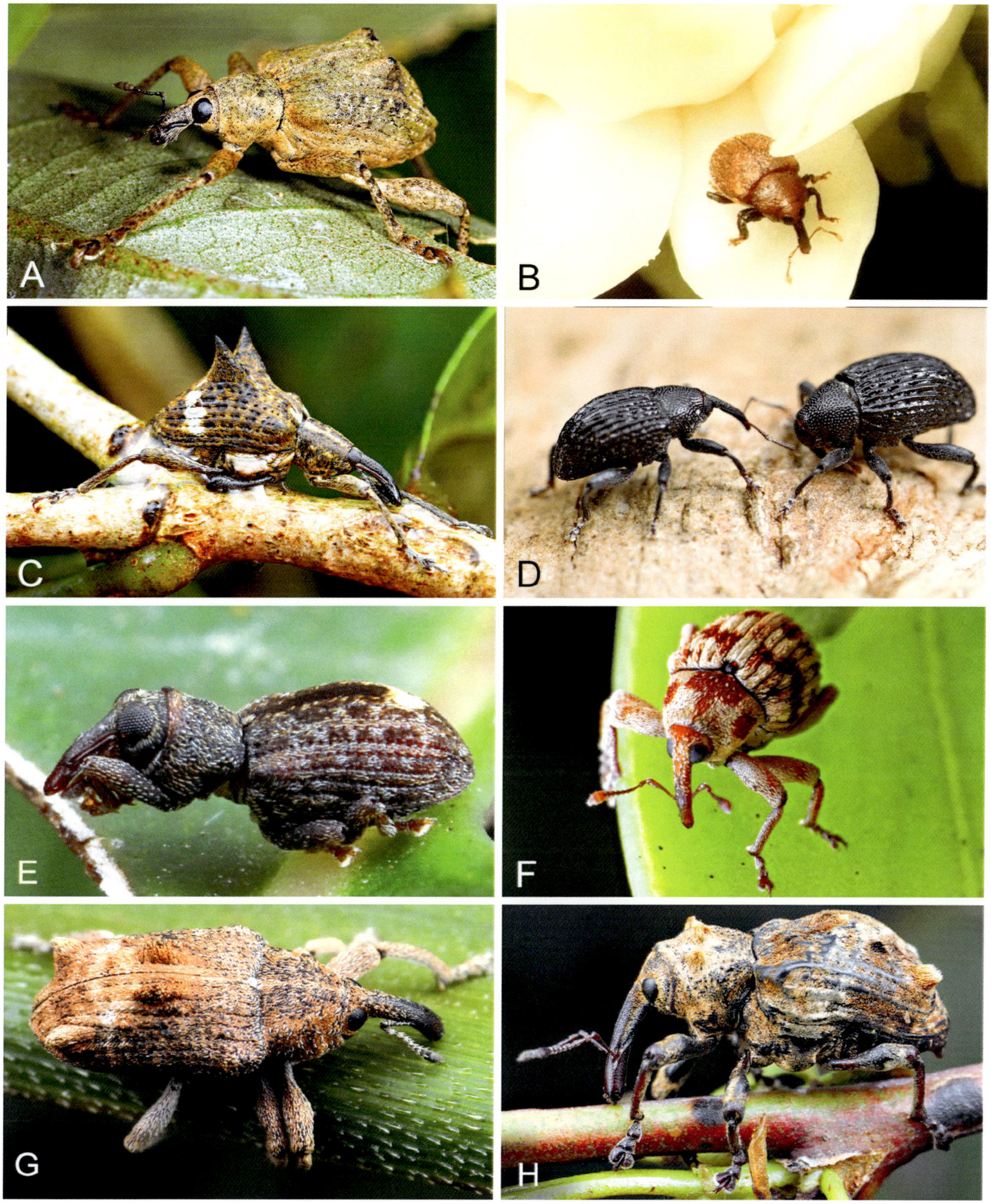

Plate 15. Curculionidae. A, *Eurychirus bituberculatus* Waterhouse (Hyperinae) © Nick Monaghan; B, *Elleschodes* sp., Mt. Glorious, QLD (Curculioninae: Storeini s.l.) © Jessa Thurman; C, *Eurychirus* sp., Lamington N.P., QLD (Hyperinae) © Nick Monaghan; D, *Melanterius* sp., Cape Otway, VIC (Curculioninae: Cleogonini) © Nick Porch; E, *Placorrhinus* sp., Cunninghams Gap, QLD (Curculioninae: Storeini s.l.) © Nick Monaghan; F, *Lybaeba* sp., Mt. Lofty, VIC (Curculioninae: Cleogonini) © Nick Monaghan; G, *Myllorhinus posticus* (Chevrolat) (Curculioninae: Storeini s.l.) © Nick Monaghan; H, *Myllorhinus bicaudatus* Boisduval (Curculioninae: Storeini s.l.) © Nick Monaghan.

Plate 16. Curculionidae (Cossoninae). A, Undet. genus, Kallista, VIC © Nick Porch; B, Undet. genus, Mt. St Gwinear, VIC © Nick Porch; C, Undet. genus, Mt. Glorious, QLD © Nick Monaghan; D, *Conlonia litoralis* Lea © Nick Porch; E, *Cossonus nitidirostris* Lea © Nick Monaghan; F, Undet. genus, Ferny Creek, VIC © Nick Porch; G, Undet. genus, Cape Conran, VIC © Nick Porch; H, *Stereomimetes* sp., Kallista, VIC © Nick Porch.

Plate 17. Curculionidae. A, *Ambrosiophilus compressus* (Lea) (Scolytinae: Xyleborini) © Hermes E. Escalona; B, *Hylurgus ligniperda* (Fabricius) (Scolytinae: Hylurgini) © Nick Porch; C, *Kuschelorhynchus macadamiae* Jennings & Oberprieler (Curculioninae: Cryptoplini) © Jessa Thurman; D, *Kuschelorhynchus macadamiae* Jennings & Oberprieler, larva (Curculioninae: Cryptoplini) © Jessa Thurman; E, *Mogulones larvatus* (Schultze) (Curculioninae: Ceutorhynchini) © Nick Monaghan; F, *Haplonyx* sp., WA (Curculioninae: Cryptoplini) © Jean and Fred Hort; G, *Haplonyx* sp., Fitzgerald River N.P., WA (Curculioninae: Cryptoplini) © Nick Monaghan; H, *Haplonyx* sp., Fitzgerald River N.P., WA (Curculioninae: Cryptoplini) © Nick Porch.

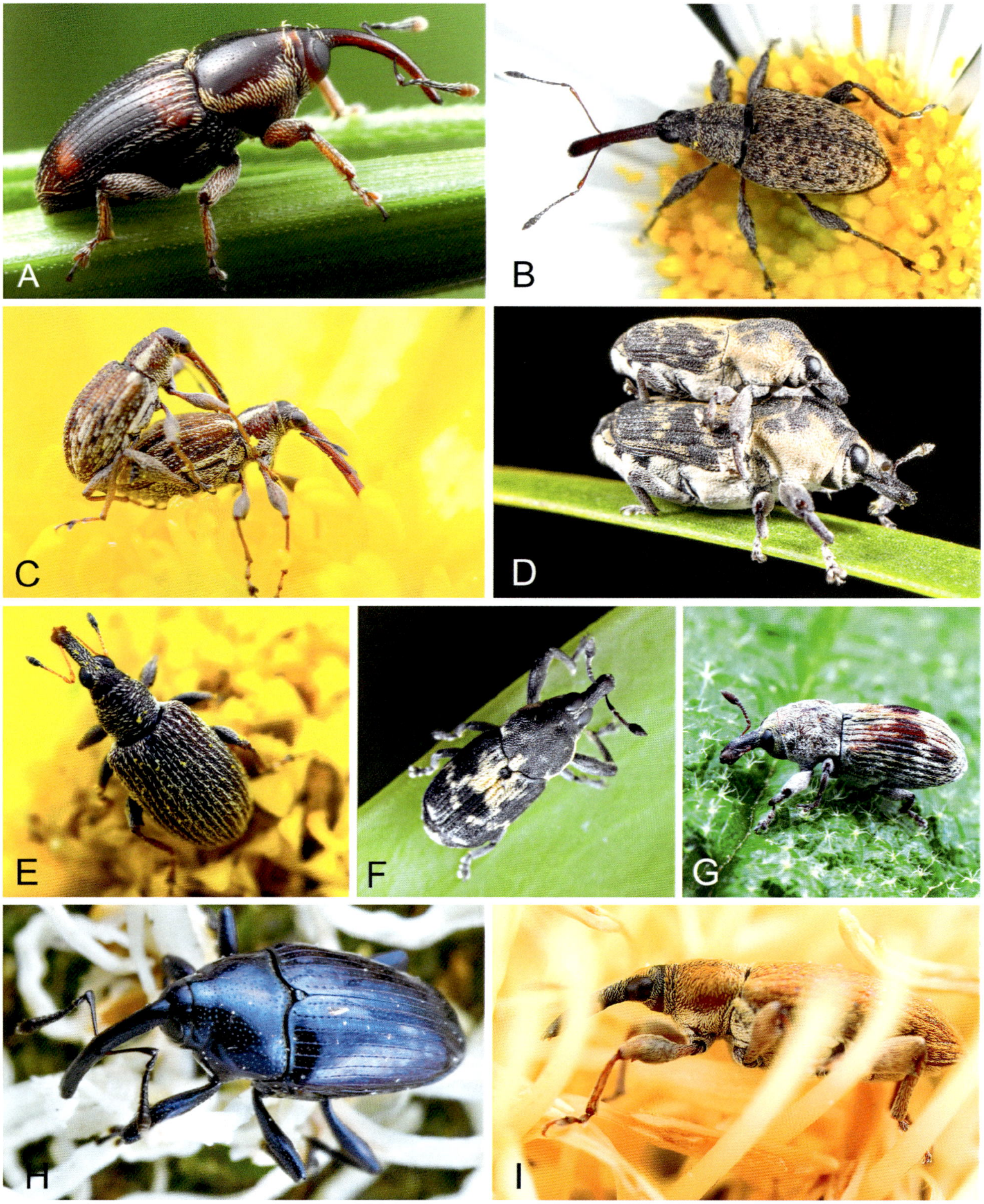

Plate 18. Curculionidae. A, *Apinocis variipennis* Lea (Conoderinae: Baridini) © Nick Monaghan; B, *Meriphus fullo* Erichson (Curculioninae: Eugnomini) © Nick Monaghan; C, *Meriphus* sp., Cape Otway, VIC (Curculioninae: Eugnomini) © Nick Porch; D, *Neolaemosaccus* sp., Mt. Ridley, WA (Molytinae: Mesoptiliini) © Jiri Lochman; E, *Ancyttalia* sp., Morwell N.P., VIC (Curculioninae: Eugnomini) © Nick Monaghan; F, *Neolaemosaccus longimanus* (Pascoe) (Molytinae: Mesoptiliini) © Nick Monaghan; G, *Notomagdalis mamillata* (Lea) (Molytinae: Mesoptiliini) © Nick Monaghan; H, *Ipsichora* sp., Noosa N.P., QLD (Conoderinae: Baridini) © Nick Monaghan; I, *Myossita* sp., Twin Creeks Reserve, WA (Curculioninae: Eugnomini) © Nick Porch.

Plate 19. Curculionidae (Molytinae). A, B, *Orthorhinus cylindrirostris* (Fabricius) (Orthorhinini) © Nick Monaghan; C, D, *Tranes* sp. on male cones of *Lepidozamia peroffskyana* Regel, Lismore, NSW (Molytinae) © Nicholas J. Fisher; E, *Lixus mastersii* Pascoe (Lixini) © Jiri Lochman; F, *Rhinocyllus conicus* (Frölich) (Lixini) © Nick Porch; G, *Paratranes zimmermani* Hsiao & Oberprieler (Molytinae) © Nick Monaghan.

Plate 20. Curculionidae. A, *Lixus tasmanicus* Germar (Molytinae: Lixini) © Nick Monaghan; B, *Curculio* sp., Kondalilla N.P., QLD (Curculioninae: Curculionini) © Nick Monaghan; C, *Sclerixus* sp., Kallista, VIC (Phrynixini) © Nick Porch; D, *Sclerixus* sp., Mt. Donna Buang, VIC (Phrynixini) © Nick Porch; E, *Poropterus antiquus* (Erichson) (Molytinae: Cryptorhynchini) © Nick Porch; F, *Mecopus* sp., Lismore, NSW (Conoderinae: Conoderini) © Nicholas J. Fisher; G, *Poropterus* sp., QLD (Molytinae: Cryptorhynchini) © Dash Huang; H, *Mormosintes rubus* Pascoe (Molytinae: Cryptorhynchini) © Jessa Thurman.

Plate 21. Curculionidae (Molytinae: Cryptorhynchini). A, *Enteles vigorsii* Gyllenhal © Nick Monaghan; B, *Tychreus* sp., Mt. Field N.P., TAS © Nick Porch; C, *Ouroporopterus* sp., Kondalilla N.P., QLD © Nick Monaghan; D, *Poropterus ellipticus* Pascoe © Bruno Bell; E, *Decilaus* sp., Lake St Clair N.P., TAS © Nick Porch; F, *Tentegia stupida* (Fabricius) © Hermes E. Escalona; G, *Drassicus* sp., Mapleton Falls N.P., QLD © Nick Monaghan; H, *Tentegia stupida* (Fabricius) © Hermes E. Escalona.

Plate 22. Curculionidae (Molytinae). A, *Onidistus* sp., Mossman Gorge, QLD (Cryptorhynchini) © Nick Monaghan; B, *Glochinorhinus* sp., ♂, Kenilworth, QLD (Cryptorhynchini) © Nick Monaghan; C, *Glochinorhinus* sp., ♀, Kenilworth, QLD (Cryptorhynchini) © Nick Monaghan; D, *Paleticus* sp., Tamborine Mountain, QLD (Cryptorhynchini) © Nick Monaghan; E, *Roptoperus* sp., Mt. Field N.P., TAS (Cryptorhynchini) © Nick Porch; F, *Atragopus caliginosus* (Lea) (Cryptorhynchini) © Nick Monaghan; G, *Camptorhinus* sp., Cape Tribulation, QLD (Ithyporini) © Nick Monaghan; H, *Tyrtaeosus* sp., McMahons Creek, VIC (Cryptorhynchini) © Nick Monaghan.

Plate 23. Cerambycidae: Cerambycinae (A, B, D), Prioninae (C, G), Lamiinae (E), Spondylidinae (F). A, *Stenoderus suturalis* (Olivier); B, *Australodon nearnsi* Escalona & Ślipiński, ♀; C, *Analophus parallelus* Waterhouse (Prioninae); D, *Katerina cantharoides* Ślipiński & Escalona; E, *Acalolepta ovina* (Pascoe); F, *Arhopalus rusticus* (Linnaeus); G, *Agrianome spinicollis* (Macleay). A © Marie Lochman; B © Nicholas J. Fischer; C, D © Roger de Keyzer; E © Pavel German; F © Reiner Richter; G © Jiri Lochman.

Plate 24. Cerambycidae: Lamiinae. A, *Zygocera metallica* (Westwood); B, *Excastra albopilosa* Tweed, Ashman & Ślipiński; C, *Rhytiphora dallasii* Pascoe; D, *Batocera frenchi* Van de Pol; E, *Microtragus arachne* Pascoe; F, *Ancita marginicollis* (Boisduval); G, *Rhytiphora pardalis* (Newman). A, C, E, F, G © Jiri Lochman; B © James Tweed; D © Jack Hasenpusch.

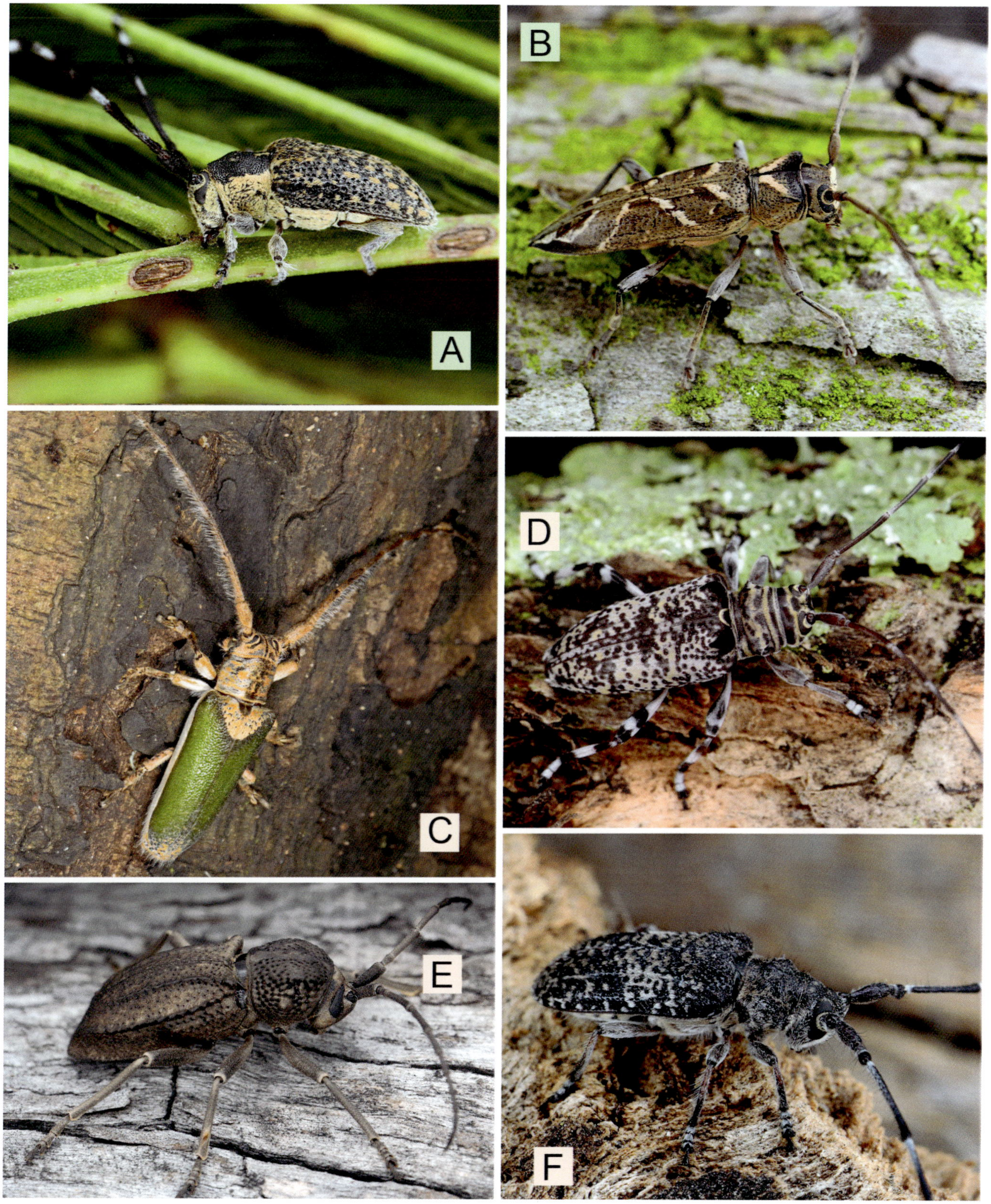

Plate 25. Cerambycidae: Lamiinae. A, *Ancita marginicollis* (Boisduval); B, *Temnosternus* sp.; C, *Rhytiphora nigrovirens* (Donovan); D, *Disterna* sp.; E, *Microtragus mormon* Pascoe; F, *Probatodes plumula* Newman © Owen Kelly; C © Living Li; remaining © Roger de Keyzer.

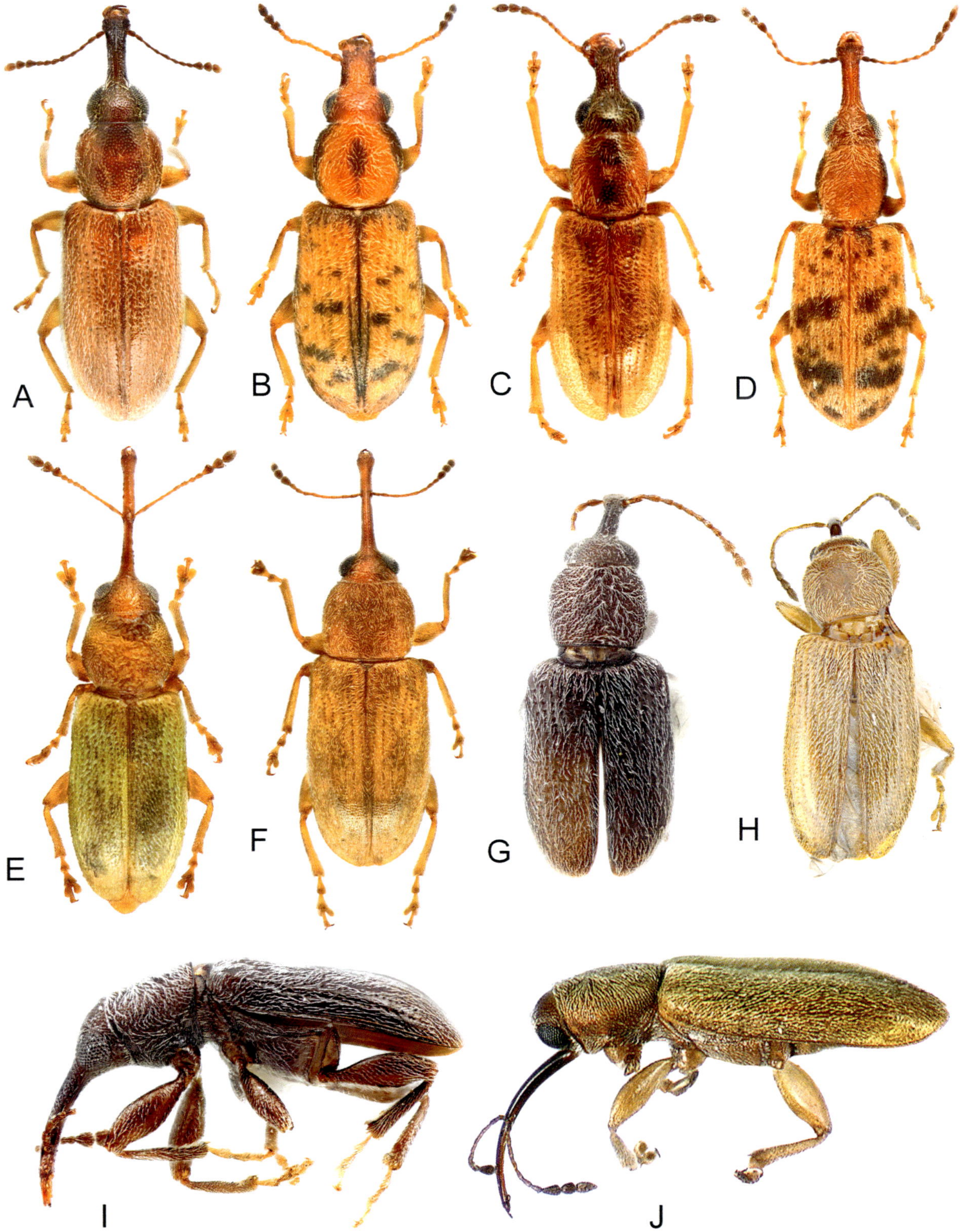

Plate 26. Nemonychidae. A, *Aragomacer uniformis* Kuschel, 4.0 mm (Mecomacerini); B, *Bunyaeus monteithi* Kuschel, 2.9 mm (Mecomacerini); C, *Eutactobius puellus* Kuschel, 2.2 mm (Mecomacerini); D, *Notomacer reginae* Kuschel, 3.4 mm (Mecomacerini); E, *Basiliogeus prasinus* Kuschel, 2.7 mm (Rhinorhynchini); F, *Basiliorhinus araucariae* Kuschel, 3.7 mm (Rhinorhynchini); G, I, *Pagomacer deceptus* Kuschel, dorsal and lateral, 2.5 mm (holotype, QMBA, Rhinorhynchini); H, J, *Zimmiellus fronto* Kuschel, dorsal and lateral, 2.8 mm (holotype, ANIC, Rhinorhynchini).

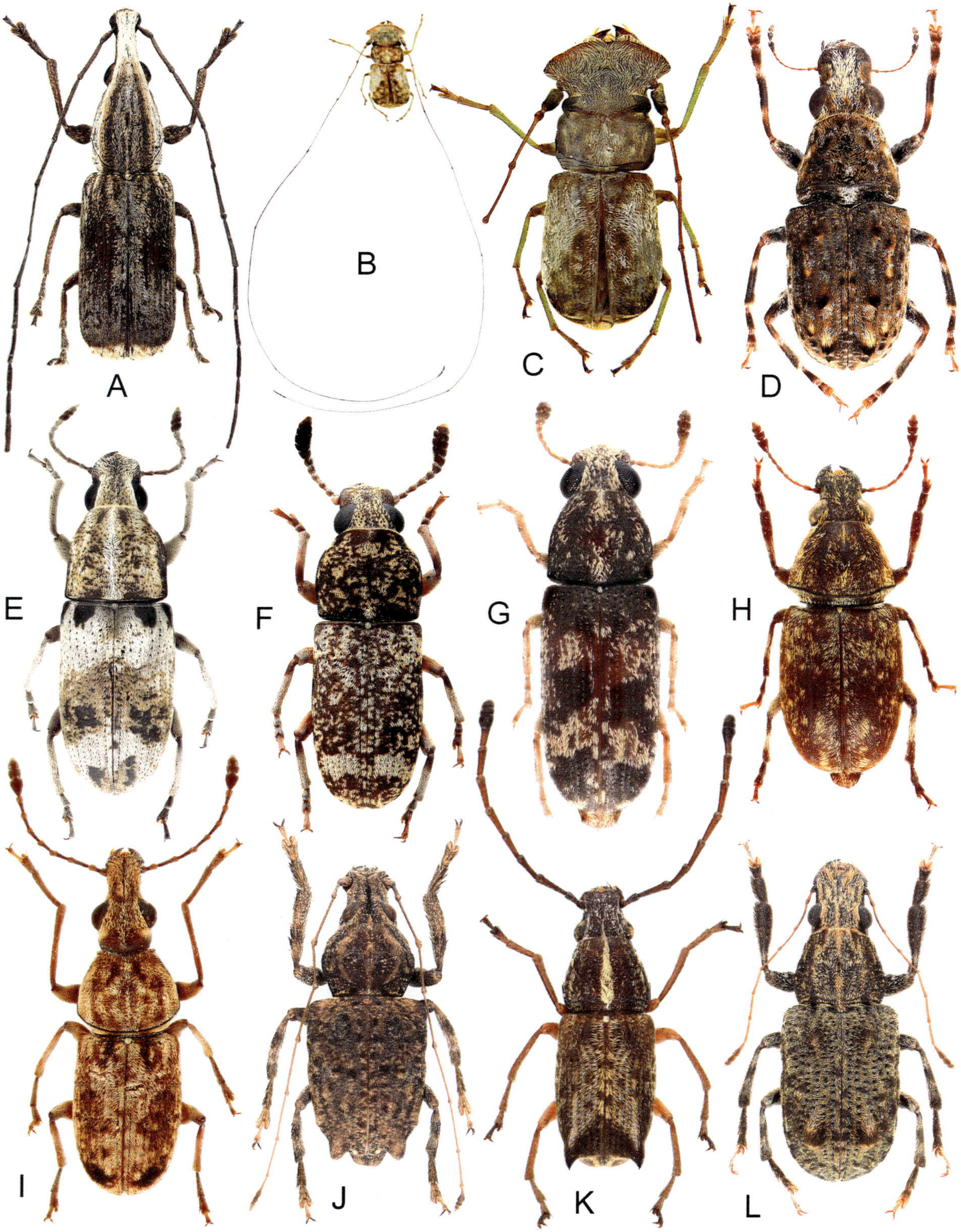

Plate 27. Anthribidae (Anthribinae). A, *Ancylotropis waterhousei* Jekel, 14.0 mm; B, C, *Antribisomus latifrons* Kuschel, ♂, 3.3 mm, habitus and main body detail; D, *Apatenia insignis* Jordan, 8.0 mm; E, *Basitropis pallida* Blackburn, 9.9 mm; F, *Basitropis peregrina* Pascoe, 7.2 mm; G, *Basitropis solitaria* Pascoe, 4.4 mm; H, *Cacephatus* sp., Dorrigo, NSW, 5.3 mm; I, *Cacephatus* sp., Norfolk Island, 4.0 mm; J, *Commista baccula* Jordan, 12.6 mm; K, *Commista bispina* (Erichson), 3.9 mm; L, *Commista latifrons* Jordan, 8.8 mm.

Plate 28. Anthribidae (Anthribinae). A, *Dendropemon albopictus* (Pascoe), 13.3 mm; B, *Dendropemon marmoratus* (Montrouzier), 10.9 mm; C, *Enspondus bigibbosus* Blackburn, 4.6 mm; D, *Entromus dorsoplagiatus* Blackburn, 4.8 mm; E, *Epargemus marmoratus* Blackburn, 4.5 mm; F, *Ethneca bakewellii* Pascoe, 4.8 mm; G–H, *Euciodes suturalis* Pascoe, G, ♀, 2.7 mm, H, ♂, 2.7 mm; I, *Erichsonocis albuginosus* (Erichson), 2.5 mm; J, *Erichsonocis* sp., Sugarloaf Creek, NSW, 3.4 mm; K, *Eczesaris australis* Frieser, 4.7 mm; L, *Eucorynus crassicornis* (Fabricius), 7.8 mm.

Plate 29. Anthribidae (Anthribinae). A, *Eupanteos bifasciatus* (Lea), 2.9 mm; B, *Eupanteos montanus* (Oke), 3.8 mm; C, *Eupanteos ornatus* Jordan, 5.4 mm; D, *Eupanteos* sp., Springbrook, QLD, 3.8 mm; E, *Euparius lateripictus* (Jordan), 9.2 mm; F, *Exillis australis* Frieser, ♀, 3.5 mm; G, *Exillis australis* Frieser, ♂, 3.9 mm; H, *Genethila retusa* Pascoe, 5.3 mm; I, Undet. genus sp., Mareeba, QLD, 2.1 mm; J–K, *Howeanthribus bufo* Zimmerman, J, ♂, 3.3 mm, K, ♀, 4.7 mm; L, *Hucus persimilis* Jordan, 4.1 mm.

Plate 30. Anthribidae (Anthribinae). A, *Mauia subnotata* (Boheman), 3.1 mm; B, *Mauia* sp., Mt. Finnigan, QLD, 2.6 mm; C, *Mecocerinopis amabilis* (Pascoe), ♂, 7.7 mm; D, *Mecocerinopis amabilis* (Pascoe), ♀, 7.4 mm; E, *Mecocerinopis balli* (Olliff), 11.4 mm; F, *Leaoanthribus crucifer* (Lea), 2.7 mm; G, *Ozotomerus bipunctatus* (Montrouzier), ♀, 9.0 mm; H, *Ozotomerus bipunctatus* (Montrouzier), ♂, 8.2 mm; I, *Litocerus macrophthalmus* (Montrouzier), 7.4 mm; J, *Phaulimia albirostris* (Frieser), 3.8 mm; K, *Phaulimia australiaca* (Jordan), 4.2 mm; L, *Mucronianus lunulatus* Frieser, 8.3 mm.

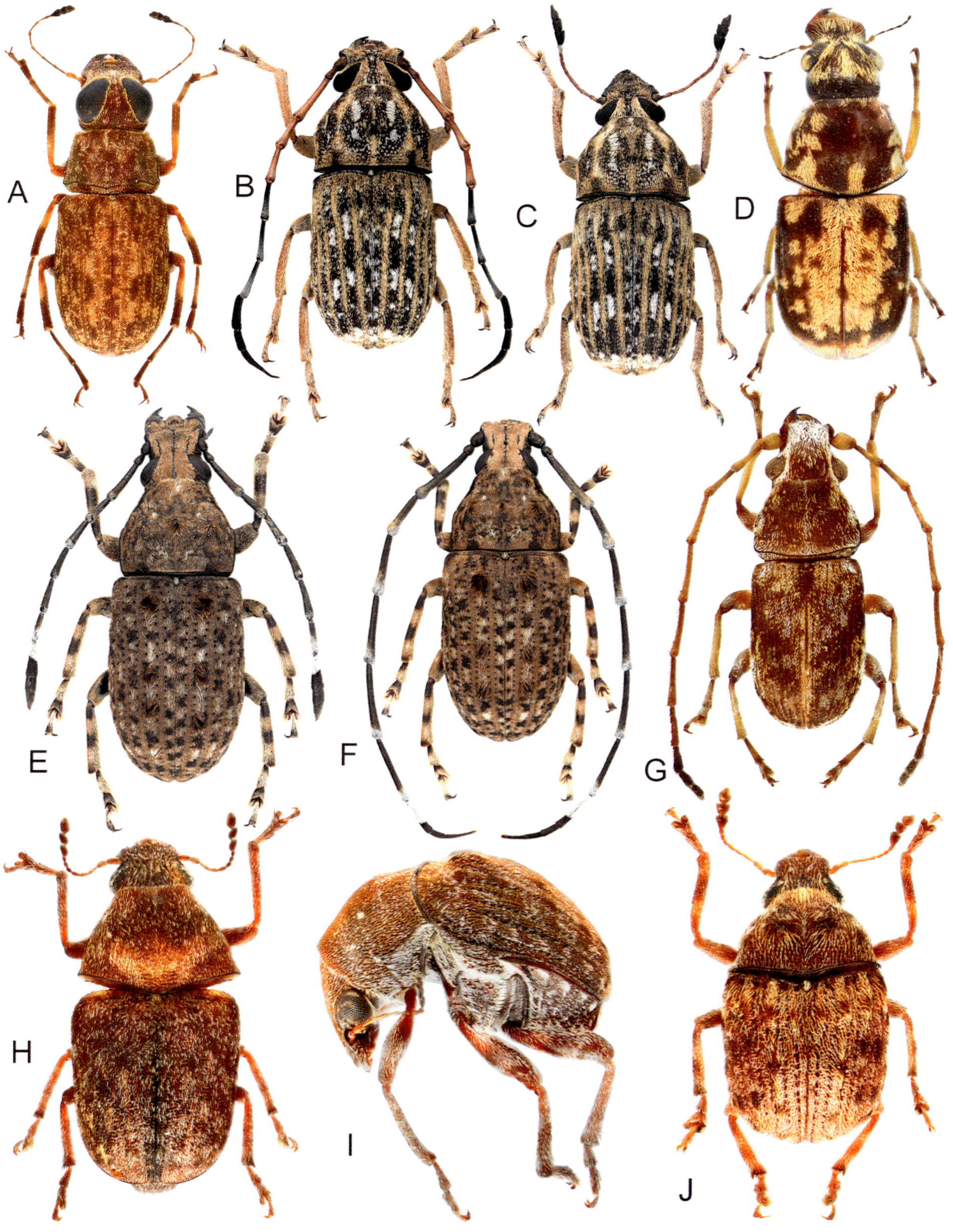

Plate 31. Anthribidae (Anthribinae). A, *Notoecia reticulata* Blackburn, 3.2 mm; B, *Phloeobius gigas* (Fabricius), ♂, 9.3 mm; C, *Phloeobius gigas* (Fabricius), ♀, 10.0 mm; D, *Plesiobasis* sp., Atherton, QLD, 2.8 mm; E–F, *Platystomos wallacei* (Pascoe), E, ♀, 12.4 mm, F, ♂, 17.8 mm; G, *Protaedus bryanti* Jordan, 2.9 mm; H–I, *Streneoderma contemptum* Blackburn, H, dorsal, I, lateral, 3.0 mm; J, *Streneoderma planatum* Blackburn, 2.1 mm.

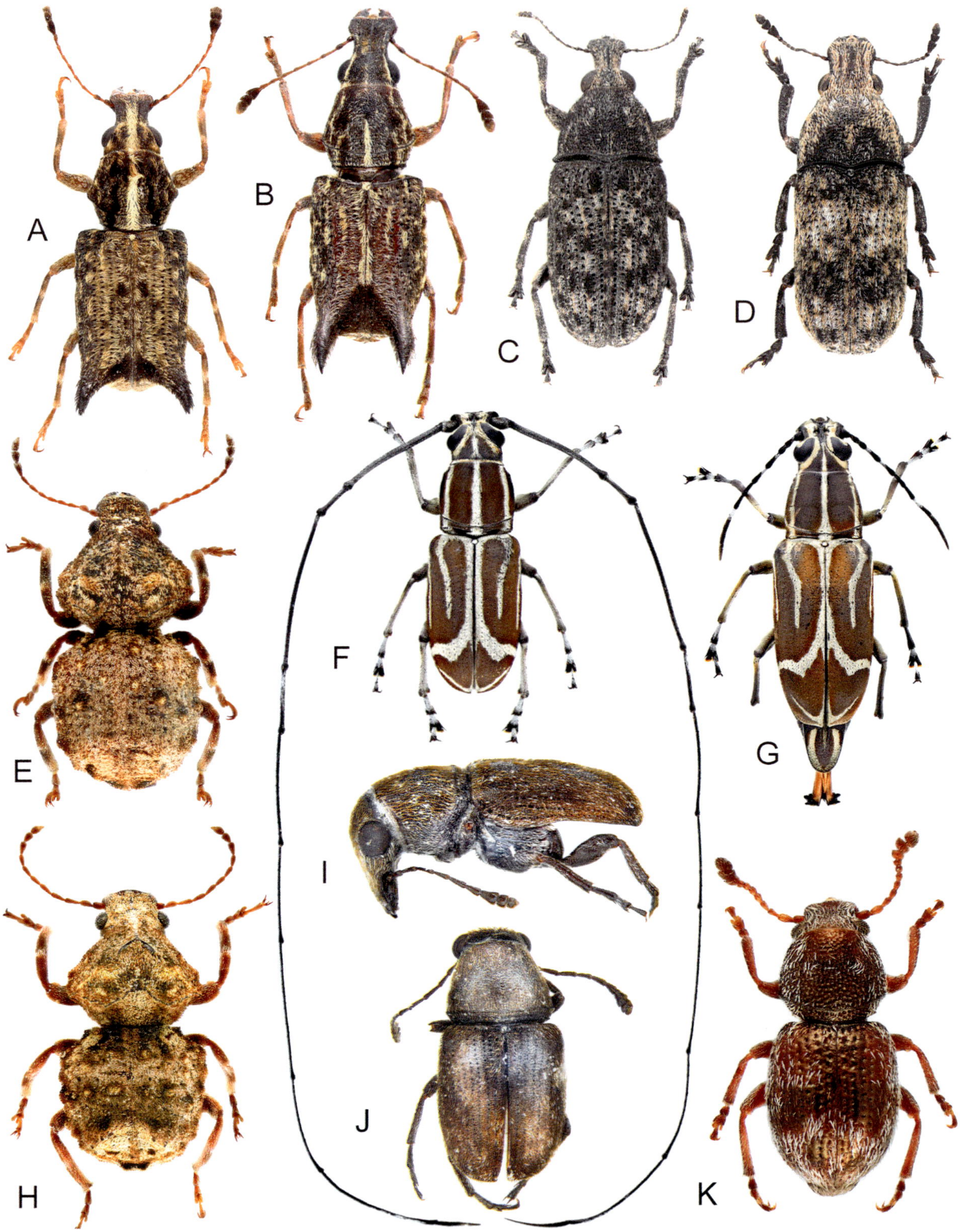

Plate 32. Anthribidae (Anthribinae). A, *Telala armigera* Jordan, 6.5 mm; B, *Telala* sp., Illawarra, NSW, 6.9 mm; C, *Taburnus neglectus* Zimmerman, 12.8 mm; D, *Taburnus breviceps* Zimmerman, 7.4 mm; E, *Teratanthribus* sp., St. Lucia, QLD, 4.2 mm; F, *Xenocerus speracerus* Montrouzier, ♂, 12.1 mm; G, *Xenocerus speracerus* Montrouzier, ♀, 19.4 mm; H, *Teratanthribus monstrosus* Zimmerman, 4.1 mm; I, J, *Atropideres evanescens* (Blackburn), 2.6 mm, lateral and dorsal; K, *Xynotropis micans* Blackburn, 1.7 mm.

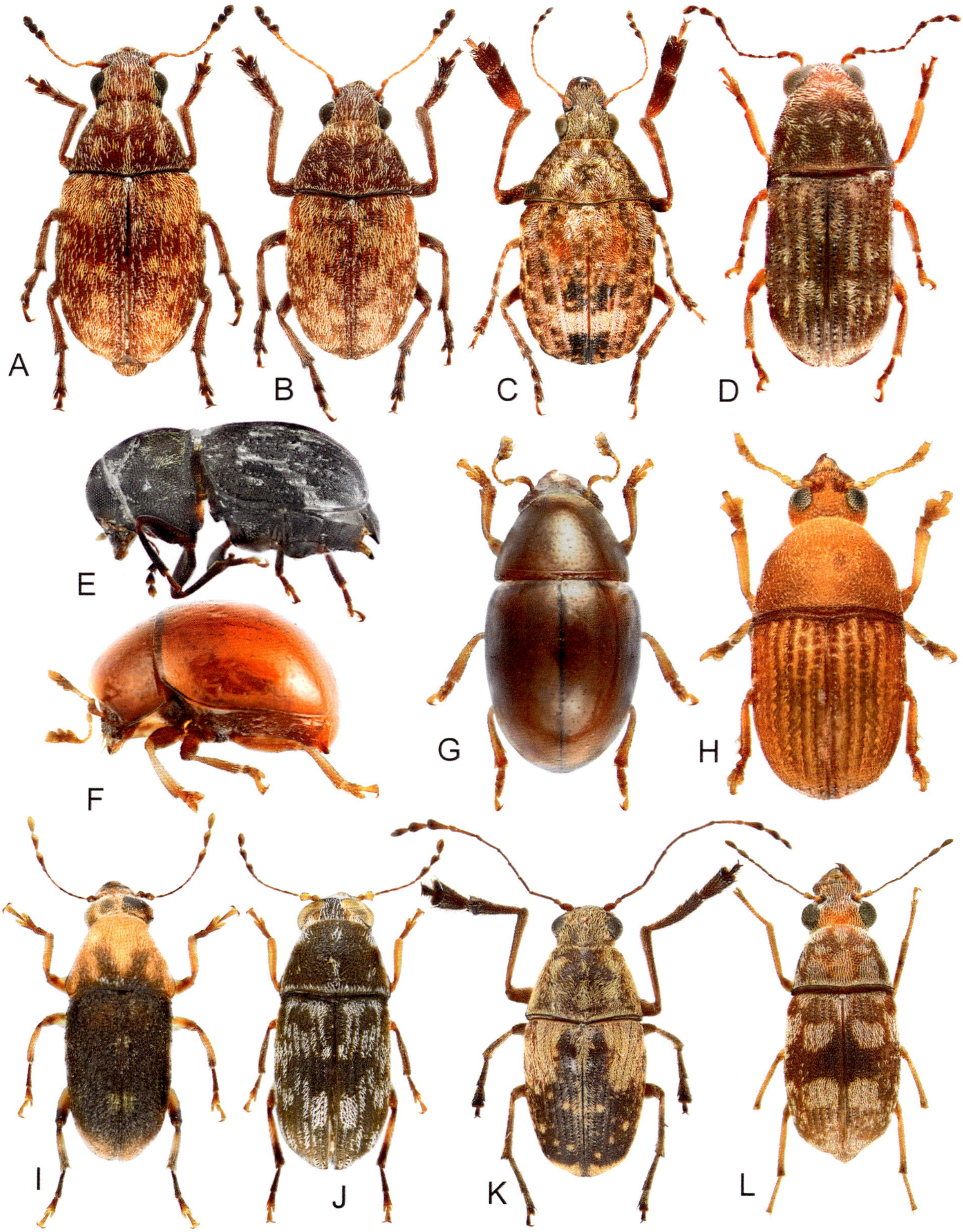

Plate 33. Anthribidae (Choraginae). A, *Araecerus lindensis* Blackburn, ♀, 2.5 mm; B, *Araecerus lindensis* Blackburn, ♂, 2.6 mm; C, *Araecerus palmaris* (Pascoe), ♂, 4.9 mm; D, *Araeocerodes* sp., Atherton, QLD, 2.2 mm; E, *Araeocerodes* sp., Daintree, QLD, 2.5 mm, lateral; F, *Cisanthribus* sp., Lord Howe Island, NSW, 1 mm, lateral; G, *Cisanthribus* sp., Norfolk Island, 1.4 mm; H, *Stenorhis* sp., Daintree, QLD, 1.4 mm; I, *Deropygus* sp., Mt. Tozer, QLD, 2.3 mm (paler colour); J, *Deropygus* sp., Mt. Tozer, QLD, 2.5 mm (darker colour); K, *Misthosima* sp., ♂, Christmas Island, 4.0 mm; L, *Misthosima dorsonotata* Blackburn, 2.9 mm, ♀.

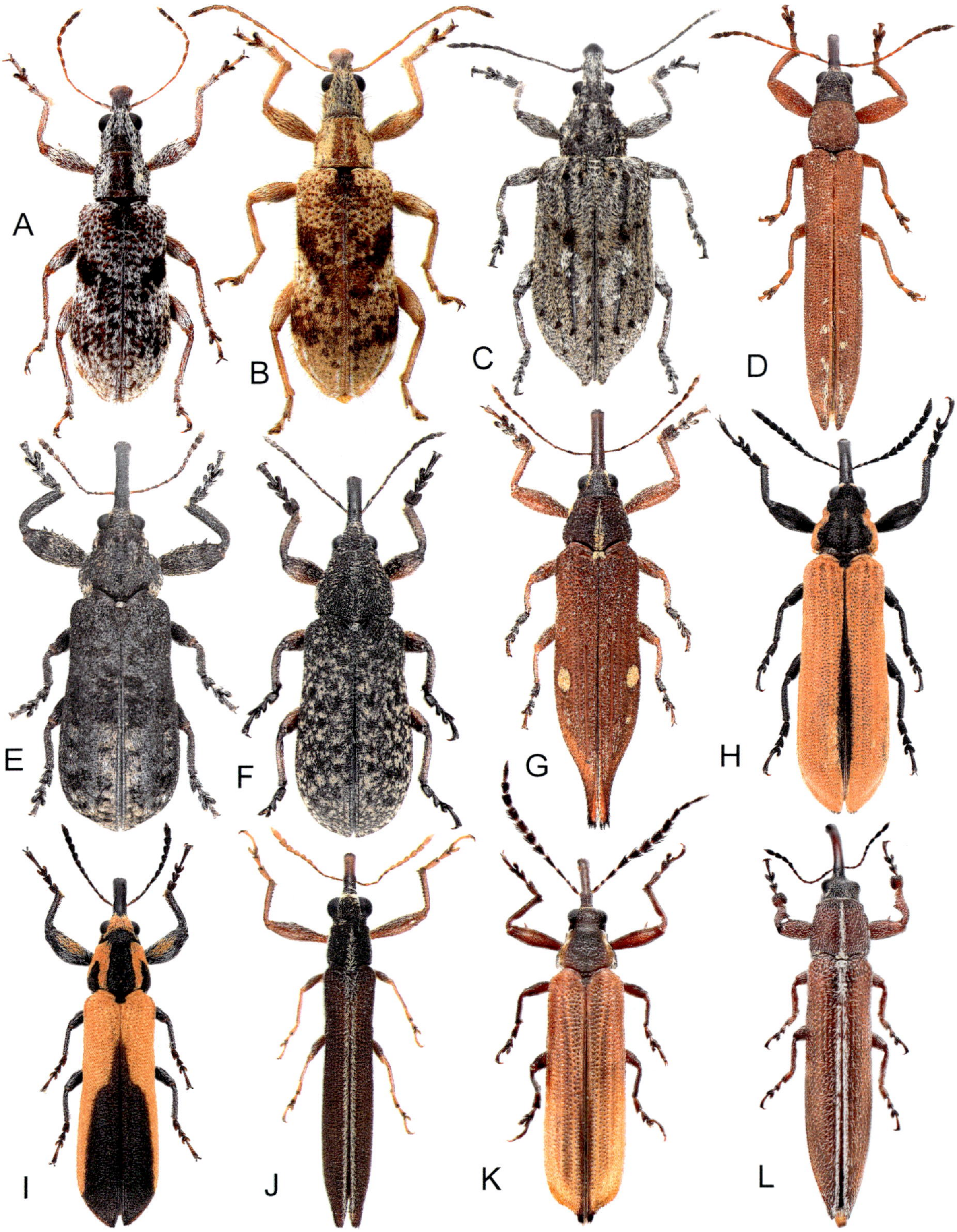

Plate 34. Belidae (Belinae). A–B, *Agnesiotis pilosula* Pascoe, 7.9 mm and 8.6 mm respectively (Agnesiotidini); C, *Cyrotyphus fascicularis* Pascoe, 14.6 mm (Agnesiotidini); D, *Araiobelus filum* (Jekel), 9.3 mm (Belini); E, *Isacantha dermestiventris* (Boisduval), 12.3 mm (Belini); F, *Isacanthodes monilis* (Newman), 11.8 mm (Belini); G, *Rhinotia bidentata* (Donovan), 15.6 mm (Belini); H, *Rhinotia haemoptera* Kirby, 16.9 mm (Belini); I, *Rhinotia princeps* Zimmerman, 11.0 mm (Belini); J, *Rhinotia tenuis* (Lea), 7.0 mm (Belini); K, *Rhinotiodes spinipennis* (Lacordaire), 10.2 mm (Belini); L, *Stenobelus tibialis* (Blackburn), 10.4 mm (Belini).

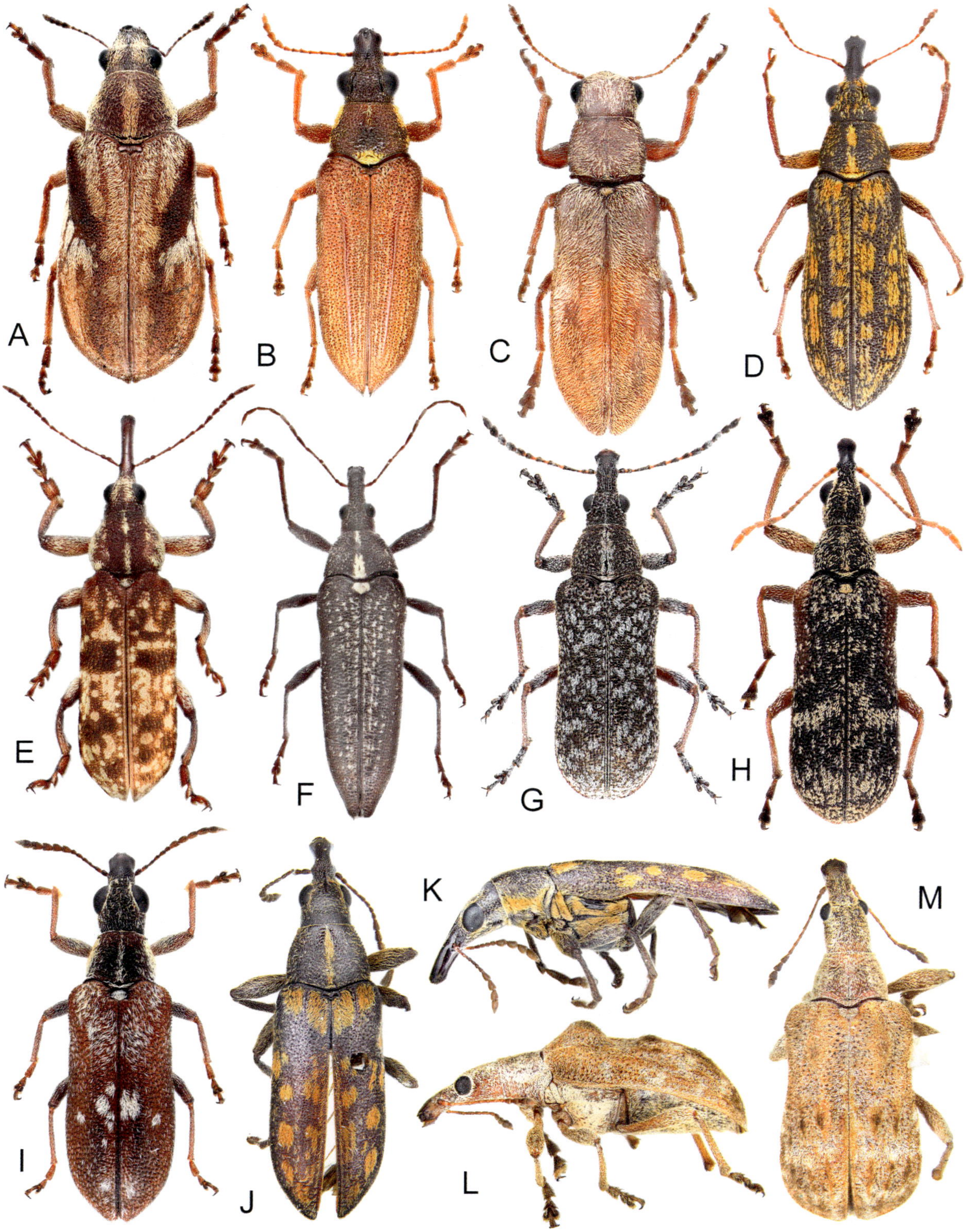

Plate 35. Belidae (Belinae, Pachyurini). A, *Agathobelus bivittatus* Zimmerman, 8.3 mm; B, *Apagobelus brevirostris* (Lea), 8.5 mm; C, *Arhinobelus agathophagus* Zimmerman, 4.3 mm; D, *Basiliobelus flavovittatus* Zimmerman, 8.0 mm; E, *Hadrobelus undulatus* (Zimmerman), 12.4 mm; F, *Macrobelus insignis* (Lea), 15.3 mm; G, *Pachyura australis* Hope, 10.0 mm; H, *Sphinctobelus cinereus* (Blanchard), 4.8 mm; I, *Sphinctobelus quadrimaculatus* (Lea), 5.5 mm; J, K, *Habrobelus maculatus* Zimmerman, dorsal and lateral, 8.0 mm (holotype, ANIC); L, M, *Pachybelus tuberculatus* Zimmerman, 8.5 mm, lateral and dorsal (holotype, ANIC).

Plate 36. Attelabidae A–H, Caridae I–L. A, *Euops bakewelli* Jekel, 3.0 mm (Attelabinae); B, *Euops eucalypti* Pascoe, 3.0 mm (Attelabinae); C, *Euops falcatus* (Guérin-Méneville), ♂, 3.5 mm (Attelabinae); D, *Euops rudis* Lea, ♀, 5.3 mm (Attelabinae); E, *Euops* sp., Mt. Tamborine, QLD, 3.7 mm (Attelabinae); F, *Metopum densum* (Lea), 2.7 mm (Rhynchitinae); G, *Metopum kingi* (Lea), 1.8 mm (Rhynchitinae); H, *Australotobius zimmermani* (Legalov), 4.3 mm (Rhynchitinae); I, *Car condensatus* Blackburn, 3.6 mm; J, *Car pini* Lea, 2.1 mm; K, *Car* sp., Monarto, SA, 4.5 mm; L, *Carodes revelatus* Zimmerman, 4.3 mm.

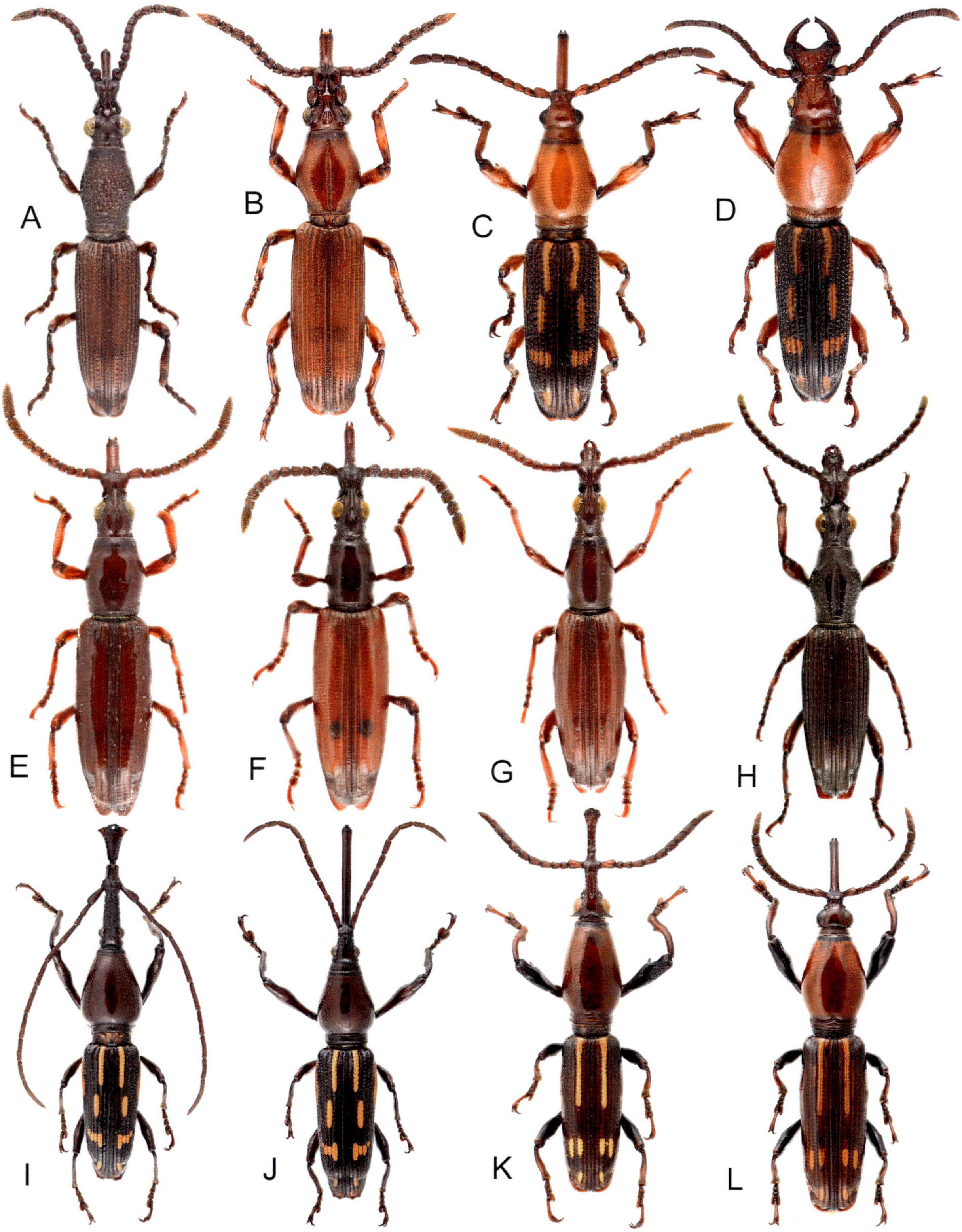

Plate 37. Brentidae (Brentinae: Brentini). A, *Ankleineella australis* (Lacordaire), 9.1 mm; B, *Ankleineella sulcicollis* (Pascoe), 6.3 mm; C, *Baryrhynchus lineicollis* Power, ♀, 15.2 mm; D, *Baryrhynchus lineicollis* Power, ♂, 11.8 mm; E, *Cordus demarzi* Damoiseau, 5.8 mm; F, *Cordus hospes* Schoenherr, 7.7 mm; G, *Cordus schoenherri* Power, 7.0 mm; H, *Kleineella barbata* (Kleine), 7.7 mm; I, *Ectocemus decemmaculatus* (Montrouzier), ♂, 18.8 mm; J, *Ectocemus decemmaculatus* (Montrouzier), ♀, 16.2 mm; K, *Orychodes digramma* (Boisduval), ♂, 10.2 mm; L, *Orychodes digramma* (Boisduval), ♀ 11.6 mm.

Plate 38. Brentidae (Brentinae). A, B, *Schizoeupsalis promissa* (Pascoe), major and minor, ♂, 10.3 mm, 7.0 mm (Brentini); C, *Cylas formicarius* (Fabricius), ♂, 4.8 mm (Cyladini); D, *Ancylobrentus australicola* Damoiseau, (Cyphagogini), 4.2 mm; E, *Allaeometrus breviceps* Senna, 5.6 mm (Cyphagogini); F, *Callipareius planitarsus* (Perroud & Montrouzier), 8.1 mm (Cyphagogini); G, *Catagogus diorymerus* (Lea), 6.2 mm (Cyphagogini); H, *Cerobates laevipennis* Senna, 5.9 mm (Cyphagogini); I, *Cyphagogus crassitarsus* Damoiseau, 7.5 mm (Cyphagogini); J, *Cyphagogus delicatus* Lea, 4.9 mm; K, *Hoplopisthius kolbei* Senna, 6.1 mm (Cyphagogini); L, *Isomorphus* sp., Mt. Tozer, QLD, 2.8 mm (Cyphagogini).

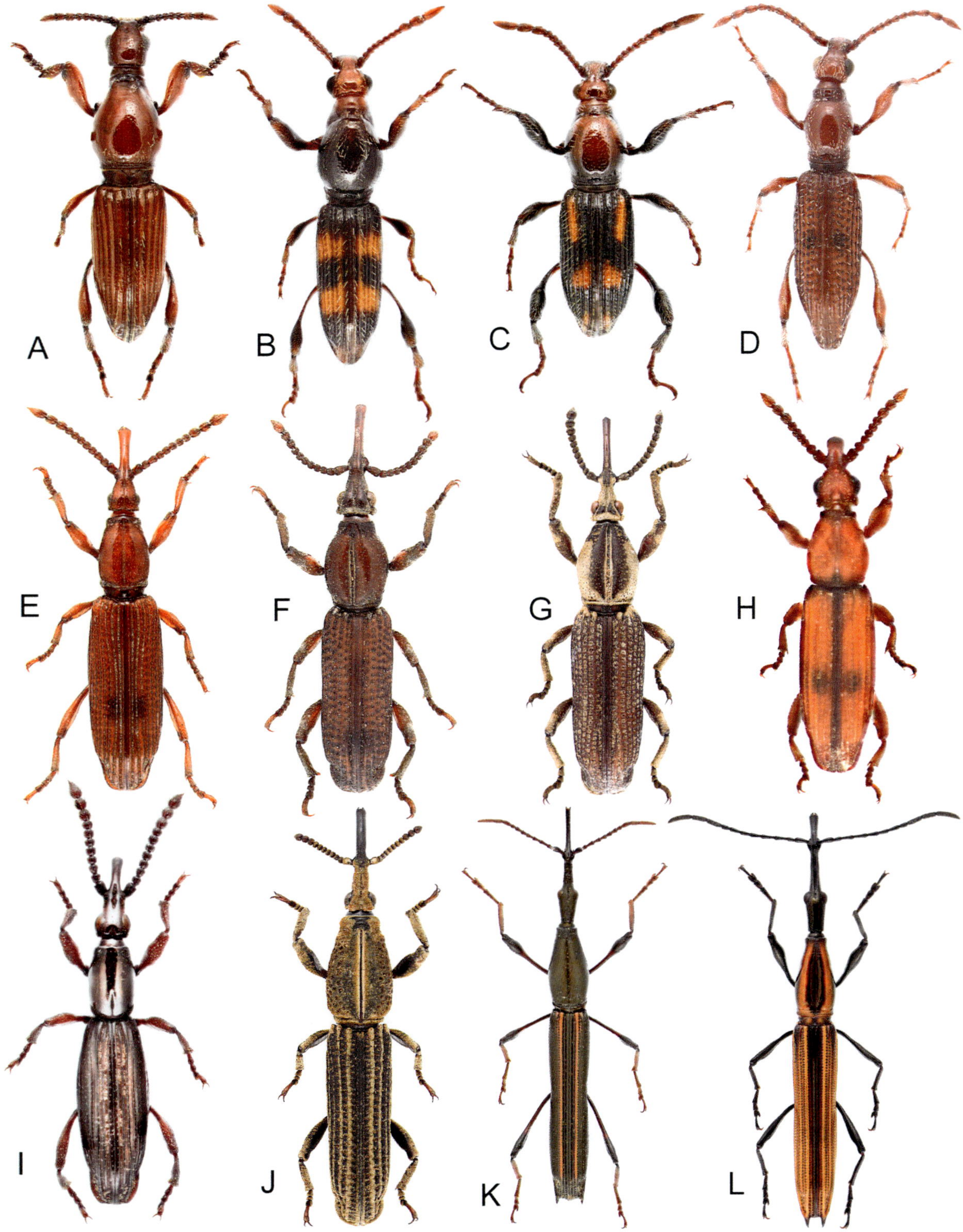

Plate 39. Brentidae (Brentinae). A, *Neosebus* sp., Iron Range, QLD, 5.3 mm (Cyphagogini); B, *Mesoderes guttatus* (Kleine), ♂, 5.4 mm (Cyphagogini); C, *Mesoderes guttatus* (Kleine), ♀, 3.9 mm (Cyphagogini); D, *Opisthenoxys ochraceus* Kleine, Indonesia, 4.6 mm (Cyphagogini); E, *Araiorrhinus howittii* (Pascoe), 4.8 mm (Cyphagogini); F, *Euschizus internatus* (Pascoe), 6.3 mm (Trachelizini); G, *Euschizus dictatorius* Kleine, 10.2 mm (Trachelizini); H, *Higonius novenarius* Damoiseau, 2.5 mm (Cyphagogini); I, *Howeius micropterus* Mantilleri, 4.7 mm (Cyphagogini); J, *Hormocerus reticulatus* (Lund), 16.2 mm (Trachelizini); K, *Ithystenus hollandiae* (Boisduval), 18.7 mm (Trachelizini); L, *Mesetia amoena* Blackburn, 14.2 mm (Trachelizini).

Plate 40. Brentidae (Brentinae A–G, Eurhynchinae H–K, Nanophyinae L). A, *Microtrachelizus queenslandicus* Damoiseau, 3.0 mm (Cyphagogini); B, *Miolispa australiana* Senna, 8.7 mm (Trachelizini); C, *Stenobrentus lineatus* Damoiseau, ♂, 9.1 mm (Trachelizini); D, *Stenobrentus lineatus* Damoiseau, ♀, 10.3 mm (Trachelizini); E, *Trachelizus bisulcatus* (Lund), 9.9 mm (Trachelizini); F, *Tracheloschizus dichrous* (Lacordaire), 13.8 mm (Trachelizini); G, *Uropteroides gestroi* (Senna), 23.5 mm (Trachelizini); H, *Aporhina australis* (Heller), 9.3 mm; I, *Ctenaphides porcellus* Pascoe, 9.6 mm; J, *Eurhynchus acanthopterus* Boisduval, 10.1 mm; K, *Eurhynchus* sp., Cairns, QLD, 15.2 mm; L, *Austronanodes nigrovarius* (Lea), 2.2 mm (Nanophyini).

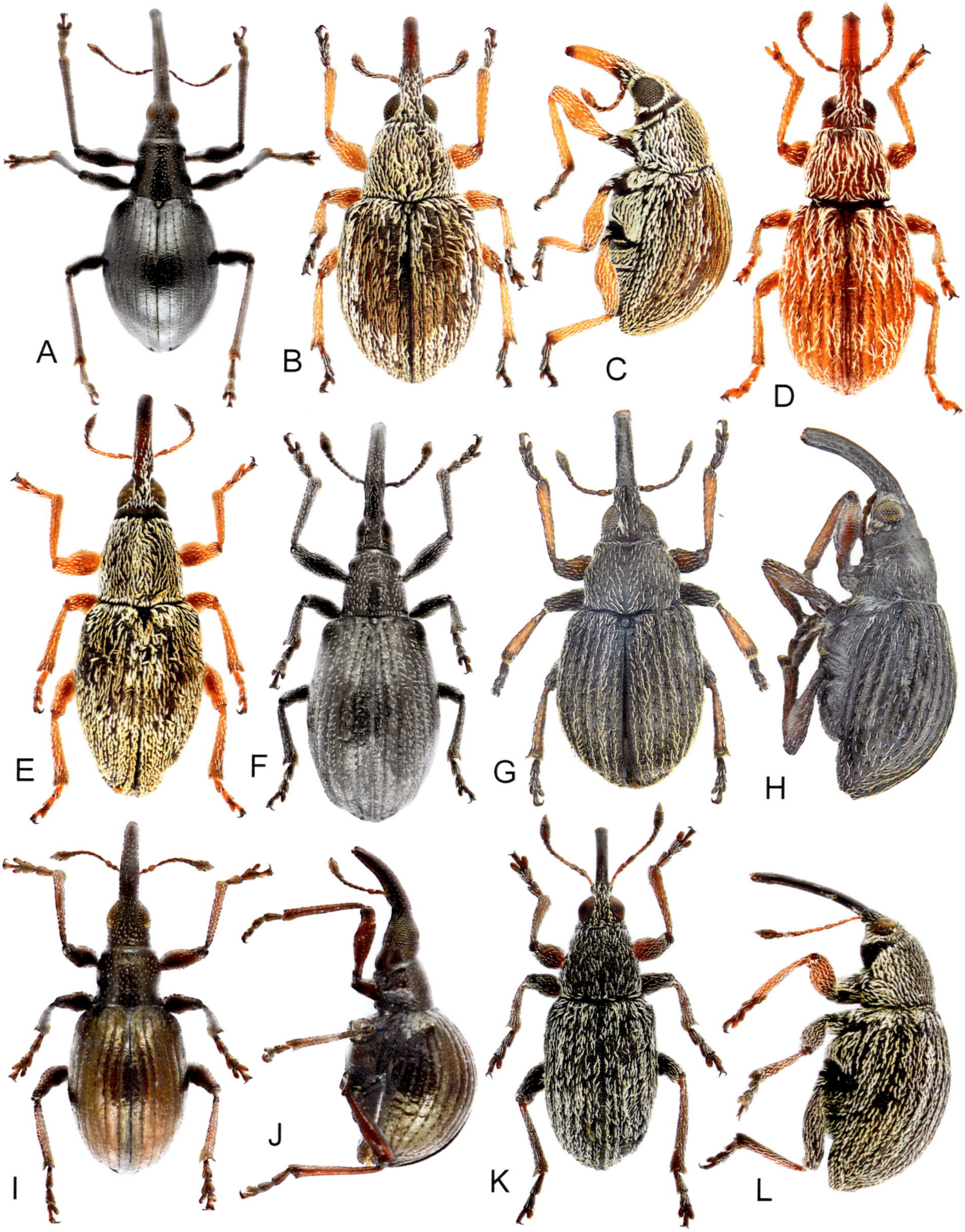

Plate 41. Brentidae (Apioninae: Apionini). A, *Pseudopiezotrachelus convexipennis* (Lea), 2.7 mm; B, C, *Pseudapion hoblerae* (Lea), dorsal and lateral, 1.7 mm; D, *Apiotherium microscopicum* (Lea), 1.3 mm; E, *Pseudapion philanthum* (Lea), 1.9 mm; F, *Pseudoconapion terraereginae* (Blackburn), 3.1 mm; G, H, *Coelocephalapion pigrae* Kissinger, dorsal and lateral, 1.3 mm; I, J, *Conapium argutulum* (Pascoe), dorsal and lateral, 1.4 mm; K, L, *Exapion ulicis* (Forster), dorsal and lateral, 2.0 mm.

Plate 42. Brentidae (Apioninae). A, B, *Perapion antiquum* (Gyllenhal), dorsal and lateral, 2.1 mm (Apionini); C, D, *Pseudopiezotrachelus macleayensis* (Lea), dorsal and lateral, 2.1 mm (Apionini); E, *Pseudapion aemulum* (Lea), 1.5 mm (Apionini); F, *Pseudorhinapion agonis* (Lea), 2.2 mm (Apionini); G, *Apiotherium commersoniae* (Rheinheimer), 3.2 mm (Apionini); H, *Rhynolaccus exsertus* (Pascoe), 3.2 mm (Myrmacicelini); I, *Alissapion zimmermani* Wanat, 1.8 mm (Rhadinocybini); J, *Allonotapion subulirostre* Wanat, 1.5 mm (Rhadinocybini); K, *Ganoapion rhadinocyboides* Wanat, 1.9 mm (Rhadinocybini); L, *Lissapion varistriatum* (Lea), 1.8 mm (Rhadinocybini).

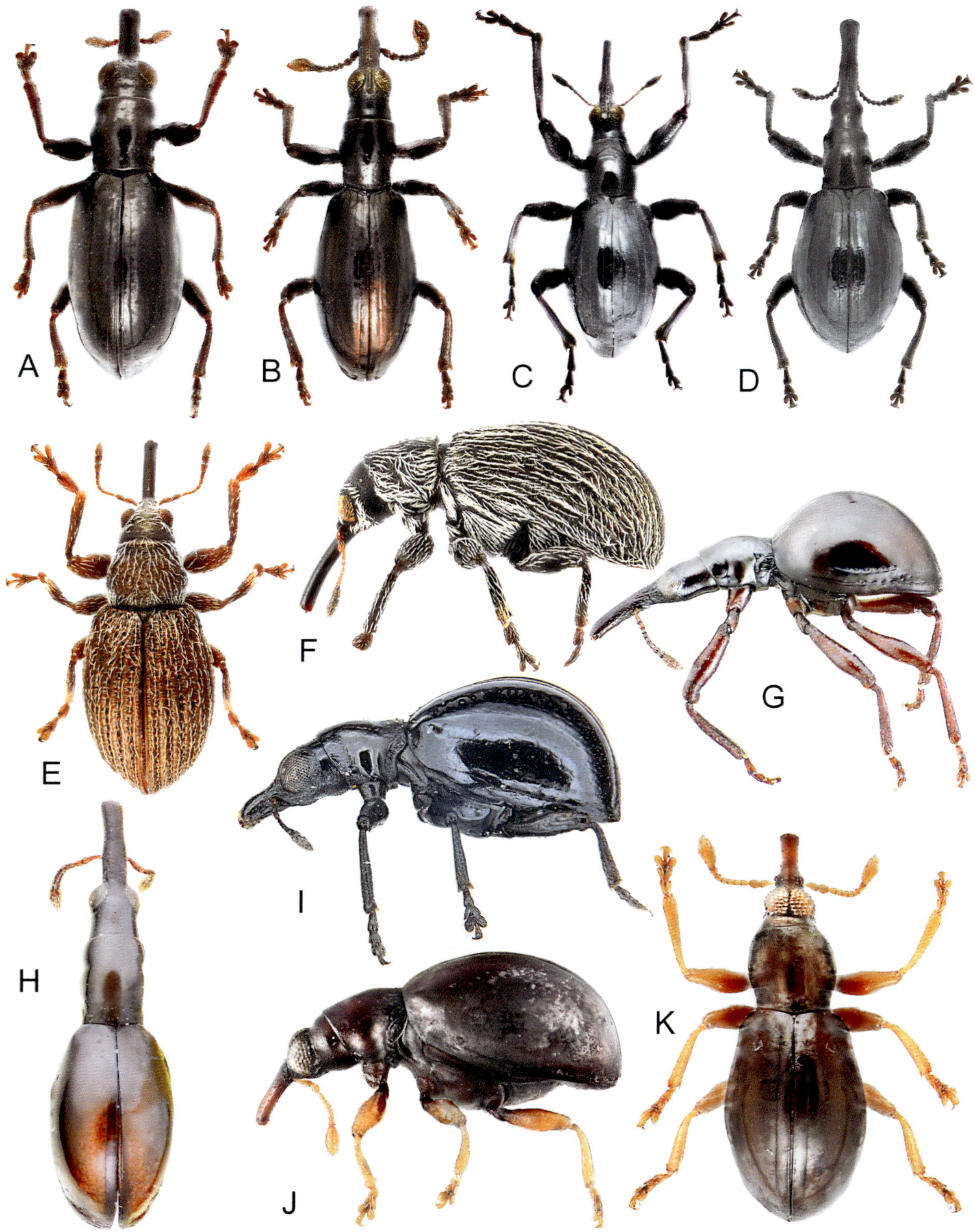

Plate 43. Brentidae (Apioninae). A, *Micronotapion gibbiceps* Wanat, 1.7 mm (Rhadinocybini); B, *Myrmecapion australicum* Wanat, 1.8 mm (Rhadinocybini); C, *Nanomyrmacyba illawarrensis* (Lea), 2.5 mm (Rhadinocybini); D, *Notapion tenuistriatum* (Lea), 2.9 mm (Rhadinocybini); E, F, *Rhinorhynchidius australasiae* (Lea), dorsal and lateral, 1.7 mm (Rhinorhynchidiini); G, H, *Coeliapion mednonstriatum* (Lea), dorsal and lateral, 2.5 mm (type SAM, Rhadinocybini); I, *Brachycybus* sp., Taree, NSW, 1.8 mm (Rhadinocybini); J, *Ommatocybus* sp., Lake Eacham N.P., QLD, 1.3 mm (Rhadinocybini); K, *Ommatocybus* sp., Wongabel S.F., QLD, 1.3 mm (Rhadinocybini).

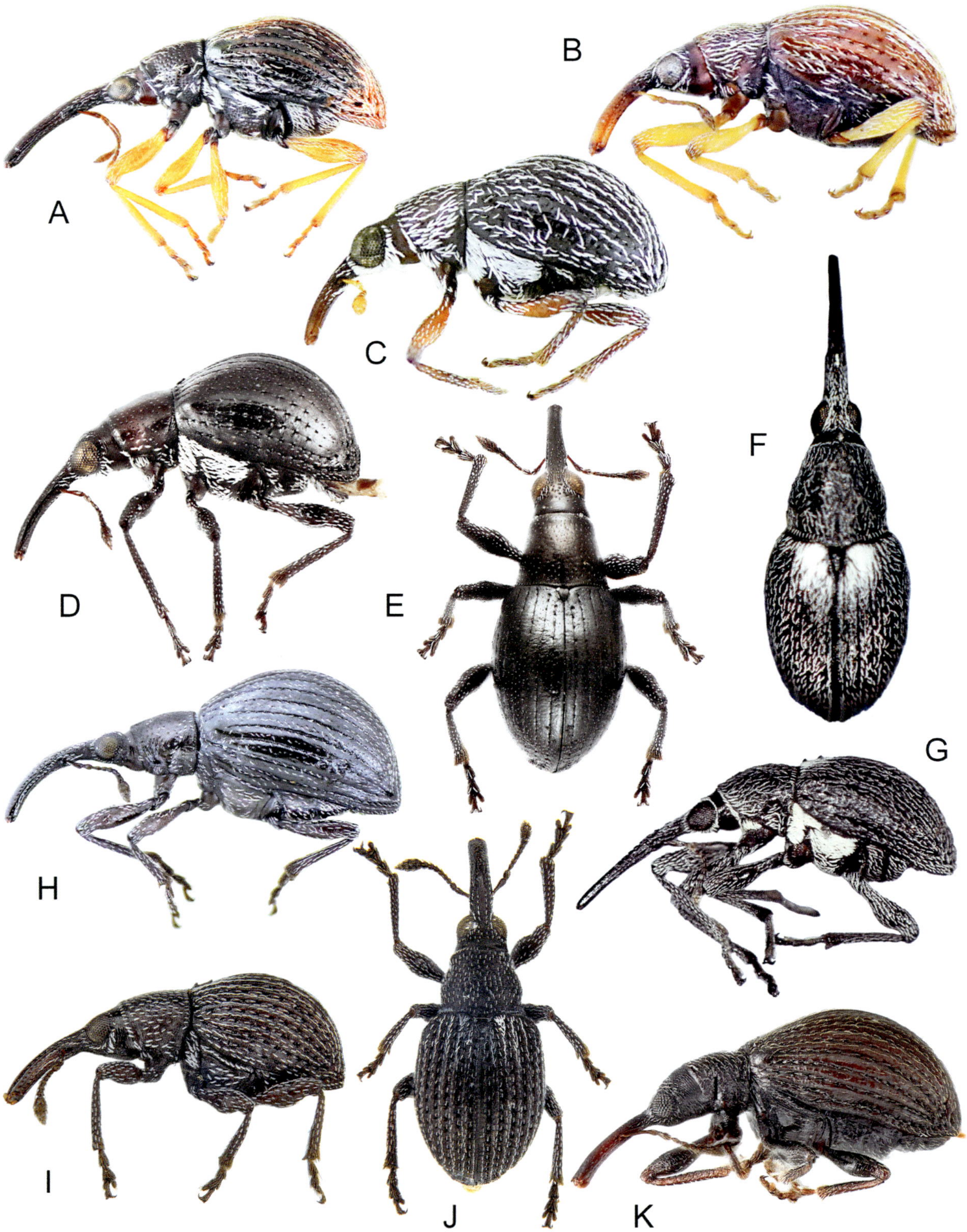

Plate 44. Brentidae (Apioninae). A, *Apiotherium amabile* (Lea), 4.2 mm, (syntype, SAM) (Apionini); B, *Apiotherium condensatum* (Lea), 2.6 mm (type, SAM) (Apionini); C, *Apiotherium melvillense* (Lea), 1.9 mm (type, SAM) (Apionini); D, E, *Apiotherium niveodispersum* (Lea), dorsal and lateral, 1.9 mm (Apionini); F, G, *Lopatinapion* sp., Drysdale River, WA, dorsal and lateral, 3.9 mm (Apionini); H, *Pseudoconapion terrareginae* (Blackburn), 4.4 mm (Apionini); I, J, *Pseudaspidapion subopacum* (Lea), 1.6 mm, lateral and dorsal (Apionini) K, *Diselmobius lechi* Wanat, 1.6 mm (*incertae sedis*).

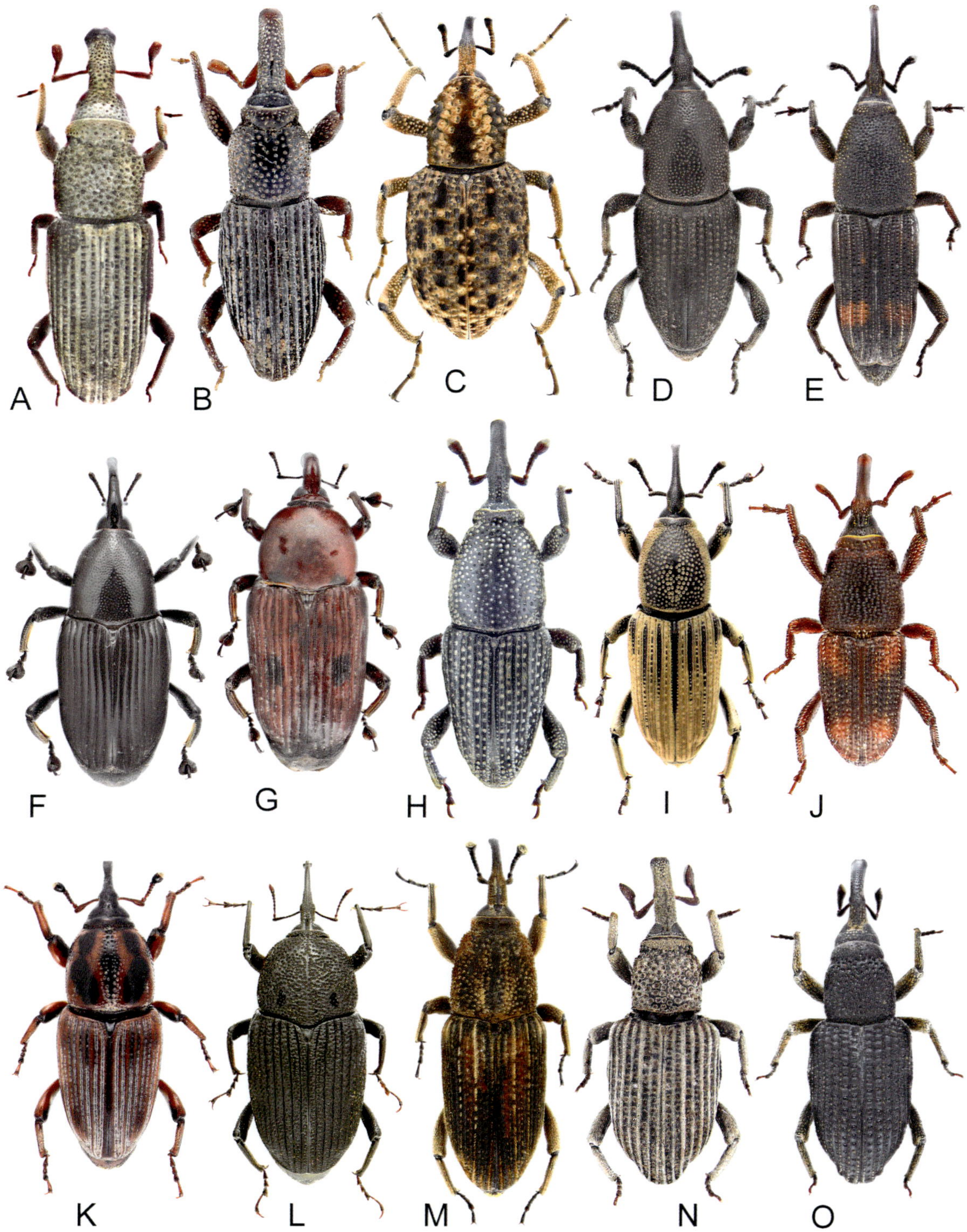

Plate 45. Curculionidae (Dryophthorinae). A, *Dryophthorus assimilis* Gahan, 2.4 mm (Dryophthorini); B, *Stenommatus* sp., South Johnstone, QLD, 1.5 mm (Dryophthorini); C, *Sipalinus gigas* (Fabricius), 19.8 mm (Orthognathini); D, *Cosmopolites sordidus* (Germar), 11.4 mm (Rhynchophorini); E, *Diocalandra frumenti* (Fabricius), 5.6 mm (Rhynchophorini); F, *Diathetes morio* Pascoe, 15.0 mm (Rhynchophorini); G, *Iphthimorhinus australasiae* Roelofs, 33.0 mm (Rhynchophorini); H, *Polytus mellerborgii* (Boheman), 3.8 mm (Rhynchophorini); I, *Rhabdoscelus interstitialis* (Boheman), 10.6 mm (Rhynchophorini); J, *Sitophilus oryzae* (Linnaeus), 2.8 mm (Rhynchophorini); K, *Sphenophorus brunnipennis* (Germar), 7.6 mm (Rhynchophorini); L, *Trigonotarsus calandroides* Guérin-Méneville, 31.3 mm (Rhynchophorini); M, *Trochorhopalus strangulatus* (Gyllenhal), 6.7 mm (Rhynchophorini); N, *Allaeotes* sp., Mt. Tozer, QLD, 3.6 mm (Stromboscerini); O, *Dryophthoroides* sp., Mulgrave River, QLD, 5.4 mm (Stromboscerini).

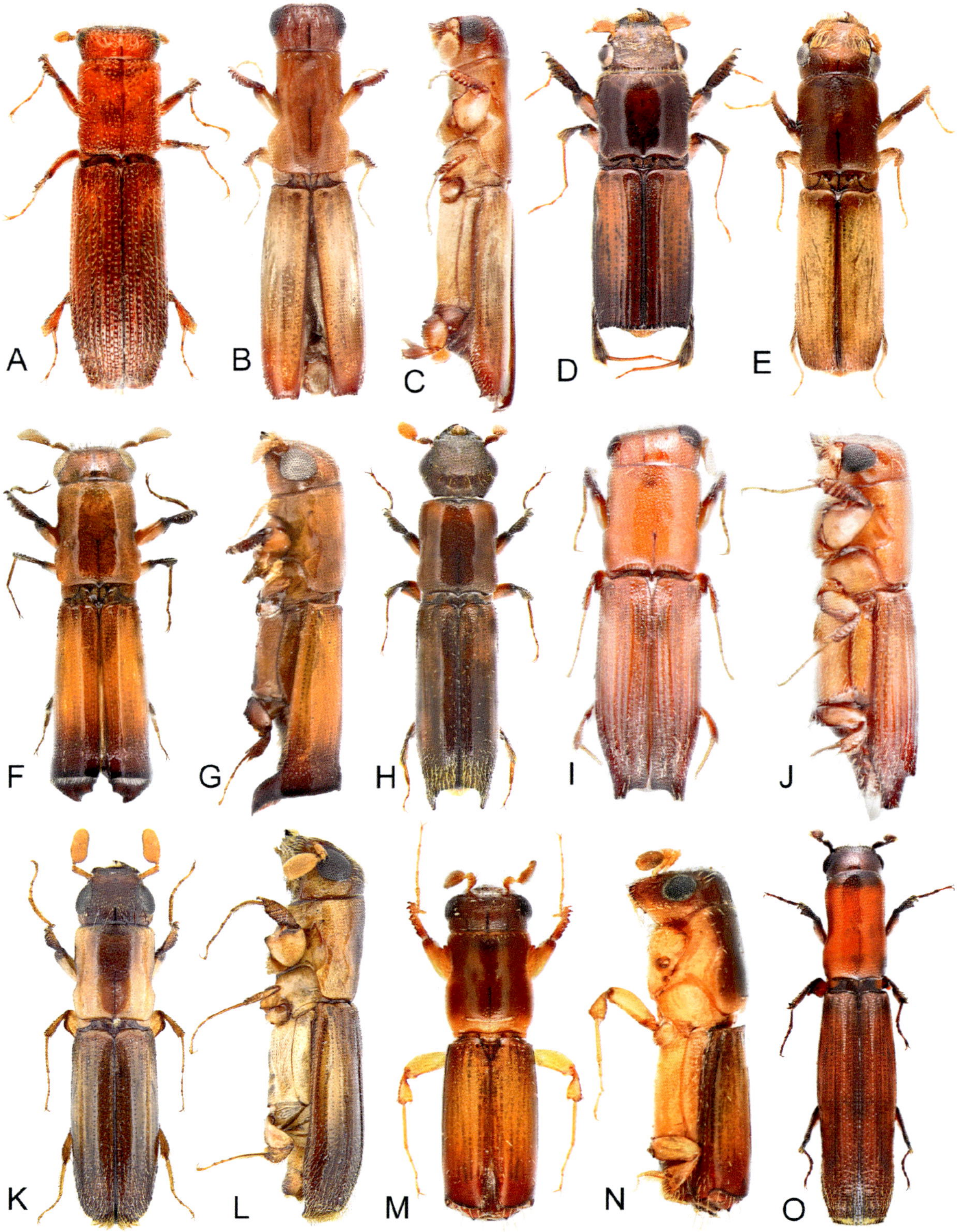

Plate 46. Curculionidae (Platypodinae). A, *Austroplatypus incompertus* (Schedl), ♂, 4.4 mm (Platypodini); B, C, *Baiocis pernanulus* (Schedl), dorsal and lateral, ♂, 2.5 mm (Platypodini); D, *Crossotarsus mniszechi* Chapuis, ♂, 6.1 mm (Platypodini); E, *Crossotarsus subpellucidus* Lea, ♂, 3.6 mm (Platypodini); F, G, *Dinoplatypus pallidus* (Chapuis), dorsal and lateral, ♂, 3.4 mm (Platypodini); H, *Platypus norfolkensis* Schedl, ♂, 5.0 mm (Platypodini); I, J, *Euplatypus parallelus* (Fabricius), dorsal and lateral, ♂, 4.0 mm (Platypodini); K, L, *Treptoplatypus crenatus* (Chapuis), dorsal and lateral, ♀, 5.0 mm (Platypodini); M, N, *Diapus pusillimus* Chapuis, ♂, 1.8 mm (Tesserocerini); O, *Notoplatypus elongatus* Lea, ♀, 6.3 mm (Tesserocerini).

Plate 47. Curculionidae (Bagoinae, Brachycerinae). A, B, *Azollaebagous clarenciensis* (Blackburn), dorsal and lateral, 1.5 mm (Bagoinae); C, *Bagous australasiae* Blackburn, 3.0 mm (Bagoinae); D, *Bagous utriculariae* O'Brien, 2.8 mm (Bagoinae); E, *Aonychus hopei* Boheman, 5.2 mm (Erirhinini); F, *Aonychus lituratus* Lea, 5.3 mm (Erirhinini); G, '*Bagoopsis*' sp., Regans Ford, WA, 3.4 mm (Erirhinini); H, *Echinocnemus exsul* Faust, 3.5 mm (Erirhinini); I, *Echinocnemus latus* (Lea), 4.6 mm (Erirhinini); J, *Tadius barbatus* Lea, 4.7 mm (Erirhinini); K, *Hexonymus reginalis* Kuschel, 0.9 mm (Myrtonymini); L, *Myrtonymus moorei* Kuschel, 0.9 mm (Myrtonymini).

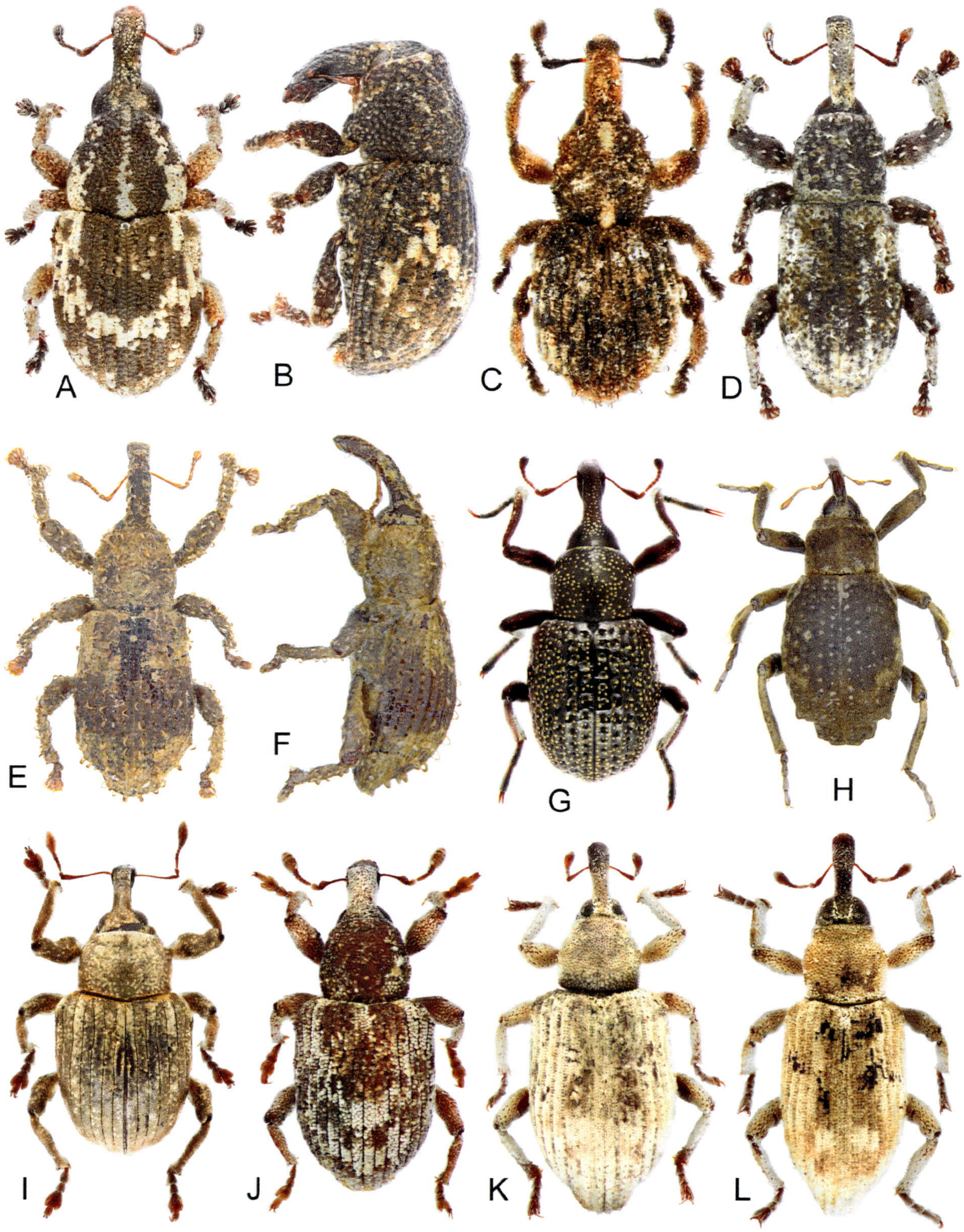

Plate 48. Curculionidae (Brachycerinae: Tanysphyrini). A, B, *Athor* sp., Black Mountain, ACT, dorsal and lateral, 2.0 mm; C, *Baeosomus uvidus* (Oke), 1.4 mm; D, *Cenchrena* sp., Green Island, NSW, 2.3 mm; E, F, *Cenchrena alternata* Lea, dorsal and lateral, 2.5 mm; G, *Cyrtobagous salviniae* Calder & Sands, 2.1 mm; H, *Hydrotimetes natans* Kolbe, 3.2 mm; I, *Neochetina eichhorniae* Warner, 3.7 mm; J, *Neohydronomus affinis* Hustache, 2.1 mm; K, *Niphobolus deceptor* Blackburn, 3.0 mm; L, *Niphobolus* sp., Borroloola, NT, 2.7 mm.

Plate 49. Curculionidae (Cyclominae: Amycterini). A, *Acantholophus alpicola* Ferguson, 19.8 mm; B, *Acantholophus amycteroides* Macleay, 21.6 mm; C, *Acantholophus aureolus* (Boheman), 12.2 mm; D, *Acantholophus niveovittatus* Blackburn, 17.6 mm; E, *Acherres mamillatus* Pascoe, 9.8 mm; F, *Achorostoma hystricosum* (Boheman), 9.5 mm; G, *Aedriodes fastigiatus* Pascoe, 10.3 mm; H, *Alexirhea aurita* Pascoe, 10.7 mm; I, *Amorphorhinus australis* (Germar), 10.6 mm; J, *Amorphorhinus polyacanthus* Pascoe, 6.9 mm; K, *Amorphorhinus rugicollis* (Lea), 9.3 mm; L, *Anascoptes obliteratus* Ferguson, 7.4 mm.

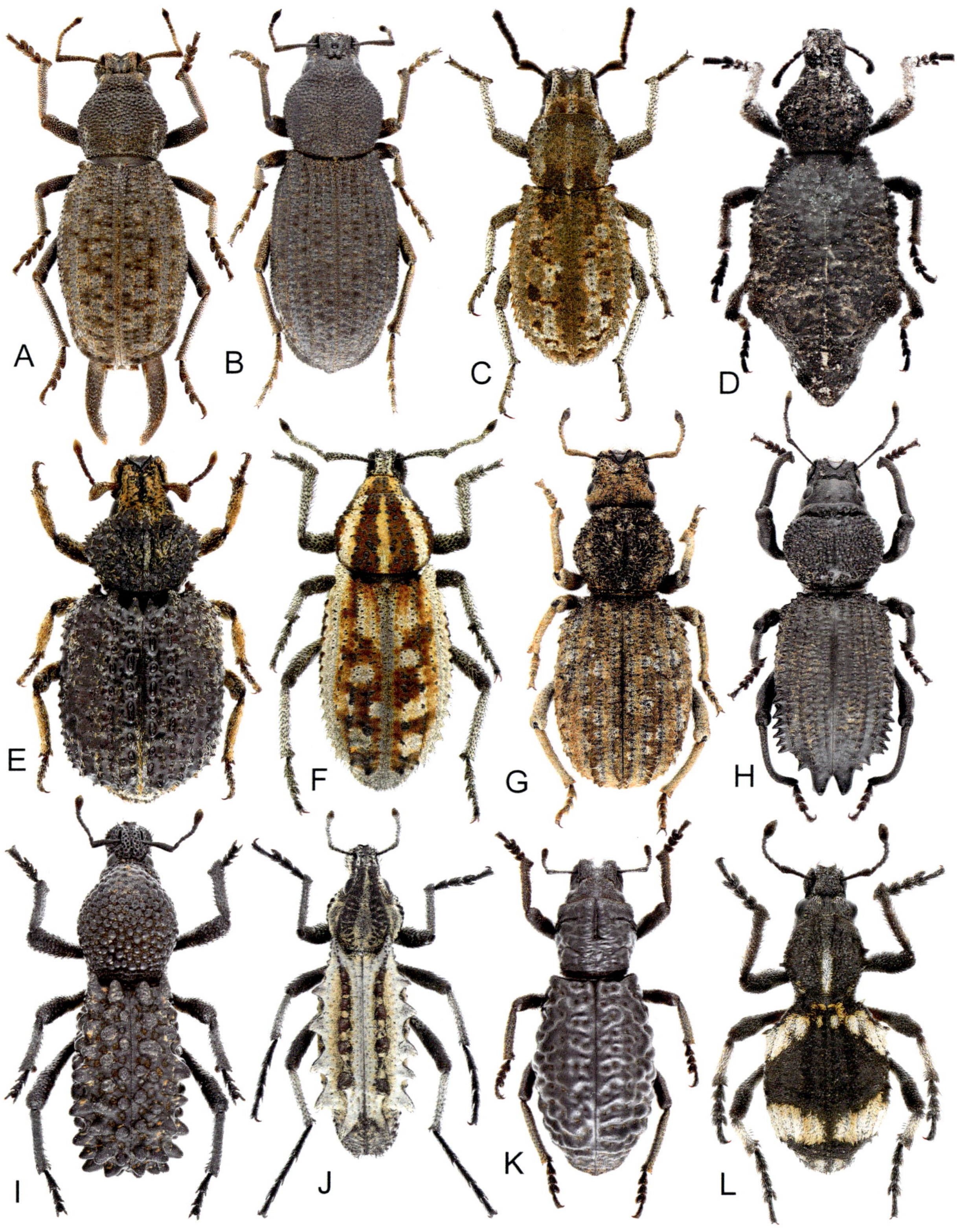

Plate 50. Curculionidae (Cyclominae: Amycterini). A, *Amycterus frenchi* (Ferguson), ♂, 21.8 mm; B, *Amycterus mirabilis* (Kirby), 21.2 mm; C, *Antalaurinus pulverulentus* (Macleay), 10.3 mm; D, *Atychoria funesta* Pascoe, 9.7 mm; E, *Brachyrothus auritus* (Ferguson), 6.1 mm; F, *Chriotyphus acromialis* Pascoe, 12.6 mm; G, *Cubicorhynchus calcaratus* Macleay, 14.5 mm; H, *Cubicorhynchus dohrnii* (Waterhouse), 16.6 mm; I, *Cucullothorax horridus* Ferguson, 10.3 mm; J, *Dialeptopus collaris* (Boheman), 13.9 mm; K, *Dicherotropis damelii* (Macleay), 11.5 mm; L, *Ennothus fallax* Pascoe, 6.3 mm.

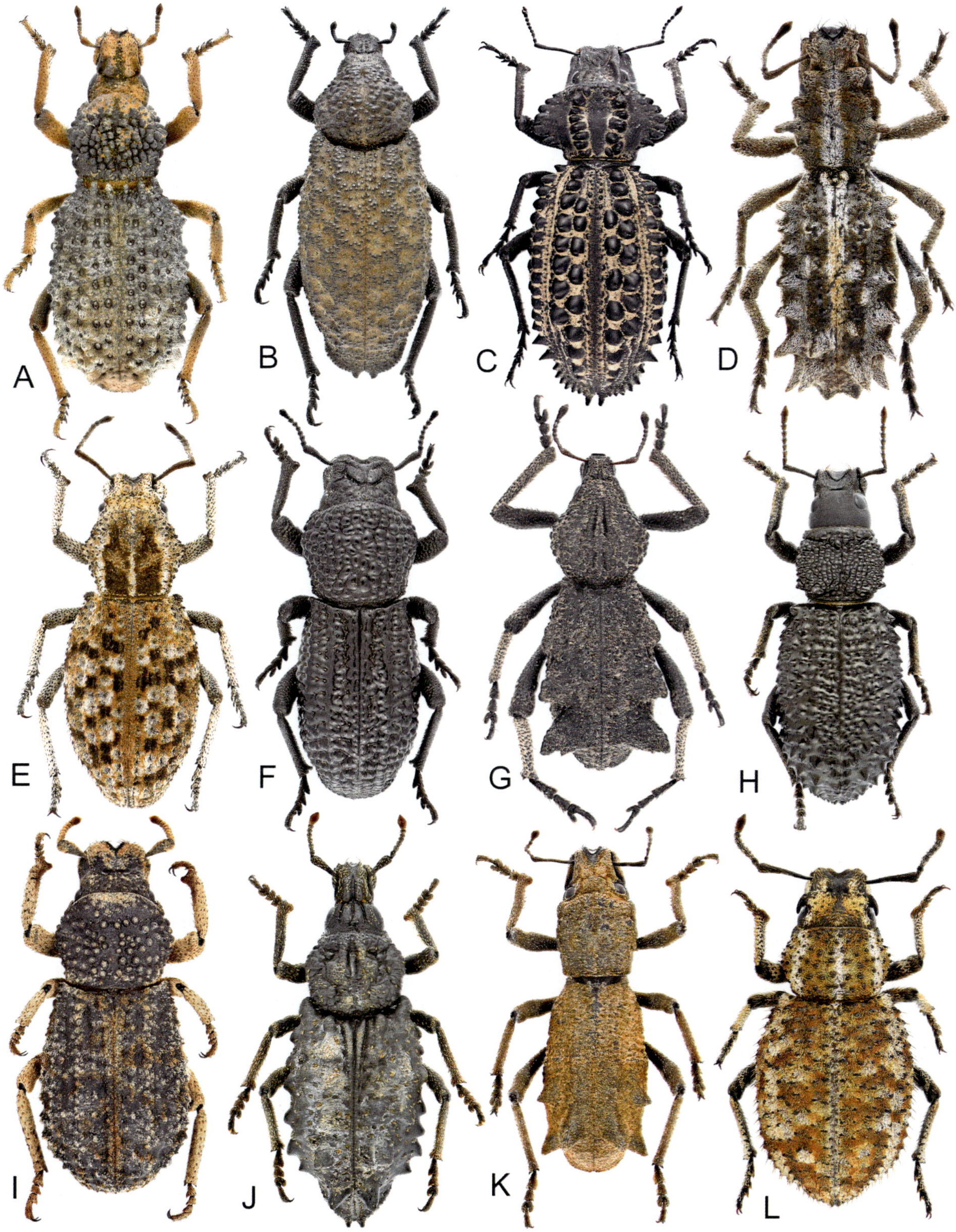

Plate 51. Curculionidae (Cyclominae: Amycterini). A, *Euomella retusa* (Pascoe), 12.2 mm; B, *Euomus insculptus* (Boheman), 23.9 mm; C, *Gagatophorus draco* (Macleay), 28.4 mm; D, *Hyborrhinus bicornutus* (Macleay), 8.4 mm; E, *Hypotomops angulicollis* (Lea), 15.2 mm; F, *Lataurinus rugiceps* (Macleay), 19.7 mm; G, *Melanegis stygius* Pascoe, 9.2 mm; H, *Molochtus gagates* Pascoe, 15.7 mm; I, *Myotrotus obtusus* Pascoe, 11.0 mm; J, *Mythites tuberculatus* Schoenherr, 13.8 mm; K, *Neohyborrhynchus rugosus* (Macleay), 10.9 mm; L, *Notonophes cichlodes* (Pascoe), 11.6 mm.

Plate 52. Curculionidae (Cyclominae: Amycterini). A, *Oditesus indutus* Pascoe, 7.8 mm; B, *Ophthalamycterus laticeps* (Macleay), 18.0 mm; C, *Parahyborrhynchus convexiusculus* (Macleay), 9.1 mm; D, *Pseudonotonophes lemmus* (Pascoe), 6.8 mm; E, *Sclerorrhinella manglesii* (Boheman), 17.9 mm; F, *Sclerorinus albovittatus* Ferguson, 16.0 mm; G, *Sclerorinus adelaidae* Macleay, 17.9 mm; H, *Sosytelus pithecius* (Pascoe), 11.2 mm; I, *Talaurinus suttoni* Carter, 21.6 mm; J, *Talaurinellus sulciventris* (Ferguson), 10.4 mm; K, *Tetralophus sculpturatus* (Waterhouse), 10.1 mm; L, *Xenommamycterus capito* (Pascoe), 16.0 mm.

Plate 53. Curculionidae (Cyclominae: Aterpini). A, *Aades cultratus* (Fabricius), 8.4 mm; B, *Acalonoma pusilla* Blackburn, 2.8 mm; C, *Aesiotes notabilis* Pascoe, ♂, 15.0 mm; D, *Aesiotes notabilis* Pascoe, ♀, 17.1 mm; E, *Alphitopis nivea* Pascoe, 14.3 mm; F, *Anomocis apicalis* Lea, 14.8 mm; G, *Aoplocnemis rufipes* Boheman, 7.4 mm; H, *Aoplocnemis* sp., Labertouche, VIC, 9.6 mm; I, *Atelicus atrophus* Pascoe, 5.5 mm; J, *Atelicus* sp., Glen Innes, NSW, 9.9 mm; K, *Aterpodes seriatus* (Boisduval), 15 mm; L, *Aterpodes tuberculatus* (Gyllenhal), 13.2 mm.

Plate 54. Curculionidae (Cyclominae: Aterpini). A, *Aromagis echinata* Pascoe, 6.6 mm; B, *Cechides amoenus* Pascoe, 11.8 mm; C, *Chrysolopus detritus* Chevrolat, 14.7 mm; D, *Chrysolopus spectabilis* (Fabricius), 17.4 mm; E, *Chrysolopus spectabilis* (Fabricius), 18.7 mm; F, *Chrysophoracis pulcher* Lea, 12.8 mm; G, *Cyllorhamphus mimicus* Lea, 7.4 mm; H, *Cyllorhamphus tuberosus* Erichson, 5.2 mm; I, *Dixoncis pictus* Oke, 7.2 mm; J, *Euthyphasis acuta* Pascoe, 7.4 mm; K, *Euthyphasis funerea* Blackburn, 4.7 mm; L, *Iphisaxus asper* Pascoe, 11.6 mm.

Plate 55. Curculionidae (Cyclominae: Aterpini). A, *Kershawcis cylindricus* Lea, 8.7 mm; B, *Nemestra incerta* Pascoe, 5.4 mm; C, *Oenopus guttatus* (Boheman), 4.2 mm; D, *Ophthalmorychus angustus* Blackburn, 3.8 mm; E, *Pelororhinus argentosus* Gyllenhal, 13.7 mm; F, *Pelororhinus grandis* (Lea), 14.0 mm; G, *Pelolorhinus transversus* (Boisduval), 10.6 mm; H, *Rhadinosomus lacordairei* Pascoe, 7.4 mm; I, *Rhinaria cristata* Kirby, 10.5 mm; J, *Rhinaria granulosa* (Fahraeus), 12.4 mm; K, *Rhinoplethes foveatus* Pascoe, 8.9 mm; L, *Strongylorhinus clarki* Marshall, 11.3 mm.

Plate 56. Curculionidae (Cyclominae). A, *Anorthorhinus pictipes* Blackburn, 3.5 mm (Listroderini); B, *Listroderes foveatus* (Lea), 7.0 mm (Listroderini); C, *Listronotus bonariensis* (Kuschel), 2.8 mm (Listroderini); D, *Methypora postica* Pascoe, 5.9 mm (Listroderini); E, *Steriphus cordatus* (Pascoe), 6.3 mm (Listroderini); F, *Steriphus sericeus* (Blackburn), 6.5 mm (Listroderini); G, *Aphela helopoides* Pascoe, 5.4 mm (Notiomimetini); H, *Mandalotina atranotata* Oke, 1.8 mm (Listroderini); I, *Neosyagrius cordipennis* Lea, 2.8 mm (Listroderini); J, *Notiomimetes pascoei* Wollaston, 1.2 mm (Notiomimetini); K, *Psaldus liosomoides* Pascoe, 3.0 mm (Notiomimetini); L, *Tasmanica* sp., Kettering, TAS, 1.7 mm (Listroderini).

Plate 57. Curculionidae (Cyclominae). A, *Wollastonicis minutus* Lea, 1.5 mm (Notiomimetini); B, *Aparete palpebrosa* Pascoe, 6.7 mm (Rhythirrinini); C, *Ethemaia sellata* Pascoe, 6.1 mm (Rhythirrinini); D, *Medicasta leucura* Pascoe, 3.8 mm (Rhythirrinini); E, *Ophryota nodosa* (Blackburn), 7.3 mm (Rhythirrinini); F, *Phrenozemia lyproides* Pascoe, 3.7 mm (Rhythirrinini); G, *Pocius parallelus* (Lea), 3.8 mm (Rhythirrinini); H, Undet. genus, 38.1 km WSW of Pannawonica WA, 3.2 mm (WAM); I, Undet. genus, 89 km SE of Wittenoom WA, 4.5 mm (WAM); J, Undet. genus, 25 km NE of Wittenoom WA, 4.2 mm (WAM); K, Undet. genus, 41 km WNW of Wittenoom WA, 3.7 mm (WAM); L, Undet. genus, 10.2 km SW of Pannawonica WA, 3.4 mm (WAM).

Plate 58. Curculionidae (Gonipterini). A, *Bryachus squamicollis* Pascoe, 10.9 mm; B, *Gonipterus notographus* Boisduval, 7.6 mm; C, *Gonipterus pulverulentus* Lea, 8.2 mm; D, *Iptergonus cionoides* (Pascoe), 4.6 mm; E, *Oxyops bilunaris* Pascoe, 10.0 mm; F, *Oxyops fasciculatus* Redtenbacher, 7.0 mm; G, *Oxyops scabrosus* (Boisduval), 18.0 mm; H, *Pantoreites trivirgatus* Lea, 5.4 mm; I, *Pantoreites virgatus* Pascoe, 5.5 mm; J, *Prophaesia tenuirostris* Lea, 4.2 mm; K, *Syarbis pachypus* Pascoe, 4.2 mm; L, *Syarbis subnitida* (Roelofs), 6.0 mm.

Plate 59. Curculionidae (Hyperinae). A, *Agriochaeta crinita* Pascoe, ♀, 4.7 mm; B, *Agriochaeta crinita* Pascoe, ♂, 4.6 mm; C, *Byrsodes binodipennis* (Lea), 6.9 mm; D, *Byrsodes ceratus* (Pascoe), 4.9 mm; E, *Cassythicola* sp., Paluma, QLD, 3.2 mm; F, *Eniopea viridisquama* Lea, 2.3 mm; G, *Euhackeria insignis* Lea, 5.0 mm; H, *Eurychirus bituberculatus* Waterhouse, 9.2 mm; I, *Gerynassa nodulosa* Pascoe, 3.8 mm; J, *Glaucopela acaciae* Marshall, 2.1 mm; K, *Glaucopela unicolor* Pascoe, 3.3 mm; L, *Lexithia rufipennis* Pascoe, 2.5 mm.

Plate 60. Curculionidae, unplaced to subfamily. A, *Lycosura bispinosa* Pascoe, 7.1 mm (Hyperinae); B, *Nedyleda semiusta* Pascoe, 2.9 mm (Hyperinae); C, *Nothyperus cionodes* Marshall, 4.2 mm (Hyperinae); D, *Olanaea nigricollis* Pascoe, 2.8 mm (Hyperinae); E, *Xeda notabilis* Blackburn, 4.0 mm (Hyperinae); F, *Cisolea umbratilis* Oke, 4.3 mm (Phrynixini); G, *Crossixus punctatus* Kuschel, 4.1 mm (Phrynixini); H, *Dinichus terreus* Pascoe, 7.7 mm (Phrynixini); I, *Lophixus major* (Oke), 9.3 mm (Phrynixini); J, *Mallixus sylvicola* (Oke), 4.5 mm (Phrynixini); K, *Sclerixus victoriae* (Oke), 5.5 mm (Phrynixini); L, *Geochus* sp., Lord Howe Island, NSW, 1.8 mm (Viticiini).

Plate 61. Curculionidae (unplaced to subfamily, Molytinae). A, *Austrocis bicaudatus* Zimmerman & Oberprieler, 3.9 mm (Viticiini); B, *Eupholocis dentipes* Lea, 2.3 mm (Viticiini); C, *Peliocis subcylindricus* Lea, 3.1 mm (Viticiini); D, *Platynotocis pyriformis* Lea, 3.0 mm (Viticiini); E, *Tivicis* sp., Cairns, QLD, 2.7 mm (Viticiini); F, *Ethadomorpha australis* (Erichson), 2.3 mm; G, *Hypsomus* sp., Tenant Ck, NT, 7.0 mm; H, *Symbothynus* sp., Victoria River District, NT, 3.2 mm; I, *Symbothynus* sp., Pebbly Beach, NSW, 2.6 mm; J, *Thechia pygmaea* Pascoe, 2.0 mm; K, *Syagrius intrudens* Waterhouse, 6.5 mm (Molytinae); L, *Thaumastophasis oculata* Wollaston, 3.5 mm (Molytinae).

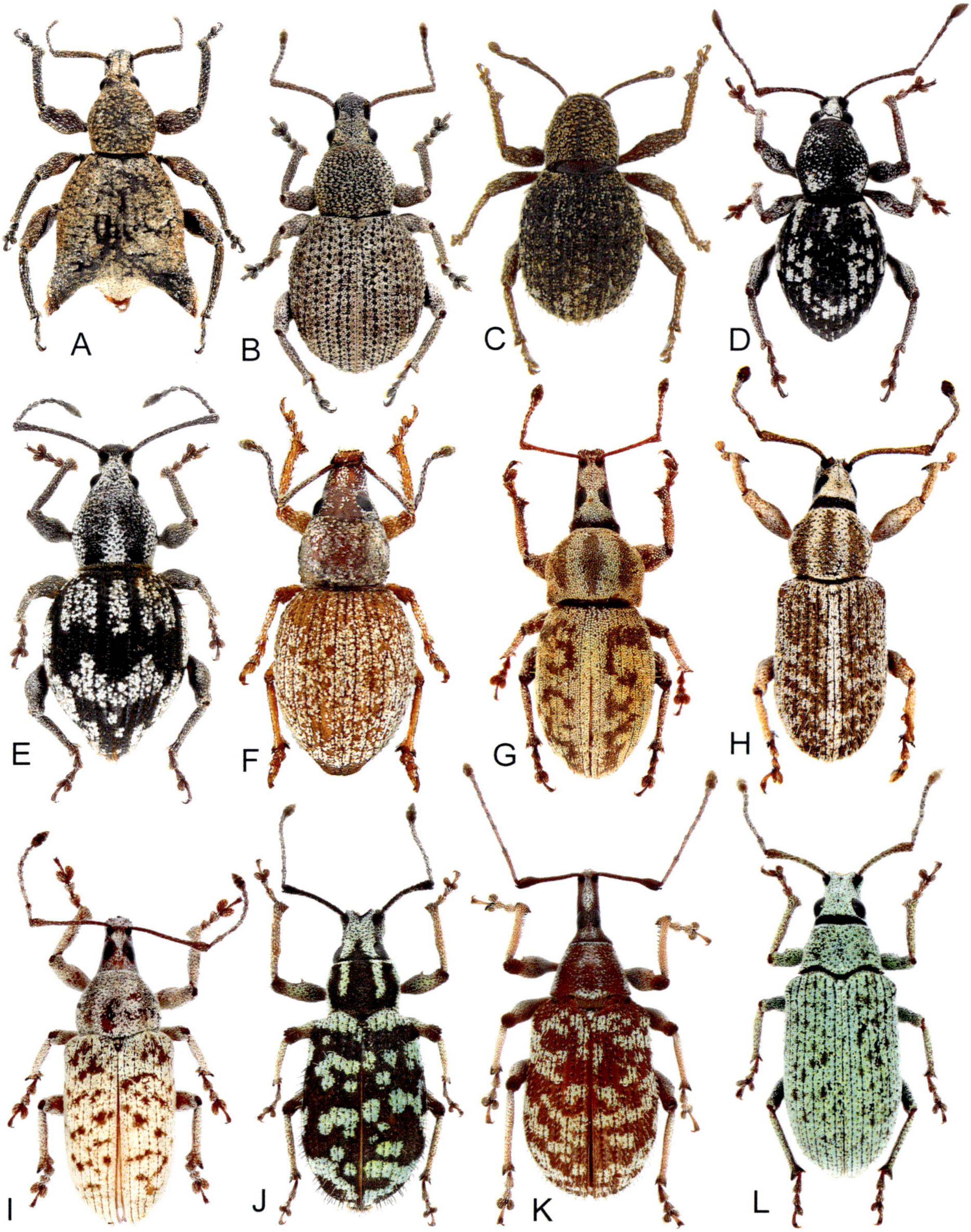

Plate 62. Curculionidae (Entiminae: Celeuthetini, Cyphicerini). A, *Apirocalus cornutus* Pascoe, 8.6 mm (Celeuthetini); B, *Celeuthetes australis* (Gyllenhal), 6.0 mm (Celeuthetini); C, *Celeuthetinus gnomus* Oberprieler & Zimmerman, 3.0 mm (Celeuthetini); D, *Idiopsodes ocularis* (Lea), 4.2 mm (Celeuthetini); E, *Oribius gestroi* (Pascoe), 5.3 mm (Celeuthetini); F, *Apomyllocerus ovatus* Oberprieler & Zimmerman, 2.6 mm (Cyphicerini); G, *Atmesia marginata* Pascoe, 3.6 mm (Cyphicerini); H, *Ecmyllocerus trivitticollis* (Lea), 4.2 mm (Cyphicerini); I, *Euphalia pardalis* Pascoe, 5.3 mm (Cyphicerini); J, *Hackeria viridivaria* Lea, 5.0 mm (Cyphicerini); K, *Matesia maculata* Lea, 5.4 mm (Cyphicerini); L, *Myllocerus* cf. *M. australis* Boisduval, 4.7 mm (Cyphicerini).

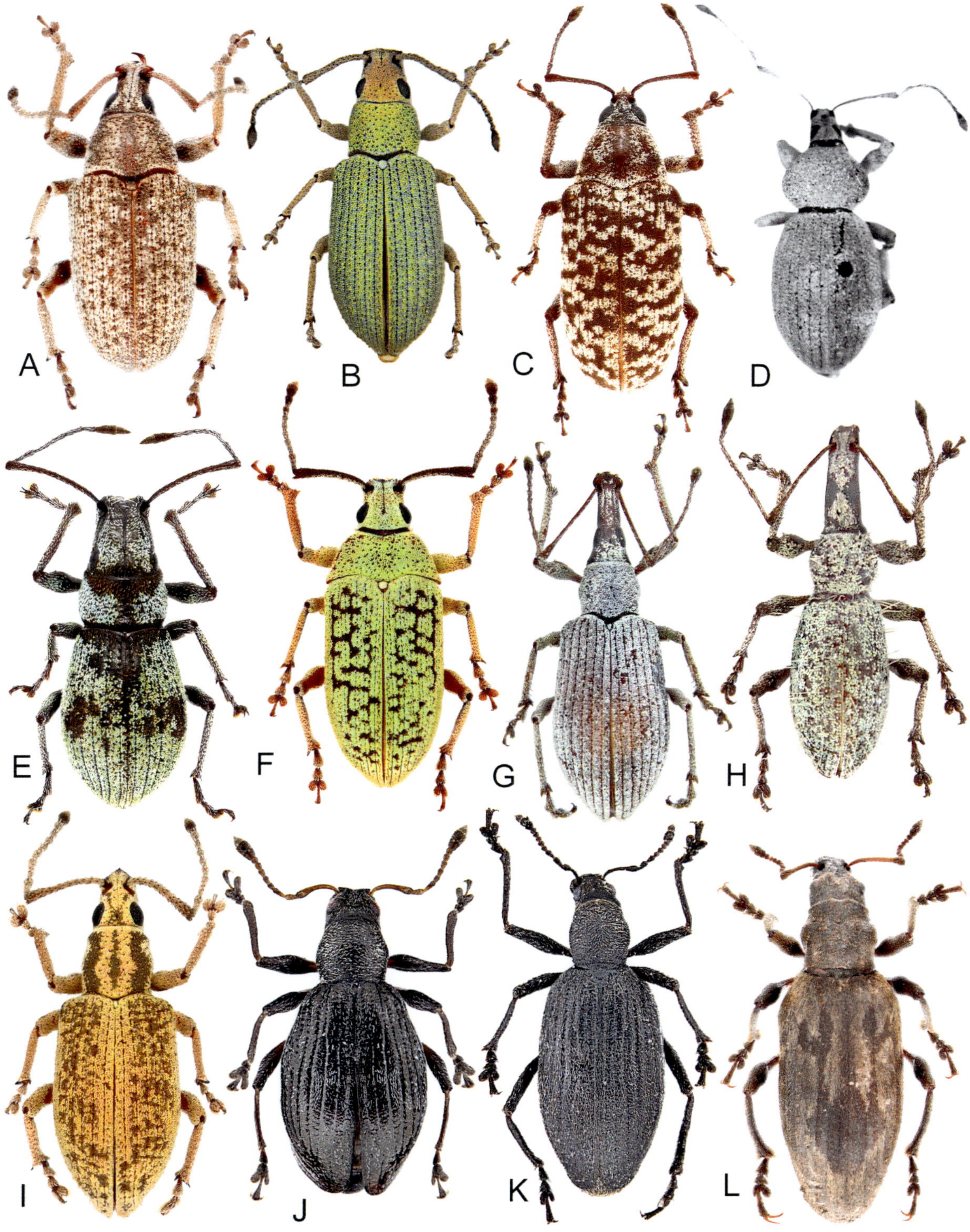

Plate 63. Curculionidae (Entiminae: Cyphicerini, Ectemnorhinini). A, *Myllocerus obscurus* Lea, 6.8 mm (Cyphicerini); B, *Myllocerus viridimicans* Lea, 3.6 mm (Cyphicerini); C, *Paratitinia ceratorhina* (Lea), 5.7 mm (Cyphicerini); D, *Sematia glaucina* (Pascoe), ♀ (holotype, NHMUK) (Cyphicerini); E, *Synomus* sp., Mitchell Plateau, WA, 4.4 mm (Cyphicerini); F, *Syrphax bovilli* (Blackburn), 4.5 mm (Cyphicerini); G, *Tanymyllocerus tenuirostris* Oberprieler & Zimmerman, 4.2 mm (Cyphicerini); H, *Tesiama hystricosa* Oberprieler & Zimmerman, 3.0 mm (Cyphicerini); I, *Titinia grata* (Pascoe), 3.5 mm (Cyphicerini); J, *Bothrometopus brevis* (C. O. Waterhouse), 3.9 mm (Ectemnorhinini); K, *Bothrometopus gracilipes* (C. O. Waterhouse), 4.5 mm (Ectemnorhinini); L, *Canonopsis sericea* C. O. Waterhouse, 9.6 mm (Ectemnorhinini).

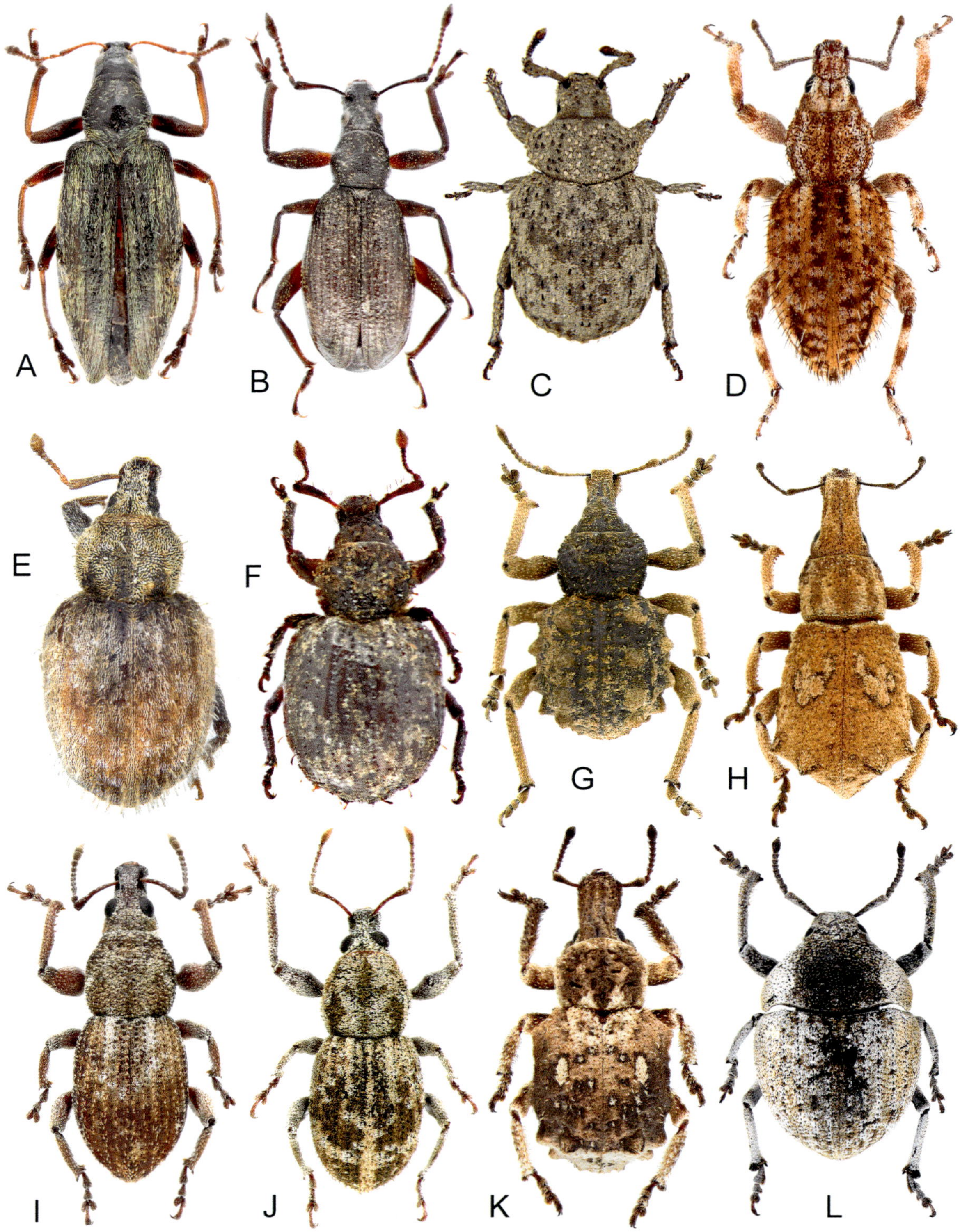

Plate 64. Curculionidae (Entiminae: Ectemnorhinini, Embrithini, Leptopiini). A, *Ectemnorhinus viridis* G. R. Waterhouse, 6.7 mm (Ectemnorhinini); B, *Palirhoeus eatoni* (C. O. Waterhouse), 4.2 mm (Ectemnorhinini); C, *Afrophloeus squamifer* (Boheman), 2.8 mm (Embrithini); D, *Agroicus ateropterus* (Boheman), 6.7 mm (Leptopiini); E, *Allotimareta setistriata* (Lea), 2.8 mm (type) (Leptopiini); F, *Amandalotus cordipennis* (Lea), 2.0 mm (Leptopiini); G, *Amisallus* sp., Gosford, NSW, 9.3 mm (Leptopiini); H, *Amnemus quadrituberculatus* (Boheman), 8.8 mm (Leptopiini); I, *Asceparnodes duplicatus* (Lea), 5.4 mm (Leptopiini); J, *Asceparnus subfasciatus* (Blackburn), 5.4 mm (Leptopiini); K, *Baryopadus corrugatus* Pascoe, 6.9 mm (Leptopiini); L, *Basedowia basicollis* Lea, 7.2 mm (Leptopiini).

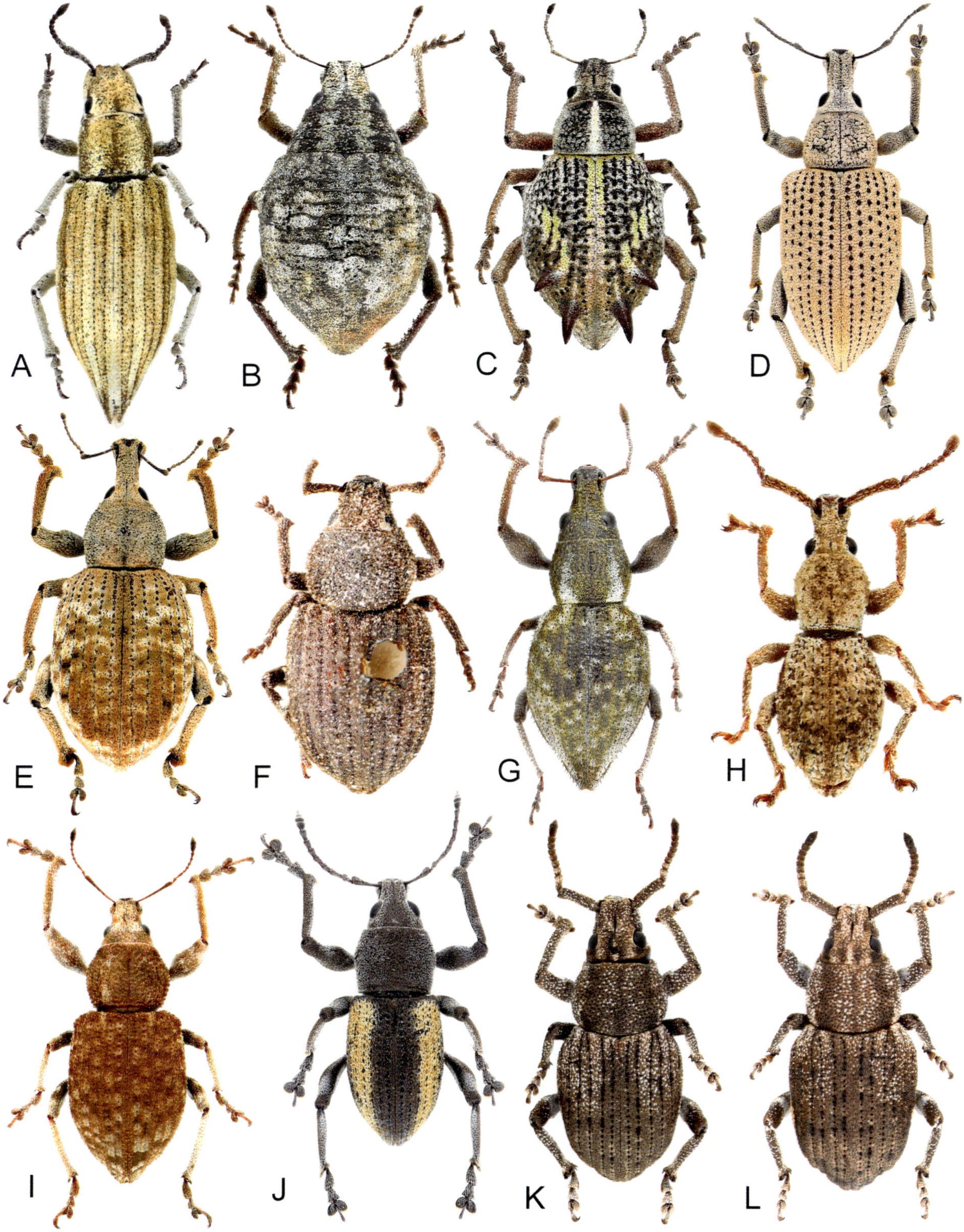

Plate 65. Curculionidae (Entiminae: Leptopiini). A, *Bothynorhynchus decorum* (Blackburn), 7.7 mm; B, *Catasarcus bakeri* Thompson, 6.7 mm; C, *Catasarcus carbo* Pascoe, 12.4 mm; D, *Catastygnus* sp., No Locality (Brock collection, ANIC), 12.2 mm; E, *Centyres* sp., Windsor Tablelands, QLD, 16.3 mm; F, *Chaodius nigrescens* Pascoe, 4.5 mm (holotype NHMUK); G, *Decienus sphasodes* Pascoe, 6.5 mm; H, *Ecrizothis inaequalis* Blackburn, ♂, 3.6 mm; I, *Enchymus punctonotatus* Pascoe, 10.8 mm; J, *Esmelina flavovittata* Pascoe, 6.6 mm; K, *Essolithna pluviata* Pascoe, ♂, 7.9 mm; L, *Essolithna pluviata* Pascoe, ♀, 8.3 mm.

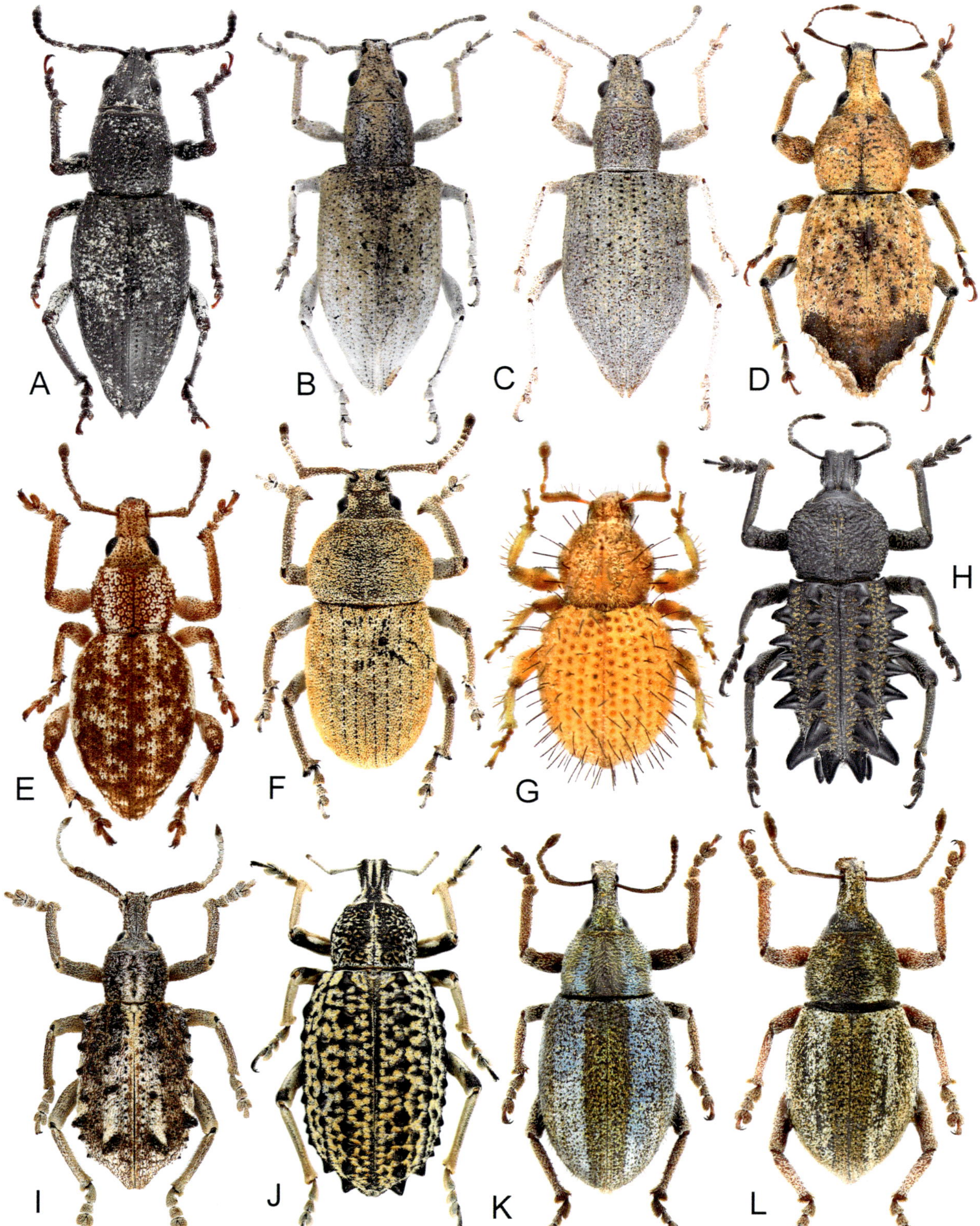

Plate 66. Curculionidae (Entiminae: Leptopiini). A, *Evadomorpha everardensis* (Blackburn), 9.7 mm; B, *Evas argenteiventris* Pascoe, 8.2 mm; C, *Evas latipennis* Lea, 7.5 mm; D, *Gastrocis montanus* Lea, 9.3 mm; E, *Geosomus macleayi* Faust, 4.1 mm; F, *Hadrorhinus lepidopterus* Gyllenhal, 7.4 mm; G, *Howeocis setosus* Lea, 2.4 mm; H, *Leptopius biordinatus* (Blackburn), 23.2 mm; I, *Leptopius clavus* (Fabricius), 17.4 mm; J, *Leptopius gravis* (Blackburn), 32.0 mm; K, *Lipothyrea chloris* Pascoe, ♀, 6.8 mm; L, *Lipothyrea chloris* Pascoe, ♂, 5.2 mm.

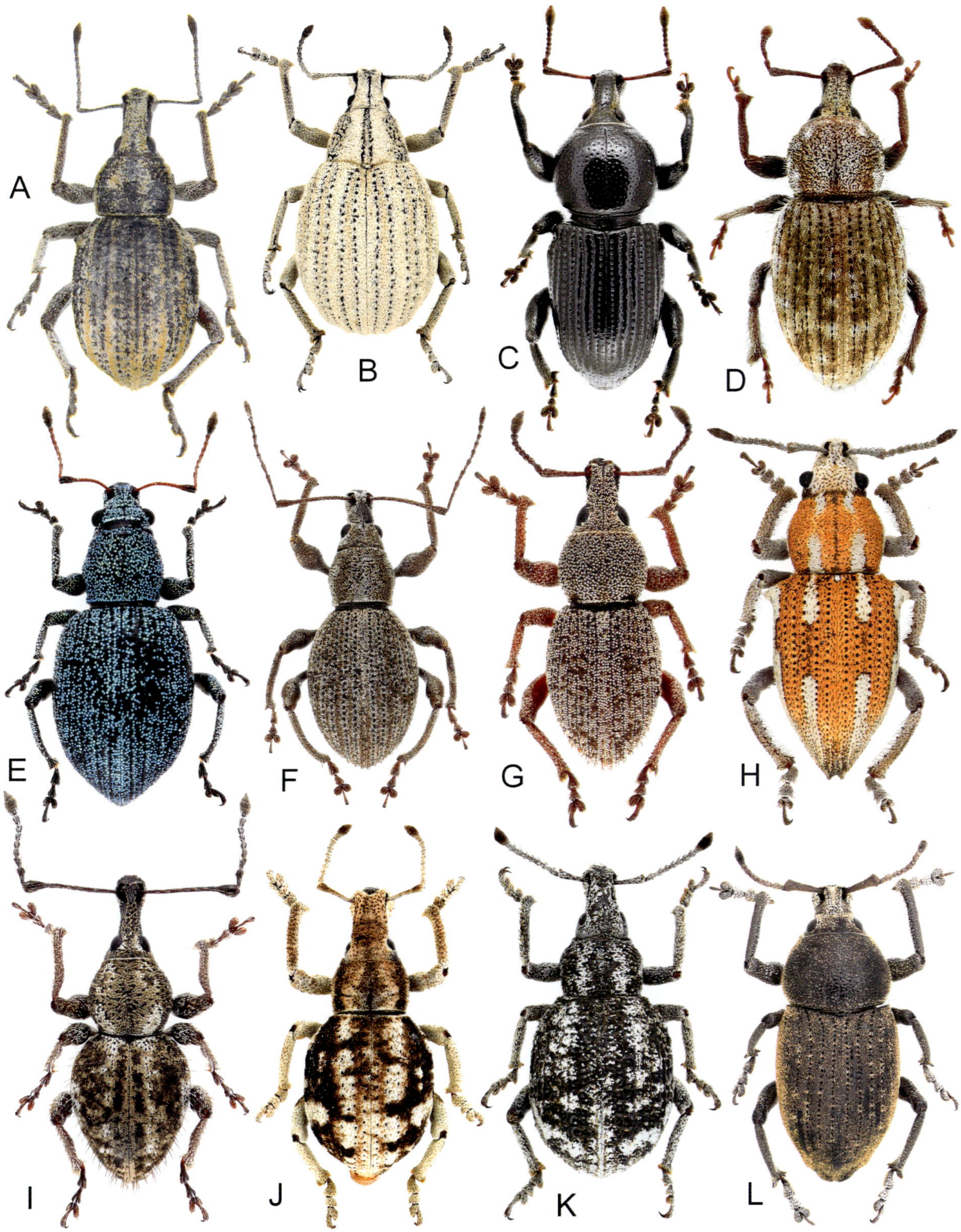

Plate 67. Curculionidae (Entiminae: Leptopiini). A, *Lysizone conifer* (Lea), ♂, 9.8 mm; B, Undet. genus, Darwin NT, ♀, 11.5 mm; C, *Mandalotus glaber* (Blackburn), 6.4 mm; D, *Mandalotus sabulosus* Lea, 4.9 mm; E, *Merimnetes celmisiae* Lea, 4.1 mm; F, *Merimnetes uniformis* Boheman, 6.5 mm; G, *Neomerimnetes* sp., Brindabella Ranges, ACT, 3.6 mm; H, *Ochrometa amoena* Pascoe, 8.5 mm; I, *Ocynoma antennatum* Pascoe, 3.6 mm; J, *Onesorus maculosus* Pascoe, 7.1 mm; K, *Onesorus tarsalis* Lea, 8.0 mm; L, *Oophthalmus pistor* (Germar), 9.2 mm.

Plate 68. Curculionidae (Entiminae: Leptopiini). A, *Pascoellus echimys* (Pascoe), 4.8 mm; B, *Peripagis rufipes* Pascoe, 11.3 mm; C, *Perperus innocuus* Boheman, 4.5 mm; D, *Perperus lateralis* (Boisduval), 10.3 mm; E, *Platypterocis paradoxus* Lea, 5.8 mm; F, *Polyphrades fortis* Blackburn, 13.7 mm; G, *Polyphrades laticollis* Fahraeus, 5.7 mm; H, *Polyphrades* sp., Whyalla, SA, 9.8 mm; I, *Prosayleus hopei* Boheman, 7.3 mm; J, *Prostomus scutellaris* (Fabricius), 17.7 mm; K, *Prypnus quinquenodosus* Schoenherr, 16.6 mm; L, *Prypnus tibialis* Lea, 13.7 mm.

Plate 69. Curculionidae (Entiminae: Leptopiini). A, *Psapharus infaustus* (Olivier), 13.8 mm; B, *Psapharus mistothes* (Herbst), 17.0 mm; C, *Psapharus oesalon* (Pascoe), 7.7 mm; D, *Pseudotimareta subterranea* (Lea), 4.3 mm; E, *Rhinoscapha doriae* Pascoe, 15.8 mm; F, *Rhyparophilus howensis* (Lea), ♂, 4.3 mm; G, *Rhyparophilus howensis* (Lea), ♀, 4.8 mm; H, *Scotasmus litoralis* (Lea), 4.8 mm; I, *Scotasmus parvicornis* (Lea), 11.5 mm; J, *Stenocorynus aridus* Pascoe, 8.2 mm; K, *Synaptonyx ovatus* G. R. Waterhouse, 5.6 mm; L, *Telenica nebulosa* Pascoe (holotype, NHMUK).

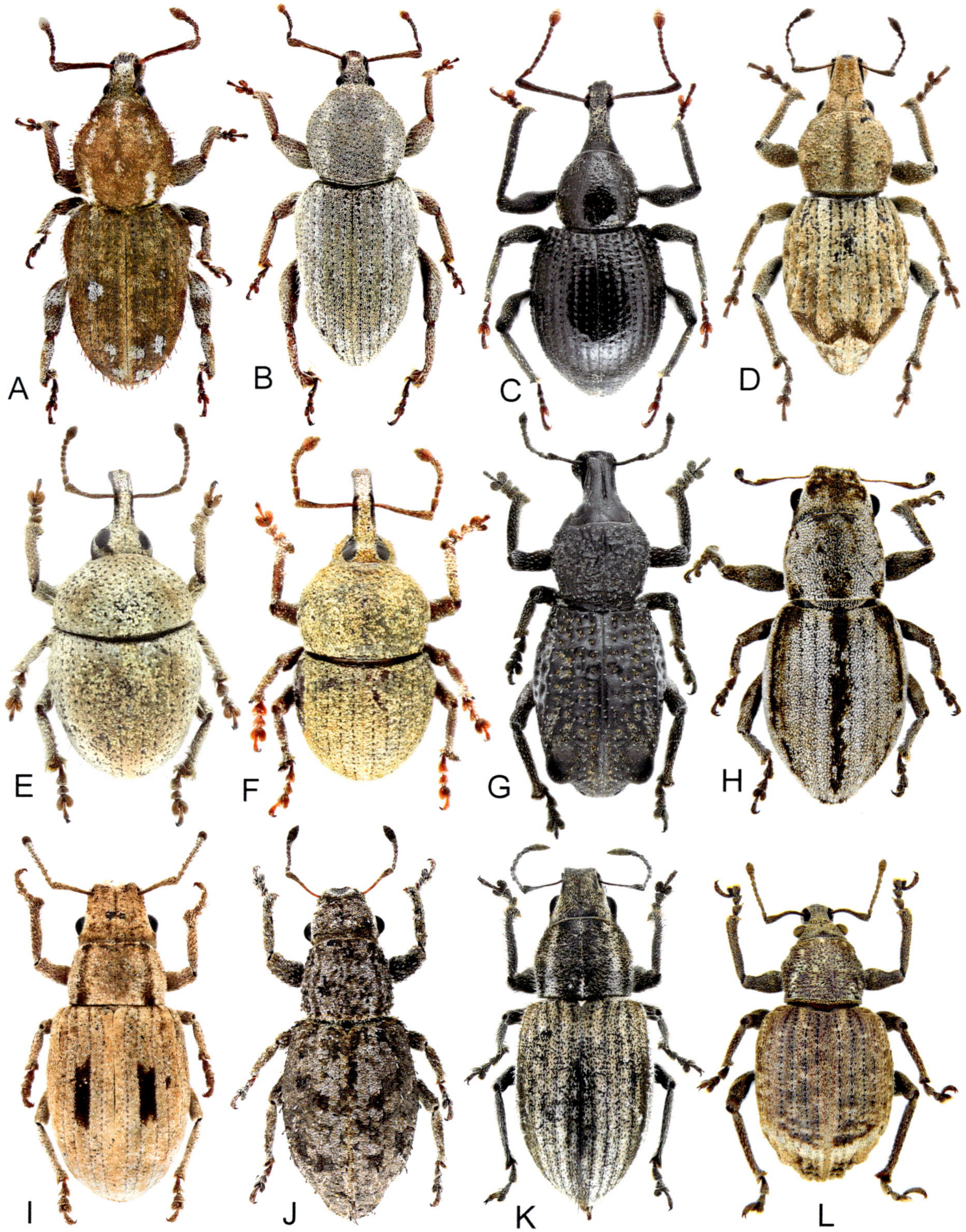

Plate 70. Curculionidae (Entiminae: Leptopiini, Naupactini, Oosomini). A, *Timareta figurata* Pascoe, 3.9 mm (Leptopiini); B, *Timareta pilosa* (Blackburn), 6.5 mm (Leptopiini); C, *Undarobius howarthi* Escalona & Oberprieler, 4.0 mm (abraded specimen) (Leptopiini); D, *Uroleptops impar* Lea, 7.0 mm (Leptopiini); E, *Xynaea saginata* Pascoe, 4.0 mm (Leptopiini); F, *Xyneella uniformis* (Lea), 3.0 mm (Leptopiini); G, *Zymaus binodosus* Pascoe, 11.6 mm (Leptopiini); H, *Atrichonotus taeniatulus* (Berg), 6.3 mm (Naupactini); I, *Eurymetopus birabeni* Kuschel, 6.4 mm (Naupactini); J, *Floresianus sordidus* Hustache, 5.1 mm (Naupactini); K, *Naupactus leucoloma* Boheman, 10.6 mm (Naupactini); L, *Phlyctinus callosus* (Schoenherr), 6.8 mm (Oosomini).

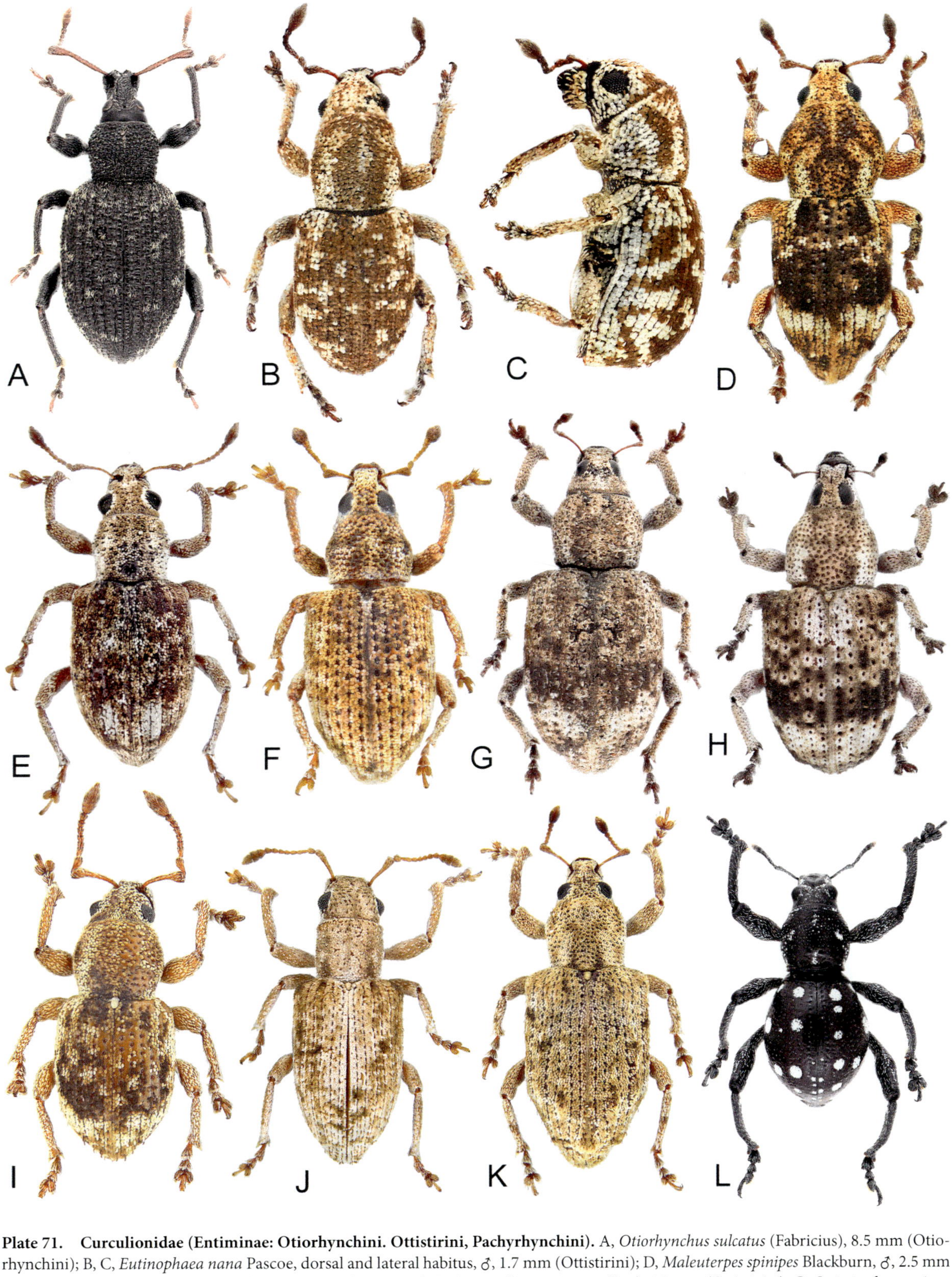

Plate 71. Curculionidae (Entiminae: Otiorhynchini. Ottistirini, Pachyrhynchini). A, *Otiorhynchus sulcatus* (Fabricius), 8.5 mm (Otiorhynchini); B, C, *Eutinophaea nana* Pascoe, dorsal and lateral habitus, ♂, 1.7 mm (Ottistirini); D, *Maleuterpes spinipes* Blackburn, ♂, 2.5 mm (Ottistirini); E, *Monteithus falcatus* (Lea), 2.4 mm (Ottistirini); F, *Ottistirellus variegatus* (Lea), 1.9 mm (Ottistirini); G, *Ottistirodes murinus* (Lea), ♀, 5.8 mm (Ottistirini); H, *Ottistirodes murinus* (Lea), ♂, 5.8 mm (Ottistirini); I, *Reidius fasciculatus* (Lea), 2.1 mm (Ottistirini); J, *Tregoleus griseus* Oberprieler & Zimmerman, 2.9 mm (Ottistirini); K, *Weirius suturalis* (Lea), 3.4 mm (Ottistirini); L, *Pantorhytes stanleyanus* (White), 11.3 mm (Pachyrhynchini).

Plate 72. Curculionidae (Entiminae: Sitonini, Tanymecini; Molytinae). A, *Sitona discoideus* Gyllenhal, 3.7 mm (Sitonini); B, *Homoeotrachelus australasiae* Faust, ♂, 4.4 mm (Tanymecini); C, *Homoeotrachelus australasiae* Faust, ♀, 5.3 mm (Tanymecini); D, *Hypomeces obscurus* (Macleay), ♂, 15.8 mm (Tanymecini); E, *Jenningsius debbiae* Oberprieler & Zimmerman, 4.1 mm (Tanymecini); F, *Pullenius kimberii* Oberprieler & Zimmerman, 8.8 mm (Tanymecini); G, *Tanymecus* near *T. boettcheri* Voss, 6.3 mm (Tanymecini); H, *Rhyncholobus andrewsi* Gahan, 12.4 mm (Entiminae); I, *Eristinus eucalypti* Lea, 1.6 mm (Molytinae); J, *Euthebus troglodytes* Pascoe, 5.6 mm (Molytinae); K, *Exeiratus carinatus* (Lea), 4.6 mm (Molytinae); L, *Opsittis atomaria* Pascoe, 7.2 mm (Molytinae).

Plate 73. Curculionidae (Curculioninae). A, *Epamoebus ziczac* Lea, 2.3 mm (Acalyptini); B, *Eudelodes bicolor* (Faust), 3.0 mm (Acalyptini); C, *Tithene linospadicus* (Rheinheimer), 2.3 mm (Acalyptini); D, *Tithene* sp., Iron Range, QLD, 3.6 mm (Acalyptini); E, *Hypohypurus medvedevi* (Korotyaev), 1.5 mm (Ceutorhynchini); F, *Hypurus bertrandi* (Perris), 2.2 mm (Ceutorhynchini); G, *Mecysmoderes* sp., Mt. Webb, N.P., QLD, 1.5 mm (Ceutorhynchini); H, *Mogulones geographicus* (Goeze), 4.7 mm (Ceutorhynchini); I, *Mogulones larvatus* (Schultze), 3.5 mm (Ceutorhynchini); J, *Rhinoncus nigriventris* Pascoe, 2.2 mm (Ceutorhynchini); K, *Trichosirocalus briesei* Alonso-Zarazaga & Sánchez-Ruiz, 4.2 mm (Ceutorhynchini); L, *Stephanorhynchus crassus* Broun, 3.5 mm (Eugnomini).

Plate 74. Curculionidae (Curculioninae). A, *Cranoides* sp., Kuranda, QLD, 3.3 mm (Cranopoeini); B, *Cranopoeus* sp., Gosford, NSW, 1.4 mm (Cranopoeini); C, *Cratoscelocis foveicollis* Lea, 2.0 mm (Cranopoeini) specimen abraded; D, *Enneaeus bituberculatus* (Lea), 1.7 mm (Cranopoeini); E, *Fergusoniella cristata* (Lea), 2.0 mm (Cranopoeini); F, *Fergusoniella* sp., Julatten, QLD, 2.9 mm (Cranopoeini); G, *Cryptoplus rubiginosus* (Pascoe), 3.3 mm (Cryptoplini); H, *Haplonyx maximus* Lea, 10.7 mm (Cryptoplini); I, *Kuschelorhynchus macadamiae* Jennings & Oberprieler, 7.3 mm (Cryptoplini); J, *Menechirus fuscodorsalis* (Heller), 7.3 mm (Cryptoplini); K, *Sigastus fascicularis* Pascoe, 9.8 mm (Cryptoplini); L, *Zeopus* sp., Madura, WA, 2.7 mm (Cryptoplini).

Plate 75. Curculionidae (Curculioninae). A, *Curculio simulator* Zimmerman & Oberprieler, 3.3 mm (Curculionini); B, *Pycnochirus fuscovarius* (Faust), 7.0 mm (Curculionini); C, *Ancyttalia* sp., Millgrove, VIC, 2.2 mm (Eugnomini); D, *Aptilonotus ater* (Lea), 2.0 mm (Eugnomini); E, *Bothrophasis* sp., Mt. Tozer, QLD, 1.6 mm (Eugnomini); F, *Meripherinus fimbriatus* Lea, 5.3 mm (Eugnomini); G, *Meriphus australis* (Boisduval), 4.0 mm (Eugnomini); H, *Myossita melanocephala* Pascoe, 4.9 mm (Eugnomini); I, *Orpha flavicornis* Pascoe, 3.9 mm (Eugnomini); J, *Rhopalomerus piceosetosus* (Lea), 3.2 mm (Eugnomini); K, *Orchestes australiae* Lea, 2.6 mm (Rhamphini); L, *Rhamphus* sp., Albany, WA, 1.1 mm (Rhamphini).

Plate 76. Curculionidae (Curculioninae). A, *Smicronyx longirostris* (Lea), 3.1 mm (Smicronychini); B, *Emplesis* sp., Mt. Glorious, QLD, 3.4 mm (Storeini s.str.); C, *Leucomelacis quadrinotatus* Lea, 2.5 mm (Storeini s.str.); D, *Microberosiris albus* Lea, 1.7 mm (Storeini s.str.); E, *Placorrhinus* sp., Beaury S.F., NSW, 2.4 mm (Storeini s.str.); F, *Pseudostoreus placitus* Lea, 3.5 mm (Storeini s.str.); G, *Storeus maximus* Lea, 7.5 mm (Storeini s.str.); H, *Abethas varians* (Blackburn), 2.8 mm (Storeini s.l.); I, *Agestra horridae* Rheinheimer, 2.5 mm (Storeini s.l.); J, *Anarciarthrum viride* Blackburn, 2.3 mm (Storeini s.l.); K, *Cisowhitea longicollis* Lea, 2.0 mm (Storeini s.l.); L, *Cydmaea bimaculata* Pascoe, 2.7 mm (Storeini s.l.).

Plate 77. Curculionidae (Curculioninae: Storeini s.l.). A, *Dicomada* sp., Oallen, NSW, 3.1 mm; B, *Dyschoenium flavum* Blackburn, 1.8 mm; C, *Elleschodes hamiltoni* Blackburn, 2.6 mm; D, *Elleschodes* sp., Atherton, QLD, 2.9 mm; E, *Empira* sp., Sydney, NSW, 2.8 mm; F, *Empolis granulatus* Lea, 4.1 mm; G, *Encosmia cornuta* Blackburn, 4.6 mm; H, *Epacticus nigrirostris* Blackburn, 2.1 mm; I, *Erytenna consputa* Pascoe, 4.6 mm; J, *Hibberticola punctulatus* (Blackburn), 2.7 mm; K, *Imathia* sp., Mt. Tozer, QLD, 2.5 mm; L, *Ixamine* sp., Coomberdale, WA, 2.7 mm.

Plate 78. Curculionidae (Curculioninae: Storeini s.l.). A, *Micraonychus* sp., Caloundra, QLD, 2.1 mm; B, *Misophrice dispar* Blackburn, 1.9 mm; C, *Misophricoides* sp., Cocklebiddy, WA, 1.2 mm; D, *Myllorhinus bicaudatus* Boisduval, 12.2 mm; E, *Myllorhinus insignis* (Lea), 7.3 mm; F, *Ochrophoebe uniformis* Pascoe, 2.7 mm; G, *Olbiodorus hirsutus* Blackburn, 4.1 mm; H, *Paryzeta musiva* Pascoe, 3.1 mm; I, *Orichora trivirgata* Pascoe, ♀, 2.2 mm; J, *Orichora trivirgata* Pascoe, ♂, 2.6 mm; K, *Phaunaeus longirostris* Lea, 2.5 mm; L, *Phaunaeus medioalbus* Lea, 2.3 mm.

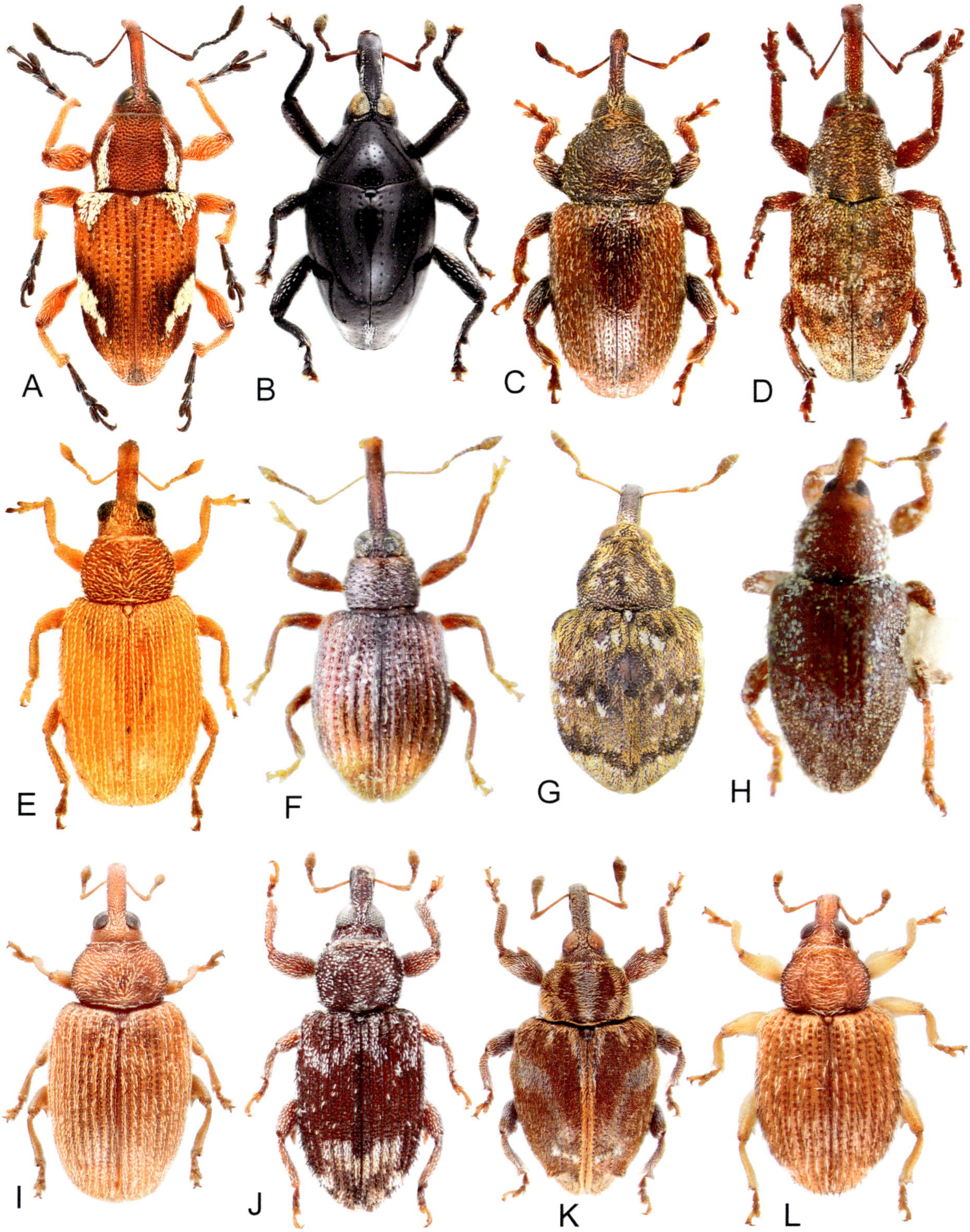

Plate 79. Curculionidae (Curculioninae). A, *Plaesiorhinus notatus* Blackburn, 4.6 mm (Storeini s.l.); B, *Rhinidotasia suturalis* Lea, 2.4 mm (Storeini s.l.); C, *Sellechus tibialis* Lea, 2.7 mm (Storeini s.l.); D, *Themelia inconspicua* Blackburn, 2.6 mm (Storeini s.l.); E, '*Tychius*' *clavivarius* Lea, 2.5 mm (Ochyromerini); F, *Meripherellus apicalis* Lea, 2.4 mm (syntype, SAM, Storeini s.l.); G, *Mesembrinocis variegatus* Lea, 4.6 mm (holotype SAM, Storeini s.l.); H, *Minia opalescens* Pascoe, 2.6 mm (holotype NHMUK, Storeini s.l.); I, '*Eudela*' *castanea* Lea, 2.6 mm (Ochyromerini); J, Undet. genus, NT, 2.6 mm (Ochyromerini); K, Undet. genus, Mt. Tamborine, QLD, 3.5 mm (Ochyromerini); L, Undet. genus, Bunya Mts QLD, 2.7 mm (Ochyromerini).

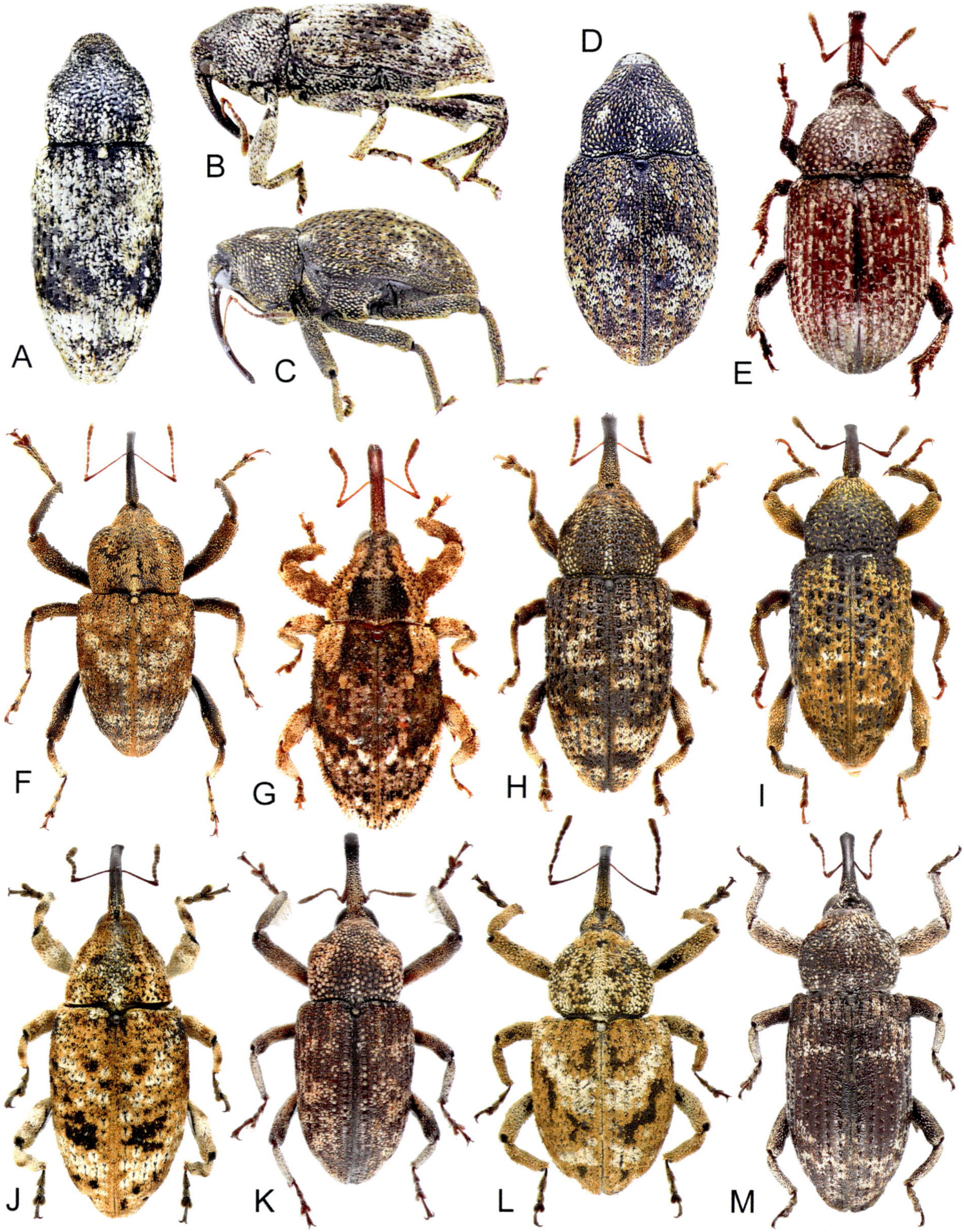

Plate 80. Curculionidae (Molytinae: Aedemonini). A, B, *Cincius triangulifer* (Lea), dorsal and lateral, 8.0 mm (holotype SAM); C, D, *Damippus tenuirostris* (Lea), 9.5 mm, lateral and dorsal; E, *Derbyiella laminata* (Lea), 5.6 mm; F, *Diatassa denticulata* (Pascoe), 10.6 mm; G, *Dystropicus* sp., QLD, 6.9 mm; H, *Epirus maerens* (Lea), 9.2 mm; I, *Magius cancellatus* (Lea), 6.2 mm; J, *Mechistocerus calidris* (Pascoe), 11.6 mm; K, *Parendymia pilipes* Kirsch, 7.4 mm; L, *Rhadinomerus mastersi* (Pascoe), 11.2 mm; M, *Roecus vulneratus* (Lea), 9.0 mm.

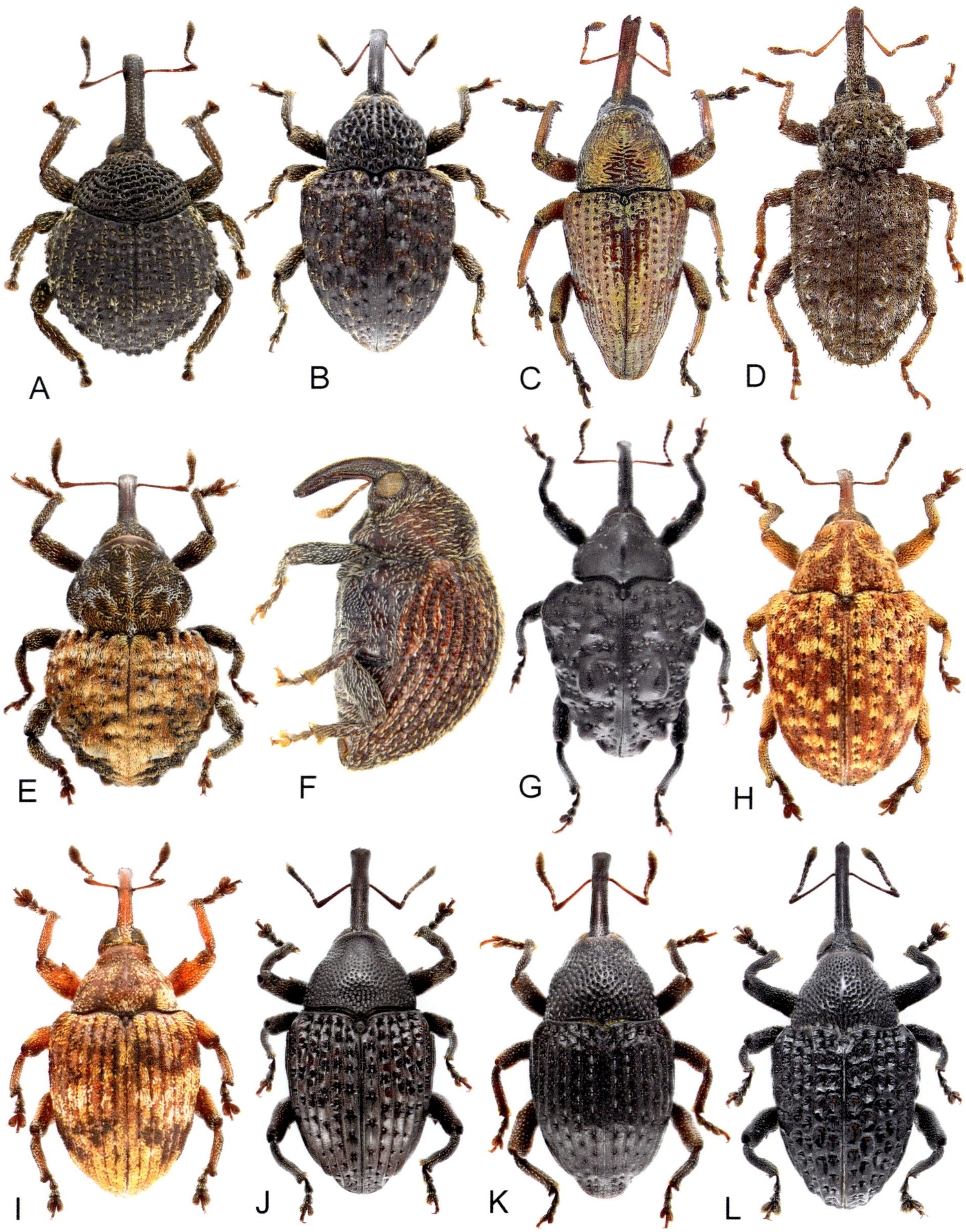

Plate 81. Curculionidae (Molytinae: Cleogonini). A, *Achelocis rudis* Lea, 3.0 mm; B, *Arthriticosoma* sp., Mt. Tozer, QLD, 4.6 mm; C, *Chalcodermus serripes* Fahraeus, 4.6 mm; D. *Conotrachelus albocinereus* Fiedler, 4.1 mm; E, *Cycloporopterus mysticus* Lea, 3.8 mm; F, *Eurymelanterius monteithi* Zimmerman & Oberprieler, 5.4 mm, (holotype ANIC); G, *Hybophorus rufotuberosus* Waterhouse, 7.5 mm; H, *Lybaeba squamivaria* Lea, 3.8 mm; I, *Lybaeba subfasciata* Pascoe, 3.4 mm; (J, *Melanterius bidentatus* Lea, 6.9 mm; K, *Melanterius corosus* (Boisduval), 3.6 mm; L, *Melanterius porcatus* Erichson, 6.3 mm.

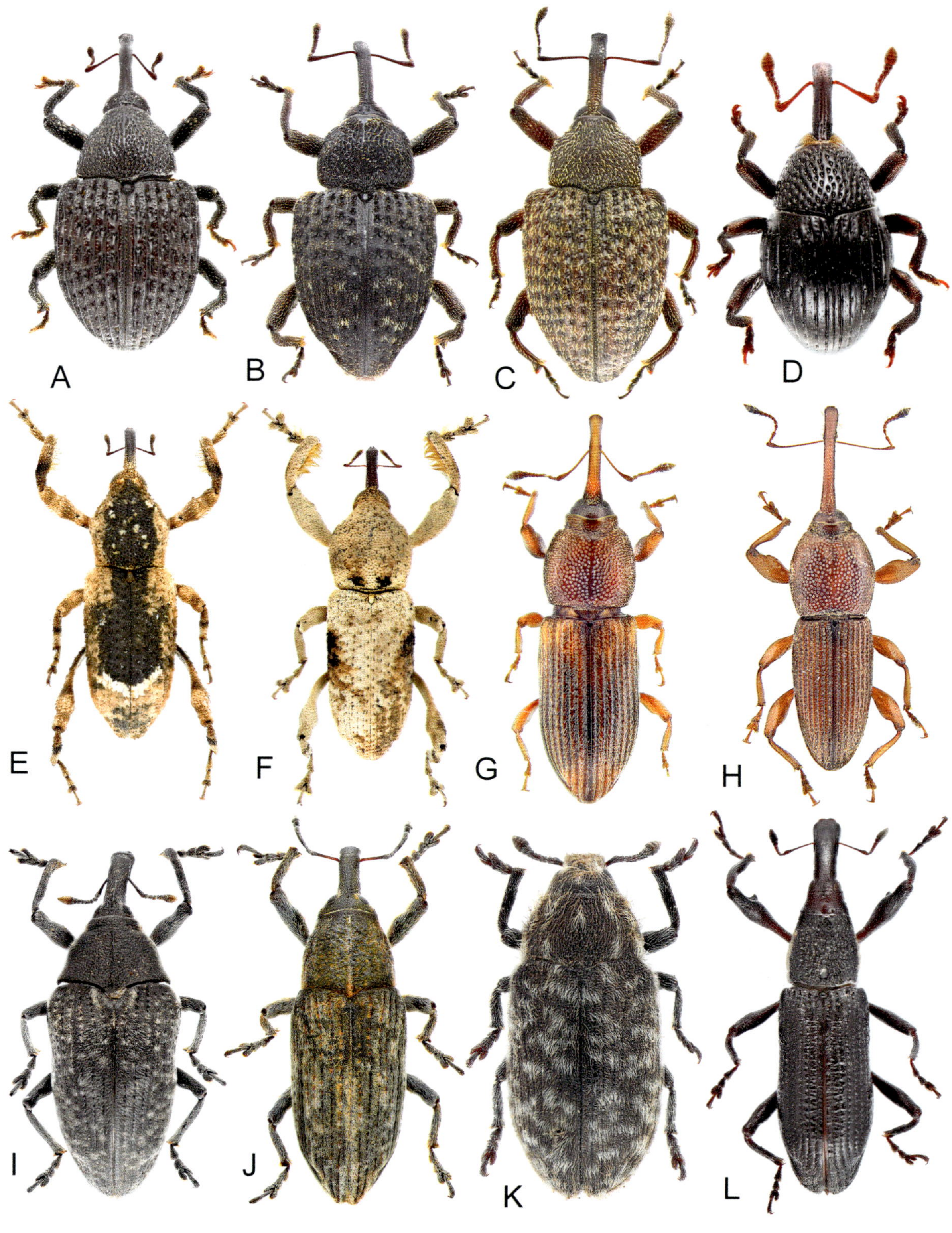

Plate 82. Curculionidae (Molytinae). A, *Moechius anaglyptus* Pascoe, 5.8 mm (Cleogonini); B, *Neomelanterius carinicollis* (Pascoe), 6.9 mm (Cleogonini); C, *Neomelanterius longirostris* Lea, 6.2 mm (Cleogonini); D, *Teutheria insculpta* Pascoe, 2.5 mm (Cleogonini); E, *Camptorhinus dorsalis* (Boisduval), 7.1 mm (Ithyporini); F, *Camptorhinus inornatus* Lea, 10.3 mm (Ithyporini); G, *Arecophaga* sp., ♂, Lord Howe Island, NSW, 3.4 mm (Juanorhinini); H, *Arecophaga* sp., ♀, Norfolk Island, 4.8 mm (Juanorhinini); I, *Larinus latus* (Herbst), 15.8 mm (Lixini); J, *Lixus tasmanicus* Germar, 10.8 mm (Lixini); K, *Rhinocyllus conicus* (Frölich), 5.2 mm (Lixini); L, *Peribleptus dealbatus* (Boisduval), 10.7 mm (Paipalesomini) specimen abraded.

Plate 83. Curculionidae (Molytinae: Cryptorhynchini). A, *Achopera* sp., Mt. Barker, WA, 4.3 mm; B, *Achoperinus infulatus* (Erichson), 4.1 mm; C, *Acidinus sedecimtuberculatus* (Lea), 10.0 mm; D, *Acrotychreus fasciculatus* (Lea), 4.4 mm; E, *Aethreus cicatricosus* Pascoe, 10.9 mm; F, *Agathicis distinctus* Lea, 7.2 mm; G, *Agenopus agricola* Pascoe, 5.3 mm; H, *Alatidotasia rubiventris* Lea, 2.9 mm; I, *Amleanus bicrucicollis* (Lea), 5.7 mm; J, *Ampagia abdominalis* (Pascoe), 12.4 mm; K, *Ampagia erinacea* Pascoe, 2.7 mm; L, *Andracalles fasciculatus* (Lea), 1.6 mm.

Plate 84. Curculionidae (Molytinae: Cryptorhynchini). A, *Athanasius latipennis* (Lea), 8.5 mm; B, *Atragopus caliginosus* (Lea), 5.5 mm; C, *Axionicus insignis* Pascoe, 16.7 mm; D, *Baeodontocis megapholus* Lea, 6.5 mm; E, *Bepharus ellipticus* Pascoe, 3.7 mm; F, *Blepiarda undulata* Pascoe, 9.0 mm; G, *Bleptocis collaceratus* Lea, 6.2 mm; H, *Brachyphyes thoracicus* (Lea), 6.4 mm; I, *Brachyporopterus montanus* Lea, 6.4 mm; J, *Cairnsicis opalescens* Lea, 5.9 mm; K, *Chaetectetorus bifasciatus* (Boheman), 7.5 mm; L, *Chimades lanosus* Pascoe, 6.3 mm.

Plate 85. Curculionidae (Molytinae: Cryptorhynchini). A, *Cratomerocis flavonotatus* Lea, 5.2 mm; B, *Critomerus iliacus* (Pascoe), 7.2 mm; C, *Cyphoderocis dorsalis* Lea, 8.3 mm; D, *Decilaus squamosus* Pascoe, 3.8 mm; E, *Decilaus suturaelevata* (Rheinheimer), 1.6 mm; F, *Drassicus nigricornis* Pascoe, 5.2 mm; G, *Dysopirhinus grandis* Lea, 14.9 mm; H, *Ecildaus* sp., Patchewollock, VIC, 5.2 mm; I, *Elaeagna squamibunda* Pascoe, 6.8 mm; J, *Embaphiodes phyxidatus* Pascoe, 7.8 mm; K, *Emydica brevisetosa* Lea, 4.8 mm; L, *Enteles vigorsii* Gyllenhal, 8.5 mm.

Plate 86. Curculionidae (Molytinae: Cryptorhynchini). A, *Ephrycinus pilistratus* Lea, 4.2 mm; B, *Ephrycus obliquus* Pascoe, 2.9 mm; C, *Episodiocis* sp., Nimbin, NSW, 5.7 mm; D, *Ethocis bifasciatus* Lea, 3.8 mm; E, *Eufaustia mirabilis* Lea, 7.2 mm; F, *Euryporopterus funereus* Lea, 8.1 mm; G, *Eutyrhinus meditabundus* (Fabricius), 10.3 mm; H, *Eutinobothrus pilosellus* (Boheman), 3.4 mm; I, *Exithius cariosus* (Erichson), 5 mm; J, *Evaniocis ellipticus* Lea, 5.9 mm; K, *Genuacalles* sp., Dorrigo N.P., NSW, 1.3 mm; L, *Glochinorhinus evanidus* Lea, 12.1 mm.

Plate 87. Curculionidae (Molytinae: Cryptorhynchini). A, *Griffithia anomala* Lea, 4.3 mm; B, *Gygaeus prodigus* Pascoe, 11.2 mm; C, *Gymnoporopterus pictipes* Lea, 2.5 mm; D, *Hexymus australis* (Boisduval), 7.9 mm; E, *Hoplidotasia torresensis* Lea, 4.0 mm; F, *Hybomorphus melanosomus* Saunders & Jekel, 15.4 mm; G, *Hyparinus dispar* Pascoe, 11.2 mm; H, *Hyposcolyphrus semipunctatus* (Lea), 6.3 mm; I, *Idiopterocis trilineatus* Lea, 3.7 mm; J, *Imaliodes subfasciatus* Pascoe, 6.5 mm; K, *Hypsophorus dromedarius* (Boisduval), 16.9 mm; L, *Imalithus patella* Pascoe, 5.5 mm.

Plate 88. Curculionidae (Molytinae: Cryptorhynchini). A, *Isax planipennis* Lea, 8.2 mm; B, *Leptidotasia macrops* (Lea), 3.6 mm; C, *Lophocheirus cruciger* (Pascoe), 11.7 mm; D, *Mallus costatus* Marshall, 4.6 mm; E, *Mecryptorhynchus* sp., 9.2 mm; F, *Meniomorpha inconstans* Lea, 7.0 mm; G, *Menios internatus* Pascoe, 4.9 mm; H, *Mesalcus* sp., Nangur S.F., QLD, 3.3 mm; I, *Metacymia marmorea* Pascoe, 7.4 mm; J, *Methidrysis afflicta* Pascoe, 6.4 mm; K, *Metraniomorpha entima* Lea, 10.0 mm; L, *Metyrus albicollis* (Germar), 7.8 mm.

Plate 89. Curculionidae (Molytinae: Cryptorhynchini). A, *Metyrculus bimaculatus* Lea, ♂, 3.7 mm; B, *Metyrculus bimaculatus* Lea, ♀, 4.4 mm; C, *Microcryptorhynchus oreas* Lea, 1.5 mm; D, *Microporopterus regularis* Lea, 6.4 mm; E, *Miotus crux* (Marshall), 3.4 mm; F, *Miotus foveicollis* (Lea), 5.7 mm; G, *Mitrastethus baridioides* Redtenbacher, 5.3 mm; H, *Mormosintes rubus* Pascoe, 18.7 mm; I, *Myrtesis* sp., Bunya Mts N.P., QLD, 5.8 mm; J, *Nechyrus incomptus* Pascoe, 8.5 mm; K, *Neodecilaus picus* Lea, 3.7 mm; L, *Neomystocis squamiventris* Lea, 8.9 mm.

Plate 90. Curculionidae (Molytinae: Cryptorhynchini). A, *Neoporopterus lithodermus* (Boisduval), 10.8 mm; B, *Niconotus stenotarsus* Lea, 6.7 mm; C, *Notocalviceps rarus* Lea, 8.2 mm; D, *Notocryptorhynchus sinuatus* Lea, 9.0 mm; E, *Nototragopus plagiatus* (Pascoe), 11.8 mm; F, *Nototragopus* sp., Kuranda, QLD, 8.8 mm; G, *Oemethylus triangularis* Lea, 9.8 mm; H, *Omydaus impressicollis* Lea, 9.9 mm; I, *Onidistus nodipennis* Pascoe, 6.8 mm; J, *Ophrythyreocis valgus* (Lea), 4.9 mm; K, *Orochlesis annularis* Pascoe, 6.1 mm; L, *Orphanistes eustictus* Pascoe, 18.9 mm.

Plate 91. Curculionidae (Molytinae: Cryptorhynchini). A, *Orthoporopterus elongatus* Lea, 5.4 mm; B, *Osaces naso* Pascoe, 6.5 mm; C, *Ouroporopterus diurus* Lea, 7.3 mm; D, *Pachyporopterus satyrus* (Pascoe), 16.2 mm; E, *Paleticus pedestris* Pascoe, 10.6 mm; F, *Paletonidistus trisinuatus* Lea, 7.2 mm; G, *Parasalcus latissimus* (Pascoe), 5.5 mm; H, *Perissops ocellatus* (Redtenbacher), 11.6 mm; I, *Phloeoglymma alternans* Pascoe, 4.8 mm; J, *Phloeoglymma dorsale* (Pascoe), 4.6 mm; K, *Pezichus binotatus* Lea, 11.2 mm; L, *Plagiocorynus quadrituberculatus* Waterhouse, 12.8 mm.

Plate 92. Curculionidae (Molytinae: Cryptorhynchini). A, *Platyporopterus* sp., Brindabella Ranges, ACT, 8.6 mm; B, *Platytenes varius* Pascoe, 6.3 mm; C, *Poropteroides dichotomus* Lea, 8.9 mm; D, *Poropterus antiquus* (Erichson), 11.6 mm; E, *Poropterus fasciculatus* Lea, 10.7 mm; F, *Bothynacrum sterioides* Lea, 5.3 mm; G, *Pseudapries corticalis* Lea, 4.3 mm; H, *Pseudometyrus antares* (Erichson), 9.0 mm; I, *Pseudonidistus cordatus* Lea, 4.8 mm; J, *Pseudoporopterus cruciatus* (Fabricius), 8.0 mm; K, *Pteroporopterus lacunosus* Lea, 6.1 mm; L, *Salcus* sp., Cardstone, QLD, 6.8 mm.

Plate 93. Curculionidae (Molytinae: Cryptorhynchini). A, *Roptoperus basalis* Lea, 4.6 mm; B, *Roptoperus tasmaniensis* Lea, 4.6 mm; C, *Scolyphrus* sp., Newcastle Range, QLD, 14.4 mm; D, *Stenoporopterus canaliculatus* Lea, 6.9 mm; E, *Sternochetus mangiferae* (Fabricius), 7.9 mm; F, *Sybulus yorkensis* Lea, 5.8 mm; G, *Tamphilus amplicollis* (Faimaire), 4.3 mm; H, *Tapinocis abundans* Lea, 4.8 mm; I, *Tapinosomus exitiosus* (Pascoe), 13.2 mm; J, *Tentegia amplipennis* Lea, 10.2 mm; K, *Tentegia stupida* (Fabricius), 12.8 mm; L, *Tepperia sterculiae* Lea, 8.7 mm.

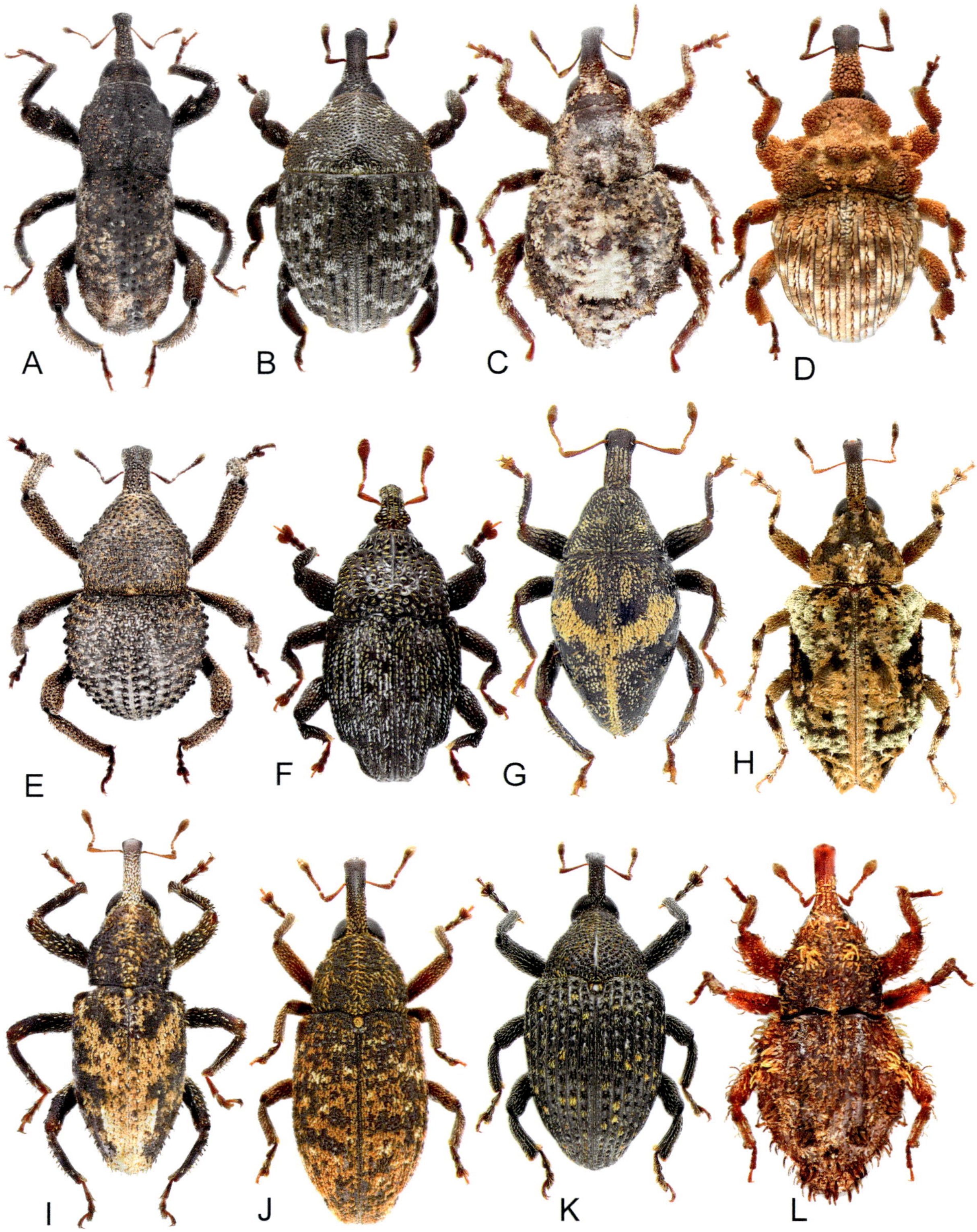

Plate 94. Curculionidae (Molytinae: Cryptorhynchini). A, *Terporopus tenuicornis* Lea, 5.6 mm; B, *Tetengia solenopa* Lea, 4.6 mm; C, *Tituacus ostracion* (Pascoe), 3.5 mm; D, *Tomweirius mirus* Zimmerman & Oberprieler, 3.2 mm; E, *Tragopus muticus* (Lea), 6.0 mm; F, *Trigonopterus lewisensis* Riedel, 3 mm; G, *Tropidotasia femoralis* Lea, 3.1 mm; H, *Tychreus camelus* Pascoe, 6.7 mm; I, *Tyrtaeosellus coxalis* Lea, 3.8 mm; J, *Tyrtaeosus aequus* Lea, 4.6 mm; K, *Tyrtaeosus flavonotatus* Lea, 6.3 mm; L, *Zenoporopterus mirus* Lea, 2.9 mm.

Plate 95. Curculionidae (Molytinae: Cryptorhynchini). A, B, *Albius humifer* (Boheman), dorsal and lateral, 7.0 mm (OUMNH); C, *Amorphocis mirus* Lea, 4.2 mm (holotype SAM); D, *Ampagiosoma convexum* Lea, 4.2 mm; E, F, *Ancocis collaris* Lea, dorsal and lateral, 8.5 mm (holotype SAM); G, *Anepigraphocis basiventris* Lea, 8.5 mm (holotype SAM); H, *Anilaus sordidus* Pascoe, 5.0 mm (lectotype NHMUK); I, J, *Anoplocis ferrugineus* Lea, dorsal and lateral, 6.5 mm (holotype SAM); K, *Athyreocis triangulifer* Lea, 3.8 mm; L, *Athyreocis tarsalis* Lea, 3.3 mm (SAM).

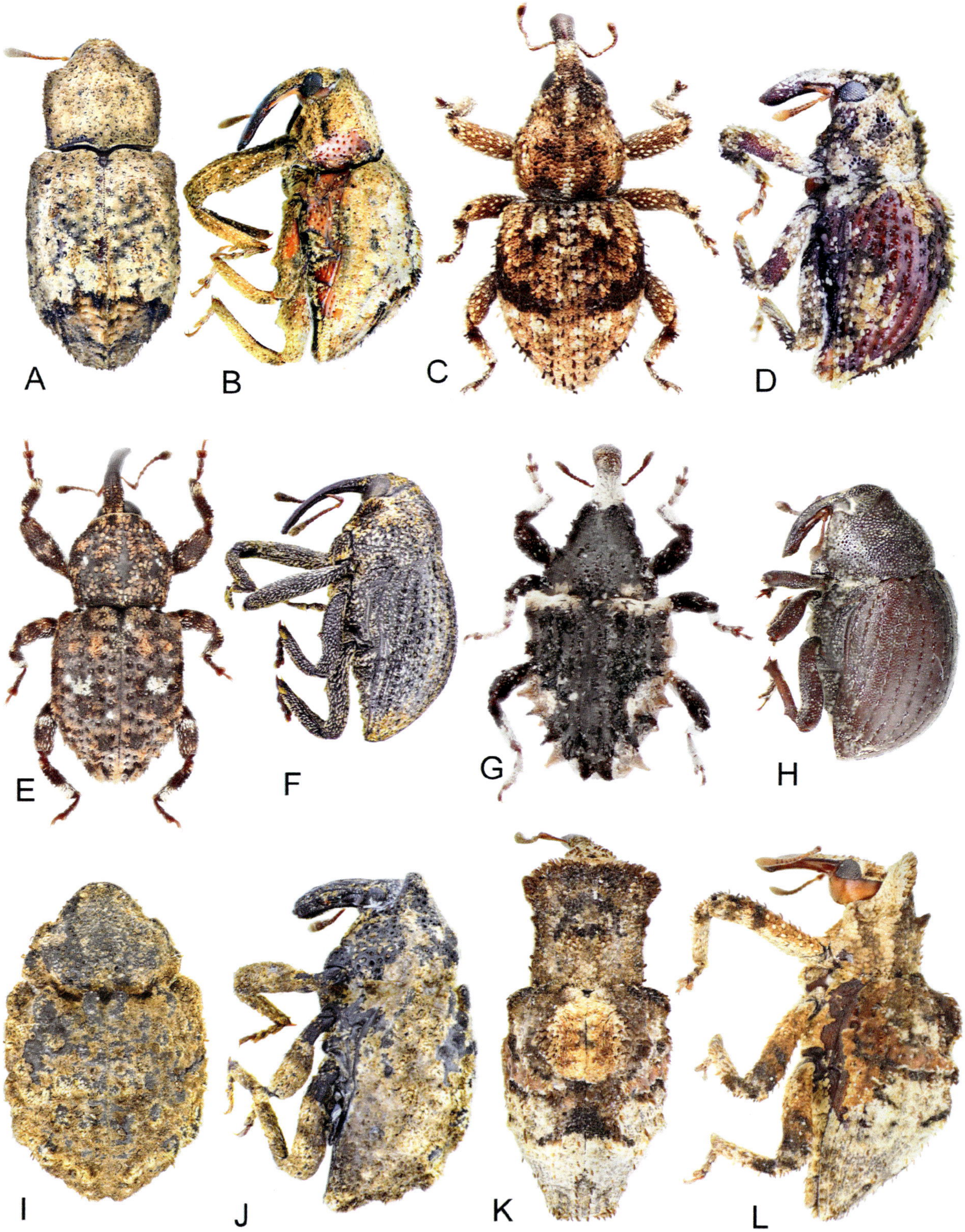

Plate 96. Curculionidae (Molytinae: Cryptorhynchini). A, B, *Austrectopsis oblonga* Lea, dorsal and lateral, 6.5 mm (holotype SAM); C, *Cardiopterocis variegatus* Lea, 3.1 mm; D, *Cardiopterocis variegatus* Lea, 3 mm (holotype SAM); E, *Diaphorocis variegatus* Lea, 6.1 mm; F, *Didorus pictifrons* (Lea), 8.5 mm (syntype, SAM); G, *Eucalyptocis fasciculatus* Lea, 2.9 mm; H, *Euoropis castanea* Pascoe, 5.4 mm (holotype NHMUK); I, J, *Eurycis multinodosus* Lea, dorsal and lateral, 7.0 mm (lectotype SAM); K, L, *Eurymetopocis bilobus* Lea, dorsal and lateral, 6.5 mm (holotype SAM).

Plate 97. Curculionidae (Molytinae: Cryptorhynchini). A, B, *Exithioides punctatus* Lea, dorsal and lateral, 5.0 mm (holotype SAM); C, D, *Gymnocis impunctatus* Lea, dorsal and lateral, 4.9 mm; E, F, *Hoplodecilaus marmoratus* Lea, dorsal and lateral, 2.8 mm (holotype SAM); G, H, *Leucothyreocis convexus* Lea, dorsal and lateral, 2.0 mm (syntype, SAM); I, *Hyperiosoma* sp., Kuranda, QLD, 4.4 mm; J, *Odosyllis scutellaris* Lea, 10.5 mm; K, L, *Poropterculus subnitidus* Lea, dorsal and lateral, 2.6 mm (holotype SAM).

Plate 98. Curculionidae (Molytinae: Cryptorhynchini). A, *Paratituacia dorsosignata* Lea, 3.7 mm; B, *Poropterellus intercoxalis* Lea, 3.7 mm; C, *Poropterinus trilobus* Lea, 4.7 mm; D, *Pseudomydaus tenuis* Lea, 6.5 mm; E, Undet. genus, Mt. Trafalgar WA, 2.8 mm; F, G, *Pseudotepperia compta* Lea, 7.7 mm, dorsal and lateral (holotype ANIC); H, *Rhaebocnemocis posterus* Lea, 4.9 mm (syntype, SAM); I, *Scleropoides triangu-liferus* Lea, 5.8 mm (MV); J, K, *Scotinocis sagittifer* Lea, 2.7 mm, dorsal and lateral (type SAM); L, *Sympediosoma albifrons* Lea, 7.7 mm.

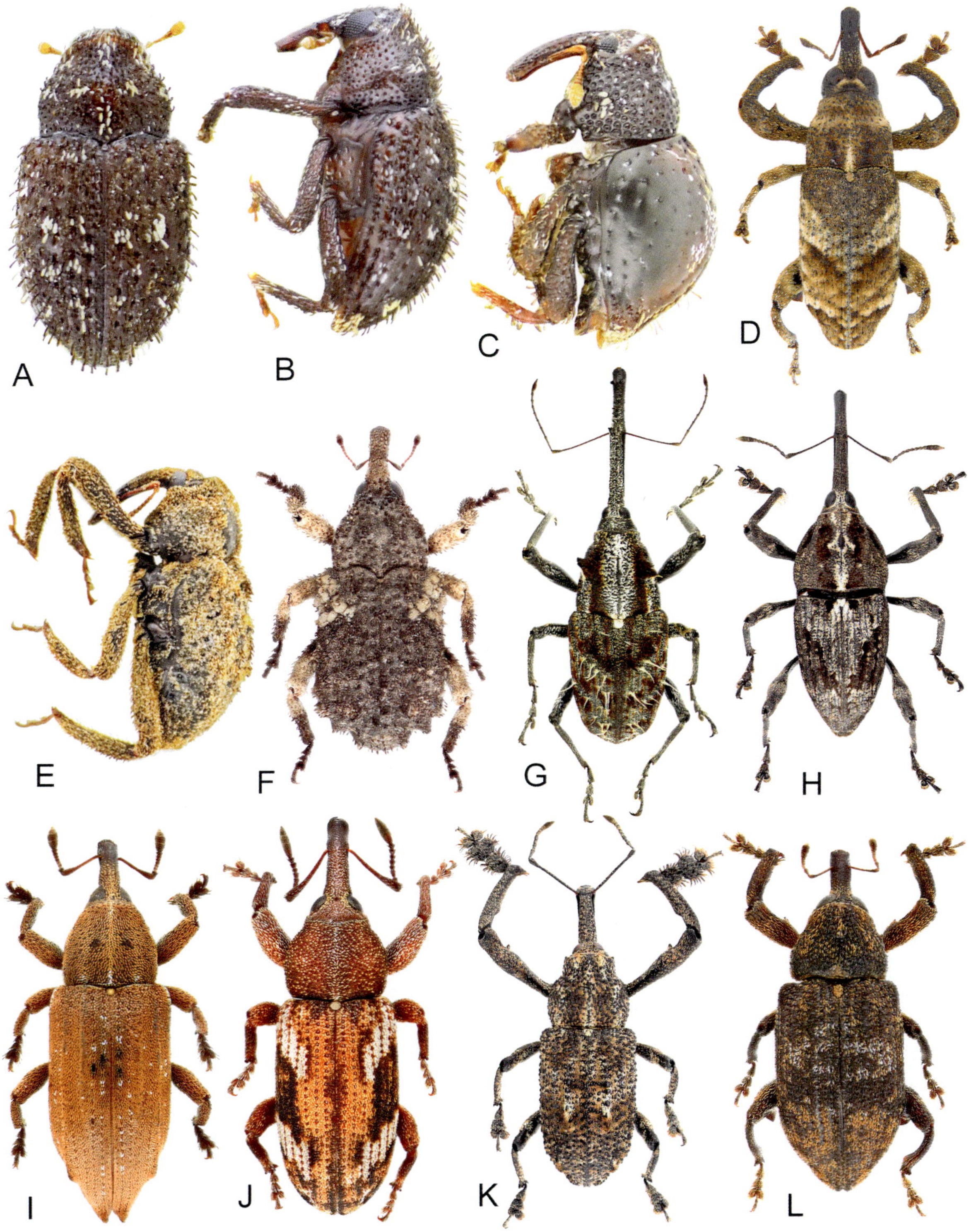

Plate 99. Curculionidae (Molytinae). A, B, *Sophronocis albonotatus* Lea, dorsal and lateral, 3.1 mm (holotype SAM, Cryptorhynchini); C, *Triptocis puncticollis* Lea, dorsal and lateral, 2.4 mm (holotype SAM, Cryptorhynchini); D, *Homorthorrhinus* sp., Cairns, QLD, 6.4 mm (Orthorhinini); E, *Tepalicus semicalvus* Lea, 7.5 mm (syntype, SAM, Cryptorhynchini); F, *Tylocis squamibundus* Lea, 6.2 mm (Cryptorhynchini); G, *Eurhamphus fasiculatus* Shuckard, 26.6 mm (Orthorhinini); H, *Ilacuris laticollis* Pascoe, 15.5 mm (Orthorhinini); I, *Imbilius araucariae* Marshall, 3.9 mm (Orthorhinini); J, *Notopissodes pictus* Zimmerman & Oberprieler, 5.9 mm (Orthorhinini); K, *Orthorhinus cylindrirostris* (Fabricius), 18.3 mm (Orthorhinini); L, *Parorthorhinus aethiops* (Boisduval), 6.3 mm (Orthorhinini).

Plate 100. Curculionidae (Molytinae). A, *Cylindralcides bubo* (Fabricius), 5.0 mm (Mecysolobini); B, *Sternuchopsis pentasticta* (Ancey), 7.4 mm (Mecysolobini); C, *Aethemagdalis instabilis* (Lea), 2.3 mm (Mesoptiliini); D, *Allolaemosaccus latirostris* (Lea), 1.7 mm (paratype ANIC, Mesoptiliini); E, *Atopomagdalis renoculus* Zimmerman, 1.5 mm (Mesoptiliini); F, *Laemosaccellus festivus* (Lea), 2.9 mm (Mesoptiliini); G, *Melaemosaccus angulatus* (Janczyk), 5.7 mm (Mesoptiliini); H, *Neolaemosaccus chadwicki* (Janczyk), 4.9 mm (Mesoptiliini); I, *Neolaemosaccus rufipennis* (Lea), 3.1 mm (Mesoptiliini); J, *Notomagdalis rufimanus* (Lea), 2.9 mm (Mesoptiliini); K, *Zimmermanianthus judaicus* (Lea), 4.1 mm (Mesoptiliini); L, *Aclees* sp., Boar Pocket Rd, QLD, 14.7 mm (Molytini).

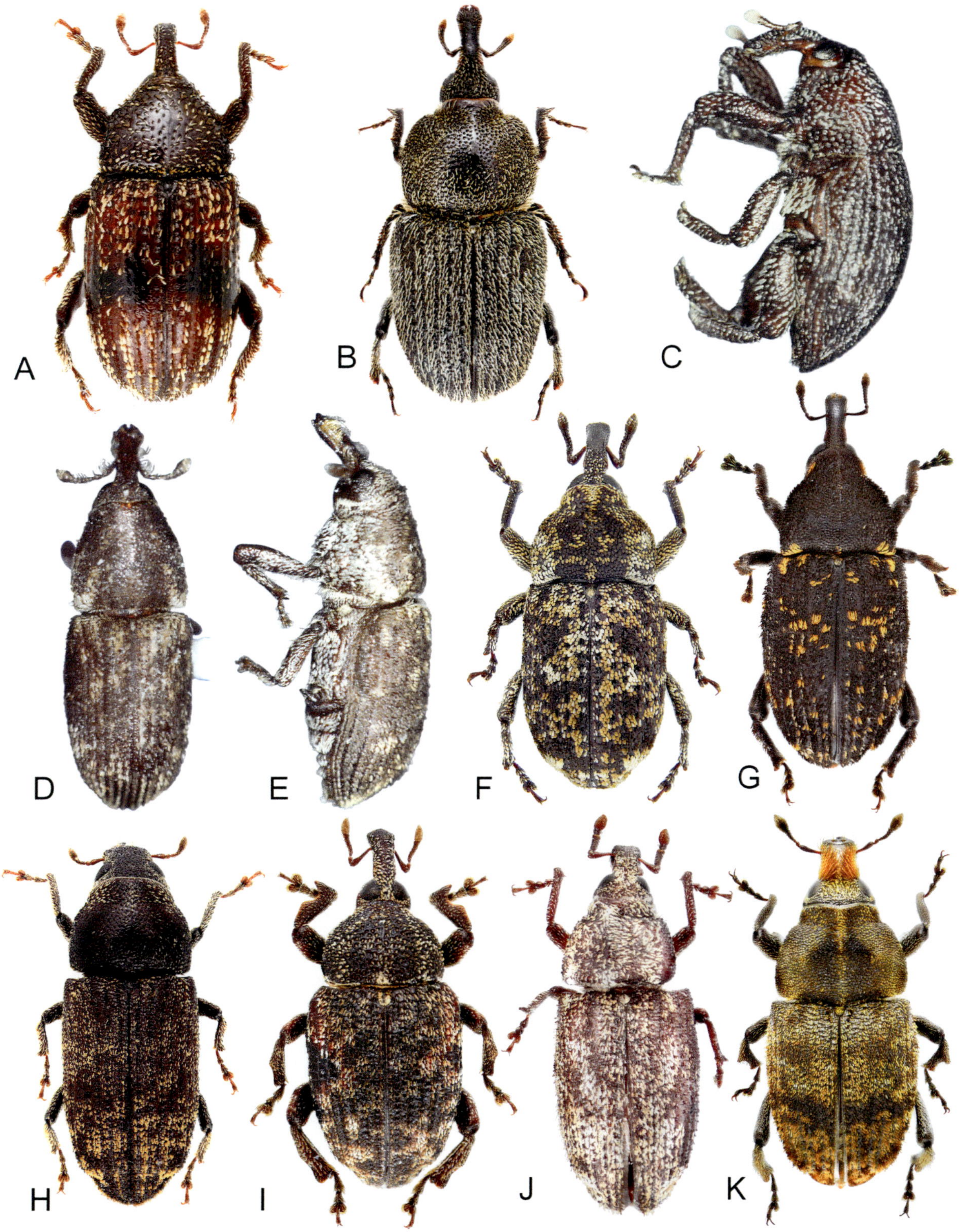

Plate 101. Curculionidae (Molytinae: Psepholacini). A, *Calpurnius niger* (Lea), 4.7 mm; B, *Catocalephe minans* Blackburn, 6.1 mm; C, *Celidaus michaelsoni* Lea, 5.0 mm; D, E, *Glechinus trichoceras* (Montrouzier), dorsal and lateral, 5.3 mm; F, *Neozeneudes* sp., Maryborough, QLD, 5.1 mm; G, *Oreda notata* White, 10.1 mm; H, *Protohylastes* sp., Atherton Tablelands, QLD, 8.6 mm; I, *Psepholacipus fossilis* Lea, 6.5 mm; J, *Psepholasoma rostrale* Lea, 5.9 mm; K, *Psepholax egerius* Pascoe, 7.5 mm.

Plate 102. Curculionidae (Molytinae). A, *Psepholasoma cylindricum* (Lea), 6.8 mm (Psepholacini); B, *Pseudotherebus sculptipennis* Lea, 6.9 mm (Psepholacini); C, *Sympiezoscelus spencei* Waterhouse, 7.8 mm (Psepholacini); D, *Therebiosoma variegatum* Lea, 8.1 mm (Psepholacini); E, *Therebus cepuroides* Pascoe, 8.6 mm (Psepholacini); F, *Wiburdia scrobiculata* Lea, 10.3 mm (Psepholacini); G, *Zeneudes sterculiae* Pascoe, 7.6 mm (Psepholacini); H, *Colobodes* sp., Mt. Tozer, QLD, 8 mm (Sophrorhinini); I, *Eprias aspratilis* (Lea), 4.8 mm (Sophrorhinini); J, K, *Perrhaebius tibialis* (Lea), dorsal and lateral, 10.0 mm (holotype SAM, Sophrorhinini); L, *Deretiosus aridus* Pascoe, 6.6 mm (Sophrorhinini).

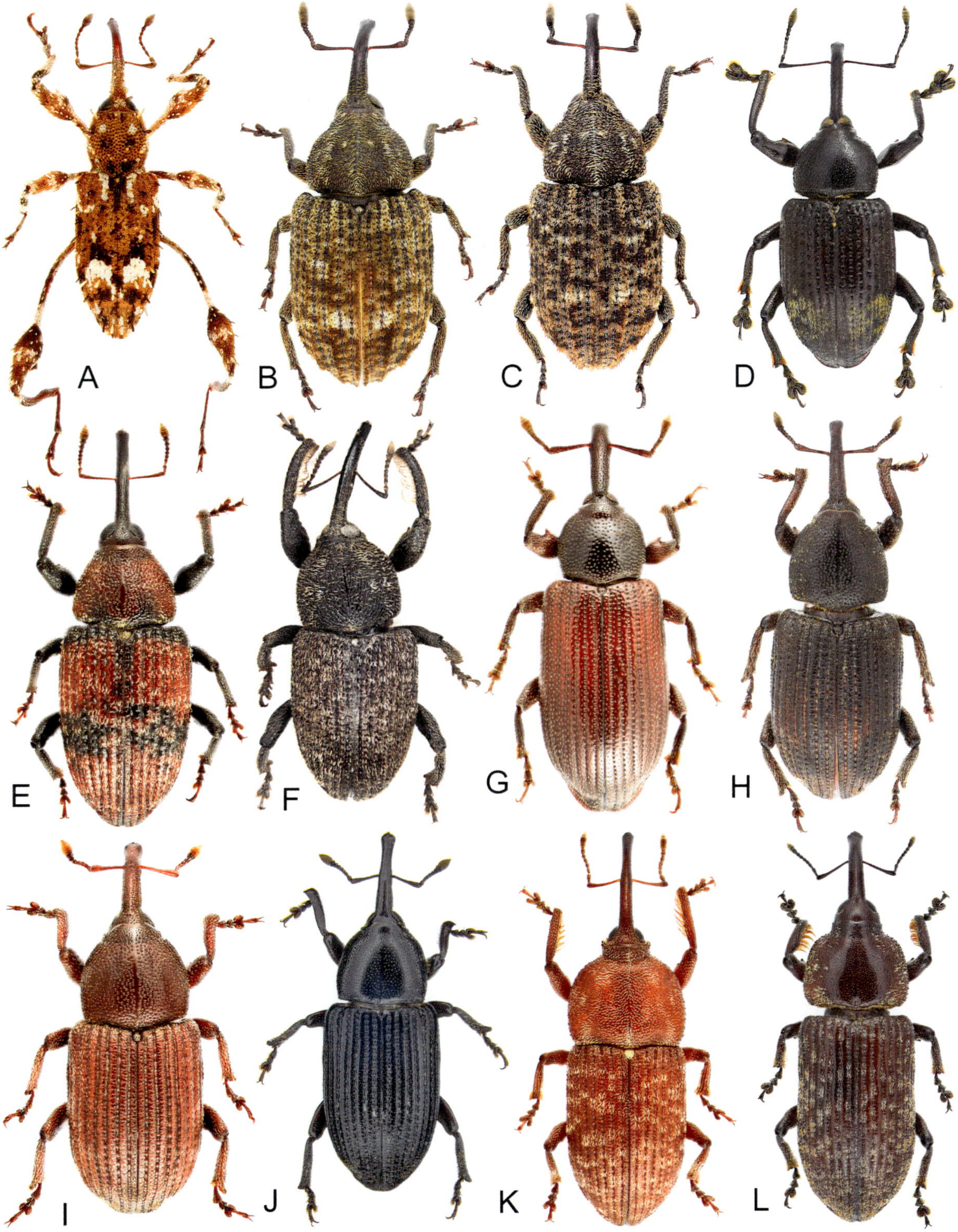

Plate 103. Curculionidae (Molytinae). A, *Acicnemis meriones* Pascoe, 3.3 mm (Trachodini); B, *Demyrsus meleoides* Pascoe, 10.8 mm (*Tranes* group); C, *Demyrsus digmon* Hsiao & Oberprieler, 11.4 mm (*Tranes* group); D, *Howeotranes insularis* (Pascoe), 10.9 mm (*Tranes* group); E, *Miltotranes prosternalis* (Lea), 6.1 mm (*Tranes* group); F, *Miltotranes subopacus* (Lea), 7.4 mm (*Tranes* group); G, *Paratranes monopticus* (Pascoe), 7.6 mm (*Tranes* group); H, *Siraton internatus* (Pascoe), ♂, 7.9 m (*Tranes* group); I, *Siraton internatus* (Pascoe), ♀, 9.1 mm (*Tranes* group); J, *Siraton roei* (Boheman), 10.1 mm (*Tranes* group); K, *Tranes lyterioides* (Pascoe), 6.8 mm (*Tranes* group); L, *Tranes vigorsii* Boheman, 15.5 mm (*Tranes* group).

Plate 104. Curculionidae (Conoderinae: Baridini). A, *Apinocis* sp., Hunter Valley, NSW, 3.1 mm; B, *Baris* sp., 3.4 mm; C, *Baris sublaminata* Lea, 3.1 mm; D, *Cosmobaris discolor* (Boheman), 2.9 mm; E, *Gymnobaris polita* Lea, 3.6 mm (syntype, ANIC); F, *Ipsichora desiderabilis* Lea, 4.2 mm; G, *Linogeraeus urbanus* (Boheman), 3.9 mm; H, *Myctides imberbis* Lea, 5.5 mm; I, *Nyella tuberculata* Oke, 3.2 mm; J, *Orchidophilus aterrimus* (Waterhouse), 5.4 mm; K, *Solenobaris decipiens* Lea, 3.8 mm.

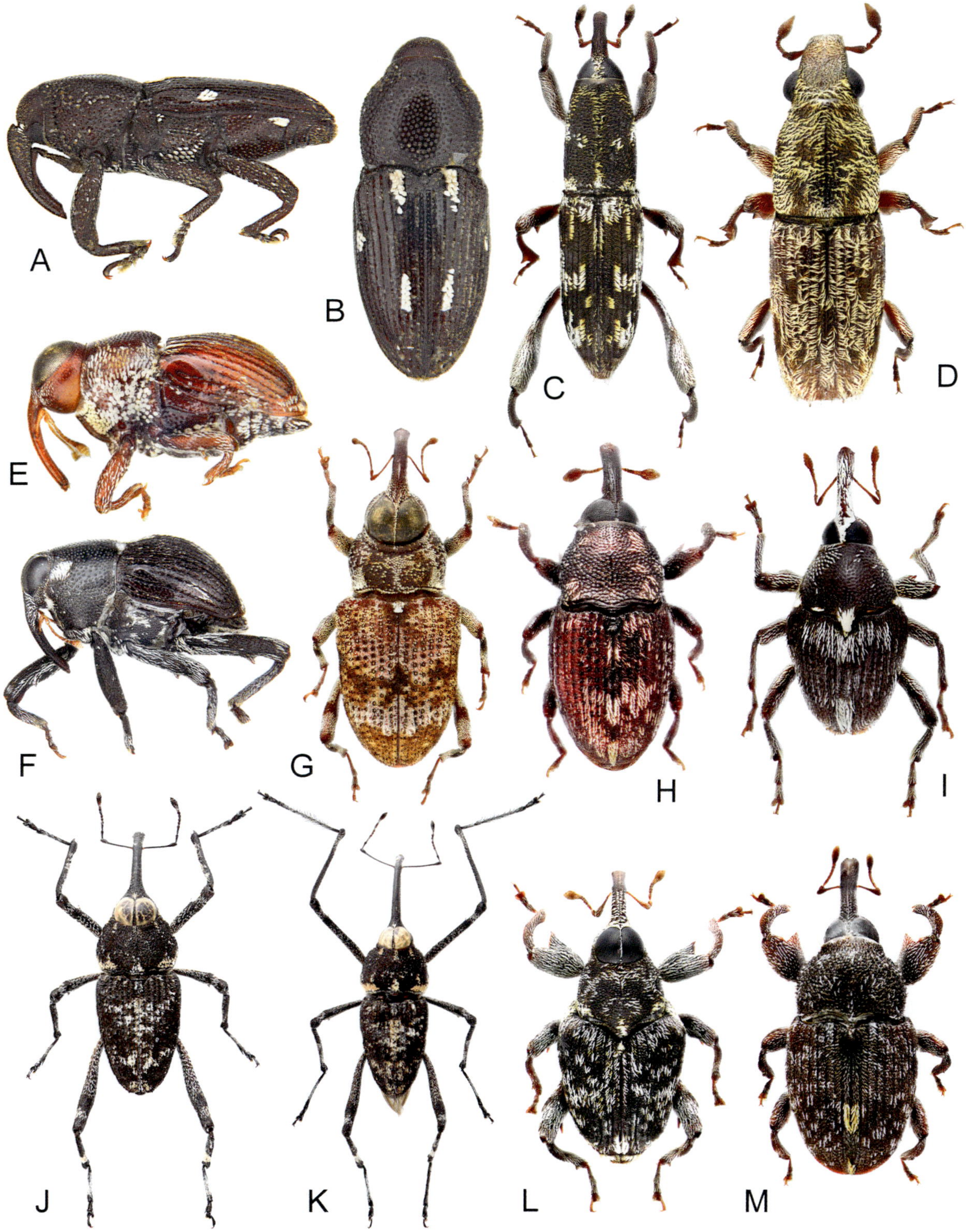

Plate 105. Curculionidae (Conoderinae). A–B, *Acythopeus bigeminatus* Pascoe, 3.3 mm (Baridini), lateral and dorsal respectively; C, *Phaenomerus auriceps* Thompson, 3.7 mm (Campyloscelini); D, *Tomicoproctus* sp., Wyong, NSW, 3.3 mm (Campyloscelini); E. *Brimoides exilis* (Pascoe), 1.6 mm (paralectotype ANIC, Conoderini); F. *Othippia guttula* Pascoe, 3.3 mm (Conoderini); G, *Agametis bifasciata* Lea, 4.3 mm (Conoderini); H, *Lamitema decipiens* Lea, 2.5 mm (syntype, ANIC, Conoderini); I, *Lobotrachelus stigma* Pascoe, 2.2 mm (Conoderini); J–K, *Mecopus tipularius* Pascoe, (Conoderini), J, ♀, 5.2 mm, K, ♂, 6.8 mm; L, *Metialma australiae* Lea, 3.4 mm (Conoderini); M, *Temialma suturalis* Lea, 3.5 mm (Conoderini).

Plate 106. Curculionidae (Cossoninae). A, *Agrilochilus* sp., Lord Howe Island, NSW, 2.5 mm; B, *Allopentarthrum elumbe* (Boheman), 2.6 mm; C, *Aphanocorynes depressus* Wollaston, 4.4 mm; D, *Catolethrobius* sp., Bellenden Ker, QLD, 2.5 mm; E, *Choerorhinus* sp., Humpty Doo, NT, 2.3 mm; F, *Conlonia litoralis* Lea, 2.0 mm; G, *Coptocorynus araucariae* Marshall, 1.6 mm; H, *Cossonideus* sp., Eneabba, WA, 4.3 mm; I, *Cossonus vicarius* Lea, 5.6 mm; J, *Dryotribus mimeticus* Horn, 2.8 mm; K, *Eiratus* sp., Mt. Bartle Frere, QLD, 1.8 mm; L, *Entium* sp., Lord Howe Island, NSW, 1.4 mm.

Plate 107. Curculionidae (Cossoninae). A, *Exomesites* sp., Lord Howe Island, NSW, 5.0 mm; B, *Halorhynchus caecus* Wollaston, 2.1 mm; C, *Himatium* sp., Bellenden Ker, QLD, 3.2 mm; D, *Himatium* sp., Christmas Island, 2.2 mm; E, *Hoplocossonus lethargicus* (Olliff), 5.2 mm; F, *Macrorhyncolus* sp., Christmas Island, 2.2 mm; G, *Macrorhyncolus* sp., ♀, Bunya Mts, QLD, 3.1 mm; H, *Mastersinella octoarticulata* Lea, 2.5 mm; I, *Microtribus* sp., Bunya Mts, QLD, 1.8 mm; J, *Morronella lawsoni* Wollaston, 2.0 mm, NZ; K, *Ochronanus* sp., Mt. Lewis, QLD, 2.6 mm; L, *Omeretes onychialis* Marshall, 3.3 mm.

Plate 108. Curculionidae (Cossoninae). A, *Orthotemnus polixus* (Erichson), 3.3 mm; B, *Oxydema procer* (Boheman), 4.5 mm; C, *Pachyops insularis* (Lea), 4.0 mm; D, *Pachyops* sp., Christmas Island, 3.7 mm; E, *Pentamimus australis* (Erichson), 3.6 mm; F, *Pentarthrocis ammophilus* Lea, 2.6 mm; G, *Pentarthrum* sp., Christmas Island, 2.0 mm; H, *Pselactus spadix* (Herbst) 3.3 mm; I, J, *Rhyncolosoma dubium* (Gahan), dorsal and lateral, 3.0 mm; K, *Rhyncolus ater* (Linnaeus), USA Cal, 2.7 mm; L, *Sericotrogus subaenescens* Wollaston, 2.6 mm.

Plate 109. Curculionidae (Cossoninae). A, *Stenotrupis* sp., Lord Howe Island, NSW, 2.9 mm; B, *Stenotrupis* sp., Christmas Island, 2.7 mm; C, *Stereoborus inductus* Pascoe, 4.8 mm; D, *Stilbocara* sp., Lord Howe Island, NSW, 2.3 mm; E, *Temnorrhamphus* sp., Christmas Island, 2.3 mm; F, *Toura* sp., Clyde Mt., NSW, 2.6 mm; G, *Unas* sp., VIC, 3.6 mm; H, *Xenocnema spinipes* Wollaston, 3.7 mm; I, *Xenotrupis* sp., PNG, 3.8 mm; J, *Hexarthroides punctulatum* Lea; K, *Stereomimetes crassicornis* Wollaston (holotype NHMUK); L, *Stereoderus macleayi* Lea, 4.6 mm.

Plate 110. Curculionidae (Scolytinae). A, *Cryphalus mangiferae* Stebbing, 1.9 mm (Cryphalini); B, *Cryphalus* sp., Christmas Island, 1.5 mm (Cryphalini); C, *Acacicis* sp., Canberra, ACT, ♀, 1.3 mm (Diamerini); D, *Diamerus interstitialis* (Lea), ♀, 4.7 mm (Diamerini); E, F, *Coccotrypes carpophagus* (Hornung), dorsal and lateral, ♀, 2.4 mm (Dryocoetini); G, *Cyrtogenius nitidus* (Hagedorn), ♀, 2.3 mm (Dryocoetini); H, I, *Dryocoetiops moestus* (Blandford), dorsal and lateral, 3.6 mm (Dryocoetini); J, K, *Peridryocoetes queenslandi* Schedl, dorsal and lateral, 2.1 mm (Dryocoetini).

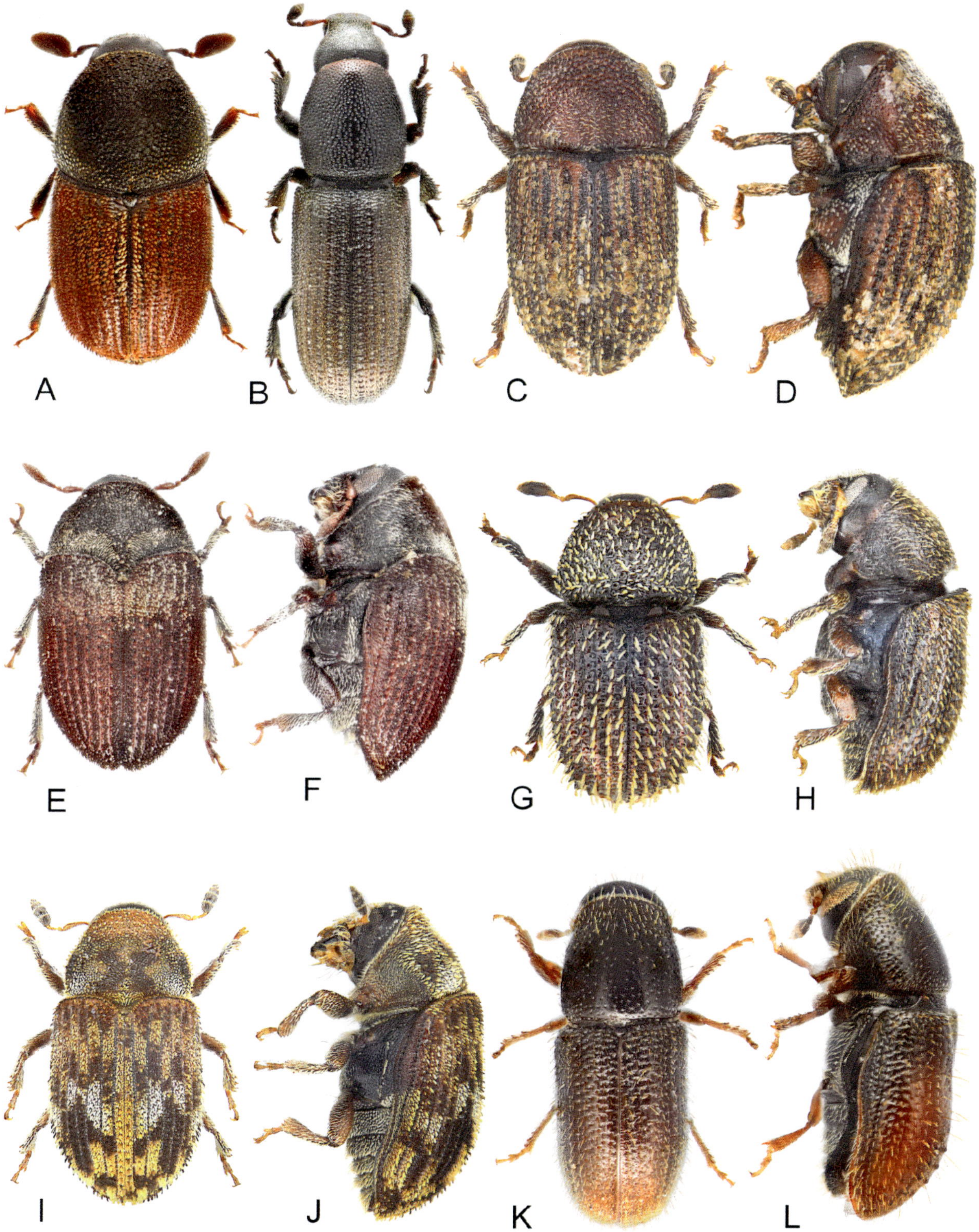

Plate 111. Curculionidae (Scolytinae). A, *Eidophelus darwini* Eichhoff, 2.6 mm (Ernoporini); B, *Hylastes ater* (Paykull), 4.3 mm (Hylastini); C, D, *Ficicis porcatus* (Chapuis), dorsal and lateral, ♀, 2.4 mm (Hylesinini); E, F, *Hylesinus cordipennis* Lea, dorsal and lateral, ♂, 3.2 mm (Hylesinini); G, *Zygophloeus australis* Schedl, 1.6 mm (Hylesinini); H, *Zygophloeus* sp., Warburton, VIC, 1.7 mm (Hylesinini);I, J, *Chaetoptelius tricolor* (Schedl), dorsal and lateral, 2.8 mm (Hylurgini); K, L, *Hylurdrectonus piniarius* Schedl, dorsal and lateral, ♂, 1.4 mm (Hylurgini).

Plate 112. Curculionidae (Scolytinae). A, *Hylurgus ligniperda* (Fabricius), 5.0 mm (Hylurgini); B, *Pachycotes kuscheli* Schedl, ♀, 3.9 mm (Hylurgini); C, D, *Sueus niisimai* (Eggers), dorsal and lateral, ♀, 1.6 mm (Hyorrhynchini); E, *Liparthrum* sp., Kakadu N.P., NT, 1.9 mm (Hypoborini); F, *Acanthotomicus australis* (Schedl), 2.2 mm (Ipini); G, *Ips grandicollis* (Eichhoff), dorsal, 3.5 mm (Ipini);); H, *Ips grandicollis* (Eichhoff), lateral, ♀, 3.5 mm (Ipini); I, J, *Hyledius cribratus* (Blandford), dorsal and lateral, 2.5 mm (Phloeosinini); K, *Hyleops glabratus* Schedl, ♀, 4.0 mm (Phloeosinini); L, *Phloeosinopsioides leai* Wood, ♂, 2.6 mm (Phloeosinini).

Plate 113. Curculionidae (Scolytinae). A, *Dactylipalpus transversus* Chapuis, ♀, 8.2 mm (Hylesinini); B, *Phloeosinus cupressi* Hopkins, 2.7 mm (Phloeosinini); C, *Aricerus eichhoffi* Blandford, ♂, 3.9 mm (Phloeotribini); D, *Phloeotribus acaciae* (Lea), ♂, 1.9 mm (Phloeotribini); E, *Scolytus multistriatus* (Marsham), ♂, 3.2 mm (Scolytini); F, G, *Cosmoderes* sp., dorsal and lateral, Mt. Tozer, QLD, 1.7 mm (Trypophloeini); H, *Hypothenemus birmanus* (Eichhoff), 2.3 mm (Trypophloeini); I, *Amasa truncata* (Erichson), ♀, 3.1 mm (Xyleborini); J, K, *Ambrosiodmus asperatus* (Blandford), dorsal and lateral, ♀, 2.4 mm (Xyleborini); L, *Arixyleborus malayensis* (Schedl), ♀, 2.2 mm (Xyleborini).

Plate 114. Curculionidae (Scolytinae: Xyleborini). A, *Ambrosiophilus compressus* (Lea), ♀, 3.6 mm; B, *Ambrosiophilus pityogenes* (Schedl), ♀, 2.7 mm; C, *Beaverium insulindicus* (Eggers), ♀, 5.0 mm; D, *Cnestus solidus* (Eichhoff), ♀, 4.1 mm; E, *Cryptoxyleborus subnaevus* Schedl, ♀, 2.5 mm; F, *Debus emarginatus* (Eichhoff), ♀, 3.7 mm; G, *Fraudatrix pileatula* (Schedl), ♀, 2.3 mm (PNG, Roger Beaver's collection, © Sarah Smith); H, *Diuncus haberkorni* (Eggers), ♀, 2.4 mm; I, *Eccoptopterus spinosus* (Olivier), ♀, 2.5 mm; J, *Euwallacea destruens* (Blandford), ♀, 4.4 mm; K, L, *Euwallacea fornicatus* (Eichhoff), dorsal and lateral, ♀, 2.5 mm.

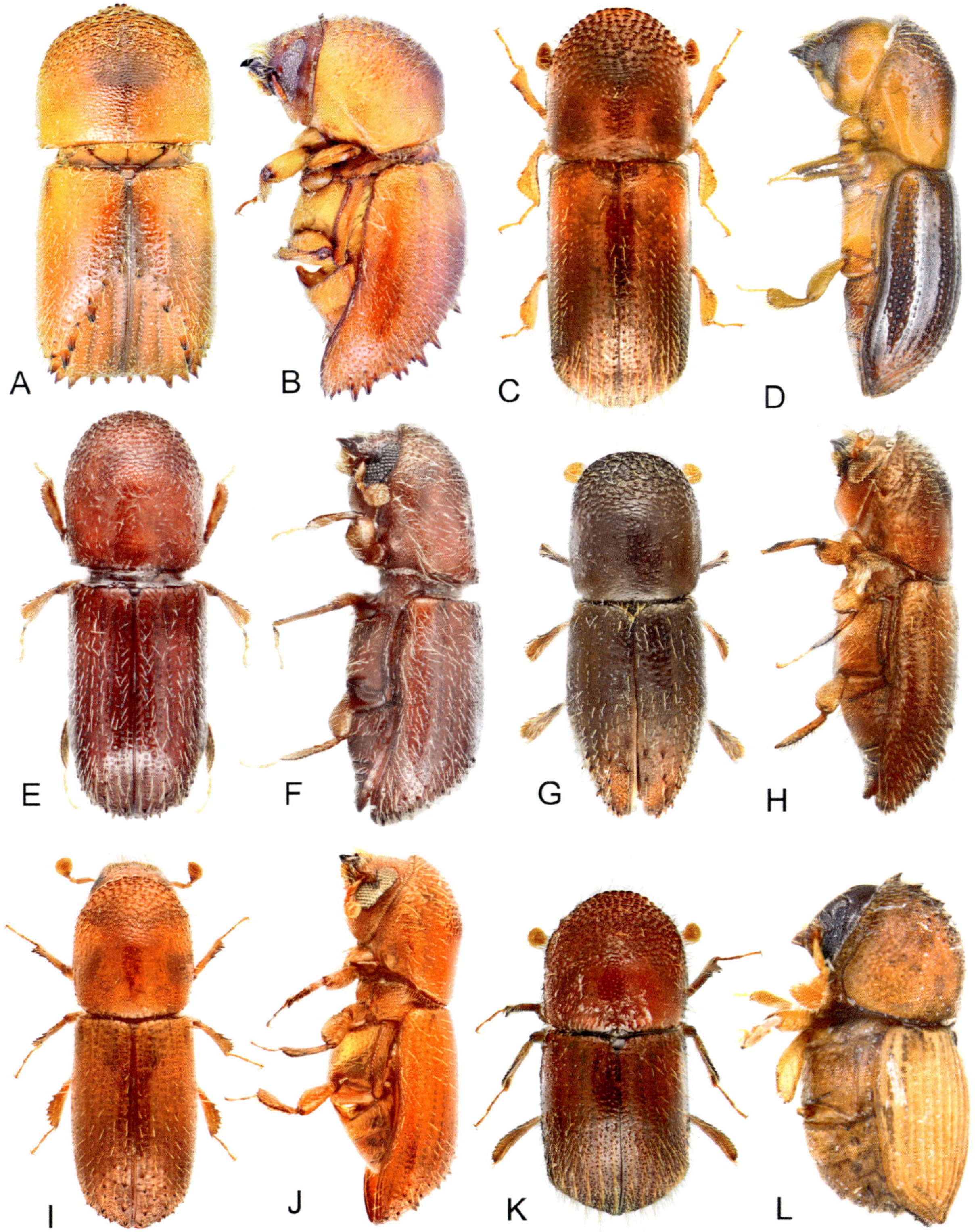

Plate 115. Curculionidae (Scolytinae). A, B, *Immanus acanthurus* (Lea), dorsal and lateral, ♀, 7.5 mm (Xyleborini); C, *Microperus eucalypticus* (Schedl), ♀, 1.9 mm (Xyleborini); D, *Planiculus* sp., Iron Range, QLD, ♀,1.9 mm (Xyleborini); E, F, *Truncaudum agnatum* (Eggers), dorsal and lateral, ♀, 2.3 mm (Xyleborini); G, H, *Xyleborinus andrewesi* (Blandford), dorsal and lateral, ♀, 1.8 mm (Xyleborini); I, J, *Xyleborus perforans* (Wollaston), ♀, 2.5 mm (Xyleborini); K, *Xylosandrus crassiusculus* (Motschulsky), ♀, 2.5 mm (Xyleborini); L, *Scolytomimus pusillus* (Eggers), 1.3 mm (Xyloctonini).

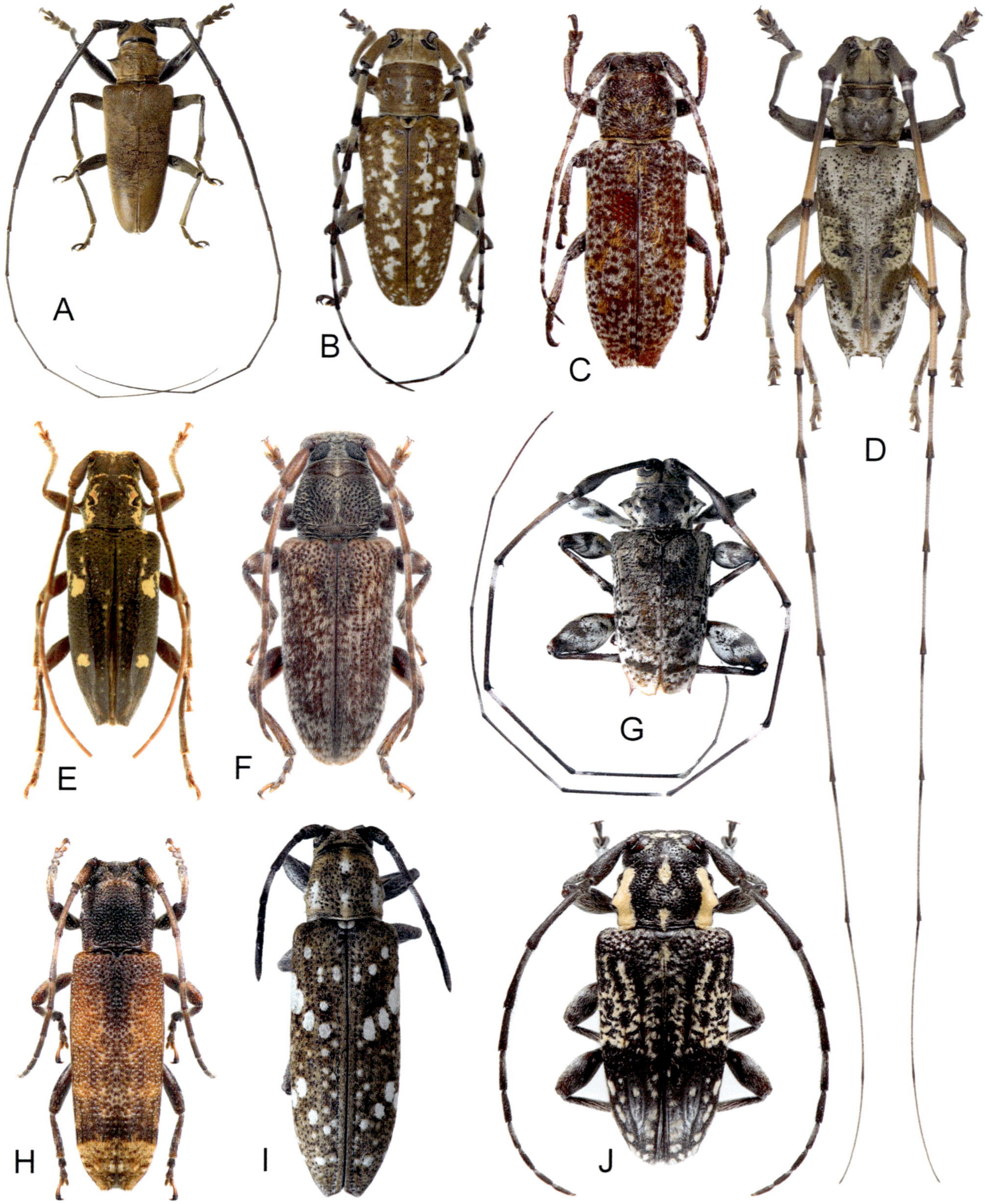

Plate 116. Cerambycidae (Lamiinae). A, *Acalolepta aesthetica* (Olliff); B, *A. ovina* (Pascoe); C, *Rhytiphora subovata* Ashman & Ślipiński; D, *Acalolepta* sp.; E, *Amblymora australica* Breuning; F, *Amblymoropsis fusca* Breuning; G, *Acanista alphoides* Pascoe; H, *Atimura terminata* Pascoe; I, *Apomecyna histrio* (Fabricius); J, *Ancita crocogaster* (Boisduval). After Ślipiński & Escalona (2013).

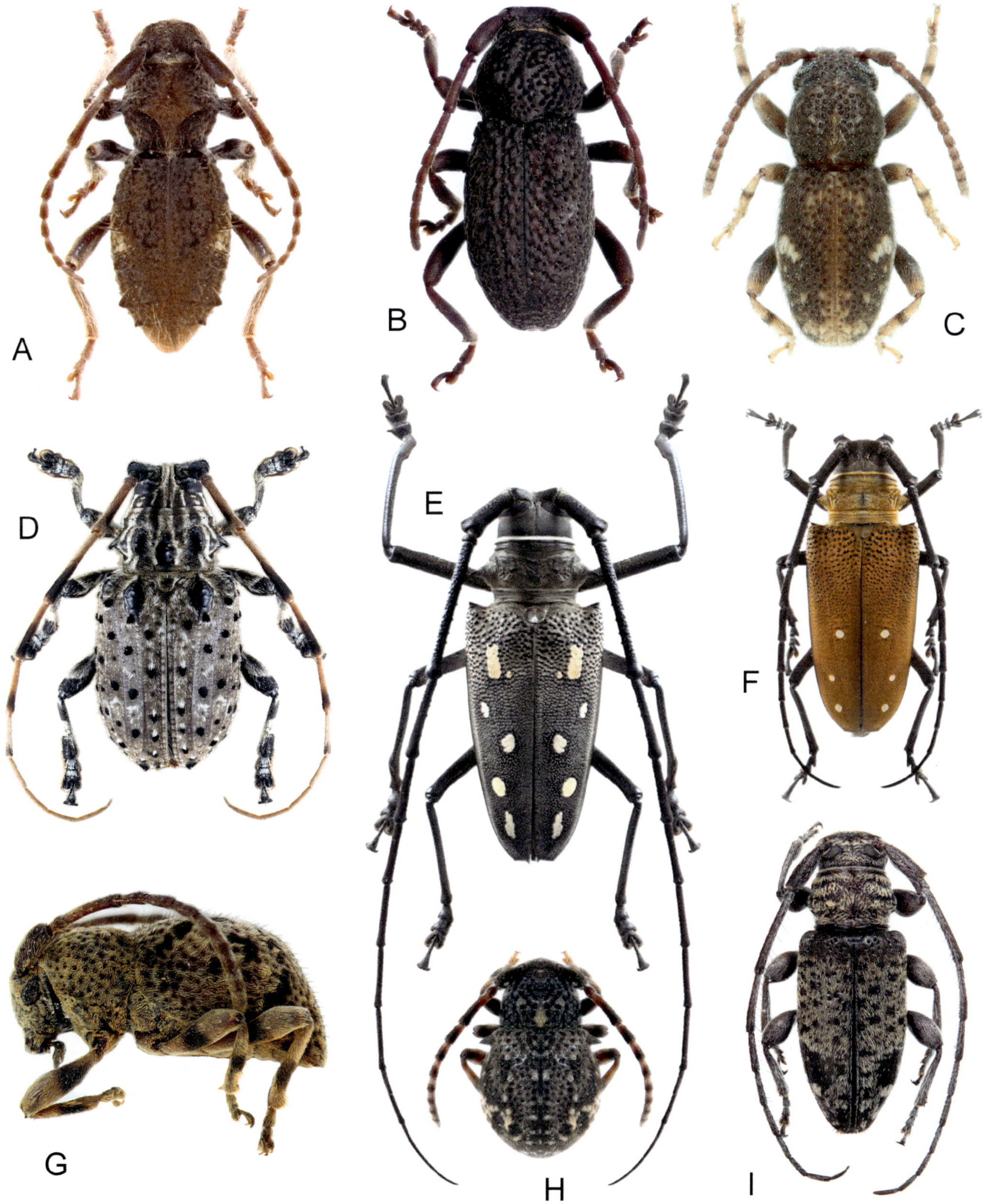

Plate 117. Cerambycidae (Lamiinae). A, B, *Athemistus* spp.; C, G, H, *Austrosomatidia* spp.; D, *Callipyrga turrita* Newman; E, *Batocera frenchi* Van de Pol; F, *Batocera laena* Thomson; I, *Bucynthia spiloptera* (Pascoe). After Ślipiński & Escalona (2013).

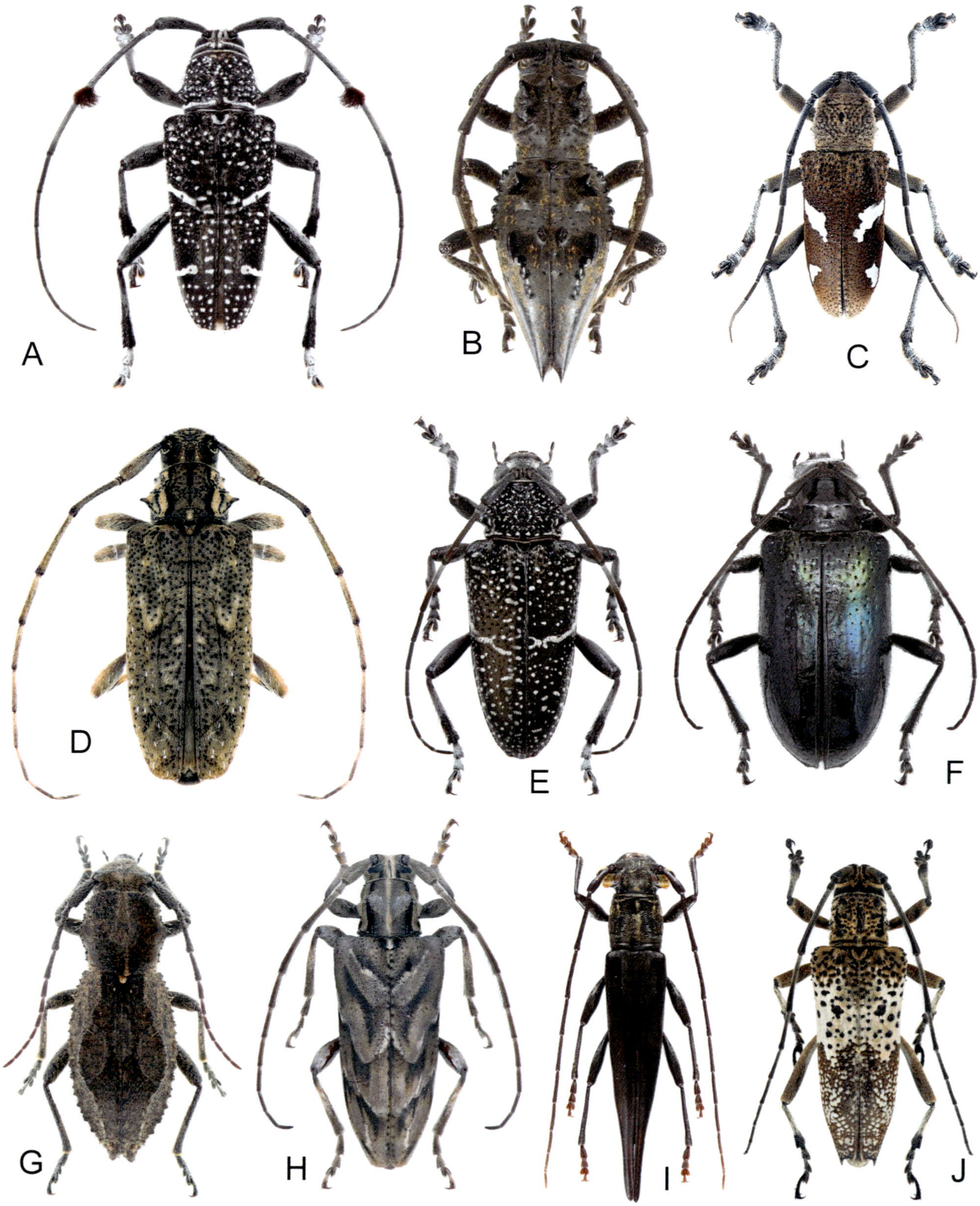

Plate 118. Cerambycidae (Lamiinae). A, *Zygocera bivittata* (Breuning); B, *Ceraegidion horrens* Boisduval; C, *Cyocyphax praonethoides* Thomson; D, *Didymocentrotus foveatus* (Aurivillius); E, *Zygocera pruinosa* (Boisduval); F, *Zygocera metallica* (Westwood); G, *Dorcadida bilocularis* White; H, *Dysthaeta anomala* Pascoe; I, *Enicodes fichteli* (Schreibers); J, *Demonassa capitalis* (Blackburn). After Ślipiński & Escalona (2013).

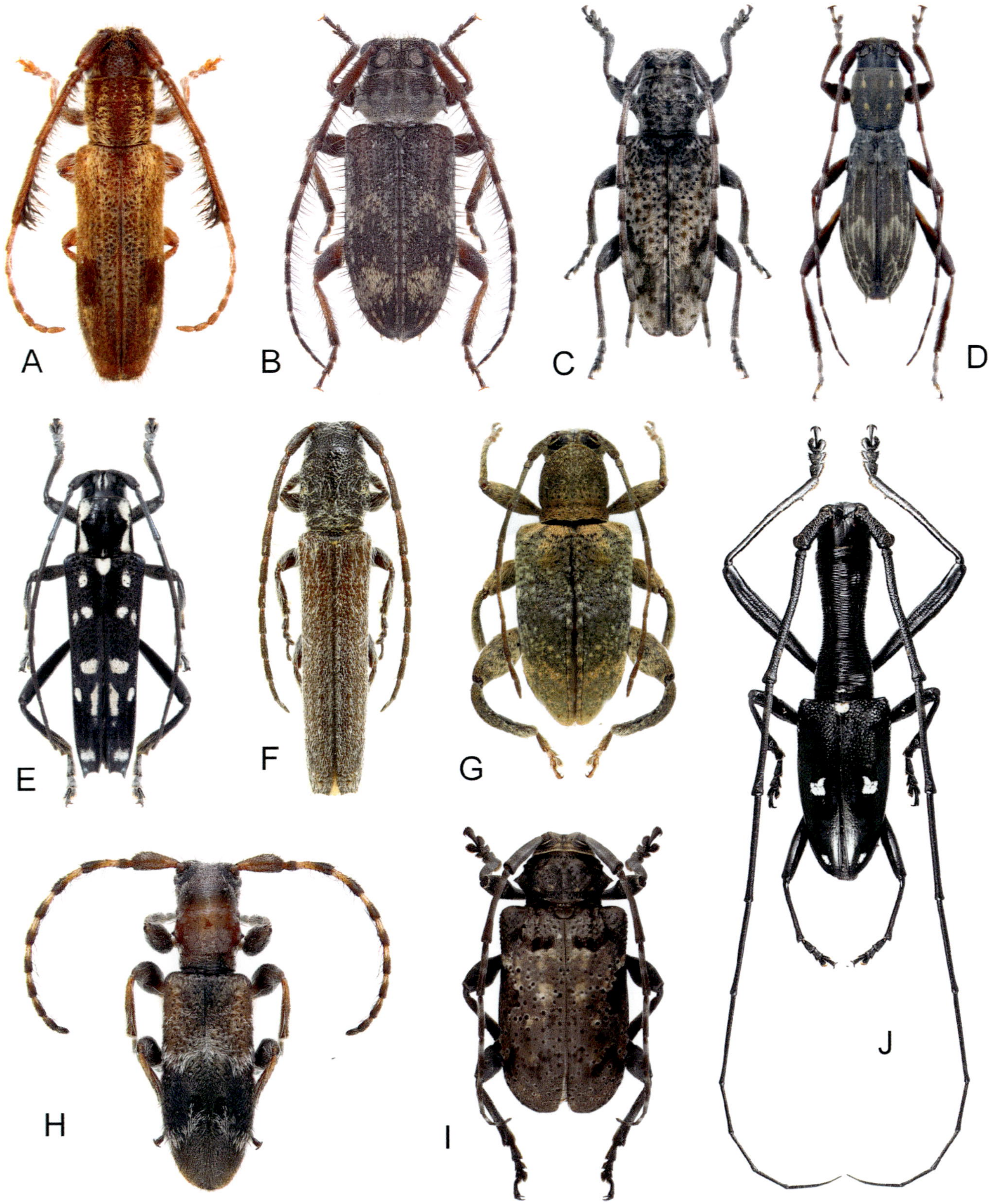

Plate 119. Cerambycidae (Lamiinae). A, *Essisus dispar* Pascoe; B, *Exocentrus* sp.; C, *Carteridion wilsonensis* Ślipiński & Escalona; D, *Mesolita lineolata* Pascoe; E, *Glenea spinifera* (Voet); F, *Itheum alboscutellare* Breuning; G, *Alice dekeyzeri* Ślipiński & Escalona; H, *Illaena* sp.; I, *Lagocheirus funestus* Thomson; J, *Gnoma australis* Schwarzer. After Ślipiński & Escalona (2013).

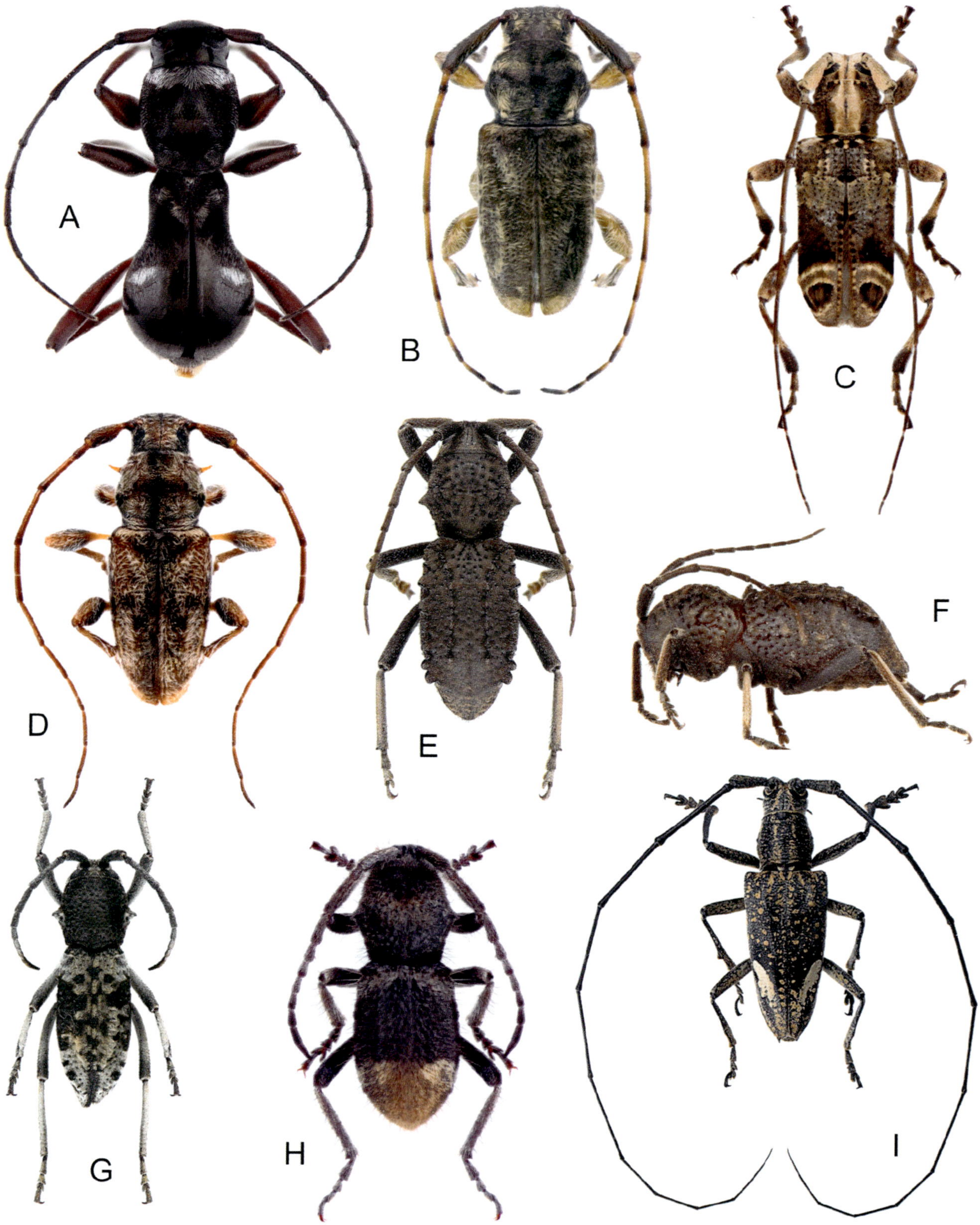

Plate 120. Cerambycidae (Lamiinae). A, *Mesolita myrmecophila* Lea; B, *Microlamia viridis* Ślipiński & Escalona; C, *Meton digglesii* Pascoe; D, *Microlamia* sp.; E, F, *Microtragus* sp.; G, *Microtragus arachne* Pascoe; H, *Microtragus bifasciatus* Lea; I, *Pelargoderus rubropunctatus* (Guérin-Méneville). After Ślipiński & Escalona (2013).

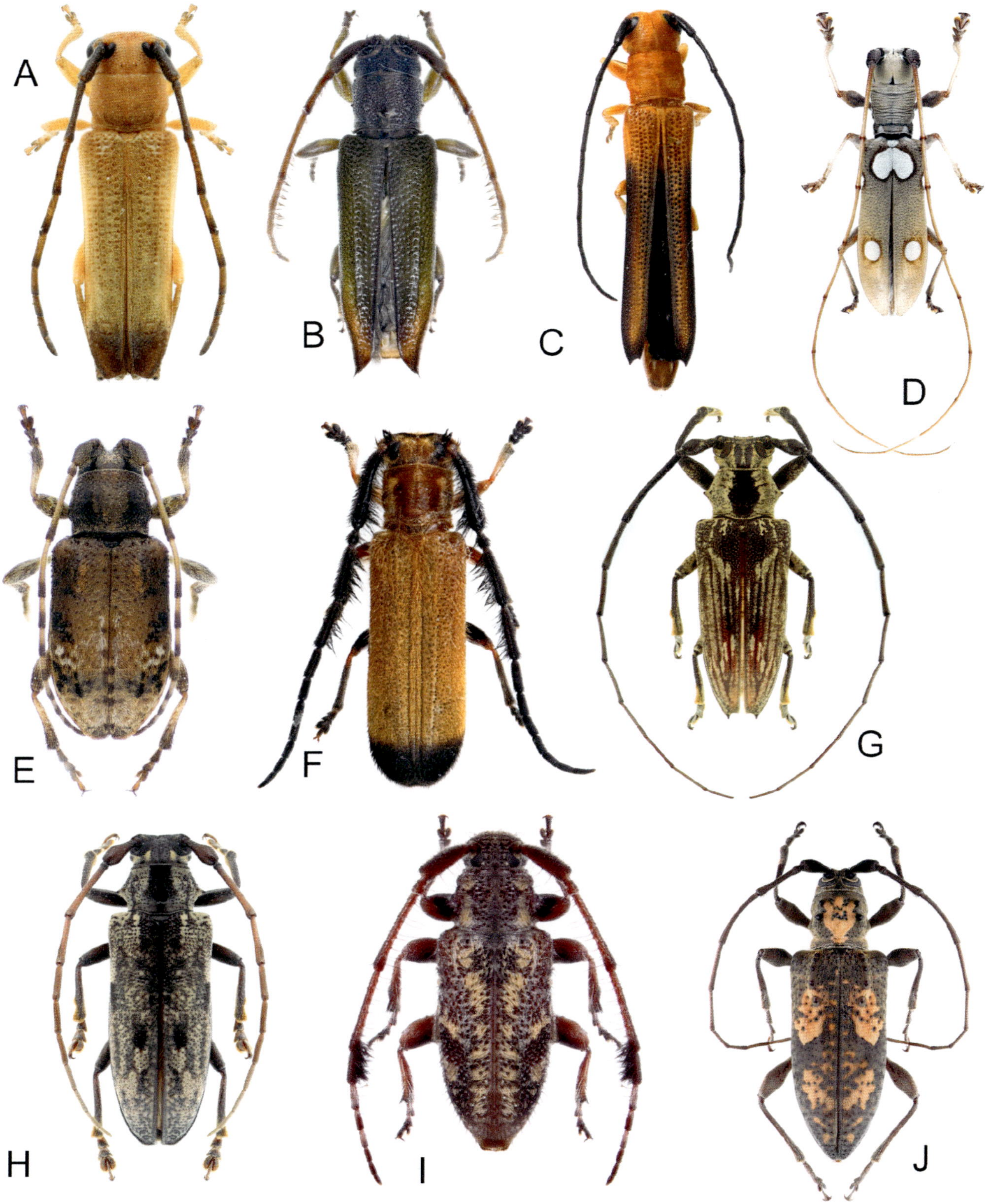

Plate 121. Cerambycidae (Lamiinae). A, *Nupserha vexator* Pascoe; B, *Paressisus viridipennis* Aurivillius; C, *Oberea nefasta* Pascoe; D, *Olenecamptus bilobus* (Fabricius); E, *Oricopis* sp.; F, *Momisis melanura* Gahan, female; G, *Mulciber maculosus* Breuning; H, *Notomulciber* sp.; I, *Pentacosmia cacioides* (Breuning); J, *Oricopis maculiventris* Lea. After Ślipiński & Escalona (2013).

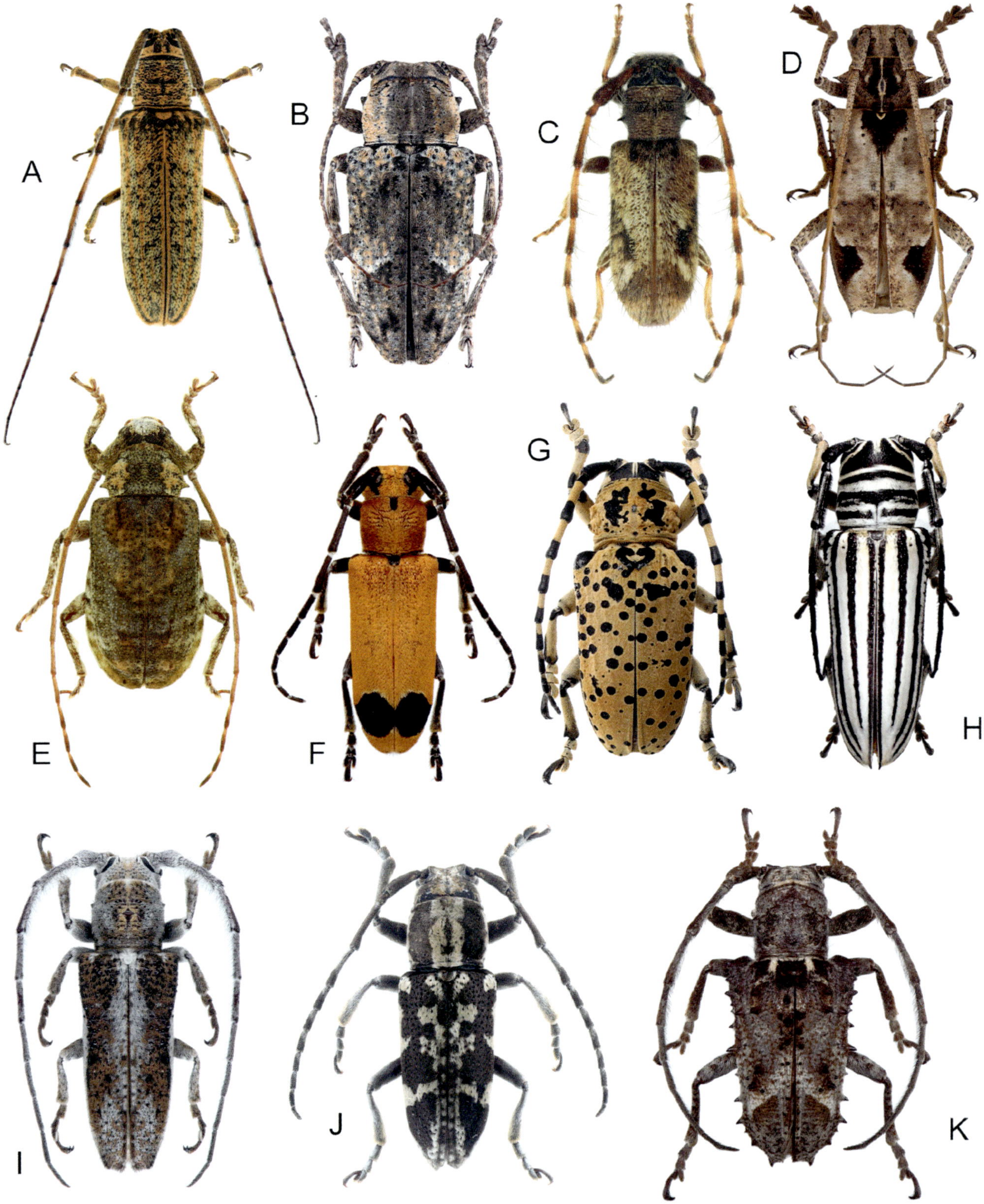

Plate 122. Cerambycidae (Lamiinae). A, *Phelipara subvittata* Blair; B, *Rhytiphora bakewelli* (Pascoe); C, *Neissa* sp.; D, *Potemnemus scabrosus* (Olivier); E, *Phyxium* sp.; F, *Rhytiphora diva* (Thomson); G, *Rhytiphora pardalis* (Newman); H, *Rhytiphora dallasii* Pascoe; I, *Rhytiphora lateralis* (Pascoe); J, *Pterolophia luctuosa* (Pascoe); K, *Rhytiphora obliqua* (Donovan). After Ślipiński & Escalona (2013).

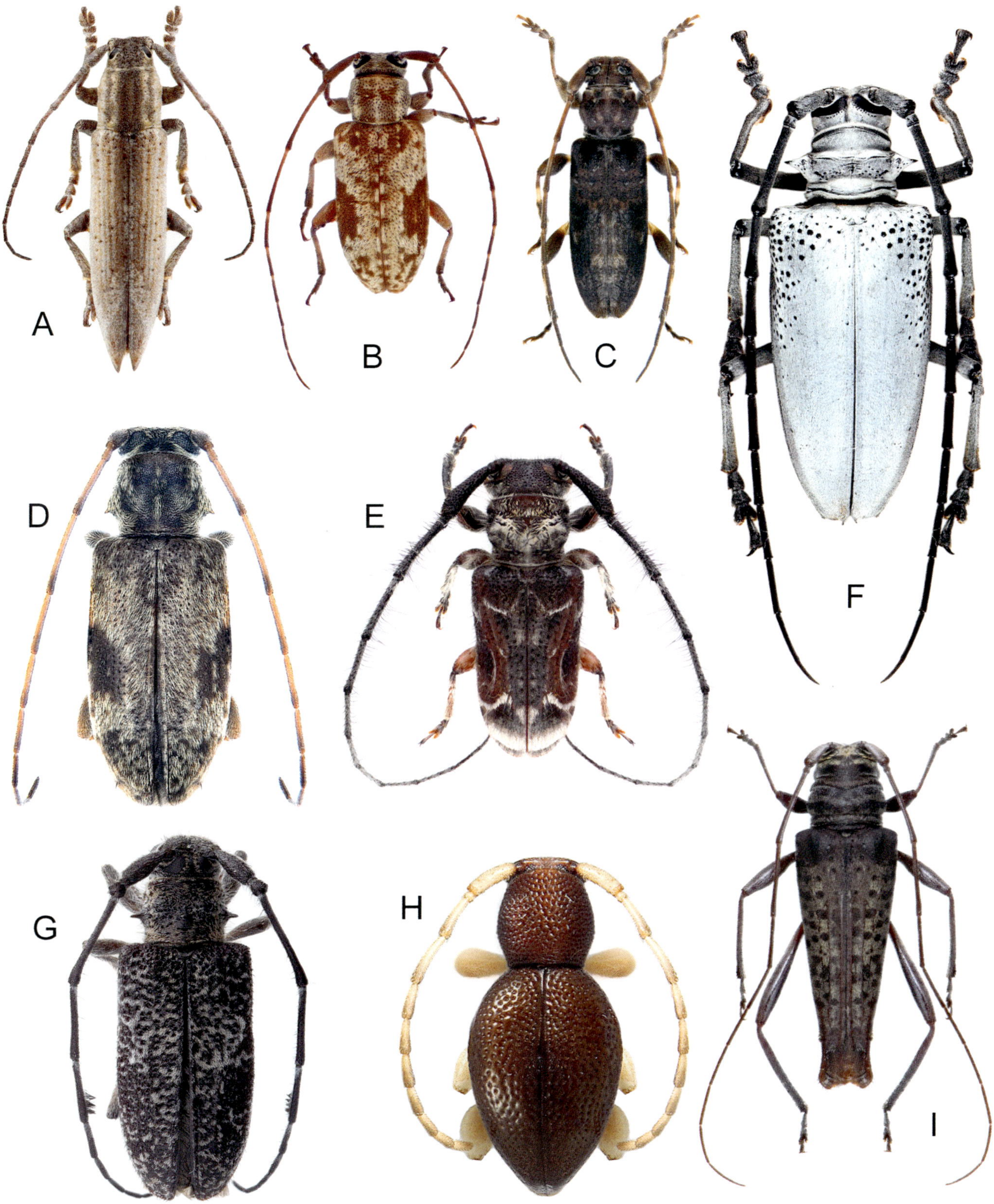

Plate 123. Cerambycidae (Lamiinae). A, *Rhytiphora lobata* (Breuning); B, *Mimostenellipsis* sp.; C, *Stenellipsis* sp.; D, *Sciades exiguua* (Gahan); E, *Similosodus venosus* (Pascoe); F, *Rosenbergia candidipennis* Heyne & Taschenberg; G, *Probatodes plumula* (Newman); H, *Somatidia aranea* Olliff; I, *Annemabel elongata* (Breuning). After Ślipiński & Escalona (2013).

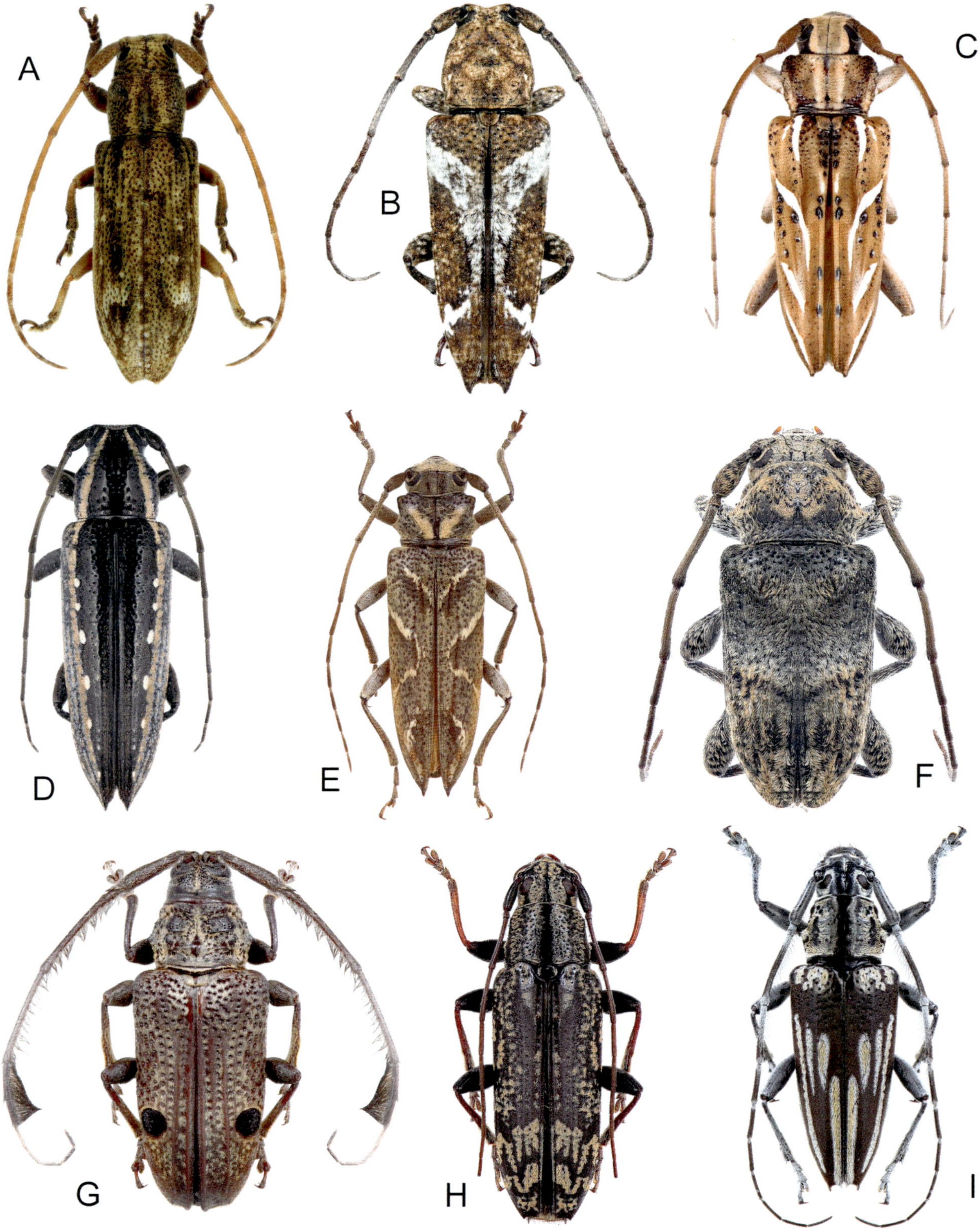

Plate 124. Cerambycidae (Lamiinae). A, *Ropica elongatula* Breuning; B, *Sybra mastersi* Blackburn; C, *Temnosternus niveoscriptus* McKeown; D, *Sybra truncata* (Aurivillius); E, *Temnosternus* sp.; F, *Velora sordida* (Pascoe); G, *Thyada barbicornis* (Pascoe); H, *Tmesisternus monteithi* Gressitt; I, *Trigonoptera vittata* (Gestro). After Ślipiński & Escalona (2013).

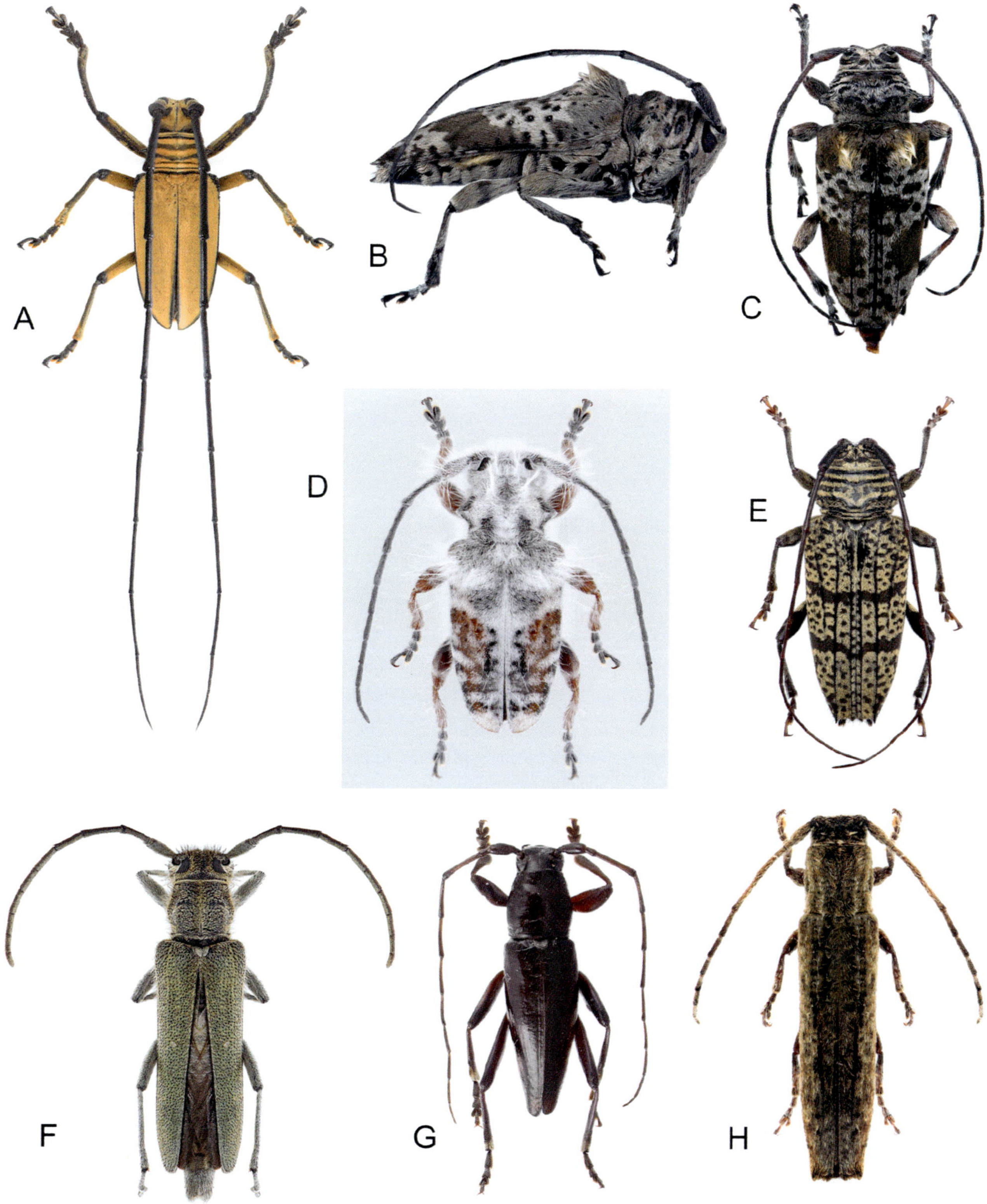

Plate 125. Cerambycidae (Lamiinae). A, *Nemophas tomentosus* (Buquet); B, C, *Disterna plumifera* (Pascoe); D, *Excastra albopilosa* Tweed, Ashman & Ślipiński; E, *Disterna bifasciata* (Pascoe); F, *Phytoecia coerulescens* (Scopoli); G, *Xylotoles selwyni* Olliff; H, *Zorilispe albosetosa* Breuning. After Ślipiński & Escalona (2013), except A, F and D.

46. CURCULIONIDAE: SCOLYTINAE LATREILLE, 1806

James R. M. Bickerstaff, Sarah M. Smith, Anthony I. Cognato, Debbie Jennings, Rolf G. Oberprieler and Hermes E. Escalona

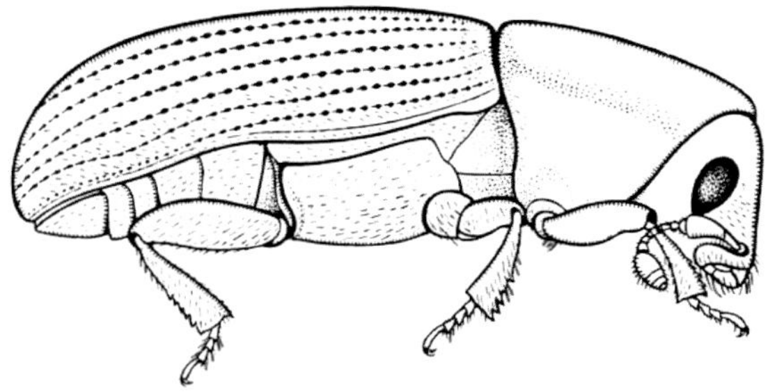

Fig. 46.1. *Hylastes ater* (Paykull).

Introduction. The subfamily Scolytinae comprises more than 6000 species in about 280 extant genera, which are grouped into 26 tribes (Oberprieler *et al.* 2007; Hulcr *et al.* 2015; Pistone *et al.* 2018). The species are commonly found in forested ecosystems on most continents, with most of the diversity occurring in the tropical regions of the world. Scolytinae are commonly referred to as bark and ambrosia beetles, which is reflective of their life styles, but the name 'ambrosia beetles' encompasses several lineages of Scolytinae and Platypodinae associated with symbiotic fungi, which are critical for their development and from which all life stages obtain nutrition (Hulcr & Stelinski 2017). Scolytinae not associated with ambrosia fungi are generally referred to as bark beetles, although some of these are also associated with fungi but do not depend entirely on them for their development (Beaver 1989; Hulcr & Stelinski 2017).

The taxonomic position of Scolytinae in Curculionoidea has been highly contentious historically. In the past the subfamily was usually regarded as closely related to Platypodinae and classified either inside and outside of Curculionidae (Wood 1984; Oberprieler *et al.* 2007; Jordal 2014; Jordal *et al.* 2014), but recent phylogenomic analyses have shown that it falls inside the CCCMS clade of 'higher' Curculionidae (Gunter *et al.* 2016), whereas Platypodinae form a more basal lineage of Curculionidae related to Dryophthorinae (e.g. Mugu *et al.* 2018; Shin *et al.* 2018; Smit *et al.* 2022). The monophyly of Scolytinae is also not yet resolved, as the tribe Scolytini generally does not cluster together with other scolytines in phylogenetic analyses spanning a wider range of CCCMS taxa (Gillett *et al.* 2014; Mugu *et al.* 2018; Pistone *et al.* 2018; Shin *et al.* 2018), although some species of *Scolytus* do so (Smit *et al.* 2022).

The classification of Scolytinae into natural tribes and genera is also still unsettled. More than half of all scolytine species have been described by three authors, Hans Eggers, Karl E. Schedl and Stephen L. Wood (Hulcr *et al.* 2015). Schedl alone described 1607 species, including 203 Australian ones (Pullen *et al.* 2014). Despite these impressive numbers, many of their names have since sunk into synonymy, due to over splitting species based on minor morphological variations (Wood & Bright 1992) or not verifying species described by other taxonomists (Hulcr *et al.* 2015). Much of the synthesis of Scolytinae into a modern classification system is a result of the work by Wood and co-workers. The world catalogue of Scolytinae by Wood & Bright (1992) and its supplements (Bright & Skidmore 1997, 2002; Bright 2014, 2021), together with Wood's (1984) key to the subfamilies and tribes of 'Scolytidae', constitute a major and admirable attempt of bringing some degree of order to the numerous genera of the subfamily and are still regarded as the cornerstone of the current knowledge of its taxonomy. However, modern phylogenetic analyses have revealed many dissents with Wood's classification. For instance, a study by Pistone *et al.* (2018), analysing more than 10 kb of nucleotide data from 182 species of Scolytinae representing 24 tribes, showed that several of the tribes, specifically Hylurgini, Hylastini and Hylesinini, are not ostensibly monophyletic and that many other nodes of relationships remain unresolved, thus indicating that much further work is needed before a natural division and classification of the subfamily can be attained.

The evolutionary age of Scolytinae is also debated, with molecular phylogenetic dating appearing to conflict with the fossil record, specifically with the age of the two oldest described scolytine fossils, *Cylindrobrotus pectinatus* Kirejtshuk *et al.* (Kirejtshuk *et al.* 2009) and *Microborus inertus* Cognato & Grimaldi (Cognato & Grimaldi 2009), both found in Cretaceous amber dated as about 100 Ma in age and predating the minimum ages for the subfamily derived from phylogenetic analyses. However, the single specimen of *C. pectinatus* displays several characters conflicting with an assignment to Scolytinae and even to Curculionoidea (Gunter *et al.* 2016) and its identity remains to be vetted and confirmed, and the origin and age of the *M. inertus* fossil are in doubt as its amber does not fluoresce under UV light as Burmese amber does (Clarke *et al.* 2018). As a result, neither of these fossils can be used to calibrate any lineage of Scolytinae, and dating estimates that have done so cannot be relied for the reconstruction of the evolutionary history of the subfamily. More credible age estimates for Scolytinae range from 86 Ma and 82 Ma (McKenna *et al.* 2009; Gunter *et al.* 2016) to 70 Ma and even 60 Ma (Shin *et al.* 2018; McKenna *et al.* 2019).

Biology. Scolytinae typically live and breed in trunks of dead or dying trees, but some develop in woody herbs and ferns as well as in twigs, petioles and seedlings and even fruits and seeds (Jordal 2014; Hulcr & Stelinski 2017). Most bark beetles are phloeophagous, obtaining all their nutrition from phloem tissue, in which they typically construct a single vertical tunnel that is extended into perpendicular branches by the larvae (Kirkendall *et al.* 2015; Raffa *et al.* 2015). However, the larvae of several species develop in other plant parts, such as xylem tissues (xylophagy), seeds, cones and fruits (spermatophagy), e.g. the Coffee Berry Borer, *Hypothenemus hampei* (Ferrari), and even in free-living fungi (mycetophagy) (Beaver 1989; Kirkendall *et al.* 2015). A number of species attack live, healthy trees (Beaver 1989; Hulcr & Stelinski 2017) and are then usually regarded as economically important pests able to decimate entire plantations and forests, such as *Dendroctonus ponderosae* (Hopkins) and *Ips typographus* (Linnaeus) (Cognato 2015; Six & Bracewell 2015). Several instances of associations between bark beetles and fungi have been documented, e.g. the European species *Tomicus minor* (Hartig), whose larvae initially feed on phloem but later migrate to the xylem of their host tree to feed on inoculated symbiotic fungi (Kirkendall *et al.* 2015). Such species are xylomycetophagous but not true ambrosia beetles, as not all their life stages are obligately dependent on cultivated fungi.

True ambrosia beetles are also xylomycetophagous and require nutrients from symbiotic fungi in all their life stages (Jordal 2014; Hulcr & Stelinski 2017). In this symbiosis, the beetles benefit by having a constant source of food and protection against host tree defences, and the fungi benefit by being propagated and dispersed to new trees by the beetles (Beaver 1989; Kirkendall *et al.* 2015). Symbiotic relationships of scolytine ambrosia beetles exist with a diverse range of fungus families. The prevalent fungal symbionts belong to the order Ophiostomatales of Ascomycota, typically to the genera *Harringtonia*, *Leptographium, Ophiostoma* and *Raffaelea* (Hulcr & Stelinski 2017; De Beer *et al.* 2022; Mayers *et al.* 2022), but species of Ceratocystidaceae (order Microascales) are also common symbionts of these beetles, with *Ambrosiella* a common and well-studied symbiont of Xyleborini (Hulcr & Stelinski 2017; Mayers *et al.* 2022; Bickerstaff *et al.* 2023). Symbioses also exist with less common fungi, such as between *Euwallacea* and *Fusarium* (Ascomycota: Hypocreales), a well-studied genus of phytopathogens. A few species of ambrosia beetles are associated with fungi inhabiting decaying wood, which can be detrimental to the longevity of gallery systems. To date, three species of *Ambrosiodmus* (Xyleborini) are known to be associated with *Flavodon ambrosius*, a white-rot basidiomycote fungus of decaying timber (You *et al.* 2015; Kasson *et al.* 2016; Simmons *et al.* 2016). Other genera of Scolytinae (e.g. *Ambrosiophilus*, *Diuncus* and *Xylosandrus*) have independently evolved mycetoclepsia (fungus-stealing; also incorrectly termed 'mycocleptism') (Hulcr & Cognato 2010; Kirkendall *et al.* 2015; Skelton *et al.* 2018). As recorded for *Diuncus*, species construct their galleries in close proximity to those of fungus-cultivating species and feed on the fungal hyphae penetrating through the xylem (Hulcr & Cognato 2010).

Scolytinae can transport fungal symbionts in a variety of ways, for example *Xyleborinus saxesenii* (Ratzeburg) can carry fungal spores in their gut (Beaver 1989), but their most common way to carry fungal spores is in pocket-like structures broadly known as mycangia (Francke-Grosmann 1956; Giese 1967; Mayers *et al.* 2022), although the correct term is mycetangia (Vega & Biedermann 2020). The mycangia are highly variable in their structure and location on the body, but typically they are tubes, pouches or cuticular invaginations of various sizes, often surrounded by glandular cells to reduce desiccation of spores or fungal propagules, and located on or under the pronotum, near the scutellum, on the mesonotum, in the head capsule, in the mandibles or in the pre-oral cavity (Francke-Grosmann 1956; Batra 1963; Hulcr & Stelinski 2017; Mayers *et al.* 2022). The structure of the mycangia is typically genus-specific, with some variation between species (Mayers *et al.* 2020, 2022). There can also be variation in their structure and even their presence between the sexes, as in being present in females only in *Dactylipalpus* (Hylesinini) and Xyleborini. They can also vary in the diversity of symbiont fungi they house, e.g. the mandibular mycangia of *Xyleborus* have been found to contain more fungal spores of various species than the pronotal mycangia of *Xylosandrus* (Kostovcik *et al.* 2015).

Scolytines display a large variety of social behaviours, many of which have apparently evolved repeatedly. All species are considered to be social to some degree, with the beetles engaging in gregarious feeding, parental investment and alloparental care (Kirkendall *et al.* 2015). Parental investment in offspring commences with adults excavating galleries in host plant tissues for reproduction. Following oviposition, either one or both parents remain in the gallery to protect their offspring, maintain gallery hygiene and, in the case of ambrosia beetles, tend to the fungal symbionts (Kirkendall *et al.* 2015). Alloparental care, in the form of delayed dispersal of offspring, is also common in Scolytinae. Adults delaying their dispersal typically display behaviours they would otherwise only undertake if they were in their own galleries, such as frass removal, gallery defence and, in the case of ambrosia beetles, fungus cultivation (Biedermann *et al.* 2011; Kirkendall *et al.* 2015; Biedermann 2020). Apparently, facultative eusociality has also evolved in several species of xyleborine ambrosia beetles, where females practice reproductive altruism without suffering sterility (Kirkendall *et al.* 2015).

Scolytinae also have a large variety of mating and reproductive systems, which have also evolved multiple times independently in different lineages. The most common mating system is female-initiated monogamy. Females begin the construction of a gallery in a suitable host plant, to which the males are then attracted either by host plant volatiles or pheromones. Copulation typically follows shortly after the males arrive in the gallery, and males typically remain in the gallery after oviposition and contribute to gallery hygiene or defence (Kirkendall *et al.* 2015). Bigyny, characterised by males mating with two females, has also evolved multiple times, having been recorded in 19 genera across eight tribes (Kirkendall *et al.* 2015). Similar to bigyny is harem polygyny, in which multiple females join one male, either unmated or already mated, and mate with him, approximately 26 genera in eight tribes display this behaviour (Kirkendall *et al.* 2015). An extension of harem polygyny is colonial polygyny, in which multiple males and multiple females breed in the same gallery system. This system has been described in *Crypturgus* and *Cyrtogenius* (Kirkendall *et al.* 2015), with a colony of *Cyrtogenius brevior* (Eggers) in one tree branch consisting of 12 galleries containing 20 males and 32 females (Beaver 1976). Inbreeding polygyny, defined as mating between siblings of

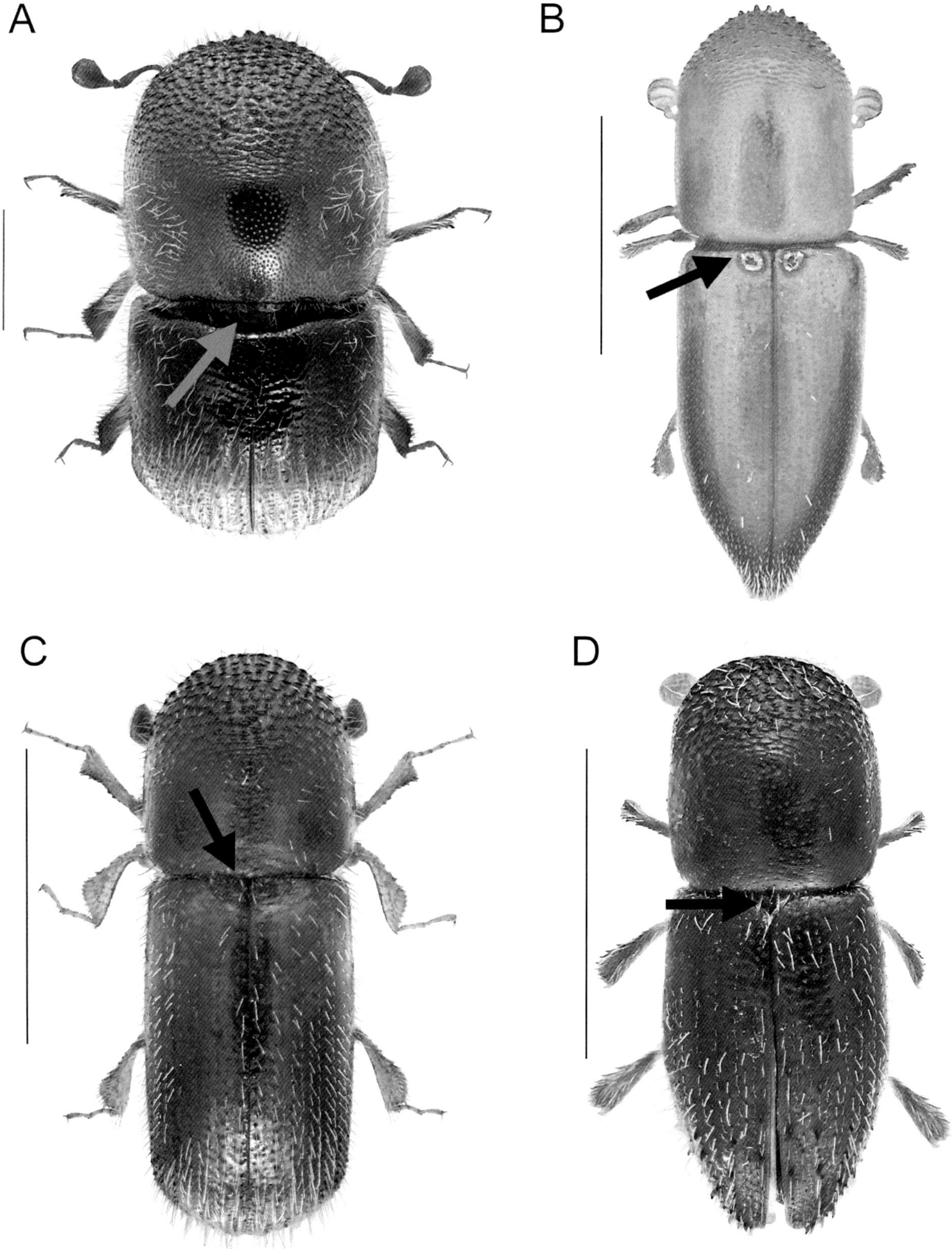

Fig. 46.2. Mycangia in some species of Xyleborini ambrosia beetles. A, *Cnestus solidus* (Eichhoff); B, *Cryptoxyleborus subnaevus* Schedl; C, *Microperus eucalypticus* (Schedl); D, *Xyleborinus andrewsei* (Blandford). Scale bar 1 mm.

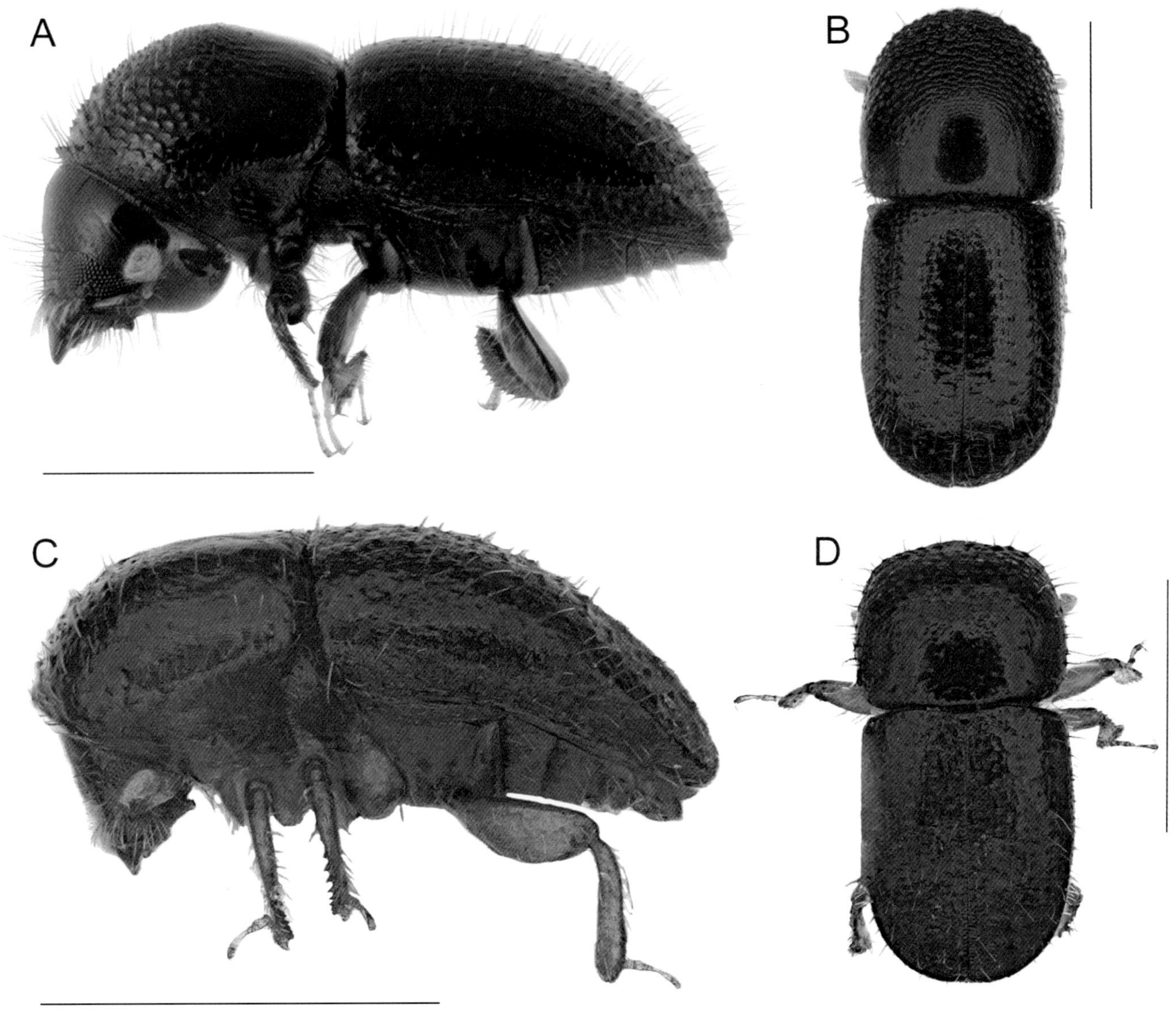

Fig. 46.3. Structural variation between the female (A–B) and male (C–D) of the haplodiploid *Euwallacea fornicatus* (Eichhoff) (Xyleborini). Scale bar 1 mm.

a brood and commonly also referred to as sibling mating or permanent inbreeding (Kirkendall *et al.* 2015; Johnson *et al.* 2018), has evolved multiple times and occurs in approximately 27 % of all species, with about 97 % of all inbreeding species belonging to the tribes Dryocoetini and Xyleborini. Inbreeding also occurs in Bothrosternini, Corthylini, Hyorrhynchini, Ipini, Trypophloeini and Xyloterini (Kirkendall *et al.* 2015). Partial inbreeding has been described in at least two species of *Dendroctonus*, with genetic data suggesting that populations have reduced levels of heterozygosity (Six & Bracewell 2015; Kegley *et al.* 1997).

Several modes of parthenogenesis have been described for Scolytinae. Thelytoky, clonal production of females without reproduction, has been identified in *Pityophthorus puberulus* (LeConte) (Smith & Hulcr 2015). Pseudogamy, parthenogenetic reproduction of clonal offspring that requires fertilisation for oocyte development, has been recorded in *Ips* species (Lanier & Kirkendall 1986). Arrhenotoky, sexual production of diploid females and asexual reproduction of haploid males, is predominant in the haplodiploid Xyleborini and Dryocoetini (Kirkendall *et al.* 2015), and pseudo-arrhenotoky, fertilisation of the egg followed by paternal genome elimination resulting in offspring produced by meiosis without passing on the paternal genotype, has been identified only in *Hypothenemus hampei* (Vega *et al.* 2015).

Australian fauna. The Australian Scolytinae fauna as currently known comprises 156 species in 50 genera and 16 tribes

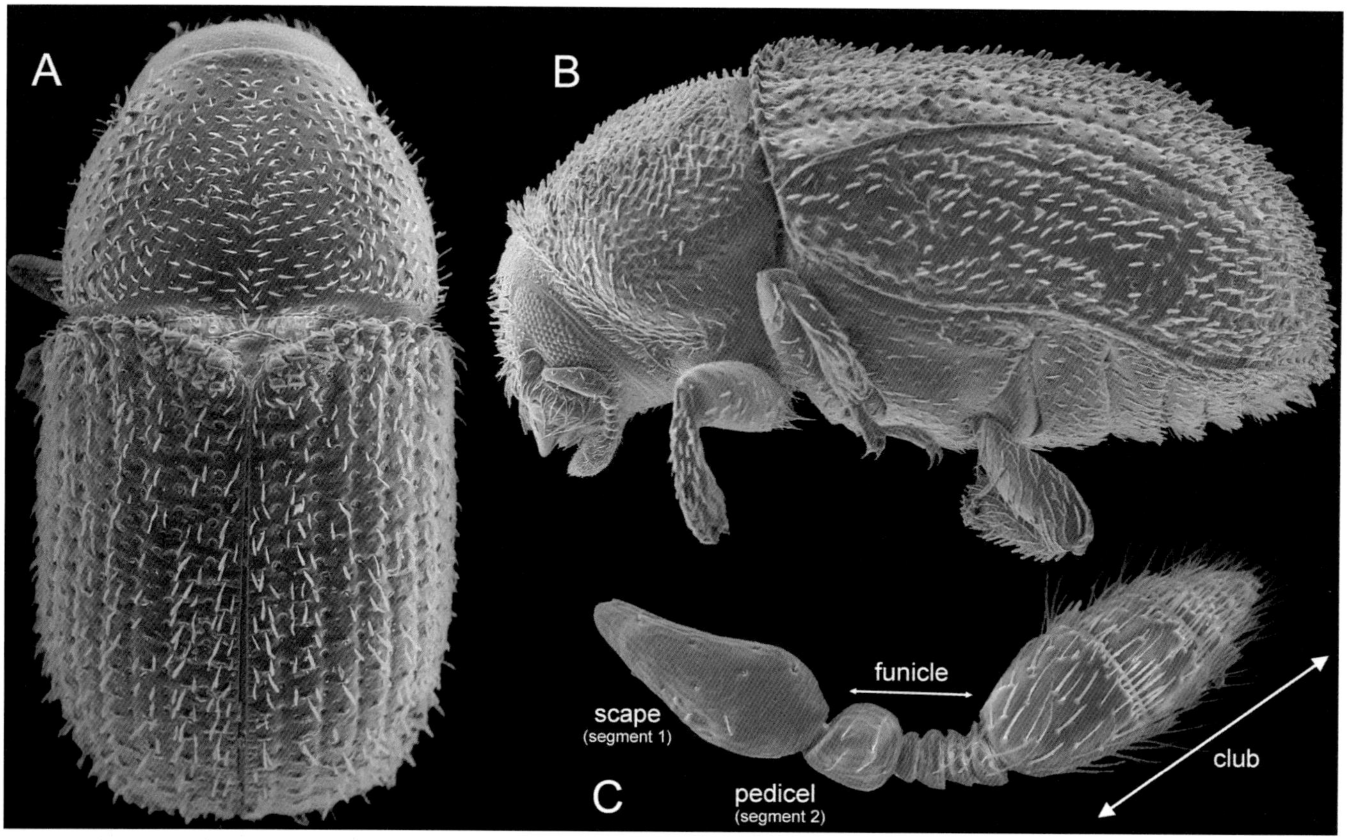

Fig. 46.4. *Phloeotribus* sp. (Phloeotribini). A–B, habitus: A, dorsal; B, lateral; C, antenna.

(Pl. 17A–B, Table 46.1), but its taxonomy has not been addressed adequately. The number of species per genus ranges from 1 to 19, with an average of three species per genus. The largest tribe in Australia are the Xyleborini, represented by 59 species and 19 genera and by approximately 1300 species in 43 genera globally (Knížek & Smith 2024). The second-largest tribe in Australia are the Cryphalini, with 19 species of *Cryphalus*, of which 17 are native or endemic. This large genus is globally distributed and contains approximately 252 species (Johnson *et al.* 2020b).

The scolytinae fauna in Australia occurs predominantly in mesic forests along the eastern coast, only few species having been described from Western Australia (e.g. *Amasa banksiae* Schedl) and the Northern Territory (e.g. *Amasa darwini* Schedl). However, collecting in the Kimberley rainforests of northern Western Australia (Naumann *et al.* 1991) and jarrah and karri forests of southern Western Australia (Bickerstaff & Escalona, personal observations) has yielded several other scolytinae taxa as well, and further targeted sampling in these areas may discover additional taxa. Scolytinae also occur on islands around Australia. Extensive sampling on Christmas Island yielded a number of species of *Coccotrypes*, *Cryphalus*, *Hypothenemus* and Xyleborini (in ANIC), several species of inbreeding polygynous *Coccotrypes* and *Xyleborus* are known from Lord Howe Island (in ANIC) and some endemic species occur on Norfolk Island, including *Pachycotes kuscheli* Schedl and *Xyleborus norfolkensis* Schedl.

From an ecological perspective, several broad life history traits are represented among the Scolytinae in Australia. Some lineages of fungus-farming ambrosia beetles are present, such as those of *Dactylipalpus*, *Sueus* and the Xyleborini, whereas all other lineages are predominantly those of phloeophagous bark beetles. Similarly, several inbreeding lineages are represented in Australia, such as those of *Coccotrypes*, *Hypothenemus* and *Sueus* and the Xyleborini, but most lineages have the monogynous mating system. Several genera are known to have definite host plant preferences but with varying levels of specificity. At a genus level according to Wood & Bright (1992), *Araucaria* species are common obligate hosts of *Hyleops*, *Hylurdrectonus pinarius* Schedl and *Pachycotes*, *Acacia* obligate hosts of *Acacicis* and *Zygophloeus* and *Ficus* trees are specific hosts of *Acanthotomicus* and *Aricerus*. Some introduced species of bark beetles are also host-specific, such as species of *Hylastes*, *Hylurgus* and *Ips* regularly attacking *Pinus* trees in plantations (Neumann 1987; Yousuf *et al.* 2014) and *Scolytus multistriatus* (Marsham) ornamental *Ulmus* trees (Neumann & Minko 1985).

Ecological, economic and invasive species impacts in Australia. Scolytines are considered among the most ecologically and economically important forest insect pests that contribute to tree mortality, either through host tree infestation or the spread of symbiotic phytopathogenic fungi (Grégoire *et al.* 2015; Biedermann *et al.* 2019). Their wood-

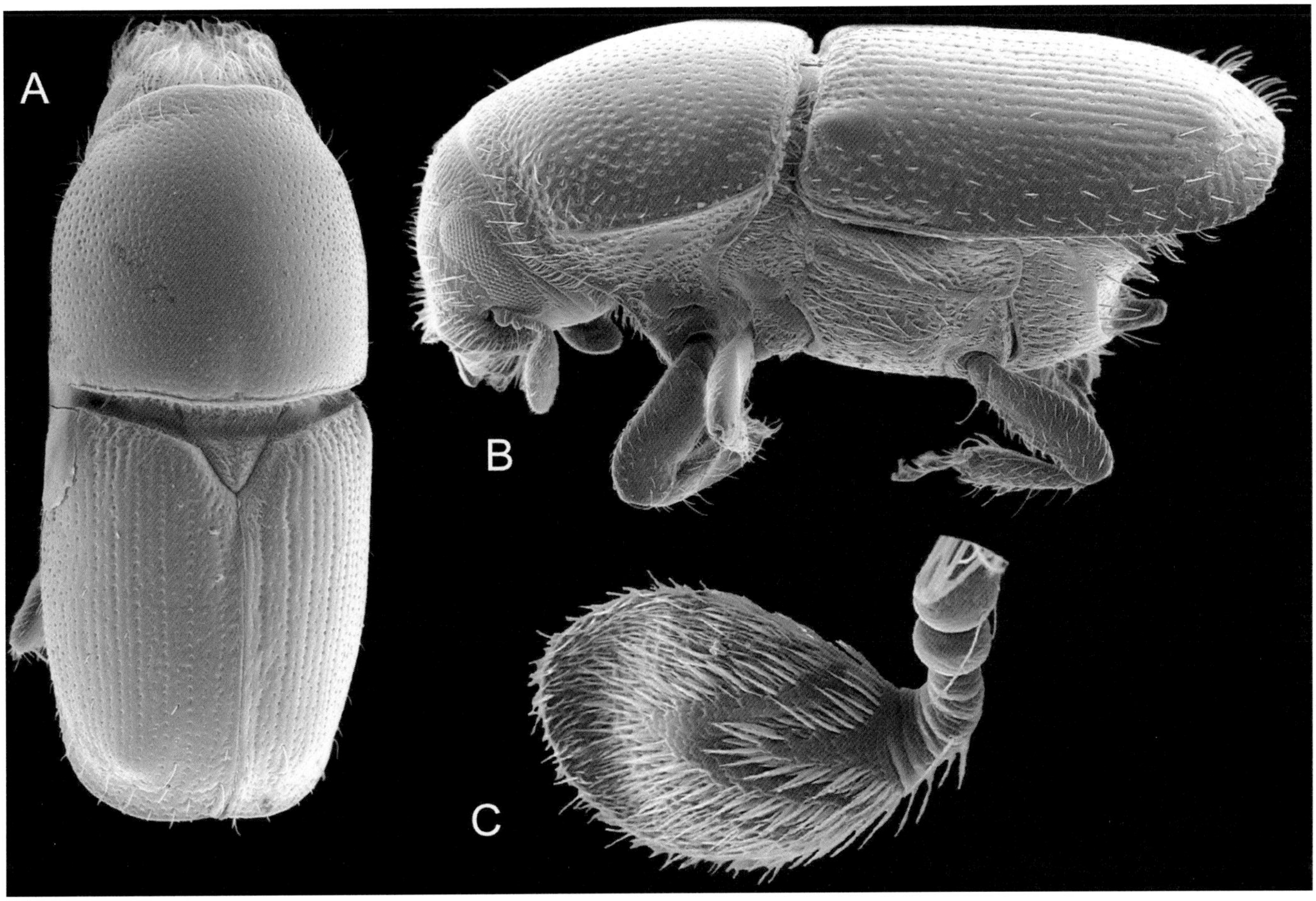

Fig. 46.5. *Scolytus multistriatus* (Marsham) (Scolytini). A–B, habitus: A, dorsal; B, lateral; C, antenna.

boring habits and small sizes predispose them to be readily introduced to novel areas through human-mediated transport of raw wood, wood products, wood packaging materials and plant trade (Brockerhoff *et al.* 2006; Rassati *et al.* 2015; Meurisse *et al.* 2019). Additionally, life history traits such as polyphagy, inbreeding polygyny and fungus farming may also contribute to the establishment of introduced species (Lantschner *et al.* 2020; Vilardo *et al.* 2022). For example, a study exploring the colonisation success on tropical islands found that inbreeding polygyny facilitates successful establishment over outbreeding systems (Jordal *et al.* 2001). Modelling scolytine invasions, however, has identified that human trade (Lantschner *et al.* 2020) and the number of individuals that arrive in novel areas (Vilardo *et al.* 2022) are the strongest predictors of successful species invasion. Despite the global distribution of several economically important scolytine species, only approximately 2.17 % of all species have successfully established in areas outside of their native range (Lantschner *et al.* 2020). In North America, bark beetle impacts are considered to be more destructive to forest ecosystems than hurricanes, tornadoes and fires, leading to an average annual impact of 2 040 000 ha and an estimated US$ 1 billion per year (Grégoire *et al.* 2015). Similarly, in Europe bark beetles were ranked between the years 1950 and 2000 as the third-most destructive agent in forest ecosystems, after storms and fire, and have led to the destruction of 288 hectares per year of forest between 1958 and 2001 (Seidl *et al.* 2011; Grégoire *et al.* 2015).

The ecological and economic impacts that scolytines have on Australian natural and managed forest ecosystems are unknown. However, their impact on Australian forests and agriculture has not gone unnoticed. Scolytines introduced to Australia can be very common in Australian forests (J. Bickerstaff, personal observation). Static-panel trapping throughout multiple plantation forests has regularly found that invasive species dominate the abundance of scolytines in Queensland (Wylie *et al.* 2008), New South Wales (Stone *et al.* 2010) and Tasmania (Bashford 2008; Wotherspoon 2008). Similarly, in natural ecosystems such as national parks, invasive species typically outnumber native species in trapping experiments (J. Bickerstaff, personal observation). Invasive species are known to cause economic impacts in Australia as well, such as *Ips grandicollis* (Eichhoff) in pine plantations (Yousuf *et al.* 2014; Nahrung & Carnegie 2020), *Hypothenemus* in macadamia orchards (Mitchell & Maddox 2010) and *Euwallacea fornicatus* (Eichhoff) among ornamental trees in Perth, Western Australia (Cook & Broughton 2023).

One of the most speciose scolytine genera is *Hypothenemus*, which is common in tropical and subtropical regions of the world. Its species are typically seed and twig borers, very small and difficult to distinguish morphologically. Several species are significant economic and ecological pests, such as the Coffee Berry Borer, *Hypothenemus hampei* (Vega *et al.* 2015), which has become established in many coffee-growing regions across the world though not thus far in Australia. In Brazil, it is estimated to cause losses of US$ 215–358 million per year (Oliveira *et al.* 2013), and globally it has impacted several million hectares of coffee plantations (Vega *et al.* 2015). Several species of *Hypothenemus* have been introduced to Australia, of which *H. eruditus* (Westwood) and *H. seriatus* (Eichhoff) have been recorded to destroy macadamia nuts (Mitchell & Maddox 2010).

Scolytus multistriatus is a potentially destructive species that was introduced to Australia from Europe, first found in elm tree (*Ulmus*) branches in Melbourne in 1974 (Rosel & French 1975; Neumann & Minko 1985). It is now established in southeastern Australia and attacks ornamental elm trees. In other regions of the world it transmits the highly phytopathogenic fungus *Ophiostoma ulmi*, which causes Dutch Elm Disease and can lead to significant tree death (Smith & Hulcr 2015; Bernier 2022). In Australia, *S. multistriatus* is not known to transmit this disease, but it does so in New Zealand (Neumann & Minko 1985; Gadgil *et al.* 2000).

The invasive scolytines *Ips grandicollis*, *Hylastes ater* (Paykull) and *Hylurgus ligniperda* attack managed pine plantations in Australia. *Ips grandicollis* was introduced from America by 1943 (Brimblecombe 1953; Marchant 1976), whereas *H. ater* and *H. lingiperda* were introduced from Europe in 1939 and 1942, respectively (Swan 1942; Brimblecombe 1953). All three species obligately infest pine trees, but the position and timing of their attacks differs. *Ips* beetles typically constructs galleries under the bark of trunks of weakened, damaged or stressed trees (Yousuf *et al.*, 2014). In contrast, *H. ater* and *H. ligniperda* beetles attack the logs and stumps of felled pine trees, those of *H. ater* first colonising the xylem tissue, followed by *H. ligniperda* beetles (Neumann 1987; Griggs 1998). In southeastern South Australia by 1985, the population of *H. ligniperda* increased due to a build-up of logs and stumps that were damaged by extensive bushfires in February 1983. Populations built up over a 12–18 months period, and by late 1985, beetles had dispersed to nearby areas, resulting in the death of several hundred managed pine trees (Neumann 1987). All three pine borers are known to be associated with ophiostomatoid fungi as well, *Ips grandicollis* with the blue-stain fungus *Ophiostoma ips* (Yousuf *et al.* 2014; Trollip *et al.* 2021), which does not directly lead to host tree mortality but causes discoloration and reduces the structural integrity of the softwood timber, thus causing economic damage (Stone & Simpson 1987). Ophiostomatoid fungi of the genera of *Grosmannia, Hyalorhinocladiella, Leptographium, Ophiostoma* and *Sporothrix* are known the be associated with *Hylurgus* (Kim *et al.* 2011) and *Graphium*, *Leptographium* and *Ophiostoma* with *Hylastes* (Griggs 1998).

The tribe Xyleborini also includes several introduced species in Australia. Members of the tribe are characterised by inbreeding polygyny, haplodiploidy and fungal farming, which are thought to facilitate the establishment of the species in novel areas (Jordal *et al.* 2001; Gohli & Jordal 2017). A substantial number of xyleborine species has been introduced to Australia, with roughly 33% of all introduced species belonging to this tribe. One of the most recent introductions is that of the Polyphagous Shot-Hole Borer (PSHB), *Euwallacea fornicatus*, in Perth in September 2021. This species is also now established in Argentina, Israel, Palestine, South Africa and the U.S.A. (Ceriani-Nakamurakare *et al.* 2023; Cook & Broughton 2023). Throughout its native, Southeast Asia, and introduced range, PSHB attacks various agricultural and ornamental trees, including avocado, cacao, citrus and tea, and can cause serious damage by heavy infestations and its phytopathogenic fungi, leading to significant economic losses (Smith & Hulcr 2015; De Wit *et al.* 2022; Cook & Broughton 2023). In South Africa, it is estimated to have caused losses of approximately Int.$ 18.45 billion over a 10-year period (De Wit *et al.* 2022). Similarly, costs to eradicate the species in Perth have been estimated at AUD$ 45 million, and, if not eradicated, its management to cost AUD$ 6.8 million per annum in 30 years (Cook & Broughton 2023). *Euwallacea fornicatus* transmits the phytopathogenic fungus *Fusarium euwallacea*, which can cause branch dieback and canker disease (Freeman *et al.* 2013; Ploetz *et al.* 2013). A closely related species, *E. perbrevis* (Schedl), occurs along the southern Queensland coast of Australia but does not cause substantial damage to its host trees (Smith *et al.* 2019).

Four invasive *Xylosandrus* species have been introduced to Australia, *X. compactus, X. crassiusculus* (Motschulsky), *X. discolor* (Blandford) and *X. morigerus* (Blandford). All are twig borers native to Southeast Asia and typically agricultural and sylvicultural pests (Dole & Cognato 2010). The first two have dispersed to all tropical and subtropical countries in the world and are some of the most widespread and abundant wood-boring beetles (Smith & Hulcr 2015). As other *Xylosandrus* species, they are associated with *Ambrosiella* fungi (Harrington *et al.* 2014; Skelton *et al.* 2018). *Xylosandrus crassiusculus* is reported to attack 200 tree species in 46 families and primarily regarded as a pest of small trees in nurseries and urban settings, but it can also damage stored timber (Smith & Hulcr 2015). Population genomics indicates that it comprises two primary genetic clusters, both introduced multiple times in various countries (Storer *et al.* 2017; Urvois *et al.* 2023). In Australia, it only reached the mainland in about 2011 (Nahrung & Carnegie 2021) and now occurs in southeastern Queensland and northern New South Wales, with both genetic clusters detected in the country (Tran *et al.* 2022). Its potential economic and ecological impacts in Australia are yet unknown. In Brazil it attacks avocado, cacao, coffee and mango and causes significant economic losses (Oliveira *et al.* 2008).

There are several other widely distributed species of Xyleborini in Australia of which little is known regarding their economic and ecological impacts. *Xyleborinus saxesenii* (Ratzeburg) is a Palearctic species first recorded in Australia in 1936 (Brimblecombe 1953) and now commonly found along the east coast from Queensland to Tasmania (Wardlaw & Lawson 2012), where it can be intercepted in large numbers in bark beetle traps (J. Bickerstaff pers. obs.). It commonly attacks fruit trees but has no significant economic impacts

Table 46.1. The diversity of Australian Scolytinae by tribe and genus.

Invasive species estimates are listed by Pullen *et al.* (2014). Larval feeding mode and reproductive mode are modified from the Appendix of Kirkendall *et al.* (2015).

Tribe	Genus	Number of species	Approx. no. of invasive species	Larval feeding mode	Reproductive mode
Cryphalini	*Cryphalus*	19	2	phloeophagous	monogynous
Diamerini	*Acacicis*	2	0	phloeophagous	monogynous
Diamerini	*Diamerus*	2	0	phloeophagous	monogynous
Dryocoetini	*Coccotrypes*	13	6	phloeophagous, spermatophagous and non-woody structures	inbreeding polygynous (haplodiploid)
Dryocoetini	*Cyrtogenius*	5	2	phloeophagous and xylomycetophagous	monogynous and colonial polygynous
Dryocoetini	*Dryocoetiops*	1	1	non-woody structures (rarely)	harem polygynous and bigynous (rarely)
Dryocoetini	*Peridryocoetes*	1	0	phloeophagous (rarely)	unknown
Ernoporini	*Eidophelus*	6	1	myelophagous, phloeophagous	bigynous (rarely), monogamous
Hylastini	*Hylastes*	1	0	phloeophagous	monogynous
Hylesinini	*Dactylipalpus*	1	0	xylomycetophagous	monogynous
Hylesinini	*Ficicis*	5	0	phloeophagous	monogynous
Hylesinini	*Hylesinus*	1	0	phloeophagous	monogynous
Hylesinini	*Zygophloeus*	1	0	unknown	unknown
Hylurgini	*Chaetoptelius*	3	0	phloeophagous	monogynous
Hylurgini	*Hylurdrectonus*	1	0	phloeophagous	monogynous
Hylurgini	*Hylurgus*	1	1	phloeophagous	monogynous
Hylurgini	*Pachycotes*	6	1	xylophagous	monogynous
Hypoborini	*Liparthrum*	1	1	phloeophagous and non-woody structures	monogynous
Hyorrhynchini	*Sueus*	1	0	xylomycetophagous	inbreeding polygynous
Ipini	*Acanthotomicus*	1	0	phloeophagous and non-woody structures (rarely)	harem polygynous and monogynous (rarely)
Ipini	*Ips*	1	1	phloeophagous	harem polygynous
Phloeosinini	*Hyledius*	1	0	phloeophagous	monogynous and harem polygynous (rarely)
Phloeosinini	*Hyleops*	1	0	phloeophagous	monogynous
Phloeosinini	*Phloeosinopsioides*	3	0	unknown	unknown
Phloeosinini	*Phloeosinus*	3	1	phloeophagous and xylophagous (rarely)	monogynous and bigynous (rarely)
Phloeotribini	*Aricerus*	2	0	phloeophagous	monogynous
Phloeotribini	*Phloeotribus*	1	0	phloeophagous	monogynous
Scolytini	*Scolytus*	1	1	phloeophagous	monogynous
Trypophloeini	*Cosmoderes*	2	0	unknown	unknown
Trypophloeini	*Hypothenemus*	8	7	phloeophagous, xylomycetophagous (rarely), spermatophagous, myelophagous, non-woody structures (rarely)	inbreeding polygynous
Xyleborini	*Amasa*	6	0	xylomycetophagous	inbreeding polygynous (haplodiploid)
Xyleborini	*Ambrosiodmus*	3	1	xylomycetophagous	inbreeding polygynous (haplodiploid)
Xyleborini	*Ambrosiophilus*	4	0	xylomycetophagous	inbreeding polygynous (haplodiploid)
Xyleborini	*Arixyleborus*	1	0	xylomycetophagous	inbreeding polygynous (haplodiploid)
Xyleborini	*Beaverium*	1	0	xylomycetophagous	inbreeding polygynous (haplodiploid)

Table 46.1. continued

Tribe	Genus	Number of species	Approx. no. of invasive species	Larval feeding mode	Reproductive mode
Xyleborini	*Cnestus*	1	0	xylomycetophagous	inbreeding polygynous (haplodiploid)
Xyleborini	*Cryptoxyleborus*	1	0	xylomycetophagous	inbreeding polygynous (haplodiploid)
Xyleborini	*Debus*	2	0	xylomycetophagous	inbreeding polygynous (haplodiploid)
Xyleborini	*Diuncus*	2	2	xylomycetophagous	inbreeding polygynous (haplodiploid)
Xyleborini	*Eccoptopterus*	1	0	xylomycetophagous	inbreeding polygynous (haplodiploid)
Xyleborini	*Euwallacea*	7	1	xylomycetophagous	inbreeding polygynous (haplodiploid)
Xyleborini	*Fraudatrix*	1	0	xylomycetophagous	inbreeding polygynous (haplodiploid)
Xyleborini	*Immanus*	1	0	xylomycetophagous	inbreeding polygynous (haplodiploid)
Xyleborini	*Microperus*	4	0	xylomycetophagous	inbreeding polygynous (haplodiploid)
Xyleborini	*Planiculus*	1	0	xylomycetophagous	inbreeding polygynous (haplodiploid)
Xyleborini	*Truncaudum*	1	0	xylomycetophagous	inbreeding polygynous (haplodiploid)
Xyleborini	*Xyleborinus*	4	1	xylomycetophagous	inbreeding polygynous (haplodiploid)
Xyleborini	*Xyleborus*	10	7	xylomycetophagous	inbreeding polygynous (haplodiploid)
Xyleborini	*Xylosandrus*	8	3	xylomycetophagous	inbreeding polygynous (haplodiploid)
Xyloctonini	*Scolytomimus*	2	0	phloeophagous	monogynous

(CABI 2022). Several species of globally distributed *Xyleborus* are also present in Australia, all of them morphologically very similar; *Xyleborus affinis* Eichhoff, *X. ferrugineus* (Fabricius), *X. perforans* (Wollaston) and *X. volvulus* (Fabricius). Recent phylogenetic analyses have indicated that these species were globally distributed prior to human expansion across the world but that recent human-aided translocation of species has accelerated their rates of colonisation (Gohli *et al.* 2016). Conversely, the recently described Australian-native *Amasa parviseta* Knížek & Smith has been introduced in *Eucalyptus* plantations in Europe and South America (Knížek & Smith 2024). This species is morphologically similar to *A. truncata* and distributed sympatrically with it throughout eastern Australia.

Characteristics. Adults. Body length 0.9–7.5 mm, 1.4–4.17 × as long as wide. Body typically rounded to elongate. Sexual dimorphism present in most species, affecting forehead, abdominal tergites or ventrites. In haplodiploid species (e.g. Xyleborini) males dwarfed, flightless and pronotum typically highly modified with horns or excavated depressions. Forehead and vertex often flat to concave in males, flat to convex in females, smooth to densely punctate, glabrous to setose. Eyes oval (Fig. 46.15B) to shallowly or deeply emarginate (Fig. 46.22C) or divided. Antennal funicles 2- to 7-segmented (Fig. 46.11A–G); clubs variable, either oval, circular, ovate, pseudolamellate, flat or truncate, segments divided by straight, curved or sinuate sutures. Postmentum more or less flattened (Fig. 46.13D) or strongly depressed (Fig. 46.25E), in ventral view as long as wide, wider than long or longer than wide. Pronotum often covering head in dorsal view, variable in shape, anterior edge serrate or rounded to subquadrate asperate, smooth to punctate, glabrous to setose. Scutellar shield at same level as elytra or not exposed, disc impressed. Elytra with anterior margin straight or curved, if curved occasionally crenulate (Fig. 46.10I, J); disc variable in length and shape, convex or straight; apical declivity variable in shape, slightly to steeply rounded, truncate, attenuate and often with granules, tubercles or spines (Fig. 46.10D–H); surface typically with strial punctures weakly to deeply impressed, occasionally forming deep striae. Metanepisternal sutures mostly lined with sclerolepidia (Fig. 46.13J). Procoxae and mesocoxae contiguous (Fig. 46.11P) to separate (Fig. 46.11R). Protibiae rounded to triangular and typically armed with socketed denticles (Fig. 46.10A). Tarsi with third segment bilobed (Fig. 46.10A, B) or cylindrical (Fig. 46.10C). Modified from Jordal (2014).

Larvae (last instar). Body length 2–7.5 mm. Head width 0.3–1.3 mm. Setae pallid to pale brown to dark brown, slender in most species, moderately long (Fig. 46.8B) on head capsule in some species. Head capsule free, hypognathous, typically hemispherical in shape, stemmata absent. Head yellowish brown to reddish brown in some species. Frontal suture variable. Frons with pair of dark brown tubercles in several genera (Fig. 46.7I). Epistoma rarely with tubercle (Fig. 46.7A) near posterior margin. Clypeus with setae of variable length (e.g. Fig. 46.8B), occasionally pigmented and often long as labrum. Labrum typically transverse, diamond-shaped in some genera, either elongate or pigmented (e.g. Fig. 46.8E). Tormae strong, subparallel, straight to curved in shape. Mandibles red brown to black in colour, variously dentate, usually with small molar tooth or projection, usually with two setae (Figs 46.6R; 46.7C, J; 46.8C, O). Antennae exposed, conical in shape, with 3 to 5 sensory hairs. Thorax legless, typically not sclerotised (Figs 46.6J). Abdomen 10-segmented. Thoracic spiracles circular to broadly ovate, abdominal spiracles usually half as large as thoracic ones, typically circular. Alimentary canal of modal form in most species. Pseudocerci, when present, short, pallid and pubescent. Modified from May (1994).

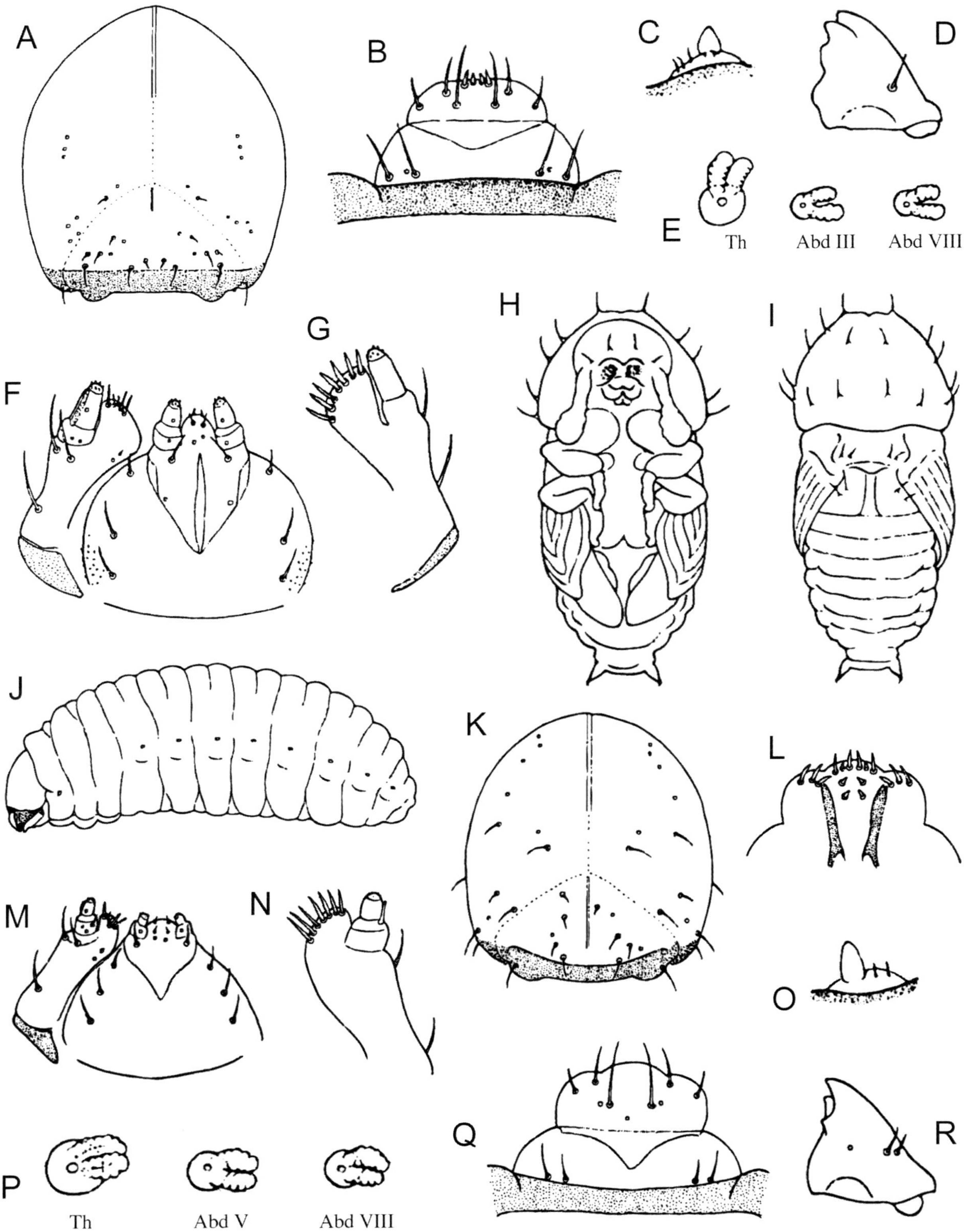

Fig. 46.6. Scolytinae, structural details of immature stages. A–G, *Cryphalus wapleri* Eichhoff (Cryphalini), larva (New Zealand): A, head, dorsal; B, labrum, clypeus and epistoma; C, antenna; D, mandible; E, spiracle, showing alignment; F, mala and labium, ventral; G, mala, dorsal; H–I, *Cryphalus wapleri* Eichhoff, pupa: H, habitus, ventral; I, habitus, dorsal; J–Q, *Coccotrypes carpophagus* (Hornung) (Dryocoetini), larva: J, habitus, lateral; K, head, dorsal; L, epipharyngeal lining; M, mala and labium, ventral; N, mala, dorsal; O, antenna; P, spiracles, showing alignment; Q, labrum, clypeus and epistoma; R, mandible. After May (1994).

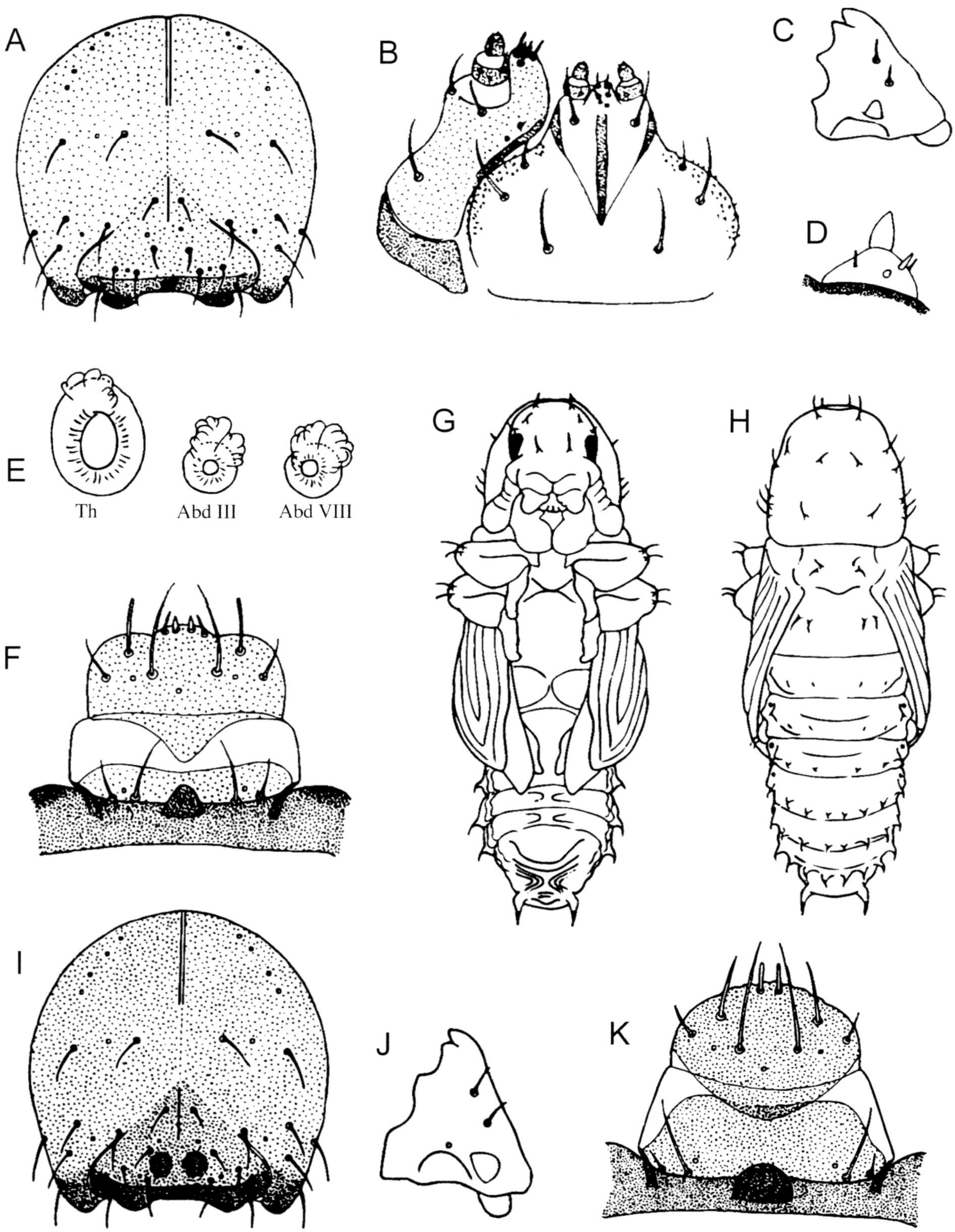

Fig. 46.7. Scolytinae, structural details of immature stages. A–F, *Hylastes ater* (Paykull) (Hylastini), larva (New Zealand): A, head, dorsal; B, mala and labium, ventral; C, mandible; D, antenna; E, spiracles, showing alignment; F, labrum, clypeus and epistoma; G–H, *Hylastes ater* (Paykull), pupa: G, habitus, ventral; H, habitus, dorsal; I–K, *Hylurgus ligniperda* (Fabricius) (Hylurgini), larva: I, head, dorsal; J, mandible; K, labrum, clypeus and epistoma. After May (1994).

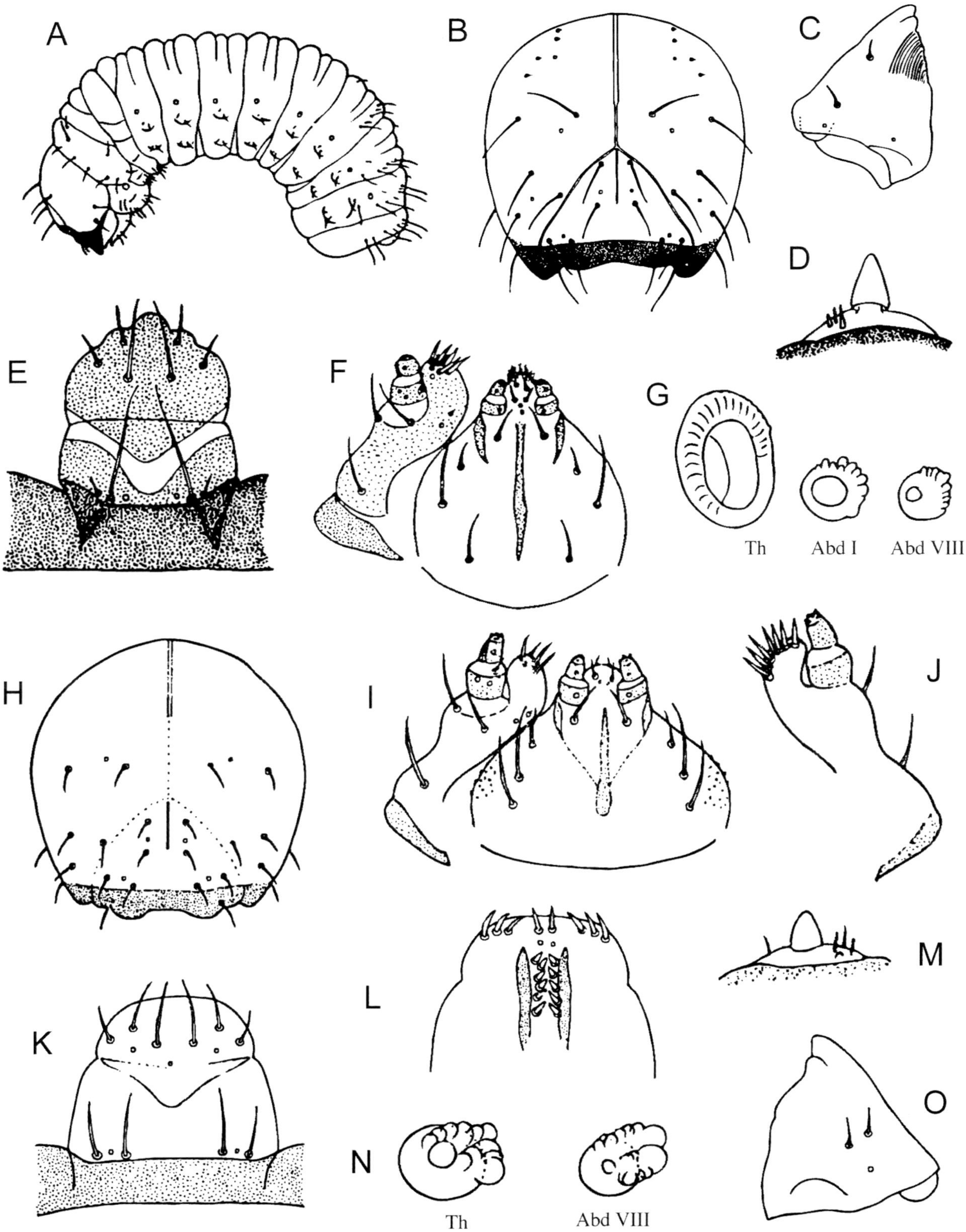

Fig. 46.8. Scolytinae, structural details of immature stages. A–G, *Pachycotes australis* Schedl (Hylurgini), larva: A, habitus, lateral; B, head, dorsal; C, mandible; D, antenna; E, labrum, clypeus and epistoma; F, mala and labium, ventral; G, spiracles, showing alignment; H–O, *Ips grandicollis* (Eichhoff) (Ipini), larva: H, head, dorsal; I, mala and labium, ventral; J, mala, dorsal; K, labrum, clypeus and epistoma; L, epipharyngeal lining; M, antenna; N, spiracles, showing alignment; O, mandible. After May (1994).

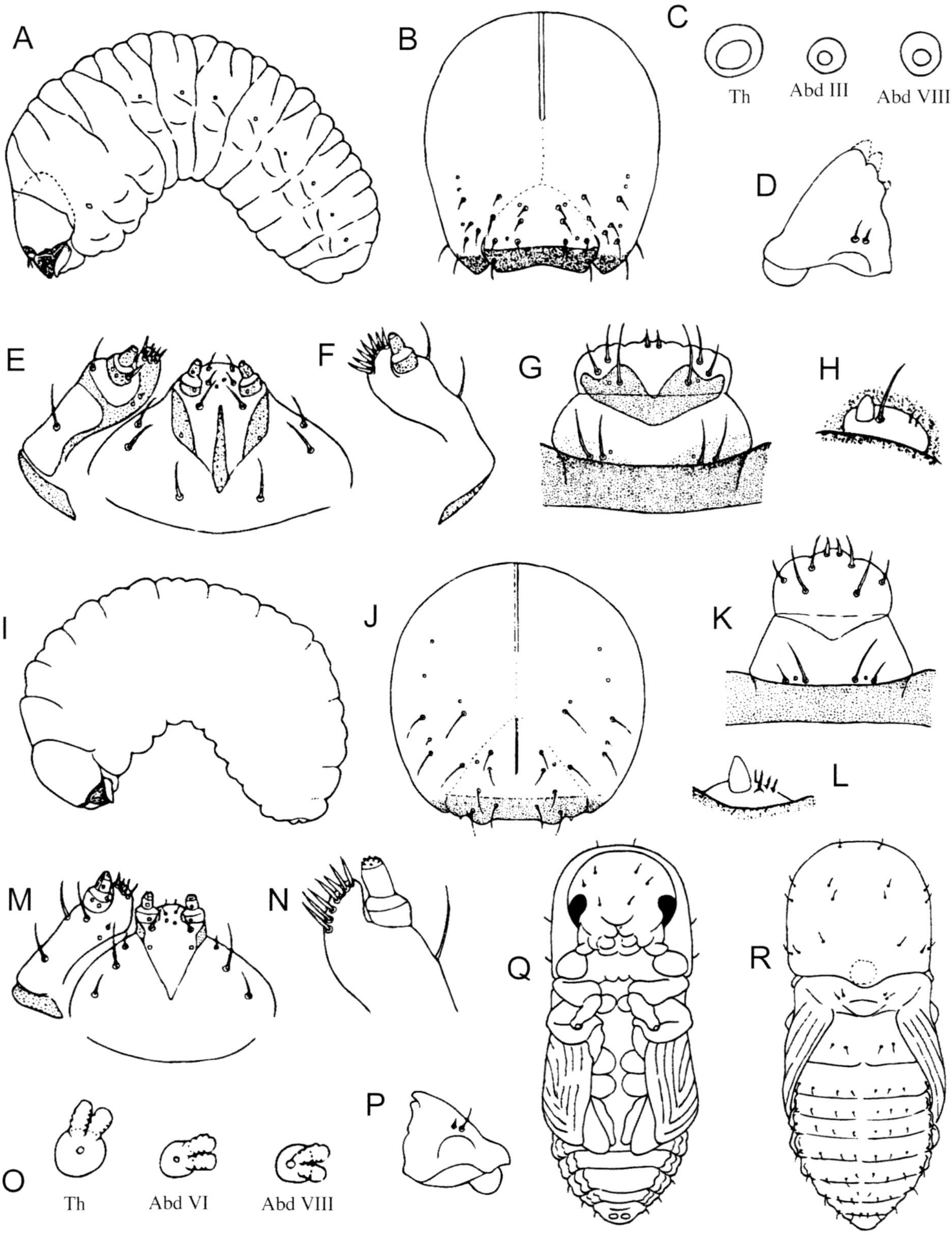

Fig. 46.9. Scolytinae, structural details of immature stages. A–H, *Phloeosinus cupressi* Hopkins (Phloeosinini), larva (New Zealand): A, habitus, lateral; B, head, dorsal; C, spiracles, showing alignment; D, mandible; E, mala and labium, ventral; F, mala, dorsal; G, labrum, clypeus and epistoma; H, antenna; I–P, *Xyleborinus saxesenii* (Ratzeburg) (Xyleborini), larva: I, habitus, lateral; J, head, dorsal; K, labrum, clypeus and epistoma; L, antenna; M, mala and labium, ventral; N, mala, dorsal; O, spiracles, showing alignment; P, mandible; Q–R, *Xyleborinus saxesenii* (Ratzeburg), pupa: Q, habitus, ventral; R, habitus, dorsal. After May (1994).

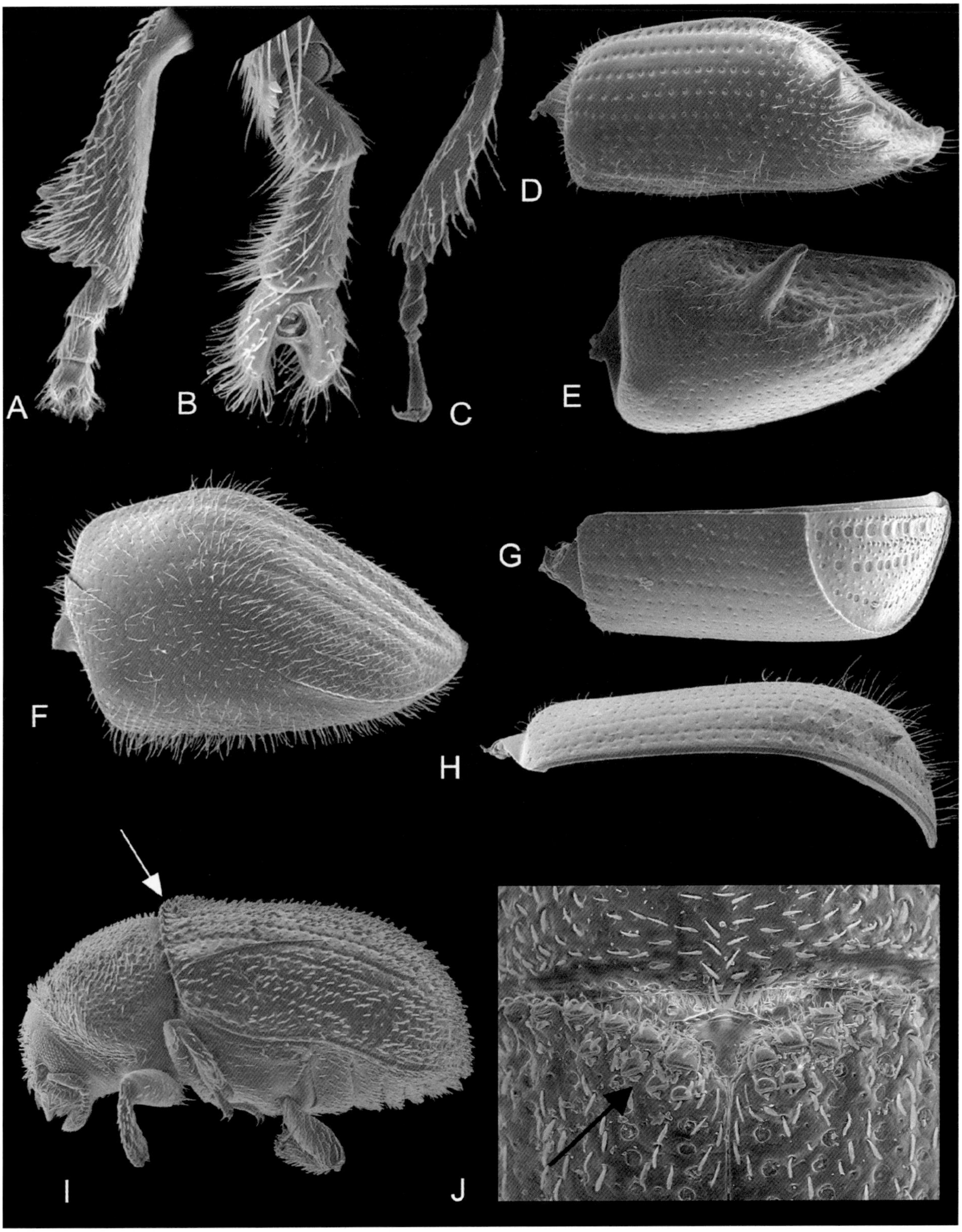

Fig. 46.10. Diagnostic characters for the key to the genera of Australian Scolytinae. A–B, bilobed third tarsal segment: A, *Phloeotribus* sp., B, *Aricerus eichhoffi* Blandford; C, cylindrical third tarsal segment, *Hypothenemus* sp.; D–E, spines and armature on elytral declivity (dorsal): D, *Ips grandicollis* (Eichhoff); E, *Eccoptopterus spinosus* (Olivier); F, obliquely truncate elytral declivity (dorsal), *Cnestus solidus* (Eichhoff); G, truncate elytral declivity (dorsal), *Amasa parviseta* Knížek & Smith; H, rounded elytral declivity (lateral), *Ambrosiophilus compressus* (Lea); I, upturned elytral basal margin (lateral), *Phloeotribus* sp.; J, crenulate elytral basal margin, *Phloeotribus* sp.

Key to the tribes and genera of Scolytinae in Australia

The presence of *Gnathotrichus* Eichhoff in Australia is doubtful, but the genus is included in the key as it has been intercepted in the past. The genera of Xyleborini are represented by females only as males are rarely collected. The species of many other tribes and genera are sexually dimorphic, but this does not affect the reliability of the key as most of the dimorphism occurs on the head (such as vestiture or tubercles) or in body size, armature of the elytral declivity and the ability to stridulate (as in *Hylurgus*).

1. Head usually visible in dorsal view; elytra with basal margin procurved dorsally, typically upturned in middle and crenulate (Fig. 46.10I–J) ... **2**

– Head usually not visible in dorsal view; elytra with basal margin dorsally straight, simple (Fig. 46.10D–G) ... **20**

2(1). Antennal clubs compact and conical (Fig. 46.11A), loosely segmented or pseudolamellate (Fig. 46.11B, D) ... **3**

– Antennal clubs rounded to oval in shape (Fig. 46.11C) ... **17**

3(2). Elytra with basal margin weakly crenulate and straight; body length 4.0–4.3 mm ... **HYLASTINI ... *Hylastes* Erichson**

– Elytra with basal margin crenulate and procurved (Fig. 46.10I–J) ... **4**

4(3). Pronotum conspicuously asperate (Fig. 46.11H, K–O) ... **HYLESININI ... 5**

– Pronotum weakly asperate or without asperities (Fig. 46.11I) ... **8**

5(4). Antennal funicles 6-segmented; body length 1.0–1.3 mm ... ***Zygophloeus* Schedl**

– Antennal funicles 7-segmented ... **6**

6(5). Pronotum quadrate, with invagination (mycangia) in female (Pl. 113A); antennal clubs without defined suture lines; body length 5.4–9.5 mm ... ***Dactylipalpus* Chapuis**

– Pronotum rounded, without mycangia; antennal clubs with well-defined suture lines ... **7**

7(6). Elytra with base strongly curved (Pl. 111E); declivity rounded to emarginate, unarmed; protibiae armed with six socketed denticles; body length 3.0–3.3 mm ... ***Hylesinus* Fabricius**

– Elytra with base not as strongly curved (Pl. 111C); declivity attenuate, granulate; protibiae armed with 2 or 3 three socketed denticles; body length 1.7–2.0 mm ... ***Ficicis* Lea**

8(4). Antennal clubs loosely segmented or pseudolamellate (Fig. 46.11B, D) ... **PHLOEOTRIBINI ... 9**

– Antennal clubs compact (Fig. 46.11A) ... **10**

9(8). Antennal clubs with basal 2 segments arranged in chevron pattern (Fig. 46.11B); pronotum weakly asperate anterolaterally (Fig. 46.11J); setae on pronotum and elytra short, fine; body length 3.6 mm ... ***Aricerus* Blandford**

– Antennal clubs with segments loose, conical, not arranged in chevron pattern (Fig. 46.11D); pronotum without asperities; setae on pronotum and elytra short, thick; body length 2 mm ... ***Phloeotribus* Latreille**

10(8). Antennal clubs circular in transverse view ... **11**

– Antennal clubs flattened in transverse view ... **PHLOEOSININI ... 14**

11(10). Eyes entire ... **HYLURGINI (part) ... 12**

– Eyes completely divided; body length 1.6 mm ... **HYORRHYNCHINI ... *Sueus* Murayama**

12(11). Antennal funicles 7-segmented ... **13**

– Antennal funicles 6-segmented; body length 4.5–5.0 mm ... ***Hylurgus* Latrielle**

13(12). Pronotum and elytra densely setose; body length 2.1–2.9 mm ... ***Chaetoptelius* Fuchs**

– Pronotum sparsely setose, densely punctate; elytra with disc sparsely setose but declivity densely setose; body length 2.3–4.2 mm ... ***Pachycotes* Sharp**

14(10). Eyes entire; antennal clubs with sutures complete ... **15**

– Eyes divided; antennal clubs with sutures incomplete; body length 2.6 mm ... ***Hyledius* Sampson**

15(14). Eyes emarginate; antennal funicles 5-segmented ... **16**

– Eyes oval; antennal funicles 7-segmented; body length ca. 3.5 mm ... ***Hyleops* Schedl**

16(15). Eyes shallowly emarginate; elytra with interstrial setae in single rows, declivity without tubercles; third tarsal segments cylindrical; body length 1.4–2.5 mm ... ***Phloeosinopsioides* Schedl**

– Eyes deeply emarginate; elytra with declivity tuberculate; third tarsal segments bilobed; body length 1.7–2.5 mm ... ***Phloeosinus* Chapuis**

17(2). Abdomen ascending towards elytral apex, ventrite 2 almost perpendicular to ventrite 1 (Fig. 46.5B) in lateral view; body length 2.6–3 mm ... **SCOLYTINI ... *Scolytus* Geoffroy**

– Ventrites horizontal in lateral view ... **18**

18(17). Antennal funicles 7-segmented ... **DIAMERINI ... 19**

– Antennal funicles 4-segmented; body length 1.0–1.3 mm ... **HYPOBORINI ... *Liparthrum* Wollaston**

19(18). Body stout, 1.4 × as long as wide; forehead, pronotum and elytra densely covered with setae; pronotum anteriorly asperate; body length 1.1–2.5 mm ... ***Acacicis* Lea**

– Body elongate, 1.8 × as long as wide; forehead, pronotum and elytral disc densely punctate, without setae; pronotum without asperities; setae on elytral declivity restricted to interstriae; body length 3–5 mm ... ***Diamerus* Erichson**

20(1). Third tarsal segments bilobed (Fig. 46.10A, B) ... **21**

– Third tarsal segments cylindrical (Fig. 46.10C) ... **22**

21(20). Antennal clubs circular to oval in shape (Fig. 46.11E); pronotum coarsely asperate; body length 1.1–2 mm ... **CRYPHALINI ... *Cryphalus* Erichson**

– Antennal clubs ovate; pronotum lacking asperities; body length 1.4–1.6 mm ... **HYLURGINI ... *Hylurdrectonus* Schedl**

22(20). Metanepisterna concealed by elytra ... **CORTHYLINI ... *Gnathotrichus* Eichhoff**

– Metanepisterna exposed ... **23**

23(22). Setae thick, scale like, with rounded apex, typically uniformly covering elytra and pronotum ... **24**

– Setae erect, slender, with acute apex, coverage variable ... **26**

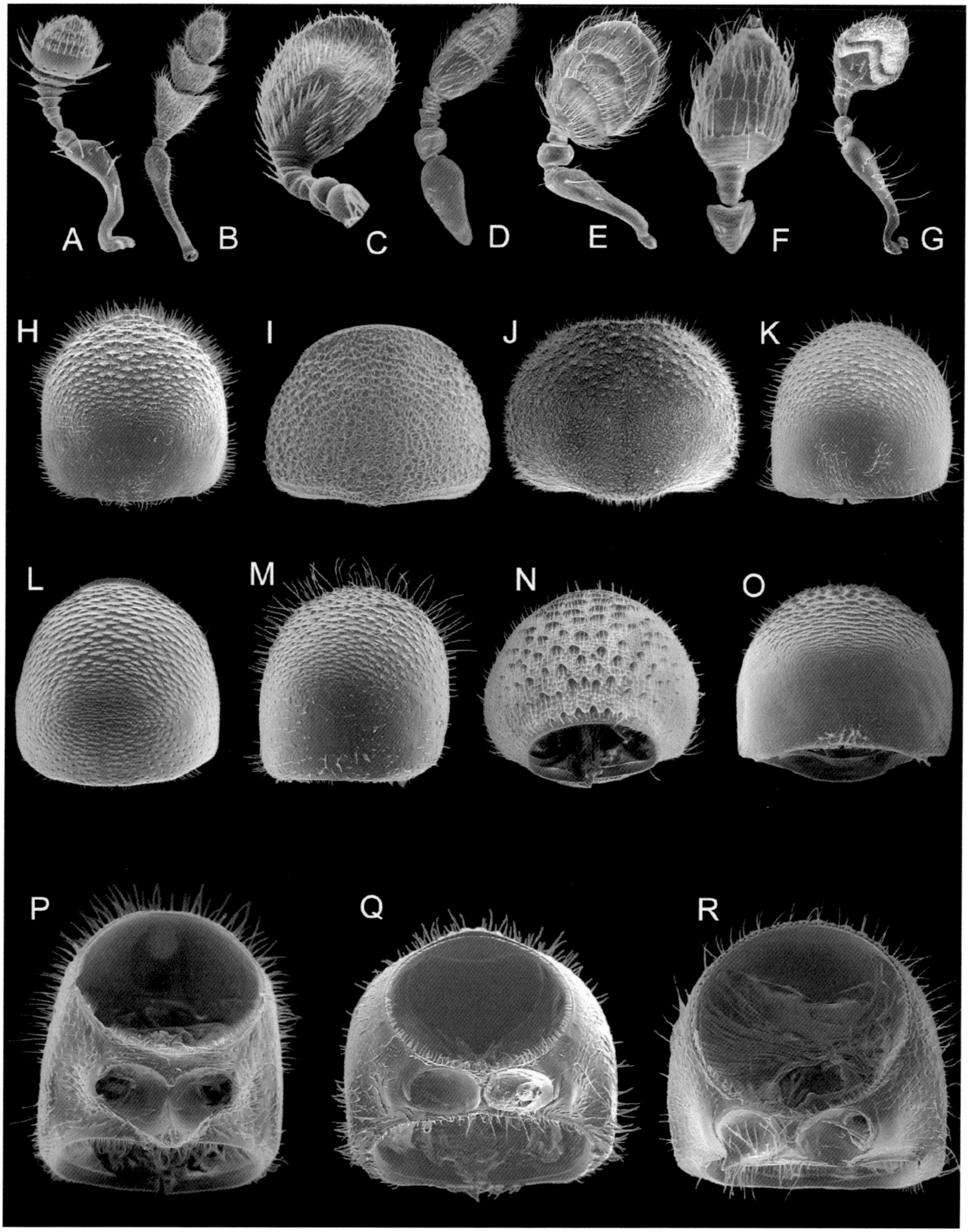

Fig. 46.11. Diagnostic characters for Australian Scolytinae. A–G, antenna: A, *Hylurgus ligniperda* (Fabricius); B, *Aricerus eichhoffi* Blandford; C, *Scolytus multistriatus* (Marsham); D, *Phloeotribus* sp.; E, *Cryphalus* sp.; F, *Hypothenemus* sp.; G, *Ips grandicollis* (Eichhoff); H–O, pronotum, dorsal: H, *Cnestus solidus* (Eichhoff); I, *Phloeosinus* sp.; J, *Aricerus eichhoffi* Blandford; K, *Xylosandrus crassiusculus* (Motschulsky); L, *Beaverium insulindicus* (Eggers); M, *Ambrosiophilus compressus* (Lea); N, *Cryphalus* sp.; O, *Eccoptopterus spinosus* (Olivier); P–R, pronotum, ventral: P, contiguous procoxae, *Ips grandicollis* (Eichhoff); Q, narrowly separate procoxae, *Cnestus solidus* (Eichhoff); R, widely separate procoxae, *Xylosandrus crassiusculus* (Motschulsky).

24(23). Eyes oval, outline entire; body length 1.4–2.5 mm .... **ERNOPORINI** ... ***Eidophelus* Eichhoff**
– Eyes emarginate .... **TRYPOPHLOEINI** ... **25**
25(24). Antennal funicles 2- to 3-segmented; clubs with segments fused, without sutures; pronotum either longer than wide or as wide as long; body length 1.6–1.7 mm .... ***Cosmoderes* Eichhoff**
– Antennal funicles 4- to 5-segmented; clubs with two to three sutures (Fig. 46.11F), sometimes a partial septum; pronotum wider than long (0.7–0.9 × as long as wide); body length 0.9–1.8 mm .... ***Hypothenemus* Westwood**
26(23). Abdominal ventrites ascending toward elytral apex in lateral view (Pl. 115L); body length 1.4–2.1 mm .... **XYLOCTONINI** ... ***Scolytomimus* Blandford**
– Abdominal ventrites horizontal in lateral view **27**
27(26). Antennal funicles 5-segmented; clubs with sutures bisinuate or procurved; elytra with declivity deeply excavated and armed with 2–5 large teeth .... **IPINI** ... **28**
– Antennal funicles 2- to 5-segmented; clubs typically truncate or flat with no bisinuate sutures; elytra with declivity not as above, if dentate not as above **29**
28(27). Antennal clubs with sutures procurved; elytra with declivity armed with 2 spines; procoxae narrowly separate; body length 2.0–2.1 mm .... ***Acanthotomicus* Blandford**
– Antennal clubs with sutures bisinuate (Fig. 46.11G); elytra with declivity armed with 5 spines (Fig. 46.10D); procoxae contiguous (Fig. 46.11P); body length 3.5–3.9 mm .... ***Ips* De Geer**
29(27). Pronotum gradually descending anteriad in lateral view; middle and hind tibiae slender, abruptly narrowed apically, typically armed with 5 or fewer socketed denticles .... **DRYOCOETINI** ... **30**
– Pronotum typically steeply descending anteriad in lateral view; middle and hind tibiae typically expanded and tapering towards apex, armed with 5 or more socketed denticles .... **XYLEBORINI** ... **33**
30(29). Pronotum wider than long (0.87–0.9 × as long as wide) **31**
– Pronotum as long as wide or longer than wide (1.0–1.45 × as long as wide) **32**
31(30). Antennal clubs longer than wide; pronotum completely asperate; pronotum and elytra covered with short setae; body length 1.6–1.8 mm .... ***Peridryocoetes* Wood**
– Antennal clubs about as long as wide; pronotum asperate anteriorly, punctate on disc; pronotum and elytra covered with long setae; body length 1.9–4.0 mm .... ***Dryocoetiops* Schedl**
32(30). Forehead with surface aciculate; elytra with declivity typically rounded in lateral view, sometimes steeply so, or attenuate in dorsal view; body length 1.2–2.5 mm .... ***Coccotrypes* Eichhoff**
– Forehead with surface deeply punctate (male) or setose (female); elytra with declivity always steeply rounded in lateral view, almost truncate in dorsal view; body length 1.3–3 mm .... ***Cyrtogenius* Strohmeyer**
33(29). Antennal funicles 3-segmented; body length 2.1–2.5 mm .... ***Fraudatrix* Cognato, Smith & Beaver**
– Antennal funicles 5-segmented **34**
34(33). Scutellar shield conical, protruding, or absent **35**
– Scutellar shield at same level of elytra or lower **37**
35(34). Scutellar shield conical, surrounded by setae; body length 1.8–2.8 mm .... ***Xyleborinus* Reitter**
– Scutellar shield not exposed **36**
36(35). Elytra with large anteromedial mycangial pit (Fig 46.2B), anterior edge without setae; body length 2.3–2.7 mm .... ***Cryptoxyleborus* Wood & Bright**
– Elytra without large anteromedial mycangial pit, anterior edge with setae (Fig. 46.2C); body length 1.8–3.1 mm .... ***Microperus* Wood**
37(34). Pronotum with mycangial tufts along posterior edge **38**
– Pronotum without mycangial tufts along posterior edge **40**
38(37). Metatibiae greatly enlarged (Pl. 114I); elytra with declivity deeply excavate and armed with several large spines (Fig. 46.10E); body length 2.5–3.7 mm .... ***Eccoptopterus* Motschulsky**
– Metatibiae proportional in size to other tibiae; elytra with declivity unarmed (Fig. 46.10F) **39**
39(38). Anterior margin of pronotum with 2 flat serrations protruding anteriad (Fig. 46.11H); procoxae narrowly separate (Fig. 46.11Q); body length 3.6–4.1 mm .... ***Cnestus* Sampson**
– Anterior margin of pronotum serrate or granulate (Fig. 46.11K); procoxae separate (Fig. 46.11R); body length 1.4–3.4 mm .... ***Xylosandrus* Reitter**
40(37). Elytra with declivity emarginate or deeply excavate **41**
– Elytra with declivity entire **43**
41(40). Elytra with declivity deeply excavate and tuberculate or bearing projections **42**
– Elytra with declivity emarginate and granulate; body length 1.7–2.4 mm .... ***Planiculus* Hulcr & Cognato**
42(41). Pronotum wider than long (0.76 × as long as wide); body length 7.0–7.5 mm .... ***Immanus* Hulcr & Cognato**
– Pronotum longer than wide (1.22 × as long as wide); body length 2.5–3.6 mm .... ***Debus* Hulcr & Cognato**
43(40). Elytra with declivity truncate in lateral view **44**
– Elytra with declivity rounded or flat in lateral view **45**
44(43). Elytra with declivity surrounded by continuous carina (Fig. 46.10G); body length 2.5–3.5 mm .... ***Amasa* Lea**
– Elytra with declivity tuberculate (Pl. 115E, F); body length 2.1–2.9 mm .... ***Truncaudum* Hulcr & Cognato**
45(43). Pronotum with anterior edge carinate (Fig. 46.11L) **46**
– Pronotum with anterior edge not carinate (Fig. 46.11M) **47**
46(45). Pronotum with anterior margin without distinct serrations (Fig. 46.11L); body length 4.9–6.2 mm .... ***Beaverium* Hulcr & Cognato**
– Pronotum with anterior margin with 2 distinct serrations protruding anteriad; body length 1.4–1.9 mm .... ***Diuncus* Hulcr & Cognato**
47(45). Elytral declivity with interstriae deeply impressed, forming ridges (Pl. 113L); body length 2.0–2.1 mm .... ***Arixyleborus* Hopkins**
– Elytral declivity with interstriae not deeply impressed **48**

48(47). Pronotum completely asperate (Fig. 46.11L); body length 2.3–2.8 mm .... ***Ambrosiodmus* Hopkins**
– Pronotum only anterolaterally asperate (Fig. 46.11M) **49**
49(48). Elytral disc as long as declivity, the latter steeply rounded (Fig. 46.10H); body length 2.5–3.9 mm .... ***Ambrosiophilus* Hulcr & Cognato**
– Elytral disc longer or shorter than declivity, the latter more gradually rounded **50**
50(49). Pronotum in lateral view with indistinct summit; body length 1.9–3.1 mm .... ***Xyleborus* Eichhoff**
– Pronotum in lateral view with distinct summit; body length 2.2–4.6 mm .... ***Euwallacea* Hopkins**

Classification of the Australian genera of Scolytinae

The generic classification used follows that of Pullen *et al.* (2014) but with subsequent taxonomic changes noted. The origins of the species are given as native or introduced based on Pullen *et al.* (2014). The distribution of genera by country follows Wood & Bright (1992), with updates incorporated when available.

Tribe Cryphalini Lindemann, 1877

The tribe was revised by Johnson *et al.* (2020a), resulting in the following changes from the classification of Pullen *et al.* (2014). The genera *Cosmoderes* and *Hypothenemus* were transferred to Trypophloeini and *Eidophelus* to Ernoporini; the names *Ptilopodius* and *Scolytogenes* were synonymised with *Eidophelus* and *Hypocryphalus* with *Cryphalus*.

Cryphalus Erichson, 1836 (Pl. 110A, B; Fig. 46.12)

Type species, by subsequent designation (Thomson, 1859: 146): *Bostrichus asperatus* Gyllenhal, 1813.

Characteristics. Body length 1.1–2.0 mm, 2.1–2.4 × as long as wide. Forehead flat, punctate and setose. Postmentum flat to slightly impressed, either as long as wide or wider than long. Eyes emarginate, typically with upper portion smaller than lower but sometime equal. Antennae with 4-segmented funicle (3- to 5-segmented in non-Australian species) and oval to circular club marked with 2–3 procurved sutures. Pronotum covering head, as wide as long to wider than long; surface coarsely asperate and setose. Scutellar shield at same level as elytra. Elytra with anterior margin straight; disc typically flat, as long as declivity, densely setose, impunctate; declivity rounded, unarmed. Procoxae typically narrowly separate, mesocoxae contiguous to separate; protibiae armed with 1–8 socketed denticles; tarsi with third segment bilobed.

Australian species. Native: *C. asperulus* Schedl, 1948, *C. brimblecombei* Schedl, 1948, *C. compactus* Lea, 1910, *C. constrictus* Schedl, 1942, *C. moorei* (Schedl, 1964), *C. niger* Schedl, 1942, *C. nigrosetosus* (Schedl, 1948), *C. ovalicollis* (Schedl, 1942), *C. pilosellus* Erichson, 1842, *C. puberulus* Schedl, 1942, *C. punctipennis* Schedl, 1942, *C. sidneyanus* (Nördlinger, 1856), *C. sparsepilosus* Schedl, 1942, *C. spathulatus* (Schedl, 1938), *C. subcompactus* Lea, 1910, *C. walkeri* (Blandford, 1896), *C. wapleri* Eichhoff, 1872. Introduced: *C. mangiferae* Stebbing, 1914.

Distribution. Australia (NSW, NT, QLD, TAS, VIC, WA), Africa, Asia, Europe, Oceania, North, introduced in Central and South America.

Notes. Johnson *et al.* (2020a) subsumed the genus *Hypocryphalus* Hopkins, 1915 into *Cryphalus* and transferred all the species listed under the former genus in Pullen *et al.* (2014) to *Cryphalus*, and they also synonymised the species name *Cryphalus perminimus* Schedl, 1942 with that of *C. constrictus* Schedl, 1942. All species are phloeophagous, monogynous bark beetles, and several attack living hosts and can transmit plant pathogens (Johnson *et al.* 2017). In Australia, *C. subcompactus* has been recorded to attack branches of stressed macadamia trees, the beetles boring in the bark (Mitchell & Maddox 2010). *Cryphalus mangiferae*, a globally distributed mango pest, is also present in Australia; the beetles usually colonise branches of dead mango trees but also attack stressed or diseased trees (Johnson *et al.* 2017).

Tribe Diamerini Hagedorn, 1909

Acacicis Lea, 1910 (Pl. 110C)

Type species, by monotypy: *Acacicis abundans* Lea, 1910.

Characteristics. Body length 1.1–2.5 mm, approximately 1.4 × as long as wide. Forehead flat, setose. Postmentum flat, as wide as long. Eyes oval. Antennae with 7-segmented funicle and oval club typically devoid of sutures, if present as 1 or 2 feeble straight lines. Pronotum covering head, wider than long; surface asperate anterolaterally, densely setose. Scutellar shield not exposed. Elytra with anterior margin curved, granulate in males, granules increasing in size towards middle of elytra; disc convex, shorter than declivity, shining, densely setose, strial punctures irregular or absent; declivity rounded, broad in dorsal view. Metanepisterna covered by elytra. Pro- and mesocoxae widely separate; protibiae triangular, armed with 6–8 socketed denticles.

Australian species. Native: *A. atomarius* (Chapuis, 1869), *A. minor* Schedl, 1936.

Distribution. Australia (ACT, NSW, QLD, TAS, VIC), Africa (tropical and subtropical countries), Papua New Guinea, Southeast Asia, Sri Lanka.

Notes. All species are phloeophagous, monogynous bark beetles (Wood 1986; Kirkendall *et al.* 2015). *Acacicis atomarius* has been found to breed in *Acacia* trees (Wood & Bright 1992).

Fig. 46.12. Cryphalini, *Cryphalus* sp., Garradunga QLD, structural details. A, head, dorsal; B, antenna; C, lateral; D, mouthparts; E–F, prothorax: E, dorsal; F, ventral; G–H, elytron: G, dorsal; H, lateral; I–J, pterothorax: I, ventral; J, lateral; K, abdominal ventrites; L, scutellum; M, tergite VIII, dorsal; N, hindleg; O, foreleg; P, middle leg; Q, hindleg.

Fig. 46.13. Dryocoetini, *Coccotrypes fallax* (Eggers), structural details. A, head, dorsal; B, antenna; C, head, lateral; D, mouthparts; E–F, prothorax: E, dorsal; F, ventral; G–H, elytron: G, dorsal; H, lateral; I–J, pterothorax: I, ventral; J, lateral; K, abdominal ventrites; L, scutellum; M, tergite VIII, dorsal; N, foreleg; O, middle leg; P, hindleg.

Diamerus Erichson, 1836 (Pl. 110D)

Type species, by monotypy: *Hylesinus hispidus* Klug, 1832.

Characteristics. Body length 3.0–5.0 mm, about 1.8 × as long as wide. Forehead flat, densely punctate, sparsely setose. Eyes oval. Antennae with 7-segmented funicle and large, oval club with 1 procurved suture. Postmentum wider than long, slightly to deeply impressed. Pronotum covering head, wider than long, anterior margin curved, lateral edge carinate; surface densely punctate. Scutellar shield not exposed. Elytra with anterior margin curved, costate and granulate, laterally deeply emarginate; disc flat, longer than declivity, striae deep; declivity steeply rounded, broad, emarginate, interstriae with more than one row of setae. Metanepisterna covered by elytra. Pro- and mesocoxae widely separate; protibiae apically rounded, armed with 6–8 socketed denticles.

Australian species. Native: *D. curvifer* (Walker, 1859), *D. interstitialis* (Lea, 1910).

Distribution. Australia (NSW, QLD), Africa (tropical and subtropical countries), China, Papua New Guinea, Southeast Asia.

Notes. All species are phloeophagous, monogynous, polyphagous bark beetles (Wood 1986; Kirkendall *et al.* 2015). *Diamerus interstitialis* has been recorded from *Artocarpus anisophyllus* trees, and *D. curvifer* attacks trees of this species as well as of *Macaranga andamanica*, *Ficus* and *Symplocos* (Wood & Bright 1992).

Tribe Dryocoetini Lindemann, 1877

Coccotrypes Eichhoff, 1878 (Pl. 110E, F; Fig. 46.13)

Type species, by subsequent designation (Alonso-Zarazaga & Lyal, 2009: 52): *Bostrichus dactyliperda* Fabricius, 1801.

Characteristics. Body length 1.2–3.5 mm, 2.1–2.5 × as long as wide. Forehead flat, aciculate, occasionally sparsely to densely setose. Eyes emarginate, with the upper portion smaller than lower. Antenna with 5-segmented funicle and truncate club marked with 0–2 suture lines. Postmentum flat, as wide as long. Pronotum covering head, as wide as long or longer than wide, anterior edge sometimes serrate; surface anterolaterally to entirely asperate, often typically densely punctate and sparsely setose. Scutellar shield flat, at same level as elytra. Elytra with anterior margin straight; disc flat to slightly convex, smooth to shining, interstriae with long thin sparse setae; declivity rounded, unarmed. Procoxae contiguous to separate, mesocoxae narrowly to widely separate; protibiae typically slender, armed with 1–5 socketed denticles.

Australian species. Native: *C. carpophagus* (Hornung, 1842), *C. distinctus* (Motschulsky, 1866), *C. fallax* (Eggers, 1927), *C. fijianus* (Schedl, 1942), *C. queenslandi* (Schedl, 1942), *C. tutuilensis* (Beeson, 1929), *C. vulgaris* (Eggers, 1923). Introduced: *C. advena* Blandford, 1894, *C. cardamomi* Schaufuss, 1905, *C. cyperi* (Beeson, 1929), *C. dactyliperda* (Fabricius, 1801), *C. myristicae* (Roepke, 1919), *C. nubilus* (Blandford, 1894).

Distribution. Australia (Christmas Is., NSW, QLD), circumtropical.

Notes. The species are phloeophagous, spermatophagous, inbreeding polygynous, haplodiploid bark beetles (Kirkendall *et al.* 2015). Most excavate galleries in seeds and fruits, but some develop in leafstalks, phloem tissues, pith of twigs, in ferns or mangrove propagules (notably the Australian *C. fallax*) (Jordal *et al.* 2002; Kirkendall *et al.* 2015) and also in clothes buttons (*C. dactyliperda*). Mated females construct galleries in the host trees and lay fertilised and unfertilised eggs, the former through arrhenotokous parthenogenesis and yielding haploid sons, the latter yielding diploid daughters (Jordal *et al.* 2002; Kirkendall *et al.* 2015). Offspring engage in regular sibling-mating, and mated females leave the gallery to disperse. Many species have broad ranges and are able colonisers of novel environments. *Coccotrypes carpophagous* and *C. dactyliperda* have a near-global distribution (Lantschner *et al.* 2020) and are highly polyphagous, with *C. dactyliperda* regarded as a pest of date-palm seeds (Raffa *et al.* 2015).

Cyrtogenius Strohmeyer, 1910 (Pl. 110G; Fig. 46.14)

Type species, by monotypy: *Cyrtogenius bicolor* Strohmeyer, 1910.

Characteristics. Body length 1.3–3.0 mm, 2.9 × as long as wide. Forehead flat, densely punctate (males) or setose (females). Eyes emarginate, both portions equal in size. Antennae with 3- to 5-segmented funicle (in Australia mostly 4- or 5-segmented) and truncate club typically without sutures, rarely with 1 procurved suture. Postmentum flat, as wide as long. Pronotum covering head, longer than wide; surface anterolaterally asperate, disc densely punctate or sparsely setose. Scutellar shield at same level as elytra. Elytra with anterior margin straight; disc flat, longer than declivity, strial punctures large, interstriae setose; declivity rounded, appearing almost truncate in dorsal view, sometimes granulate. Procoxae contiguous to narrowly separate, mesocoxae narrowly separate; protibiae slender, armed with 1–5 socketed denticles.

Australian species. Native: *C. dimorphus* (Schedl, 1936), *C. nitidus* (Hagedorn, 1910), *C. rotundicollis* (Eggers, 1928).

Introduced: *C. brevior* (Eggers, 1927), *C. fijianus* (Schedl, 1951).

Distribution. Australia (NSW, NT, QLD), Africa (tropical and subtropical countries), Fiji, Hong Kong, Japan, Korea, Madagascar, New Caledonia, Papua New Guinea, Samoa, China.

Fig. 46.14. Dryocoetini, *Cyrtogenius brevior* (Eggers), structural details. A, head, dorsal; B, antenna; C, head, lateral; D, mouthparts; E–F, prothorax: E, dorsal; F, ventral; G–H, elytron: G, dorsal; H, lateral; I–J, pterothorax: I, ventral; J, lateral; K, abdominal ventrites; L, scutellum; M, metanotum; N, foreleg; O, middle leg; P, hindleg; Q, hindleg.

Notes. The species are phloeophagous or xylomycetophagous, monogynous or colonial polygynous, diplodiploid and typically outbred bark and ambrosia beetles (Browne 1961; Gómez *et al.* 2012), but courtship and family structures vary. Adults construct irregular galleries, eggs are laid in niches in the sides of the gallery tunnel, the larvae construct tunnels in the phloem tissue and the pupation typically occurs in the outer phloem or bark (Beaver 1976).

Dryocoetiops Schedl, 1957 (Pl. 110H, I)

Type species, by monotypy: *Ozopemon laevis* Strohmeyer, 1911.

Characteristics. Body 1.9–4.0 mm in size, 2.4 × as long as wide. Forehead flat, without setae and punctures, epistome setose. Eyes emarginate, both portions equal in size. Antennae with 5-segmented funicle and truncate club marked with 1 recurved suture. Postmentum flat, wider than long. Pronotum covering head, wider than long, posterior margin straight; surface anterolaterally asperate, on disc densely punctate and sparsely covered with long thin setae. Scutellar shield flat, at same level as elytra. Elytra with anterior margin straight; disc convex, longer than declivity, sparsely covered with long thin setae, interstrial punctures irregular; declivity steeply rounded, interstriae granulate. Pro- and mesocoxae narrowly separate; protibiae slender, armed with 4 socketed denticles.

Australian species. Introduced: *D. moestus* (Blandford, 1894).

Distribution. Australia (NSW, QLD, VIC), Papua New Guinea, Southeast Asia.

Notes. The names of the two species listed under the genus by Pullen *et al.* (2014), *D. australis* (Schedl) and *D. coffeae* (Eggers), have since been synonymised with that of *D. moestus* (Beaver *et al.* 2019a), which is thus the only species present in Australia. *Dryocoetiops* species are bigynous or harem polygynous and, similar to *Coccotrypes*, inbreeding and thought to be haplodiploid (Beaver *et al.* 2019a). They are polyphagous, the galleries constructed in twigs by mated females. The males are unknown, but as the spermathecae of females contain sperm, males exist and reproduction is not parthenogenic (Beaver *et al.* 2019a).

Peridryocoetes Wood, 1984 (Pl. 110J, K)

Type species, by original designation: *Ozodendron nitens* Schedl, 1964.

Characteristics. Body length 1.6–1.8 mm, 2.0–2.1 × as long as wide. Forehead flat, punctate, setose. Eyes emarginate, both upper and lower halves equal in size. Antennae with 3-, 4- or 5-segmented funicle (5-segmented in *P. queenslandi*) and elongate, truncate club typically without sutures but sometimes with 1 recurved suture line. Postmentum flat, prementum wider than long. Pronotum covering head, wider than long, posterior margin straight; surface entirely asperate, covered with short setae. Scutellar shield flat, at same level as elytra. Elytra with anterior margin straight; disc flat, shining, strial punctures large, interstriae setose; declivity shorter than disc, steeply rounded, unarmed. Procoxae contiguous, mesocoxae widely separate; protibiae slender, armed with 1–5 socketed denticles.

Australian species. Native: *P. queenslandi* Schedl, 1972.

Distribution. Australia (QLD), Papua New Guinea, Southeast Asia.

Notes. Little is known regarding the life histories of the species of this genus (Kirkendall *et al.* 2015). They are presumed to be phloeophagous and polyphagous, having been collected from galls of *Garcinia* species and *Flindersia schottiana* (Schedl 1972), but the reproductive system is unknown.

Tribe Ernoporini Nüsslin, 1911

This tribe was not recorded from Australia by Pullen *et al.* (2014) as its name was considered to be synonym of that of Cryphalini at the time (Alonso-Zarazaga & Lyal 2009). It was treated as a distinct tribe by Johnson *et al.* (2020a) and is represented in Australia by the genus *Eidophelus*. Johnson *et al.* (2020a) also transferred *Hemicryphalus* Schedl to Ernoporini, but this genus is no longer considered to occur in Australia (Roger Beaver pers. comm.).

Eidophelus Eichhoff, 1878 (Pl. 111A)

Type species, by monotypy: *Eidophelus imitans* Eichhoff, 1876.

Characteristics. Body length 1.4–2.5 mm, 2.0–2.3 × as long as wide. Forehead flat, smooth to punctate, sparsely setose. Eyes oval. Antenna with 3- to 4-segmented funicle and large, flat, oval club marked with 1 partial suture. Postmentum flat, wider than long. Pronotum covering head, wider than long, posterior margin straight; surface asperate anterolaterally or entirely, covered with short thick setae. Scutellar shield various, occasionally flat and at same level as elytra (*E. darwini*) or not exposed (*E. tricolor*). Elytra with anterior margin straight; disc convex, longer than declivity; interstriae variably setose, setae scale- or hair like, covering entire elytra or not; declivity steeply rounded to vertical, unarmed. Procoxae contiguous, mesocoxae widely separate; protibiae apically rounded, armed with 2–6 socketed denticles.

Australian species. Native: *E. agnatus* (Schedl, 1942) (Schedl 1942: 168 not Schedl 1939 as in Pullen *et al.* 2014),

E. australis (Schedl, 1942), *E. brimblecombei* (Schedl, 1972), *E. furvus* Johnson, 2020, *E. tricolor* (Lea, 1910). Introduced: *E. darwini* (Eichhoff, 1878).

Distribution. Australia (Christmas Is., QLD), all continents except Antarctica.

Notes. The species are myelophagous or phloeophagous, bigynous or monogynous bark beetles (Kirkendall *et al.* 2015) and presumably polyphagous, having been collected from *Cryptolepis buchananii* (*E. darwini*) and *Millettia megasperma* (*E. tricolor*) (Wood & Bright 1992). *Eidophelus* is a senior synonym of *Scolytogenes* Eichhoff and *Miocryphalus* Schedl. After Jordal (2021), *Eidophelus* also includes *E. agnatus* Schedl, previously placed in *Afromicracis* Schedl (Micracidini) but suspected to be misclassified there by Pullen *et al.* (2014).

Tribe Hylastini LeConte, 1876

Hylastes Erichson, 1836 (Pl. 111B)

Type species, by subsequent designation (Westwood, 1838: 39): *Bostrichus ater* Paykull, 1800.

Characteristics. Body length 4.0–4.3 mm, 3.1–3.3 × as long as wide. Forehead and vertex flat, punctate, sparsely setose. Eyes oval. Antenna with 7-segmented funicle and ovate club marked with 3 straight suture lines. Postmentum impressed, long, shallow. Pronotum not covering head, longer than wide, posterior margin straight; surface densely punctate, laterally sparsely setose. Scutellar shield small, at same level as elytra. Elytra with anterior margin straight, sparsely setose; disc flat, much longer than declivity; declivity steeply rounded, unarmed, strial punctures deep. Procoxae contiguous, mesocoxae widely separate; protibiae apically triangular to rounded, armed with 3–4 four socketed denticles; tarsi with third segment bilobed.

Australian species. Introduced: *H. ater* (Paykull, 1800).

Distribution. Australia (ACT, NSW, SA, TAS, VIC), China, Europe, Jamaica, Japan, Korea, New Zealand, North America, South Africa, South America (Argentina and Chile), Turkey.

Notes. The single species introduced into Australia is a phloeophagous, monogynous bark beetle specific to Pinacaeae (Wood & Bright 1992; Kirkendall *et al.* 2015). The beetles are primarily attracted to roots of weakened, stressed or recently killed trees, but they also breed in trunks and branches of dead trees (Neumann & Harris 1974), excavating tunnels under the bark to feed on the phloem tissues. They form monogamous pairs, and parents overlap with offspring in the galleries (Kirkendall *et al.* 2015). Sapstain fungi have been found to be associated with *Hylastes*, commonly the genera *Graphium*, *Leptographium* and *Ophiostoma* (Griggs 1998). These fungi are of economic concern in pine forestry as they cause cosmetic and, occasionally, structural damage to the trees and timber (Trollip *et al.* 2021). A notable attack by *H. ater* occurred in the Stanley Plantation in Victoria, 1962, in which all 1-year-old *Pinus radiata* seedlings were killed (Neumann 1987).

Tribe Hylesinini Erichson, 1836

Dactylipalpus Chapuis, 1869 (Pl. 113A)

Type species, by subsequent designation (Hopkins, 1914: 120): *Dactylipalpus transversus* Chapuis, 1869.

Characteristics. Body length 5.4–9.5 mm, 1.86–2.1 × as long as wide. Forehead and vertex flat and sparsely setose in females, slightly concave and glabrous in males. Eyes oval. Antennae with 7-segmented funicle and ovate club with no defined suture lines. Postmentum impressed, strongly reduced. Pronotum partially covering head, wider than long, anterior margin flat, posterior margin curved; surface densely punctate, slightly setose to glabrous, in females in anterior third with transverse slit-like mycangia. Scutellar shield small, at same level as elytra. Elytra with anterior margin curved, granulate; disc flat to slightly convex, longer than declivity, striae deep, interstriae with more than one row of setae; declivity rounded, granulate. Procoxae and mesocoxae widely separate; protibiae apically triangular, armed with 8–10 socketed denticles; tarsi with third segment bilobed.

Australian species. Native: *D. transversus* Chapuis, 1869.

Distribution. Australia (NSW, QLD), Southeast Asia, Papua New Guinea.

Notes. The species are xylomycetophagous and monogynous (Kirkendall *et al.* 2015). They are assumed to be ambrosia beetles, but their primary fungal symbionts are unknown and it is uncertain whether they practice true, obligate fungal cultivation (Hulcr & Stelinski 2017; Mayers *et al.* 2022). Observations from collecting of specimens indicate that parental galleries are stained black with fungi and larval galleries are decayed by fungi (Wood 1986). The single species in Australia is not host-specific, as specimens have been collected from tree species belonging to different families, such as *Ficus religiosa* (Moraceae), *Magnolia insignis* (Magnoliaceeae), *Mesua ferrea* (Calophyllaceae), *Myristica philippensis* (Myristicaceae) and *Neolamarckia cadamba* (Rubiaceae) (Wood & Bright 1992).

Ficicis Lea, 1910 (Pl. 111C, D)

Type species, by subsequent designation (Hopkins, 1914: 122): *Ficicis varians* Lea, 1910.

Characteristics. Body length 1.7–2.0 mm, 1.9–2 × as long as wide. Forehead and vertex variable in shape, flat or concave (males). Forehead typically smooth, sparsely punctate, occasionally with median carina and densely setose. Eyes oval. Antennae with 7-segmented funicle and ovate club marked with 3 straight suture lines. Postmentum slightly impressed, about half as wide as long. Pronotum covering head, wider than long, anterior margin flat, posterior margin curved; surface asperate, punctate on disc, densely covered with setae. Scutellar shield small, at same level as elytra. Elytra with anterior margin curved, costate, granulate, entirely or partly setose; disc variable in length, convex, striae deep, interstriae with slender or thick setae; declivity attenuate dorsally, rounded laterally, granulate. Procoxae narrowly to widely separate, mesocoxae widely separate; protibiae slender to triangular, armed with 3 or fewer socketed denticles; tarsi with third segment bilobed.

Australian species. Native: *F. elongatus* (Schedl, 1942), *F. maculipennis* (Schedl, 1975), *F. porcatus* (Chapuis, 1869), *F. robustus* (Eggers, 1939), *F. varians* Lea, 1910.

Distribution. Australia (NSW, QLD, WA), Fiji, Papua New Guinea, Samoa, Southeast Asia.

Notes. The species are phloeophagous, monogynous, polyphagous bark beetles (Wood & Bright 1992; Kirkendall *et al.* 2015).

Hylesinus Fabricius, 1801 (Pl. 111E, F)

Type species, by subsequent designation (Latreille, 1810: 431): *Hylesinus crenatus* Fabricius, 1801.

Characteristics. Body length 3.0–3.3 mm, 1.7–2 × as long as wide. Forehead and vertex variable in shape, flat or concave (in males), mostly smooth and densely covered with setae. Eyes oval. Antennae with 7-segmented funicle and ovate club marked with 3 straight suture lines. Pronotum covering head, anterior margin rounded, armed with several tubercles, posterior margin strongly curved; surface punctate, densely setose. Scutellar shield small, at same level as elytra. Elytra with anterior margin curved, costate and granulate, densely covered with short, thick setae, striae deep; disc convex, gradually rounded onto declivity; declivity slightly rounded, truncate, unarmed. Pro- and mesocoxae widely separate; protibiae apically rounded, armed with 5 socketed denticles; tarsi with third segment bilobed.

Australian species. Native: *H. cordipennis* Lea, 1910.

Distribution. Australia (QLD), cosmopolitan.

Notes. The species are phloeophagous, monogynous, polyphagous bark beetles (Wood & Bright 1992; Kirkendall *et al.* 2015). Specimens of the single Australian species have been found boring in *Fraxinus* trees, but it is unknown whether it also colonises other hosts. *Trichogramma* and *Entedon ergias* wasps are known to parasitise eggs of *Hylesinus* (Wegensteiner *et al.* 2015).

Zygophloeus Schedl, 1958 (Pl. 111G, H)

Type species, by monotypy: *Zygophloeus australis* Schedl, 1958.

Characteristics. Body length 1.0–1.3 mm, 2.0 × as long as wide. Forehead and vertex mostly smooth but with some punctures, setose. Eyes oval. Antennae with 6-segmented funicle and oval club marked with 2 straight sutures. Pronotum covering head, wider than long, posterior margin curved; surface smooth with some punctures, setose. Scutellar shield not exposed. Elytra with anterior margin curved; disc flat, steeply rounded onto declivity, setose, strial punctures deep; declivity unarmed, setose. Pro- and mesocoxae widely separate; protibiae slender to triangular, armed with up to 5 socketed denticles.

Australian species. Native: *Z. australis* Schedl, 1958.

Distribution. Australia (NSW, QLD).

Notes. The single species in Australia is assumed to be a phloeophagous, monogynous bark beetle (Wood 1986) host-specific to *Acacia* (Wood & Bright 1992).

Tribe Hylurgini Gistel, 1848

Chaetoptelius Fuchs, 1913 (Pl. 111I, J)

Type species, by monotypy: *Hylesinus vestitus* Mulsant & Rey, 1860.

Characteristics. Body length 2.1–2.9 mm, 2.0–2.6 × as long as wide. Forehead and vertex variable, flat or impressed (in males), setose. Eyes oval. Antennae with 7-segmented funicle and ovate club with 2 straight suture lines. Pronotum not covering head, setose, anterior margin feebly armed with asperities, posterior margin curved. Scutellar shield at same level as elytra. Elytra with anterior margin curved, granulate; disc longer than declivity, flat, steeply rounded onto declivity; striae deep; setae short, densely covering elytra; declivity unarmed. Pro- and mesocoxae widely separate; protibiae triangular, typically armed with 6 socketed denticles; tarsi with third segment bilobed.

Australian species. Native: *C. bimaculatus* (Schedl, 1936), *C. squamosus* (Schedl, 1942), *C. tricolor* (Schedl, 1938).

Distribution. Australia (NSW, QLD), Brazil, Chile, Equatorial Africa, New Zealand, Papua New Guinea, Southeast Asia.

Notes. The species are phloeophagous, monogynous, polyphagous bark beetles. The eggs are laid in niches in the galleries and packed in frass, and the larvae mine away from the parental tunnel (Wood 1986). *Chaetoptelius bimaculatus* has been found in lianas and *C. tricolor* in wood of *Baloghia lucida* (Euphorbiaceae). One species is regarded as a pest, *C. vestitus*, which is distributed throughout Africa and Europe and attacks pistachio trees (*Pistacia*, Anacardiaceae), infestations leading to reduced growth and eventual mortality (Yaman 1969).

Hylurdrectonus Schedl, 1938 (Pl. 111K, L)

Type species, by monotypy: *Hylurdrectonus piniarius* Schedl, 1938.

Characteristics. Body length 1.4–1.6 mm, 2.5–2.8 × as long as wide. Forehead and vertex typically impressed. Eyes oval. Antennae with 5-segmented funicle and ovate club and marked with 2 straight sutures. Postmentum slightly impressed, twice as wide as long. Pronotum not covering head, conical in shape, anterior and posterior margins straight; surface smooth. Scutellar shield at same level as elytra. Elytra with anterior margin straight; disc flat, longer than declivity, steeply rounded onto declivity, sparsely setose, striae shallow; declivity with fine (females) to large (males) granules. Pro- and mesocoxae widely separate; protibiae slender, armed with 5 socketed denticles; tarsi with third segment bilobed.

Australian species. Native: *H. piniarius* Schedl, 1938.

Distribution. Australia (QLD), Papua New Guinea.

Notes. The species are mostly phloeophagous, monogynous bark beetles, but *H. piniarius* is known to be spermatophagous (Schedl 1938). The phloeophagous species construct galleries in branchlets, whereas spermatophagous species do so in cones. All species are host specific to *Araucaria* (Araucariaceae) (Wood & Bright 1992), with *H. araucariae* Schedl regarded as a pest species of Hoop Pine plantations in Papua New Guinea, where major infestations lead to foliage dieback (Gray & Lamb 1975).

Hylurgus Latreille, 1806 (Pl. 112A; Fig. 46.15)

Type species, by monotypy: *Bostrichus ligniperda* Fabricius, 1787.

Characteristics. Body length 4.0–5.5 mm, 2.8–3.0 × as long as wide. Forehead and vertex typically flat to slightly impressed, densely punctate, slightly to densely setose, frons with tubercle above epistome in males. Eyes oval. Antennae with 6-segmented funicle and ovate club and marked with 3 straight sutures. Postmentum slightly impressed, twice as wide as long. Pronotum not covering head, conical in shape, anterior margin straight, posterior margin curved; surface densely punctate. Scutellar shield at a lower level than elytra. Elytra with anterior margin curved; disc flat, longer than declivity, steeply rounded onto declivity, sparsely setose, striae shallow; declivity unarmed, setose. Procoxae narrowly separate, mesocoxae widely separate; protibiae triangular, armed with 6 socketed denticles; tarsi with third segment bilobed.

Australian species. Introduced: *H. ligniperda* (Fabricius, 1787).

Distribution. Australia (ACT, NSW, QLD, SA, TAS, VIC); native in Asia, Atlantic islands, Europe, North Africa; introduced in Argentina, Brazil, Chile, Eswatini, Japan, New Zealand, North America, South Africa, Sri Lanka, St. Helena, Uruguay.

Notes. The species are phloeophagous, monogynous bark beetles (Kirkendall *et al.* 2015). The single species in Australia commonly attacks *Pinus radiata* trees, preferentially developing in large-diameter felled logs or stumps and excavating galleries under the bark and feeding on the phloem tissue (Neumann 1987; Mausel *et al.* 2007). It can co-exist with *Hylastes ater* in fallen pines, but it typically succeeds the former in gallery excavation and can outcompete extraspecifics (Neumann 1987). Generally, *H. ligniperda* only attacks weakened, stressed or recently dead pine trees, but it has the capacity to attack healthy hosts when present in large populations. Population numbers grow when large volumes of pine slash are built up through clear-felling operations. In South Australia, populations built up in pine plantations damaged by fires in 1985 and dispersed over 25 km to nearby plantations, killing several hundred pines trees aged between 4 and 14 years (Neumann 1987). While the species is considered as a bark beetle, specimens have been found in association with several fungal genera, including *Grosmannia, Hyalorhinocladiella, Leptographium, Ophiostoma* and *Sporothrix*, which are common blue-stain fungi that cause economic damage to trees through cosmetic staining of the timber and also some structural damage (Kim *et al.* 2011).

Pachycotes Sharp, 1877 (Pl. 112B)

Type species, by monotypy: *Pachycotes ventralis* Sharp, 1877.

Characteristics. Body length 2.3–4.2 mm, 2.1–2.4 × as long as wide. Forehead and vertex typically flat to impressed, slightly to densely punctate. Eyes oval. Antennae with 7-segmented funicle and ovate club marked with 2–3 straight sutures. Pronotum not covering head, wider than long, anterior margin straight to emarginate, posterior margin curved; surface densely punctate. Scutellar shield at same level as elytra or depressed below level of elytra. Elytra with anterior margin curved, striae deep; disc flat, steeply rounded onto declivity; declivity setose, granulate. Procoxae narrowly separate, mesocoxae widely separate; protibiae slender, armed with 4 socketed denticles; tarsi with third segment bilobed.

Australian species. Native: *P. australis* Schedl, 1938, *P. clavatus* Schedl, 1938, *P. kuscheli* Schedl, 1972, *P. minor* Wood, 1985, *P. villosus* Schedl, 1962. Introduced: *P. peregrinus* (Chapuis, 1869).

Distribution. Australia (NSW, Norfolk Is., QLD, WA), New Zealand, Papua New Guinea.

Notes. The species are xylophagous, monogynous bark beetles (Kirkendall *et al.* 2015) typically host-specific to *Araucaria* trees (Araucariaceae) (Mecke 2004), only *P. peregrinus* found to be associated with a different species and family, *Pseudotsuga menziesii* of Pinaceae (Wood & Bright 1992).

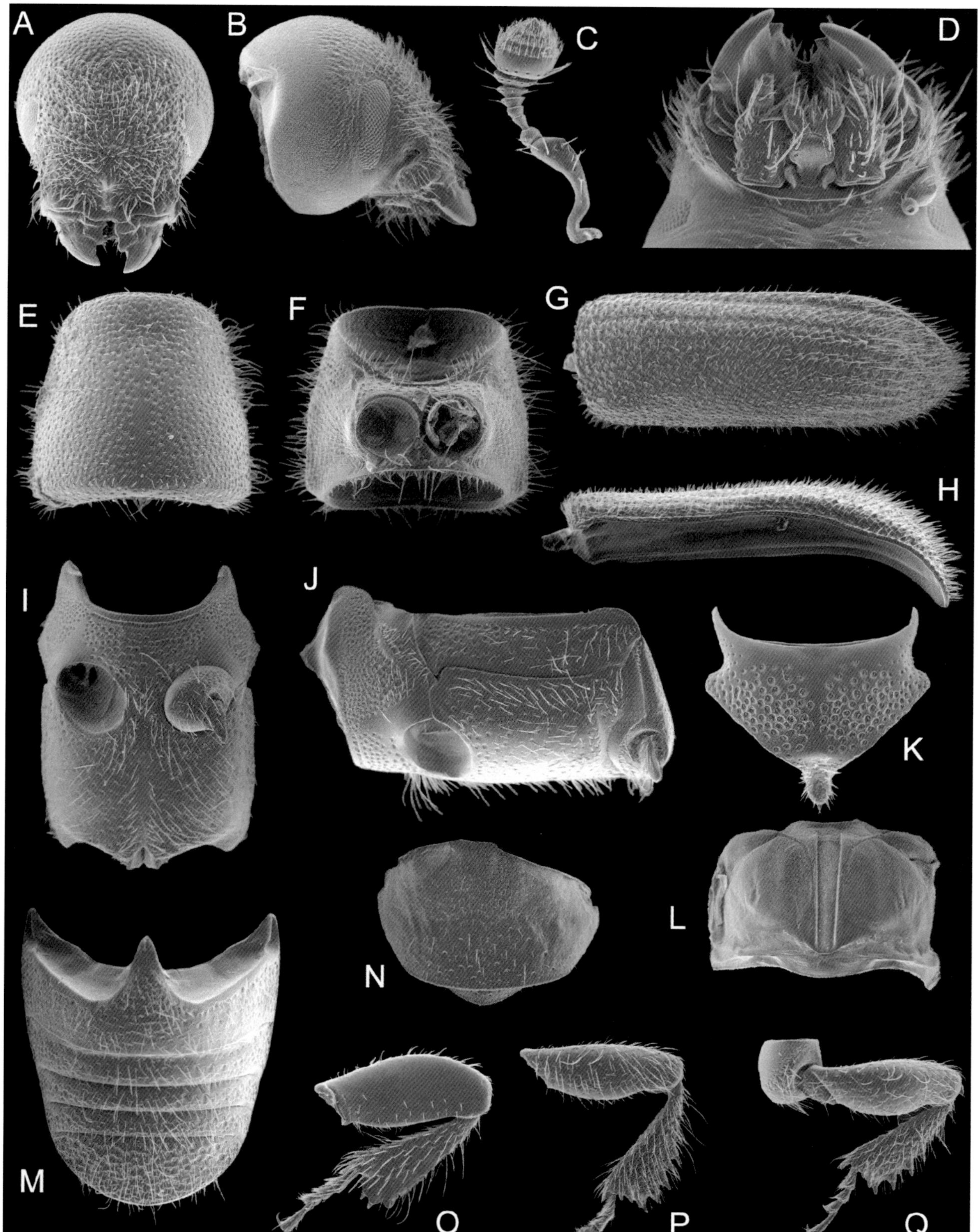

Fig. 46.15. Hylurgini, *Hylurgus ligniperda* (Fabricius), structural details (♀). A–B, head: A, dorsal; B, lateral; C, antenna; D, mouthparts; E–F, prothorax: E, dorsal; F, ventral; G–H, elytron: G, dorsal; H, lateral; I–J, pterothorax: I, ventral; J, lateral; K, scutellum; M, abdominal ventrites; N, tergite VIII, dorsal; L, metanotum; O, hindleg; P, middle leg; Q, foreleg.

Tribe Hyorrhynchini Hopkins, 1915

Sueus Murayama, 1951 (Pl. 112C, D)

Type species, by original designation: *Sueus sphaerotrypoides* Murayama, 1951.

Characteristics. Body length 1.6 mm, 2.0 × as long as wide. Forehead flat, weakly punctate, sparsely setose, with median line. Eyes completely divided, upper portion larger than lower. Antennae with 6-segmented funicle and oval club marked with 2 straight sutures. Postmentum slightly impressed, as wide as long. Pronotum not covering head, wider than long, quadratic in shape, posterior margin straight; surface densely punctate. Scutellar shield at same level as elytra, disc impressed. Elytra with anterior margin curved, costate, raised in the middle; disc flat, setose, longer than declivity and steeply rounded onto declivity, striae deep; declivity unarmed, setose. Pro- and mesocoxae widely separate; protibiae triangular, unarmed; tarsi with third segment bilobed.

Australian species. Native: *Sueus niisimai* (Eggers, 1926).

Distribution. Australia (QLD), Japan, Papua New Guinea, Southeast Asia; introduced in Fiji and Martinique.

Notes. The single species in Australia was described from Japan and occurs widely in Southeast Asia as far south as New Guinea and Australia. It was surmised by Pullen *et al.* (2014) to have possibly been introduced in Australia, but Smith *et al.* (2022a) regarded it as native in the country. It is a xylomycetophagous, inbreeding polygynous, highly polyphagous ambrosia beetle (Kirkendall *et al.* 2015), having been recorded from 39 tree species including one conifer and 24 angiosperm families (Smith *et al.* 2022a). It is sexually dimorphic, the females being larger than the dwarfed and flightless males. The females construct radial galleries in twigs, which house one female parent and her offspring, and the males do not leave their natal nest (Beaver 1984; Kirkendall *et al.* 2015; Smith *et al.* 2022b). The location of the mycangia is still unknown. The primary symbiont of the species has been identified as *Diatrypella japonica*, the Canker Disease fungus (Li *et al.* 2020), which is a pathogen of sycamore trees (*Platanus*) in Japan, and pathogenicity trials have found that Shumard Oak (*Quercus shumardii*) responded with lesion but no dieback or mortality (Li *et al.* 2020).

Tribe Hypoborini Nüsslin, 1911

Liparthrum Wollaston, 1854 (Pl. 112E)

Type species, by original designation: *Liparthrum bituberculatum* Wollaston, 1854.

Characteristics. Body length 1.7 mm, 2.1 × as long as wide. Forehead flat, mostly smooth with some punctures, densely setose. Eyes oval. Antennae with 4-segmented funicle and oval club without suture lines. Pronotum covering head, slightly longer than wide, posterior margin curved; surface densely punctate, setose. Scutellar shield not exposed. Elytra with anterior margin curved; disc flat, steeply rounded onto declivity, densely setose, strial punctures deep; declivity unarmed, setose. Procoxae narrowly separate, mesocoxae widely separate; protibiae slender, armed with 4 socketed denticles.

Australian species. Introduced: *Liparthrum* sp.

Distribution. Australia (NT), Africa, Europe, North America, South America, Southeast Asia, Western Asia.

Notes. The species are phloeophagous, monogynous bark beetles (Kirkendall *et al.* 2015). The species associated with *Euphorbia* form a monophyletic group, indicating some host specialization to exist in the group (Jordal *et al.* 2004).

Tribe Ipini Bedel, 1888

Acanthotomicus Blandford, 1894 (Pl. 112F)

Type species, by monotypy: *Acanthotomicus spinosus* Blandford, 1894.

Characteristics. Body length 2.0–2.1 mm, 2.8 × as long as wide. Forehead smooth with some punctures, sparsely setose. Eyes emarginate, both halves equal in size. Antenna with 5-segmented funicle and truncate club marked with 2 procurved sutures. Postmentum flat, as wide as long. Pronotum covering head, longer than wide, posterior margin straight; surface smooth with some punctures, asperate anteriorly. Scutellar shield depressed below level of elytra. Elytra with anterior margin straight; disc flat, longer than declivity, strial punctures shallow; declivity truncate, excavate, armed with 2 spines, margin costate, declivital face sparsely setose. Procoxae contiguous to narrowly separate, mesocoxae narrowly separate; protibiae triangular, armed with 3–4 socketed denticles.

Australian species. Native: *A. australis* (Schedl, 1972).

Distribution. Australia (QLD), Africa (tropical countries), China, Japan, North America, South America, Southeast Asia.

Notes. The species are phloeophagous, harem polygynous or monogynous bark beetles distributed throughout tropical regions, with a large number occurring in Asia. They typically construct galleries in dead and dying trees, and, unlike other Ipini species, they prefer angiosperms over gymnosperms as hosts (Cognato & Smith 2020). Little is known regarding the life history of the genus, but *A. suncei* Cognato, regarded as an emerging pest of sweetgum in China, has been well studied. It attacks American Sweetgum, *Liquidambar styraciflua*, and Chinese Sweetgum, *L. formosana*. It has killed over 30 000 trees in Shanghai, but it has not been determined whether it is

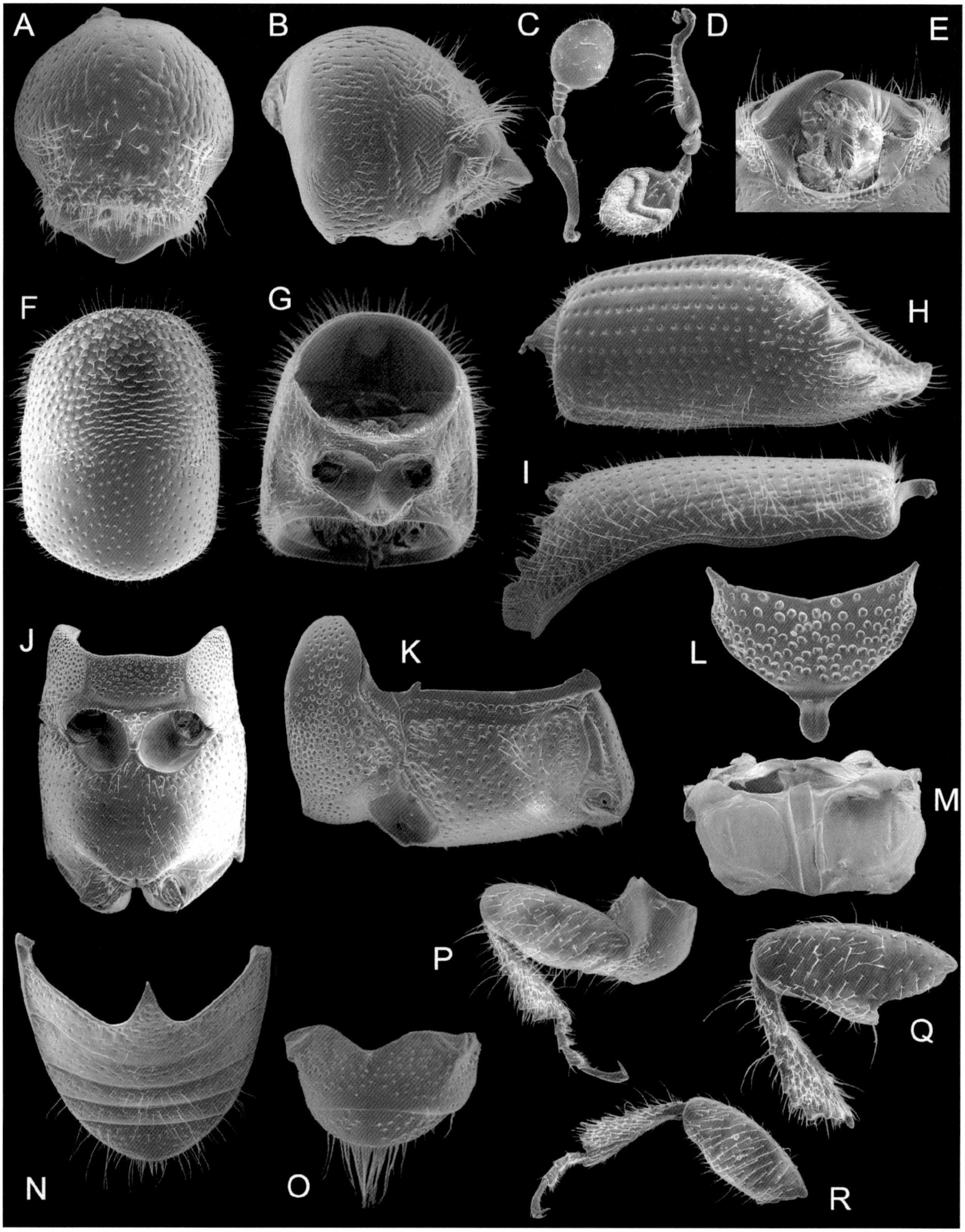

Fig. 46.16. Ipini, *Ips grandicollis* (Eichhoff), structural details. A–B, head: A, dorsal; B, lateral; C–D, antenna: C, dorsal and D, ventral; E, mouthparts; F–G, prothorax: F, dorsal; G, ventral; H–I, elytron: H, dorsal; I, lateral; J–K, pterothorax: J, ventral; K, lateral; L, scutellum; M, metanotum; N, abdominal ventrites; O, tergite VIII, ventral; P, foreleg; Q, hindleg; R, middle leg.

the beetles, the fungal associates or a combination of both that cause the host mortality. The common phytopathogenic fungi *Geosmithia* and *Phaeoacremonium tuscanum* are associates of *A. suncei* in Shanghai, but they have been experimentally shown to not harm American Sweetgum hosts (Gao *et al.* 2021). In Australia, *A. australis* has been recorded only from *Ficus macrophylla* (Wood & Bright 1992), but whether it is host-specific or polyphagous is yet to be determined.

Ips De Geer, 1775 (Pl. 112G, H; Fig. 46.16)

Type species, by subsequent designation (Crotch, 1870: 44): *Dermestes typographus* Linnaeus, 1758.

Characteristics. Body length 3.5–3.9 mm, 2.6–2.8 × as long as wide. Forehead smooth with some punctures, with central tubercles in males, sparsely setose. Eyes emarginate, halves equal in size. Antennae with 5-segmented funicle and truncate club marked with 2 bisinuate sutures. Postmentum flat, wider than long. Pronotum covering head, longer than wide, posterior margin straight; surface densely punctate, sparsely setose, asperate anteriorly. Scutellar shield at same level as elytra, disc impressed. Elytra with anterior margin straight; disc flat, longer than declivity, strial punctures deep; declivity truncate, excavate, armed with 5 spines, costate and setose. Procoxae contiguous, mesocoxae narrowly separate; protibiae triangular, armed with 4 socketed denticles.

Australian species. Introduced: *I. grandicollis* (Eichhoff, 1868).

Distribution. Australia (ACT, NSW, QLD, SA, VIC, WA), Asia, Europe, North America, Southeast Asia.

Notes. The species are phloeophagous, harem polygynous bark beetles (Kirkendall *et al.* 2015) obligately associated with *Pinus* species in Australia. The beetles usually attack weakened, damaged or moribund trees, either at the base or the mid-section of the trunk and branches (Yousuf *et al.* 2014). Typically, galleries are initiated by males by the construction of a nuptial chamber to attract mates, and males mate with multiple females. All mated individuals reside in the same gallery, with females constructing brood galleries that branch from the nuptial chamber, in which the eggs are laid (Cognato 2015). Infestation of host trees occurs by an initial attack by a small number of beetles, followed by mass colonisation. Males produce ipsenol, a pheromone that attracts both female mates and other males and can facilitate mass aggregation (Yousuf *et al.* 2014). Host trees are damaged not through the boring activity but by blue-stain fungi, *Ophiostoma ips*, which blocks xylem tissues. In Australia, *I. grandicollis* is typically a secondary pest in pine plantations, damaging stressed and moribund trees, but if populations build up following extreme and prolonged stress events on trees (such as drought), outbreaks of the species can occur and healthy trees can be subject to attack (Neumann 1987). No biological control agents have been successful in controlling *I. grandicollis* populations in Australia, and appropriate silvicultural management has best controlled this species (Yousuf *et al.* 2014).

Tribe Phloeosinini Nüsslin, 1912

Hyledius Sampson, 1921 (Pl. 112I, J)

Type species, by monotypy: *Hyledius asper* Sampson, 1921.

Characteristics. Body length 2.6 mm, 2.3 × as long as wide. Forehead and vertex punctate, setose. Eyes divided, upper portion smaller. Antenna with 5-segmented funicle and oval club marked with 2 partial straight sutures. Postmentum as wide as long, slightly impressed. Pronotum not covering head, wider than long, posterior margin curved; surface densely punctate, sparsely setose. Scutellar shield flat, at same level as elytra. Elytra with anterior margin curved, costate and granulate; disc flat, longer than declivity, sparsely setose, striae deep; declivity steeply rounded, setose, granulate. Abdominal ventrites covered with setae bifurcate. Pro- and mesocoxae widely separate; protibiae triangular, armed with 4 socketed denticles.

Australian species. Native: *H. cribratus* (Blandford, 1896).

Distribution. Australia (QLD), Papua New Guinea, Southeast Asia.

Notes. The species are phloeophagous, monogynous or harem polygynous bark beetles (Kirkendall *et al.* 2015) specific to Myristicaceae (Hulcr *et al.* 2007). The single species in Australia has been found to breed in *Knema furfuracea*, *Myristica*, *Palaquium gutta* and *Pterocymbium* (Wood & Bright 1992).

Hyleops Schedl, 1938 (Pl. 112K)

Type species, by monotypy: *Hyleops glabratus* Schedl, 1938.

Characteristics. Body length 3.5 mm, 2.0 × as long as wide. Forehead and vertex concave, densely punctate, setose. Eyes oval, entire. Antenna with 7-segmented funicle and oval club marked with 2 straight sutures. Postmentum wider than long, slightly impressed. Pronotum not covering head, wider than long, posterior margin straight; surface densely punctate, setose. Scutellar shield not exposed. Elytra with anterior margin curved, costate, granulate; disc convex, gradually rounded onto declivity, setose, strial punctures shallow; declivity setose, armed with 4 tubercles in females, granulate in males. Procoxae contiguous, mesocoxae narrowly separate; protibiae triangular, armed with up to 7 socketed denticles; tarsi with third segment bilobed.

Australian species. Native: *H. glabratus* Schedl, 1938.

Distribution. Australia (QLD).

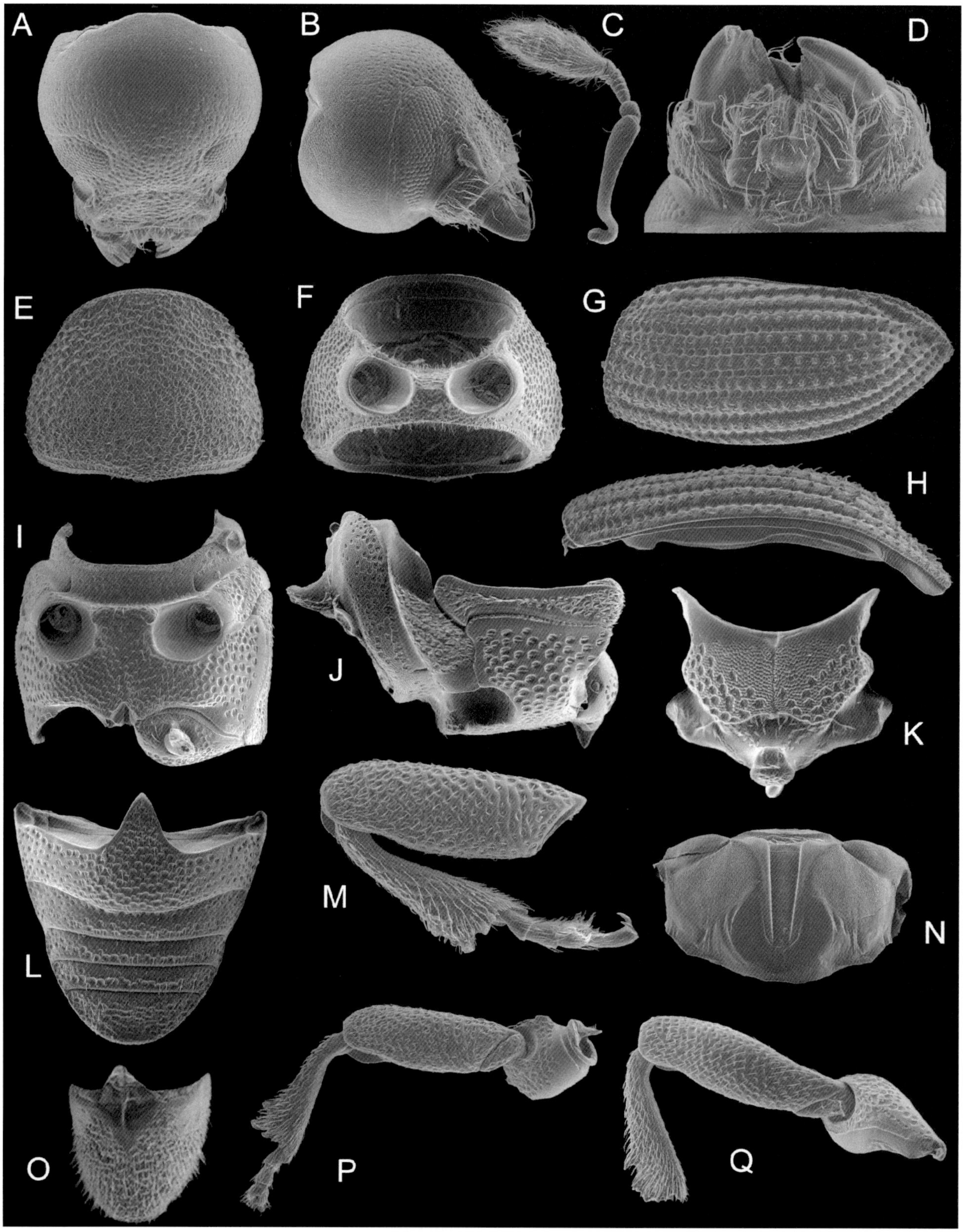

Fig. 46.17. Phloeosinini, *Phloeosinus* sp., Garradunga, QLD, structural details. A–B, head: A, dorsal; B, lateral; C, antenna; D, mouthparts; E–F, prothorax: E, dorsal; F, ventral; G–H, elytron: G, dorsal; H, lateral; I–J, pterothorax: I, ventral; J, lateral; K, scutellum; L, abdominal ventrites; M, foreleg; N, metanotum; O, terminalia, dorsal; P, middle leg; Q, hindleg.

Notes. The adults are phloeophagous, monogynous bark beetles (Kirkendall *et al.* 2015), but the larvae are xylomycetophagous (Wood 1986). The single species in Australia has only been recorded from *Araucaria cunninghamii*. The beetles excavate galleries in the branches, and larger larvae bore into the xylem tissue and feed on fungi growing in the tunnels, but the galleries constructed by the adult beetles are devoid of symbiotic fungi (Wood 1986). The identity of the symbiont fungi on which the larvae feed is unknown, as is whether the adults or the larvae cultivate the symbionts (Hulcr & Stelinski 2017).

Phloeosinopsioides Schedl, 1964 (Pl. 112L)

Type species, by original designation: *Phloeosinopsis triseriata* Schedl, 1964.

Characteristics. Body length 1.4–2.5 mm, 2.1–2.4 × as long as wide. Forehead impressed in males, convex in females, forehead and vertex densely setose, mostly smooth with some punctures. Eyes slightly emarginate, upper portion larger than lower. Antennae with 5-segmented funicle and oval club marked with 2 straight sutures. Pronotum not covering head, wider than long, posterior margin curved; surface asperate anterolaterally, densely setose. Scutellar shield small, at same level as elytra. Elytra with anterior margin curved, costate, granulate; disc flat, rounded onto declivity, striae deep, interstriae with single row of thick, short setae; declivity unarmed. Pro- and meso-coxae widely separate; protibiae triangular, armed with 6–8 socketed denticles.

Australian species. Native: *P. leai* (Schedl, 1936), *P. formosanus* (Schedl, 1935), *P. pumilus* Wood, 1985.

Distribution. Australia (NT, QLD), Papua New Guinea, Southeast Asia.

Notes. Little is known about the life history of the species. Wood (1986) stated that they are monogamous and phloeophagous, but this has not been confirmed (Beaver 2011a). The only recorded hosts are unidentified lianas.

Phloeosinus Chapuis, 1869 (Pl. 113B; Fig. 46.17)

Type species, by subsequent designation (Hopkins, 1914: 126): *Hylesinus thujae* Perris, 1855.

Characteristics. Body length 1.7–2.5 mm, 1.7–2.3 × as long as wide. Forehead and vertex punctate. Eyes emarginate, upper portion larger. Antennae with 5-segmented funicle and oval club marked with 2 straight sutures. Pronotum not covering head, wider than long, and posterior margin curved; surface densely punctate, sparsely setose. Scutellar shield at same level as elytra. Elytra with anterior margin curved, costate, granulate; disc steeply rounded onto declivity, sparsely setose, strial punctures shallow to deep; declivity armed with tubercles evenly spaced on interstriae 1 to 4, densely setose. Pro- and mesocoxae widely separate; protibiae triangular, armed with up to 8 socketed denticles; tarsi with third segment bilobed.

Australian species. Native: *P. transversarius* Schedl, 1936. Introduced: *P. cupressi* Hopkins, 1903, *P. dubiosus* Schedl, 1972 (possibly, see Pullen *et al.* 2014).

Distribution. Australia (ACT, NSW, QLD, VIC), China, Europe, Japan, North Africa, North America, Papua New Guinea, Southeast Asia.

Notes. The species are phloeophagous (rarely xylophagous), monogynous (rarely bigynous) bark beetles largely specific to Cupressaceae (Wood & Bright 1992). Several species have been introduced in regions outside their native range, with *P. cupressi* being of particular importance in Australia. This species is widely distributed in eastern Australia and attacks ornamental cypress trees (Neumann 1987), but little is known about the damage it causes (Smith & Hulcr 2015). Little is also known about the fungal symbionts associated with *Phloeosinus* species. Several species of *Geosmithia* and *Talaromyces* fungi have been identified as being associated with *P. armatus* and *P. bicolor* in Israel, of which *Talaromyces cupressi* was found to be pathogenic when inoculated into saplings of *Cupressus sempervirens* (Meshram *et al.* 2022).

Tribe Phloeotribini Chapuis, 1869

Aricerus Blandford, 1894 (Pl. 113C; Fig. 46.18)

Type species, by subsequent designation (Hopkins, 1914: 117): *Aricerus chapuisi* Blandford, 1894.

Characteristics. Body length 3.6–4.8 mm, 1.4–1.5 × as long as wide. Forehead and vertex mostly smooth, with some punctures, setose. Eyes oval. Antennae with 5-segmented funicle and pseudolaminate club with constricted, recurved sutures forming a chevron pattern. Pronotum not covering head, wider than long, posterior margin curved; surface asperate anterolaterally, otherwise smooth, sparsely punctate, densely setose. Scutellar shield at same level as elytra. Elytra with anterior margin curved, costate, granulate; disc rounded onto declivity, covered with short setae, striae deep from apical margin to declivity; declivity unarmed, similarly covered with short setae. Pro- and mesocoxae widely separate; protibiae slender, unarmed, outer margin with 2 spines; tarsi with third segment bilobed.

Australian species. Native: *A. chapuisi* Blandford, 1894, *A. eichhoffi* Blandford, 1894.

Distribution. Australia (NSW, QLD), Papua New Guinea.

Notes. The species are phloeophagous, monogynous bark beetles (Kirkendall *et al.* 2015) only recorded to construct galleries in the limbs of *Ficus* trees (Wood & Bright 1992).

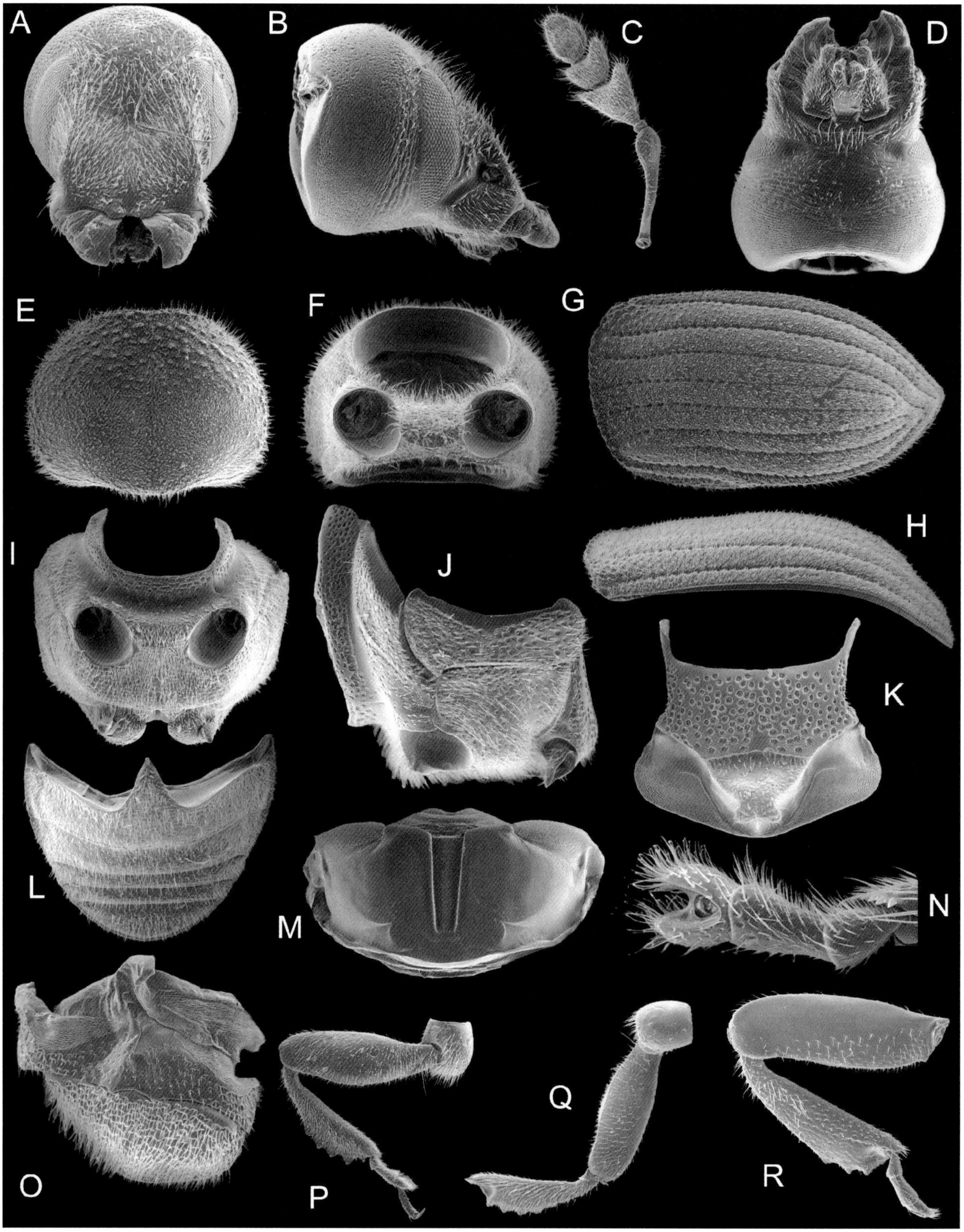

Fig. 46.18. Phloeotribini, *Aricerus eichhoffi* Blandford, structural details. A–B, head: A, dorsal; B, lateral; C, antenna; D, mouthparts; E–F, prothorax: E, dorsal; F, ventral; G–H, elytron: G, dorsal; H, lateral; I–J, pterothorax: I, ventral; J, lateral; K, scutellum; L, abdominal ventrites; M, metanotum; N, tarsi, dorsal; O, tergite VIII, dorsal; P, foreleg; Q, middle leg; R, hindleg.

Phloeotribus Latreille, 1797 (Pl. 113 D; Fig. 46.4)

Type species, by subsequent monotypy: *Bostrichus oleae* Fabricius, 1793.

Characteristics. Body length 2.0 mm, 2.0 × as long as wide. Forehead and vertex mostly smooth, densely setose. Eyes oval. Antenna with 5-segmented funicle and pseudolaminate club appearing almost ovate, first suture distinctly constricted. Pronotum not covering head, wider than long, posterior margin curved; surface punctate, covered with thick setae. Scutellar shield not exposed. Elytra with anterior margin curved, costate, granulate; disc steeply rounded onto declivity, covered with short thick setae, striae deep; declivity unarmed, similarly setose. Pro- and mesocoxae widely separate; protibiae slender, armed with up to 4 socketed denticles; tarsi with third segment bilobed.

Australian species. Native: *P. acaciae* (Lea, 1910).

Distribution. Australia (NSW, TAS, VIC), Europe, North Africa, North America, Central America, South America.

Notes. The species are phloeophagous, monogynous, polyphagous bark beetles (Kirkendall *et al.* 2015), although *P. acaciae* has only been recorded from *Acacia* species (Wood & Bright 1992). One species, *P. scarabaeoides* (Bernard), is regarded as a pest of olive trees throughout the Mediterranean region (Lozano *et al.* 1996), and *P. liminaris* (Harris) has been introduced from its native range in the Nearctic to northern Italy (Kirkendall & Faccoli 2010).

Tribe Scolytini Latreille, 1804

Scolytus Geoffroy, 1762 (Pl. 113E; Fig. 46.5)

Type species, by subsequent designation (ICZN, 1963: 416): *Bostrichus scolytus* Fabricius, 1775.

Characteristics. Body length 2.6–3.0 mm, 2.0–2.3 × as long as wide. Forehead and vertex punctate, setose. Eyes emarginate, halves equal in size. Antennae with 7-segmented funicle and obovate club marked with 2 procurved sutures. Postmentum flat, as wide as long. Pronotum not covering head, wider than long, anterior margin slightly emarginated in the middle, pre margin with a subtle groove (collar) more evident laterally, posterior margin straight; surface punctate, sparsely setose along apical margin. Scutellar shield large, at same level as elytra. Elytra evenly curved in profile, without sharply marked declivity, anterior margin bent around scutellar shield, apex truncate; surface flat, smooth, sparsely setose, strial punctures shallow. Abdomen ascending towards apex of elytra, ventrite 2 almost perpendicular to ventrite 1. Procoxae contiguous, mesocoxae widely separate; protibiae rectangular, unarmed with socketed denticles but uncinate; tarsi with third segment bilobed.

Australian species. Introduced: *S. multistriatus* (Marsham, 1802).

Distribution. Australia (ACT, NSW, VIC); native in China, Europe, Himalayan areas, Japan, Korea, Russia, North, Central and South America; some species introduced in Mexico, New Zealand, North and South America, South Africa.

Notes. The species are phloeophagous, typically monogynous bark beetles (Kirkendall *et al.* 2015), but several Neotropical species are bigamous or polygamous (Smith & Hulcr 2015). They are also polyphagous, constructing their galleries in a wide range of tree families. In monogamous species, such as *S. multistriatus*, females begin gallery construction and are later joined by males. They construct entrance tunnels, nuptial chambers and egg galleries and then lay a single egg in each egg gallery and cover it with sawdust. During this stage, the males seal the entrance tunnel with their abdomen and assist in ejecting frass. Once the eggs have been cared for, the males leave and the females seal the gallery entrance with their abdomen, where they die. The larvae construct tunnels radiating from the egg galleries and in them consume phloem tissue (Smith & Hulcr 2015). *Scolytus multistriatus* is the only species of the genus and the tribe present in Australia (Rosel & French 1975; Neumann 1987). It is regarded as a pest of elm trees (*Ulmus*, Ulmaceae) and is the primary vector of the ascomycete fungus causing Dutch Elm Disease, *Ophiostoma ulmi*. This disease is regarded as one of the world's worst tree diseases, having killed half of all European elms and up to three quarters of North American elms since 1910 (Bernier 2022). It is not present in Australia, and Australian populations of *S. multistriatus* therefore do not carry the pathogenic fungus that causes it (Neumann 1987). *Scolytus multistriatus* was accidentally introduced to Australia in 1974 and has since spread to all regions of the country in which elms are grown (Neumann 1987). It typically attacks stressed or weakened trees, rarely healthy ones. It was also introduced to New Zealand but does carry the pathogenic fungus there. Initial attempts to eradicate the beetle came close to success, but management programs are now implemented to stop the spread of disease (Gadgil *et al.* 2000).

Tribe Trypophloeini Nüsslin, 1911

Like Ernoporini, this tribe was also not recorded from Australia by Pullen *et al.* (2014) as its name was considered to be synonym of that of Cryphalini at the time (Alonso-Zarazaga & Lyal 2009). It was treated as a distinct tribe by Johnson *et al.* (2020a) and is represented in Australia by the genera *Cosmoderes* Eichhoff and *Hypothenemus* Westwood, which were previously classified in Cryphalini.

Cosmoderes Eichhoff, 1878 (Pl. 113F, G)

Type species, by monotypy: *Cosmoderes monilicollis* Eichhoff, 1878.

Characteristics. Body length 1.6–1.7 mm, 2.3–2.6 × as long as wide. Forehead mostly smooth, with some punctures, sparsely setose. Eyes emarginate, upper portion smaller than lower. Antennae with 2- to 3-segmented funicle and oval club lacking sutures. Postmentum as wide as long, slightly impressed. Pronotum covering head, as long as wide to longer than wide, lateral margins either rounded or carinate, posterior margin straight; surface punctate, asperate anterolaterally, covered with short thick setae. Scutellar shield at same level as elytra. Elytra with anterior margin straight; disc slightly convex, shining, covered with erect thick setae, strial punctures shallow; declivity rounded, unarmed, covered with erect thick setae. Pro- and mesocoxae contiguous; protibiae slender, triangular, armed with up to 6 socketed denticles.

Australian species. Native: *C. queenslandi* (Schedl, 1938), *C. setistriatus* (Lea, 1910).

Distribution. Australia (NSW, QLD, WA), China, Democratic Republic of the Congo, Fiji, Japan, Madagascar, Southeast Asia.

Notes. The life history of the genus is poorly understood. The galleries are constructed in the pith of twigs and vines, and the larvae feed on phloem tissue. The species are apparently host-specific, but no host associations are known for the two Australian species (Wood & Bright 1992).

Hypothenemus Westwood, 1834 (Pl. 113H; Fig. 46.19)

Type species, by monotypy: *Tomicus (Hypothenemus) eruditus* Westwood, 1834.

Characteristics. Body length 0.9–1.8 mm, 2.25–2.5 × as long as wide. Forehead mostly smooth, with some punctures. Eyes emarginate, upper portion smaller than lower. Antennae with 4- to 5-segmented funicle and oval club marked by 2–3 procurved to bisinuate sutures and sometimes a partial septum. Postmentum slightly impressed, variable in shape. Pronotum covering head, wider than long, lateral margins carinate, posterior margin straight; surface setose, coarsely asperate anterolaterally. Scutellar shield flat, at same level as elytra. Elytra with disc convex, shining, setose, setae short and thick or thin, strial punctures shallow to deep; declivity rounded, unarmed, covered with short and thick or thin setae. Procoxae contiguous, mesocoxae contiguous or slightly separate; protibiae slender, armed with 2–6 socketed denticles.

Australian species. Native: *H. melasomus* (Lea, 1910). Introduced: *H. areccae* (Hornung, 1842), *H. birmanus* (Eichhoff, 1878) (possibly, see Pullen *et al.* 2014), *H. californicus* Hopkins, 1915, *H. crudiae* (Panzer, 1791), *H. eruditus* (Westwood, 1834), *H. ingens* Schedl, 1942, *H. seriatus* (Eichhoff, 1872).

Distribution. Australia (NSW, NT, QLD); most tropical and subtropical regions of the world.

Notes. The species are phloeophagous, myelophagous, spermatophagous or, rarely, xylomycetophagous, inbreeding polygynous bark beetles (Kirkendall *et al.* 2015). They are highly polyphagous and globally distributed and include many invasive species of economic concern. *Hypothenemus eruditus* is one of the most polyphagous species, constructing galleries in many parts of plants as well as in galleries of other insects, fruiting bodies of fungi and even human-made products such as books (Vega *et al.* 2015). Galleries are initiated by a single mated female, which lays eggs yielding both male and female offspring. The sex determination system is pseudo-arrhenotoky, the sex ratio is biased towards females and all individuals are functionally haplodiploid (Vega *et al.* 2015).

The genus includes several economically important pest species. *Hypothenemus hampei*, the Coffee Berry Borer, is native to Africa but has been introduced in most countries that produce coffee, though not in Australia. It is specific to *Coffea* and constructs galleries in coffee berries (fruits) and seeds. *Hypothenemus obscurus* is another species of seed-boring beetles occurring in most tropical countries but not yet present in Australia. It is also highly polyphagous and constructs galleries in the seeds of commercially important agricultural products. In Hawaii it caused US$ 0.3–1.8 million in losses per year due to damage in macadamia plantations between 1998 and 2012 (Vega *et al.* 2015).

Tribe Xyleborini LeConte, 1876

Several genera of this tribe have been revised since the catalogue of Pullen *et al.* (2014) and additional taxa have also been recorded and described from Australia. The genus *Wallacellus* was subsumed into *Euwallacea* by Storer *et al.* (2015). The species *Ambrosiodmus compressus* (Lea), *A. latecompressus* (Schedl) and *Cyclorhipidion pityogenes* (Schedl) were transferred to *Ambrosiophilus* by Smith *et al.* (2022b), and several species names have been synonymised.

Amasa Lea, 1894 (Pl. 113I; Fig. 46.20)

Type species, by monotypy: *Amasa thoracica* Lea, 1894.

Characteristics. Body length 2.5–3.5 mm, 2.2–2.9 × as long as wide. Forehead punctate, sparsely setose. Eyes deeply emarginate, upper portion smaller than lower. Antennae with 5-segmented funicle and flat club marked with 2 procurved sutures, club segment 1 smaller than 2. Postmentum deeply impressed, longer than wide. Pronotum covering head, as wide as long, posterior margin straight;

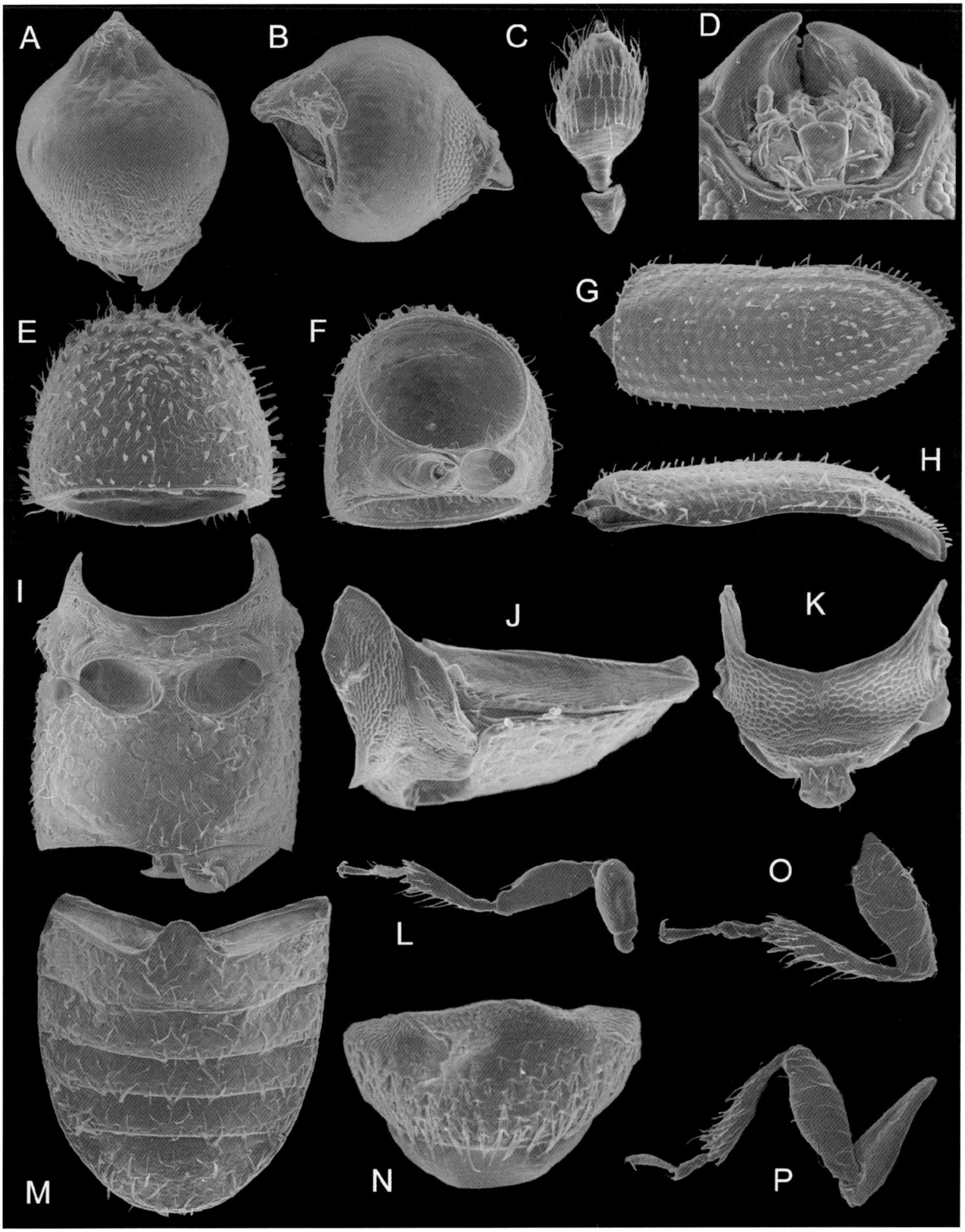

Fig. 46.19. Trypophloeini, *Hypothenemus birmanus* (Eichhoff), structural details. A–B, head: A, dorsal; B, lateral; C, antenna; D, mouthparts; E–F, prothorax: E, dorsal; F, ventral; G–H, elytron: G, dorsal; H, lateral; I–J, pterothorax: I, ventral; J, lateral; K, scutellum; L, foreleg; M, abdominal ventrites; N, tergite VIII, dorsal; O, middle leg; P, hindleg.

surface sparsely setose, asperate anterolaterally. Scutellar shield flat, at same level as elytra. Elytra with disc shining, flat, ascending posteriorly; declivity truncate, unarmed, 3 striae on declivital face. Procoxae contiguous, mesocoxae separate; protibiae slender, slightly inflated, armed with 6–8 socketed denticles.

Australian species. Native: *A. banksiae* (Schedl, 1964), *A. darwini* (Schedl, 1972), *A. exacta* (Schedl, 1964), *A. parviseta* Knížek & Smith 2024, *A. schlichii* (Stebbing, 1914), *A. truncata* (Erichson, 1842).

Distribution. Australia (ACT, NSW, NT, QLD, TAS, WA), Fiji, Papua New Guinea, China; introduced in South America and Europe.

Notes. The species are xylomycetophagous, haplodiploid, inbreeding polygynous, polyphagous ambrosia beetles (Kirkendall *et al.* 2015). The ones in Australia are widespread, with *A. truncata* commonly intercepted along the eastern Australian coastline in mesic forest habitats (Bickerstaff, pers. obs.), where it constructs galleries in Myrtaceae trees (Wood & Bright 1992). The ambrosia symbionts associated with the genus are fungi of the *Raffaelea* group, which are transported by the beetles in mandibular mycangia (Hulcr & Stelinski 2017). The recently described *Amasa parviseta* has been introduced to France, Spain, Portugal, Brazil, Uruguay and Chile (Kirkendall & Faccoli 2010; Viñolas & Verdugo 2011; Gómez *et al.* 2017; Rainho *et al.* 2018; Barnouin *et al.* 2020; Córdoba *et al.* 2023; Knížek & Smith 2024).

Ambrosiodmus Hopkins, 1915 (Pl. 113J, K)

Type species, by original designation: *Xyleborus tachygraphus* Zimmermann, 1868.

Characteristics. Body length 2.3–2.8 mm, 2.3–2.7 × as long as wide. Forehead punctate, sparsely setose. Eyes emarginate, upper portion smaller than lower. Antennae with 5-segmented funicle and flat club with 2 recurved sutures, club segment 1 smaller than 2. Postmentum flat to slightly impressed, twice as wide as long. Pronotum covering head, as wide as long, short, anteriorly rounded, posterior margin straight; surface sparsely setose, completely asperate. Scutellar shield flat, at same level as elytra. Elytra with disc longer than declivity, shining, convex, sparsely setose, strial punctures shallow; declivity rounded, armed with 3–5 tubercles, finely setose. Procoxae contiguous, mesocoxae narrowly separate; protibiae rounded, armed with 6–8 socketed denticles.

Australian species. Native: *A. asperatus* (Blandford, 1895), *A. optatus* (Schedl, 1973). Introduced: *A. rubricollis* (Eichhoff, 1876).

Distribution. Australia (NSW), near global except for Europe and Antarctica.

Notes. The species are xylomycetophagous, haplodiploid inbreeding, polygynous, polyphagous and widely distributed ambrosia beetles (Wood & Bright 1992; Kirkendall *et al.* 2015). A wood-rot fungus, *Flavodon ambrosius,* has been found to be the primary nutritional symbiont of the *Ambrosiodmus* species native to North America (You *et al.* 2015; Kasson *et al.* 2016; Simmons *et al.* 2016). This species is unique among ambrosia fungi in that it grows as hyphae in the mycangia of its hosts and is also a white-rot saprophyte in gallery ecosystems. Unlike other ambrosia beetle species that thrive in either living host tree or newly dead timber, *Ambrosiodmus* species can, by inoculating galleries with *F. ambrosius*, colonise and thrive in decaying timber that is inaccessible to other ambrosia beetle species (You *et al.* 2015).

Ambrosiophilus Hulcr & Cognato, 2009 (Pl. 114A, B; Fig. 46.21)

Type species, by original designation: *Xyleborus restrictus* Schedl, 1939.

Characteristics. Body length 2.5–3.9 mm, 2.4–2.6 × as long as wide. Forehead sparsely to densely punctate, sparsely setose. Eyes emarginate, upper portion smaller than lower. Antennae with 5-segmented funicle and flat club marked with 3 variable suture lines, club segment 1 as large as or smaller than 2. Postmentum flat to slightly impressed, twice as wide as long. Pronotum covering head, as wide as long to longer than wide, short, anteriorly rounded, posterior margin straight; surface sparsely setose, anteriorly asperate, disc densely punctate. Scutellar shield at same level as elytra, disc slightly impressed. Elytra with disc shining, convex, sparsely setose, strial punctures shallow; declivity rounded, slightly emarginate, armed with 2–5 tubercles, finely setose. Procoxae contiguous, mesocoxae narrowly separate; protibiae rounded to triangular, armed with 6–8 socketed denticles.

Australian species. Native: *A. compressus* (Lea, 1894), *A. latecompressus* (Schedl, 1936), *A. pityogenes* (Schedl, 1936), *A. restrictus* (Schedl, 1939).

Distribution. Australia (ACT, NSW, QLD, VIC), India, Japan, North Korea, Papua New Guinea, South Korea, Southeast Asia; introduced in Europe and North America (*A. atratus* (Eichhoff) and *A. osumiensis* (Murayama)).

Notes. The species are xylomycetophagous, haplodiploid and inbreeding polygynous (Kirkendall *et al.* 2015). Several are true ambrosia beetles, but others practice mycetoclepsia (Hulcr & Cognato 2010). The ambrosia species cultivate the same basidiomycete as *Ambrosiodmus* species do, *Flavodon ambrosius* (You *et al.* 2015; Kasson *et al.* 2016; Simmons *et al.* 2016), and these two genera contain the only ambrosia beetles that have formed a symbiosis with a wood-rot fungus. By associating with this symbiont, the

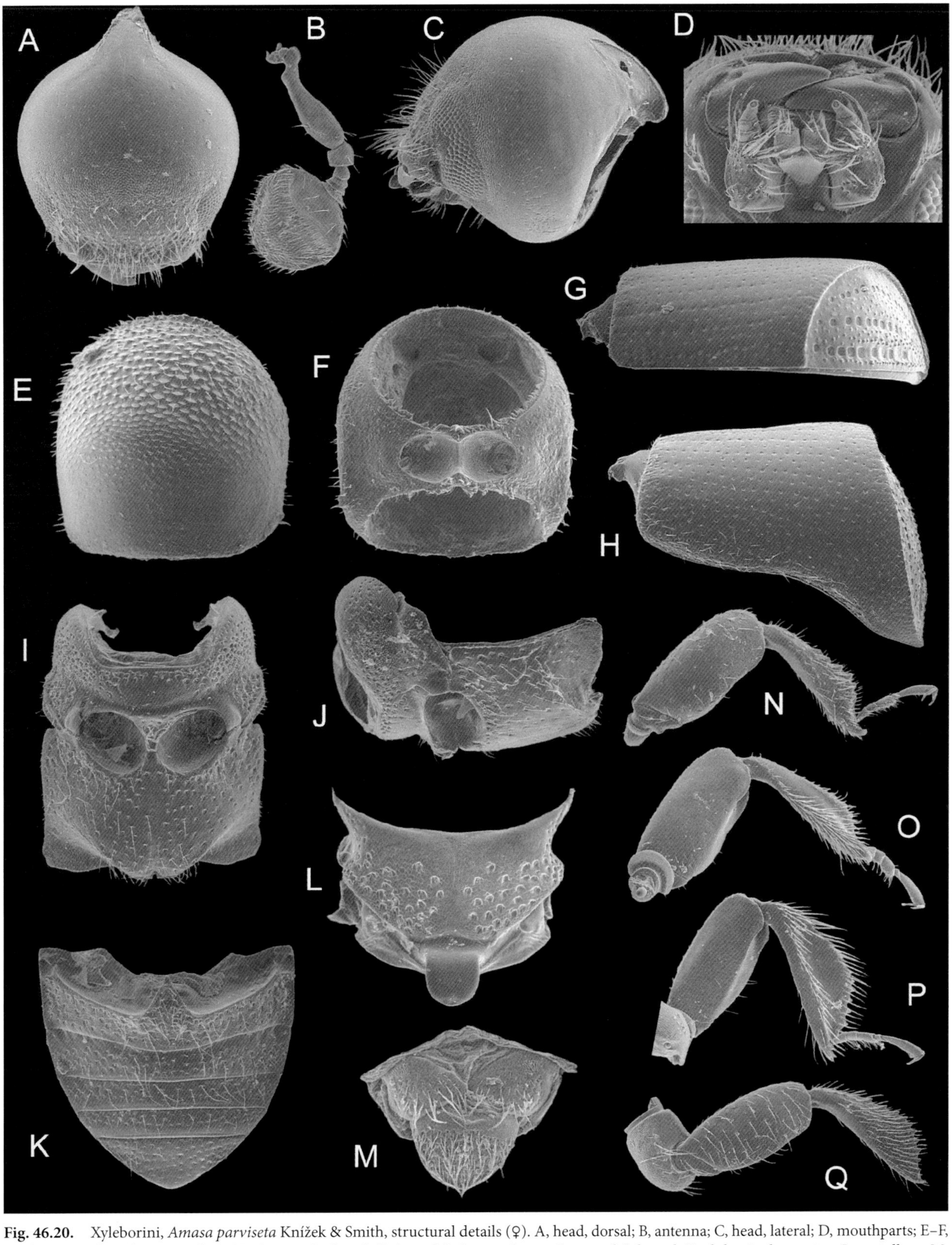

Fig. 46.20. Xyleborini, *Amasa parviseta* Knížek & Smith, structural details (♀). A, head, dorsal; B, antenna; C, head, lateral; D, mouthparts; E–F, prothorax: E, dorsal; F, ventral; G–H, elytron: G, dorsal; H, lateral; I–J, pterothorax: I, ventral; J, lateral; K, abdominal ventrites; L, scutellum; M, Tergites VII and VIII, dorsal; N, foreleg; O, middle leg; P, hindleg; Q, hindleg.

beetles' galleries can persist in wood that is typically too decayed for other ambrosia beetle species to use. Notably this is also one of a few documented cases of a single symbiont species being associated with multiple species and genera of ambrosia beetles, and across continents (Li *et al.* 2017). Several instances of mycetoclepsia have also been discovered in *Ambrosiophilus*. The females of these species construct galleries in close proximity to those of other ambrosia beetles, typically *Beaverium* species but one report of *Ambrosiodmus rubricollis* also exists, and divert mycelial growth from the host gallery into their own. Species that practice mycetoclepsia lack mycangia (Hulcr & Cognato 2009, 2010). Of the species present in Australia, *A. restrictus* has been reported to be mycetocleptic (Hulcr & Cognato 2010), but whether the others are as well remains to be determined.

Arixyleborus Hopkins, 1915 (Pl. 113L)

Type species, by original designation: *Arixyleborus rugosipes* Hopkins, 1915.

Characteristics. Body length 2.0–2.1 mm, 2.6–2.63 × as long as wide. Forehead smooth, sparsely setose. Eyes deeply emarginate, upper portion smaller than lower. Antennae with 5-segmented funicle and truncate club marked with 3 recurved sutures, edge of club segment 1 continuous on both sides and segment 2 visible on posterior face. Postmentum deeply impressed, twice as wide as long. Pronotum covering head, longer than wide, anteriorly rounded and asperate, posterior margin straight; surface smooth, sparsely setose. Scutellar shield at same level as elytra. Elytra with disc shining, convex, sparsely setose, strial punctures shallow; declivity obliquely truncate, striae deep, interstrial ridges granulate, covered with short setae. Procoxae contiguous, mesocoxae narrowly separate; protibiae rounded, armed with 6–8 socketed denticles.

Australian species. Native: *A. malayensis* (Schedl, 1954).

Distribution. Australia (Christmas Is.), Africa (tropical and subtropical countries), Japan, China, Southeast Asia.

Notes. The species are xylomycetophagous, haplodiploid, inbreeding polygynous, polyphagous ambrosia beetles (Kirkendall *et al.* 2015). *Arixyleborus grandis* has been recorded to attack mango trees (*Mangifera indica*) in Thailand (Sittichaya 2012).

Beaverium Hulcr & Cognato, 2009 (Pl. 114C; Fig. 46.22)

Type species, by original designation: *Xyleborus insulindicus* Eggers, 1923.

Characteristics. Body length 4.9–6.2 mm, 2.1–2.3 × as long as wide. Forehead punctate, sparsely setose. Eyes deeply emarginate, portions equal in size. Antennae with 5-segmented funicle and flat club marked with 3 straight to recurved sutures, club segment 1 as large as 2. Postmentum slightly impressed, twice as wide as long. Pronotum covering head, as wide as long, round, anterior edge with 2 flat denticles, posterior margin straight; surface entirely asperate, punctate. Scutellar shield at same level as elytra. Elytra with disc shining, convex, sparsely setose, strial punctures shallow; declivity unarmed, costate, slightly emarginate, subquadrate, rounded. Procoxae contiguous, mesocoxae narrowly separate; protibiae rounded, armed with 6–8 socketed denticles.

Australian species. Native: *B. insulindicus* (Eggers, 1923).

Distribution. Australia (QLD), Fiji, Papua New Guinea, Southeast Asia.

Notes. The species are xylomycetophagous, haplodiploid, inbreeding polygynous ambrosia beetles (Kirkendall *et al.* 2015). Roberts (1977) described the gallery of *B. insulindicus* as short and radial in shape, approximately 2–4 cm in length, with longitudinal galleries parallel to the stem of the host plant but without enlarged brood chambers.

Cnestus Sampson, 1911 (Pl. 114D; Fig. 46.23)

Type species, by monotypy: *Cnestus magnus* Sampson, 1911.

Characteristics. Body length 3.6–4.1 mm, 1.6–1.9 × as long as wide. Forehead punctate. Eyes emarginate, both portions equal in size. Antennae with 5-segmented funicle and truncate club marked with 3 recurved sutures, club segment 1 extending over entire posterior face. Postmentum deeply impressed, twice as wide as long. Pronotum covering head, as wide as long, rounded, anterior margin with 2 prominent denticles, posterior margin straight, medially with tufts of mycangial setae; surface asperate, anterolaterally setose, disc punctate, shining. Scutellar shield at same level as elytra. Elytra with disc shining, convex, shorter than declivity; declivity obliquely truncate, costate, setose, armed with interstrial granules. Procoxae narrowly separate, mesocoxae widely separate; protibiae triangular, armed with 4 socketed denticles.

Australian species. Native: *C. solidus* (Eichhoff, 1868).

Distribution. Australia (NSW, QLD, TAS, VIC, WA), Oceania, Southeast Asia; introduced in North and South America.

Notes. The name of *Cnestus pseudosolidus* (Schedl) was recently synonymised with that of *C. solidus* (Eichhoff) by Bickerstaff *et al.* (2023). The species are xylomycetophagous, haplodiploid, inbreeding polygynous ambrosia beetles (Kirkendall *et al.* 2015). The single species in Australia is widespread throughout the eastern parts but has also been collected in Perth, Western Australia (Froggatt 1927). In eastern Australia it is highly polyphagous and commonly found is mesic forests. Most populations are associated with one species of *Ambrosiella,* but one population in southeast New South Wales is associated with

Fig. 46. 21. Xyleborini, *Ambrosiophilus compressus* (Lea), structural details (♀). A, head, dorsal; B, antenna; C, head, lateral; D, mouthparts; E–F, prothorax: E, dorsal; F, ventral; G–H, elytron: G, dorsal; H, lateral; I–J, pterothorax: I, ventral; J, lateral; K, abdominal ventrites; L, scutellum; M, tergites, dorsal; N, foreleg; O, middle leg; P, hindleg.

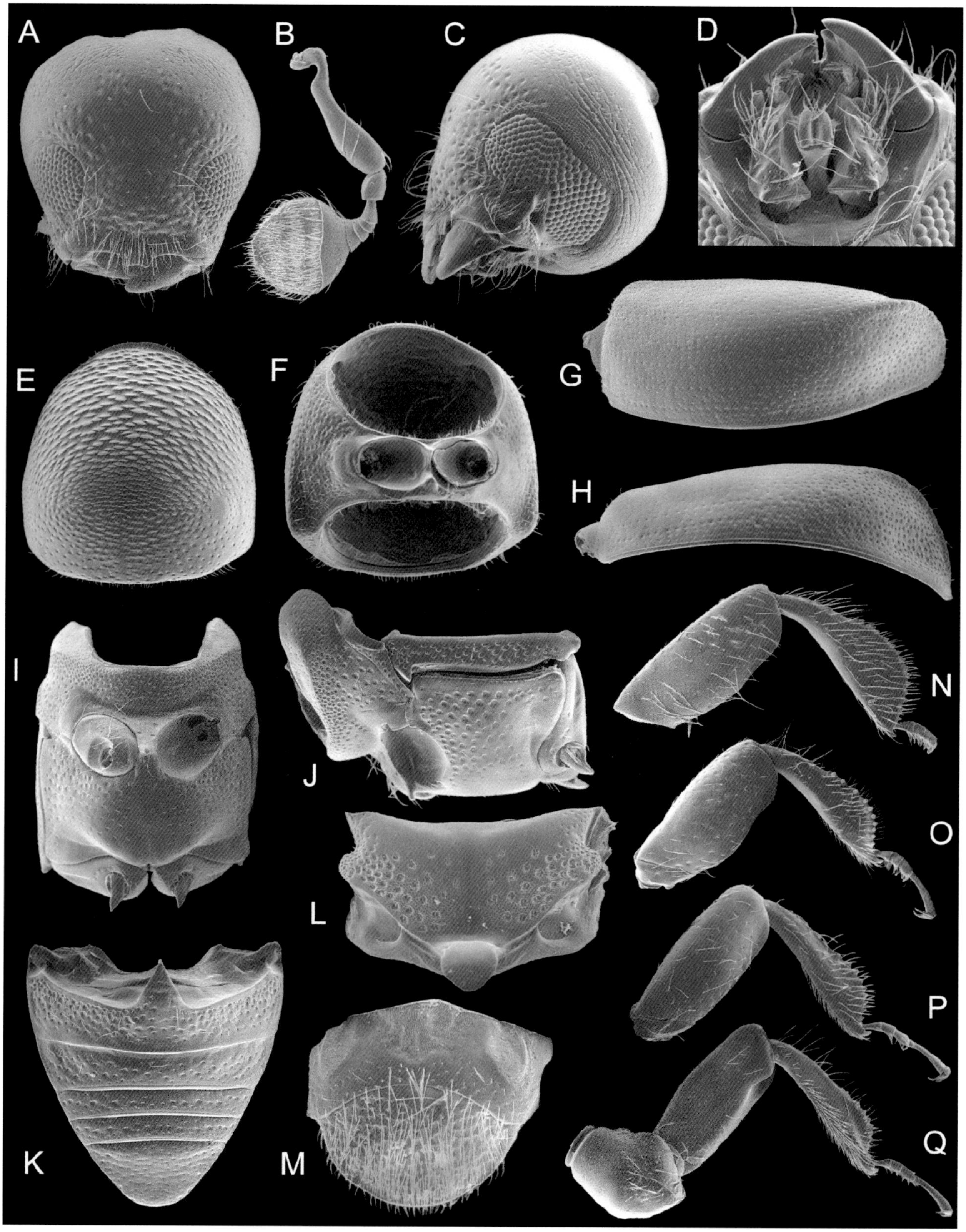

Fig. 46.22. Xyleborini, *Beaverium insulindicus* (Eggers), structural details (♀). A, head, dorsal; B, antenna; C, head, lateral; D, mouthparts; E–F, prothorax: E, dorsal; F, ventral; G–H, elytron: G, dorsal; H, lateral; I–J, pterothorax: I, ventral; J, lateral; K, abdominal ventrites; L, scutellum; M, tergites VII and VIII, dorsal; N, hindleg; O, middle leg; P, foreleg; Q, foreleg.

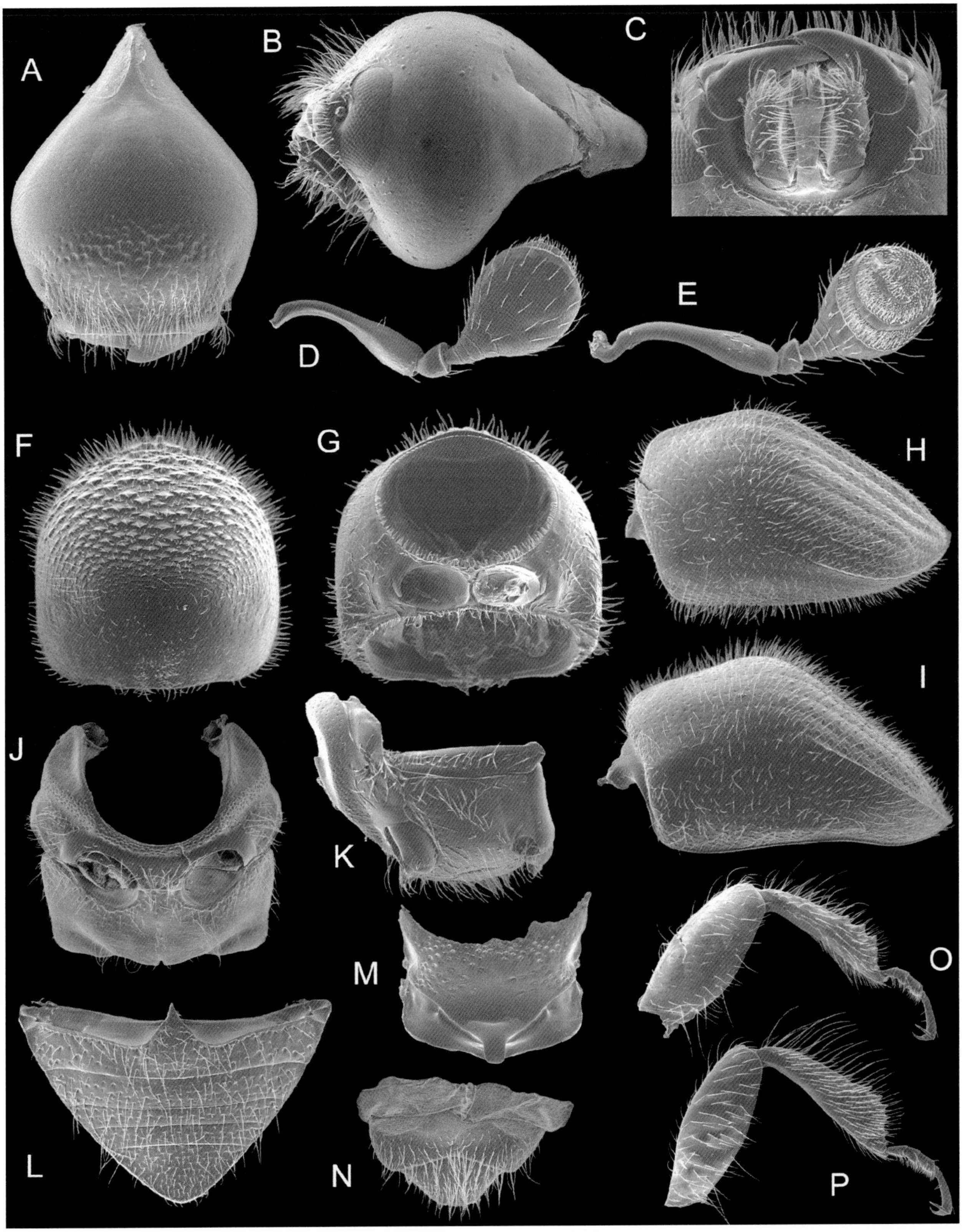

Fig. 46.23. Xyleborini, *Cnestus solidus* (Eichhoff), structural details (♀). A–B, head: A, dorsal; B, lateral; C, mouthparts; D–E, antenna: D, dorsal; E, ventral; F–G, prothorax: E, dorsal; G, ventral; H–I, elytron: H, dorsal; I, ventral; J–K, pterothorax: J, ventral; K, lateral; L, abdominal ventrites; M, scutellum; N, tergites VI–VIII, dorsal; O, foreleg; P, middle leg.

another, distantly related strain of *Ambrosiellia. Cnestus* populations are also infected with the endosymbiont *Wolbachia.* A recent study by Bickerstaff *et al.* (2023) identified two sympatric and genetically distinct populations of *C. solidus*, one of which was predominately infected with *Wolbachia* , but it was determined that this infection occurred after the populations separated, with inbreeding and *Wolbachia* infection reinforcing reproductive barriers between the two lineages following their current distribution.

Cryptoxyleborus Wood & Bright, 1992 (Pl. 114E)

Type species, by original designation: *Cryptoxyleborus naevus* Schedl, 1937.

Characteristics. Body length 2.3–2.7 mm, 3.29–4.17 × as long as wide. Forehead smooth. Eyes emarginate, both portions equal in size. Antenna with 5-segmented funicle and flat club marked with 3 procurved sutures, club segment 1 smaller than 2, all 3 segments distinct on posterior face. Postmentum deeply impressed, wider than long. Pronotum covering head, longer than wide, anteriorly rounded, sides parallel, anterior margin with 4 prominent denticles, posterior margin straight; surface anterolaterally asperate, anteriorly sparsely setose, disc smooth, shining, slightly punctate. Scutellar shield not exposed. Elytra with disc shining, convex, longer than declivity; declivity unarmed, densely setose, laterally flat, dorsally attenuate; each elytron anteromedially with large mycangial pit (*C. subnaevus*) the number of mycangial pits varying in other Southeast Asian species. Procoxae contiguous, mesocoxae narrowly separate; protibiae rounded, armed with fewer than 6 socketed denticles.

Australian species. Native: *C. subnaevus* Schedl, 1937.

Distribution. Australia (QLD), Papua New Guinea, Southeast Asia.

Notes. The species are xylomycetophagous, haplodiploid, inbreeding polygynous ambrosia beetles (Kirkendall *et al.* 2015). They are only recorded from Dipterocarpaceae (Beaver & Hulcr 2008), and their gallery systems are unbranched, with a single tunnel leading to a single brood chamber (Browne 1961; Beaver & Hulcr 2008).

Debus Hulcr & Cognato, 2010 (Pl. 114F)

Type species, by original designation: *Xyleborus emarginatus* Eichhoff, 1878.

Characteristics. Body length 2.5–3.6 mm, 2.8–3.57 × as long as wide. Forehead mostly smooth with some punctures. Eyes deeply emarginate, upper portion smaller than lower. Antennae with 5-segmented funicle and truncate club marked with 2 or 3 procurved sutures, club segment 1 continuous on both sides of anterior face, segment 2 only distinct on posterior face. Postmentum deeply impressed, wider than long to as wide as long. Pronotum covering head, longer than wide, anteriorly rounded, posterior margin is straight; surface anterolaterally asperate, shining, densely punctate. Scutellar shield flat, at same level as elytra. Elytra with disc shining, convex, longer than declivity; declivity deeply excavate, emarginate in dorsal view, rounded with apex emarginate in lateral view, on each elytron armed with up to 4 spines. Procoxae contiguous; protibiae triangular, armed with fewer than 6 socketed denticles.

Australian species. Native: *D. emarginatus* (Eichhoff, 1878), *D. pumilus* (Eggers, 1923).

Distribution. Australia (QLD), Bhutan, China, Japan, Papua New Guinea, South Korea, Southeast Asia.

Notes. The species are xylomycetophagous, haplodiploid, inbreeding polygynous, highly polyphagous ambrosia beetles (Wood & Bright 1992; Kirkendall *et al.* 2015). In *Debus emarginatus*, the beetles construct galleries between the bark and xylem tissue of host plants, unlike other ambrosia beetles, which construct galleries in xylem tissues in horizontal planes (Beaver & Browne 1978). In *Debus pumilus*, on the other hand, the beetles form short and flat galleries, with a preference for Moraceae as hosts (Beaver & Browne 1978).

Diuncus Hulcr & Cognato, 2009 (Pl. 114H)

Type species, by original designation: *Xyleborus papatrae* Schedl, 1972.

Characteristics. Body length 1.4–1.9 mm, 2.2–2.5 × as long as wide. Forehead smooth. Eyes emarginate, upper portion smaller than lower. Antennae with 5-segmented funicle and truncate club marked with 1–3 recurved sutures, club segment 1 encircling anterior face. Postmentum flat, wider than long. Pronotum covering head, as wide and long or wider than long, rounded, anterior margin with 2 flat denticles, posterior margin straight; surface asperate anterolaterally, sparsely setose, smooth with few punctures. Scutellar shield flat, at same level as elytra. Elytra with disc smooth, flat to convex, as long as or shorter than declivity; declivity carinate, laterally broad in dorsal view, gradually descending in lateral view, flat or gradually rounded; on each elytron armed with up to 2 spines. Procoxae contiguous, mesocoxae narrowly separate; protibiae triangular, armed with fewer than 6 socketed denticles.

Australian species. Native: *D. justus* (Schedl, 1931). Introduced: *D. haberkorni* (Eggers, 1920).

Distribution. Australia (NSW, QLD), India, Papua New Guinea, Southeast Asia, China; introduced in Africa.

Notes. The species are xylomycetophagous, haplodiploid, inbreeding polygynous (Kirkendall *et al.* 2015). Most are mycetocleptic, the females constructing galleries close to galleries of other xyleborine ambrosia beetles and diverting the growth of ambrosia mycelia from these galleries into their own brood chambers. As in other mycetocleptic lineages, the species have secondarily lost their mycangia. The females mostly construct galleries in dead or dying trees, and though they are polyphagous in terms of host trees, the presence of their preferred host beetles is the limiting factor in gallery construction. Gallery systems vary between the species and are dependent on the size of the host tree substrate (Hulcr & Cognato 2009, 2010, 2013).

Eccoptopterus Motschulsky, 1863 (Pl. 114I; Fig. 46.24)

Type species, by monotypy: *Eccoptopterus sexspinosus* Motschulsky, 1863.

Characteristics. Body length 2.5–3.7 mm, 2.0–2.2 × as long as wide. Forehead mostly smooth, with some punctures. Eyes emarginate, upper portion smaller than lower. Antennae with 5-segmented funicle and truncate club marked with 2 sutures, club segment 1 encircling anterior face. Postmentum deeply impressed, as wide as long to wider than long. Pronotum covering head, as wide as long, rounded, anterior margin with 2 flat denticles, posterior margin straight with medial mycangial tufts; surface asperate anteriorly, mostly smooth. Scutellar shield flat, at same level as elytra. Elytra with disc smooth, flat, shorter than declivity; declivity costate, deeply excavate and rounded in dorsal view, flat, gradually descending in lateral view, on each elytron armed with 3 spines. Procoxae contiguous, mesocoxae narrowly separate; protibiae rounded, armed with 6–8 socketed denticles; metatibiae greatly enlarged.

Australian species. Native: *E. spinosus* (Olivier, 1795).

Distribution. Australia (QLD), China, India, Papua New Guinea, Southeast Asia, Uganda.

Notes. The species are xylomycetophagous, haplodiploid, inbreeding polygynous, highly polyphagous ambrosia beetles (Kirkendall *et al.* 2015). The females construct short galleries in small branches that follow various axial planes (Hulcr & Cognato 2013).

Euwallacea Hopkins, 1915 (Pl. 114J–L; Figs 46.3, 46.25)

Type species, by original designation: *Xyleborus wallacei* Blandford, 1896.

Characteristics. Body length 2.2–4.6 mm, 2.2–3.2 × as long as wide. Forehead smooth to punctate. Eyes emarginate, upper portion smaller than lower. Antennae with 5-segmented funicle and truncate or flat club marked with 2 straight or recurved sutures, club segment 1 as large as 2 or encircling anterior face, segment 2 sometimes distinct on posterior face. Postmentum slightly impressed, twice as wide as long. Pronotum covering head, as wide as long, subquadrate or rounded, sides always parallel, posterior margin straight; surface mostly smooth but with some punctures and asperate anterolaterally. Scutellar shield flat, at same level as elytra. Elytra with disc shining, costate or carinate, longer or shorter than declivity; declivity rounded, sometimes with interstrial granules and/or tubercles. Procoxae contiguous, mesocoxae narrowly separate; protibiae rounded or triangular, mostly armed with 6–10 socketed denticles but sometimes fewer than 5.

Australian species. Native: *E. destruens* (Blandford, 1896), *E. funereus* (Lea, 1910), *E. perbrevis* (Schedl, 1951), *E. piceus* (Motschulsky, 1863), *E. similis* (Ferrari, 1867), *E. wallacei* (Blandford, 1896). Introduced: *E. fornicatus* (Eichhoff, 1868),

Distribution. Australia (NSW, QLD), Africa, Southeast Asia, Oceanic islands; some species introduced in Australia (WA), Argentina, Israel, North America.

Notes. The species are xylomycetophagous, haplodiploid, inbreeding polygynous ambrosia beetles (Kirkendall *et al.* 2015) typically associated with fungal symbionts of *Raffaelea* or the Ambrosia *Fusarium* Complex (AFC), which are carried in paired preoral mycangial pockets (Mayers *et al.* 2022). Most species are highly polyphagous, with *E. fornicatus* recorded to attack more than 680 tree species (Cook & Broughton 2023), but to reproduce in only 168 (Mendel *et al.* 2021). Galleries are typically excavated in the trunk of the host, with variable tunnel shape and structure. In *E. destruens*, galleries can reach nearly 1.4 m in length and are excavated in a single transverse plane, whereas in *E. wallacei* the galleries are long, branching tunnels without chambers (Hulcr & Cognato 2013). Several species are regarded as serious pests of managed and natural ecosystems, the best-studied being the *E. fornicatus* species complex, which comprises *E. fornicatus,* the Polyphagous Shot Hole Borer (PSHB), *E. perbrevis*, the Tea Shot Hole Borer a (TSHBa), *E. fornicatior*, the Tea Shot Hole Borer b (TSHBb) and *E. kuroshiro*, the Kuroshio Shot Hole Borer (KSHB) (Stouthamer *et al.* 2017; Gomez *et al.* 2018; Smith *et al.* 2019). *Euwallacea fornicatus* has been introduced to the U.S.A., Israel, South Africa and Palestine and was also detected in Perth, Western Australia, in September 2021 (Cook & Broughton 2023). Throughout its native and introduced range, it attacks various cultivated trees, including tea, citrus, avocado and cacao, and heavy infestations with weakly phytopathogenic fungi can cause serious damage and lead to significant economic losses (Smith & Hulcr 2015; De Wit *et al.* 2022; Cook & Broughton 2023). Both species of TSHB are also regarded as economic pests but do not cause as much damage as PSHB. Of these, only *E. perbrevis* is present in Australia. In its native range it is

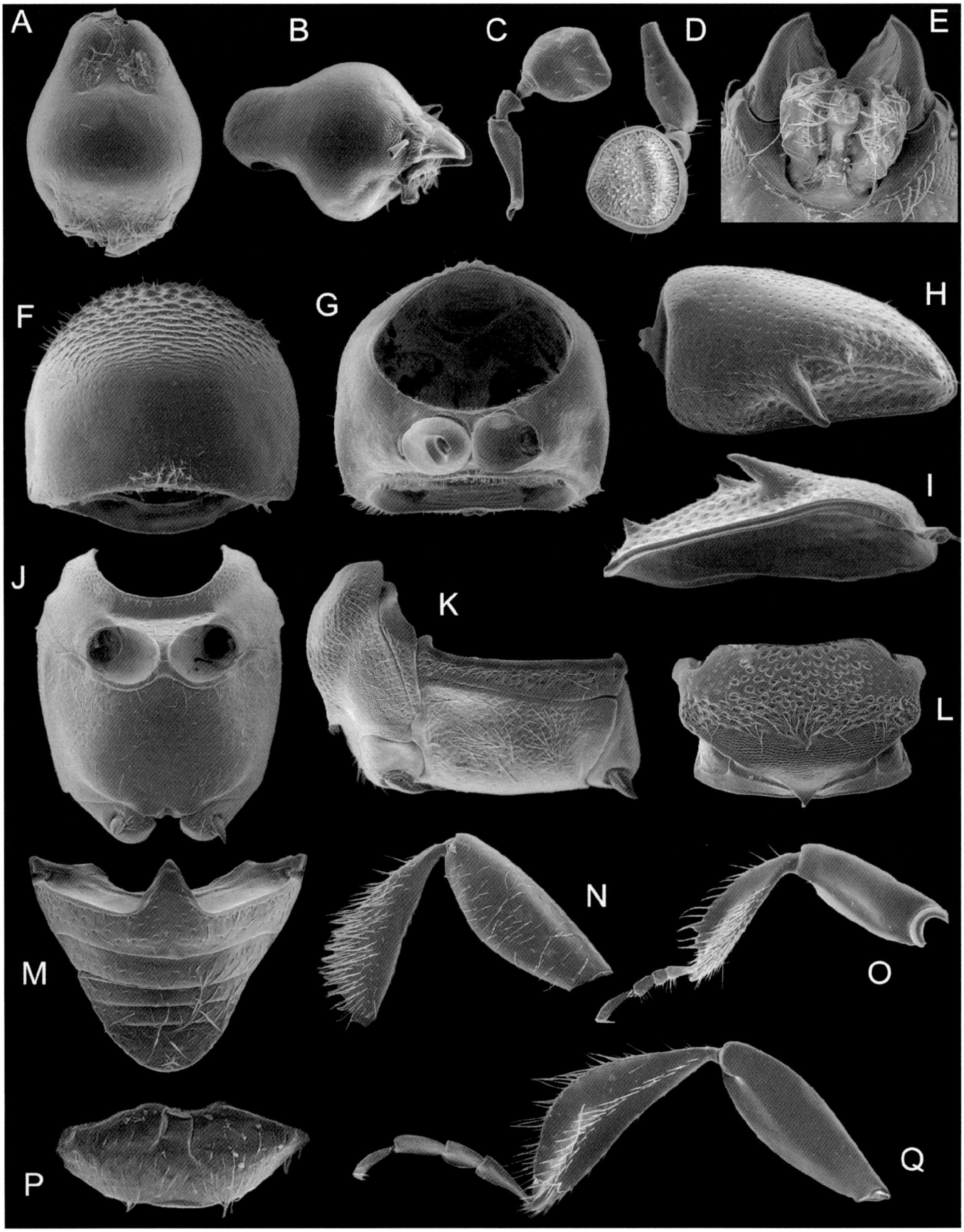

Fig. 46.24. Xyleborini, *Eccoptopterus spinosus* (Olivier), structural details (♀). A–B, head: A, dorsal; B, lateral; C–D, antenna: C, dorsal; D, ventral; E, mouthparts; F–G, prothorax: F, dorsal; G, ventral; H–I, elytron: H, dorsal; I, lateral; J–K, pterothorax: J, ventral; K, lateral; L, scutellum; M, abdominal ventrites; N, foreleg; O, middle leg; P, tergite VIII, dorsal; Q, hindleg.

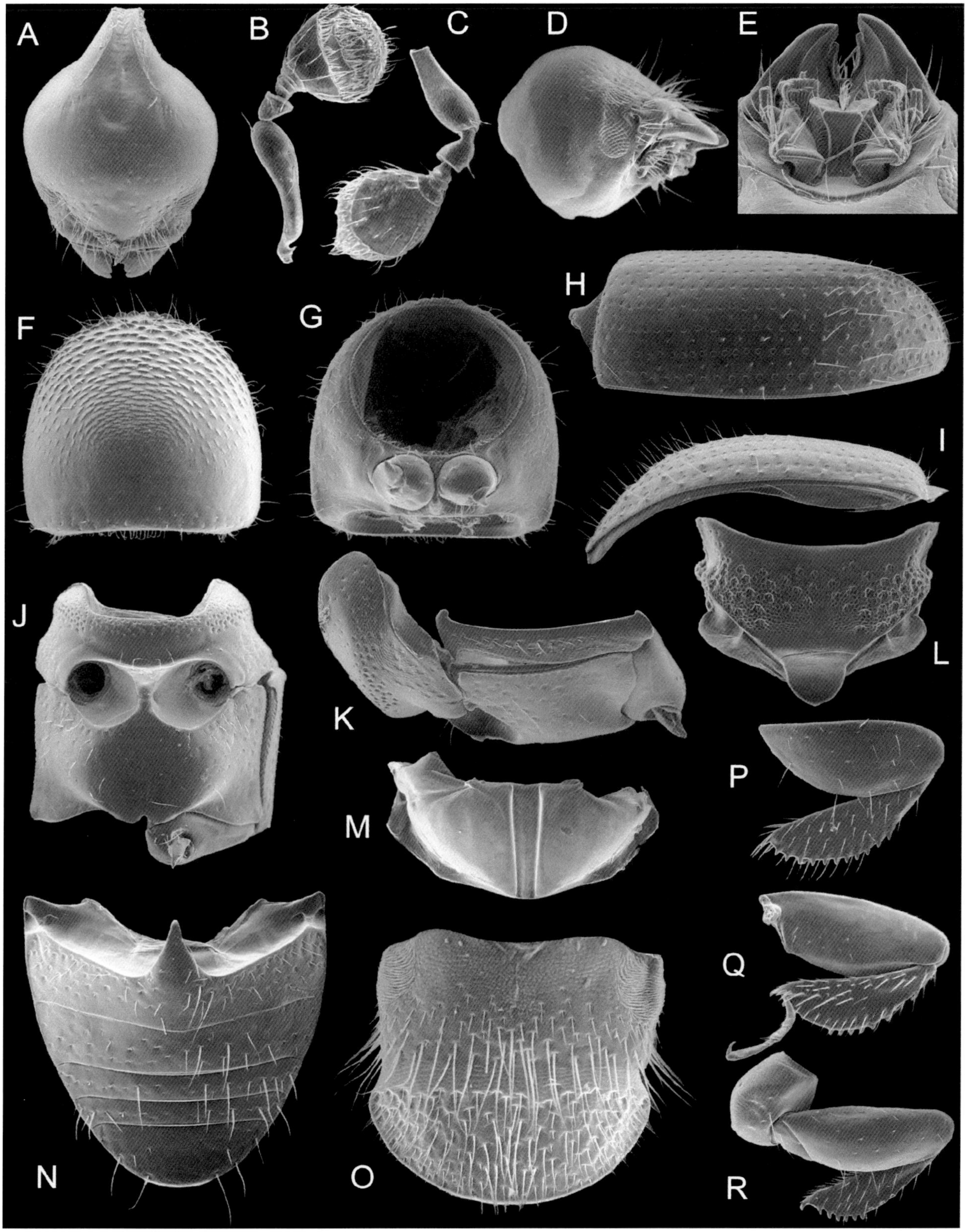

Fig. 46.25. Xyleborini, *Euwallacea fornicatus* (Eichhoff), structural details (♀). A, head, dorsal; B, antenna, ventral; C, antenna, dorsal; D, head, lateral E, mouthparts; F–G, prothorax: F, dorsal; G, ventral; H–I, elytron: H, dorsal; I, lateral; J–K, pterothorax: J, ventral; K, lateral; L, scutellum; M, metanotum; N, abdominal ventrites; O, tergite VIII, dorsal; P, foreleg; Q, middle leg; R, hindleg.

regarded as a pest of tea (*Camellia sinensis*) in Sri Lanka, but in its range of introduction it is polyphagous and attacks a range of plant species. Its impact is, however, low compared to that of PSHB (Carrillo *et al.* 2012; Cooperband *et al.* 2016).

Fraudatrix Cognato, Smith & Beaver, 2020 (Pl. 114G)

Type species, by original designation: *Xyleborus melas* Eggers, 1927.

Characteristics. Body length 2.1–2.50 mm, 3.13 × as long as wide. Forehead slightly convex, punctate. Eyes emarginate, both portions equal in size. Antennae with 3-segmented funicle and truncate club with segment 2 visible on posterior face. Postmentum slightly impressed, longer than wide. Pronotum covering head, longer than wide, sides parallel, posterior margin straight; surface smooth to slightly punctate, asperate anteriorly. Scutellar shield at same level as elytra. Elytra with disc convex, longer than declivity, interstrial punctures bearing a single seta; declivity laterally rounded, dorsally attenuate, armed with granules. Procoxae contiguous; protibiae triangular, armed with 5 or 6 socketed denticles.

Australian species. Native: *F. pileatula* (Schedl, 1975)

Distribution. Australia (QLD), China, Hong Kong, Papua New Guinea, Southeast Asia.

Notes. The species are xylomycetophagous, haplodiploid, inbreeding polygynous ambrosia beetles (Kirkendall *et al.* 2015) apparently specific to Dipterocarpaceae (Cognato *et al.* 2020). The gallery system of *F. cuneiformis* Schedl has branched tunnels in a longitudinal plane with small brood chambers (Browne 1961).

Immanus Hulcr & Cognato, 2013 (Pl. 115A, B)

Type species, by original designation: *Xyleborus colossus* Blandford, 1896.

Characteristics. Body length 7.0–7.5 mm, 2.14 × as long as wide. Forehead densely punctate. Eyes emarginate, both portions equal in size. Antennae with 5-segmented funicle and flat club with segment 1 smaller than 2. Postmentum slightly impressed, twice as wide as long. Pronotum covering head, wider than long, rounded, anterior margin with denticles, posterior margin straight; surface entirely asperate. Scutellar shield flat, at same level as elytra. Elytra with disc shining, setose, convex, smaller than declivity; declivity laterally rounded, dorsally excavate, margin typically armed with spines, sometimes also spines on declivital face. Procoxae contiguous, mesocoxae narrowly separate.

Australian species. Native: *I. acanthurus* (Lea, 1910).

Distribution. Australia (QLD), China, Japan, Papua New Guinea, Southeast Asia.

Notes. The single species in Australia is a xylomycetophagous, haplodiploid, inbreeding polygynous ambrosia beetle (Kirkendall *et al.* 2015). Its host trees and gallery systems are not recorded.

Microperus Wood, 1980 (Pl. 115C; Fig. 46.26)

Type species, by original designation: *Xyleborus theae* Eggers, 1940.

Characteristics. Body length 1.8–3.1 mm, 2.7–3.1 × as long as wide. Forehead mostly smooth, with some punctures. Eyes emarginate, upper portion smaller than lower. Antennae with 5-segmented funicle and truncate club marked with 2 straight sutures, club segment 1 encircling anterior face. Postmentum deeply impressed, twice as wide as long. Pronotum covering head, as wide as long, anteriorly rounded, sides parallel, posterior margin straight; surface anterolaterally asperate, disc smooth to punctate. Scutellar shield not exposed. Elytra with disc longer than elytral declivity, shining, convex, setose; declivity rounded, unarmed, setose. Procoxae contiguous, mesocoxae narrowly separate; protibiae triangular, armed with 6–8 socketed denticles.

Australian species. Native: *M. chimbui* (Schedl, 1973), *M. eucalypticus* (Schedl, 1938), *M. intermedius* (Eggers, 1923), *M. parvus* (Lea, 1894).

Distribution. Australia (NSW, QLD), Russia, Korea, Japan, Southeast Asia, Papua New Guinea, China, India.

Notes. The species are xylomycetophagous, haplodiploid, inbreeding polygynous, polyphagous ambrosia beetles (Kirkendall *et al.* 2015). *Microperus parvus* has been reported to create short and flat gallery tunnels (Hulcr & Cognato 2013).

Planiculus Hulcr & Cognato, 2010 (Pl. 115D)

Type species, by original designation: *Xyleborus bicolor* Blandford, 1894.

Characteristics. Body length 1.7–2.4 mm, 2.8–3.6 × as long as wide. Forehead smooth. Eyes emarginate, upper portion smaller than lower. Antennae with 5-segmented funicle and truncate club marked with 2 sutures, edge of club segment 1 continuous on both sides, segment 2 distinct on posterior face. Postmentum slightly impressed, as wide as long. Pronotum covering head, longer than wide, rounded, posterior margin straight; surface anterolaterally asperate, disc smooth. Scutellar shield at same level as elytra. Elytra with disc longer than declivity, shining, convex, sparsely setose or asetose; declivity rounded, granulate. Procoxae contiguous, mesocoxae narrowly separate; protibiae triangular, armed with 6 socketed denticles.

Australian species. Native: *P. bicolor* (Blandford, 1894).

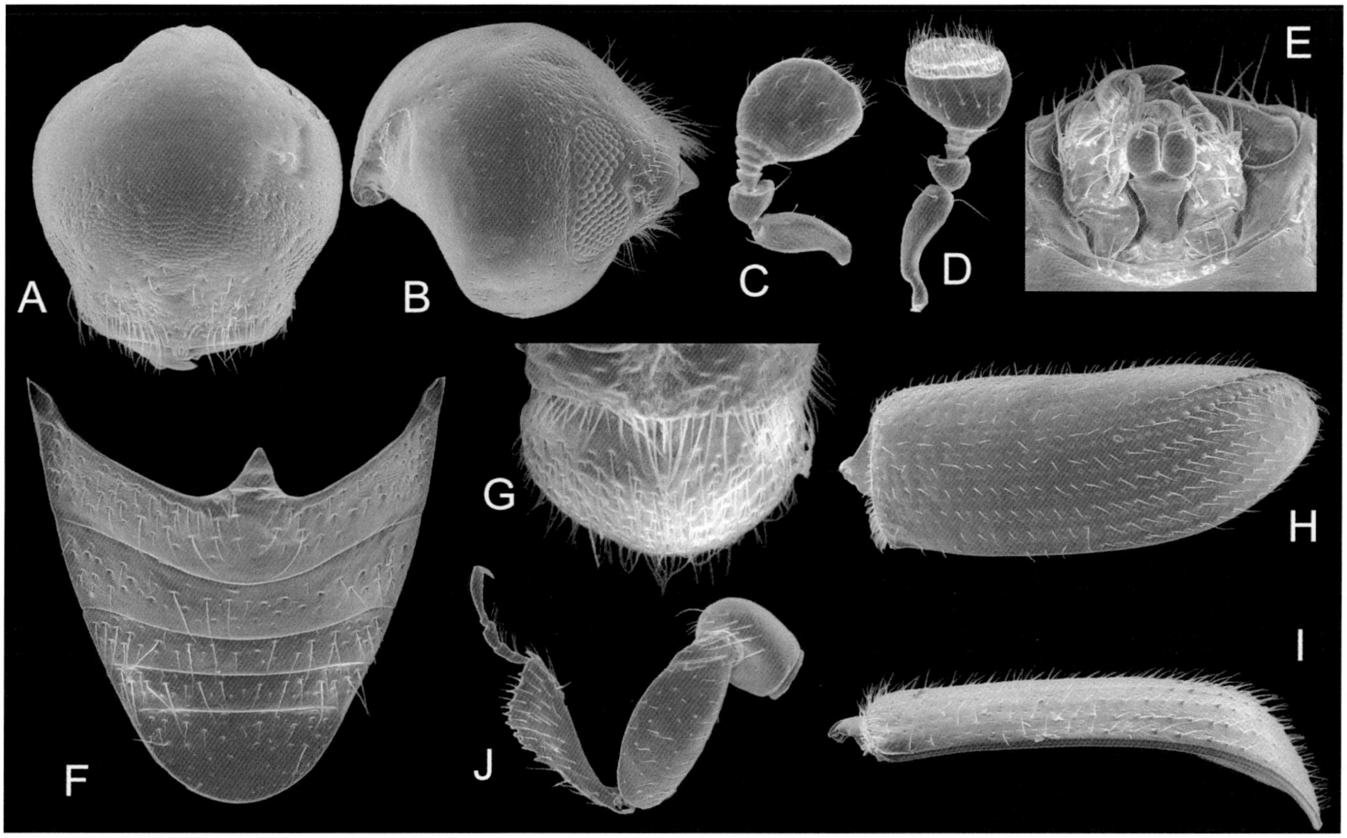

Fig. 46.26. Xyleborini, *Microperus* sp., Lamington NP, QLD, structural details (♀). A–B, head: A, dorsal; B, lateral; C–D, antenna: C, dorsal; D, ventral; E, mouthparts; F, abdominal ventrites; G, tergites VII and VIII, dorsal; H–I, elytron: H, dorsal; I, lateral; J, hindleg.

Distribution. Australia (QLD), Bangladesh, Fiji, India, Japan, Myanmar, Nepal, New Caledonia, Papua New Guinea, Samoa, Seychelles, Southeast Asia, Solomon Is., Sri Lanka.

Notes. The species are xylomycetophagous, haplodiploid, inbreeding polygynous, polyphagous ambrosia beetles (Wood & Bright 1992; Kirkendall *et al.* 2015). Galleries are constructed in one horizontal plane and contain few branches, and occasionally they are formed between phloem and sapwood (Browne 1961).

Truncaudum Hulcr & Cognato, 2010 (Pl. 115E, F; Fig. 46.27)

Type species, by original designation: *Xyleborus impexus* Schedl, 1942.

Characteristics. Body length 2.1–2.9 mm, 2.4–2.9 × as long as wide. Forehead mostly smooth. Eyes emarginate, upper portion smaller than lower. Antennae with 5-segmented funicle and truncate club, edge of club segment 1 continuous on both sides, segment 2 distinct on posterior face. Postmentum deeply impressed, wider than long. Pronotum covering head, longer than wide, anterior margin rounded, posterior margin straight; surface asperate anterolaterally, disc smooth. Scutellar shield at same level as elytra. Elytra with disc longer than declivity, shining, flat, setose; declivity truncate, tuberculate, setose. Procoxae contiguous to narrowly separate, mesocoxae narrowly separate; protibiae rounded, armed with 6–8 socketed denticles.

Australian species. Native: *T. agnatum* (Eggers, 1923).

Distribution. Australia (QLD), Africa, Southeast Asia, China, Papua New Guinea.

Biology. The species are xylomycetophagous, haplodiploid, inbreeding polygynous, polyphagous ambrosia beetles (Wood & Bright 1992; Kirkendall *et al.* 2015). Their galleries consist of a short tunnel leading to a large brood chamber (Hulcr & Cognato 2013).

Xyleborinus Reitter, 1913 (Pl. 115G, H; Fig. 46.28)

Type species, by subsequent designation (Swaine, 1918: 50): *Bostrichus saxesenii* Ratzeburg, 1837.

Characteristics. Body length 1.8–2.8 mm, 2.3–3.3 × as long as wide. Frons mostly smooth, with some punctures,

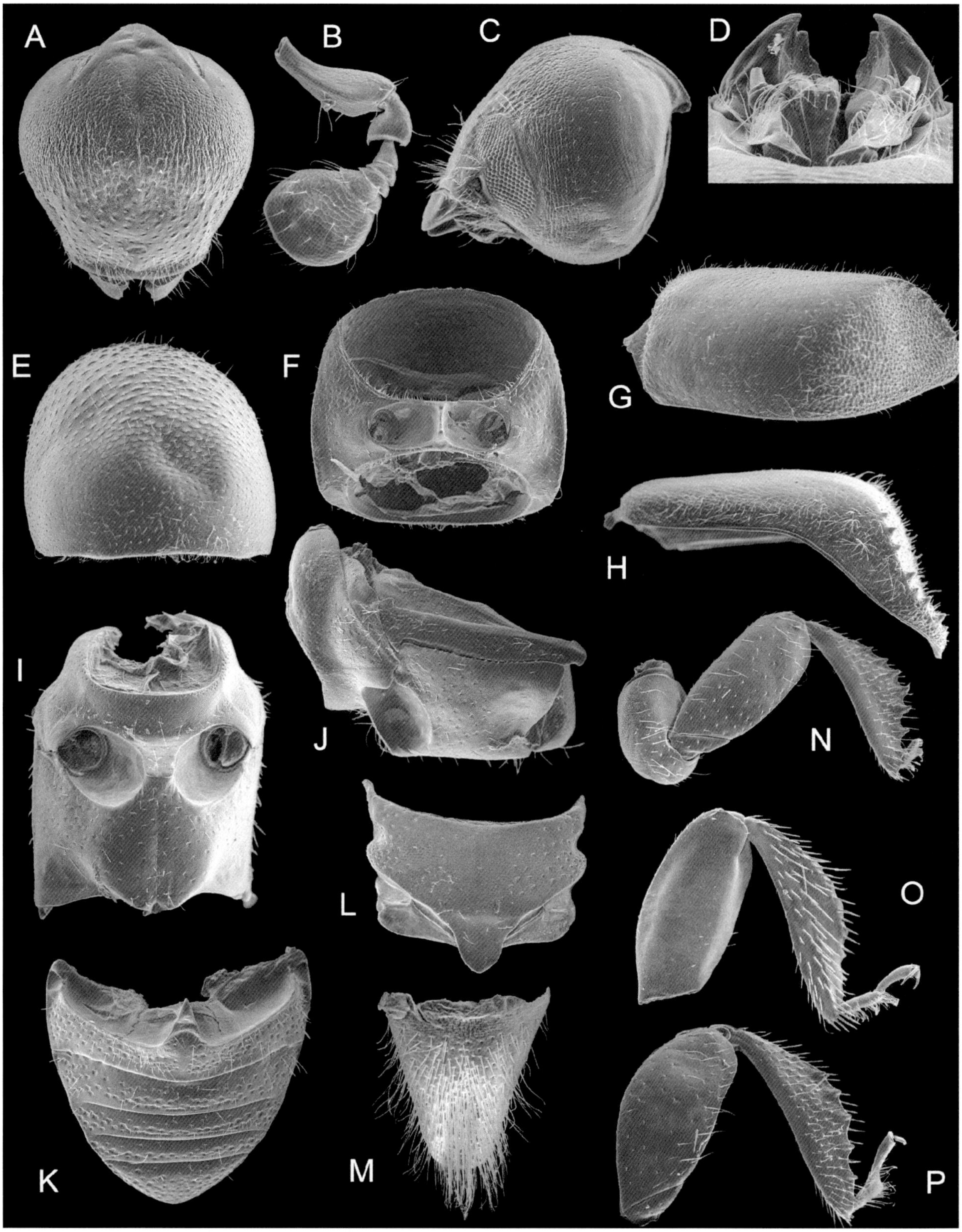

Fig. 46.27. Xyleborini, *Truncaudum* sp., East Claudie River, QLD, structural details (♀). A, head, dorsal; B, antenna; C, head, lateral; D, mouthparts; E–F, prothorax: E, dorsal; F, ventral; G–H, elytron: G, dorsal; H, lateral; I–J, pterothorax: I, ventral; J, lateral; K, abdominal ventrites; L, scutellum; M, tergite VIII, dorsal; N, foreleg; O, middle leg; P, hindleg.

sparsely setose, epistome setose. Eyes emarginate, upper portion smaller than lower. Antennae with 5-segmented funicle and truncate club marked with 2 sutures, club segment 1 encircling anterior face. Postmentum deeply impressed, as wide as long. Pronotum covering head, longer than wide, anteriorly rounded, posterior margin straight; surface asperate anterolaterally, disc smooth with some punctures. Scutellar shield conical, surrounded by setae indicating the presence of elytral mycangia. Elytra with disc longer than declivity, shining, flat; declivity rounded or attenuate dorsally, rounded laterally, tuberculate, setose. Procoxae contiguous, mesocoxae narrowly separate; protibiae triangular, armed with up to 8 socketed denticles.

Australian species. Native: *X. andrewesi* (Blandford, 1896). Introduced: *X. artestriatus* (Eichhoff, 1878) (possibly, see Pullen *et al.* 2014), *X. exiguus* (Walker, 1859) (possibly, see Pullen *et al.* 2014), *X. saxesenii* (Ratzeburg, 1837).

Distribution. Australia (NSW, QLD, WA), all continents except Antarctica, including Hawaii, Japan, Madagascar, Oceanic and Pacific Islands, Papua New Guinea, Southeast Asia.

Notes. The species are xylomycetophagous, haplodiploid, inbreeding polygynous ambrosia beetles (Kirkendall *et al.* 2015) typically associated with *Raffaelea* symbiotic fungi, which are carried in mycangia located in elytral notches (Mayers *et al.* 2022). Of all species of the genus, *Xyleborinus saxesenii* is the best studied. It is globally distributed, highly polyphagous (Wood & Bright 1992) and commonly intercepted (Bickerstaff, pers. obs.). It is commonly referred to as the Fruit-tree Pinhole Borer and regarded as a minor pest of fruit trees (CABI 2022). Females typically delay their dispersal from the natal nest when eggs and dependent siblings are present, and they do not reproduce either, suggesting that they practice cooperative breeding and alloparental care (Peer & Taborsky 2007). Microbiome analyses of *X. saxesenii* have revealed a large diversity of mutualistic and commensal fungi associated with the species, including a pathogenic biofilm fungus detrimental to larvae and adults (Biedermann *et al.* 2013). Several bacterial associates have also been identified, including cycloheximide-producing *Streptomyces* species that may inhibit parasites of the gallery ecosystem (Grubbs *et al.* 2020). Furthermore, amplicon sequencing has detected turnover of fungal symbionts in the galleries, *R. canadensis* present in stable numbers through time but *R. sulphurea* dominating in young nests and then decreasing in abundance over time (Diehl *et al.* 2023).

Xyleborus Eichhoff, 1864 (Pl. 115I, J; Fig. 46.29)

Type species, by subsequent designation (Lacordaire, 1865: 381): *Bostrichus monographus* Fabricius, 1793.

Characteristics. Body length 1.9–3.1 mm, 2.5–3.1 × as long as wide. Forehead mostly smooth, with some punctures. Eyes emarginate, upper portion smaller than lower. Antennae with 5-segmented funicle and truncate or flat club, shape of club segment 1 variable (if club truncate, edge of segment 1 continuous on both sides and segment 2 distinct on posterior face; if club flat, segment 1 as large as 2). Postmentum slightly or deeply impressed, twice as wide as long. Pronotum covering head, proportions variable, anterior margin rounded, posterior margin straight; surface asperate anterolaterally, disc smooth with some punctures. Scutellar shield at same level as elytra. Elytra with disc longer than declivity, shining, either convex or flat; declivity rounded, armed with spines, tubercles or granules. Procoxae contiguous, mesocoxae narrowly separate; protibiae triangular, armed with up to 8 socketed denticles.

Australian species. Native: *X. analis* Schedl, 1973, *X. celsoides* Hagedorn, 1908, *X. norfolkensis* Schedl, 1972. Introduced: *X. affinis* Eichhoff, 1868, *X. bidentatus* (Motschulsky, 1863), 1908, *X. ferrugineus* (Fabricius, 1801), *X. granatus* Schedl, 1979, *X. perforans* (Wollaston, 1857), *X. volvulus* (Fabricius, 1775), *X. xylographus* (Say, 1826).

Distribution. Australia (NSW, QLD), all continents except Antarctica, including Hawaii, Madagascar, Japan, Southeast Asia, Papua New Guinea, Oceanic and Pacific Islands.

Notes. The species are xylomycetophagous, haplodiploid, inbreeding polygynous ambrosia beetles (Kirkendall *et al.* 2015) typically associated with *Raffaelae* fungi carried in paired preoral mycangial pockets (Smith & Hulcr 2015; Mayers *et al.* 2022). Many species have been dispersed globally, with some carrying harmful pathogenic fungi, such as *X. glabratus* Eichhoff transmitting *Harringtonia lauricola*, a phytopathogenic fungus that causes Laurel Wilt Disease in host trees. In the southeastern U.S.A., this beetle and disease have caused the death of half a billion laurel trees and the functional elimination of tree species in natural ecosystems (Smith & Hulcr 2015). Delayed dispersal and alloparental care have been observed in *X. affinis* (Biedermann *et al.* 2011; Biedermann 2020). A rich diversity of microbial symbionts has been characterised and identified in *X. affinis* galleries, including *Raffaelea*, yeast symbionts and bacteria, and many of these microbiota were found to vary in abundance throughout the life cycle of the host beetle (Ibarra-Juarez *et al.* 2020), and one bacterium, a cycloheximide-producing *Streptomyces* species, may inhibit parasites from invading the gallery (Grubbs *et al.* 2020).

Xylosandrus Reitter, 1913 (Pl. 115K; Fig. 46.30)

Type species, by monotypy: *Xyleborus morigerus* Blandford, 1894.

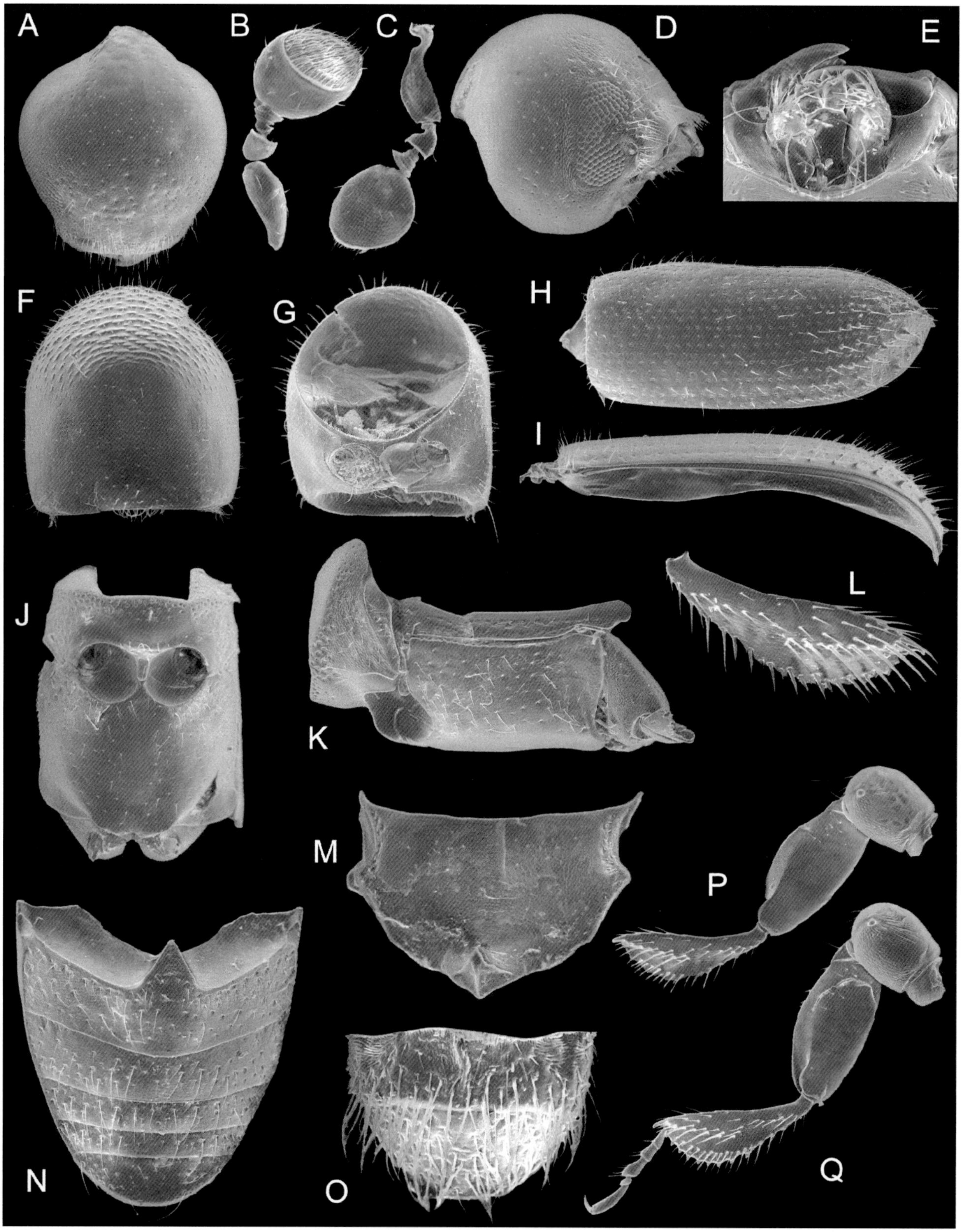

Fig. 46.28. Xyleborini, *Xyleborinus saxesenii* (Ratzeburg), structural details (♀). A, head, dorsal; B–C, antenna: B, ventral; C, dorsal; D, head, lateral; E, mouthparts; F–G, prothorax: F, dorsal; G, ventral; H–I, elytron: H, dorsal; I, lateral; J–K, pterothorax: J, ventral; K, lateral; L, foretibia; M, scutellum; N, abdominal ventrites; O, tergites VII and VIII, dorsal; P, hindleg; Q, hindleg.

Characteristics. Body length 1.4–3.4 mm, 2.0–2.5 × as long as wide. Forehead mostly smooth, with some punctures. Eyes emarginate, upper portion smaller than lower. Antennae with 5-segmented funicle and truncate club with segment 1 encircling segment 2. Postmentum slightly or deeply impressed, twice as wide as long or as wide as long. Pronotum covering head, wider than long, anterior margin rounded and serrate, posterior margin straight, with medial mycangial tuft of setae; surface asperate anterolaterally, disc smooth with some punctures. Scutellar shield at same level as elytra. Elytra with disc variable in length, shining, convex, sparsely or moderately setose; declivity steeply rounded, setose, sometimes granulate. Pro- and mesocoxae widely separate; protibiae slender, armed with up to 5 socketed denticles.

Australian species. Native: *X. abruptulus* (Schedl, 1953), *X. monteithi* Dole & Beaver, 2008, *X. queenslandi* Dole & Beaver, 2008, *X. woodi* Dole & Beaver, 2008. Introduced: *X. compactus* (Eichhoff, 1876), *X. crassiusculus* (Motschulsky, 1866), *X. discolor* (Blandford, 1898), *X. morigerus* (Blandford, 1894).

Distribution. Australia (NSW, QLD), all continents except Antarctica, including Hawaii, Indonesia, Japan, Madagascar, New Zealand and Papua New Guinea.

Notes. The species are xylomycetophagous, haplodiploid, inbreeding polygynous, polyphagous ambrosia beetle (Wood & Bright 1992; Kirkendall *et al.* 2015) that cultivate symbiotic *Ambrosiella* fungi (Skelton *et al.* 2018). *Xylosandrus compactus* has been shown to be able to feed on and complete its life cycle on various *Ambrosiella* species, but it only stores spores of closely related species that it typically cultivates in its mycangia (Skelton *et al.* 2018). *Xylosandrus crassiusculus* has been introduced from its native range in Southeast Asia to all continents except Antarctica. Population genomic analyses (Storer *et al.* 2017; Urvois *et al.* 2023) have identified two genotypes of *X. crassiusculus*, one originating in Southeast Asia including mainland China and the other in Japan. The latter genotype has been introduced to North and Central America, whereas the Southeast Asian one has been introduced to Africa and Oceania. Both genotypes have been detected in Hawaii, and mitochondrial analysis by Tran *et al.* (2022) has also identified both genotypes to be present in Australia.

Tribe Xyloctonini Eichhoff, 1878

Scolytomimus Blandford, 1895 (Pl. 115L)

Type species, by monotypy: *Scolytomimus dilutus* Blandford, 1895.

Characteristics. Body length 1.4–2.1 mm, 1.75–2.0 × as long as wide. Forehead punctate. Eyes deeply emarginate to completely divided, upper portion smaller than lower. Antennae with 6-segmented funicle and ovate to oval club marked with 1 procurved suture. Postmentum flat, as long as wide or longer than wide. Pronotum covering head, wider than long, posterior margin straight; surface mostly smooth, with punctures and asperate anterolaterally. Scutellar shield large, at same level as elytra. Elytra with anterior margin straight; disc flat; without distinct declivity. Abdomen with ventrites ascending toward elytral apex, 1 and 2 slightly raised, 3–5 perpendicular to 1 and 2. Procoxae contiguous, mesocoxae widely separate; protibiae rounded, unarmed.

Australian species. Native: *S. philippinensis* (Eggers, 1923), *S. pusillus* (Eggers, 1927).

Distribution. Australia (QLD, WA), Fiji, Papua New Guinea, Samoa, Southeast Asia.

Notes. The species are phloeophagous, monogynous, polyphagous bark beetles (Kirkendall *et al.* 2015) that construct biramous galleries under bark (Wood 1986). *Scolytomimus philippinensis* has been recorded from *Pometia pinnata* and *S. pusillus* from *Achras sapota, Mimusops littoralis, Palaquium maingayi* and *P. stellatum* (Wood & Bright 1992).

Taxa doubtfully recorded from Australia

Tribe Corthylini LeConte, 1876

Gnathotrichus Eichhoff, 1869

Type species, by monotypy: *Gnathotrichus corthyloides* Eichhoff, 1869.

Characteristics. Body length 2.9 mm. Forehead flat, punctate. Eyes emarginate, upper and lower portions equal in size. Antennae with 5-segmented funicle and obovate club (oval with a narrow base) marked with 2 procurved or straight sutures. Postmentum flat, wider than long. Pronotum covering head, longer than wide, rounded, anterior margin marked by a row of serrations, posterior margin straight; surface anteriorly asperate, disc sparsely punctate. Elytra covering metanepisterna, anterior margin straight; disc flat, smooth, longer than declivity; declivity rounded, emarginate. Pro- and mesocoxae contiguous; protibiae slender to triangular, armed with 1–5 socketed denticles.

Species. Possibly introduced: *G. retusus* (LeConte, 1868).

Distribution. North and Central America; introduced in western Europe and possibly Australia.

Notes. The species are xylomycetophagous, monogynous ambrosia beetles (Kirkendall *et al.* 2015). Males establish the galleries and are joined by females after gallery construction, then guard the gallery entrance while females lay eggs, care for the larvae and tend to fungal gardens by removing frass from galleries, which the males then expel from the entrance. The symbiont ambrosia fungi are mostly *Raffaelea* species, as known.

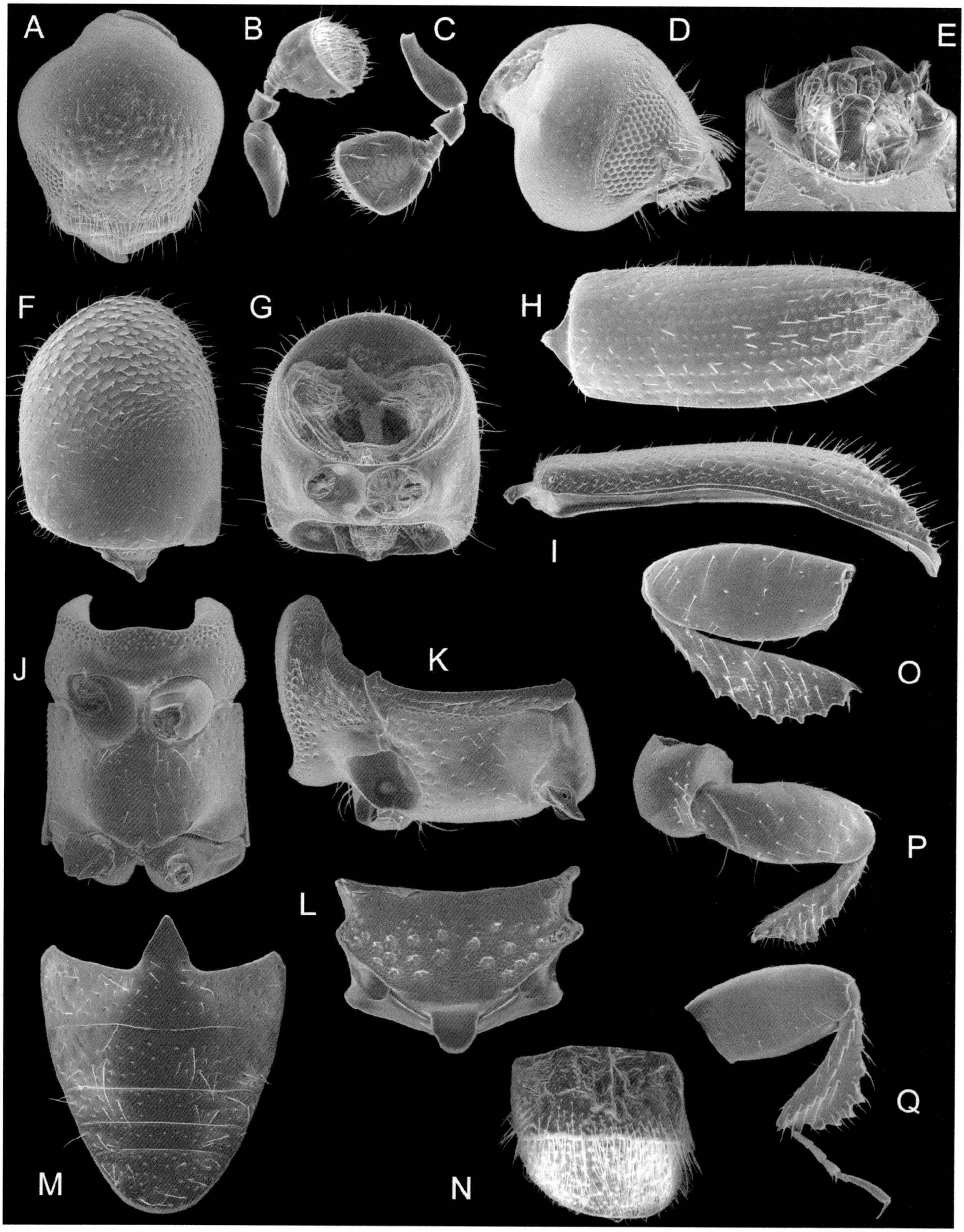

Fig. 46.29. Xyleborini, *Xyleborus perforans* (Wollaston), structural details (♀). A, head, dorsal; B–C, antenna: B, ventral; C, dorsal; D, head, lateral; E, mouthparts; F–G, prothorax: F, dorsal; G, ventral; H–I, elytron: H, dorsal; I, lateral; J–K, pterothorax: J, ventral; K, lateral; L, scutellum; M, abdominal ventrites; N, terminalia, dorsal; O, foreleg; P, middle leg; Q, hindleg.

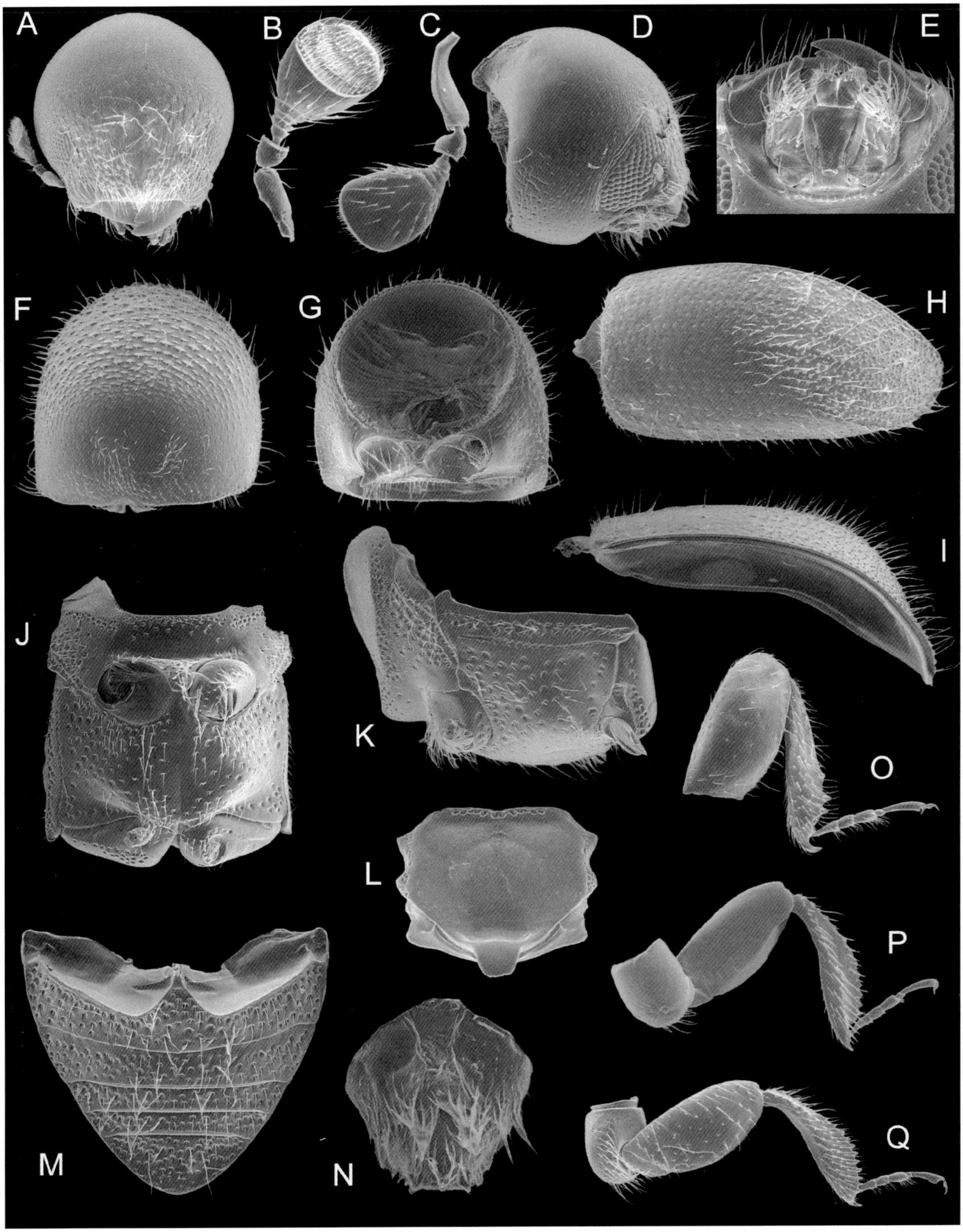

Fig. 46.30. Xyleborini, *Xylosandrus crassiusculus* (Motschulsky), structural details (♀). A, head, dorsal; B–C, antenna: B, ventral; C, dorsal; D, head, lateral; E, mouthparts; F–G, prothorax: F, dorsal; G, ventral; H–I, elytron; H, dorsal; I, lateral; J–K, pterothorax: J, ventral; K, lateral; L, scutellum; M, abdominal ventrites; N, tergite VIII, dorsal; O, foreleg; P, middle leg; Q, hindleg

Galleries are constructed in both conifers and hardwoods, those of *G. retusus* commonly found in trees of the conifer genera *Abies, Larix, Pinus, Pseudotsuga* and *Tsuga* and the angiosperm genera *Alnus, Betula* and *Populus* (Smith & Hulcr 2015). *Gnathotrichus retusus* has been frequently intercepted in Australia but its presence and distribution in the country remain uncertain (Pullen *et al.* 2014).

Tribe Crypturgini LeConte, 1876

Crypturgus Erichson, 1836

Type species, by subsequent designation (Thomson, 1859: 147): *Bostrichus pusillus* Gyllenhal, 1813.

Characteristics. Body length 0.9–1.5 mm. Forehead flat to convex, punctate, setose. Eyes deeply emarginate, upper and lower portions equal in size. Antennae with 2-segmented funicle and obovate club with 1 partial septum and no distinct suture lines. Postmentum flat, as wide as long. Pronotum covering head, longer than wide; surface punctate. Scutellar shield at same level as elytra. Elytra with anterior margin straight; disc flat, smooth, longer than declivity; declivity rounded laterally, slightly attenuate in dorsal view, unarmed, sometimes setose. Procoxae contiguous, mesocoxae separate.

Species. Possibly introduced: *Crypturgus* sp.

Distribution. Holarctic region.

Notes. The species are phloeophagous, monogynous bark beetles (Kirkendall *et al.* 2015) associated with conifers. They often utilise the entrance tunnels of other insects (Wood 1986). The possible presence of the genus in Australia is based on a 1944 record from *Araucaria* by Brimblecombe (1953), but the identification of the specimens is unconfirmed and Pullen *et al.* (2014) listed it only provisionally as occurring in Australia, where it would then be introduced.

Acknowledgements

We thank Lourdes Chamorro (Smithsonian Institution – National Museum of Natural History, U.S.A), Kylie Ireland (Department of Primary Industries and Regional Development – Western Australia), Harald Schillhammer (Naturhistorisches Museum Wien, Austria), James Buxton (Agriculture Victoria), Karin Koch (Queensland Museum), Justin Bartlett (Queensland Department of Agriculture and Forestry) and Ben Parslow (South Australia Museum) for access to specimens used for imaging. We are also grateful to Roger Beaver (Thailand) for providing insight into the distribution of Australian fauna. JRMB would like to thank Markus Riegler (Western Sydney University) and Bjarte Jordal (University of Bergen, Norway) for their support during his research on bark beetles. This study was supported, in part, by a Cooperative Agreement from USDA Forest Service (16-CA-11420004-072 [to A.I.C.]).

References

Alonso-Zarazaga MA, Lyal CHC (2009) A catalogue of family and genus group names in Scolytinae and Platypodinae with nomenclatural remarks (Coleoptera: Curculionidae). Zootaxa **2258**, 1–134.

Barnouin T, Soldati F, Roques A, Faccoli M, Kirkendall LR, *et al.* (2020) Bark beetles and pinhole borers recently or newly introduced to France (Coleoptera: Curculionidae, Scolytinae and Platypodinae). *Zootaxa* **4877**(1), 51–74. doi:10.11646/zootaxa.4877.1.2

Bashford R (2008) The development of static trapping systems to monitor for wood-boring insects in forestry plantations. *Australian Forestry* **71**(3), 236–241. doi:10.1080/00049158.2008.10675041

Bateman C, Šigut M, Skelton J, Smith KE, Hulcr J (2016) Fungal associates of the *Xylosandrus compactus* (Coleoptera: Curculionidae, Scolytinae) are spatially segregated on the insect body. *Environmental Entomology* **45**(4), 883–890. doi:10.1093/ee/nvw070

Batra LR (1963) Ecology of ambrosia fungi and their dissemination by beetles. *Transactions of the Kansas Academy of Science (1903-)* **66**(2), 213. doi:10.2307/3626562

Beaver RA (1976) The biology of Samoan bark and ambrosia beetles (Coleoptera, Scolytidae and Platypodidae). *Bulletin of Entomological Research* **65**(4), 531–548. doi:10.1017/S0007485300006210

Beaver RA (1984) The biology of the ambrosia beetle *Sueus niisimai* (Eggers) (Col. Scolytidae), in Fiji. *Entomologist's Monthly Magazine* **120**, 99–101.

Beaver RA (1989) Insect–fungus relationships in the bark and ambrosia beetles. In *Insect-fungus Interactions* (Eds N Wilding, NM Collins, PM Hammond and JF Weber) pp. 121–143. Elsevier. doi:10.1016/B978-0-12-751800-8.50011-2

Beaver RA (2011a) A review of the genus *Phloeosinopsioides* Schedl (Coleoptera: Curculionidae: Scolytinae). *Zootaxa* **3012**(1), 59. doi:10.11646/zootaxa.3012.1.5

Beaver RA (2011b) (Coleoptera: Curculionidae: Scolytinae, Platypodinae). *Koleopterologische Rundschau* **81**, 277–289.

Beaver RA, Browne FG (1978) The Scolytidae and Platypodidae (Coleoptera) of Penang, Malaysia. *Oriental Insects* **12**(4), 575–624. doi:10.1080/00305316.1978.10432538

Beaver RA, Hulcr J (2008) A review of the ambrosia beetle genus *Cryptoxyleborus* Schedl (Coleoptera: Curculionidae: Scolytinae). *The Coleopterists Bulletin* **62**(1), 133–153. doi:10.1649/1026.1

Beaver RA, Smith SM, Sanguansub S (2019a) A review of the genus *Dryocoetiops* Schedl, with new species, new synonymy and a key to species (Coleoptera: Curculionidae: Scolytinae). *Zootaxa* **4712**(2), 236–250. doi:10.11646/zootaxa.4712.2.4

Beaver RA, Sittichaya W, Liu LY (2019b) A review of the genus *Immanus* Hulcr & Cognato with a key to species (Coleoptera: Curculionidae: Scolytinae: Xyleborini). *Zootaxa* **4585**(2), 378–387. doi:10.11646/zootaxa.4585.2.11

Bernier L (2022) Dutch elm disease. In *Forest Microbiology.* (Eds FO Asiegbu and A Kovalchuk) pp. 291–309. Elsevier. doi:10.1016/B978-0-323-85042-1.00010-0

Bickerstaff JRM, Jordal BH, Riegler M (2023) Two sympatric lineages of Australian *Cnestus solidus* share *Ambrosiella* symbionts but not *Wolbachia*. *Heredity* **132**(1), 43–53. doi:10.1038/s41437-023-00659-w

Biedermann PHW (2020) Cooperative breeding in the ambrosia beetle *Xyleborus affinis* and management of its fungal symbionts. *Frontiers in Ecology and Evolution* **8**, 1–12. doi:10.3389/fevo.2020.518954

Biedermann PHW, Klepzig KD, Taborsky M (2011) Costs of delayed dispersal and alloparental care in the fungus-cultivating ambrosia beetle *Xyleborus affinis* Eichhoff (Scolytinae: Curculionidae). *Behavioral Ecology and Sociobiology* **65**(9), 1753–1761. doi:10.1007/s00265-011-1183-5

Biedermann PHW, Klepzig KD, Taborsky M, Six DL (2013) Abundance and dynamics of filamentous fungi in the complex ambrosia gardens of the primitively eusocial beetle *Xyleborinus saxesenii* Ratzeburg (Coleoptera:

Curculionidae, Scolytinae). *FEMS Microbiology Ecology* **83**(3), 711–723. doi:10.1111/1574-6941.12026

Biedermann PHW, Müller J, Grégoire JC, Gruppe A, Hagge J, *et al.* (2019) Bark beetle population dynamics in the Anthropocene: challenges and solutions. *Trends in Ecology & Evolution* **34**(10), 914–924. doi:10.1016/j.tree.2019.06.002

Bright DE (2014) A catalog of Scolytidae and Platypodidae (Coleoptera), Supplement 3 (2000–2010), with notes on subfamily and tribal reclassifications. *Insecta Mundi* **0356**, 1–336.

Bright DE (2021) *A Catalog of Scolytidae (Coleoptera), Supplement 4 (2011–2019) with an Annotated Checklist of the World Fauna (Coleoptera: Curculionoidea: Scolytidae).* Colorado State University, Fort Collins, Colorado.

Bright DE, Skidmore RE (1997) *A Catalog of Scolytidae and Platypodidae (Coleoptera): Supplement 1 (1990–1994)* (Vol. 1). NRC Research Press, Ottawa, Ontario.

Bright DE, Skidmore RE (2002) *A catalog of Scolytidae and Platypodidae (Coleoptera): Supplement 2 (1995–1999)* (Vol. 2). NRC Research Press, Ottawa, Ontario.

Brimblecombe AR (1953) An annotated list of the Scolytidae occurring in Australia. *Queensland Journal of Agricultural Science* 10(3), 167–205.

Brockerhoff EG, Bain J, Kimberley M, Knížek M (2006) Interception frequency of exotic bark and ambrosia beetles (Coleoptera: Scolytinae) and relationship with establishment in New Zealand and worldwide. *Canadian Journal of Forest Research* **36**(2), 289–298. doi:10.1139/x05-250

Browne FG (1961) The biology of Malayan Scolytidae and Platypodidae. *Malayan Forest Records* **22**(22), 1–255.

CABI (2022) *Xyleborinus saxesenii (fruit-tree pinhole borer).* CABI Compendium. doi:10.1079/cabicompendium.57038

Carrillo D, Duncan RE, Peña JE (2012) Ambrosia beetles (Coleoptera: Curculionidae: Scolytinae) that breed in avocado wood in Florida. *Florida Entomologist* **95**(3), 573–579. doi:10.1653/024.095.0306

Ceriani-Nakamurakare E, Johnson AJ, Gomez DF (2023) Uncharted territories: first report of *Euwallacea fornicatus* (Eichhoff) in South America with new reproductive hosts records. *Zootaxa* **5325**(2), 289–297. doi:10.11646/zootaxa.5325.2.10

Clarke DJ, Limaye A, McKenna DD, Oberprieler RO (2018) The weevil fauna preserved in Burmese amber – snapshot of a unique, extinct lineage (Coleoptera: Curculionoidea). *Diversity* **11**(1), 1. doi:10.3390/d11010001

Cognato AI (2015) Biology, systematics, and evolution of *Ips*. In *Bark Beetles* (Eds FE Vega and RW Hofstetter) pp. 351–370. Elsevier. doi:10.1016/B978-0-12-417156-5.00009-5

Cognato AI, Grimaldi D (2009) 100 million years of morphological conservation in bark beetles (Coleoptera: Curculionidae: Scolytinae). *Systematic Entomology* **34**(1), 93–100. doi:10.1111/j.1365-3113.2008.00441.x

Cognato AI, Smith SM (2010) Resurrection of *Dryotomicus* Wood and description of two new species from the Amazon River Basin (Coleoptera, Curculionidae, Scolytinae, Phloeotribini). *ZooKeys* **56**, 49–64. doi 10.3897/zookeys.56.518

Cognato AI, Smith SM (2020) *Acanthotomicus diaboliculus* and *A. enzoi* (Coleoptera: Curculionidae: Scolytinae: Ipini), new species from Southeast Asia. *The Coleopterists Bulletin* **74**(3), 538. doi:10.1649/0010-065X-74.3.538

Cognato AI, Smith SM, Beaver RA (2020) Two new genera of Oriental xyleborine ambrosia beetles (Coleoptera, Curculionidae: Scolytinae). *Zootaxa* **4722**(6). doi:10.11646/zootaxa.4722.6.2

Cook DC, Broughton S (2023) Economic impact of polyphagous shot hole borer *Euwallacea fornicatus* (Coleoptera: Curculionidae: Scolytinae) in Western Australia. *Agricultural and Forest Entomology* **25**(3), 449–457. doi:10.1111/afe.12566

Cooperband MF, Stouthamer R, Carrillo D, Eskalen A, Thibault T, *et al.* (2016) Biology of two members of the *Euwallacea fornicatus* species complex (Coleoptera: Curculionidae: Scolytinae), recently invasive in the U.S.A., reared on an ambrosia beetle artificial diet. *Agricultural and Forest Entomology* **18**(3), 223–237. doi:10.1111/afe.12155

Córdoba SP, Atkinson TH, Mendoza EA (2023) Checklist of the subfamily Scolytinae (Coleoptera: Curculionidae) in Tucumán province, Argentina. *Zootaxa* **5353**(6), 501–532. doi:10.11646/zootaxa.5353.6.1

Crotch GR (1870) The genera of Coleoptera studied chronologically (1735–1801). *Transactions of the Entomological Society of London* **1870**(1), 41–52.

De Beer ZW, Procter M, Wingfield MJ, Marincowitz S, Duong TA (2022) Generic boundaries in the *Ophiostomatales* reconsidered and revised. *Studies in Mycology* **101**(1), 57–120. doi:10.3114/sim.2022.101.02

De Wit MP, Crookes DJ, Blignaut JN, De Beer ZW, Paap T, *et al.* (2022) An assessment of the potential economic impacts of the invasive polyphagous shot hole borer (Coleoptera: Curculionidae) in South Africa. *Journal of Economic Entomology* **115**(4), 1076–1086. doi:10.1093/jee/toac061

Diehl JMC, Keller A, Biedermann PHW (2023) Comparing the succession of microbial communities throughout development in field and laboratory nests of the ambrosia beetle *Xyleborinus saxesenii. Frontiers in Microbiology* **14**, 1151208. doi:10.3389/fmicb.2023.1151208

Dole SA, Cognato AI (2010) Phylogenetic revision of *Xylosandrus* Reitter (Coleoptera: Curculionidae: Scolytinae: Xyleborina). *Proceedings of the California Academy of Sciences* **61**(7), 451.

Flechtmann CAH, Cognato AI (2011) First Report of *Amasa truncata* (Erichson) (Coleoptera: Curculionidae: Scolytinae) in Brazil. *The Coleopterists Bulletin* **65**(4), 417–421. doi:10.1649/072.065.0419

Francke-Grosmann H (1956) Hautdrüsen als Träger der Pilzsymbiose bei Ambrosiakäfern. *Zeitschrift Für Morphologie Und Ökologie Der Tiere* **45**(3), 275–308. doi:10.1007/BF00430256

Freeman S, Sharon M, Maymon M, Mendel Z, Protasov A, *et al.* (2013) *Fusarium euwallaceae* sp. nov. – a symbiotic fungus of *Euwallacea* sp., an invasive ambrosia beetle in Israel and California. *Mycologia* **105**(6), 1595–1606. doi:10.3852/13-066

Froggatt WW (1927) *Forest Insects and Timber Borers.* Alfred James Kent, Government Printer, Phillip-Street Sydney.

Gadgil PD, Bulman LS, Dick MA, Bajn J (2000) Dutch Elm Disease in New Zealand. In *The Elms.* (Ed. CP Dunn) pp. 189–199. Springer US. doi:10.1007/978-1-4615-4507-1_11

Gao L, Li Y, Wang Z, Zhao J, Hulcr J, *et al.* (2021) Biology and associated fungi of an emerging bark beetle pest, the sweetgum inscriber *Acanthotomicus suncei* (Coleoptera: Curculionidae). *Journal of Applied Entomology* **145**(6), 508–517. doi:10.1111/jen.12861

Giese RL (1967) The Columbian Timber Beetle, *Corthylus columbianus* (Coleoptera: Scolytidae): V. A description of the mycetangia. *The Canadian Entomologist* **99**(1), 54–58. doi:10.4039/Ent9954-1

Gillett C, Crampton-Platt A, Timmermans MJTN, Jordal BH, Emerson BC, *et al.* (2014) Bulk de novo mitogenome assembly from pooled total DNA elucidates the phylogeny of weevils (Coleoptera: Curculionoidea). *Molecular Biology and Evolution* **31**(8), 2223–2237. doi:10.1093/molbev/msu154

Gohli J, Jordal BH (2017) Explaining biogeographical range size and measuring its effect on species diversification in bark beetles. *Journal of Biogeography* **44**(9), 2132–2144. doi:10.1111/jbi.13005

Gohli J, Selvarajah T, Kirkendall LR, Jordal BH (2016) Globally distributed *Xyleborus* species reveal recurrent intercontinental dispersal in a landscape of ancient worldwide distributions. *BMC Evolutionary Biology* **16**(1), 37. doi:10.1186/s12862-016-0610-7

Gomez DF, Skelton J, Steininger MS, Stouthamer R, Rugman-Jones P, *et al.* (2018) Species delineation within the *Euwallacea fornicatus* (Coleoptera: Curculionidae) complex revealed by morphometric and phylogenetic analyses. *Insect Systematics and Diversity* **2**(6), 2. doi:10.1093/isd/ixy018

Gómez D, Martinez G, Beaver RA (2012) First record of *Cyrtogenius luteus* (Blandford) (Coleoptera: Curculionidae: Scolytinae) in the Americas and its distribution in Uruguay. *The Coleopterists Bulletin* **66**(4), 362–364. doi:10.1649/072.066.0414

Gómez D, Suárez M, Martínez G (2017) *Amasa truncata* (Erichson) (Coleoptera: Curculionidae: Scolytinae): a new exotic ambrosia beetle in Uruguay. *The Coleopterists Bulletin* **71**(4), 825. doi:10.1649/0010-065X-71.4.825

Gray B, Lamb KP (1975) Biology of *Hylurdrectonus araucariae* Schedl (Coleoptera, Scolytidae), a pest of hoop pine plantations in New Guinea. *Bulletin of Entomological Research* **65**(1), 21–32. doi:10.1017/S000748530000571X

Grégoire JC, Raffa KF, Lindgren BS (2015) Economics and politics of bark beetles. In *Bark Beetles*. (Eds FE Vega and RW Hofstetter) pp. 585–613. Elsevier. doi:10.1016/B978-0-12-417156-5.00015-0

Griggs JA (1998) *Management of Hylastes Ater (Coleoptera: Scolytidae) Attacking Pinus Radiata Seedlings*. University of Tasmania.

Grubbs KJ, Surup F, Biedermann PHW, McDonald BR, Klassen JL, *et al.* (2020) Cycloheximide-producing *Streptomyces* associated with *Xyleborinus saxesenii* and *Xyleborus affinis* fungus-farming ambrosia beetles. *Frontiers in Microbiology* **11**, 562140. doi:10.3389/fmicb.2020.562140

Gunter NL, Oberprieler RG, Cameron SL (2016) Molecular phylogenetics of Australian weevils (Coleoptera: Curculionoidea): exploring relationships in a hyperdiverse lineage through comparison of independent analyses. *Austral Entomology* **55**(2), 217–233. doi:10.1111/aen.12173

Harrington TC, McNew D, Mayers C, Fraedrich SW, Reed SE (2014) *Ambrosiella roeperi* sp. nov. is the mycangial symbiont of the granulate ambrosia beetle, *Xylosandrus crassiusculus*. *Mycologia* **106**(4), 835–845. doi:10.3852/13-354

Hopkins AD (1914) List of generic names and their type-species in the coleopterous superfamily Scolytoidea. *Proceedings of the United States National Museum* **48(2066)**, 115–136.

Hulcr J, Cognato AI (2009) Three new genera of oriental Xyleborina (Coleoptera: Curculionidae: Scolytinae). *Zootaxa* **2204**(1), 19–36. doi:10.11646/zootaxa.2204.1.2

Hulcr J, Cognato AI (2010) Repeated evolution of crop theft in fungus-farming ambrosia beetles: mycocleptism in ambrosia beetles. *Evolution* **64**(11), 3205–3212. doi:10.1111/j.1558-5646.2010.01055.x

Hulcr J, Cognato AI (2013) *Xyleborini of New Guinea, a Taxonomic Monograph (Coleoptera: Curculionidae: Scolytinae)* (Vol. 33). Entomological Society of America.

Hulcr J, Stelinski LL (2017) The ambrosia symbiosis: from evolutionary ecology to practical management. *Annual Review of Entomology* **62**(1), 285–303. doi:10.1146/annurev-ento-031616-035105

Hulcr J, Mogia M, Isua B, Novotny V (2007) Host specificity of ambrosia and bark beetles (Col., Curculionidae: Scolytinae and Platypodinae) in a New Guinea rainforest. *Ecological Entomology* **32**(6), 762–772. doi:10.1111/j.1365-2311.2007.00939.x

Hulcr J, Atkinson TH, Cognato AI, Jordal BH, McKenna DD (2015) Morphology, taxonomy, and phylogenetics of bark beetles. In *Bark Beetles*. (Eds FE Vega and RW Hofstetter) pp. 41–84. Elsevier. doi:10.1016/B978-0-12-417156-5.00002-2

Ibarra-Juarez LA, Burton MAJ, Biedermann PHW, Cruz L, Desgarennes D, *et al.* (2020) Evidence for succession and putative metabolic roles of fungi and bacteria in the farming mutualism of the ambrosia beetle *Xyleborus affinis*. *mSystems* **5**(5), e00541-20. doi:10.1128/mSystems.00541-20

Johnson AJ, Knížek M, Atkinson TH, Jordal BH, Ploetz RC, *et al.* (2017) Resolution of a global mango and fig pest identity crisis. *Insect Systematics and Diversity* **1**(2), 1–10. doi:10.1093/isd/ixx010

Johnson AJ, McKenna DD, Jordal BH, Cognato AI, Smith SM, *et al.* (2018) Phylogenomics clarifies repeated evolutionary origins of inbreeding and fungus farming in bark beetles (Curculionidae, Scolytinae). *Molecular Phylogenetics and Evolution* **127**, 229–238. doi:10.1016/j.ympev.2018.05.028

Johnson AJ, Hulcr J, Knížek M, Atkinson TH, Mandelshtam MY, *et al.* (2020a) Revision of the bark beetle genera within the former Cryphalini (Curculionidae: Scolytinae). *Insect Systematics and Diversity* **4**(3), 1–81. doi:10.1093/isd/ixaa002

Johnson AJ, Li Y, Mandelshtam M Yu, Park S, Lin CS, *et al.* (2020b). East Asian *Cryphalus* Erichson (Curculionidae, Scolytinae): new species, new synonymy and redescriptions of species. *ZooKeys* **995**, 15–66. doi:10.3897/zookeys.995.55981

Jordal BH (2014) 3.7. 12 Scolytinae Latreille, 1806. In *Coleoptera, Beetles. Volume 3: Morphology and Systematics (Phytophaga)*. (Eds RAB Leschen and RG Beutel) pp. 633–642. Walter deGruyter, Berlin.

Jordal BH (2021) Molecular and morphological revision of Afrotropical Hypoborini (Coleoptera: Curculionidae: Scolytinae) revealed novel bark beetle taxa with narrow geographical distributions. *European Journal of Entomology* **118**, 90–110. doi:10.14411/eje.2021.011

Jordal BH, Beaver RA, Kirkendall LR (2001) Breaking taboos in the tropics: incest promotes colonization by wood-boring beetles. *Global Ecology and Biogeography* **10**(4), 345–357. doi:10.1046/j.1466-822X.2001.00242.x

Jordal BH, Kirkendall LR, Harkestad K (2004) Phylogeny of a Macaronesian radiation: host-plant use and possible cryptic speciation in *Liparthrum* bark beetles. *Molecular Phylogenetics and Evolution* **31**(2), 554–571. doi:10.1016/j.ympev.2003.09.008

Jordal BH, Smith SM, Cognato AI (2014) Classification of weevils as a data-driven science: leaving opinion behind. *ZooKeys* **439**, 1–18. doi:10.3897/zookeys.439.8391

Jordal BH, Normark BB, Farrell BD, Kirkendall LR (2002) Extraordinary haplotype diversity in haplodiploid inbreeders: phylogenetics and evolution of the bark beetle genus *Coccotrypes*. *Molecular Phylogenetics and Evolution* **23**(2), 171–188. doi:10.1016/S1055-7903(02)00013-1

Kasson MT, Wickert KL, Stauder CM, Macias AM, Berger MC, *et al.* (2016) Mutualism with aggressive wood-degrading *Flavodon ambrosius* (Polyporales) facilitates niche expansion and communal social structure in *Ambrosiophilus* ambrosia beetles. *Fungal Ecology* **23**, 86–96. doi:10.1016/j.funeco.2016.07.002

Kegley SJ, Furniss MM, Grégoire JC (1997) Electrophoretic comparison of *Dendroctonus punctatus* LeConte and *D. micans* (Kugelann) (Coleoptera: Scolytidae). *Pan-Pacific Entomologist* **73**(1), 40–45.

Kim S, Harrington TC, Lee JC, Seybold SJ (2011) *Leptographium tereforme* sp. nov. and other Ophiostomatales isolated from the root-feeding bark beetle *Hylurgus ligniperda* in California. *Mycologia* **103**(1), 152–163. doi:10.3852/10-096

Kirejtshuk AG, Azar D, Beaver RA, Mandelshtam MY, Nel A (2009) The most ancient bark beetle known: a new tribe, genus and species from Lebanese amber (Coleoptera, Curculionidae, Scolytinae). *Systematic Entomology* **34**(1), 101–112. doi:10.1111/j.1365-3113.2008.00442.x

Kirkendall LR, Faccoli M (2010) Bark beetles and pinhole borers (Curculionidae, Scolytinae, Platypodinae) alien to Europe. *ZooKeys* **56**, 227–251. doi:10.3897/zookeys.56.529

Kirkendall LR, Biedermann PHW, Jordal BH (2015) Evolution and diversity of bark and ambrosia beetles. In *Bark Beetles*. (Eds. FE Vega and RW Hofstetter) pp. 85–156. Elsevier. doi:10.1016/B978-0-12-417156-5.00003-4

Knížek M, Smith SM (2024) A new widely distributed invasive alien species of *Amasa* ambrosia beetles (Coleoptera: Curculionidae: Scolytinae: Xyleborini). *Zootaxa* **5403**(3), 385–390. doi:10.11646/zootaxa.5403.3.8

Kostovcik M, Bateman CC, Kolarik M, Stelinski LL, Jordal BH, *et al.* (2015) The ambrosia symbiosis is specific in some species and promiscuous in others: evidence from community pyrosequencing. *The ISME Journal* **9**(1), 126–138. doi:10.1038/ismej.2014.115

Kuschel G, Leschen RAB, Zimmerman EC (2000) Platypodidae under scrutiny. *Invertebrate Systematics* **14**(6), 771–805. doi:10.1071/IT00024

Lacordaire T (1865[1866]) *Histoire naturelle des Insectes. Genera des Coléoptères ou exposé méthodique et critique de tous les genres proposés jusqu'ici dans cet ordre d'insectes*. Tome septième. Roret, Paris, 620 pp. doi:10.5962/bhl.title.67686

Lanier GN, Kirkendall LR (1986) Karyology of pseudogamous *Ips* bark beetles. *Hereditas* **105**(1), 87–96. doi:10.1111/j.1601-5223.1986.tb00646.x

Lantschner MV, Corley JC, Liebhold AM (2020) Drivers of global Scolytinae invasion patterns. *Ecological Applications* **30**(5), e02103. doi:10.1002/eap.2103

Latreille PA (1810) *Considérations générales sur l'ordre naturel des animaux, composant les classes des Crustacés, des Arachnides, et des Insectes; avec un tableau méthodique de leurs genres, disposés en familles*. F. Schoell, Paris, 444 pp. doi:10.5962/bhl.title.39620

Lea AM (1910) On Australian and Tasmanian Coleoptera, with descriptions of new species. Part I. *Proceedings of the Royal Society of Victoria* **22**(2), 113–152.

Li Y, Bateman CC, Skelton J, Jusino MA, Nolen ZJ, *et al.* (2017) Wood decay fungus *Flavodon ambrosius* (Basidiomycota: Polyporales) is widely farmed by two genera of ambrosia beetles. *Fungal Biology* **121**(11), 984–989. doi:10.1016/j.funbio.2017.08.004

Li Y, Skelton J, Adams S, Hattori Y, Smith ME, *et al.* (2020) The ambrosia beetle *Sueus niisimai* (Scolytinae: Hyorrhynchini) is associated with the canker disease dungus *Diatrypella japonica* (Xylariales). *Plant Disease* **104**(12), 3143–3150. doi:10.1094/PDIS-03-20-0482-RE

Lin W, Li Y, Meng LZ (2023) First record of the genus *Immanus* Hulcr & Cognato (Coleoptera: Curculionidae: Scolytinae: Xyleborini) from China, with description of a new species. *Zootaxa* **5352**(3), 433–438. doi:10.11646/zootaxa.5352.3.7

Lozano C, Kidd NAC, Campos M (1996) The population dynamics of the bark beetle *Phloeotribus scarabaeoides* (Col., Scolytidae) on European olive (*Olea europaea*). *Journal of Applied Entomology* **120**(1–5), 193–197. doi:10.1111/j.1439-0418.1996.tb01590.x

Marchant KR (1976) *Worldwide Introduction and Establishment of Bark and Timber Beetles (Coleoptera: Scolytidae and Platypodidae)*. Simon Fraser University.

Mausel DL, Gara RI, Lanfranco D, Ruiz C, Ide S, *et al.* (2007) The introduced bark beetles *Hylurgus ligniperda* and *Hylastes ater* (Coleoptera: Scolytidae) in Chile: seasonal flight and effect of *Pinus radiata* log placement on colonization. *Canadian Journal of Forest Research* **37**(1), 156–169. doi:10.1139/x06-215

May BM (1994) An introduction to the immature stages of Australian Curculionoidea. In *Australian Weevils. Volume II. Brentidae, Eurhynchidae and Apionidae and a Chapter on Immature Stages by Brenda May*. (Ed. EC Zimmerman) pp. 365–728. CSIRO, Australia, x + 751 pp.

Mayers CG, Harrington TC, Biedermann PHW (2022) Mycangia define the diverse ambrosia beetle–fungus symbioses. In *The Convergent Evolution of Agriculture in Humans and Insects*. (Eds TR Schultz, R Gawne, PN Peregrine) pp. 105–142. The MIT Press. doi:10.7551/mitpress/13600.003.0013

Mayers CG, Harrington TC, Masuya H, Jordal BH, McNew DL, *et al.* (2020) Patterns of coevolution between ambrosia beetle mycangia and the *Ceratocystidaceae*, with five new fungal genera and seven new species. *Persoonia – Molecular Phylogeny and Evolution of Fungi* **44**(1), 41–66. doi:10.3767/persoonia.2020.44.02

McKenna DD, Sequeira AS, Marvaldi AE, Farrell BD (2009) Temporal lags and overlap in the diversification of weevils and flowering plants. *Proceedings of the National Academy of Sciences* **106**(17), 7083–7088. doi:10.1073/pnas.0810618106

McKenna DD, Shin S, Ahrens D, Balke M, Beza-Beza C, *et al.*(2019) The evolution and genomic basis of beetle diversity. *Proceedings of the National Academy of Sciences* **116**(49), 24729–24737. doi:10.1073/pnas.1909655116

Mecke R (2004) New species of *Pachycotes* Sharp and *Xylechinus* Chapuis (Coleoptera: Curculionidae: Scolytinae, Tomicini) from New Caledonian *Araucaria* spp. (Coniferales: Araucariaceae). *New Zealand Journal of Zoology* **31**(4), 343–349. doi:10.1080/03014223.2004.9518387

Mendel Z, Lynch SC, Eskalen A, Protasov A, Maymon M, *et al.* (2021) What determines host range and reproductive performance of an invasive ambrosia beetle *Euwallacea fornicatus*; lessons from Israel and California. *Frontiers in Forests and Global Change* **4**, 654702. doi:10.3389/ffgc.2021.654702

Meshram V, Sharma G, Maymon M, Protasov A, Mendel Z, *et al.* (2022) Symbiosis and pathogenicity of *Geosmithia* and *Talaromyces* spp. associated with the cypress bark beetles *Phloeosinus* spp. and their parasitoids. *Environmental Microbiology* **24**(8), 3369–3389. doi:10.1111/1462-2920.16016

Meurisse N, Rassati D, Hurley BP, Brockerhoff EG, Haack RA (2019) Common pathways by which non-native forest insects move internationally and domestically. *Journal of Pest Science* **92**(1), 13–27. doi:10.1007/s10340-018-0990-0

Mitchell A, Maddox C (2010) Bark beetles (Coleoptera: Curculionidae: Scolytinae) of importance to the Australian macadamia industry: an integrative taxonomic approach to species diagnostics. *Australian Journal of Entomology* **49**(2), 104–113. doi:10.1111/j.1440-6055.2010.00746.x

Nahrung HF, Carnegie AJ (2020) Non-native forest insects and pathogens in Australia: establishment, spread, and impact. *Frontiers in Forests and Global Change* **3**, 1–12. doi:10.3389/ffgc.2020.00037

Nahrung HF, Carnegie AJ (2021) Border interceptions of forest insects established in Australia: intercepted invaders travel early and often. *NeoBiota* **64**, 69–86. doi:10.3897/neobiota.64.60424

Naumann ID, Weir TA, Edwards ED (1991) Insects of Kimberley rainforests. In *Kimberley Rainforests of Australia*. (Eds NL McKenzie, RB Johnston and PG Kendrick) pp. 299–332. Surrey Beatty & Sons in association with Dept. of Conservation and Land Management, Western Australia and Dept. of Arts, Heritage and Environment, Canberra Chipping, Norton, NSW.

Neumann FG (1987) Introduced bark beetles on exotic trees in Australia with special reference to infestations of *Ips grandicollis* in pine plantations. *Australian Forestry* **50**(3), 166–178. doi:10.1080/00049158.1987.10674513

Neumann FG, Harris JA (1974) Pinhole borers in green timber. *Australian Forestry* **37**(2), 132–141. doi:10.1080/00049158.1974.10675605

Neumann FG, Minko G (1985) Studies on the introduced smaller European elm bark beetle, *Scolytus multistriatus*, a potential vector of Dutch elm disease in Victoria. *Australian Forestry* **48**(2), 116–126. doi:10.1080/00049158.1985.10674432

Oberprieler RG, Marvaldi AE, Anderson RS (2007) Weevils, weevils, weevils everywhere*. *Zootaxa* **1668**(1), 491–520. doi:10.11646/zootaxa.1668.1.24

Oliveira CM, Auad AM, Mendes SM, Frizzas MR (2013) Economic impact of exotic insect pests in Brazilian agriculture. *Journal of Applied Entomology* **137**(1–2), 1–15. doi:10.1111/jen.12018

Oliveira CM, Flechtmann CAH, Frizzas MR (2008) First record of *Xylosandrus compactus* (Eichhoff) (Coleoptera: Curculionidae: Scolytinae) on soursop, *Annona muricata* L. (Annonaceae) in Brazil, with a list of host plants. *The Coleopterists Bulletin* **62**(1), 45–48. doi:10.1649/1039.1

Peer K, Taborsky M (2007) Delayed dispersal as a potential route to cooperative breeding in ambrosia beetles. *Behavioral Ecology and Sociobiology* **61**(5), 729–739. doi:10.1007/s00265-006-0303-0

Pistone D, Gohli J, Jordal BH (2018) Molecular phylogeny of bark and ambrosia beetles (Curculionidae: Scolytinae) based on 18 molecular markers. *Systematic Entomology* **43**(2), 387–406. doi:10.1111/syen.12281

Ploetz RC, Hulcr J, Wingfield MJ, de Beer ZW (2013) Destructive tree diseases associated with ambrosia and bark beetles: Black Swan events in tree pathology? *Plant Disease* **97**(7), 856–872. doi:10.1094/PDIS-01-13-0056-FE

Pullen KR, Jennings D, Oberprieler RG (2014) Annotated catalogue of Australian weevils (Coleoptera: Curculionoidea). *Zootaxa* **3896**(1), 1–481. doi:10.11646/zootaxa.3896.1.1

Raffa KF, Aukema BH, Bentz BJ, Carroll AL, Hicke JA, *et al.* (2008) Cross-scale drivers of natural disturbances prone to anthropogenic amplification: the dynamics of bark beetle eruptions. *BioScience* **58**(6), 501–517. doi:10.1641/B580607

Raffa KF, Grégoire JC, Staffan Lindgren B (2015) Natural history and ecology of bark beetles. In *Bark Beetles* (Eds FE Vega and RW Hofstetter) pp. 1–40. Elsevier. doi:10.1016/B978-0-12-417156-5.00001-0

Rainho HL, Silva WD, Leite MOG, Bento JMS (2018) Notes on the distribution of the exotic ambrosia beetle *Amasa truncata* (Erichson) (Coleoptera: Curculionidae: Scolytinae) in southeastern Brazil. *The Coleopterists Bulletin* **72**(4), 870–872. doi:10.1649/0010-065X-72.4.870

Rassati D, Faccoli M, Petrucco Toffolo E, Battisti A, Marini L (2015) Improving the early detection of alien wood-boring beetles in ports and surrounding forests. *Journal of Applied Ecology* **52**(1), 50–58. doi:10.1111/1365-2664.12347

Roberts H (1977) Observations on the biology of some tropical rain forest Scolytidae (Coleoptera) from Fiji: II. Subfamily Ipinae – tribe Xyleborini. *Journal of Natural History* **11**(3), 251–272. doi:10.1080/00222937700770181

Rosel A, French J (1975) Elm bark disease in Australia. *Nature* **253**(5490), 305–305. doi:10.1038/253305b0

Schedl KE (1938) Scolytidae and Platypodidae. Contribution 49. New species from Australia and the Fiji Islands with some revisional notes. *Transactions of the Royal Society of South Australia* **62**(July), 34–52.

Schedl KE (1942) Interessante und neue Scolytiden und Platypodiden aus der australischen Region. 79. Beitrag zur Morphologie und Systematik der Scolytoidea. *Mitteilungen der Münchener Entomologischen Gesellschaft* **32**, 162–201.

Schedl KE (1971) Bark and timber beetles from Australia (Coleoptera: Scolytidae and Platypodidae). *Australian Journal of Entomology* **11**(2), 143–149. doi:10.1111/j.1440-6055.1972.tb01615.x

Schedl KE (1972) Bark and timber beetles from Australia (Coleoptera: Scolytidae and Platypodidae). *Australian Journal of Entomology* **11**(2), 143–149. doi:10.1111/j.1440-6055.1972.tb01615.x

Seidl R, Schelhaas M, Lexer MJ (2011) Unraveling the drivers of intensifying forest disturbance regimes in Europe. *Global Change Biology* **17**(9), 2842–2852. doi:10.1111/j.1365-2486.2011.02452.x

Shin S, Clarke DJ, Lemmon AR, Lemmon ME, Aitken AL, *et al.* (2018) Phylogenomic data yield new and robust insights into the phylogeny and evolution of weevils. *Molecular Biology and Evolution* **35**(4), 823–836. doi:10.1093/molbev/msx324

Simmons DR, Li Y, Bateman CC, Hulcr J (2016) *Flavodon ambrosius* sp. nov., a basidiomycetous mycosymbiont of *Ambrosiodmus* ambrosia beetles. *Mycotaxon* **131**(2), 277–285. doi:10.5248/131.277

Sittichaya W (2012) Bark and ambrosia beetles (Coleoptera: Curculionidae: Scolytinae and Platypodinae) infesting mango trees (*Mangifera indica* L.) in Southern Thailand, with two new species recorded for Thailand. *Songklanakarin Journal of Science and Technology* **34**(2), 153–155.

Sittichaya W, Smith SM, Beaver RA, Thaochan N. (2021) Revision of the xyleborine ambrosia beetle genus *Microperus* Wood, 1980 (Curculionidae, Scolytinae, Xyleborini) of Thailand with four new species and four newly recorded species. *ZooKeys* **1074**, 191–214. doi:10.3897/zookeys.1074.76235

Six DL, Bracewell R (2015) *Dendroctonus*. In *Bark Beetles*. (Eds FE Vega and RW Hofstetter) pp. 305–350. Elsevier. doi:10.1016/B978-0-12-417156-5.00008-3

Skelton J, Johnson AJ, Jusino MA, Bateman CC, Li Y, *et al.* (2018) A selective fungal transport organ (mycangium) maintains coarse phylogenetic congruence between fungus-farming ambrosia beetles and their symbionts. *Proceedings of the Royal Society B: Biological Sciences* **286**(1894), 20182127. doi:10.1098/rspb.2018.2127

Smit SJ, Allsopp E, Nethavhani Z, Caleca V, Oberprieler RG, van Asch B (2022) Mitogenomics of the Olive Seed Weevil, *Anchonocranus oleae* Marshall and implications for its phylogenetic position in Curculionidae. *Insects* **13**(7), 607.

Smith SM, Cognato AI (2014) A taxonomic monograph of Nearctic *Scolytus* Geoffroy (Coleoptera, Curculionidae, Scolytinae). *ZooKeys* **450**, 1–182. doi:10.3897/zookeys.450.7452

Smith SM, Hulcr J (2015) *Scolytus* and other economically important bark and ambrosia Beetles. In *Bark Beetles*. (Eds FE Vega and RW Hofstetter) pp. 495–531. Elsevier. doi:10.1016/B978-0-12-417156-5.00012-5

Smith SM, Beaver RA, Cognato AI (2020) A monograph of the Xyleborini (Coleoptera, Curculionidae, Scolytinae) of the Indochinese Peninsula (except Malaysia) and China. *ZooKeys* **983**, 1–442. doi:10.3897/zookeys.983.52630

Smith SM, Touroult J, Cognato AI (2022a) The first report of *Sueus niisimai* (Eggers, 1923) (Coleoptera: Curculionidae: Scolytinae: Hyorrhynchini) from the Western Hemisphere, from the Caribbean Island of Martinique. *The Coleopterists Bulletin* **76**(3). doi:10.1649/0010-065X-76.3.364

Smith SM, Beaver RA, Sittichaya W, Cognato AI (2022b) One hundred eighteen taxonomic changes among Xyleborine ambrosia beetles (Coleoptera: Curculionidae: Scolytinae). *Zootaxa* **5194**(2), 151–175. doi:10.11646/zootaxa.5194.2.1

Smith SM, Gomez DF, Beaver RA, Hulcr J, Cognato AI (2019) Reassessment of the species in the *Euwallacea fornicatus* (Coleoptera: Curculionidae: Scolytinae) complex after the rediscovery of the "Lost" type specimen. *Insects* **10**(9), 261. doi:10.3390/insects10090261

Stone C, Simpson JA (1987) Influence of *Ips grandicollis* on the incidence and spread of bluestain fungi in *Pinus elliottii* billets in north-eastern New South Wales. *Australian Forestry* **50**(2), 86–94. doi:10.1080/00049158.1987.10674500

Stone C, Goodyer G, Sims K, Penman T, Carnegie A (2010) Beetle assemblages captured using static panel traps within New South Wales pine plantations. *Australian Journal of Entomology* **49**(4), 304–316. doi:10.1111/j.1440-6055.2010.00769.x

Storer CG, Breinholt JW, Hulcr J (2015) *Wallacellus* is *Euwallacea*: molecular phylogenetics settles generic relationships (Coleoptera: Curculionidae: Scolytinae: Xyleborini). *Zootaxa* **3974**(3), 391–400. doi:10.11646/zootaxa.3974.3.6

Storer C, Payton A, McDaniel S, Jordal BH, Hulcr J (2017) Cryptic genetic variation in an inbreeding and cosmopolitan pest, *Xylosandrus crassiusculus*, revealed using dd RAD seq. *Ecology and Evolution* **7**(24), 10974–10986. doi:10.1002/ece3.3625

Stouthamer R, Rugman-Jones P, Thu PQ, Eskalen A, Thibault T, *et al.* (2017) Tracing the origin of a cryptic invader: phylogeography of the *Euwallacea fornicatus* (Coleoptera: Curculionidae: Scolytinae) species complex. *Agricultural and Forest Entomology* **19**(4), 366–375. doi:10.1111/afe.12215

Swaine JM (1918) Canadian bark-beetles. Part II. A preliminary classification, with an account of the habits and means of control. *Bulletin, Dominion of Canada Department of Agriculture, Entomological Branch* **14**(2), 1–143.

Swan D (1942) The bark beetle – *Hylastes ater* Payk (Coleoptera: Scolytidae) attacking pines in South Australia. *Journal of the Department of Agriculture of South Australia* **12**, 86–90.

Thomson CG (1859) *Skandinaviens Coleoptera synoptiskt bearbetade. 1. Tom.* Berlingska Boktryckeriet, Lund, 394 pp.

Tran HX, Doland Nichols J, Li D, Le NH, Lawson SA (2022) Seasonal flight and genetic distinction among *Xylosandrus crassiusculus* populations invasive in Australia. *Australian Forestry* **85**(4), 224–231. doi:10.1080/00049158.2022.2151722

Trollip C, Carnegie AJ, Dinh Q, Kaur J, Smith D, *et al.* (2021) Ophiostomatoid fungi associated with pine bark beetles and infested pines in south-eastern Australia, including *Graphilbum ipis-grandicollis* sp. nov. *IMA Fungus* **12**(24), 1–27. doi:10.1186/s43008-021-00076-w

Urvois T, Perrier C, Roques A, Sauné L, Courtin C, *et al.* (2023) The worldwide invasion history of a pest ambrosia beetle inferred using population genomics. *Molecular Ecology*, **32**(15), 4381–4400. doi:10.1111/mec.16993

Vega FE, Infante F, Johnson AJ (2015) The Genus *Hypothenemus*, with emphasis on *H. hampei*, the coffee berry borer. In *Bark Beetles*. (Eds FE Vega and RW Hofstetter) pp. 427–494. Elsevier. doi:10.1016/B978-0-12-417156-5.00011-3

Vega FE, Biedermann PH (2020) On interactions, associations, mycetangia, mutualists and symbiotes in insect-fungus symbioses. *Fungal Ecology* **44**, 100909.

Vilardo G, Faccoli M, Corley JC, Lantschner MV (2022) Factors driving historic intercontinental invasions of European pine bark beetles. *Biological Invasions* **24**(9), 2973–2991. doi:10.1007/s10530-022-02818-2

Viñolas A, Verdugo A (2011) Nuevas especies de coleópteros para la Península Ibérica. Familias Zopheridae, Corylophidae y Curculionidae. *Orsis* **25**, 131–139.

Wardlaw T, Lawson S (2012) Establishing pest detection systems in South Pacific countries and Australia. In *Australian Centre for International Agricultural Research* (April; pp. 1–61). Australian Centre for International Agricultural Research.

Wegensteiner R, Wermelinger B, Herrmann M (2015) Natural enemies of bark beetles. In *Bark Beetles*. (Eds FE Vega and RW Hofstetter) pp. 247–304. Elsevier. doi:10.1016/B978-0-12-417156-5.00007-1

Wood SL (1984) New generic synonymy and new genera of Scolytidae (Coleoptera). *Great Basin Naturalist* **44**(2), 223–230.

Wood SL (1986) A reclassification of the genera of Scolytidae (Coleoptera). *Great Basin Naturalist Memoirs* **10**(1), 2.

Wood SL, Bright DE (1992) A catalog of Scolytidae and Platypodidae (Coleoptera). Part 2: Taxonomic index. *Great Basin Naturalist Memoirs* **13**, 1–1553.

Wotherspoon KP (2008) Forest health surveillance in Tasmania. *Australian Forestry* **71**(3), 182–187. doi:10.1080/00049158.2008.10675033

Wylie FR, Griffiths M, King J (2008) Development of hazard site surveillance programs for forest invasive species: a case study from Brisbane, Australia. *Australian Forestry* **71**(3), 229–235. doi:10.1080/00049158.2008.10675040

Yaman IKA (1969) Biology and control studies of the Pistachio Beetle, *Hylesinus vistitus* Muls., in Iraq. *Zeitschrift Für Angewandte Entomologie* **64**(1–4), 426–436. doi:10.1111/j.1439-0418.1969.tb03055.x

You L, Simmons DR, Bateman CC, Short DPG, Kasson MT, *et al.* (2015) New fungus-insect symbiosis: culturing, molecular, and histological methods determine saprophytic Polyporales mutualists of *Ambrosiodmus* ambrosia beetles. *PLOS ONE* **10**(9), e0137689. doi:10.1371/journal.pone.0137689

Yousuf F, Gurr GM, Carnegie AJ, Bedding RA, Bashford R, *et al.* (2014) Biology of the bark beetle *Ips grandicollis* Eichhoff (Coleoptera: Scolytinae) and its arthropod, nematode and microbial associates: a review of management opportunities for Australia. *Austral Entomology* **53**(3), 298–316. doi:10.1111/aen.12077

47. CERAMBYCIDAE LATREILLE, 1802

Adam Ślipiński, Roger de Keyzer and Hermes E. Escalona

Fig. 47.1. *Australodon nearnsi* Escalona & Ślipiński.

Common name. Longhorn Beetles

Introduction. Cerambycidae are exceptionally speciose, they include over 35 000 species worldwide (Švácha & Lawrence 2014). About 1500 species have been described from Australia (Lawrence & Ślipiński 2013). Longhorn Beetles are one of the most popular groups among the beetle collectors and are known to include the largest world beetle, *Titanus giganteus* (Linnaeus) from South America and the largest Australian beetle, *Batocera wallacei* Thomson from northern Queensland.

Long antennae and rather elongate bodies are used as general diagnostic characteristics for these beetles. However, longhorn beetles come in all sizes, shapes, and colours, even commonly mimicking unpalatable Lycidae or Cantharidae beetles, stinging ants or wasps (Pl. 23). Larvae of Cerambycidae develop exclusively in living or dead plant tissues. They often utilise damaged or dead trees for their development and, through feeding on rotten wood form an important element of the saproxylic fauna, speeding wood decay and energy circulation in woodland habitats. Many species are pests of quarantine concern, attacking and killing forest or ornamental trees, spreading injurious nematodes or, by developing in seasoned wood, causing structural damage to timber constructions.

Cerambycidae are members of the Phytophaga, a highly derived group of polyphagous beetles that feed primarily on vascular plants as larvae and adults. This clade (Reid 2014; McKenna *et al.* 2015, 2019; Haddad & McKenna 2016; Zhang *et al.* 2018) includes two large superfamilies, Curculionoidea (weevils) and Chrysomeloidea. Chrysomeloidea in a traditional sense (Crowson 1955) used to contain just three families, Chrysomelidae (leaf beetles), Bruchidae (seed beetles) and Cerambycidae (longhorn beetles) but that has drastically changed in recent years with Cerambycidae split into Vesperidae, Oxypeltidae, Disteniidae and Cerambycidae, and Megalopodidae and Orsodacnidae removed from Chrysomelidae (Reid 2014; Haddad & McKenna 2016). In Australia Chrysomeloidea is represented only by three families: Chrysomelidae, Megalopodidae and Cerambycidae (Lawrence & Ślipiński 2013).

The subfamily level classification of Longhorn beetles has varied over time and the most recent classification of Cerambycidae by Švácha & Lawrence (2014) recognised eight subfamilies: Prioninae, Parandrinae, Dorcasominae, Cerambycinae, Spondylidinae, Necydalinae, Lepturinae, and Lamiinae. Subsequent molecular research involving various groups of Cerambycidae provided further refinement of the subfamily classification by supporting the inclusion of Dorcasominae in Cerambycinae and of Parandrinae in Prioninae (Haddad *et al.* 2018; Ashman *et al.* 2022; Jin *et al.* 2022, 2023). This classification is followed in the Cerambycidae chapters.

The Australian Cerambycidae has been recently revised at the generic level in three comprehensive volumes of the Australian Longhorn Beetles series (Ślipiński & Escalona 2013, 2016; Ślipiński *et al.* 2023) and consequently, it will be treated only very briefly in this series.

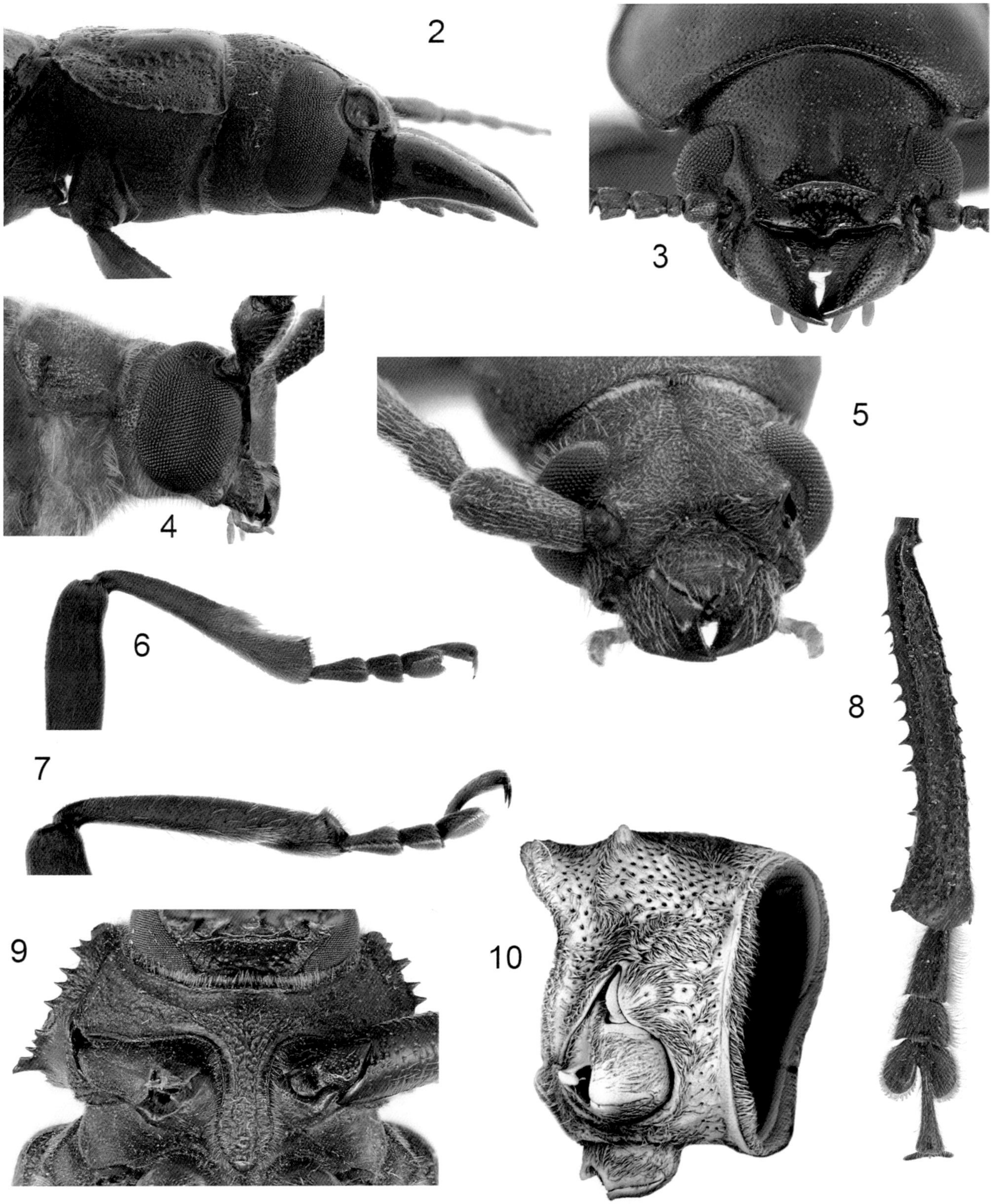

Figs 47.2–47.10. Cerambycidae, adults. 2, *Eurynassa* sp. (Prioninae), head and prothorax, lateral; 3, *Storeyanda frenchi* (Blackburn), head frontal view; 4, *Rhipidocerus weiri* (Jin *et al.*), male head lateral; 5, *Arhopalus rusticus* (Linnaeus), head frontal view; 6, *Athemistus* sp., mesotibia and tarsus; 7, *Athemistus* sp., foretibia and tarsus; 8, *Cnemoplites australis* (Erichson), foretibia and tarsus; 9, *Cnemoplites edulis* Newman, prothorax; 10, *Ancita* sp., prothorax lateral. 2, 3, 5–7, 10 after Ślipiński & Escalona (2013); 4, 8, 9 after Ślipiński *et al.* (2023).

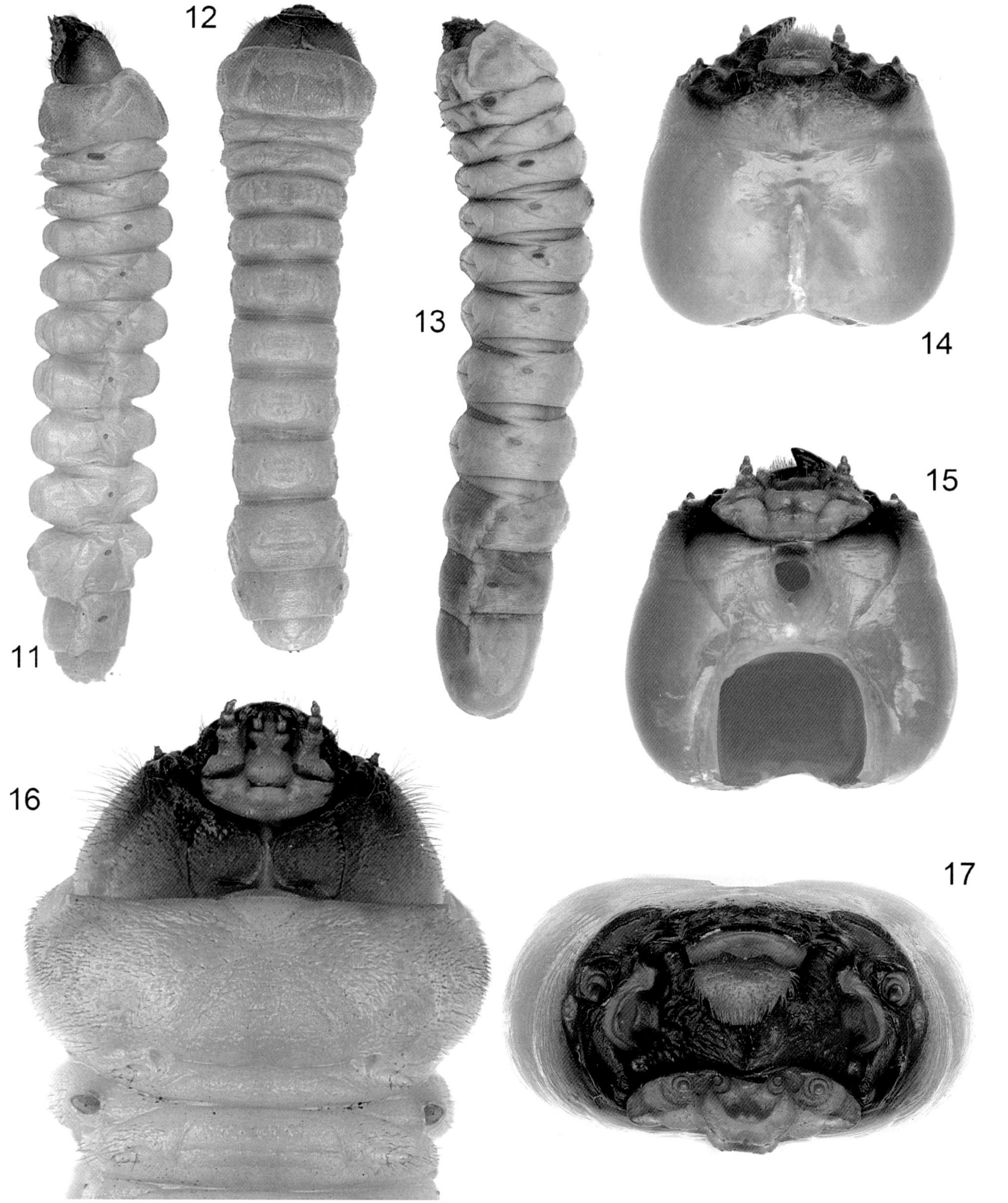

Figs 47.11–47.17. Cerambycidae, larvae. 11, 12, *Arhopalus rusticus* (Linnaeus) habitus lateral and dorsal. 13–15, 17, *Cnemoplites edulis* Newman, 13, habitus lateral; 14, head dorsal; 15, head ventral, 16, *Arhopalus rusticus* (Linnaeus), head, prothorax and mesothorax, ventral; 17, head frontal. After Ślipiński & Escalona (2013).

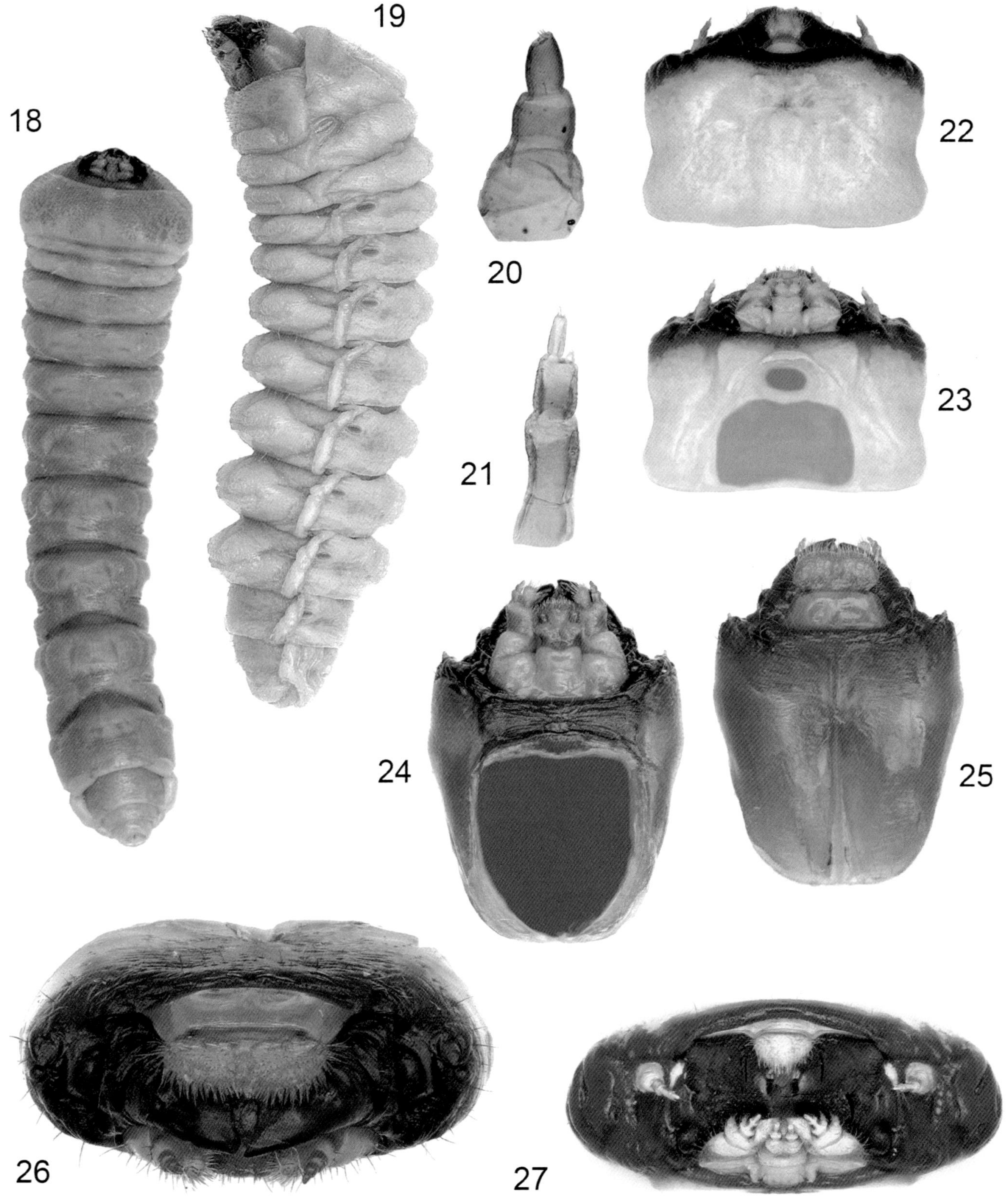

Figs 47.18–47.27. Cerambycidae, larvae. 18, *Coptocercus biguttatus* (Donovan), habitus dorsal; 19, *Rhytiphora solandri* (Fabricius), habitus lateral; 20, *Cnemoplites* sp., antenna. 21–23, 27 *Phoracantha* sp., 21, antenna; 22, head dorsal; 23, head ventral; 24–26, *Acalolepta* sp., 24, head ventral; 25, head dorsal; 26, head frontal; 27, head frontal. After Ślipiński & Escalona (2013).

Key to subfamilies of Australian Cerambycidae

ADULTS

1. Lateral pronotal carinae present, often incomplete and dentate (Figs 47.2, 47.3, 47.4). Procoxae strongly transverse with large exposed trochantins (Fig. 47.9). Protibia often expanded and serrate along external edge (Fig. 47.8) .... **PRIONINAE**

– Lateral pronotal carinae absent (Fig. 47.10) (sometimes pronotum irregularly crenulate laterally). Procoxae oval or weakly transverse at most with small trochantins visible in procoxal extensions (Fig. 47.10). Protibia never expanded and serrate along external edge **2**

2(1). Protibia with inner grooved antenna cleaner (Fig. 47.7); mesotibia with sulcate or comb-like antenna cleaner (Fig. 47.6). Head usually strongly deflexed. Procoxal cavities almost always externally closed .... **LAMIINAE**

– Protibia and mesotibia without antenna cleaners. Head usually prognathous or weakly deflexed. Procoxal cavities externally closed or open **3**

3(2). Antennal pedicel 1.5 times longer than wide and usually about 0.5 times as long as scape or antennomere 3 (Fig. 47.5). Mandible externally densely setose, without distinct tuft of long setae (Fig. 47.5) .... **SPONDYLIDINAE**

– Antennal pedicel subquadrate or transverse, ≤0.3 times as long as scape or antennomere 3; antennae variable but usually extending beyond middle of elytra. Mandible externally glabrous or with distinct tuft of long setae .... **CERAMBYCINAE**

LARVAE

1. Head distinctly elongate (Fig. 47.25). Clypeus broad (Fig. 47.26). Maxillary cardines fused with labium to form single plate; occipital foramen undivided (Fig. 47.24). Legs absent (Fig. 47.19) .... **LAMIINAE**

– Head subquadrate to transverse (Figs 47.14, 47.22). Clypeus narrow or broad (Figs 47.17, 47.27). Maxillary cardines and labium separate; occipital foramen divided or not. Legs usually present (Figs 47.11, 47.13) **2**

2(1). Occipital foramen not divided by tentorial bridge, gula exposed, narrow and long (Fig. 47.16). Abdomen with segment IX bearing simple, short urogomphi (Fig. 47.12) .... **SPONDYLIDINAE**

– Occipital foramen divided into two parts by tentorial bridge (Figs 47.15, 47.23). Abdomen with segment IX simple (Fig. 47.18), rarely with complex armature **3**

3(2). Clypeus strongly narrowing anteriorly, not filling space between mandibular articulations (Fig. 47.27). Mandible with incisor area concave (gouge-like); maxillary mala with setae sparse and short. Antenna 3-segmented (Fig. 47.21) .... **CERAMBYCINAE**

– Clypeus broad and only slightly narrowing anteriorly, entirely filling space between mandibular articulations (Fig. 47.17). Mandible with incisor area grooved; maxillary mala with setae distinct and dense. Antenna 2- or 3-segmented (Fig. 47.20) .... **PRIONINAE**

Acknowledgements

Nicholas J. Fischer, Marie and Jiri Lochman and Reiner Richter have provided photographs of living longhorn beetles.

References

Ashman LG, Shin S, Zwick A, Ślipiński A, McKenna DD (2022) The first phylogeny of Australasian Lamiinae longhorn beetles (Coleoptera: Cerambycidae) reveals poor tribal classification and a complex biogeographic history. *Systematic Entomology* **47**, 213–230.

Crowson RA (1955) *The Natural Classification of the Families of Coleoptera*. Nathaniel Lloyd, London.

Haddad S, McKenna DD (2016) Phylogeny and evolution of the superfamily Chrysomeloidea (Coleoptera: Cucujiformia). *Systematic Entomology* **41**, 697–716.

Haddad S, Shin S, Lemmon AR, Lemmon EM, Svacha P, *et al.* (2018) Anchored hybrid enrichment provides new insights into the phylogeny and evolution of longhorned beetles (Cerambycidae). *Systematic Entomology* **43**, 68–89.

Jin M, De Keyzer R, Ashman LG, Zwick A, Ślipiński A (2023) Phylogeny and Tribal Classification of Australian Prioninae (Coleoptera: Cerambycidae). *Annales Zoologici* **73**, 499–512.

Jin M, Shin S, Ashman LG, Leschen RAB, Zwick A, *et al.* (2022) Phylogenomics resolves timing and patterns in the evolution of Australasian Cerambycinae (Coleoptera: Cerambycidae), and reveals new insights into the subfamily-level classification and historical biogeography of longhorn beetles. *Molecular Phylogenetics and Evolution* **172**, 107486.

Lawrence JF, Ślipiński A (2013) *Australian Beetles. Morphology, Classification and Keys.* Volume 1. CSIRO Publishing, Collingwood, Victoria.

McKenna DD, Shin S, Ahrens D, Balke M, Beza-Beza C, *et al.* (2019) The evolution and genomic basis of beetle diversity. *Proceedings of the National Academy of Sciences* **116**(49), 24729–24737.

McKenna DD, Wild AL, Kanda K, Bellamy CL, Beutel RG, *et al.* (2015) Beetles survived the end Permian mass extinction to diversify during the Cretaceous terrestrial revolution. *Systematic Entomology* **40**, 835–880.

Reid CAM. (2014) 2. Chrysomeloidea Latreille, 1802. In *Handbuch der Zoologie/Handbook of Zoology. Band/Volume IV Arthropoda: Insecta Teilband/Part 38. Coleoptera, Beetles. Volume 3. Morphology and Systematics (Polyphaga).* (Eds RAB Leschen and RG Beutel) pp. 11–15. W. DeGruyter, Berlin.

Ślipiński A, Escalona HE (2013) *Australian Longicorn Beetles (Coleoptera). Volume 1. Introduction and Subfamily Lamiinae.* CSIRO Publishing, Collingwood, Victoria.

Ślipiński A, Escalona HE (2016) *Australian Longicorn Beetles (Coleoptera). Volume 2. Subfamily Cerambycinae.* CSIRO Publishing, Collingwood, Victoria.

Ślipiński A, de Keyzer R, Jin M (2023) *Australian Longhorn Beetles (Coleoptera: Cerambycidae). Volume 3. Subfamily Prioninae of the Australo-Pacific Region.* CSIRO Publishing and Australian Biological Resources Study.

Švácha P, Lawrence JF (2014) 2.4 Cerambycidae Latreille, 1802. In *Handbuch der Zoologie/Handbook of Zoology. Band/Volume IV Arthropoda: Insecta Teilband/Part 38. Coleoptera, Beetles. Volume 3. Morphology and Systematics (Polyphaga).* (Eds RAB Leschen and RG Beutel) pp. 77–177. W DeGruyter, Berlin.

Zhang SQ, Che LH, Li Y, Liang D, Pang H, *et al.* (2018) Evolutionary history of Coleoptera revealed by extensive sampling of genes and species. *Nature Communications* **9**(1), 205.

48. CERAMBYCIDAE: LAMIINAE LATREILLE, 1825

Adam Ślipiński, Lauren G. Ashman, Roger de Keyzer and Hermes E. Escalona

Fig. 48.1. *Rhytiphora pardalis* (Newman).

Common name. Flat-faced Longhorn Beetles.

Introduction. The Lamiinae with over 3000 genera and 20 000 species are the largest subfamily of the Cerambycidae (Švácha & Lawrence 2014). The subfamily is cosmopolitan, but it is particularly diverse in the tropics. The Australian fauna with about 540 described species of Lamiinae (Pls 24, 25) is one of few continents where they are distinctly outnumbered by a much more diverse Cerambycinae.

The history of classification and research on Australian Lamiinae have been summarised by Ślipiński & Escalona (2013), who rejected the existing tribal classification as largely artificial and impractical. This has been confirmed by the subsequent molecular research of Souza *et al.* (2020) and Ashman *et al.* (2022a), who exposed para- and polyphyly in most of the larger tribes. However, both papers have covered only a small representation of world genera and the monophyly of most of the tribes have not been tested due to the lack of material for molecular analysis from type genera of many tribes. In consequence, many studied Australian taxa have been placed as Lamiinae *incertae sedis* (Ashman *et al.* 2022a). In the absence of a practical tribal classification of the Lamiinae, we follow Ślipiński & Escalona (2013) and record the Australian genera alphabetically.

The genera *Temnolamnia* Breuning (1961) and *Tetroeopsis* Breuning (1940a) are treated as taxa 'incertae sedis' pending discoveries of their type material (Ślipiński & Escalona 2013). *Pseudocalamobius incertus* Breuning (1940b), described from Queensland, is removed from the Australian fauna as it is based on a badly damaged Asian specimen. Thanks to the information received from Dr. Antonio Santos Silva (Universidade de São Paulo, Brazil), two Neotropical genera are also removed from the Australian fauna: *Australoleiopus* Breuning 1970: 487 and *A. marmoratus* Breuning, 1970: 488 described from Sydney, NSW are synonymised with the Neotropical genus *Nyssodrysternum* Gilmour, 1960 and *N. reticulatum* (Melzer, 1934) **Syn. Nov.** *Elongatocontoderus* Breuning, 1977: 273 and *E. fuscofasciatus* Breuning, 1977: 273 described from Adelaide, SA, are synonymous with the genus *Neoeutrypanus* Monne, 1977 and *N. decorus* (Bates, 1881) **Syn. Nov.**

Biology. The Lamiinae, unlike most of the other Cerambycidae, are known for their universal and indispensable adult feeding, called maturation feeding, on leaves or fresh bark of the plant that usually serves as the larval host. Their females almost universally use their mandibles to make a deep incision (egg scar) in the bark or in stems of an herbaceous plant, then turn around and lay a single egg in each cavity with their long ovipositor. Females of many Australian *Rhytiphora* girdle living branches of the acacias or eucalypts in which they lay eggs (Froggatt 1923), providing their larvae with weakened, but still living, host material. The species that attack healthy and vigorous plants can develop in stems of herbaceous plants and in twigs of woody plants, often inducing gall-like structures on the host plants.

Highly polyphagous and broadly distributed species of the lamiine genus *Acalolepta* are known to attack passion vines, grapevines and mango trees (Goodwin & Pettit 1994; Goodwin *et al.* 1994; Smith 1996). Two Australian species (*Rhytiphora diva* and *R. stigmatica* – Lucerne Crown Borers) can attack healthy woody and herbaceous plants and are regarded as minor pests of soybeans, lucerne and summer pulses in tropical Queensland. Three Asian species of Lamiinae are serious pests of quarantine concern: Asian Longhorn Beetle (*Anoplophora glabripennis*), Citrus Longhorn Beetle (*A. chinensis*) and Pine Sawyer Beetle (*Monochamus alternatus*). Their accidental introduction would be very damaging to the native and ornamental trees in Australia.

Characteristics. Adults. Moderately small to large longhorns, 2 to 60 mm in length. Body colour very variable, but most of the species are densely setose with matt cryptic mottled colouration, with rare exceptions. Eyes almost always deeply emarginate or divided into upper and lower parts (Figs 48.11–48.16). Antennae usually 11-segmented and filiform of variable length. Scape usually much longer than pedicel; pedicel very short. Terminal maxillary palpomere fusiform (triangular in males of some genera). Lateral pronotal carinae always absent. Prosternal process (Figs 48.17–48.21, 28.23) almost always complete and procoxal cavities closed externally. Procoxae often projecting well below prosternum (Fig. 48.19) usually with lateral extension and exposed trochantin. Elytra always completely covering abdomen. Tibial spurs usually 2-2-2, well developed, rarely reduced to 1 or absent on some pairs of legs. Protibia with oblique band of dense setae, usually within pubescent groove (antenna cleaner) on inner face, and mesotibia in many genera with similar structure on outer face (Figs 48.85, 48.86).

Larvae. Head prognathous, usually elongate and slightly retracted in prothorax; posterior edge of head capsule not, or only slightly, emarginate. Epicranial stem absent or concealed beneath head retractor muscles. Median endocarina long, extending to clypeus. Stemmata 3 or none in late instars. Frontoclypeal suture present and associated with a sclerotised internal transverse bar. Clypeus broad, trapezoidal, occupying most of the area between dorsal mandibular articulations. Labrum free, transverse and broadly rounded apically. Antennae minute, 3-segmented with terminal antennomere reduced and sensorium conical. Ventral mouthparts slightly retracted; maxillary articulating area fused with cardines and submentum; maxillary palp usually 3-segmented with reduced and fixed palpifer. Hypostomal rods moderately long to strongly reduced. Tentorial bridge rudimentary. Legs always absent, reduced to calli. Abdomen with tergum IX usually simple, rarely with median sclerotised tubercle or spine.

Key to the Australian genera

1. Procoxal cavities broadly open externally (Fig. 48.17); prosternal process narrowing to apex; large beetles, more than 30 mm long with very long antennae **2**

– Procoxal cavities closed externally; prosternal process usually expanded apically (Figs 48.18–48.20); body and antennae size variable but usually under 30 mm **3**

2(1). Antennal scape without apical carina or coarse punctures; mesoventrite anteriorly with median projection .... ***Rosenbergia* Ritsema**

– Antennal scape with apical carina and coarse punctures; mesoventrite without median projection .... ***Batocera* Dejean**

3(1). Pretarsal claws appendiculate **4**

– Pretarsal claws simple **7**

4(3). Antennal scape very long, reaching elytral bases, body densely setose; male head strongly modified (Fig. 48.15) .... ***Momisis* Pascoe**

– Antennal scape not extending beyond middle of pronotum, body sparsely setose; male head not modified **5**

5(4). Eyes with upper and lower lobes completely separate; body colour grey or green .... ***Phytoecia* Dejean**

– Eyes emarginate but not completely divided; body colours black and orange **6**

6(5). Abdomen longer than head and thorax combined (Fig. 48.50); antennal scape flattened ventrally; scutellum emarginate apically .... ***Oberea* Dejean**

– Abdomen shorter than head and thorax combined (Fig. 48.51); antennal scape cylindrical; scutellum rounded apically .... ***Nupserha* Thomson**

7(3). Antennal tubercles prominent and very close to each other, their inner surfaces separated by less than one diameter of antennal foramen (Figs 48.7, 48.8) **8**

– Antennal tubercles usually flat and directed laterally, their inner surfaces separated at least by 1.5 diameter of antennal foramen (Figs 48.9, 48.10) **17**

8(7). Metaventrite very short, about as long as or shorter than mesoventrite (Fig. 48.28) **9**

– Metaventrite of normal length, 1.7–2.0 times as long as mesoventrite (Figs 48.22, 48.27) **10**

9(8). Pronotum and elytra with large and pointed tubercles (Fig. 48.31); antennal tubercles situated above upper margin of eye (Fig. 48.36) .... ***Ceraegidion* Boisduval**

– Pronotum and elytra with granulations or weak blunt tubercles (Fig. 48.32); antennal tubercles situated below or at the upper level of eyes (Fig. 48.37) .... ***Microtragus* White**

10(8). Prothorax cylindrical without dorsal or lateral projections or coarse granulations **11**

– Prothorax variable but always with at least small lateral projection and usually with dorsal tubercles **12**

11(10). Antenna very long 12-segmented; larger beetle, length 15–18 mm .... ***Phelipara* Pascoe**

– Antenna shorter, 11-segmented; smaller beetle, less than 10 mm long .... ***Essisus* Pascoe**

12(10) Prosternal process forming broad, flat, anteriorly angulate surface overlapping entire length of procoxae from below (Figs 48.23, 48.41) **13**

– Prosternal process strongly arcuate, sometimes overlapping procoxae posteriorly **14**

13(12). Scape without apical carina (Fig. 48.2); antennae with dense and darker setae on antennomeres 4–6 (Fig. 48.34) .... ***Thyada* Pascoe**

– Antennal scape with apical carina (Figs 48.3, 48.44); antenna sparsely setose (Fig. 48.43) .... ***Callipyrga* Newman**

14(13) Prosternal process moderately broad, angulate medially forming articulation point with procoxa (Fig. 48.38) and overlaps posterior part of procoxae from below .... ***Zygocera* Dejean**

– Prosternal process narrow between procoxae (Fig. 48.39) **15**

15(14). Antennal scape distinctly pyriform; antennal tubercles very close but separated from each other .... ***Oricopis* Pascoe** [part]

– Antennal scape subcylindrical; antennal tubercles touching each other apically (Fig. 48.40) **16**

16(15). Pronotum with complex tubercles (Fig. 48.45); elytra bispinose apically (Fig. 48.46) .... ***Demonassa* Thomson**

– Pronotum granulose with small simple tubercles (Fig. 48.42); elytra blunt apically (Fig. 48.47) .... ***Cyocyphax* Thomson**

17(7). Prothorax strongly elongate, cylindrical, medially narrow, forming a long, transversely ringed neck (Fig. 48.30) that is about as long as elytra (Fig. 48.48) .... ***Gnoma* Fabricius**

– Prothorax variable but never forming a long cylindrical medially narrow neck **18**

18(17). Antennal scape with apical carina at least partially visible (Figs 48.3, 48.4) **19**

– Antennal scape without apical carina (Figs 48.2, 48.5) **26**

19(18). Prosternal process broad, overlapping procoxae from below (Fig. 48.33); antennal scape with apical carina reduced and close to apical margin, area between carina and scape edge smooth .... ***Disterna* Thomson** [part]

– Prosternal process narrow between coxae; antennal scape with apical carina separated from anterior margin, area between carina and apical edge coarsely punctate and setose **20**

20(19). Pronotum subquadrate, lateral margins with scarcely visible median tubercle (Fig. 48.29), pronotal disc transversely striate .... ***Pelargoderus* Audinet-Serville**
– Pronotum transverse, lateral margin with clearly visible median tubercle (Figs 48.21, 48.25), pronotal disc without transverse striation **21**
21(20). Large beetles, above 15.0 mm; tarsi 4-4-4; protibia bent inwards in apical third (Fig. 48.49) **22**
– Size variable but usually less than 15.0 mm; tarsi 5-5-5; protibia rarely bent apically **24**
22(21). Mesoventrite with median projection anteriorly; elytra often with sharp and serrate lateral carina and short spines on dorsal surface (Fig. 48.49) .... ***Potemnemus* Thomson**
– Mesoventrite anteriorly rounded, without median projection; elytra with lateral edges blunt, without lateral carina, dorsal surfaces smooth or weakly granulose at bases **23**
23(22). Humeral part of elytra not granulate (Fig. 48.113); head, pronotum and elytra covered with light brown or variably coloured setae .... ***Acalolepta* Pascoe**
– Humeral part of elytra granulate; head, pronotum and elytra covered in dense yellow to orange setae .... ***Nemophas* Thomson**
24(21). Antennal scape expanded at apex with complete, semicircular apical carina (Fig. 48.3); elytra with strong tubercles at base of third interval .... ***Meton* Pascoe**
– Antennal scape subcylindrical, thickening apically with apical carina incomplete, arcuate or oblique (Fig. 48.4), rarely complete and elytral base smooth or with weak tubercles **25**
25(24). Antennal scape with apical carina oblique (Fig. 48.4) or rarely complete; female last ventrite bearing dense setae .... ***Ancita* Thomson**
– Antennal scape with apical carina arcuate and incomplete; female with last ventrite without dense setae .... ***Bucynthia* Pascoe**
26(18). Body subcylindrical, colour dark blue and iridescent with white spots (Fig. 48.55); labrum apex thickened and truncate .... ***Glenea* Newman**
– Body shape and colour different from above; labrum apex usually flat **27**
27(26) Eye broadly divided, lobes separated by wide area bearing dense yellow setae (Fig. 48.11); head transverse, strongly expanded in males; Norfolk and Philip Islands .... ***Enicodes* Gray**
– Eye lobes never separated by broad yellow setose area; head in males never expanded as above **28**
28(27). Antennomere 3 very short, slightly longer than 2 (Fig. 48.54) .... ***Probatodes* Thomson**
– Antennomere 3 clearly longer than 2 .... **29**
29(28). Antennal scape very long, extending to or beyond posterior margin of pronotum (Fig. 48.59); antennal length about twice body length .... ***Acanista* Pascoe**
– Antennal scape never reaching posterior margin of pronotum; antennae usually not as long as above **30**
30(29). Habitus as in Fig. 48.58; inner surfaces of tibiae keeled; posterior tibia strongly arched (Fig. 48.52) .... ***Alice* Ślipiński & Escalona**
– Habitus not as above; inner surfaces of tibiae rounded or flat; posterior tibia more or less straight **31**
31(30). Membranous anteclypeus not visible; antennomeres 5–11 short and compact (Fig. 48.53) .... ***Apomecyna* Dejean**
– Membranous anteclypeus exposed; antennomeres 5–11 usually longer and loosely articulated **32**
32(31). Dorsal surfaces apparently glabrous; body elongate-oval (Fig. 48.57); Lord Howe and Norfolk Islands .... ***Xylotoles* Newman [part]**
– Dorsal surfaces variably setose (very sparsely so in some species) **33**
33(32). Lateral extension of procoxal cavities closed (Fig. 48.20) protrochantin not visible externally **34**
– Lateral extension of procoxal cavities open (Figs 48.19, 48.21, 48.26), opening very narrow in *Microlamia* or very short in *Somatidia*, at least slightly exposing protrochantin **37**
34(33). Lateral margin of prothorax with spine or tubercle (Fig. 48.56); large beetles, length about 30 mm .... ***Lagocheirus* Dejean**
– Lateral margin of prothorax straight to arcuate without tubercle (Figs 48.61, 48.62); smaller beetles, length less than 25 mm **35**
35(34). Elytra at base as broad as prothorax, humeral angles blunt or rounded (Fig. 48.62) .... ***Mesolita* Pascoe**
– Elytra at base broader than prothorax, humeral angles prominent (Fig. 48.61) **36**
36(35). Prosternum in front of coxae as long or shorter than coxal longitudinal diameter; smaller beetles, length less than 6 mm .... ***Stenellipsis* Bates**
– Prosternum in front of coxae longer than coxal longitudinal diameter; larger beetles, length more than 8 mm; Lord Howe Island .... ***Xylotoles* Newman [part]**
37(33). Metaventrite very short about as long as mesoventrite (Fig. 48.28) **38**
– Metaventrite distinctly longer than mesoventrite (Fig. 48.27) **41**
38(37). Lateral margin of pronotum rounded; scutellum not visible or scarcely visible externally **39**
– Lateral margin of pronotum with sharp angulate projection, spine or tubercle; scutellum visible, sometimes reduced **40**
39(38). Mesofemur weakly broadened at middle (Fig. 48.67), not apically clavate .... ***Austrosomatidia* McKeown**
– Mesofemur abruptly expanded, apically clavate (Fig. 48.63) .... ***Somatidia* Thomson**
40(38). Elytra with serrate lateral carina (Fig. 48.65) .... ***Dorcadida* White**
– Elytra tuberculate, blunt laterally without lateral carina (Fig. 48.64) .... ***Athemistus* Pascoe**
41(37). Abdominal tergites VII–VIII projecting beyond apices of elytra (Fig. 48.75) .... ***Annemabel* Ślipiński & Escalona**
– Abdominal tergites VII–VIII never projecting beyond apices of elytra **42**
42(41). Prosternal process broad and flat, projecting beyond procoxae and fitting into mesoventrital cavity (Fig. 48.68) **43**
– Prosternal process variable, usually arcuate, never strongly projecting posteriorly to be received in mesoventrital cavity **44**
43(42). Scutellum large and projecting forward over posterior edge of pronotum; eye lobes separate (Fig. 48.13); frontoclypeal suture visible; antennomere 3 slightly longer than 4; ventrite 5 with two apical spines .... ***Tmesisternus* Latreille**
– Scutellum smaller and not overlapping posterior edge of pronotum; eye deeply emarginate but eyes lobes not separate; frontoclypeal suture not visible; antennomere 3 clearly longer than 4; ventrite 5 without apical spines .... ***Trigonoptera* Perroud**
44(42). Pretarsal claws narrowly divergent (Fig. 48.72) **45**
– Pretarsal claws broadly divergent (Fig. 48.73) **54**

45(44). Prothorax with broad median projection (Figs 48.101, 48.102); antenna scape pear-shaped (Fig. 48.5) 46
– Prothorax without median projection, at most with small projection or spine close to anterior angles (Fig. 48.24); antennal scape not modified . 47
46(45). Elytral apex with external spine (Fig. 48.69); eye lobes separate .... ***Mulciber* Thomson**
– Elytral apex rounded (Fig. 48.71); eyes deeply emarginate but not divided .... ***Notomulciber* Blackburn**
47(45). Body very slender, elongate with prothorax and elytra almost parallel sided (Figs 48.66, 48.76); small species, length almost always less than 10 mm 48
– Body always broader and often broadly oval, in more elongate species at least elytra distinctly narrowing posteriorly and body length usually more than 10 mm 49
48(47). Labrum anteriorly thickened and carinate; elytral intervals always weakly convex .... ***Zorilispe* Pascoe**
– Labrum flat without preapical carina; elytra usually without raised intervals .... ***Itheum* Pascoe**
49(47). Antennal scape flat ventrally with distinct lateral edges (Fig. 48.90) 50
– Antennal scape ventrally convex without distinct lateral edges 51
50(49). Mesotibia antennal cleaner sulcate (Fig. 48.85); tarsi 4-segmented .... ***Ropica* Pascoe**
– Mesotibia antennal cleaner comb-like (Fig. 48.86); tarsi 5-segmented .... ***Pterolophia* Newman**
51(49). Mesotibia antennal cleaner sulcate (Fig. 48.85) 52
– Mesotibia antennal cleaner not sulcate, comb-like (Fig. 48.86) 53
52(51). Apical third of elytra moderately obliquely truncate; some elytral intervals weakly costate (Fig. 48.79); procoxa usually with short spine .... ***Atimura* Pascoe**
– Apical third of elytra gradually descending to apex (Fig. 48.77); elytral intervals flat or weakly convex; procoxa without spine .... ***Sybra* Pascoe**
53(51). Antennal scape somewhat clubbed and dorsally densely and coarsely punctate and setose (Fig. 48.84); most antennomeres densely and equally setose on all sides; pretarsal claws separation apically broader (Fig. 48.73); abdominal ventrite 2 in male with paired sex patches, covered by hind margin of ventrite 1 .... ***Similosodus* McKeown**
– Antennal scape variable, finely punctate and setose; antennae at most with ventral fringe of longer setae; pretarsal claws separation apically narrower; male abdominal ventrite 2 usually with paired sex patches, visible or covered by long setae along hind margin of ventrite 1 (Figs 48.87, 48.88) .... ***Rhytiphora* Audinet-Serville**
54(44). Prothorax with lateral margins smooth or with short spine near anterior angle 55
– Prothorax with median or postmedian lateral tubercle or projection 58
55(54). Dorsal surface of antennal scape rough, asperate (Fig. 48.6); antennae and legs very long; body appearance distinctive (Fig. 48.81) .... ***Olenecamptus* Chevrolat**
– Dorsal surface of antennal scape with regular punctures and setae; antennae and legs shorter 56
56(55). Prosternal and mesoventrital processes flat, prosternum and mesoventrite at the same level; Lord Howe and Norfolk Islands 57
– Prosternal and mesoventrital process on different levels; mainland Australia .... ***Mimostenellipsis* Breuning**
57(56). Eye divided into distinct separate lobes (Fig. 48.92); antennal scape longer than antennomere 3 (Fig. 48.93); length less than 6 mm; Lord Howe and Norfolk Islands .... ***Microlamia* Bates**
– Eye lobes connected posteriorly by one row of ommatidia; antennal scape as shorter than antennomere 3; length over 10 mm. Lord Howe Island .... ***Xylotoles* Newman** [part]
58(54). Male abdominal ventrites 3–4 with median setose patches each bearing central opening (Fig. 48.91) .... ***Carteridion* Ślipiński & Escalona**
– Male abdominal ventrites 3–4 without setose patches or openings (Fig. 48.89) 59
59(58). Antennomeres 3–4 distinctly longer than 5–11 combined (Fig. 48.99) .... ***Pentacosmia* Newman**
– Antennomeres 3–4 distinctly shorter than 5–11 combined 60
60(59). Body shape characteristic, as in Fig. 48.98, mostly green; antennomeres 3–6 with ventral fringe of dense setae, elytra sparsely setose, green and shiny with apices pointed bearing brushes of setae .... ***Paressisus* Aurivillius**
– Body colour and shape different from above 61
61(60). Elytral epipleura ventral, at abdominal level very broad and complete to apex (Fig. 48.100) .... ***Phyxium* Pascoe**
– Elytral epipleura oblique, at abdominal level narrow and incomplete apically 62
62(61). Dorsum, including antennae covered by erect, white setae some combined forming spires (Fig. 48.103) .... ***Excastra* Tweed, Ashman et Ślipiński**
– Dorsum and antennae different 63
63(62). Prothorax distinctly constricted at base; length 2.5–3.5 mm .... ***Illaena* Erichson**
– Prothorax not distinctly constricted at base or if somewhat constricted at base body much larger, 10 mm or above 64
64(63). Mesoventrite distinctly angulate often with median tubercle or projection 65
– Mesoventrite relatively flat, or weakly arcuate without angle or projection 68
65(64). Frontoclypeus strongly transverse (Fig. 48.105); prosternum in front of coxae as long as or longer than coxal length; prosternal process with median projection (Fig. 48.26) meeting angulate mesoventrital surface; antennal tubercles flat, broadly separated and directed laterally .... ***Temnosternus* White**
– Frontoclypeus square or trapezoidal (Fig. 48.104), rarely weakly transverse; prosternum in front of coxa shorter than a coxal length (Fig. 48.33); prosternal process without projection, angulate; antennal tubercles variable but often much closer together and prominent 66
66(65). Pronotum with anterior and posterior impressed lines or grooves but without distinct dorsal tubercles (Fig. 48.95), constricted posteriorly; eyes finely facetted; frontoclypeus trapezoidal or subquadrate, gena taller than lower eye lobe (Fig. 48.104) .... ***Disterna* Thomson**
– Pronotum without transverse ridges with or without dorsal tubercles; eyes usually coarsely facetted; frontoclypeus elongate, square or slightly transverse; gena as tall as or shorter than lower eye lobe 67
67(66). Pronotum and elytral bases with various tubercles (Fig. 48.35) .... ***Oricopis* Pascoe** [part]
– Pronotum without tubercles, rarely with small nodules; base of elytra without tubercles but often with raised area joined to broadly convex interval 3 .... ***Amblymora* Pascoe**
68(64). Larger beetles, over 10 mm long with broad based median straight prothoracic projection (Figs 48.101, 48.102) 69
– Smaller beetles, less than 7 mm long with prothorax bearing sharp posteriorly directed or short posterior spine (Fig. 48.111) 71

69(68). Antennal tubercles distinctly elevated, moderately separated; femora not clavate (Fig. 48.106) .... ***Dysthaeta*** **Pascoe**
– Antennal tubercles weakly prominent; femora distinctly clavate (Fig. 48.97) **70**
70(69). Pronotum distinctly uneven, tuberculate; lateral projection broad at base (Fig. 48.102) .... ***Velora*** **Thomson**
– Pronotum relatively even, without tubercles; lateral projection with narrower base (Fig. 48.101) .... ***Didymocentrotus*** **McKeown**
71(68). Antennal scape clavate and roughly punctate along external surfaces .... ***Neissa*** **Pascoe**
– Antennal scape gradually thickening apically, regularly punctate and setose **72**
72(71). Prothorax strongly transverse, lateral margin with broad-based spine (Fig. 48.107); antennae and dorsum with long tactile setae .... ***Exocentrus*** **Dejean**
– Pronotum subquadrate or weakly transverse, lateral margin with spine usually narrow-based (Fig. 48.110); antennae and dorsum with short mostly adpressed setae **73**
73(72). Smaller, less than 5 mm long, lightly sclerotised beetle; prothoracic spines broader based (Fig. 48.111); lower eye lobe distinctly separated from mandibular articulation .... ***Sciades*** **Pascoe**
– Larger, over 5 mm long, heavily sclerotised beetles; prothoracic spines smaller and more narrow-based (Fig. 48.110); lower eye lobe almost touching mandibular articulation .... ***Amblymoropsis*** **Breuning**

Classification of Australian genera

Acalolepta Pascoe, 1858 (Pl. 116A, B, D; Fig. 48.113)

Type species. *Acalolepta pusio* Pascoe, 1858.
Cypriola Thomson, 1864
Dihammus Thomson, 1864
Neanthes Pascoe, 1878
Haplohammus Bates, 1884
Astynoscelis Pic, 1904
Niphohammus Matsushita, 1932
Saitoa Matsushita, 1937
Australian species. 13.
Distribution. *Acalolepta* is distributed from Japan, China through Southeast Asia to New Guinea, Australia and many Pacific Islands. It includes approximately 250 species and many subspecies (Vitali 2011) that are often poorly defined. Several common and broadly distributed species occur mostly in northern and eastern Australia but at least one species (Fig Longicorn – *A. mixta*) has been recorded also from Victoria and South Australia. *Acalolepta fasciatus* and *A. mixta* have been recorded from Lord Howe Island; *A. nativitatis* Gahan was described from Christmas Island.
Biology. Several widely distributed species are extremely polyphagous and show remarkably wide host ranges that include many unrelated native and introduced species of trees, shrubs and vines (Vitali 2011). This genus has been recorded occasionally as a serious pest of grapevines in the Hunter Valley (NSW), and as a pest of mango trees in Northern Territory (Smith 1996).
Notes. The species *Acalolepta vastator* (Newman, 1847) seems to be valid and not a synonym of *A. mixta* (Hope, 1841) according to Vitali (2017), although its type species has not been examined.

Acanista Pascoe, 1864 (Pl. 116G; Fig. 48.59)

Type species. *Acanista alphoides* Pascoe, 1864.
Australian species. One.
Distribution. A single species classified in *Acanista* is distributed from the southeastern part of Indonesia and New Guinea to Cape York Peninsula (QLD).

Alice Ślipiński & Escalona, 2013 (Pl. 119G; Fig. 48.58)

Type species. *Alice dekeyzeri* Ślipiński & Escalona, 2013
Australian species. One.
Distribution. *Alice dekeyzeri* is endemic to northern NSW and southern QLD.
References. Ślipiński & Escalona (2013).

Amblymora Pascoe, 1867 (Pl. 116E)

Type species. *Amblymora instabilis* Pascoe, 1867
Australothelais Breuning, 1963
Australian species. Three.
Distribution. About 25 species of this genus are widely distributed in Australia, New Guinea, Philippines and Indonesia. In Australia it is common on the Atherton Tableland and in the Cairns District (QLD).

Amblymoropsis Breuning, 1958 (Pl. 116F; Fig. 48.110)

Type species. *Amblymoropsis fusca* Breuning, 1958.
Setosophroniella Breuning, 1961
Australian species. Two.
Distribution. Four described species are known from New Guinea and Australia. The Australian specimens have been collected mostly around Cairns and Kuranda in QLD.

Ancita Thomson, 1864 (Pls 24F, 25A, 116J; Fig. 48.114)

Type species. *Ancita crossotoides* Thomson, 1864.

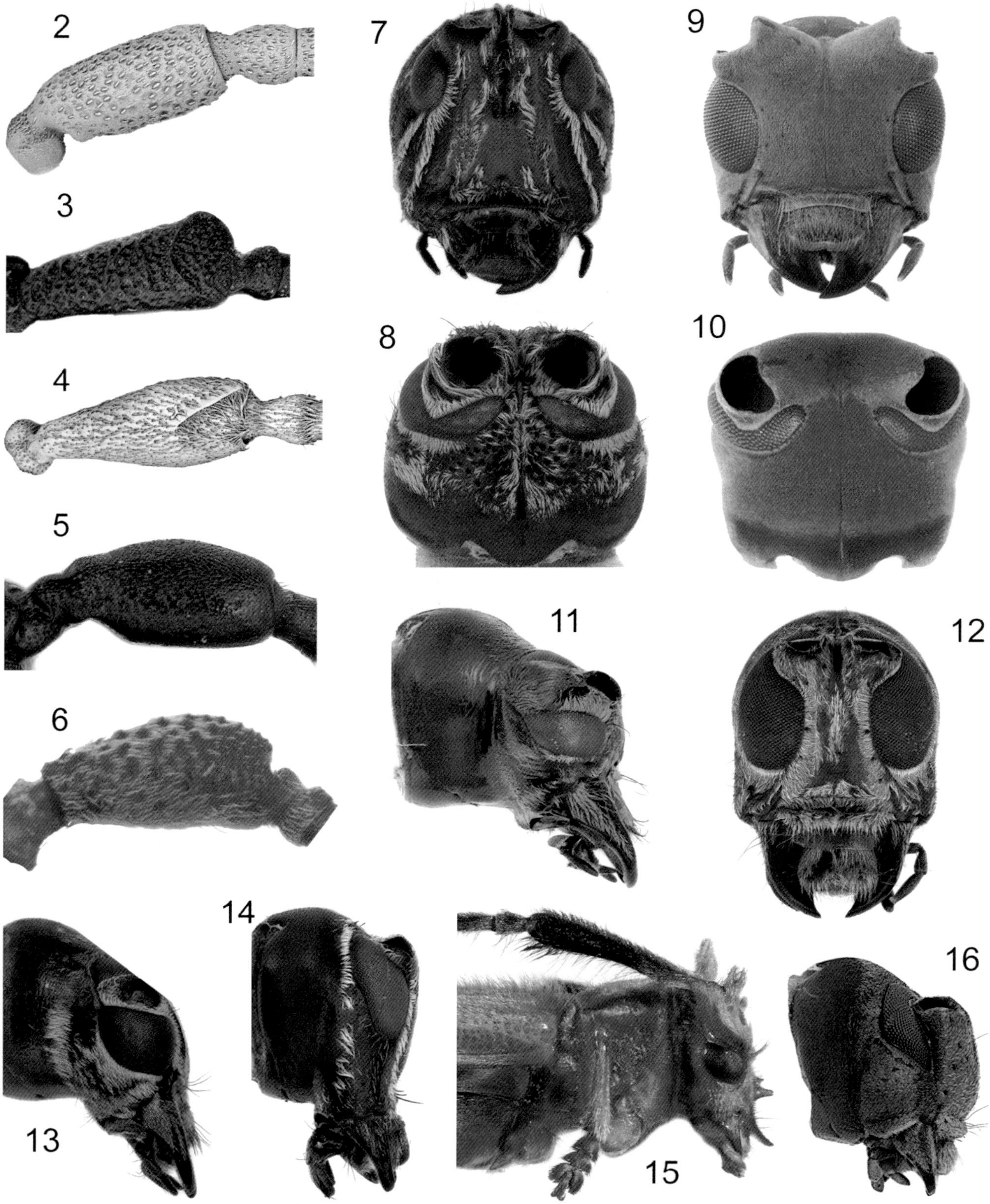

Figs 48.2–48.16. Cerambycidae: Lamiinae. 2–6, Antennal scape; 7, 9, 12, head frontal view; 8, 10, head dorsal; 11, 13, 14, 16, head lateral view. 2, *Sybra* sp.; 3, *Gnoma* sp.; 4, *Ancita* sp.; 5, *Xylotoles* sp.; 6, *Olenecamptus* sp.; 7, *Callipyrga* sp.; 8, *Thyada* sp.; 9, 10, *Acalolepta* sp.; 11, *Enicodes* sp.; 12, *Demonassa* sp.; 13, *Temnosternus* sp.; 14, *Zygocera* sp.; 15, *Momisis* sp., male; 16, *Microtragus* sp.

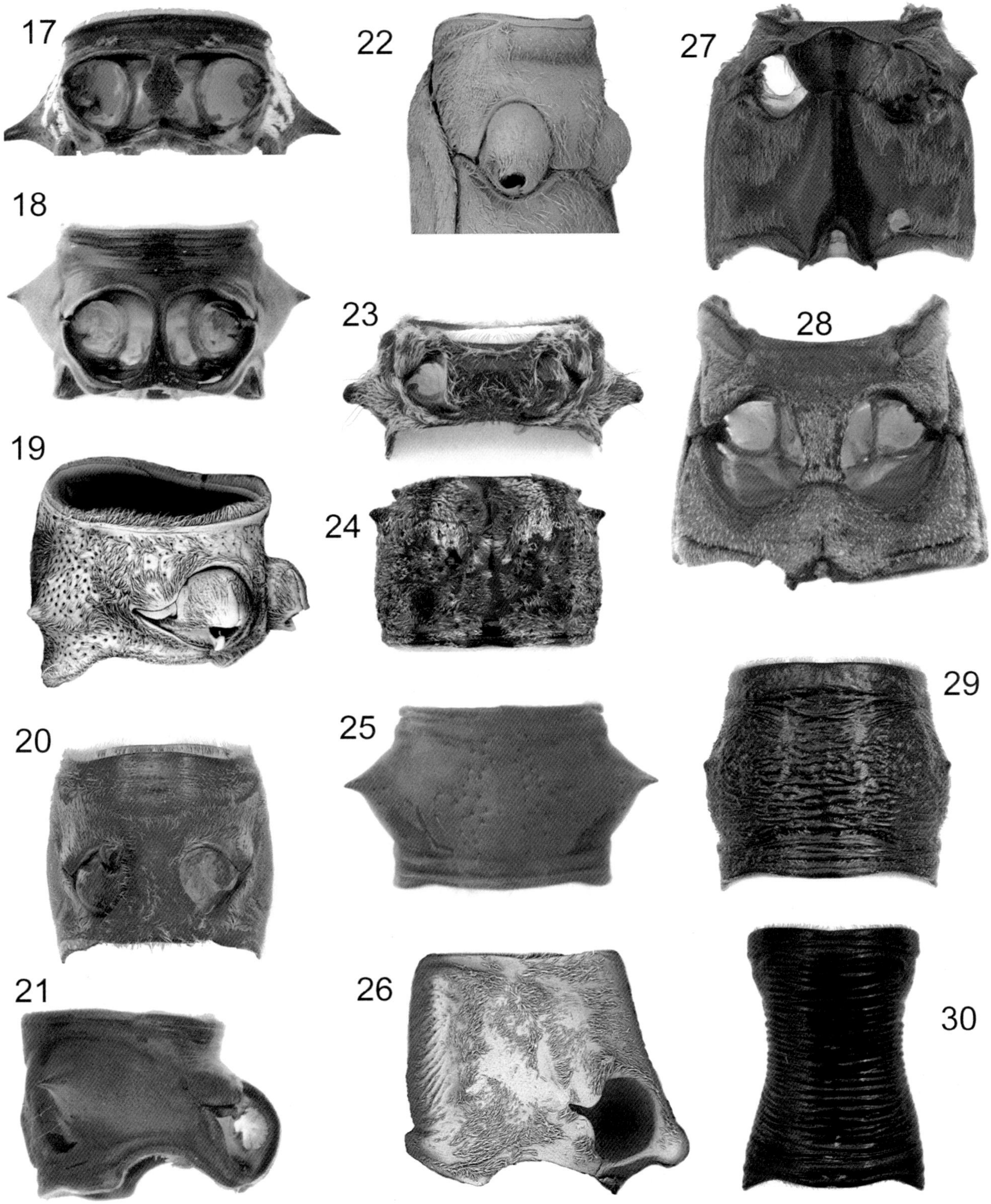

Figs 48.17–48.30. Cerambycidae: Lamiinae, prothorax. 17, *Batocera* sp.; 18, 21, 25, *Acalolepta* sp.; 19, *Ancita* sp.; 20, *Enicodes* sp.; 22, *Essisus* sp.; 23, 27, *Disterna* sp.; 24, *Rhytiphora* sp.; 26, *Temnosternus* sp.; 28, *Dorcadida* sp.; 29, *Pelargoderus* sp.; 30, *Gnoma* sp.

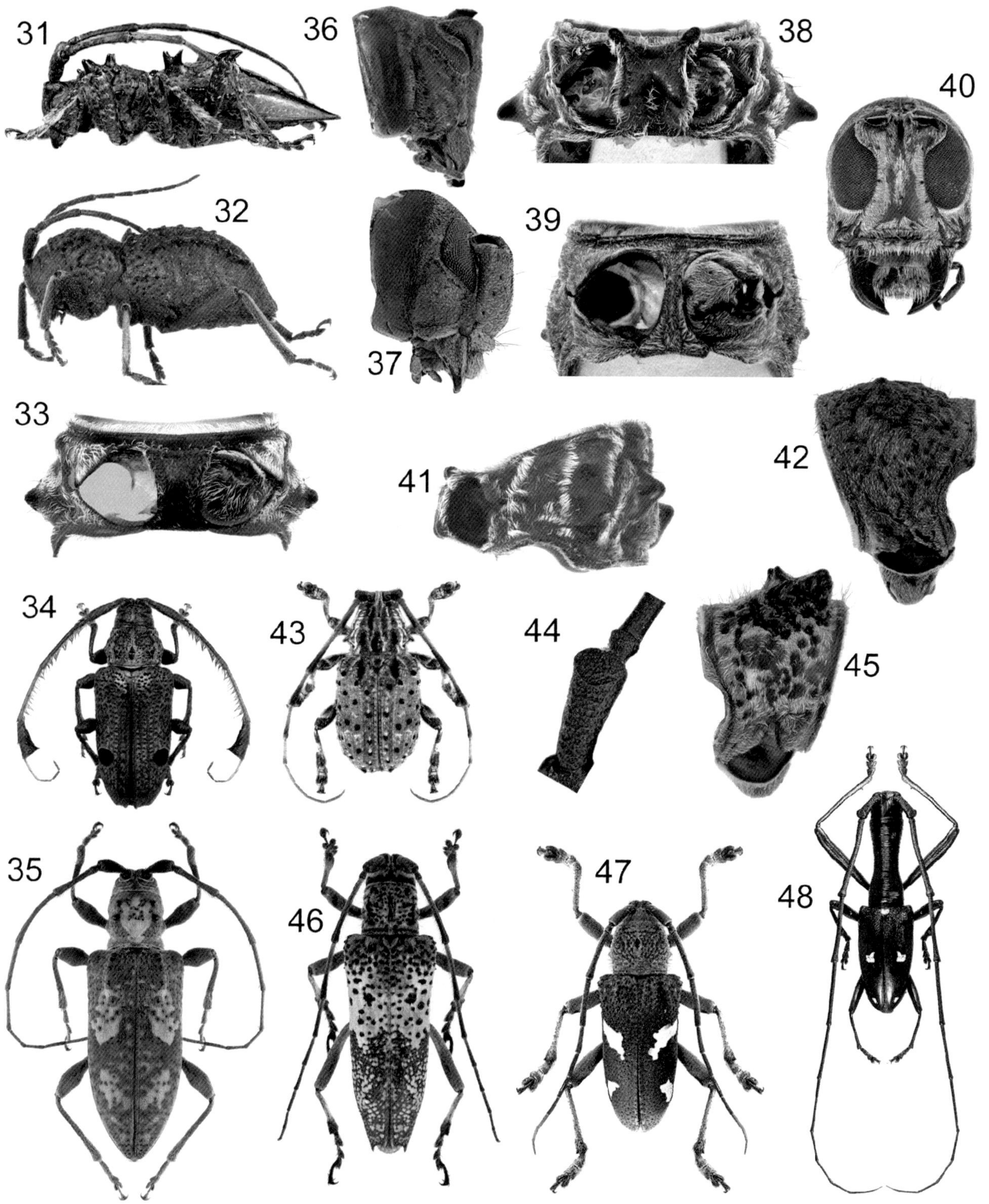

Figs 48.31–48.48. Cerambycidae: Lamiinae. 31, *Ceraegidion* sp., body lateral; 32, *Microtragus* sp., body lateral; 33, *Disterna* sp., prothorax, ventral; 34, *Thyada barbicornis* (Pascoe); 35, *Oricopis maculiventris* Lea; 36, *Ceraegidion* sp., head lateral; 37, *Microtragus* sp., head lateral; 38, *Callipyrga* sp., prothorax ventral; 39, *Microtragus* sp., prothorax ventral; 40, *Demonassa* sp., head frontal; 41, *Callipyrga* sp., prothorax lateral; 42, *Cyocyphax* sp., prothorax lateral; 43, *Callipyrga turrita* Newman; 44, *Callipyrga* sp., antennal scape; 45, *Demonassa* sp., prothorax lateral; 46, *Demonassa capitalis* Blackburn; 47, *Cyocyphax praonethoides* Newman; 48, *Gnoma australis* Schweizer.

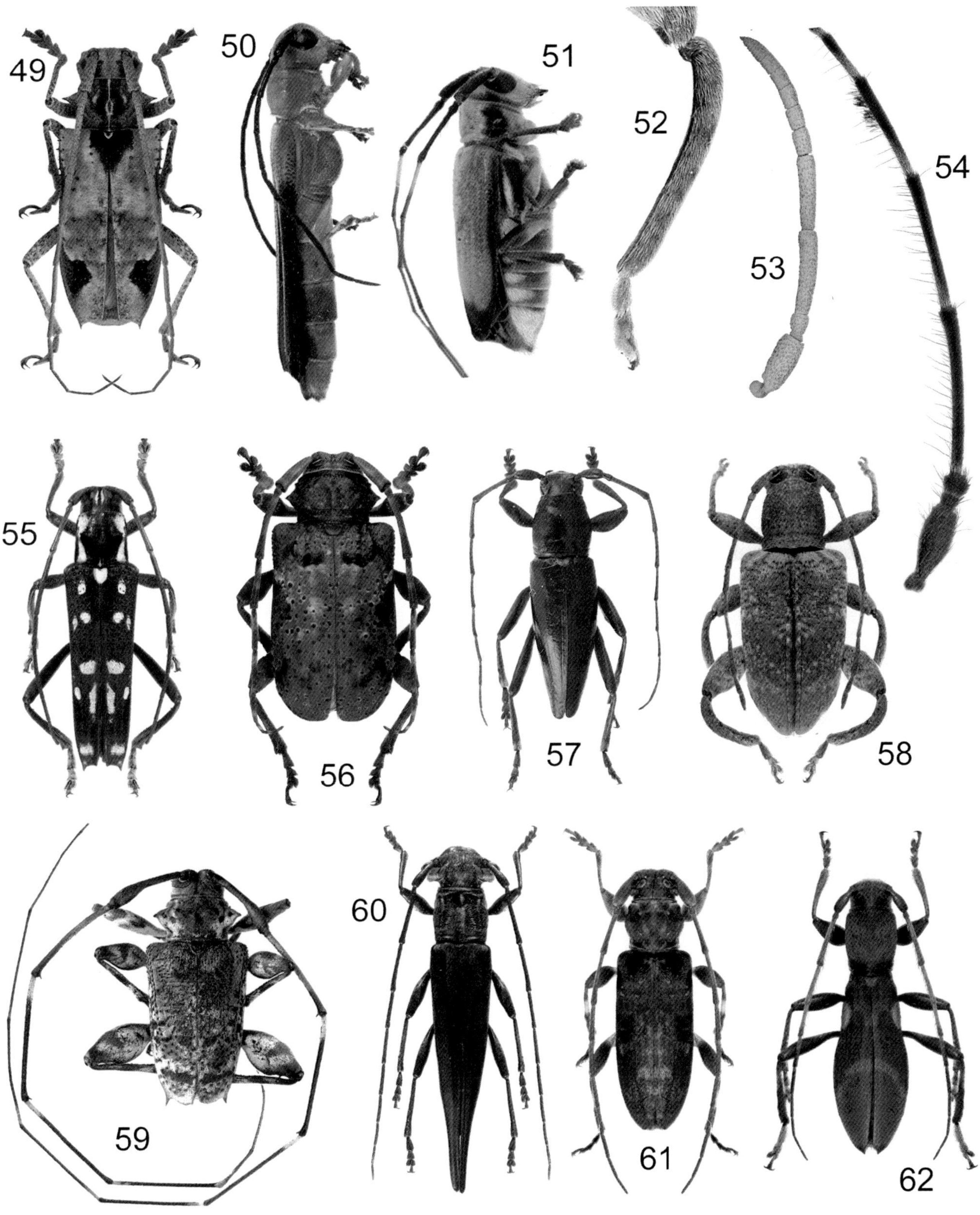

Figs 48.49–48.62. Cerambycidae: Lamiinae. 49, *Potemnemus scabrosus* (Olivier); 50, *Oberea nefasta* Pascoe; 51, *Nupserha* sp.; 52, *Alice* sp., hind tibia; 53, *Apomecyna* sp., antenna; 54, *Probatodes* sp., antenna; 55, *Glenea spinifera* (Voet); 56, *Lagocheirus funestus* Thomson; 57, *Xylotoles selwyni* Oliff; 58, *Alice dekeyzeri* Ślipiński & Escalona; 59, *Acanista alphoides* Pascoe; 60, *Enicodes fichteli* (Schreibers); 61, *Stenellipsis* sp.; 62, *Mesolita* sp.

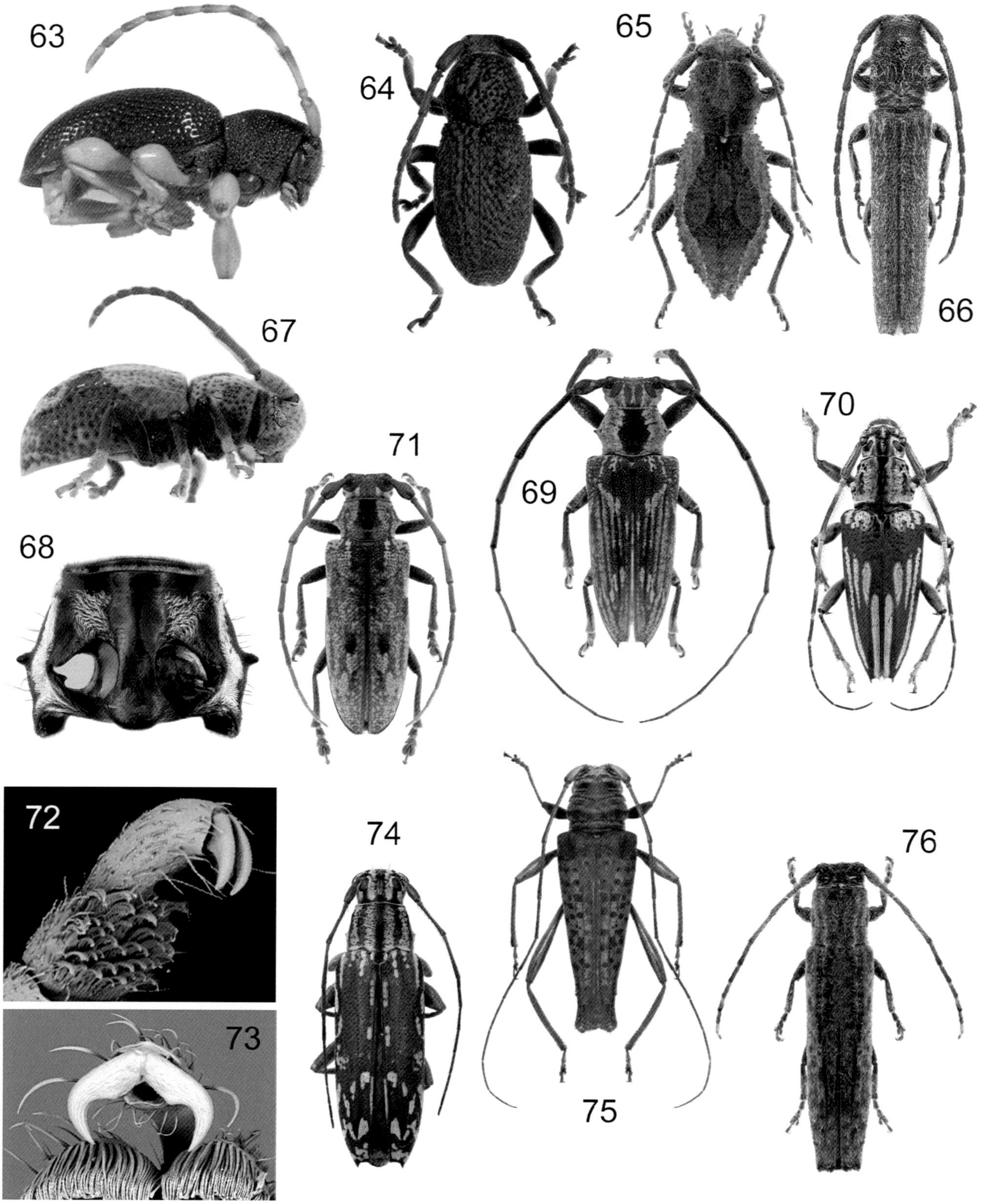

Figs 48.63–48.76. Cerambycidae: Lamiinae. 63, *Somatidia aranea* Oliff; 64, *Athemistus* sp.; 65, *Dorcadida bilocularis* White; 66, *Itheum alboscutellare* Breuning; 67, *Austrosomatidia* sp.; 68, *Trigonoptera* sp., prothorax, ventral; 69, *Mulciber maculosus* Breuning; 70, *Trigonoptera vittata* (Gestro); 71, *Notomulciber* sp.; 72, *Ropica* sp.; 73, *Somatidia* sp.; 74, *Tmesisternus* sp.; 75, *Annemabel elongata* (Breuning); 76, *Zorilispe albosetosa* Breuning.

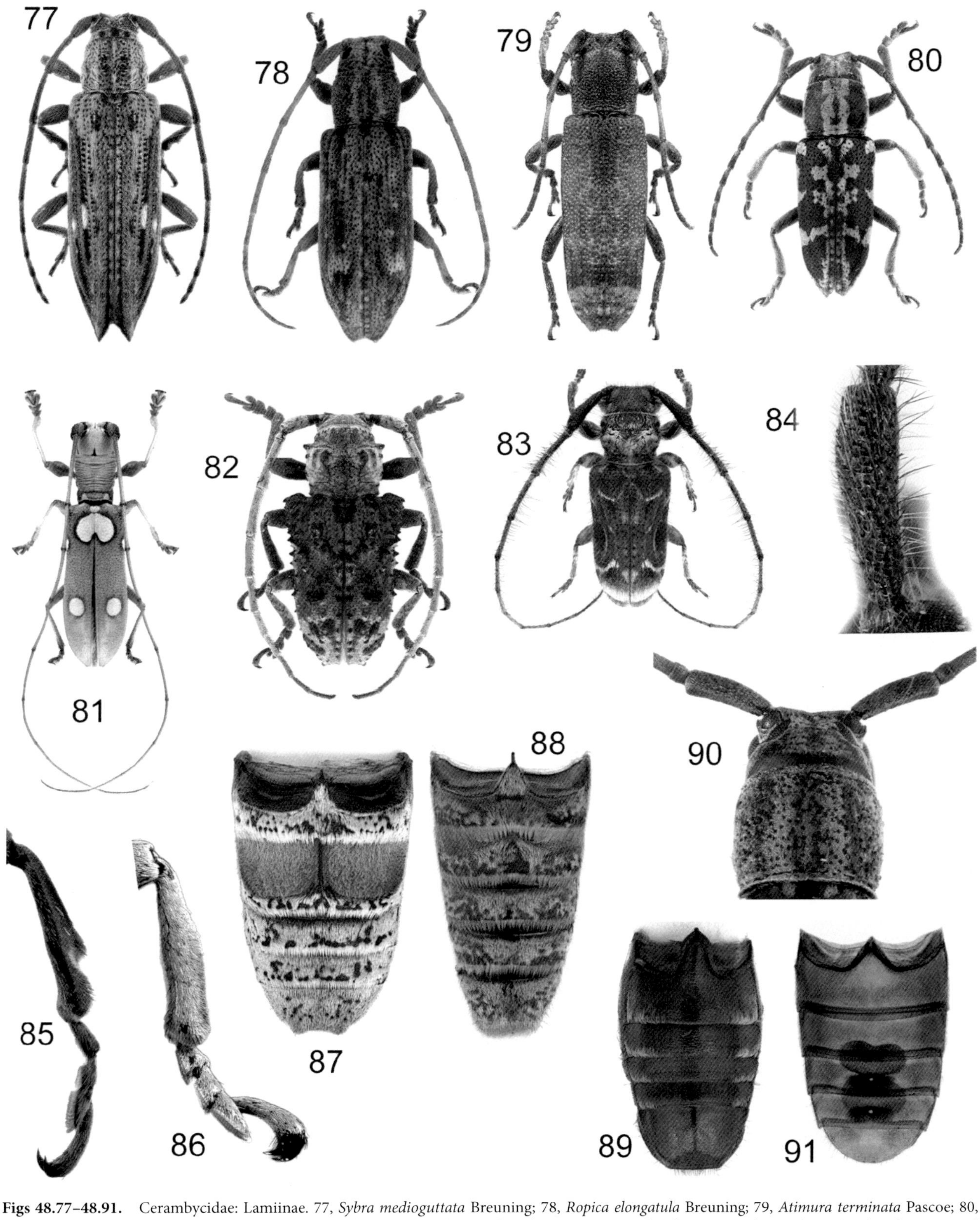

Figs 48.77–48.91. Cerambycidae: Lamiinae. 77, *Sybra medioguttata* Breuning; 78, *Ropica elongatula* Breuning; 79, *Atimura terminata* Pascoe; 80, *Pterolophia luctuosa* (Pascoe); 81, *Olenecamptus bilobus* (Fabricius); 82, *Rhytiphora* sp.; 83, *Similosodus venosus* (Pascoe) 84, *Similosodus*, antennal scape; 85, *Sybra* sp., mid tibia; 86, *Rhytiphora* sp., mid tibia; 87, 88, *Rhytiphora* spp., male abdomen with sex patches; 89, *Mesolita* sp.; 90, *Ropica* sp., head and pronotum; 91, *Carteridion* sp., male abdomen ventral showing glands and openings.

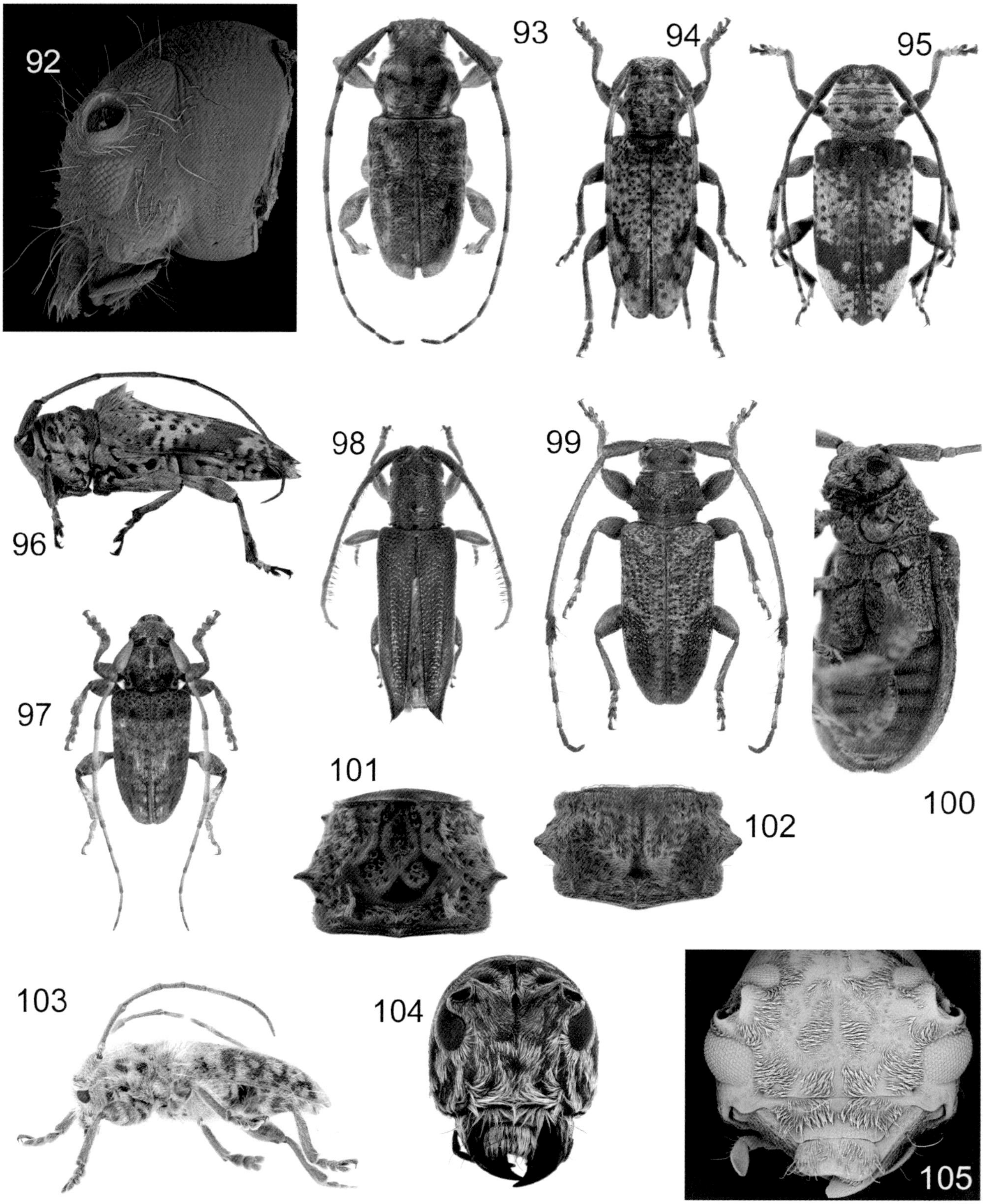

Figs 48.92–48.105. Cerambycidae: Lamiinae. 92, *Microlamia* sp., head lateral; 93, *Microlamia viridis* Ślipiński & Escalona; 94, *Carteridion wilsonensis* Ślipiński & Escalona; 95, *Disterna* sp.; 96, *Disterna plumifera* (Pascoe); 97, *Velora* sp.; 98, *Paressisus viridipennis* Aurivillius; 99, *Pentacosmia cacioides* (Breuning); 100, *Phyxium* sp., ventral; 101, *Didymocentrotus* sp., prothorax; 102, *Velora*, prothorax; 103, *Excastra albopilosa* Tweed *et al.*; 104, *Disterna* sp., head frontal; 105, *Temnosternus*, head frontal.

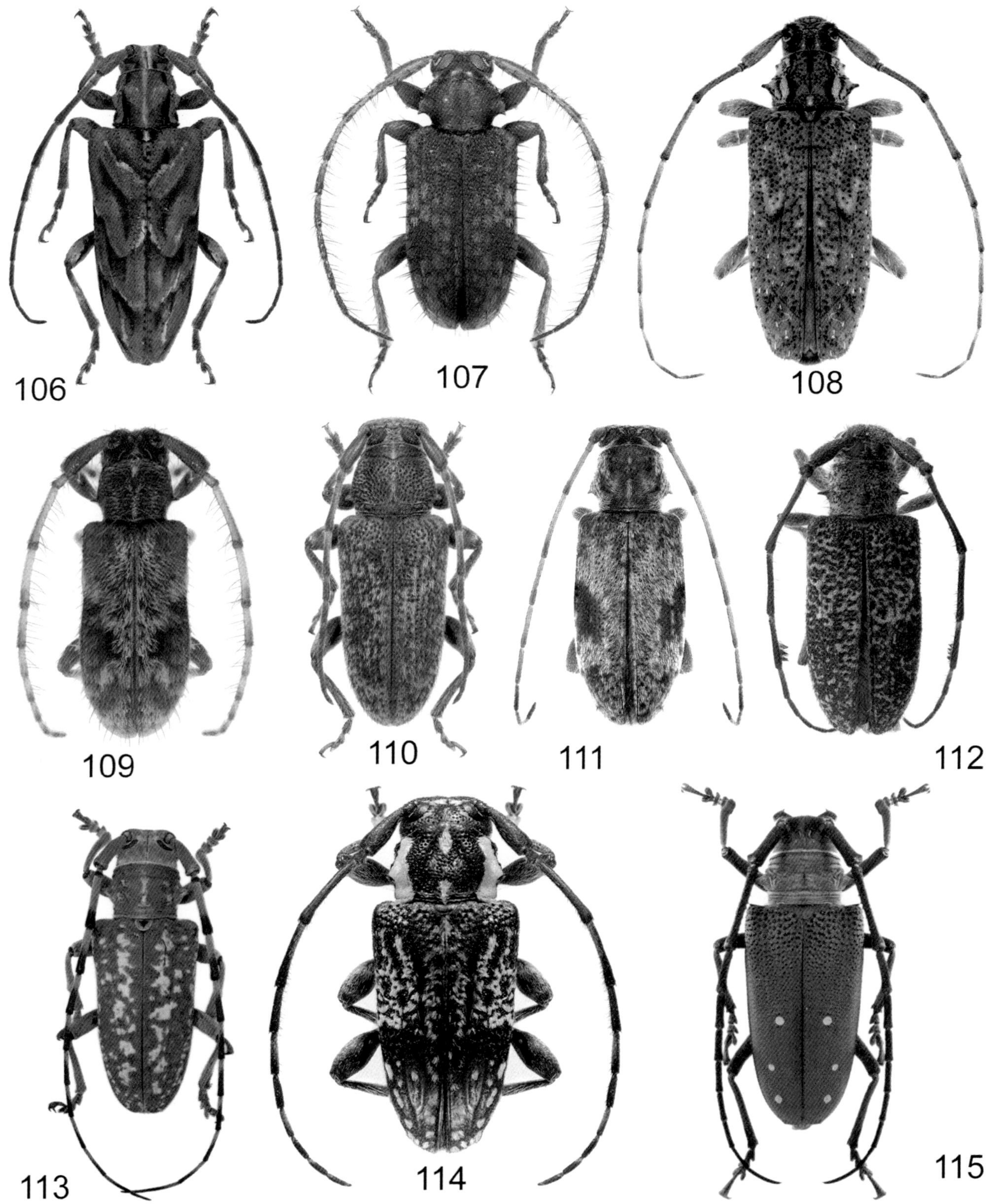

Figs 48.106–48.115. Cerambycidae: Lamiinae. 106, *Dysthaeta anomala* Pascoe; 107, *Exocentrus miselloides* Breuning; 108, *Didymocentrotus foveatus* (Aurivillius); 109, *Neissa* sp.; 110, *Amblymoropsis fusca* Breuning; 111, *Sciades exiguua* (Gahan); 112, *Probatodes plumula* (Newman); 113, *Acalolepta ovina* (Pascoe); 114, *Ancita crocogaster* (Boisduval); 115, *Batocera laena* Thomson

Hebecerus Dejean, 1835
Hebesecis Pascoe, 1865
Australiorondonia Breuning, 1982
Australian species. 25.
Distribution. Australia and New Guinea.
Biology. Adults of *Ancita* are frequently collected on *Acacia* trees and shrubs during the summer months, often when plants are flowering. Adults feed on bark of branches; the eggs are deposited in dead or dying wood, and larvae develop in the branches.

Annemabel Ślipiński & Escalona, 2013 (Pl. 123I; Fig. 48.75)

Type species. *Zygocera elongata* Breuning, 1939.
Australian species. One.
Distribution. Endemic to eastern Australia, *Annemabel elongata* is known from QLD and NSW.

Apomecyna Dejean, 1821 (Pl. 116I)

Type species. *Saperda alboguttata* Megerle, 1803.
Anapomecyna Pic, 1925
Crassapomecyna Breuning, 1958
Mecynapus Thomson, 1858
Pseudalbana Pic, 1895
Australian species. Two.
Distribution. Twenty-two species have been described in *Apomecyna* from the tropics of Africa, Madagascar, Asia and New Guinea. The widely distributed *A. histrio* occurs from northern Queensland to as far south as Sydney, while *A. nigritarsis* was described from Christmas Island (Indian Ocean).
Biology. Several species of *Apomecyna* are pests of cucumbers, pumpkins, squashes, watermelons, and gourds.

Athemistus Pascoe, 1859 (Pl. 117A, B; Fig. 48.64)

Type species. *Lamia rugosula* Guérin-Méneville, 1831.
Mimoparmena Breuning, 1940
Parmenomorpha Blackburn, 1889
Hoplathemistus Aurivillius, 1917
Parathemistus Breuning, 1950
Mesathemistus Breuning, 1950
Microtragoides Breuning, 1950
Spinhoplathemistus Breuning, 1973
Australian species. 37.
Distribution. Endemic to eastern Australia from VIC to northern QLD.
Biology. All species are wingless and nocturnal. Specimens have been taken crawling into the troughs of flight intercept traps, pitfall traps, under and in timber, extracted from litter in Berlese funnels or found wandering around at night.

Atimura Pascoe, 1863 (Pl. 116H; Fig. 48.79)

Type species. *Atimura terminata* Pascoe, 1863.
Australian species. One.
Distribution. Sixteen described species are known from Asia and northern Australia (Cape York Peninsula, QLD).

Austrosomatidia McKeown, 1945 (Pl. 117C, G, H; Fig. 48.67)

Type species. *Austrosomatidia pulleni* McKeown, 1945.
Australian species. One.
Distribution. *Austrosomatidia* is endemic to higher elevations along the eastern coast of NSW (New England N.P.) to southern QLD (Binna Burra, Eungella N.P.). It includes one described and several undescribed wingless species.

Batocera Dejean, 1835 (Pls 24D, 117E, F; Fig. 48.115)

Type species. *Cerambyx rubus* Linnaeus, 1758.
Megacriodes Pascoe, 1866
Australian species. Four, including *Batocera wallacei* Thomson, the largest Australian beetle.
Distribution. *Batocera* is widespread throughout the Old World tropics and subtropics in the Afrotropical, Oriental and Australian Regions.
Biology. The widely distributed Australian species, *B. boisduvali*, utilises a large variety of unrelated trees as its hosts, but most often it has been recorded from various figs (*Ficus* spp.) and Milky Pine (*Alstonia scholaris*).

Bucynthia Pascoe, 1866 (Pl. 117I)

Type species. *Zygocera spiloptera* Pascoe, 1866.
Spinostenellipsis Breuning, 1959.
Mimozygocera Breuning, 1963.
Australian species. Three.
Distribution. Endemic to eastern Australia, *Bucynthia* species occur in coastal NSW and QLD.

Carteridion Ślipiński & Escalona, 2013 (Pl. 119C; Fig. 48.94)

Type species. *Carteridion wilsonensis* Ślipiński & Escalona, 2013.
Australian species. One.
Distribution. Australian endemic genus. *Carteridion wilsonensis* is known from central and southern NSW, and an undescribed species is known from southern QLD.
References. Ślipiński & Escalona (2013).

Callipyrga Newman, 1842 (Pl. 117D; Fig. 48.43)

Type species. *Callipyrga turrita* Newman, 1842.
Australian species. One.
Distribution. *Callipyrga turrita* is endemic to eastern Australia.

Ceraegidion Boisduval, 1835 (Pl. 118B; Fig. 48.31)

Type species. *Ceraegidion horrens* Boisduval, 1835.
Australian species. Two.
Distribution. The two known species of *Ceraegidion* are endemic to higher elevations in subtropical and temperate rainforests of eastern NSW and VIC.

Cyocyphax Thomson, 1878 (Pl. 118C; Fig. 48.47)

Type species. *Cyocyphax praonethoides* Thomson, 1878.
Australian species. One.
Distribution. *Cyocyphax* is monotypic and endemic to northern QLD.

Demonassa Thomson, 1864 (Pl. 118J; Fig. 48.46)

Type species. *Demonassa funeraria* Thomson, 1864.
Tuberozygocera Breuning, 1974
Australian species. Four.
Distribution. *Demonassa* is an Australian endemic genus and includes four species distributed along the eastern coast of NSW and QLD.

Didymocentrotus McKeown, 1945 (Pl. 118D; Fig. 48.108)

Type species. *Didymocentrus foveatus* Aurivillius, 1917.
Didymocentrus Aurivillius, 1917
Mimotemnosternus Breuning, 1949
Australian species. Two.
Distribution. Two endemic Australian species are found along the eastern coast of NSW and QLD.

Disterna Thomson, 1864 (Pls 25D, 125B, C, E; Figs 48.95, 48.96)

Type species. *Zygocera bifasciata* Pascoe, 1859.
Paradisterna Breuning, 1959
Zygoceropsis Breuning, 1960
Parahybolasius Breuning, 1982
Australian species. 18.
Distribution. *Disterna* is widely distributed in Australia (including Lord Howe Island) and New Guinea.
Biology. Adults have been collected mostly from the foliage or branches of various shrubs or trees (*Eucalyptus*, *Acacia*) or reared from timber.

Dorcadida White, 1846 (Pl. 118G; Fig. 48.65)

Type species. *Dorcadida bilocularis* White, 1846.
Falsoceraegidion Breuning, 1950
Australian species. Two.
Distribution. Endemic to Tasmania.

Dysthaeta Pascoe, 1859 (Pl. 118H; Fig. 48.106)

Type species. *Dysthaeta anomala* Pascoe, 1859.
Australian species. Three.
Distribution. *Dysthaeta anomala* is endemic to eastern Australia, ranging from northern NSW to central QLD. *Dysthaeta naevia* was described from Norfolk Island.
Biology. Several specimens of *D. anomala* have been reared from *Araucaria cunninghamii*.

Enicodes Gray, 1832 (Pl. 118I; Fig. 48.60)

Type species. *Cerambyx fichteli* Schreibers, 1802.
Australian species. One.
Distribution. A single Australian species is known from Norfolk Island, the remaining two species occur in New Caledonia.
Biology. Recorded from vines and many shrubs of unrelated families, including passion vine (*Passiflora* sp.) and coffee bean (*Coffea* spp.).
References. Sudre *et al.* 2010.

Essisus Pascoe, 1866 (Pl. 119A)

Type species. *Essisus dispar* Pascoe, 1866.
Australian species. One.
Distribution. *Essisus dispar* is endemic to eastern Australia.

Excastra Tweed et al., 2024 (Pls 24B, 125D; Fig. 48.103)

Type species. *Excastra albopilosa* Tweed *et al.*, 2024.
Australian species. One.
Distribution. This recently described genus includes one species known from a single specimen collected in Binna Burra in Lamington N.P. (QLD).
References. Tweed *et al.* 2024.

Exocentrus Dejean, 1835 (Pl. 119B; Fig. 48.107)

Type species. *Cerambyx lusitanus* Linnaeus, 1767.
Formosexocentrus Breuning, 1958
Bicolorihirtus Kusama & Tahira, 1978
Australian species. Five.
Distribution. This genus includes about 260 species and is widespread throughout the world, being found in the Palaearctic, Oriental, Afrotropical and Neotropical Regions. Australian species are mostly confined to the northern part of QLD and NT.

Glenea Newman, 1842 (Pl. 119E; Fig. 48.55)

Type species. *Saperda novemguttata* Guérin-Méneville, 1831.
Sphenura Dejean, 1835
Macroglenea Aurivillius, 1920

Australian species. One.
Distribution. This large genus includes about 850 species from Africa, Asia, Australia and Melanesia. The widespread Indo-Pacific species, *G. spinifera* (Voet), reaches Cape York Peninsula (QLD).

Gnoma Fabricius, 1801 (Pl. 119J; Fig. 48.48)

Type species. *Cerambyx longicollis* Fabricius, 1801.
Australian species. One.
Distribution. *Gnoma* is widely distributed from Malaysia through Indonesia, Philippines to New Guinea and Pacific Islands and northern Australia.

Illaena Erichson, 1842 (Pl. 119H)

Type species. *Illaena exilis* Erichson, 1842.
Allomicrus Gahan, 1893
Australian species. Three.
Distribution. *Illaena* is endemic to Australia and is distributed in NSW, SA and WA. Only the holotype of *I. exillis* is known from TAS.

Itheum Pascoe, 1864 (Pl. 119F; Fig. 48.66)

Type species. *Itheum vittigerum* Pascoe, 1864.
Australian species. Four.
Distribution. *Itheum* is endemic to coastal and inland regions of NT, SA, VIC, NSW and QLD.

Lagocheirus Dejean, 1835 (Pl. 119I; Fig. 48.56)

Type species. *Cerambyx araneiformis* Linnaeus, 1767
Karadinia McKeown, 1942
Sternocheirus Dillon, 1957
Archlagocheirus Dillon, 1957
Australian species. One.
Distribution. *Lagocheirus funestus* Thomson was introduced to Australia from Mexico in 1935 to control Velvety Tree Pear (*Opuntia tomentosa*). Currently known only from around Toowoomba (QLD).
References. Delfosse 1985.

Mesolita Pascoe, 1862 (Pls 119D, 120A; Fig. 48.62)

Type species. *Mesolita transversa* Pascoe, 1862.
Corestetha Pascoe, 1875
Cyrtillus Aurivillius, 1917
Australian species. 13.
Distribution. *Mesolita* is known from New Zealand and Australia. In Australia it is distributed along the eastern coast extending from central NSW to northern QLD.

Meton Pascoe, 1862 (Pl. 120C)

Type species. *Meton granulicolle* Pascoe, 1859
Dulichium Thomson, 1864
Australian species. Three.
Distribution. Eastern Indonesia (Aru Islands) to Australia. In Australia the species are widespread across northern NSW, QLD and NT.

Microlamia Bates, 1874 (Pl. 120B, D; Fig. 48.93)

Type species. *Microlamia pygmaea* Bates, 1874.
Parapteridotelus Breuning, 1962
Mimoxylotoles Breuning, 1962
Australian species. Two.
Distribution. New Zealand, Norfolk Island, Lord Howe Island, Fiji and Solomon Islands.

Microtragus White, 1846 (Pls 24E, 25E, 120E–H; Fig. 48.32)

Type species. *Microtragus senex* White, 1846.
Stychus Lacordaire, 1869
Australian species. 15.
Distribution. The genus is widespread in arid areas of Australia and has been recorded from all states except for TAS.

Mimostenellipsis Breuning, 1956 (Pl. 123B)

Type species. *Mimostenellipsis albertisi* Breuning, 1956.
Subexocentrus Breuning, 1963
Striomiaena Breuning, 1963
Australian species. Three.
Distribution. Three species of this genus are known from Yule Island (Papua New Guinea) and the wet tropics of North Australia, from Cairns to Mount Tozer in QLD.

Momisis Pascoe, 1867 (Pl. 121F; Fig. 48.15)

Type species. *Momisis aegrota* Pascoe, 1867.
Australian species. One.
Distribution. Seven *Momisis* species are distributed in Asia and Australia. A single Australian species is known from a few localities in northern QLD.
References. Vives & Heffern 2012.

Mulciber Thomson, 1864 (Pl. 121G; Fig. 48.69)

Type species. *Mulciber linnei* Thomson, 1864.
Australian species. One.
Distribution. Species of *Mulciber* are mostly found on the Moluccas, New Guinea and surrounding islands. *Mulciber maculosus* Breuning has been described from Thursday Island in the Torres Strait Islands.

Neissa Pascoe, 1866 (Pl. 122C; Fig. 48.109)

Type species. *Neissa inconspicua* Pascoe, 1866.

Australian species. Four.
Distribution. The genus is widespread in Australia.

Nemophas Thomson, 1864 (Pl. 125A)

Type species. *Nemophas batoceroides* Thomson, 1864.
Australian species. One.
Distribution. *Nemophas* contains about 20 species widely distributed in the Australo-Pacific region, one species *N. tomentosus* (Buquet, 1860) has been recently found in Iron Range (QLD).
References. Vitali 2014; Wallin *et al.* 2014.

Notomulciber Blackburn, 1894 (Pl. 121H; Fig. 48.71)

Type species. *Notomulciber carpentariae* Blackburn, 1894.
Australian species. One.
Distribution. Approximately 22 species of *Notomulciber* have been described from South and Southeast Asia, Pacific Islands and northern Australia.
Biology. Larvae develop in fronds of various palms.

Nupserha Thomson, 1860 (Pl. 121A; Fig. 48.51)

Type species. *Stibara cosmopolita* Thomson, 1860.
Australian species. One.
Distribution. About 125 species of *Nupserha* have been described from Africa, Australia, China, South Asia and Southeast Asia. A single female of an undescribed species is known from Iron Range (QLD).

Oberea Dejean, 1835 (Pl. 121C; Fig. 48.50)

Type species. *Cerambyx linearis* Linnaeus, 1761.
Australian species. Two.
Distribution. *Oberea* is a large genus with about 250 (Breuning 1967) species distributed in North America, Europe, Asia, Africa, and Melanesia. At least two uncommon species of *Oberea* are found in northern Australia.

Olenecamptus Chevrolat, 1835 (Pl. 121D; Fig. 48.81)

Type species. *Olenecamptus serratus* Chevrolat, 1835.
Authades Thomson, 1857
Ibidiomorphum Motschulsky, 1860
Australian species. One.
Distribution. *Olenecamptus* includes about 60 species distributed in Africa and Asia to New Guinea, the Pacific Islands and Australia. The widely distributed Asian-Papuan species, *O. bilobus* (F.), occurs on Cape York Peninsula (QLD), and *O. basalis* Gahan occurs on Christmas Island.
Biology. *Olenecamptus bilobus* develops in various figs (*Ficus* sp.) and is sometimes considered to be an occasional pest.

Oricopis Pascoe, 1863 (Pl. 121E, J; Fig. 48.35)

Type species. *Oricopis umbrosus* Pascoe, 1863.
Elasmostoma Olliff, 1890
Elasmotena McKeown, 1945
Cristhybolasius Breuning, 1959
Mimoricopis Breuning, 1969
Australian species. Nine.
Distribution. Widely distributed along the eastern coast of Australia.

Paressisus Aurivillius, 1917 (Pl. 121B; Fig. 48.98)

Type species. *Paressisus viridipennis* Aurivillius, 1917.
Australian species. One.
Distribution. *Paressisus viridipennis* is endemic to coastal eastern Australia, extending from as far south as Armidale (NSW) to Lamington N.P. and Brisbane, QLD.

Pelargoderus Audinet-Serville, 1835 (Pl. 120I)

Type species. *Pelargoderus vittatus* Audinet-Serville, 1835.
Nanyohammus Matsushita, 1935
Paragnoma Blanchard, 1853
Rhamses Thomson, 1857
Australian species. One.
Distribution. *Pelargoderus* is widely distributed from Japan and China through Southeast Asia to New Guinea and Australia. A single Australian species, *P. rubropunctatus*, occurs in northern QLD (Cape Tribulation, Torres Strait and Iron Range).

Pentacosmia Newman, 1842 (Pl. 121I; Fig. 48.99)

Type species. *Pentacosmia scoparia* Newman, 1842.
Mimancita Breuning, 1938
Australian species. Four.
Distribution. *Pentacosmia* is an endemic Australian genus distributed mostly along the eastern and southern coastal area in QLD to TAS and southern WA.

Phelipara Pascoe, 1866 (Pl. 122A)

Type species. *Phelipara marmorata* Pascoe, 1866.
Oncideropsis Aurivillius, 1922
Falsopothyne Pic, 1931
Paraphelipara Breuning, 1940
Australian species. Two.
Distribution. Approximately 18 described species of *Phelipara* are known from Asia and northern Australia. Two Australian species are from Christmas Island (*P. subvittata*) and from northern QLD (probably an undescribed species).

Phytoecia Dejean, 1835 (Pl. 125F)

Type species. *Saperda cylindrica* Fabricius, 1775.

Australian species. One.

Distribution. Large genus of herbaceous stem boring beetles distributed worldwide. One species, *P. coerulescens* (Scopoli), has been introduced to Australia as a biological control agent against the weed Paterson's curse (*Echium plantagineum*) and can be sporadically found in ACT and NSW.

Phyxium Pascoe, 1864 (Pl. 122E; Fig. 48.100)

Type species. *Phyxium bufonicum* Pascoe, 1864.

Australian species. Three (undescribed).

Distribution. Two or three unnamed species have been collected in northern QLD. Six species are recorded from eastern Indonesia, Papua New Guinea, and New Caledonia.

Potemnemus Thomson, 1864 (Pl. 122D; Fig. 48.49)

Type species. *Cerambyx scabrosus* Olivier, 1790.

Periaptodes Pascoe, 1866

Spinopotemnemus Breuning, 1973

Australian species. Four.

Distribution. *Potemnemus* includes 17 described species from Indonesia, New Guinea, Solomon Islands and northern Australia (Cape York, QLD).

References. Wallin & Kvamme 2015.

Probatodes Thomson, 1864 (Pls 25F, 123G; Fig. 48.112)

Type species. *Acanthocinus plumula* Newman, 1851.

Ameipsis Pascoe, 1865

Australian species. Two.

Distribution. *Probatodes* is endemic to Australia. Two geographically separated species are known from TAS, VIC and NSW (*P. plumula*) and northern QLD (*P. marginicollis*).

Pterolophia Newman, 1842 (Pl. 122J; Fig. 48.80)

Type species. *Mesosa bigibbera* Newman, 1842.

Praonetha Dejean, 1835

Praonetha Pascoe, 1862

Prioneta Blanchard, 1853

Lychrosis Pascoe, 1866

Chaetostigme Pascoe, 1875

Australian species. 13.

Distribution. Africa, Pacific Islands, East Asia, Melanesia, India, Moluccas, New Guinea and Australia.

Rhytiphora Audinet-Serville, 1835 (Pls 24C, G, 25C, 116C, 122B, F–I, K, 123A; Figs 48.1, 48.82)

Type species. *Lamia rugicollis* Dalman, 1817.

Achriotypa Pascoe, 1875.

Australian species. 163.

Distribution. *Rhytiphora* is found across Australia in all major biomes. The majority of species diversity is in Australia; there are also at least 50 species described from New Guinea, Pacific Islands and Southeast Asia.

Biology. Australian *Rhytiphora* species are found on many different host plants, both woody and herbaceous; the most common host plant genus is *Acacia*, with Fabaceae and Myrtaceae as the most common host plant families (Ashman *et al.* 2023). Some *Rhytiphora* species have been recorded from a wide range of host plants, while *R. solandri* is exclusively found in the flower stalks of grass trees (*Xanthorrhoea* sp.) (Froggatt 1894). *Rhytiphora diva* (= *Zygrita diva*, Lucerne Crownborer) and *R. stigmatica* (= *Corrhenes stigmatica*) are minor pests of Lucerne (Jarvis & Smith 1946; Duffy 1963). Serious infestations of *Rhytiphora pedicornis* on the legume *Stylosanthes scabra* have been recorded in the Gulf of Carpentaria, Queensland (Hall 1980).

References. Ashman *et al.* 2022a, b, 2023.

Notes. Other synonyms: *see* Ślipiński & Escalona 2013: 196–197.

Ropica Pascoe, 1858 (Pl. 124A; Fig. 48.78)

Type species. *Ropica piperata* Pascoe, 1858.

Australian species. Six.

Distribution. About 140 species in this genus have been described from Asia, Pacific Islands and along the eastern coast of Australia.

Rosenbergia Ritsema, 1881 (Pl. 123F)

Type species. *Rosenbergia mandibularis* Ritsema, 1881.

Mimorosenbergia Breuning, 1980

Australian species. Three.

Distribution. About 20 species of *Rosenbergia* have been described from New Guinea and northern Australia.

References. Rigout 1992; Wallin & Nylander 2006, 2007.

Sciades Pascoe, 1864 (Pl. 123D; Fig. 48.111)

Type species. *Leiopus suffusus* Pascoe, 1859

Miaenia Pascoe, 1864

Australian species. Two.

Distribution. About 20 species are distributed predominantly in Asia with a few reaching New Guinea and northern Queensland. *Sciades exigua* was described from Christmas Island (Indian Ocean).

Similosodus McKeown, 1945 (Pl. 123E; Fig. 48.83)

Type species. *Sodus verticallis* Pascoe, 1865.

Sodus Pascoe, 1864
Venosodus Breuning 1961
Australian species. One.
Distribution. About 20 species of *Similosodus* have been described from South and Southeast Asia, New Guinea and northern Queensland.

Somatidia Thomson, 1864 (Pl. 123H; Fig. 48.63)

Type species. *Parmena antarctica* White, 1846.
Araneosoma Breuning, 1950
Tenebrosoma Breuning, 1950
Australian species. Six.
Distribution. About 50 species of this genus have been described from New Zealand, New Caledonia and Australia.

Stenellipsis Bates, 1874 (Pl. 123C; Fig. 48.61)

Type species. *Xylotoles bimaculatus* White, 1846.
Brachyrhabdus Aurivillius, 1917
Australian species. Nine.
Distribution. *Stenellipsis* includes many species described from New Zealand, New Caledonia and the eastern coast of Australia.

Sybra Pascoe, 1865 (Pl. 124B, D; Fig. 48.77)

Type species. *Ropica stigmatica* Pascoe, 1859.
Australian species. 38.
Distribution. *Sybra* is widespread throughout the eastern Palaearctic, Oriental and Australian Regions. It is found Australia-wide, occurring in all states and territories except Tasmania.
References. Weigel & Skale 2009, 2011.
Notes. Synonyms: *see* Ślipiński & Escalona 2013: 226.

Temnosternus White, 1855 (Pls 25B; 124C, E)

Type species. *Temnosternus planiusculus* White, 1855.
Temnosternopsis Breuning, 1939
Tuberothelais Breuning, 1963
Falsapolia Breuning, 1945
Australian species. 14.
Distribution. *Temnosternus* is the only genus of Tmesisternini that is endemic to Australia. Fourteen described species are confined to the eastern part of the continent, ranging from northern QLD to VIC.
References. Gressitt 1984: 23–25.

Thyada Pascoe, 1863 (Pl. 124G; Fig. 48.34)

Type species. *Zygocera barbicornis* Pascoe, 1859.
Australian species. One.
Distribution. *Thyada barbicornis* occurs in northern NSW and southern QLD, with a single specimen collected in WA and on Lord Howe Island.

Tmesisternus Latreille, 1829 (Pl. 124H; Fig. 48.74)

Type species. *Tmesisternus bizonulatus* Guérin-Méneville, 1831.
Icthyosoma Boisduval, 1835
Arrhenotus Pascoe, 1858
Polyxo Thomson, 1864
Apolia Thomson, 1864
Mneside Thomson, 1864
Tmesisternopsis Breuning, 1945
Australian species. Two.
Distribution. Over 300 described species of *Tmesisternus* are distributed primarily in Indonesia and Melanesia, with few species recorded outside of this area, including species from Cape York Peninsula (QLD).
References. Gressitt 1984; Weigel 2003.

Trigonoptera Perroud, 1855 (Pl. 124I; Fig. 48.70)

Type species. *Trigonoptera maculata* Perroud, 1855.
Arsysia Pascoe, 1867
Australian species. Two.
Distribution. Most of the *Trigonoptera* species occur in Melanesia (principally New Guinea). Two species are found in northeastern Australia.
References. Gressitt 1984.

Velora Thomson, 1864 (Pl. 124F; Fig. 48.97)

Type species. *Velora australis* Thomson, 1864.
Pseudancita Breuning, 1949
Australian species. Four.
Distribution. The genus is distributed in New Guinea, northern QLD reaching northern NSW.

Xylotoles Newman, 1840 (Pl. 125G; Fig. 48.57)

Type species. *Saperda grisea* Fabricius, 1775.
Blaxotes McKeown, 1945
Trichoxylotoles Breuning, 1950
Australian species. Four.
Distribution. *Xylotoles* is predominantly a New Zealand genus. Described Australian species are known from Norfolk Island and Lord Howe Island, and an unnamed species is known from QLD.

Zorilispe Pascoe, 1865 (Pl. 125H; Fig. 48.76)

Type species. *Zorilispe fulvisparsa* Pascoe, 1865.

Australian species. Two.

Distribution. *Zorilispe* is widespread through the eastern Palaearctic, Oriental and Australian Regions. In Australia it is found only in coastal NSW.

Zygocera Dejean, 1835 (Pls 24A; 118A, E, F)

Type species. *Acanthocinus pruinosus* Boisduval, 1835.

Disternopsis Breuning, 1939

Plocamocera Breuning, 1939

Pseudodisterna Breuning, 1953

Trichozygocera Breuning, 1956

Australian species. Seven.

Distribution. *Zygocera* is found Australia-wide but mostly in the south, along the east coast from SA to northern QLD; one species is found in the southern part of WA.

Acknowledgements

Our research on Australian Cerambycidae has been supported by multiple grants from the Australian Biological Resources Study (ABRS) and the Department of Agriculture, Fisheries and Forestry. Cate Lemann and Lingzi Zhou (ANIC) are thanked for their technical support. Marie and Jiří Lochman, Owen Kelly, Jack Hasenpusch and James Tweed have provided photographs of living longhorn beetles.

References

Ashman LG, De Keyzer R, Ślipiński A (2023) The Australian genus *Rhytiphora* (Coleoptera: Cerambycidae: Lamiinae) with a revision of the *Rhytiphora collaris* group. *Zootaxa* **5312**(1), 1–62. doi:10.11646/zootaxa.5312.1.1

Ashman LG, Shin S, Zwick A, Ślipiński A, McKenna DD (2022a) The first phylogeny of Australasian Lamiinae longhorn beetles (Coleoptera: Cerambycidae) reveals poor tribal classification and a complex biogeographic history. *Systematic Entomology* **47**(24), 213–230. doi:10.1111/syen.12526

Ashman LG, Hartley D, Jin M, Rowell DM, Teasdale L, *et al.* (2022b) *Rhytiphora*: a phylogenetic and morphological study of Australia's largest longhorn beetle genus (Coleoptera: Cerambycidae). *Invertebrate Systematics* **36**(6), 493–505. doi:10.1071/IS21071

Bates HW (1881) Insecta, Coleoptera, Cerambycidae. *Biologia Centrali-Americana* **5**, 153–224, pls. XII–XV.

Breuning S (1940a) Novae species Cerambycidarum VIII. *Folia Zoologica et Hydrobiologica* **10**(1), 37–85.

Breuning S (1940b) Novae species Cerambycidarum X. *Folia Zoologica et Hydrobiologica* **10**(1), 407–437.

Breuning S (1961) Nouvellles formes de Lamiaires (Treizième Partie). *Bulletin de l'Institut Royal des Sciences Naturelles de Belgique* **37**, 1–44.

Breuning S (1967) *Catalogue des Lamiaires du Monde (Col. Céramb.).* Museum G. Frey, Tutzing bei München, pp. 771–864.

Breuning S (1970) Nouveaux Coléoptères Cerambycidae Lamiinae des collections du Muséum de Paris. *Bulletin du Muséum National Histoire Naturelle* (2) **42**(2), 460–488.

Breuning S (1977) Révision de la tribu des Acanthocinini de la région Asiato-Australienne (Coleoptera: Cerambycidae) (deuxième partie). *Mitteilungen aus dem Zoologische Museum in Berlin* **53**(1), 199–276. doi:10.1002/mmnz.19770530203

Delfosse ES (1985) *Echium plantagineum* in Australia: effects of a major conflict of interest. In *Proceedings of VI International Symposium of Biological Control of Weeds, 19–25 August 1984, Vancouver Canada.* (Ed. ES Delfosse) pp. 293–299. Agriculture Canada.

Duffy EAJ (1963) *A Monograph of the Immature Stages of Australasian Timber Beetles (Cerambycidae).* British Museum (Natural History), London, 235 pp. 11 pls.

Froggatt WW (1923) *Forest Insects of Australia.* Government Printer, Sydney, 171 pp.

Gilmour EF (1960) Some new genera and species of the neotropical Acanthocinini (Coleoptera, Cerambycidae, Lamiinae). *Acta Biologica Venezuelica* **3**(2), 43–65.

Goodwin S, Pettit MA (1994) *Acalolepta vastator* (Newman) (Coleoptera: Cerambycidae) infesting grapevines in the Hunter Valley, New South Wales 2, Biology and Ecology. *Australian Journal of Entomology* **33**(4), 391–397. doi:10.1111/j.1440-6055.1994.tb01258.x

Goodwin S, Pettit MA, Spohr LJ (1994) *Acalolepta vastator* (Newman) (Coleoptera: Cerambycidae) infesting grapevines in the Hunter Valley, New South Wales 1, Distribution and Dispersion. *Australian Journal of Entomology* **33**(4), 385–390. doi:10.1111/j.1440-6055.1994.tb01257.x

Gressitt JL (1984) Systematics and biogeography of the longicorn beetle tribe Tmesisternini. *Pacific Insects Monograph* **41**, 1–263.

Jarvis H, Smith JH (1946) Lucerne pests. *Queensland Agricultural Journal* **62**(2), 79–89.

Melzer J (1934) Novos subsídios para o conhecimento dos Cerambycídeos neotrópicos (Col.). *Revista de Entomologia* **4**(1), 70–110.

Monné MÁ (1977) Contribuição ao conhecimento dos Acanthocinini. VI (Coleoptera, Cerambycidae, Lamiinae). *Revista Brasileira de Biologia* **37**, 693–711.

Rigout J (1992) Rosenbergia nouveaux, récents ou peu connus. *Bulletin de la Société Sciences Nat* **75–76**, 78.

Ślipiński A, Escalona HE (2013) *Australian Longicorn Beetles (Coleoptera). Volume 1. Introduction and Subfamily Lamiinae.* CSIRO Publishing, Collingwood, Victoria, 484 pp.

Smith ESC (1996) Control of longicorn beetles in mango trees. Northern Territory Department of Primary Industries, Darwin. *Agnote* **685**, 1–2.

Souza DS, Marinoni L, Monné ML, Gómez-Zurita J (2020) Molecular phylogenetic assessment of the tribal classification of Lamiinae (Coleoptera: Cerambycidae). *Molecular Phylogenetics and Evolution* **145**, 106736. doi:10.1016/j.ympev.2020.106736

Sudre J, Vives E, Cazères S, Mille C (2010) Contribution à l'étude des Cerambycidae (Coleoptera) de la Nouvelle-Calé,donie – 1e partie: sous-famille des Lamiinae. *Mémoires de la Société linnéenne de Lyon* **1**, 1–76.

Švácha P, Lawrence JF (2014) 2.4 Cerambycidae Latreille, 1802. In *Handbuch der Zoologie/Handbook of Zoology. Band/Volume IV Arthropoda: Insecta Teilband/Part 38. Coleoptera, Beetles. Volume 3. Morphology and Systematics (Polyphaga).* (Eds RAB Leschen and RG Beutel) pp. 77–177. W DeGruyter, Berlin.

Tweed JMH, Ashman LG, Ślipiński A (2024) *Excastra albopilosa*, a remarkable new genus and species of Lamiinae (Coleoptera: Cerambycidae) from South East Queensland, Australia. *Australian Journal of Taxonomy* **54**, 1–8. doi:10.54102/ajt.iv1x5

Vitali F (2011) Notes on the genus *Acalolepta* Pascoe, 1858 (Coleoptera: Cerambycidae) from Indonesian Papua and the Moluccas. In *Biodiversity, Biogeography and Nature Conservation in Wallacea and New Guinea. Vol. 1.* (Ed. D Telnov) pp. 291–297. The Entomological Society of Latvia, Riga.

Vitali F (2014) Taxonomic revision of the genus *Nemophas* Thomson, 1864 (Coleoptera, Cerambycidae). *Lambillionea* **113** [2013], 255–266.

Vitali F (2017) Taxonomic notes on *Acalolepta mixta* (Hope, 1841) and *Acalolepta vastator* (Newman, 1847) (Coleoptera: Cerambycidae). In *Biogeography and Nature Conservation in Wallacea and New Guinea, III.*

Vol. 3. (Eds D Telnov, MVL Barclay, OSG Pauwels, GW Beccaloni) pp. 291–297. The Entomological Society of Latvia.

Wallin H, Kvamme T (2015) A review of the genus *Potemnemus* Thomson, 1864 (Coleoptera: Cerambycidae) with description of a new species from Papua New Guinea. *Zootaxa* **4040**(5), 501–529.

Wallin H, Nylander U (2006) Description of the genitalia characteristics of species within the *Rosenbergia xenium* Gilmour, 1959 complex and related species. A taxonomic revision. Part 1. (Coleoptera Cerambycidae). *Lambillionea* **106**(3), 1–24.

Wallin H, Nylander U (2007). Description of the genitalia characteristics of species within the *Rosenbergia weiskei* Heller, 1902 complex and related species, including the description of a new species of *Rosenbergia*. A taxonomic revision. Part 2. (Coleoptera Cerambycidae). *Lambillionea* **107**, 1–29.

Wallin H, Kvamme T, Nylander U (2014) A revision of the genus *Nemophas* Thomson, 1864 (Coleoptera: Cerambycidae), with descriptions of a new subgenus *Pilomophas* and a new genus *Nemoplophora*. In *Biodiversity, Biogeography and Nature Conservation in Wallacea and New Guinea. Vol. 2.* (Ed. D Telnov) pp. 407–436. The Entomological Society of Latvia, Riga.

Weigel A (2003) Zur taxonomie, synonymie und faunistik der gattung *Tmesisternus* Latreille, 1829 mit beschreibung zweier neuer Arten (Coleoptera: Cerambycidae, Tmesisternini). *Veröffentlichungen Naturkundemuseum Erfurt* **22,** 127–136.

Weigel A, Skale A (2009) Zur Systematik, Taxonomie und Faunistik der Apomecynini der orientalischen und australischen Region (Coleoptera: Cerambycidae: Lamiinae). Revision der Gattung *Sybra* Pascoe, 1865. Teil 1. *Vernate* **28**, 421–450.

Weigel A, Skale A (2011) Revision der Gattung Sybra Pascoe, 1865 und Anmerkungen zu weiteren Gattungen, Teil 2. [Systematik, Taxonomie und Faunistik der Apomecynini der orientalischen und australischen Region (Coleoptera: Cerambycidae: Lamiinae)]. In *Biodiversity, Biogeography and Nature Conservation in Wallacea and New Guinea. Vol. 1.* (Ed. D Telnov) pp. 335–350, pls 78–83. The Entomological Society of Latvia, Riga.

INDEX OF GENERIC AND FAMILY GROUP NAMES